BIOLOGY

FOR A CHANGING WORLD

WITH PHYSIOLOGY

Fourth Edition

BIOLOGY

FOR A CHANGING WORLD

WITH PHYSIOLOGY

Michèle Shuster/Janet Vigna/Matthew Tontonoz

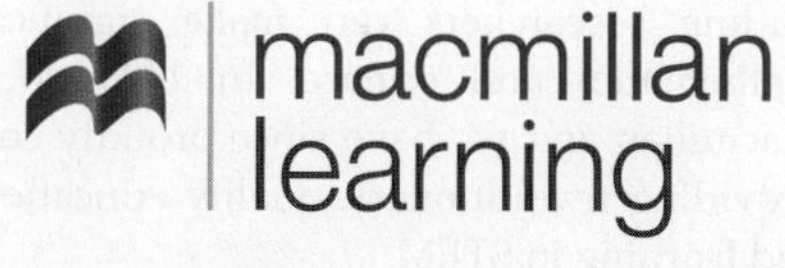

Austin · Boston · New York · Plymouth

Senior Vice President, STEM: Daryl Fox
Executive Program Director, Life Sciences: Sandy Lindelof
Program Manager: Shannon Howard
Executive Marketing Manager: Will Moore
Executive Content Development Manager, STEM: Debbie Hardin
Senior Development Editor: Susan Moran
Executive Project Manager, Content, STEM: Katrina Mangold
Editorial Project Manager: Karen Misler
Director of Content, Life Sciences: Jennifer Driscoll Hollis
Senior Media Editor: Mary N. Tibbets
Media Editor: Jennifer Compton
Learning Solutions Specialist: Jim Zubricky
Assistant Editor: Casey Blanchard
Marketing Assistant: Morgan Psiuk
Director of Content Management Enhancement: Tracey Kuehn
Senior Managing Editor: Lisa Kinne
Senior Content Project Manager: Martha Emry
Assistant Director, Process Workflow: Susan Wein
Senior Workflow Manager: Jennifer Wetzel
Director of Design, Content Management: Diana Blume
Design Services Manager: Natasha Wolfe
Cover Design Manager: John Callahan
Text Designer: Gary Hespenheide
Art Manager: Matthew McAdams
Illustrations: Eli Ensor
Director of Digital Production: Keri deManigold
Media Project Manager: Daniel Comstock
Media Permissions Manager: Christine Buese
Photo Editor: Krystyna Borgen, Lumina Datamatics, Inc.
Composition: Lumina Datamatics, Inc.
Printing and Binding: LSC Communications
Cover Image: Cora Rosenhaft/Getty Images

Library of Congress Control Number: 2020939799

Student Edition Paperback:
ISBN-13: 978-1-319-27096-4
ISBN-10: 1-319-27096-4

Student Edition Loose-leaf:
ISBN-13: 978-1-319-36338-3
ISBN-10: 1-319-36338-5

© 2021, 2018, 2014, 2012 by W. H. Freeman and Company

All rights reserved.

Printed in the United States of America

1 2 3 4 5 6 25 24 23 22 21 20

Macmillan Learning
One New York Plaza
Suite 4600
New York, NY 10004-1562
www.macmillanlearning.com

In 1946, William Freeman founded W. H. Freeman and Company and published Linus Pauling's *General Chemistry*, which revolutionized the chemistry curriculum and established the prototype for a Freeman text. W. H. Freeman quickly became a publishing house where leading researchers can make significant contributions to mathematics and science. In 1996, W. H. Freeman joined Macmillan and we have since proudly continued the legacy of providing revolutionary, quality educational tools for teaching and learning in STEM.

To our teachers and students

You are our inspiration

About the Authors

© 2021 Macmillan, Photo by Shannon Howard.

Janet Vigna, Michèle Shuster, and Matthew Tontonoz

MICHÈLE SHUSTER, PH.D., is a Professor in the Biology Department at New Mexico State University in Las Cruces, New Mexico. She focuses on biology teaching and learning and teaches introductory biology, microbiology, and cancer biology classes at the undergraduate level, as well as working on several K–12 science education programs. Michèle is involved in mentoring graduate students and postdoctoral fellows in effective teaching, thereby preparing the next generation of undergraduate educators. She is the recipient of numerous teaching awards, including the Westhafer Award for Teaching Excellence at NMSU. Michèle received her Ph.D. from the Graduate School of Biomedical Sciences at Tufts University School of Medicine, where she studied meiotic chromosome segregation in yeast.

JANET VIGNA, PH.D., is Professor of Biology and chair of the Biology Department at Grand Valley State University in Allendale, Michigan. She is a science education specialist in the Integrated Science Program, with a passion for training and mentoring K–12 science teachers and teaching nonmajors biology. Her scholarly interests include biology curriculum development, retention strategies for academically at-risk students, and research on the effects of biological pesticides on amphibian communities. She has been recognized with several teaching awards, including the Michigan Science Teachers Association College Science Teacher of the Year. She received her B.S. in biology from the University of Michigan and her Ph.D. in microbiology from the University of Iowa.

MATTHEW TONTONOZ is a science writer living in Brooklyn, New York. His writing has appeared in *Scientific American, Popular Science, OnCancer, History of Psychology, Science as Culture,* and *Cancer Immunology Research.* For ten years, he was a development editor for textbooks in biology before shifting his focus to writing. He is currently senior science writer at Memorial Sloan Kettering Cancer Center. Matt received his B.A. in biology from Wesleyan University and his M.A. in the history and sociology of science from the University of Pennsylvania.

Dear Student,

Thank you for opening this book! We hope that your journey through it will be as rewarding for you as our journey in writing it has been. When we first came together to collaborate on the development of this text, our biggest overarching goal was to get students interested in biology by showing its relevance to daily life. We wanted to create a textbook that students would actually want to read. Our model and partner in this process has been *Scientific American,* a visually stunning magazine that's been successfully bringing science to the public for more than 150 years. The result is a unique textbook that takes a novel approach to teaching biology, one that we think has the potential to greatly improve learning. We hope that this brief introduction will serve as a road map of the book, so that you can get the most out of your experience with it and be as captivated by the wonders of life as we are.

The main approach of each chapter is to present key science concepts within the context of a relevant and engaging story—a story of discovery, of determination, of human interest, of adventure. From the search for life on Mars to the problem of antibiotic-resistant bacteria, we use stories to bring science to life and to show scientists in action. After all, science is not just a collection of facts, so why would we present it that way? We ask you, our students, to study biology so you can use knowledge to make choices in the real world. We value those stories that will lead you to ask questions about life and how it works and to see the relevance of biology to daily activity. We have seen how stories engage students in our classrooms, and we hope you will be similarly intrigued.

While gripped by a story, you may not even realize how much you are learning. To reinforce the basic learning process, we rely on several strategies:

- Each story is prefaced by a set of **Driving Questions**. By keeping these questions in mind as you navigate the story, you will have a good framework for learning the key science concepts.
- Eye-catching **Infographics** highlight and drill down into the science of each story. The set of Infographics in a chapter provides a science storyboard for that chapter, illustrating the key scientific concepts and linking them to the story.
- Each Infographic has a **question** to help you ensure that you have grasped the concept illustrated.
- **Key terms** are defined in the margins, making it easy to check a definition without having to leave the story.
- **Chapter summaries** provide a concise set of bullet points that distill the key scientific concepts.
- **Test Your Knowledge** questions at the end of each chapter reinforce basic facts and allow you to apply these facts through data interpretation and mini cases.

By taking full advantage of these resources, you will be better able to appreciate how biology affects each and every one of us as well as our relatives—both close and distant—on this planet. We hope that you will talk about biology with your friends and family, and that what you learn here will be applicable to your life. We hope that you will think as critically about choices you make outside the classroom as we will ask you to do here in these pages.

Welcome to ***Biology for a Changing World.*** We hope that you enjoy your journey, and complete it more prepared for your life in a changing world.

Michèle Shuster

Janet Vigna

Matthew Tontonoz

Real stories about real people and real science

From the groundbreaking partnership of Macmillan Learning and *Scientific American* comes this one-of-a-kind introduction to the science of biology and its impact on the way we live.

Now supported in **Achieve, *Biology for a Changing World*** explores the core ideas of biology through chapters written and illustrated in the style of a *Scientific American* article. Chapters don't just feature compelling stories of real people—each chapter is a newsworthy story that serves as a context for covering the standard non-majors biology curriculum.

Overview of key features

MEDIA AND ASSESSMENT

The Fourth Edition of ***Biology for a Changing World*** is now supported in **Achieve,** Macmillan's new online learning system. Achieve supports educators and students throughout the full range of instruction, including assets suitable for pre-class preparation, in-class active learning, and post-class study and assessment. This simple yet powerful platform both supports instructors and is packed with engaging assessments, multimedia assets, and self-study tools to support students in their journey.

New media and assessment features in Achieve include:

- **Simulation activities** focus on the most challenging concepts in a non-majors biology course. These scaffolded, interactive lessons help students apply the principles they have learned from the text or lecture in the context of a relevant "real life" question or familiar scenario.
- **Tutorials** give students an opportunity to work through a concept step by step, with support provided through visuals, hints, feedback, and links to the e-book. Tutorials were developed to work as preparations for the simulations, but are also effective as standalone assignments.
- **Pre-built, curated homework assignments** are easy to assign and can be customized by the instructor. Homework questions feature hints, error-specific feedback, and solutions to provide a powerful formative assessment experience for students.
- **Animations** are embedded directly into associated quizzes so students can reference these powerful visualization tools as they work through the questions. These animations are also available in the instructor resources for use in the classroom.

THE TEXTBOOK

Biology for a Changing World is a unique text for non-majors biology students. By teaching science through the context of an engaging and magazine-like story in each chapter, it encourages students to keep reading and to appreciate why biology matters to them and to their world.

Features of the text include:

- Each chapter of ***Biology for a Changing World*** is written in the style of a *Scientific American* article. This **story-based approach** grabs students' interest and teaches the relevance of biology and why it matters to students.
- Engaging and informative **Infographics** appear throughout the book. These powerful pieces of art teach students how to learn from charts, graphs, and images, and add visual appeal to the science. Animations of many of the infographics are available in Achieve, accompanied by quiz questions.
- **Milestone mini-chapters** highlight historically important discoveries in biology. These features present biology as a living science by teaching students how we know what we know, prompting them to consider how future research will expand our understanding of biology.
- **Driving Questions** provide the pedagogical framework for the chapter material by prompting students to consider the questions they need to be able to answer to have a full understanding of the material.
- **End-of-chapter questions,** each written by Michèle Shuster, are framed around the chapter's Driving Questions; each question set includes Interpreting Data, Mini-Case, and Bring It Home questions to help students develop higher-order thinking skills.

Achieve is the culmination of years of development work put toward creating the most powerful online learning tool for biology students. It houses all of our renowned assessments, multimedia assets, e-books, and instructor resources in a powerful new platform.

Achieve supports educators and students throughout the full range of instruction, including assets suitable for pre-class preparation, in-class active learning, and post-class study and assessment. The pairing of a powerful new platform with outstanding biology content provides an unrivaled learning experience.

For more information or to sign up for a demonstration of Achieve, contact your local Macmillan representative or visit **macmillanlearning.com/achieve**

Full Learning Path

Achieve includes a full learning path of content for pre-class preparation, in-class active learning, and post-class engagement and assessment. It's the perfect solution for engaging all students in an inclusive teaching experience.

LearningCurve Adaptive Quizzing

LearningCurve's game-like quizzing motivates each student to engage with the course content, and reporting tools help teachers get a handle on what their class needs.

Book-specific quizzes adapt their level of difficulty based on individual student performance.

Students earn points for correct answers ... and receive immediate feedback on incorrect answers.

Questions are tagged to sections of the e-book to provide comprehensive coverage and easy-to-find help.

Upon completion, each student receives a study plan with links to additional study tools.

The quiz adapts to each student's needs based on performance, automatically providing more questions on topics where the student is struggling.

Performance is reported by Learning Objectives and can also be tracked to give instructors actionable information about topics that need extra emphasis.

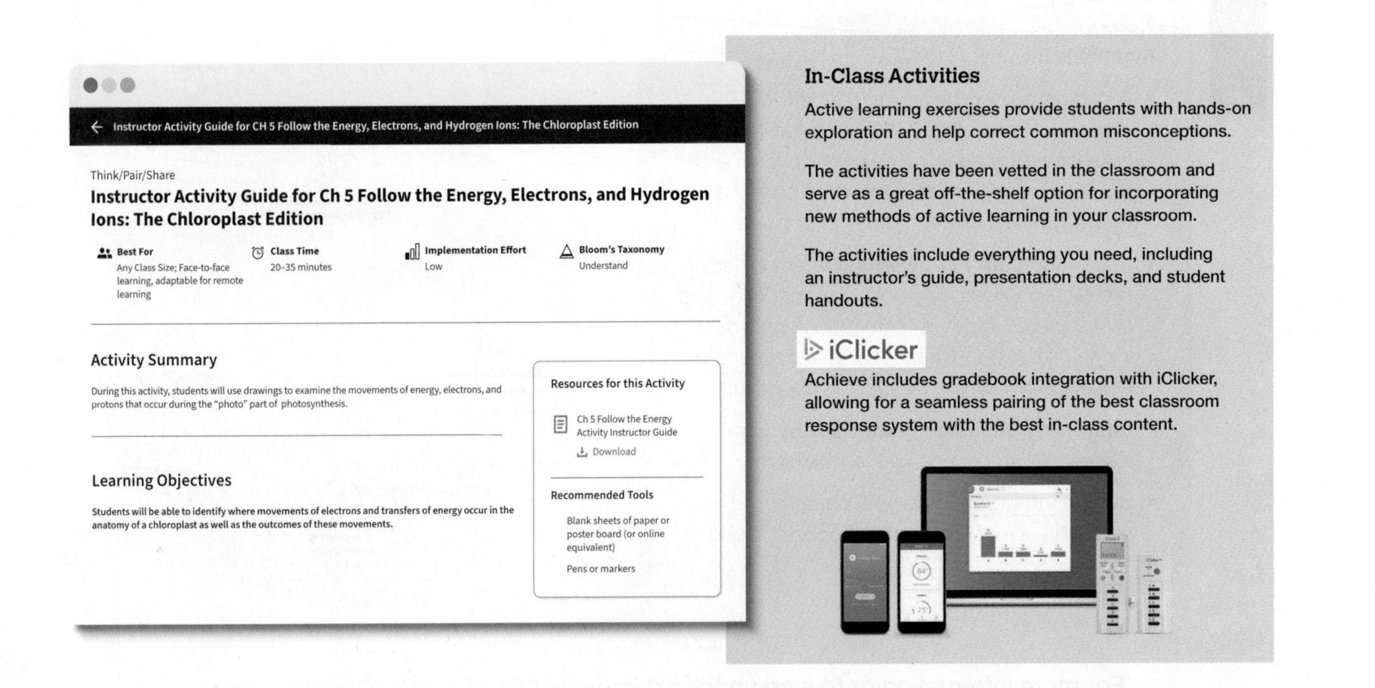

In-Class Activities

Active learning exercises provide students with hands-on exploration and help correct common misconceptions.

The activities have been vetted in the classroom and serve as a great off-the-shelf option for incorporating new methods of active learning in your classroom.

The activities include everything you need, including an instructor's guide, presentation decks, and student handouts.

Achieve includes gradebook integration with iClicker, allowing for a seamless pairing of the best classroom response system with the best in-class content.

Powerful Content for an Engaging Learning Experience

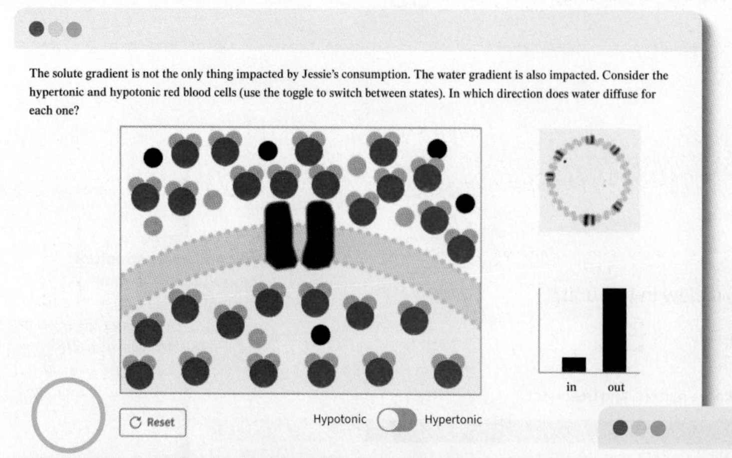

Simulation Activities focus on the most challenging concepts in a non-majors biology course. These scaffolded, interactive lessons help students apply the principles they have learned from the text or lecture in the context of a real-life question or familiar scenario.

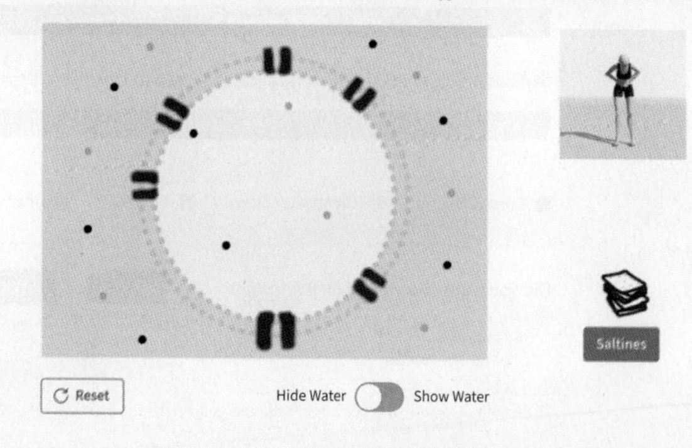

Tutorials give students an opportunity to work through a concept step by step, with support provided through visuals, hints, feedback, and links to the e-book. Tutorials were developed to work as preparations for the simulations, but are also effective as standalone assignments.

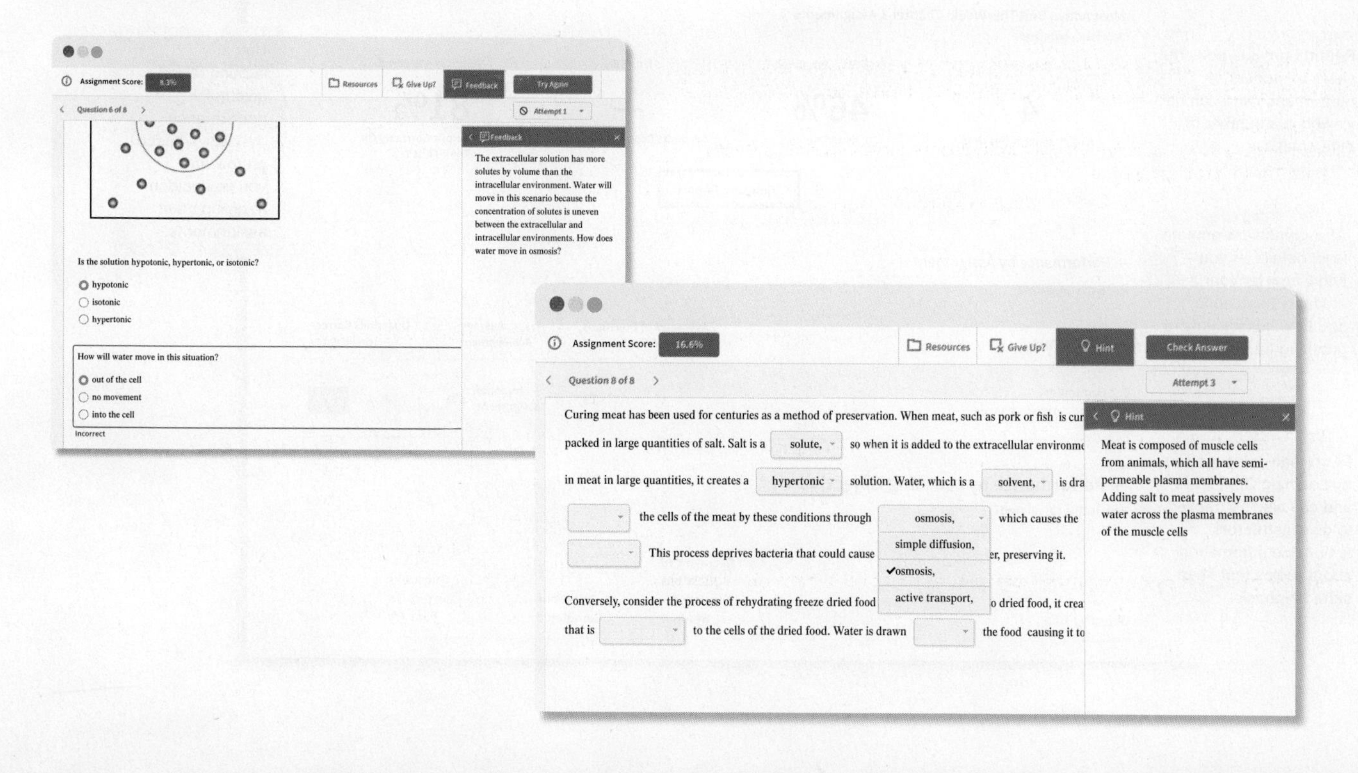

Powerful analytics, viewable in an elegant dashboard, offer instructors a window into student progress. Achieve gives you the insight to address trouble areas and misconceptions before students struggle on a test.

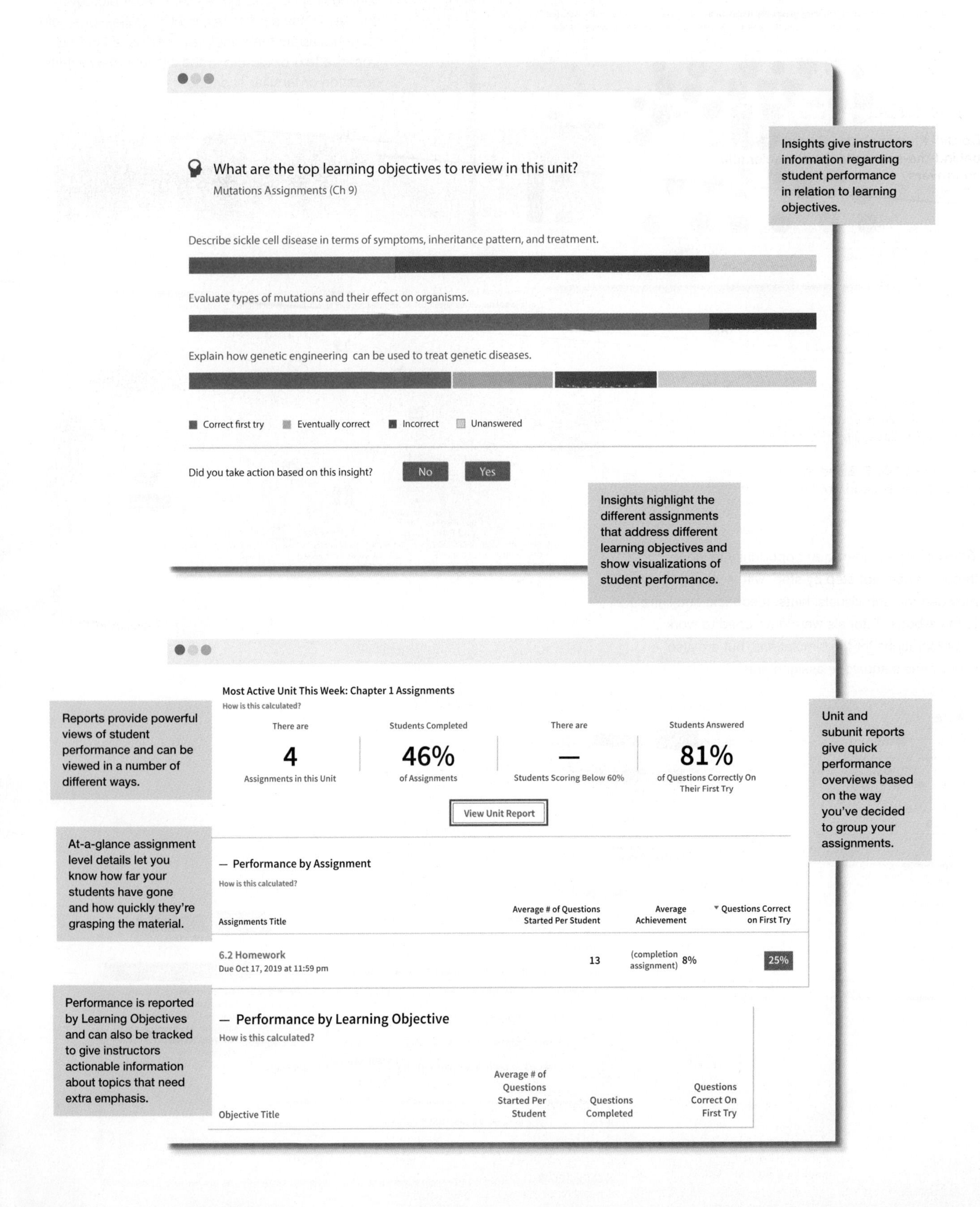

Science through Stories

Each chapter of ***Biology for a Changing World*** is written in the style of a *Scientific American* article. This story-based approach captures students' interest immediately and teaches not only the fundamental concepts of the discipline but also why understanding those concepts matters to students' lives and the world in general.

NEW CHAPTER STORIES

Do Cell Phones Cause Cancer? A behind-the-scenes looks at a scientific controversy
(Chapter 1, Process of Science)

New Gene, New Me: Gene therapy offers hope to people with debilitating genetic conditions
(Chapter 9, Mutations and Genetic Engineering)

Invisible You: A hidden world of microbes helps make you who you are
(Chapter 17, Prokaryotic Diversity)

Bringing Bison Back: A controversial plan to rewild the American Prairie
(Chapter 22, Ecosystem Ecology)

NEW MILESTONE! Pandora's Dish: The power, promise, and politics of stem cells

INFOGRAPHIC 9.2

Mutations Can Alter Protein Shape and Function

Mutations alter the nucleotide sequence of DNA. If a mutation changes the coding region of a gene, the resulting protein may have an altered structure and function. In this case, altered hemoglobin causes cells to take on a sickled shape, and interferes with the ability of red blood cells to carry oxygen to tissues.

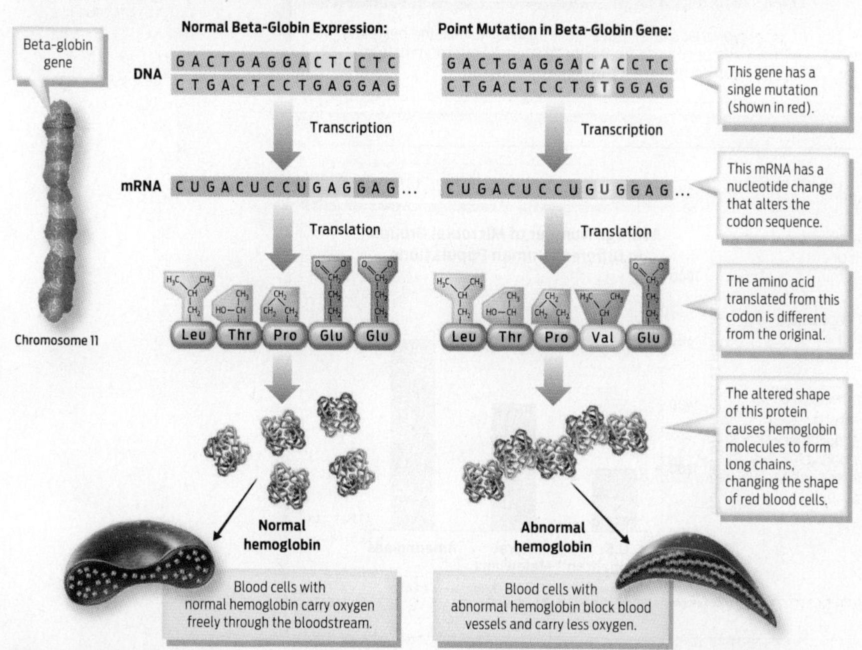

? Which of the following molecules are altered due to the sickle cell mutation in the beta-globin gene: beta-globin DNA, beta-globin mRNA, or hemoglobin protein?

Science through Infographics

Engaging and informative **Infographics** are used throughout the book to provide careful and complete explanations of the key scientific concepts. These powerful pieces of art teach students how to learn from charts, graphs, and images, and add visual appeal to the science. Animated Infographics in Achieve are accompanied by assignable quiz questions.

Each infographic includes a thought-provoking **Infographic Question** at the end to encourage students to think about the information presented in the figure.

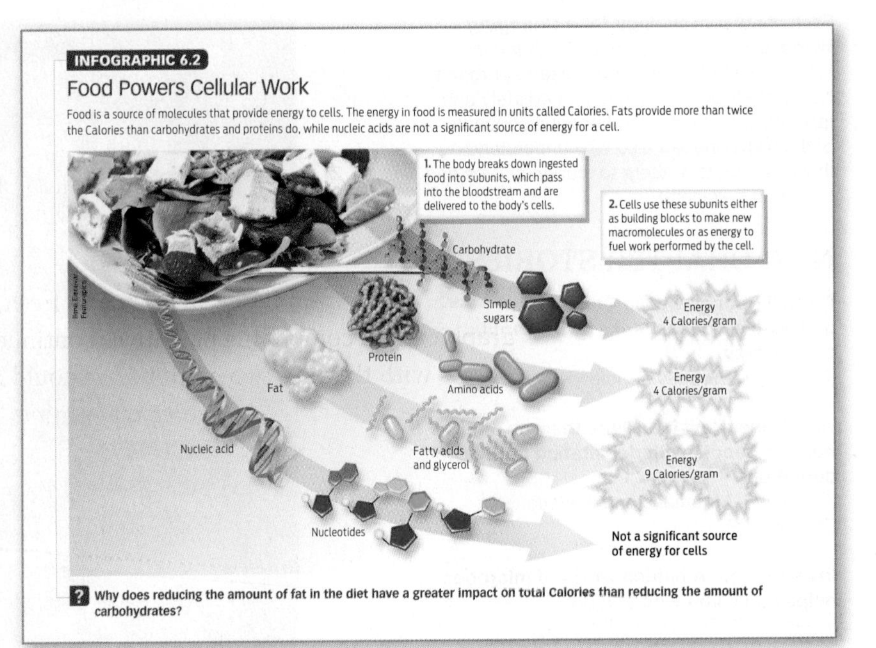

ADDITIONAL PEDAGOGICAL FEATURES

Driving Questions provide the pedagogical framework for the chapter content by prompting students to consider the questions they need to be able to answer to have a full understanding of the material.

End-of-chapter questions, written by Michèle Shuster, are framed around the chapter's **Driving Questions.** Each question set includes **Interpreting Data, Mini Case,** and **Bring It Home** questions to help students develop higher-order thinking skills. Selected questions are also assignable through Achieve.

Mini Case

Apply Your Knowledge

23. Your friend has had a virus-caused cold for 3 days and is still so stuffy and hoarse that he is hard to understand. He seems to be telling you that his doctor called in a prescription for an antibiotic for him to pick up at his pharmacy. You hope that you misunderstood him, but you realize that you heard him perfectly well.

a. Will the antibiotic help your friend's cold?

b. What are the risks to your friend if he takes the antibiotic? (Think about what might happen if he should develop a wound infection in the future.)

c. Your friend is a wrestler. What are the risks to his teammates or competitors if he takes the antibiotic?

Bring It Home

Apply Your Knowledge

24. Your roommate has been prescribed an antibiotic for bacterial pneumonia. She is feeling better and stops taking her antibiotic before finishing the prescribed dose, telling you that she will save the remainder to take the next time she becomes sick. What can you tell your roommate to convince her that this is not a good plan?

Interpreting Data

Apply Your Knowledge

20. It has been suggested that increasing rates of immune disorders, such as asthma and some food allergies, are the result of a reduced exposure to a diversity of "friendly" bacteria in the human microbiome.

a. Based on the data shown in the bar chart, which population would you predict to have the lowest rates of asthma and allergies? Note that the Amerindians are from rural communities in Venezuela.

b. If your prediction is correct, is that evidence of correlation or causation?

c. Design an epidemiological study to test the hypothesis that reduced diversity of the human microbiome is responsible for increasing rates of immune disorders. (Hint: Refer back to Chapter 1 for observational and epidemiological studies.)

Average Number of Microbial Groups in Different Human Populations

Number of groups of microbes: 1000, 1200, 1400, 1600, 1800

U.S. metropolitan | Rural Malawians | Amerindians

Data from Yatsunenko et al. 2012. Human gut microbiome viewed across age and geography. *Nature* 486: 222–228

Acknowledgments

We are thrilled to introduce the fourth edition of *Scientific American Biology for a Changing World.* In addition to updating every chapter, we have replaced four stories and added one new Milestone to keep this book at the forefront of current issues in biology. A fresh, updated design showcases our unparalleled Infographics, and compelling narratives continue to reveal how biology is relevant to daily life.

As with the first two editions, we could not have completed this project without the help of a fantastic team at Macmillan Learning. The authors would like to thank Susan Winslow, Daryl Fox, Sandy Lindelof, and the folks at Macmillan Learning and *Scientific American* for continuing to support this vision for biology education. They recognized our diverse strengths and brought us together to make this vision a reality. We continue to learn so much from one another on this challenging and rewarding professional journey, and none of us has likely worked so hard and so passionately on a project as we all have on this one.

We would like to thank all those who were interviewed and who generously contributed information for the chapters in this edition. Their stories are central to the impact that this book will have on the students we teach. They serve as authentic examples of biology in a changing world, and they bring this book to life.

A special thank you is required for our Program Manager, Shannon Howard, for her unwavering encouragement and ability to bring stable direction and support to the project. The Development Editor Susan Moran spent many hours in the pages of this book, editing the details, managing our chaos, and smoothing our rough edges, while Karen Misler managed the chaos of the schedule. We thank them for their dedication, patience, experience, and expertise. We are grateful to the skilled media team and the expertise they have brought to this edition: Jennifer Driscoll Hollis, Jennifer Compton, Mary Tibbets, and Daniel Comstock. Additional thanks goes to Learning Solutions Specialist, Jim Zubricky, whose work training and supporting book adopters helped steer the media and resources.

Many thanks to Martha Emry, Jill Hobbs, Diana Blume, Natasha Wolfe, Matthew McAdams, Christine Buese, Krystyna Borgen, Susan Wein, Jennifer Wetzel, and all the people behind the scenes at Macmillan Learning for translating our ideas into a beautiful, cohesive product. We would like to thank Eli Ensor for his outstanding work on the Infographics. We appreciate his patience with the many edits and quick timelines throughout the project. He does amazing work. We also wish to acknowledge the skillful work of designer Gary Hespenheide, who created the design and did page layout for this edition.

We would like to thank Will Moore for his enthusiasm and hard work in promoting this book within the biology education community. We thank the enthusiastic group of salespeople who connect with biology educators across the country and do a wonderful job representing this book.

We would like to thank our families and friends who have been close to us during this process. They have been our consultants, served as sounding boards about challenges, celebrated our successes, shared our passions, and supported the extended time and energy we often diverted away from them to this project. We are grateful for their patience and unending support.

And finally, a sincere thank you to our many teachers, mentors, and students over the years who have shaped our views of biology and the world, and how best to teach about one in the context of the other. You are our inspiration.

Contributing Authors, Media and Assessment

J. Michelle Cawthorn, Georgia Southern University, *Student Study Guide*
Georgianne Connell, Western Washington University, *Biology Simulations*
Pratima Darr, Georgia Gwinnett College, *Instructor Activity Guides*
Juwen DuBois, *LearningCurve Adaptive Quizzing*
Ching-Yu Huang, Virginia Commonwealth University, *Infographic Animation Assessment update*
Barbekka C. Hurtt, University of Denver, *Test Bank*
Karl Jarvis, Southern Utah University, *Lecture Slides*
Melissa Lail-Trecker, Western New England University, *LearningCurve Adaptive Quizzing, Test Bank*
Barbara Salvo, Carthage College, *Media Learning Objectives, Instructor Activity Guides*
David Steinberg, Bucknell University, *Clicker Questions*
JodiAnne Wood, *LearningCurve Adaptive Quizzing*

Faculty Advisory Board

Tonya Bates, University of North Carolina at Charlotte
Tiffany Bensen, University of Mississippi
Ruth Birch, St. Louis Community College–Florissant Valley
Lisa L. Boggs, Southwestern Oklahoma State University–Weatherford
Matthew Cook, Washburn University
Elizabeth Deimeke, Clark Atlanta University
Tamar Goulet, University of Mississippi
Ching-Yu Huang, Virginia Commonwealth University
Molly Kucera, College of DuPage
Anica D. Lee, Texas State University–San Marcos
Mark Manteuffel, St. Louis Community College–Florissant Valley
Crystal McAlvin, University of Tennessee, Knoxville
Catarino Morales, Southwest Texas Junior College
Dan Porter, Amarillo College
Jaime L. Sabel, University of Memphis
Wendy E. Sera, Houston Community College–Northwest
Elizabeth Shaffer-McCarthy, College of DuPage
Wendy Stankovich, University of Wisconsin–Platteville
Mary Staton, Wake Technical Community College
Melissa J. Walsh, University of Texas at Arlington
Heather Wilson-Ashworth, Utah Valley University

Chapter Reviewers

Natalie Abram, Kentucky Community and Technical College System
Karen Alvarez-Delfin, Miami Dade College–Hialeah
Christy C. Andrade, Gonzaga University
Kristin Andrud, University of Denver–Denver
Jennifer Bandura, University of Pennsylvania–Lockhaven
Ronald C. Barwick, King University
Greg Beaulieu, University of Victoria
Stacy Bennetts, Augusta University
Lisa L. Boggs, Southwestern Oklahoma State University–Weatherford
Greg Burchett, Riverside City College
Suparna Chatterjee, New Mexico State University
Kimberly Cline-Brown, University of Northern Iowa
Carin M. Cruz, Southwest Texas Junior College
Brian E. Dalton, Western State College–Gunnison
Pratima Chakrabarti Darr, Georgia Gwinnett College
Leigh Delaney-Tucker, University of South Alabama
Robert Dillon, Amarillo College
Amy L. S. Donovan, Columbia Basin College
Lisa M. Farmer, University of Houston
Christa Florea, Gateway Community and Technical College–New Haven
Reza Forough, Bellevue College
Amie Mazzoni Frazer, Fresno City College
Leah C. Freeman, Central New Mexico Community College
Larry M. Frolich, Miami Dade College–Wolfson
Rebecca Gehringer, Ozarks Technical Community College
Olivia L. George, University of Hawai'i—West O'ahu

Allison J. Gong, Cabrillo College
Tamar Goulet, University of Mississippi
Kristin Hennessy-McDonald, University of Memphis
Tina T. Hopper, Missouri State University–Springfield
Carina Endres Howell, Lock Haven University
Barbekka Hurtt, University of Denver
Karl Jarvis, Southern Utah University
Kristy Y. Johnson, The Citadel Military College of South Carolina
Jonathan Kniss, State University of New York–Fredonia
Melissa Lail-Trecker, Western New England University
Nancy Lane, Bellevue College
Anica D. Lee, Texas State University
Holly A. Little, Saginaw Valley State University
Qinglan Liu, Eastern Washington University
Paul Lonquich, California State University–Northridge
Jeffery Masters, University of Louisville
Crystal McAlvin, University of Tennessee, Knoxville
Wendy McBride, Coconino Community College
Kendra Merchant, Miami Dade College
Soma Mukhopadhyay, Augusta State University
Fran Norflus, Clayton State University
Matthew Nusnbaum, Georgia State University–Atlanta
Akinyele Oni, Morgan State University
M. Pantastico-Caldas, Los Angeles Trade Technical College
Dan Porter, Amarillo College–Amarillo
Samiksha Raut, University of Alabama at Birmingham
Theus Rogers, Georgia Gwinnett College
Amy Ryan, State University of New York–Plattsburgh
Jaime L. Sabel, University of Memphis
Helen D. Sarantopoulos, East Los Angeles College
Wendy E. Sera, Houston Community College–Northwest
A. K. M. Shahjahan, Baton Rouge Community College
Oliver Starks, Jefferson Community and Technical College–Downtown
John Starnes, Southcentral Kentucky Community and Technical College
C. Michael Stinson, Southside Virginia Community College
Dennis M. Toback, Broward College
Cristy Tower-Gilchrist, Emory University
Chris Trzepacz, Murray State University
Melissa J. Walsh, University of Texas at Arlington
Cholani Kumari Weebadde, Michigan State University
Sandra L. Whisler, Central Texas College
Kelly E. Williams, University of Hawai'i—West O'ahu
Charlie Willis, University of Minnesota
Heather Wilson-Ashworth, Utah Valley University
Gary T. ZeRuth, Murray State University
Brenda Zink, Northeastern Junior College

Media and Resources Reviewers

Sylvia Bonner, University of Memphis
Eunice Chin, Capilano University
Kimberly Cline-Brown, University of Northern Iowa
Brian E. Dalton, Western Colorado University
Alejandro D'Brot, Southern Methodist University
Robert Dillon, Amarillo College
Lisa M. Farmer, University of Houston
Olivia George, University of Hawai'i—West O'ahu
Jennifer Harrell, Community College of Aurora
Karl Jarvis, Southern Utah University
Melissa Lail-Trecker, Western New England University
Holly A. Little, Saginaw Valley State University
Helen Liu, Eastern Washington University
Crystal McAlvin, University of Tennessee, Knoxville
Erin McNally-Goward, Grand Valley State University
Fran Norflus, Clayton State University
Robert Pressley, Delta State University
Sylvia Rabacchi, Seton Hall University
Amy Ryan, State University of New York—Plattsburgh
Elizabeth Shaffer-McCarthy, College of DuPage
Alka Sharma, University of Memphis
Ilse Silva-Krott, Central Texas College
Heidi Tarus, Minnesota West Community and Technical College
Jennifer Trusty, Georgia State University, Perimeter College
Ramey Wauer, Central Texas College
Cholani Kumari Weebadde, Michigan State University
Daniel B. Williams, Brooklyn College, City University of New York
Kelly E. Williams, University of Hawai'i—West O'ahu
Charlie Willis, University of Minnesota

About *Scientific American*:

Scientific American is the authority on science and technology for a general audience, with coverage that explains how research changes our understanding of the world and shapes our lives. First published in 1845, *Scientific American* is the longest continuously published magazine in the United States. The magazine has published articles by more than 150 Nobel Prize-winning scientists and built a loyal following of influential and forward-thinking readers. With daily coverage in digital media, 12 issues per year of *Scientific American*, 6 issues of *Scientific American Mind* and more than 170 years of archives, the magazine continues to be the leading source for business and policy leaders, education professionals, and science enthusiasts of all kinds.

Scientific American is published by Springer Nature, a leading global research, educational, and professional publisher, home to an array of respected and trusted brands providing quality content through a range of innovative products and services.

www.ScientificAmerican.com

Contents

Unit 1: **What Is Life Made of? Chemistry, Cells, Energy**

Unit 2: **How Does Life Reproduce? Cell Division and Inheritance**

menonsstocks/E+/Getty Images

James D. Watson Collection, Cold Spring Harbor Laboratory Archives

mycteria/Shutterstock

Courtesy Jennelle Stephenson

Julie DeRoche/All Canada Photos/Alamy

PHYSIOLOGY

Unit 6: **How Do Animals Work? Physiology**

Harry Kikstra/Getty Images

Digital Property Ltd/Image Source/Corbis

Photos © Copyright Eli Lilly and Company. All Rights Reserved. Photos courtesy of Eli Lilly and Company Archives.

1 Process of Science

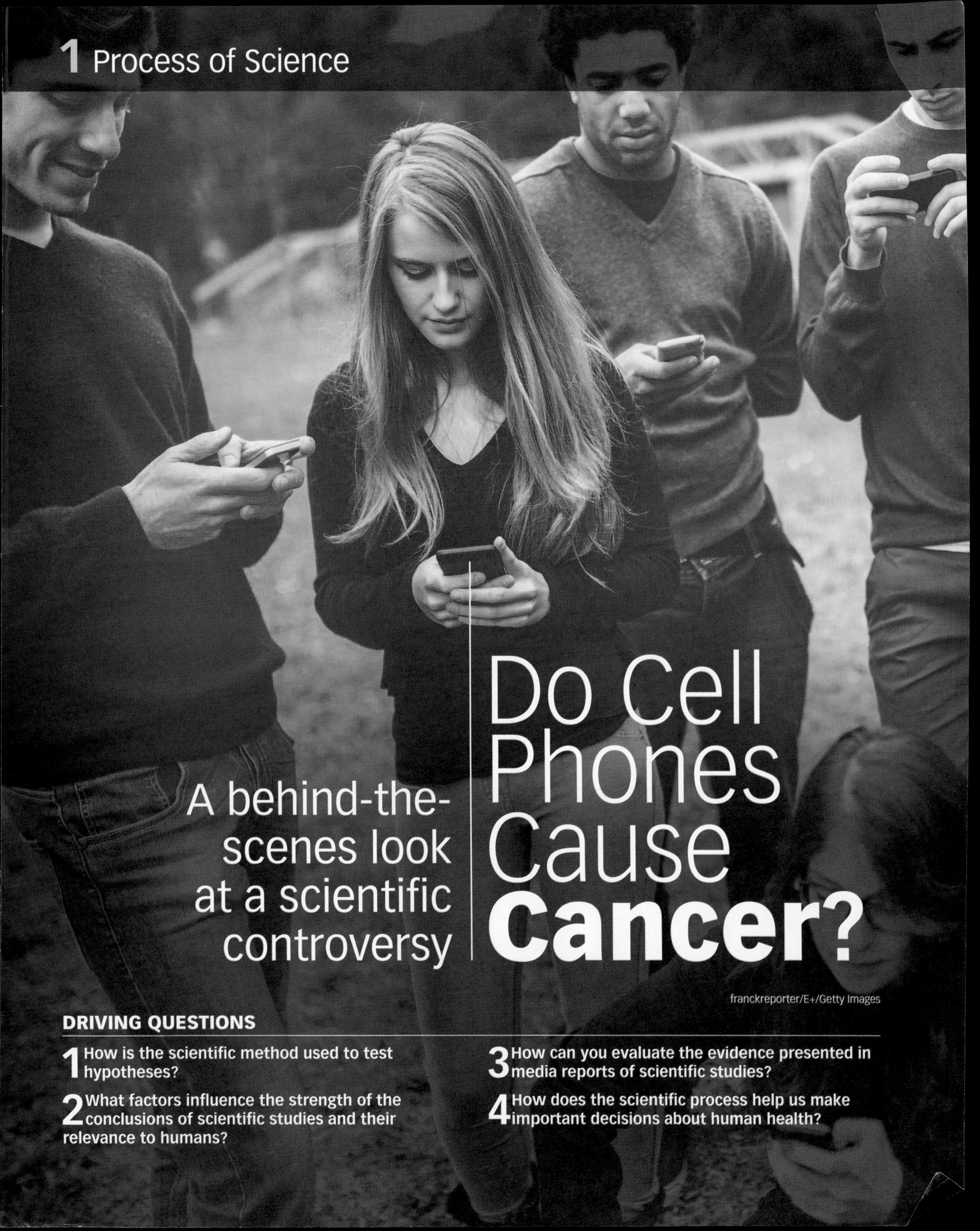

franckreporter/E+/Getty Images

Do Cell Phones Cause **Cancer?**

A behind-the-scenes look at a scientific controversy

DRIVING QUESTIONS

1. How is the scientific method used to test hypotheses?
2. What factors influence the strength of the conclusions of scientific studies and their relevance to humans?
3. How can you evaluate the evidence presented in media reports of scientific studies?
4. How does the scientific process help us make important decisions about human health?

In January 1993, David Reynard, a resident of St. Petersburg, Florida, appeared on CNN's *Larry King Live* program to raise alarm bells about the dangers of cell phone radiation. Mr. Reynard's wife, Susan, had recently died of a rare brain cancer called astrocytoma. She was a heavy cell phone user, commonly holding it up to her ear for hours at a time.

"It appeared that the tumor was in the location directly next to the antenna [where first-generation cell phones received their signal] and seemed to be growing inward from that direction," Reynard told King's audience.

Surgeons removed the tumor, but it did little good. Mrs. Reynard died in 1992 at age 33.

This photo from a research group in Italy shows rats being exposed to radiofrequency radiation equivalent to that emitted by 3G cell phone towers. The researchers will autopsy rats at the end of their lives to look for diseases such as cancer.

Mauro Fermariello/Science Source

Convinced that radiation from the cell phone had caused his wife's cancer, Mr. Reynard filed a lawsuit against the cellphone manufacturer and the carrier. It was the first legal case to allege a link between cell phones and cancer. Since then, many other such cases have been filed.

The Florida Circuit Court that heard Mr. Reynard's case ultimately found that the evidence was too weak to support his claim and dismissed the case. But that ruling did little to calm public debate about the potential dangers of cell phones, which has flared up repeatedly over the past two decades.

The latest conflagration was ignited by a study conducted by the U.S. National Toxicology Program, a division of the National Institutes of Health (NIH). Final results of the study were released in early 2018. This study entailed exposing a large number of rodents, both rats and mice, to various levels of cell phone radiation for 9 hours per day for as long as two years. To the worry of many, the researchers concluded that there is "clear evidence" that radiation from cell phones causes cancer—specifically, a type of nerve cell cancer called a schwannoma. They also found "some evidence" that this radiation causes glioma, a type of brain tumor.

Yet even these definitive-sounding results did not settle the issue once and for all in the minds of scientists. "I am unable to accept the authors' conclusions," said NIH Deputy Director Michael Lauer when the findings were released.

The director of the Food and Drug Administration's (FDA's) Center for Devices and Radiological Health, Jeffrey Shuren, chimed in as well. "We disagree with the conclusions of [the NTP's] final report," he said.

Decades of study and millions of dollars later, we might wonder why it is so difficult to answer the question of whether cell phones cause cancer. In some ways, that's just the nature of science: it can take a while to establish a firm conclusion. But this issue also illustrates why some questions are harder to answer than others.

Gut Check

▶ Evidence and scientific methods

When many people think about science, they think of a body of facts to be memorized: water boils at 100°C; the nucleus is a part of a cell. But that's not the whole story. **Science** is a *way* of knowing, a *method* of seeking answers to questions on the basis of observation and experimentation. Scientists draw conclusions from the best evidence they have at a given moment, but the process is not always easy or straightforward. Conclusions based on today's evidence may be modified in the future as scientists uncover additional data or make better observations. Improved technology may support more-refined data gathering. This new information can cast a different light on old conclusions. Science is a never-ending process.

In the quest to draw sound conclusions, scientists rely on different types of evidence. Susan Reynard's cancer is an example of **anecdotal evidence.** This kind of evidence is based on personal, often first-hand observations, but is often unreliable. Because only one or a few data points are involved, it's not possible to draw firm conclusions.

Without data from other cell phone users, their time spent holding their phones to their head, and their incidence of cancer, it's impossible to know for sure whether Susan Reynard's cancer had anything to do with her cell phone use, or whether it would have developed anyway even if she had never used a cell phone.

Anecdotal evidence can feel very persuasive, especially if it seems to corroborate what we have "seen with our own eyes" or feels right in our gut. This sort of evidence often fuels many scientific controversies you read about in the news.

A child develops signs of autism after receiving a vaccine shot, so parents think the vaccine caused the autism. A woman is hit with a baseball in her chest and later develops breast cancer, so she thinks this physical trauma led to her cancer. In many ways, the whole point of science is to use more reliable evidence to draw conclusions freed from the limitations of our own—often biased or limited—perspective **(TABLE 1.1)**.

SCIENCE
The process of using observations and experiments to draw conclusions based on evidence.

ANECDOTAL EVIDENCE
An informal observation that has not been systematically tested.

TABLE 1.1 Types of Evidence

Anecdotal evidence can be very persuasive, as it is based on personal, first-hand observations. Scientific evidence is generated using a process intended to reduce personal bias.

Anecdotal Evidence	Scientific Evidence
• Anecdotal evidence is derived from personal observation and experience. • Ideas are inspired by evidence, but are not systemically tested. • One or a few points of data are considered. • Data are analyzed from the perspective of personal experience. • Conclusions are informally vetted in conversation with family and acquaintances, based on personal experiences. • Conclusions are published on social media. • Conclusions are not verified through independent testing.	• Scientific evidence is derived from systematic observation and experimentation. • Hypotheses that are testable and falsifiable are examined. • Hundreds to thousands of data points are considered. • Data are subjected to statistical analysis. • Conclusions are formally vetted in a peer-review process by other scientists who have expertise in the area. • Conclusions are published in scientific journals. • Conclusions are supported or disproved through further scientific experimentation.

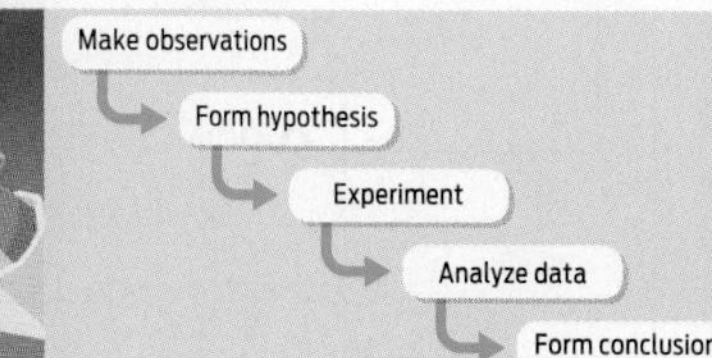

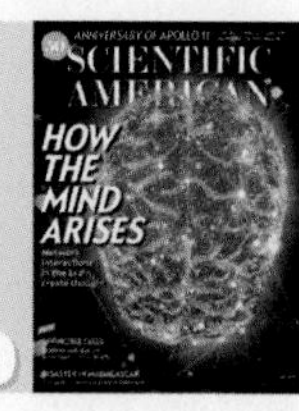

Although anecdotal evidence is often unreliable, it can nevertheless be a starting point for a scientific investigation. As an observation, it can lead you to formulate a question that you want to investigate further. For example, one question might be: Is cell phone radiation dangerous?

Typically, when scientists seek to answer a question, they begin by reviewing existing literature on the topic. This information is available in online databases of journal articles and in university libraries. As part of your background research, for example, you might read articles comparing the strength of cell phone radiation to other types of radiation we experience in our lives, such as that coming from the sun.

Generally, the information in scientific journals is reliable because it has been subjected to **peer review,** a process in which experts in the same field as the investigator review an article before it is published. The aim of peer review is to weed out sloppy research as well as overstated claims, thereby ensuring the integrity of the journal and the scientific findings it publishes. To further reduce the chance of bias, authors must declare any possible conflicts of interest and name all funding sources (for example, pharmaceutical or soft drink companies). Armed with this information, reviewers and readers can view the study with a more critical eye.

The next step of the scientific method is to generate a **hypothesis,** a possible answer to the question under investigation. For example, one hypothesis about cell phones and cancer might be that cell phone radiation damages DNA, the molecule that contains our genetic information and that is often altered in cancer. Another hypothesis might be that the more hours someone talks on the phone, the more likely that person is to get cancer.

PEER REVIEW
A process in which independent experts read scientific studies before they are published to ensure that the authors have appropriately designed and interpreted the study.

HYPOTHESIS
A tentative explanation for a scientific observation or question.

INFOGRAPHIC 1.1

Science Is a Process: Narrowing Down the Possibilities

Multiple scientists doing multiple experiments narrow down the pool of possible hypotheses. Those that are rigorously tested and supported by other experiments emerge with greatest confidence.

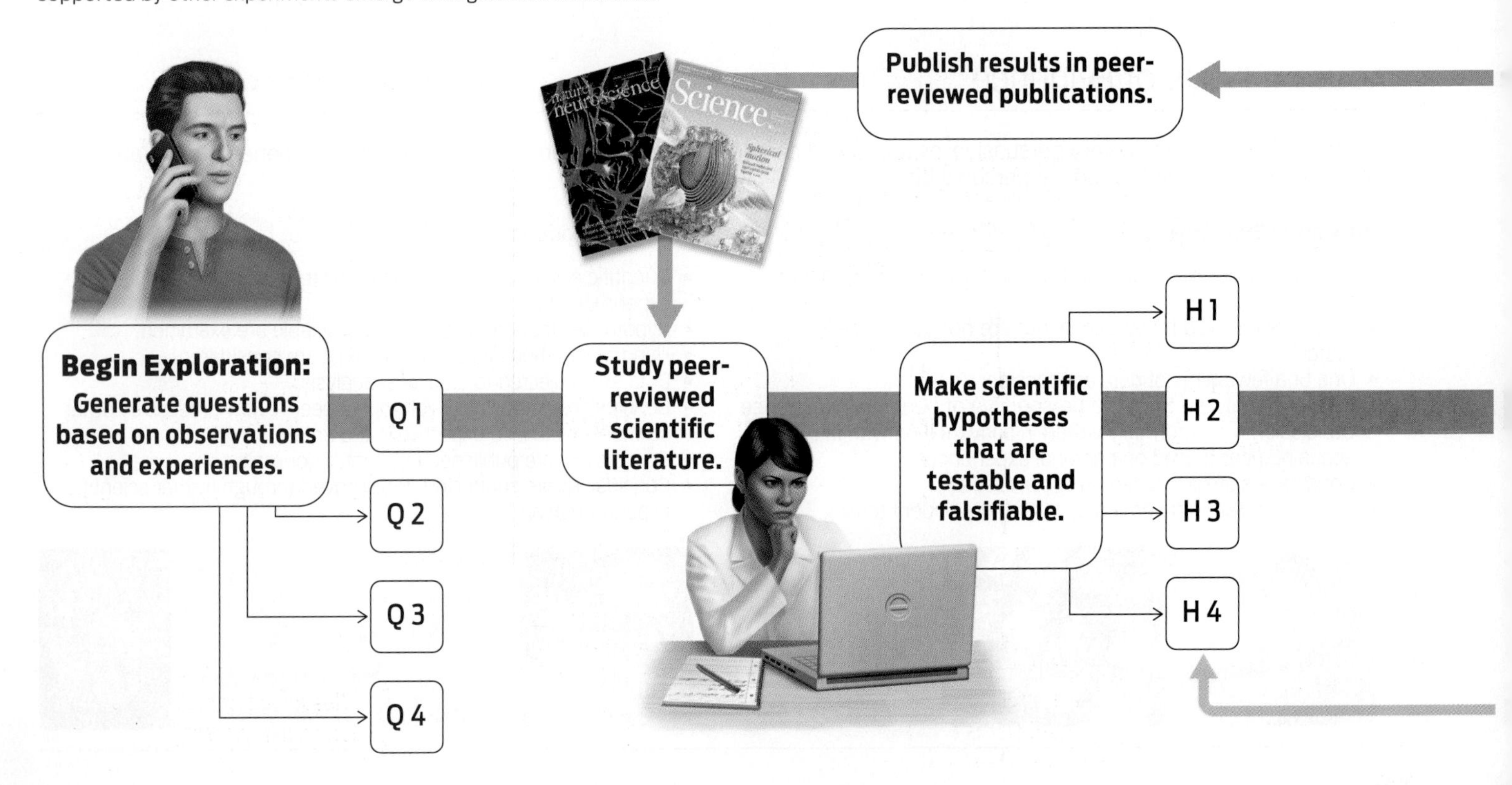

A scientific hypothesis must be **testable** and **falsifiable**—that is, it can be supported or rejected by evidence. Hypotheses that are supported by evidence gain credibility, while ones that are not supported are rejected. A scientific hypothesis can never be proven to be correct with absolute certainty, because no one can test it in every possible scenario. But it can be proven wrong—which is what it means for a hypothesis to be falsifiable. Explanations that depend on supernatural forces (God or ghosts, for example) are not scientific hypotheses because they cannot be tested or refuted.

With a clear scientific hypothesis in hand—"cell phone radiation damages DNA"—the next step is to test it, generating evidence for or against the idea. On the one hand, if a hypothesis is shown to be false—cell phone radiation does not damage DNA—the hypothesis can be rejected and removed from the list of possible answers to the original question. As the scientist, you are then forced to consider other hypotheses. On the other hand, if data support the hypothesis, it will be accepted, at least until further testing and data show otherwise **(INFOGRAPHIC 1.1)**.

Getting (Closer) to the Truth

▸ Controlled experiments to test hypotheses

The first scientific studies of cell phones and cancer began almost immediately after David Reynard took to CNN to make his case that cell phones were dangerous. Share prices for cell phone company stocks plummeted in the days following his appearance on the Larry King talk show. Fearful that the bad publicity would threaten their growing industry, a telecommunications trade group, the Cellular Telecommunications and Internet Association (CTIA), hired a scientist with a law degree, named George Carlo, to spearhead an investigation.

With a budget of $20 million, Carlo commissioned scientists from around the country to conduct studies on the potential biological effects of cell phone radiation. These investigations included tests of cell phone radiation on DNA integrity; studies of the relationship between handedness and brain tumor site; and effects of cell phone radiation on the blood–brain barrier, which helps protect the brain by keeping out potentially harmful substances in the blood. Carlo's team also

TESTABLE
Describes a hypothesis that can be supported or rejected by carefully designed experiments or observational studies.

FALSIFIABLE
Describes a hypothesis that can be ruled out by data that show the hypothesis does not explain the observation.

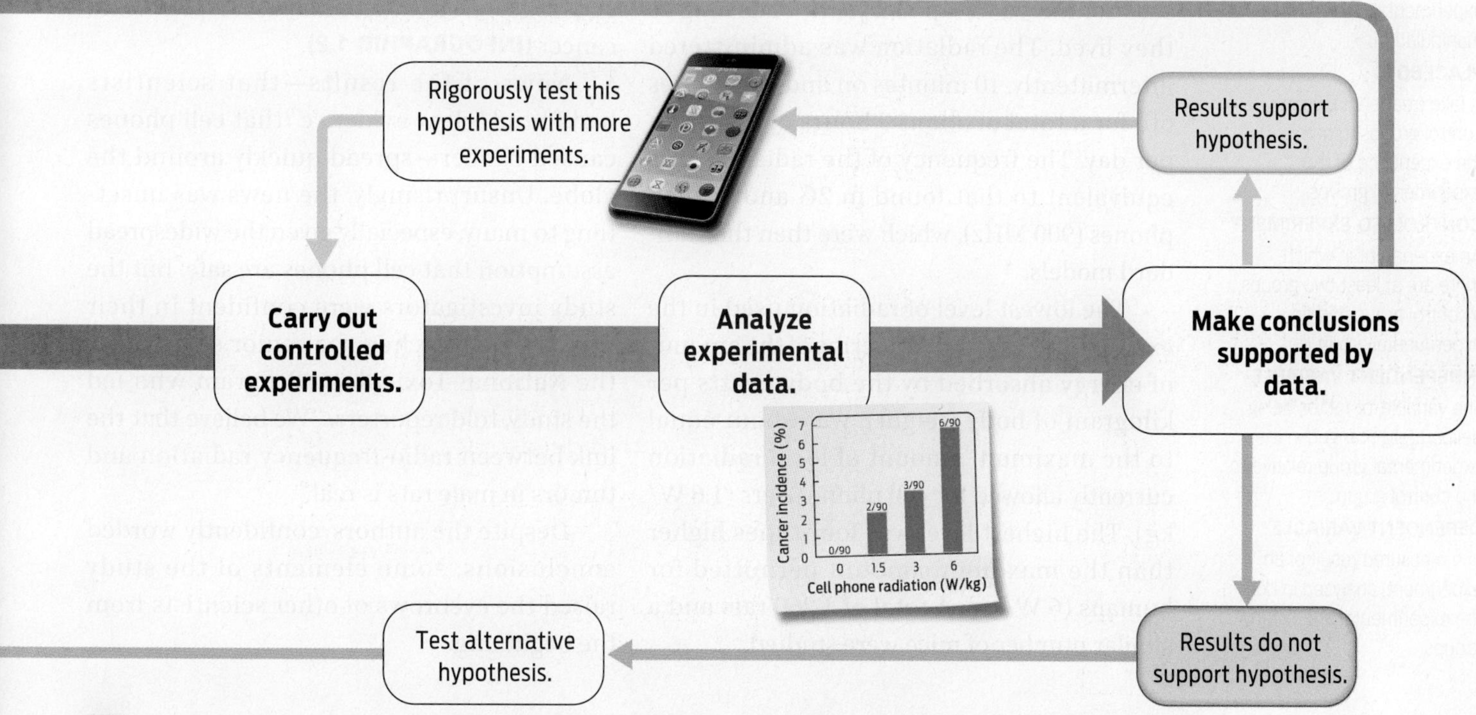

? **If the data from an experiment appear to support a hypothesis, what are the next steps?**

reviewed the existing literature on cell phone radiation.

By 1999, Carlo's investigation had uncovered some evidence that seemed to suggest cell phone radiation could cause genetic damage and increase the risk of certain brain tumors. Carlo suggested that more research was needed to confirm the findings. Industry spokespeople downplayed the potential risks, however, and insisted that their products were safe. (Their behavior would lead some journalists to accuse "Big Wireless" of covering up known cell phone dangers.)

With the safety of cell phones in question, and controversy heating up like an overused cell phone, the FDA asked scientists at the National Toxicology Program to study the issue with greater rigor. They obliged, and it was this study whose results were reported in 2018.

Let's take a close look at the study, including its design, its results, and its conclusions. The study was an **experiment,** a scientific procedure designed to test a specific hypothesis. It involved male and female rats and mice. The animals were housed in chambers and exposed to cell phone radiation starting while the rats were in their mother's womb, or when mice were 5 weeks old. The exposure continued for up to 2 years, or as long as they lived. The radiation was administered intermittently, 10 minutes on and 10 minutes off, for a total of about 9 hours of radiation per day. The frequency of the radiation was equivalent to that found in 2G and 3G cell phones (900 MHz), which were then the standard models.

The lowest level of radiation used in the experiment, measured in terms of the amount of energy absorbed by the body (watts per kilogram of body weight), was about equal to the maximum amount of local radiation currently allowed for cell phone users (1.6 W/kg). The highest level was four times higher than the maximum amount permitted for humans (6 W/kg). A total of 1,260 rats and a similar number of mice were studied.

Like all well-designed experiments, this one had certain key components. First, it included a **control group.** This group of animals did not receive radiation but was otherwise treated exactly the same as the **experimental group,** the rodents that received radiation. A control group in an experiment will sometimes receive an inactive treatment, called a **placebo.** An experiment that includes both a control group and an experimental group is called a **controlled experiment.**

In this experiment, radiation was the **independent variable**—the factor that is being changed in a deliberate way. The occurrence of cancer was the **dependent variable**—the outcome that may "depend" on radiation.

The National Toxicology Program study took more than 10 years to complete, at a cost of $30 million. The results were released in February 2018. The study found that male rats exposed to the highest level of cell phone radiation had a higher rate of occurrence, or incidence, of two types of cancer when compared with control rats. About 6% of the rats exposed to the highest level of radiation developed a schwannoma, a type of cancer, in their hearts. Roughly 3% of the exposed rats developed glioma, a type of brain cancer. None of the control rats developed either cancer **(INFOGRAPHIC 1.2).**

News of the results—that scientists had found "clear evidence" that cell phones caused cancer—spread quickly around the globe. Unsurprisingly, the news was unsettling to many, especially given the widespread assumption that cell phones are safe. But the study investigators were confident in their results. John Bucher, the senior scientist at the National Toxicology Program who led the study, told reporters: "We believe that the link between radio-frequency radiation and tumors in male rats is real."

Despite the authors' confidently worded conclusions, some elements of the study raised the eyebrows of other scientists from the beginning.

EXPERIMENT
A carefully designed test, the results of which will either support or rule out a hypothesis.

CONTROL GROUP
The group in an experiment that experiences no experimental intervention or manipulation.

EXPERIMENTAL GROUP
The group in an experiment that experiences the experimental intervention or manipulation.

PLACEBO
A fake treatment given to control groups to mimic the experience of the experimental groups.

CONTROLLED EXPERIMENT
An experiment in which there are at least two groups, a control group and an experimental group.

INDEPENDENT VARIABLE
The variable, or factor, being deliberately changed in the experimental group relative to the control group.

DEPENDENT VARIABLE
The measured result of an experiment, analyzed in both the experimental and control groups.

For one, the increased incidence of cancer occurred only in male rats. Female rats exhibited no such increase. What's more, when looking at the results for the mice, no difference in cancer incidence was seen in either males or females.

Another oddity was the fact that the rats exposed to radiation lived longer than the controls. One commenter joked that headlines should read, "Cell phone radiation makes rats live longer."

Finally, the study found an increased incidence of schwannomas in rats exposed to cell phone radiation, but only in their heart tissue. Other tissues where schwannomas can form did not have a higher incidence.

Geoffrey Kabat, a cancer epidemiologist and author of *Hyping Health Risks,* thinks that these internal inconsistencies cast doubt on the strength of the findings. "Properly interpreted, this study should actually reassure those who are worried about the cancer risks

INFOGRAPHIC 1.2

Anatomy of an Experiment

There are many ways to approach a scientific problem. Controlled experiments are one way. As illustrated here, controlled experiments have two groups—the control group and the experimental group—that differ only in the independent variable.

Hypothesis: Exposure to cell phone radiation increases the incidence of cancer.

Male and female rats

	Control Group	Experimental Group
Random placement into equivalent groups (with respect to age, gender, health, activity level, etc.)		
Independent variable (the variable that is changed in a systematic way)	**Control Treatment** No cell phone exposure	**Test Treatment** Cell phone radiation: 10 min on/ 10 min off for 9 hrs each day
Dependent variable (the variable that is measured in the experiment)	**Incidence of Cancer** Monitored over 2 years or until death	**Incidence of Cancer** Monitored over 2 years or until death
Results from data	**Male rats** had **lower** incidence of certain cancers compared to the experimental group.	**Male rats** had **higher** incidence of certain cancers compared to the control group.
Evidence-based conclusion	Cell phone radiation increases the incidence of certain cancers in male rats, but not in female rats.	

? What are the dependent and independent variables in this experiment? Which one is intentionally changed between the control group and the experimental group?

of cell phones," he says. "But the reasons are subtle and difficult to convey in a media sound bite."

Results You Can Trust

▶ Statistical significance; everyday theory versus scientific theory

So let's dig deeper. When researchers observe a difference between the experimental group and the control group—say, in cancer incidence—they need to know if this difference is "real" or if it could have occurred simply by chance. To find out, they conduct statistical tests. Various types of tests can be performed, but they share the common purpose of using math to understand probabilities and gauge **statistical significance.** A result has statistical significance when it is unlikely to have occurred by chance alone.

A greater sample size tends to reduce large swings in the data due to chance alone.

By convention, when statistical testing shows that the likelihood, or probability, of the same result occurring by chance is less than 5%, we say that the difference is "statistically significant"—that is, it is unlikely to be due to chance. This number means that if we were to carry out the same experiment 100 times, we could be confident that our result was real more than 95% of the time. Statistical tests give us confidence that our results aren't fooling us.

Note that we can never be 100% sure. At the 95% level of statistical certainty, we could expect to see our results occur maybe five times out of 100 just as a result of chance.

One thing that can contribute to statistical significance, and therefore strengthen our confidence in the results of a scientific study, is **sample size.** Sample size is the number of individuals participating in a study, or the number of times an experiment or set of observations is repeated. The larger the sample size, the less likely it is that chance will skew the results in a meaningful way.

To see why this is so, consider a familiar example: flipping a coin. We know that it should be heads half the time and tails half the time. However, if you flip a coin 10 times, you wouldn't be too surprised if you got 4 heads and 6 tails, or even 3 heads and 7 tails. However, it would be very surprising if you flipped a coin 100 times and got 30 heads and 70 tails, and even more surprising if you flipped the coin 1,000 times and got 300 heads and 700 tails. That's because random variations are averaged out with more flips. The same is true in science: a greater sample size tends to reduce large swings in the data due to chance alone **(INFOGRAPHIC 1.3).**

Now consider the cell phone data. The National Toxicology Program experiment involved 1,260 rats. That may sound like a lot, but keep in mind that this was the total number. Each individual group of rats given a particular dose of radiation contained 90 rats. The investigators found that 6% of male rats given the highest dosage of cell phone radiation developed heart schwannomas. (This was the strongest finding of the study, the one for which there was "clear" evidence.) Six percent of 90 is just 5 rats. That isn't very many. Given the small sample size of each group, and the much smaller number of rats that developed cancer, it is hard to be sure that this result is real and not due to chance.

The difficulty is compounded by what cancer epidemiologist Kabat calls the "multiple comparisons" problem. "The researchers didn't just look for the evidence of heart schwannomas. They looked for many types of cancer in many different organs. The more tissues you look in, the higher the odds that you will find something unusual simply by chance," he says.

He continues: "If you conduct 100 different analyses looking for tumors in different organs in male and female rodents, just by chance alone, you would expect to

STATISTICAL SIGNIFICANCE
A measure of confidence that the results obtained are "real" and not due to chance.

SAMPLE SIZE
The number of experimental subjects or the number of times an experiment is repeated. In human studies, sample size is the number of participants.

find tumors in 5 of those analyses." For that reason, he explains, researchers are supposed to adjust their tests of significance to take these multiple comparisons into account. The NTP researchers failed to do this.

Another unusual aspect of the study: the control rats did not live as long as the experimental rats. It's possible that if they lived longer, some of them would have developed these cancers, too. Taken together, these oddities of the study make the results hard to interpret.

Let's assume for the sake of argument that the results had been consistent. Could we then assume that the results automatically apply to humans? Definitely not.

Although laboratory animals like rats and mice are extremely useful for testing hypotheses, they are usually only a starting point for

INFOGRAPHIC 1.3

Sample Size Influences the Strength of a Conclusion

The larger the sample analyzed in an experiment, the more you can trust the conclusions.

One vs. Multiple Small Size Experiments

Results of experiments that have a small sample size can vary widely depending on which individuals are chosen and the rarity of the condition being measured. Repeating the experiment multiple times can reveal variability.

Small vs. Large Size Experiments

Experiments using small sample sizes can be inaccurate due to chance. Increasing the sample size can give results that more accurately represent the population under study.

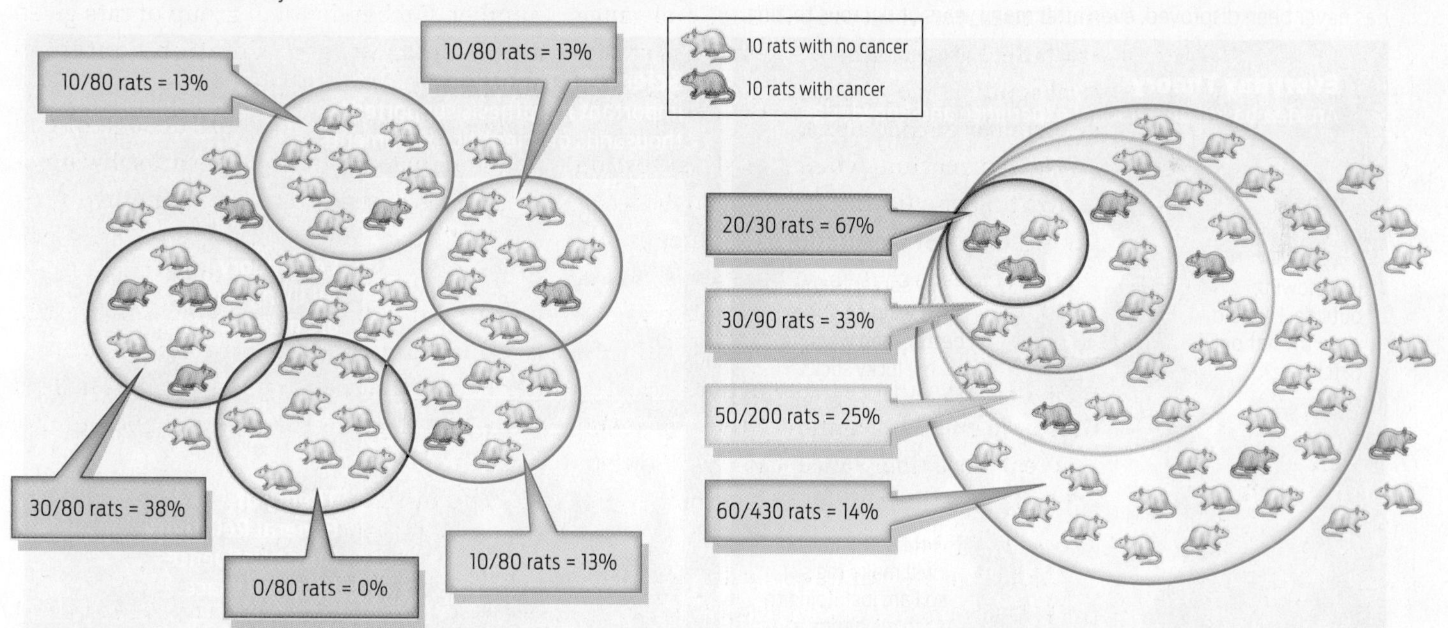

Individual small size experiments:	**0 to 30/80**	**= 0% to 38%**
Average of 5 small size experiments:	**60/400**	**= 15%**
Actual total for entire population:	**70/500**	**= 14%**

Smallest sample size experiment:	**20/30**	**= 67%**
Largest sample size experiment:	**60/430**	**= 14%**
Actual total for entire population:	**70/500**	**= 14%**

? In the small-sample experiments (left), what is the range of percentages of affected rats? How do the high and low percentages compare to the actual percentage in the total population? In the right panel, which sample gives the percentage that most closely matches the percentage in the total population?

research. Just because something is true in rodents, that does not automatically mean it will be true in people.

In addition, the rodents received radiation over their whole bodies, for a much longer time than people normally spend on their phones, and at a much higher dose. Given these differences from the radiation associated with typical cell phone use, it's not possible to extrapolate the results to the human situation.

Finally, the NTP investigation was just one study. In science, it's important for unusual findings to be confirmed by other experiments and other investigators. Before that is done, it pays to be a bit skeptical.

SCIENTIFIC THEORY
An explanation of the natural world that is supported by a large body of evidence and has never been disproved.

Generally, the more experiments that support a hypothesis, the more confident you can be that it is true. Additional experiments, conducted by other researchers, decrease the likelihood that a particular result is a fluke. They also help counter individual biases that any particular researcher or research group may have. In this way, the scientific process is self-correcting.

Hypotheses that have been confirmed many, many times may eventually attain the status of a **scientific theory,** a high point of scientific knowledge. The word "theory" in science means something very different from its everyday meaning. In casual conversation, we may say something is "just a theory,"

INFOGRAPHIC 1.4

Everyday Theory versus Scientific Theory

In daily life, people use the word "theory" to refer to an idea that explains an everyday event. In science, a theory is a hypothesis that has never been disproved, even after many years of rigorous testing.

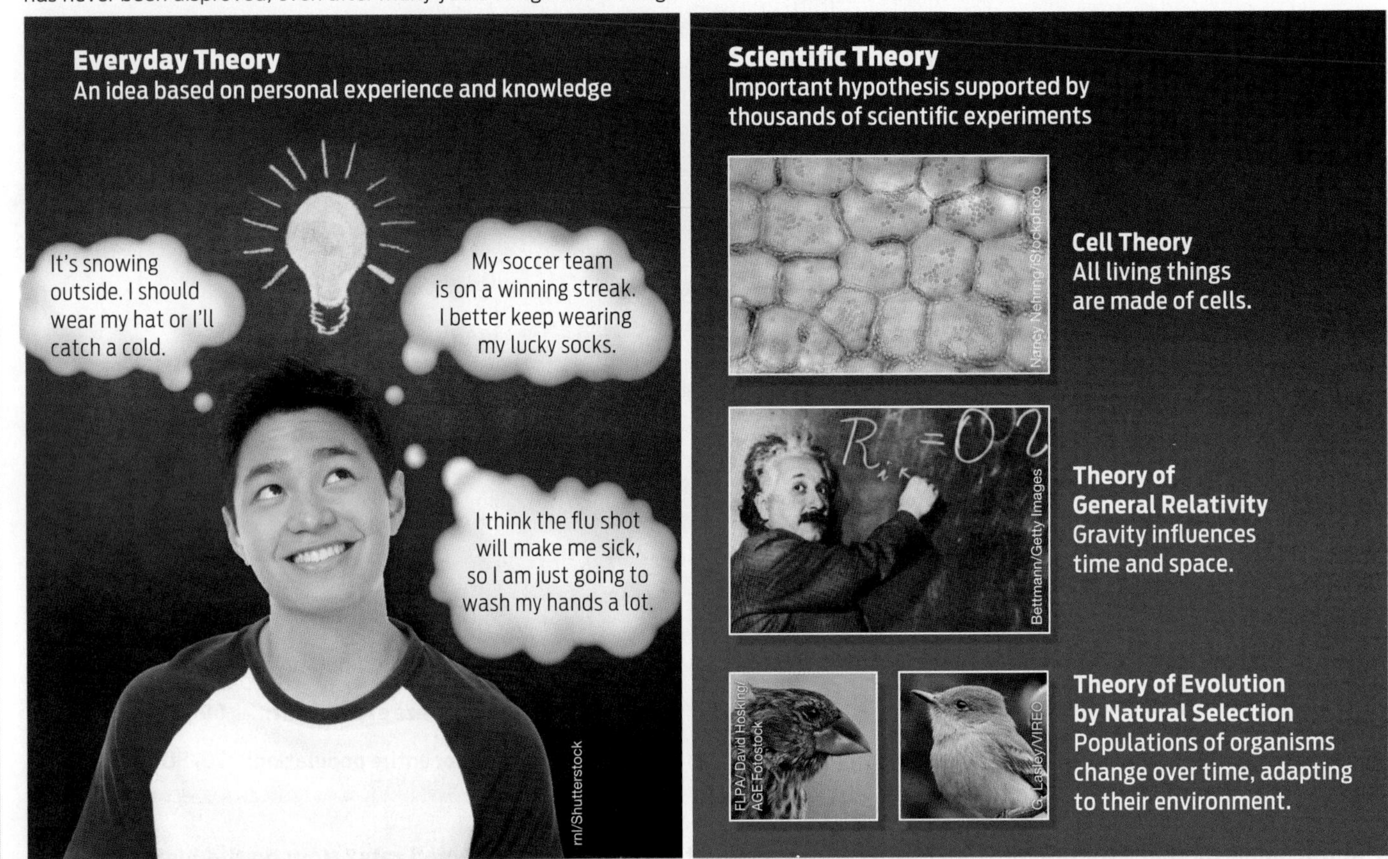

? **An article stated that experiments have been carried out to test the theory that cell phones cause cancer. What is wrong with this statement?**

meaning that it hasn't been proved. In contrast, in science, a theory is an explanation of the natural world that is supported by a large body of evidence compiled over time by numerous researchers. Far from being a fuzzy or unsubstantiated claim, a theory is a scientific explanation that has been extensively tested and has never been disproved **(INFOGRAPHIC 1.4)**.

Not All Radiation Is the Same

▸ Evaluating biological mechanisms

The word "radiation" can sound scary, conjuring up images of nuclear fallout or mutant-creating rays. In reality, radiation encompasses many commonly encountered forms of energy that we experience every day, such as sunlight and radio waves.

Like these forms of radiation, cell phone radiation is a part of the electromagnetic spectrum. It falls on the low end of intensity of the spectrum, between microwaves and FM radio waves. These waves, in the radiofrequency range, are relatively low energy. What has troubled many scientists who have investigated the cell phone claims is the lack of a plausible biological mechanism for how this type of radiation might promote cancer.

Radiation that falls at the high energy end of the electromagnetic spectrum is referred to as ionizing because it's powerful enough to directly damage molecules in cells, including the DNA that contains our genetic instructions. DNA-damaging agents—both chemicals and high-energy radiation—have indeed been shown to cause cancer. But non-ionizing radiation, of the sort emitted by cell phones, is not known to damage DNA, making it difficult to explain how it could cause cancer **(INFOGRAPHIC 1.5)**.

Heating can be a result of cell phone radiation, but not at the levels that phones produce. When your phone gets hot, that's because of the battery—not the radiation.

Of course, cell phone radiation might potentially interact with body tissues in other ways. Indeed, one recent study found that cell phone radiation could affect cells' use of glucose, a type of fuel. When people held a cell phone to their ear for 50 minutes, brain tissues on the same side of the head as the phone used more glucose than did tissues on the other side of the brain. It's unclear what relevance, if any, this finding may have to the question of whether cell phone use causes cancer.

The word "theory" in science means something very different from its everyday meaning.

No Epidemic of Brain Cancer

▸ Epidemiological studies; correlation versus causation

It can be easy to lose sight of the big picture when you're looking at the details of rat experiments. But it pays to keep that big picture in focus because some of the strongest evidence against a cell phone–cancer link comes from studies that look at the rates of diseases in population. This area of research is called **epidemiology.** Epidemiological studies can provide important big-picture context.

If cell phones truly caused brain cancer, you would expect to see an increase in the incidence of brain cancer in the population over time. But, according to the best numbers we have, the rates of brain cancer have remained constant over the past few decades.

Given how rare brain cancer is—only about 7 cases are diagnosed per 100,000 individuals in the United States each year—it should be very clear whether a potent new cancer cause has emerged in our midst. The latest data we have tend to suggest that no such cause exists. Of course, it's always possible that not enough time has passed for the effects of cell phone radiation to

EPIDEMIOLOGY
The study of patterns of disease in populations, including risk factors.

INFOGRAPHIC 1.5

The Electromagnetic Spectrum

Electromagnetic radiation is characterized by its wavelength and its frequency. Shorter wavelengths have higher frequencies and higher energy. Radiofrequency radiation (emitted by cell phones) is low frequency, in the range from 10^8 hertz (one megahertz, MHz) to 10^9 hertz (one gigahertz, GHz). The number of hertz equals the number of waves per second.

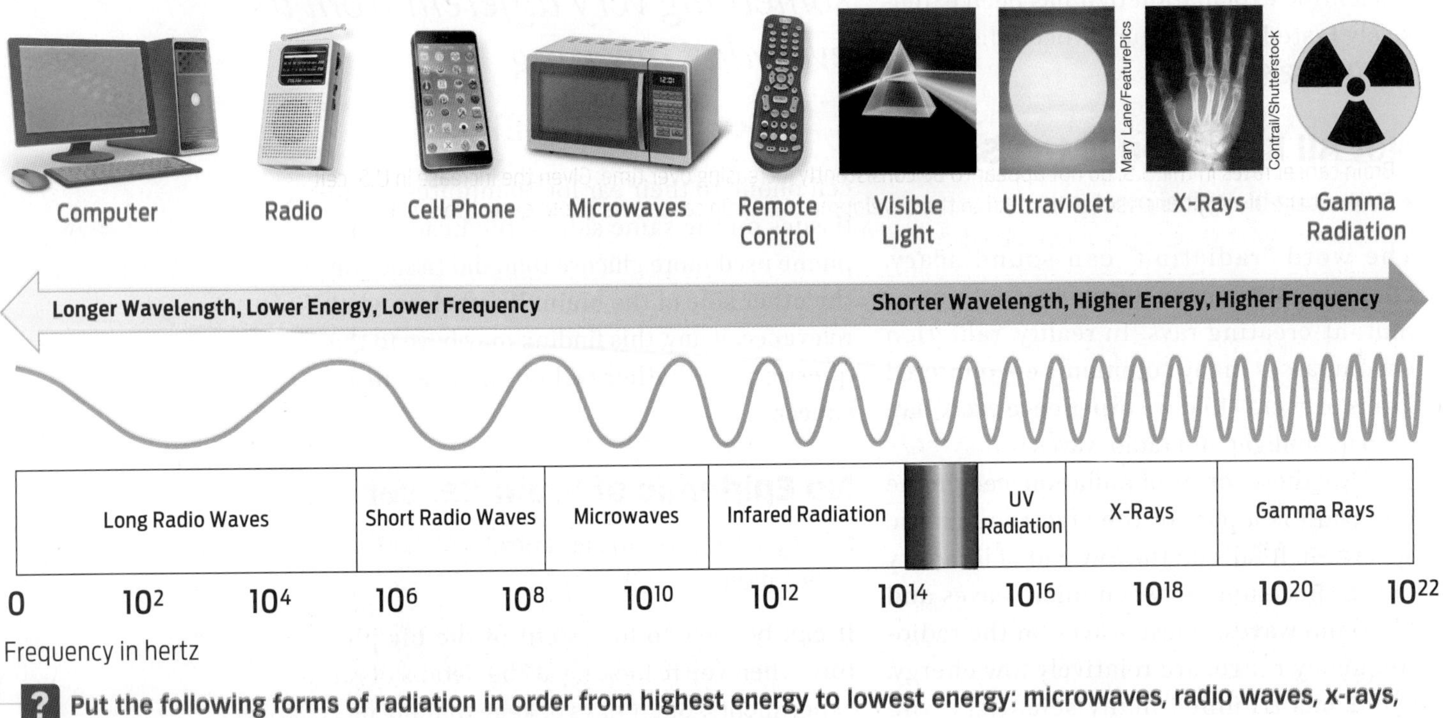

? Put the following forms of radiation in order from highest energy to lowest energy: microwaves, radio waves, x-rays, ultraviolet (UV) radiation.

have fully influenced the development of brain cancer in a way that can be measured **(INFOGRAPHIC 1.6)**.

Epidemiological studies that look at specific slices of the population have also largely come up negative. A common type of epidemiological study is called a **case-control study.** In this type of study, two groups that differ in a disease state or condition, such as having brain cancer or not, are compared based on a variable of interest. Case-control studies are a type of observational study: the researchers do not manipulate the variable, but instead merely track, or observe, it. In the cell phone–cancer situation, the variable of interest is cell phone use—so investigators compare cell phone use among people with brain tumors (cases) to cell phone use among people without brain tumors (controls). If cell phone radiation caused cancer, you would expect the group with tumors (the cases) to have used their cell phones more than the controls. Several studies of this type have been conducted.

The largest case-control study done to date, called INTERPHONE, looked at cell phone use among more than 5,000 people from 13 countries who had brain tumors and a similar group of people without cancer. Overall, the study uncovered no link between brain tumor risk and cell phone use. There did appear to be an increased risk of a type of brain tumor called a glioma in the 10% of people who used their cell phones the most. However, the investigators deemed this finding unreliable because people in this segment of respondents reported implausibly high levels of cell phone use—more than 12 hours per day.

Likewise, researchers have not found a consistent **dose–response relationship.** If cell phone radiation causes cancer, you might expect that the likelihood of cancer would increase with more time spent on the phone. In other words, a greater percentage of the

CASE-CONTROL STUDY
A type of epidemiological study to assess an association between an exposure and an outcome.

DOSE–RESPONSE RELATIONSHIP
A relationship between the amount of a chemical or physical (e.g., radiation) exposure, and the risk of a specific outcome in an exposed organism.

individuals who spend a lot of time on their cell phone would develop cancer. But no such relationship was found.

Like many case-control studies, INTERPHONE relied upon a retrospective questionnaire, in which people were asked to estimate the length and frequency of their cell phone calls. Such questionnaires have known problems, one of which is **recall bias.** This type of bias arises when a person's memory of past behavior is influenced by present circumstances. It is quite conceivable that people

RECALL BIAS
A type of error resulting from inaccurate recollection or reporting of past events.

INFOGRAPHIC 1.6

Brain Cancer Diagnoses Do Not Appear to Be Rising

Brain cancer rates in the U.S. do not appear to be consistently increasing over time. Given the increase in U.S. cell phone usage, if cell phone use contributed to the development of brain cancer, we would expect to see a corresponding increase in new brain cancer diagnoses.

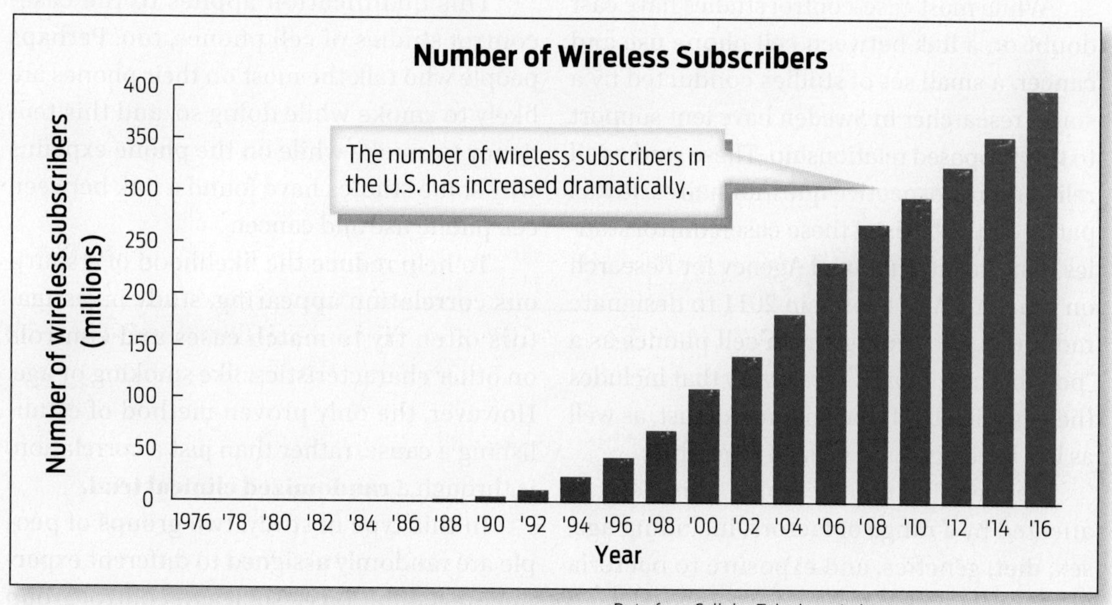

Data from Cellular Telephone Industry Association, www.ctia.org

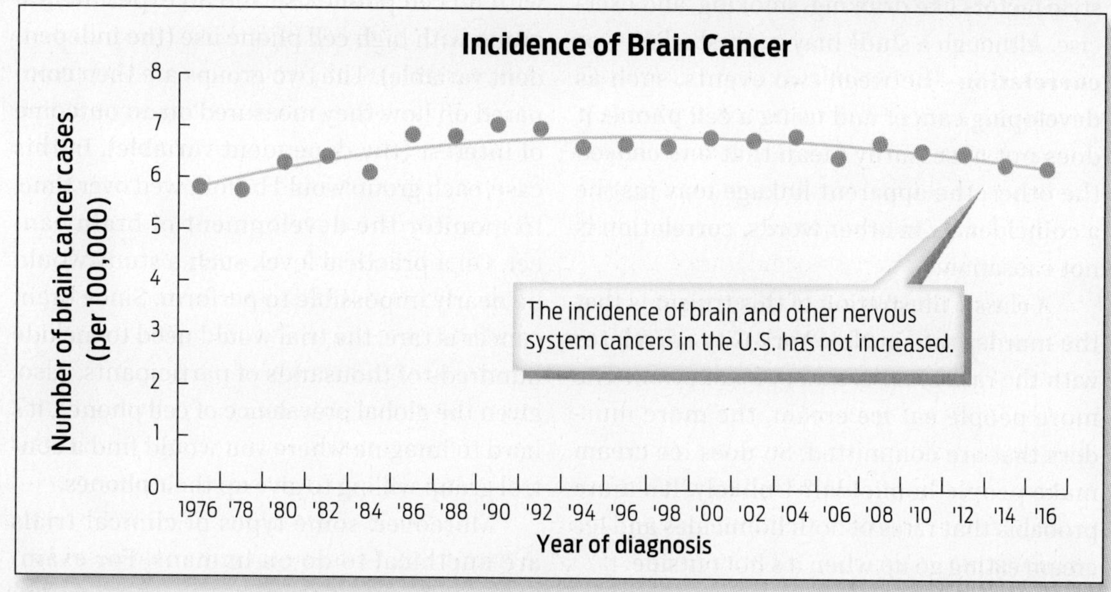

Data from SEER 9

? Given the percentage increase in cell phone subscribers between 2000 and 2006, if cell phone use was an immediate cause of brain cancer, what would we have expected for the incidence of brain cancer diagnoses in 2006?

diagnosed with brain cancers might inflate the amount of time they spent on the phone as a way to make sense of their present condition. Seeking answers or explanations for bad luck is a very human behavior.

Investigators in Denmark carried out a case-control study that did not depend on a retrospective questionnaire. Specifically, they compared all the people in Denmark who had a cell phone subscription between 1982 and 1995 (about 400,000 people) to those without a subscription to look for a possible increase in brain tumors. They didn't find one.

While most case-control studies have cast doubt on a link between cell phone use and cancer, a small set of studies conducted by a single researcher in Sweden have lent support to this proposed relationship. These studies all relied on retrospective questionnaires. It was partly on the basis of these case-control studies that the International Agency for Research on Cancer (IARC) voted in 2011 to designate radiofrequency energy from cell phones as a "possible carcinogen," a category that includes the pesticide DDT and engine exhaust, as well as hot beverages and pickled vegetables.

Cancer, like many complex diseases, is affected by a range of factors, including age, sex, diet, genetics, and exposure to bacteria and environmental chemicals, as well as lifestyle factors like drinking, smoking, and exercise. Although a study may suggest a link—or **correlation**—between two events, such as developing cancer and using a cell phone, it does not necessarily mean that one caused the other; the apparent linkage may just be a coincidence. In other words, correlation is not causation.

A classic illustration of this truism is that the murder rate in New York City correlates with the rate of ice cream consumption. The more people eat ice cream, the more murders that are committed. So does ice cream make people homicidal? Unlikely. It's more probable that rates of both homicides and ice cream eating go up when it's hot outside.

Most of the scientific results you hear about in the news, particularly in the area of nutrition, are based on correlations. One study published in 2018 found that people who buy organic food tend to have lower rates of cancer. One possible explanation for this finding is that organic food contains fewer cancer-causing pesticides, so eating these pesticide-free foods caused the lower cancer rate. But we can't say that for sure. Perhaps buying organic food correlates with other aspects of a person's life that play a role in lowering cancer risk. For example, people who eat organic foods may be more health conscious in general or exercise more.

This qualification applies to the case-control studies of cell phones, too. Perhaps people who talk the most on their phones are likely to smoke while doing so, and this tendency to smoke while on the phone explains why some studies have found a link between cell phone use and cancer.

To help reduce the likelihood of a spurious correlation appearing, study investigators often try to match cases and controls on other characteristics, like smoking or age. However, the only proven method of establishing a cause, rather than just a correlation, is through a **randomized clinical trial.**

In this type of study, two groups of people are randomly assigned to different experimental conditions—such as a control group with no cell phone use and an experimental group with high cell phone use (the independent variable). The two groups are then compared on how they measured on an outcome of interest (the dependent variable). In this case, each group would be followed over time, to monitor the development of brain cancer. On a practical level, such a study would be nearly impossible to perform. Since brain cancer is rare, the trial would need to include hundreds of thousands of participants. Also, given the global prevalence of cell phones, it's hard to imagine where you would find a control group willing to give up their phones.

Moreover, some types of clinical trials are unethical to do on humans. For example, you could not ethically conduct a trial in which participants were asked to consume a

CORRELATION
A consistent relationship between two variables.

RANDOMIZED CLINICAL TRIAL
A controlled medical experiment in which subjects are randomly chosen to receive either an experimental treatment or a standard treatment (or a placebo).

potential poison as a way to find out if that poison caused disease. It would be highly unethical to subject them to this risk **(INFOGRAPHIC 1.7)**. In this case, we would have to rely on other types of scientific evidence—for example, monitoring disease rates in people who had been unintentionally exposed to the poison, such as during

INFOGRAPHIC 1.7

How to Evaluate the Impact of Cell Phone Use

A randomized clinical trial is considered the "gold standard" to reach a convincing conclusion (for example, about cause and effect). Epidemiological studies that look back in time, asking people with and without a disease about their personal and medical histories, are subject to recall bias, and cannot confirm cause and effect.

Randomized Clinical Trial

Participants are randomly assigned to control and experimental groups. The two groups vary based on the independent variable.

Control Group:
Does not use cell phones

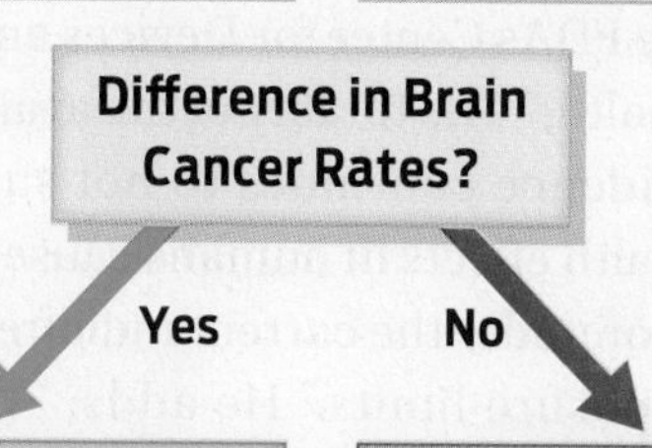

Experimental Group:
Uses cell phones

Difference in Brain Cancer Rates?

Yes → **Hypothesis Supported:** Cell phones may cause cancer

No → **Hypothesis Rejected:** Cell phones do not cause cancer

Cell Phone Research Results:

- To date, no such trial has been carried out in humans.
- Similar experiments in animals have shown no overall link between brain tumor risk and cell phone exposure.

Challenges:

- A large sample size is necessary to accurately measure rare brain tumor events.
- In the case of a human clinical trial, it is unlikely that human participants randomly assigned to the control group (no cell phone usage) would actually give up their cell phones.

Epidemiological Case-Control Study

Two otherwise comparable groups are identified that differ based on current status (e.g., brain cancer or no brain cancer). A questionnaire and/or medical or other records are used to evaluate past exposures that may explain the current disease status.

Control Group:
No brain cancer

Case Group:
Has brain cancer

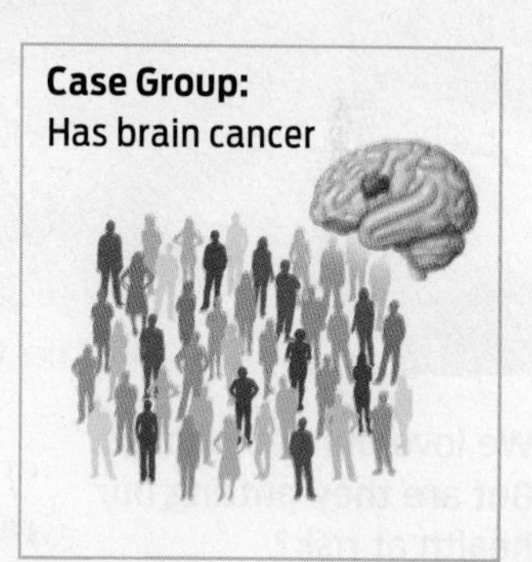

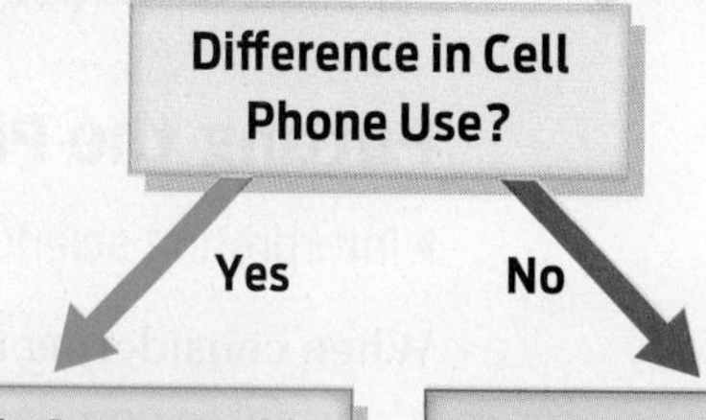

Difference in Cell Phone Use?

Yes → **Hypothesis Supported:** Cell phones may cause cancer

No → **Hypothesis Rejected:** Cell phones do not cause cancer

Cell Phone Research Results:

- To date, no conclusive link between brain tumor risk and cell phone use has been found in humans.
- No dose–response relationship between cell phone exposure and brain tumor incidence has been found.

Challenges:

- Participants have recall bias when filling out surveys to record behaviors like cell phone use.
- Medical or cell phone usage records are often incomplete.
- Even when a correlation is found in these types of studies, it is not possible to conclude that a particular behavior or exposure is directly responsible for the current condition.

? **Of the two types of experiments described here, which is most similar to the NTP rodent study? Explain your answer.**

Sidekick/Getty Images

We love our cell phones. But are they putting our health at risk?

an industrial accident, and comparing those rates to the disease rates in people with no exposure to the poison.

Putting the Pieces Together

▸ Interpreting science in the news

When considering the risk of cell phone use, it's useful to compare the collective data from a confirmed cause of cancer: cigarette smoking. When the U.S. Surgeon General issued his 1964 report on the dangers of smoking, he had a large body of evidence backing him up. Not only did epidemiological studies strongly link smoking with lung cancer, but animal studies clearly showed that the chemicals in tobacco smoke could cause tumors in laboratory animals. It was the abundance, consistency, and coherence of the data that clinched the case against tobacco. By comparison, the case against cell phones is much weaker.

You wouldn't know that from some of the media reports, however. A headline in *Mother Jones* read: "'Game-Changing' Study Links Cellphone Radiation to Cancer." "Cellphone–Cancer Link Found in Government Study," announced the *Wall Street Journal*. "Massive government study concludes cell phone radiation causes brain cancer," pronounced the website *NaturalNews*. The online health site STAT was more restrained, adding a pinch of skepticism: "Major US study links cellphone exposure to cancer—at least in rats."

It can be challenging for news headlines to accurately capture the subtleties of a scientific study. The body of the article should, in principle, add the details that the headline omits. Unfortunately, that is frequently not the case. Even if the details are there, many people do not read the whole article. The headline is satisfying enough, and pronounced with enough confidence, that people walk away believing it. The result is that the findings of scientific studies are often "lost in translation" when they move from the lab to the media **(INFOGRAPHIC 1.8)**.

So what's the upshot of all this? Do cell phones cause cancer?

According to Jeffrey Shuren, Director of the FDA's Center for Devices and Radiological Health, "The totality of the available scientific evidence continues to not support adverse health effects in humans caused by exposures at or under the current radiofrequency energy exposure limits." He adds: "We believe the existing safety limits for cell phones remain acceptable for protecting the public health."

With all the conflicting messages we read in the media about the results of scientific studies, how are ordinary people expected to judge for themselves what to believe?

There are a few strategies that people can use, including being on the lookout for certain "red flags." For example, is the newsworthy claim based on a small or large study? Was the study conducted in mice or in people? Was it funded by a company or industry that stands to gain financially from the conclusions? These and other tips can help put media claims in context (see **Tips for Interpreting Science in the News**).

Cancer epidemiologist Kabat argues that we also need to get better at distinguishing false problems from real problems. "By our nature, we are disposed to want to find external causes to account for diseases we don't understand," he says. "This helps explain the enormous appetite for stories about what are extremely low-level exposures in the environment."

But the biggest influencers on disease, he says, are things that are actually under our control: things like smoking, heavy sun

INFOGRAPHIC 1.8

From the Lab to the Media: Lost in Translation

The data as reported in peer-reviewed journals are often very complex. Scientists interpret these data in lengthy discussions, but the public receives them as isolated media headlines.

Data from scientific studies provide a large amount of technical information that's difficult to interpret

Data shown only for male rats

Table 5. Incidence of schwannomas in male Hsd: Sprague Dawley®SD® (Harlan) rats exposed to GSM- or CDMA-modulated RFR§,¶

	Control	GSM			CDMA		
	0 W/kg	1.5 W/kg	3 W/kg	6 W/kg	1.5 W/kg	3 W/kg	6 W/kg
Number examined	90	90	90	90	90	90	90
Heart‡	0*	2 (2.2%)	1 (1.1%)	5 (5.5%)	2 (2.2%)	3 (3.3%)	6 (6.6%)**
Other sites†	3 (3.3%)	1 (1.1%)	4 (4.4%)	2 (2.2%)	2 (2.2%)	1 (1.1%)	2 (2.2%)
All sites (total)	3 (3.3%)	3 (3.3%)	5 (5.5%)	7 (7.7%)	4 (4.4%)	4 (4.4%)	8 (8.8%)

§ Data presented as number of animals per group with tumors (percentage of animals per group with tumors).
¶ February 2, 2018 update adds one additional tumor in the 6 W/kg CDMA group, identified as a result of additional pathology reviews of other sites.
* Significant SAR level-dependent trend for GSM and CDMA, poly 3 test ($p < 0.05$)
** Significantly higher than controls, poly 3 test ($p < 0.05$)
‡ Historical control incidence in NTP studies: 9/699 (1.3%), range 0–6%
† Mediastinum, thymus, and fat

National Institutes of Health, National Toxicology Program Technical Report 595, 2018

Small sample size

Cancers at other sites observed even in the absence of radiation (control rats)

Confusing medical jargon

Rats were exposed to full body irradiation, not just focused near the head and brain

Heart cancers at highest rates in rats exposed to highest radiation levels

Cancers at other sites don't change with increasing radiation levels

Complicated statistical notations

Media reports greatly simplify the results for the public, often leading to misleading or sensationalized information

? **What key information may be missing from media reports of scientific studies?**

Tips for Interpreting Science in the News

Study size	How big was the study? Were 10 people included or 1,000? The larger the study, the more confident you can be in the results.
Model system	Was the study conducted in a lab dish? In mice? In humans? The greater the distance between the animal studied (model system) and humans, the less confident we should be in extrapolating results from the study to humans.
Replication	Is this the first study to report this result, or have others repeated and confirmed it?
Hyped language	Beware of terms like "cure," "breakthrough," and "revolutionize," especially in news headlines. These are often overused to generate "clicks" when the underlying data don't support the claim.
Outside input	Are other experts, who weren't involved in the study, cited in the article? Getting outside feedback can help counter any biases that the study authors might have about their own work.
Funding source	Who funded the study? Was it an industry that stands to gain financially from the results? Or was it a nonprofit foundation or government? Although funding sources don't necessarily compromise results, it pays to be skeptical of industry-funded research.
Publication reputation	Where are you reading about this study? In the *New York Times*, *Slate*, or *Chicago Tribune*? Or a personal blog? More established news organizations have departments devoted to fact checking that smaller outfits may not.

exposure, excess body weight, poor diet, lack of physical activity, and exposure to certain microorganisms. "Unfortunately, these mundane factors do not inspire anywhere near the kind of concern that is inspired by trace exposures to chemicals and radiation in the environment," he says. In the case of cell phones, people who would like to reduce their risk can, for example, use wireless headphones or keep the phone away from their bed at night.

But it's also important not to lose perspective. Cell phones have far more obvious dangers—namely, the risk that arises from texting while driving. And unlike the risk of cancer, this one is not up for debate. ■

CHAPTER 1 Summary

Driving Question 1 How is the scientific method used to test hypotheses?

- Science is a process—a way of seeking answers to questions using observations and experiments.
- A potential answer to a scientific question is called a hypothesis. Hypotheses are tested in controlled experiments or in observational studies, whose results can either support or rule out a hypothesis. Hypotheses can be supported by data but cannot be proved absolutely, as future studies may provide new findings.

Driving Question 2 What factors influence the strength of the conclusions of scientific studies and their relevance to humans?

- Scientists rely on peer-reviewed scientific reports to learn about new advances in the field. Peer review helps ensure that the scientific results are valid as well as accurately and fairly presented.
- Every experiment should have a control—a group that is identical in every way to the experimental group except for one factor: the independent variable.
- The independent variable in an experiment is the one being deliberately changed in the experimental group (e.g., radiation exposure). The dependent variable is the measured result of the experiment (e.g., cancer incidence).
- Often, a control group takes a placebo, a fake treatment that mimics the experience of the experimental group.
- The strength of the conclusions of a scientific study depends on, among other factors, the type of study carried out and the sample size.

Driving Question 3 How can you evaluate the evidence presented in media reports of scientific studies?

- Scientific theories differ from everyday theories. A scientific theory has undergone extensive testing, is supported by a significant body of evidence, and has never been disproved.
- Most of the general public relies on media reports for scientific information. Media reports are not always completely accurate in how they portray the conclusions of scientific studies.
- To understand a study properly, it is often necessary to look at how the study was designed and to evaluate the strength of the study conclusions for yourself.

Driving Question 4 How does the scientific process help us make important decisions about human health?

- Case control studies are observational studies in which cases (who have a disease of interest) are compared to controls (who do not have the disease of interest) to identify an exposure that is responsible for the disease in the cases.
- In epidemiological studies, a relationship between an independent variable (such as cell phone use) and a dependent variable (such as development of cancer) does not necessarily mean one caused the other; in other words, correlation does not equal causation.
- In a randomized clinical trial, test participants are randomly chosen to receive either a standard treatment (or a placebo) or an experimental treatment (e.g., cell phone use).

More to Explore

- PubMed.gov [Use the search bar at the top of the page to search for, say, "cell phones and cancer"; abstracts (summaries) of scientific publications will be returned.]
- National Cancer Institute, *Cell Phones and Cancer Risk*, https://www.cancer.gov/about-cancer/causes-prevention/risk/radiation/cell-phones-fact-sheet
- Kabat, Geoffrey. (2016). *Getting Risk Right: Understanding the Science of Elusive Health Risks*. New York, NY: Columbia University Press.
- Wyde, Michael, Mark Cesta, Chad Blystone, et al. (2018). *Report of Partial Findings from the National Toxicology Program Carcinogenesis Studies of Cell Phone Radiofrequency Radiation in Hsd: Sprague Dawley® SD Rats (Whole Body Exposures)*, https://www.biorxiv.org/content/10.1101/055699v3.abstract
- Oreskes, Naomi (2019). *Why Trust Science?* Princeton, NJ: Princeton University Press.

  See also: Oreskes, Naomi (2014). TED Talks: Why we should trust scientists, https://www.ted.com/talks/naomi_oreskes_why_we_should_believe_in_science/discussion

CHAPTER 1 Test Your Knowledge

Driving Question 1 How is the scientific method used to test hypotheses?

By answering the questions below and studying Infographics 1.1, 1.2, and 1.7, you should be able to generate an answer for this broader Driving Question.

Know It

1. When scientists carry out an experiment, they are testing a

a. theory.
b. question.
c. hypothesis.
d. control.
e. variable.

2. Of the following, which is the earliest step in the scientific process?

a. generating a hypothesis
b. analyzing data
c. conducting an experiment
d. drawing a conclusion
e. asking a question about an observation

3. In a controlled experiment, which group receives a placebo?

a. the experimental group
b. the control group
c. the scientist group
d. the independent group
e. all groups

4. In the rat experiments discussed in this chapter, the independent variable is ______ and the dependent variable is ______.

a. radiation; no radiation
b. cancer; radiation
c. radiation; cancer
d. cancer; no cancer
e. rats; radiation

Use It

5. You carry out a clinical trial to test whether a new drug relieves the symptoms of arthritis better than a placebo does. You have four groups of participants, all of whom have mildly painful arthritis (rated 6 on a scale of 1 to 10). Each group receives a daily pill as follows: group 1 (control), placebo; group 2, 15 mg; group 3, 25 mg; group 4, 50 mg. At the end of 2 weeks, participants in each group are asked to rate their pain on a scale of 1 to 10. What is the independent variable in this experiment?

- **a.** the amount of pain experienced at the start of the experiment
- **b.** the amount of pain experienced at the end of the experiment
- **c.** the degree to which pain symptoms changed between the start and the end of the experiment
- **d.** the drug dosage
- **e.** The independent variable could be a, b, or c.

6. You are working on an experiment to test whether a specific drug reduces the risk of breast cancer in postmenopausal women. Describe your control and experimental groups with respect to age, gender, and breast cancer status.

7. Design a randomized clinical trial to test the effects of caffeinated coffee on brain activity. Design your study so that the results will apply to as many people in as many scenarios as possible.

Driving Question 2 What factors influence the strength of the conclusions of scientific studies and their relevance to humans?

By answering the questions below and studying Infographics 1.3, 1.4, and 1.7, you should be able to generate an answer for this broader Driving Question.

Know It

8. In which of the following would you have the most confidence?

- **a.** a randomized clinical trial with 15,000 subjects
- **b.** a randomized clinical trial with 5,000 subjects
- **c.** a case-control study with 5,000 subjects
- **d.** an endorsement of a product by a movie star
- **e.** a report on a study presented by a news organization

9. What is the importance of statistical tests?

- **a.** They can reveal whether the data have been fabricated.
- **b.** They can be used to support or reject the hypothesis.
- **c.** They can be used to determine whether any observed differences between two groups are real or a result of chance.
- **d.** all of the above
- **e.** both b and c

10. Can an epidemiologist who finds a correlation between the use of tanning beds and melanoma (an aggressive form of skin cancer) in college-age women conclude that tanning beds cause skin cancer?

- **a.** yes, as long as the correlation was statistically significant
- **b.** yes, but only for college-age women
- **c.** yes, but only melanoma skin cancer, not other forms of skin cancer
- **d.** no; the study would have to be done with a wider range of participants (males and females of different ages) before it can be concluded that tanning beds cause melanoma
- **e.** no; correlation is not proof of causation

Use It

11. You carry out a clinical trial to test whether a new drug relieves the symptoms of arthritis better than a placebo does. You have four groups of participants, all of whom have mildly painful arthritis (rated 6 on a scale of 1 to 10). Each group receives a daily pill as follows: group 1 (control), placebo; group 2, 15 mg; group 3, 25 mg; group 4, 50 mg. At the end of 2 weeks, participants in each group are asked to rate their pain on a scale of 1 to 10. The mean pain rating of the participants was 6.5 for the placebo, 6.0 for 15 mg of the drug, 4.5 for 25 mg of the drug, and 4.5 for 50 mg of the drug. What is your next step?

- **a.** Invest in the drug company.
- **b.** Conclude that the drug relieves arthritis pain.
- **c.** Conduct a statistical test to determine if the differences are significant.
- **d.** Conclude that the drug doesn't work very well (even the placebo group went down on the pain scale, and there was no difference in results between doses of 25 mg and 50 mg of the drug).
- **e.** both a and b

12. Looking at Infographic 1.3 (Sample Size Influences the Strength of a Conclusion), the left panel shows several small studies (each with 80 rats drawn from a larger population). Explain why each of these experiments has a different percentage of rats with cancer. Looking at the right side of Infographic 1.3, explain why an experiment with a larger sample size gives a result that is very similar to that for the entire population.

Bring It Home

Apply Your Knowledge

13. From what you have read in this chapter, which experiments, if any, will cause you to alter your cell phone use? Explain why you will or will not alter your cell phone use, referring to specific studies.

Interpreting Data

Apply Your Knowledge

14. Most statistical tests report a *p* value that determines whether the results are statistically significant (i.e., not produced by chance). Usually the cutoff for *p* values is 0.05: if the *p* value is less than 0.05, the results are considered to be statistically significant. The graph below shows data from a 2012 study published in the *New England Journal of Medicine*. The study examined the impact of the drug tofacitinib on ulcerative colitis. From the data shown, what dose(s) of tofacitinib is/are significantly better than the placebo in treating ulcerative colitis?

a. 0.5 mg
b. 3 mg
c. 10 mg
d. 15 mg
e. both 10 mg and 15 mg
f. All doses are more effective than the placebo.

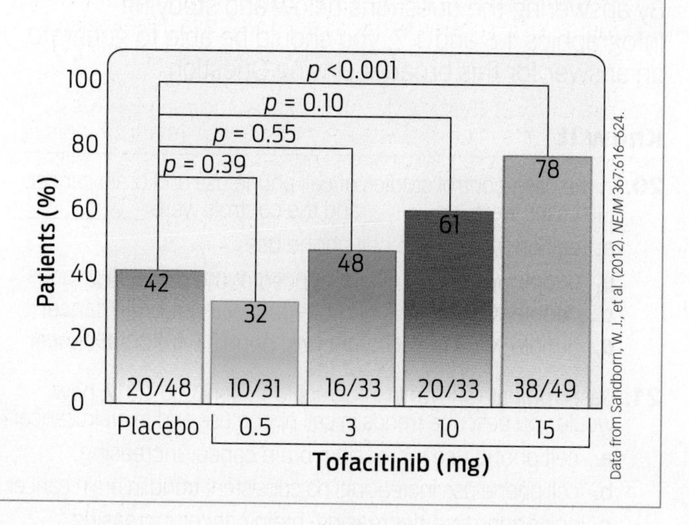

Driving Question 3 How can you evaluate the evidence presented in media reports of scientific studies?

By answering the questions below and studying Infographics 1.7 and 1.8, you should be able to generate an answer for this broader Driving Question.

Know It

15. You hear a news report about a new asthma treatment. What would you want to know before you asked your doctor if this treatment was right for you?

a. Was the drug tested in a randomized clinical trial?
b. How many participants were in the trial?
c. Was there a significant difference between the effect of the new drug and the treatment used in the control group?
d. Did any of the researchers have financial ties to the manufacturer of the new asthma drug?
e. all of the above

16. You are listening to a news report that claims a new study has found convincing evidence that a particular weight-loss product is much more effective than diet and exercise. What can you infer about the "convincing evidence" in this case?

a. It agrees with the hypothesis.
b. Statistical tests showed significantly more weight loss in the participants who used the weight-loss product than those who relied on diet and exercise.
c. All the participants lost at least 10 pounds.
d. Only the participants who used the weight-loss product lost weight.
e. The participants who used the weight-loss product lost an average of 3 pounds, while the participants who used diet and exercise lost an average of 2 pounds.

Use It

17. How can two different studies investigating the same thing (e.g., the relationship, if any, between caffeinated coffee and memory) come to different conclusions?

a. They may have had different sample sizes.
b. They may have used different types of participants (e.g., participants of different ages or professions).
c. They may have used different amounts of caffeine.
d. They may have evaluated memory differently (e.g., long-term versus short-term memory).
e. all of the above

18. A scientist who reads an article in a scientific or medical journal can be confident that the report has been peer reviewed.

a. What is a "peer-reviewed" report? Is an article in a daily newspaper a peer-reviewed article?
b. What is the role of a peer reviewer of a scientific article?
c. Why do scientists place so much value on the peer-review process?

19. You may have seen advertisements on television that show beautiful people with clear skin who claim that a specific skin care product is "scientifically proven" to reduce acne. The product reportedly gave these people glowing, clear skin.

a. Is their testimony alone strong enough evidence for you to act on? Why or why not?
b. What kind of scientific evidence would persuade you to spend money on this product? Explain your answer.

Driving Question 4 How does the scientific process help us make important decisions about human health?

By answering the questions below and studying Infographics 1.6 and 1.7, you should be able to generate an answer for this broader Driving Question.

Know It

20. In the case-control studies of cell phone use and brain cancer, the cases were ________ and the controls were ___________.

a. cell phone use; no cell phone use
b. people with brain cancer; people who used cell phones
c. people with brain cancer; people without brain cancer
d. people who used cell phones; people with brain cancer

21. Based on data shown in Infographic 1.6 for 2000–2017, how would you describe trends in cell phone use and in brain cancer?

a. cell phone use increasing; brain cancer increasing
b. cell phone use increasing; no consistent trend in brain cancer
c. cell phone use decreasing; brain cancer increasing
d. cell phone use decreasing; brain cancer decreasing
e. cell phone use steady; no consistent trend in brain cancer

"If you're worried about cell phone microwaves, stick a piece of popcorn in your ear. When it pops, it's time to hang up."

Use It

22. Based on the data shown in Infographic 1.6, does there appear to be a correlation between cell phone use and brain cancer? Explain your answer.

23. Following the prompts below, design a clinical trial to test the impact of a particular intervention on a specific aspect of human health. You will need to use everything you have learned in this chapter to do this.

a. From scientific articles or press releases from health organizations you have read, or from your own experiences, what observation(s) can you start with?
b. Do some reading and research to generate a testable hypothesis.
c. Design the trial. Consider sample size, whether you will use a placebo, and possible independent and dependent variables.

Apply Your Knowledge

Mini Case

24. There are many misconceptions about breast cancer and its causes. In the late 1990s, rumors suggested that antiperspirants cause breast cancer. Even today, some retail sources continue to offer alternative underarm hygiene products that claim to reduce the risk of breast cancer. One viral e-mail claimed that by blocking perspiration, antiperspirants prevent the body from purging toxins, instead forcing the body to store the toxins in lymph nodes in the underarm area near breast tissue. The e-mail stated that men were less likely to develop breast cancer from antiperspirants because their underarm hair trapped most of the product away from direct contact with skin. And as men are less likely to shave their underarms, they are less likely to have shaving nicks through which antiperspirants can enter the body.

a. Read the abstracts of the two articles for which URLs are provided below.

Darbre, 2005: http://is.gd/pPLxwZ

Darbre et al., 2004: https://www.ncbi.nlm.nih.gov/pubmed/14745841

From the abstracts, and from any other investigation you do, name the components of underarm deodorants and antiperspirants that have been identified as possible culprits in causing breast cancer.

b. Briefly comment on the strengths and weaknesses of each study (consider sample size, control groups, and overall study design).

c. Based on what you read in the abstracts and any other research you do (cite any additional reliable sources that you consulted), do you think that use of antiperspirants or deodorants, or both, is a consistent risk factor for breast cancer? Has your opinion about underarm hygiene changed? Explain how and why your opinion has either changed or remained consistent, referring to the abstracts that you have reviewed.

Mission *to* Mars

Prospecting for life on the red planet

Stocktrek Images/Getty Images

DRIVING QUESTIONS

1. How do we define life?
2. How is matter organized into the molecules of living organisms?
3. What is the basic structural unit of life, and what are its properties?
4. Why is water so important for life and living organisms?

"For the landing to succeed, hundreds of events will need to go right, many with split-second timing and all controlled autonomously by the spacecraft."

—Pete Theisinger

Seven minutes of terror. That's how NASA scientists described the anticipated final moments of their 2012 Mars landing. In that harrowing interval, a speeding spacecraft carrying a massive cargo would need to slow from about 13,200 mph to less than 2 mph as it dropped like a stone through the thin Martian atmosphere.

"Those seven minutes are the most challenging part of this entire mission," said Pete Theisinger, project manager at NASA. "For the landing to succeed, hundreds of events will need to go right, many with split-second timing and all controlled autonomously by the spacecraft."

The most frightening part was the last few moments of the descent, when—if all went well—the spacecraft would release its cargo—a 1-ton, SUV-size rover—down a floating "sky crane," essentially three nylon cables suspended from a rocket-powered backpack. The maneuver had never been attempted before, and even NASA's own scientists had their doubts about it, dubbing it "rover on a rope."

***Curiosity* is lowered to Mars down the floating sky crane.**
NASA/JPL-Caltech

But it worked. On August 6, 2012, at 1:32 A.M. Eastern Standard Time, NASA's *Mars Science Laboratory,* also known as *Curiosity,* landed successfully on the surface of the red planet. "Touchdown confirmed," announced NASA engineer Allen Chen. "We're safe on Mars!"

News of the successful landing sent NASA's Jet Propulsion Laboratory into loud cheers as people hugged and high-fived one another. Moments later, *Curiosity* began sending back grainy black-and-white images of its landing site to a planet full of eager witnesses.

Curiosity isn't the first rover to explore Mars. It follows on the heels of *Spirit* and *Opportunity,* which arrived in 2004. Nor will it be the last. NASA is already gearing up for a 2020 mission, which will send another rover to the red planet. But *Curiosity* represents a significant milestone. With an entire laboratory built into its sleek frame and the ability to travel for miles over rocky terrain, *Curiosity* is unquestionably the most technologically advanced rover to date.

It's also the most socially connected. Shortly after touching down, *Curiosity* began tweeting news of its progress to its more than 1 million followers, even checking into Mars on Foursquare: "One check-in closer to being Mayor of Mars!" it chirped.

The purpose of *Curiosity*'s daring trip to Mars, which is part of NASA's larger Mars

Exploration Program, is to find out whether the planet could have once supported life—and might support it still. NASA calls the mission the "prospecting" stage of its search for life on Mars. What *Curiosity* discovers will help scientists begin to answer some fundamental questions not only about life on Mars, but also about life on Earth. Questions like: How did life begin? Is there more than one type of life? Could life have arrived here on a meteorite from outer space?

These are big questions, but NASA's high-tech rover is nothing if not curious.

BRIAN VAN DER BRUG/United Press International (UPI)/Newscom

Relief and excitement at NASA's Jet Propulsion Laboratory when *Curiosity* landed safely.

Will We Know It When We See It?

▸ Challenges of defining "life"

Answering the question of whether there is life on Mars seems as if it should be pretty straightforward: look and see if anything is growing, or running around, or texting. By these measures, clearly, there is no life on Mars. The earliest pictures of Mars obtained by NASA in 1965 revealed a dry, rocky landscape—more reminiscent of our lifeless moon than the lush, blue marble we call home. But what if Mars harbors microscopic life, invisible to the naked eye? Could life be lurking in the Martian soil? And what if life looks different on Mars than it does on Earth—would we ever recognize it?

Biology is the study of life, so naturally it's important for biologists to know which things fall under that heading. Mosses, for example, are living—but the rocks they grow on are not. Yet coming up with a strict definition of life can be tricky.

Biologists generally agree that—on Earth, at least—living things have five life-defining properties. First, living things grow—they increase in size or cell number. Second, they reproduce, by producing offspring that are similar, if not quite identical, to themselves. Third, they maintain a relatively stable internal environment in the face of changing external circumstances—producing heat when they're cold, for example—a phenomenon called **homeostasis.** Fourth, they sense and respond to their environment, as when a plant grows toward sunlight. Fifth, to carry out these various activities, they obtain and use **energy,** the power to do work **(INFOGRAPHIC 2.1).**

If scientists found an alien being with all five of these properties, they could make a good case for having found life. But what if the alien specimen had only some of these properties? Would it be alive? No one can say for sure, and we won't know until we have a candidate life form to consider.

Even on Earth, our definitions of life don't always hold. For example, a mule (the offspring of a female horse and a male donkey) is clearly alive, but it is sterile and cannot reproduce. Likewise, fire grows, reproduces, and uses energy, but most people would not say fire is alive. By what criteria, then, should NASA evaluate evidence of life, past or present, on Mars? One option—and the one that *Curiosity* is using—is chemistry.

BIOLOGY
The study of life.

HOMEOSTASIS
The maintenance of a relatively constant internal environment.

ENERGY
The ability to do work. Living organisms obtain energy either directly from sunlight (through photosynthesis) or from food they consume.

INFOGRAPHIC 2.1

Properties That Define Life

Living organisms share five properties that define them as "alive."

All living organisms...

Africa Studio/Shutterstock

Grow
For unicellular (one-celled) organisms, growth is an increase in cell size before reproduction. For multicellular organisms, growth refers to an increase in an organism's size as the number of cells making up the organism increases.

JUNIORS BILDARCHIV/AGE Fotostock

Reproduce
Reproduction is the process of producing new organisms. Offspring are similar, but not necessarily identical, to their parents in general structure, function, and properties.

globestock/iStockphoto

Maintain Homeostasis
Organisms maintain a stable internal environment, even when the external environment changes.

© Kazuo Ogawa/AGE Fotostock

Sense and Respond to Stimuli
Organisms respond to stimuli in many ways. For example, they may move toward a food source or move away from a threatening predator.

Andres Rodriguez/AGE Fotostock

Obtain and Use Energy
All living organisms require an input of energy to power their activities. Organisms obtain energy from food (which they either produce themselves by photosynthesis or consume from the environment). Chemical reactions convert that energy into usable forms. The sum total of all these reactions is metabolism.

? **Consider an avocado tree. How does it demonstrate each of the five properties that define it as living?**

Curious about Chemistry

▸ Elements and atoms

In his lab at NASA's Goddard Space Flight Center in Greenbelt, Maryland, chemist Paul Mahaffy monitors a device the size of a microwave oven, hairy with cords and wires. When it boots up, the device declares: "Sam I am, I am Sam!" in playful homage to Dr. Seuss. SAM stands for "Sample Analysis on Mars." It is an exact replica of the tool that *Curiosity* is using to check for life's chemical building blocks. Mahaffy is SAM's principal investigator.

SAM is mounted on *Curiosity* like a backpack. If the powerful cameras on board *Curiosity* are its eyes, then SAM is its extremely sensitive nose. On Mars, says Mahaffy, SAM is doing a lot of sniffing.

NASA

Meet SAM, *Curiosity*'s organic molecule detector.

The sniffing takes place in a series of interconnected chambers. Once *Curiosity* scoops up a sample of rock or dirt, it is loaded into SAM, where it is baked to a high temperature. The gases given off from this high-tech oven are then analyzed to determine their precise chemical makeup. The whole operation takes about a day, but interpreting the results can take much longer. With SAM's assistance, *Curiosity* is analyzing the chemical components of the rocks, soil, and atmosphere it encounters on Mars.

Chemical composition is a reasonable stand-in marker for life because of how consistent life's recipe is on Earth. All life we know of—from amoebas to zebras—uses the same basic chemical recipe: a stew of carbon-based ingredients floating in a broth of water. Carbon (C) is one of 118 known elements in the universe. **Elements** are substances that cannot be broken down by chemical reactions into smaller substances. They are themselves considered the fundamental components of anything that takes up space and has mass—the **matter** in the universe. Elements make up both living and nonliving things. The rocky surface of Mars, for example, appears red because of an abundance of the element iron (Fe) that has long since rusted. Your sweat contains the elements sodium and chloride, which is why it tastes salty.

The smallest unit of an element that still retains the property of that element is an **atom.** Atoms are made up of three types of subatomic particles: positively charged **protons,** negatively charged **electrons,** and neutral **neutrons.** The relatively heavy and large protons and neutrons are packed into the atom's dense core, or **nucleus,** while the light and tiny electrons orbit it in electron shells defined by their energy. The number of subatomic particles in an atom determines its physical and chemical properties—its heaviness, for example, and how it reacts with atoms of other elements. The number of protons in an atom is called the **atomic number;** by convention, this number determines the atom's identity. A carbon atom, for example, is carbon because it has six protons. It also has six electrons and six neutrons **(INFOGRAPHIC 2.2).**

Carbon is the fourth most common element in the universe, and the second most common element in the human body. In fact, just six elements make up the bulk of your body: oxygen (65%), carbon (18%), hydrogen (10%), nitrogen (3%), calcium (1.5%), and phosphorus (1%).

ELEMENT
A pure substance that cannot be chemically broken down; each element is made up of and defined by a single type of atom.

MATTER
Anything that takes up space and has mass.

ATOM
The smallest unit of an element that still retains the property of the element.

PROTON
A positively charged subatomic particle in the nucleus of an atom.

ELECTRON
A negatively charged subatomic particle with negligible mass in an atom.

NEUTRON
An electrically uncharged subatomic particle in the nucleus of an atom.

NUCLEUS
The dense core of an atom.

ATOMIC NUMBER
The number of protons in an atom, which determines the atom's identity.

INFOGRAPHIC 2.2

All Matter Is Made of Elements

Elements are the fundamental building blocks of matter. The periodic table of elements represents all known elements in the universe. Each element is placed in order on the table by its atomic number, the number of protons found in the nucleus of its corresponding atom.

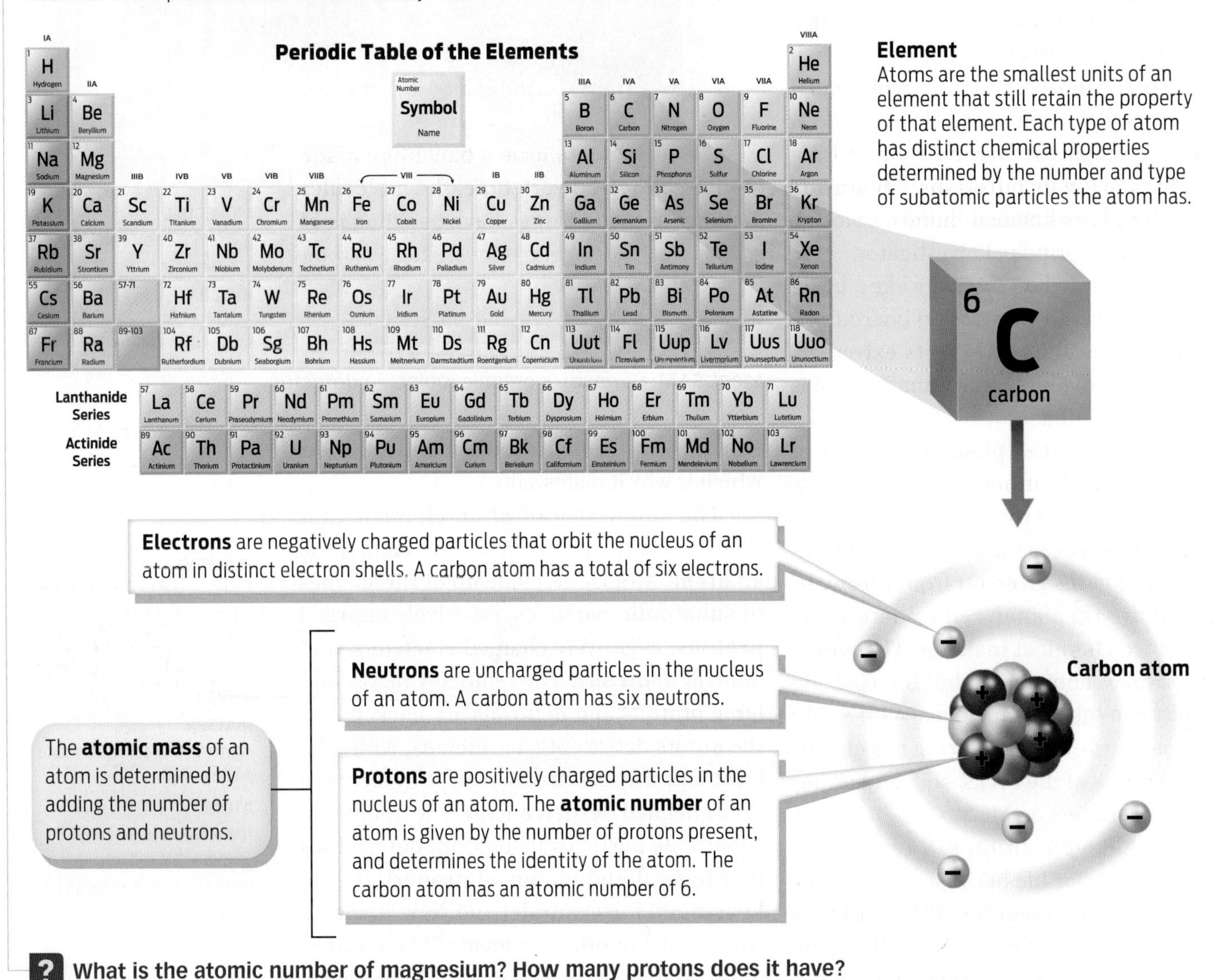

? **What is the atomic number of magnesium? How many protons does it have?**

Life's Backbone

▶ Chemical bonds and organic molecules

On Mars, NASA set *Curiosity* down in a large canyon called Gale Crater, at the center of which is an elevated region dubbed Mount Sharp. Geologically, this region is similar to Earth's Grand Canyon, with exposed strata of past eons layered one on top of the other, like a many-layered cake. NASA chose this as the best spot to try to reconstruct the past history of Mars, including whether the ancient atmosphere and soil of the planet could have supported life.

Curiosity is not actually testing for the presence of living organisms in Martian soil; that will be the goal of future missions. Instead, it is looking for evidence of the chemical building blocks of life. If life were ever present, it may have left a chemical signature that *Curiosity* can detect, maybe even telltale carbon compounds.

When astrobiologists (and science fiction writers) talk about "carbon-based life forms," they are referring to the fact that carbon forms the backbone of nearly every molecule making up living things. A **molecule** is defined as two or more atoms joined together by a **covalent bond,** a strong attraction formed between two atoms that share a pair of electrons, one from each atom.

Different atoms can form different numbers of covalent bonds, depending on the number of electrons in their outer shell. Carbon atoms have four outer-shell electrons with which to form covalent bonds, so a single carbon atom can form bonds with four other atoms, giving the element enormous versatility in forming molecules.

"Carbon is very cool, very flexible, very useful," says Chris McKay, an astrobiologist with NASA's Ames Research Center in California who studies the evolution of life and helped pack *Curiosity*'s chemical toolkit. "We don't see any other element that has the sort of flexibility and utility that carbon has."

> *"We don't see any other element that has the sort of flexibility and utility that carbon has."*
> **—Chris McKay**

Just as humans have a backbone made of interconnected vertebrae, the molecules making up living things have backbones of interconnected carbon atoms. These carbon backbones can be linear, like our spine, or circular, in which case the first carbon in the chain binds to the last carbon in the chain.

Molecules made with a backbone of interconnected carbon atoms and at least one carbon atom attached to a hydrogen atom are called **organic** molecules. Most organic

MOLECULE
Atoms linked by covalent bonds.

COVALENT BOND
A strong interaction resulting from the sharing of a pair of electrons between two atoms.

ORGANIC
Describes a molecule with a carbon-based backbone and at least one C–H bond.

***Curiosity* safely landed in Gale Crater.**
It has been navigating toward and up Mount Sharp, collecting and analyzing soil and rock samples, and capturing important images along the way.

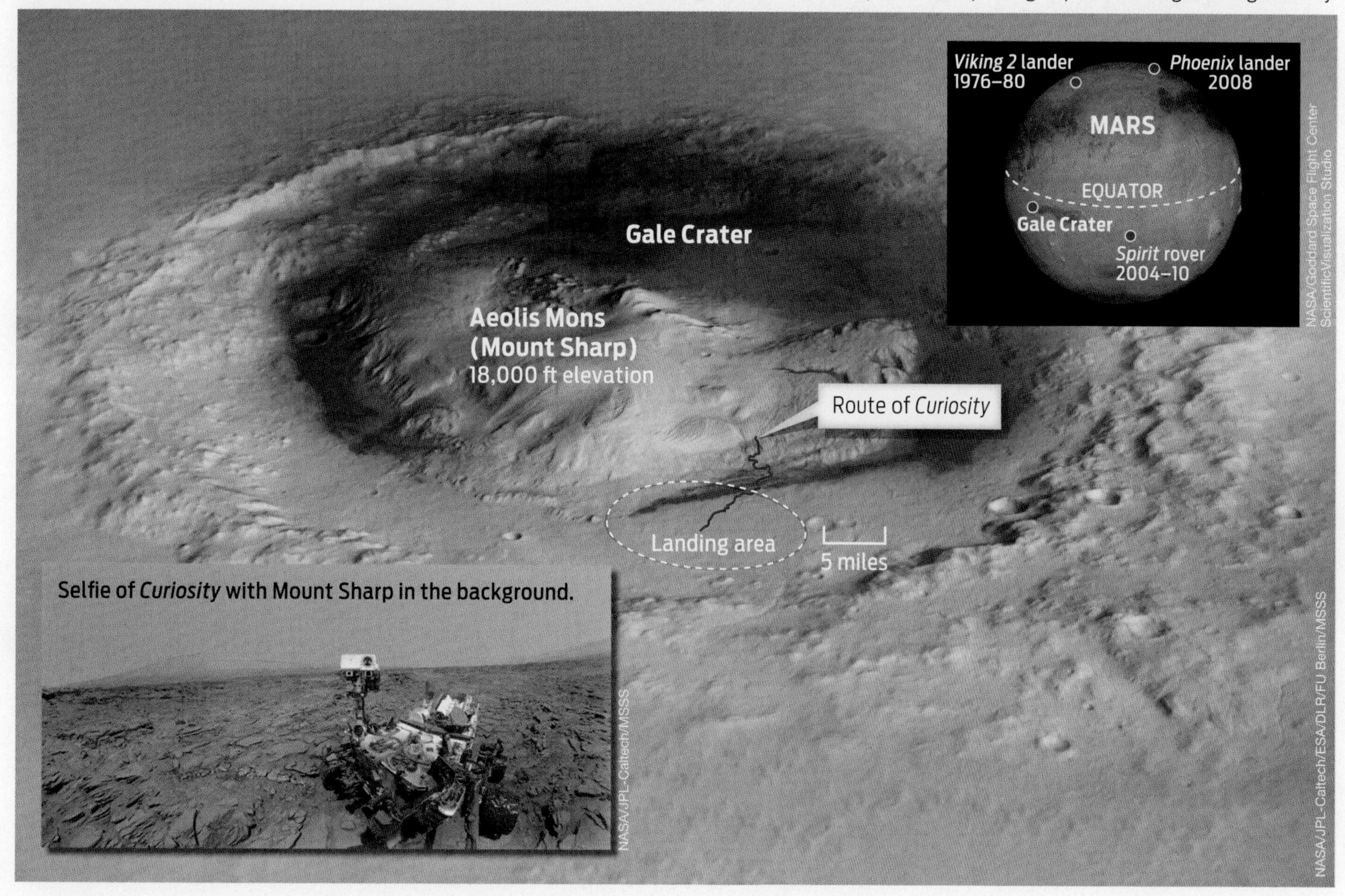

INORGANIC
Describes a molecule that lacks a carbon-based backbone and C–H bonds.

molecules require living things to make them, which is why they are often telltale signs of life. An example of a simple organic molecule is glucose, a type of sugar. Its molecular formula (a shorthand for representing the composition of a molecule) is $C_6H_{12}O_6$. This formula indicates that each molecule of glucose has 6 carbon atoms, 12 hydrogen atoms, and 6 oxygen atoms. Glucose is a ring-shaped molecule, with the carbon atoms forming the backbone of the ring.

Nonliving things can also contain carbon, but this carbon is **inorganic:** inorganic molecules do not have a carbon–carbon backbone and a carbon–hydrogen bond. Carbon dioxide (CO_2), for example, is an inorganic molecule, one found in the atmospheres of both Mars and Earth **(INFOGRAPHIC 2.3).**

INFOGRAPHIC 2.3

Carbon Is a Key Component of Life's Molecules

Molecules are atoms linked by covalent bonds. An atom of the element carbon can form four covalent bonds with other atoms. This leads to a great deal of diversity in the shapes and compositions of biologically important molecules. Carbon is found in both organic and inorganic molecules.

Carbon atoms can form four covalent bonds with other atoms:

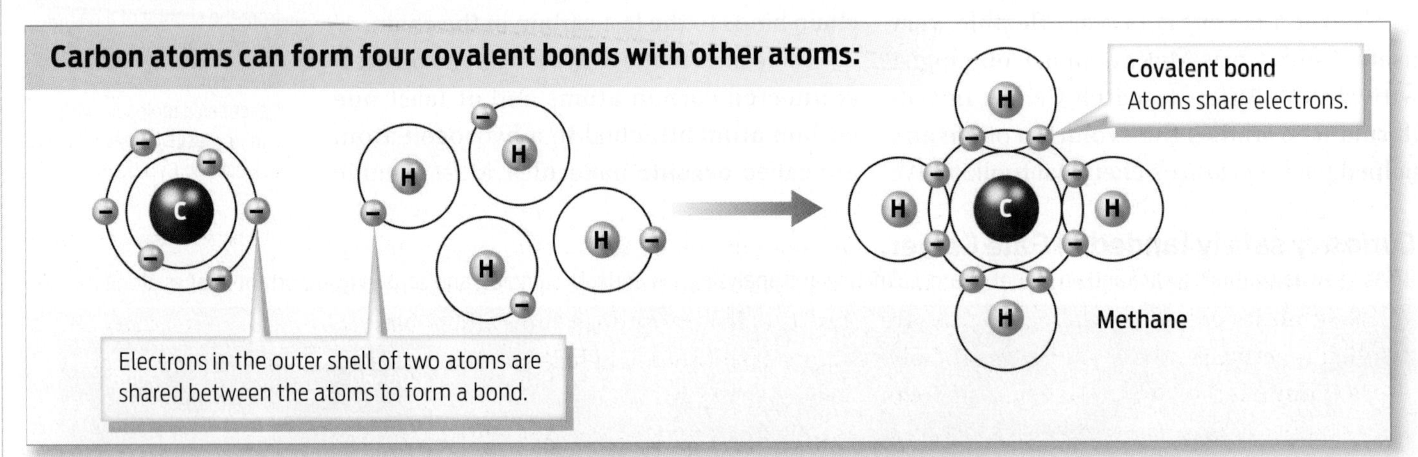

Organic molecules contain C–C and C–H bonds:

Organic molecules form when carbon atoms are covalently bound to other carbon and hydrogen atoms to form a diversity of chained, branching, and ring structures found in the molecules of life.

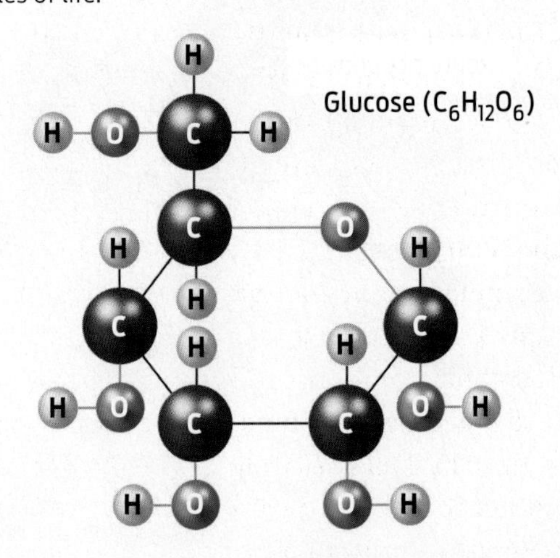

Inorganic molecules do not have C–C or C–H bonds:

Carbon atoms may also be found in inorganic molecules bound to non-carbon atoms. Inorganic molecules cannot form the large chained polymers found in organic molecules.

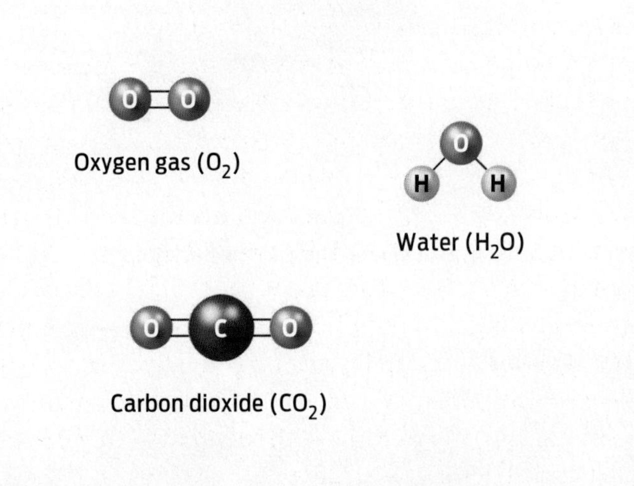

? **How many covalent bonds does every carbon in glucose have? How many bonds does the carbon of CO_2 have?**

Living things on Earth are made up of just four types of organic molecules: **carbohydrates, proteins, lipids,** and **nucleic acids.** Every molecule in the human body can be classified as one of these organic molecules. Your skin, for example, is composed of the proteins collagen and elastin, and the protein hemoglobin carries oxygen in your blood. The padding in your soft spots is composed of lipids called triglycerides, also known as fats. And in your liver and muscle cells, a carbohydrate called glycogen helps store energy. All of these organic molecules have a backbone of interlinked carbon atoms.

Organic molecules can be quite large and, therefore, are considered **macromolecules.** Macromolecules share a similar organization, in that they are composed of subunits called **monomers** linked together in a chain. When two or more monomers join, they form a **polymer.** Carbohydrates, for example, are polymers made up of linked monomers called **monosaccharides;** proteins are made up of subunits called **amino acids** that are bonded together; and nucleic acids are polymers composed of **nucleotides** that form long chains. Lipids, a more diverse group of molecules, are not classified as polymers **(INFOGRAPHIC 2.4).**

As far as we know, these large organic macromolecules are made only by living organisms. Therefore, if they are found on Mars, they would strongly suggest that life is present, or had been present at one time. "Organic compounds are fundamental to our search for life," Mahaffy, SAM's creator, says.

Since *Curiosity* landed in August 2012, it has been searching for chemical markers of life in a region of Gale Crater right around the landing site, an area called Yellowknife Bay, as well as in a higher ridge, the Vera Rubin Ridge. By drilling into the rocks and analyzing the samples in SAM, the rover has found evidence of hydrogen, oxygen, carbon, nitrogen, sulfur, and phosphorus—all elements that are critical for life as we know it.

In 2013, SAM made an even more startling discovery. In the gases given off when

JPL/NASA

Mars 2020 logo with mission objectives.

rock samples were baked, SAM sniffed a variety of carbon-containing compounds—the first time that such organic molecules were definitively detected on Mars. At this point, it's not possible to say whether the molecules were produced by living or nonliving means, but the discovery does show that Mars has many of the chemical building blocks necessary to make life.

"A fundamental question for this mission is whether Mars could have supported a habitable environment," said Michael Meyer, the lead scientist of NASA's Mars Science Laboratory, when the discovery of organic molecules was made. "From what we know now, the answer is yes."

As of late 2019, *Curiosity* had been on Mars for 7 years, traveling more than 13 miles and ascending more than 1,200 feet up Mount Sharp. Its journey is far from over, however. It is roughly halfway through a clay-rich region on the side of Mount Sharp and has just drilled its 22nd soil sample. It has enough energy in its nuclear power pack to last several more years, at which point its power will be carefully budgeted so that it can still function for years to come. In 2020, if all goes well, it will be joined by another rover tasked with looking for evidence of past or present Martian life, including the presence of a defining feature of life on Earth: cells.

CARBOHYDRATE
An organic molecule made up of one or more sugars.

PROTEIN
An organic molecule made up of linked amino acid subunits.

LIPIDS
Organic molecules that generally repel water.

NUCLEIC ACIDS
Organic molecules made up of linked nucleotide subunits; DNA (deoxyribonucleic acid) and RNA (ribonucleic acid) are examples.

MACROMOLECULES
Very large organic molecules that make up living organisms; they include carbohydrates, proteins, and nucleic acids.

MONOMER
One chemical subunit of a polymer.

POLYMER
A molecule made up of individual subunits, called monomers, linked together in a chain.

MONOSACCHARIDE
The building block, or monomer, of a carbohydrate.

AMINO ACID
The building block, or monomer, of a protein.

NUCLEOTIDE
The building block, or monomer, of a nucleic acid.

INFOGRAPHIC 2.4 UP CLOSE

Molecules of Life

a. Carbohydrates Are Made of Monosaccharides

Carbohydrates are made up of repeating subunits of simple sugars known as monosaccharides. Carbohydrates act as energy-storing molecules in many organisms. Other carbohydrates provide structural support for cells.

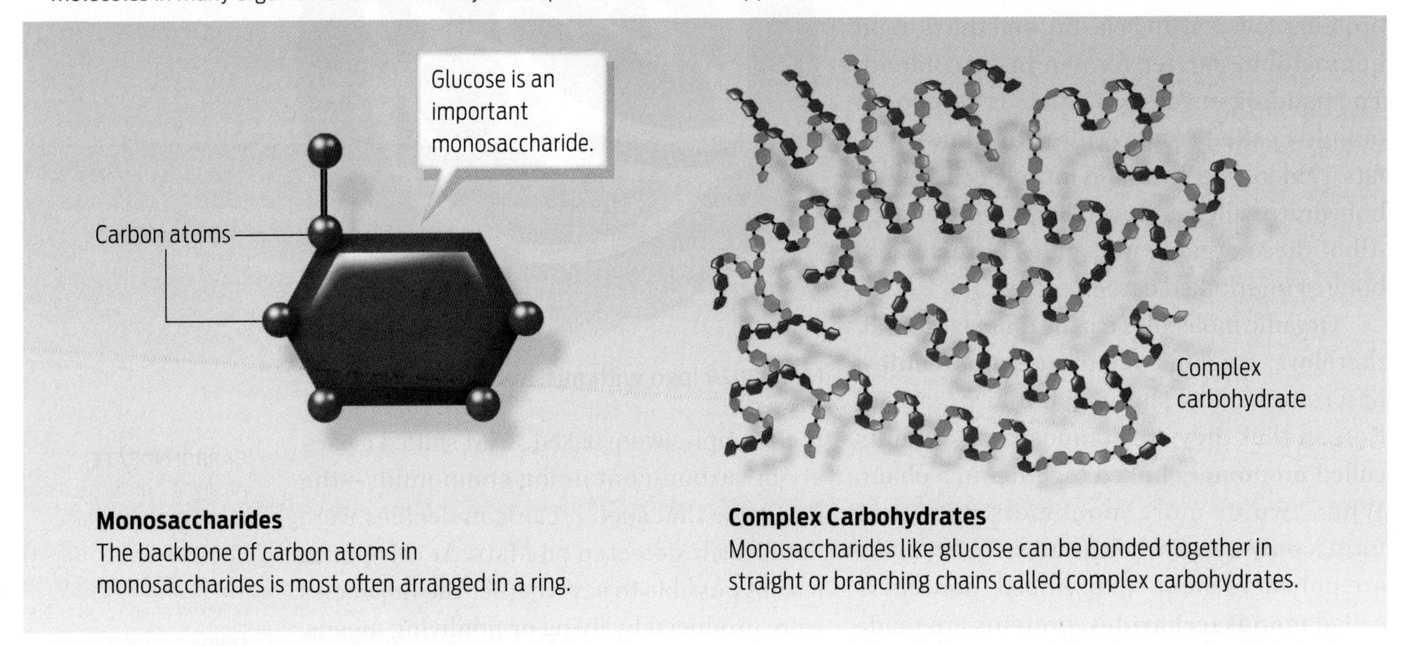

Monosaccharides
The backbone of carbon atoms in monosaccharides is most often arranged in a ring.

Complex Carbohydrates
Monosaccharides like glucose can be bonded together in straight or branching chains called complex carbohydrates.

b. Proteins Are Made of Amino Acids

Proteins are folded polymers of small repeating units called amino acids. Proteins carry out many functions in cells. They help speed up the rate of chemical reactions. They also move things through and around cells and even help entire cells move.

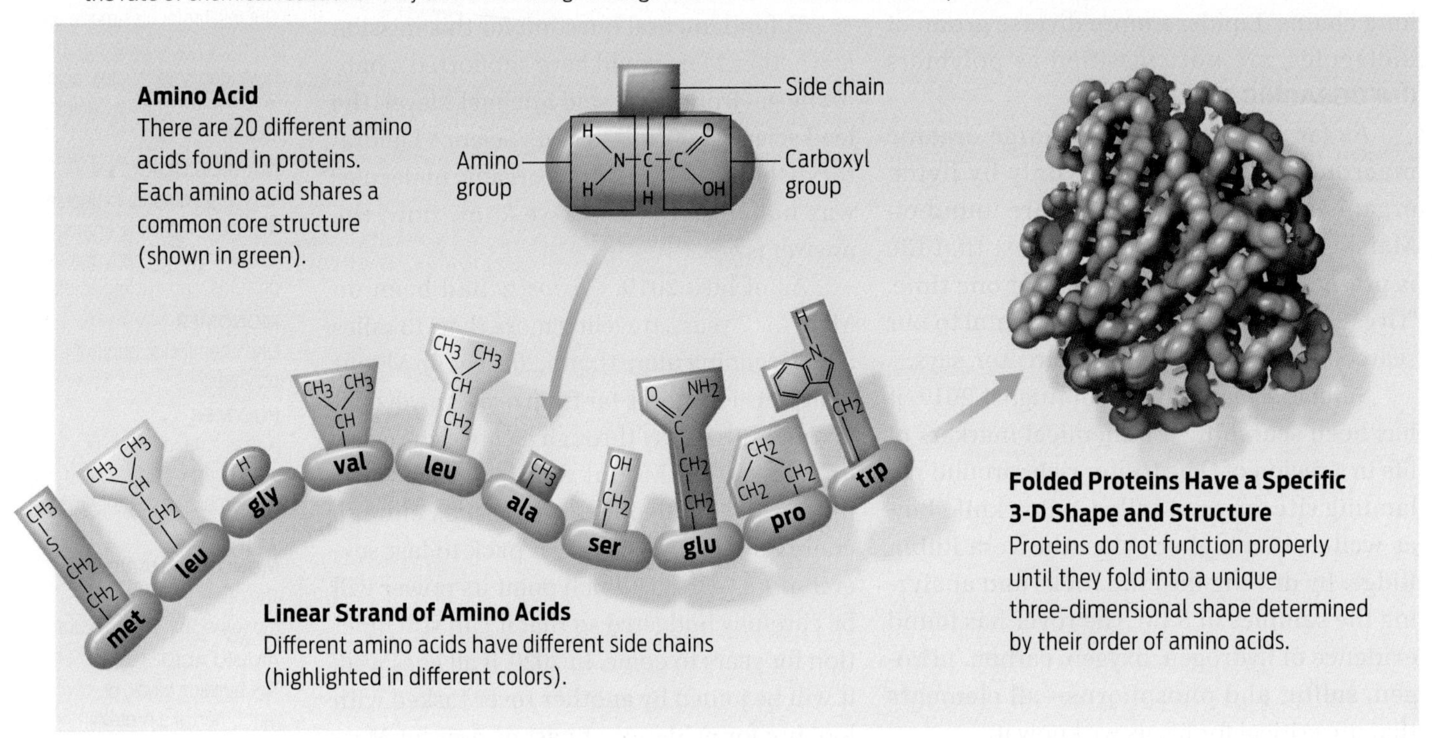

Amino Acid
There are 20 different amino acids found in proteins. Each amino acid shares a common core structure (shown in green).

Linear Strand of Amino Acids
Different amino acids have different side chains (highlighted in different colors).

Folded Proteins Have a Specific 3-D Shape and Structure
Proteins do not function properly until they fold into a unique three-dimensional shape determined by their order of amino acids.

c. Lipids Are Hydrophobic Molecules

There are different types of lipids, each with a distinct structure and function. Lipids are not polymers of repeating subunits like the other molecules of life, but they are all organic molecules. They are also all hydrophobic molecules, meaning they don't mix with water.

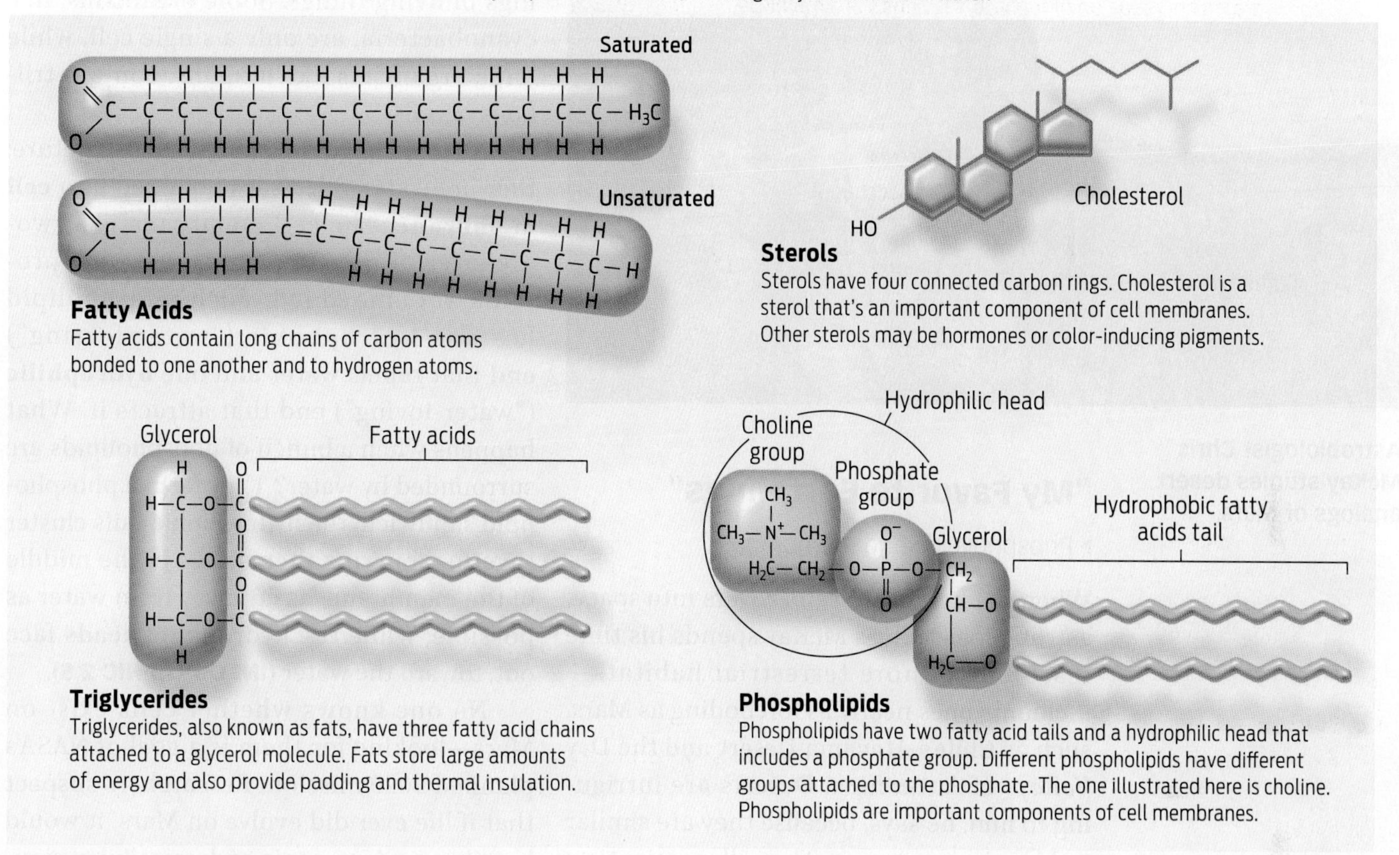

Fatty Acids
Fatty acids contain long chains of carbon atoms bonded to one another and to hydrogen atoms.

Sterols
Sterols have four connected carbon rings. Cholesterol is a sterol that's an important component of cell membranes. Other sterols may be hormones or color-inducing pigments.

Triglycerides
Triglycerides, also known as fats, have three fatty acid chains attached to a glycerol molecule. Fats store large amounts of energy and also provide padding and thermal insulation.

Phospholipids
Phospholipids have two fatty acid tails and a hydrophilic head that includes a phosphate group. Different phospholipids have different groups attached to the phosphate. The one illustrated here is choline. Phospholipids are important components of cell membranes.

d. Nucleic Acids Are Made of Nucleotides

Nucleic acids are polymers of repeating subunits known as nucleotides. There are two types of nucleic acids, DNA and RNA, each of which is made up of slightly different types of nucleotides. DNA and RNA are critical for the storage, transmission, and execution of genetic instructions.

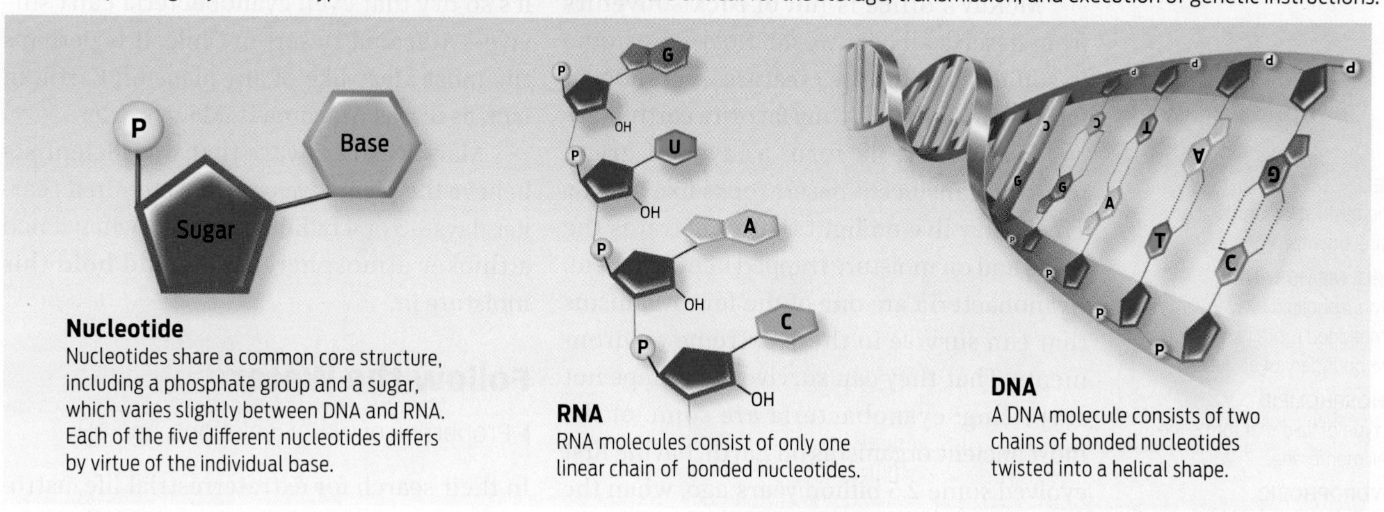

Nucleotide
Nucleotides share a common core structure, including a phosphate group and a sugar, which varies slightly between DNA and RNA. Each of the five different nucleotides differs by virtue of the individual base.

RNA
RNA molecules consist of only one linear chain of bonded nucleotides.

DNA
A DNA molecule consists of two chains of bonded nucleotides twisted into a helical shape.

NASA

Astrobiologist Chris McKay studies desert analogs of Mars.

Not bad for an organism made up of only a single **cell.** Cells are the basic structural unit of life on Earth. They are like microscopic bricks making up the walls and ceilings of living things. Some organisms, like cyanobacteria, are only a single cell, while large organisms like humans, contain trillions of cells.

All cells have the same basic structure: they are water-filled sacs bounded by a **cell membrane.** The cell membrane is a two-ply layer of **phospholipids** in which proteins are embedded. Each phospholipid has one **hydrophobic** ("water-fearing") end that repels water and one **hydrophilic** ("water-loving") end that attracts it. What happens when a bunch of phospholipids are surrounded by water? They form a phospholipid sandwich: the hydrophobic tails cluster together, burying themselves in the middle of the membrane, as far away from water as possible, while the hydrophilic heads face out, toward the water **(INFOGRAPHIC 2.5).**

No one knows whether cells exist on Mars—looking for them is a goal of NASA's planned 2020 mission. Researchers suspect that if life ever did evolve on Mars, it would have been microscopic and unicellular, similar to the earliest life forms on Earth.

McKay knows of only one desert where it's so dry that even cyanobacteria can't survive—Atacama Desert in Chile. It is perhaps the most Mars-like of any place on Earth. In fact, as dry as Atacama is, Mars is drier.

Mars wasn't always that way. Scientists believe the planet was much wetter in its earlier days—3 or 4 billion years ago, when it had a thicker atmosphere, and could hold this moisture in.

"My Favorite Earthlings"

▸ Phospholipids and membranes

When not helping to send rovers into space, astrobiologist Chris McKay spends his time researching more terrestrial habitats—including ones nearly as foreboding as Mars, such as Chile's Atacama Desert and the Dry Valleys of Antarctica. Deserts are intriguing to him, he says, because they are similar to Mars in many ways. They allow scientists to ask questions about the Martian environment without actually traveling there.

McKay's office is full of rock souvenirs from deserts all over world. Rocks are home to some of his favorite creatures—cyanobacteria—which he calls "my favorite earthlings."

Cyanobacteria form a layer of green beneath translucent desert rocks like quartz. There they live on light that penetrates the rocks and on moisture trapped beneath them. Cyanobacteria are one of the few organisms that can survive in these extreme environments. That they can survive is perhaps not surprising: cyanobacteria are some of the most ancient organisms on Earth, having first evolved some 2.5 billion years ago, when the world was a much more hostile place. Over time, through photosynthesis (see Chapter 5), they filled the Earth's atmosphere with oxygen, making it habitable for the rest of life.

CELL
The basic structural unit of living organisms.

CELL MEMBRANE
A phospholipid bilayer with embedded proteins that forms the boundary of all cells.

PHOSPHOLIPID
A type of lipid that forms the cell membrane.

HYDROPHOBIC
"Water-fearing"; hydrophobic molecules repel water.

HYDROPHILIC
"Water-loving"; hydrophilic molecules attract water.

Follow the Water

▸ Properties of water, solutions, and pH

In their search for extraterrestrial life, astrobiologists use a rule of thumb: "follow the water." Water is viewed as a proxy for life because it is so crucial to life on Earth. Water makes up more than 75% of a cell's weight.

INFOGRAPHIC 2.5

Membranes Rich in Phospholipids Define Cell Boundaries

Cells are the basic structural unit of life. They have a water-based interior that is separated from a chemically distinct water-based exterior by a cell membrane. The cell membrane is a lipid bilayer with embedded proteins.

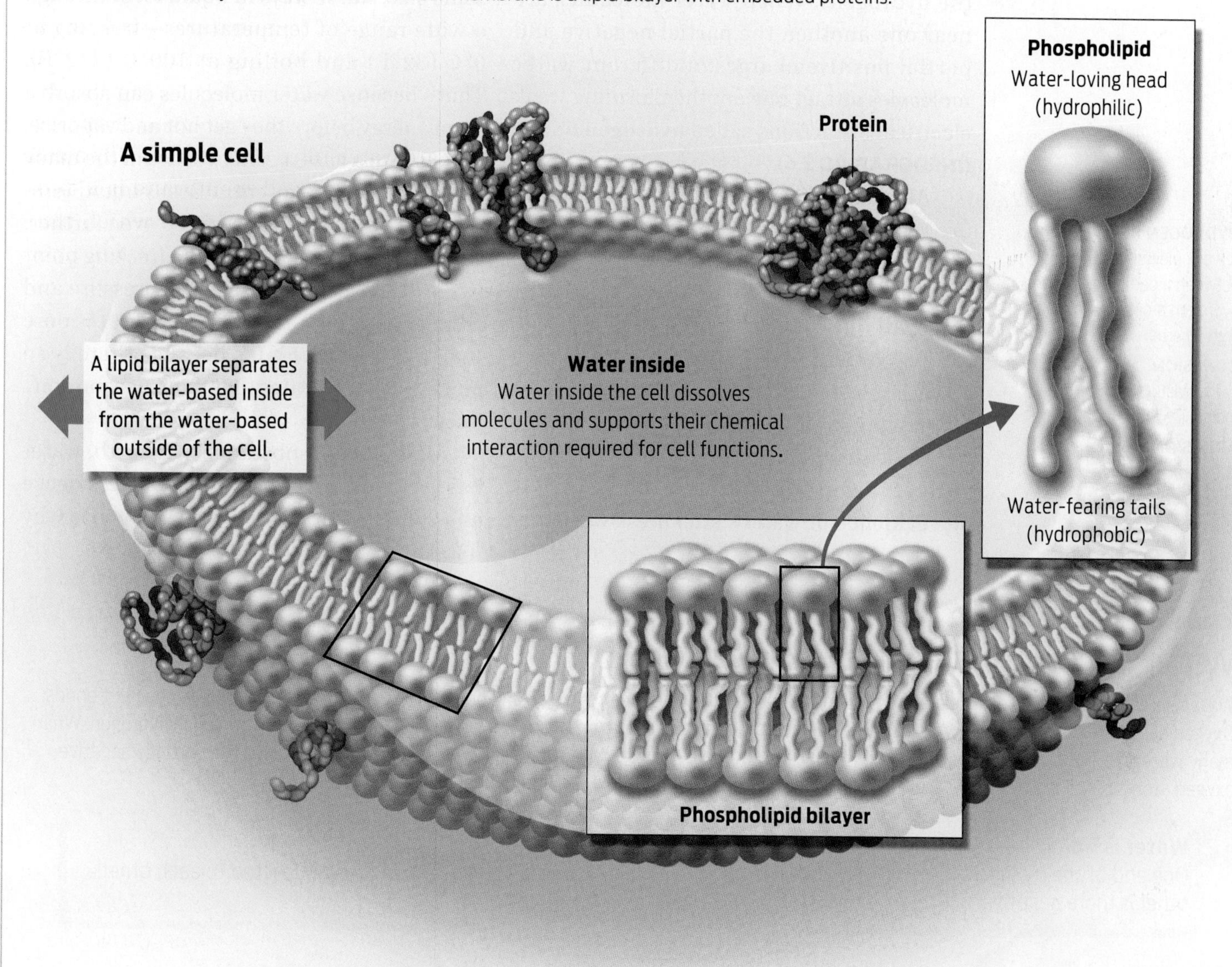

? **What two types of molecules make up cell membranes?**

All of life's chemical reactions take place in water, and many living things can survive only a few days without it.

A water molecule, H_2O, is shaped like Mickey Mouse's head: the oxygen atom is the face, and the two hydrogens are the ears. Many of water's life-conducive properties result from this simple shape. Water is a **polar molecule,** meaning that the electrons in the bonds between the oxygen atom and the hydrogen atoms are shared unequally. The oxygen atom has a stronger pull on electrons than the hydrogen atoms do, so the negatively charged electrons are often

POLAR MOLECULE
A molecule in which electrons are not shared equally between atoms, causing a partial negative charge at one end and a partial positive charge at the other. Water is a polar molecule.

closer to the oxygen atom than to the hydrogen atoms. As a result, a water molecule has a partial negative charge on the oxygen side and a partial positive charge on the side with the hydrogens. When water molecules are near one another, the partial negative and partial positive charges of different water molecules attract one another, forming weak electrical attractions called **hydrogen bonds** **(INFOGRAPHIC 2.6)**.

Although individual hydrogen bonds are relatively weak, there are a large number of hydrogen bonds in a volume of water, and they add up to a significant attractive force. All those hydrogen bonds make water "sticky," by acting as a kind of glue holding water molecules together. The stickiness of water allows water molecules to cling to one another (demonstrating **cohesion**) or to a surface (demonstrating **adhesion**). You can see evidence of water's stickiness wherever you look: a drop of water clinging to a leaf despite the downward pull of gravity, or an insect able to walk on the surface of a pond.

HYDROGEN BOND
A weak electrical attraction between a partially positive hydrogen atom and an atom with a partial negative charge.

COHESION
The attraction between molecules (or other particles).

ADHESION
The attraction between molecules (or other particles) and a surface.

Compared to other molecules of the same size, water stays in liquid form through a wide range of temperatures—freezing at 0°C (32°F) and boiling at 100°C (212°F). That's because water molecules can absorb a lot of energy before they get hot and vaporize, or turn into a gas—again because of the many hydrogen bonds present. Water's liquid temperature range can be extended even further: add salt and you can lower the freezing point to −46°C (−50°F); increase the pressure and you can bump up the boiling point to more than 343°C (650°F). It's because there is so much salt in seawater that most oceans don't freeze in winter.

Unlike most substances on Earth, water has the unusual property of being less dense as a solid than it is as a liquid—which is why

INFOGRAPHIC 2.6

Water Is Polar and Forms Hydrogen Bonds

Water is a polar molecule because electrons are not shared equally between the oxygen and the hydrogen atoms in each of its covalent bonds. Oxygen has a stronger pull on electrons, resulting in a slight negative charge on the oxygen and a slight positive charge on each hydrogen. When many water molecules are near one another, the partially positive hydrogen atoms of some molecules are attracted to the partially negative oxygen atoms of nearby water molecules. These weak electrical attractions are hydrogen bonds.

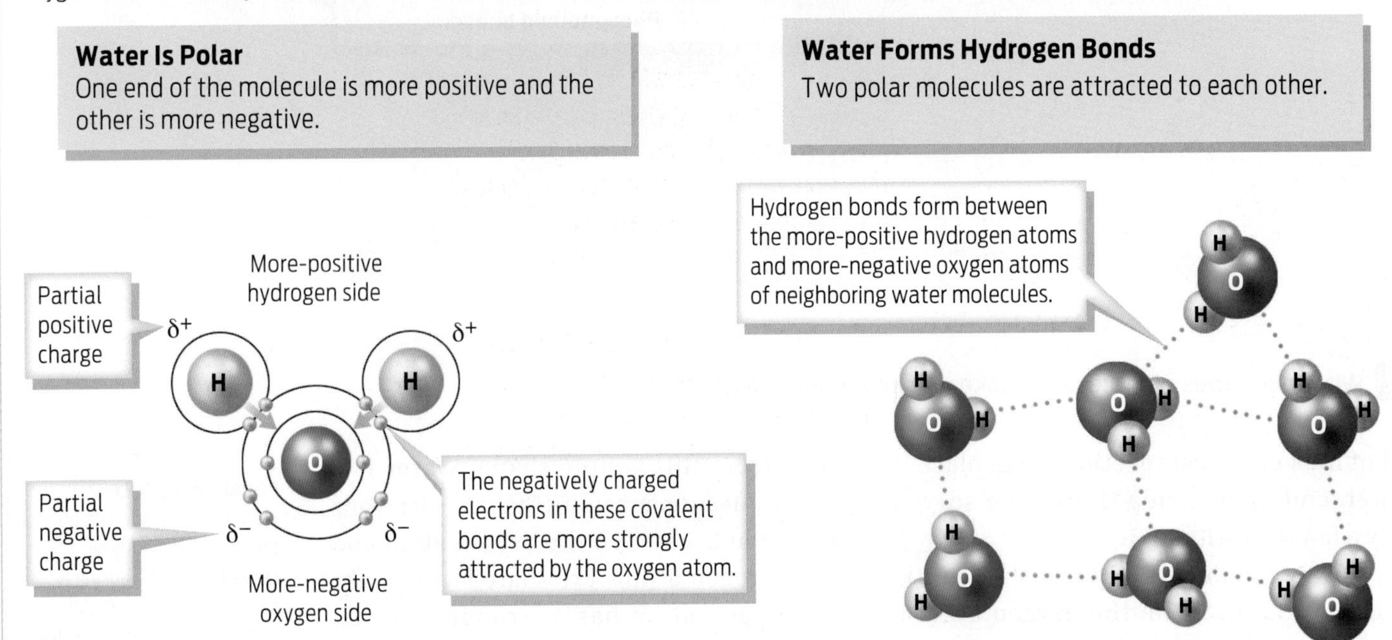

? **Look at the water molecule on the bottom right of the right panel. Draw another water molecule hydrogen bonded to this one.**

ice floats. When water freezes, the water molecules form an ordered array, with individual water molecules spaced at an equal and fixed "arm's length" from one another. This increased separation between water molecules decreases the density of ice, allowing ice to float on water. And because ice floats, fish can live beneath frozen lakes in winter and not turn into ice cubes **(INFOGRAPHIC 2.7)**.

Perhaps the most important life-conducive property of water is that it is a good **solvent:** its polar structure makes it capable of dissolving many substances on which life depends. Water dissolves these substances by coating, or surrounding, the individual molecules (or ions, in the case of ionic substances, discussed below) making up the substance. These individual components then separate from one another and disperse through the liquid. Water transports all of life's dissolved molecules, or **solutes,** from place to place—whether through a cell, a body, or an ecosystem. Life, in essence, is a water-based **solution,** a mixture of solvent and solutes. Many biological molecules, such as proteins and DNA, have the specific shapes they do

SOLVENT
A substance in which other substances can dissolve. Water is a good solvent.

SOLUTE
A dissolved substance.

SOLUTION
The mixture of solute and solvent.

INFOGRAPHIC 2.7

Hydrogen Bonds Give Water Its Unique Properties

Because so many hydrogen bonds can form in a body of water (or ice), even these weak bonds can collectively give water its unique properties.

AGE Fotostock/Superstock

Water Is Sticky
Hydrogen bonding allows water molecules to stick to one another (cohesion) and to other surfaces (adhesion), making them wet.

iStockphoto/Thinkstock

Water Can Absorb a Lot of Energy
It takes a lot of heat energy to disrupt the many hydrogen bonds between water molecules. That is why water is a liquid at a wide range of temperatures.

igorad1/Shutterstock

Ice Is Less Dense Than Liquid Water
When water freezes, the bonds between the water molecules become rigid and expand, increasing the overall volume of the water. This makes ice less dense than liquid water. That is why ice floats.

? **These properties are all the result of hydrogen bonding between water molecules. Draw two water molecules that are hydrogen bonded to each other.**

only because of the way that their atoms interact with water. Hydrophobic regions tend to cluster together in the interior of the molecule, whereas hydrophilic regions tend to be exposed to, and interact with, the water molecules surrounding them.

Water is an excellent solvent for other polar molecules with partial charges and for substances that contain **ionic bonds,** which are the strong attractions formed between atoms that have opposite electrical charges (positive and negative). Such charged atoms, or **ions,** form when one atom loses a negatively charged electron (becoming a positively charged ion) and another atom gains that electron (becoming a negatively charged ion). These oppositely charged ions then form an ionic bond. Water dissolves substances containing ionic bonds by surrounding the substance and breaking the ionic bonds within the substance. Table salt, or sodium chloride, is an example of a substance that contains ionic bonds and is easily dissolved by water and is therefore hydrophilic. Of course, even water cannot dissolve everything; you have probably seen firsthand what happens when you try to mix fats, such as butter or oils in salad dressing, with water. That is because these hydrophobic substances repel water and so clump together rather than being dispersed **(INFOGRAPHIC 2.8).**

IONIC BOND
A strong electrical attraction between oppositely charged ions formed by the transfer of one or more electrons from one atom to another.

ION
An electrically charged atom, the charge resulting from the loss or gain of electrons.

When astrobiologists speak about the importance of water for life, they make an important qualification: *liquid* water. Frozen water is found throughout the universe; there are abundant quantities on Mars and on other planets and moons in our solar system, for example. But only on Earth does water exist primarily in its liquid form, because of its moderate temperatures. "Liquid water is the key requirement in the search for life," says McKay. "The other worlds of the solar system have enough light, enough carbon, and enough of the other key elements for life. Water in the liquid form is rare."

Though liquid water is not present on the surface of Mars today, scientists suspect that liquid water—lots of it—may have once

Evidence for an ancient freshwater lake on Mars.

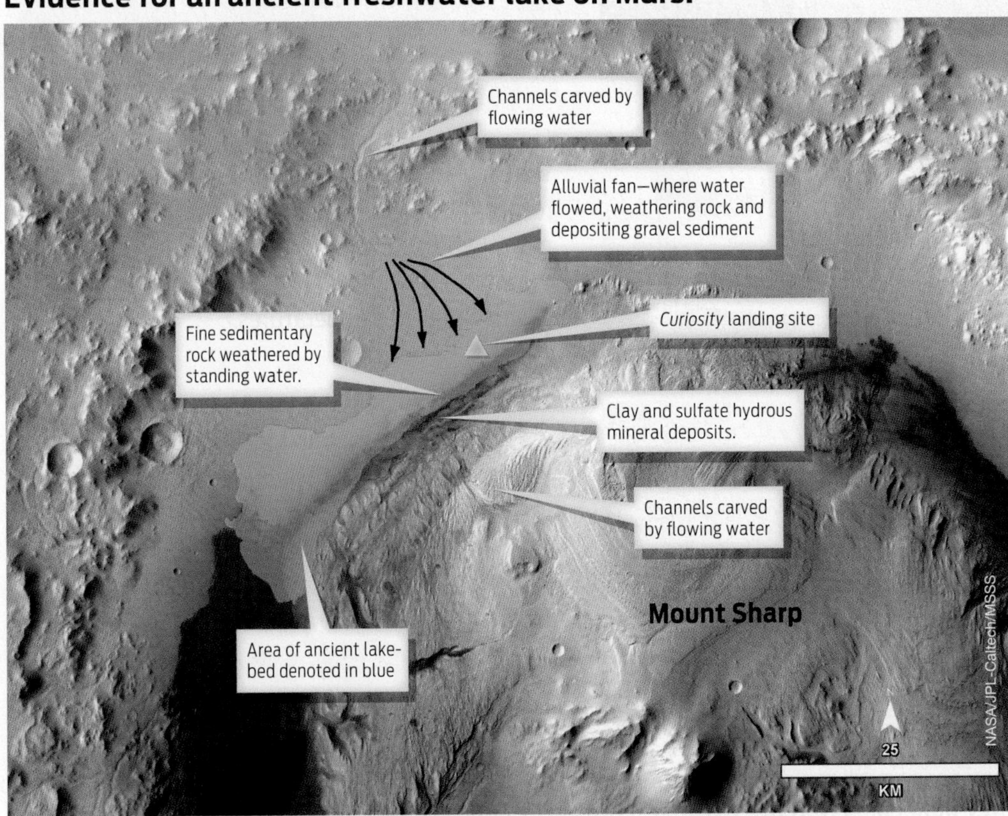

INFOGRAPHIC 2.8

Water Is a Good Solvent

Because water molecules have partial charges, they can interact with charged ions and other hydrophilic molecules, allowing water to coat and then dissolve these solutes.

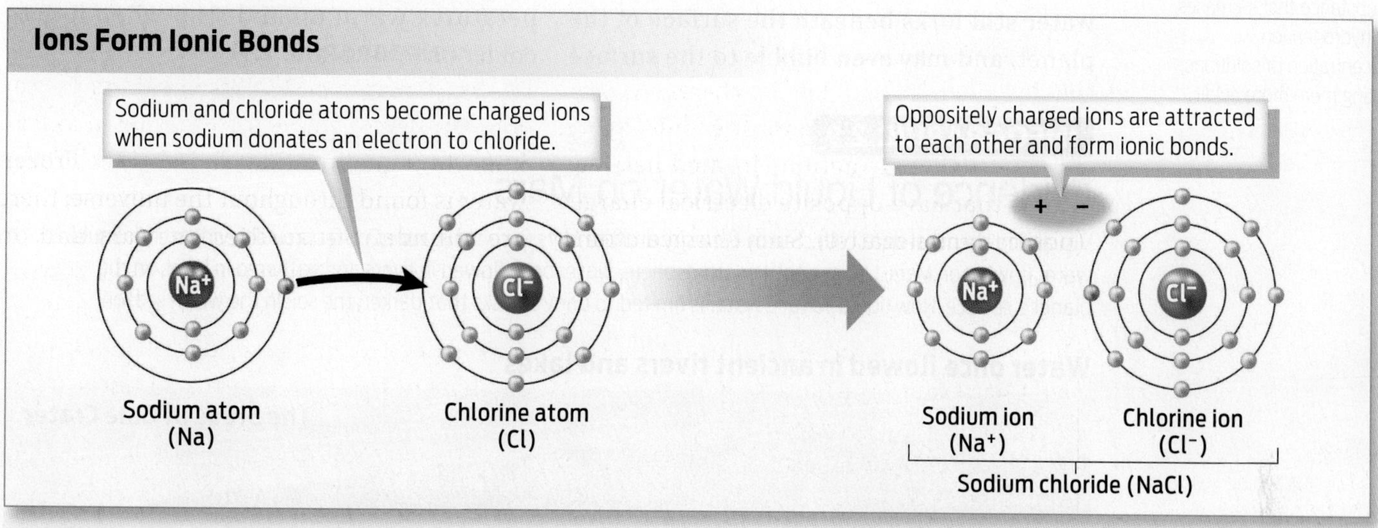

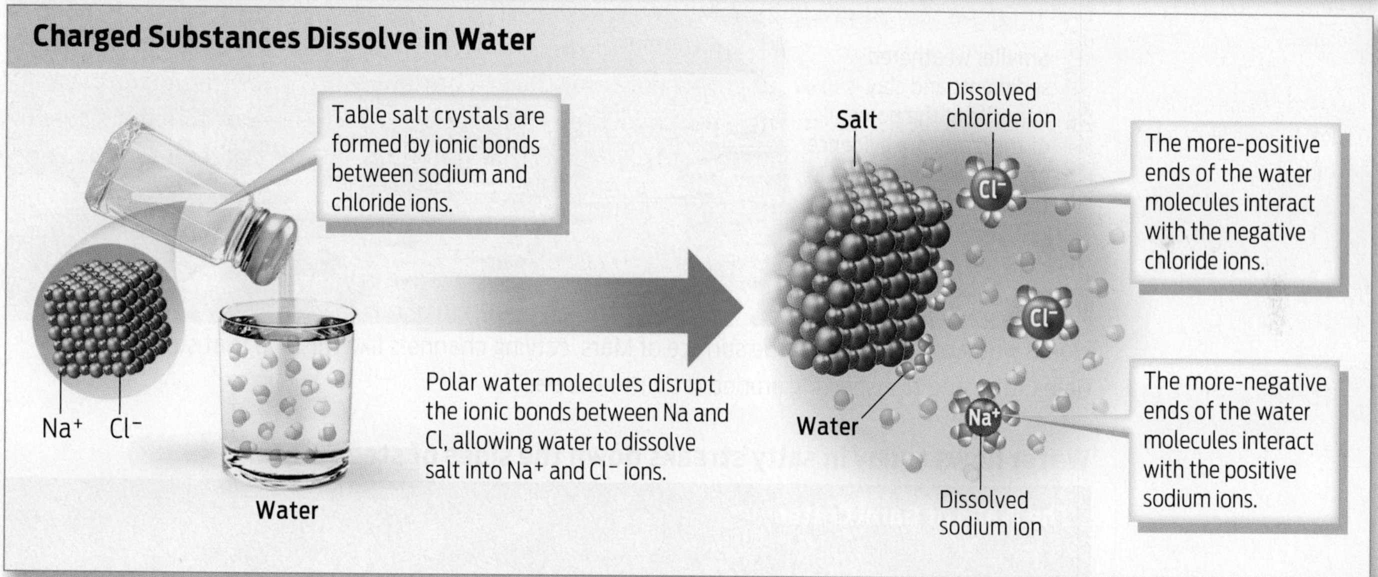

? **Why is a sodium ion positively charged? Why is a chloride ion negatively charged?**

covered the planet. Clues to this ancient water can be seen all over the surface of the planet, which in many places is carved out like sections of the Grand Canyon. There are also salt deposits like those you can see when seawater evaporates, another telltale sign of liquid water's past on the surface of Mars. Gale Crater, where *Curiosity* landed, has clay deposits and rock formations that suggest it was once the site of a large lake. "The case for water is tight," McKay says.

"From the *Curiosity* rover, we now know that Mars once was a planet very much like Earth, with warm, salty seas, with freshwater lakes, probably snow-capped peaks, and clouds and a water cycle," NASA scientist John Grunsfeld said at a September 2015 press conference. Where all this water went, no one knows. For flowing water to have once existed on Mars, the planet's atmosphere must have been much thicker and warmer than it is today. Something happened to the

pH
A measure of the concentration of hydrogen ions, H^+, in a solution.

ACID
A substance that increases the hydrogen ion concentration of solutions, making them more acidic.

atmosphere over time such that now the planet's atmosphere is thin and blisteringly cold. Exactly how this happened remains a mystery.

Some evidence suggests that liquid water still lurks beneath the surface of the planet, and may even bubble to the surface periodically. Photographs taken by NASA's *Mars Reconnaissance Orbiter*, which has been circling the planet since 2006, show what appear to be rivulets of water flowing down rocky slopes. These flows occur when temperatures warm up and stop when it gets colder **(INFOGRAPHIC 2.9)**.

INFOGRAPHIC 2.9

Evidence of Liquid Water on Mars

Visual and chemical data collected by *Curiosity* and the *Mars Reconnaissance Orbiter* provide evidence for liquid water flowing on Mars. Under a thicker atmosphere, water once flowed in rivers and gathered in lakes on the planet's surface. Now liquid surface water is limited to briny streaks that darken the soil in the warm season.

Water once flowed in ancient rivers and lakes

Liquid water once flowed on the surface of Mars, carving channels like this one that slope into the Gale Crater and transporting eroded rock into the basin.

Water flows today in salty streaks down the sides of steep craters

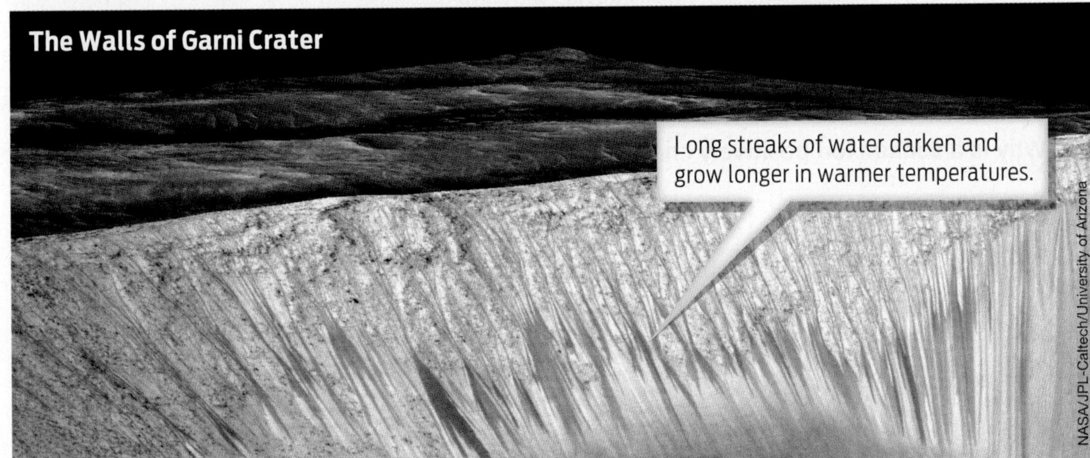

Minerals detected in these locations suggest that salty water flows when the temperature in these areas climbs above −10°F (−23°C).

? Why is the discovery of liquid water on Mars important?

In 2018, scientists with the European Space Agency announced that they had used radar to discover a subterranean lake of liquid water located nearly a mile beneath the icy surface of Mars's south pole. This finding is tantalizing because microscopic life grows in similar subglacial lakes in Antarctica on Earth.

Even with evidence of past and present water on Mars, questions still remain about whether it has (or had) the same properties as Earth water, and could therefore support life. Depending on what's dissolved in it, water can have a wide range of characteristics—from caustic drain cleaner to calming chamomile tea. The different chemical properties of water-based solutions reflect their **pH,** the concentration of hydrogen ions (H^+) in a solution, denoted as a range from 0 to 14. In every water-based solution, water molecules (H_2O) split briefly into separate hydrogen (H^+) and hydroxide (OH^-) ions. In pure water, the number of separated H^+ ions is, by definition, exactly equal to the number of separated OH^- ions; the pH is therefore 7, or neutral—the dead center of the 0 to 14 scale. Acidic solutions, or **acids,** have a higher concentration of hydrogen ions (H^+), a lower concentration of hydroxide ions (OH^-), and a pH closer to 0. When acids are added to water, they increase the concentration of hydrogen ions and make the solution more acidic. By contrast, basic solutions, or **bases,** have a lower concentration of H^+ ions and a pH closer to 14. Bases remove H^+ ions from a solution, thereby increasing the proportion of OH^- ions.

Strong acids and bases are highly reactive with other substances, which makes them destructive to the molecules in a cell. Also, many biochemical reactions take place only at a certain pH. As a consequence, living things are extremely sensitive to changes in pH, and most function best when their pH stays within a specific range. The pH of human blood is about 7.4. If that pH were to fall even slightly, to 7, our biochemistry would malfunction and we would die.

Previous missions to Mars have calculated the pH of Martian soil as 7.7—mild enough to grow asparagus, as one NASA chemist put it **(INFOGRAPHIC 2.10).**

The planned landing site for NASA's 2020 Mars mission is a place called Jezero Crater,

BASE
A substance that reduces the hydrogen ion concentration of solutions, making them more basic.

INFOGRAPHIC 2.10

Solutions Have a Characteristic pH

The pH of a solution is a measure of the concentration of hydrogen ions (H^+) in it. Solutions with a low concentration of H^+ ions have a basic pH (greater than pH 7). Solutions with a high concentration of H^+ ions have an acidic pH (a pH of less than 7). Both acids and bases can be damaging because they are highly reactive with other substances. A neutral solution has a pH of 7.

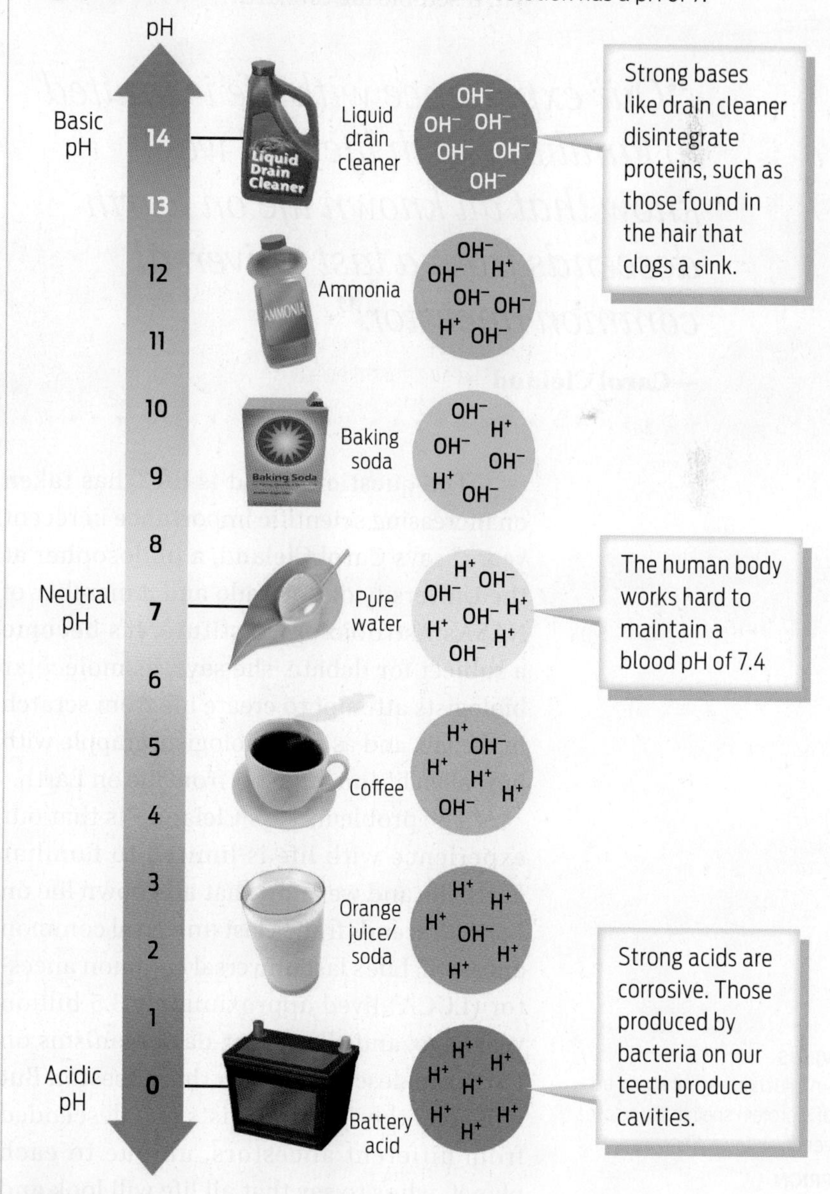

? What is the normal pH of human blood?

which was chosen because it's the site of an ancient lake. It was determined to be the most likely place to find evidence of cellular life.

"Weird Life"

▸ Viruses and other entities that bend life's rules

So far, NASA's search for life on Mars has stuck very closely to our understanding of life on Earth, where living things seem to share certain chemical and structural properties, like carbon-based molecules and cells. But how can we be sure that life on other planets will resemble life on Earth?

"Our experience with life is limited to familiar Earth life and we know that all known life on Earth descends from a last universal common ancestor."

—Carol Cleland

"The question 'What is life?' has taken on increasing scientific importance in recent years," says Carol Cleland, a philosopher at the University of Colorado and a member of NASA's Astrobiology Institute. It's become a subject for debate, she says, as molecular biologists attempt to create life from scratch in the lab, and as astrobiologists grapple with how alien life might differ from life on Earth.

"The problem," says Cleland, "is that our experience with life is limited to familiar Earth life and we know that all known life on Earth descends from a last universal common ancestor." Life's last universal common ancestor (LUCA) lived approximately 3.5 billion years ago, and all present-day organisms on Earth are descended from this ancestor. But if life on other planets exists and descended from different ancestors, unique to each planet, who's to say that all life will look and behave just as it does on this planet?

VIRUS
An infectious agent made up of a protein shell that encloses genetic information.

PRION
An infectious agent composed only of protein.

More and more, scientists are accepting that they need to keep an open mind about what life on other planets might look like. In 2007, the National Academy of Sciences issued a "weird life" report suggesting that NASA should not be so narrowly focused on water and organic molecules in its search for life on other planets. Although water may be crucial to life on Earth, perhaps other solvents—ammonia or methane, for example—might support life elsewhere. The report also urged the space agency to avoid being "fixated on carbon," even though carbon forms the scaffold of life on Earth. Other elements, such as silicon, could hypothetically provide a functional scaffold for life on other planets.

Even on Earth, a few exceptions, or boundary cases, bend the "rules" of life as they are currently conceived. **Viruses,** for example, reproduce and pass their genetic information on to new viruses, but they are not made of cells. Instead, they consist of a protein shell that encloses one or more DNA or RNA molecules containing genetic information. A virus reproduces by infecting a host cell and hijacking its cellular machinery to make copies of itself. Other noncellular, self-reproducing entities include **prions,** infectious proteins that are responsible for "mad cow disease" and related illnesses. Whether viruses and prions are alive is a point hotly debated among scientists.

Mars may seem an inhospitable place to find life, but microscopic life has been found thriving in some of the most extreme places on Earth, from radioactive waste and deep-sea vents to frozen Antarctic lakes submerged under miles of ice. Such examples reveal that life is nothing if not adaptive **(INFOGRAPHIC 2.11)**.

Could similarly adaptive organisms have once inhabited Mars? Might they still? NASA's Chris McKay, for one, is cautiously optimistic: "I spend my time and energy in the search for evidence of life on Mars," he says. "Obviously, this is because I think there must have been life there and we have a good chance of finding evidence of it." ■

INFOGRAPHIC 2.11

Life on the Fringe

Several entities on Earth are not made of cells and therefore defy our criteria for "life," yet are still able to reproduce (and in some cases cause disease). Other organisms on Earth stretch our notions of where life can survive.

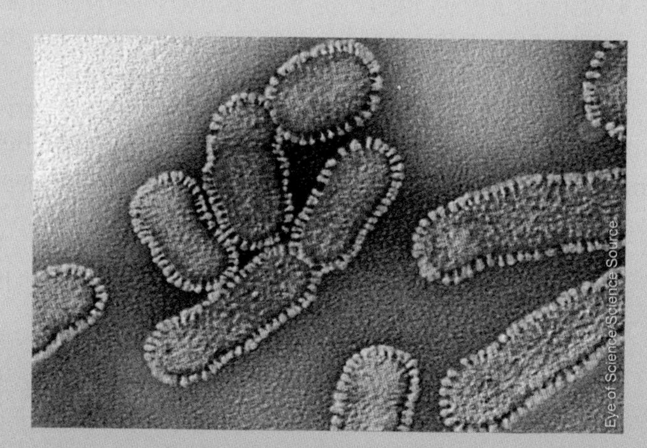

Viruses are not cellular. They infect cells and use host cell machinery to replicate.

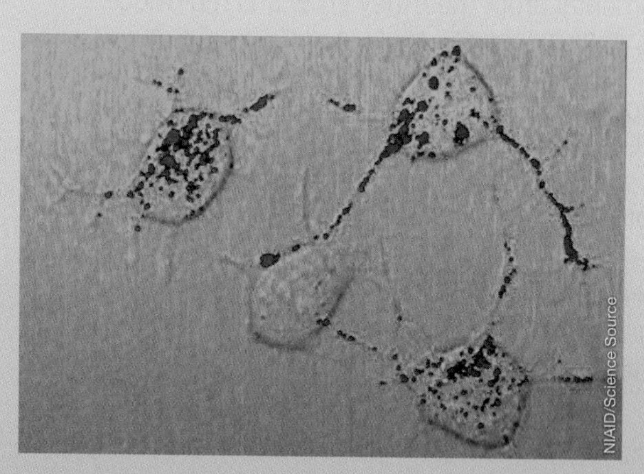

Prions are not cellular. They are infectious proteins that replicate in cells, causing disease.

These bacteria live in a high concentration of arsenic, a chemical that is toxic to most organisms.

These bacteria live in conditions of extreme heat and pressure that would kill most organisms.

? Compare and contrast viruses and prions.

CHAPTER 2 Summary

Driving Question 1 How do we define life?

- On Earth, living organisms share a number of fundamental properties: they grow and reproduce, maintain homeostasis, sense and respond to their environment, and rely on energy to carry out their functions.

Driving Question 2 How is matter organized into the molecules of living organisms?

- All matter is composed of elements; there are 118 known elements in the universe. Each element has a unique atomic structure, with a particular number of protons, neutrons, and electrons.
- When atoms share pairs of electrons, they form covalent bonds, building molecules.
- On Earth, living organisms are made up of organic molecules, which contain a backbone composed of interconnected atoms of the element carbon and with one or more carbon–hydrogen bonds.
- Four types of carbon-based organic molecules make up living things: carbohydrates, proteins, nucleic acids, and lipids.

Driving Question 3 What is the basic structural unit of life, and what are its properties?

- Living organisms on Earth are made of cells, which contain water and are bounded by a cell membrane; cells are the smallest unit of life.
- Cell membranes are made of phospholipids, which have a hydrophilic (water-loving) head group and a hydrophobic (water-fearing) tail. When surrounded by water, they form a lipid bilayer.

Driving Question 4 Why is water so important for life and living organisms?

- Water is a polar molecule, with its two hydrogen atoms carrying a partial positive charge and its single oxygen atom carrying a partial negative charge.
- Because of its partial charges, a water molecule can form hydrogen bonds (weak attractions between these opposite partial charges) with other water molecules and interact with other charged molecules.
- Water has many properties that make it a crucial component of life on Earth: it is "sticky," it regulates heat well, it floats when frozen, and it is a good solvent.
- When atoms lose or gain electrons, they become ions. Oppositely charged ions can form ionic bonds—strong electrical attractions. Water is a good solvent of substances with ionic bonds.
- The concentration of hydrogen ions (H^+) ions in a solution determines its pH, which is measured on a scale that ranges from 0 to 14. Most chemical reactions in cells take place at a nearly neutral pH (approximately 7).

More to Explore

- NASA, *Mars Science Laboratory* (*Curiosity*) mission page: http://www.nasa.gov/mission_pages/msl
- Video: *Curiosity*'s Seven Minutes of Terror: http://www.jpl.nasa.gov/video/details.php?id=1090
- Follow *Curiosity* on Twitter @MarsCuriosity.
- Special issue of the journal *Science* devoted to *Curiosity*: http://www.sciencemag.org/site/extra/curiosity/
- McKay, C. P. (2004). What is life—and how do we search for it in other worlds? *PLoS Biol* 2(9):1260–1263.
- *The Limits of Organic Life in Planetary Systems* [the "Weird Life" Report]. (2007). Washington, DC: National Academies Press.

CHAPTER 2 Test Your Knowledge

Driving Question 1 How do we define life?

By answering the questions below and studying Infographic 2.1, you should be able to generate an answer for this broader Driving Question.

Know It

1. Which of the following is not a generally recognized characteristic of most (if not all) living organisms?
   a. the ability to reproduce
   b. the ability to maintain homeostasis
   c. the ability to obtain energy directly from sunlight
   d. the ability to sense and respond to the environment
   e. the ability to grow

2. What is homeostasis? Why is it important to living organisms?

Use It

3. How would you assess whether a possibly living organism from another planet was truly alive?

4. Which of the characteristics of living organisms (if any) allow you to distinguish between living and formerly living (that is, dead) organisms? Explain your answer.

Driving Question 2 How is matter organized into the molecules of living organisms?

By answering the questions below and studying Infographics 2.2, 2.3, and 2.4, you should be able to generate an answer for this broader Driving Question.

Know It

5. What subatomic particles are located in the nucleus of an atom?
   a. protons
   b. neutrons
   c. electrons
   d. protons, neutrons, and electrons
   e. protons and neutrons

6. When an atom gains an electron, what happens?
   a. It becomes positively charged.
   b. It becomes negatively charged.
   c. It becomes neutral.
   d. Nothing happens.
   e. Atoms cannot gain an electron because atoms have a defined number of electrons.

7. Glucose (a monosaccharide) has the molecular formula $C_6H_{12}O_6$. How many carbon atoms are in each glucose molecule?

8. What does it mean to say that a macromolecule is a polymer? Give an example.

9. A collection of nucleotides could be used to build a
   a. protein.
   b. complex carbohydrate.
   c. triglyceride.
   d. nucleic acid.
   e. cell.

Use It

10. Consider the types of lipid.
    a. How does a sterol, such as cholesterol, differ from a triglyceride?
    b. Structurally, what do triglycerides and phospholipids have in common?

11. If a cell were unable to take up or make sugars, which class of molecules would it be unable to make?
    a. carbohydrates
    b. proteins
    c. lipids
    d. nucleic acids
    e. all of the above
    f. a and d

12. Phenylketonuria (PKU) is an inherited condition in which consumption of the amino acid phenylalanine results in a toxic buildup of that amino acid. What macromolecule should be limited in the diet of a person with PKU? Explain your answer.

Driving Question 3 What is the basic structural unit of life, and what are its properties?

By answering the questions below and studying Infographics 2.5 and 2.11, you should be able to generate an answer for this broader Driving Question.

Know It

13. The basic building blocks of life are
    a. DNA molecules.
    b. cells.
    c. proteins.
    d. phospholipids.
    e. inorganic molecules.

14. The cell membrane is made of
    a. water.
    b. proteins.
    c. phospholipids.
    d. nucleotides.
    e. b and c

Use It

15. What are the arguments for and against considering viruses to be living organisms?

16. Why do phospholipids form a bilayer in water-based solutions?

Driving Question 4 Why is water so important for life and living organisms?

By answering the questions below and studying Infographics 2.6, 2.7, 2.8, and 2.10, you should be able to generate an answer for this broader Driving Question.

Know It

17. Is olive oil hydrophobic or hydrophilic? What about salt? Explain your answer.

18. The "stickiness" of water results from the ____________ bonding of water molecules.

a. hydrogen
b. ionic
c. covalent
d. acidic
e. hydrophobic

19. Coffee or tea with sugar dissolved in it is an example of a water-based solution.

a. What is the solvent in such a beverage?
b. What is the solute in such a beverage?
c. Given that the sugar has dissolved in the beverage, are sugar molecules hydrophobic or hydrophilic?

20. As an acidic compound dissolves in water, the pH of the water

a. increases.
b. remains neutral.
c. decreases.
d. doesn't change.
e. becomes basic.

21. In a water molecule, the bond between the oxygen atom and a hydrogen atom is a(n) ____________ bond.

a. covalent
b. hydrogen
c. ionic
d. hydrophobic
e. noncovalent

22. How do ionic bonds compare to hydrogen bonds? What are the similarities? What are the differences?

Use It

23. Why do olive oil and vinegar (a water-based solution) tend to separate in salad dressing? Will added salt dissolve in the oil or in the vinegar? Explain your answer.

24. Which of the following is/are most likely to dissolve in olive oil?

a. a polar molecule
b. a nonpolar molecule
c. a hydrophilic molecule
d. a and c
e. b and c

Interpreting Data

Apply Your Knowledge

25. Look at Infographic 2.10. For the substances drain cleaner, coffee, and soda, answer the following questions: Is the substance an acid or a base? What is the concentration of hydrogen ions relative to a solution with a neutral pH?

Mini Case

Apply Your Knowledge

26. One approach to finding out if there is life on Mars is to bring Martian dirt samples to Earth for analysis. What are possible considerations for science and society if a Martian life form is released on Earth? Given that *Curiosity* has landed on Mars, what are the possible consequences if an Earth life form is released on Mars? What steps can NASA take to minimize these risks?

Bring It Home

Apply Your Knowledge

27. Your tax dollars are being invested in projects such as the *Curiosity* rover project. Investigate the NASA website to learn more about NASA's rationale for the investment in this mission. Now draft a letter to your congressional representative that expresses your opinion about this expenditure of taxpayer dollars. If you agree, state specific reasons why you think this a good investment of your money. If you disagree, state your reasons, and describe at least two other scientific programs that you would prefer to see funded, providing a rationale for why these are more important.

Wonder Drug

How a chance discovery in a London laboratory revolutionized medicine

Wim Van Egmond/Science Source

DRIVING QUESTIONS

1. What structural features are shared by all cells, and what are the key differences between prokaryotic and eukaryotic cells?
2. How do solutes and water cross cell membranes, and what determines the direction of movement of solutes and water in different situations?
3. How do different antibiotics target bacteria?
4. What are the key eukaryotic organelles and their functions?

On a September morning in 1928, biologist Alexander Fleming returned to his laboratory at St. Mary's Hospital in London after a short summer vacation. As usual, the place was a mess. Flasks were scattered everywhere, and his workbench was strewn with the petri plates on which he was growing bacteria. On this day, as Fleming sorted through the plates, he noticed that one was growing a patch of fluffy white mold. It had been contaminated, likely by a rogue mold spore that had drifted in from a neighboring laboratory.

Alexander Fleming in his lab.

Bettmann/Getty Images

Fleming was about to toss the plate in the sink when he noticed something unusual: wherever mold was growing, there was a zone around the mold where the bacteria did not seem to grow. Curious, he looked under a microscope and saw that the bacterial cells near the mold had lysed, or burst. Something in the mold was killing the bacteria.

Fleming scooped a bit of mold from the plate, grew it in a broth of nutrients, and then tested the liquid in a battery of additional experiments. The results were clear and dramatic: even when diluted 800 times, this "mold juice"—Fleming's term—was a potent killer of many different kinds of bacteria. What's more, no other fungus that he tested—including one obtained from a

Penicillium mold

Bacteria cannot grow near the mold due to antibiotics produced by *Penicillium*.

Staphylococcus bacterium

The original plate on which Alexander Fleming observed *Penicillium* mold inhibiting the growth of *Staphylococcus aureus* bacteria.

Mary Evans Picture Library/Alamy

pair of moldy old shoes—had this remarkable killing power. Fleming published his results in 1929 in the *British Journal of Experimental Pathology.* He named the antibacterial substance "penicillin," after the fungus that produces it, *Penicillium notatum.* His discovery marked the birth of the first **antibiotic** **(INFOGRAPHIC 3.1).**

Fleming was not the first to notice the bacteria-killing property of *Penicillium,* but he was the first to study it scientifically and publish the results. "Penicillin started as a chance observation," Fleming said many years later. "My only merit is that I did not neglect the observation and that I pursued the subject as a bacteriologist."

Before penicillin, doctors could do little to help a patient with a serious bacterial infection like pneumonia, syphilis, gonorrhea, or meningitis. Now, they had a powerful weapon on their side. As physician and author Lewis Thomas recalled in his 1992 memoir, "We could hardly believe our eyes on seeing that bacteria could be killed off without at the same time killing the patient. It was not just amazement, it was a revolution."

Bug Bullet

▶ Prokaryotic versus eukaryotic cells

What makes antibiotics special is not just their ability to kill bacteria. After all, bleach kills bacteria just fine. The important thing about antibiotics is that they are selective—they destroy bacteria without (typically) harming the human or animal host. Although Fleming didn't know it at the time, penicillin and other antibiotics selectively kill bacteria because they target something unique about bacterial cells.

ANTIBIOTIC
A chemical that can slow or stop the growth of bacteria; many antibiotics are produced by living organisms.

INFOGRAPHIC 3.1

How Penicillin Was Discovered

A fortuitous observation by Fleming led to the discovery of the first antibiotic. He realized that the fungus on his culture plate was somehow inhibiting the reproduction of bacteria.

1. A single bacterial cell lands on the surface of a nutrient-rich plate.

2. Nutrients in the plate support the growth and division of the bacterial cells.

3. After many rounds of cell division, enough cells accumulate to be visualized as noticeable bacterial growth on the plate.

Both bacteria and mold can grow and divide on the nutrient-rich petri plate. However, bacteria cannot reproduce in the area surrounding the mold.

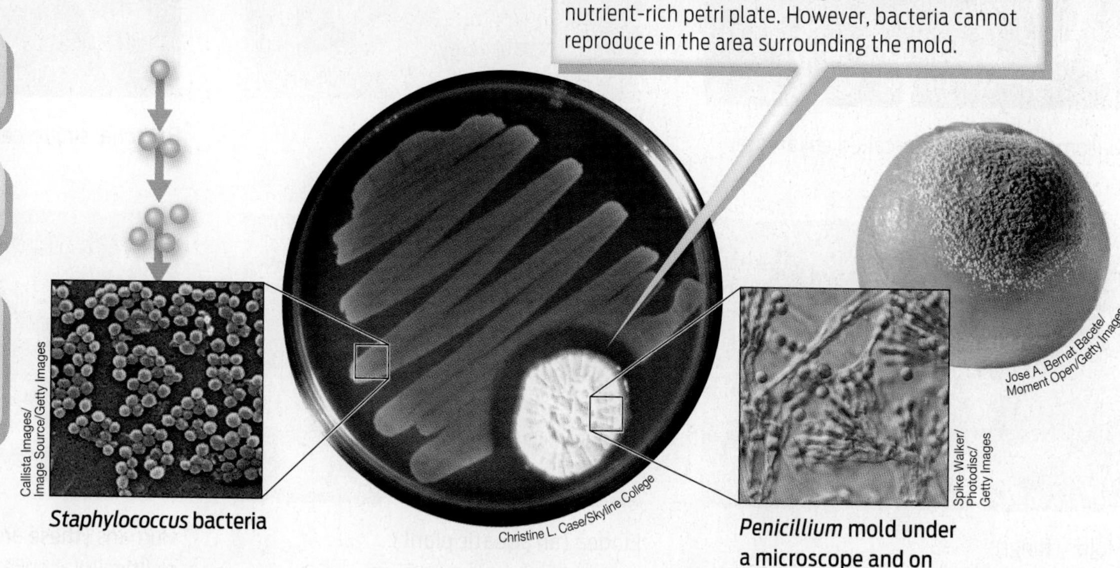

? **What would the plate look like if the fungus did not inhibit the growth of bacteria?**

CELL THEORY
The concept that all living organisms are made of cells and that cells are formed by the division of existing cells.

PROKARYOTIC CELLS
Cells that lack internal membrane-bound organelles.

EUKARYOTIC CELLS
Cells that contain membrane-bound organelles, including a central nucleus.

ORGANELLES
The membrane-bound compartments of eukaryotic cells that carry out specific functions.

CELL MEMBRANE
A phospholipid bilayer with embedded proteins that forms the boundary of all cells.

Cells are the basic building blocks of life. According to the **cell theory,** all living things are made of cells, and every new cell comes from the division of a preexisting one. But not all cells are alike. Cells come in many shapes and sizes and perform a variety of functions. Moreover, they fall into one of two fundamentally different categories: **prokaryotic** or **eukaryotic.** Prokaryotic cells are relatively small and lack internal membrane-bound compartments, called **organelles.** Eukaryotic cells, by contrast, are much larger and contain many different organelles.

Like all prokaryotic organisms, known as prokaryotes, bacteria exist as single cells. They include the harmless *Escherichia coli* that live in our gut and the *Lactobacillus acidophilus* that make our yogurt, as well as the many bacteria that make us sick. Eukaryotes may be single-celled or composed of multiple cells. Plants, animals, tiny single-celled organisms like amoeba, and fungi are all examples of eukaryotic organisms **(INFOGRAPHIC 3.2).**

To understand why antibiotics affect bacterial and human cells differently, it helps to first understand what these cells have in common. All cells, both prokaryotic and eukaryotic, are surrounded by a **cell membrane** composed of phospholipids and proteins. This flexible yet sturdy structure forms a boundary between the external environment and the cell's watery interior; it separates inside from outside and literally defines the cell. The watery solution inside every cell, enclosed by the cell membrane,

INFOGRAPHIC 3.2

Cell Theory: All Living Things Are Made of Cells

All living organisms are composed of cells. These cells arise from the reproduction of existing cells. Different cells have different structures and functions.

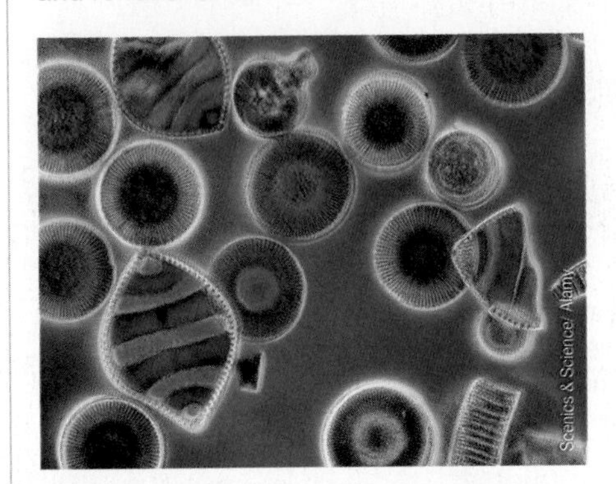

Diatoms (algae): single-celled eukaryotes

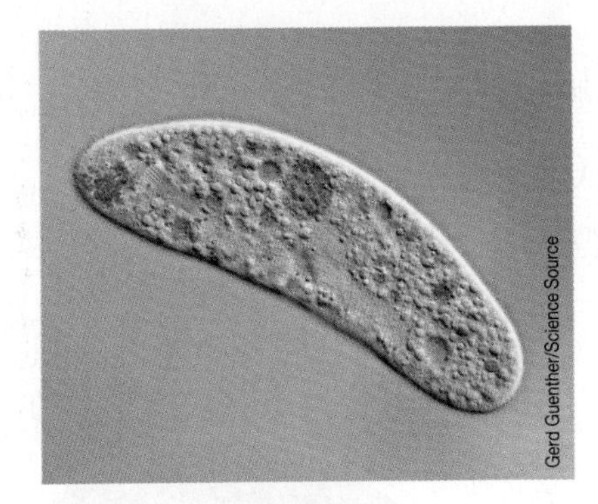

Amoeba (a protist): a single-celled eukaryote

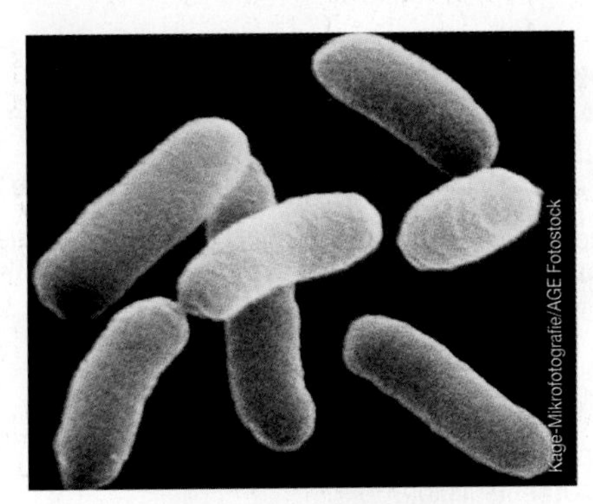

Bacteria: single-celled prokaryotes

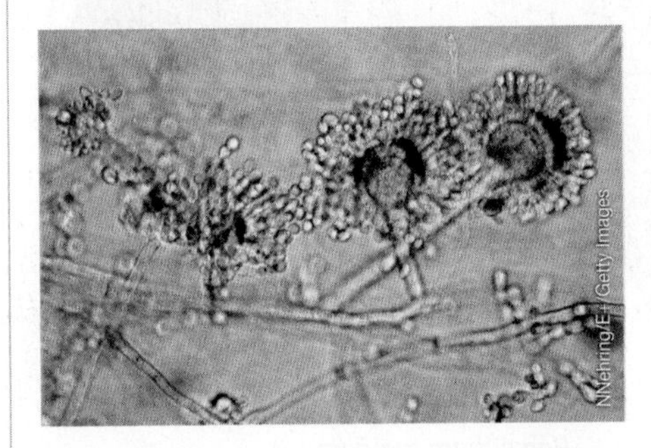

Molds (fungi): single and multicellular eukaryotic cells

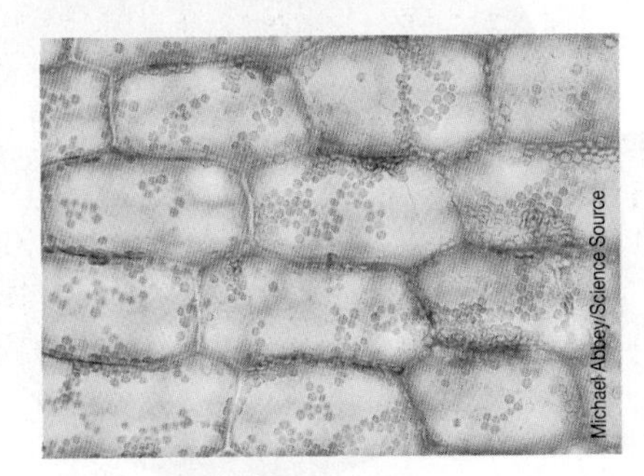

Elodea (an aquatic plant): a multicellular eukaryote

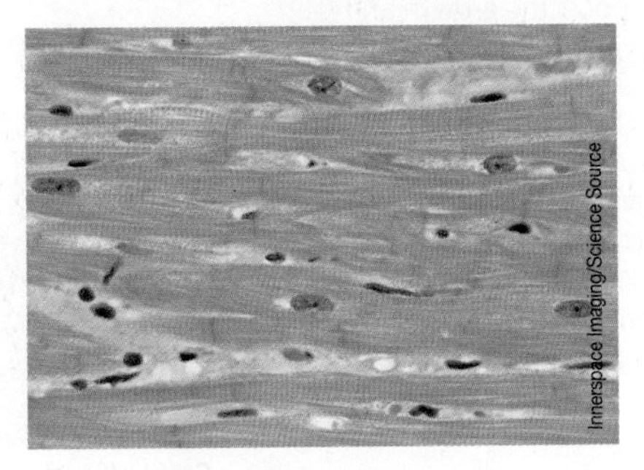

Humans (these are heart cells): multicellular eukaryotes

Which of the nonhuman organisms shown have cells of the same structural type as human cells?

is the **cytoplasm.** In addition, all cells have **ribosomes,** structures that synthesize the proteins crucial to cell function. Finally, cells contain DNA, the molecule of heredity.

However, beyond these common features—cell membrane, cytoplasm, ribosomes, and DNA—the two cell types are structurally quite different. For one thing, the DNA in a prokaryotic cell floats freely within the cell's cytoplasm, while in a eukaryotic cell it is housed within a central "command center" called the **nucleus.** The nucleus is one of many organelles found within eukaryotic cells but not in their simpler prokaryotic cousins **(INFOGRAPHIC 3.3).**

Water World

▸ Cell wall and osmosis

Penicillin selectively kills bacteria because of an important difference between prokaryotic and eukaryotic cells. Most prokaryotic cells, including bacteria, are surrounded by a **cell wall,** a rigid structure that encloses the cell membrane. Many eukaryotic cells, including those of humans and other animals, lack this structure. The cell wall is what allows bacteria to survive in watery environments—for example, in the intestines, in blood, or in a pond. Without it, water would tend to rush into the cell and cause it to burst.

CYTOPLASM
The gelatinous, aqueous interior of all cells.

RIBOSOME
A complex of RNA and proteins that carries out protein synthesis in all cells.

NUCLEUS
The organelle in eukaryotic cells that contains the genetic material.

CELL WALL
A rigid structure present in some cells that encloses the cell membrane and helps the cell maintain its integrity.

INFOGRAPHIC 3.3

Features of Prokaryotic and Eukaryotic Cells

While all cells have a cell membrane, cytoplasm, ribosomes, and DNA, there are specific structural differences between prokaryotic and eukaryotic cells. Eukaryotic cells contain a variety of membrane-enclosed organelles, while prokaryotic cells do not. All prokaryotic cells have a cell wall surrounding the cell membrane, while many eukaryotic cells do not.

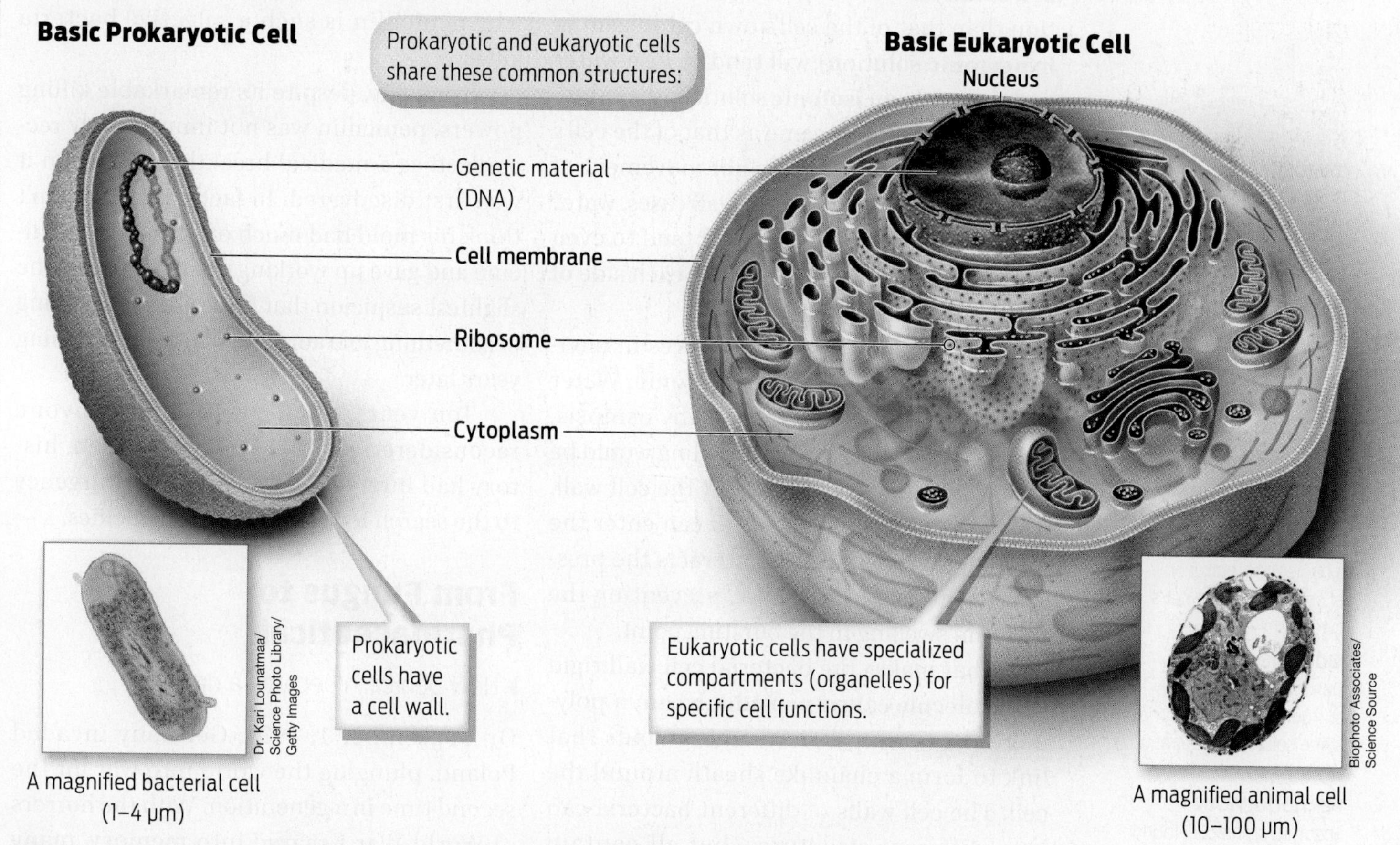

? Describe at least one difference between eukaryotic and prokaryotic cells.

Water's tendency to move across cell membranes is called **osmosis.** Water moves across a cell membrane by osmosis when the concentration of dissolved substances, or solutes, on one side of a membrane is different from that on the other side. Water moves from the solution with the lower solute concentration to the solution with the higher solute concentration so as to balance the solute concentrations on each side. As a result, a cell placed in a solution with a lower solute concentration than that of the cell's own cytoplasm (a **hypotonic** solution) will tend to take up water and swell. Conversely, a cell placed in a solution with a higher solute concentration than that of the cell's own cytoplasm (a **hypertonic** solution) will tend to lose water and shrivel. In an **isotonic** solution, the solute concentration is the same as that of the cell's cytoplasm, and there is no net movement of water into or out of the cell. In all cases, water moves in the direction that will tend to even out the solute concentrations on each side of the membrane.

"I had not the slightest suspicion that I was at the beginning of something extraordinary".

—Alexander Fleming

Many bacteria find themselves in environments that tend to be hypotonic. Water then enters the bacterial cells by osmosis, causing them to swell. This swelling would be fatal to bacteria were it not for the cell wall, which limits how much water can enter the cell. The rigid cell wall counteracts the pressure of the incoming water, preventing the cell from swelling to the bursting point.

What makes the bacterial cell wall rigid is a molecule called **peptidoglycan,** a polymer made of sugars and amino acids that link to form a chainlike sheath around the cell. The cell walls of different bacteria can have different structures, but all contain peptidoglycan, and peptidoglycan is found only in bacteria. And here's where penicillin comes in: by interfering with the synthesis of peptidoglycan, penicillin weakens the bacterial cell wall, so that it can no longer withstand the pressure of the incoming water. Eventually, the bacterial cell bursts **(INFOGRAPHIC 3.4).**

Eukaryotic cells lacking cell walls have other ways of protecting themselves from water pressure. For example, many single-celled eukaryotic organisms, called protists, live in freshwater ponds. These organisms have a structure called a contractile vacuole that acts as a water pump, continually pumping excess water out of the cell. In multicellular eukaryotic organisms, the environments surrounding cells (e.g., blood) are often maintained as an isotonic solution with respect to the cells' cytoplasm.

Some eukaryotic cells, including those of plants and some fungi, have cell walls. However, bacteria are the only ones that have a cell wall made of peptidoglycan—which is why penicillin is such a selective bacteria killer.

Ironically, despite its remarkable killing powers, penicillin was not immediately recognized as a medical breakthrough when it was first discovered. In fact, Fleming didn't think his mold had much of a future in medicine and gave up working on it. "I had not the slightest suspicion that I was at the beginning of something extraordinary," recalled Fleming years later.

Ten years would pass before anyone reconsidered Fleming's mold. By then, history had intervened and given new urgency to the search for antibacterial medicines.

From Fungus to Pharmaceutical

▶ How penicillin became a drug

On September 1, 1939, Germany invaded Poland, plunging the world into war for the second time in a generation. With the horrors of World War I seared into memory, many feared the death toll that would result from

OSMOSIS
The diffusion of water across a membrane from an area of lower solute concentration to an area of higher solute concentration.

HYPOTONIC
Describes a solution surrounding a cell that has a lower concentration of solutes than the cell's cytoplasm.

HYPERTONIC
Describes a solution surrounding a cell that has a higher concentration of solutes than the cell's cytoplasm.

ISOTONIC
Describes a solution surrounding a cell that has the same solute concentration as the cell's cytoplasm.

PEPTIDOGLYCAN
The macromolecule found in all bacterial cell walls that gives the cell wall its rigidity.

INFOGRAPHIC 3.4

Water Flows across Cell Membranes by Osmosis

The direction of water movement across the cell membrane is determined by the solute concentration on either side. Water always moves toward the side with the higher solute concentration. In a hypotonic solution, water will flow into the cell, and in a hypertonic solution, water will flow out of a cell. The bacterial cell wall helps protect the cell from lysing in a hypotonic environment. In the presence of some antibiotics the cell wall is disrupted, leaving the cell susceptible to lysis.

Hypotonic Solution	**Isotonic Solution**	**Hypertonic Solution**
▪ Higher solute concentration inside cell ▪ Water flows into the cell.	▪ Equal solute concentration in and out of cell ▪ Water flows equally in both directions.	▪ Higher solute concentration outside cell ▪ Water flows out of the cell.
Water flows in. The cell swells.	Water flows equally in both directions.	Water flows out. The cell shrivels.

In the presence of penicillin:

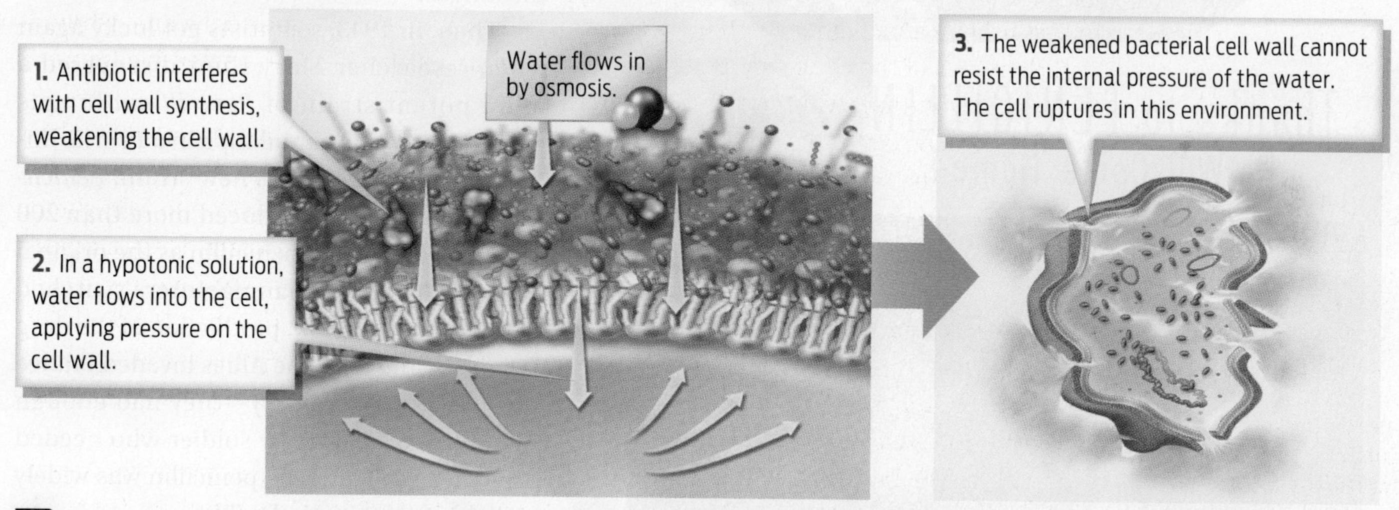

? **Human cells do not have cell walls. What will happen to a human cell placed in a hypotonic environment?**

the hostilities. Millions of soldiers and civilians had died in World War I—many not as a direct result of combat injuries, but instead from infected wounds. With few other antibacterial medicines available, penicillin suddenly became the focus of research during World War II.

In 1938, Ernst Chain, a German-Jewish biochemist, was working in Harold Florey's pathology lab at Oxford University, having fled Germany for England in 1933 when the Nazis came to power. Both Chain and Florey were interested in the biochemistry of antibacterial substances. Chain stumbled across Fleming's 1929 paper on penicillin and decided to try to isolate and concentrate the active ingredient from the mold. By 1940, he had succeeded. His breakthrough allowed Florey's group to begin testing the drug's clinical performance. They injected the purified chemical into bacteria-infected mice and found that the mice were quickly rid of their infection. Human trials followed next, in 1941, with the same remarkable result.

As encouraging as these results were, there was one nagging problem: it took as much as 2,000 liters (more than 500 gallons) of mold growing in liquid culture to obtain enough pure penicillin to treat one person. The Oxford doctors used almost their whole supply of the drug treating their first patient, a policeman ravaged by a staphylococcal infection.

Manufacturing penicillin in 1943: culture flasks are filled with the nutrient solution in which penicillin mold is grown.

Daily Herald Archive/Getty Images

Fortunately, U.S. government scientists stepped in at that point. They devised a growth method that took advantage of something the United States had in abundance: corn. Scientists were eventually able to produce penicillin in much greater quantities using a by-product of large-scale corn processing as a culture medium in which to grow the fungus.

Then, in 1943, scientists got lucky again when researcher Mary Hunt discovered a more potent strain of *Penicillium* fungus growing on a ripe cantaloupe in a Peoria, Illinois, supermarket. This new strain, *Penicillium chrysogenum,* produced more than 200 times the amount of penicillin as the original strain. When drug manufacturers switched to use *P. chrysogenum,* production of the drug soared. By the time the Allies invaded France on June 6, 1944—D-Day—they had enough penicillin to treat every soldier who needed it. By the following year, penicillin was widely available to the general public.

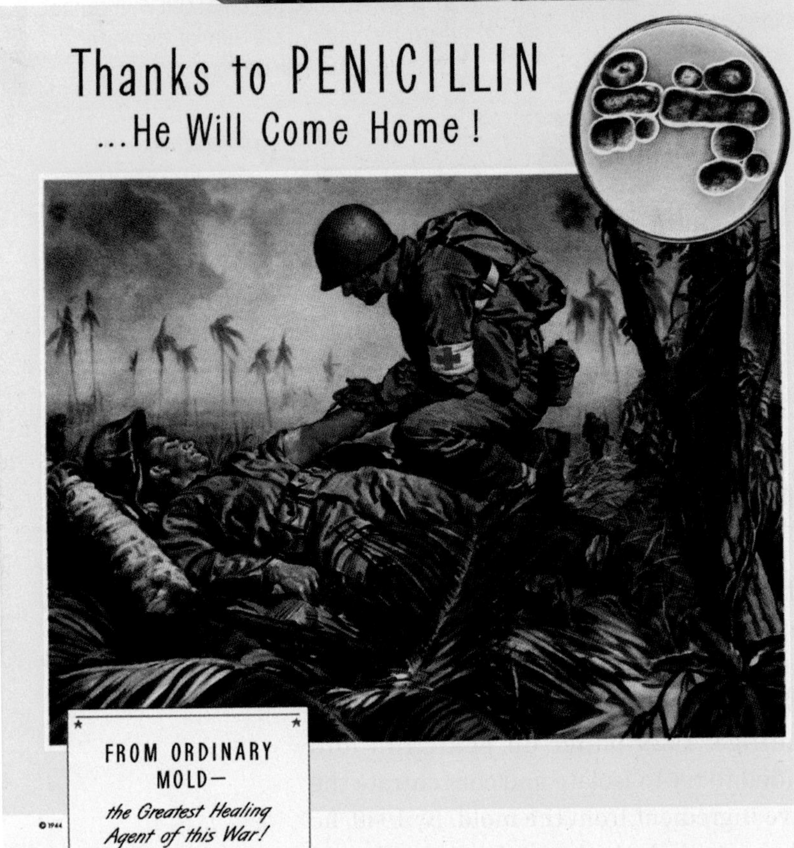

Penicillin and the war effort: feelings of wartime patriotism were enlisted to support production of the drug.

Advertising Archive/Courtesy Everett Collection

The optimism with which patients and doctors greeted the new bacteria-killer cannot be overstated. "Penicillin seemed to justify a carefree attitude to infection," says science historian Robert Bud, principal curator of the Science Museum in London. "In Western countries, for the first time in human history, most people felt that infectious disease was ceasing to be a threat, and sexually infectious disease had already been conquered. For many it seemed cure would be easier than prevention."

"Penicillin seemed to justify a carefree attitude to infection. In Western countries, for the first time in human history, most people felt that infectious disease was ceasing to be a threat."

—Robert Bud

Yet, as effective as penicillin was, it was effective only against certain types of bacteria. Against others, it was powerless.

Stockpiling the Antibiotic Arsenal

▶ How different antibiotics work

As Fleming knew, most of the bacterial world falls into one of two categories, **Gram-positive** or **Gram-negative.** These names reflect the way bacterial cell walls trap a dye known as Gram stain (after its discoverer, the Danish scientist Hans Christian Gram). Gram-positive bacteria retain the dye, whereas Gram-negative bacteria do not.

Fleming found that while penicillin easily killed Gram-positive bacteria like *Staphylococcus* and *Streptococcus*—the microbes that cause staph infections and strep throat, respectively—it had little effect on Gram-negative bacteria like *E. coli.* and *Salmonella.* That's because the cell wall of Gram-negative bacteria includes an extra layer of lipids surrounding the peptidoglycan layer. This extra lipid layer prevents penicillin from reaching the peptidoglycan underneath.

Researchers in the 1940s began to look for antibiotics that could kill Gram-negative bacteria. The first such antibiotic was streptomycin, discovered in 1943 by researchers at Rutgers University. In addition to killing Gram-negative bacteria, streptomycin was the first effective treatment for the deadly bacterial disease tuberculosis. The reason for its effectiveness? Streptomycin has

GRAM-POSITIVE
Describes bacteria with a cell wall that includes a thick layer of peptidoglycan that retains the Gram stain.

GRAM-NEGATIVE
Describes bacteria with a cell wall that includes a thin layer of peptidoglycan surrounded by an outer lipid membrane that does not retain the Gram stain.

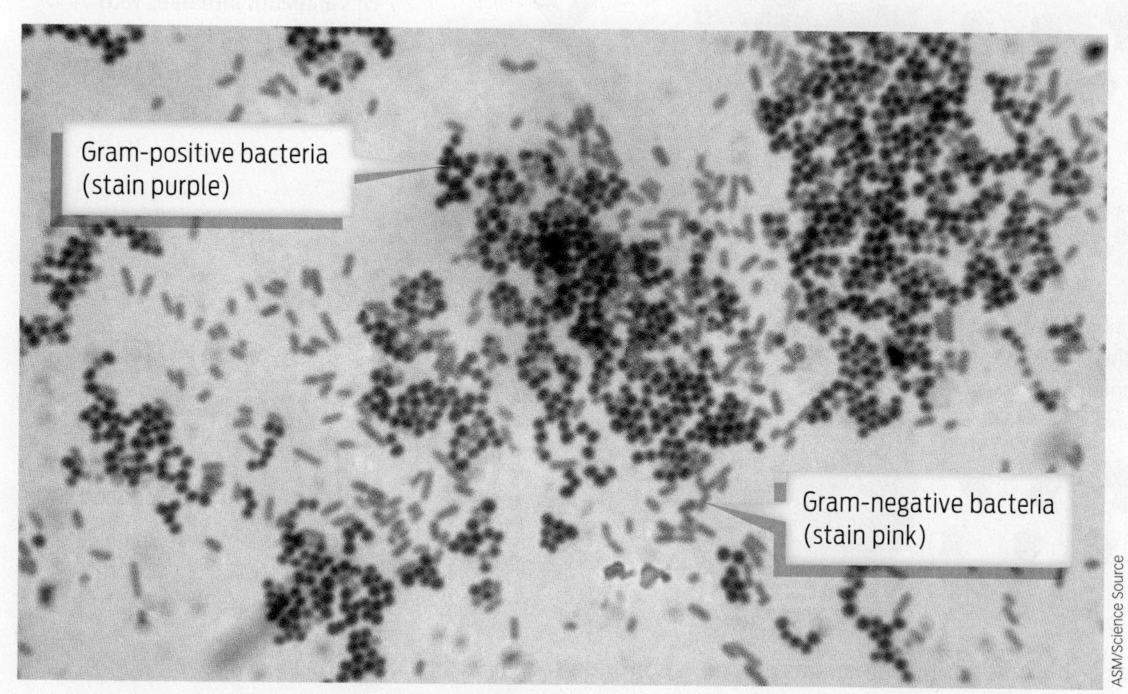

To identify bacteria, researchers treat them with a violet Gram stain. Gram-positive bacteria retain the dye well and so appear purple or blue under a microscope. Gram-negative bacteria do not retain the dye as well, but do retain a second dye, and so appear red or pink.

a chemical structure that allows it to pass more easily through the outer lipid layer of the Gram-negative bacterial cell wall. Streptomycin is classified as a broad-spectrum antibiotic because it is effective against both Gram-positive and Gram-negative bacteria.

Once inside the cell, streptomycin works by interfering with protein synthesis by bacterial ribosomes. Ribosomes are the molecular machines that assemble a cell's proteins. While both eukaryotic and prokaryotic cells have ribosomes, their ribosomes are of different sizes and have different structures. Because streptomycin targets features specific to bacterial ribosomes, it doesn't harm the eukaryotic cells of the human who is taking it **(INFOGRAPHIC 3.5).**

Other antibiotics work in different ways—by inhibiting a bacterium's ability to make a critical vitamin or to copy its DNA before dividing, for example. When either of these mechanisms are inhibited, the bacterium dies instead of reproducing.

Crossing Enemy Lines

▶ Membrane structure and transport

For any drug to be effective, it has to reach its designated target. In the case of many antibiotics, that means getting inside the cell to do their work. But the cell membrane, which surrounds all cells, acts as a barrier to the free flow of substances into the cell. How do antibiotics get past this barrier?

INFOGRAPHIC 3.5

Some Antibiotics Inhibit Prokaryotic Ribosomes

Ribosomes are responsible for the synthesis of proteins in both prokaryotic and eukaryotic cells, but their structure is slightly different in the two types of cells. Antibiotics that interfere with prokaryotic ribosomes leave eukaryotic ribosomes unaffected.

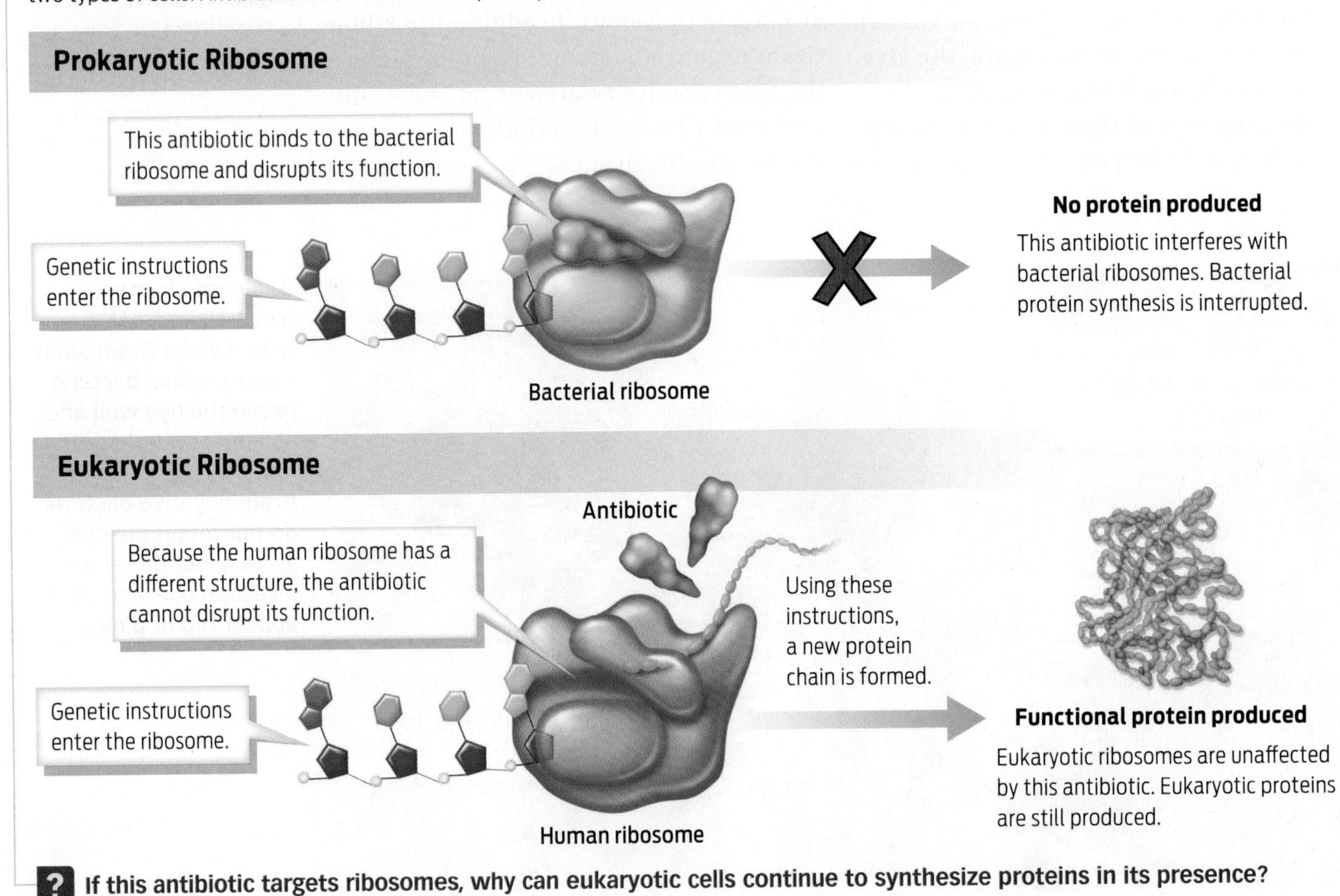

? If this antibiotic targets ribosomes, why can eukaryotic cells continue to synthesize proteins in its presence?

To understand how substances move across the cell membrane, it helps to know a bit more about its structure. The cell membrane is a flexible yet durable structure composed of phospholipids and proteins. A phospholipid has two main parts: a hydrophilic "head" and a hydrophobic "tail." In the watery context of a cell, the hydrophobic tails of phospholipids cluster together in the middle of the membrane away from water, while the hydrophilic heads face outward, toward water. When arranged in this way, the phospholipids form a two-ply structure called a bilayer. Proteins sit nestled in the lipid bilayer, where they perform a variety of functions such as relaying signals across the membrane and transporting nutrients in and wastes out of the cell **(INFOGRAPHIC 3.6).**

INFOGRAPHIC 3.6

Membranes: All Cells Have Them

All cells are enclosed by a cell membrane made of phospholipids and proteins. The hydrophilic heads of the phospholipids face "out" (on either side of the membrane) and interact with water, while the hydrophobic tails cluster together inside the membrane away from water.

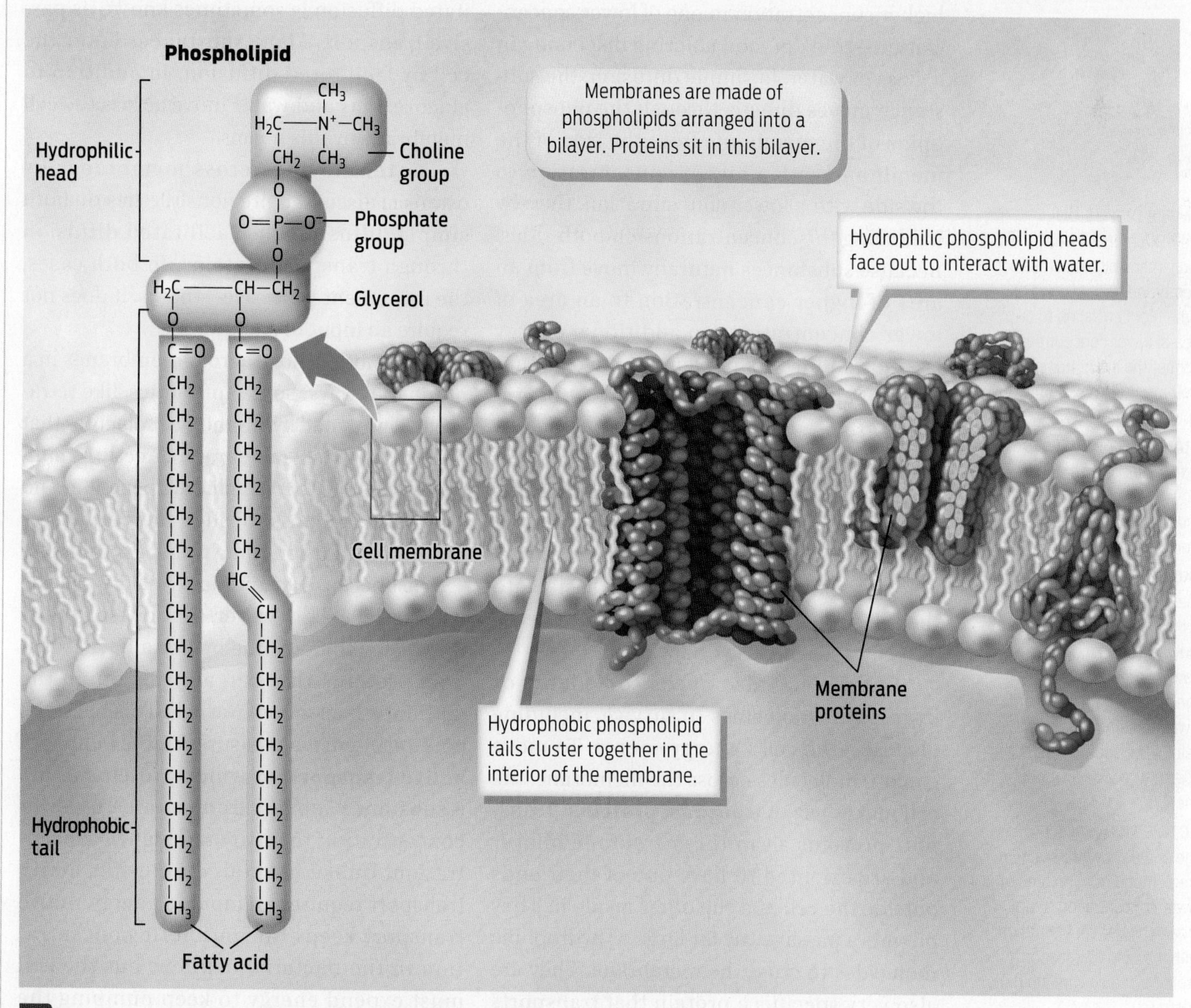

? Which portion of a phospholipid is found in the middle (interior) of a membrane?

The cell membrane is semipermeable, meaning that only substances with certain characteristics can cross it easily without help. Because of its densely packed collection of hydrophobic phospholipid tails, the lipid bilayer is largely impermeable to large molecules, like glucose, and hydrophilic or charged substances, like sodium ions; it is only weakly permeable to water. In fact, the only things that cross the lipid bilayer easily and without assistance are small, uncharged molecules like oxygen (O_2) and carbon dioxide (CO_2) gas, which cross by **simple diffusion.** Diffusion is the natural tendency of dissolved substances to move from an area of higher concentration to one of lower concentration—think of food coloring dispersing in a glass of water. In simple diffusion, the substance moves directly through the phospholipids of the membrane from the side of the membrane with a higher concentration to the side with a lower concentration, thereby balancing the concentrations on both sides. Because substances naturally move from an area of higher concentration to an area of lower concentration, no additional energy is required for this movement beyond that stored in the concentration difference, or gradient, itself.

Take oxygen, for example. The concentration of oxygen molecules, which are small and uncharged, is often higher outside the cell and lower inside it. This concentration gradient allows oxygen to diffuse easily into the cell—a good thing, because the cell needs oxygen to survive (see Chapter 6).

But the cell also needs some large or hydrophilic molecules to survive, including glucose—the cell's energy source. To move glucose molecules across the membrane, the cell makes use of **transport proteins.** Transport proteins sit in the membrane bilayer and are oriented to have one of their ends outside the cell and the other inside it. They provide a passageway for large or hydrophilic molecules to cross the membrane. They are also very specific: a protein that transports glucose will not transport calcium ions, for example. The cells of the body contain hundreds of different transport proteins, each specific for a different type of cargo.

Transport proteins can move substances either with or against their concentration gradient—either "downhill" or "uphill" across the membrane. When a substance moves "downhill" by a transport protein from an area of higher concentration to an area of lower concentration, the process is called **facilitated diffusion.** Like simple diffusion, facilitated diffusion requires no additional energy besides that stored naturally in the concentration gradient. For this reason, facilitated diffusion is sometimes known as passive transport. Many substances enter the cell by facilitated diffusion. In addition to glucose, ions and water move across the cell membrane by this means.

Water moving across membranes by osmosis, discussed previously, relies on both simple diffusion and facilitated diffusion through transport proteins. In both cases, the movement is passive—that is, it does not require an input of energy.

Antibiotics move across membranes in a number of ways. Some antibiotics, like tetracycline, are small hydrophobic molecules that can cross the cell membrane directly by simple diffusion. Others, including penicillin and streptomycin, pass through membranes by facilitated diffusion using transport proteins.

Just because an antibiotic makes it inside a bacterial cell, however, that doesn't mean it will stay there. Some bacteria have transport proteins that can actively pump the antibiotic back out of the cell. This bacterial counteroffensive measure is an example of **active transport,** in which proteins pump a substance "uphill" from an area of lower concentration to an area of higher concentration. Unlike facilitated diffusion, active transport requires an input of energy. Active transport keeps the antibiotic concentration in the bacterial cell low, but the cell must expend energy to keep pumping the

SIMPLE DIFFUSION
The movement of small, uncharged solutes across a membrane from an area of higher concentration to an area of lower concentration without the aid of transport proteins; does not require an input of energy.

TRANSPORT PROTEINS
Proteins involved in the movement of molecules and ions across the cell membrane.

FACILITATED DIFFUSION
The process by which large, hydrophilic, or charged solutes move across a membrane from an area of higher concentration to an area of lower concentration with the help of transport proteins; does not require an input of energy.

ACTIVE TRANSPORT
The process by which solutes are pumped from an area of lower concentration to an area of higher concentration with the help of transport proteins; requires an input of energy.

INFOGRAPHIC 3.7

Molecules Move across the Cell Membrane

The cell membrane is semipermeable: only a few substances can cross it unassisted. Solutes move across the membranes depending on solute size and charge, the relative concentration on each side of the membrane, and the presence or absence of transport proteins.

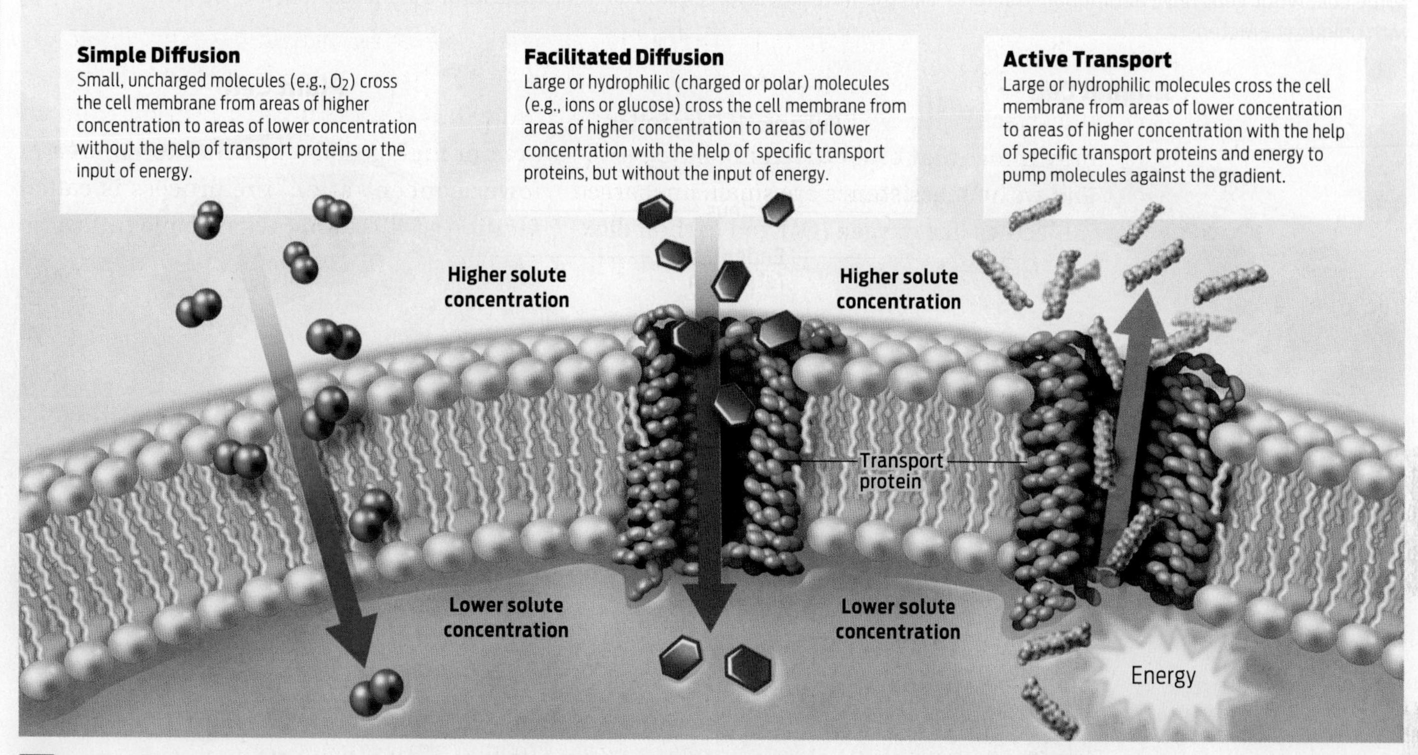

? **What is a major difference between diffusion and active transport?**

antibiotic out against its concentration gradient. Eukaryotic cells use active transport for several important purposes—for example, to maintain ion gradients across the membranes of nerve cells and to import certain nutrients "uphill" against their concentration gradient **(INFOGRAPHIC 3.7).**

Pumping antibiotics out of the bacterial cell is one way in which bacteria can resist the destructive power of an antibiotic. Others include chemically breaking down the antibiotic with enzymes. Why would bacteria have such built-in mechanisms for counteracting or resisting these drugs? Remember that penicillin was originally isolated from a living organism, a fungus. Streptomycin comes from microorganisms living in soil. Penicillin and streptomycin are chemical defenses evolved by these organisms that protect them from bacteria. In turn, their combatants have evolved countermeasures that give them resistance to these antibiotics. Humans thus find themselves embroiled in a battle originally waged solely between microorganisms.

Your Inner Bacterium

▸ Eukaryotic organelles

Antibiotics kill bacteria but leave humans unharmed because bacterial and human cells have different structures. Of all the ways that prokaryotic and eukaryotic cells differ, the most obvious is that eukaryotic cells are more complex than their smaller prokaryotic cousins. In particular, eukaryotic cells

INFOGRAPHIC 3.8

Eukaryotic Cells Have Organelles

Humans and other animals, as well as plants, fungi, and protists, are eukaryotes—they are made up of eukaryotic cells that contain internal organelles. All eukaryotic cells have a nucleus, endoplasmic reticulum, ribosomes, mitochondria, and Golgi apparatus. Some eukaryotic cells have additional structures, like the water vacuole, cell wall, and chloroplasts of a plant cell, that provide them with unique characteristics.

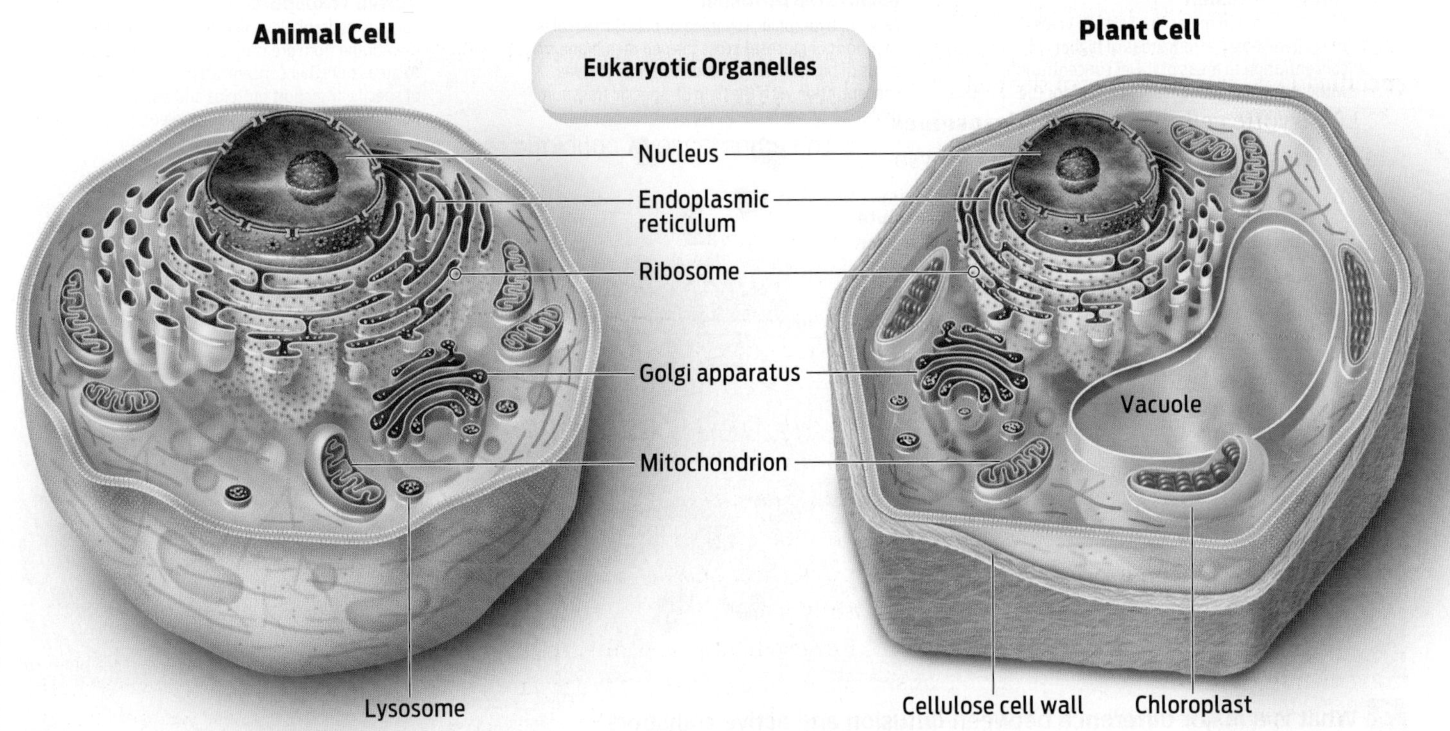

? Name and give the functions of at least three organelles present in both plant and animal cells.

contain multiple, distinct membrane-bound organelles **(INFOGRAPHIC 3.8).**

You can think of a eukaryotic cell as a miniature factory with an efficient division of labor. Each organelle is separated from the cell's cytoplasm by a membrane similar to the cell's outer membrane, and each performs a distinct function.

The nucleus is the defining organelle of eukaryotic cells (whose name is derived from the Greek *eu,* meaning "good" or "true," and *karyon,* meaning "nut" or "kernel"). It is surrounded by the **nuclear envelope,** a double membrane made of two lipid bilayers dotted by small openings called pores. The nucleus encloses the cell's DNA and acts as a kind of control center. Important reactions for interpreting the genetic instructions contained in DNA take place in the nucleus.

Other organelles in a eukaryotic cell perform other specialized tasks. **Mitochondria** are the cell's "power plants": they use oxygen to extract energy from food and convert that energy into a useful form. All eukaryotes—including plants—have mitochondria. Humans who inherit or develop defects in their mitochondria usually die—an indication of just how important these organelles are.

The **endoplasmic reticulum (ER)** is a vast network of membranes that serves as a kind of assembly line for the manufacture of proteins and lipids. The "rough" ER is studded with ribosomes making proteins; the

NUCLEAR ENVELOPE
The double membrane surrounding the nucleus of a eukaryotic cell.

MITOCHONDRIA (SINGULAR: MITOCHONDRION)
Membrane-bound organelles responsible for important energy-conversion reactions in eukaryotes.

ENDOPLASMIC RETICULUM (ER)
A network of membranes in eukaryotic cells where proteins and lipids are synthesized.

"smooth" ER makes lipids. Newly made proteins travel from the ER to the **Golgi apparatus,** an organelle that packs the protein "cargo" into small membrane-enclosed sacs called vesicles and then ships them to specific destinations, such as the cell membrane, other organelles, and the bloodstream. The nucleus, ER, and Golgi apparatus work together to make and transport proteins to specific locations in and out of the cell.

Eukaryotic cells also contain **lysosomes,** which digest and repurpose molecules. Lysosomes can be thought of as the cell's recycling centers.

In addition to these membrane-bound structures, a vast network of protein fibers called the **cytoskeleton** allows cells to move and maintain their shape, much the same way the human skeleton does.

Finally, in addition to the previously described organelles, plant cells contain **chloroplasts,** which carry out photosynthesis. They also have a cell wall made of cellulose **(INFOGRAPHIC 3.9).**

Prokaryotic cells carry out similar functions of energy conversion and protein transport, but they don't confine these processes within separate organelles. Instead, everything occurs in the cytoplasm.

How did eukaryotic cells develop their factory-like compartments? That question has long intrigued biologists. One fascinating hypothesis was proposed in the 1960s by biologist Lynn Margulis. She argued that eukaryotic organelles such as mitochondria and chloroplasts were once free-living prokaryotic cells that became incorporated—engulfed—by other free-living prokaryotic cells.

Although many considered Margulis's hypothesis of **endosymbiosis** a crazy idea at first, a wealth of evidence now supports it (see **Milestone 1: Scientific Rebel**). Mitochondria and chloroplasts are about the same size and shape as bacteria. Both mitochondria and chloroplasts have circular strands of DNA, just like prokaryotic cells. They also contain ribosomes that are similar in structure to prokaryotic ribosomes—so similar, in fact, that some antibiotics that target prokaryotic ribosomes can affect the ribosomes in eukaryotic mitochondria, which accounts for both the toxicity and the side effects of these antibiotics.

Winning the Battle, Losing the War

▸ The emergence of antibiotic resistance

To the patients who first benefited from its healing powers, penicillin truly seemed to be a wonder drug. A once-lethal bacterial infection could now be cleared in a matter of days with a course of an antibiotic. Today, antibiotics are some of the most commonly prescribed medications.

Antibiotics are so common, in fact, that many people routinely take them when they catch a cold or the flu. But antibiotics are powerless against these illnesses, because they are caused by viruses, not bacteria. Since viruses are not made of cells—and according to the cell theory are not even considered to be alive—they can't be killed with an antibiotic.

But that hasn't stopped people from trying. The Centers for Disease Control and Prevention (CDC) estimates that as many as 50% of the antibiotics prescribed for upper respiratory illnesses are unnecessary, since they are being used (ineffectively) to treat colds and other viral infections. What's more, many patients who are prescribed antibiotics for bacterial infections use them improperly. Taking only part of a prescribed dose, for example, can spare some harmful bacteria living in the body, and those bacteria that survive are often heartier and more resistant to the antibiotic than the ones that were killed. These patterns of overuse and misuse of antibiotics have led to an epidemic of antibiotic resistance, which the CDC calls "one of the world's most pressing public health problems."

GOLGI APPARATUS
An organelle made up of stacked membrane-enclosed discs that packages proteins and prepares them for transport.

LYSOSOME
An organelle in eukaryotic cells that is filled with enzymes that can degrade worn-out cellular structures.

CYTOSKELETON
A network of protein fibers in eukaryotic cells that provides structure and facilitates cell movement.

CHLOROPLAST
An organelle in plant and algal cells that is the site of photosynthesis.

ENDOSYMBIOSIS
The scientific theory that free-living prokaryotic cells engulfed other free-living prokaryotic cells billions of years ago, forming eukaryotic organelles such as mitochondria and chloroplasts.

INFOGRAPHIC 3.9 UP CLOSE

Eukaryotic Organelles

Nucleus

The nucleus is the defining organelle of eukaryotic cells. The nucleus is separated from the cytoplasm by a double membrane (two phospholipid bilayers) known as the nuclear envelope. The nuclear envelope controls the passage of molecules between the nucleus and cytoplasm. The nucleus contains the DNA, the stored genetic instructions of each cell. In addition, important reactions for interpreting these genetic instructions occur in the nucleus.

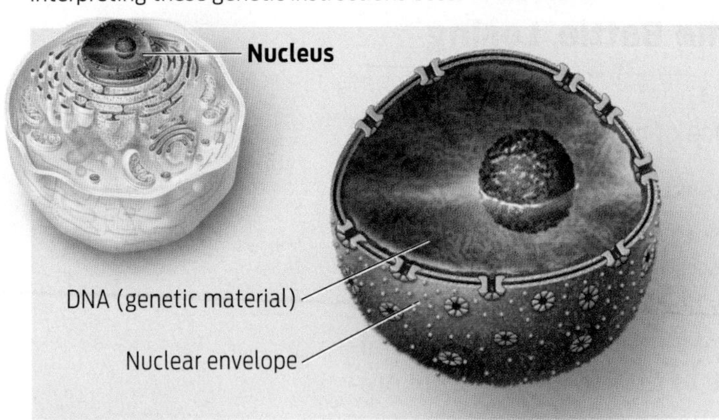

Endoplasmic Reticulum

The endoplasmic reticulum (ER) is an extensive, membranous intracellular "plumbing" system that is critical for the production of new proteins. The "rough" ER has a rough appearance because it is studded with ribosomes that are making proteins. The rough ER is contiguous with the "smooth" ER, the site of lipid production.

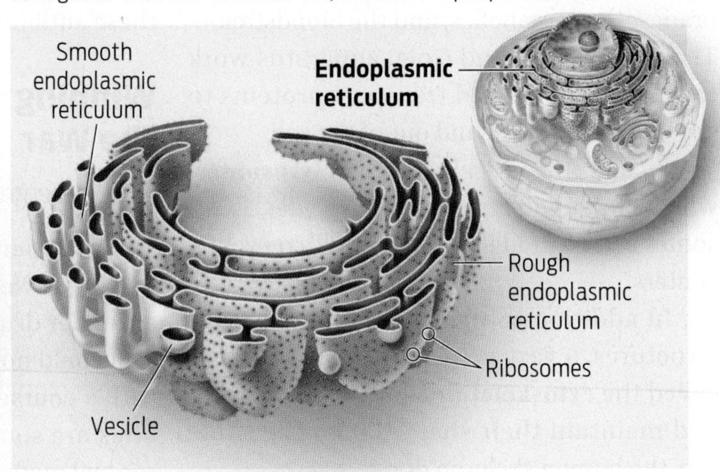

Golgi Apparatus

The Golgi apparatus is a series of flattened membrane compartments, whose purpose is to process and package proteins produced in the rough endoplasmic reticulum.

The Nucleus, Endoplasmic Reticulum, and Golgi Apparatus Work Together to Produce and Transport Proteins

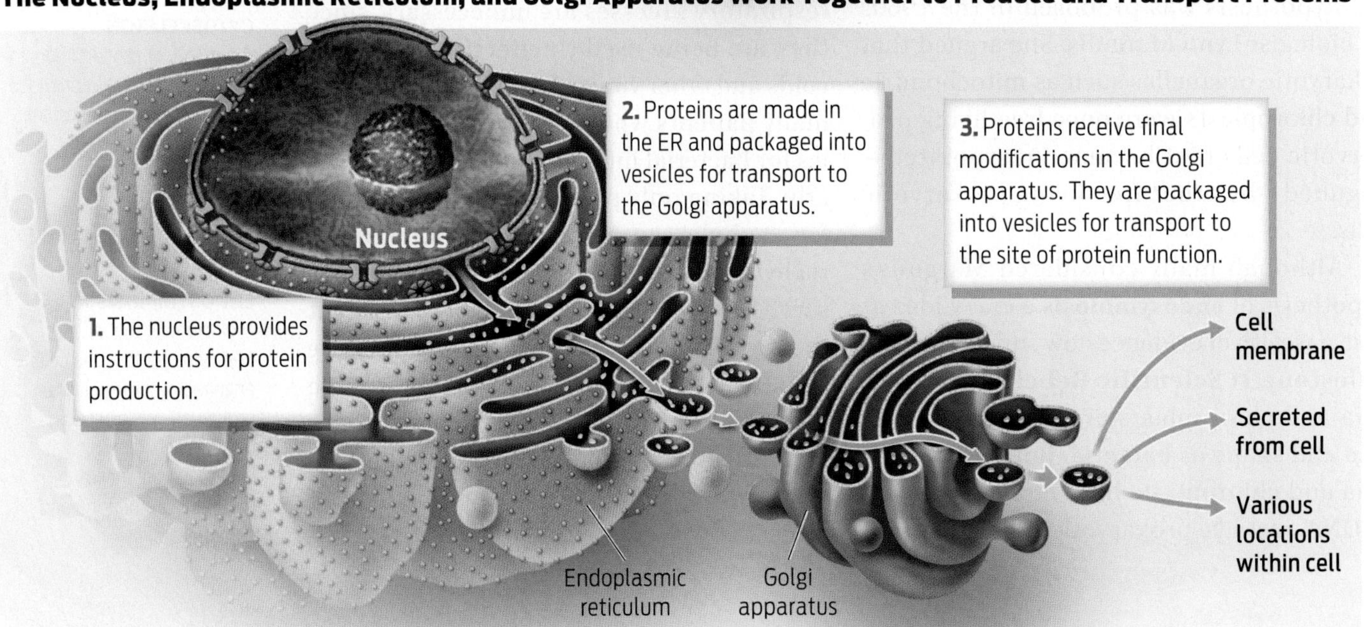

Mitochondria

Mitochondria are found in almost all eukaryotes, including plants. Mitochondria have two membranes surrounding them. The inner one is highly folded. Mitochondria carry out critical steps in the extraction of energy from food, and the conversion of that "trapped" energy to a useful form. They are the cell's "power plants."

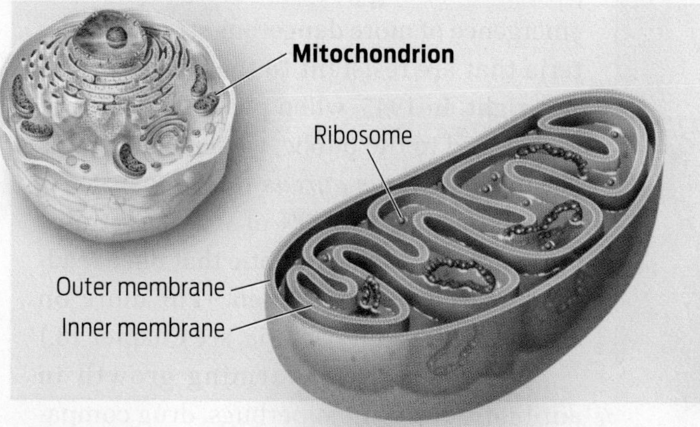

Chloroplast

Chloroplasts are organelles found in algae and in the green parts of plants. Chloroplasts have two membranes surrounding them, as well as an internal system of stacked membrane discs. Chloroplasts are the sites of photosynthesis, the reactions by which plants capture the energy of sunlight and convert it to a usable form.

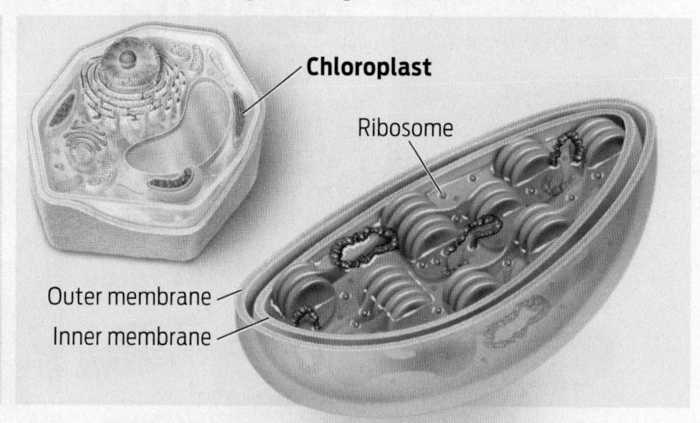

Lysosome

Lysosomes are the cell's digestive and recycling centers. They break down food and other molecules taken into the cell. The digested parts are then used in the cell for various function. Lysosomes also recycle worn-out organelles and other cellular components, so their subunits can be reused for new cell structures.

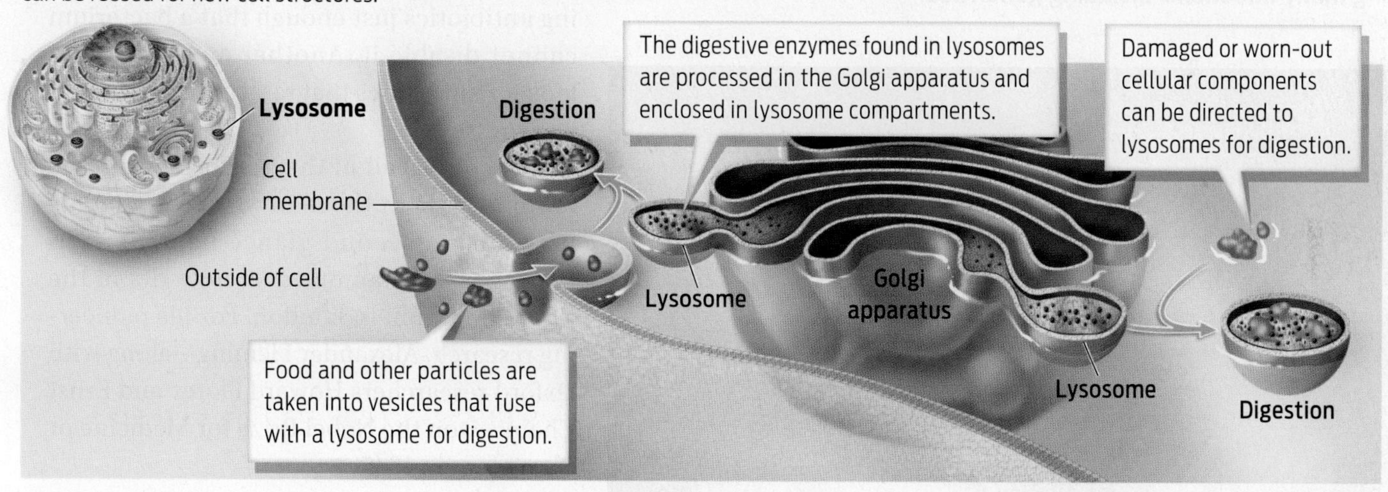

Cytoskeleton

The cytoskeleton is a meshwork of protein fibers that carry out a variety of functions, including cell support, cell movement, and movement of structures within cells. There are three main types of cytoskeletal fibers: microfilaments, intermediate filaments, and microtubules.

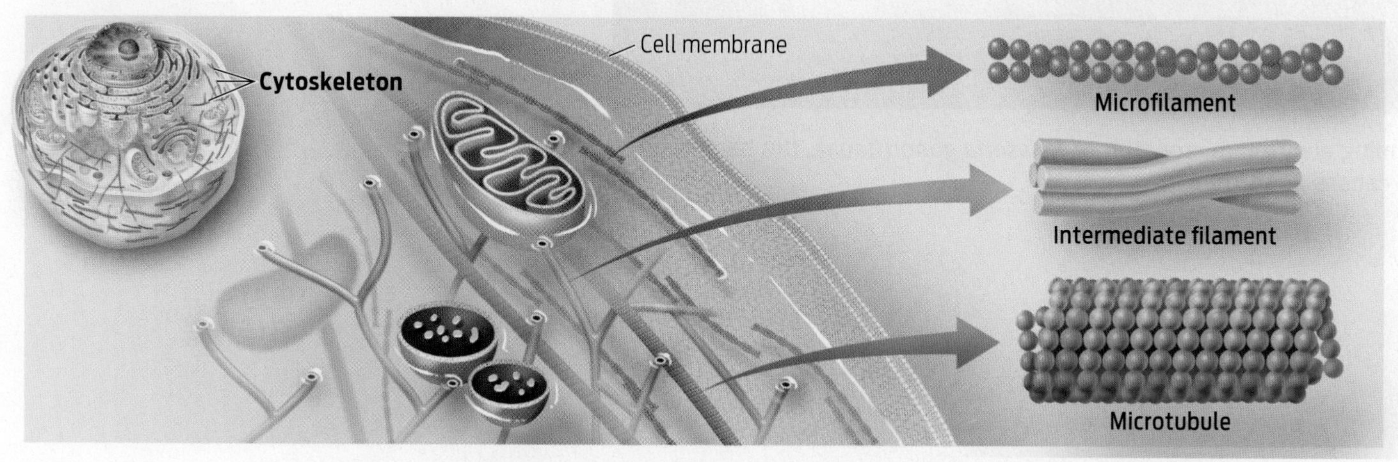

National Library of Medicine

Once perceived as a magic bullet, penicillin is now largely ineffective at treating many infections, including gonorrhea.

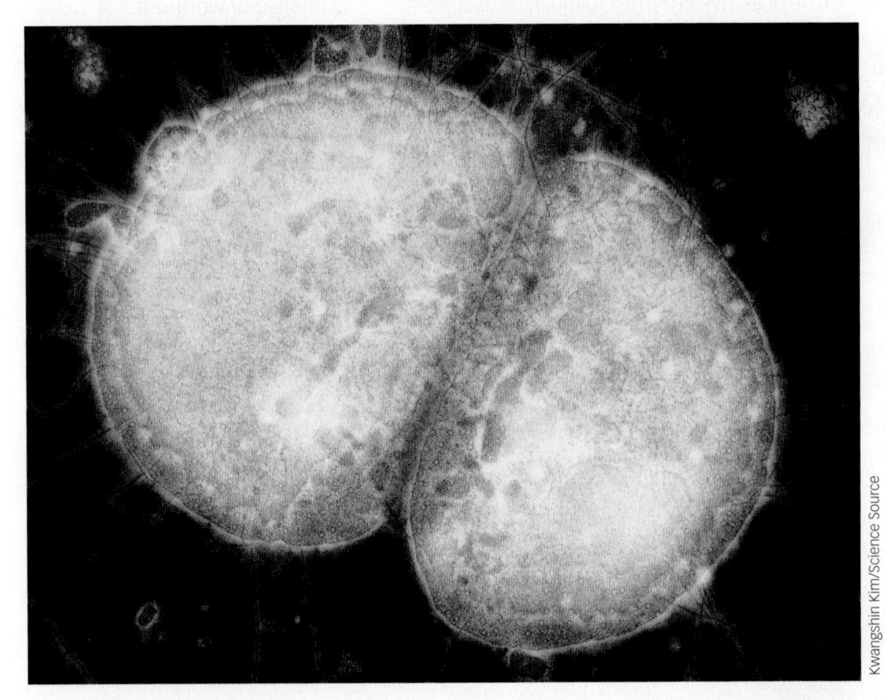

Kwangshin Kim/Science Source

Scanning electron micrograph of *Neisseria gonorrhoeae,* the bacterium that causes gonorrhea.

Fleming foresaw this very danger. In his own research, he found that whenever he used too little penicillin or when he used it for too short a time, populations of bacteria emerged that were resistant to the antibiotic. Fleming warned that improper use of penicillin among patients could lead to the emergence of more dangerous strains of bacteria that are resistant to the drug—and he was right. In 1945, when penicillin was first introduced to the public, virtually all strains of *Staphylococcus aureus* were sensitive to it. Today, more than 90% of *S. aureus* strains are resistant to the antibiotic that once readily defeated this pathogen. (For more on antibiotic-resistant bacteria, see Chapter 13.)

Because of the alarming growth in antibiotic-resistant superbugs, drug companies and researchers are working to develop new antibiotics. One strategy they employ is to tweak the chemical structure of existing antibiotics just enough that a bacterium cannot disable it. Another approach is to look for antibiotics that target other bacterial weaknesses.

Of course, all of these efforts would be nothing without the man who gave a moldy petri dish a second glance nearly a century ago. That famous dish now sits in the British Museum in London. For his pioneering research, Alexander Fleming—along with Oxford researchers Howard Florey and Ernst Chain—won the Nobel Prize for Medicine or Physiology in 1945. ■

CHAPTER 3 Summary

Driving Question 1 What structural features are shared by all cells, and what are the key differences between prokaryotic and eukaryotic cells?

- According to the cell theory, all living organisms are made of cells. New cells are formed when an existing cell divides.
- The two fundamental types of cells—prokaryotic and eukaryotic—are distinguished by their structures. Prokaryotic cells lack membrane-bound organelles; eukaryotic cells have a variety of membrane-bound organelles, including a central nucleus.
- All cells are enclosed by a cell membrane made up of phospholipids and proteins. Some cells also have a cell wall surrounding the cell membrane that imparts additional strength.
- All cells contain DNA and ribosomes, the molecular machines for making proteins.

Driving Question 2 How do solutes and water cross cell membranes, and what determines the direction of movement of solutes and water in different situations?

- Water has a tendency to cross cell membranes by osmosis in the direction that will balance the solutes on each side. In a hypotonic solution, water moves into the cell, causing it to swell; in a hypertonic solution, water leaves the cell, causing it to shrivel.
- The cell membrane is semipermeable: only substances with certain characteristics can cross it freely without help.
- Small hydrophobic molecules can cross cell membranes by simple diffusion, a process that does not require an input of energy.
- Large, hydrophilic, or charged molecules are transported across the membrane with the help of membrane transport proteins.
- Facilitated diffusion is the transport of molecules down a concentration gradient through a transport protein; it does not require an input of energy.
- Active transport is the transport of molecules up a concentration gradient through a transport protein; it requires an input of energy.

Driving Question 3 How do different antibiotics target bacteria?

- Antibiotics are chemicals, originally produced by living organisms, that selectively target and kill bacteria.
- Bacteria are surrounded by a cell wall containing peptidoglycan, a molecule not found in eukaryotes. Some antibiotics, like penicillin, work by preventing peptidoglycan synthesis, weakening the cell wall.
- Other antibiotics, like streptomycin, work by interfering with prokaryotic ribosomes and therefore protein synthesis in the cell.
- Increased and sometimes inappropriate use of antibiotics has led to the emergence of antibiotic-resistant bacteria. Infections caused by these bacteria are very hard to treat.

Driving Question 4 What are the key eukaryotic organelles and their functions?

- Eukaryotic cells contain a number of specialized organelles, including a nucleus, endoplasmic reticulum, Golgi apparatus, mitochondria, lysosomes, and chloroplasts. Each of these organelles carries out a distinct function.
- Eukaryotic cells likely evolved as a result of endosymbiosis, the engulfing of one single-cell prokaryote by another.

More to Explore

- Video: Penicillin Acting on Bacteria: http://www.hhmi.org/biointeractive/penicillin-acting-bacteria
- Centers for Disease Control and Prevention, Antibiotic Resistance: http://www.cdc.gov/drugresistance/
- Fleming, A. (1929). On the antibacterial action of cultures of a penicillium. *Br J Exper Pathol* 10:226–236.
- Lax, R. (2005). *The Mold in Dr. Florey's Coat: The Story of the Penicillin Miracle.* New York: Owl Books.
- Williams, Michelle. (2019). We can't despair about our antibiotic crisis. *Washington Post*. https://www.washingtonpost.com/opinions/dont-despair-our-antibiotic-resistance-crisis/2019/07/08/b164ea4e-9f62-11e9-b27f-ed2942f73d70_story.html

CHAPTER 3 Test Your Knowledge

Driving Question 1 What structural features are shared by all cells, and what are the key differences between prokaryotic and eukaryotic cells?

By answering the questions below and studying Infographics 3.2, 3.3, 3.5, 3.6, and 3.8, you should be able to generate an answer for this broader Driving Question.

Know It

1. What does the cell theory state?

2. Which of the following statements best explains why bacteria are considered living organisms?

- **a.** They can cause disease.
- **b.** They are made up of biological macromolecules.
- **c.** They move around.
- **d.** They are made of cells.
- **e.** They contain organelles.

3. What are the two main types of cells found in organisms?

4. Which of the following is/are *not* associated with human cells?

- **a.** cell membrane
- **b.** ribosomes
- **c.** DNA
- **d.** cell wall
- **e.** All of the above are associated with human cells.

5. Bacteria are ____________ cells, defined by the ____________.

- **a.** prokaryotic; presence of a cell wall
- **b.** eukaryotic; presence of organelles
- **c.** eukaryotic; absence of a cell wall
- **d.** prokaryotic; absence of organelles
- **e.** eukaryotic; absence of organelles

6. Which of the following is/are associated with eukaryotic cells but not with prokaryotic cells?

- **a.** cell membrane
- **b.** cell wall
- **c.** DNA
- **d.** ribosomes
- **e.** nucleus

Use It

7. According to the cell theory, all living organisms are made of cells. More specifically, what do all living organisms have in common? For example, do all living organisms carry genetic instructions? Do their cells all have a nucleus? What other features do they have in common?

8. In the soil of a forest, you find a single-celled organism with a cell wall—could this organism be an animal? Why or why not? Which of the following facts would convince you that the organism is a bacterium and not a plant?

- **a.** The cell wall is made of cellulose.
- **b.** The DNA is contained in a nucleus.
- **c.** The cell wall is made of peptidoglycan.
- **d.** a and b
- **e.** b and c

Driving Question 2 How do solutes and water cross cell membranes, and what determines the direction of movement of solutes and water in different situations?

By answering the questions below and studying Infographics 3.4 and 3.7, you should be able to generate an answer for this broader Driving Question.

Know It

9. The two major components of cell membranes are ____________ and ____________.

- **a.** phospholipids; DNA
- **b.** DNA; proteins
- **c.** peptidoglycan; phospholipids
- **d.** peptidoglycan; proteins
- **e.** phospholipids; proteins

10. If a solute is moving through a phospholipid bilayer from an area of higher concentration to an area of lower concentration without the assistance of a protein, the manner of transport must be

- **a.** active transport.
- **b.** facilitated diffusion.
- **c.** simple diffusion.
- **d.** any of the above, depending on the solute.
- **e.** Solutes cannot cross phospholipid bilayers.

11. Consider the movement of molecules across the cell membrane.

- **a.** What do simple diffusion and facilitated diffusion have in common?
- **b.** What do active transport and facilitated diffusion have in common?

12. Water is moving across a membrane from solution A into solution B. What can you infer?

- **a.** Solution A must be pure water.
- **b.** Solution A must have a lower solute concentration than Solution B does.
- **c.** Solution A must have a higher solute concentration than Solution B does.
- **d.** Solution A and Solution B must have the same concentration of solutes.
- **e.** Solution B must be pure water.

Use It

13. Why does facilitated diffusion require membrane transport proteins while simple diffusion does not?

14. Sugars are large, hydrophilic molecules that are important energy sources for cells. How can they enter cells from an environment with a very high concentration of sugar?

- **a.** by simple diffusion
- **b.** by osmosis
- **c.** by facilitated diffusion
- **d.** by active transport
- **e.** by using ribosomes

15. Many foods—for example, bacon and salt cod—are preserved by applying high concentrations of salt. How can high concentrations of salt inhibit the growth of bacteria? (Think about the high solute concentration of the salty food relative to the solute concentration in the bacterial cells. What will happen to the bacterial cells under these conditions?)

Mini Case

Apply Your Knowledge

16. Marc, a first-year college student, starts out on a backpacking trip in southern New Mexico. It is September, so the daytime temperatures are quite high, and the desert air is very dry. He has a portable water filter to treat river and stream water that he finds on his planned route through the Gila wilderness. On the second day of his week-long trip, Marc's water filter breaks. He is afraid of contracting giardiasis (a protozoal disease spread through water contaminated by animal feces), so he drinks only the small amount of water that he can boil on his camp stove at night. By the fifth night, Marc is feeling weak and thirsty, and starts to hike out. He makes it to a local highway and collapses. A passing motorist calls 911 for an ambulance.

- **a.** Given that Marc has sweated a lot, and that sweating causes the loss of more water than solutes, what has happened to the solute concentration of Marc's blood as a result of his dehydration?
- **b.** From the solute concentration of Marc's blood, what is likely to be happening to his body cells that are in contact with his blood and related fluids (e.g., lymph and cerebral spinal fluid)?
- **c.** The paramedics have three saline solutions available. The first is a "normal" isotonic saline solution (0.9% NaCl), the second is a hypertonic saline solution (3% NaCl), and the third is a "half normal" saline solution (0.45% NaCl). Which one would you use to treat Marc? Why?

Driving Question 3 How do different antibiotics target bacteria?

By answering the questions below and studying Infographics 3.4 (bottom) and 3.5, you should be able to generate an answer for this broader Driving Question.

Know It

17. Penicillin interferes with the synthesis of

- **a.** bacterial cell membranes.
- **b.** peptidoglycan.
- **c.** the nuclear envelope.
- **d.** membrane proteins.
- **e.** ribosomes.

18. Would phospholipids of the cell membrane be a good target for an antibiotic? Explain your answer.

Use It

19. A bacterial infection is being treated with an antibiotic that stops bacterial reproduction by blocking DNA replication (bacterial cells cannot reproduce if they cannot replicate their DNA). The physician decides to add penicillin, which inhibits the production of new peptidoglycan. Would this use of penicillin be effective (i.e., will it have any additional impact on treating the infection)? Explain your answer.

20. If bacterial cells were placed in a nutrient-containing solution (one that supports their growth) that had the same solute concentration as the cytoplasm but also contained penicillin, would the cells burst? Explain your answer. What if the same experiment were repeated with lysozyme, an enzyme that degrades intact peptidoglycan? What if the two experiments were repeated in solutions that have lower solute concentrations than the cytoplasm, and did *not* contain growth-supporting nutrients?

21. Fungi are eukaryotic organisms. Scientists have found it more challenging to develop treatments for fungal infections (e.g., yeast infections, athlete's foot, and certain nail infections) than for bacterial infections. Why is this so?

Apply Your Knowledge

Interpreting Data

22. Bacteria can be characterized as sensitive, intermediately resistant, or fully resistant to different antibiotics. If a strain of bacteria is sensitive to an antibiotic, we can prescribe that antibiotic to treat an infection caused by that strain and have confidence it will work. If the strain is fully resistant to an antibiotic, that antibiotic cannot treat that infection. In cases of intermediate resistance, it is better to try to find an antibiotic to which the strain is sensitive, as the infection may not respond to antibiotics to which it has intermediate resistance.

The table shows the concentrations of antibiotics that determine how a bacterial species will respond to those antibiotics. A sensitive strain will be killed by the concentration of antibiotic shown in the "sensitive" column. A strain with intermediate resistance will be killed only by concentrations in the range indicated in the "intermediately resistant" column. A fully resistant strain requires concentrations shown in the "fully resistant" column to be killed.

A hospital patient has a *Staphylococcus aureus* infection. As part of laboratory testing, the *S. aureus* from the patient was grown in different concentrations of various antibiotics. For oxacillin, the lowest concentration that inhibited the growth of the strain was 8 µg/ml; for vancomycin, 4 µg/ml; for erythromycin, 16 µg/ml; for tetracycline, 32 µg/ml; and for levofloxacin, 8 µg/ml. Which antibiotic should be used to treat the infection in this patient?

Antibiotic	Effective Dose for Sensitive Bacterial Strains (µg/ml)	Effective Dose for Intermediately Resistant Bacterial Strains (µg/ml)	Effective Dose for Fully Resistant Bacterial Strains (µg/ml)
Oxacillin	≤2 µg/ml		≥4 µg/ml
Vancomycin		8–16 µg/ml	≥32 µg/ml
Erythromycin	≤0.5 µg/ml	1–4 µg/ml	≥8 µg/ml
Tetracycline	≤4 µg/ml		≥16 µg/ml
Levofloxacin	≤2 µg/ml	4 µg/ml	≥8 µg/ml

Apply Your Knowledge

Bring It Home

23. Many patients pressure their physicians to prescribe antibiotics for colds. Is this a good idea? Why or why not?

Driving Question 4 What are the key eukaryotic organelles and their functions?

By answering the questions below and studying Infographics 3.8 and 3.9, you should be able to generate an answer for this broader Driving Question.

Know It

24. Briefly describe the structure and function of each of the following eukaryotic organelles:

a. mitochondrion
b. nucleus
c. endoplasmic reticulum
d. chloroplast

25. Which of the following is *not* a cytoskeletal fiber in eukaryotic cells?

a. macrotubules
b. intermediate filaments
c. microfilaments
d. microtubules
e. All of the above are cytoskeletal fibers.

26. Insulin is a protein hormone secreted by certain pancreatic cells into the bloodstream. Which of the following organelles is/are involved in the synthesis and secretion of insulin?

a. rough ER
b. Golgi apparatus
c. ribosomes
d. all of the above
e. a and c

Use It

27. Some inherited syndromes, such as Tay-Sachs disease and MERRF (myoclonic epilepsy with ragged red fibers), interfere with the function of specific organelles. MERRF disrupts mitochondrial function. From what you know about mitochondria, why do you think the muscles and the nervous system are the predominant tissues affected in MERRF? (Think about the activity of these tissues compared to, say, skin.)

28. Which organelle would cause the most damage to cytoskeletal fibers in the cytoplasm if its contents were to leak into the cytoplasm?

a. smooth ER
b. nucleus
c. lysosome
d. Golgi apparatus
e. rough ER

29. Cystic fibrosis is an inherited condition that affects the lungs and digestive tract (see Chapter 11). In many people with cystic fibrosis, a membrane channel protein is found in the rough endoplasmic reticulum instead of the cell membrane. How could a cell membrane protein end up in the rough endoplasmic reticulum? (*Hint:* Look at the box about the cooperation of the nucleus, endoplasmic reticulum, and Golgi apparatus in Infographic 3.9, Eukaryotic Organelles.)

MILESTONES IN BIOLOGY 1

An immune cell (green) engulfs a cell of thrush fungus (orange). ▼

Scientific Rebel

Lynn Margulis and the theory of endosymbiosis

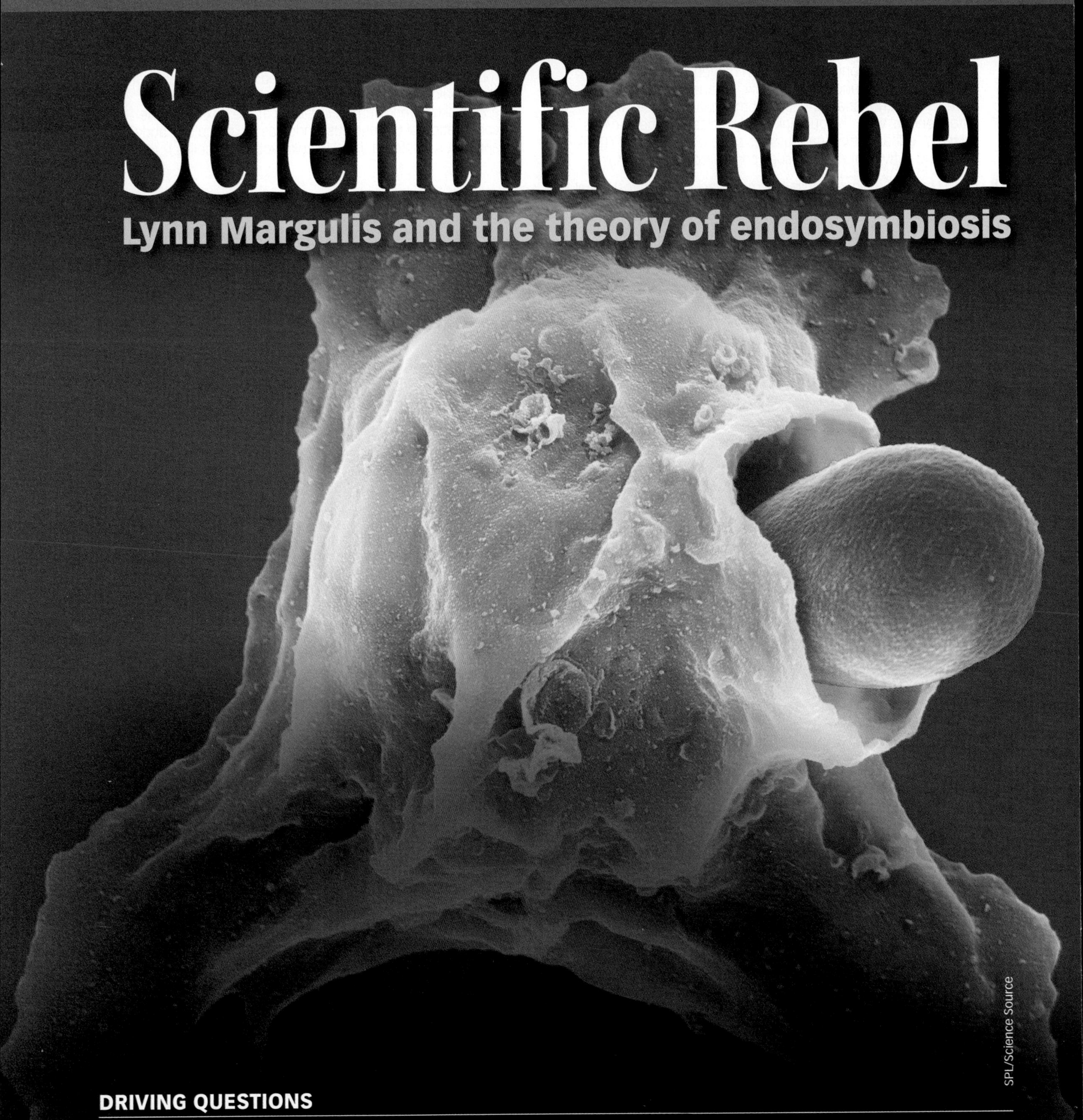

SPL/Science Source

DRIVING QUESTIONS

1. What does endosymbiosis say about the origin of two organelles, the mitochondrion and the chloroplast?

2. What evidence supports the proposed origin of mitochondria and chloroplasts?

Lynn Margulis never met a microbe she didn't like. From the time she first peered through a microscope and witnessed a world of one-celled organisms swimming in a drop of pond water, she was hooked. Other biologists were impressed by the adaptations of plants and animals, but Margulis was smitten with microscopic life. Long before there were animals on the planet, she pointed out, there were microbes, and if it weren't for them, we humans wouldn't be here.

"Life on Earth is such a good story you can't afford to miss the beginning," Margulis once said. "Do historians begin their study of civilization with the founding of Los Angeles? This is what studying natural history is like if we ignore the microcosm."

In 1966, when she was 28, Margulis used her knowledge of microbial life to propose a radical hypothesis about how cells had come to be. What distinguishes eukaryotic cells from prokaryotic cells, Margulis knew, was the presence in eukaryotic cells of internal membrane-bound organelles (see Chapter 3). Margulis proposed that eukaryotic cells, with their internal organelles, had formed when one prokaryotic cell engulfed—ate—another. Instead of being digested, the ingested cell survived and took up residence in its new host. Eventually, the two cells formed a mutually beneficial relationship, and the result was a complex eukaryotic cell. Margulis called her idea endosymbiosis (from the Greek *endo,* meaning "within," and *symbiosis,* meaning "living together").

Many scientists dismissed this idea outright as a crackpot notion. The paper Margulis wrote proposing the idea, which she titled "On the Origin of Mitosing Cells," was rejected by about 15 scientific journals before finally being accepted by the *Journal of Theoretical Biology* in 1967. But an interesting thing happened after it was published: people couldn't stop talking about it.

Microscopic Clues

▸ Evidence for endosymbiosis

In one sense, the idea of endosymbiosis was not entirely new. A few biologists in the late 19th century and early 20th century had suggested it after observing the remarkable resemblance between free-living bacteria and certain organelles of the eukaryotic cell. But in those early days it was not possible to test the idea, so it was largely ignored. In 1925, Edmund Wilson, a prominent cell biologist, wrote, "To many, no doubt, such speculations may appear too fantastic for present mention in polite biological society." But he went on to suggest that those speculations might "someday call for serious consideration."

Margulis first became intrigued by the notion of endosymbiosis in 1960, while working on her master's degree in biology at the University of Wisconsin. Her advisors, Walter Plaut and Hans Ris, had recently made the startling discovery that chloroplasts—the tiny green organelles inside plant cells that carry out photosynthesis—have their own DNA, the molecule of heredity. Most DNA in a eukaryotic cell is housed in the nucleus, where it serves as the genetic blueprint for life. What was DNA doing in a chloroplast? To Margulis, the discovery suggested that chloroplasts had once led a separate existence as independent, free-living cells. Those free-living cells must have needed DNA to reproduce, much as bacteria do.

> *"Life on Earth is such a good story you can't afford to miss the beginning."*
>
> **—Lynn Margulis**

Margulis pursued this idea further as part of her Ph.D. work at the University of California at Berkeley. In particular, she studied the small unicellular eukaryote called euglena, which lives in water and contains numerous chloroplasts. She used radioactively labeled nucleotides, the building blocks of DNA, to verify that the little circular squiggle inside a euglena chloroplast was indeed DNA, as her advisors had claimed. (Cells incorporate the nucleotides into their DNA when they divide, and the radioactivity can be detected with photographic film.) This work lent additional support to the idea of endosymbiosis.

The more Margulis looked, the more evidence she found. Not only did chloroplasts

J. Theoret. Biol. (1967) **14**, 225–274

On the Origin of Mitosing Cells

LYNN SAGAN

Department of Biology, Boston University
Boston, Massachusetts, U.S.A.

(*Received* 8 *June* 1966)

A theory of the origin of eukaryotic cells ("higher" cells which divide by classical mitosis) is presented. By hypothesis, three fundamental organelles: the mitochondria, the photosynthetic plastids and the (9+2) basal bodies of flagella were themselves once free-living (prokaryotic) cells. The evolution of photosynthesis under the anaerobic conditions of the early atmosphere to form anaerobic bacteria, photosynthetic bacteria and eventually blue-green algae (and protoplastids) is described. The subsequent evolution of aerobic metabolism in prokaryotes to form aerobic bacteria (protoflagella and protomitochondria) presumably occurred during the transition to the oxidizing atmosphere. Classical mitosis evolved in protozoan-type cells millions of years after the evolution of photosynthesis. A plausible scheme for the origin of classical mitosis in primitive amoeboflagellates is presented. During the course of the evolution of mitosis, photosynthetic plastids (themselves derived from prokaryotes) were symbiotically acquired by some of these protozoans to form the eukaryotic algae and the green plants.

The cytological, biochemical and paleontological evidence for this theory is presented, along with suggestions for further possible experimental verification. The implications of this scheme for the systematics of the lower organisms is discussed.

1. Introduction

All free-living organisms are cells or are made of cells. There are two basic cell types: *prokaryotic* and *eukaryotic*. Prokaryotic cells include the eubacteria, the blue-green algae, the gliding bacteria, the budding bacteria, the pleuropneumonia-like organisms, the spirochaetes and rickettsias, etc. Eukaryotic cells, of course, are the familiar components of plants and animals, molds and protozoans, and all other "higher" organisms. They contain subcellular organelles such as mitochondria and membrane-bounded nuclei and have many other features in common.

"The numerous and fundamental differences between the eukaryotic and prokaryotic cell which have been described in this chapter have been fully recognized only in the past few years. In fact, this basic divergence in cellular structure which separates the bacteria and blue-green algae

T.B. 225 16

Above: Lynn Margulis in her greenhouse. Right: The 1967 paper that made her famous, published under her married name at the time, Lynn Sagan.

Nancy R. Schiff/Getty Images (Lynn Margulis); Republished with permission of Elsevier, from Sagan, L., "On the origin of mitosing cells," *Journal of Theoretical Biology*, (1967), volume 3, 255–274; permission conveyed though Copyright Clearance Center, Inc.

INFOGRAPHIC M1.1

Chloroplasts and Mitochondria Share Traits with Bacteria

Margulis observed that chloroplasts and mitochondria shared several traits with free-living bacteria.

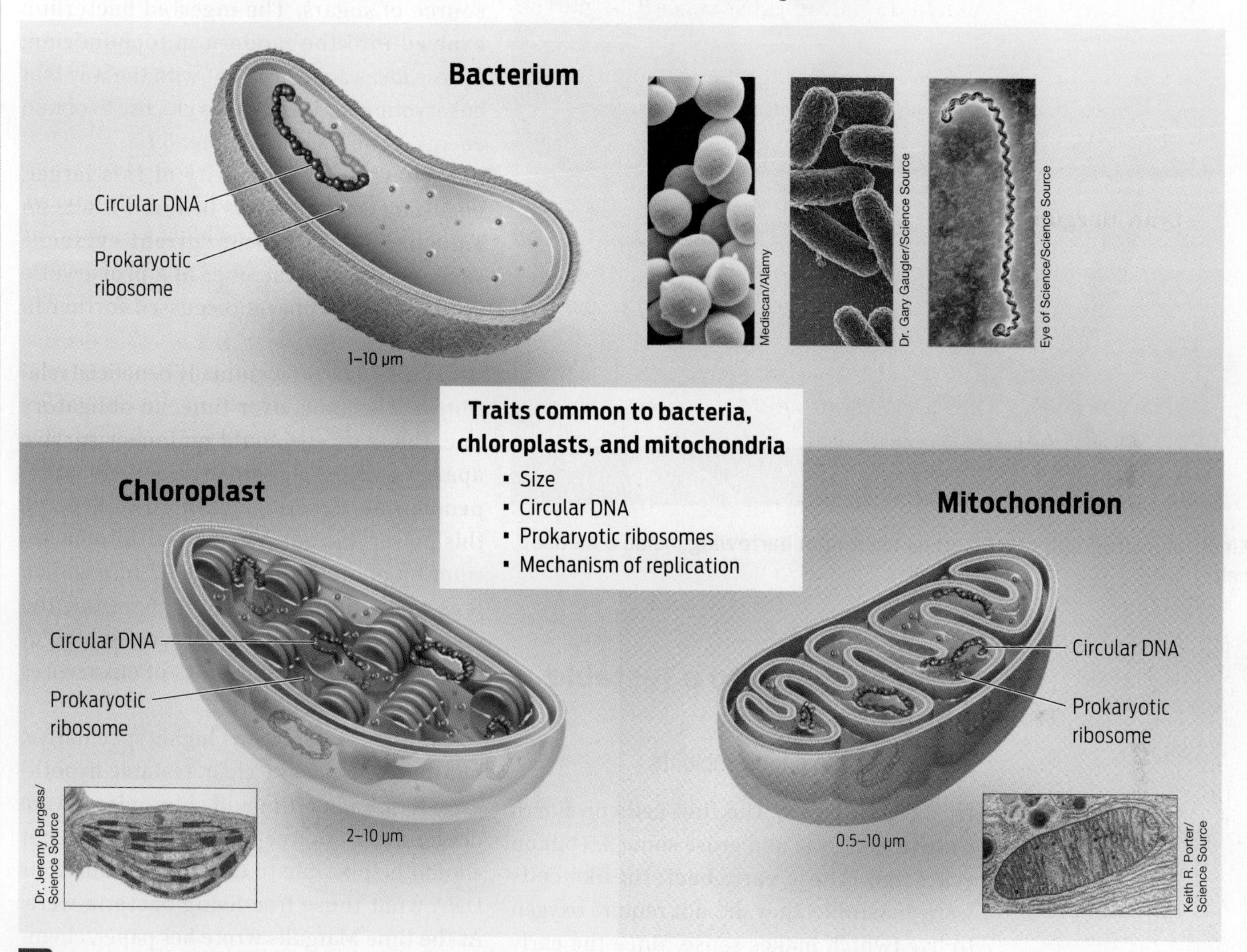

? **What is the shape of the DNA molecule in bacteria, chloroplasts, and mitochondria?**

have their own circular DNA, but they also had ribosomes—the structures that, in the cytoplasm of both prokaryotic and eukaryotic cells, synthesize proteins—and these ribosomes were essentially the same size as those in prokaryotic cells. Chloroplasts were also roughly the same size as bacteria, and to reproduce they divided much the same way bacteria divide. Further investigation revealed that mitochondria—the organelles that serve as the cell's "power plants"—also contained circular DNA and prokaryotic-size ribosomes, and divided in a fashion similar to bacteria. This evidence suggested that mitochondria, too, had once been free-living cells **(INFOGRAPHIC M1.1)**.

In her 1967 paper, Margulis gave credit to the biologists who had previously suggested endosymbiosis, but she did more than simply rehash old ideas. For the first time, she brought together all the existing evidence from cell biology and biochemistry, weaving it together into a coherent account. She also put the idea into evolutionary context and offered a rough timeline of when these events happened.

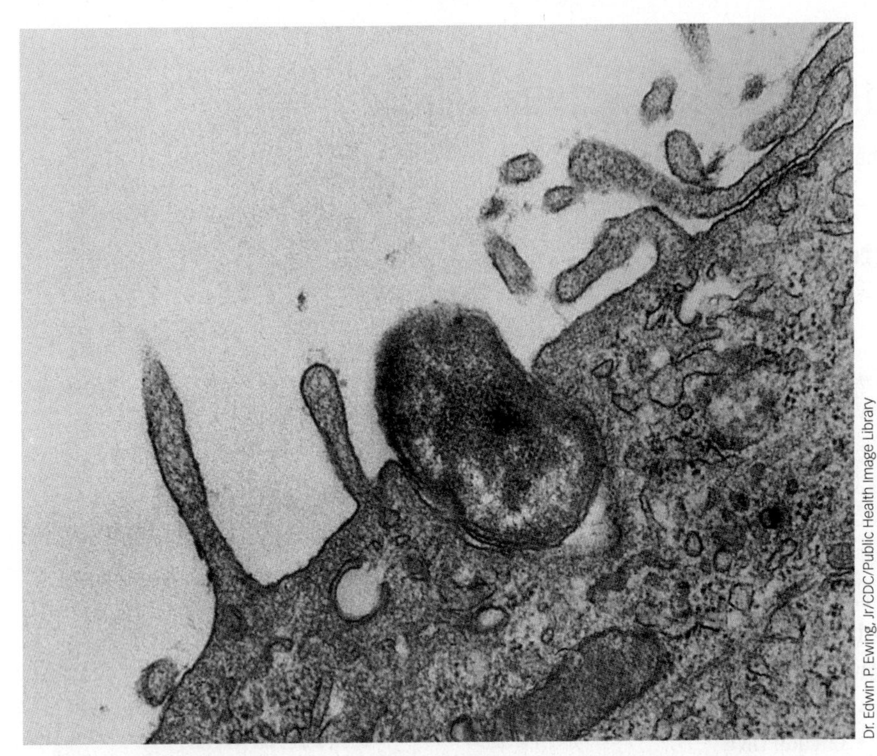
Dr. Edwin P. Ewing, Jr/CDC/Public Health Image Library

Electron micrograph of a Rikettsia bacterium burrowing inside a mouse cell.

From Evidence to a Testable Model

▸ Identifying the endosymbionts

In Margulis's view, the first cells on Earth were prokaryotic and arose some 3.5 billion years ago. These early bacteria-like cells were anaerobic: they did not require oxygen to live (which makes sense, since the early Earth had no substantial oxygen). These cells evolved and diversified for a billion years until, about 2.5 billion years ago, some developed the capacity to harvest the energy of sunlight while also splitting water molecules and releasing oxygen gas as a by-product. As oxygen built up in the atmosphere as a result of this photosynthesis, it produced an environment that strongly favored bacteria that could use oxygen to extract energy from food.

At that point, according to Margulis, one such oxygen-using, or aerobic, bacterium was engulfed by a larger anaerobic prokaryotic cell. The ingested bacterium was not destroyed; instead, the two cells formed a symbiotic, mutually beneficial relationship. The smaller aerobic bacterium enabled the larger anaerobic cell to use oxygen to obtain energy from food, and the larger anaerobic cell provided the smaller bacterium with a source of sugars. The ingested bacterium evolved into the modern mitochondrion. (These ideas are consistent with the way that eukaryotic cells break down glucose to obtain energy, as we'll see in Chapter 6.)

What was the identity of this larger, engulfing cell? Margulis believed it was an amoeba-like organism; current evidence suggests it was a member of a prokaryotic group called Archaea, discussed further in Chapter 17.

What began as a mutually beneficial relationship became, over time, an obligatory one: the two cells could no longer survive apart, as each became increasingly codependent on the other. At some point later, this power duo got cozy with a third bacterium—a photosynthetic one—and engulfed it as well. This third bacterium became the chloroplast. These composite cells, Margulis argued, are the ancestors of eukaryotes **(INFOGRAPHIC M1.2)**.

Margulis's paper was highly speculative, but it provided some clear, testable hypotheses. If mitochondria and chloroplasts were descended from free-living bacteria, then it should be possible to determine from their DNA what those free-living bacteria were. At the time Margulis wrote her paper, chemical analysis of DNA was in its infancy, so she couldn't use this technique. But by the mid-1970s, analyzing DNA had become routine, and it was DNA evidence that clinched her case: by comparing the sequences of mitochondrial and chloroplast DNA to a wide range of bacterial DNA, researchers discovered that mitochondrial DNA closely resembled the DNA from a small bacterium called Rikettsia. Interestingly, Rikettsia is a type of intracellular parasite that burrows inside other cells to live. Chloroplast DNA, the researchers discovered, is essentially the same as cyanobacterial DNA.

Cyanobacteria were the first oxygen-generating photosynthesizers on Earth, evolving some 2.5 billion years ago. By taking up residence in a larger host cell, these smaller bacteria endowed the host cell with the capacity to photosynthesize, paving the way for the evolution of green plants. "Plants are something that hold up cyanobacteria. That's all plants are," Margulis has stated. "Fundamentally, if you cut them out of the plant cell, and throw away the rest of the plant cell, the little green dot is the only thing

INFOGRAPHIC M1.2

The First Eukaryotes Were Products of Endosymbiosis

All eukaryotic cells are characterized by the presence of a membrane-enclosed nucleus and other organelles. While the origin of the nucleus is not yet completely understood, there is good evidence that mitochondria and chloroplasts arose when ancient eukaryotic ancestors engulfed bacteria.

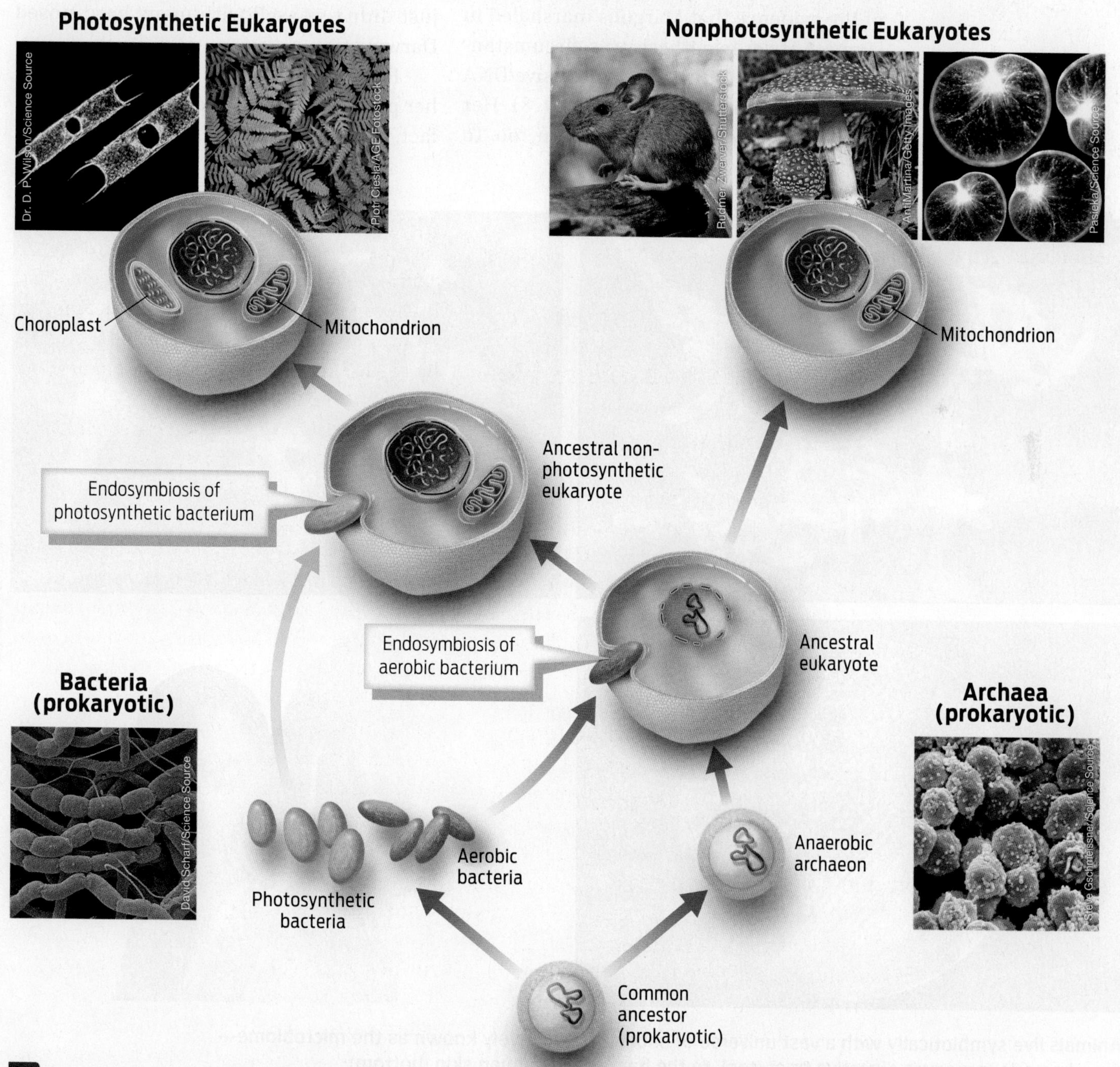

? **Are mitochondria most closely related to archaea, photosynthetic bacteria, or nonphotosynthetic bacteria?**

that can do that oxygen production. That is the greatest achievement of life on Earth, and it occurred extremely early in the history of life."

From Heresy to Orthodoxy

▶ A new scientific theory is born

As soon as her 1967 paper was published, criticism rolled in. Many of Margulis's colleagues were skeptical, even dismissive, citing a lack of supporting evidence. Most of the evidence that Margulis marshaled in support of her hypothesis was circumstantial rather than direct (the conclusive DNA evidence did not emerge until 1978). Her ideas faced philosophical opposition, too. To many people, bacteria were "bad" because they were known to cause disease; the idea that we have bacteria to thank for our very existence was difficult for many to accept. In addition, the notion of endosymbiosis seemed to go against evolutionary dogma, which held that evolution occurred in small steps as a result of an individualistic "struggle for existence." Endosymbiosis was not a small step—it was a huge one. And it wasn't about competition as much as cooperation. For these reasons, Margulis's hypothesis just didn't sit well with many hard-nosed Darwinists.

For her part, Margulis never shrank from her position or gave up pushing her case. In fact, the resistance she encountered seemed

Animals live symbiotically with a vast universe of bacteria, collectively known as the microbiome—from bacteria in a cow's digestive tract (top), to the bacteria on human skin (bottom).

Clockwise from upper left: Janice Carr/CDC/Public Health Image Library; Peter Zijlstra/Shutterstock; DWaschnig/Shutterstock; CNRI/Science Source.

almost to embolden her, and she spent years uncovering many other examples of symbiotic relationships at work in nature.

"Look at a cow," she said in a 2011 interview with *Discover* magazine. "It is a 40-gallon fermentation tank on four legs. It cannot digest grass and needs a whole mess of symbiotic organisms in its overgrown esophagus to digest it."

Or look at your own body. "There are hundreds of ways your body wouldn't work without bacteria," she pointed out. "Between your toes is a jungle; under your arms is a jungle. There are bacteria in your mouth, lots of spirochetes, and other bacteria in your intestines. We take for granted their influence. Bacteria are our ancestors."

Much current research is focused on understanding our human microbiome—the population of bacteria, archaea, and fungi that lives on and in our bodies and influences many aspects of our health (see Chapter 17). In addition to helping us digest food and shaping our immune system, our unique microbiome may even influence our susceptibility to conditions such as diabetes and obesity. The idea that these microbes are not passive freeloaders, but rather crucial constituents of our bodies is a relatively new way of looking at things—but it would hardly have come as a surprise to Margulis.

When Margulis died in 2011 at the age of 73, she was remembered as the person who most fundamentally changed our view of cells. These days, the idea that mitochondria and chloroplasts started as free-living organisms is accepted as fact by the scientific mainstream, and we now refer to the "theory of endosymbiosis," acknowledging the abundant evidence there is to support it.

"The evolution of the eukaryotic cells was the single most important event in the history of the organic world," said Ernst Mayr, the grandfather of modern evolutionary studies. "Margulis's contribution to our understanding the symbiotic factors was of enormous importance." Richard Dawkins, a prominent British evolutionary biologist, described the theory of endosymbiosis as "one of the great achievements of twentieth-century evolutionary biology." Botanist Peter Raven said the idea caused "nothing less than a revolution" in our thinking about the cell.

Not all of Margulis's ideas have gained widespread acceptance. In particular, her claim that the whiplike tail of a sperm cell derives from a formerly free-living bacterium called a spirochete lacks convincing evidence and is not accepted as fact by the scientific establishment. But on one of the most important questions of modern biology, her intellectual daring paid off.

> *"We take for granted their influence. Bacteria are our ancestors."*
>
> **—Lynn Margulis**

Asked by the *Discover* interviewer how she felt about being the source of so many controversial ideas over the years, Margulis responded in typical fashion: "I don't consider my ideas controversial. I consider them right." ■

PAUL HOSEFROS/The New York Times/Redux Picture

Margulis receiving the National Medal of Science from President Bill Clinton in 1999.

More to Explore

- Sagan, L. (1967). On the origin of mitosing cells. *Journal of Theoretical Biology* 14:225–274.
- Margulis, L., and Sagan, D. (1986). *Microcosmos: Four Billion Years of Evolution from Our Microbial Ancestors.* New York: Summit Books.
- Teresi, D. (2011). Interview with Lynn Margulis. *Discover.* http://discovermagazine.com/2011/apr/16-interview-lynn-margulis-not-controversial-right
- Sapp, J. (2012). Too fantastic for polite society: A brief history of symbiosis theory. In *Lynn Margulis: The Life and Legacy of a Scientific Rebel,* edited by Dorian Sagan (pp. 54–67). White River Junction, VT: Chelsea Green Publishing.

MILESTONES IN BIOLOGY 1 Test Your Knowledge

1. What is the function of mitochondria? Of chloroplasts?

2. What evidence did Margulis present to support her hypothesis that organelles had once been free-living prokaryotic organisms?

3. On the basis of DNA sequence analysis:

a. Which bacteria are likely the closest relatives of the chloroplast?

b. Which bacteria are likely the closest relatives of the mitochondria?

4. Which of the following is *not* a trait shared by chloroplasts and prokaryotic cells?

a. the size of their ribosomes

b. the shape of their DNA molecule

c. the presence of a nucleus

d. their mechanism of replication

5. From what you have read here about endosymbiosis:

a. Could you live without your endosymbiotic organelles? Why or why not?

b. Could you live if plants did not have their endosymbiotic organelles? Explain your answer.

The PEANUT BUTTER Project

One doctor's crusade to end malnutrition in Africa, a spoonful at a time

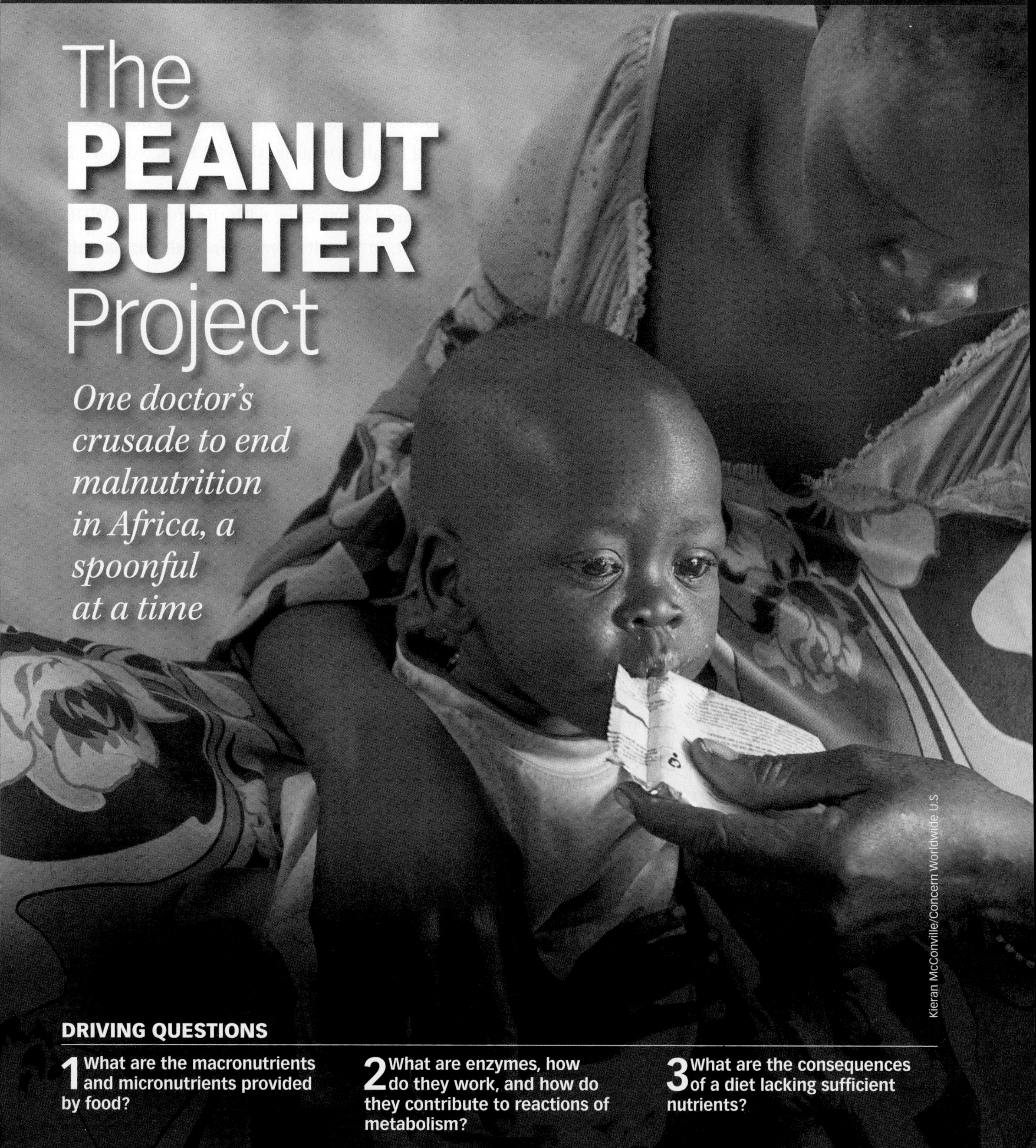

Kieran McConville/Concern Worldwide U.S

DRIVING QUESTIONS

1. What are the macronutrients and micronutrients provided by food?
2. What are enzymes, how do they work, and how do they contribute to reactions of metabolism?
3. What are the consequences of a diet lacking sufficient nutrients?

At a medical conference in 2003, during a panel on malnutrition, someone in the audience stood up and shouted at the presenter, "You're killing children!" Mark Manary was discussing a new method of treating malnutrition in children that he had used successfully in some of the poorest countries in Africa. But it flew in the face of guidelines endorsed by the World Health Organization (WHO). Now, leaders in the field were angry.

Manary, a pediatrician from St. Louis, Missouri, had firsthand experience with the standard WHO treatment. For much of the 1990s, he had devoted his life to it: in hospital wards in Malawi, a small, landlocked country in southeastern Africa, he faithfully administered fortified milk solutions to sickly children in accordance with WHO guidelines for treating malnutrition. Unfortunately, even with this treatment, children continued to die, and only a small percentage ever got better. "On our best days only 10% of the kids would die. But still only 25% would recover," he says.

Manary decided to find a better way. After searching around for potential alternatives, he settled on a surprising solution: peanut butter.

Peanut butter, he found, is well suited to the purpose. It's packed with nutrients, it doesn't require cooking, and—most important—it doesn't spoil easily and can be kept unrefrigerated in tropical climates for as long as 3 months.

Once criticized by the international aid community, Manary's method is now being hailed as a kind of miracle cure.

Manary began testing his peanut butter treatment in 2001. He gave his patients an ample supply of peanut butter and released them from the hospital. "Basically, what we did was empty the place out and send everybody home," Manary says.

The hospital staff was appalled by his brashness, but the gamble paid off. Within weeks, 95% of the children eating peanut butter had fully recovered.

Manary is now the chief executive officer of Project Peanut Butter, a nonprofit organization using the American pantry staple to end malnutrition in Africa. Spin-offs of the organization have sprung up all around the globe, including in Haiti and South Asia.

Once criticized by the international aid community, Manary's method is now being hailed as a kind of miracle cure. Supporters of the approach have even likened it to the discovery of penicillin, praising this treatment as having the potential to save millions of lives every year.

The Elephant in the Room

▸ Malnutrition and macronutrients

In 1994, Manary became a visiting faculty member at the University of Malawi School of Medicine. When he arrived, he asked the hospital director, "What's your biggest problem?" The director replied, "The malnutrition ward." So that's where Manary chose to work.

The ward was a dismal and discouraging place—a large room where 50 starving children lay close to death on crowded cots, their bodies little more than skin and bone. Nearly one-third of the children would die, despite the doctors' best efforts at rehabilitating them. This poor response rate was, Manary says, "the elephant in the room."

Severe acute **malnutrition** is the number one killer of children in the world. An estimated 3.5 million children die from malnutrition every year—more than the number who die from AIDS, tuberculosis, and malaria combined. Most of these deaths occur in sub-Saharan Africa, where grinding poverty is a regular part of life, and food is scarce for large portions of the year **(INFOGRAPHIC 4.1)**.

In Malawi, the crisis is particularly acute. The Food and Agriculture Organization of the United Nations estimates that 3.3 million people, or about 18% of the country's population, lacked adequate food between 2016 and 2018. One measure of chronic childhood malnutrition is stunting—being shorter than expected based on age. In 2004, more than half of all Malawian children under five were stunted. While this statistic has since improved, the latest numbers indicate that 37% of all children under five in Malawi are stunted. The mortality rate for children under five in Malawi was 50 per 1,000 in 2018, compared to 7 per 1,000 in the United States in the same year.

Why is food so scarce? Malawi is one of the poorest countries in the world, with a population mostly made up of subsistence farmers. The primary agricultural crops are corn and soybeans. Farmers also grow

MALNUTRITION
A medical condition resulting from a lack of essential nutrients in the diet. Malnutrition is often, but not always, associated with starvation.

INFOGRAPHIC 4.1

Percentage of Undernourished People Worldwide

Food is the most basic of human requirements for survival, but at least 1 in 9 people goes to bed hungry each night.

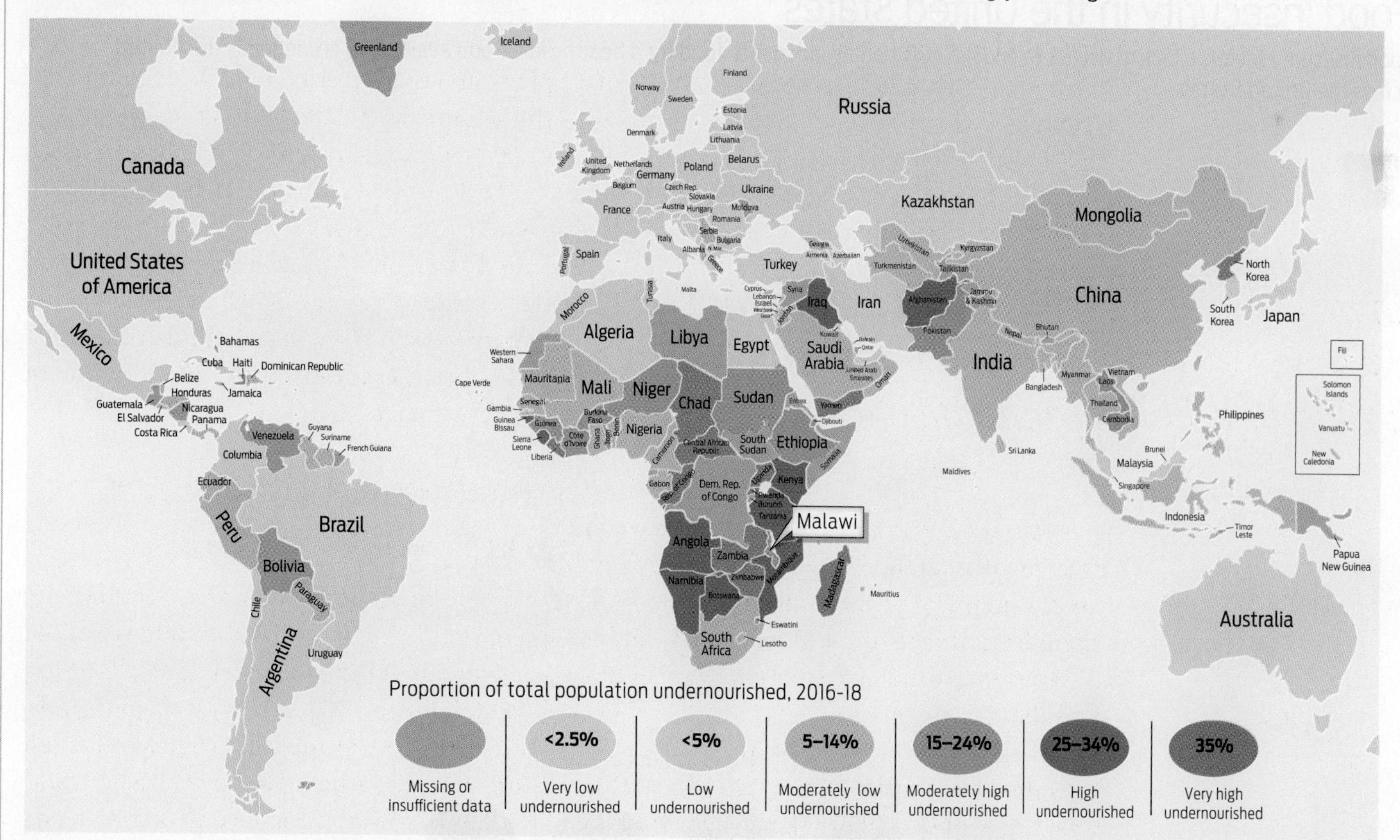

Data courtesy World Food Programme, 2019

? **Find two countries in which more than 35% of the population is undernourished. Try to find a country that meets this description and is not part of Africa.**

commercial cash crops for export, including tobacco, sugar cane, coffee, and tea. But agriculture is not easy in Malawi. Rain falls only during the December to March time span, when it might rain every day. New crops are planted during this rainy season, but the harvest must wait until March. Quite often, supplies from the previous year's harvest run out before the next crop is in. Locals call this time of food insecurity "the hungry season."

Lack of access to sufficient food to support a healthy and active life, or food insecurity, is by no means restricted to developing nations like Malawi. Even in the United States, many families do not know where their next meal is coming from. In 2017, the U.S. Department of Agriculture estimated that 11.8% of American households were food insecure at some point in the previous year, a percentage representing roughly 15 million households. Unlike in the developing world, in the United States the problem is not so much food shortages as poverty that prevents access to food **(INFOGRAPHIC 4.2)**.

Without enough food, people lack adequate **nutrients**, the chemical components in the diet that our bodies need to live, grow, and repair themselves. All organisms are made up of chemical building blocks such as organic molecules, water, and ions (see Chapter 2). Because humans (and other animals) can't make these components in their bodies, we need to obtain them from our diet. These nutrients also provide us with the **energy** needed to power essential life activities. (Both building blocks and energy are crucial components of the nutrients in food, but for simplicity we will discuss them

NUTRIENTS
Components in food that the body needs to grow, develop, and repair itself.

ENERGY
The ability to do work, including the work of building complex molecules.

INFOGRAPHIC 4.2

Food Insecurity in the United States

Food insecurity means a lack of access to enough nutritionally adequate food for a healthy life. Food availability is unequally distributed across the United States.

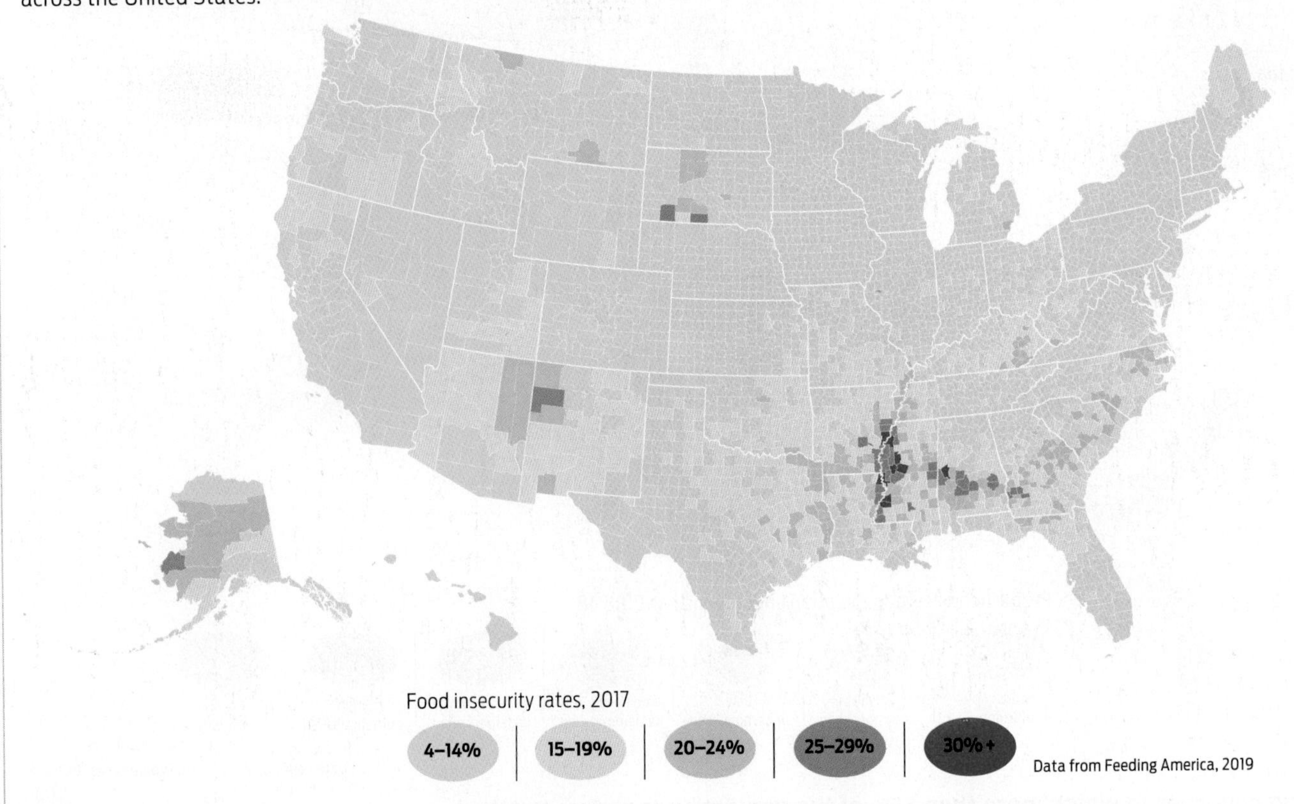

? What is the food insecurity rate in your home state and county?

separately. This chapter focuses on food as a source of chemical building blocks; Chapters 5 and 6 consider energy in more detail.)

When experts talk about a nutritious diet, they mean one that provides all the nutrients our bodies need for health, in the appropriate amounts. Nutrients that the body requires in large amounts are called **macronutrients.** The macronutrients in our diet include carbohydrates, proteins, and fats (a type of lipid)—three of the four organic macromolecules discussed in Chapter 2. Because most foods contain mixtures of these macronutrients, people who eat a varied diet that includes vegetables, oils, grains, meat, and dairy products can easily obtain all the macronutrients their bodies need **(INFOGRAPHIC 4.3).**

MACRONUTRIENTS
Nutrients, including carbohydrates, proteins, and fats, that organisms must ingest in large amounts to maintain health.

INFOGRAPHIC 4.3

Food Is a Source of Macronutrients

The most important dietary macronutrients are carbohydrates, proteins, and fats (a type of lipid). While most foods contain all of these, one or two macronutrients predominate in each food type. A well-balanced diet is one that includes a variety of foods to ensure that the body gets enough of each macronutrient to grow and remain healthy.

? Which macronutrient would not be well represented in a diet rich in rice and beans?

Macronutrients from the diet cannot be used directly by our bodies, in part because they are too large to be absorbed into the bloodstream from the digestive tract. To be useful, macronutrients must first be broken down into smaller subunits by digestion (see Chapter 27). These subunits are small enough to be absorbed from the digestive tract, taken up by cells, and used to build the macromolecules our cells need. For example, dietary carbohydrates from bread and pasta are broken down into simple sugars, which our bodies use to build an energy-storing carbohydrate called glycogen in the liver and muscle tissue. Proteins from a steak are broken down into amino acids, which can be taken up and used to build new proteins, like those making up our muscles. Fats consumed as part of the diet are broken down into fatty acids and glycerol, which are used to assemble the phospholipids that make up cell membranes.

The food we eat also contains nucleic acids, the fourth type of macromolecules making up cells. Although not considered macronutrients (because we need them in smaller amounts), nucleic acids are also

INFOGRAPHIC 4.4

Macronutrients Build and Maintain Cells

The four macromolecules that make up cells are carbohydrates, proteins, lipids (including fats), and nucleic acids. Cells synthesize these macromolecules from the subunits released by the digestion of macronutrients in food.

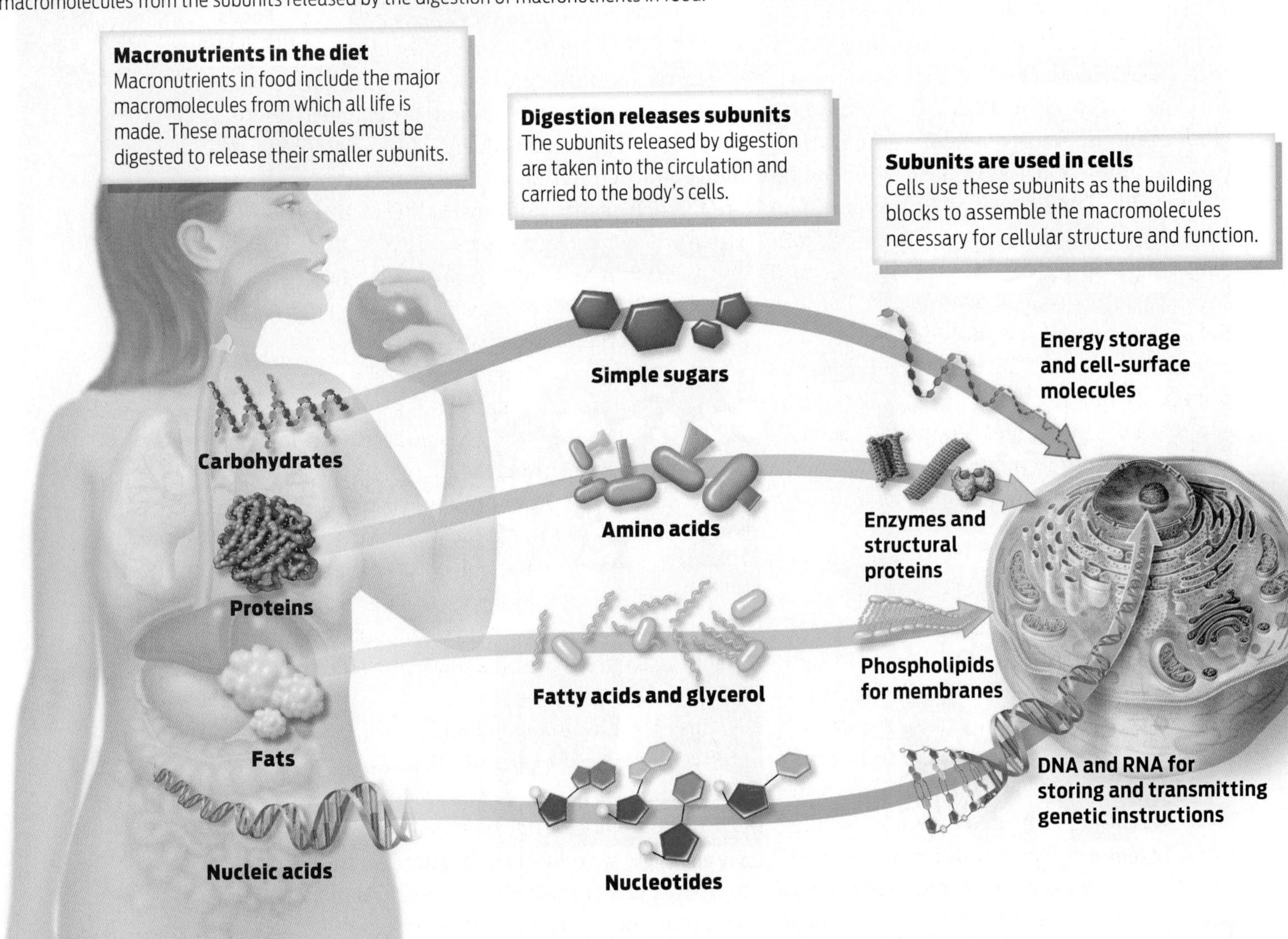

? Which subunits are released by the digestion of proteins?

broken down into smaller subunits called nucleotides, which cells use to build DNA and RNA. Breaking down food to build up our bodies means that, quite literally, we are what we eat **(INFOGRAPHIC 4.4)**.

To a certain extent, our bodies can compensate for a deficiency in a specific nutrient by synthesizing it from other chemical components. For example, if a particular amino acid is in short supply, cells may be able to make it from another amino acid that is present in excess. But some nutrients can't be manufactured by our bodies, but instead must be obtained in a pre-assembled form from our diet. These nutrients are called **essential nutrients.**

In Malawi, most families subsist on a single crop—corn. While corn is a good source of complex carbohydrates, like **starch,** it is not a significant source of protein or fat. A corn-based diet lacks many essential nutrients needed for a healthful diet, including vital amino acids. From starting materials in food, adults can synthesize 11 of the 20 amino acids they need to make proteins. The other 9 must be obtained pre-assembled from our diet. Because our body can't manufacture them, these 9 amino acids are called **essential amino acids.** A few more amino acids are considered essential for infants and children. Animal products such as meat, eggs, fish, and dairy are the richest sources of essential amino acids, but in many places around the globe, these foods are luxuries people can't afford.

In 1999, when Manary first began looking for an alternative treatment to combat severe hunger, he considered sending kids home with a bag of ingredients, such as corn flour and soy, to make traditional meals. He quickly realized that this approach would not work, for several reasons. First, such foods would need to be cooked, and cooking is extra work for an already overburdened family. Second, severely malnourished children would need to eat many bowls of these foods per day to obtain enough nutrients to recover. Third, the food wouldn't keep.

"If you eat RUTF, you don't need to eat anything else. You're getting everything you need—period."

—Mark Manary

In contrast, peanut butter—especially when it has been fortified with additional nutrients—is a highly effective way of delivering the most nutrients per spoonful of food. Technically, it's known as a ready-to-use therapeutic food (RUTF), which means it is a complete source of nutrition. "If you eat RUTF, you don't need to eat anything else," says Manary. "You're getting everything you need—period."

The RUTF that Manary uses consists of four main ingredients: full-fat milk powder, sugar, vegetable oil, and peanut butter. Peanut butter is a useful treatment for malnutrition partly because it is energy-dense, which helps children put on weight fast. "Peanuts are the only plant I know of that has 50% fat, when you get rid of the water," says Manary. But the fattening quality is only part of peanut butter's appeal. Because it contains very little water and has a pasty consistency that keeps out air, peanut butter is naturally resistant to spoiling; bacteria can't grow without water. It also doesn't require cooking, so it's ready to eat at any time. Finally, peanut butter is full of protein, a crucial macronutrient for growing children. All these things make peanut butter a near-perfect therapeutic food, supplying children with the necessary fats, carbohydrates, proteins, and essential amino acids (from the milk powder) that they would otherwise lack.

Incidentally, many people in industrialized countries have peanut allergies. But in developing countries, such allergies are rare or nonexistent. That difference has more to do with the way we train our immune systems in our hyperscrubbed world, Manary explains, than with any intrinsic quality of peanuts.

ESSENTIAL NUTRIENTS
Nutrients that can't be made by the body, and so must be obtained from the diet.

STARCH
A complex plant carbohydrate made of linked chains of glucose molecules; a source of stored energy.

ESSENTIAL AMINO ACIDS
Amino acids that can't be made by the body, and so must be obtained in pre-assembled form from the diet.

Life in the Village

▶ Chemical reactions and enzymes

In 1999, having spent 5 years working in a hospital in Malawi, Manary took an unusual step for an American-trained doctor: he left the hospital and went to live in a rural village for 10 weeks. By making this move, Manary sought to get a firsthand view of the way people lived and the obstacles they faced.

The thing that struck him the most was the sheer monotony of village life. "Every day is the same," he says. "It starts with going to the water pump, hauling water. Then it's finding firewood. Then it's cooking food. It's work all day, 365 days a year."

Instantly, he understood why childhood malnutrition is such a terrible trap for families. Families live perilously close to the edge of sufficiency, and just a slight change in their circumstances can push them over into disaster. Even one extra mouth to feed can create food shortages. Moreover, for many rural populations, the nearest hospital is miles away. Taking a sick child to the hospital is a major disruption to family life, so most mothers do it only as a last resort, when the child is already quite sick. Lastly, telling mothers—as they routinely were told by hospital doctors—that they needed to feed their malnourished child seven times a day was not a workable solution. As Manary explains, "It'd be like me telling you 'Hey, you need to walk back and forth from New York to Connecticut every day.'" Put simply, the doctors' advice wasn't realistic.

Malnutrition-related problems usually start when children are 6 months old. Babies younger than 6 months are sustained easily by breast-feeding alone. Breast milk is a perfect food for children at this age, explains Manary. It is nutritionally appropriate for what infants' bodies can handle, and it provides protection against infections and disease. Since 2001, WHO has recommended that women breast-feed exclusively until a child is 6 months old (and they are encouraged to continue for 2 years and beyond, even after infants begin to eat other food). Virtually all women in Malawi breast-feed their babies.

At 6 months, children's growing bodies require more nourishment than breast milk can provide, but no other food is as plentiful or accessible. Over a period of time, then, children begin to develop malnutrition due to chronic undereating, usually between ages 1 and 3.

© Colin Carson/AGE Fotostock

Dr. Mark Manary

In Malawi, villages are sometimes located many miles from the nearest health facility. Right: Physician Mark Manary tends to a malnourished child.

Scientists sometimes refer to the first 1,000 days of life, from gestation to age 2, as the "golden interval." "Babies go through a huge amount of physical and cognitive development during this period of 1,000 days," says Mary Arimond, a nutrition expert at the University of California, Davis, who studies nutrition in the developing world. "If deficits occur in this window, they are difficult to reverse, especially in continued conditions of poverty."

Growth and development are essentially a series of chemical reactions requiring nutrients as starting materials. A **chemical reaction** is a process that transforms one set of chemical substances into another by forming or breaking bonds between atoms. For example, the organic molecules that make up food are transformed by chemical reactions during the process of digestion. Reactions that break down larger structures into smaller ones are **catabolic reactions.** Reactions that build new structures from smaller subunits are **anabolic reactions.** Together, all the chemical reactions occurring in the body constitute the body's **metabolism.**

To proceed normally, the chemical reactions of metabolism require the assistance of helper proteins called **enzymes.** Nearly every chemical reaction in the body requires enzymes for that reaction to happen at a rate fast enough to keep up with the needs of a living organism. For example, digestive enzymes made by cells in the digestive tract help us digest food molecules into their constituent subunits. In the absence of such digestive enzymes, food molecules would remain stable and intact; they would not break down spontaneously at a rate that would suit our metabolic needs. When cells divide—for instance, in bone marrow to generate new blood cells, or in skin to replace cells lost to death or injury—they rely on enzymes to copy the DNA in cells. Similarly, enzymes produced by cells in bone carry out anabolic reactions that contribute to the formation of new bone.

"Babies go through a huge amount of physical and cognitive development during this period of 1,000 days."

—Mary Arimond

Enzymes work by speeding up, or catalyzing, chemical reactions—a process called **catalysis.** To accelerate a chemical reaction, an enzyme must bind to the molecules involved in the reaction. The molecules that enzymes bind to are called **substrates.** The part of the enzyme that binds to a substrate is called its **active site.** Each enzyme has an active site that fits only one particular substrate molecule or molecules, making each enzyme specific for the reaction that it catalyzes. Given the thousands of reactions occurring in organisms, organisms have to produce and rely on the activity of thousands of different enzymes, each catalyzing a specific reaction. The digestion of a cheeseburger, for example, requires a variety of digestive enzymes—some that specifically digest protein, some that digest carbohydrates, and one that digests fat.

Enzymes speed up reactions by facilitating the breaking or formation of chemical bonds in substrates. In the case of catabolic reactions, the binding of the enzyme to its substrate puts stress on a chemical bond, causing it to break. In anabolic reactions, the enzyme brings substrate molecules into closer proximity, increasing the likelihood that a bond will form between them **(INFOGRAPHIC 4.5).**

In effect, enzymes catalyze reactions by lowering the amount of energy required to nudge a chemical reaction into motion. Enzymes substantially reduce this **activation energy,** allowing the reaction to occur more easily. You can think of activation energy as the amount of energy an athlete needs to surmount a hurdle: the steeper the hurdle,

CHEMICAL REACTION
A rearrangement of atoms in molecules to form different molecules by forming or breaking bonds.

CATABOLIC REACTION
Any chemical reaction that breaks down complex molecules into simpler molecules.

ANABOLIC REACTION
Any chemical reaction that combines simple molecules to build more complex molecules.

METABOLISM
All biochemical reactions occurring in an organism, including reactions that break down food molecules and reactions that build new cell structures.

ENZYME
A protein that speeds up the rate of a chemical reaction.

CATALYSIS
The process of speeding up the rate of a chemical reaction (e.g., by an enzyme).

SUBSTRATE
A molecule to which an enzyme binds and on which the enzyme acts.

ACTIVE SITE
The part of an enzyme that binds to a substrate.

ACTIVATION ENERGY
The energy required for a chemical reaction to proceed. Enzymes accelerate reactions by reducing their activation energy.

the more energy required. Enzyme-catalyzed reactions involve leaping over a much lower bar, so less energy is required to clear it **(INFOGRAPHIC 4.6)**.

If a child doesn't get enough to eat, the enzymes that support the anabolic reactions necessary for growth lack the substrates they need to act upon, so those reactions can't occur. What's more, in a malnourished child, catabolic reactions will begin to break down muscle protein to obtain the amino acids needed for other purposes. The results are

INFOGRAPHIC 4.5

Enzymes Facilitate Chemical Reactions

Cells require enzymes to break down and build up macromolecules. Enzymes are proteins that speed up chemical reactions.

Catabolic Reactions: Bonds are broken

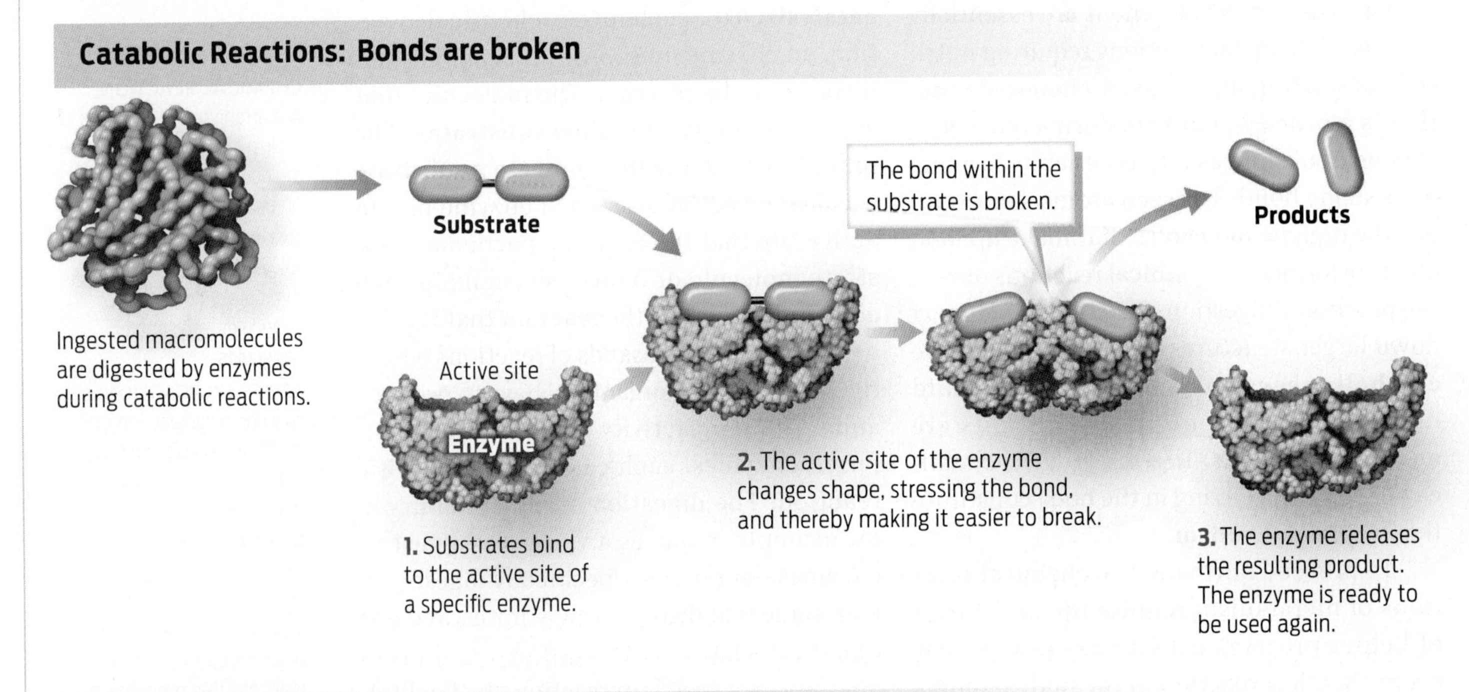

Anabolic Reactions: Bonds are created

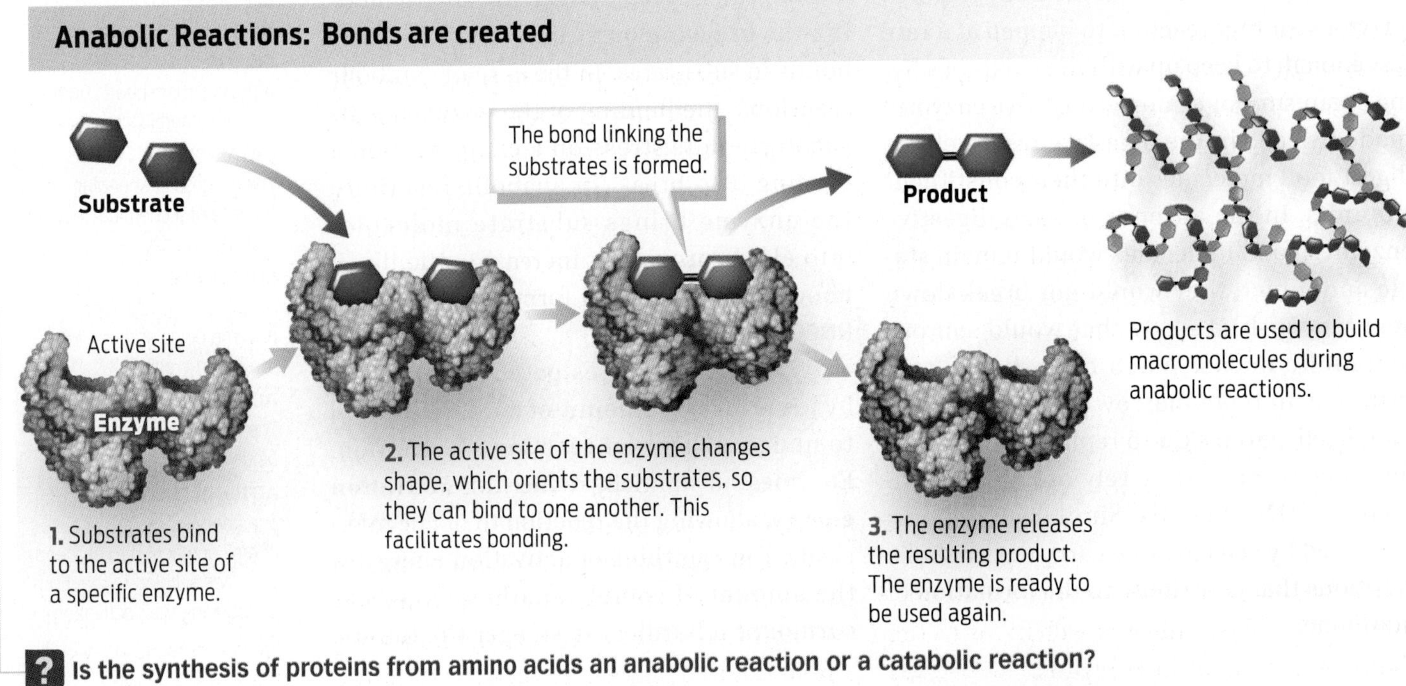

? Is the synthesis of proteins from amino acids an anabolic reaction or a catabolic reaction?

INFOGRAPHIC 4.6

Enzymes Catalyze Reactions by Lowering Activation Energy

The activation energy is the energy that must be put into a reaction to make it "go." Enzymes reduce the activation energy in both anabolic and catabolic reactions, making them occur more rapidly.

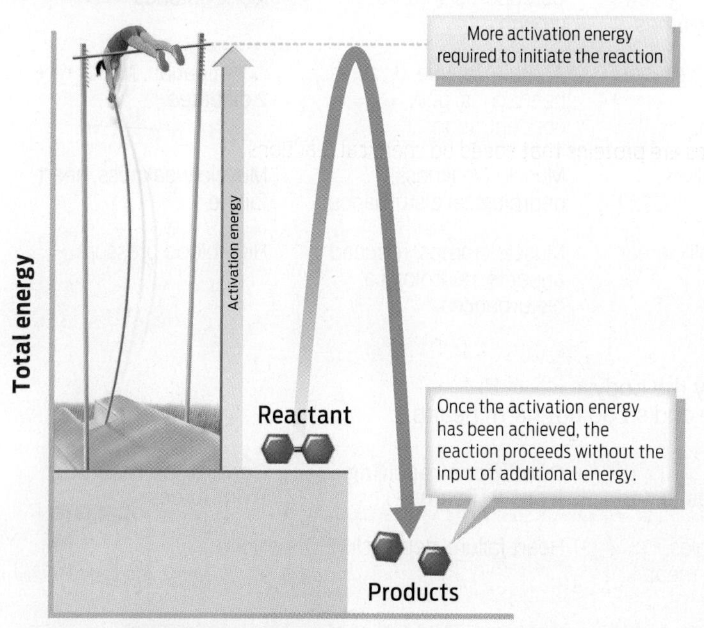

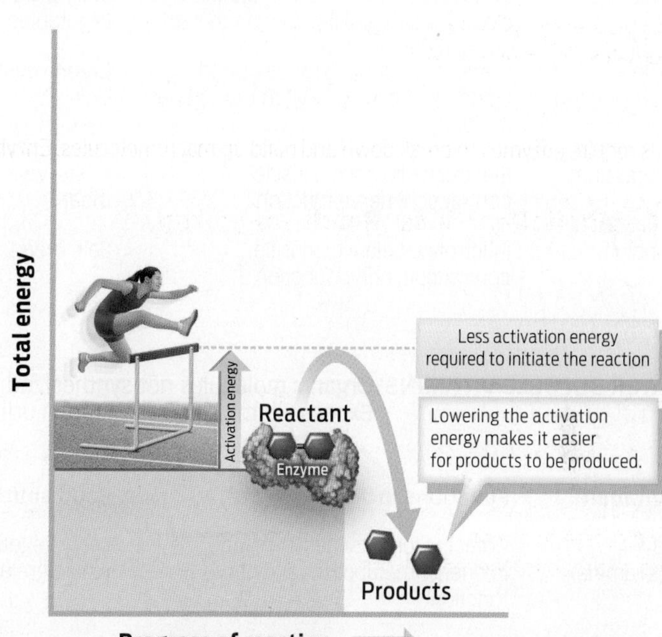

? **How does an enzyme affect the energy of the reactants, the energy of the products, and the activation energy?**

the telltale signs of malnutrition: thin arms with skin wrinkling over wasted muscle; painful swelling of the legs and feet caused by a buildup of fluid; blond or rust-colored hair resulting from a deficiency of protein; and a distended, bloated stomach.

Because a healthful diet requires a balance of many different kinds of nutrients, it's not just the sheer quantity of food that is important: the specific type of food consumed matters, too. Some nutrients can be missing from a diet, even if a child is otherwise well fed.

Hidden Hunger

▶ Vitamins, cofactors, and other micronutrients

When Manary first arrived at the Malawi hospital in 1994, he made a few changes to the then-standard treatment regimen of fortified milk. One thing he did right away was to add extra potassium to the milk the children were drinking. Immediately, the fatality rates dropped—from 33% to 10%. Potassium is a chemical element required for proper muscle contraction and nerve function. Our bodies need many such chemical elements, called **minerals,** to stay healthy.

Most minerals are required only in small amounts—hence, they are known as **micronutrients** (as opposed to macronutrients, which we need in much larger quantities). Organic molecules called **vitamins** are another kind of micronutrient. Of course, just because our bodies require only small amounts of micronutrients, that doesn't mean they aren't important. In fact, micronutrient deficiency can have serious health consequences. Iron deficiency can impair the blood's ability to carry oxygen, causing anemia, for example, and lack of vitamin C

MINERAL
An inorganic chemical element required by organisms for normal growth, reproduction, and tissue maintenance; examples include calcium, iron, potassium, and zinc.

MICRONUTRIENTS
Nutrients, including vitamins and minerals, that organisms must ingest in small amounts to maintain health.

VITAMIN
An organic molecule required in small amounts for normal growth, reproduction, and tissue maintenance.

TABLE 4.1 A Sample of Micronutrients in Your Diet

MINERALS: Inorganic elements not synthesized by the body.

Mineral	Functions include	Food sources	Conditions resulting from deficiency	Conditions resulting from excess
Calcium	Bone and teeth formation, blood clotting, cell signaling, nerve function	Dairy products, green vegetables, legumes	Osteoporosis, stunted growth	Kidney stones
Iron	Component of hemoglobin in red blood cells; carries oxygen throughout the body	Green vegetables, beef, liver	Anemia, fatigue, dizziness, headaches, poor concentration	Constipation, risk of type 2 diabetes
Potassium	Electrolyte balance, muscle contraction, nerve function	Fruits, vegetables, meat	Muscle weakness, neurological disturbances	Muscle weakness, heart failure
Sodium	Electrolyte balance, muscle contraction, nerve function	Salt, bread, milk, meat	Muscle cramps, reduced appetite, neurological disturbances	High blood pressure

WATER-SOLUBLE VITAMINS: Organic molecules not synthesized by the body.
Excess vitamins are excreted in urine and so do not harm health.

Vitamin	Functions include	Food sources	Conditions resulting from deficiency	Conditions resulting from excess
B_1 (thiamine)	Cofactor for enzymes involved in energy metabolism and nerve function	Leafy vegetables, whole grains, meat	Heart failure, depression	None
Folate	Cofactor for enzymes involved in DNA synthesis and cell production	Dark green vegetables, nuts, legumes, whole grains	Neural tube defects, anemia	None
B_{12}	Cofactor for enzymes involved in the breakdown of fatty acids and amino acids and nerve cell maintenance	Meat, milk, eggs	Anemia, neurological disturbances	None
C	Cofactor for enzymes involved in collagen synthesis; improves iron absorption and immunity	Citrus fruits	Scurvy, poor wound healing	None

FAT-SOLUBLE VITAMINS: Organic molecules not synthesized by the body (except vitamin D).
Excess vitamins are stored in fat cells and can harm health.

Vitamin	Functions include	Food sources	Conditions resulting from deficiency	Conditions resulting from excess
A (retinol)	Component of eye pigment; supports skin, bone, and tooth growth; supports immunity and reproduction	Fruits and vegetables, liver, egg yolk	Skin problems, blindness	Headaches, intestinal pain, bone pain
D	Calcium absorption, bone growth	Fish, dairy products, eggs	Bone deformities	Kidney damage
E	Antioxidant, supports cell membrane integrity	Green leafy vegetables, legumes, nuts, whole grains	Neural tube defects, anemia, digestive-health problems	Fatigue, headaches, blurred vision, diarrhea
K	Supports synthesis of blood clotting factors	Green leafy vegetables, cabbage, liver	Abnormal blood clotting, bruising	Liver damage, anemia

causes a tissue-deteriorating disease called scurvy.

In Malawi, as in many parts of the developing world, people often suffer deficiencies of vitamin A and zinc. Vitamin A deficiency can cause impairment of vision, while zinc deficiency can prevent the immune system from working properly. Such micronutrient deficiencies are sometimes known as “hidden hunger” because the problem is not lack of food per se, but rather a lack of necessary micronutrients. Manary says the problem is widespread because most of the world’s staple crops—foods like rice, corn, wheat, and cassava—do not contain adequate micronutrients.

Food producers routinely fortify foods with some micronutrients that are hard to obtain from natural sources. The mineral element iodine, for example, is added to table salt (in “iodized” salt) to prevent goiter, an abnormal thickening of the neck caused by an enlarged thyroid gland **(TABLE 4.1)**.

Vitamins and minerals play numerous roles in the body. Some play structural roles—the mineral calcium, for example, is a primary component of bones and teeth. Others play functional roles, helping other molecules to act. Perhaps their most critical role is serving as **cofactors** that assist enzymes.

Cofactors are accessory or “helper” substances that enable enzymes to function. Cofactors include inorganic metals such as zinc, copper, and iron. Cofactors can also be organic molecules, in which case they are called **coenzymes.** Most vitamins are important coenzymes. Without cofactors and coenzymes that bind to enzymes and enable them to bind to substrates, cell metabolism would grind to a halt **(INFOGRAPHIC 4.7)**.

By adding extra vitamins and minerals to the milk that malnourished children were receiving in the hospital, Manary was able to get the death rate to drop dramatically. Yet, the recovery rate still wouldn’t budge—so children still needed to be hospitalized. In the face of this dilemma, Manary knew it was time to try a new approach.

Emptying the Wards

▶ Testing peanut butter–based RUTF

By the time Manary began seriously thinking about home-based therapy, he had been working in Malawi for about 5 years, had witnessed the failure of the standard WHO treatment, and had spent time in villages. In his head, he’d been tossing around the idea of something new and different. Out of the blue, he got an e-mail from a doctor in France, André Briend, who was also thinking about home-based therapy as a treatment for malnutrition.

The two scientists corresponded for about a year by e-mail, weighing the pros and cons of various foods that could be eaten at home yet would still pack a nutritional wallop. After considering various options—including biscuits, pancakes, and even Nutella—the two researchers eventually decided on peanut butter as the best choice. The peanut butter RUTF that they devised contains a mineral and vitamin mix that makes up 1.6% of the paste by weight. Although peanut butter naturally contains many micronutrients, the amounts that malnourished children need are greater than those normally found in most foods. To provide the extra quantities, the peanut paste is deliberately enriched with micronutrients.

COFACTOR
An inorganic substance, such as a metal ion, required to activate an enzyme.

COENZYME
A small organic molecule, such as a vitamin, required to activate an enzyme.

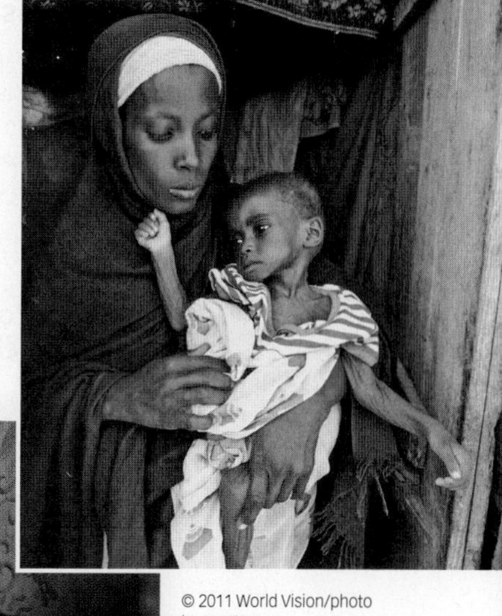

© 2011 World Vision/photo by Jon Warren

© 2011 World Vision

For many starving children, peanut butter RUTF has been a life-saver. Top right shows a malnourished child before receiving RUTF.

INFOGRAPHIC 4.7

Vitamins and Minerals Have Essential Functions

Vitamins and minerals are micronutrients, nutrients that are essential for health but required in far smaller amounts than macronutrients.

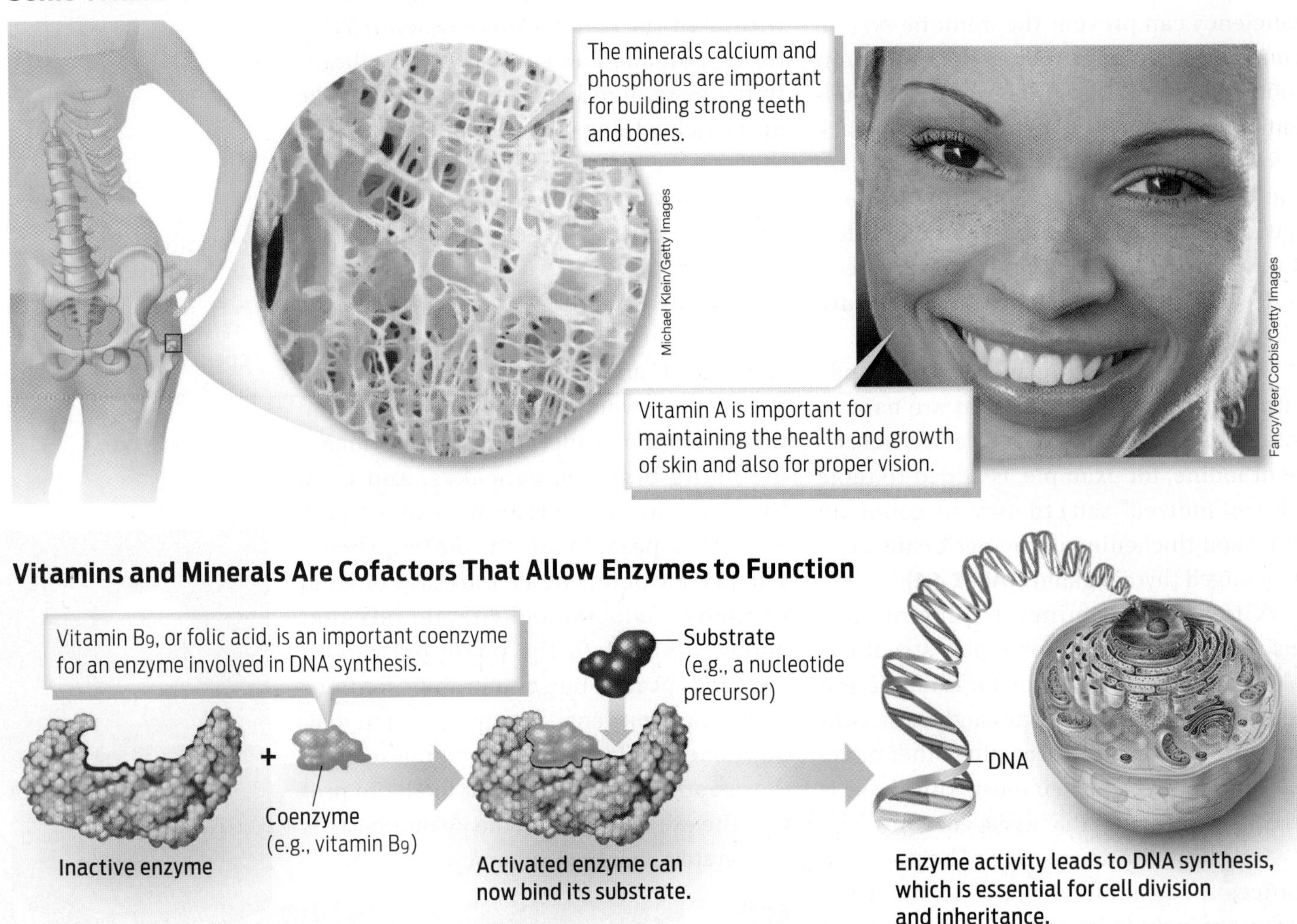

? Does the fact that vitamins and minerals are micronutrients make them less important than macronutrients? Explain why or why not.

In 2001, Manary and Briend were ready to test the idea. Briend had some of the peanut butter RUTF made up in France, sealed in foil packets, and shipped to Malawi. The two scientists then carefully designed a scientific study to test the product. After a brief stabilization phase in the hospital (during which antibiotics were given, if necessary), the children were discharged and sent home on one of three different treatment regimens: (1) ample amounts of traditional food—corn flour and soy; (2) a small amount of peanut butter–based RUTF, to be used as a supplement to the normal diet at home; and (3) the full dose of peanut butter–based RUTF containing sufficient nutrients to meet the children's total nutritional needs. The goal of the study was to see which treatment regimen was most effective.

Within a few months, the results were clear—and impressive: 95% of the children who received the full peanut butter–based RUTF recovered. Those who received traditional food or supplemental RUTF also did pretty well—about 75% of them recovered—but the full peanut butter–based RUTF worked even better. Notably, all of these home-based treatments were significantly better than the standard hospital-based milk therapy, which historically had a 25% to 40% recovery rate.

Manary couldn't quite believe how effective the treatment was. "I said, 'Damn, this stuff really works!'" **(INFOGRAPHIC 4.8).**

One of the main reasons that peanut butter RUTF—which locals call chiponde, or "nutpaste"—is more effective than standard therapy is that it can be administered safely at home. When children are malnourished, their immune systems aren't functioning at optimal levels, which means that hospitals are often the worst place for them to be because of the risk of infection from other sick patients. Peanut butter RUTF is

INFOGRAPHIC 4.8

Peanut Butter–Based RUTF Saves More Children

Studies show that significantly more children recover when treated at home with peanut butter–based RUTF compared to a corn/soy flour diet or their regular diet supplemented with a small amount of the RUTF supplement.

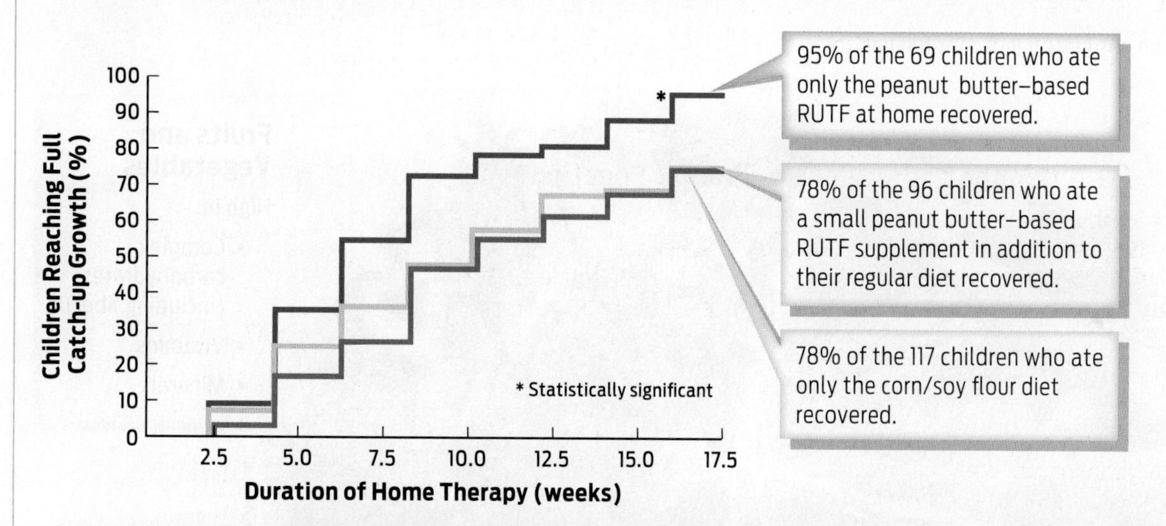

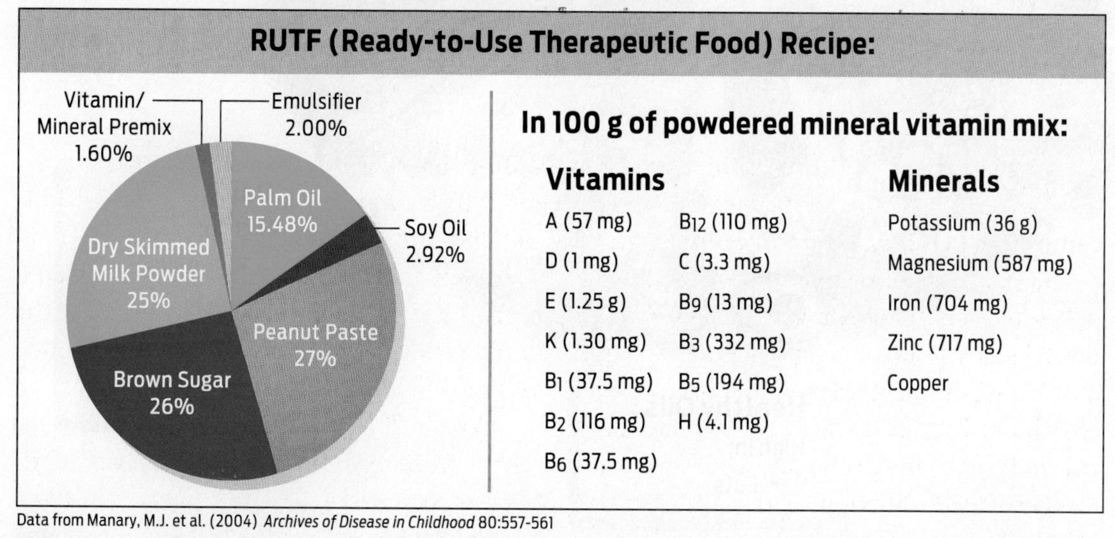

Data from Manary, M.J. et al. (2004) *Archives of Disease in Childhood* 80:557-561

? In this study, did the RUTF work better as a full dietary replacement or as a supplement to the normal diet?

also something that children can eat on their own, without help from their parents. And, because they clearly like the taste of it (one doctor described it as tasting like the inside of a Reese's Peanut Butter Cup), they gobble it up.

As dramatic as Manary's initial results were, however, not everyone was convinced. In fact, most leaders in nutrition science were vehemently opposed to the approach, which is why Manary was heckled at the conference in 2003.

Manary wasn't the only person in the field to face a backlash. Other doctors working in this area faced similar opposition. "It was pretty nasty," says Steve Collins, a physician and early advocate of home-based treatment whose work was also criticized.

To raise awareness of his peanut butter approach and to start making the treatment available on a wider basis, in 2004 Manary started Project Peanut Butter, which would produce the product locally and distribute it to families in Malawi. So far, he says, more than 500,000 children have been helped by peanut butter RUTF—not only in Malawi, but all across the world, including in Haiti and Sierra Leone.

INFOGRAPHIC 4.9

A Balanced Diet

A balanced diet includes all the nutrients needed for full health.

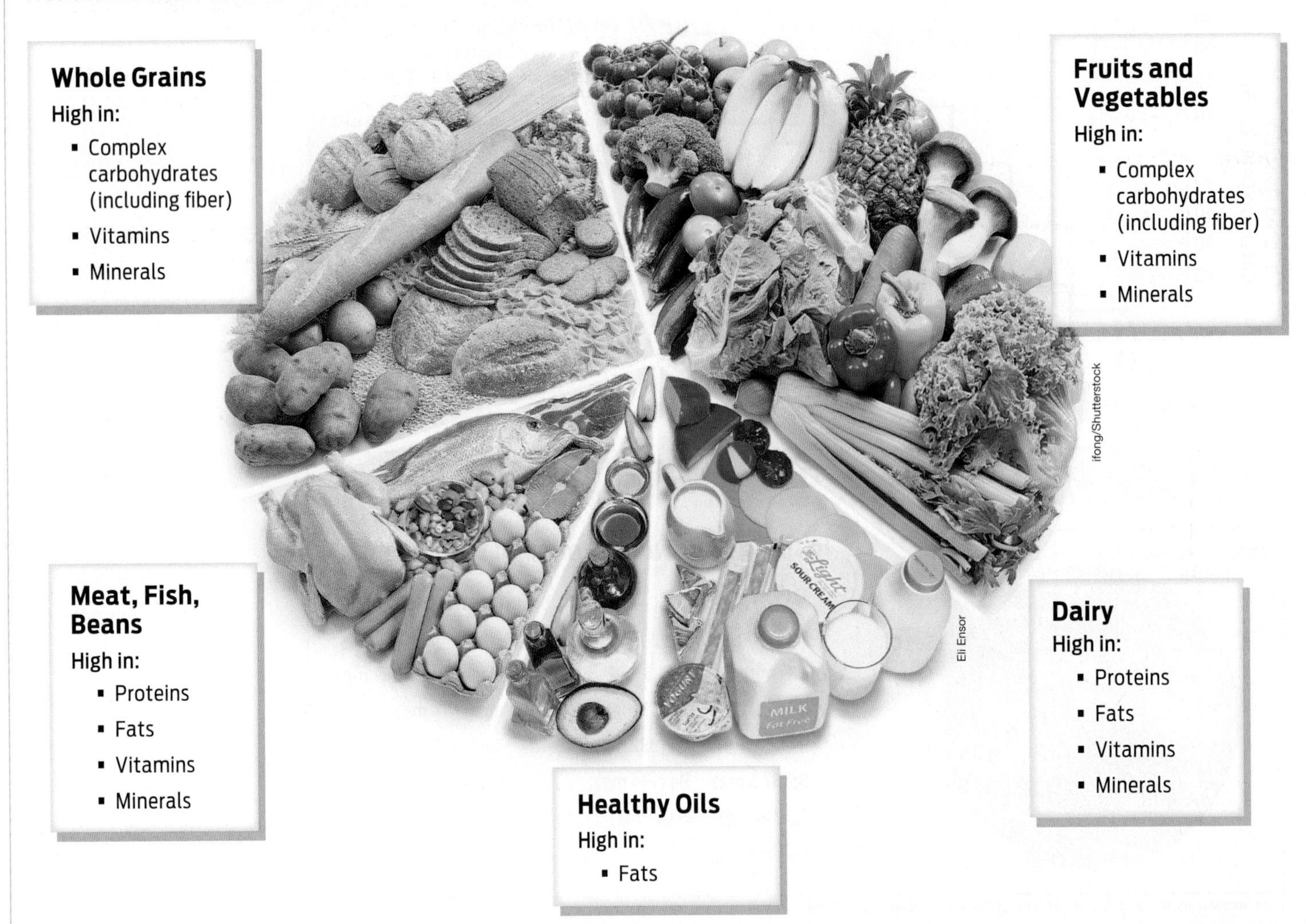

? **Some diet plans restrict or eliminate entire food groups. How easily will these diet plans fit into a balanced diet?**

Thanks in part to Project Peanut Butter, and the extensive body of evidence that Manary and others have collected, the world aid community eventually came around to Manary's view: in 2007, in a dramatic about-face, the joint UN relief agencies—UNICEF, WHO, and the World Food Programme—issued an official statement saying that home-based therapy with peanut butter RUTF is the preferred way to treat acute malnutrition. Of course, eating a balanced diet is still the ideal way to prevent malnutrition **(INFOGRAPHIC 4.9)**.

Manary believes that the ultimate solution to malnutrition is prevention through improved agriculture. To that end, he is also working on a project sponsored by the Bill and Melinda Gates Foundation that aims to use genetic engineering to improve the nutritional quality of subsistence crop plants in Africa. Manary has high hopes for the project, and envisions a day when children in this region do not routinely die from hunger. "My professional goal is to fix malnutrition for kids in Africa," he says. ■

CHAPTER 4 Summary

Driving Question 1 What are the macronutrients and micronutrients provided by food?

- Food is a source of nutrients. Nutrients provide the chemicals required to build and maintain cells and the energy that cells need to function.
- Macronutrients are nutrients required in large amounts. Micronutrients are nutrients required in smaller amounts. Both are important for good health.
- Macronutrients include carbohydrates, proteins, and lipids, including dietary fats; these are among the organic macromolecules that make up our cells.

Driving Question 2 What are enzymes, how do they work, and how do they contribute to reactions of metabolism?

- Enzymes are proteins that accelerate the rates of chemical reactions. Nearly all reactions in the body require enzymes, including those reactions required for growth and development.
- Enzymes speed up reactions by binding specifically to substrates and reducing the activation energy necessary for a reaction to occur. Enzymes mediate both bond-breaking (catabolic) and bond-building (anabolic) reactions.
- Many enzymes require small "helper" chemicals called cofactors to function. Micronutrients such as minerals and vitamins, which are found abundantly in fruits and vegetables, are important cofactors (referred to as coenzymes in the case of vitamins).

Driving Question 3 What are the consequences of a diet lacking sufficient nutrients?

- Eating a balanced diet that contains abundant quantities of fruits and vegetables is the best way to ensure proper nutrition.
- Malnutrition results when a person's diet lacks adequate macronutrients or micronutrients.
- Malnutrition is especially dangerous for children, whose bodies are, or should be, growing rapidly.

More to Explore

- Project Peanut Butter: http://www.projectpeanutbutter.org
- UN World Food Programme, Malawi: https://www.wfp.org/countries/malawi
- U.S. Department of Agriculture, Food Security in the United States: http://www.ers.usda.gov/topics/food-nutrition-assistance/food-security-in-the-us.aspx
- Film: *A Place at the Table:* http://www.magpictures.com/aplaceatthetable/.
- Manary, M. J., et al. (2004). Home based therapy for severe malnutrition with ready-to-use food. *Arch Dis Child 89*:557–561.
- Diamond, J. (1999) *Guns, Germs, and Steel: The Fates of Human Societies.* New York: W. W. Norton.

CHAPTER 4 Test Your Knowledge

Driving Question 1 What are the macronutrients and micronutrients provided by food?

By answering the questions below and studying Infographics 4.3, 4.4, 4.7, and 4.9 and Table 4.1, you should be able to generate an answer for this broader Driving Question

Know It

1. A macronutrient is
- **a.** a nutrient with a large molecular weight.
- **b.** a nutrient that is abundant in the diet.
- **c.** a nutrient that is required in large amounts.
- **d.** a nutrient that is stored in large amounts in the body.
- **e.** a nutrient that the body makes in large quantities.

2. Which of the following is/are macronutrient(s)?
- **a.** protein
- **b.** iodine
- **c.** vitamin C
- **d.** fats
- **e.** all of the above
- **f.** a and d

3. A multivitamin supplement is a(n) __________ supplement.
- **a.** macronutrient
- **b.** micronutrient
- **c.** mineral
- **d.** enzyme
- **e.** a and b

4. Which of the following foods is a rich source of protein?
- **a.** lean meat, such as chicken breast
- **b.** whole grains (e.g., whole-wheat bread)
- **c.** olive oil
- **d.** leafy greens
- **e.** berries (e.g., blueberries and raspberries)

5. Proteins are broken down into which subunits during digestion?
- **a.** fatty acids
- **b.** amino acids
- **c.** glycerol
- **d.** nucleotides
- **e.** simple sugars

6. Where (or how) do we obtain essential amino acids?
- **a.** from carbohydrates in our diet
- **b.** by synthesizing them from other amino acids
- **c.** from oils in our diet
- **d.** from bright orange fruits and vegetables
- **e.** from protein in our diet

Use It

7. Explain the difference between macronutrients and micronutrients.

8. A typical multivitamin supplement contains vitamin A, vitamin C, vitamin D, vitamin E, vitamin K, vitamin B_1, vitamin B_2, vitamin B_6, biotin, calcium, iron, magnesium, zinc, selenium, copper, manganese, and chromium. Explain your answers to the following questions.
- **a.** Are all of these vitamins? If there are ingredients that are not vitamins, what are they?
- **b.** Are all of these micronutrients?

9. Our bodies cannot synthesize vitamin C, but require it. Therefore, vitamin C is
- **a.** an essential micronutrient.
- **b.** an essential mineral.
- **c.** an essential macronutrient.
- **d.** a nonessential vitamin.
- **e.** a nonessential amino acid.
- **f.** a and c

10. Corn lacks the essential amino acids isoleucine and lysine. Beans lack the essential amino acids tryptophan and methionine. Soy contains all the essential amino acids.
- **a.** Could someone survive on a diet containing only a corn-based protein? Why or why not?
- **b.** Why do many traditional diets combine corn (e.g., in tortillas) with beans?
- **c.** Why did one of the home-based feeding therapies in Malawi combine soy flour with corn flour?

Driving Question 2 What are enzymes, how do they work, and how do they contribute to reactions of metabolism?

By answering the questions below and studying Infographics 4.5, 4.6, and 4.7, you should be able to generate an answer for this broader Driving Question.

Know It

11. The substrate of an enzyme is
- **a.** an organic accessory molecule.
- **b.** the molecule(s) released at the end of an enzyme-facilitated reaction.
- **c.** the shape of the enzyme.
- **d.** one of the amino acids that makes up the enzyme.
- **e.** what the enzyme acts on.

12. Compare and contrast enzyme cofactors and coenzymes.

13. Enzymes speed up chemical reactions by
- **a.** increasing the activation energy.
- **b.** decreasing the activation energy.
- **c.** breaking bonds.
- **d.** forming bonds.
- **e.** releasing energy.

14. How is folate (folic acid) best described?
- **a.** as a substrate of an enzyme
- **b.** as a nucleotide
- **c.** as an organic cofactor (coenzyme)
- **d.** as an enzyme
- **e.** a and b

Use It

15. If the shape of an enzyme's active site were to change, what would happen to the reaction that the enzyme usually speeds up?

16. Considering the function of folate (folic acid) given in Infographic 4.7, why would you say pregnant women (and women who could become pregnant) should ensure that they have adequate levels of folate in their diets?

Driving Question 3 What are the consequences of a diet lacking sufficient nutrients?

By answering the questions below and studying Infographics 4.7 and 4.8 and Table 4.1, you should be able to generate an answer for this broader Driving Question.

Know It

17. When vitamins are consumed:
- **a.** Why does excess vitamin E cause problems, but excess vitamin C does not?
- **b.** If you were to take a supplement containing a high amount of vitamin C, what would happen to all that vitamin C? Would it all be used? Would some of it be stored in your body?

18. Which ingredient(s) in peanut butter–based RUTF specifically help bone growth? (*Hint:* Refer to Table 4.1.)
- **a.** calcium
- **b.** vitamin D
- **c.** potassium
- **d.** all of the above
- **e.** a and b

Interpreting Data

Apply Your Knowledge

19. Infographic 4.8 shows the results of a study examining three different home-based therapies for malnourished children in Malawi.
- **a.** From the data shown, how many of all the children in the study reached full catch-up growth?
- **b.** What percentage of the children in the study does this number represent? How does this compare to previous recovery rates of 25% to 40% for children who had received standard hospital therapy?

The children who received the RUTF were given enough of it to supply 730 kJ of energy per kilogram of body weight. This is sufficient energy to meet their needs.
- **c.** Suppose that a malnourished 2-year-old girl weighs a mere 6 kg (about 13 pounds; an average 2-year-old American girl weighs approximately 28 pounds). If she had been in the RUTF group in the study, how many daily kilojoules (kJ) of energy would she have obtained from the RUTF?
- **d.** If the same malnourished 2-year-old had been in the RUTF supplement group, she would have received 2,100 kJ per day from the supplement. What percentage of her daily energy needs would this represent? (*Hint:* Use your answer to part c.)
- **e.** Children in the RUTF supplement group ate a traditional diet of corn/soy flour to make up the rest of their diet. Corn/soy flour contains 4 kJ of energy per gram. How many grams of the traditional mix would this 2-year-old need to consume (on top of the RUTF supplement) to meet her daily needs?

Mini Case

Apply Your Knowledge

20. A college student returns home at the end of the school year. His mother is shocked by the large number of unhealed scrapes and sores on his knees and arms. She also notices that he has put on a few pounds. The student tells his mother that the scrapes are just left over from a skateboarding mishap a few weeks ago and that he guesses he could cut back on some of his snacks. A week after coming home, he goes to the dentist for his yearly checkup. The dentist is alarmed by his bleeding and swollen gums. When asked about his diet, the student notes that he and some of his friends challenged one another to see who could go for the longest time eating nothing but eggs, mac 'n' cheese, and toast with butter. He proudly announces that he had stayed on this diet for 6 months.

a. Could this student be suffering from malnutrition? Explain your answer.

b. What mineral(s) or vitamin(s) (or both) are you most concerned about, given the symptoms noted by the dentist?

c. What dietary recommendations would you make for this student?

Bring It Home

Apply Your Knowledge

21. Is malnutrition a problem solely in developing countries? Do some investigative research on at least two food-aid programs operating at either a local, federal, or international scale. What criteria would you consider before deciding to donate money to a food-aid program? Explain your answer.

The *Future* of Fuel?

Scientists seek to make algae the next alternative energy source

Sinclair Stammers/Science Source

DRIVING QUESTIONS

1 Why are photosynthetic organisms like algae so important?

2 What are the different types of energy, and what transformations of energy do organisms carry out?

3 How do plants and algae convert the energy in sunlight into energy-rich organic molecules? (And why can't other organisms, including humans, do this?)

4 How do algal biofuels compare to other fuels in terms of costs, benefits, and sustainability?

As an engineer working for the Navy Seals in 1978, Jim Sears took a nighttime scuba dive off the coast of Panama City, Florida—one of many he performed to do underwater research. The dive started out routinely, but then glowing phosphorescent algae suddenly appeared as if out of nowhere. When Sears put his hands out in front of him, sparkling streamers of microbes trickled off his fingertips. "It was magical," he recalls.

Sears is an inventor with many interesting devices to his credit. In the 1970s and 1980s, he built an underwater speech descrambler and a portable mine detector, among other gadgets. Later, he moved on to more creative technologies, including a "hump-o-meter" that could tell farmers when their animals were in heat or mating.

But the seeds of his real claim to fame weren't sown until 2004, when Sears was working in agricultural electronics. That's when he turned his attention toward what he felt was the world's biggest problem: dwindling fossil fuel reserves. After he did some thinking and a little research, the tiny, glowing organisms that had wowed Sears during his nighttime dive more than two decades earlier came to mind. He realized suddenly that they might be able to help.

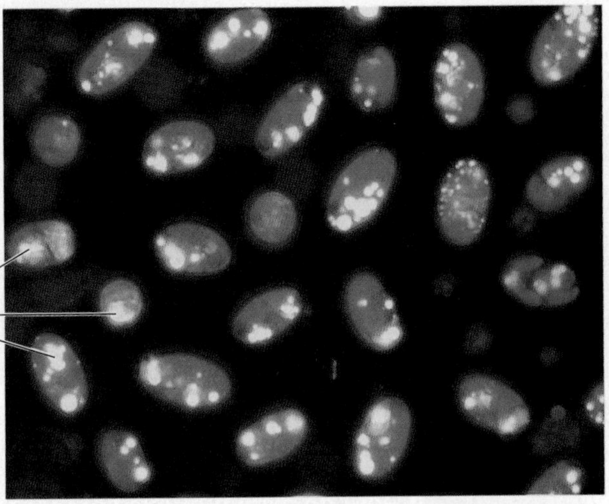

Lipid-filled compartments appear yellow

Courtesy Dr. Ann Wilkie, University of Florida-IFAS

Some algae capture energy from sunlight and convert a portion of it into oil.

Algae are perhaps best known for the layer of green scum they create on the surface of ponds and swimming pools, but they have more impressive qualities, too. Like plants, algae can capture the **energy** of sunlight and convert it into a form that other organisms can use. Even more remarkable, algae trap much of this energy in the form of oils ideally suited to making fuel. The oil that algae produce is very similar to common vegetable oil. It accumulates inside the microbes' cells, and once extracted, it can be processed to make biodiesel, gasoline, or jet fuel. "The more I looked into them, the more amazing they were," Sears says of algae.

That's good news, because America is desperate for new fuels. Americans burn through nearly 400 million gallons of gasoline a day, enough to fill roughly 600 Olympic-size swimming pools. And despite the fact that our fuel demand will likely increase over the next 25 years, the sources of our precious gasoline—oil reserves buried deep underground—are finite, take millions of years to replenish, and largely lie outside U.S. borders **(INFOGRAPHIC 5.1)**.

Confronted with this looming crisis, scientists and politicians are increasingly turning toward alternative energy sources such as **biofuels**—renewable fuels made from living organisms. In an effort to reduce our dependence on oil, in 2007 President George W. Bush signed the Energy Independence and Security Act, which requires the United States to produce 36 billion gallons of biofuels by 2022. Among other things, biofuels include biodiesel made from plant scraps, ethanol made from sugar cane, and natural gas harvested from municipal wastewater—as well as fuels made from algae.

Convinced of the promise of algae biofuels, in 2006, Sears founded Solix, one of the first biotechnology companies to produce biodiesel from algae. In 2009, the company began commercial production of its algae-based fuel, with the goal of making 3,000

INFOGRAPHIC 5.1

Distribution of Recoverable Oil Reserves

The gasoline and diesel used to power cars and trucks begins as oil formed deep in the ground over millions of years. The United States depends heavily on oil recovered from other countries for its fuel supply.

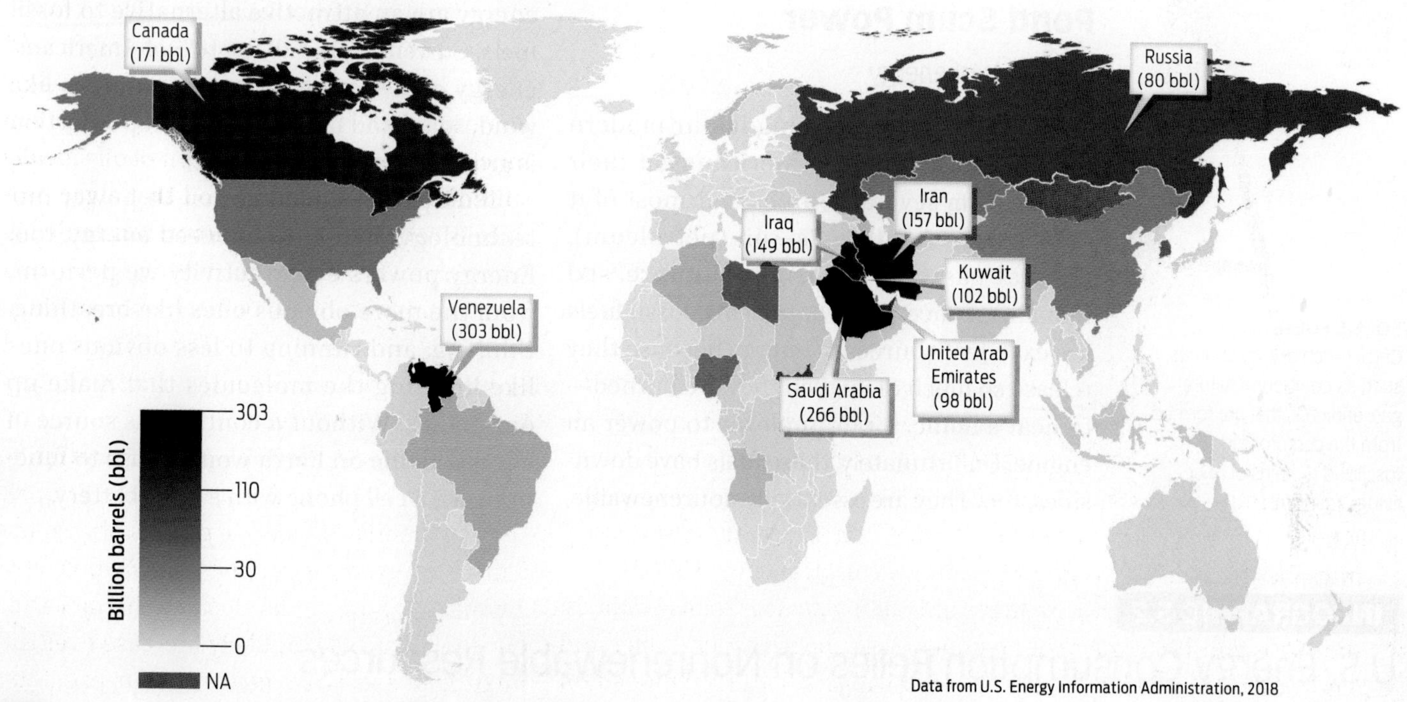

Data from U.S. Energy Information Administration, 2018

? What are the top three oil-producing countries?

gallons of biodiesel per acre of cultivated algae per year. Other companies soon joined the race, including San Diego–based Sapphire Energy and San Francisco Bay Area–based Solazyme. In September 2009, a modified Toyota Prius dubbed Algaeus drove 3,750 miles across the country powered in part by Sapphire's algae-based gasoline. Two years later, in November 2011, United Airlines flew a passenger plane fueled by Solazyme's algae-based jet fuel.

Despite the initial excitement, algae biofuels face an uphill climb in becoming commercially viable. Several of the early ventures—which benefited from the 2009 stimulus package approved by the Barack Obama administration—ultimately could not compete with low gas prices and are no longer in the algae biofuel business. Nevertheless, this alternative fuel source continues to draw attention. In fact, ExxonMobil, the largest oil company in the world, is now one of the biggest investors in

ENERGY
The capacity to do work. Cellular work includes processes such as building complex molecules and moving substances into and out of the cell.

BIOFUELS
Renewable fuels made from living organisms (e.g., plants and algae).

MCT/Getty Images

The Toyota Algaeus, powered by algae biofuel, gasoline, and batteries.

algae biofuel science. It seems to be betting that, along with other alternative energy sources, algae will be a fuel source of the future.

Pond Scum Power

▶ Sources of energy

To power nearly everything in our modern lives, we need energy. Americans get their energy from several sources, but most of it comes from **fossil fuels**—oil (petroleum), natural gas, and coal. The compressed remains of once-living organisms, fossil fuels are excellent sources of energy because they release so much of it when they are burned—to heat a home, for example, or to power an engine. Unfortunately, these fuels have downsides, too. They are essentially nonrenewable, meaning they won't last forever. In addition, extracting them from the earth and burning them can cause environmental damage, including oil spills and climate change. Those are two big reasons why renewable sources of energy are an attractive alternative to fossil fuels. Currently only about 10% of Americans' energy comes from renewable sources like wind, solar, and biofuels, but that proportion is growing **(INFOGRAPHIC 5.2)**.

FOSSIL FUELS
Carbon-rich energy sources, such as petroleum, natural gas, and coal, that are formed from the compressed, fossilized remains of once-living organisms.

Energy isn't needed just to power our technology: living things need energy, too. Energy powers every activity we perform, from the more obvious ones like breathing, thinking, and running to less obvious ones like building the molecules that make up our bodies. Without a continuous source of energy, all life on Earth would cease to function, like a cell phone with a dead battery.

INFOGRAPHIC 5.2

U.S. Energy Consumption Relies on Nonrenewable Resources

The United States leads the world in petroleum consumption, and is the second largest net consumer of fossil fuels behind China. Fossil fuels are considered nonrenewable because they take millions of years to create by natural processes. As we continue to deplete fossil fuels, new energy sources are being developed that reduce our demand on petroleum and other fossil fuels.

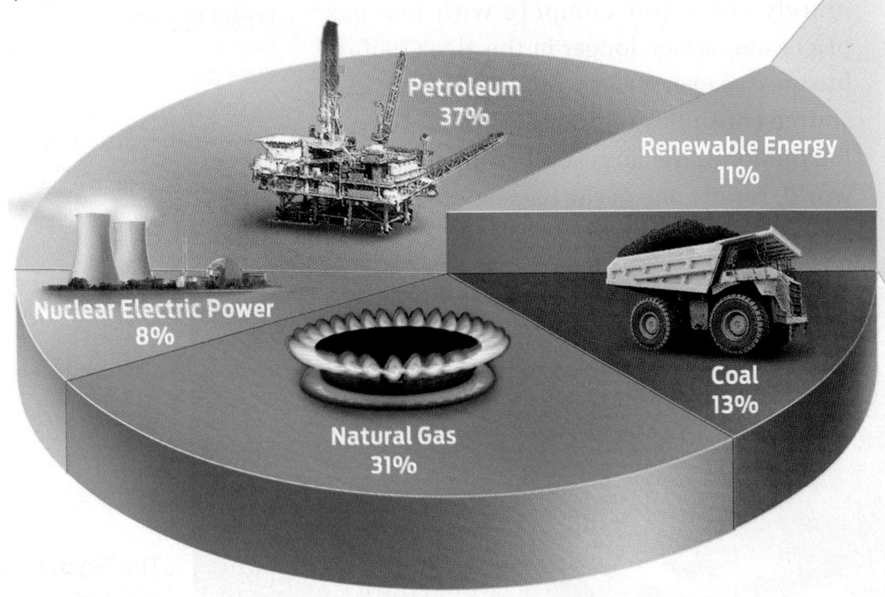

2% Geothermal
8% Solar
21% Biofuels
22% Wind
23% Hydroelectric
23% Biomass Waste and Wood

Russ Munn/AGE Fotostock; Ellen Isaacs/AGE Fotostock; Zeljko Radojko/Shutterstock; CSP_Mokai/AGE Fotostock

Data from U.S. Energy Information Administration, 2018

? Of the energy sources shown, which are fossil fuels?

Humans and other animals obtain the energy they need by eating food. We've already seen that food is a source of organic molecules—the carbohydrates, fats, and proteins that our digestive systems break down into smaller subunits (see Chapter 4). The chemical bonds in these subunits represent a form of stored **chemical energy** that can be used to power cell functions.

In contrast, algae and plants get their energy from the sun. They trap the energy of sunlight and store it in the form of molecules inside their cells. Algae are a promising source of biofuels because they are very good at trapping and storing energy—the oil they produce is rich in chemical energy. Even more impressive, all they need to make this oil is sunlight, carbon dioxide, and water (plus a few nutrients). When provided with these basic ingredients, algae grow rapidly—some strains double their volume in 12 hours—while accumulating gobs of oil inside their cells. This oil can be used to make biodiesel. Algae also make carbohydrates (sugar), which can be converted into other biofuels like ethanol and butanol. As a bonus, other algae molecules can be used for a variety of purposes **(INFOGRAPHIC 5.3)**.

Algae come in many different forms. There are large multicellular varieties, such as the seaweed that washes up on a beach after a storm, and there are microscopic single-celled varieties, such as the ones that turn ponds green and scummy. For making biofuels, the most useful algae are the single-celled pond scum variety. These organisms produce some of highest amounts of oil per cell—as much as 70% of the cell's dry weight.

CHEMICAL ENERGY
Potential energy stored in the bonds of biological molecules.

INFOGRAPHIC 5.3

Algae Capture Energy in Their Molecules

Algae can use sunlight, carbon dioxide, water, and nutrients to produce a high volume of oil readily available to produce biofuel, in addition to carbohydrates and proteins that can be useful as additional energy sources.

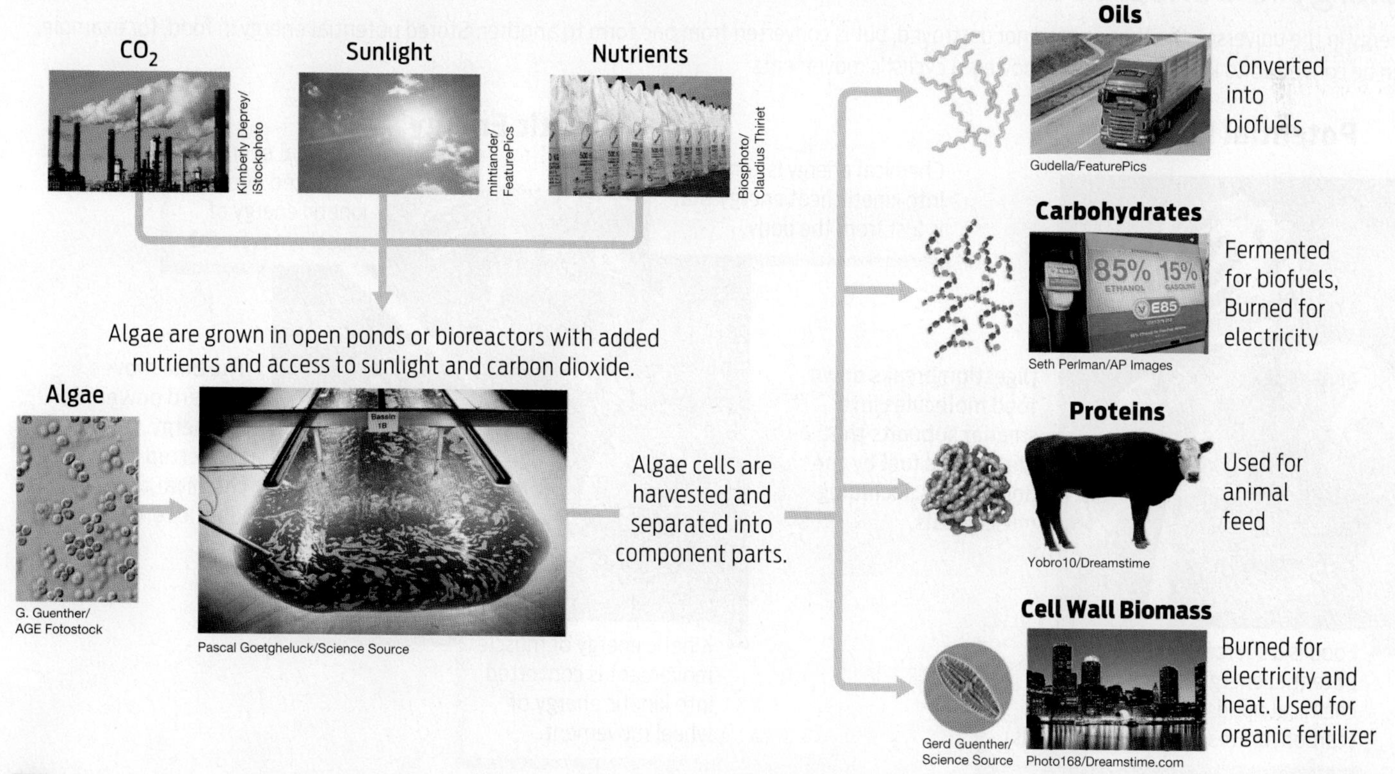

? Which algal component is used to produce biofuels? Which can be used to generate electricity?

Putting Energy to Use

▸ Conservation of energy

With the topic of dwindling energy reserves frequently in the news, it's tempting to think that energy is something that we use up over time. In reality, energy cannot be created or destroyed. When energy is used to power our cars—or our brain cells—that energy is not destroyed but rather changes form, a principle known as the **conservation of energy.** One of the laws of thermodynamics, this principle is key to understanding energy use, as applicable to people as it is to cars and cell phones.

CONSERVATION OF ENERGY
The principle that energy cannot be created or destroyed, but can be transformed from one form to another.

POTENTIAL ENERGY
Stored energy.

KINETIC ENERGY
The energy of motion or movement.

HEAT
The kinetic energy generated by random movements of molecules or atoms.

Consider a cyclist who eats a power bar before going on a ride. The bar contains chemical energy in the form of covalent bonds in the carbohydrate, lipid, and protein molecules that make up the bar. Chemical energy is **potential energy,** energy that is stored and waiting to be used. Digestion breaks these molecules into smaller subunits for use as a fuel (discussed further in Chapter 6). As the cyclist begins to pedal, his body converts this chemical potential energy into the **kinetic energy** of motion. Initially, chemical energy is converted into the kinetic energy of muscle contraction and **heat.** This kinetic energy of muscle movement is then converted into the kinetic energy of moving wheels and heat. From start to finish, from power bar to spinning wheels, energy is converted from one form into another, but is never destroyed **(INFOGRAPHIC 5.4).**

Energy from biofuel has a similar life story. Oil from algae is rich in chemical energy: the lipids in the oil store energy in their bonds. When these lipids undergo chemical reactions that break these bonds—for example, when they are burned—large amounts of energy are made available to power machines. In a car's combustion engine, the chemical energy in biofuel is

INFOGRAPHIC 5.4

Energy Is Conserved

Energy in the universe is neither created nor destroyed, but is converted from one form to another. Stored potential energy in food, for example, can be converted to kinetic energy as it powers a cyclist's movements.

Potential Energy

Food is a source of chemical potential energy, stored in the chemical bonds of food molecules (carbohydrates, proteins, and lipids).

Digestion breaks down food molecules into smaller subunits that are used as fuel by the body's cells, including muscle cells.

Kinetic Energy

Chemical energy is converted into kinetic heat energy that is lost from the body.

Chemical energy is converted into the kinetic energy of muscle movement.

Kinetic energy of muscle movement is converted into kinetic energy of wheel movement.

KENZO TRIBOUILLARD/Getty Images

The bike moves forward powered by energy converted from chemical to kinetic form.

? **Into what forms of energy is the chemical energy in the food converted as the cyclist completes his training ride?**

INFOGRAPHIC 5.5

Energy Conversion Is Not Efficient

As energy is converted from one form to another, only some of the available energy is fully converted to the next form. Some of it is converted into heat that escapes into the environment, and some energy is not converted at all.

Unconverted fuel energy is removed in exhaust.

Heat energy (kinetic) escapes

Heat energy (kinetic) escapes

Heat energy (kinetic) escapes

Pixelgnome/Dreamstime.com

Fuel Is Chemical Energy: The chemical bonds of biofuel molecules store potential energy.

Vasily Smirnov/Dreamstime.com

Fuel Combustion: Chemical energy is converted to kinetic energy of heated gas molecules.

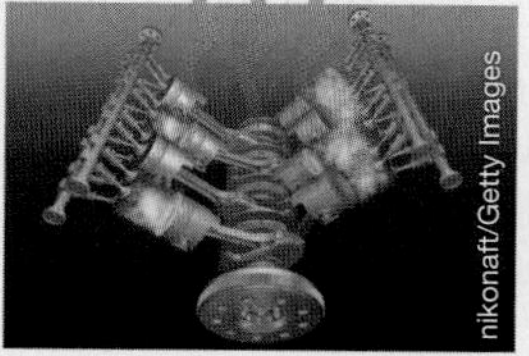
nikonaft/Getty Images

Pistons Fire: Kinetic energy of heated gas molecules is converted to kinetic energy of piston movement.

Motoring Picture Library/Alamy

Tires Roll: Kinetic energy of pistons is converted to kinetic energy of tire movement.

Arjuna Kodisinghe/Shutterstock

Energy in Car Is Depleted: More chemical energy is required to keep tires rolling.

? When the vehicle is fueled, what form of energy is being added to the tank?

rapidly and explosively converted to the kinetic energy of heat within a chamber. Gas molecules in the chamber are warmed, causing them to move faster and the gas to expand. The expanding gas pushes against the pistons, causing the wheels to move. Through this process, the chemical potential energy of biofuel is converted into the kinetic energy of car movement.

If energy is never destroyed, only converted, why do we need to keep filling our cars' gas tanks? Essentially it's because energy doesn't stay in the car system (or in your body). Energy flows from the fuel to the engine to the tires and eventually leaves the car system as heat, which can no longer do the work of running the car. What's more, the conversion of energy from one form to another isn't 100% efficient. With every energy conversion, a bit of energy is "lost" to the environment as heat. That explains why car engines are warm after being driven (and why our bodies heat up when we exercise). Just how well your car converts the chemical energy of gas into the kinetic energy of car speed determines how many miles per gallon you get. If an engine doesn't combust efficiently, some of the fuel molecules will undergo chemical reactions and be converted to other molecules—like pollutants—rather than generating heat to power the pistons. If the pistons can't use the heat efficiently, the heat will leave the car without powering the wheels. If you drive a heavy SUV, that vehicle will require more of the energy in gasoline to reach a given speed than will a lighter compact model. At each step of energy transformation, energy is lost from the car system and into the environment, sending us back to the fuel pump once more **(INFOGRAPHIC 5.5)**.

Solar-Powered Cells

▸ Autotrophs versus heterotrophs

To grow algae for biofuel, scientists are experimenting with several different production

Mark Boster/Getty Images

Algae farms like this one in San Diego, California, circulate algae as they are grown for biofuel production.

techniques. One involves growing algae in large open-air tanks—the method closest to the natural process of algae growing on ponds. This approach is the one that Sears's company Solix used and is also the one currently being used by ExxonMobil. Another method uses closed bioreactors in which algae are not exposed to outside air, but instead the necessary carbon dioxide is pumped in. It offers the advantage that individual tanks can be stacked vertically to reduce the overall footprint of the operation and better prevent contamination. A third approach involves using algae to convert plant sugars into oil in large vats. In all cases, the goal is the same: convert as much energy from sunlight as possible into biofuel.

Plentiful and free, sunlight is a biofuel maker's dream. These qualities are the main reason why so much research has gone into developing technologies to capture solar energy. Yet figuring out how to tap and store this energy cheaply has proved challenging. So far, even with our best technology, humans can't beat what plants and algae do naturally—which is why many biofuels rely on these organisms as their starting point.

The secret to their success is **photosynthesis**—the process through which plants, algae, and a few other organisms capture the energy of sunlight and convert it into the chemical energy of sugar molecules. That may sound simple, but photosynthesis surely ranks as one of the most sophisticated and consequential achievements of evolution. By performing this energy conversion, photo synthesis makes possible nearly all life on Earth **(INFOGRAPHIC 5.6)**.

There's another name for the energy-rich products of photosynthesis: food. Organisms such as plants, algae, and certain bacteria that can make food from inorganic (nonliving) starting materials—from carbon dioxide, water, and sunlight, for example—are called **autotrophs.** (Their name literally means "self-feeders.") Autotrophs include not just

PHOTOSYNTHESIS
The process by which plants and algae harness the energy of sunlight to make sugar from carbon dioxide and water.

AUTOTROPHS
Organisms such as plants, algae, and certain bacteria that can make their own food from inorganic starting materials (e.g., CO_2).

INFOGRAPHIC 5.6

Photosynthesis Converts Light Energy into Chemical Energy

Through the process of photosynthesis, plants, algae, and some bacteria are able to convert light energy from the sun into chemical energy stored in glucose.

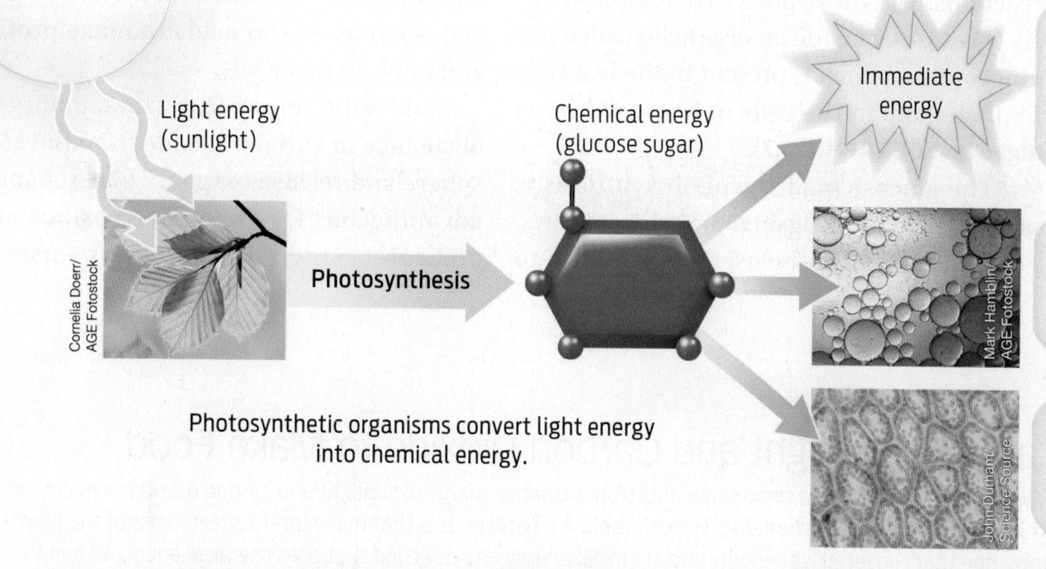

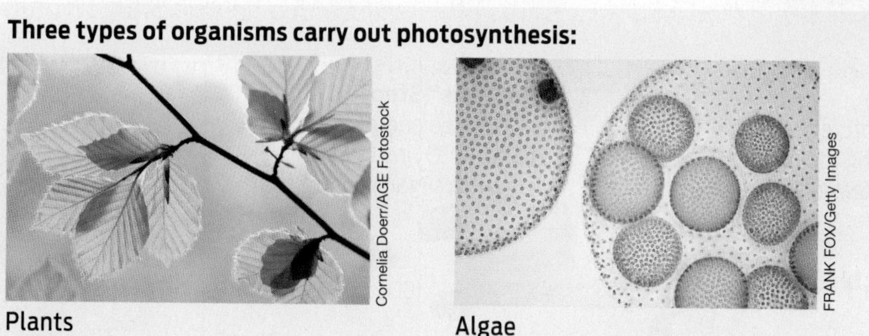

? How can the glucose produced in photosynthesis be used by the plant?

crop plants like wheat, corn, and soybeans, but also all flowering plants, trees, and bushes. Oceans and lakes have an abundance of autotrophs, too, including much of the planet's algae and photosynthetic bacteria.

Organisms that can't make their own food and must consume organic molecules produced by other living organisms to obtain energy are called **heterotrophs** ("other-feeders"). This group includes all animals, fungi, and most bacteria. Photosynthesis, in other words, is what makes life possible for the rest of us.

Using Sunlight to Make Food

▶ Steps of photosynthesis

So how do plants and algae accomplish this unique feat of creating their own food? The process of photosynthesis can be divided into two main steps: a "photo" step and a "synthesis" step. First, during the "photo" step, light energy is captured in chemical form. In this process, water molecules are split and oxygen is released as a by-product. Then, during the "synthesis" step, the chemical energy captured from light powers the formation of

HETEROTROPHS
Organisms, such as humans and other animals, that obtain energy by consuming organic molecules that were produced by other organisms.

glucose molecules, using the carbon atoms of carbon dioxide to form the carbon backbone of the glucose molecule. This synthesis phase does not directly require sunlight, but it does require the products of the "photo" reactions. The entire process of photosynthesis takes place inside an organelle called the **chloroplast,** which is present in the leaf cells of plants and in the cells of photosynthetic algae **(INFOGRAPHIC 5.7).**

CHLOROPLAST
The organelle in plant and algae cells where photosynthesis occurs.

The glucose made by photosynthesis is used by plants (or algae) in a variety of ways. For example, it can be used for growth—to build new plant parts, like stems and fruit—or it can be used as an energy source to power cellular reactions. Ultimately, glucose and other smaller sugars produced during the "synthesis" reactions provide the building materials for many different molecules in the cell, such as amino acids to make proteins and lipids to make oils.

Like all photosynthesizers, plants and algae take in carbon dioxide from the atmosphere and release oxygen. This exchange is advantageous for life on Earth, since many creatures—including humans—use this

INFOGRAPHIC 5.7

Photosynthesis Captures Sunlight and Carbon Dioxide to Make Food

Photosynthesis is the process by which plants, algae, and some other autotrophs use the energy of sunlight and carbon dioxide to make food. In plant and algae cells, photosynthesis occurs in an organelle called the chloroplast (found in cells that make up the green parts of the plant). Photosynthesis has two main steps, one that converts light energy into chemical energy and a second that uses chemical energy to build sugar molecules from carbon dioxide.

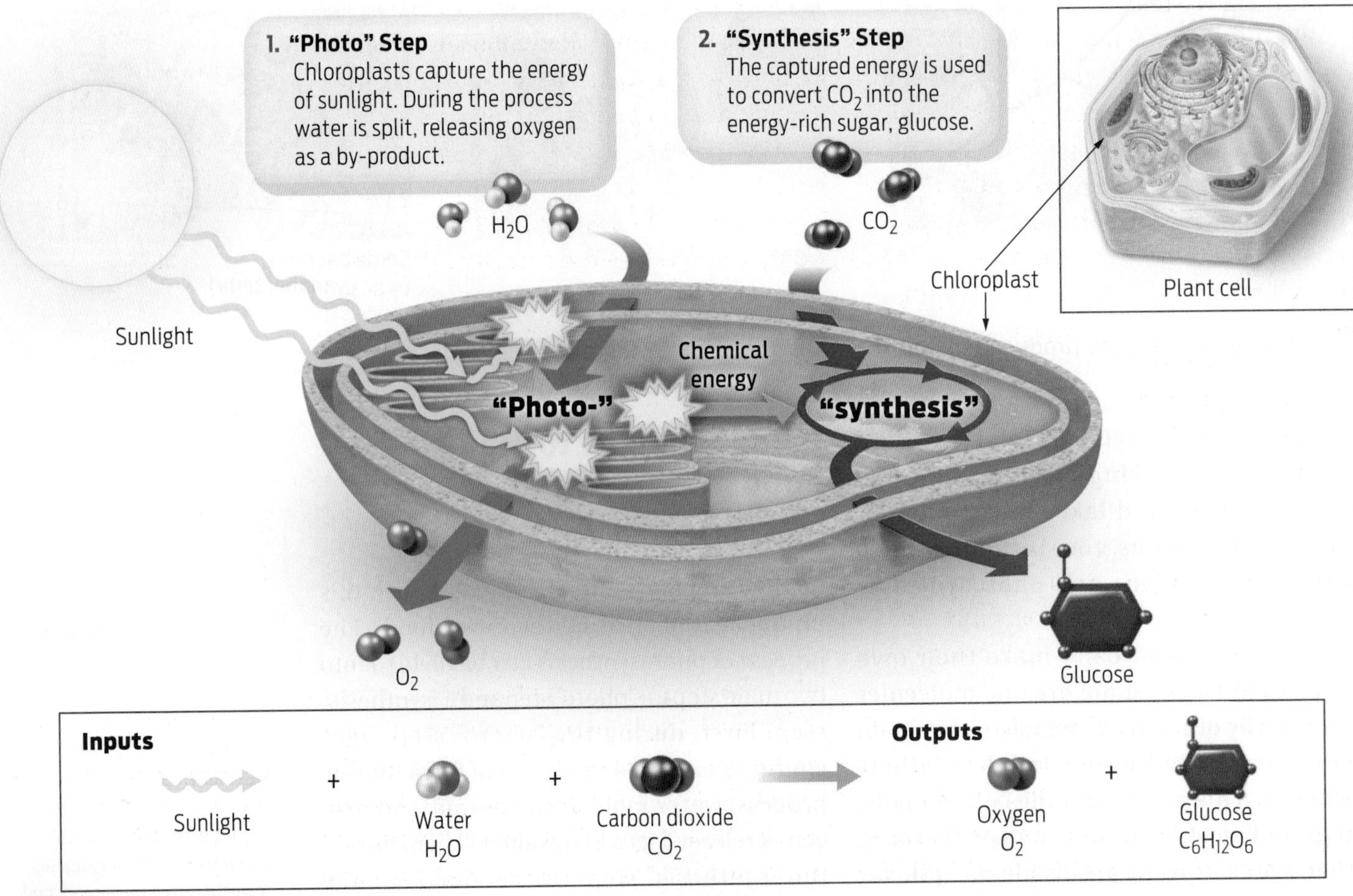

? What is the source of the carbon atoms in the glucose and of the oxygen atoms in the O_2 produced by photosynthesis?

oxygen to breathe. Nearly all the breathable oxygen that exists on our planet comes from photosynthesis. In turn, by giving off carbon dioxide (CO_2) as a waste product when we breathe out, humans and other organisms provide the raw material for photosynthesis to continue: a win–win situation for life on Earth.

Carbon dioxide, like sunshine, is free and readily available to power photosynthesis, but the amount required to make biofuel can exceed that found in the surrounding air. As a result, biofuel producers sometimes have to obtain their supply of carbon dioxide from another source, which can be costly. Sears's company, Solix, set up its first biofuel production plant next to a beer manufacturer that produces carbon dioxide as a by-product of brewing. The company could siphon off this carbon dioxide and feed it to its algae, helping them grow. Other companies have considered placing their algae bioreactors next to carbon dioxide–emitting power plants—thus turning waste into profit.

Photosynthesizers' ability to suck up so much carbon dioxide from the atmosphere has several benefits for our planet—and, by extension, for us. Carbon dioxide (CO_2) is not just the gas that plants and algae take in during photosynthesis, but also the gas that is released by burning fossil fuels. If you think about it, this makes perfect sense: fossil fuels such as coal, petroleum, and natural gas are the compressed remains of once-living photosynthetic organisms that have formed over millions of years; burning these fuels releases their stored carbon dioxide, sending it back into the atmosphere. Carbon dioxide is a greenhouse gas that accumulates in the atmosphere and is partly responsible for the increasing temperatures around the globe and other signs of climate change (see Chapter 23).

In contrast, when plant and algae products are converted to fuel and burned, the CO_2 released is the same CO_2 that they took up by photosynthesis. This means that they are not contributing additional CO_2 to the atmosphere when they are burned; in other words, they are "carbon neutral." By allowing more fossil fuels to stay in the ground, algae biofuels could help with the problem of climate change, too.

From Sun to Fuel

▶ Light energy and chlorophyll

You may be wondering how something as intangible as sunlight can carry energy. If you've ever walked barefoot across a sandy beach on a hot summer day, you know that sunlight is a potent source of heat energy. You may also have a sense that certain colors absorb or reflect sunlight better than others: on a sunny day, wearing a reflective white shirt keeps you cooler than wearing a black one that absorbs more of the sun's rays.

These properties of sunlight reflect the nature of **light energy**, a type of electromagnetic radiation (which also includes x-rays, microwaves, and radio waves). Light energy from the sun travels to Earth in waves. These waves of light are made up of discrete packets of energy called **photons.** Photons with different wavelengths contain different amounts of energy, and different objects on Earth absorb and reflect different wavelengths of light. Some of these wavelengths of light energy, when viewed by the human eye and interpreted by the human brain, appear to us as different colors. They form the visible light portion of the electromagnetic spectrum **(INFOGRAPHIC 5.8)**.

When sunlight hits a green plant, for example, its leaves absorb red and blue wavelengths and reflect green wavelengths, which is why plants appear green to our eyes. The molecule within the plant cells that absorbs and reflects these wavelengths of light is the pigment **chlorophyll**—a crucial player in photosynthesis. It is chlorophyll that actually captures the energy of sunlight. During the "photo" reactions, chlorophyll molecules within chloroplasts absorb energy from the red and blue wavelengths of sunlight. In addition to chlorophyll, plants and algae contain other pigment molecules that absorb and reflect other wavelengths of light, giving them their distinctive colors. Chlorophyll, however,

LIGHT ENERGY
A type of electromagnetic radiation that includes visible light.

PHOTONS
Packets of light energy, each with a specific wavelength and quantity of energy.

CHLOROPHYLL
The pigment present in the green parts of plants that absorbs photons of light energy during the "photo" reactions of photosynthesis.

INFOGRAPHIC 5.8

The Energy in Sunlight Travels in Waves

The sun emits electromagnetic radiation with a spectrum of wavelengths. The majority of the electromagnetic radiation emitted by the sun is ultraviolet (UV), visible, and infared (IR), each with a specific wavelength.

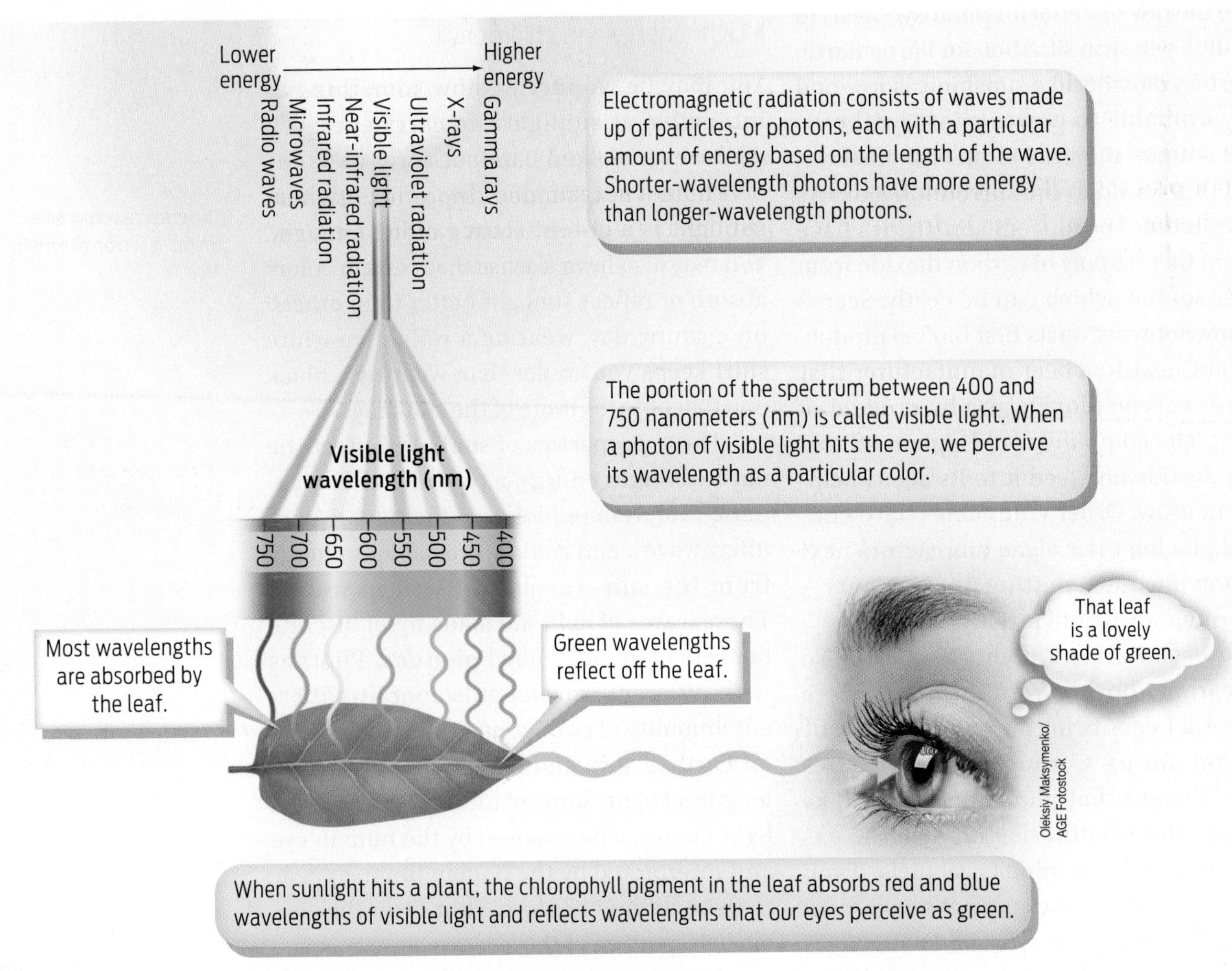

? **Which color of visible light has the most energetic photons?**

is the main pigment involved in photosynthesis. When red and blue photons of sunlight hit chlorophyll, the electrons in its atoms become excited—boosted to a higher energy level. The excitement of these electrons is the initial step in the conversion of light energy into chemical energy during photosynthesis.

Next, the excited electrons in chlorophyll are put to use, powering the gears of the photosynthesis machinery. Eventually, they are knocked off the molecule completely. They go on to power the formation of an energy-carrying molecule called **adenosine triphosphate (ATP),** which is used in the "synthesis" part of photosynthesis as an energy source to make sugar. (We'll talk more about ATP in Chapter 6.) Meanwhile, the electrons lost by chlorophyll are replaced when water is split into oxygen (O_2) and hydrogen (and electrons). At the end of the "photo" reactions, the excited electrons from chlorophyll are captured by a molecule called NADPH. NADPH, an electron carrier, brings the electrons to the "synthesis" reactions. Energy from these excited electrons, as well as from the breakdown of ATP, is then used in the synthesis reactions to incorporate carbon dioxide molecules into sugar.

ADENOSINE TRIPHOSPHATE (ATP)
The molecule in cells that powers energy-requiring functions.

Capturing Carbon

▶ Carbon fixation

Photosynthesis accomplishes two main things. First, it converts light energy from the sun into chemical energy that can be used to make food and fuel for plants and animals. Second, it captures inorganic carbon dioxide gas from the air and incorporates those carbon atoms into organic sugar molecules in a process called **carbon fixation.** What is being "fixed" is the inorganic carbon in CO_2, which is essentially useless to nonphotosynthetic organisms. We must rely on plants, algae, and other photosynthesizers to "fix" carbon into organic sugars that our bodies can use **(INFOGRAPHIC 5.9)**.

The protein that does most of the heavy lifting of incorporating CO_2 into an organic molecule is an enzyme called **Rubisco.** Located inside chloroplasts in the green parts of plants, it is estimated to be the most abundant enzyme on Earth.

Photosynthesis does more for us than just create food—as if that weren't enough. By continuously pulling carbon dioxide out of

CARBON FIXATION
The conversion of inorganic carbon (e.g., CO_2) into organic forms (e.g., sugars like glucose, $C_6H_{12}O_6$).

RUBISCO
The enzyme responsible for the first step of carbon fixation.

INFOGRAPHIC 5.9

Photosynthesis: A Closer Look

1. Light "Photo" Reactions
Chlorophyll pigments within internal chloroplast membranes absorb photons. Chlorophyll electrons (e^-) become excited and enter a series of reactions that generate energy-carrying molecules called ATP and electron-carrying molecules called NADPH, which are used in the synthesis reactions.

2. Carbon "Synthesis" Reactions
Energy from the breakdown of ATP and electrons from NADPH are used in the carbon reactions to fix carbon dioxide into organic sugar molecules, a form of stored chemical energy.

Water (H_2O)
Water is split during the light reactions. Split water molecules release electrons that replace electrons lost by excited chlorophyll molecules.

Carbon dioxide (CO_2)
CO_2 gas enters plant cells from the atmosphere. The carbon atoms are incorporated into organic sugar molecules.

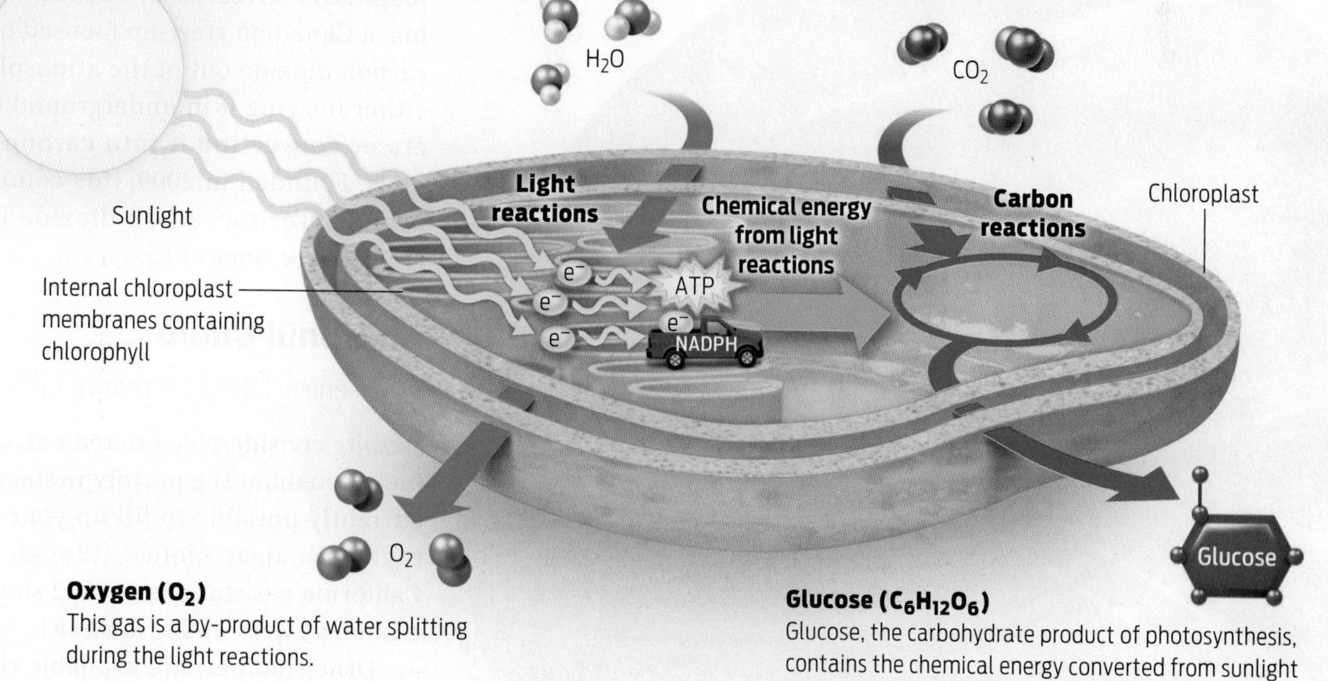

Oxygen (O_2)
This gas is a by-product of water splitting during the light reactions.

Glucose ($C_6H_{12}O_6$)
Glucose, the carbohydrate product of photosynthesis, contains the chemical energy converted from sunlight and the fixed carbon from atmospheric CO_2.

? **Which molecules provide the energy to fix carbon dioxide in the carbon reactions?**

the atmosphere and fixing it into organic sugars, plants, algae, and other photosynthetic organisms perform an important service in maintaining our habitable climate. In fact, scientists are looking for ways to enhance

Believe it or not, the healthy soil in your backyard is actually photosynthetic.

this natural process through various carbon capture systems.

When Jim Sears left Solix in 2007, he became president of A2BE Carbon Capture, a company based in Boulder, Colorado. This company's aim was to reduce carbon dioxide levels in the atmosphere by augmenting the natural process of carbon fixation.

Believe it or not, says Sears, the healthy soil in your backyard is actually photosynthetic. Tiny organisms called cyanobacteria thrive in healthy soil, and they perform photosynthesis, taking in carbon dioxide and fixing carbon into sugar. One square meter of healthy, undisturbed soil can remove 30 grams of atmospheric carbon per year, according to Sears.

A spirulina cyanobacteria farm seen from the air.

Pete Atkinson/Getty Images

The problem is that approximately 2 billion out of Earth's 13 billion total hectares of landmass have been damaged by human activity, with construction and fires among the biggest culprits. According to Sears, it can take anywhere from 30 to 3,000 years for soil microorganisms to regenerate after being destroyed. In the meantime, the damaged soil is unable to remove carbon dioxide from the atmosphere.

Sears's idea was to take small samples of microorganisms from healthy soil, grow them in a contained facility, and then transplant them to damaged soil, where they can spread out and thrive. He estimates that if 1 billion hectares of land were restored in this way, one-seventh of the world's greenhouse gas problem would be solved because of the vast amounts of carbon dioxide that would be pulled out of the atmosphere by the photosynthetic cyanobacteria in the regenerated soil.

Other companies are taking a more direct approach to carbon capture and storage. Chevron and Occidental Petroleum have invested in Carbon Engineering, a Canadian start-up focused on taking carbon dioxide out of the atmosphere and either burying it in underground containers or converting it into carbon-neutral fuels. Founded in 2009, this company has been capturing carbon dioxide from the atmosphere since 2015.

An Uphill Climb

▸ Challenges facing alternative fuels

Despite considerable excitement, algae biofuels remain at the prototype stage. It's not currently possible to fill up your car's gas tank with algae biofuel (though a test at California gas stations in 2012 showed that customers were willing to buy it).

Other biofuels are available right now. Ethanol derived from corn is routinely added to gasoline in the United States, where it helps to offset carbon emissions from fossil fuels. Soybeans and rapeseed are used to make biodiesel, which is blended with regular diesel for use in diesel vehicles.

The trouble with both of these biofuels is that the plants used to make them are grown on land that might otherwise be used for food crops. This means they compete with food production, so the use of this land for biofuels instead of food can raise the price of food. Algae don't present this problem. They can be grown on land that is unsuitable for growing food crops because they don't require fertile soil.

Compared to corn, soybeans, and rapeseed, algae have the potential to produce much more fuel for the amount of space they take up. By some estimates, more than three times the total area of cropland in the United States would need to be devoted to soybean cultivation to meet just 50% of U.S. transport fuel needs, but only 3% of total U.S. cropland would be required if algae were grown instead.

Algae do require significant amounts of water to grow, but they aren't as picky as crop plants and can grow in brackish (salty) water, which is not suitable for agriculture. Researchers with NASA are even exploring the use of wastewater from toilets as a source of nutrient-rich water for growing algae **(INFOGRAPHIC 5.10)**.

INFOGRAPHIC 5.10

How Green Are Biofuels?

Every fuel source has a unique set of environmental impacts and benefits.

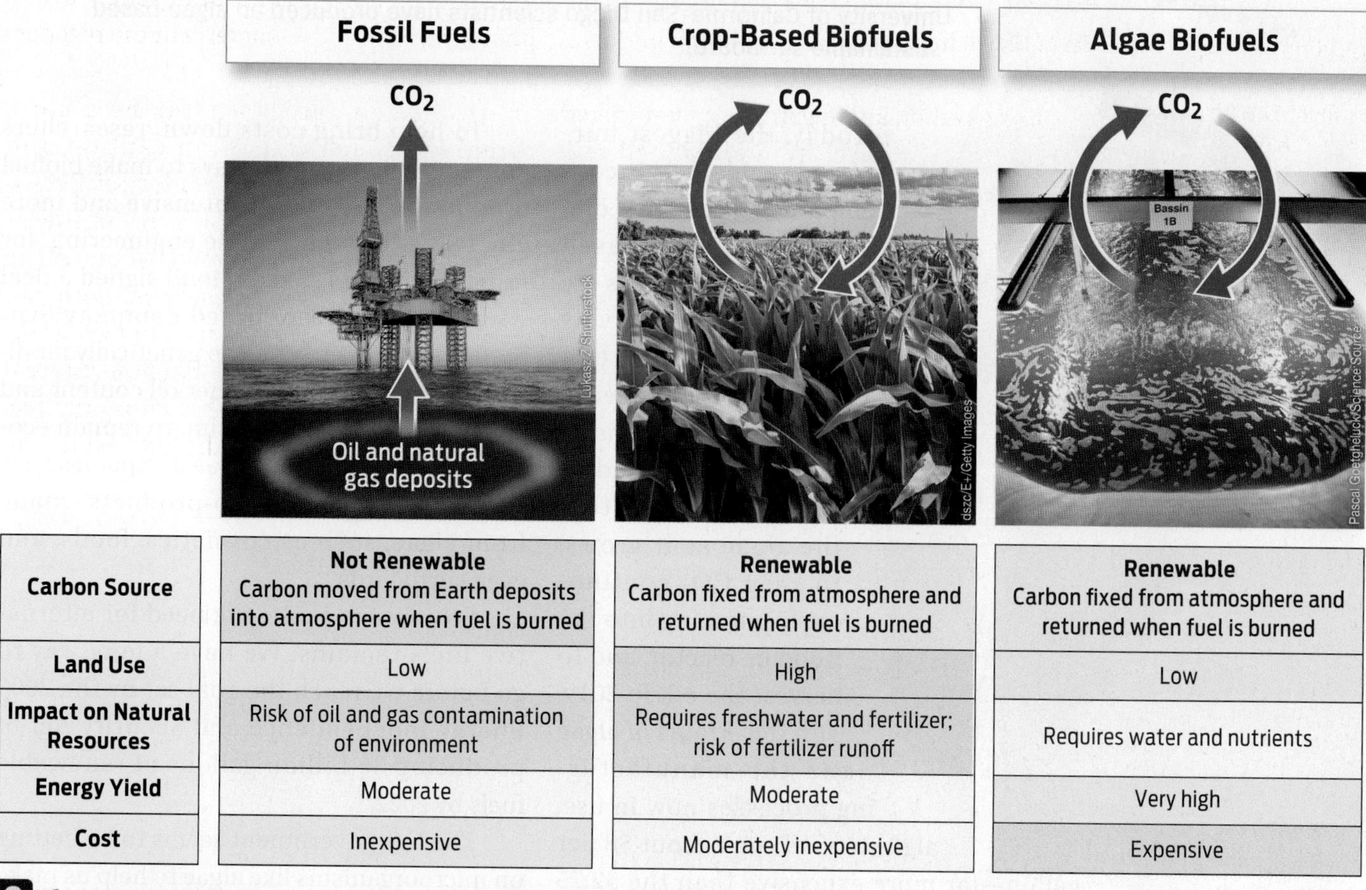

	Fossil Fuels	Crop-Based Biofuels	Algae Biofuels
Carbon Source	**Not Renewable** Carbon moved from Earth deposits into atmosphere when fuel is burned	**Renewable** Carbon fixed from atmosphere and returned when fuel is burned	**Renewable** Carbon fixed from atmosphere and returned when fuel is burned
Land Use	Low	High	Low
Impact on Natural Resources	Risk of oil and gas contamination of environment	Requires freshwater and fertilizer; risk of fertilizer runoff	Requires water and nutrients
Energy Yield	Moderate	Moderate	Very high
Cost	Inexpensive	Moderately inexpensive	Expensive

? If a green box is given a score of "3," a yellow box a score of "2," and a pink box a score of "1," which fuel has the highest score? If you were trying to "score" fuels, would you give each row equal weight (make each row equally important)? Why or why not?

California Center for Algae Biotechnology, UC San Diego

Photo by Erik Jepsen, UC San Diego

University of California–San Diego scientists have produced an algae-based sustainable surfboard.

Today, the biggest hurdle for algae biofuels is cost. Compared to gasoline, diesel, or even corn-based ethanol, algae biofuels are currently prohibitively expensive. Their high cost is mostly a consequence of the energy-intensive technology needed to contain and protect the algae as it grows, to keep CO_2 continuously pumping into the pond or reactor, and to harvest the oil. In 2019, with the strains of algae and the manufacturing processes now in use, algae biofuel runs about $8 per gallon—far more expensive than the $2.75 per gallon that we paid for gasoline in 2019.

To help bring costs down, researchers are experimenting with ways to make biofuel production less energy intensive and more efficient—through genetic engineering, for example. In 2011, ExxonMobil signed a deal with the San Diego–based company Synthetic Genomics to develop genetically modified algae strains with higher oil content and faster growth. In the interim, to remain economically viable, many algae companies have turned to producing "co-products" made from algae, such as cosmetics, foods, and even surfboards.

Cost issues aside, the need for alternative fuels remains. We have a long way to go before we reach the goal set by the 2007 Energy Independence and Security Act of producing 36 billion gallons of renewable fuels by 2022.

The U.S. government seems to be betting on microorganisms like algae to help us make up the difference. In 2018, the Department of

Energy awarded $40 million for 31 projects that are exploring ways to make algae and other biofuels cheaper and more productive.

That seemingly simple organisms like algae are so vital to life on Earth is surprising enough. To think that they may one day be an important alternative fuel source gives new meaning to the term "green energy." And all this promise comes from single-celled organisms that have just one major claim to fame: they soak up sunlight and carbon dioxide to make sugar. "Algae truly are the foundations of our entire planet," Sears says. ■

"Algae truly are the foundations of our entire planet."

—Jim Sears

CHAPTER 5 Summary

Driving Question 1 Why are photosynthetic organisms like algae so important?

- All living organisms require energy to live and grow. The ultimate source of energy on Earth is the sun.
- Photosynthetic organisms convert the energy of sunlight into energy-rich food molecules.

Driving Question 2 What are the different types of energy, and what transformations of energy do organisms carry out?

- Energy is neither created nor destroyed, but is converted from one form into another; this principle is known as the conservation of energy.
- Kinetic energy is the energy of motion and includes heat energy. Potential energy is stored energy and includes chemical energy.
- Energy flows from the sun, is captured and transferred through living organisms, and then flows back into the environment as heat.
- Energy conversions are inefficient. With every conversion of energy, some energy is lost to the environment as heat.

Driving Question 3 How do plants and algae convert the energy in sunlight into energy-rich organic molecules? (And why can't other organisms, including humans, do this?)

- Photosynthesis is a series of chemical reactions that captures the energy of sunlight and converts it into chemical energy in the form of sugar. This energy is used by all living organisms to fuel cellular processes.
- Photosynthesis can be divided into two main parts: a "photo" part, during which the pigment chlorophyll captures light energy and water is split to produce high-energy electrons, and a "synthesis" part, during which captured energy and electrons are used to fix carbon dioxide into glucose.
- Photosynthetic organisms are known as autotrophs ("self-feeders"); they include plants, algae, and some bacteria. Animals do not photosynthesize; they are known as heterotrophs ("other-feeders").

Driving Question 4 How do algal biofuels compare to other fuels in terms of costs, benefits, and sustainability?

- Photosynthetic algae convert glucose into energy-rich oils that can be used as fuel to power automobiles and aircraft. These biofuels show promise as alternatives to fossil fuels.

More to Explore

- Energy 101: Algae-to-Fuels: https://www.energy.gov/eere/videos/energy-101-algae-fuel
- PBS, NOVA, *From Pond Scum to Power*: https://www.pbs.org/wgbh/nova/tech/algae-biodiesel.html
- Can Algae Feed the World and Fuel the Planet? A Q&A with Craig Venter: http://www.scientificamerican.com/article/can-algae-feed-the-world-and-fuel-the-planet/
- Saad, M. G., Dosoky, N. S., Zoromba, M. S., and Shafik, H. M. (2019). Algal biofuels: Current status and key challenges. *Energies*, *12*(10):1920.
- Hunter-Cevera, J., et al. (2012). *Sustainable Development of Algal Biofuels. National Academy of Sciences Report.* Washington, DC: National Academies Press. https://www.nap.edu/resource/13437/Sustainable-Devel-Algal-Biofuels.pdf

CHAPTER 5 Test Your Knowledge

Driving Question 1 Why are photosynthetic organisms like algae so important?

By answering the questions below and studying Infographics 5.3, 5.6, and 5.7, you should be able to generate an answer for this broader Driving Question.

Know It

1. What do algae, cyanobacteria, and plants have in common?
2. Can animals directly use the energy of sunlight to make their own food (in their own bodies)? Briefly explain your answer.
3. What organelle(s) would a nonphotosynthetic alga need to obtain to carry out photosynthesis?
   a. mitochondria
   b. nucleus
   c. chloroplast
   d. solar transformer
   e. cell membrane
4. Why do many species of algae appear green?
5. Compare and contrast the ways photosynthetic algae and animals obtain and use energy.

Use It

6. What would happen to humans and other animals if algae, cyanobacteria, and plants were wiped out? Would we only lose a food source (e.g., plants), or would there be other repercussions?
7. Why would a dark dust cloud that prevented sunlight from reaching Earth's surface be potentially devastating to animal life?

Driving Question 2 What are the different types of energy, and what transformations of energy do organisms carry out?

By answering the questions below and studying Infographics 5.4 and 5.5, you should be able to generate an answer for this broader Driving Question.

Know It

8. The fuel energy you provide your car is best described as
   a. kinetic.
   b. chemical.
   c. heat.
   d. potential.
   e. both chemical and potential.
9. The energy in a cereal bar is __________ energy. The energy of a cyclist pedaling is __________ energy.
   a. light; chemical
   b. potential; chemical
   c. chemical; kinetic
   d. potential; potential
   e. kinetic; potential
10. Kinetic energy is best described as
   a. stored energy.
   b. light energy.
   c. the energy of movement.
   d. heat energy.
   e. any of the above, depending on the situation

Use It

11. If you wanted to get the most possible energy from photosynthetic algae, should you eat algae directly or feed algae to a cow and then eat a burger made from that cow? Explain your answer.

Driving Question 3 How do plants and algae convert the energy in sunlight into energy-rich organic molecules? (And why can't other organisms, including humans, do this?)

By answering the questions below and studying Infographics 5.6, 5.7, 5.8, and 5.9, you should be able to generate an answer for this broader Driving Question.

Know It

12. Which of the following photon wavelengths contains the greatest amount of energy?
   a. violet
   b. red
   c. green
   d. yellow
   e. blue
13. Glucose is a product of photosynthesis. Where do the carbon atoms in glucose come from?
   a. starch
   b. cow manure
   c. molecules in air
   d. water
   e. soil

14. Mark each of the following as an input (I) or an output (O) of photosynthesis.
Oxygen _____
Carbon dioxide_____
Photons _____
Glucose _____
Water _____

15. Photosynthetic algae are
a. eukaryotic autotrophs.
b. prokaryotic autotrophs.
c. eukaryotic heterotrophs.
d. prokaryotic heterotrophs.

Use It

16. Global warming is linked to elevated atmospheric carbon dioxide levels. How might a rise in carbon dioxide levels affect photosynthesis? If global warming causes ocean levels to rise, in turn causing forests to be immersed in water, how would photosynthesis be affected?

17. Why are energy-rich lipids from algae more useful as a fuel than energy-rich sugars and other carbohydrates produced by photosynthetic organisms like corn and wheat?

18. Outline (with words or a diagram) the process of photosynthesis. Include the following forms of energy and molecules: sunlight; carbon dioxide; glucose (stored chemical energy); water; ATP; heat.

Driving Question 4 How do algal biofuels compare to other fuels in terms of cost, benefits, and sustainability?

By answering the questions below and studying Infographics 5.2 and 5.10, you should be able to generate an answer for this broader Driving Question.

Know It

19. Which of the following is/are necessary for biofuel production by photosynthetic algae?
a. sunlight
b. sugar
c. CO_2
d. soil
e. all of the above
f. a and b
g. a and c

20. Why are algae considered more valuable for biofuel than are plants (such as corn)?
a. because their photosynthetic products are an oil
b. because they are cheaper to grow
c. because they do not require as much CO_2
d. because they do not require as much fertilizer
e. all of the above

Use It

21. Many types of algae can divert the sugars they make by photosynthesis into lipids that can be used to make biodiesel. Biodiesel is a promising replacement for fossil fuels. Describe the energy conversions required to make algal lipids for biodiesel and explain why biodiesel might be a more promising fuel than fuel from lipids extracted from animals.

22. What do you think are some of the advantages and disadvantages of growing algae in enclosed tubes or bags compared to growing them in open vats? Make a table listing the advantages and disadvantages of each approach and explain your reasoning.

23. Many biofuels, such as corn-derived ethanol, require arable land (land that is suitable for agriculture) for their production. Discuss competing needs for arable land in the context of human requirements for food and fuel, and describe how algae may alleviate this tension.

Mini Case

Apply Your Knowledge

24. A CEO of a new algal biofuel company is trying to select the site for a production facility. There are three possible options:

- The desert of southern New Mexico (sunny, hot, mild winters, nonarable land, remote)
- Denver, Colorado (sunny, cold winters, urban area with CO_2 emissions from factories and cars)
- Central Washington State (sunny, hot, a rich agricultural zone)

Discuss the pros and cons of each site and make a recommendation.

Apply Your Knowledge

Bring It Home

25. The airline Virgin Atlantic has committed to using a "green" fuel produced by microbes that use carbon monoxide (CO) from industrial emissions (such as from steel factories) as its carbon and energy source. Through a fermentation process that occurs in a reactor chamber, the microbes convert CO into usable ethanol, a viable "green" fuel that can be converted to jet fuel. This fuel is predicted to reduce CO_2 emissions by 60% relative to conventional jet fuel. Compare and contrast this fuel with algal biofuel and corn ethanol. If this "green" fuel venture is successful, would Virgin Atlantic's use of "green" fuel influence your decision to choose this airline over another air carrier? Why or why not?

Apply Your Knowledge

Interpreting Data

26. The United States currently uses approximately 19 million barrels of petroleum per day. Of this amount, approximately 8 million barrels per day is imported, and the rest comes from U.S. sources (which includes biofuels). The table shows the production cost estimated for different petroleum sources. (The actual cost is driven by a variety of market and geopolitical factors, so we will use production cost as a substitute for actual cost.)

a. Using the data for cost per barrel from various sources, calculate the production cost to meet current U.S. daily use. Assume that approximately 40% of the imports come from Canada, approximately 49% come from OPEC and Persian Gulf countries with production costs similar to Saudi Arabia, and 11% are split evenly between Venezuela and Iraq.

b. Suppose the United States replaces half of its current oil imports from OPEC and Persian Gulf countries with domestically produced algal biofuel. How will this change affect the cost of production to meet our daily needs?

c. How could algal biofuel companies work to decrease the production costs for their product?

Petroleum Source	Production Cost per Barrel (Average)
United States	$65.00
Canada	$91.00
Saudi Arabia	$3.00
Venezuela	$20.00
Iraq	$6.00
Algal biofuel	~$273

Data from U.S. Energy Information Administration (2019, October 4). Frequently asked questions: How much petroleum does the United States import and export? Retrieved from https://www.eia.gov/tools/faqs/faq.php?id=727&t=6; and Knoema (2019, January 18). Cost of oil production by country. Retrieved from https://knoema.com/vyronoe/cost-of-oil-production-by-country

A "NEAT" Path to

WEIGHT MANAGEMENT?

Burning calories through everyday living

DRIVING QUESTIONS

1. How does the body use the energy in food?
2. How does aerobic respiration extract useful energy from food?
3. When does fermentation occur, and why can't a human survive strictly on fermentation?
4. What factors influence weight gain and weight loss?

"In a mere 150 years, Homo sapiens *has become addicted to the chair."*

—James Levine

When James Levine was 11 years old, he began a science experiment in his bedroom. His test subjects? Snails.

In glass-walled fish tanks he built himself, Levine collected pond snails from nearby Regents Park in his native London. He then methodically monitored and recorded their movements.

"My idea was that every snail has a built-in hard-wired style of movement," explains Levine. "One snail will do swirly-whirly-whirly, while another snail will always move in a straight line."

To test this hypothesis, every night, between 9:00 P.M. and 5:00 A.M., he'd wake up hourly and mark on the glass where the snail had moved. At the end of the night, he'd trace the snail's journey.

"Not surprisingly, I was constantly asleep at school," Levine confesses.

By the time he finished his experiment, Levine had 270 snail tracings. What he discovered was that snails do indeed move in stereotypical ways.

"Joanna"—he named the snails—"always does ziggidy-zaggidy-ziggidy-zaggidy. And John always moves in a smooth way."

The snail experiment made a lasting impression on Levine, who eventually went on to medical and graduate school. Now, as a physician-scientist, Levine has turned his number-crunching talents to another slow-moving creature: the human couch potato.

Levine is a professor of endocrinology and nutrition research at the Mayo Clinic in Phoenix, Arizona. He is an expert on the physiology of weight gain and loss, with a special focus on obesity. **Obesity**—having an unhealthy amount of body fat—has been called America's number one health crisis. Obesity has been linked to a whole host of health problems, including heart disease, diabetes, stroke, Alzheimer's disease, hypertension, and even cancer. It is also a major killer: in the United States, only tobacco use causes more premature deaths.

What does it mean to be obese, exactly? To assess whether someone is obese, doctors and other health professionals often use a tool called the **body mass index (BMI),** which provides an indirect estimate of body fat based on a person's height and weight. The BMI is calculated by dividing one's weight in kilograms by one's height in meters squared (kg/m^2). A BMI between 19 and 24 is considered to be in the healthy range; a BMI between 25 and 29 is considered overweight; and a BMI of 30 or higher is in the obese range.

The BMI is a useful tool for quickly estimating healthy and unhealthy weight in individuals, but it can sometimes be misleading. Athletes and other people with more muscle mass, for example, will sometimes register as overweight when, in fact, they are a perfectly healthy weight (muscle weighs more than fat). Nonetheless, the BMI accurately predicts body fat more than 80% of the time, which is why doctors continue to use it.

OBESITY
Having more body fat than is considered healthy.

BODY MASS INDEX (BMI)
An estimate of body fat based on height and weight.

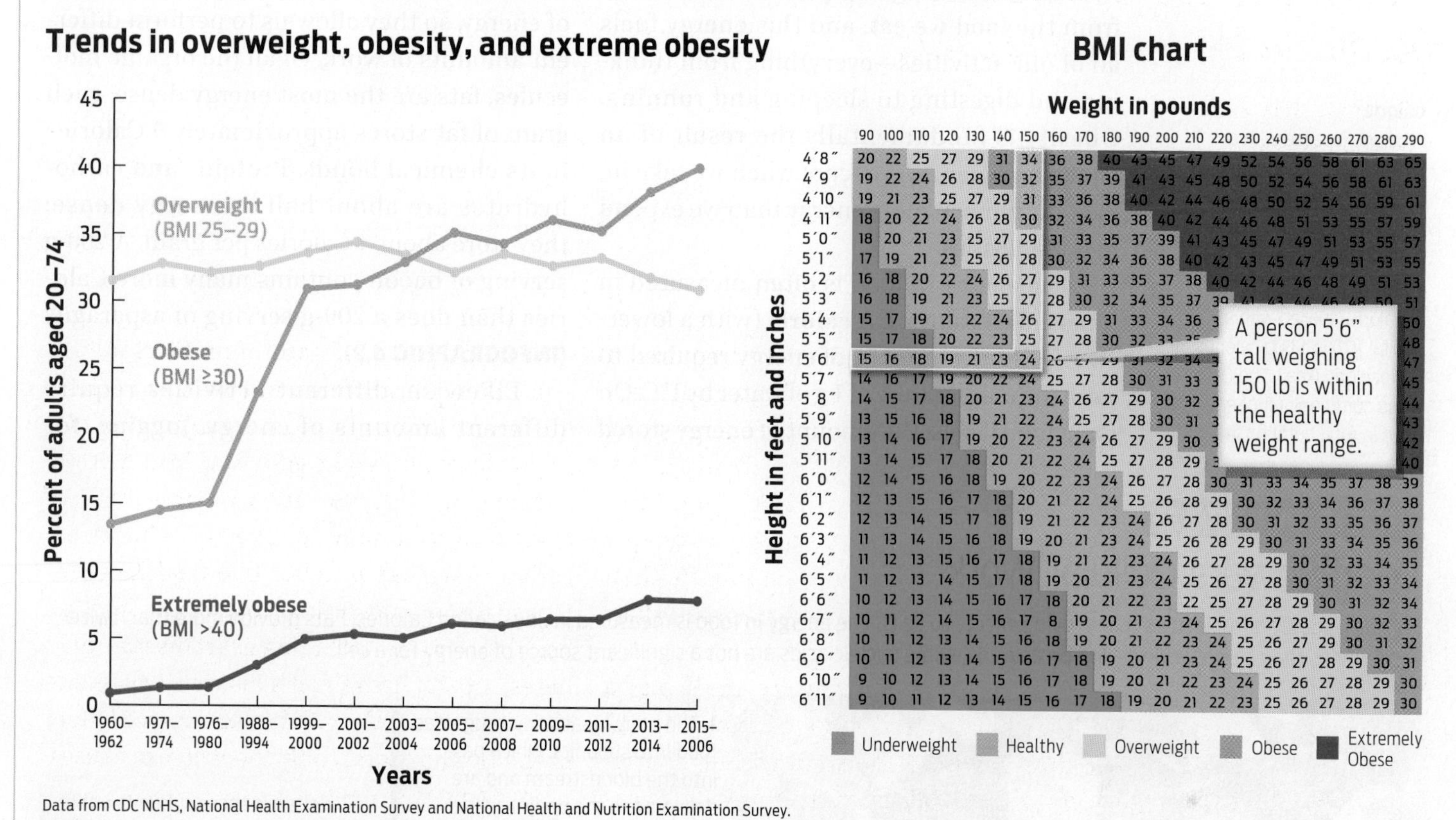

INFOGRAPHIC 6.1

Body Mass Index and Increasing Obesity Rates

A body mass index (BMI) chart provides an indirect measure of body fat based on the ratio of body height to weight. Using the BMI to categorize the U.S. population, studies show that the percentage of U.S. adults who are obese has increased since 1960.

? Is someone who is 5′10″ tall and weighs 210 pounds considered underweight, of normal weight, overweight, or obese?

Obesity rates have risen sharply over the past five decades, leading many to refer to an "obesity epidemic." Today, Americans weigh 25 pounds more, on average, than they did in 1970. As of 2019, nearly 40% of U.S. adults were obese, compared with 15% in 1970. Rates in children are not far behind **(INFOGRAPHIC 6.1)**.

While it's clear that obesity is on the rise in the United States and around the world, scientists are divided about the underlying causes of these trends. Is increased food intake—linked perhaps to larger portion sizes on menus and in supermarkets—mostly to blame? Or is decreased energy expenditure—a result of our increasingly sedentary lifestyle—the bigger player? No one can say for sure. But now, with the help of some clever experiments using sophisticated undergarments, Levine thinks he's found an important piece of the puzzle.

A Creeping Problem

▶ Energy intake and expenditure

To understand the rise in obesity, says Levine, you have to realize that it didn't happen overnight. "Obesity doesn't occur over minutes and hours," says Levine. "Obesity occurs over years, decades, and generations."

A person doesn't necessarily need to be a serious overeater to develop a weight problem. Even small changes in the way we live can seriously add up. Something as simple as the amount of time a person spends sitting or standing—if multiplied consistently across time—can profoundly affect how much energy that individual uses and, therefore, how much weight they gain.

Recall that energy is defined as the capacity to do work (Chapter 5). We obtain energy from the food we eat, and this energy fuels all of our activities—everything from thinking and digesting to sleeping and running. Obesity is fundamentally the result of an energy imbalance: it occurs when we take in, over time, much more energy than we expend in our activities.

calorie
A calorie (spelled with a lowercase "c") is the amount of energy required to raise the temperature of 1 gram of water by 1°C.

Calorie
A Calorie (spelled with a capital "C") is 1,000 calories or 1 kilocalorie (kcal). The Calorie is the common unit of energy used in food nutrition labels.

The energy in food is often measured in units called calories. A **calorie** (with a lowercase "c") is the amount of energy required to raise the temperature of 1 g of water by 1°C. On most food labels, the amount of energy stored is listed in kilocalories, which are also referred to as kcals or Calories (with a capital "C"). One **Calorie** is equal to 1,000 calories, which is equal to 1 kcal.

Different foods contain different amounts of energy, so they allow us to perform different amounts of work. Of all the organic molecules, fats are the most energy dense: each gram of fat stores approximately 9 Calories in its chemical bonds. Proteins and carbohydrates are about half as energy dense: they store about 4 Calories per gram. A 200-g serving of bacon contains many more Calories than does a 200-g serving of asparagus **(INFOGRAPHIC 6.2)**.

Likewise, different activities require different amounts of energy. Jogging, for

INFOGRAPHIC 6.2

Food Powers Cellular Work

Food is a source of molecules that provide energy to cells. The energy in food is measured in units called Calories. Fats provide more than twice the Calories than carbohydrates and proteins do, while nucleic acids are not a significant source of energy for a cell.

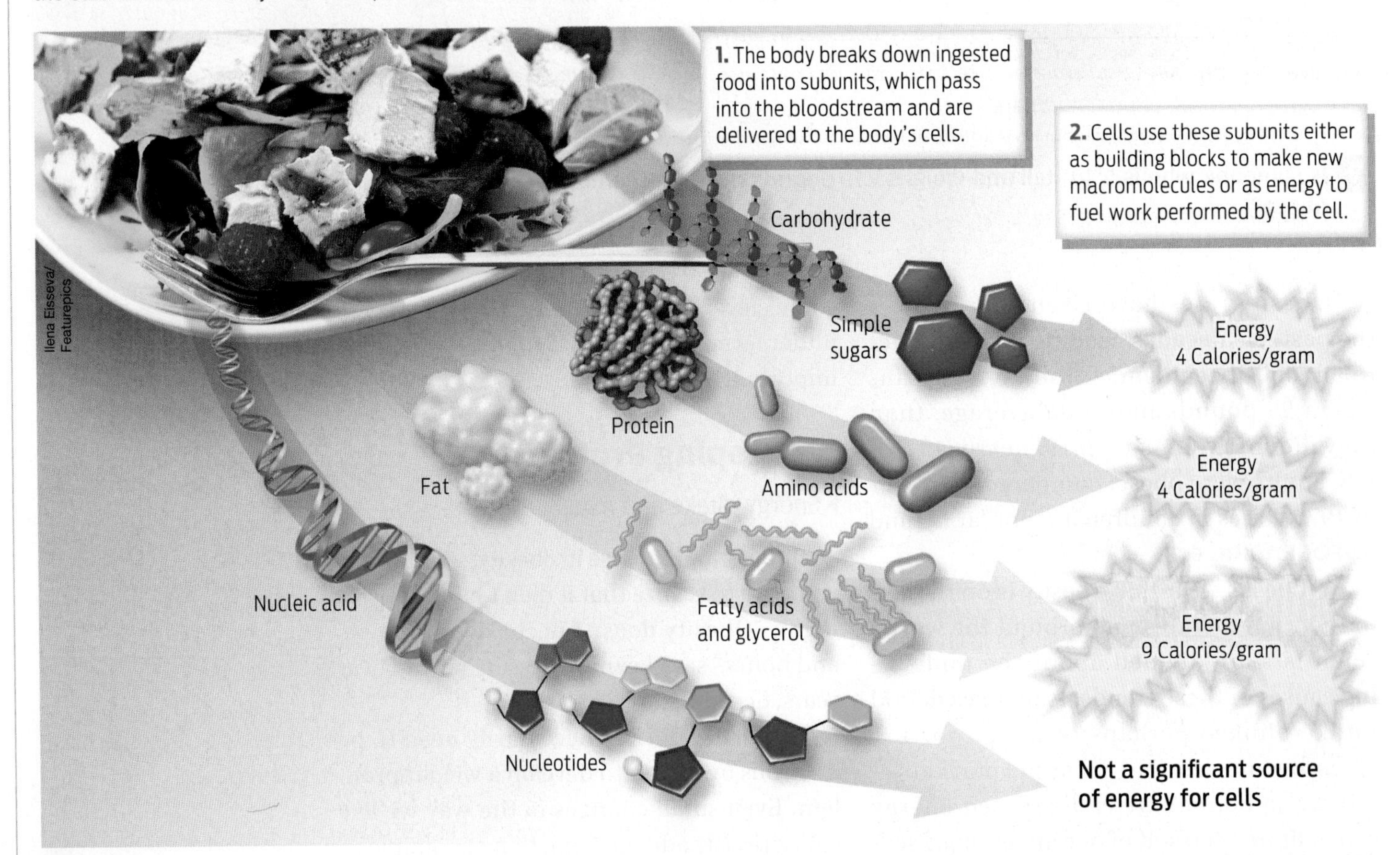

? **Why does reducing the amount of fat in the diet have a greater impact on total Calories than reducing the amount of carbohydrates?**

INFOGRAPHIC 6.3

Balancing Energy In with Energy Out

Different foods provide different numbers of Calories, and different activities expend different numbers of Calories. Ultimately, the balance between Calorie input and expenditure determines whether a person gains, maintains, or loses weight.

Food	Calories (kcal)
Medium apple (125 g)	65
1 hard-boiled egg	70
1 slice whole-wheat bread	79
¼ cup cooked white rice	102
4 oz chicken breast	120
12 oz nonfat milk	120
1 slice thick-crust pizza	256
4 oz sirloin steak	280
Subway Turkey Breast Sandwich (6 in.)	282
1 Starbucks Grande Mocha Frappucino with whipped cream	360
Burger King Whopper	670

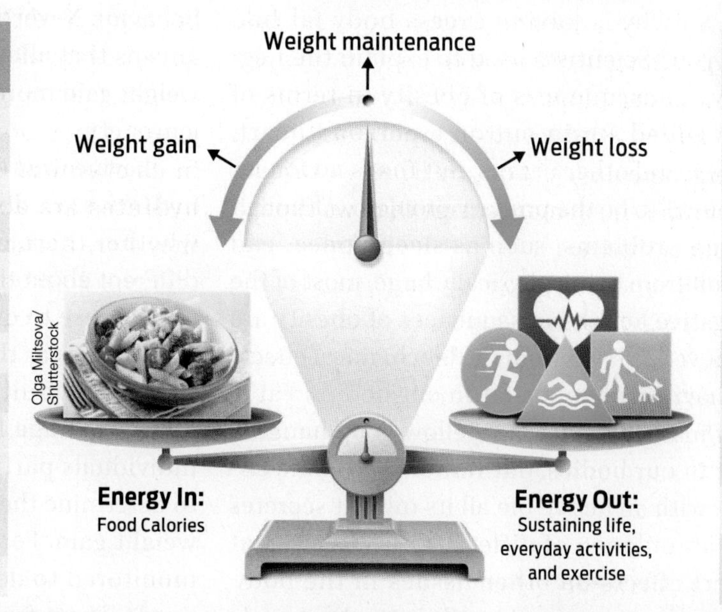

Activity	Calories (kcal)/hr
Sitting	5
Standing	15
Gum chewing	20
Walking (2 mph)	120
Stair climbing	200
Biking (moderate)	450+
Jogging (5 mph)	500+
Swimming (active)	500+
Hiking	500+
Cycling (stationary bike)	650

? How long would a person have to stand to burn off the Calories in a slice of whole-wheat bread?

example, burns about 500 Calories an hour, whereas sitting burns only about 5 Calories per hour.

There are three main ways our bodies expend energy each day. The first is our basal metabolism—the many thousands of chemical reactions that keep our cells and organs functioning, and us alive. Basal metabolism is, by far, the largest portion of the energy we expend each day; it constitutes about 60% of the total. Another 10% of our daily energy expenditure goes to digesting food. The final 30% is claimed by the physical activities we engage in—everything from walking and doing chores to running and playing tennis. Deliberate exercise makes up a relatively small portion of this third category; for most people, it's less than 10% of our total energy expenditure.

How many Calories a person needs to eat to meet daily energy needs depends on many factors, including gender, age, genetics, body type, and physical activity levels. A sedentary college-age average-size male, for example, would need to consume anywhere from 2,200 and 2,400 Calories per day to power his activities and maintain his weight, whereas a football player would need more than 3,200 Calories each day to carry out his activities and maintain his weight **(INFOGRAPHIC 6.3)**.

The only way to gain weight is by taking in more Calories than we expend through all of our activities, including basal metabolism. If we take more food energy into our bodies than we use to power cellular reactions and physical movement and generate heat, the excess is stored as fat. In other words, our waistlines obey the principle of conservation of energy: energy is neither created nor destroyed but merely converted from one form into another.

Even with a few extra pounds here and there, most people still fall within a healthy weight range. Only when our total body fat passes a certain point do the scales tip toward unhealthy. The exact tipping point is the subject of much current research, and many questions and controversies remain about the exact relationship between health and body fat. For example, is it possible to be obese and relatively healthy? Probably. Does being skinny necessarily mean you are in good cardiovascular health? No. Nevertheless, the preponderance of evidence suggests that there are real risks to obesity.

So why is having excess body fat bad for you? Scientists used to explain the negative consequences of obesity in terms of the added strain put on a person's heart, lungs, and other organs. But that's no longer thought to be the primary problem. Although some problems, such as sleep apnea, can result from being physically large, most of the negative health consequences of obesity are believed to stem from the biochemical effects of having too much fat in our bodies. Fat is not just a blob of inert yellow goop hanging out in our bodies, but rather a dynamic tissue with an active life all its own. It secretes an abundance of different hormones that exert effects on other tissues in the body. These hormones cause a low-grade chronic inflammatory state throughout the body that may be the primary means by which obesity causes health problems and contributes to the development of many chronic diseases.

The Secret to Weight Control?

▸ Non-exercise activity thermogenesis (NEAT)

Most of us are surrounded by an abundance of easy-to-obtain, tasty foods. Food manufacturers also spend billions of dollars designing foods that will trigger our "bliss point" and marketing these products to us. So does overeating explain the rising rates of obesity in the United States? Are we simply being goaded into eating more food than our ancestors did and gaining weight as a result? While that may be part of the problem, Levine and his colleagues think an even more important factor is what's happening on the other side of the equation: the amount of energy we expend.

We all have observed people who seem to eat anything they like and never gain weight; in fact, they may even struggle to maintain their weight. Other people, it seems, merely have to look at food to put on pounds. It turns out there is scientific support for these subjective observations. Studies that have looked at weight gain in response to overeating have shown that people vary greatly in how much weight they accumulate as a result of this behavior. Nevertheless, the biological mechanisms that allow some individuals to resist weight gain more than others are still largely a mystery.

Levine wanted to get to the bottom of this mystery, so he designed an experiment to test whether there was something measurably different about those who do not gain weight in response to overeating. He and a team of researchers at the Mayo Clinic recruited 16 nonobese adults: 12 males and 4 females, ranging in age from 25 to 36 years. These individuals participated in a 10-week study to determine the effect of daily overeating on weight gain. For 2 weeks, the subjects were monitored to determine their daily caloric needs. Then, for the next 8 weeks, each participant consumed 1,000 Calories each day over the amount required to maintain his or her current weight. Meals were prepared and consumed in a research facility at the clinic, and activity levels were strictly monitored through daily interviews and accelerometers worn by the study participants.

Because each person in the study was overeating to the same extent, you might expect that each person would gain the same amount of weight. However, the researchers found that the amount of body fat a person gained varied 10-fold among individuals in the study, ranging from a gain of about a pound of fat to a gain of nearly 10 pounds of fat. Where was all the extra energy provided by overeating going if not into the bodies of the study participants? One obvious possibility is different amounts of energy expenditure. In Levine's study, he found only very minor increases in participants' basal metabolic rate and the energy expended to digest food, which were not enough to account for the 10-fold variance in fat gain among them. By contrast, levels of physical activity varied markedly between study participants and could account for the differences in weight gain. Interestingly, intentional exercise was not the crucial difference in activity level.

Levine designed his study so that intentional exercise was kept at a constant and minimum level across all test subjects. Therefore, the differences in physical activity were principally not related to exercise. Levine has a name for this type of activity: **NEAT,** short for "non-exercise activity thermogenesis." NEAT includes all sorts of activities of daily living that are not deliberate exercise, such as walking up stairs, doing chores, gardening, and playing a musical instrument. It also includes spontaneous movements such as fidgeting, pacing, and even chewing gum.

In his study, Levine found that different participants increased their NEAT by different amounts in response to overeating. The amounts differed by as much as 800 Calories per day. Changes in NEAT were inversely correlated with weight gain: participants who increased their NEAT the most gained the smallest amount of weight. In other words, NEAT accounted for how these individuals resisted weight gain. "People who are resistant to weight gain and can stay thin are people who can switch on their NEAT in response to overfeeding and never gain a pound," Levine says **(INFOGRAPHIC 6.4).**

NEAT
Non-exercise activity thermogenesis; the amount of energy expended in everyday activities.

INFOGRAPHIC 6.4

NEAT Activities Influence Weight

When fed 1,000 Calories per day more than needed to maintain weight, participants who increased their NEAT to a greater extent gained less fat. In another study, lean people had higher levels of NEAT (for example, higher levels of standing and moving around).

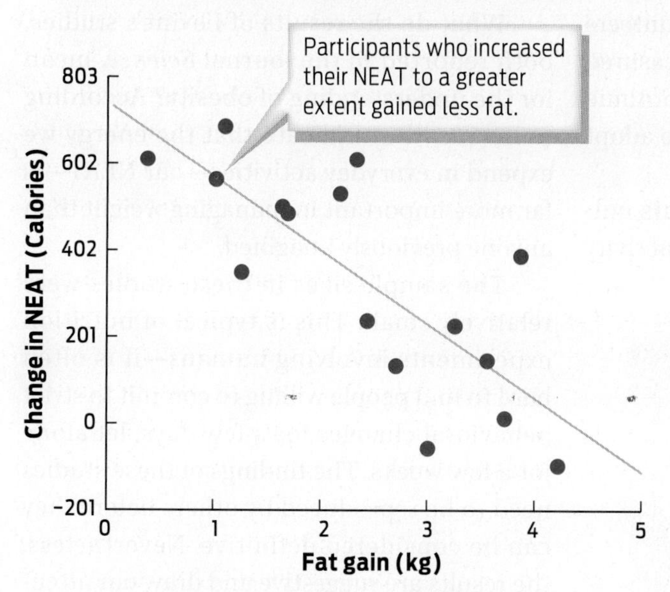

Data from Levine JA, Eberhardt NL, Jensen MD. *Science* 1999; 283: 212–4.

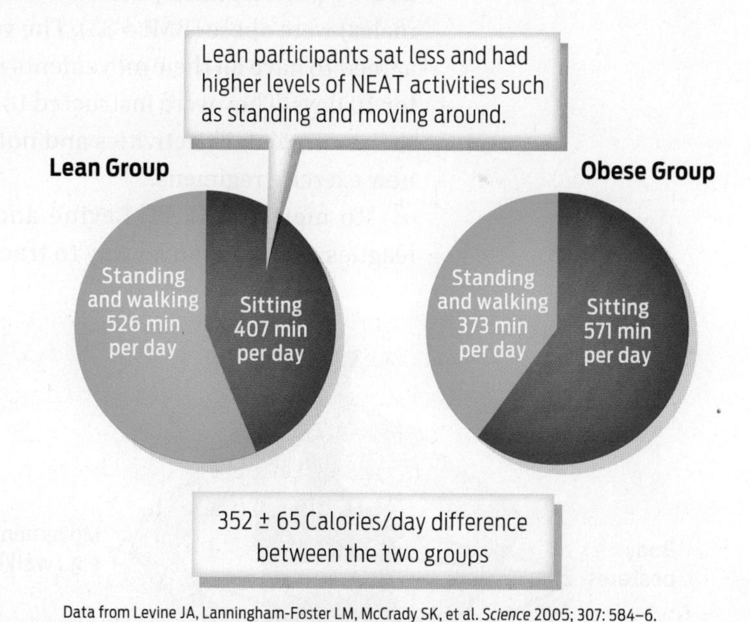

Data from Levine JA, Lanningham-Foster LM, McCrady SK, et al. *Science* 2005; 307: 584–6.

NEAT includes non-exercise activities such as walking the dog, shopping, dishwashing, social activities, and fidgeting.

? Consider someone who cannot handle a kickboxing class but who walks the dog and weeds the garden. What evidence suggests that these activities may help this person avoid weight gain?

Next, Levine wanted to know whether NEAT plays a role in obesity. Do lean and obese people differ in their average levels of NEAT, for example? To get at this question, Levine and his colleagues conducted another experiment in which they recruited 20 healthy volunteers who were self-proclaimed "couch potatoes." Ten participants (five females and five males) were lean (BMI ≈ 23) and 10 participants (five females and five males) were obese (BMI ≈ 33). The volunteers agreed to have all their movements measured for 10 days. They were instructed to continue their normal daily activities and not to adopt new exercise regimens.

> *"People who are resistant to weight gain and can stay thin are people who can switch on their NEAT in response to overfeeding and never gain a pound."*
>
> **—James Levine**

To measure NEAT, Levine and his colleagues devised a novel way to track activity levels in test participants. They built a special kind of undergarment outfitted with electronic sensors that detect movement. The undergarment, which Levine calls "magic underwear," was designed to allow people to wear it essentially all the time—even while going to the bathroom and having sex. The magic underwear collects data "every half second of every day for days and days and days on end," Levine says.

Over the 10-day period, Levine's team collected 25 million data points on NEAT for each individual. From those measurements, they calculated how much time each person spent standing, walking, sitting, or sleeping per day, and how many calories each person expended as a result.

The main takeaway? Obese individuals in the study sat, on average, 2.25 hours longer per day than their leaner counterparts. By sitting less, the lean people burned an additional 350 Calories per day.

What do the results of Levine's studies, both reported in the journal *Science,* mean for the understanding of obesity? According to Levine, they indicate that the energy we expend in everyday activities—our NEAT—is far more important in managing weight than anyone previously imagined.

The sample sizes in these studies were relatively small. This is typical of nutrition experiments involving humans—it is often hard to find people willing to commit to strict behavioral changes for a few days, let alone for a few weeks. The findings of these studies need to be reproduced by others before they can be considered definitive. Nevertheless, the results are suggestive and draw our attention to an underappreciated source of energy expenditure.

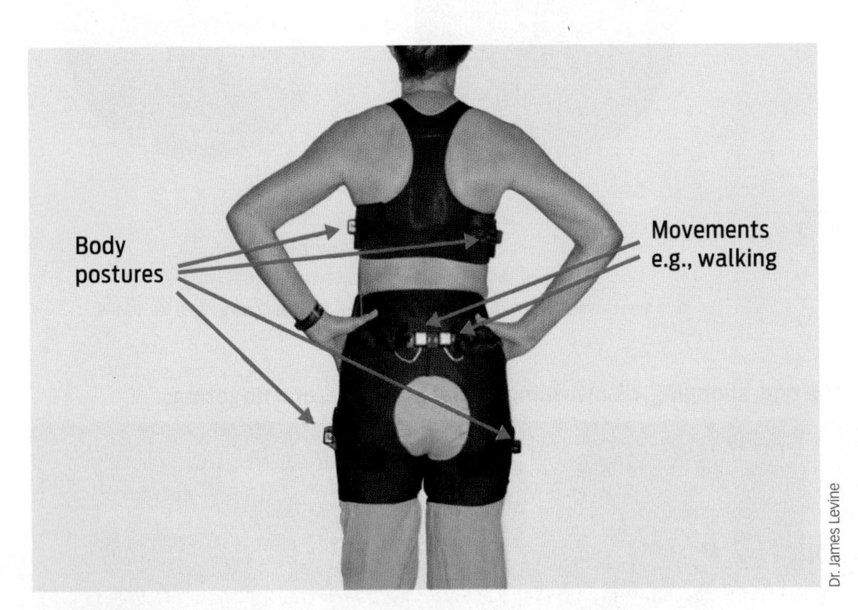

To study NEAT, Levine employs what he calls "magic underwear"—specially designed undergarments with sensors that detect movement.

The Cell's Energy Currency

▸ Adenosine triphosphate (ATP)

All physical activity—both NEAT and deliberate exercise—burns energy. Much in the same way that an automobile burns gasoline to power its pistons, the human body burns food molecules like sugars and fats to carry out its many chemical reactions. Both processes are a kind of controlled combustion, but they differ in important ways, too.

For the energy released from food molecules to be useful to the body, it has to be captured in a form that can participate in the cell's chemical reactions. That form is a molecule called **adenosine triphosphate (ATP).**

To make ATP, our bodies first break down food molecules into their smaller subunits by digestion: carbohydrates into sugars, fats into fatty acids and glycerol, proteins into amino acids. Once released from food, these subunits leave the small intestine and enter the bloodstream, which transports them to the body's cells. Inside cells, enzymes break apart the bonds holding these subunits together. The energy stored in those bonds is then captured and transferred into the chemical bonds that make up ATP. When cells need energy, they break these bonds in ATP, releasing the stored energy, which can then participate in chemical reactions.

You can think of food as being like a bar of gold: it has a great deal of value, but if you brought that gold bar to your local convenience store, you wouldn't be able to buy even a cup of coffee with it. Instead, you would first have to convert your gold bar into bills and coins. ATP is the energetic equivalent of bills and coins; it's currency that your body can actually spend **(INFOGRAPHIC 6.5).**

ADENOSINE TRIPHOSPHATE (ATP)
The molecule that cells use to power energy-requiring functions.

INFOGRAPHIC 6.5

ATP: The Energy Currency Cells

Just as a gold bar must be converted to currency to buy merchandise, the energy in food must be converted to ATP before it can be used by the cell.

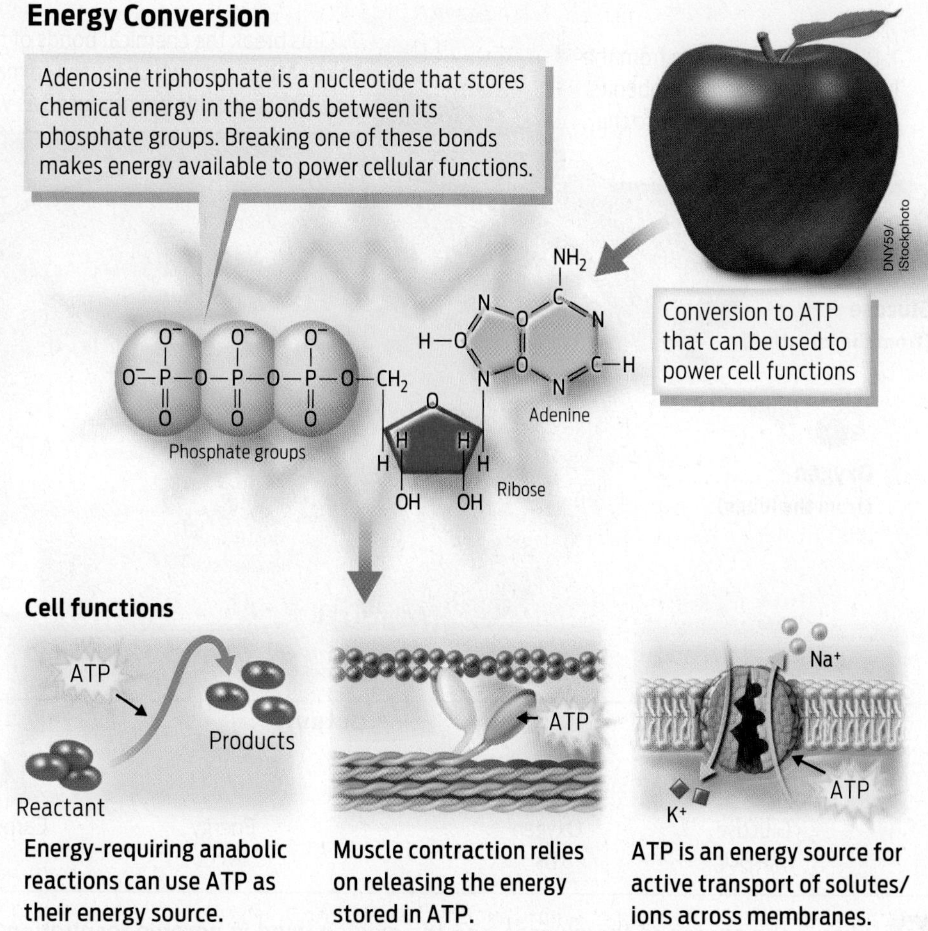

? What do muscle contraction and active transport have in common?

ATP is spent anytime a muscle contracts or a neuron fires. When we go for a run, or even when we sit in a chair but tap our feet, our muscle cells break the bonds in ATP. The energy released allows muscle fibers to contract, powering movement. In our large brain, which runs principally on glucose, ATP is used to move ions across cell membranes, enabling our neurons to fire. When we die, rigor mortis sets in because ATP is required for muscle proteins to slide past one another; without it, they stay locked in place, and become stiff (and so do you). Ensuring a steady supply of ATP is therefore one of life's most critical activities.

AEROBIC RESPIRATION
A series of reactions that occurs in the presence of oxygen and converts energy stored in food into ATP.

Burning Food for Energy

▸ Steps of aerobic respiration

The primary process that eukaryotic organisms—animals, plants, fungi, and protists—use to convert food energy into ATP is called aerobic cellular respiration, or just **aerobic respiration.** "Aerobic" means "in the presence of oxygen"; thus, as the term suggests, aerobic respiration requires a continual source of oxygen. Sugars, fats, and amino acids from our diets can all be burned in this process to make ATP. During aerobic respiration, oxygen is consumed, energy is captured in the bonds of ATP, and carbon dioxide and water are given off as waste products. As we saw in

INFOGRAPHIC 6.6

Aerobic Respiration Transfers Food Energy to ATP

During aerobic respiration, our cells use the oxygen we inhale to help extract energy from food. Cells convert the energy stored in food molecules into the bonds of ATP, the cell's energy currency.

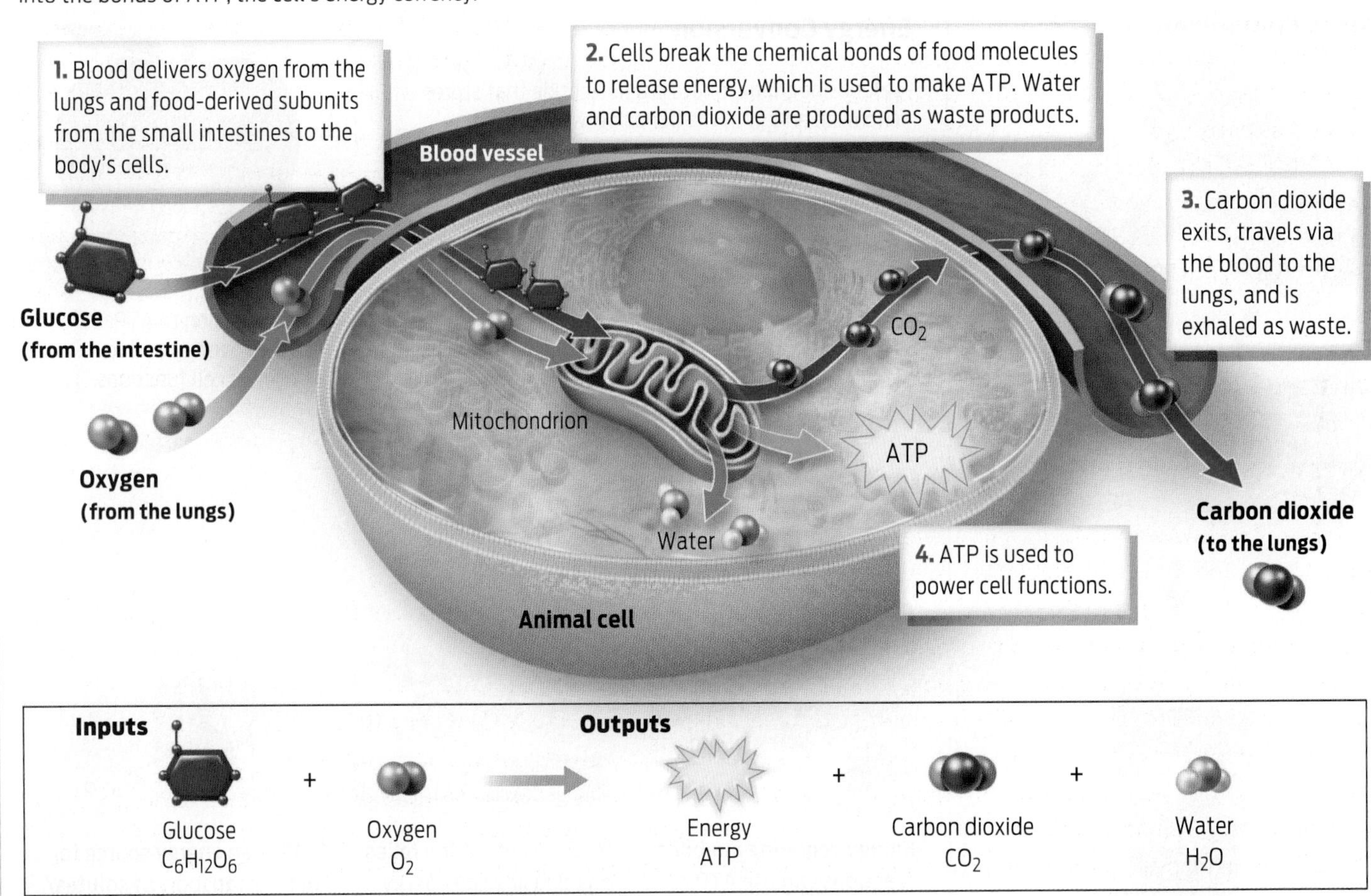

? **What is the source of the glucose and the oxygen used in aerobic respiration?**

Chapter 5, energy conversion reactions are not 100% efficient, so some of the energy is also released as heat. For simplicity, we focus here on glucose, which is the most common fuel source for living organisms, from bacteria to humans **(INFOGRAPHIC 6.6)**.

Aerobic respiration proceeds through three stages that take place in different parts of the cell. The first stage, **glycolysis,** takes place in the cell's cytoplasm. Glycolysis is a series of chemical reactions that splits glucose into two smaller molecules of pyruvate. The pyruvate molecules then enter the cell's mitochondria. Glycolysis produces a small amount of ATP, but this amount makes only a small contribution to the total amount of ATP produced by aerobic respiration.

The second stage, the **citric acid cycle,** takes place in the mitochondria. A series of reactions strips electrons from the bonds between the carbon and hydrogen atoms that were originally in glucose and are now in pyruvate. In the process, the pyruvate is broken down into smaller and smaller carbon-based molecules, and eventually exhaled as carbon dioxide from the lungs.

As the energy-rich bonds in glucose and pyruvate are broken, some of the energy released is used to make a small amount of ATP. The rest of the energy is stored in electrons released from the broken bonds. These electrons are picked up by a molecule called **NAD^+.** When NAD^+ picks up electrons, it becomes NADH (the electron-carrying form of the molecule). NADH then carries the electrons to the inner membrane of the mitochondria, where NADH gives them up (reverting to NAD^+). The electrons then go through the third and last stage of aerobic respiration: the **electron transport chain.**

During electron transport, the energetic electrons are passed like hot potatoes down a chain of molecules in the inner mitochondrial membrane. As electrons pass down the chain, they release their stored energy, which is used to power reactions that form many molecules of ATP. Eventually the electrons are passed to oxygen molecules, which combine with hydrogen atoms to produce water **(INFOGRAPHIC 6.7)**.

The complete aerobic respiration of one molecule of glucose produces approximately 30–32 molecules of ATP—enough to power the contraction of a medium-sized muscle.

When Oxygen Is Scarce

▸ Fermentation

Aerobic respiration requires a steady supply of oxygen, which is transported to cells of the body in blood traveling from the lungs. Occasionally, when we perform very strenuous activities, the rate at which oxygen can be delivered to muscles is lower than the rate at which oxygen is consumed in the chemical reactions of aerobic respiration. Without oxygen to accept electrons, the electron transport chain stops, and aerobic respiration grinds to a halt. When this happens, another form of metabolism, **fermentation,** comes into play. Fermentation is an anaerobic process, which means that it occurs without oxygen. During this process, the products of glycolysis do not go through the citric acid cycle and the electron transport chain. Instead, they are shunted into a different set of reactions, which take place in the cell's cytoplasm and produce lactic acid as a by-product.

Fermentation does not actually produce any more ATP beyond the amount produced by glycolysis. So you might wonder: why do cells do it? In essence, fermentation is a way to keep glycolysis running. As glucose is converted to pyruvate in glycolysis, glucose gives up electrons to NAD^+. In the absence of oxygen, NADH cannot unload its electrons to the electron transport chain, so after a while no NAD^+ is available to pick up electrons from glucose (since all the NAD will be in the NADH form). Soon glycolysis will stop making ATP—that is, unless there is some way for the cell to regenerate NAD^+. This is precisely what fermentation does. In fermentation, NADH unloads its electrons onto pyruvate (making lactic acid), thereby regenerating

GLYCOLYSIS
A series of reactions that breaks down sugar into smaller units; glycolysis takes place in the cytoplasm and is the first stage of both aerobic respiration and fermentation.

CITRIC ACID CYCLE
A set of reactions that takes place in mitochondria and helps extract energy (in the form of high-energy electrons) from food; the second stage of aerobic respiration.

NAD^+
An electron carrier. NAD^+ can accept electrons, becoming NADH in the process.

ELECTRON TRANSPORT CHAIN
The transfer of electrons that takes place in mitochondria and produces the bulk of ATP during aerobic respiration; the third stage of aerobic respiration.

FERMENTATION
A series of chemical reactions beginning with glycolysis and taking place in the absence of oxygen. Fermentation produces far less ATP than does aerobic respiration.

INFOGRAPHIC 6.7

Aerobic Respiration: A Closer Look

Nearly all eukaryotic organisms carry out aerobic respiration. The three main stages of aerobic respiration occur in specific locations within the cell and yield distinct products.

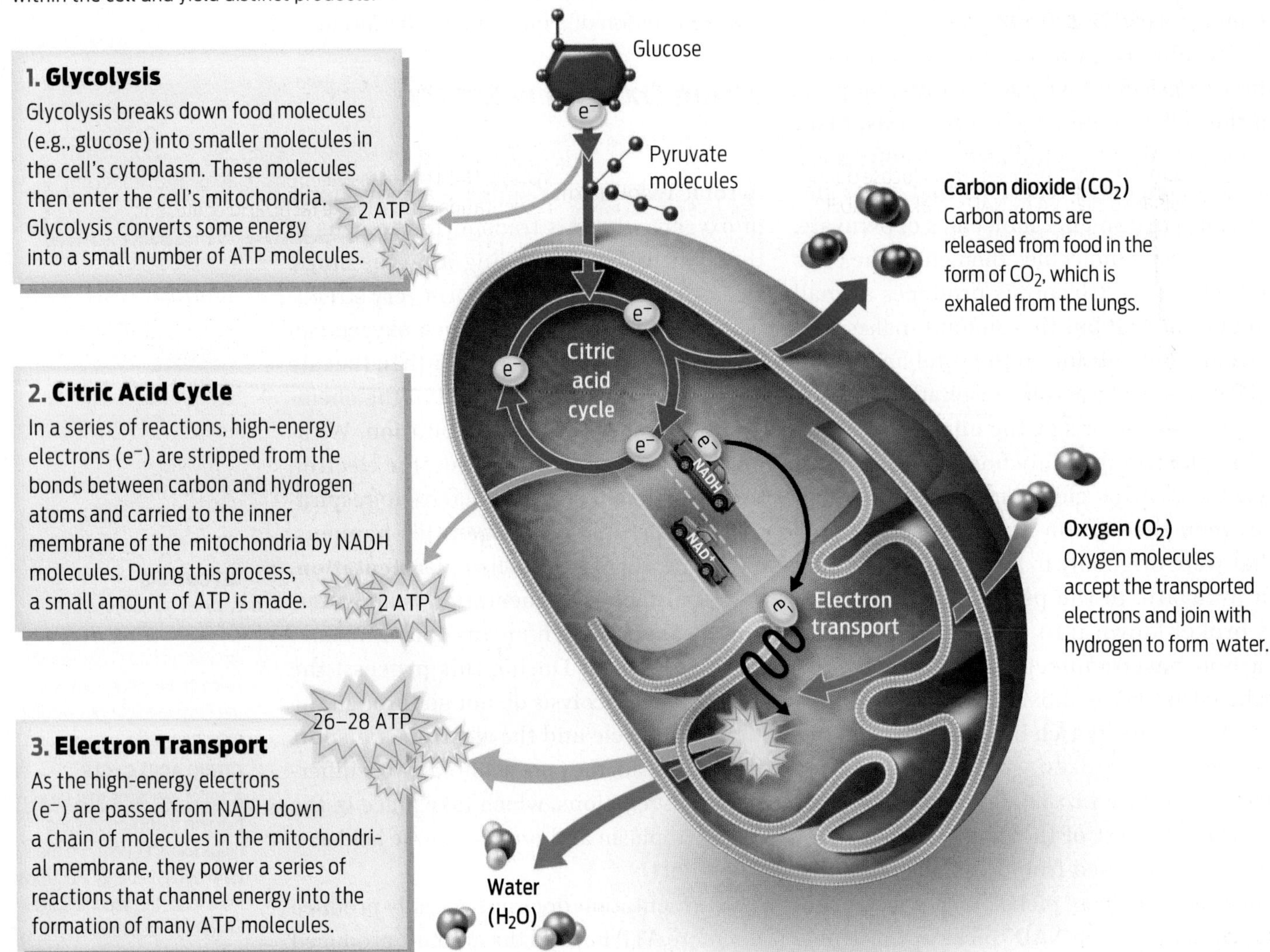

? **List the three main steps of aerobic respiration. Where in the cell does each step occur?**

NAD^+ and allowing glycolysis to continue **(INFOGRAPHIC 6.8)**.

In humans, fermentation takes place primarily during bursts of energy-intensive activities, such as sprinting, when oxygen in muscles is scarce. Red blood cells—which carry lots of oxygen but lack mitochondria—also must rely on glycolysis and fermentation to produce ATP.

In some organisms, fermentation produces alcohol rather than lactic acid as a by-product. Brewer's yeast, for example, is a fungus that ferments sugar, producing alcohol as a result. Humans use brewer's yeast to make beer and wine.

Since fermentation does not break down glucose as completely as does aerobic respiration, there is still quite a bit of carbohydrate energy left in such beverages as beer and wine, about 7 Calories per gram. Not surprisingly, then, most weight-loss diets eliminate alcohol.

INFOGRAPHIC 6.8

Fermentation Occurs When Oxygen Is Scarce

Glycolysis occurs whether or not oxygen is present. When oxygen is not available, aerobic respiration cannot be completed, and NADH cannot drop off electrons to the electron transport chain. Fermentation occurs in the cytoplasm, and allows NADH to drop off electrons to pyruvate. This regenerates NAD+, which can keep glycolysis running.

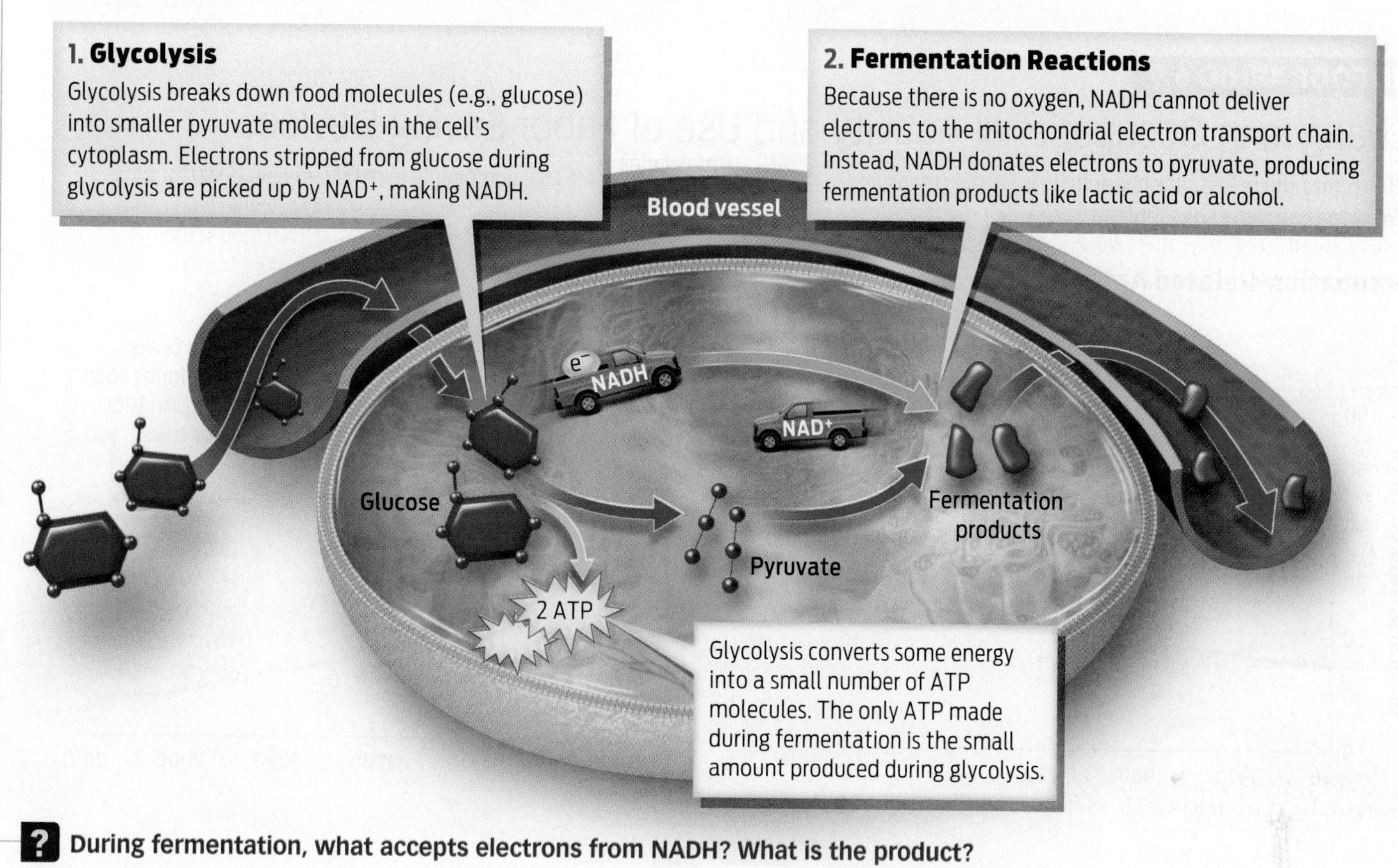

? **During fermentation, what accepts electrons from NADH? What is the product?**

Modern Times, Modern Problems

▸ Energy balance and imbalance

When Levine gives presentations at scientific conferences, he asks the audience to consider how different our modern lifestyles are from those of our not-so-distant ancestors. A hundred and fifty years ago, 90% of the world's population lived in agricultural regions. They walked to work, performed manual labor, and walked home at the end of the day. They prepared their own food and washed their clothes by hand.

Today, in industrialized countries, most people live in cities or suburbs and work behind a computer. They sit during their drive to work, sit all day at work, sit to drive home, and sit in the evening watching television, surfing the Internet, or playing video games. They even sit as the washing machine washes, rinses, and spins their laundry.

"In a mere 150 years," says Levine, "*Homo sapiens* has become addicted to the chair."

In adopting this sedentary lifestyle, humans have decreased their NEAT by approximately 1,500 Calories per day, according to Levine. At the same time, we have easier access to high-caloric foods and can pretty much eat whenever we want. As a result, we are experiencing a population-wide

GLYCOGEN
A complex animal carbohydrate, made up of linked chains of glucose molecules, that stores energy for short-term use.

positive energy balance: more Calories in than Calories out **(INFOGRAPHIC 6.9)**.

When we eat Calories beyond what our bodies expend, the extra energy is stored in one of two forms: as glycogen in muscle and liver cells, or as triglycerides (fat) in fat cells. **Glycogen,** a polymer of glucose, is the energy-storing carbohydrate found in animal cells. It's essentially a short-term storage system. When we require short bursts of

INFOGRAPHIC 6.9

Changes in Occupational Activity and Use of Labor-Saving Devices

Jobs today require less physical activity intensity (measured in metabolic equivalents: METS), and people expend less energy performing their jobs than they did in 1960. At the same time, Americans have access to more labor-saving devices like washing machines and cars.

Occupation-Related Activity

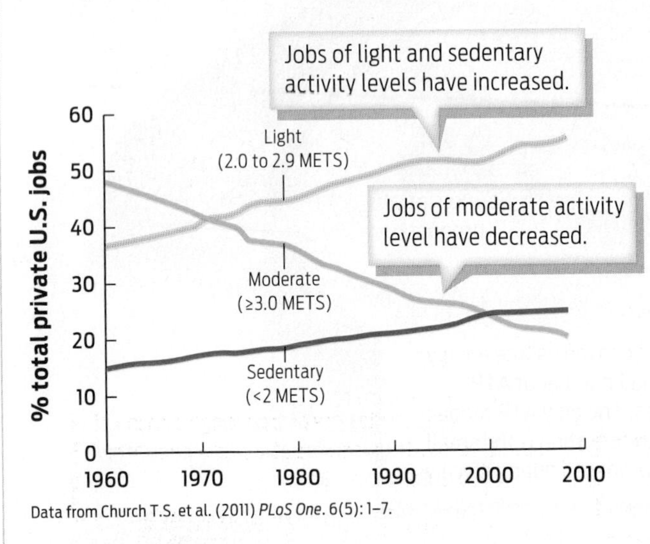

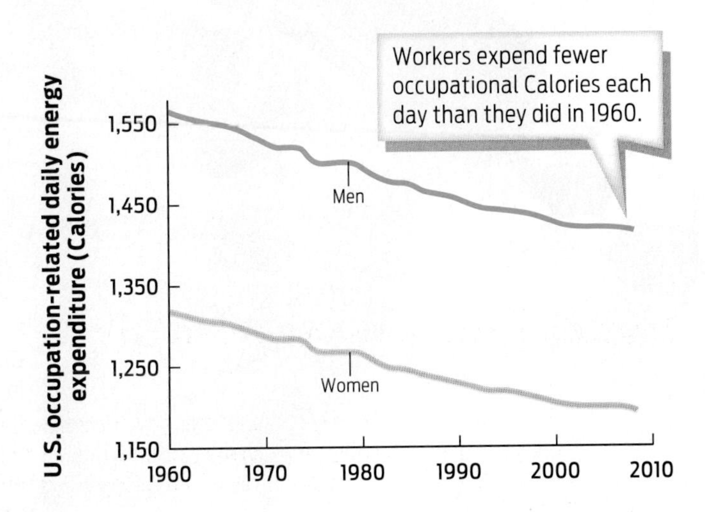

Data from Church T.S. et al. (2011) *PLoS One*. 6(5): 1–7.

Domestic Machine Useage

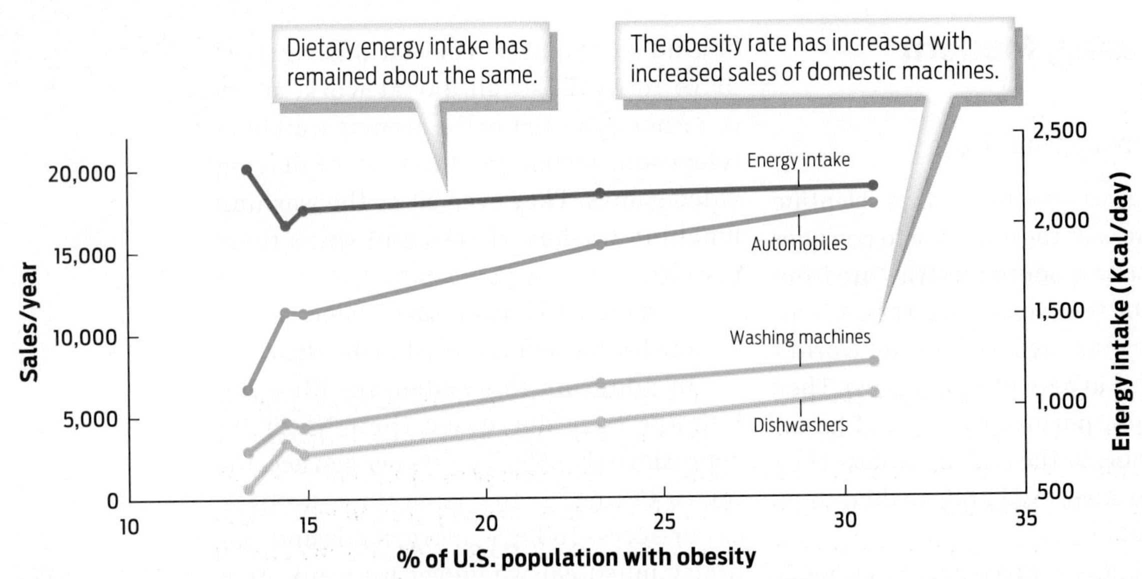

Data from Levine, J., et al. (2006). Non-exercise activity thermogenesis: The crouching tiger hidden dragon of societal weight gain. *Arteriosclerosis, Thrombosis, and Vascular Biology*. 26: 729–736.

? In what year did light-activity jobs overtake moderate-activity jobs as a percentage of the total number of jobs?

INFOGRAPHIC 6.10

Glycogen and Fat Store Excess Calories

When we ingest more Calories than our bodies need, they are stored as glycogen molecules in muscle and liver cells. Once the body's glycogen stores have been replenished, any excess Calories are stored as triglyceride molecules in fat cells.

Glycogen carbohydrates provide short-term energy storage.

Excess simple sugars not immediately used for energy or cell structures are bound together in branching chains called glycogen.

Glycogen is stored short term in muscle and liver tissue. Glycogen stores approximately 4 Calories per gram and is a quick source of energy when food molecules are unavailable in the bloodstream.

Simple sugars

Glycogen

Triglyceride fats provide long-term energy storage.

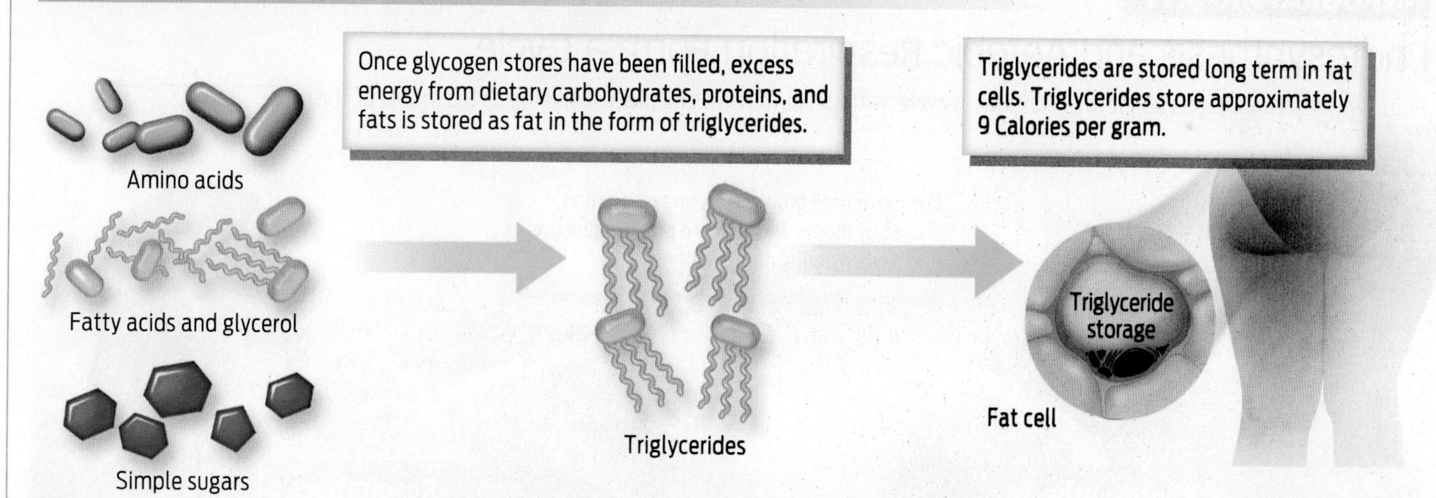

? Which molecule (a triglyceride or glycogen) stores more energy per gram? Where is this molecule stored?

energy—in a sprint, for example—the body breaks down glycogen into its component glucose molecules, and uses those molecules in aerobic respiration to obtain usable energy (ATP). However, because a gram of glycogen stores only half as many Calories as a gram of fat (about 4 Calories per gram versus 9), our bodies would have to carry around twice as much glycogen to store the same amount of energy. So our bodies store most excess energy as **triglycerides** in fat cells, which actually allows us to carry around less weight overall. The body burns this fat only after it has already used up both food molecules in the bloodstream and its stored glycogen **(INFOGRAPHIC 6.10)**.

TRIGLYCERIDES
A type of lipid found in fat cells that stores excess energy for long-term use.

The average adult American today weighs 25 pounds more than the average person did in the 1970s. Clearly, as a population we are taking in more energy than we are using. Scientists generally agree that some combination of consuming more Calories and expending less energy is at work, but pinning down the exact contributions has proved difficult.

"There is a huge debate about this," says Eric Ravussin, director of the Nutrition Obesity Research Center at the Pennington Biomedical Research Center at Louisiana State University. "Portion sizes have increased, the number of meals, the amount of snacking," he notes, "but also sitting more in front of computers, not walking as much to go to work." Separating out these influences can be challenging.

Many chroniclers of the obesity epidemic have noted that our collective weight has crept up in tandem with a change in *what* we are eating. Dietary guidelines issued by the U.S. government shifted in the late 1970s, for the first time recommending that Americans cut back on the amount of fat (particularly saturated fat) that they consume. The goal of this recommendation was to lower the risk of heart disease, which some studies had linked to consuming certain types of fats. While a reasonable idea in principle, the problem is that people responded by replacing fats in their diet with carbohydrates—particularly starchy, refined carbohydrates like those in bread, pasta, and sugary drinks. Food companies even began creating and marketing fat-free cookies—still loaded with sugar—as a healthful alternative to eating fat.

While a Calorie is a Calorie regardless of whether it comes from fat or sugar (or protein), some evidence indicates that eating lots of refined carbohydrates leaves us

INFOGRAPHIC 6.11

Photosynthesis and Aerobic Respiration Form a Cycle

Photosynthesis and respiration form a continuous cycle, with the outputs of one process serving as the inputs of the other.

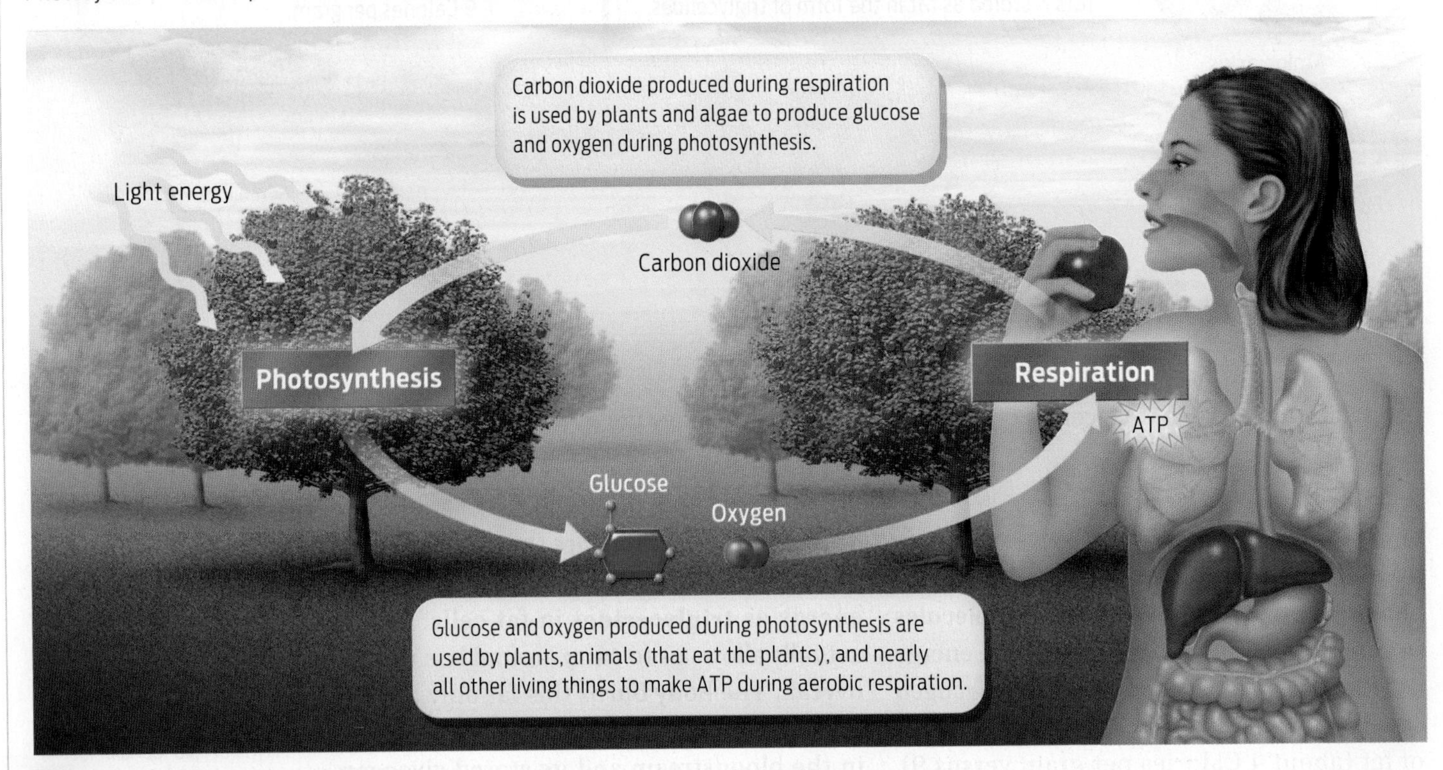

? **What is the role of carbon dioxide in photosynthesis and in cellular respiration?**

"What's really cool about NEAT is that everyone can do it."

—James Levine

less satiated and more hungry than eating foods with fat. As a result, we consume more Calories overall. This tendency, say some experts, could be a contributing factor to the obesity epidemic.

Regardless of what we eat, there is no escaping the first law of thermodynamics, which states that energy is neither created nor destroyed. All the food we eat—whether burger, broccoli, or bread—is a source of energy that originally came from the sun. During photosynthesis, plants capture the energy of sunlight and convert it into chemical energy stored in sugar (see Chapter 5). We then eat this sugar (or eat animals that have eaten this sugar), and that stored energy becomes available to us. Because energy is never destroyed, only converted from one form into another, excess energy in our diet can find its way onto our belly, hips, and thighs.

What happens when we expend more Calories than we take in? Stored fat molecules are burned by aerobic respiration. The energy stored in fat is captured by ATP, while the carbon atoms that made up the backbone of the fat molecules are given off as carbon dioxide when we exhale. Plants take up this carbon dioxide during photosynthesis, and the cycle then repeats **(INFOGRAPHIC 6.11)**.

The only way to lose weight is to shift our energy balance to the negative—that is, to expend more energy than we take in. While cutting back on how much we eat and increasing our deliberate exercise are two ways of achieving this goal, many people find doing either to be difficult—especially over the long term. Levine's work suggests a powerful third way—through increasing our NEAT.

Although different people may have different baseline levels of NEAT, based in part on their genetics, we may be able to change our patterns. Through deliberate practice and conscious choice, Levine says, we can retrain our brains to become "NEATer." We can also change our environment in ways that encourage us to increase our NEAT, which over time will add up to big Calorie differences.

A Moratorium on the Chair

▶ Strategies for increasing NEAT

When we first caught up with Levine to discuss NEAT, we found him in transit, walking to his office in downtown Phoenix, Arizona. Levine walks *a lot*. He routinely conducts meetings, interviews, and many other tasks on the go. "Contrary to popular belief, I do sit from time to time," he says.

Asked just how bad sitting is for you, Levine rattles off a list of 16 associated health risks, including obesity, diabetes, hypertension, high cholesterol, cardiovascular disease, depression, swollen ankles, joint problems, back pain, depression, and cancer. Even a person's creativity, he suggests, may be dulled from sitting too much.

That's why Levine has called for a "moratorium on the chair." He believes it is time to fundamentally redesign our environments so that higher NEAT is the norm. Toward that end, he is working with the mayor of Phoenix on initiatives to encourage more walking among commuters, and also works with businesses and school systems to make workplaces and

Christina Paolucci/AP Images

James Levine, working at his computer in his office at the Mayo Clinic, uses a specially designed elevated desk that includes a treadmill.

INFOGRAPHIC 6.12

Small Changes in NEAT Behavior Add Up to Big Calorie Rewards

Small, consistent changes in lifestyle can tip the scales considerably toward more Calories burned per week. Over time this may be all that is necessary for some to maintain a healthy weight.

NEAT Calories Burned per Hour
(Calories burned above those burned at rest for a 150-pound person)

0–50 Calories	50–100 Calories	100–200 Calories
Sitting Activities: Driving, Desk work	Standing Activities: Cooking, Stretching, Ironing	Walking Activities: Dog walking, House cleaning, Gardening

Sedentary Activities	Calories burned in one week
Park by the door and take the elevator to and from your third floor office or apartment.	210
Relax with your feet up while eating during your 45-minute lunch break.	175
Spend an hour sitting while talking on the phone.	105
Drive to and from the convenience store for milk.	70
Spend an hour on your favorite social media sites.	70
Order pizza and sit while waiting for it.	53
Watch TV for 30 minutes after dinner.	53
Hire a maid.	20
Total Calories:	756

NEAT Activities	Calories burned in one week
Park several blocks away, walk and take the stairs to and from your third floor office or apartment.	840
Get out for a 30-minute walk, then sit down for a quick 15-minute lunch.	875
Use a headset to walk or pace while making phone calls.	910
Walk 15 minutes to and from the convenience store and carry the milk	350
Spend an hour playing cards or working a puzzle.	210
Move about the kitchen making dinner for 30 minutes.	490
Take a 30-minute walk after dinner.	700
Clean one room each day for 15 minutes.	210
Total Calories:	4,585

Data from Levine, J. and Yeager, S. *Move a Little, Lose a Lot*. Three Rivers Press, 2009 and *The Compendium of Physical Activities Tracking Guide*, 2011.

? What three NEAT activities will allow you to burn approximately 1,500 Calories in a week?

classrooms more active—or, at Levine put it, "NEATer."

"What's really cool about NEAT is that everyone can do it," says Levine **(INFOGRAPHIC 6.12)**.

The degree of energy imbalance that has produced our obesity epidemic has been termed our "energy gap." By some estimates, a lifestyle change that reduced energy intake or increased energy expenditure by 100 Calories per day would completely abolish the energy gap, and permit weight maintenance, for most of the population. Walking just one extra mile each day can increase a person's energy expenditure by about 100 Calories per day. Similarly, taking just a few less bites of food at each meal can reduce energy intake by 100 Calories per day. The problem, of course, is that our current environment strongly discourages both of these good habits.

Levine once referred to NEAT as the "crouching tiger, hidden dragon" of societal weight gain. By that phrase, he meant that the behaviors that lead to obesity may be both sneakier and more deadly than we ever imagined.

"Whoever thinks about the amount of time they spend sitting?" Levine asks. "No one does." ■

CHAPTER 6 Summary

Driving Question 1 How does the body use the energy in food?

- The macronutrients in our food (proteins, carbohydrates, and fats) are sources of dietary energy.
- Fats are the most energy-rich organic molecules in our diet. Fats contain twice as many Calories per gram as do carbohydrates and proteins.
- When we consume more Calories than we use, our bodies store the excess energy in the bonds of glycogen in muscle and liver cells, and the bonds of triglycerides in fat cells.
- During exercise, glycogen is used first. Stored fats are tapped only when glycogen stores have been depleted, which may happen during long periods of exercise.

Driving Question 2 How does aerobic respiration extract useful energy from food?

- Cells carry out chemical reactions that break down food to obtain usable energy in the form of ATP.
- In the presence of oxygen, aerobic respiration produces large amounts of ATP from the energy stored in food.
- Aerobic respiration occurs in three stages: (1) glycolysis, (2) the citric acid cycle, and (3) electron transport. The first stage occurs in the cytoplasm, and the latter two in the mitochondria. Electron transport produces the bulk of ATP.
- The ultimate source of energy in food is the sun. Photosynthesizers such as plants trap the energy of sunlight and convert it into the chemical energy of sugar. Animals then eat this sugar either directly or indirectly. Both plants and animals can use this sugar as an energy source for aerobic cellular respiration.
- Photosynthesis and respiration form a cycle: the carbon dioxide given off by animals, plants, and all organisms that perform aerobic respiration is used by photosynthesizers to make glucose and oxygen during photosynthesis.

Driving Question 3 When does fermentation occur, and why can't a human survive strictly on fermentation?

- In the absence of oxygen, fermentation follows glycolysis and produces lactic acid in animals (or, in some organisms, alcohol). Fermentation produces far less ATP than does aerobic respiration.

Driving Question 4 What factors influence weight gain and weight loss?

- The balance between Calorie input and expenditures determines whether a person gains, maintains, or loses weight.
- Obesity—having an unhealthy amount of body fat—is the result of a persistent energy imbalance: more energy in than calories out. Obesity is associated with serious health risks, including heart disease, diabetes, stroke, Alzheimer's disease, hypertension, and cancer.
- Scientists don't understand fully what is causing obesity rates to rise. NEAT is one hypothesis that focuses on the "energy out" side of the equation.
- Activity—both intentional exercise and NEAT—helps burn stored Calories. A combination of eating fewer Calories and being active will result in weight loss.

More To Explore

- NOVA's Secret Life of Scientists and Engineers. (2014). James Levine: "I came alive as a person": https://www.youtube.com/watch?v=fLgGf0BO4tw
- Levine, J. (2014). *Get Up!: Why Your Chair Is Killing You and What You Can Do About It.* New York: St. Martin's Griffin.
- Ludwig, D. (2018). *Always Hungry? Conquer Cravings, Retrain Your Fat Cells, and Lose Weight Permanently.* London: Orion Books. And: www.youtube.com/watch?v=OK1zePxBJu4
- Mayor, S. (2015). Prolonged sitting increases risk of serious illness and death regardless of exercise, study finds. *BMJ* 350:306.
- Miller, D. (August 30, 2019). Could you skip your workout if you built more activity into your day? *Washington Post.*

CHAPTER 6 Test Your Knowledge

Driving Question 1 How does the body use the energy in food?

By answering the questions below and studying Infographics 6.2, 6.3, 6.5, and 6.10, you should be able to generate an answer for this broader Driving Question.

Know It

1. Which snack will provide the lowest number of Calories?
   - **a.** 25 g sugar, 5 g protein, 0 g fat
   - **b.** 30 g sugar, 0 g protein, 5 g fat
   - **c.** 10 g sugar, 10 g protein, 10 g fat
   - **d.** 0 g sugar, 15 g protein, 15 g fat
   - **e.** 10 g sugar, 25 g protein, 0 g fat

2. Which type of organic molecule serves as long-term energy storage in humans?
   - **a.** proteins
   - **b.** starch
   - **c.** nucleic acid
   - **d.** fats (triglycerides)
   - **e.** b and d

3. Which part of an ATP molecule is associated with its ability to store energy?
   - **a.** the adenine
   - **b.** the ribose
   - **c.** the phosphate groups
   - **d.** all of the above

4. If you exercise for an extended period of time, you will use energy first from ____________, then from ____________.
   - **a.** fats; glycogen
   - **b.** proteins; fats
   - **c.** glycogen; proteins
   - **d.** fats; proteins
   - **e.** glycogen; fats

Use It

5. Compare the weight of 1,000 Calories stored as glycogen with the weight of 1,000 Calories stored as fat.

Apply Your Knowledge

Interpreting Data

6. Consider a well-trained 130-pound female marathon runner. She has just loaded up on a carbohydrate meal and has the maximum amount of stored glycogen (6.8 g of glycogen per pound of body weight).
   - **a.** How many grams of glycogen is she storing?
   - **b.** How many Calories does she have stored as glycogen?
   - **c.** If this same number of Calories were stored as fat, how much would it weigh?
   - **d.** Suppose this athlete decides to go for a run at a pace of 9 mph (she will be running 6.5-minute miles). Given her weight, she will burn 885 Calories per hour at this pace. How long will it take her to deplete her glycogen stores? How many miles can she run before her glycogen supplies run out? Will she be able to complete a 26.2-mile marathon?
   - **e.** Once her glycogen supplies run out, what has to happen if she wants to keep running?

Driving Question 2 How does aerobic respiration extract useful energy from food?

By answering the questions below and studying Infographics 6.5, 6.6, 6.7, and 6.11, you should be able to generate an answer for this broader Driving Question.

Know It

7. Which process is *not* correctly matched with its cellular location?
   - **a.** glycolysis—cytoplasm
   - **b.** citric acid cycle—mitochondria
   - **c.** glycolysis—mitochondria
   - **d.** electron transport—mitochondria
   - **e.** none of the above; they are all correctly matched

8. In the presence of oxygen, humans use ____________ to fuel ATP production, and plants use ____________ to fuel ATP production from their stored sugars.
   - **a.** aerobic respiration; photosynthesis
   - **b.** aerobic respiration; aerobic respiration
   - **c.** fermentation; aerobic respiration
   - **d.** fermentation; photosynthesis
   - **e.** glycolysis; photosynthesis

9. Given 1 g of each of the following, which would yield the greatest amount of ATP by aerobic respiration?
   - **a.** fat
   - **b.** protein
   - **c.** carbohydrate
   - **d.** nucleic acid
   - **e.** alcohol

10. During aerobic respiration, what molecule has (and carries) electrons stripped from food molecules?

a. NAD^+
b. NADH
c. O_2
d. H_2O
e. pyruvate

11. During aerobic respiration, how does NADH give up electrons to regenerate NAD^+?

a. by giving electrons to O_2
b. by giving electrons to pyruvate
c. by giving electrons to glucose
d. by giving electrons to the electron transport chain
e. by giving electrons to another NAD^+

Use It

12. Draw a carbon atom that is part of a CO_2 molecule such as you just exhaled. In a written description or a diagram, trace what happens to that carbon atom as it is absorbed by the leaf of a spinach plant and then what happens to the carbon atom when you eat that leaf in a salad.

13. If you ingest carbon in the form of sugar and use it to generate ATP via cellular respiration, how is that carbon released from your body?

a. as sugar
b. as fat
c. as CO_2
d. as protein
e. in urine

Mini Case

Apply Your Knowledge

14. Begin by reviewing the number of ATP molecules generated at each step of aerobic respiration. Now consider a molecule that interferes with the third stage of aerobic respiration (electron transport). This molecule does not prevent the actual passing of electrons along the chain of molecules in the inner mitochondrial membrane, but it does prevent the capture of the energy as ATP.

What do you predict the impact of this molecule to be on overall ATP production?

Now consider what happens to the energy that was not captured as ATP. (*Hint:* Consider the law of conservation of energy and review Infographics 5.4 and 5.5. In what other form can that energy leave the system?) In human babies, a particular type of fat tissue (known as brown fat) is thermogenic and helps keep infants warm. It does so because it has highly specialized mitochondria. Based on all the information provided, what do you think is unique about the mitochondria in brown fat?

Driving Question 3 When does fermentation occur, and why can't a human survive strictly by fermentation?

By answering the questions below and studying Infographic 6.8, you should be able to generate an answer for this broader Driving Question.

Know It

15. Compared to aerobic respiration, fermentation produces ____________ ATP.

a. much more
b. the same amount of
c. a little less
d. much less
e. no

16. Which process is most directly prevented in the absence of adequate oxygen?

a. citric acid cycle
b. glycolysis
c. electron transport chain
d. a, b, and c
e. a and b

17. During fermentation, how does NADH give up electrons to regenerate NAD^+?

a. by giving electrons to O_2
b. by giving electrons to pyruvate
c. by giving electrons to glucose
d. by giving electrons to the electron transport chain
e. by giving electrons to another NAD^+

18. Where in the cell does fermentation take place?

a. cytoplasm
b. mitochondria
c. nucleus
d. cytoplasm and mitochondria
e. Fermentation doesn't occur in cells, but rather in the liquid portion of blood.

Use It

19. Explain how the presence or absence of oxygen affects ATP production. (The terms *aerobic respiration* and *fermentation* should be in your answer.)

20. Consider fermentation.

a. How much ATP is generated during fermentation?
b. How does the amount of ATP generated by fermentation compare to the amount generated by aerobic respiration?
c. In humans, why can't fermentation sustain life? (*Hint:* Think of two reasons—one is related to the product of fermentation and what happens if it accumulates.)

Apply Your Knowledge

Bring It Home

21. A 60-year-old CEO wants to lose some weight, but he has bad knees, so he can't work out as he once did. He is skeptical about a weight-loss plan not based on exercise. What specific activities can you suggest to him, and what information can you provide to persuade him to give a NEAT plan a try?

Driving Question 4 What factors influence weight gain and weight loss?

By answering the questions below and studying Infographics 6.1, 6.3, 6.4, 6.9 and 6.12, you should be able to generate an answer for this broader Driving Question.

Know It

22. A 6′0″ male weighs 230 pounds. Use Infographic 6.1 to determine his BMI. Would he be considered underweight, of normal weight, overweight, or obese?

23. If a person wants to gain weight, which of the following will contribute to the necessary Calorie imbalance?

a. doing more housecleaning
b. eating more
c. exercising more
d. all of the above

Use It

24. Consider the 6′0″, 230-pound male from question 22. If you learned that he was an NFL quarterback, would you reconsider how to interpret his BMI? Explain your answer. (*Hint:* Muscle is denser than fat, so a given volume of muscle will weigh more than the same volume of fat.)

25. If an average person wanted to balance Calories consumed in a snack of an apple and a hard-boiled egg, how long would the person have to stand? How long would the person have to walk (at 2 mph)?

26. Two friends are trying to lose weight. They adopt the same diet, but one briskly walks her dog for 30 minutes twice a day at about 2 mph, while the other jogs at about 5 mph for 15 minutes a day. Predict the outcomes in 90 days. Will there be a substantial difference in weight loss? Why or why not?

27. Use the data in the top right graph of Infographic 6.9 (that show occupation-related daily energy expenditure over time) to complete the table below.

	Males	Females
Calories expended per day 1960		
Calories expended per day 2009		
Change in Calories expended		
Change as a % of the 1960 amount		

28. Look at Infographic 6.12. The data are shown for one week. If you adopted all the NEAT activities shown for an entire month (4 weeks), how many additional Calories would you burn compared to a month (4 weeks) of sedentary activities? Assuming that it takes approximately 3,500 Calories to lose 1 pound of fat, how many additional pounds of fat would you lose during a month of NEAT activities?

Biologically Unique

How DNA helped free an innocent man

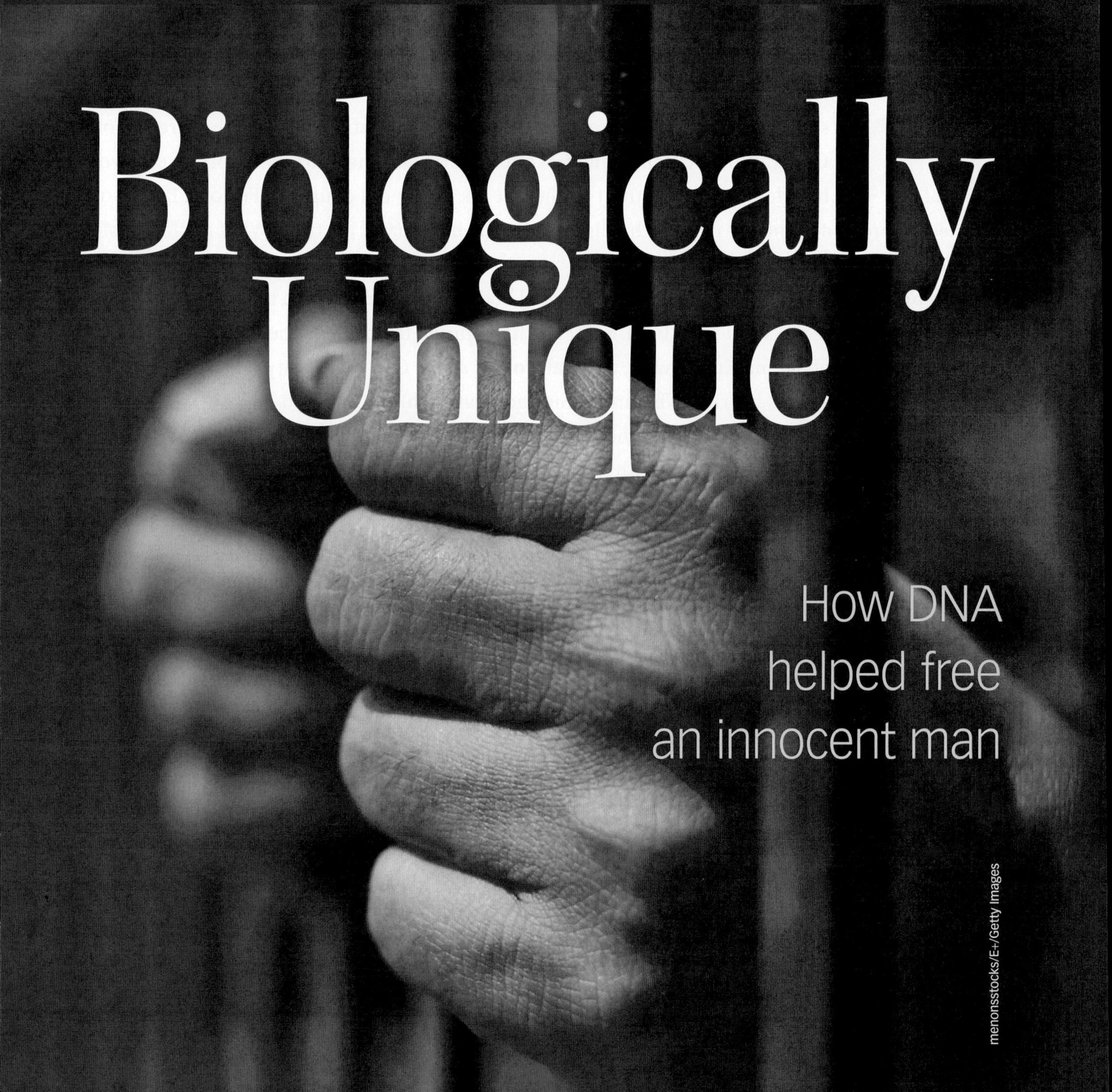

menonsstocks/E+/Getty Images

DRIVING QUESTIONS

1. What is the function of DNA, and how is DNA organized in cells?
2. What is the structure of DNA, and why is each person's DNA unique?
3. How is DNA copied in living cells, and how can DNA be amplified for forensics?
4. How does DNA profiling make use of genetic variation in DNA sequences?

Roy Brown thought the police were just checking up on him when an officer knocked on his door one day in May 1991. Only a week earlier, Brown, a self-professed hard drinker who earned a living selling magazine subscriptions, had been released after serving an 8-month prison term. His crime: threatening to kill the director of the Cayuga County Department of Social Services in upstate New York. A caseworker had deemed Brown unfit to care for his 7-year-old daughter. Furious, Brown made a series of threatening phone calls to the director. But he had served his time. What could the officer want from him now?

Three days earlier, police had found the battered body of a woman lying in the grass about 300 feet from the farmhouse where she had lived. Someone had burned the place to the ground. The body was identified as that of Sabina Kulakowski, a social worker at the Cayuga County Department of Social Services. The crime was horrific. The murderer had beaten the 49-year-old Kulakowski, bitten her several times, dragged her outside, and then stabbed and strangled her. It was obvious that Kulakowski had struggled; her body was covered with defensive wounds.

Although Kulakowski had not been involved in Brown's case, officers arrested Brown that day on suspicion of murder. Eight months later, a jury found Brown guilty of homicide and sentenced him to prison for 25 years to life. The prosecution argued that Brown's motive was revenge against the Department of Social Services. But what really nailed the case was testimony from an expert who stated that bite marks on the victim's body matched Brown's teeth.

Brown, however, maintained his innocence. "I never knew Ms. Kulakowski, and I had nothing to do with that woman's death . . . I am truly innocent," he told the court and onlookers after the verdict had been announced.

Even from prison, Brown never stopped trying to prove his innocence. He repeatedly petitioned, in vain, for a retrial.

Then something unexpected happened: Brown uncovered additional evidence that strongly suggested he was not the perpetrator. The evidence was so compelling, in fact, that in late 2004, after Brown had spent 12 years in prison, his lawyers decided to contact the Innocence Project, a nonprofit organization founded in 1992. Its mission: to use DNA evidence to free people wrongly convicted of crimes.

Dr. Frank Wright, a forensic dentist, uses tooth molds like this one (which is not Roy Brown's) for bite-mark analysis.

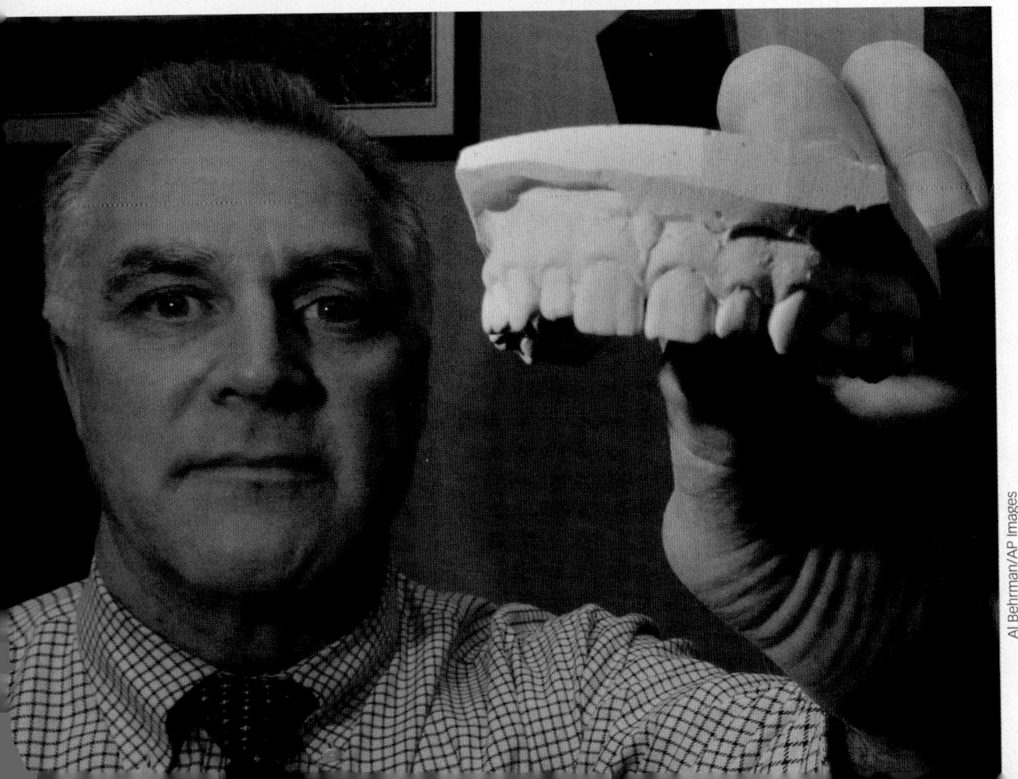

Al Behrman/AP Images

Life's Blueprint

▶ Function and location of DNA

When the jury convicted Brown in 1992, DNA testing was still in its infancy, so DNA was rarely used as evidence in criminal cases. But over the next decade, DNA testing became a standard part of court cases, as science increasingly showed that it was an extremely accurate way to match crime scene evidence to perpetrators.

How can scientists use DNA to identify a person? The answer lies in the chemical makeup of this molecule, often referred to as the blueprint of life. **Deoxyribonucleic acid (DNA)** stores biological information and serves as the instruction manual from which we are built. It is passed from parents to offspring during reproduction, and is the reason children resemble their biological relatives. All life forms—from bacteria to plants to humans—use DNA as the storehouse of their biological information and transmit it to their offspring. DNA is the hereditary molecule of life, establishing the uniqueness of each person.

Where can you find DNA? In eukaryotic cells, DNA is found inside the nucleus, where it exists in the form of chromosomes. A **chromosome** is a single, large DNA molecule wound around proteins. Human cells have 23 pairs of chromosomes; we inherit one chromosome of each pair from our mother and the other from our father, for a total of 46 chromosomes. The 23rd chromosome pair consists of the sex chromosomes, X and Y, which determine a person's sex. Men have an X and a Y, and pass on one or the other during reproduction. Women have two Xs, and therefore can only pass on an X. Children who inherit a Y from their dad and an X from their mom are males, while children who inherit one X from each parent are females. Since nearly all cells contain DNA, scientists can collect evidence such as blood, skin, semen, saliva, or hair from a crime scene and extract DNA from it to identify a perpetrator **(INFOGRAPHIC 7.1)**.

Courtesy The Innocence Project

Barry Scheck and Peter Neufeld, the lawyers who founded the Innocence Project.

DNA testing has helped the Innocence Project free more than 350 people from prison since 1992, including 18 who served time on death row. This technology has not only given these people their lives back, but has also shown a spotlight on flaws in our criminal justice system. Why were people wrongly convicted and placed on death row? Innocence Project lawyers have found several culprits: dishonest witnesses, unscrupulous police officers, apathetic or overburdened lawyers, and mistakes in eyewitness identification. Fortunately, new developments such as those in DNA technology are helping to improve the system.

"DNA is only one example of how advances in science have made the criminal justice system more reliable," says Peter Neufeld, who, along with Barry Scheck, both from the Benjamin N. Cardozo School of Law in New York City, founded the Innocence Project. "But what we really hope to do now is use DNA as the gold standard of reliability to weed out junk science."

DEOXYRIBONUCLEIC ACID (DNA)
The molecule of heredity, common to all life forms, which is passed from parents to offspring.

CHROMOSOME
A single, large DNA molecule wrapped around proteins. Chromosomes are located in the nuclei of eukaryotic cells.

INFOGRAPHIC 7.1

What Is DNA and Where Is It Found?

Deoxyribonucleic acid, or DNA, is the hereditary molecule common to all living organisms. It is the instruction manual from which an organism is built.

DNA is found in the nucleus of eukaryotic cells.
DNA is coiled around proteins into discrete structures called chromosomes.

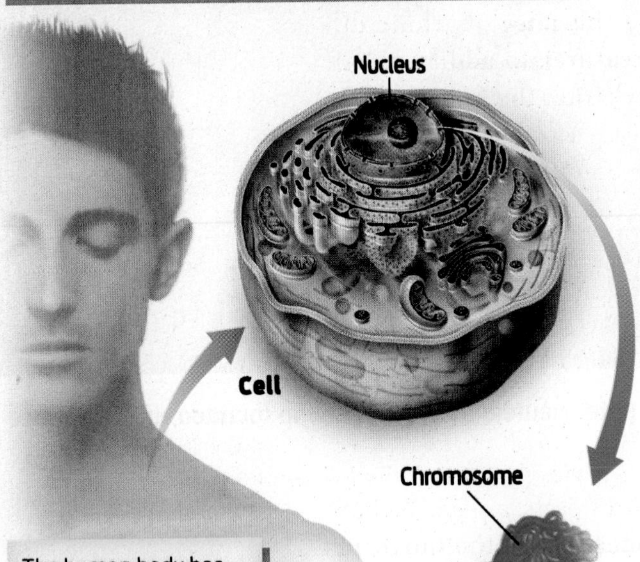

The human body has trillions of cells, each containing DNA with the instructions for cell structure and function.

Human cells have 23 pairs of chromosomes.
One chromosome of each pair is inherited from the mother, the other from the father. The 23rd chromosome pair determines a person's sex.

Human Male Chromosomes
(23 chromosome pairs, XY)

23rd chromosome pair determines a person's sex.

Each chromosome consists of a single, long DNA molecule wrapped around proteins.

? **How many DNA molecules are in the nucleus of each human cell?**

More-Reliable Evidence

▶ The DNA double helix

Indeed, it was "junk science" that convicted Roy Brown. The only physical evidence linking Brown to the case was his teeth. A dentist hired by the prosecution testified that the bite marks on Kulakowski's body matched Brown's teeth. But as the defense pointed out, the bite marks came from someone with a full set of upper front teeth—and Brown was missing two of his. The prosecution's witness argued that Brown could have twisted Kulakowski's skin while biting her and filled in the gaps—an argument that ultimately proved convincing to the jury.

Bite-mark analysis is a particularly troubling form of evidence. No widely accepted rules or standards govern its use, and no government or outside scientific commission has ever validated its claims. In fact, studies show error rates—the rate at which experts have falsely identified bite marks as belonging to a particular person—as high as 91%.

Hair analysis, another common type of evidence, can be equally unreliable. In dozens of cases, Innocence Project lawyers found that forensic scientists had testified that hairs from crime scenes matched the accused, explains Neufeld. But when scientists subsequently tested the DNA inside the follicle cells from those hairs, the DNA didn't match.

The problem is that hair analysis, when performed under a microscope, can reveal only certain characteristics: it can determine whether hair is human, or show a person's ancestry (because of ethnic differences in hair texture). In addition, hair analysis can tell whether the hair has been dyed, cut in a certain way or pulled out, and where on the body it came from. Hair samples can exclude a suspect, but not positively identify one.

By contrast, each person's DNA is unique. To understand how DNA varies from person to person, consider its structure. A DNA molecule is made up of two strands of subunits linked together in long chains. As we learned in Chapter 2, each subunit—called a **nucleotide**—has three parts: a sugar, a phosphate group, and a base. In each DNA strand, the phosphate group of one nucleotide binds to the sugar of the next nucleotide to form a

NUCLEOTIDES
The building blocks of DNA. Each nucleotide consists of a sugar, a phosphate group, and a base. The sequence of nucleotides (As, Cs, Gs, Ts) along a DNA strand is unique to each person.

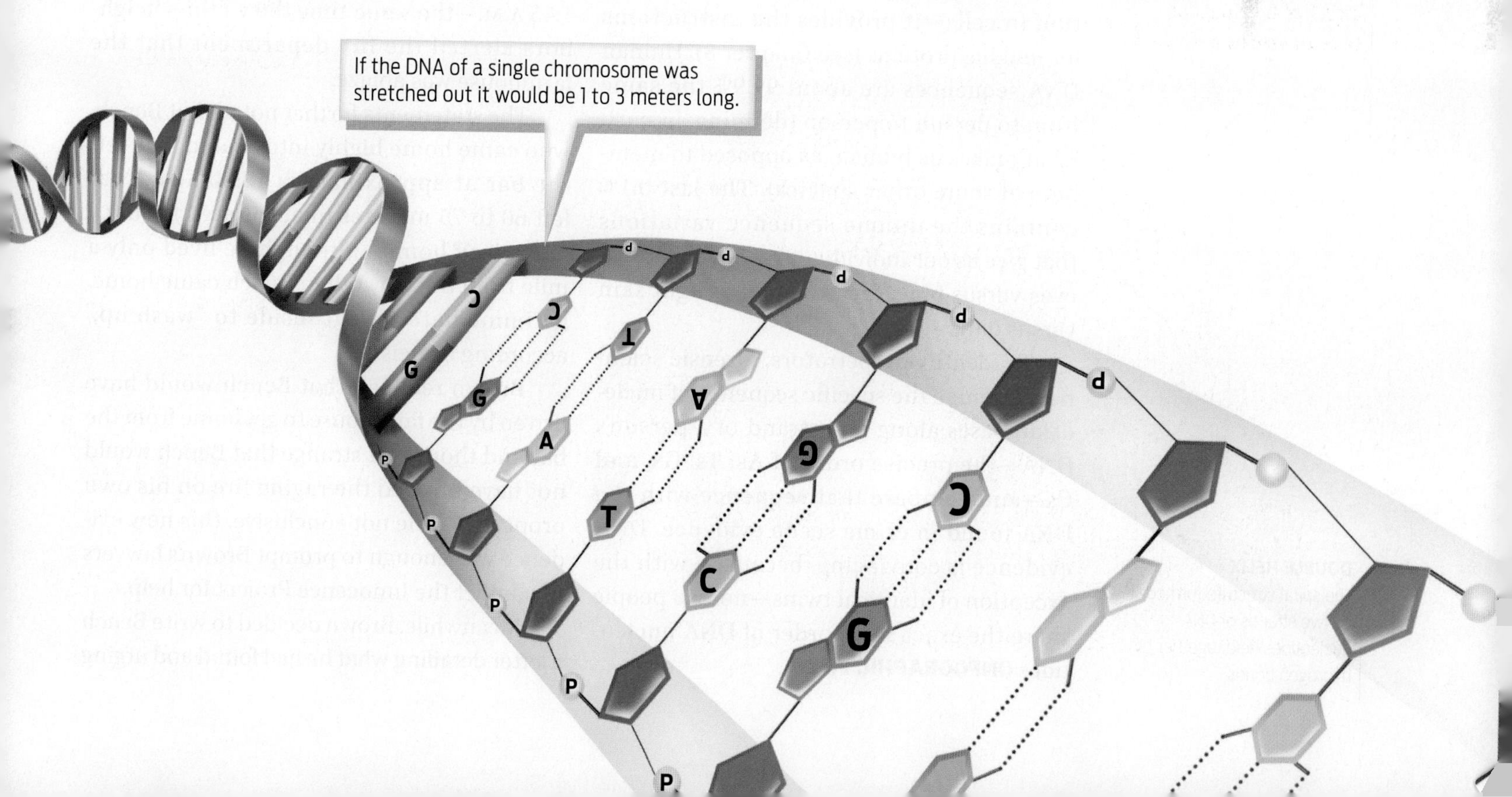

chain of linked nucleotides. The two strands of linked nucleotides pair up and twist around each other to form a spiral-shaped **double helix.** The sugars and phosphates form the outside "backbone" of the helix, while the bases point toward its center, forming internal "rungs," like steps on a twisting ladder. The bases in one strand associate with bases from the other strand through hydrogen bonds, which hold the two strands of the double helix together.

The rungs of the DNA ladder, made up of the bases, are the most useful part of DNA for profiling. There are four possible nucleotide bases: adenine (A), thymine (T), guanine (G), and cytosine (C). In DNA, these four nucleotide bases are repeated over and over,

To identify perpetrators, forensic scientists examine the specific sequence of nucleotide bases along one strand of a person's DNA.

billions of times, in different orders along a DNA strand. The order of nucleotide bases in a strand is a key form of genetic information in cells—it provides the instructions for making proteins (see Chapter 8). Human DNA sequences are about 99.9% the same from to person to person (defining, in part, what makes us human, as opposed to members of some other species). The last 0.1% contains the unique sequence variations that give us our individuality—having brown eyes versus blue, for example, or light skin versus dark.

To identify perpetrators, forensic scientists examine the specific sequence of nucleotide bases along one strand of a person's DNA—the precise order of As, Ts, Gs, and Cs—and compare that sequence with the DNA found in crime scene evidence. DNA evidence is convincing, because—with the exception of identical twins—no two people share the exact same order of DNA nucleotides **(INFOGRAPHIC 7.2)**.

DOUBLE HELIX
The spiral structure formed by two strands of DNA nucleotides held together by hydrogen bonds.

Brown Gets a Break

While Brown sat in prison, he never stopped trying to prove his innocence. For 11 years, he petitioned for an appeal and requested specifically that DNA tests be performed on evidence collected at the crime scene. On each occasion, the judge denied his request. Finally, in 2003, Brown filed a Freedom of Information Act request to obtain copies of all documents relating to his case. That's when he made a surprising discovery.

The additional evidence Brown obtained included four affidavits collected by the Cayuga County Sheriff's Department the day after the murder—documents that neither Brown nor his lawyers had ever seen. In the affidavits, four people described the suspicious behavior of another man: Barry Bench, the brother of Kulakowski's former boyfriend. The Bench family owned the farmhouse in which Kulakowski had been living.

The affidavits included sworn testimony from neighbors as well as from Bench's then-girlfriend, Tamara Heisner. They stated that on the day of Kulakowski's murder, Bench argued with Heisner, went to a local bar, and returned home between 1:30 and 1:45 A.M.—the same time the victim's neighbors alerted the fire department that the farmhouse was ablaze.

The statements further noted that Bench, who came home highly intoxicated, had left the bar at approximately 12:30 A.M. That left 60 to 75 minutes unaccounted for until he arrived home—although he lived only a mile from the bar. When Bench came home, he immediately went inside to "wash up," according to Heisner.

Brown realized that Bench would have driven by the farmhouse to get home from the bar and thought it strange that Bench would not have noticed the raging fire on his own property. While not conclusive, this new evidence was enough to prompt Brown's lawyers to contact the Innocence Project for help.

Meanwhile, Brown decided to write Bench a letter detailing what he had found and urging him to confess. He warned him of his intent to obtain a DNA test on evidence from the murder. "Judges can be fooled and juries make mistakes," he wrote, "[but] when it comes to DNA testing there's no mistakes. DNA is God's creation and God makes no mistakes."

INFOGRAPHIC 7.2

DNA Is Made of Two Strands of Nucleotides

DNA is a double-stranded molecule. Each of the two strands consists of a chain of subunits called nucleotides that are bonded together. There are four types of nucleotides: adenine (A), thymine (T), guanine (G), and cytosine (C). Nucleotides are the building blocks of DNA.

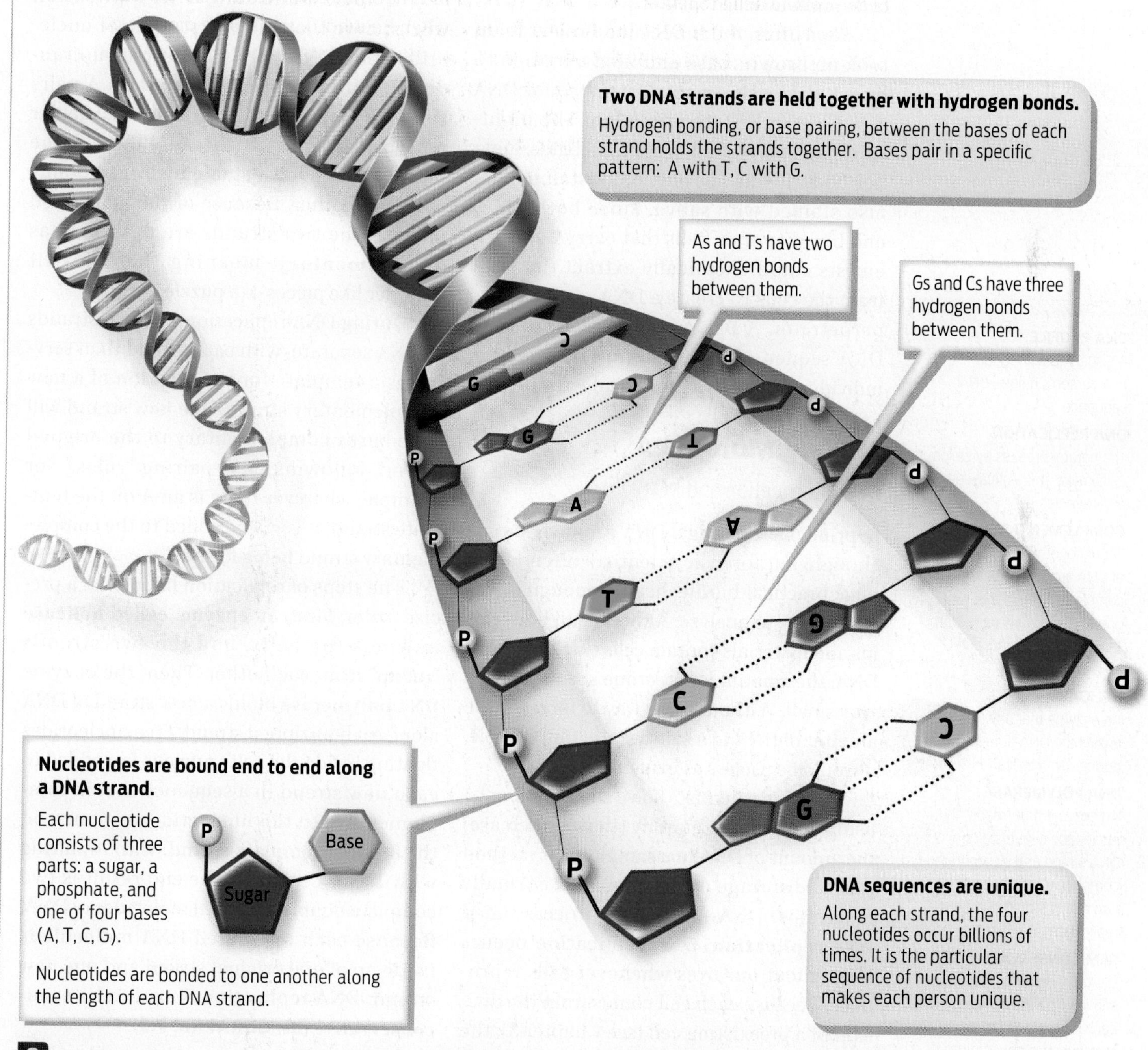

? One strand of a DNA molecule has the sequence ATGCTGA. What is the corresponding sequence on the second strand?

"Judges can be fooled and juries make mistakes, [but] when it comes to DNA testing there's no mistakes."

—Roy Brown

Five days after Brown mailed his letter, Bench threw himself in front of an Amtrak train and was killed instantly.

Soon after, the Innocence Project team took on Brown's case and filed a motion to have Kulakowski's nightshirt tested for DNA at a New York State crime lab. The nightshirt had been found in some tall grass near her body. It was not only bloodstained, but also stained with saliva. Since both saliva and blood contain cells that carry DNA, scientists could chemically extract the DNA from the cells to create a DNA profile of the perpetrator. A **DNA profile** is a readout of DNA sequences that is unique to a single individual—akin to a genetic fingerprint.

DNA PROFILE
A visual representation of a person's unique DNA sequence.

DNA REPLICATION
The natural process by which cells make an identical copy of a DNA molecule.

COMPLEMENTARY
Fitting together; two strands of DNA are said to be complementary in that A in one strand always pairs with T in the other strand, and G always pairs with C.

HELICASE
An enzyme that unwinds and unzips the DNA double helix during DNA replication.

DNA POLYMERASE
An enzyme that "reads" the nucleotide sequence of a DNA strand and incorporates complementary nucleotides into a new strand during DNA replication.

SEMICONSERVATIVE
DNA replication is said to be semiconservative because each newly made DNA molecule has one original DNA strand and one new DNA strand.

Making More DNA

▶ DNA replication and PCR

In principle, creating a DNA profile is simple enough, but forensic scientists often face a huge practical hurdle: having enough crime scene DNA to analyze. Although all body fluids and materials contain cells that house our DNA, the amount left at crime scenes may be very small. Without some way to increase the amount of DNA in a saliva stain, for example, it would be useless as evidence.

Forensic scientists solve this problem by using a method to amplify (that is, increase) the amount of DNA in a sample. This method takes advantage of the way cells normally make more DNA—through a process called **DNA replication.** DNA replication occurs throughout our lives whenever cells reproduce. Because each cell comes from the division of a preexisting cell (see Chapter 3), the DNA of the parent cell must be replicated so that each daughter cell has one copy. The process of DNA replication works essentially the same way in all organisms.

To understand how DNA replication works, note that the two strands of nucleotides in a DNA helix do not pair up randomly, but in a consistent pattern: A pairs with T, and G pairs with C. These particular nucleotides form pairs because they are the right shape to form stable hydrogen bonds with each other. Because of this patterned pairing, the two strands are described as **complementary,** meaning that they fit together like pieces of a puzzle.

During DNA replication, the two strands of DNA separate, with each strand then serving as a template for the creation of a new complementary strand. The new strand will have bases complementary to the original strand, following base-pairing "rules." For example, wherever there is an A on the template strand, a T will be added to the complementary strand being formed.

The steps of replication happen in a precise order. First, an enzyme called **helicase** unwinds the helix, and the two strands "unzip" from each other. Then, the enzyme **DNA polymerase** builds a new strand of DNA along each unzipped strand. Free nucleotides floating inside the cell's nucleus are added to each new strand in a sequence that is complementary to the nucleotide sequence on the original template strand, with A pairing with T and C with G. The end result is two complete double-stranded molecules of DNA. Because each replicated DNA molecule is made up of one original strand and one new strand, DNA replication is said to be **semiconservative (INFOGRAPHIC 7.3).**

DNA replication is a remarkably accurate process that happens at mind-boggling

INFOGRAPHIC 7.3

DNA Structure Provides a Mechanism for DNA Replication

When cells reproduce, they must first replicate their DNA so that each new cell contains a copy of the original DNA molecule. Complementary base pairing between nucleotides guides replication of two new strands.

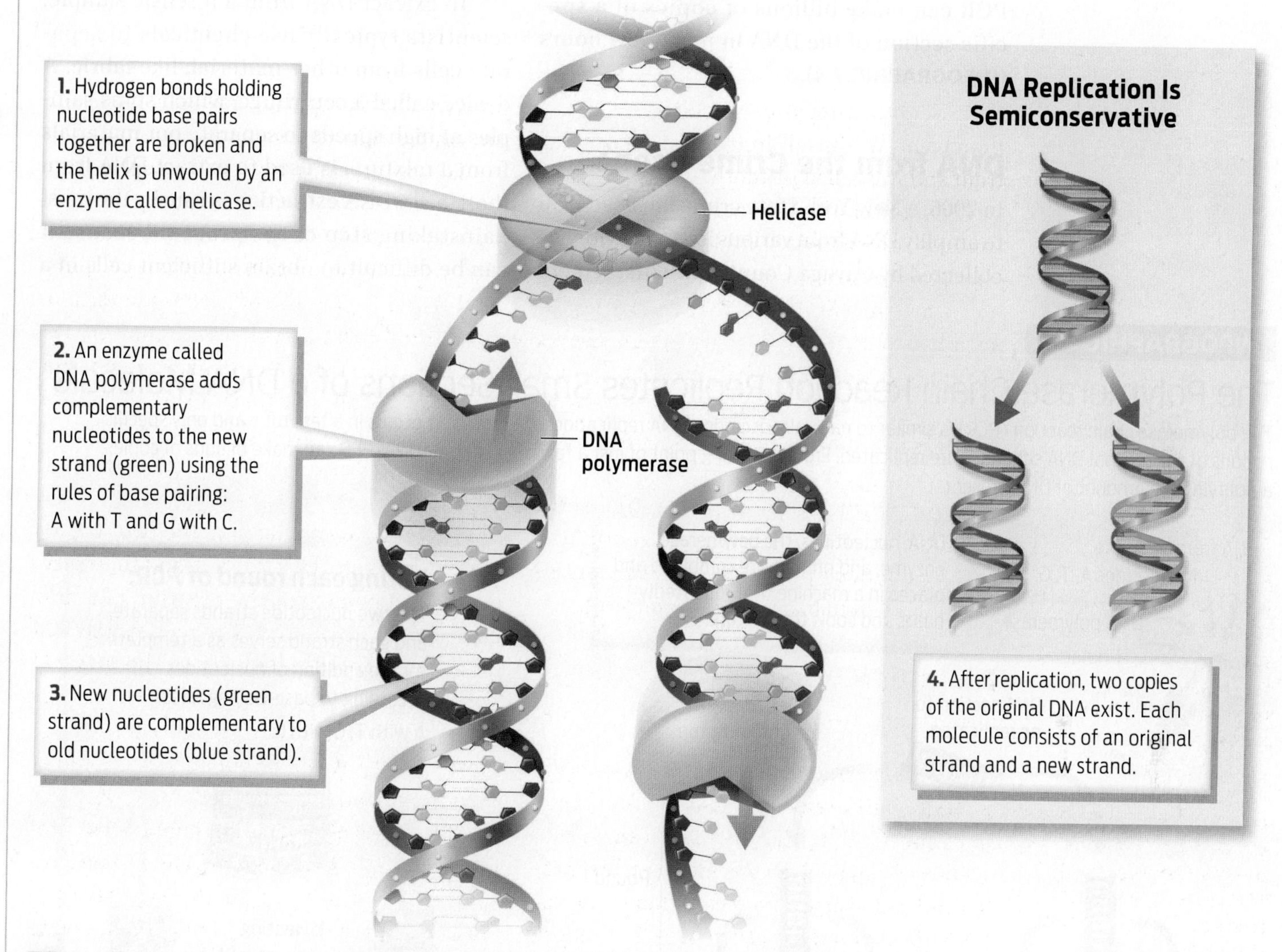

? **Which enzyme adds complementary nucleotides during DNA replication?**

speeds. In fact, the polymerase enzyme adds about 1,000 nucleotides per second and rarely makes a mistake (though the mistakes it does make often have important consequences; see Chapter 9). On a human scale, that's like a car speeding down the highway at 300 miles per hour, weaving in and out of traffic without hitting any other cars.

Forensic scientists make use of this natural cellular process when they need to amplify the DNA in a crime scene sample. The method they use, the **polymerase chain reaction (PCR),** is similar to DNA replication in cells—but it takes place in a test tube.

Here's how it works. To a small sample of DNA, scientists add nucleotides, the DNA polymerase enzyme, and primers—short segments of DNA that act as guideposts. The primers bind to complementary segments of each DNA template, and their locations flag the section to which DNA polymerase should bind to begin replication. The DNA

POLYMERASE CHAIN REACTION (PCR)
A laboratory technique used to replicate, and thereby amplify, a specific DNA segment.

is first heated to separate the strands, and then cooled to allow the primers to associate with the DNA and to allow the DNA polymerase to add new nucleotides. From a starting sample of just a few DNA molecules, PCR can make billions of copies of a specific section of the DNA in just a few hours **(INFOGRAPHIC 7.4)**.

DNA from the Crime Scene

In 2006, a New York State crime lab used PCR to amplify DNA from various items of evidence collected by Cayuga County law enforcement officials during their original investigation of Kulakowski's murder. The evidence included remnants of cotton swabs used to sample bite marks on the victim and Kulakowski's blood- and saliva-stained nightshirt.

To extract DNA from a forensic sample, scientists typically use chemicals to separate cells from other material, like fabric. A device called a centrifuge, which spins samples at high speeds to separate out materials from a mixture, is used to extract DNA from those cells. DNA extraction is usually the most painstaking step of the process because it can be difficult to obtain sufficient cells in a

INFOGRAPHIC 7.4

The Polymerase Chain Reaction Replicates Small Sections of a DNA Molecule

The polymerase chain reaction (PCR) is similar to naturally occurring DNA replication, except that it occurs in a test tube and only specific regions of the original DNA sequence are replicated. From a starting point of just a few molecules of DNA, PCR can make billions of copies, amplifying the amount of DNA present.

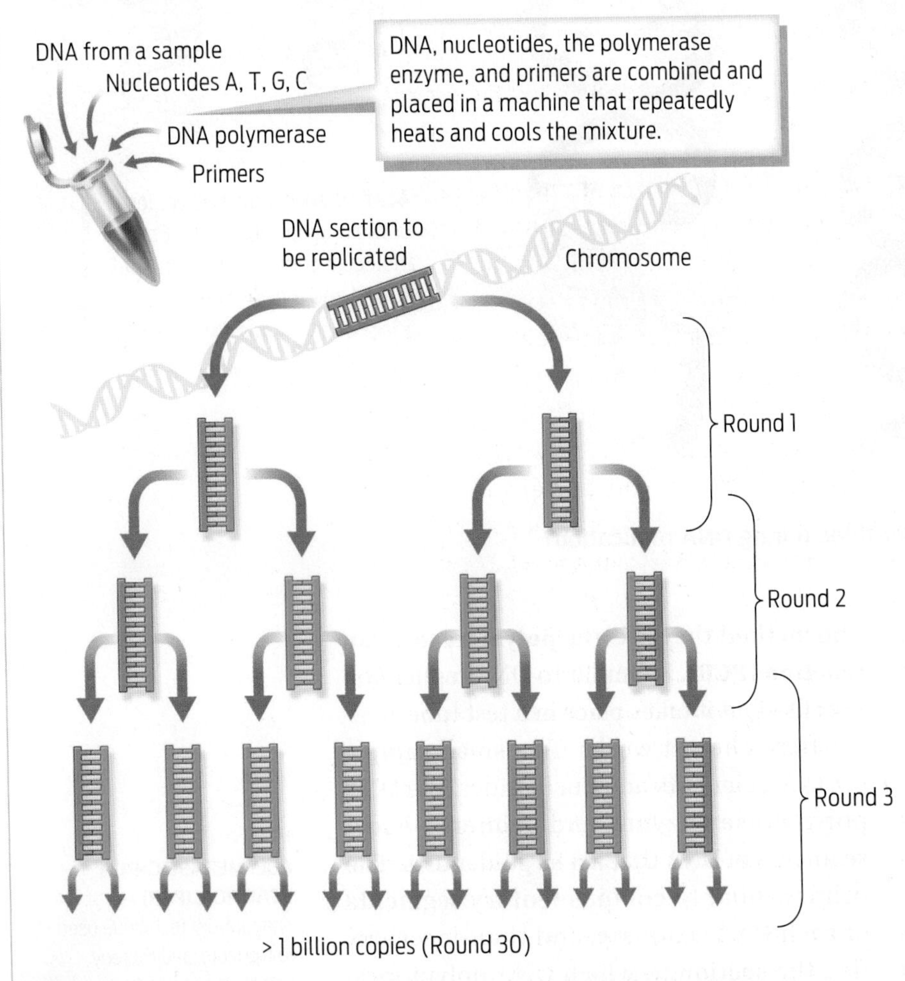

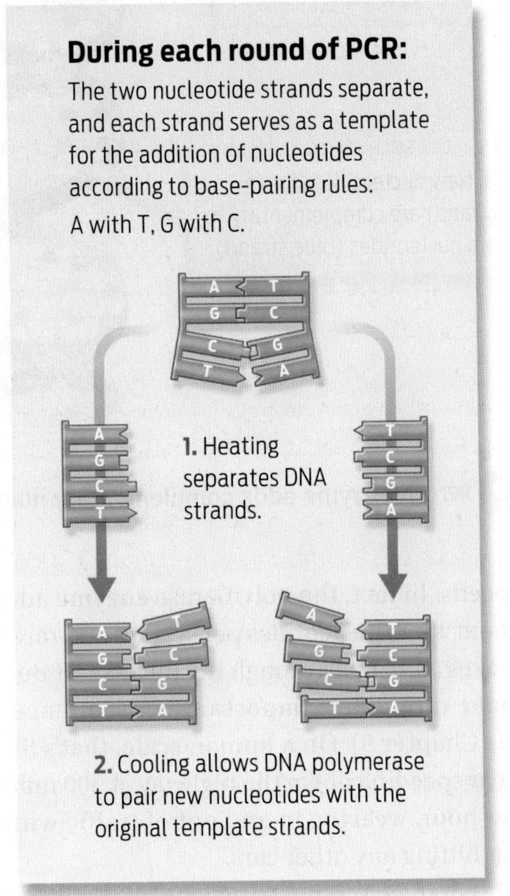

? **Which enzyme adds complementary nucleotides during PCR?**

Each person's DNA sequence is unique and can be used as a kind of molecular "fingerprint."

forensic sample to yield enough DNA for PCR. Furthermore, improperly stored samples can degrade too much to be useful. Samples can also become contaminated with foreign DNA from improper handling, which would render results useless.

In Brown's case, the laboratory's first report on the tested evidence was disappointing. Technicians hadn't been able to obtain any DNA from the bite-mark swab. The lab's second report was more conclusive: seven different pieces of the victim's nightshirt contained DNA. Moreover, the report went on to state that six of the pieces contained DNA from two different people, the victim and another person who was male.

How DNA Profiling Works

▶ STRs and gel electrophoresis

Once DNA from crime scene evidence is obtained, the next step is to analyze it. Human cells contain vast amounts of DNA—on the order of 3 billion nucleotide base pairs within the full stretch of the human **genome,** or one complete set of genetic instructions. Figuring out the sequence of every nucleotide in the genome would be extremely time consuming and expensive. So forensic scientists use a shortcut: they employ PCR to amplify specific segments of DNA and analyze just these segments. These segments are a type of DNA sequence known as short tandem repeats.

Short tandem repeats (STRs) are blocks of repeated DNA sequences found at points along our chromosomes. For example, a single STR might contain the sequence AAAAG repeated over and over again. All of us have STRs in the same places along our chromosomes, but the exact number of repeats in each STR can vary from person to person. For example, at a single STR site, one person may carry the AAAAG sequence repeated six times while another person may carry the AAAAG sequence repeated four times. Also, since we inherit two copies of every

GENOME
One complete set of genetic instructions encoded in the DNA of an organism.

SHORT TANDEM REPEATS (STRs)
Sections of a chromosome in which short DNA sequences are repeated.

chromosome, every person has two copies of each STR, and they can have two different numbers of repeats. An STR with a higher number of repeats will be longer in length. Forensic scientists use these differences in STR lengths to distinguish between individuals **(INFOGRAPHIC 7.5).**

To create a DNA profile, scientists first use PCR to increase the amount of DNA at multiple STR regions. Next, they use a method called **gel electrophoresis** to separate the amplified STRs based on their length. The segments of DNA separated by length create a specific pattern of bands on a gel that is unique to each person. This pattern is a person's DNA profile. Scientists can then compare band patterns of DNA found at a crime scene to band patterns of DNA from a suspect. DNA profiles also have other applications, including in paternity and ancestry testing **(INFOGRAPHIC 7.6).**

GEL ELECTROPHORESIS
A laboratory technique that separates fragments of DNA by size.

DNA Profiling and the Law

▶ The CODIS database

Since 1994, the federal government has been collecting DNA profiles of criminal offenders and storing this information in a computer database, the Combined DNA Index System (CODIS). This database contains more than 10 million profiles from criminals convicted of specific crimes in all 50 states. Each profile consists of a banding pattern that represents 15 specific STR regions scattered throughout

INFOGRAPHIC 7.5

DNA Profiling Uses STR Regions of Varying Length

No two people have the same exact nucleotide sequence. The specific regions of DNA that forensic scientists analyze are those that contain short tandem repeats (STRs). STRs are short stretches of repeated DNA sequences. People differ in the number of copies of an STR sequence found along their chromosomes.

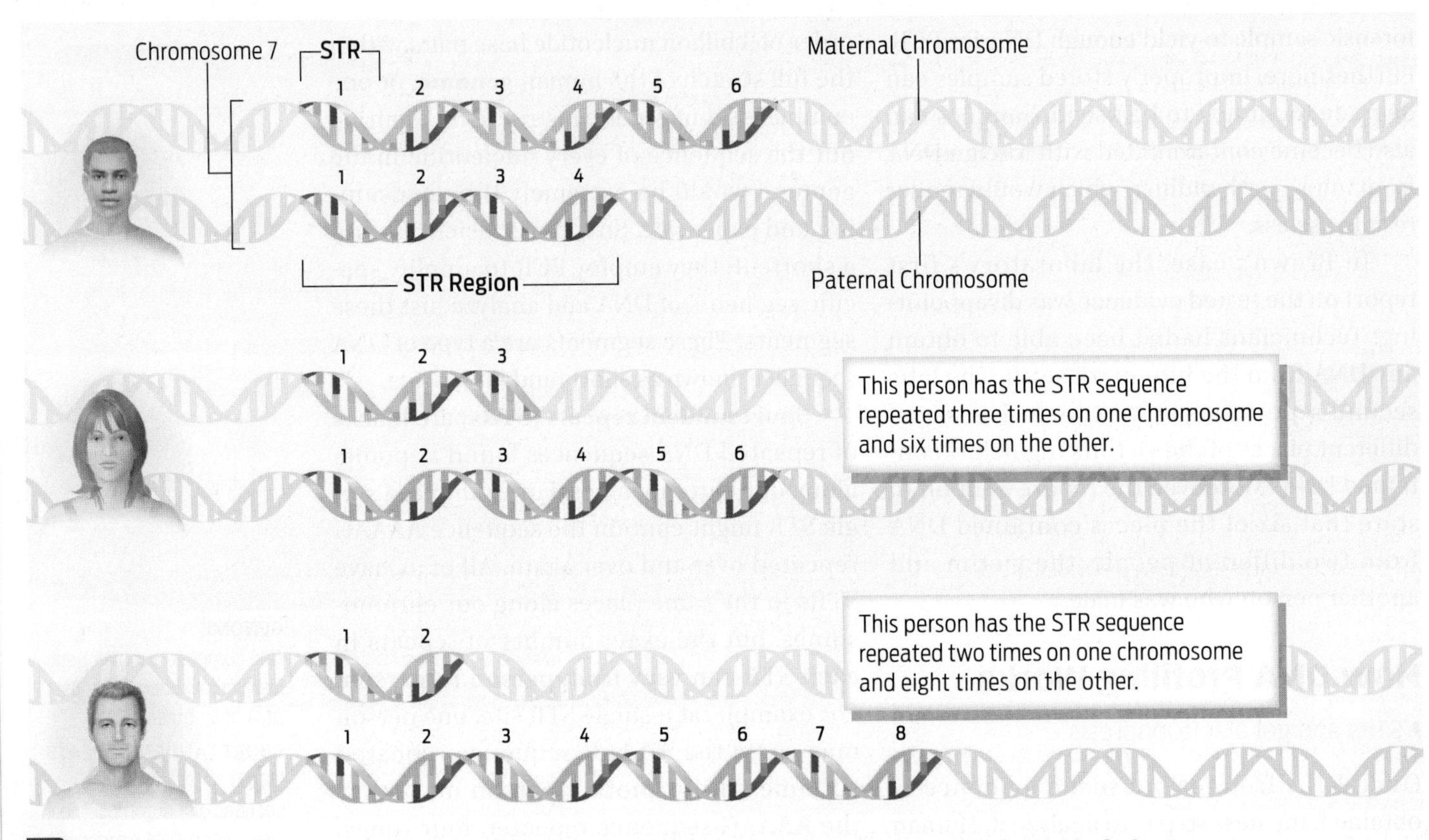

? **How many STR copies did the top person inherit from his mother and how many from his father?**

INFOGRAPHIC 7.6

Creating a DNA Profile

Cells left at crime scenes contain DNA that can be profiled. The DNA is isolated from the cells (evidence). STR regions are replicated by PCR, and their lengths are compared to those of possible suspects.

1. Collect cells from crime scene evidence and extract DNA.

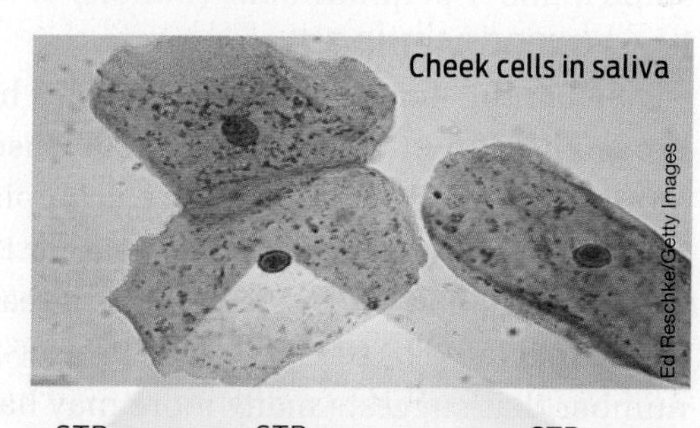

2. Use PCR to replicate multiple STR regions.

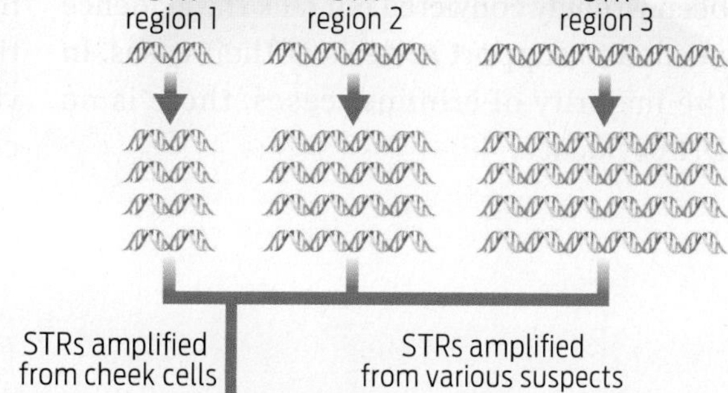

3. Separate STRs by gel electrophoresis.

PCR products are loaded into chambers at the top of a gel. An electric current applied to the gel causes the charged DNA to migrate through it. Shorter fragments of DNA travel faster—and thus farther—than longer fragments in a given amount of time. All DNA fragments of the same length will travel together, forming a band of DNA that can be visualized on the gel.

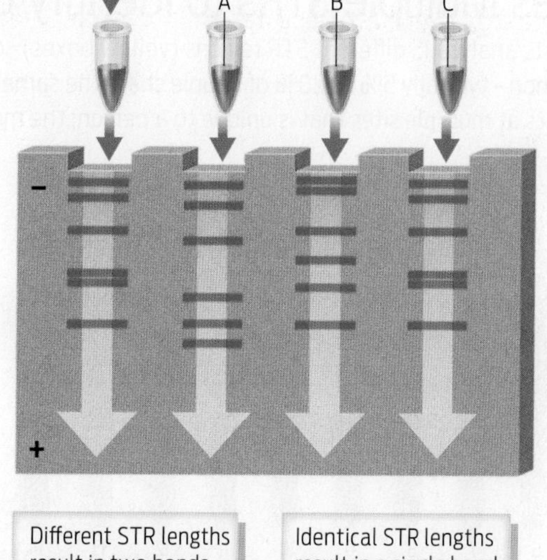

4. Compare STR banding patterns.

The gel shows the results of three different STR regions (green, red, and blue bands) analyzed from the DNA in a crime scene sample and in three suspects.

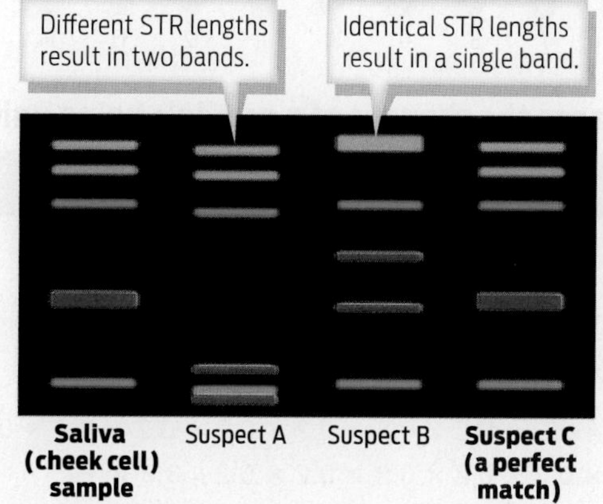

? In a paternity test, how many of the child's STR bands should match bands from the father?

our genomes. Forensic scientists typically describe the likelihood that any two unrelated people will have the same number of repeated sequences at all 15 regions as 1 in some number of quintillions (billions of billions) **(INFOGRAPHIC 7.7)**.

So far, the database of DNA profiles has proved helpful in more than 190,000 cases. More significantly, DNA evidence is helping to change the criminal justice system for the better. More than 350 prisoners have already been exonerated by the Innocence Project—a number that suggests many more may have been wrongly convicted but lack the evidence needed to support reviews of their cases. In the majority of criminal cases, there is no DNA evidence.

"How many more wrongful convictions will it take for New York to begin addressing the systemic problems that lead to such miscarriages of justice?" asked Neufeld in 2007, when Brown's case was being reviewed.

Recognizing the flaws in our criminal justice system, Innocence Project lawyers are working with several states to change the way law enforcement operates. For example, studies have found that witnesses more accurately identify perpetrators if they are shown suspects one at a time instead of in a group line-up. Project lawyers are also helping to force changes in the way interrogations are conducted, calling for them to be videotaped to reduce the possibility of forced confessions. In addition, they are lobbying

INFOGRAPHIC 7.7

DNA Profiling Uses Multiple STRs to Identify Genetically Unique Individuals

To create a DNA profile, scientists analyze 15 different STR regions (yellow boxes) scattered among our chromosomes. Sharing an STR of similar length is relatively common—typically 5% to 20% of people share the same number of repeats at any one STR site. But it is the combined pattern of STR repeats at multiple sites that is unique to a person; the more STRs tested, the more discriminating the test becomes.

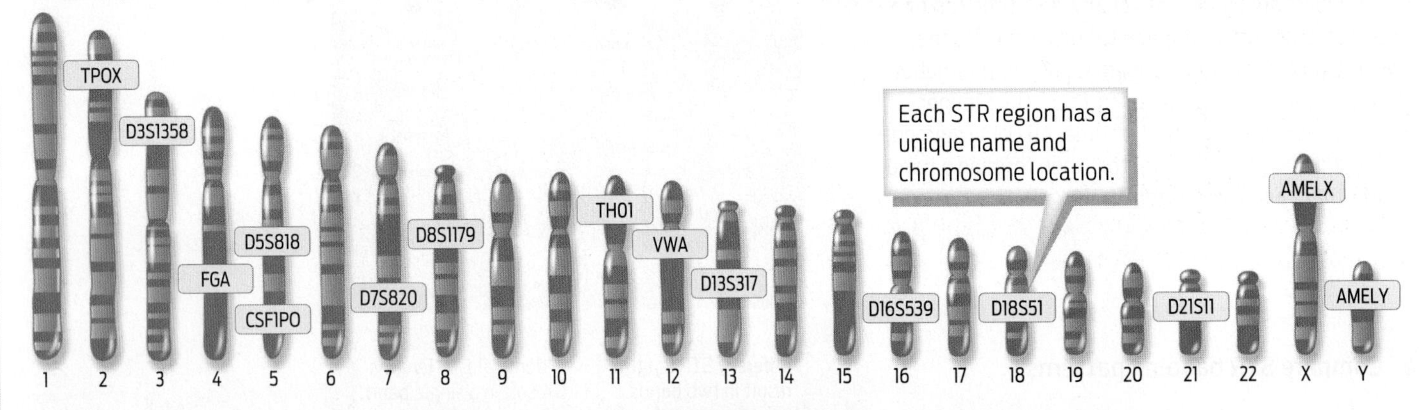

Using multiple STRs increases the chances of a profile's being unique to a single individual:

Analysis	Calculation	People sharing this profile
1 STR region	0.2	1 in 5
2 STR regions	0.2 × 0.2 = 0.04	1 in 25
5 STR regions	0.2 × 0.2 × 0.2 × 0.2 × 0.2 = 0.00032	1 in 3,125
15 STR regions	0.2 × 0.2 × 0.2 × 0.2 × 0.2 × 0.2 × 0.2 × 0.2 × 0.2 × 0.2 × 0.2 × 0.2 × 0.2 × 0.2 × 0.2 = 3.3×10^{-11}	1 in several quintillion

? If an STR analysis were carried out with three STRs, what proportion of people would be expected to share that profile (if the probability of matching at each STR is 0.2)?

for legislation to ensure that evidence from crime scenes is properly collected and maintained, since DNA evidence can be ruined or contaminated during collection. They also advocate that anyone convicted of a crime be able to gain access to DNA testing.

"The key is that DNA really gives us an opportunity to start making the other institutions in the system more scientific and reliable as well," says Neufeld.

Vindication

The DNA that the New York State crime lab extracted from Sabina Kulakowski's nightshirt contained a mixture of DNA from the victim and from another person, who was male. Analysis showed that this male DNA did not match Roy Brown's; in other words, DNA evidence excluded him as Kulakowski's murderer.

Additional testing eventually linked that DNA evidence to Barry Bench. After Bench's suicide, of course, he couldn't provide DNA directly. So lawyers pursued the next best option: a DNA sample voluntarily donated by Bench's biological daughter, Katherine Eckstadt. Because we all receive one set of chromosomes from our mother and one set from our father, half of Eckstadt's DNA would have come from her father, and therefore would show great similarity to his. The test yielded dramatic results—a 99.99% probability that the man who deposited his saliva on Kulakowski's nightshirt was Eckstadt's father, Barry Bench.

Kevin Rivoli/AP Images

Roy Brown leaves the courthouse a free man in 2007 with his lawyers, Peter Neufeld and Nina Morrison.

To clinch the case, Cayuga County prosecutors eventually agreed to have Bench's body exhumed for DNA tests. Those tests definitively matched Bench's DNA with the DNA from the saliva stains.

"We've had a lot of crazy cases," says Nina Morrison, the Innocence Project attorney who handled Brown's case, "but this is really up there with the best of them . . . the client solving his own case . . . it's insane." After serving 16 years in prison for a crime he didn't commit, Brown was ultimately cleared of all charges and released from prison in 2007. ■

CHAPTER 7 Summary

Driving Question 1 What is the function of DNA, and how is DNA organized in cells?

- DNA is the hereditary molecule of all living organisms. DNA contains instructions for building an organism.
- Humans have 23 pairs of chromosomes in their cells—one chromosome of each pair is inherited from the mother, the other from the father.
- The DNA in a eukaryotic cell is packaged into chromosomes located in the nucleus.

Driving Question 2 What is the structure of DNA, and why is each person's DNA unique?

- DNA sequences determine the genetic uniqueness and relatedness of individuals.
- DNA is a double-stranded molecule that forms a spiral structure known as a double helix.
- Each strand of DNA is made of nucleotides bonded together in a linear sequence.
- There are four distinct nucleotides in DNA: adenine (A), thymine (T), guanine (G), and cytosine (C).
- The two linear strands of a DNA molecule are bound together by hydrogen bonds between base pairs. Base pairing is complementary: A always pairs with T, and G always pairs with C.

Driving Question 3 How is DNA copied in living cells, and how can DNA be amplified for forensics?

- DNA replication is semiconservative—each original strand serves as a template to build a new strand.
- During DNA replication, helicase separates the DNA strands, and DNA polymerase adds complementary nucleotides.
- PCR enables scientists to vastly increase the number of copies of specific DNA sequences.

Driving Question 4 How does DNA profiling make use of genetic variation in DNA sequences?

- Forensic scientists use noncoding DNA sequences known as STRs to create a DNA profile.
- STRs are blocks of repeated sequences of DNA; the number of repeats varies in different people.
- STRs vary in their length, and those variations can be visualized by gel electrophoresis.
- A DNA profile is more accurate and reliable than many other forms of forensic evidence.

MORE TO EXPLORE

- The Innocence Project: http://www.innocenceproject.org/
- National Human Genome Research Institute: http://www.genome.gov
- Scheck, B., et al. (2003). *Actual Innocence: When Justice Goes Wrong and How to Make It Right.* New York: NAL Trade Paperbacks.
- NOVA, Forensics on trial, October 17, 2012: http://www.pbs.org/wgbh/nova/tech/forensics-on-trial.html
- Cold Spring Harbor Laboratory: DNA from the beginning: An animated primer of 75 experiments that made modern genetics. http://www.dnaftb.org

CHAPTER 7 Test Your Knowledge

Driving Question 1 What is the function of DNA, and how is DNA organized in cells?

By answering the questions below and studying Infographic 7.1, you should be able to generate an answer for this broader Driving Question.

Know It

1. The chromosomes shown in Infographic 7.1 are from a male individual. What would be different if the chromosomes were from a female individual?

a. There would be no Y chromosome.
b. There would be a second X chromosome.
c. There would be an additional pair of chromosomes.
d. There would be no difference.
e. a and b

2. The ____________ chromosomes in a typical human cell are found in the ____________.

a. 46; cytoplasm
b. 23; nucleus
c. 24; cytoplasm
d. 46; nucleus
e. 22; nucleus

3. Each chromosome contains

a. DNA only.
b. proteins only.
c. DNA and proteins.
d. the same number of genes and STRs.
e. the entire genome of a cell.

Use It

4. You can detect DNA that is specifically from the X chromosome in a DNA sample from a person. Can you definitively determine the sex of that person (male or female) from the presence of the X chromosome? Explain your answer.

5. Human red blood cells are enucleated; that is, they do not have nuclei. Is it possible to isolate DNA from red blood cells? Why or why not?

Driving Question 2 What is the structure of DNA, and why is each person's DNA unique?

By answering the questions below and studying Infographic 7.2, you should be able to generate an answer for this broader Driving Question.

Know It

6. Which of the following is *not* a nucleotide found in DNA?

a. adenine (A)
b. thymine (T)
c. cytosine (C)
d. guanine (G)
e. uracil (U)

7. If the sequence of one strand of DNA is AGTCTAGC, what is the sequence of the complementary strand?

a. AGTCTAGC
b. CGATCTGA
c. TCAGATCG
d. GTCGACGC
e. GCTAGACT

8. In addition to the base, what are the other components of a nucleotide?

a. sugar and polymerase
b. phosphate group and sugar
c. phosphate group and polymerase
d. phosphate group and helix
e. helix and sugar

Use It

9. Which of the following double-stranded DNA molecules do you predict will be held together the most tightly?

a. AATTAAA
TTAATTT
b. AATTAGC
TTAATCG
c. AATTGGC
TTAACCG
d. GCTTGGC
CGAACCG
e. GCGCGGC
CGCGCCG

10. A single strand of DNA has the sequence ATCG. Which of the following can you predict with certainty? Explain your answer.

a. the nucleotide sequence of the complementary strand of DNA in a double-stranded DNA molecule
b. the nucleotide before the A
c. the nucleotide after the G

Driving Question 3 How is DNA copied in living cells, and how can DNA be amplified for forensics?

By answering the questions below and studying Infographics 7.3, and 7.4, you should be able to generate an answer for this broader Driving Question.

Know It

11. DNA replication is said to be semiconservative because a newly replicated, double-stranded DNA molecule consists of

a. two old strands.
b. two new strands.
c. one old strand and one new strand.
d. two strands, each containing a mixture of old and new DNA.
e. any of the above, depending on the cell type

12. Which of the following statements about PCR is true?

a. DNA polymerase is the enzyme that copies DNA in PCR.
b. Primers are not necessary for PCR.
c. PCR does not require nucleotides.
d. PCR does not generate a complementary DNA strand.
e. PCR can make only a few copies of a DNA molecule.

Use It

13. Complete the statements below, and then number them to indicate the order of these two major steps necessary to copy a DNA sequence during PCR.

Step # ____________ The enzyme ____________ "reads" each template strand and adds complementary nucleotides to make a new strand.

Step # ____________ The two original strands of the DNA molecule can be separated by ____________.

14. Given this segment of a double-stranded DNA molecule, draw the two major steps involved in DNA replication:

```
ATCGGCTAGCTACGGCTATTTACGGCATAT
TAGCCGATCGATGCCGATAAATGCCGTATA
```

Driving Question 4 How does DNA profiling make use of genetic variation in DNA sequences?

By answering the questions below and studying Infographics 7.5, 7.6, and 7.7, you should be able to generate an answer for this broader Driving Question.

Know It

15. A person has an STR with 30 nucleotides on one chromosome and 40 nucleotides on the other chromosome. The STR is amplified by PCR, and the PCR product is run on a gel. Which lane (A–D) in this gel shows the banding pattern you would expect to see? The marker lane (M) has DNA fragments starting at 10 nucleotides (at the bottom) and increasing in 10-nucleotide increments.

M A B C D

Gel for Question 15

16. This gel shows the DNA profile of STRs from four sources: blood from crime scene evidence (E), suspect A, suspect B, and the victim (V). An eyewitness identified suspect A as fleeing the apartment building where the crime occurred. Suspect B was picked up at a local convenience store after using bloodstained money.

a. From the DNA profiles shown, can you draw any conclusions about where the crime scene DNA came from?
b. Can you draw any conclusions about relationships among the people profiled? Explain your reasoning.

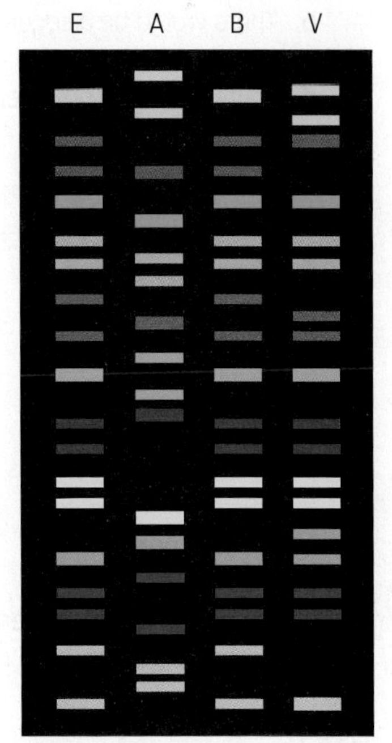

Gel for Question 16

17. Which STR will have migrated farthest through an electrophoresis gel?

a. an STR with a length of 8 nucleotides
b. an STR with a length of 12 nucleotides
c. an STR with a length of 20 nucleotides
d. an STR with a length of 24 nucleotides
e. an STR with a length of 28 nucleotides

18. Which of the following represents genetic variation between individuals?

a. whether or not G pairs with C or T

b. the presence of STRs in their genomes

c. the number of chromosomes in the nucleus

d. the sequence of nucleotides along the length of each chromosome

e. the number of chromosomes received from each parent

Use It

19. A series of statements is presented below. Mark each statement as true (T) or false (F).

a. ____________ G pairs with T.

b. ____________ Genetic information is passed on to the next generation in the form of DNA molecules.

c. ____________ All DNA sequences encode information to produce proteins.

d. ____________ Each person carries the same number of repeats in STRs on both maternal and paternal chromosomes.

e. ____________ DNA evidence can be obtained from saliva left in a bite mark.

20. Explain why the statements that you marked as true in Question 19 are in fact true.

21. Rewrite the statements that you marked as false in Question 19 to make them true.

22. Look at Infographic 7.7. From the STRs used in forensic investigations, which STRs on which chromosomes would be particularly useful in determining whether crime scene evidence was left by a female or a male?

23. Explain your response to Question 22, stating the number of STR copies you would expect to see if the perpetrator was female and if the perpetrator was male.

24. This gel shows a DNA profile using five STRs. The lane labeled W is a mother and the lane labeled C is her child. The lanes labeled M1 and M2 are two men, either of whom, according to the mother, could be the father of the child.

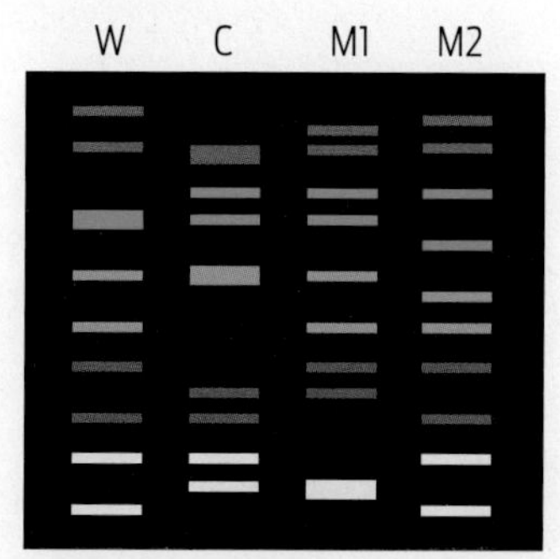

Gel for Question 24

a. Circle the STR bands that the child (C) inherited from its mother (W).

b. Use the DNA profiles to determine which man is the father of the child.

Interpreting Data

Apply Your Knowledge

25. The table on the next page shows the frequencies of STR lengths (versions) in different U.S. populations. Since we inherit two copies of every chromosome, one copy from each parent, every person has two copies of each STR. You can determine the probability of a particular combination of STRs by multiplying the frequencies.

If the STR lengths on each chromosome are identical, the probability of this pattern can be determined by multiplying the frequency by itself. For example, the probability of a Hispanic person having D3S1358 version A (0.079) on both chromosomes is $0.079 \times 0.079 = 0.0062$.

If the STR lengths on each chromosome are different, the probability of this pattern can be determined by multiplying the two frequencies together. Then, because there are two ways of inheriting this pattern (getting version A from dad and version E from mom, or getting version E from dad and version A from mom), this number must be multiplied by 2. For example, the probability of a Hispanic person having D3S1358 version A (0.079) and D3S1358 version E (0.125) is $0.079 \times 0.125 \times 2 = 0.0198$.

a. What is the probability of a Caucasian American having a C, D combination for D3S1358?

b. What is the probability of an African American having a C, D combination for D3S1358?

c. What is the probability of a Hispanic American having a C, D combination for D3S1358?

d. What is the probability of a Caucasian American having a C, D combination for D3S1358, and an A, E combination for TH01?

e. What is the probability of a Caucasian American having a C, D combination for D3S1358, an A, E combination for TH01, and a B, E combination for D18S51?

f. Consider your answers to a, d, and e. Why does forensic analysis use many CODIS STRs (and not just one or two)?

CODIS* STR	Version	Frequency (Caucasian)	Frequency (African American)	Frequency (Hispanic)
D3S1358	A	0.103	0.089	0.079
	B	0.262	0.186	0.293
	C	0.253	0.248	0.286
	D	0.215	0.242	0.204
	E	0.152	0.155	0.125
TH01	A	0.002	0.004	0
	B	0.232	0.124	0.214
	C	0.190	0.421	0.096
	D	0.084	0.194	0.096
	E	0.114	0.151	0.150
D18S51	A	0.008	0.006	0.004
	B	0.017	0.002	0.011
	C	0.127	0.078	0.118
	D	0.132	0.053	0.111
	E	0.137	0.072	0.139

Data from Butler, J. M., et al. (2003). Allele frequencies for 15 autosomal STR loci on U.S. Caucasian, African American, and Hispanic populations. *Journal of Forensic Sciences* 48(4):908–911.

*CODIS stands for "Combined DNA Index System," a government database of DNA profiles from offenders, crime scenes, and missing persons.

Mini Case

Apply Your Knowledge

26. A female eyewitness has identified a Hispanic American male as the man who stole her car. The eyewitness stated that the man was bleeding profusely from a head wound. Her car was recovered, and male blood with a C, D combination for D3S1358, an A, E combination for TH01, and a B, E combination for D18S51 was found on the driver's seat and steering wheel. Does this finding call the eyewitness evidence into question? Explain your answer.

Bring It Home

Apply Your Knowledge

27. Scientists used DNA from Barry Bench's daughter to pinpoint Bench as a possible suspect, as his DNA was not on file anywhere. Similarly, in cases of disasters, DNA evidence is sometimes required to identify victims. If a victim doesn't have a DNA profile on file, his or her identity must be reconstructed by comparing the victim's DNA profile to that of relatives. These situations illustrate that a DNA profile database has the potential to be useful in cases in which DNA-based identification is required. However, maintenance of such a database is controversial. What arguments can you make for and against banking DNA profiles in a database? If such a database existed, what restrictions would you place on it? Would you choose to register your DNA or your child's DNA in such a database?

The Model Makers

Watson, Crick, and the structure of DNA

James D. Watson Collection, Cold Spring Harbor Laboratory Archives

DRIVING QUESTIONS

1 **Whose work contributed to the discovery of the structure of DNA?**

2 **What pieces of scientific knowledge were assembled to elucidate the structure of DNA?**

In 1953, James Watson and Francis Crick announced to a crowd in their favorite pub in Cambridge, England, that they had found "the secret of life." They had every right to boast. In solving the structure of DNA, they had revealed the chemical basis of inheritance. Their discovery would revolutionize biology and push forward the study of anthropology, evolution, and medicine.

But Watson and Crick's success wasn't merely the result of a marriage between two great minds. Science is a collective enterprise and progresses through the work of many people. Breakthroughs happen when scientists are well positioned to build upon foundations laid down by other scientists. And so it was with DNA: in addition to their own insight, Watson and Crick built on the discoveries of others. They also had luck on their side—they were in the right place at the right time.

Watson and Crick met in 1951 at Cambridge University, in England. Watson was an American scientist who had just accepted a research position at the university. Crick was then a Ph.D. student, studying protein structure with a technique called x-ray crystallography.

The men didn't appear obvious collaborators. Watson was a prodigy: twelve years younger than Crick, he had earned his Ph.D. at age 22. Crick, by contrast, was a late bloomer: he was 38 years old by the time he had his Ph.D. What they did share was intellectual curiosity. Both had changed their research focus several times. By their own admission, both were more interested in solving current hot topics in science—like the structure of DNA—than pursuing the more obscure science that each had trained to do.

Although DNA was first observed in cell nuclei in the late 1860s, it took almost a century before scientists realized its importance. For a long time, the prevailing belief was that proteins carried the genetic information. But by the 1950s, scientists had pegged DNA as the more likely candidate. A key experiment carried out in 1952 convincingly demonstrated that DNA, not protein, was the genetic material. This set the stage for a race among scientists to understand the structure of DNA and how it carried genetic information.

When Crick and Watson came to the problem, they already knew, from the work of other scientists, that DNA was made up of nucleotides. Each of these nucleotides contained a sugar, a phosphate group, and one of four bases: adenine (A), thymine (T), cytosine (C), or guanine (G). But how did these components fit together? To answer this question, the pair took inspiration from the chemist Linus Pauling. Pauling had been studying the structure of proteins and had built a molecular model showing that some proteins exist as a single-stranded, twisting helix. He backed up his model with lab experiments to prove his structure was correct. If an eminent scientist like Pauling could model a protein structure without first conducting laboratory experiments, Watson and Crick thought they might be able to do the same with DNA. And, since Pauling was one of several scientists now chasing after DNA's structure,

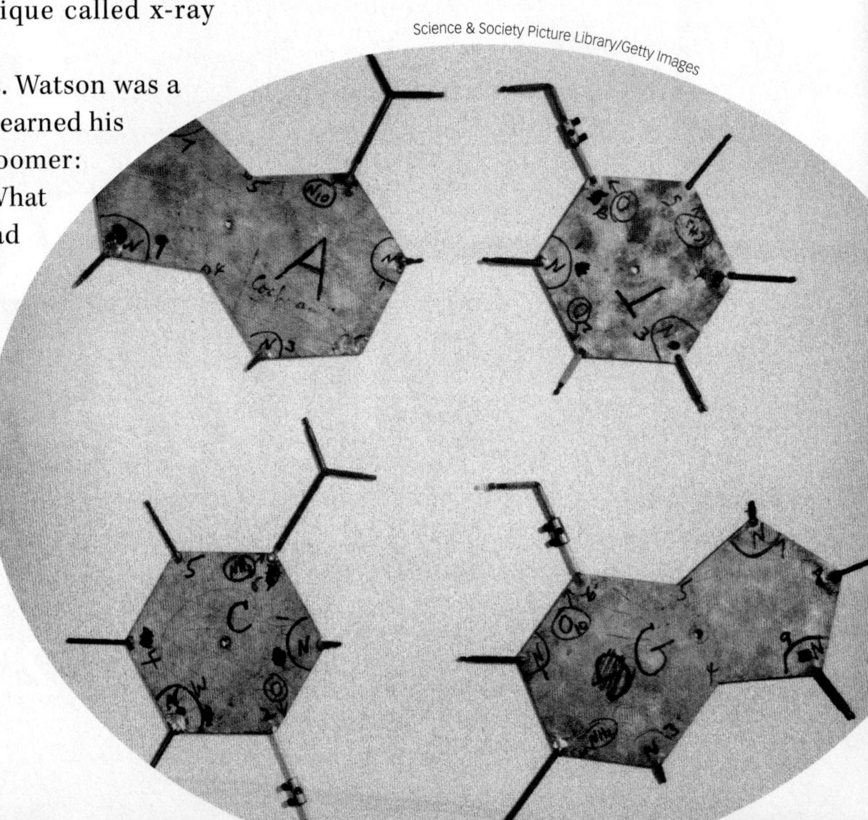

Science & Society Picture Library/Getty Images

The aluminum bases that Watson and Crick used in their DNA model.

INFOGRAPHIC M2.1

The DNA Puzzle

It was known that DNA was made of nucleotides that included a deoxyribose sugar, a phosphate group, and one of four nitrogenous bases. But no one had yet figured out how the nucleotides fit together to produce a DNA molecule.

Phosphate
Sugar
Adenine
Phosphate
Sugar
Cytosine
Phosphate
Sugar
Guanine
Phosphate
Sugar
Thymine

? What do all nucleotides have in common?

they hoped to—as Watson put it—"imitate Linus Pauling and beat him at his own game" **(INFOGRAPHIC M2.1)**.

Using wire and metal, Watson and Crick began building scale models of DNA on the basis of existing evidence about the chemical structure of nucleotides. They initially built a three-helix model with the phosphate groups on the inside and the bases radiating outward. Colleagues who analyzed the structure, however, deemed it chemically unstable.

Then came a crucial finding. In 1951, Watson attended a lecture by a 31-year-old scientist named Rosalind Franklin. In her laboratory at King's College London, she had been making x-ray diffraction pictures of DNA. The x-ray diffraction technique analyzes the way x-rays bounce off molecules to determine their chemical structure. Franklin had observed that increasing the humidity of DNA caused it to elongate. To Franklin, this suggested that water molecules were being attracted to the DNA molecule, coating it and causing it to stretch out. Since water is a polar molecule (see Chapter 2) that is attracted to charged (hydrophilic) molecules, Franklin realized that the charged, water-loving phosphate groups of DNA must therefore be on the *outside* of the structure—not, as others had suggested, on the inside.

Franklin's contribution didn't end there—she also discovered other important facts about the structure of DNA. Working with a graduate student, Raymond Gosling, she found that her x-ray diffractions confirmed that the elongated form of DNA had all the characteristics of a twisting helix **(INFOGRAPHIC M2.2)**.

Maurice Wilkins, who was Franklin's peer and worked in the same laboratory, was also studying DNA structure at the time. In early 1953, Wilkins saw Franklin's best unpublished x-ray picture of DNA and showed it to Watson without Franklin's knowledge. "The instant I saw the picture my mouth fell open and my pulse began to race," Watson recalled in his memoir of the discovery, *The Double Helix*, published in 1968. The sneak preview "gave several of the vital helical parameters."

With that clue in hand, Watson and Crick then took a crucial conceptual step and suggested that the molecule was made of two chains of nucleotides. Each formed a helix, as Franklin's data suggested, but because the DNA molecule was symmetrical—it looked the same when flipped upside down and backward—they realized that the two chains of nucleotides must be oriented in opposite directions.

To construct the model, Watson and Crick also built on a discovery made a few

INFOGRAPHIC M2.2

Rosalind Franklin and the Shape of DNA

Franklin's 1951 x-ray diffraction studies of DNA showed that the structure was likely helical, involving two strands that run in opposite directions, and that the phosphate groups were on the outside of the molecule.

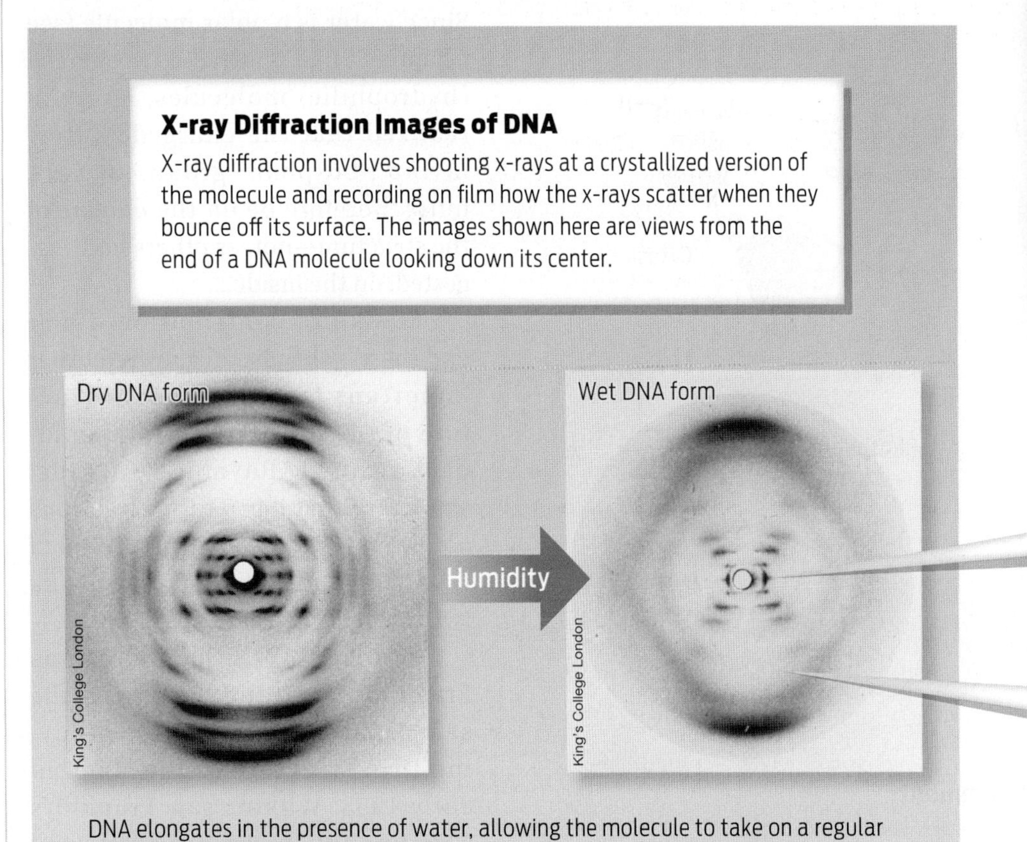

Rosalind Franklin
July 25, 1920–April 16, 1958

The signature "X" in the middle of this picture suggests a double-stranded helical structure.

The symmetry of the image suggests that the molecule is uniform in width.

? Which part of a nucleotide (the sugar, phosphate, or base) can interact with water?

months earlier. In 1952, Erwin Chargaff had found that any given DNA sample, no matter the organism, always contained equal amounts of adenine and thymine, and equal amounts of guanine and cytosine. This information was critical. As the helix had a smooth shape and a uniform thickness, and as the bases had to point toward the inside of the helix, the different-size bases somehow had to fit together in a way that allowed the helix to have a consistent width. Following a tip about the structure of bases, Watson and Crick were able to construct a model that showed that A–T pairs and G–C pairs were exactly the same width, explaining the consistent helix shape, and also accounting for Chargaff's finding that the amounts of A and T were equal to each other, as were the amounts of G and C **(INFOGRAPHIC M2.3)**.

The final double-helix model so perfectly fit the experimental data that the scientific community accepted it almost immediately. Watson and Crick published their paper on the structure of DNA in April 1953 in the prominent journal *Nature*. In it, with considerable understatement, they wrote: "It has

INFOGRAPHIC M2.3

Erwin Chargaff's Work Provided a Clue to Base Pairing

Erwin Chargaff studied the nitrogenous bases of DNA. He found that no matter which DNA molecule he analyzed, it always contained equal amounts of adenine and thymine bases and equal amounts of cytosine and guanine bases. Additionally, the width of an A–T pair is the same as the width of a C–G pair. These data suggest that in a double-stranded DNA molecule, adenine must pair with thymine and cytosine must pair with guanine. This pairing is consistent with a uniform base-pair width down the length of the DNA molecule, as Franklin's data suggested.

Chargaff's Rule of Base Pairing
Within any DNA molecule:
% adenine = % thymine
% cytosine = % guanine

Courtesy of Columbia University Medical Center

Erwin Chargaff
August 11, 1905–June 20, 2002

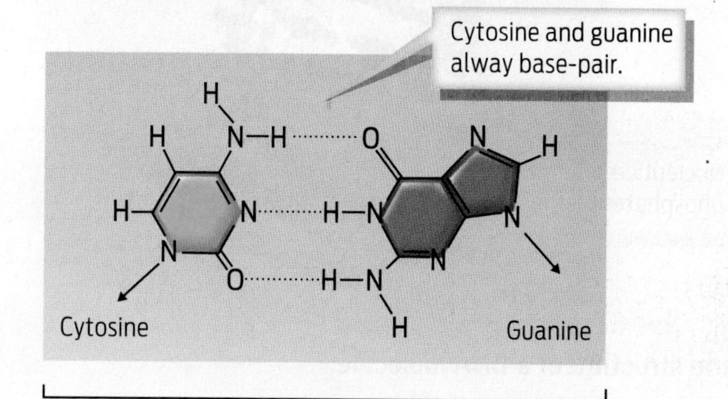

? **If a DNA molecule is 20% G, what is the %C? What is the %A?**

not escaped our notice that the specific pairing we have postulated immediately suggests a possible copying mechanism for the genetic material." Indeed, the model that Watson and Crick proposed solved at once both the structure of DNA and its mechanism of replication: each strand of an unzipped helix serves as the template for the creation of a complementary strand, thus reproducing the original pattern (see Chapter 7). The mystery of heredity had finally been unlocked.

Alongside their paper, in the same issue, individual papers by Wilkins and Franklin discussed their respective x-ray diffraction results, which supported the Watson–Crick model. In 1962, Watson, Crick, and Wilkins shared the Nobel Prize in Physiology or Medicine **(INFOGRAPHIC M2.4)**.

But what about Franklin? She had died of ovarian cancer in 1958, at the age of 37. Under the Nobel Foundation's rules, she was ineligible for nomination because prizes are not awarded posthumously.

Controversy over whether Franklin has been adequately recognized continues. Although Watson and Crick acknowledged her input in their article in *Nature*, the extent to which her x-ray pictures helped them build their DNA model was revealed only much later in Watson's 1968 book, published 10 years after Franklin's death. For example, at the time *Nature* published the papers on DNA

INFOGRAPHIC M2.4

The Structure Is Finally Known: The DNA Double Helix

The structure of DNA that Watson and Crick proposed fit all the experimental evidence—the x-ray crystallography data, the base-composition evidence, and the placement of the phosphates on the exterior of the double helix.

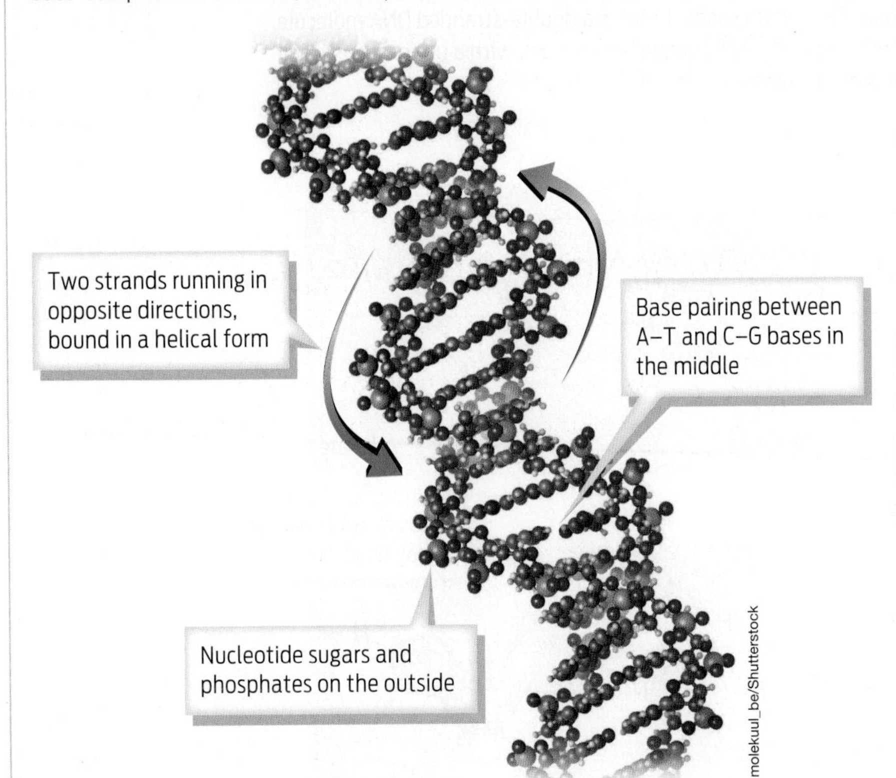

? Describe the structure of a DNA molecule.

structure, Franklin's paper was perceived as merely supporting evidence. But it was her data that helped Watson and Crick clinch the structure. Some historians argue that sexist attitudes prevented Franklin from receiving the acclaim she deserved before she died. At the time, female scientists in the biomedical sciences were few and were frequently confronted by negative attitudes from their male peers. "I'm afraid we always used to adopt—let's say, a patronizing attitude towards her," Crick publicly commented after Watson's book was published. He added that if Franklin had lived, "It would have been impossible to give the prize to Maurice [Wilkins] and not to her" because "she did the key experimental work."

Although it is quite normal for colleagues to share data, some have even argued that Wilkins showed Franklin's critical x-ray diffraction photos to Watson out of jealousy or disdain. But despite controversy, Franklin's contribution to the discovery has never been completely ignored, and she is now recognized as having been a top-notch scientist. Her notebooks show that without her thorough scientific research and original ideas, we would have had to wait much longer for what is still considered to be one of the most important discoveries in biology. ■

"It would have been impossible to give the prize to Maurice [Wilkins] and not to [Franklin because] she did the key experimental work."

—Francis Crick

More to Explore

- Watson, J. D., and Crick, F. H. C. (1953). A structure for deoxyribose nucleic acid. *Nature* 171:737–738.
- Watson, J. D. (2001 [1968]). *The Double Helix: A Personal Account of the Discovery of the Structure of DNA.* New York: Touchstone Books.
- Maddox, B. (2003). *Rosalind Franklin: The Dark Lady of DNA.* New York: Harper Perennial.
- American Masters: Decoding Watson (2019): https://www.pbs.org/video/decoding-watson-ua6jjx/

MILESTONES IN BIOLOGY 2 Test Your Knowledge

1. What technique did Rosalind Franklin use to examine the structure of DNA?

a. mass spectrometry
b. gel electrophoresis
c. x-ray diffraction
d. model building
e. all of the above

2. Which of the following statements about DNA structure is *not* true?

a. DNA is a double helix.
b. The phosphate groups are on the outside of the helix.
c. The two strands run in opposite orientations.
d. A pairs with A, T with T, C with C, and G with G.
e. The helix has a constant diameter along its length.

3. Describe the scientific contributions of Watson, Crick, Franklin, and Wilkins to the discovery of the structure of DNA.

4. How did differing amounts of water in the DNA crystals help explain the x-ray diffraction patterns?

5. Summarize the structure of a DNA double helix.

8 Genes to Proteins

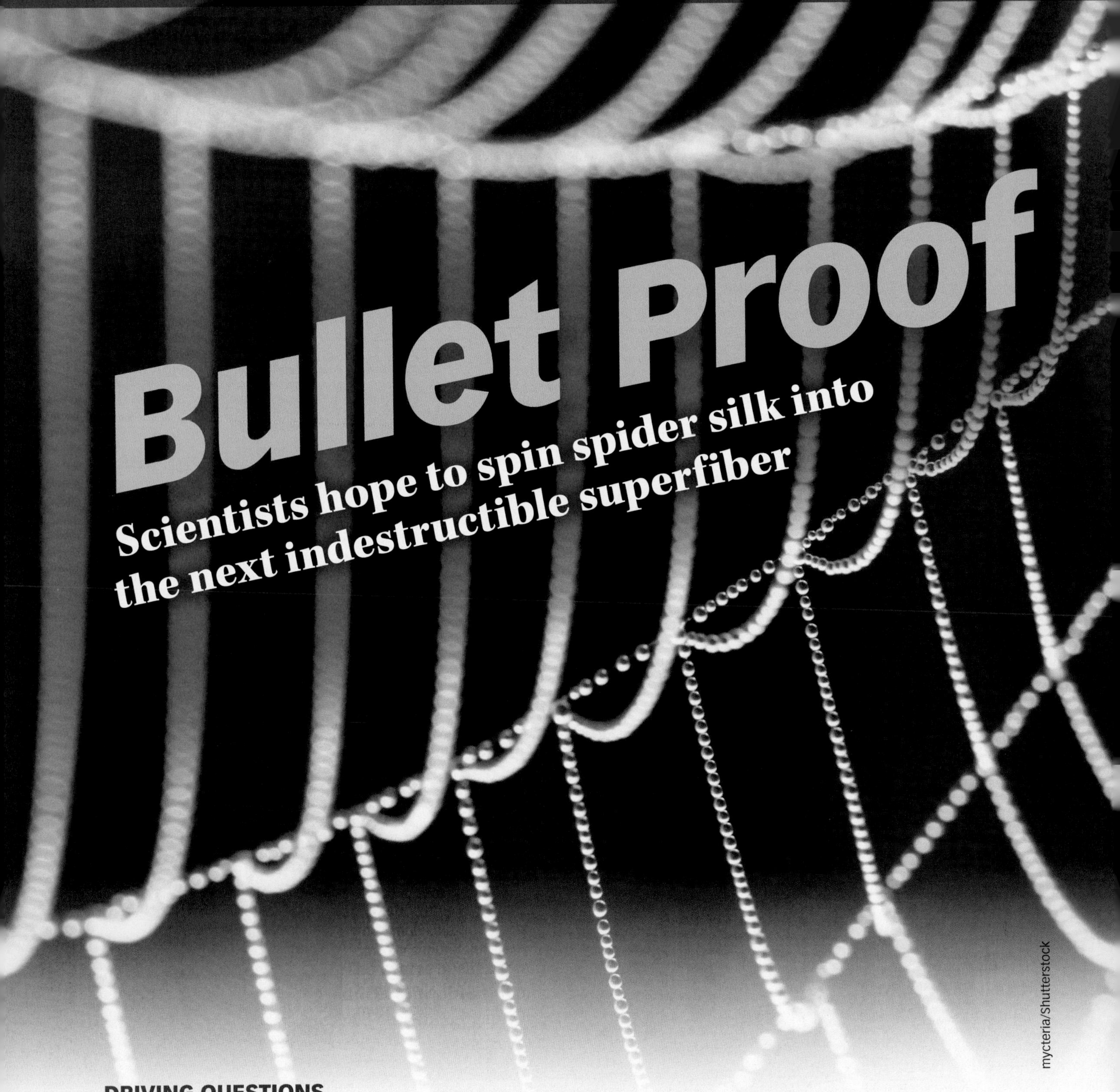

mycteria/Shutterstock

Bullet Proof

Scientists hope to spin spider silk into the next indestructible superfiber

DRIVING QUESTIONS

1. What determines the shape of a protein molecule, and why is its shape important?
2. What are the steps of gene expression, and where in the cell do they occur?
3. How can organisms be genetically modified to produce recombinant proteins?
4. What are some pros and cons of genetically modified organisms?

"On an equal weight basis, spider silk has a higher toughness than steel and Kevlar."

—David Kaplan

Peter Parker's spidey sense is tingling from the latest news out of biotech: genetically engineered spider silk, produced from spider DNA but assembled inside an entirely different organism.

For more than two decades, scientists have sought ways to harness the unique properties of spider silk—a near-miracle fiber that is pound for pound stronger than steel but also lightweight and flexible. Yet harvesting silk from spiders is challenging.

"Spiders don't like neighbors," says David Kaplan, a professor of biomedical engineering at Tufts University. "They are territorial and cannibalistic."

This makes them much harder to work with than, say, silkworms—another source of silk fibers. "If I put a hundred silkworms together on a table with enough food, I can come back in 30 days and find a hundred cocoons," Kaplan explains. "If I put a hundred spiders together with enough food and come back 30 days later, there'd be only one spider."

The other approach to obtaining spider silk—collecting it from webs in the wild—is exceedingly tedious. It would take about 1 million spider webs to make a single spider silk garment. Which is why, recently, scientists have turned to putting spider DNA into other, more congenial organisms.

The potential applications of genetically engineered spider silk are vast. They include more-durable, environmentally friendly clothing; safer medical products such as biocompatible bone screws and sutures that are not toxic and are not rejected by the body; and lightweight military vests that deflect bullets better than Kevlar. While some of these products are years away, others debuted in 2016.

Stronger than Steel

▸ Protein structure and function

If you think about it, a spider's web is, in essence, a device for stopping a speeding projectile—such as a flying insect. It's no wonder, then, that the fibers making up the web are uncommonly strong **(TABLE 8.1)**.

"On an equal weight basis, spider silk has a higher toughness than steel and Kevlar," says Kaplan, whose lab is using genetically engineered spider silks to build better medical devices. But it's also flexible and elastic, which allows it to absorb the energy of whatever hits it. A rope of spider silk the diameter of a few pencils could theoretically stop a jet landing on an aircraft carrier, though no one has tried that yet **(INFOGRAPHIC 8.1)**.

The physical properties of spider silk reflect the structure of the **proteins** making

PROTEIN
A macromolecule made up of repeating subunits called amino acids, which determine the shape and function of a protein. Proteins play many critical roles in living organisms.

TABLE 8.1 Mechanical Properties of Spider Silk

Spider silk exhibits a unique combination of strength and elasticity, enabling silk fibers to absorb a lot of energy before breaking (toughness).

Material	**Strength** (MPa) Spider silk is as strong as high-grade alloy steel and about half as strong as Kevlar.	**Extensibility** (%) Silk is extremely elastic, and some silks are able to stretch up to four times their relaxed length.	**Toughness** (MJ/m^3) The combination of strength and extensibility make silks tougher than many engineered polymers like Kevlar.
Spider dragline silk	880–1,500	21%–27%	136–194
Silkworm silk	600	18%	70
Nylon	950	18%	80
Kevlar 49 fiber	3,600	2.7%	50
High-tensile steel	1,500	0.8%	6

Data from Anna Rising and Jan Johansson. 2015. *Nature Chem Biol*. 11:309–315. Lin Römer and Thomas Scheibel. 2008. *Prion* 2(4):154–161.

it up. Recall from Chapter 2 that proteins are one of the four main macromolecules that make up cells, along with carbohydrates, nucleic acids, and lipids. Proteins are the cell's workhorse molecules. They perform myriad functions inside cells, and in that way help our bodies perform countless tasks—everything from contracting our muscles and sensing light to regulating blood sugar and fighting infections. Spiders use their silk proteins to build webs for trapping prey, sacs to protect eggs, and bungee cord–like draglines that catch them when they fall.

Tina Carvalho/MicroAngela

A colored scanning electron micrograph (SEM) of silk fibers emerging from a spider.

All proteins are made of the same building blocks, which are called **amino acids.** There are 20 different amino acids in cells. All amino acids have the same basic core structure, but each of the 20 also has a unique chemical side chain that distinguishes it from all the others. To form proteins, amino acids link together in linear chains. Spider silk proteins, called spidroins, are chains of about 3,500 amino acids. This is longer than many animal proteins, which average around 400 amino acids in length. But proteins can be much longer or much shorter. The longest human protein, titin (involved in muscle contraction), is a single chain of 34,350 amino acids. By comparison, insulin, a protein that helps regulate blood sugar, has only 51 amino acids.

In cells, the amino acid chain of a protein folds into a distinct three-dimensional shape, or conformation, that underlies a protein's function. Some proteins, like insulin,

INFOGRAPHIC 8.1

Spider Silk Characteristics and Applications

Spider silk is tough, flexible, and elastic, making it well suited for military and industrial uses. It is also biocompatible, making it ideal for a variety of surgical applications, and better for the environment, too.

Bulletproof vests

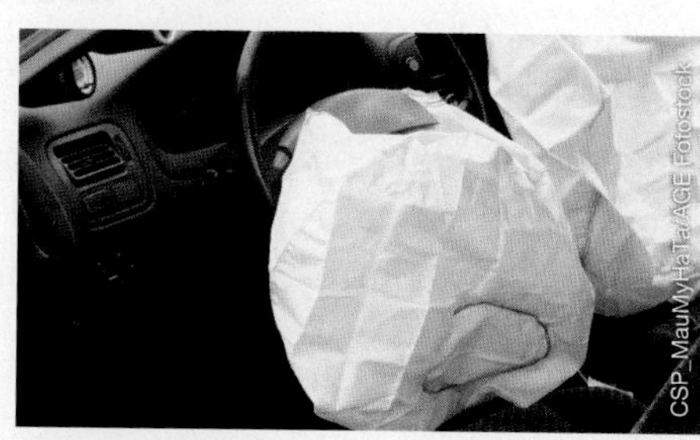

Strong, elastic airbags

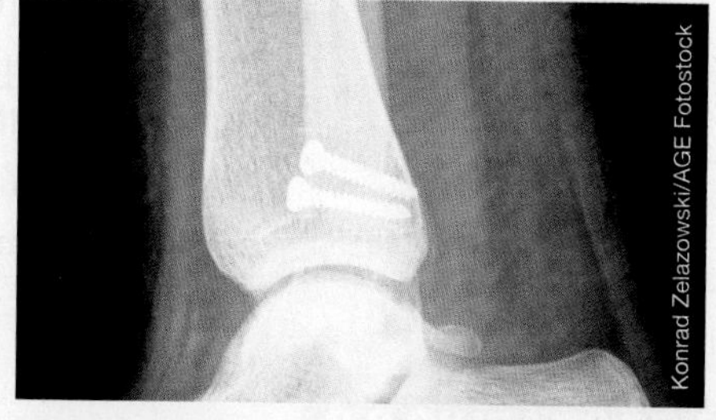

Biocompatible and biodegradable screws and sutures

Stronger skin grafts

Durable ecotextiles

Scaffolds for growing tissues

? Why are both strength and elasticity important for surgical sutures?

are made up of just one folded chain. Other proteins consist of multiple folded amino acid chains bound together—examples are the antibodies of our immune system and the hemoglobin that carries oxygen in our red blood cells. A spider silk thread is made up of many spidroin proteins linked and packed together to form a fiber.

The particular sequence of amino acids in a chain determines how the chain will fold. Interactions between amino acid side chains, and between these side chains and the surrounding water, influence the precise folding pattern. Hydrophobic amino acid side chains tend to clump together, away from water, while hydrophilic amino acids face out toward water. The distinct three-dimensional shape that forms as a result is what ultimately determines how a protein functions.

According to Kaplan, the first thing you'd notice about the amino acid sequence of a spider silk protein is its repetitive nature. Most of the protein—roughly 90%—is composed of repeated sequences of relatively hydrophobic amino acids. These repetitive sequences are responsible for the toughness of spider silks. Regions rich in the hydrophobic amino acid alanine pack closely together, away from water, and form flat, interlocking structures called beta sheets (a folded structure found in many proteins but present at a much higher frequency in silk). These regions impart strength to the protein. Other regions, rich in the amino acid glycine, form more flexible parts of the protein, which confer elasticity. Nonrepetitive regions that flank these core repeated sequences are made of charged amino acids. These hydrophilic end regions keep the spidroin proteins dissolved in the silk gland's watery environment and prevent silk proteins from crystalizing spontaneously

AMINO ACIDS
The building blocks of proteins. There are 20 different amino acids.

into fibers inside the spider, which would kill the animal **(INFOGRAPHIC 8.2)**.

Silk proteins lack the structural complexity you would see in an enzyme or an antibody. An enzyme, for example, might have a deep pocket into which other molecules fit like pieces of a puzzle. By bringing these molecules together in close proximity, the enzyme facilitates a chemical reaction between them (Chapter 4). By contrast, a silk protein is a much more uniform structure that resembles a synthetic polymer such as nylon or polyester.

Molecular Recipes

▸ Genes and gene expression

Where do spiders get the information to build such unique proteins? As with all organisms, the instructions to make proteins

INFOGRAPHIC 8.2

Amino Acid Sequence Determines Protein Shape and Function

The sequence of amino acids determines how a chain will fold into a three-dimensional shape and potentially interact with other chains to establish the final shape (and function) of that protein.

Amino Acid Structure

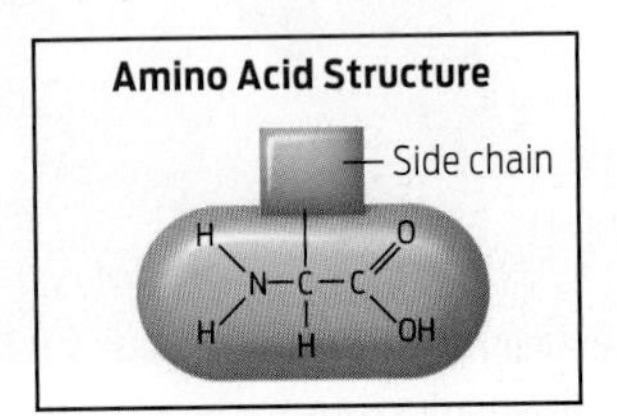

The repeated sequences are important for determining how different regions of the protein fold.

Protein ends are important for stability in water, secretion from the cell, and assembly into long, silk-fiber polymers.

Spidroin linear structure
3,000–4,000 amino acids long

Repeated amino acid sequences

1. Linear Amino Acid Chain
Amino acids bind together in linear chains. In this linear form, a chain of amino acids does not yet have a specific function.

Poly-Alanine sequences

Glycine-rich sequences

ala ala ala ala ala ala ala ala ala gly gly gln gly gln gly gly tyr gly

2. Three-Dimensional Protein Folding
Interactions among amino acid core structures, among amino acid side chains, and between side chains and water all direct three-dimensional folding. The overall shape of the protein, including placement of its side chains, determines its ultimate function.

Spidroin Three-Dimensional Protein
(Beta-sheet crystals)

Poly-Alanine sequences

Glycine-rich sequences

Silk Fiber Polymer

Spidroin beta-sheet crystals

Flexible linking regions

3. Polymer Assembly
Within the protein, beta-sheet crystalline regions, which confer strength, alternate with more flexible regions, which confer flexibility. Overall, the protein is strong and flexible.

? Leucine is a hydrophobic amino acid. Where would you expect to find it: interacting with other hydrophobic amino acids in the interior of a protein, or on the surface of a protein, interacting with water?

INFOGRAPHIC 8.3

Chromosomes Include Gene Sequences that Code for Proteins

Chromosomes have many genes along their length. Each gene contains instructions to make at least one protein. Depending on the needs of the cell at any given time, each gene may be expressed (making protein) or silenced.

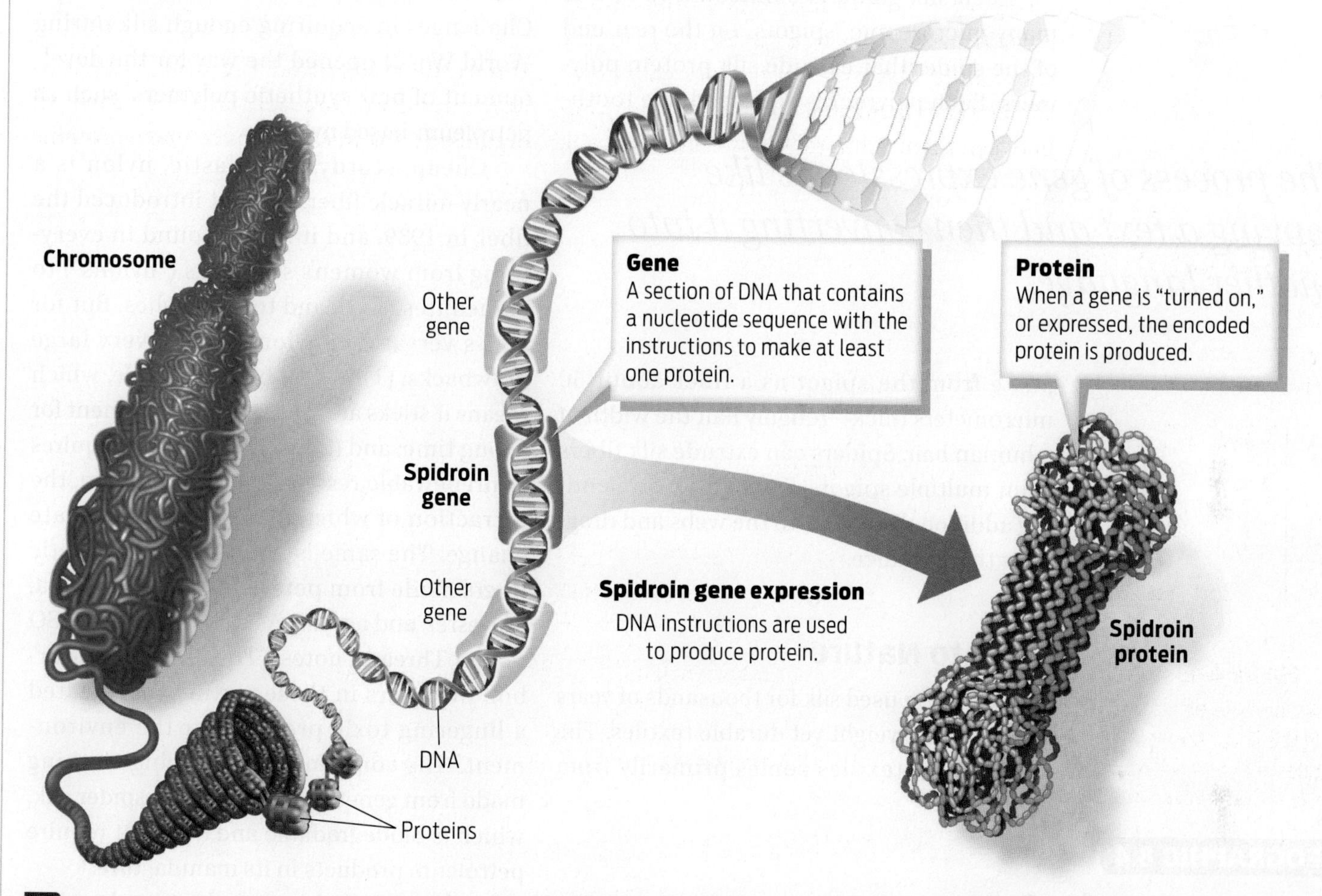

? A cell has a high concentration of protein A, but no detectable protein B. Does this mean that the cell does not have the gene encoding protein B? Explain your answer.

are encoded in the DNA, in genes. A **gene** is a sequence of DNA that provides instructions for making one or more proteins. These instructions come in the form of the particular DNA nucleotide sequence making up the gene. Genes are found along the length of chromosomes, with each specific chromosome carrying a unique set of genes.

The synthesis of a protein from the information encoded in a gene is called **gene expression.** When a cell makes the protein encoded by that gene, the gene is said to be "expressed" **(INFOGRAPHIC 8.3).**

Spiders don't express the silk gene in every cell in their body, just in the cells of their silk glands, located in their abdomen. You might wonder how this gene, which is present in every cell in the spider's body, can be turned on only in some cells. It's because every gene has two parts: a regulatory sequence and a coding sequence. A **regulatory sequence** is like an on–off switch for a gene: it determines when, where, and how much protein is produced from a gene. A **coding sequence** is the sequence of nucleotides in a gene that determine the identity of a protein: it specifies the

GENE
A sequence of DNA that contains the information to make at least one protein.

GENE EXPRESSION
The process of using DNA instructionas to make proteins.

REGULATORY SEQUENCE
The part of a gene that determines the timing, amount, and location of protein production.

CODING SEQUENCE
The part of a gene that specifies the amino acid sequence of a protein. Coding sequences determine the identity, shape, and function of proteins.

order, or sequence, of amino acids. The regulatory sequence of the silk gene specifies that it should be expressed only in a silk gland. That means the silk protein is produced only in these cells **(INFOGRAPHIC 8.4)**.

Each silk gland is connected to one of many microscopic "spigots" on the rear end of the spider that extrude silk protein polymers. Each polymer is squeezed like toothpaste from the spigot as a fiber about 50 micrometers thick—roughly half the width of a human hair. Spiders can extrude silk fibers from multiple spigots simultaneously, lending additional strength to the webs and draglines they produce.

The process of gene expression is like copying a text and then converting it into another language.

Back to Nature

Humans have used silk for thousands of years to make lightweight yet durable textiles. The silk in these textiles comes primarily from silkworms—the larvae of silk moths. During their development, silkworms produce protective cocoons made of silk, which humans can then harvest. Silk has been used to make clothing, bandages, and medical sutures, as well as military parachutes and flak vests. Challenges in acquiring enough silk during World War II opened the way for the development of new synthetic polymers, such as petroleum-based nylon.

Cheap, sturdy, and elastic, nylon is a nearly miracle fiber. DuPont introduced the fiber in 1939, and it is now found in everything from women's stockings ("nylons") to parachutes, tires, and toothbrushes. But for all its versatility, nylon has two very large drawbacks: (1) it's not biodegradable, which means it sticks around in the environment for a long time; and (2) its manufacture requires nonrenewable resources like petroleum, the extraction of which contributes to climate change. The same is true of other synthetic fibers made from petroleum, such as Lycra, polyester, and acrylic. As Dan Widmaier, CEO of Bolt Threads, notes, "The use of hydrocarbon polymers in these textiles has created a lingering toxic problem for the environment." His company is developing clothing made from genetically engineered spider silk, which is biodegradable and does not require petroleum products in its manufacture.

A former graduate student in chemical biology at the University of California, San Francisco, Widmaier started Bolt Threads in 2009 with the help of three friends and fellow scientists. Their original goal was to make lightweight bulletproof vests that would appeal to the defense industry. But Widmaier—whose wife is a fashion designer at Old Navy—eventually decided to shift the company's focus to the consumer textile industry, which represents a much larger market and has a much greater impact on the environment.

To understand how companies like Bolt are creating genetically engineered spider silk, it helps to know how spiders do it naturally.

INFOGRAPHIC 8.4

The Two Parts of a Gene

Genes are organized into two parts. Regulatory sequences determine when a protein is made from a gene and in which cells, and how much protein a gene makes. Coding sequences determine the amino acid sequence of the encoded protein, which determines its shape and function.

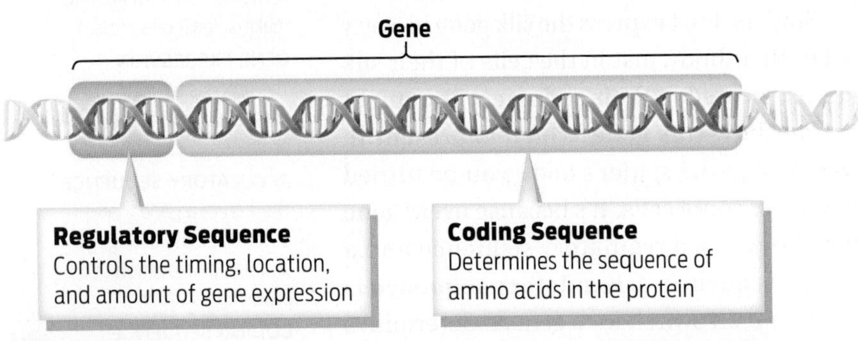

? Why does a milk protein gene (expressed in mammary glands) have both a different regulatory sequence and a different coding sequence from the insulin gene (expressed in the pancreas)?

Making Proteins, or How Genes Are Expressed

▸ Transcription, translation, and the genetic code

Inside a spider's silk gland are cells that assemble the spidroin protein from the information encoded in the spidroin gene. To get from the spidroin gene to spidroin protein, the cells carry out two major steps: transcription and translation. **Transcription** is the process of using DNA to make a **messenger RNA (mRNA)** copy of the gene. Recall from Chapter 2 that RNA is another type of nucleic acid, but one that is single stranded rather than double stranded. **Translation** is the process of using this mRNA copy as a set of instructions to assemble amino acids into a protein **(INFOGRAPHIC 8.5)**.

Why two separate steps? As the names "transcription" and "translation" imply, the process of gene expression is like copying a text and then converting it into another language. In this case, the text to be translated

TRANSCRIPTION
The first stage of gene expression, during which cells produce molecules of messenger RNA (mRNA) from the instructions encoded within genes in DNA.

MESSENGER RNA (mRNA)
The RNA copy of an original DNA sequence made during transcription.

TRANSLATION
The second stage of gene expression, during which mRNA sequences are used to assemble the corresponding amino acids to make a protein.

INFOGRAPHIC 8.5

Gene Expression: An Overview

Gene expression is the process of converting the genetic information of DNA into the amino acid sequence of a protein. Gene expression has two main steps: transcription and translation.

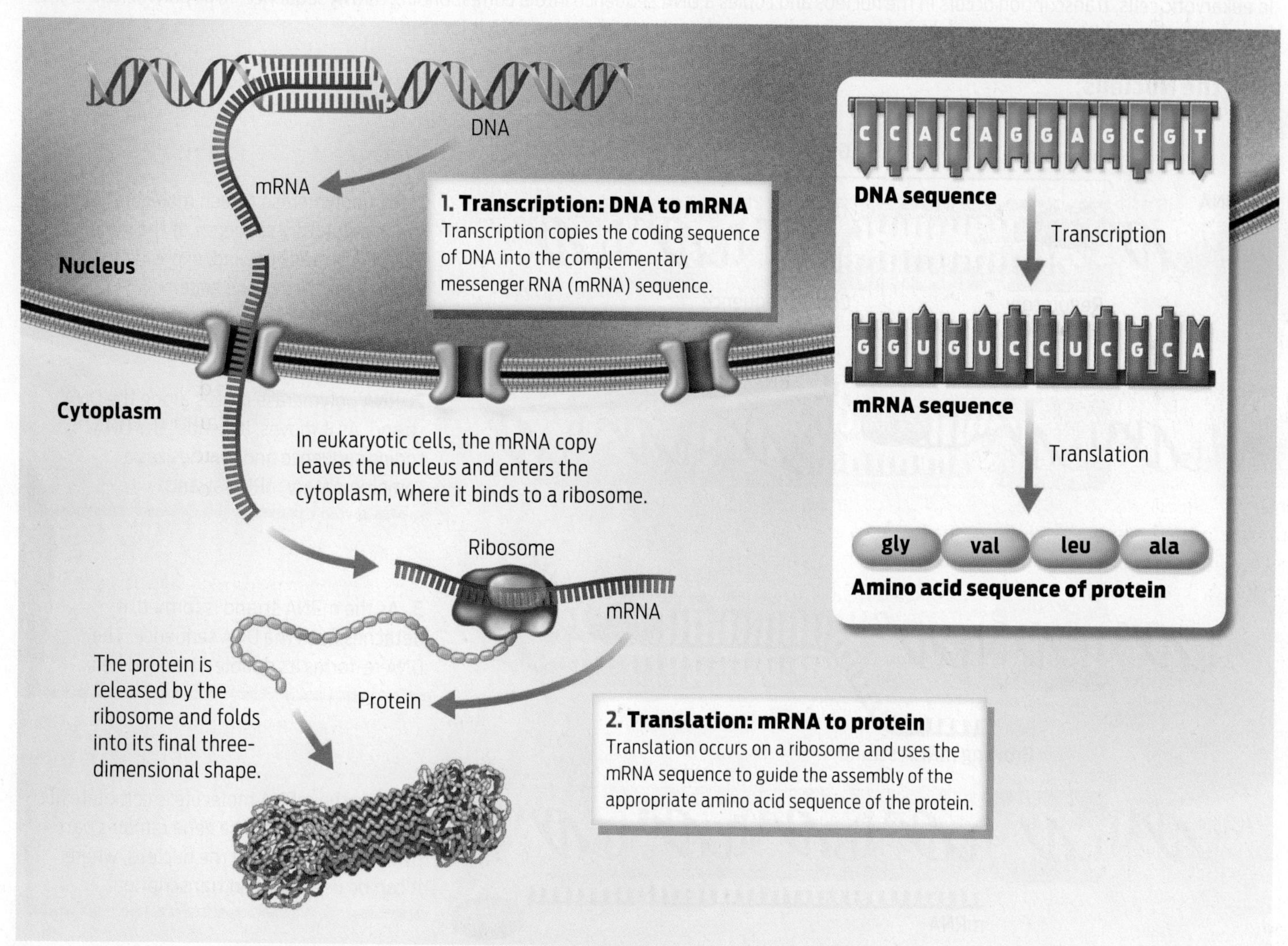

? **What are the products of transcription, and what are the products of translation?**

is a valuable, one-of-a-kind document: DNA. Just as you would be forbidden to borrow a rare manuscript from the library at school and would instead have to copy the text into your notebook or laptop, the cell cannot take DNA out of its "library"—the nucleus. Instead, it must first make a copy—the mRNA. The cell can then take this mRNA copy into the cytoplasm, where it is translated into a new language: protein.

Transcription begins in the nucleus of a cell when an enzyme called **RNA polymerase** binds to DNA at a gene's regulatory sequence, located just ahead of the coding sequence. At that site, RNA polymerase separates the two strands of the DNA double helix and begins moving along one DNA strand. As it moves, the RNA polymerase reads the DNA sequence and synthesizes a complementary mRNA strand according to the rules of complementary base pairing. The same rules that govern DNA base pairing apply here, with one difference: RNA nucleotides are made with the base uracil (U) instead of thymine (T). So the complementary base pairs are C with G and A with U **(INFOGRAPHIC 8.6)**.

RNA POLYMERASE
The enzyme that carries out transcription. RNA polymerase copies a strand of DNA into a complementary strand of mRNA.

INFOGRAPHIC 8.6

Transcription: A Closer Look

In eukaryotic cells, transcription occurs in the nucleus and copies a DNA sequence into a corresponding mRNA sequence. RNA polymerase is the key enzyme involved. In prokaryotic cells, transcription occurs in the cytoplasm, where DNA is located.

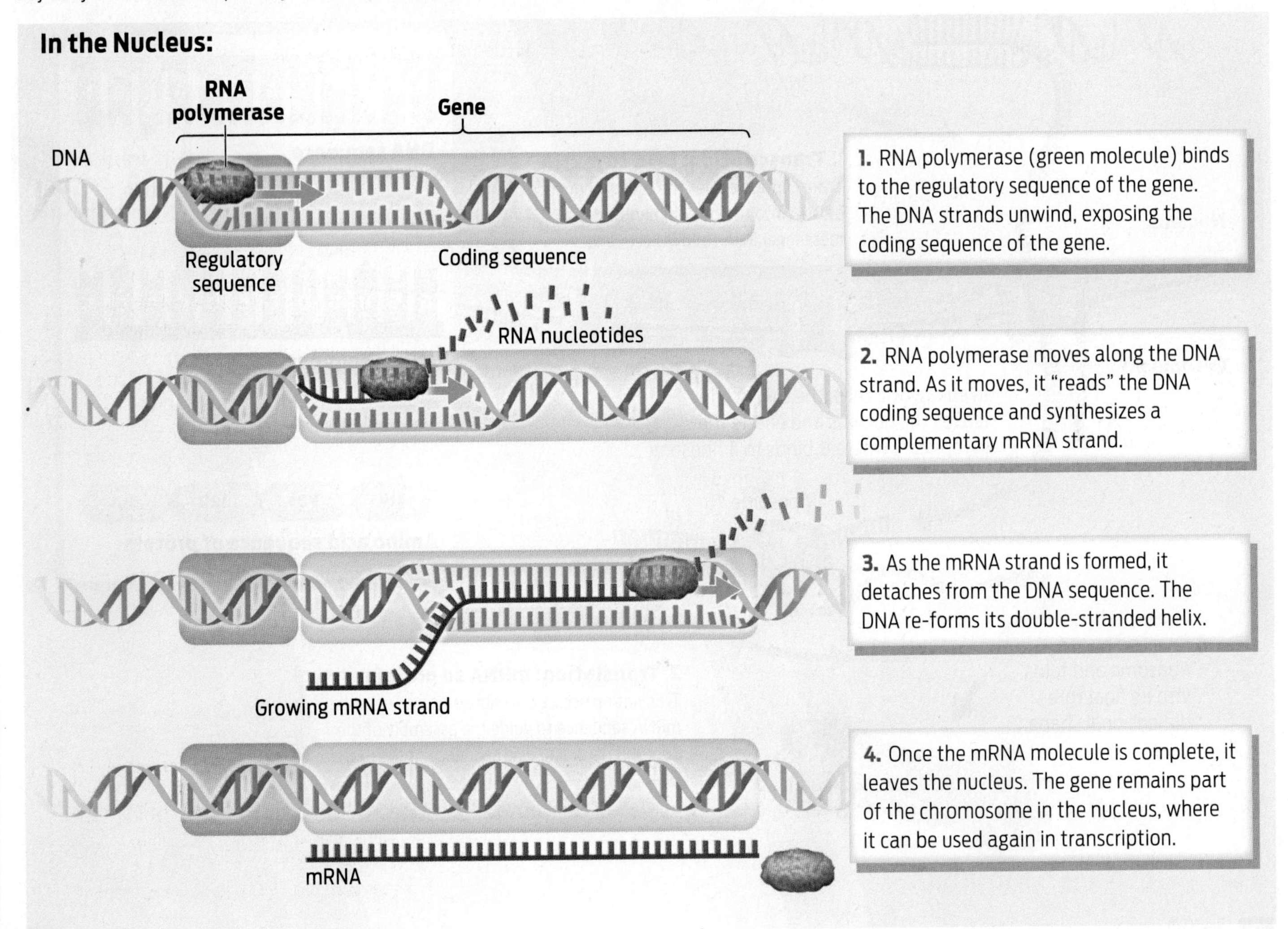

? What enzyme is responsible for transcription, and what, specifically, does it do?

As its name implies, messenger RNA serves to relay information. Once the mRNA copy, or transcript, is made, it leaves the nucleus and attaches to a complex piece of cellular machinery in the cytoplasm called the ribosome. This event marks the start of translation.

During translation, the **ribosome** reads the mRNA transcript and translates it into a chain of amino acids. The sequence of nucleotides in the mRNA transcript specifies which amino acids should be joined together in the newly forming protein chain. Each of the 20 amino acids that occur in proteins is specified by a group of three mRNA nucleotides called a **codon** that functions like a word: for example, the codon GGU specifies the amino acid glycine.

The actual building blocks of proteins—amino acids—are physically delivered to the ribosome by another type of RNA, called **transfer RNA (tRNA).** Each tRNA molecule serves as a kind of adaptor, with one end binding to a specific amino acid and the other end binding to the mRNA codon for that particular amino acid. The part that binds mRNA is called the **anticodon** because it base-pairs in a complementary fashion with the mRNA codon. When the amino acid–toting tRNA finds its mRNA codon match, the ribosome adds the amino acid to the growing protein chain **(INFOGRAPHIC 8.7).**

RIBOSOME
The cellular machinery that assembles proteins during translation.

CODON
A sequence of three mRNA nucleotides that specifies a particular amino acid.

TRANSFER RNA (tRNA)
A type of RNA that transports amino acids to the ribosome during translation.

ANTICODON
The part of a tRNA molecule that binds to a complementary mRNA codon.

INFOGRAPHIC 8.7

Translation: A Closer Look

In the cytoplasm, the ribosome reads the mRNA sequence and translates it into a chain of amino acids to make a protein.

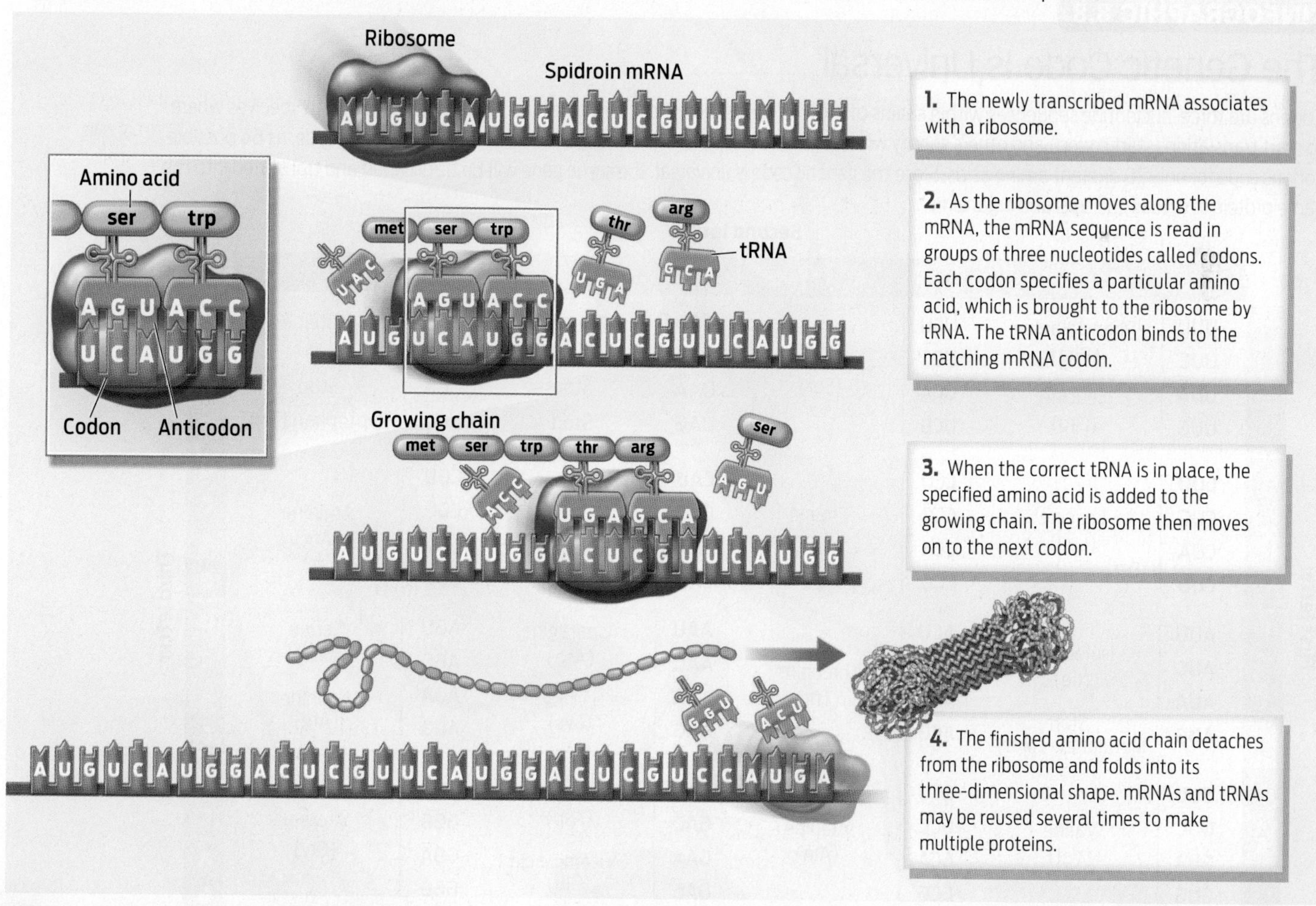

? What molecule transports amino acids to the growing protein? How does it match a specific amino acid to the corresponding mRNA codon?

The vast majority of mRNA codons specify a specific amino acid, but a few have other functions. The "start codon," which in eukaryotes codes for the amino acid methionine, is the first codon of a coding sequence; it tells the ribosome to start translating and add amino acids. "Stop codons" (there are three) tell the ribosome to stop translating and not add any more amino acids to the growing chain.

Although the human genome encodes many thousands of different proteins, each one is pieced together from the starting set of just 20 amino acids. In the same way that the 26 letters in our alphabet can spell hundreds of thousands of words, the basic set of amino acids can make hundreds of thousands of proteins. The set of rules dictating which mRNA codons specify which amino acid is called the **genetic code.** Scientists have pieced together this code by systematically studying how changes to the letters of a codon alter the specified amino acid, so that we now know what amino acid each codon stands for. Each codon specifies one and only amino acid.

GENETIC CODE
The set of rules relating particular mRNA codons to particular amino acids.

Two additional features of the genetic code stand out. First, the code is redundant: multiple codons specify the same amino acid. In many cases, a codon will differ at the third nucleotide position without changing the amino acid that is specified. (Note that while the code is redundant, it is not ambiguous—the same codon will not specify more than one amino acid.) Second, the genetic code is universal, which means that it is the same in all living organisms. This shared nature is why other organisms carrying a spider gene can express that gene and produce spider silk **(INFOGRAPHIC 8.8).**

INFOGRAPHIC 8.8

The Genetic Code Is Universal

Codons are three-nucleotide sequences within chains of mRNA. Most codons specify a particular amino acid. One codon speciÿes where to start translation (start codon) and others specify where to end it (stop codons). There is redundancy in the genetic code, as 64 possible codons code for only 20 di‹erent amino acids. Since the genetic code is universal, the same gene will be transcribed and translated into the same protein in virtually all cells and organisms.

Second letter

First letter	U		C		A		G		Third letter
U	UUU	Phenylalanine (Phe)	UCU		UAU	Tyrosine (Tyr)	UGU	Cysteine (Cys)	U
	UUC		UCC	Serine (Ser)	UAC		UGC		C
	UUA	Leucine (Leu)	UCA		UAA	Stop	UGA	Stop	A
	UUG		UCG		UAG	Stop	UGG	Tryptophan (Trp)	G
C	CUU		CCU		CAU	Histidine (His)	CGU		U
	CUC	Leucine (Leu)	CCC	Proline (Pro)	CAC		CGC	Arginine (Arg)	C
	CUA		CCA		CAA	Glutamine (Gln)	CGA		A
	CUG		CCG		CAG		CGG		G
A	AUU		ACU		AAU	Asparagine (Asn)	AGU	Serine (Ser)	U
	AUC	Isoleucine (Ile)	ACC		AAC		AGC		C
	AUA		ACA	Threonine (Thr)	AAA	Lysine (Lys)	AGA	Arginine (Arg)	A
	AUG	Start Methionine (Met)	ACG		AAG		AGG		G
G	GUU		GCU		GAU	Aspartic acid (Asp)	GGU		U
	GUC	Valine (Val)	GCC	Alanine (Ala)	GAC		GGC	Glycine (Gly)	C
	GUA		GCA		GAA	Glutamic acid (Glu)	GGA		A
	GUG		GCG		GAG		GGG		G

? What is the sequence of the start codon, and what amino acid does it specify?

Spider Silk Factories

▶ Making transgenic organisms

Scientists have tried making genetically engineered spider silk in a variety of organisms, including bacteria, insects, plants, and even goats. (The goats can be engineered to produce the silk in their milk.) But each of these organisms has drawbacks. Silks are large proteins and can be tricky to produce in the prokaryotic cells of bacteria. Animals like goats reproduce slowly and take up a lot of space and resources. And it's hard to scale up from insects like silkworms.

Bolt Threads seeks to overcome these hurdles by relying on a different unicellular organism: yeast. Though perhaps better known for their role in making bread and beer, yeasts have several attributes that make them good silk producers. Specifically, they are eukaryotic cells, which means their protein production machinery resembles that of a spider (another eukaryote). But yeasts are easier to house and cheaper to feed than more complex multicellular animals like silkworms and goats. When fed a simple diet of sugar, they grow and divide plentifully. In the process, yeasts synthesize new proteins, including—if they have been engineered to contain the silk gene—an abundance of silk. By conducting the entire process in large industrial vats, scientists can make mountains of silk protein this way.

The process of making genetically engineered silk begins with isolating spider DNA from spider cells or synthesizing it from scratch. This bit of DNA is inserted into a cell of a new organism, which is then coaxed to make the protein encoded by the spider gene. Organisms that have been genetically modified to contain genes from other species are called **transgenic** (*trans* means "across"—in this case, across species, from one to another).

To make a transgenic yeast, for example, scientists first fuse the coding sequence of a spider silk gene to the regulatory sequence of a yeast gene. The resulting combination is called a **recombinant gene,** since it mixes and matches segments of genes that weren't naturally found together. Next, using **genetic engineering** techniques, which manipulate DNA, scientists insert the recombinant gene into a piece of DNA that can carry the recombinant gene into the yeast cell, and ultimately into a yeast chromosome. The carrier DNA molecule is called a **vector.** The final step is gene expression, when the yeast ribosome reads the instructions in the recombinant gene and translates it into spider silk protein **(INFOGRAPHIC 8.9).**

As the yeast grow, they make abundant quantities of silk protein, which they secrete into the surrounding culture media. The scientists then harvest these proteins. Once the silk proteins are in hand, the next step in the manufacturing process is spinning them into fibers, using a process that mimics what happens naturally inside spiders. The fibers are then collected and can be woven together to make fabrics.

By tweaking the specific sequence of nucleotides in the spider genes they introduce into yeast, scientists can produce spider proteins with unique amino acid sequences and

Moon Parka, a prototype created by Spiber and The North Face Japan, is the first apparel product to integrate synthetic spider silk with existing industrial manufacturing technology. It is made using Spiber's Qmonos fiber.

Spiber, Inc.

The Adidas Biosteel shoe.

TRANSGENIC
Refers to an organism that carries one or more genes from a different species.

RECOMBINANT GENE
A genetically engineered gene that contains portions of genes not naturally found together.

GENETIC ENGINEERING
Altering or manipulating the DNA of organisms by modern laboratory techniques.

VECTOR
A DNA molecule used to deliver a recombinant gene to a host cell.

INFOGRAPHIC 8.9

Making a Transgenic Organism

Transgenic organisms contain genes from other organisms. For the foreign gene to be expressed in the new host, it needs to be modified. The modified recombinant gene contains a regulatory sequence from the host organism and the coding sequence from the gene of interest. The recombinant gene is then inserted into and expressed in the host.

1. Create a Recombinant Gene

The yeast regulatory sequence and spider spidroin coding sequence are cut out of donor cell chromosomes and joined together using specialized enzymes.

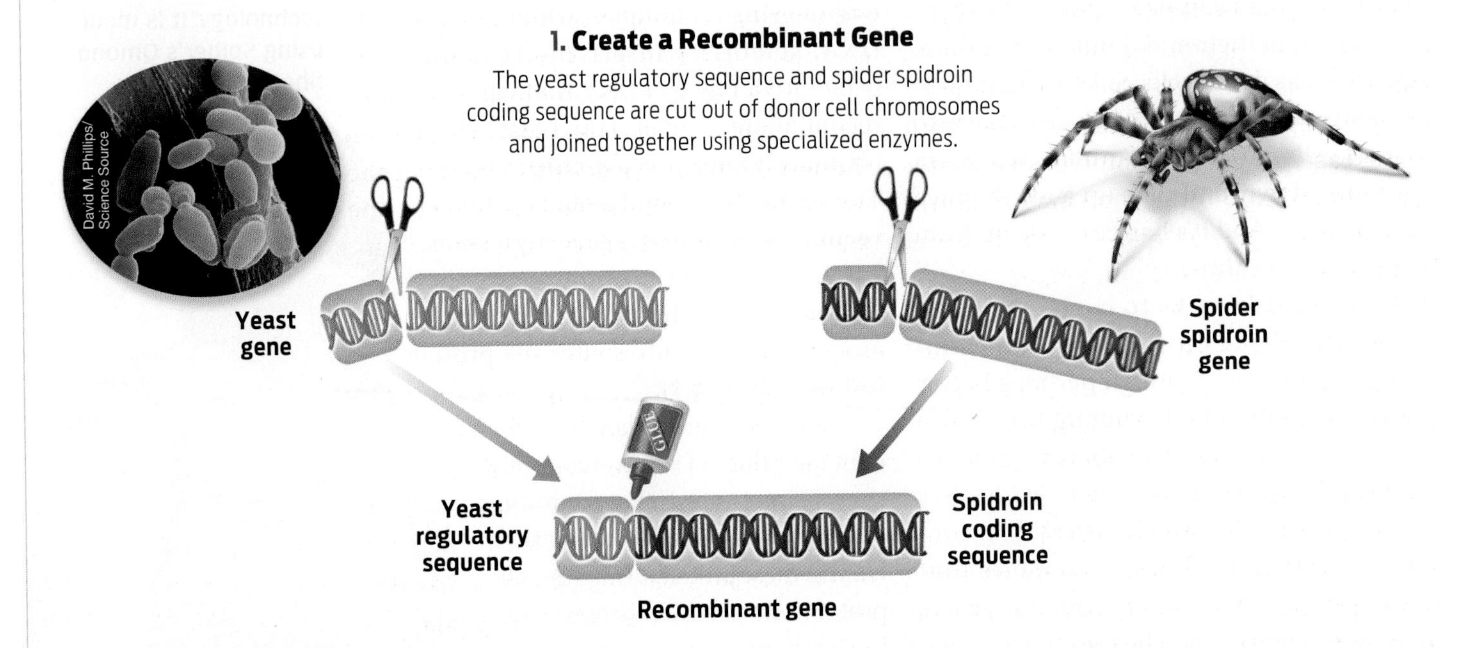

2. Insert the Recombinant Gene into a Yeast Chromosome

The recombinant gene is inserted into a small piece of DNA. This recombinant DNA is added to yeast nuclei, where it becomes part of a yeast chromosome.

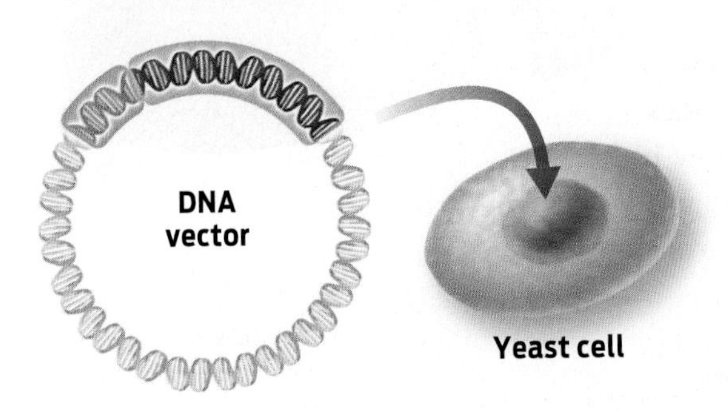

3. Spidroin Secretion and Harvest

Yeast cells are grown in a large tank under conditions that allow them to express and secrete spidroin protein.

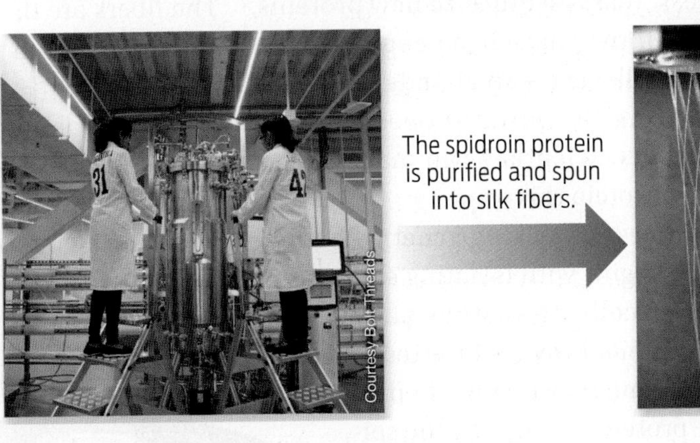

? Why did the scientists give the recombinant spidroin gene a yeast regulatory sequence?

therefore unique properties—perhaps silk that is sturdier than the native form, or stickier, or more elastic. With genetic engineering techniques, the possibilities for producing designer proteins are nearly endless.

Besides Bolt Threads, several other biotech companies are hoping to snag a piece of the spider silk action, including Spiber Inc. in Japan and Kraig Biocraft Laboratories in Ann Arbor, Michigan.

Brave New World?

▶ Challenges and opportunities of genetic engineering

In May 2016, Bolt Threads announced that it was partnering with clothing manufacturer Patagonia to develop a new line of eco-friendly clothing, which it plans to bring to market in the coming years. The Japan-based company Spiber, which has partnered with North Face, unveiled a prototype of a winter parka in 2016. And the shoe manufacturer Adidas recently announced the creation of a running shoe made with biodegradable silk fibers, dubbed Biosteel.

Yet these are just the tip of the iceberg when it comes to spider silk products in development. "Nature provides a great starting point and then once we're outside of the spider we can expand the set of materials and uses," says bioengineer Kaplan. His lab at Tufts University is focused on medical applications—like genetically engineered spider fibers that can substitute for bone and ligaments.

Other researchers have their eyes set on industrial uses, like a less bruising material for car air bags. Eventually, the goal is to make super-strong products that substitute spider silk for Kevlar, such as in bulletproof vests. Even superhuman tissues might not be out of the question. In 2013, Dutch researchers grew human skin cells together with spider silk proteins to make what they call "bulletproof skin," which can stop a bullet traveling at half the normal speed of a typical bullet.

Although genetically engineered spider silks may capture headlines, other uses of genetic engineering are already common in daily life. Much of the corn we eat today is transgenic, as are the soybeans that we feed to farm animals. Transgenic organisms are examples of **genetically modified organisms (GMOs)**—organisms whose genomes have been altered through modern genetic engineering techniques, sometimes to contain new genes. Transgenic crops such as corn and soybeans usually contain genes for natural pesticides, which help the plants fight pests and reduce the amount of pesticides a farmer must use (see Chapter 24).

Genetic engineering also has important medical applications. The drug insulin, used for treating diabetes, is commonly produced inside a genetically engineered bacterium—one into which the (human) insulin gene has been inserted. Through **gene therapy**—replacing a defective human gene with a healthy one—scientists hope to one day be able to treat, cure, or even prevent several inherited genetic disorders, including cystic fibrosis, Huntington's disease, and hemophilia.

Despite the many actual and potential benefits of genetic engineering, the practice has inspired debate among scientists, environmentalists, and the general public alike. Some groups object to humans' meddling with the biology of organisms that have evolved naturally because they are afraid that eating GMOs might have negative effects on health. Others worry about the consequences to our environment if, for example, genetically engineered insects were to spread in wild populations. And for many, the idea of tampering with human genes to build "better" people raises the specter of eugenics—the early 20th century practice of trying to weed out the "unfit" from society.

Disquieting or not, genetic engineering appears to be speeding ahead. In November 2018, a Chinese scientist announced the birth of twins whose genomes had been altered by a gene editing technique known as CRISPR (see Chapter 9) while they were still embryos. According to the scientist, he performed the experiment to make the children resistant to HIV infection (by editing a particular protein that HIV uses to enter cells). The announcement caused a worldwide uproar in the scientific community. The editing of human embryos has long been an ethical red line not to be crossed. In response, some have

GENETICALLY MODIFIED ORGANISM (GMO)
An organism whose genome has been altered through genetic engineering techniques, often to contain a gene from another species.

GENE THERAPY
A treatment that aims to cure, treat, or prevent human disease by replacing defective genes with functional ones.

since called for a complete moratorium on gene editing of human embryos. But the case shows the difficulty of enforcing compliance with ethical standards.

Charged with both hope and fear, debates about the ethical use of genetic engineering are unlikely to go away anytime soon. And, given the rapid progress being made in the related field of bioengineering, opportunities to remake our world—perhaps even ourselves—are sure to proliferate. Even so, it will be at least a few years before Spiderman has any real competition. ■

CHAPTER 8 Summary

Driving Question 1 What determines the shape of a protein molecule, and why is its shape important?

- Proteins are folded chains of amino acids that perform many functions in cells, such as transmitting signals, catalyzing chemical reactions, and generating force for movement.
- The order and identity of amino acids in a protein chain determine the shape and function of the protein.

Driving Question 2 What are the steps of gene expression, and where in the cell do they occur?

- Genes provide instructions to make proteins. The process of using the information in genes to make proteins is called gene expression.
- Every gene has two parts: a coding sequence and a regulatory sequence. The coding sequence determines the identity of a protein; the regulatory sequence determines where, when, and how much of the protein is produced.
- Gene expression occurs in two stages, transcription and translation, which take place in separate compartments in eukaryotic cells.
- Transcription is the first step of gene expression, in which the information stored in DNA is copied into mRNA. Transcription occurs in the nucleus.
- Translation, the second step of gene expression, uses the information carried in mRNA to assemble a protein. Translation occurs in the cytoplasm.
- Proteins are assembled by ribosomes with the help of tRNA molecules, which deliver amino acids to the ribosome.
- The genetic code is the set of rules by which mRNA sequences are translated into protein sequences; the code is redundant and universal—it is shared by all living organisms.

Driving Question 3 How can organisms be genetically modified to produce recombinant proteins?

- Through genetic engineering, genes from one species of organism can be inserted into the genome of another species of organism to make a transgenic organism.

Driving Question 4 What are some pros and cons of genetically modified organisms?

- Transgenic organisms have numerous uses in biotechnology and health.
- There are also potential risks and ethical questions around transgenic organisms.

MORE TO EXPLORE

- Prasad, A. (2019). The future of silk. *Scientific American:* blogs.scientificamerican.com/observations/the-future-of-silk/.
- Hayashi, C. (2010). TED Talk: The magnificence of spider silk: https://www.ted.com/talks/cheryl_hayashi_the_magnificence_of_spider_silk
- Tokareva, O., et al. (2013). Recombinant DNA production of spider silk proteins. *Microbial Biotechnology* 6(6), 651–663.
- Center for Genetics and Society: http://www.geneticsandsociety.org
- Kevles, D. J. (1995). *In the Name of Eugenics: Genetics and the Uses of Human Heredity*. Cambridge, MA: Harvard University Press.

CHAPTER 8 Test Your Knowledge

Driving Question 1 What determines the shape of a protein molecule, and why is its shape important?

By answering the questions below and studying Infographic 8.2, you should be able to generate an answer for this broader Driving Question.

Know It

1. A protein is made up of a chain of
- **a.** nucleotides.
- **b.** amino acids.
- **c.** lipids.
- **d.** fatty acids.
- **e.** simple sugars.

2. What determines a protein's function?
- **a.** the sequence of amino acids
- **b.** the three-dimensional shape of the folded protein
- **c.** the location of its gene on the chromosome
- **d.** all of the above
- **e.** a and b

Use It

3. Spidroin proteins are in an unfolded state in the spider's silk gland before they are extruded from the spider. In their unfolded state, will they have the same properties as spider fibers in a web? Explain your answer.

4. If the repeated alanines in spidroin were changed to amino acids with hydrophilic side chains, would they still cluster together away from water? Explain your answer.

Driving Question 2 What are the steps of gene expression, and where in the cell do they occur?

By answering the questions below and studying Infographics 8.3, 8.4, 8.5, 8.6, 8.7, and 8.8, you should be able to generate an answer for this broader Driving Question.

Know It

5. "A gene contains many chromosomes. Each chromosome encodes a protein." Is this statement accurate? If not, explain why not, and rewrite the statement to make it correct.

6. What is the final product of gene expression?
- **a.** a DNA molecule
- **b.** an RNA molecule
- **c.** a protein
- **d.** a ribosome
- **e.** an amino acid

7. For each structure or enzyme listed, indicate by N (nucleus) or C (cytoplasm) where it acts in the process of gene expression in a eukaryotic cell.

_____________ RNA polymerase
_____________ Ribosome
_____________ tRNA
_____________ mRNA

8. What is encoded by a single codon?
- **a.** a single protein
- **b.** an RNA nucleotide
- **c.** a DNA nucleotide
- **d.** an amino acid
- **e.** any of the above, depending on the organism

9. A gene has the sequence ATCGATTG. What is the sequence of the complementary RNA?
- **a.** ATCGATTG
- **b.** TAGCTAAC
- **c.** GTTAGCTA
- **d.** UAGCUAAC
- **e.** CAAUCGAU

Use It

10. If a spider wasn't making the normal amount of its spidroin protein, would you suspect a problem in the regulatory or coding sequence of the spidroin gene? Explain your answer.

11. If you wanted to try to increase the amount of spidroin protein a spider produces, would you modify the regulatory sequence or the coding sequence? Explain your answer.

12. A change in DNA sequence can affect gene expression and protein function. What would be the impact of each of the following changes? How, specifically, would each change affect protein or mRNA structure, function, and levels?
- **a.** a change that prevents RNA polymerase from binding to a gene's regulatory sequence
- **b.** a change in the coding sequence that changes the amino acid sequence of the protein
- **c.** a change in the regulatory sequence that allows transcription to occur at much higher levels
- **d.** a combination of the changes in b and c

13. On the one hand, the insulin gene is normally expressed in specific cells in the pancreas, but not in a type of immune cell known as a B cell. On the other hand, B cells express large amounts of antibody proteins. What would you have to do get a B cell to express insulin? (*Hint:* Remember that all cells in an organism have the same set of chromosomes and associated genes.)

Driving Question 3 How can organisms be genetically modified to produce recombinant proteins?

By answering the questions below and studying Infographic 8.9, you should be able to generate an answer for this broader Driving Question.

Know It

14. Why is recombinant protein production in yeast an efficient strategy?

- **a.** because yeast can easily be grown in large quantities
- **b.** because yeast can secrete large amounts of recombinant proteins into their growth medium
- **c.** because yeast are multicellular, so they have a variety of cell types for recombinant gene expression
- **d.** all of the above
- **e.** a and b

15. What is the purpose of the vector in generating a transgenic organism?

16. Describe the recombinant gene that would be needed to create a transgenic spider that produces a yeast protein in its silk glands.

Use It

17. Why is it important that the transgenic yeast expressing recombinant spidroin proteins secrete the protein into their culture (growth) medium? (*Hint:* What has to happen to spidroin to convert it into actual silk?)

18. Keratin 10 is a protein expressed at high levels in human skin cell; it is an important component of the cytoskeleton of skin cells. If you wanted to express a different gene in skin cells, which part of the keratin 10 gene would you use? Why? If you wanted to produce keratin 10 in yeast cells, what part of the keratin 10 gene would you use? Why?

19. Lysozyme is a protein secreted in tears and saliva in all mammals. Amylase is a protein secreted in mammalian saliva.

- **a.** Describe the recombinant gene that you would assemble to express recombinant human lysozyme in the tears of goats.
- **b.** Describe the recombinant gene that you would assemble to express recombinant human amylase in goat saliva.

Driving Question 4 What are some pros and cons of genetically modified organisms?

By answering the questions below and studying Infographics 8.1 and 8.9, you should be able to generate an answer for this broader Driving Question.

Know It

20. Why is transgenic technology needed to produce large quantities of spider silk?

21. Why is spider silk such a valuable product?

Use It

22. Type 1 diabetes results from a loss of insulin production from the pancreas. People with diabetes take recombinant human insulin expressed in bacteria.

- **a.** Describe the gene construct necessary for expression of human insulin in bacteria.
- **b.** Describe the gene construct necessary to produce human insulin in goat's milk.
- **c.** If you were to attempt gene therapy (genetically modifying the human's genome so that insulin would be produced in the human's pancreas), would you need a recombinant form of the insulin gene? Explain your answer.

Apply Your Knowledge

Interpreting Data

23. A biotechnology company has created a number of strains of transgenic yeast with a recombinant spidroin gene. Each strain is grown in 1 L of culture medium. The cells are separated from the culture medium. All of the spidroin protein present in the culture medium is isolated and quantified. Similarly, all the cells are lysed (broken open) and all the spidroin present within the cells is quantified. The results are shown in the table below.

- **a.** How much protein (total) is produced by each strain in 1 L of culture?
- **b.** What are the differences in spidroin production in the different strains?
- **c.** Which strain should the company use to commercialize spidroin production? Explain your answer.

Strain	Spidroin protein isolated from cells (mg)	Spidroin protein present in culture medium (mg)
Nontransgenic yeast	0	0
Transgenic strain 1	80	0
Transgenic strain 2	40	40
Transgenic strain 3	0	80
Transgenic strain 4	60	20

Mini Case

Apply Your Knowledge

24. A physician is stationed in a military hospital in Iraq. She often has to treat severe wounds caused by sniper shots. Infection is always a concern, and the bandages currently used are not always flexible enough to permit movement of the affected area as the wound heals. Often, the treated soldiers need to return to duty before their stitches are ready to be removed.

Given the scenario described, what case could a representative of a spider silk biotechnology company make to the army to support spider silk research?

Bring It Home

Apply Your Knowledge

25. A number of concerns have been expressed about GMOs. Search the Internet for reliable sources about a particular GMO that you have heard of or in which you are interested (e.g., Golden Rice or genetically modified salmon). List what you consider to be the pros and cons of at least two GMOs. Has what you have read in this chapter about other genetically modified organisms and the transgenic yeast changed your opinion about GMOs? What restrictions (if any) would you place on GMOs?

New Gene, New Me

Gene therapy offers hope to people with debilitating genetic conditions

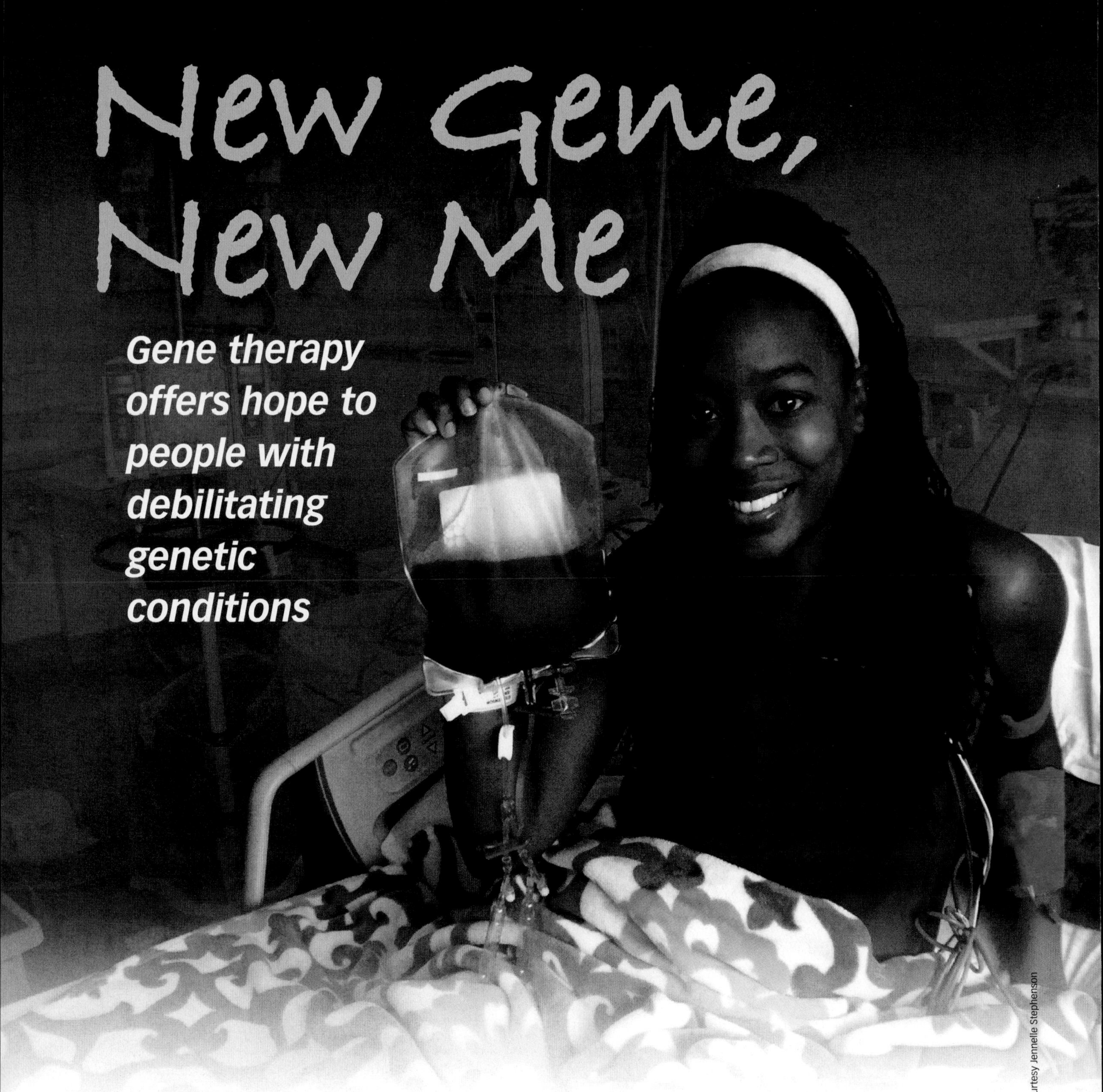

Courtesy Jennelle Stephenson

DRIVING QUESTIONS

1. What are mutations, what is their impact, and how do they occur?
2. How can genetic engineering be used to treat genetic diseases?
3. Are all mutations harmful?

For Jennelle Stephenson, a graduate student from Kissimmee, Florida, the pain had finally become unbearable. "Imagine there's an incredibly heavy object crushing you," she says. "You can't breathe, there's no air, and you feel like your bones are cracking from the pressure."

The pain comes on without warning, often beginning with a tingling sensation in her back. Then it spreads to her arms, her legs, and even her cheekbones.

When it's not the crushing sort of pain, it's sharp and localized—"like being stabbed over and over and over in same spot," she says.

During these pain crises, only the strongest narcotics will help. To obtain these, Jennelle must travel to the emergency room. But doctors and nurses there are typically reluctant to give her the medications, concerned she might be faking it. In the meantime, she must endure the excruciating pain.

Jennelle's experience, while dramatic, is all too typical for people with sickle cell disease. Pain is a constant threat, restricting their life choices.

And it's not just pain. There are also serious health dangers of the condition.

By the time Manny Johnson of Boston, Massachusetts, was 4 years old, he'd already had a stroke. A blood clot formed in a vessel in his brain, blocking the flow of oxygen. Since then, he's received a blood transfusion nearly every month to prevent a dangerous clot from occurring again. Like Jennelle, he has suffered pain his whole life.

Sickle cell disease is an inherited genetic disorder. It primarily affects people who can trace their ancestors to equatorial regions of the globe. In the United States, most sufferers of the condition are African American.

The disease gets its name from the characteristic shape of red blood cells in people with the disease. Red blood cells carry oxygen throughout the body—oxygen that is critical for cells to carry out cellular respiration (Chapter 6). Normally, a red blood cell is shaped like a jelly doughnut: round and squishy with a depression in the middle. In people with sickle cell disease, red blood cells become long and bent, like a sickle or a crescent moon.

Sickled red blood cells are less effective at carrying oxygen than normal red blood cells are, and they do not survive as long as normal blood cells. As a result, people with this condition often develop anemia, a shortage of red blood cells capable of delivering oxygen to tissues.

In addition, sickled cells tend to get stuck in the tiny blood vessels that feed tissues throughout the body, leading to blood clots and intense pain. The clots can also cause stroke, blockages in the lungs, organ failure, and, ultimately, death **(INFOGRAPHIC 9.1)**.

Few effective treatments for the disease exist. Narcotics and blood transfusions, given at the time of a pain crisis, can alleviate pain but will not prevent it. Drugs can help reduce the frequency of episodes, though these medicines don't work for everyone. The only real potential cure is a bone marrow transplant, a physically grueling and risky procedure. With such a transplant, a person's entire blood-making system is wiped out with toxic chemicals and then replaced with that of a genetically compatible donor—assuming one can be found.

GENE THERAPY
Correcting or replacing mutated genes as a treatment for a genetic disease.

But thanks to advances in genetic engineering, this dismal picture is starting to improve. Scientists at several medical centers in the United States are developing forms of **gene therapy** designed to fix the genetic mistake that these individuals are born with. Manny and Jennelle are two of the first people with sickle cell disease to receive these new therapies. Doctors and patients around the world are eagerly watching, with fingers crossed, to see if those treatments work.

Small Change, Big Effect

▸ Mutations and their consequences

Scientists have known since the 1950s that sickle cell disease is the result of a single change in the nucleotide sequence of one

INFOGRAPHIC 9.1

Sickle Cell Disease Affects Red Blood Cells

Sickle cell disease is an inherited genetic disorder that causes red blood cells to take on a sickled shape and carry less oxygen. People with sickle cell disease experience anemia, episodes of severe pain, and other complications due to clots and blockages in blood vessels.

No Sickle Cell Disease

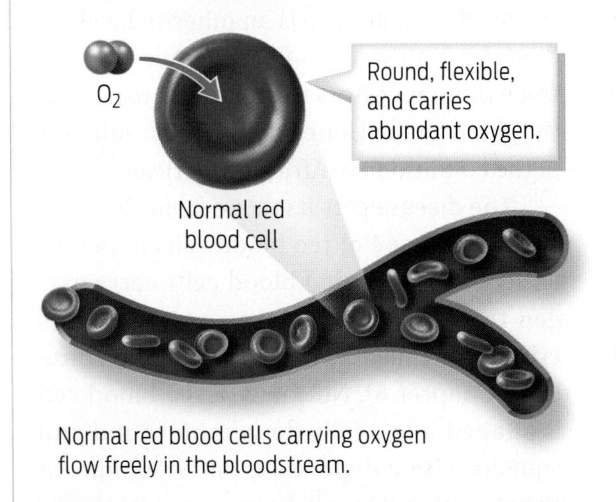

Normal red blood cells carrying oxygen flow freely in the bloodstream.

Sickle Cell Disease

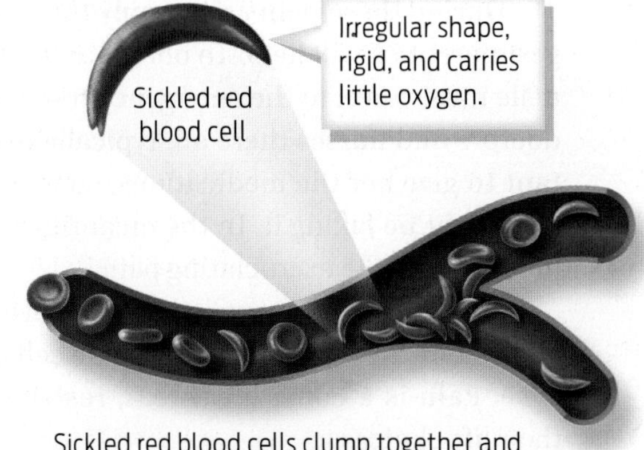

Complications of Sickle Cell Disease:
- Anemia
- Extreme pain
- Blood clots
- Stroke
- Death

Sickled red blood cells clump together and block the flow of blood and oxygen. They also do not survive as long as normal red blood cells.

? **How does the shape and function of red blood cells differ in people with and without sickle cell disease?**

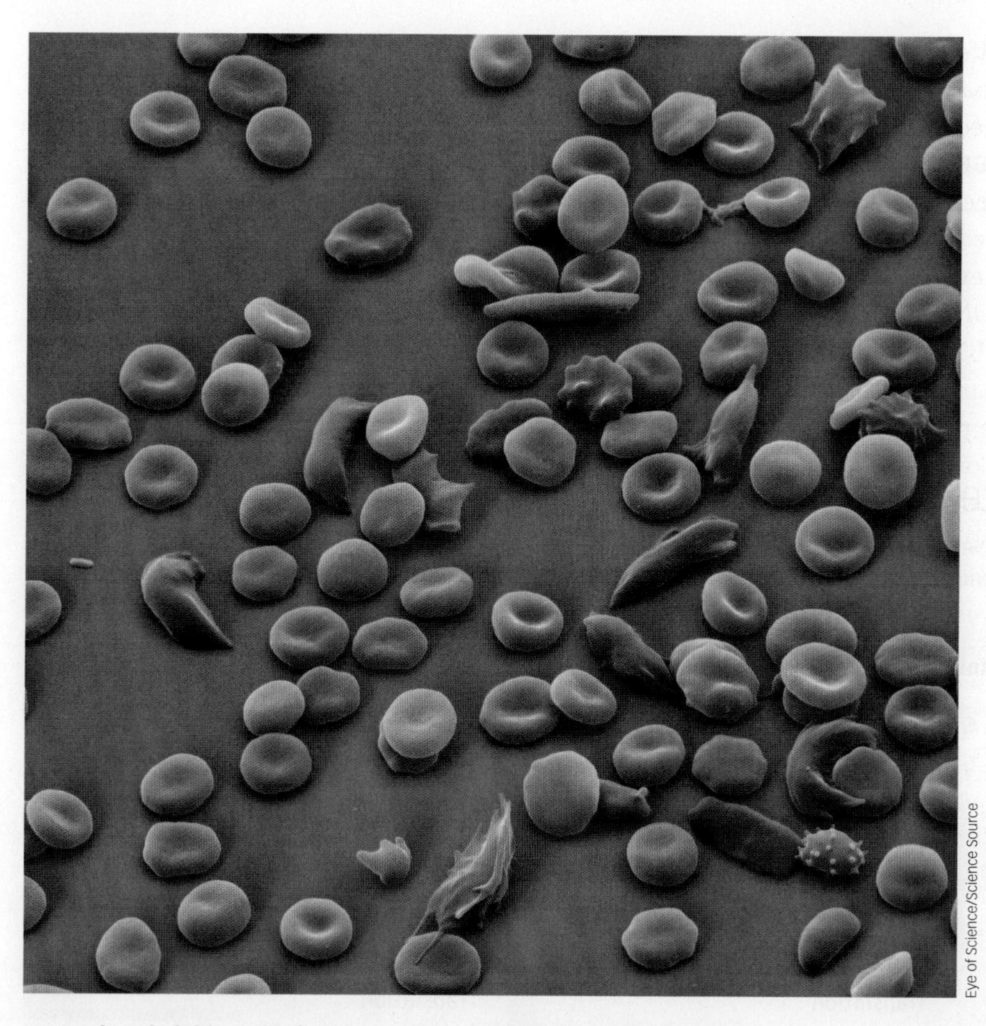
Eye of Science/Science Source

Normal and sickle cells from someone with sickle cell disease.

gene. That change alters the shape of **hemoglobin,** the oxygen-carrying molecule inside of red blood cells. Red blood cells are essentially bags of hemoglobin.

The particular gene affected encodes a protein called **beta-globin** that makes up one part of the hemoglobin molecule. What was an A-T base pair in the original version of the beta-globin gene is changed to a T-A base pair. A change in the nucleotide sequence of a DNA molecule is called a **mutation.**

Recall from Chapter 8 that during gene expression DNA is first transcribed into mRNA, which is then translated into protein. Groups of three mRNA nucleotides, called codons, specify particular amino acids according to the genetic code. In people with sickle cell disease, the original codon in the beta-globin mRNA, GAG, is changed to GUG. As a result, when the mRNA is translated into protein, glutamic acid (Glu) in the normal protein becomes a valine (Val). This amino acid change alters the physical shape and chemical properties of the protein.

Normally, hemoglobin is a compact protein molecule made up of four interacting subunits: two alpha-globin subunits and two beta-globin subunits. In sickle cell disease, the amino acid change alters the physical shape and chemical properties of the beta-globin subunit, causing it to link up with other beta-globin subunits in neighboring hemoglobin molecules. Since a protein's shape determines its function, hemoglobin molecules with these altered shapes are

HEMOGLOBIN
The oxygen-carrying protein in red blood cells.

BETA-GLOBIN
One of the proteins that makes up hemoglobin.

MUTATION
A change in the nucleotide sequence of a DNA molecule.

unable to effectively carry oxygen. Moreover, the long chains of hemoglobin molecules stretch the cell into its characteristic sickled form **(INFOGRAPHIC 9.2)**.

Sickle cell disease was the first inherited genetic disease to be understood on a molecular level. The discovery that a change in a single DNA nucleotide could wreak such havoc on the shape and function of a protein came as a surprise to most scientists at the time. But it provided a valuable lesson in understanding the importance of mutations in biology.

Mutations are central to both genetics and evolution—they are responsible for the diversity of traits that we see among

INFOGRAPHIC 9.2

Mutations Can Alter Protein Shape and Function

Mutations alter the nucleotide sequence of DNA. If a mutation changes the coding region of a gene, the resulting protein may have an altered structure and function. In this case, altered hemoglobin causes cells to take on a sickled shape, and interferes with the ability of red blood cells to carry oxygen to tissues.

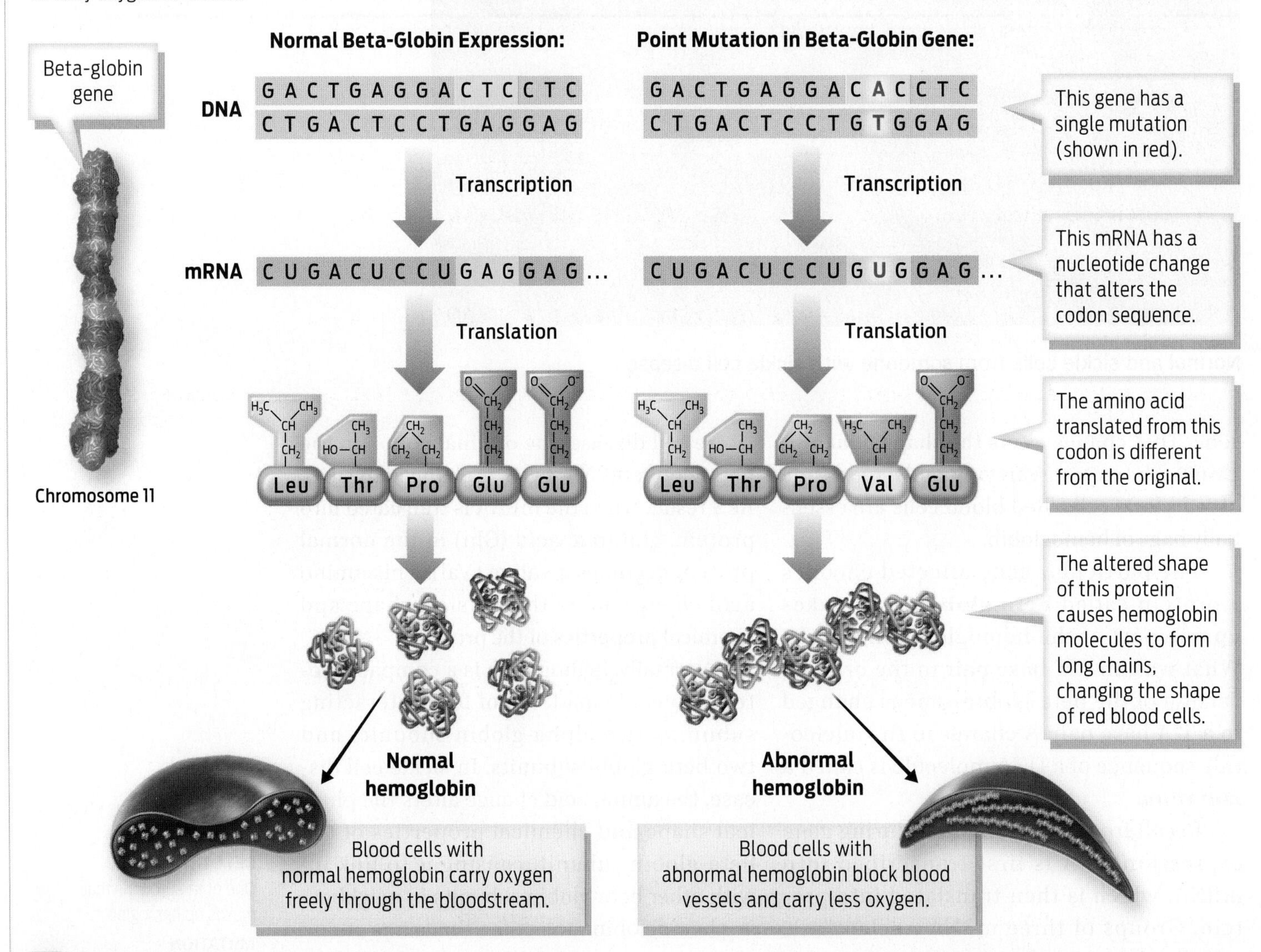

? Which of the following molecules are altered due to the sickle cell mutation in the beta-globin gene: beta-globin DNA, beta-globin mRNA, or hemoglobin protein?

individuals and among species. Mutations come in a variety of forms and can have many different effects. The type of mutation that causes sickle cell disease is called a **point mutation;** it alters a single DNA nucleotide. Depending on where a point mutation occurs in a codon, it may—or may not—change the amino acid sequence of a protein. Point mutations that change the amino acid sequence of a protein are called **missense mutations;** those that do not change the protein sequence are called **silent mutations.**

In other cases, one or more DNA nucleotides may be inserted or deleted from genes, shifting the reading frame of that gene—that is, changing where a codon begins and ends. These types of mutations are known as **frameshift mutations.**

Whole blocks of DNA can be rearranged as a result of mutation. A segment of DNA can "flip" within its normal chromosomal location (a change called an inversion), or segments of DNA can trade places between different chromosomes (translocations). Large inversions and translocations can fuse portions of different genes together, creating new proteins with novel activity **(TABLE 9.1).**

Ultimately, the impact of a mutation depends on how it affects the shape of a protein. In many cases, the shape of the protein is altered in a way that makes it nonfunctional. In other cases, the mutation changes the shape of the protein in a way that makes

POINT MUTATION
A mutation that alters a single DNA nucleotide.

MISSENSE MUTATION
A point mutation that changes the amino acid sequence of the encoded protein.

SILENT MUTATION
A point mutation that does not change the amino acid sequence of the encoded protein.

FRAMESHIFT MUTATION
A shift in the reading frame, such that codons start and end at an alternative position.

TABLE 9.1 Types of Mutations and Their Effects

Type of Mutation	Example	Effect on Protein Function
Original DNA sequence: No mutation	DNA HAS ALL YOU CAN ASK FOR	
Point Mutations		
Silent mutation: Change one nucleotide to another; no change in amino acid sequence	DNA HAS ALL YOO CAN ASK FOR	No change; normal function
Missense mutation: Change one nucleotide to another; different amino acid sequence in this location	DNA HAS ALL LOU CAN ASK FOR	Change in protein shape and function
Nonsense mutation: Change one nucleotide; introduces early stop codon	DNA HAS ALL YOU	Protein is too short and therefore not functional
Frameshift Mutations		
Insertion mutation: Insert one or more nucleotides; shifts reading frame of every codon after the insertion	DNA HAS ALL YYO UCA NAS KFO R	Severely modified sequence makes the protein not functional
Deletion mutation: Delete one or more nucleotides; shifts reading frame of every codon after the deletion	DNA HAS ALY OUC ANA SKF OR	Severely modified sequence makes the protein not functional
Rearranged DNA Mutations		
Inversion mutation: A group of DNA nucleotides are flipped to read in reverse order; different amino acid sequence in this location	DNA HAS ALL YOC UAN ASK FOR	Change in protein shape and function
Translocation mutation: Move segments of DNA from one chromosome to another, fusing portions of different genes together	DNA HAS ALL YOU CAN EAT THE DOG AND CAT ASK FOR	Significant change in protein shape and function

it overly active. In the case of sickle cell disease, the mutation causes the beta-globin subunits of hemoglobin to become "sticky" and attract the beta-globin subunits from other hemoglobin molecules.

This stickiness of hemoglobin molecules is what causes red blood cells to sickle. The degree of sickling depends on how much mutated hemoglobin is present in a red blood cell, relative to the amount of normal hemoglobin. As discussed in Chapter 7, we all have two copies of every gene—one copy from our biological father and one copy from our biological mother. People can have one, two, or no mutated beta-globin genes, depending on which versions they inherit from their parents. People with one copy of the mutated beta-globin gene and one copy of a normal beta-globin gene are carriers of the condition: they typically do not experience sickling of their cells and are said to have sickle cell trait. People with two mutated copies of the beta-globin gene, however, will have sickle cell disease.

Where do mutations like those that cause sickle cell disease come from? In the case of sickle cell disease, the mutations are inherited, meaning they are passed from parents to children and are present in every cell of the child's body at birth.

Other mutations occur during our lifetime. In this case, cells that develop a mutation will pass that mutation on to their daughter cells every time they divide (Chapter 10). As a result, only some cells in the body will have that mutation. One way that a cell can develop a new mutation is during DNA replication. Each time our cells replicate their DNA (Chapter 7), there is a small chance that a mistake will occur—say, an A nucleotide is paired with a G, instead of a T. If this mistake is not corrected, it will lead to a permanent change in the DNA sequence—a mutation. This mutation will then be passed on as the cell divides and reproduces.

MUTAGEN
Any chemical or physical agent that can damage DNA by changing its nucleotide sequence.

Our DNA is continually being bombarded by environmental factors that can also damage DNA and cause mutations. These factors include chemicals, ultraviolet light, radiation, and infectious agents like viruses. Physical or chemical agents that cause mutations are called **mutagens.**

Not all mutagens originate outside the body. For example, some of the reactions that occur in the mitochondria during cellular respiration (see Chapter 6) produce DNA-damaging molecules called free radicals. When cells attempt to repair this damage, the repairs may be carried out incorrectly, leading to a mutation **(INFOGRAPHIC 9.3)**.

One of the main existing treatments for sickle cell disease is a drug called hydroxyurea, which can reduce the amount of sickling that occurs in a person's body. Unfortunately, it doesn't work for everyone. Indeed, it didn't work for Jennelle or Manny. They also had no compatible sibling donor to provide bone marrow for a transplant. (The probability of finding a match in donor databases for African Americans is typically low.) So they needed another option.

The turning point for Jennelle came after a particularly bad pain crisis in August 2016. "I had just been discharged from the hospital and my boyfriend was encouraging me to do something because I was getting progressively worse," she says. "We had a long talk and he really convinced me that there's got to be something."

That's when she went online and discovered a number of clinical trials being offered for people like herself. She sent emails to 11 places, and heard back from one: the National Institutes of Health (NIH) in Bethesda, Maryland.

A Therapy Decades in the Making

▶ Genetic engineering techniques

Gene therapy has been a goal of medical science since the 1970s. Ever since scientists learned how to cut and splice together pieces of DNA in the lab, they have been tantalized by the prospect of using such techniques to

INFOGRAPHIC 9.3

What Causes Mutations?

Mutations are changes in the nucleotide sequence of DNA. There are several ways that a person can end up with a mutation: it may have been inherited; it may have occurred randomly during DNA replication; or it may have been the result of environmental insult.

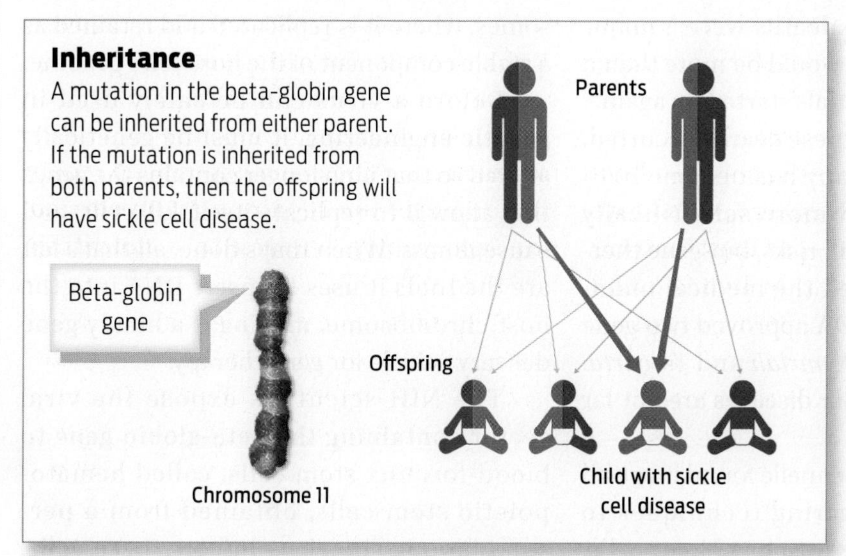

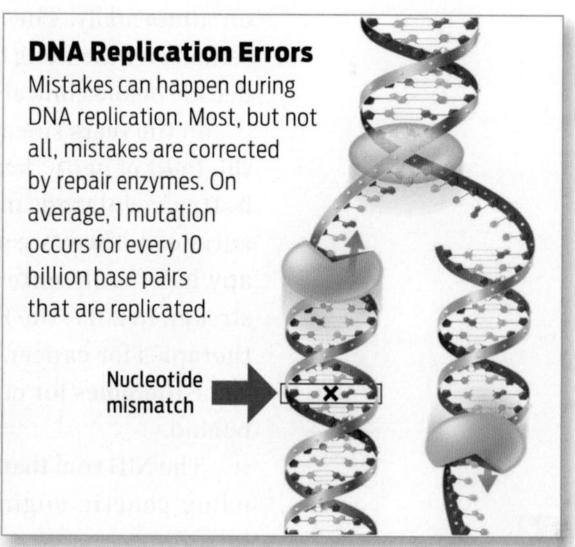

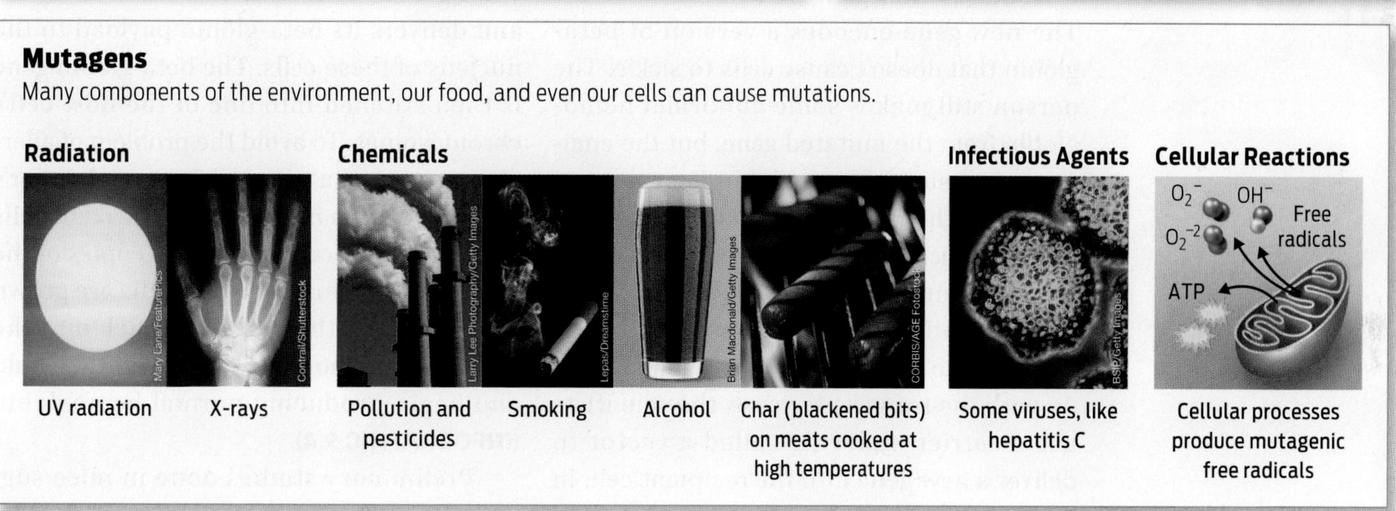

? List three things that you could do to decrease the risk of developing a cancer-causing mutation in a skin cell.

fix genetic errors and cure genetic diseases in people.

The first trials of gene therapy approaches in people were conducted in the early 1990s. A 4-year-old girl with a severe immune deficiency became the first person to receive successful gene therapy: she was cured of her disease.

Yet the field has had its share of failures as well. Most memorably, in 1999, a gene therapy trial ended in disaster when the patient, 18-year-old Jesse Gelsinger, died as a result of the procedure. The gene therapy was designed to fix a rare liver disease. Jesse developed a severe allergic reaction to the virus that was used to deliver the new gene to his liver cells. The scientist who led the trial admitted to not following his own procedures, and he shouldered some of the blame for the tragic result.

A few years later, in 2003, several people who received gene therapy developed cancer, and one died from the disease. In this case, the virus used to deliver the gene ended up landing next to a cancer gene, causing it to turn on abnormally. These deaths were a major setback for the field; it would be more than a decade before clinical trials started up again.

In the years since these deaths occurred, the field of gene therapy has become both better regulated and more scientifically advanced. There are still risks, but gene therapy has finally entered the medical mainstream. In 2017, the FDA approved two gene therapies for cancer, *Kymriah* and *Yescarta*. Gene therapies for other diseases are not far behind.

The NIH trial that Jennelle found involved using genetic engineering techniques to introduce a new gene into her blood cells. The new gene encodes a version of beta-globin that doesn't cause cells to sickle. The person still makes some abnormal hemoglobin from the mutated gene, but the engineered version of beta-globin is present in high enough amounts to prevent red blood cells from sickling.

The gene therapy the NIH team developed is similar to the process used to engineer yeast to produce spider silk (Chapter 8). As with the transgenic yeast, the scientists use a carrier molecule called a vector to deliver a new gene into the recipient cell. In this case, the vector is a genetically modified virus, and the gene is a beta-globin gene with the correct nucleotide sequence.

Recall that viruses are noncellular entities consisting of a protein "coat" surrounding genetic material (Chapter 1). Viruses make good vectors because they naturally inject genetic material—their own—into host cells as part of their normal life cycle. Specifically, viruses reproduce by infecting host cells and delivering viral genes into the infected cells. An infected host cell transcribes and translates these viral genes as if they are its own, and the virus then uses these components to make even more copies of itself.

Similarly, the engineered viral vector carrying the new beta-globin gene naturally infects host cells, delivering the beta-globin gene into those cells. The beta-globin gene is inserted into one of the host cell's chromosomes, where it is replicated and retained as a stable component of the host cell's genome.

Before a virus can be safely used in genetic engineering, it must be genetically altered so that it no longer contains the genes that allow it to replicate itself, kill cells, and cause illness. When that's done, all that's left are the tools it uses to insert DNA into the host chromosome, making it a handy gene delivery vehicle for gene therapy.

The NIH scientists expose the viral vector containing the beta-globin gene to blood-forming stem cells, called hematopoietic stem cells, obtained from a person's blood. The virus infects these cells, and delivers its beta-globin payload to the nucleus of these cells. The beta-globin gene is then stitched into one of the host cell's chromosomes. To avoid the problem of allergic reaction that caused Jesse Gelsinger's death, the virus is exposed to the stem cells outside of the body. Millions of copies of the genetically engineered stem cells are grown in the lab and then inserted back into the patient in the hope that they will take hold and begin producing normal hemoglobin **(INFOGRAPHIC 9.4)**.

Preliminary studies done in mice suggested that this approach had promise. Only with clinical trials in humans, however, could the scientists know whether it would be safe and effective in people.

Gearing Up for the Big Day

Jennelle enrolled in the clinical trial in November 2016. Over the next few months, she received several rounds of a drug designed to boost the production of her blood stem cells. Then, in December 2017, doctors took a vial of Jennelle's blood and isolated some of her blood-forming stem cells. In a lab, they exposed these cells to the modified

INFOGRAPHIC 9.4

Gene Therapy for Sickle Cell Disease: One Approach

A modified virus serves as the vector to introduce a normal beta-globin gene to hematopoietic (blood-forming) stem cells isolated from a patient with sickle cell disease. Once stem cells have been obtained from the patient, they are infected with the engineered virus. The beta-globin gene carried by the virus is incorporated into a host cell's chromosome. The genetically modified cells are then grown in the lab, so that a large number of them can be introduced back into the patient, where they will produce normal beta-globin, which will be incorporated into normal hemoglobin in red blood cells that will not sickle.

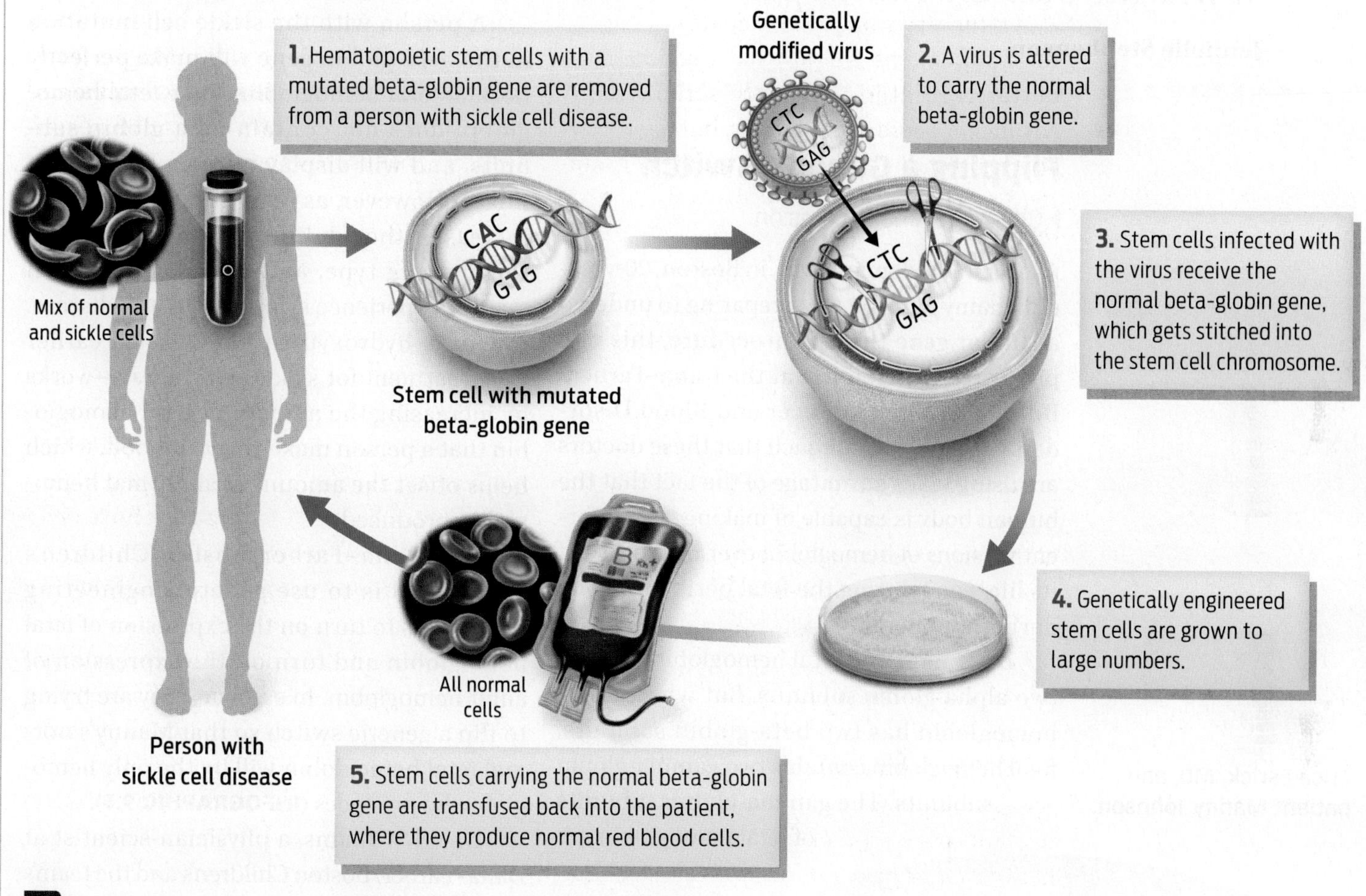

? What is the role of the virus in this gene therapy approach?

virus vector containing the new hemoglobin gene. Then, they waited.

If all went as planned, the viral vector would infect Jennelle's blood stem cells and introduce its contents into her cells. The viral enzymes would then go to work cutting and pasting the new gene into Jennelle's DNA. After growing the cells into millions of copies, the doctors would introduce the genetically altered stem cells back into Jennelle's body.

To make room for all those new cells, doctors gave Jennelle a few rounds of intense chemotherapy to kill off her existing blood stem cells, which contained the mutated version of the hemoglobin. During this time, she says, she felt awful. She lost her hair, her mouth filled with sores, and she couldn't eat. But this step was necessary to ensure the new stem cells could take hold.

Jennelle received her infusion of cells on December 26, 2017. "It was the best Christmas present ever," she says. Both her dad and her brother were there by her side, along with her team of doctors and nurses. Everyone held their breath as the bag began to drip its contents into her vein.

It would be a few months before she and her medical team would know whether the procedure worked.

"It was the best Christmas present ever."

—Jennelle Stephenson

Flipping a Genetic Switch

▸ Changing gene expression

A few hundred miles away, in Boston, 20-year-old Manny Johnson was preparing to undergo a similar gene therapy procedure, this one pioneered by scientists at the Dana–Farber/Boston Children's Cancer and Blood Disorders Center. The approach that these doctors are using takes advantage of the fact that the human body is capable of making two different versions of hemoglobin over the course of its life—one during the fetal period and one during adulthood.

Both adult and fetal hemoglobin contain two alpha-globin subunits. But where adult hemoglobin has two beta-globin subunits, fetal hemoglobin contains two gamma-globin subunits. The gamma-globin subunits of fetal hemoglobin are expressed during prenatal development, with that expression beginning to taper off around the time of birth. Beta-globin expression starts a few months before birth and continues into infancy and adulthood. The result of this gene expression pattern is a gradual loss of fetal hemoglobin after birth, and its replacement by adult hemoglobin.

Erica Esrick, MD, and patient Manny Johnson.

Dana-Farber Cancer Institute

A person with the sickle cell mutation in the beta-globin gene will make perfectly normal fetal hemoglobin, since fetal hemoglobin does not contain beta-globin subunits, and will display no symptoms as an infant. However, as beta-globin expression increases, the adult hemoglobin will be of the sickling type, and the individual will start to experience symptoms of the disease. The drug hydroxyurea—mentioned earlier as a treatment for sickle cell disease—works by increasing the amount of fetal hemoglobin that a person makes in adulthood, which helps offset the amount of abnormal hemoglobin produced.

The Dana–Farber/Boston Children's team's goal is to use genetic engineering techniques to turn on the expression of fetal hemoglobin and turn off the expression of adult hemoglobin. In essence, they are trying to flip a genetic switch so that Manny's normal fetal hemoglobin will be the only hemoglobin he produces **(INFOGRAPHIC 9.5)**.

David Williams, a physician-scientist at Dana–Farber/Boston Children's and the team's lead scientific investigator, says his group favors this "switching" approach because it takes advantage of a process that the body uses anyway. "Other trials are adding genes that encode fetal hemoglobin or corrected, non-sickling adult hemoglobin, without directly decreasing expression of the sickle hemoglobin gene," he says. "We predict our strategy is a more effective way to reduce or even eliminate the sickling of cells."

Manny was the first patient enrolled on the trial at Dana–Farber/Boston Children's. He was motivated to participate, he says, because his 7-year-old brother, Aiden,

INFOGRAPHIC 9.5

Using Gene Therapy to Switch Genes On or Off

Because fetal hemoglobin does not contain beta-globin subunits, it will not cause sickling, even in people with sickle cell disease. Genetic engineering approaches are being used to express fetal hemoglobin and shut down expression of sickle hemoglobin in people with the disease.

Types of Hemoglobin in a Person with Sickle Cell Disease, Before and After Gene Therapy

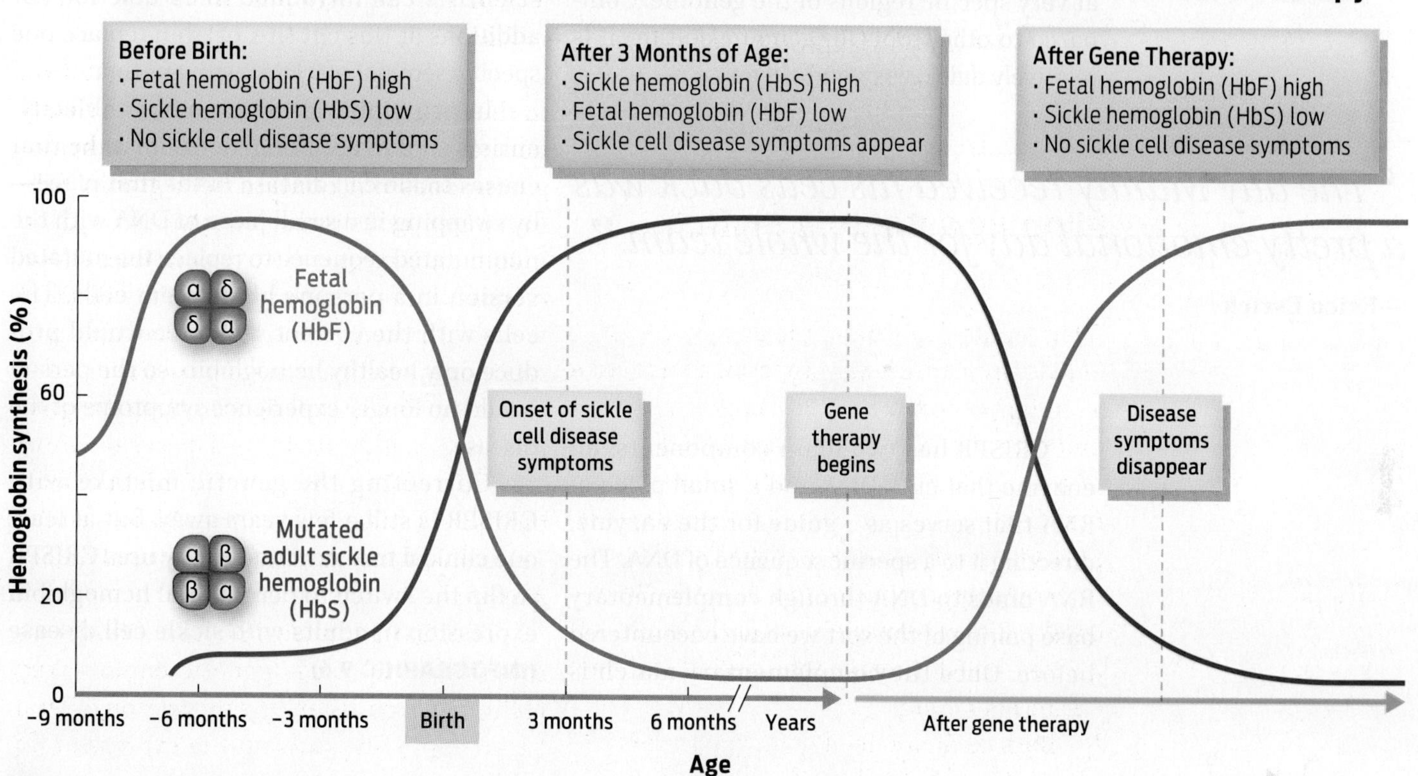

? **Which type of hemoglobin is expressed after this gene therapy? What subunits does it contain?**

also has sickle cell disease. "I wanted to do something to help him," Manny says. He has a tattoo of a sickle cell ribbon with Aiden's name in it.

Manny received his infusion of genetically modified blood stem cells in May 2018. "The day Manny received his cells back was a pretty emotional day for the whole team," says Erica Esrick, a pediatric hematologist-oncologist at Dana–Farber/Boston Children's and the co-principal investigator on the clinical trial. "It was the pinnacle of many, many years of many, many people's hard work."

Only time would tell if the massive effort would ultimately benefit Manny.

Correcting the Mistake

▶ Genome editing with CRISPR

Scientists sometimes make a distinction between gene therapy and **gene editing.** In gene therapy, entire genes are inserted or removed from a recipient cell. But what if it were possible to simply correct a genetic error in the original gene—for example, to edit out the mistaken "T" and replace it with the correct "A"? In fact, that goal may one day be possible thanks to the genome-editing tool CRISPR.

CRISPR (which stands for "clustered regularly interspaced short palindromic

GENE EDITING
A way to change the sequence of a gene.

CRISPR
A genome-editing tool based on a natural defense system in bacteria.

repeats") is a kind of molecular scissors that bacteria use to chop up viruses and thereby defend themselves from infection. Scientists have adapted this set of bacterial scissors to make it an exquisite tool for genetic engineering. CRISPR allows scientists to edit DNA at very specific regions of the genome. Compared to other tools for genome editing, it is relatively quick, easy, and cheap.

"The day Manny received his cells back was a pretty emotional day for the whole team."

—Erica Esrick

CRISPR has two main components: an enzyme that cuts DNA and a small piece of RNA that serves as a guide for the enzyme, directing it to a specific sequence of DNA. The RNA binds to DNA through complementary base pairing of the sort we have encountered before. Once the complementary match is found, the enzyme cuts the DNA in that precise location. The cell's DNA repair machinery then takes over to fill in the gap. If a supplementary piece of DNA is supplied along with the CRISPR enzyme, the cell will stitch this piece of DNA into the gap. Using CRISPR, scientists can introduce small deletions or additions at this cut site, or even replace one specific sequence of DNA with another.

In principle, CRISPR could enable scientists to correct the genetic mistake that causes sickle cell disease in the first place—by swapping in a small piece of DNA with the nonmutated sequence to replace the mutated version in a person's blood stem cells. The cells with the correct sequence would produce only healthy hemoglobin, so the person would no longer experience symptoms of the disease.

Correcting the genetic mistake with CRISPR is still a few years away. But at least one clinical trial now under way uses CRISPR to flip the switch to permit fetal hemoglobin expression in adults with sickle cell disease **(INFOGRAPHIC 9.6)**.

Red Blood Cells Before Treatment

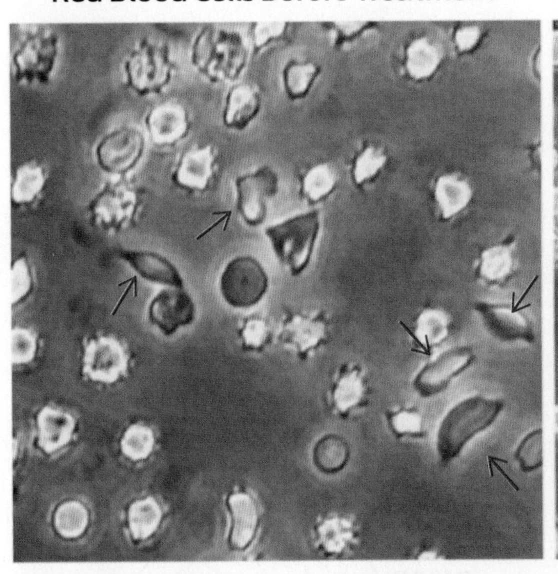

Red Blood Cells After Treatment

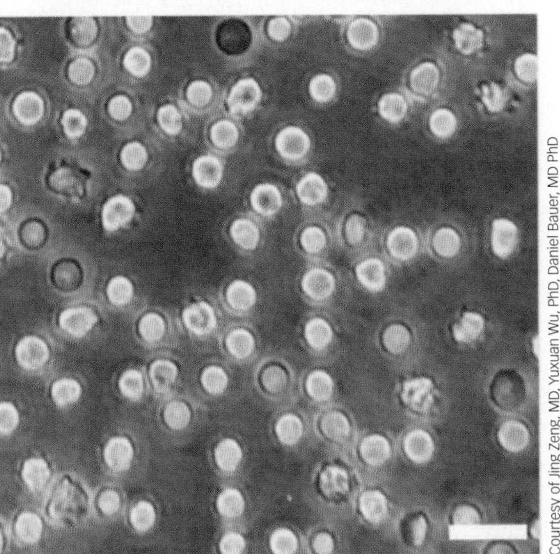

Courtesy of Jing Zeng, MD, Yuxuan Wu, PhD, Daniel Bauer, MD PhD

Left: In sickle cell disease, red blood cells sickle (red arrows). Blood stem cells from a person with sickle cell disease were edited with CRISPR with the goal of turning on fetal hemoglobin expression. Right: After genetic engineering, almost no red blood cells are sickled.

INFOGRAPHIC 9.6

CRISPR Adds Precision to Genetic Engineering

CRISPR is a genetic engineering method that can precisely modify specific gene sequences. Molecular tools target a specific DNA sequence, which can then be used to insert, delete, or alter a DNA sequence at that site.

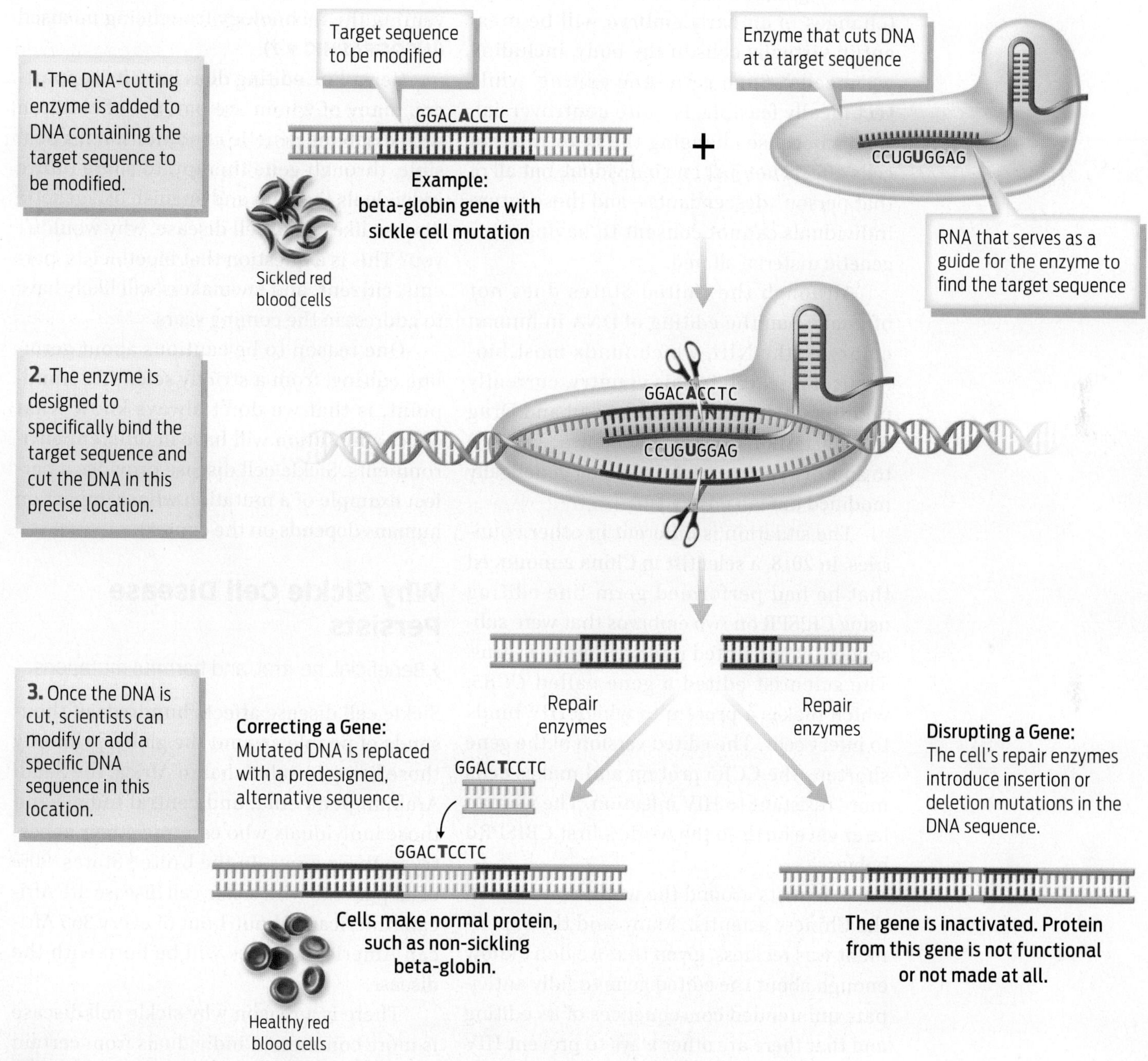

? **How does CRISPR target specific sites in the genome for modification?**

A Permanent Fix?

▸ Somatic versus germ-line editing

No matter how much Jennelle and Manny may personally benefit from their genetically engineered cells, they will not be free from all consequences of their genetic disease. That's because the changes are being made in only a subset of the **somatic cells** making up their body. They will still have the mutation

SOMATIC CELLS
Nonreproductive cells of the body.

in their **germ cells**—those that develop into sperm or eggs. That means they can still pass the mutation on to their children.

The only way to prevent that possibility would be to correct the genetic mistake in their germ cells or in an early embryo (changes in an early embryo will be present in virtually cells in the body, including germ cells). Such germ-line editing, while technically feasible, is quite controversial. That's because changing the DNA in germ cells affects not just one individual, but all of that person's descendants—and those future individuals cannot consent to having their genetic material altered.

Although the United States does not officially ban the editing of DNA in human embryos, the NIH, which funds most biomedical research in this country, currently prohibits it. In addition, the Food and Drug Administration (FDA) is not allowed (by law) to approve clinical trials involving genetically modified human embryos.

The situation is different in other countries. In 2018, a scientist in China announced that he had performed germ-line editing using CRISPR on two embryos that were subsequently implanted into a woman's uterus. The scientist edited a gene called *CCR5*, which makes a protein to which HIV binds to infect cells. The edited version of the gene shortens the CCR5 protein and makes cells more resistant to HIV infection. The woman later gave birth to the world's first CRISPR'd babies.

Scientists around the world condemned the Chinese scientist. Many said the experiment was reckless, given that we don't know enough about the edited gene to fully anticipate unintended consequences of its editing and that there are other ways to prevent HIV infection. CRISPR is such a new technology that there is not yet a lot of experience from which to learn. One concern with CRISPR technology is that it could make "off target" cuts in the genome, not just at the desired location. These unplanned changes could have serious consequences if they occur in genes that encode essential proteins.

These concerns have led some scientists to call for a complete moratorium on the editing of human embryos. The Chinese case, however, highlights the difficulty of preventing the technology from being misused **(INFOGRAPHIC 9.7)**.

Germ-line editing does have its supporters, many of whom are parents of children with disabling genetic conditions. If it's possible, through gene therapy, to spare future individuals the pain and anguish of a genetic disease like sickle cell disease, why wouldn't you? This is a question that bioethicists, parents, citizens, and lawmakers will likely have to address in the coming years.

One reason to be cautious about germ-line editing, from a strictly scientific standpoint, is that we don't always know what effect a mutation will have in different environments. Sickle cell disease provides a perfect example of a mutation whose impact on humans depends on the context.

Why Sickle Cell Disease Persists

▶ Beneficial, neutral, and harmful mutations

Sickle cell disease affects hundreds of thousands of people around the globe, primarily those living in sub-Saharan Africa, the Saudi Arabian peninsula, and central India—and those individuals who can trace their ancestry to these areas. In the United States, 90% of all patients with sickle cell disease are African American. About 1 out of every 365 African American babies will be born with the disease.

There is a reason why sickle cell disease is more common in individuals from certain geographic areas. In these environments, having one copy of the sickle cell mutation confers a benefit. The sickle cell mutation is found at high frequencies wherever the disease malaria is or has been common. Malaria is caused by a parasite that infects red blood

GERM CELLS
Reproductive cells of the body.

INFOGRAPHIC 9.7

Mutations Can Be Hereditary or Nonhereditary

Mutations that occur in sperm or egg cells are germ-line mutations that can be passed on to offspring. These inherited mutations are then found in every cell of the offspring, including egg or sperm cells, and can be passed on to subsequent generations. Somatic mutations are those that occur in any cell in the body other than sperm or egg cells. These mutations are not inherited but may cause disease in the individual that acquires them.

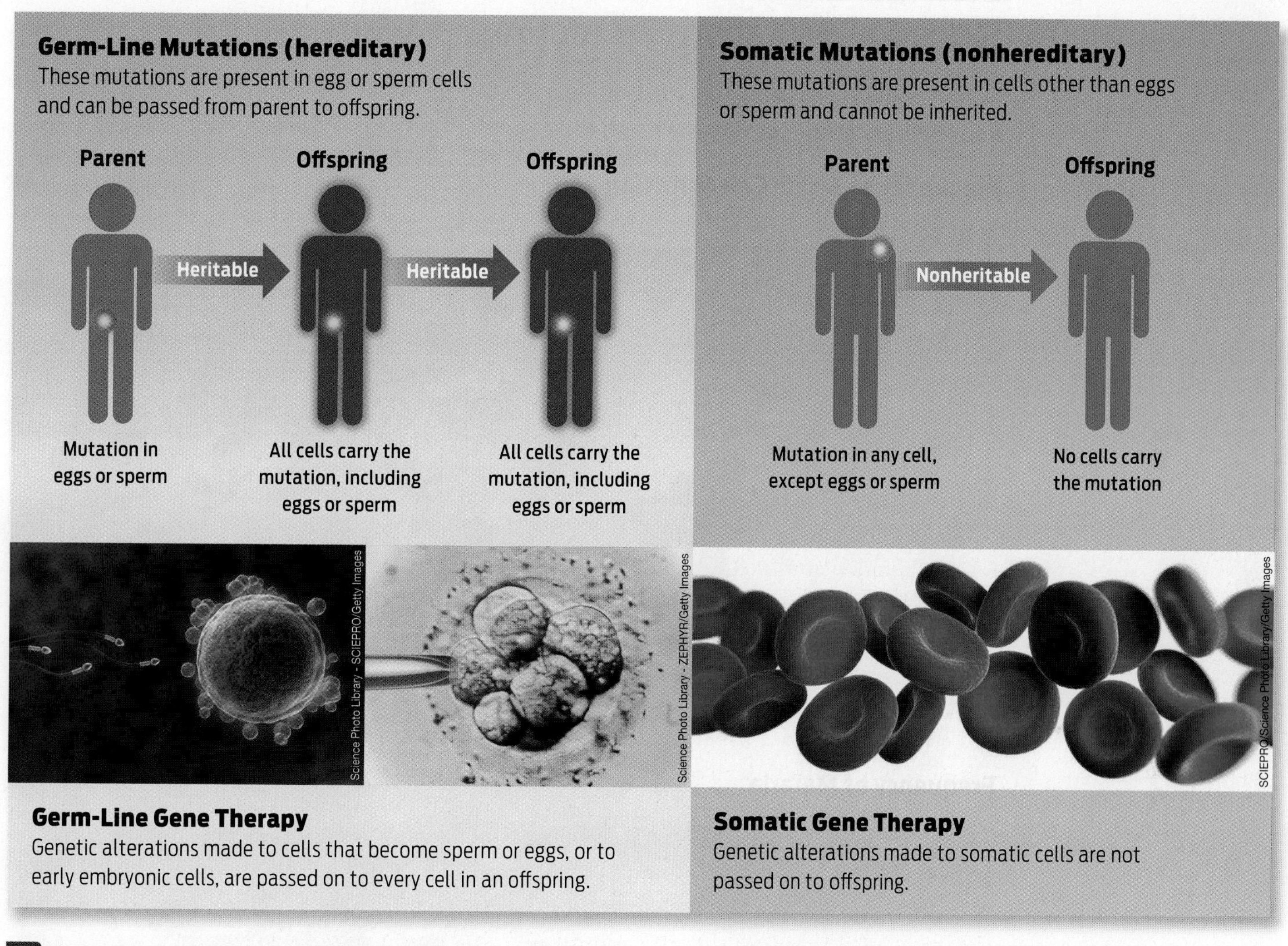

? Is the genetic alteration of hematopoietic stem cells a type of germ-line or somatic gene therapy? Explain your answer.

cells (see Milestone 5). In environments where malaria is common, having just one sickle cell mutation—the condition called sickle cell trait—provides protection against malaria infection. This is why the mutation continues to occur at high frequency in populations of people living in those areas.

In places without malaria, the mutation has no such protective effect. In addition, those who happen to inherit two copies of it suffer severe, painful consequences in the form of sickle cell disease.

In other words, whether a particular mutation provides a survival advantage, or is neutral or harmful, often depends on the environment where a person with that mutation lives. We'll have more to say about the relationship between mutation and evolutionary fitness in Chapter 13. Without mutations, there would be no evolution at all **(INFOGRAPHIC 9.8).**

Testing the Boundaries

Although gene therapy is not new, only a relatively small number of people have been

INFOGRAPHIC 9.8

Sickle Cell Mutations Protect Against Malaria

The frequency of the sickle cell mutation is higher in areas where malaria is or has been common. While having two copies of the sickle cell mutation is harmful, having one copy protects against malaria, which explains why this mutation is more frequent in certain parts of the world.

Frequency of Sickle Cell Mutation

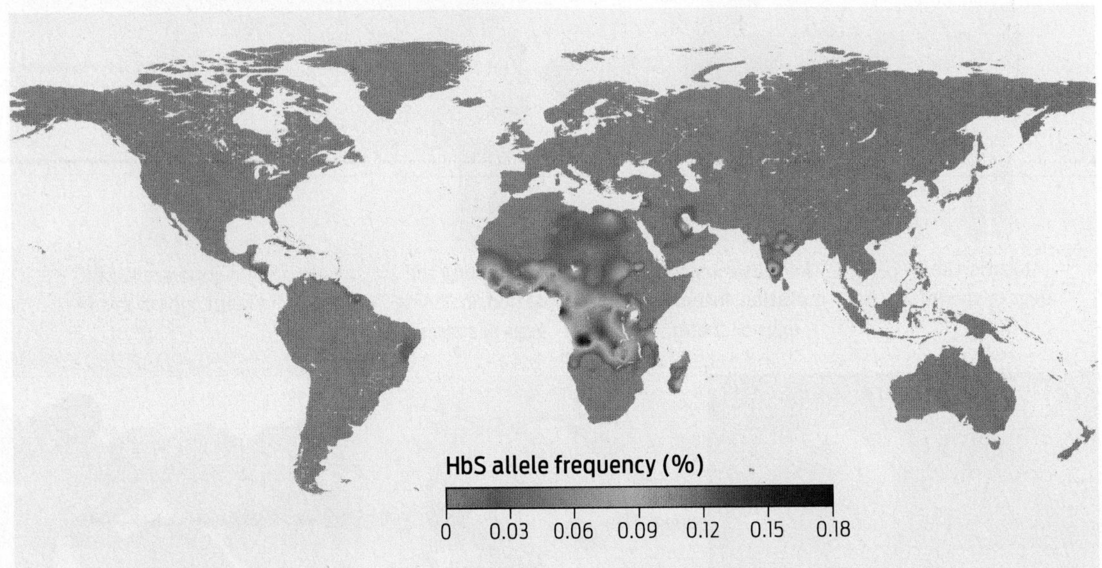

Piel, F.B., Patil, A.P., et al., Global epidemiology of sickle haemoglobin in neonates: a contemporary geostatistical model-based map and population estimates. *The Lancet*, 2013; 381: 142–151.

Frequency of Malaria

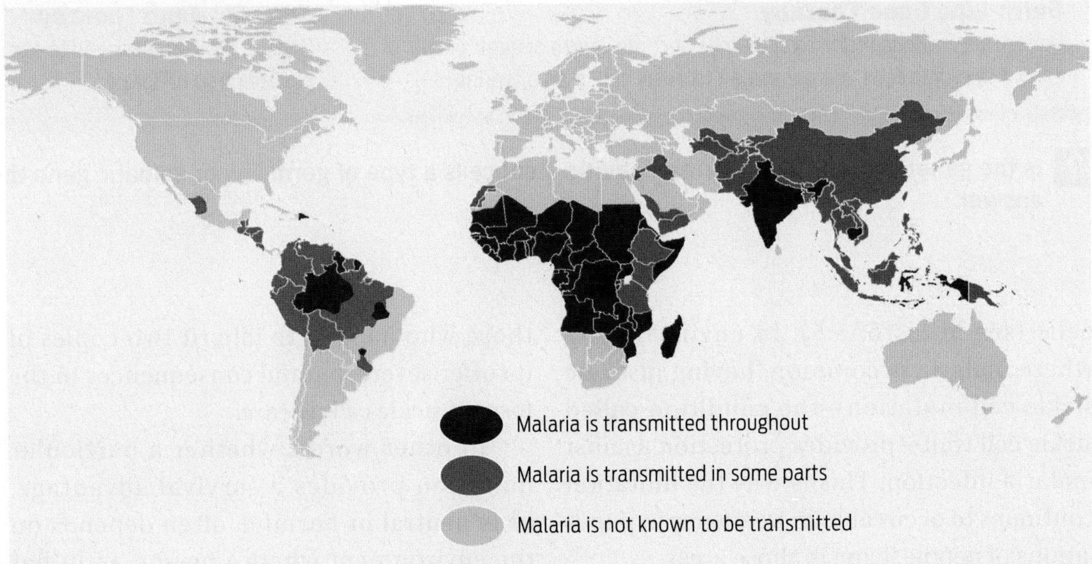

Data from the CDC.

? **Why is the sickle cell mutation present at high frequency in populations in Africa and South America?**

treated, and their short- and long-term outcomes are being closely monitored. Given past mistakes, scientists are understandably cautious about moving too fast with these powerful techniques. That's why news about the experiences of people like Jennelle and Manny is so eagerly awaited.

So far, both gene therapy approaches seem to be working. More than a year after his treatment, Manny hasn't had any pain crises or needed a transfusion. Doctors can see that his blood is full of round, healthy red blood cells. You wouldn't be able to tell his blood from someone who doesn't have sickle cell disease. Now 21, Manny is thinking about applying to college. "I feel great, I'm looking great, I'm trying to be great," he says.

It's been more than two years since Jennelle received her infusion of genetically modified blood cells. She could tell something was different within a few months.

At first it was little things, like being able to run up stairs without getting completely out of breath, or sitting on a cold train and not being in pain. Over time, she began to notice a dramatic change in herself. "I really felt like I wanted to do something to test my boundaries," Jennelle says.

She landed on the martial art of jujitsu. She enjoys the confidence she gets from being able to kick and fall on the mat without worrying that she's going to be in pain. She also feels more in synch with herself: "In my head, I've always been kind of a bubbly, smiley, outgoing person. But my body, my actions, my physical health was never able to match that."

"Now everything matches," she says. "This is me. This is who I was supposed to be." ■

Courtesy Jennelle Stephenson

Jennelle Stephenson practicing jujitsu after gene therapy for sickle cell disease.

CHAPTER 9 SUMMARY

Driving Question 1 What are mutations, what is their impact, and how do they occur?

- Mutations are changes in the nucleotide sequence of DNA.
- The different types of mutations include point mutations, insertions and deletions, translocations, and inversions.
- The effect of a mutation depends on where it occurs in the genome, and whether it changes the amino acid sequence of a protein.
- Mutations can occur spontaneously during DNA replication. They can also be caused by environmental triggers such as tobacco, ultraviolet radiation, chemicals, and viruses, and by chemicals naturally produced by the body.

Driving Question 2 How can genetic engineering be used to treat genetic diseases?

- Viruses are useful tools in genetic engineering because they make good vectors.
- CRISPR is a genome-editing tool adapted from enzymes found in bacteria. It can be used to make changes to DNA at specific locations.
- Mutations that occur in body (somatic) cells will be found only in the descendants of that particular cell. Mutations that occur in germ cells (sperm and eggs) will be inherited by offspring and therefore will be present in all the cells of that offspring's body.
- Using CRISPR to modify germ cells is not a currently accepted therapy for genetic diseases.

Driving Question 3 Are all mutations harmful?

- The impact of a mutation may vary, depending on the environment in which it's found.
- Mutations can be beneficial, harmful, or neutral in terms of the effect they have on survival and reproduction.

More to Explore

- Sickle Cell Disease, National Heart, Lung, and Blood Institute: https://www.nhlbi.nih.gov/health-topics/education-and-awareness/sickle-cell
- NIH Director's Blog: https://directorsblog.nih.gov/2019/04/02/a-crispr-approach-to-treating-sickle-cell/
- Sickle Gene Therapy Timeline: https://vector.childrenshospital.org/2018/01/sickle-cell-gene-therapy-bcl11a-timeline/
- Pauling, L., et al. (1949). Sickle-cell anemia, a molecular disease. *Science* 110:543–548.
- Wailoo, Keith. (2001). *Dying in the City of Blues: Sickle Cell Anemia and the Politics of Race and Health.* Chapel Hill, NC: University of North Carolina Press.

CHAPTER 9 Test Your Knowledge

Driving Question 1 What are mutations, what is their impact, and how do they occur?

By answering the questions below and studying Infographics 9.1, 9.2, 9.3, and 9.7 and Table 9.1, you should be able to generate an answer for this broader Driving Question.

Know It

1. What is a mutation?
  a. a cause of a disease
  b. a change in the structure of a protein
  c. a change in the function of a protein
  d. a change in the nucleotide sequence of DNA

2. At the DNA level, what is the sickle cell disease mutation?
  a. a change of a single codon in the mRNA
  b. a change of a single base pair in the gene sequence
  c. a change of a single amino acid in the protein sequence
  d. all of the above

3. Which type of cells are impacted in sickle cell disease?
  a. white blood cells
  b. all blood cells
  c. red blood cells
  d. heart cells

4. A mutation causes a substitution of one amino acid for another in the encoded protein. Which type of mutation is this?
  a. silent
  b. nonsense
  c. missense
  d. shift in reading frame caused by an insertion
  e. shift in reading frame caused by a deletion

5. Which of the following is a known mutagen?
  a. cigarette smoke
  b. sunlight
  c. charred meat cooked at high temperatures
  d. x-rays
  e. all of the above

Use It

6. How likely (very, potentially, not very) is each of the following forms of hemoglobin to cause symptoms of sickle cell disease? Explain your answer for each.
  a. hemoglobin A (2 alpha-globin subunits and 2 beta-globin subunits)
  b. hemoglobin S (2 alpha-globin subunits and 2 mutant beta-globin subunits)
  c. fetal hemoglobin (2 alpha-globin subunits and 2 gamma-globin subunits)

7. The mutation illustrated in Infographic 9.2 substituted a U for an A in the fourth codon of the mRNA shown.

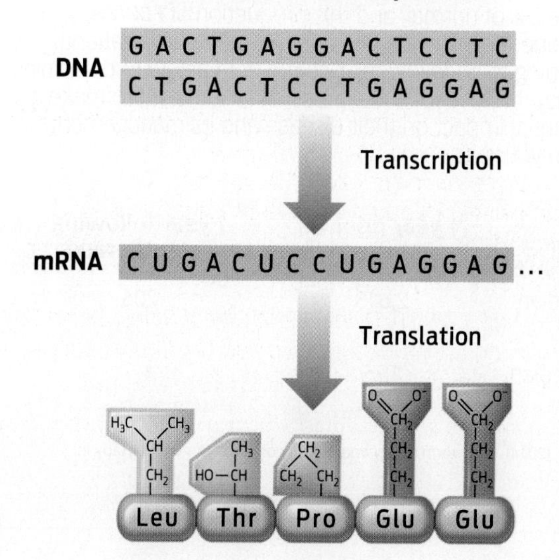

Below is a short list of mutations of the normal mRNA sequence shown in this infographic. Use the genetic code (Infographic 8.8, p. 178) to match each of the following mutations with both its effect on the protein and the type of mutation. For each mutation, put a check mark next to the corresponding effect and type of mutation.

Mutation	Effect of Mutation	Type of Mutation
Substitution of a C for the U in the third codon	_______ The protein will have an incorrect amino acid in its sequence. _______ No impact on the protein _______ The protein will be shorter than normal.	_______ Nonsense _______ Missense _______ Silent
Substitution of a U for the first G in the fifth codon	_______ The protein will have an incorrect amino acid in its sequence. _______ No impact on the protein _______ The protein will be shorter than normal.	_______ Nonsense _______ Missense _______ Silent
Substitution of an A for the C in the second codon	_______ The protein will have an incorrect amino acid in its sequence. _______ No impact on the protein _______ The protein will be shorter than normal.	_______ Nonsense _______ Missense _______ Silent

8. A mutation in codon #42 of a gene encoding a 97-amino-acid protein causes a shift in the reading frame. The shift causes different amino acids to be encoded by codons #42 and #43 and then produces a stop codon at codon #44. Based on this information, describe

a. the length (number of amino acids) of the mutant protein.

b. the number of identical amino acids between the normal (unmutated) and mutant protein.

c. how likely the mutant protein is to function properly.

Driving Question 2 How can genetic engineering be used to treat genetic diseases?

By answering the questions below and studying Infographics 9.4, 9.5, and 9.6, you should be able to generate an answer for this broader Driving Question.

Know It

9. Modifications of blood stem cells are

a. somatic changes, so they can be passed on to offspring.

b. somatic changes, so they will not be passed on to offspring.

c. germ-line changes, so they can be passed on to offspring.

d. germ-line changes, so they will not be passed on to offspring.

10. Why is a virus a good vector to deliver a new gene (like non-sickling beta-globin) into blood stem cells?

11. Manny's treatment uses CRISPR technology to edit the genome of blood stem cells, while Jennelle's treatment involves introducing a new beta-globin gene into her blood stem cells. What do their treatments have in common?

a. Both rely on genetic engineering techniques.

b. Both are designed to restore fetal hemoglobin production.

c. Both are designed to precisely correct the original beta-globin mutation.

d. Both cause permanent genetic alterations to germ-line cells.

Use It

12. Assume that Jennelle's treatment is successful (that the introduced gene continues to express a non-sickling form of beta-globin and she no longer has symptoms of sickle cell disease). If she were to have children in the future, would they inherit her mutated beta-globin and/or the therapeutic non-sickling beta-globin that she was treated with? Explain your answer.

Interpreting Data

Apply Your Knowledge

13. Hemophilia B is a genetic disease caused by a mutation in the gene (*F9*) encoding coagulation factor IX (FIX). People with hemophilia B have very low levels of FIX and are at high risk for episodes of spontaneous bleeding that can result in death. To maintain FIX levels of at least 1% of the normal level (the minimum needed to prevent spontaneous bleeding), patients must receive intravenous infusions of FIX every two to three days. A 2015 report estimated the cost of these infusions to be more than $270,000 per year for a single patient. Gene therapy (inserting a healthy *F9* gene) has been considered as a possible alternative treatment. Because the *F9* gene is expressed by liver cells, the *F9* gene must be introduced through a vector that can deliver it to liver cells.

One study of 10 patients used an "adeno-associated viral vector" known as AAV8. This viral vector has a DNA genome that is not inserted into the host cell's DNA.

All 10 patients stably produced FIX at levels between 1% and 6% of normal, and this production of FIX has been maintained for between 1.5 and 4.3 years (with patients continuing to be followed). Use the data below to determine whether this level of FIX expression is sufficient to make a substantial impact on their disease and its management. Explain your answer.

	1 year prior to gene therapy	1 year following gene therapy
Annual use of FIX (median units/kg per year)	2,613	206
Bleeding episodes (median number per year)	15.5	1.5

Data from Nathwani, A. C., et al. (2014). Long-term safety and efficacy of factor IX gene therapy in hemophilia B. *N Engl J Med* 371:21.

Bring It Home

Apply Your Knowledge

14. The annual cost to treat a patient with sickle cell disease is estimated to be more than $10,000 per year for patients younger than 10, and more than $30,000 per year for patients who are 50–64 years old. By age 45, total lifetime costs approach $1 million. The cost of gene therapy is not yet determined, but is estimated to be between $500,000 and $700,000 (as a one-time expense). Make arguments to justify the coverage (or not) of gene therapy by insurance providers.

Driving Question 3 Are all mutations harmful?

By answering the questions below and studying Infographic 9.8, you should be able to generate an answer for this broader Driving Question.

Use It

15. What is sickle cell trait?

a. a mild form of sickle cell disease
b. having a single copy of the mutated beta-globin gene
c. having sickle cell disease due to environmental rather than genetic traits
d. having abnormally shaped red blood cells in the absence of sickle cell disease

16. What disease does sickle cell trait protect against?

a. sickle cell disease
b. anemia
c. abnormal bleeding
d. abnormal blood clotting
e. malaria

Mini Case

Apply Your Knowledge

17. People who have two copies of a mutation in the *CCR5* gene are resistant to HIV infection—the virus cannot enter target cells when the protein encoded by the *CCR5* gene is mutated.

A 40-year-old male had been infected with HIV when he was 30. He had been treated with an advanced antiretroviral therapy drug cocktail for the past 4 years. He had not developed any AIDS-associated illnesses, and there was no HIV RNA detectable in his blood. It appeared that the anti-HIV therapy was successfully controlling the infection.

This patient developed a form of leukemia (a cancer of the blood). The treatment for most leukemias (as well as sickle cell disease) is a bone marrow transplant. In this case, several "matched" donors were identified. Having a matched donor minimizes the chances of transplant rejection. In this case, the transplant team was able to find a matched donor who also had two mutant copies of the *CCR5* gene. After the patient underwent chemotherapy to destroy his existing bone marrow, bone marrow stem cells from the donor were transfused into him. No anti-HIV therapy was used after the transplant, and the viral RNA levels remained undetectable (despite not taking the drug treatment).

a. Explain why this patient no longer needs anti-HIV therapy.

b. How practical is this approach as a general strategy for curing HIV?

18. When looking for mutations that protect against malaria, it made sense to look at populations living in regions where malaria is common.

a. Where should researchers look for mutations that protect against Ebola?

b. How should researchers look for mutations that increase life span (i.e., mutations that promote longevity)? (*Hint:* Think beyond geography; think about the types of populations researchers might study.)

c. How should researchers look for mutations that protect against type 2 diabetes in people who have risk factors for developing type 2 diabetes? (*Hint:* Think beyond geography; think about the types of populations researchers might study.)

10 Cell Division and Cancer

Fighting Fate

When cancer runs in the family, ordinary measures are not enough

© DR P. MARAZZI/SCIENCE SOURCE

DRIVING QUESTIONS

1. How and why do cells divide, and how is cell division regulated?
2. What factors contribute to the development of cancer?
3. How is cancer treated?

"Half of me was expecting it all my life and part of me was saying, 'No, this won't happen to me.'"

—Lorene Ahern

Lorene Ahern of Twinsburg, Ohio, wasn't totally surprised when she was diagnosed with breast cancer. "Half of me was expecting it all my life and part of me was saying, 'No, this won't happen to me,'" says the 47-year-old mother of two. She knew that her risk of cancer might be higher than average—her mother had died of cancer at 49. But until the day she was diagnosed, Ahern, who took good care of herself and had a healthy lifestyle, had never fully believed she would develop **cancer**, a disease in which cells divide repeatedly and without restraint.

There was more bad news in store for Ahern. About a year after she received the diagnosis of breast cancer, Ahern had DNA extracted from her blood and tested for mutations in two genes—*BRCA1,* located on chromosome 17, and *BRCA2,* located on chromosome 13 (*BRCA* stands for "***br****east* ***ca****ncer*"). These genes encode proteins that are important for repairing damaged DNA. Women who are born with mutations in one copy of either of these two genes have a much higher risk of developing breast and ovarian cancers.

The test was positive: Ahern had a mutation in one of her copies of the *BRCA1* gene. This meant that her breast cancer was likely caused by this mutation and could put her at high risk for developing a second cancer. Moreover, her children might also be at risk, if they inherited the mutation. When genetics are to blame, cancer can run in the family.

Aside from nonmelanoma skin cancer, breast cancer is the most common cancer affecting women worldwide. More than 250,000 women in the United States are diagnosed with breast cancer each year, according to the National Cancer Institute. For most women, the lifetime risk of developing breast cancer is about 12%, or 1 in every 8 women. For women with a mutation in *BRCA1* or *BRCA2,* however, the risk is much higher: on average, these women have a 45% to 65% lifetime risk of developing breast cancer and a 10% to 40% risk of developing ovarian cancer. In some families with particular *BRCA1* mutations, the risk of getting breast cancer can run as high as 80%.

Men are also affected by mutations in the two *BRCA* genes. Breast cancer is typically very rare in men, occurring in fewer than 1 in 1,000 men. But in men with a *BRCA* mutation, the risk rises to 5% to 10%. For most men, the risk of developing prostate cancer is about one in eight, or 12%. But for men with a *BRCA* mutation, that risk is one in four, or 25%. And for both men and women, *BRCA* mutations more than quadruple the risk of developing

CANCER
A disease in which cells divide repeatedly and without restraint, in some cases forming a tumor.

pancreatic cancer, the deadliest of all cancers **(INFOGRAPHIC 10.1)**.

The good news is that studies have shown that diet and lifestyle changes can dramatically cut a person's risk of developing many types of cancer—just quitting smoking reduces the risk by 30%. The bad news is that prevention is not that simple for people with inherited predispositions to cancer. For this group, diet and lifestyle changes make less of a difference. As scientists have learned, a genetic mistake stacks the deck against them by creating biological conditions that make developing cancer more likely. What's more, even after receiving treatment for cancer, people with mutations in one of the *BRCA* genes are at risk of developing a second cancer, such as in the other breast.

INFOGRAPHIC 10.1

Mutations in the *BRCA* Genes Increase the Risk of Cancer

People born with certain mutations in the genes *BRCA1* or *BRCA2* have a higher risk of developing cancer than people born with normal versions of the *BRCA* genes. A genetic test can ascertain whether a person carries any of the high-risk *BRCA* mutations.

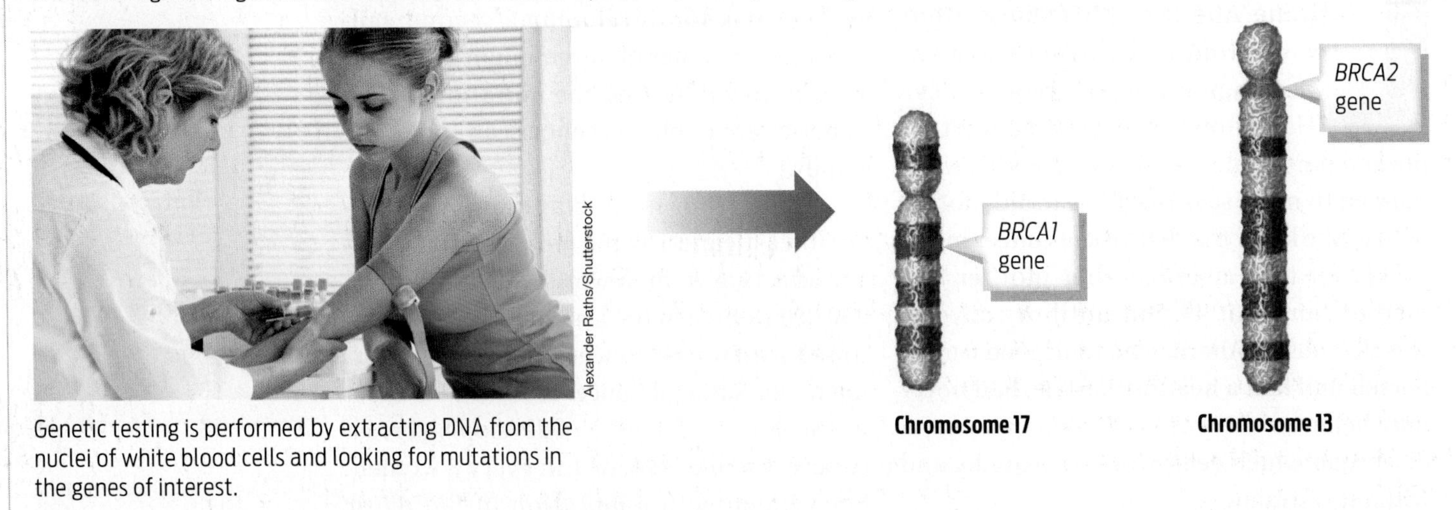

Genetic testing is performed by extracting DNA from the nuclei of white blood cells and looking for mutations in the genes of interest.

Women with one copy of a mutated version of either *BRCA* gene are at higher risk of developing breast and ovarian cancer in their lifetime. Men with these mutations are also at higher risk of developing breast and other cancers, including prostate cancer.

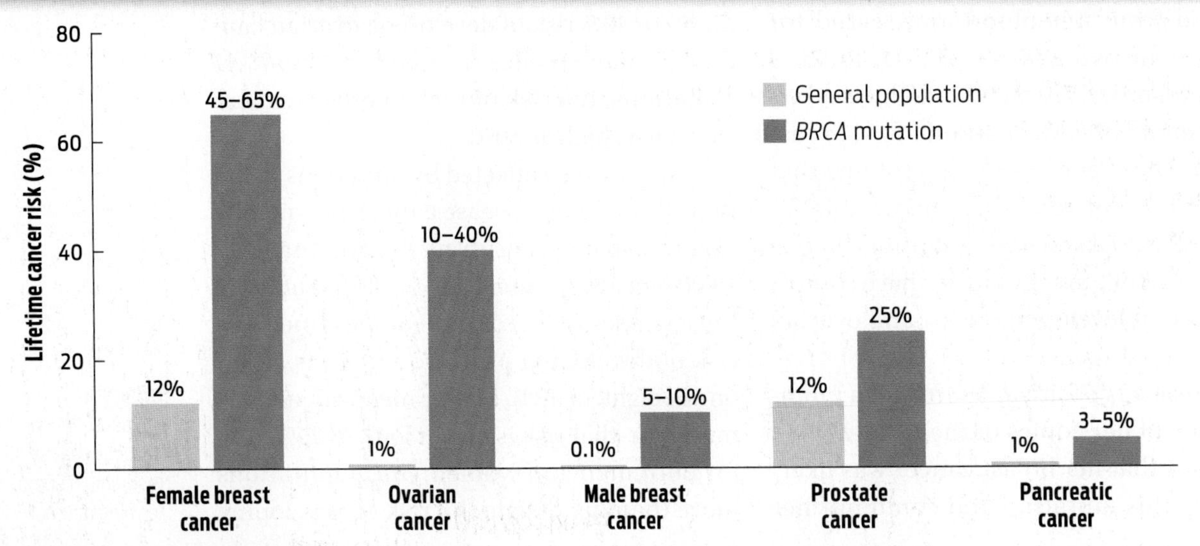

? Which cancers have a higher risk in both males and females with a *BRCA* mutation?

But some options can drastically reduce a person's risk of getting cancer, and other options are available for treating cancer if it occurs. In addition, the more scientists learn about how *BRCA* mutations increase cancer risk in some people, the better they are getting at treating cancer in all people.

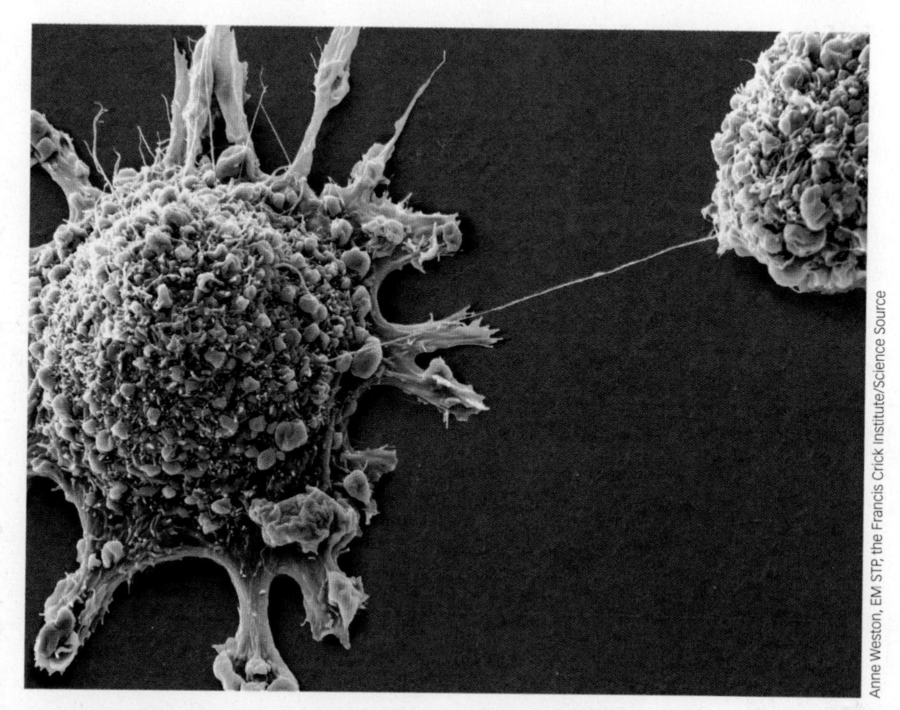

Anne Weston, EM STP, the Francis Crick Institute/Science Source

Breast cancer cells have an irregular shape, often with membrane extensions (green) that enable rapid crawling.

When Good Cells Go Bad: Cancer

▶ Cell division and the cell cycle

Cancer is a scary word. Most of us know someone who has had cancer, or maybe we have even lost a loved one to the disease. In the United States, one in two men and one in three women will develop cancer in their lifetimes. Cancer is the second leading cause of death among both men and women in the United States, right after heart disease.

Cancer has been in scientists'—and politicians'—crosshairs for decades. President Richard Nixon declared a "war on cancer" in 1971. Vice President Joe Biden launched a "cancer moonshot" in 2015. And yet, even after decades of research, cancer is still with us. One reason why cancer is so hard to beat is that it is not one disease, but many. Different tissues develop cancer as a result of different mutations, and the body has more than 200 different tissues. But all cancers do share something in common: they take advantage of what our bodies have evolved to do normally—grow and repair themselves through **cell division**.

Though we tend to think of our bodies as relatively fixed structures, most of our tissues are in constant flux as cells divide periodically to replace those cells that have reached the end of their life span. Cell division is a normal part of the development, growth, maintenance, and repair of the body. In fact, cell division in our bodies begins long before we are born. A single fertilized egg cell divides, and its daughter cells divide again and again, eventually forming trillions of cells by the time a baby is born. During childhood, cell division helps us grow larger. As we age, our tissues continually discard old cells and generate new ones in their place. And when we cut or injure ourselves, cells in the area divide to heal the wound. For nearly all cells—with the possible exceptions of adult nerve cells and heart muscle cells—cell division is a regular part of life **(INFOGRAPHIC 10.2)**.

To produce new cells by cell division, cells pass through a series of stages collectively known as the **cell cycle.** During the cell cycle, one cell becomes two. A cell doesn't simply split in half, however. If it did, each resulting cell would be smaller than the original, and with each division, each cell would lose half its contents. Instead, before a cell divides, it first makes a copy of its contents so that each new cell has the same amount of organelles, DNA, and cytoplasm as the original cell. This preparatory stage of the cell cycle, called **interphase,** is divided into three subphases: G_1, S, and G_2. During the G_1 phase, the cell grows larger and begins to produce more cytoplasm and organelles. During the synthesis phase (S), the cell's DNA is replicated—which means that its chromosomes are

CELL DIVISION
The process by which a cell reproduces itself; cell division is important for normal growth, development, maintenance, and repair of an organism.

CELL CYCLE
The ordered sequence of stages through which a cell progresses to divide. The stages include preparatory phases (G_1, S, G_2) and division phases (mitosis and cytokinesis).

INTERPHASE
The stage of the cell cycle in which dividing cells spend most of their time, preparing for cell division. There are three distinct subphases: G_1, S, and G_2.

duplicated. A replicated chromosome consists of two identical DNA molecules, known as **sister chromatids;** initially, this set of two sister chromatids is linked at their midpoint. During the G_2 phase, the cell prepares to enter the division phases. Different cell types divide at different rates—some very fast, some very slow. In a cell that takes approximately 24

INFOGRAPHIC 10.2

Why Do Cells Divide?

Cell division is a normal part of development, growth, maintenance, and repair of the body. A fertilized egg cell and its descendants divide repeatedly to generate the cells that make up a newborn. Specialized cells in tissues divide to replace cells that have reached the end of their lifespan. Cell division also acts to replace damaged cells in wounds.

Embryonic Development

A fertilized egg and its daughter cells continue to divide to create the trillions of cells that make up the human body.

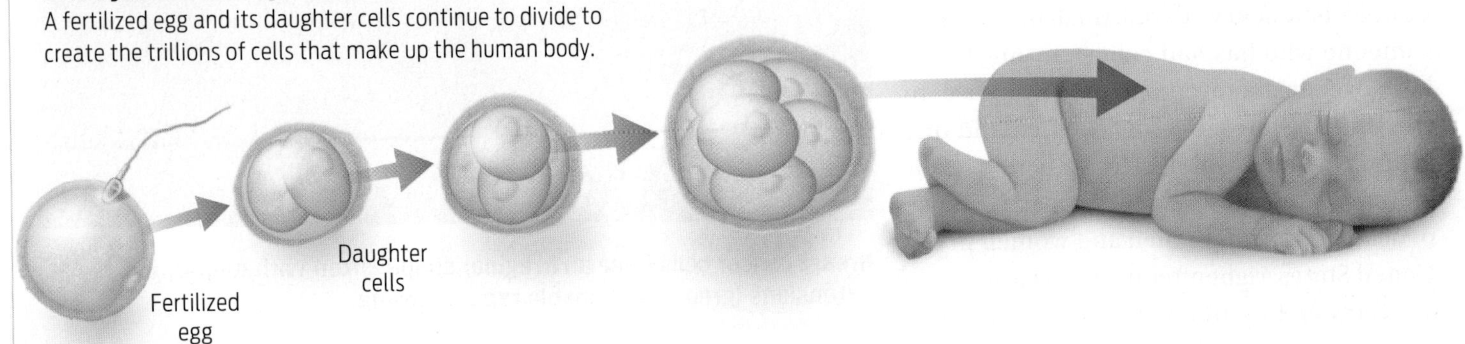

Cell Replacement

Most cells have a finite life span. Cell division within tissues regularly replaces the dying cells and maintains healthy tissues.

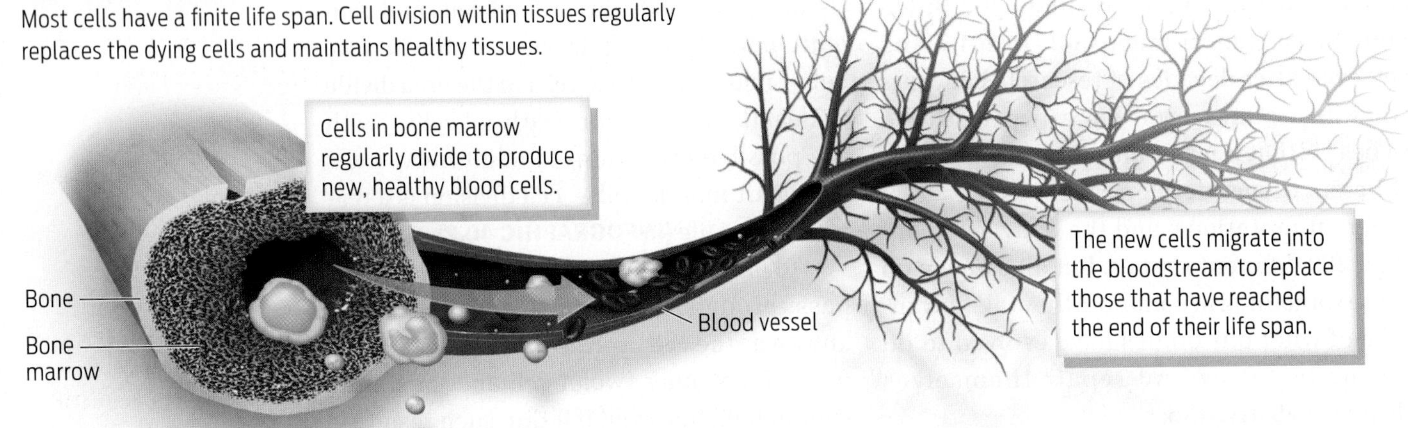

Repairing Damaged Tissue

Injury triggers cell division to replace damaged cells.

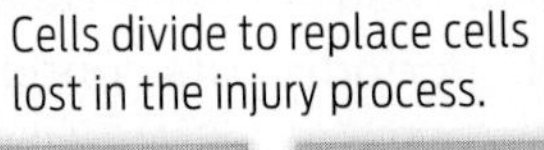

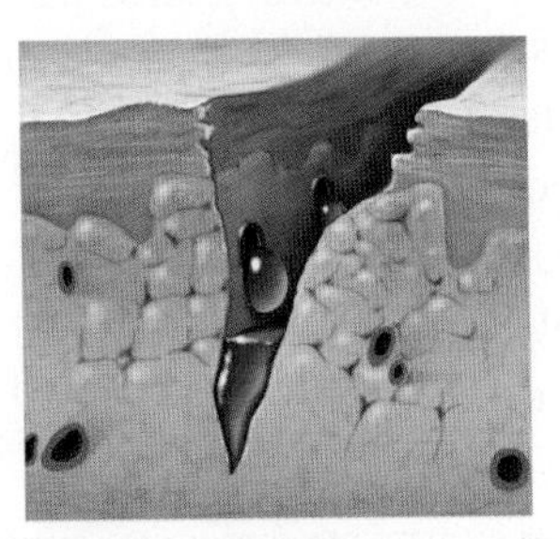

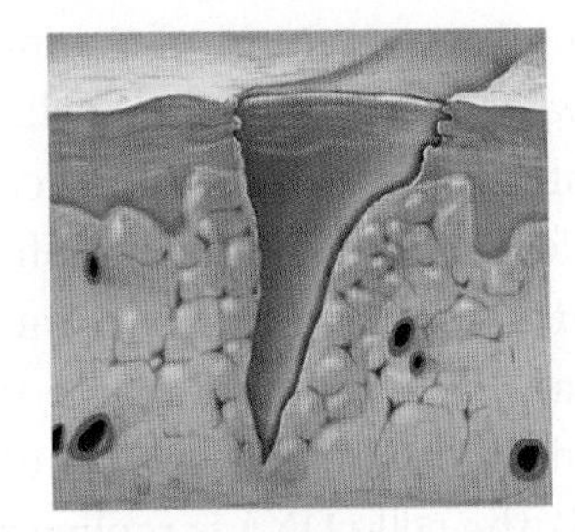

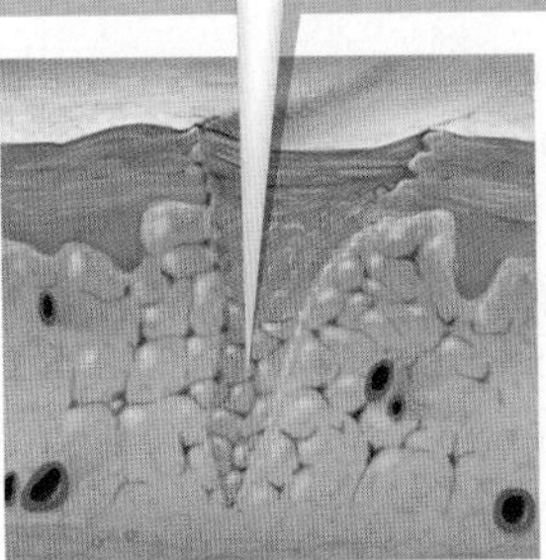

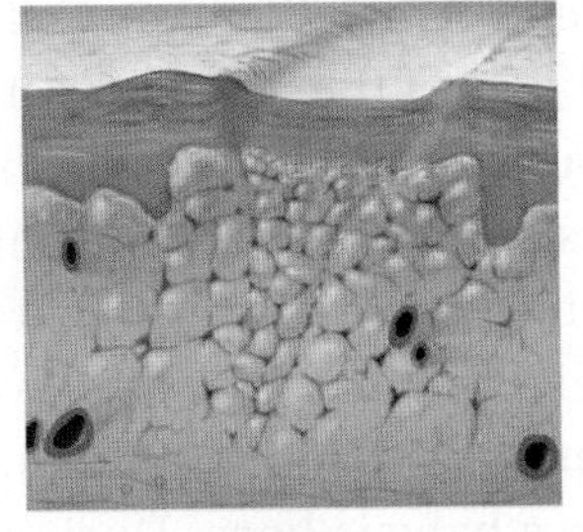

 What type of cells are produced by division of cells in the bone marrow?

hours to divide, interphase takes about 22 hours to complete.

Once the cell duplicates its contents, it enters the two division phases of the cell cycle: **mitosis,** in which the sister chromatids of each replicated chromosome become evenly divided between the two daughter cells; and **cytokinesis,** in which the two daughter cells physically separate. Once the cell is finished dividing, it may enter a resting phase of the cell cycle, called G_0. During G_0, the cell goes about its business as usual, but is not preparing to divide. Some cells in the body, like adult nerve cells, are in a permanent state of G_0 **(INFOGRAPHIC 10.3).**

Normally, cells divide only in response to appropriate growth signals relayed by molecules known as growth factors. Growth factors are critical in regulating cell division. For example, when tissue is damaged, growth factors are released. These growth factors signal cells in the area that they need to divide to repair the damage. When cells no longer need to divide—for example, when a wound has healed or when worn-out tissues have been replaced—the growth signals are turned

SISTER CHROMATIDS
The two identical DNA molecules that result from the replication of a chromosome during the S phase.

MITOSIS
The segregation and separation of replicated chromosomes during cell division.

CYTOKINESIS
The physical division of a cell into two daughter cells.

INFOGRAPHIC 10.3

The Cell Cycle: How Cells Reproduce

The purpose of the cell cycle is to replicate cells, creating two new daughter cells that are genetically identical to the original parent cell. The cell cycle consists of preparatory phases collectively known as interphase, as well as the division phases, mitosis and cytokinesis.

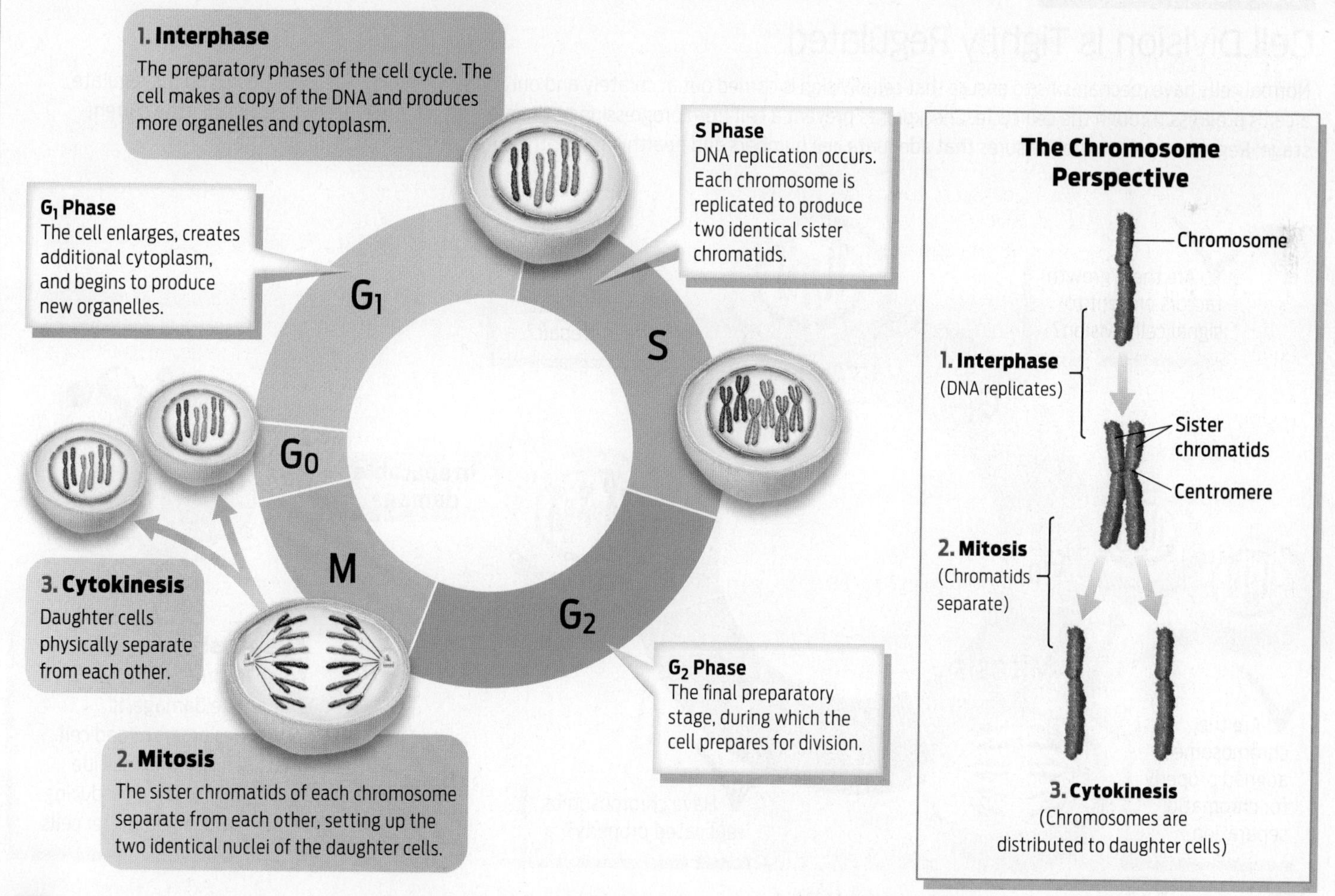

? **What is the general function of interphase, and what stages of the cell cycle are part of interphase?**

off and cells take a break from dividing. Cancer cells, unlike normal cells, divide even in the absence of growth factors and without stopping. In essence, cancer is cell division run amok.

A System of Checks and Balances

▶ Mutations and cell cycle checkpoints

What causes cells to "go rogue"? Cancer results when cells accumulate DNA sequence changes—mutations—that interfere with the orderly steps of the cell cycle. As we saw in Chapter 9, mutations can happen in a variety of ways. Every time a cell replicates its DNA, for example, there is a small chance that it will make a mistake—insert the wrong nucleotide, for instance. Environmental insults, like smoking, infections, or a bad sunburn, can also damage DNA. Substances that are known to cause cancer, or increase its risk of forming, are called **carcinogens.** Most carcinogens are also mutagens—physical or chemical agents that cause mutations in DNA.

Cells have a natural barrier against the accumulation of mutations: a system of **cell cycle checkpoints** that monitors the cell's progress through the cell cycle and prevents cells from moving into the next phase if problems arise. At one checkpoint, for example, proteins scan DNA for damage or incorrect base pairing. Another checks that chromosomes are properly aligned in mitosis. If problems are detected, one of two things happens: either the cell pauses the cycle and gives itself time to fix DNA mistakes, or, in the case of

CARCINOGEN
Any substance that causes cancer. Most carcinogens are mutagens.

CELL CYCLE CHECKPOINT
A cellular mechanism that ensures that a stage of the cell cycle is completed accurately.

INFOGRAPHIC 10.4

Cell Division Is Tightly Regulated

Normal cells have mechanisms to ensure that cell division is carried out accurately and only when necessary. Cell cycle checkpoints regulate a cell's progress through the cell cycle. Checkpoints prevent a cell from progressing to the next stage until it accurately finishes the current stage. Regulated cell division ensures that adequate cell numbers and healthy tissue structure are maintained in the body.

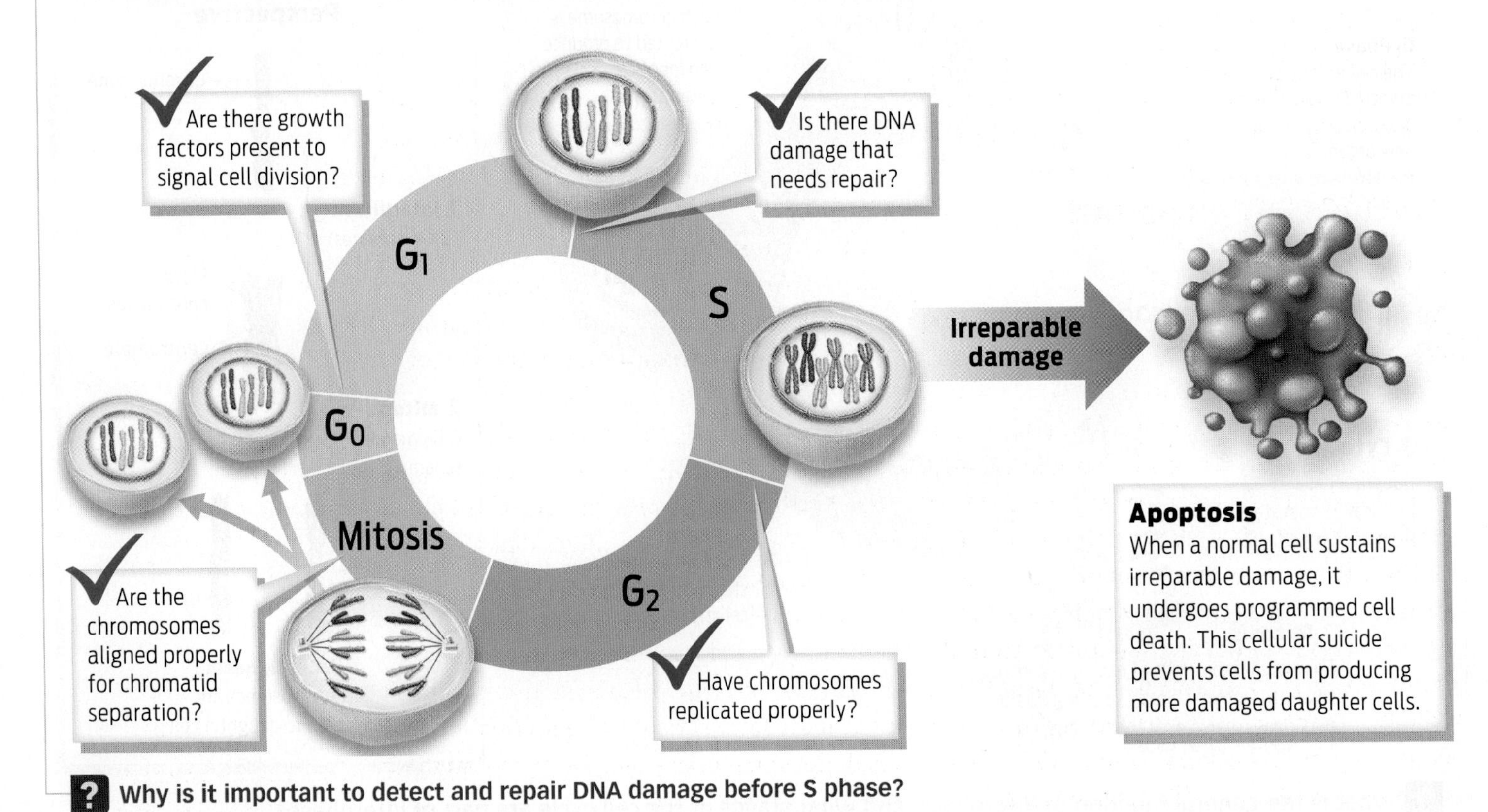

? **Why is it important to detect and repair DNA damage before S phase?**

severe and irreparable damage, the checkpoints direct a cell to commit suicide in a process called **apoptosis.** At least one checkpoint is found in each phase of the cell cycle: G_1, G_2, S, and mitosis **(INFOGRAPHIC 10.4)**.

Though checkpoint mechanisms ensure that the vast majority of DNA replication mistakes are fixed, on average one uncorrected mistake occurs for every 10 billion DNA base pairs that are replicated. That may not seem like a lot, but consider this: each human cell has 6 billion total base pairs, and some body cells divide trillions of times in our lifetimes. With so many opportunities for something to go wrong, the number of mutations that can occur in any given cell quickly adds up.

Equally important, mutations will accumulate faster if the proteins that monitor and repair DNA damage or trigger apoptosis are themselves damaged by mutation. Defective checkpoint proteins don't do their jobs, so additional mistakes continue to accumulate in these cells.

When cells accumulate enough mutations to interfere with multiple checkpoints, the result is cancer. Cancer cells plow through the cell cycle unimpeded, divide uncontrollably, and in many cases eventually form a mass of cells called a **tumor.**

Sorting Chromosomes

▶ Phases of mitosis

Mistakes can happen at any point in the cell cycle, but those that occur during mitosis can be especially catastrophic. When a cell divides, each daughter cell must receive the correct number of chromosomes—in humans, that means no more and no less than a complete set of 46 chromosomes. Mitosis is the complex and carefully orchestrated process by which a dividing cell imparts the correct number of chromosomes to each daughter cell. If any chromosome is left out of a daughter cell, all the genetic information contained in that chromosome will be missing, which could spell disaster for the cell.

Environmental insults, like smoking, infections, or a bad sunburn, can also damage DNA.

You can think of mitosis as a carefully choreographed dance of chromosomes. At the beginning of the dance, the replicated chromosomes, each with two sister chromatids, are present in the nucleus as loosely gathered threads. To avoid becoming entangled, the chromosomes condense into tightly wrapped, rod-shaped structures that can move about the cell more easily. The two identical sister chromatids of a replicated chromosome are held together at a region called the **centromere.** As mitosis progresses, these linked sister chromatids move across the cell and line up along the midline of the cell. Once they are aligned, the two sister chromatids of each chromosome will eventually be pulled apart and dragged to different sides of the cell. As a result, when the cell goes through cytokinesis, each of the two daughter cells will have one of each sister chromatid.

The cellular machine that orchestrates all this pushing and pulling is called the **mitotic spindle.** The mitotic spindle is made up of microtubules, the same protein fibers that make up part of the cell's cytoskeleton. At the beginning of mitosis, microtubule spindle fibers extend from each end of the cell and snag the centromere region of each chromatid, such that each chromatid in a chromosome is attached to a spindle fiber extending from the opposite end of the cell. By alternating between lengthening and shortening, the spindle fibers move the attached chromosomes until they are aligned at the midpoint of the cell. Toward the end of mitosis, the spindle fibers then shorten to pull each sister chromatid away from its partner to the opposite side of the cell. Each chromatid will eventually form one of two genetically identical chromosomes—one for each daughter cell **(INFOGRAPHIC 10.5)**.

APOPTOSIS
A type of cell death; often referred to as cellular suicide.

TUMOR
A mass of cells resulting from uncontrolled cell division.

CENTROMERE
The specialized region of a chromosome where the sister chromatids are joined; it is critical for proper alignment and separation of sister chromatids during mitosis.

MITOTIC SPINDLE
The microtubule-based structure that separates sister chromatids during mitosis.

INFOGRAPHIC 10.5 UP CLOSE

Phases of Mitosis

Interphase

- Each chromosome replicates in S phase, resulting in two sister chromatids connected at the centromere.
- Chromosomes are loosely gathered in the nucleus.

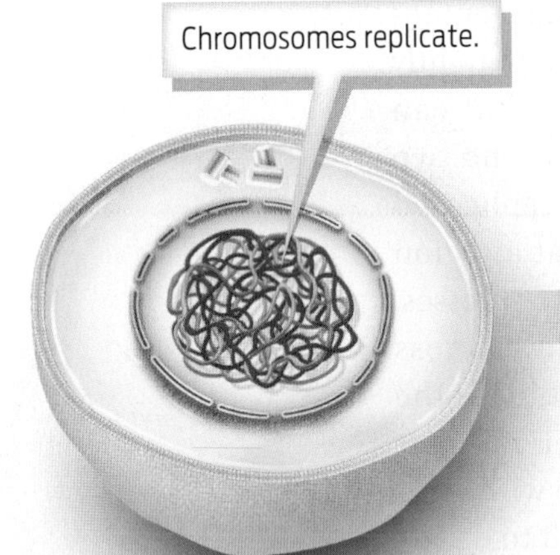

Mitosis

Prophase

- Replicated chromosomes begin to coil up.
- The nuclear membrane begins to disassemble.
- Microtubule fibers begin to form the mitotic spindle.

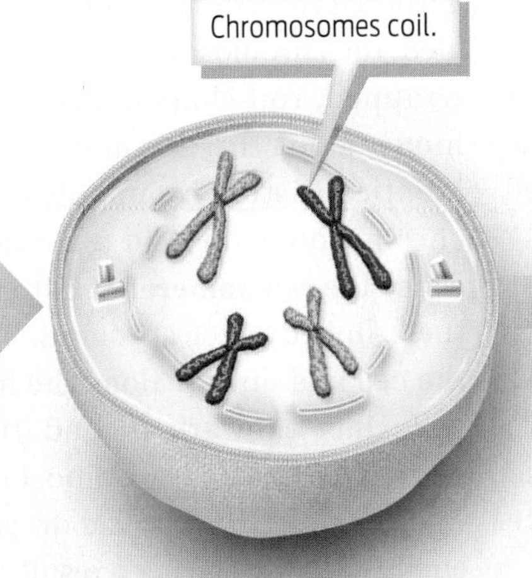

Metaphase

- Microtubule spindle fibers from opposite ends of the cell attach to the sister chromatids of each chromosome.
- Replicated chromosomes become aligned along the middle of the cell.

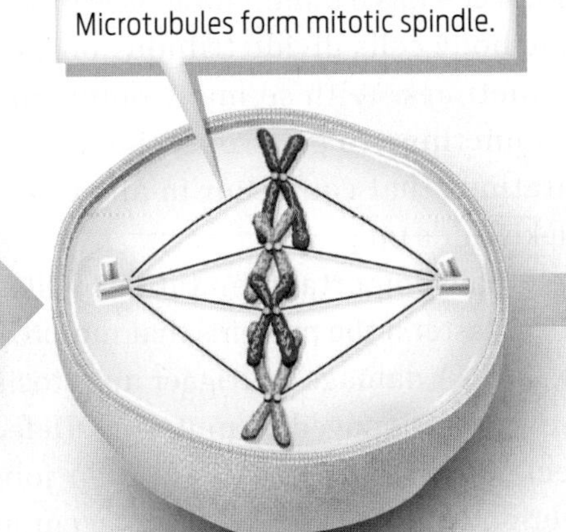

Animal Cells:

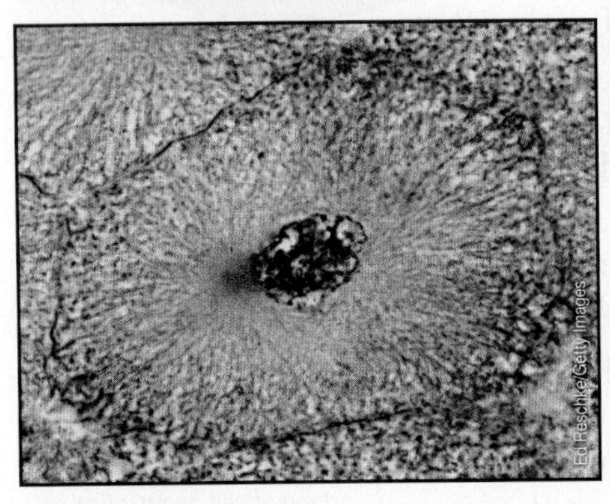

Ed Reschke/Getty Images

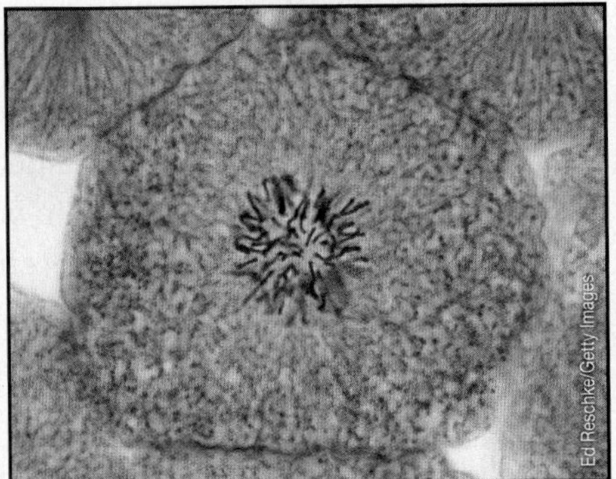

Ed Reschke/Getty Images

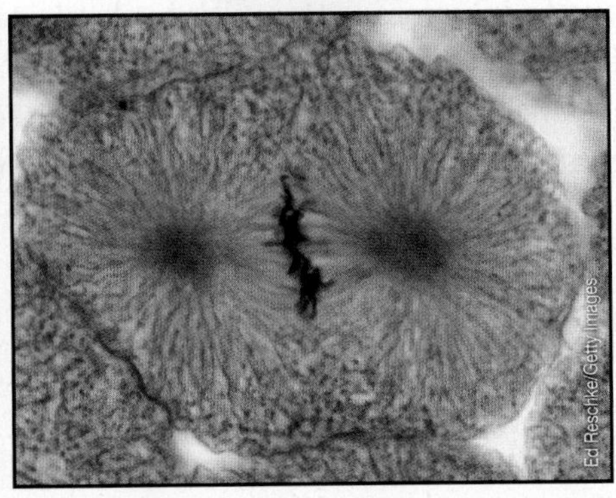

Ed Reschke/Getty Images

Plant Cells:

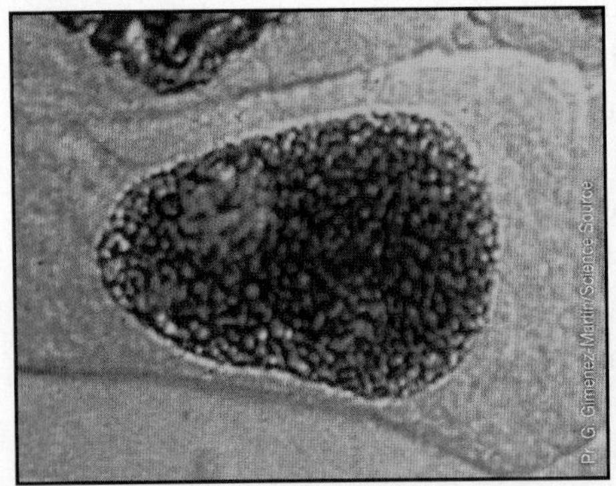

P. G. Gimenez-Martin/Science Source

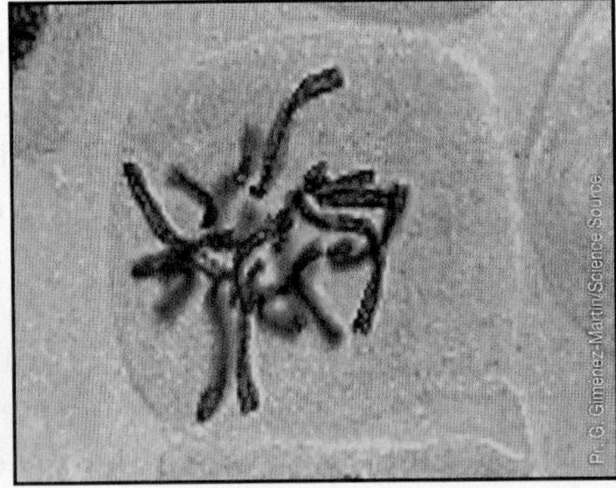

P. G. Gimenez-Martin/Science Source

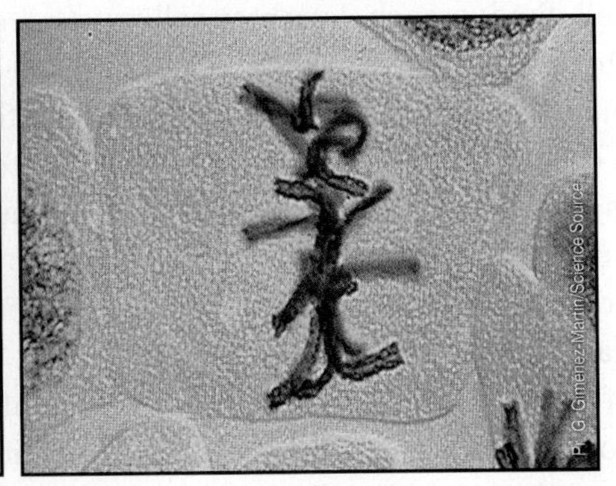

P. G. Gimenez-Martin/Science Source

Anaphase

- Microtubules shorten, pulling the sister chromatids to opposite ends of the cell.

Telophase

- Identical sets of chromosomes reach each pole.
- Microtubule spindle fibers disassemble.
- A nuclear membrane forms around each set of chromosomes, forming the daughter cell nuclei.

Cytokinesis

- Cytoplasm divides.
- Two nuclei become separated into daughter cells.

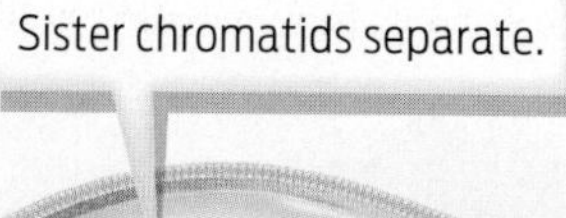

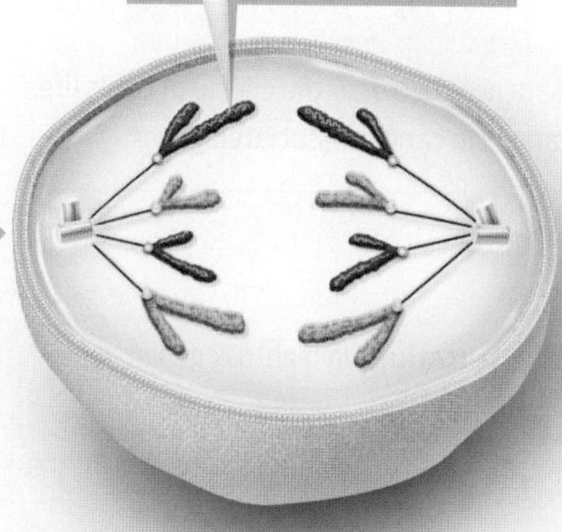

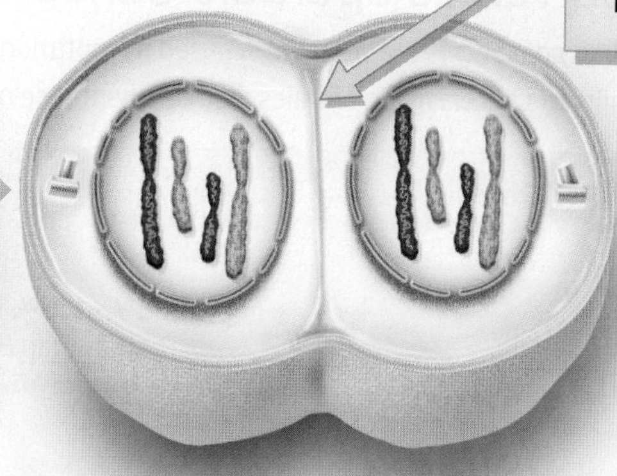

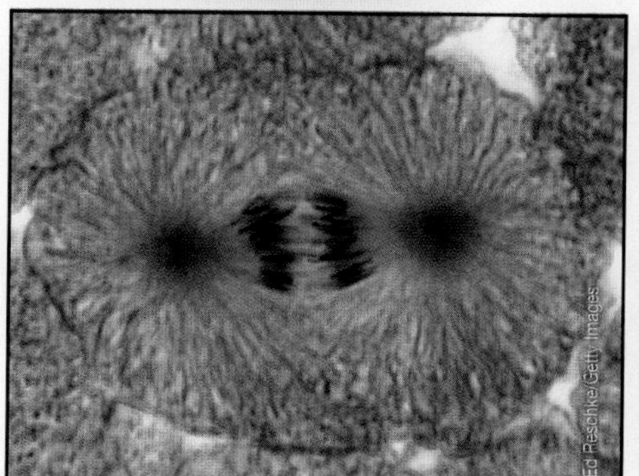

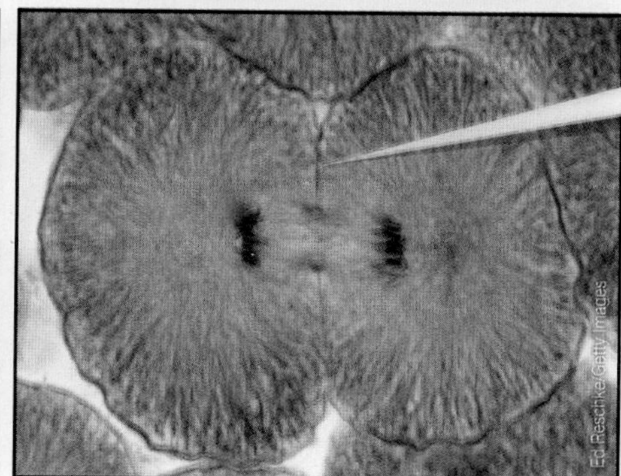

During cytokinesis in animal cells, the cell membrane pinches in to separate the daughter cells.

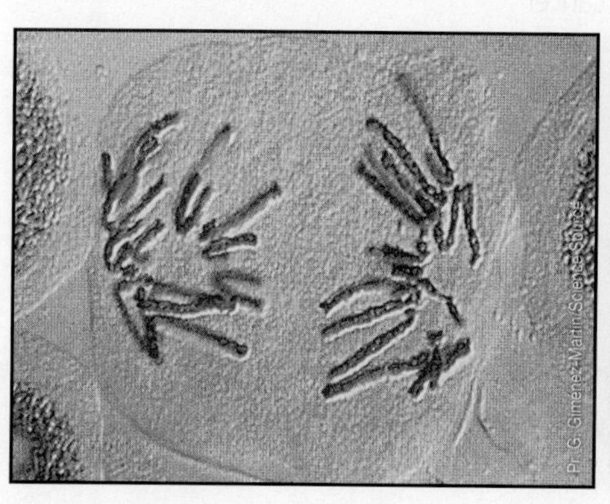

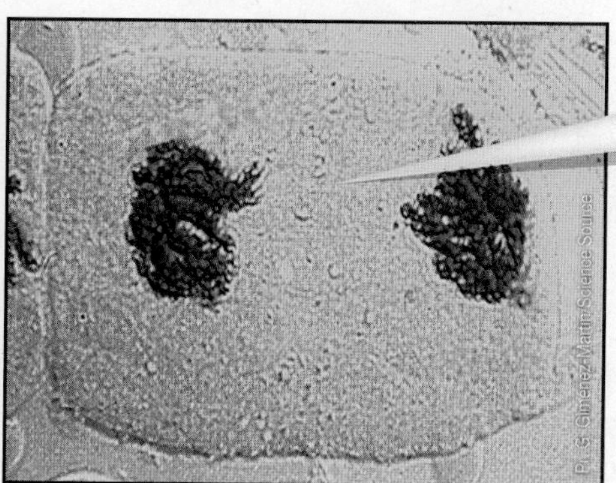

During cytokinesis in plant cells, a new cell wall is synthesized between the daughter cells.

Inherited Cancer

▶ The *BRCA* genes and DNA repair

Women who, like Ahern, have a genetic predisposition to getting cancer are often said to have "hereditary" or "inherited" cancer. This does not mean that their cancer was passed from parent to child, the way that eye or hair color is. Instead, it means that they have inherited a genetic mutation, from one or both parents, that makes the development of cancer much more likely. In other words, the cancer itself is not inherited, but the risk of getting it is.

Doctors have known for decades that breast cancer runs in certain families: a woman's chance of getting the disease is greatly increased if she has a sister or mother who has it. But it wasn't until the 1990s that researchers homed in on the reason:

INFOGRAPHIC 10.6

Inherited versus Sporadic Cancer

The vast majority of cancers are sporadic—caused by mutations that are acquired over the course of one's life. A small minority of cancers run in families, due to inheritance of mutations in cancer-associated genes.

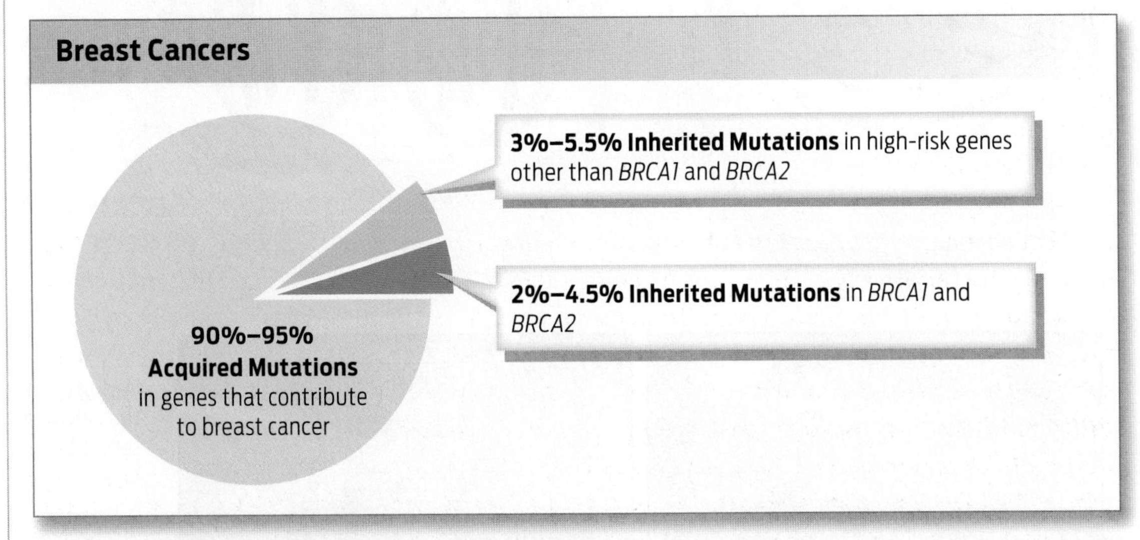

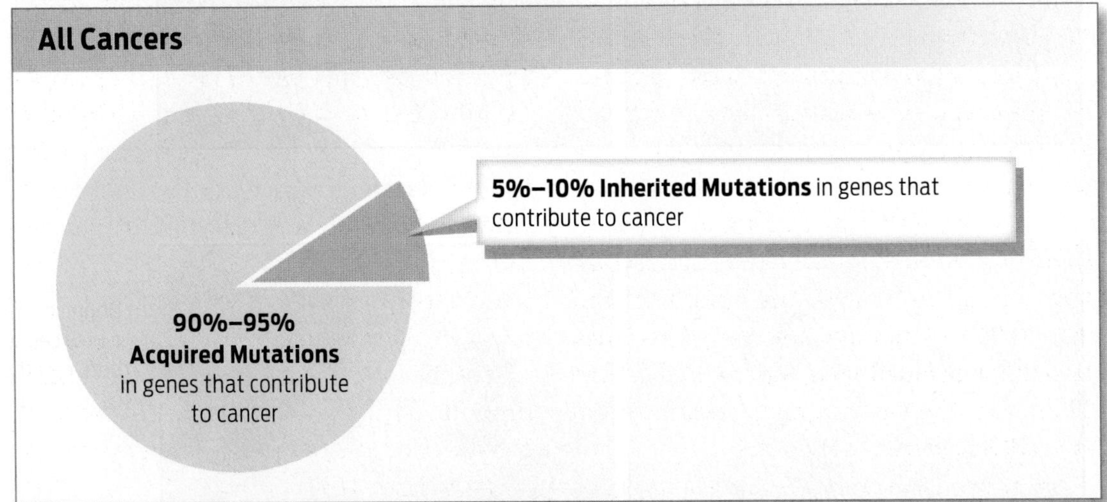

? Are most cancers sporadic or inherited? Be sure you can explain the difference between sporadic and inherited cancers.

mutations in a particular region of chromosome 17 are unusually common in these families. By the middle of the decade, scientists had isolated the specific mutated gene, which they called *BRCA1,* and determined its function. Identification of *BRCA2* occurred soon thereafter.

In their normal form, the BRCA1 and BRCA2 proteins help repair DNA damage—specifically, double-strand breaks, in which both sides of the DNA "ladder" are broken. They spring into action when a checkpoint detects this type of DNA damage and help correct it. When these proteins are mutated, cells can no longer properly repair this damage adequately. As a result, additional genetic mistakes—more mutations—accumulate in these cells, eventually leading to cancer.

When someone is born with a mutation in one copy of either *BRCA* gene—inherits it from mom or dad—that person's DNA repair system is compromised for their entire life. Having only one functional copy of either *BRCA* gene means that the person has a lower capacity to repair DNA. As a result, cells throughout their body are prone to accumulating DNA mistakes at faster rates than the cells of someone who did not inherit a *BRCA* mutation and has two working copies of the BRCA proteins. The accumulation of unrepaired DNA mistakes greatly increases an individual's risk of developing cancer. Why mutations in *BRCA* genes tend to cause only certain types of cancer—such as breast, ovarian, and prostate cancers, for example, but not lung or colon cancer—is not yet understood.

Scientists estimate that inherited mutations in genes like *BRCA1* or *BRCA2* account for about 5% to 10% of all cancers. The other 90% to 95% of cancers are caused by mutations that occur during a person's lifetime. Therefore, the vast majority of cancers are **sporadic,** rather than inherited.

However, the same gene mutations that can predispose a person to cancer when they are inherited are often found in sporadic cancers, too. What scientists are learning about inherited mutations may also apply to those who acquire mutations over their lifetime **(INFOGRAPHIC 10.6).**

A Numbers Game

▸ Oncogenes, tumor suppressors, and the multi-hit model

Mutations in *BRCA* genes are an important driver of both inherited and sporadic cancer, but they aren't the only ones. There are hundreds of known genes that can promote cancer when they are mutated. Most of these genes can be categorized into one of two types: **proto-oncogenes** and **tumor suppressor genes.** Normal proto-oncogenes activate cell division, but only in response to appropriate signals from growth factors. When proto-oncogenes are mutated, they can become permanently turned on, stimulating cells to divide all the time, even in the absence of appropriate signals from growth factors. In this state they are called **oncogenes**—literally, "genes that cause cancer." In other words, oncogenes are proto-oncogenes that have mutated to become permanently activated. *Her2,* a gene frequently mutated in certain types of breast cancer, is an example of a proto-oncogene. It provides a "go" signal for cell division when a growth signal is present. Cancer cells have more copies of the HER2 protein, which becomes active even without a growth signal. The drug Herceptin blocks this overactive HER2.

Tumor suppressor genes encode proteins that normally prevent cell division when

Scientists estimate that inherited mutations in genes like BRCA1 *or* BRCA2 *account for about 5% to 10% of all cancers.*

SPORADIC
Cancers that are caused by non-inherited (acquired) mutations.

PROTO-ONCOGENE
A gene that codes for a protein that helps cells divide normally.

TUMOR SUPPRESSOR GENE
A gene that codes for a protein that monitors and checks cell cycle progression. When these genes mutate, tumor suppressor proteins lose normal function.

ONCOGENE
A mutated and overactive form of a proto-oncogene. Oncogenes drive cells to divide continually.

problems arise that need to be corrected. These "stop" signals may pause cell division, repair damaged DNA, or initiate apoptosis. Tumor suppressor genes cause cancer when they are inactivated by mutation. More than 50% of all cancers have a mutation in the *p53* tumor suppressor gene, for example. Without a properly functioning p53 protein, cells cannot commit suicide in response to DNA damage, and mutations accumulate in cells as a result. The BRCA proteins are also tumor suppressors—they help repair DNA damage.

INFOGRAPHIC 10.7

Mutations in Two Types of Genes Cause Most Types of Cancer

Proto-oncogenes push the cell to divide when the appropriate signals are present. Mutations in these genes push the cell to divide, even in the absence of these signals. Tumor suppressor genes act to detect problems (such as DNA damage) and pause the cell cycle, or cause the cell to carry out apoptosis. Mutated tumor suppressor genes allow the cell to continue to divide, even if there is DNA damage.

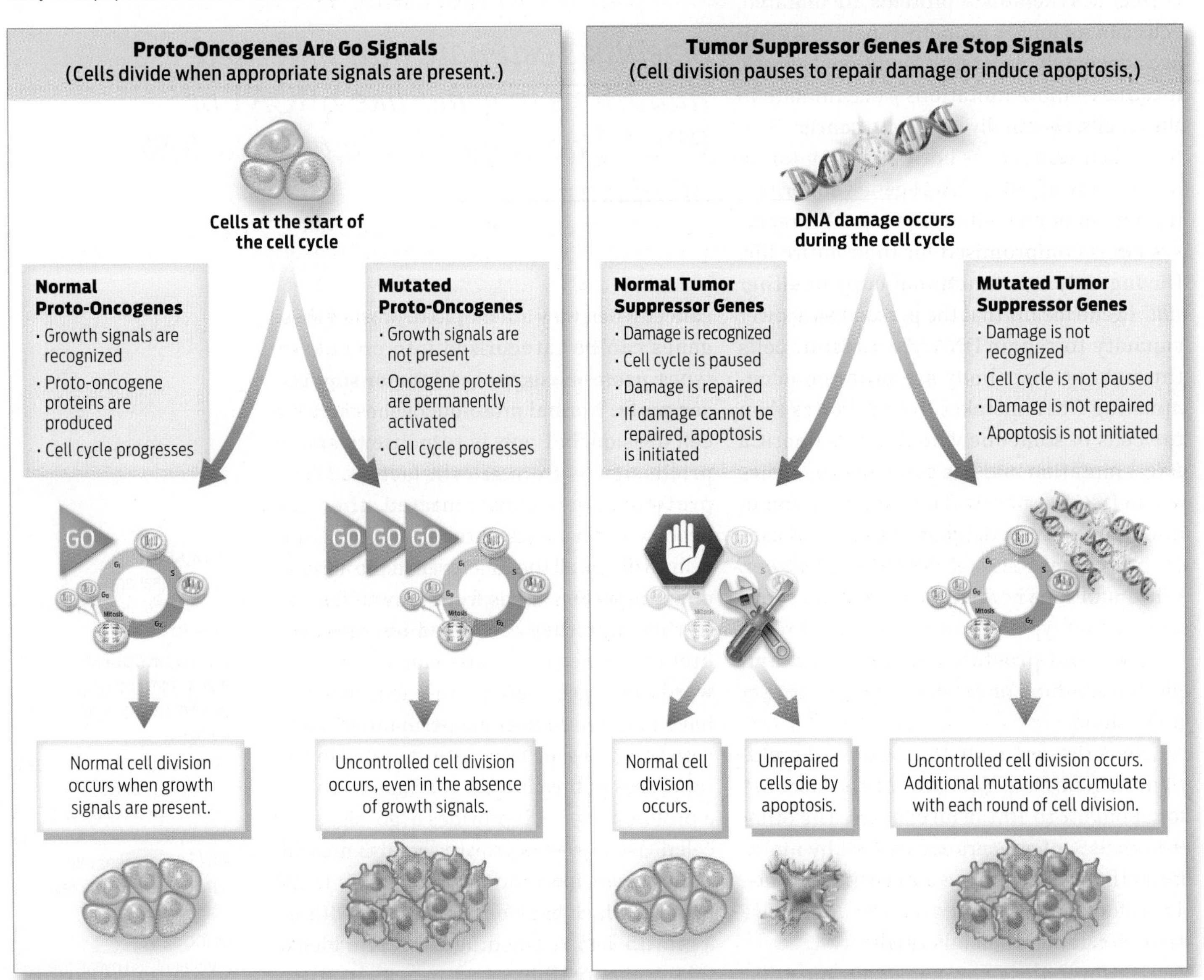

? A mutation has inactivated a protein involved in detecting DNA damage. Will this cell divide or not divide after exposure to DNA-damaging radiation?

"You can think of tumor suppressors as brakes and the oncogenes as the accelerators," says Thomas Sellers, a cancer biologist at the H. Lee Moffitt Cancer Center and Research Institute in Tampa, Florida. "They are sort of the yin and yang of each other." If a proto-oncogene is mutated, it's as if the accelerator is stuck down and cell division keeps going, even if the cell should not be dividing; if a tumor suppressor gene is mutated, it's as if the brakes don't work and the cell division cannot be stopped, even if problems occur **(INFOGRAPHIC 10.7)**.

It usually takes more than a single mutation in a cell to cause cancer. In most cases, a cell will become cancerous only after it has acquired multiple mutations in several genes that regulate the cell cycle or repair DNA damage. This is known as the multi-hit model of cancer, in which each "hit" is a mutation, and multiple hits are needed to cause the disease.

The collection of mutated genes can include a combination of tumor suppressor genes that have lost their function and proto-oncogenes that have been activated to become oncogenes. The need for multiple hits is the main reason cancer affects people more as they age: as cells accumulate mutations over time through exposure to carcinogens and repeated rounds of cell division, the chances increase that a cell will accumulate enough mutations to become cancerous.

A person who inherits a mutation in one copy of the *BRCA1* or *BRCA2* gene starts life with at least one cancer-predisposing mutation. As a result, he or she requires fewer additional mutations to develop cancer. Moreover, because it disrupts DNA repair, a single mutation in a *BRCA* gene is enough to increase the chances that additional mutations will occur in that cell: DNA damage caused by a replication error or exposure to carcinogens will go unrepaired—or be repaired incorrectly—leading to a permanent change in the nucleotide sequence of DNA. Daughter cells will accumulate mutations at a faster rate, which is why hereditary cancers often strike people who are in their 30s and 40s—much younger than people who have no inherited predisposition to cancer. Typically, by the time a person with an inherited predisposition develops cancer, their other (normal) copy of the *BRCA* gene will have been mutated as well.

> *"You can think of tumor suppressors as brakes and the oncogenes as the accelerators. They are sort of the yin and yang of each other."*
>
> **—Thomas Sellers**

Although any family may carry mutated *BRCA1* or *BRCA2* genes, certain ethnic groups are more likely than others to have them. Ahern's background is Ashkenazi Jewish, meaning that her ancestors were Jews of German and Eastern European descent. Ashkenazi Jews are more likely than the general population to carry mutations in *BRCA1* and *BRCA2*. Some studies have found that approximately 3% of Ashkenazi women carry a mutated *BRCA1* gene, compared to 0.2% of other women. For that reason, doctors may recommend that Ashkenazi Jewish women get genetic testing to determine if they carry a *BRCA* mutation.

The Road to Cancer

▶ Steps of cancer progression

Though cancers originate from a single cell that has "gone rogue," cancerous tumors do not emerge in a full-blown state overnight. Instead, they tend to develop in stages, becoming more aggressive over time. After one or two genetic hits or mutations, a dividing cell may form clump of cells called a

BENIGN TUMOR
A noncancerous tumor whose cells will not spread throughout the body.

MALIGNANT TUMOR
A cancerous tumor whose cells can spread throughout the body.

METASTASIS
The spread of cancer cells from one location in the body to another.

CONTACT INHIBITION
A characteristic of normal cells that prevents them from dividing once they have filled a space and are in contact with their neighbors.

benign tumor. Cells in a benign tumor divide more frequently than do cells in normal tissues, but they do not spread to other areas and so are usually much less dangerous. Some benign tumors remain benign, but others can turn cancerous. After several more mutations occur in this population of genetically abnormal cells, a **malignant tumor** may result. Malignant tumors have the capacity to **metastasize**—that is, they can spread to other locations in the body and form new tumors there. Metastasis is the main way that cancer kills.

Cancer cells have other distinctive properties that set them apart and contribute to malignancy. Normal cells stop dividing once they come into contact with neighboring cells, a property called **contact inhibition.** As genetic mutations accumulate, cancerous cells usually lose this property. The result is a pile of cells growing on top of each other. Normal cells also usually require connections to the tissue to which they belong, property called **anchorage dependence.** Cancer cells typically lose this constraint as well, so they are able to detach from their neighboring cells and spread. Finally, cancer cells promote the growth of new blood vessels, or **angiogenesis,** as a means of acquiring oxygen and nutrients for growth **(INFOGRAPHIC 10.8).**

The Cancer Therapy Toolbox

▶ Types of cancer treatments

For many types of cancer, the first line of treatment is surgery to remove the tumor. In the case of breast cancer, that can mean

INFOGRAPHIC 10.8

Tumors Develop in Stages as Mutations Accumulate in a Cell

It takes more than a single mutation to cause cancer. Individuals who have inherited high-risk mutations require fewer additional mutations to get to cancer, and therefore develop cancer at a much earlier age.

A. Inherit *BRCA1* mutation.
DNA damage will not be efficiently repaired. Additional mutations are more likely to occur and be passed to additional cells during cell division (green cells).

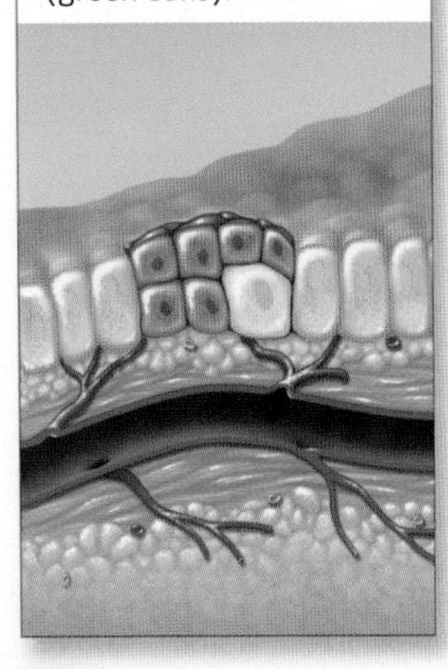

B. A carcinogen damages DNA and produces an oncogene.
The oncogene is overactive, allowing the cell to divide more often without normal checks. Cells begin to pile up into a tumor.

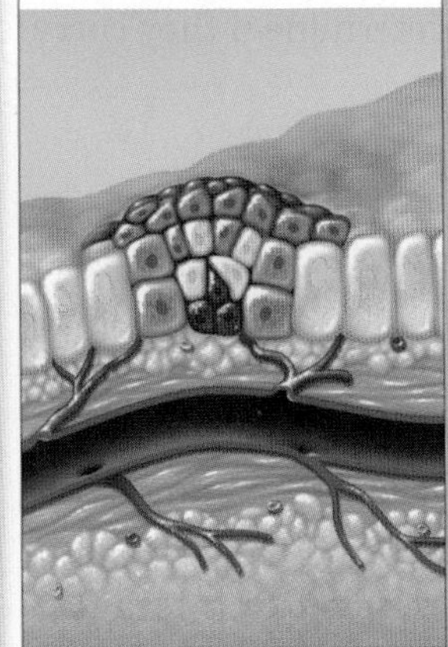

C. Smoking mutates tumor suppressor gene.
Cells fail to correct DNA damage or enter apoptosis. Cells have lost contact inhibition, dividing even when crowded.

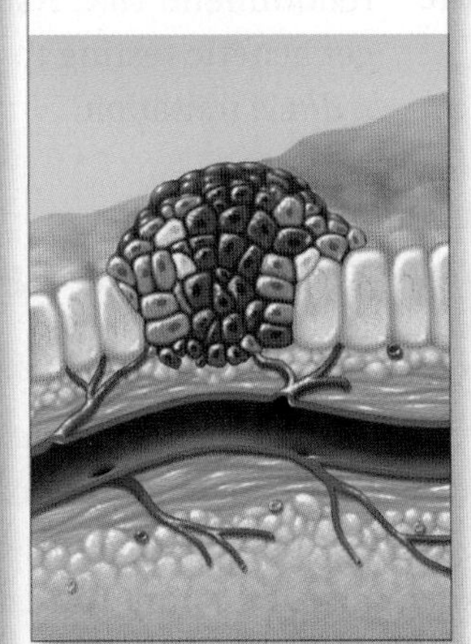

D. Additional mutations permit metastasis and new blood vessel growth.
The tumor contains malignant cells that lose their connections with the tissue, allowing them to invade surrounding tissues and spread to nearby and distant locations in a process called metastasis. The cells of the tumor also promote new blood vessel growth, allowing delivery of oxygen and nutrients.

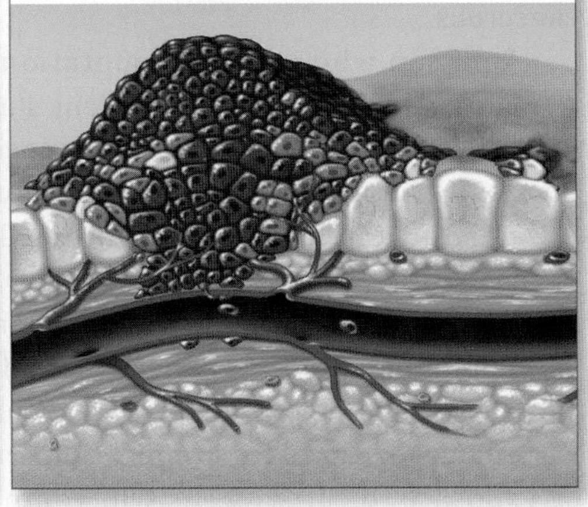

? If a person with an inherited *BRCA1* mutation develops breast cancer, is the *BRCA1* mutation likely to be the only mutation in the malignant cells? Why or why not?

complete removal of the breast, called a mastectomy. Alternatively, a surgeon may remove just the tumor plus a small amount of surrounding tissue, a procedure called a lumpectomy. Surgery is generally effective for solid tumors that are diagnosed early, but not for cancers that have spread to other parts of the body, or for cancers that do not produce tumors, like leukemia. In these cases, the best option is usually **chemotherapy**—using toxic chemicals to kill the cancer cells in the body. Most chemotherapy drugs work by interfering with one or more steps of the cell cycle in dividing cells.

Doctors may also treat a tumor with radiation. In **radiation therapy,** beams of ionizing (high-energy) radiation are focused on a tumor. The radiation damages DNA in dividing cells. Cancer cells are less able to repair this DNA damage than are normal cells, so they die.

The downside of both chemotherapy and radiation is that they can cause severe side effects. That's because neither therapy is very specific—both treatments damage all dividing cells in their path, including healthy ones. In particular, chemotherapy routinely kills healthy cells lining the intestinal tract, cells in hair follicles, and cells in the bone marrow. These cells share a common property: they all normally divide frequently. The side effects of these unintended cell deaths can include vomiting, bruising, hair loss, and susceptibility to infection.

In recent years, new forms of cancer treatment have become more common, including targeted therapy and immunotherapy. **Targeted therapies** are designed to kill cancer cells specifically. Often they combat a specific defect in the cancer cell, such as the mutated protein made from an oncogene or tumor suppressor. For example, drugs called PARP inhibitors are an effective treatment for *BRCA*-mutated cancers, including breast, ovarian, prostate, and pancreatic cancers. PARP inhibitors block the activity of an enzyme that contributes to DNA repair.

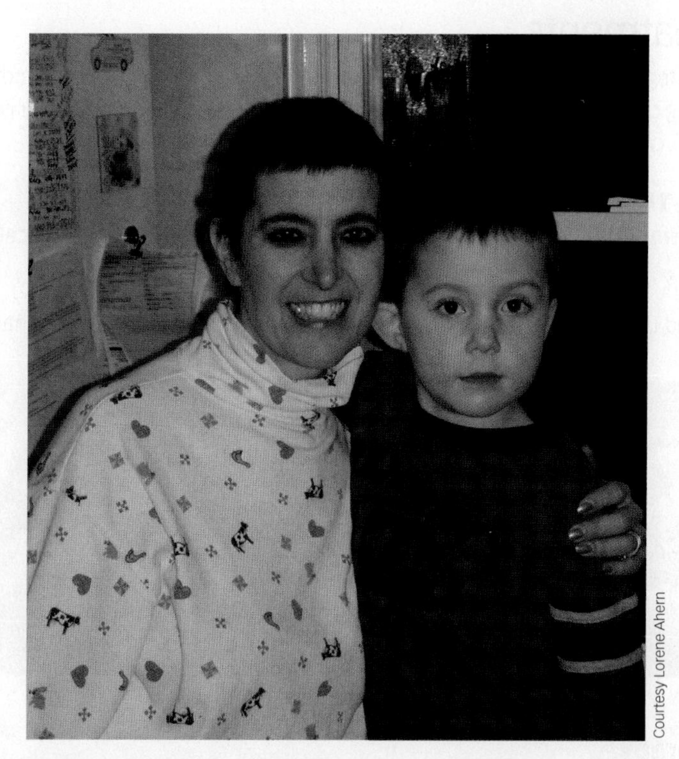
Courtesy Lorene Ahern

Lorene Ahern with her son. Her hair was just growing back after chemotherapy.

Normal cells with functioning BRCA proteins can repair DNA damage in the presence of PARP inhibitors, but *BRCA*-mutant cells cannot and consequently die.

Immunotherapies are drugs or treatments that stimulate the body's own immune system to find and fight cancer within the body. For some types of cancer—including melanoma, lung, and kidney cancer—this approach has proved very effective. Several forms of immunotherapy are available. In one common approach, drugs called checkpoint inhibitors are used to temporarily release a natural brake on immune cells that keeps the immune system in check. Once this brake is released, the immune cells can find and attack the cancer. Some cancers actively engage this brake to avoid being killed by immune cells. By releasing the brake, the checkpoint inhibitor robs the cancer of its defenses. These drugs can sometimes have serious immune-related side effects, and not everyone responds to them, but

ANCHORAGE DEPENDENCE
The need for normal cells to be in physical contact with another layer of cells or a surface.

ANGIOGENESIS
The growth of new blood vessels.

CHEMOTHERAPY
Treatment using toxic chemicals that kill cancer by interfering with cell division.

RADIATION THERAPY
The use of ionizing (high-energy) radiation to treat cancer.

TARGETED THERAPY
A cancer therapy that is specific for cancer cells and not harmful to normal cells.

IMMUNOTHERAPY
A cancer therapy that uses the immune system to recognize and destroy cancer cells.

INFOGRAPHIC 10.9

Cancer Treatments

There are many strategies to treat cancer. Surgery removes a tumor or at-risk tissue; chemotherapy and radiation therapy kill rapidly dividing cells in the body, including cancer cells. Newer treatment approaches specifically target cancer cells or improve the ability of the immune system to recognize and attack cancer cells.

Conventional Therapy

Conventional therapies are generally nonspecific, in that they can harm normal cells as well as cancer cells.

Surgery
removes diseased tissues.

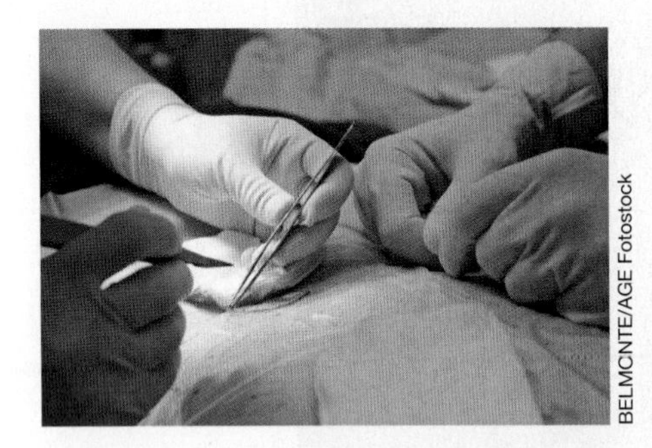

BELMONTE/AGE Fotostock

Chemotherapy
is a full-body chemical treatment that kills all rapidly dividing cells.

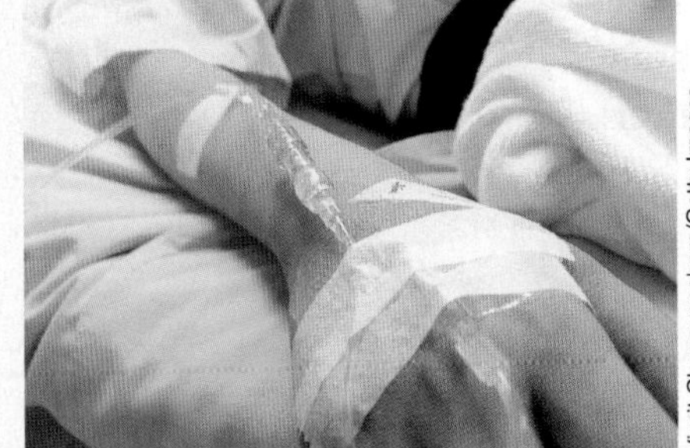

Virojt Changyencham/Getty Images

Radiation
kills exposed cells by severely damaging DNA.

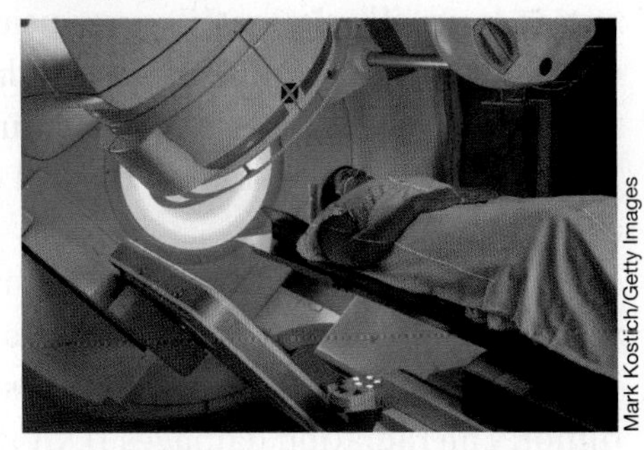

Mark Kostich/Getty Images

Targeted Therapy

Targeted therapies kill cancer cells specifically by exploiting weaknesses caused by oncogenes or mutated tumor suppressor genes.

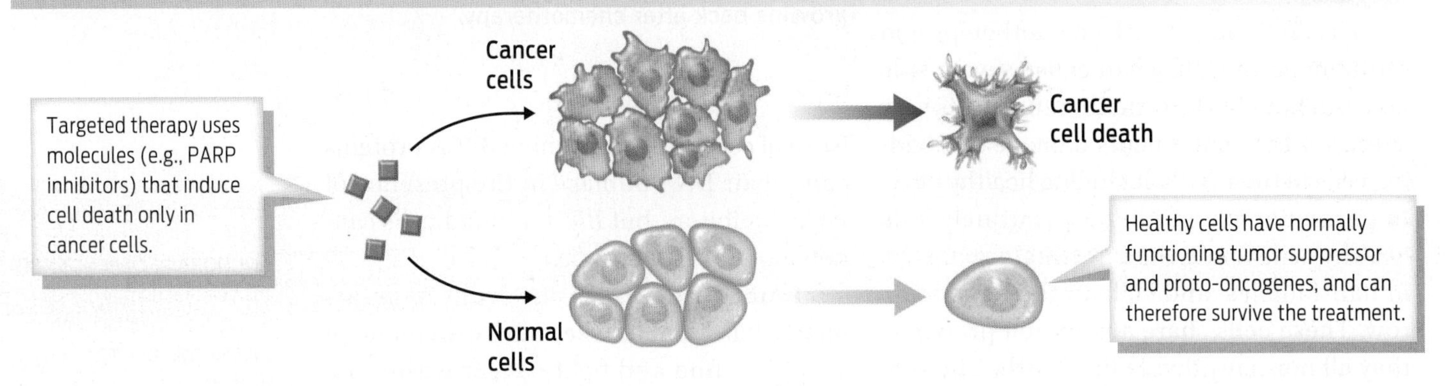

Immunotherapy

Immunotherapies stimulate the immune system to find and attack cancer.

Without Immunotherapy:

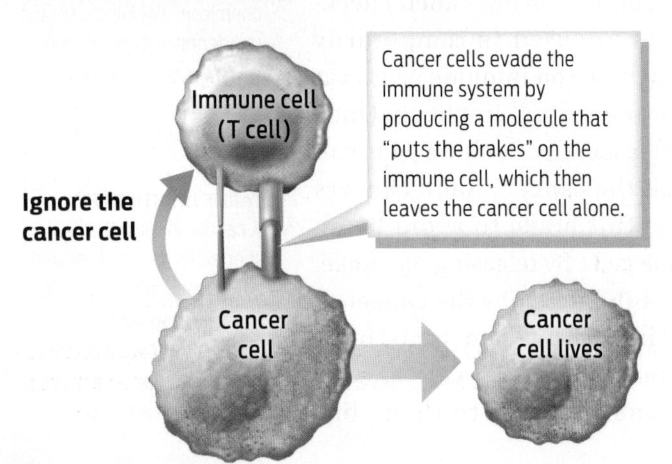

With Immunotherapy:

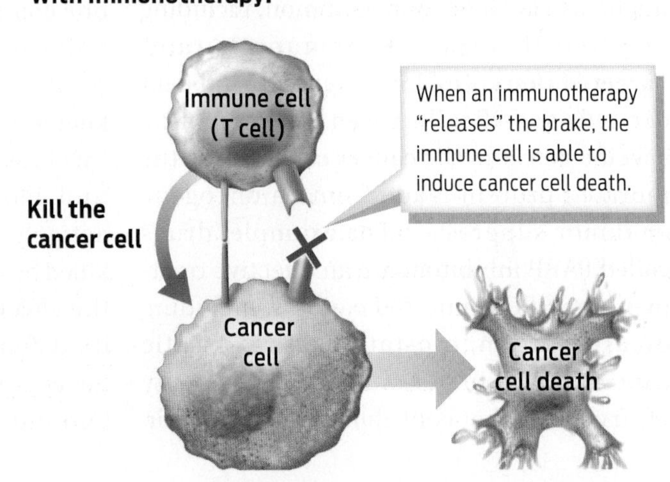

? **Compare and contrast chemotherapy and targeted therapy. Consider how each is delivered (whole body or specific location) and how it impacts healthy and cancer cells.**

individuals who do can sometimes experience dramatic improvements in their cancer **(INFOGRAPHIC 10.9)**.

When it came time for treatment, Ahern chose a lumpectomy to remove her breast tumor, followed by chemotherapy and radiation. But she also did something less conventional: she had her ovaries removed. The ovaries are a major source of estrogen in the female body. Removing this source of estrogen cuts down on cell division in breast tissue, thereby reducing the likelihood that mutations in breast cells will occur. As estrogen is also produced in small amounts by other organs, Ahern takes an estrogen-blocking drug called tamoxifen as well. All of these treatments are designed to reduce the chance that the cancer will recur in her body.

An Ounce of Prevention

▶ Cancer screening and prevention methods

If inherited mutations account for only a small fraction of cancers (about 5% to 10% of the total), that means that most cancers are the result of mutations we acquire during our life. Some of these mutations can be avoided by engaging in a healthy lifestyle—in other words, we have some control over them.

According to the American Association for Cancer Research, more than half of all cancer deaths each year are related to preventable causes. At the top of the list is tobacco use, followed closely by obesity and cancer-causing infections. For that reason, your best bet for avoiding cancer is to quit smoking, maintain a healthy weight, and get vaccinated. The human papillomavirus (HPV) vaccine, for example, prevents the infections that cause a large proportion of genital cancers and head and neck cancers—provided people get the vaccine when they are children, before they are exposed to the virus.

Screening is also very important to detect cancers early. Because tumors in the breast are hard to detect by a breast self-exam

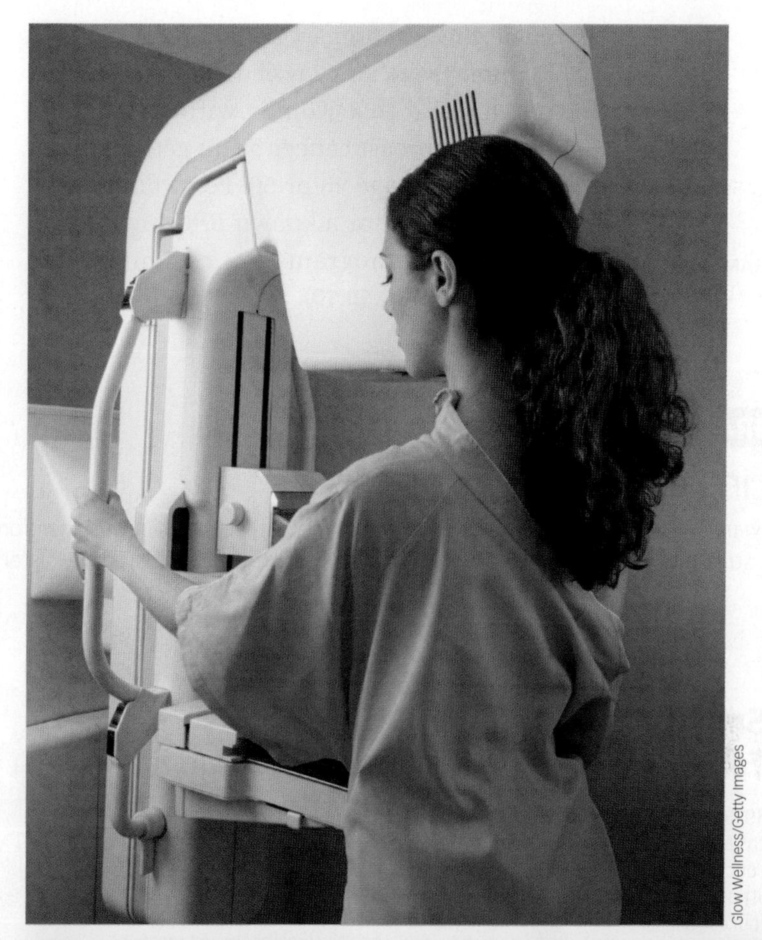

Glow Wellness/Getty Images

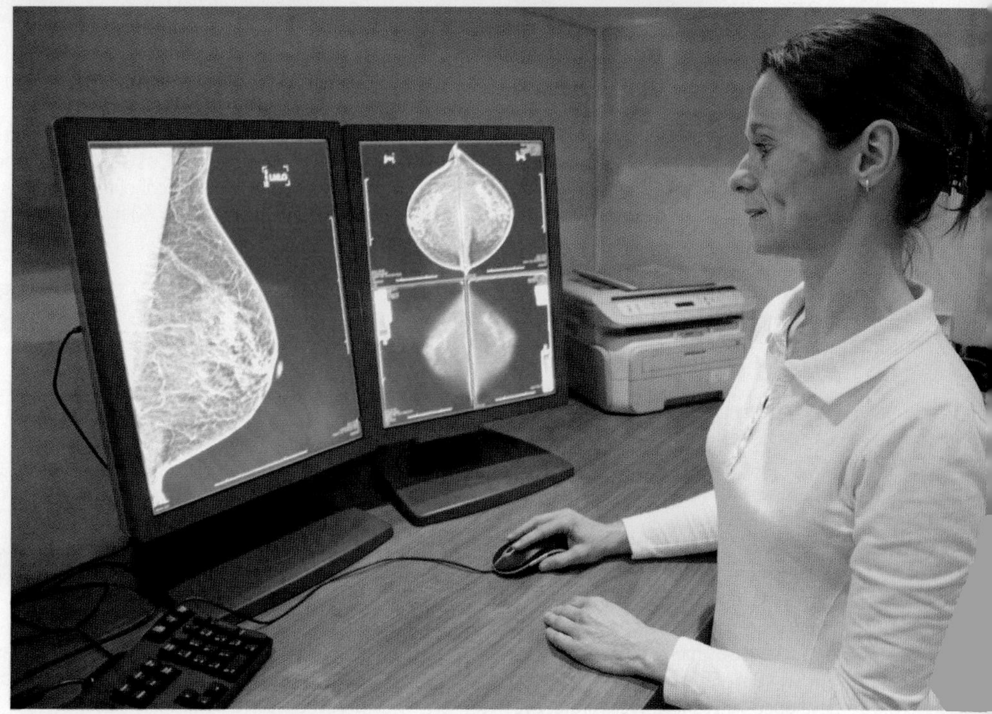

Mammograms are made by x-rays that image the breasts. Pathologists analyze the images, looking for abnormalities that could indicate cancer.

Picture Partners/AGE Fotostock

alone, the best primary screening method is a **mammogram,** an x-ray of the breast. Doctors recommend that women without a family history of breast cancer begin getting mammograms at age 45 or 50. For women with such a history or a known *BRCA* mutation, earlier mammograms are recommended **(INFOGRAPHIC 10.10).**

MAMMOGRAM
An x-ray of the breast.

Proactive Measures

▸ Debating risk-reduction surgery

With genetic testing for *BRCA* and other mutations becoming increasingly common, more and more people find themselves facing a difficult choice: wait to see if they develop cancer or take proactive action. Actress

INFOGRAPHIC 10.10

Reducing Your Risk of Cancer

By being aware of risk factors for cancer, people can minimize their exposure and therefore reduce their risk of developing cancer. Other preventive strategies include screening and genetic testing for people with a family history of cancer.

1 Don't Smoke or Chew Tobacco

More than 30% of all cancer deaths are caused by smoking or chewing tobacco. Cutting these activities reduces your cancer risk. Nonsmokers exposed to secondhand smoke are more at risk for lung cancer and heart disease, so eliminating smoking increases the health of all members of a household.

2 Eat Healthy and Maintain a Healthful Weight

4%–7% of new cancer cases in the U.S. are related to obesity. Maintaining a healthy weight can decrease this risk. Eat fruits, vegetables, and whole-grain foods. Avoid processed foods, and limit the amount of red meat in your diet. Avoid processed foods high in sugar, salt, and fat, and limit consumption of red meat. Avoid alcohol, as 3.5% of all cancer deaths in the U.S. are alcohol related. Be physically active at least 30 minutes each day.

3 Get Vaccinated

Many cancers caused by viruses and bacteria can be prevented by vaccination. Vaccinate newborns for hepatitis B and boys and girls age 11–12 for human papillomavirus (HPV).

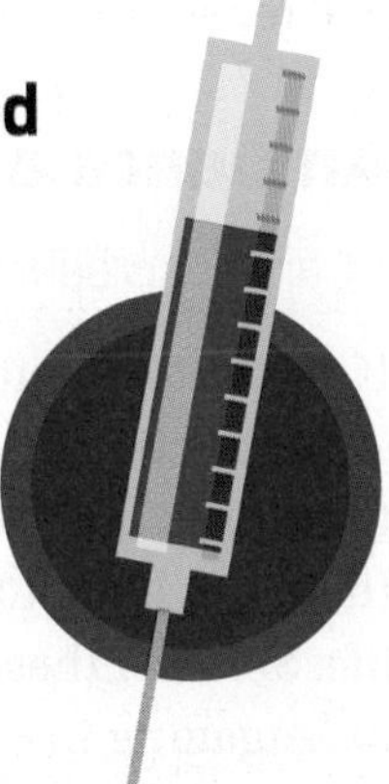

4 Avoid Harmful Environments

Environmental pollutants are responsible for 4%–19% of all cancers. Avoid pollutants in the workplace by following safety procedures. Minimize radiation exposure in your house by monitoring radon levels. By age 65, 40%–50% of Americans will have developed skin cancer. Wear protective clothing, and sunscreen daily, even when the sun is not shining brightly.

5 Screen Regularly for Cancer

Screening for cancer or conditions that could lead to cancer increases the odds of successful treatment. Regular Pap tests screen for cervical cancer, and mammograms for women over 45 check for breast cancer. Colon cancer screening is recommended after age 50. Check with your doctor for screening programs near you.

6 Get Tested for Hereditary Cancers

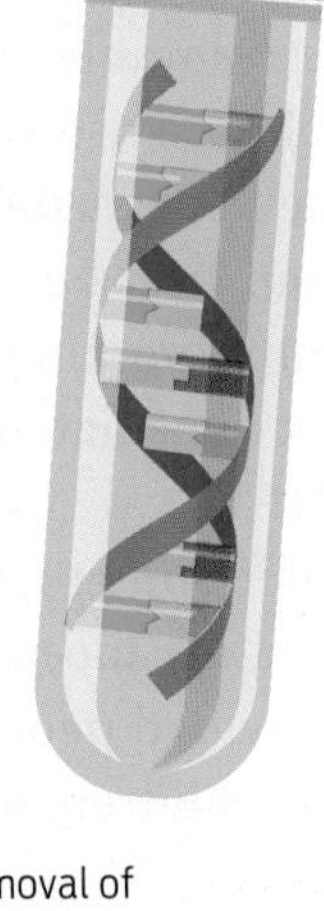

Genetic testing for high-risk cancer genes is a good idea if cancer runs in your family. Testing allows you to plan ahead for prevention and treatment options. Those who've inherited high-risk cancer genes may consider surgery to remove the high-risk tissue, as in the removal of breast and ovarian tissue for those carrying mutated *BRCA* genes.

? What is the recommended screening method for breast cancer?

Angelina Jolie made headlines in 2013 when she elected to undergo a double mastectomy as a preventive measure. Jolie had tested positive for a *BRCA1* mutation, and her mother, grandmother, and aunt had all died of either breast or ovarian cancer. In 2015, Jolie went further and had her ovaries removed as well.

A decade ago, such operations would have seemed drastic and unwarranted. But over the years, the value of surgery to prevent future cancers has been validated by science. Studies in women with mutated *BRCA* genes who have not yet developed cancer show that ovary removal cuts the risk of breast cancer by 50% and the risk of ovarian cancer by 90%. Double mastectomy cuts the risk of breast cancer by 95%. These surgeries also lower the risk of cancer recurring in women, like Ahern, who have already had cancer, although not quite as much.

Doctors admit that surgery isn't the most palatable treatment. "Surgery cuts your risk substantially, but it's still pretty traumatic," says Sellers. "It would be nice to say we've got a medication you can take and you'll have the same effect. But we just don't have that kind of treatment right now." ■

CHAPTER 10 SUMMARY

Driving Question 1 How and why do cells divide, and how is cell division regulated?

- Cell division is a fundamental feature of life, necessary for normal development, growth, maintenance, and repair of the body.
- The cell cycle is the sequence of steps that a cell undergoes to divide. During the cell cycle, one parent cell becomes two daughter cells.
- Stages of the cell cycle include interphase, in which the cell's contents, including organelles and DNA, are replicated; mitosis, in which the replicated chromosomes (in the form of sister chromatids) are pulled apart; and cytokinesis, in which the cell physically divides into two daughter cells.
- Cell cycle checkpoints ensure accurate progression through the cell cycle; repair mechanisms at each checkpoint can fix mistakes that occur, such as improper base pairing or DNA damage. If a checkpoint protein is impaired, cells may fail to properly repair DNA mistakes, leading to mutations that are passed on to daughter cells.
- Mitosis is a critical part of cell division and takes place in several phases, each of which is important to properly segregate chromosomes into daughter cells. Mistakes in mitosis can lead to abnormal chromosome numbers.

Driving Question 2 What factors contribute to the development of cancer?

- Cancer cells have lost the ability to regulate cell division and so reproduce uncontrollably, in most cases eventually forming a tumor.
- Mutations in two types of genes, proto-oncogenes and tumor suppressors, cause most cancers. Mutated proto-oncogenes (called oncogenes) cause cells to divide even in the absence of growth signals. Mutated tumor suppressors fail to pause growth in the face of DNA damage or other problems.
- Multiple mutations must occur in the same cell for the cell to become cancerous. People with "hereditary" cancer inherit specific genetic mutations that predispose them to develop the disease. These mutations are present in all body cells and can serve as the first genetic "hit."

Driving Question 3 How is cancer treated?

- Cancer is often treated with a combination of surgery, chemotherapy, and radiation. New and promising cancer treatments include targeted therapy and immunotherapy.
- Women who test positive for *BRCA* mutations may consider surgical removal of the ovaries or breasts to prevent cancer from developing in these organs.

MORE TO EXPLORE

- FORCE: http://www.facingourrisk.org/
- King, M.-C. (2014). "The race" to clone *BRCA1*. *Science* 343(6178):1462–1465.
- Skloot, R. (2010). *The Immortal Life of Henrietta Lacks.* New York: Random House.
- *Cancer: The Emperor of All Maladies* (2015), a PBS documentary: http://www.pbs.org/kenburns/cancer-emperor-of-all-maladies/home/
- Graeber, Charles. (2018). *The Breakthrough: Immunotherapy and the Race to Cure Cancer.* New York: Twelve Press.

CHAPTER 10 Test Your Knowledge

Driving Question 1 How and why do cells divide, and how is cell division regulated?

By answering the questions below and studying Infographics 10.2, 10.3, 10.4, and 10.5, you should be able to generate an answer for this broader Driving Question.

Know It

1. What process is critical for embryonic development, wound healing, and replacement of blood cells? (Hint: All these processes require new cells).

2. In the cell cycle, DNA is replicated during

- **a.** mitosis.
- **b.** G_1.
- **c.** S.
- **d.** G_2.
- **e.** cytokinesis.

3. Following mitosis and cytokinesis of a human cell, daughter cells

- **a.** are genetically unique.
- **b.** are genetically identical to each other.
- **c.** each have 23 chromosomes.
- **d.** each have 92 chromosomes.

4. During which stage of the cell cycle do sister chromatids separate from each other?

5. During which stage of the cell cycle are sister chromatids initially produced?

6. In an otherwise normal cell, what happens if one mistake is made during DNA replication?

- **a.** Nothing; mistakes just happen.
- **b.** A cell cycle checkpoint detects the error and pauses the cell cycle so that the error can be corrected.
- **c.** The cell begins to divide out of control, forming a malignant tumor.
- **d.** A checkpoint forces the cell to carry out apoptosis, a form of cellular suicide.
- **e.** The mutation is inherited by the individual's offspring.

Use It

7. A chemical is added to cells growing in a culture dish. This chemical blocks completion of DNA replication. In what stage of the cell cycle will the cells be stuck?

8. Many drugs interfere with cell division. Why shouldn't pregnant women take these drugs?

9. What would be the result if a cell completed interphase and mitosis but failed to complete cytokinesis? How many cells would there be at that point, and how many chromosomes would those cells have in comparison to the parent cell?

Driving Question 2 What factors contribute to the development of cancer?

By answering the questions below and studying Infographics 10.1, 10.6, 10.7, 10.8, and 10.10, you should be able to generate an answer for this broader Driving Question.

Know It

10. A woman with a *BRCA1* mutation

- **a.** will definitely develop breast cancer.
- **b.** is at increased risk of developing breast cancer.
- **c.** must have inherited the mutation from her mother because of the link to breast cancer.
- **d.** will also have a mutation in *BRCA2*.
- **e.** both b and c

11. Which of the following can cause cancer to develop and progress?

- **a.** a proto-oncogene
- **b.** an oncogene
- **c.** a tumor suppressor gene
- **d.** a mutated tumor suppressor gene
- **e.** both b and d
- **f.** both b and c

12. What are some differences and some similarities between tumor suppressor genes and oncogenes?

13. What is the role of *BRCA1* in normal cells?

14. Which of the following family histories most strongly suggests a risk of inherited breast cancer due to *BRCA1* mutations?

- **a.** Many female relatives were diagnosed with breast cancer in their 70s.
- **b.** Many relatives have had skin cancer.
- **c.** Many relatives were diagnosed with skin cancer at an early age.
- **d.** Many female relatives were diagnosed with breast cancer at an early age.
- **e.** Many female relatives were diagnosed with both early breast cancer and ovarian cancer.

Use It

15. What would you say to a niece if she asked you how she could reduce her risk of breast cancer? (Assume there is no family history of breast cancer.) How might each of your suggestions reduce her risk?

16. A 28-year-old male graduate student was born with an inherited predisposition to colon cancer due to a mutation in a DNA repair gene called *MLH1*. He has recently been diagnosed with colon cancer. At the cellular and genetic level, was he born with colon cancer? Was he born with a predisposition to colon cancer? At birth, were cells in his colon genetically identical to cells in his liver? Now that he has colon cancer, are his cancer cells genetically identical to his normal colon cells? Explain your answers.

17. Which of the following women would be most likely to benefit from genetic testing for breast cancer?

a. a 25-year-old woman whose mother, aunt, and grandmother had breast cancer

b. a healthy 75-year-old woman with no family history of breast cancer

c. a 40-year-old woman who has a cousin with breast cancer

d. a 55-year-old woman whose older sister was just diagnosed with breast cancer

e. All women can benefit from genetic testing for breast cancer.

18. People like Lorene Ahern have inherited a mutated version of *BRCA1*. Why does this mutation pose a problem? Why are these people at high risk of developing breast cancer when they still have a functional *BRCA1* allele? Describe how the protein encoded by normal *BRCA1* compares to that encoded by mutant alleles of *BRCA1*.

Mini Case

Apply Your Knowledge

19. Nellie has a family history similar to Lorene Ahern's. Nellie's mother died at an early age from breast cancer, as did her maternal aunt (her mother's sister). Nellie is not yet 35 but has started having annual mammograms. She has also been tested for *BRCA1* and *BRCA2* mutations. She has a *BRCA2* mutation and is considering surgery to remove the majority of her breast tissue. Her younger sister, Anne, doesn't want to know the results of Nellie's genetic testing because if Nellie has a *BRCA2* mutation, then there is a chance that Anne could have inherited the same mutation from their mother. Does Nellie or Nellie's doctor have an obligation to tell Anne about the test results? What about Nellie's older brother? Should he be told? There are personal and medical benefits and risks to consider here.

Bring It Home

Apply Your Knowledge

20. If you wanted to change your lifestyle to reduce your risk of developing cancer, what specific steps could you take with respect to each of the following factors? Be as specific as you can. Take your age and gender into consideration as you consider each factor.

a. alcohol consumption

b. sun exposure

c. tobacco use

d. exposure to pesticides

e. meat preparation (cooking method)

Driving Question 3 How is cancer treated?

By answering the questions below and studying Infographic 10.9, you should be able to generate an answer for this broader Driving Question.

Know It

21. Which of the following is a targeted cancer therapy?

a. surgery

b. radiation

c. a PARP inhibitor

d. chemotherapy

e. all of the above

22. Why is chemotherapy more effective for metastatic cancer than either surgery or radiation?

Interpreting Data

Apply Your Knowledge

23. José is a 32-year old attorney who was born and raised in Phoenix (and still lives there). He played a lot of soccer growing up and remains active: he hikes almost every weekend. José and his 64-year-old father, Ray, were both diagnosed with metastatic melanoma within 2 months of each other. Both had their tumors biopsied to look for two different potential targets for targeted therapy: the *BRAF* proto-oncogene and PD-L1, which is an immune system braking molecule on some cancer cells that can be targeted by immunotherapy. In the *BRAF* analysis, the *BRAF* proto-oncogene from each of their tumors was sequenced. In the PD-L1 analysis, cancer cells were analyzed for expression of this molecule. The data are shown in the table below.

	***BRAF* gene sequence (corresponding to amino acids 598–602; note that the DNA groups shown below correspond to mRNA codons)**	**PD-L1 expression**
José	CGA TGT CTC TTT AGA	No
Ray	CGA TGT CAC TTT AGA	Yes

a. Transcribe and translate the *BRAF* gene sequences from José's and Ray's tumors. Which amino acid is at position 600 in each?

b. Zelboraf is a drug that stops cell division in metastatic melanomas by blocking the activity of a mutant (oncogenic) *BRAF* protein that has a glutamic acid at position 600 (the proto-oncogene has a valine at position 600). Given their *BRAF* sequences, is Zelboraf a viable treatment for either José or Ray? Explain your answer.

c. Keytruda is a drug that blocks the interaction between PD-L1 on cancer cells and its binding partner (PD-1) on immune cells. This interaction acts as the "brake" on the immune cells (see Infographic 10.9): when it is disrupted, the immune cell attacks the cancer cell. Given the PD-L1 expression results, is Keytruda a potential treatment for either José or Ray? Explain your answer.

d. In addition to Zelboraf and Keytruda, what other treatment(s) are available for José and Ray? Explain your answer.

e. From the information presented, do you think Jose's and Ray's cancers are more likely to be cases of inherited cancer, or two cancers that just happened to occur in these two family members? Explain your answer, and consider other risk factors that may be involved for José and Ray.

Use It

24. What is the difference between preventive mastectomy and surgical removal of a tumor (given that both involve surgery to remove tissue from the breast)?

11 Simple Inheritance and Meiosis

Catching BREATH

One woman's mission to outrun a genetic disease

© Martin Vecchio

DRIVING QUESTIONS

1. How does the organization of chromosomes, genes, and their alleles contribute to human traits?

2. How does meiosis produce genetically diverse gametes?

3. Why do different traits have different patterns of inheritance?

4. What are some practical applications of understanding the genetic basis of human disease?

Emily Schaller never thought she'd be running marathons one day. From the time she was a teenager, the Detroit, Michigan, native has struggled just to take a deep breath without coughing. Emily spent her youth in and out of hospitals, getting treatment for the recurrent lung infections that plague people with her condition.

Emily has cystic fibrosis (CF), a genetic disorder she inherited from her parents. CF has many symptoms. The most dangerous is that mucus clogs airways in the lungs and makes it difficult to breathe. But people with CF also have trouble digesting food—mucus blocks the passageways through which digestive enzymes travel to the intestines. To overcome this problem and make sure they get enough nutrients, they must swallow enzymes before each meal.

Emily's struggle with CF reached a tipping point in 2007. Her health had declined to the point that she qualified for Social Security benefits and disability insurance at age 25. "I thought to myself, 'I'm sick of being sick. I have to do something.'"

So Emily started running—slowly and painfully at first, but eventually with more ease. "It took me 3 to 4 months just to run 2 miles—coughing and spitting every block," she says. Eventually, through running, Emily managed to raise her lung function from 50% to 75% of the normal level. "Running totally changed my life," she says.

Emily's successes haven't come from sheer determination alone. She's also benefited greatly from progress in medical science. Through research, scientists have come to understand CF better. They now know what genes are affected and are learning why some people with CF have more severe forms of the disease. This research is leading to new drugs that specifically target the defect that occurs in CF. These drugs make it possible for someone with the disease to live and breathe more easily.

Genetic Typos

▸ Mutations and alleles

Every year, approximately 2,500 babies in the United States are born with CF, making it the most common life-threatening genetic disease in this country. The current life expectancy for someone with CF is about 44 years of age. When Emily's CF was diagnosed in 1983, it was less than 25 years **(INFOGRAPHIC 11.1)**.

It wasn't until 1989 that scientists understood what caused the disease. That year,

INFOGRAPHIC 11.1

Cystic Fibrosis Life Expectancy

As a result of medical research and the development of effective treatments, the median life expectancy of people living with cystic fibrosis has increased by about 15 years since 1986. For those born between 2013 to 2017, the median predicted survival age is 43.6 years.

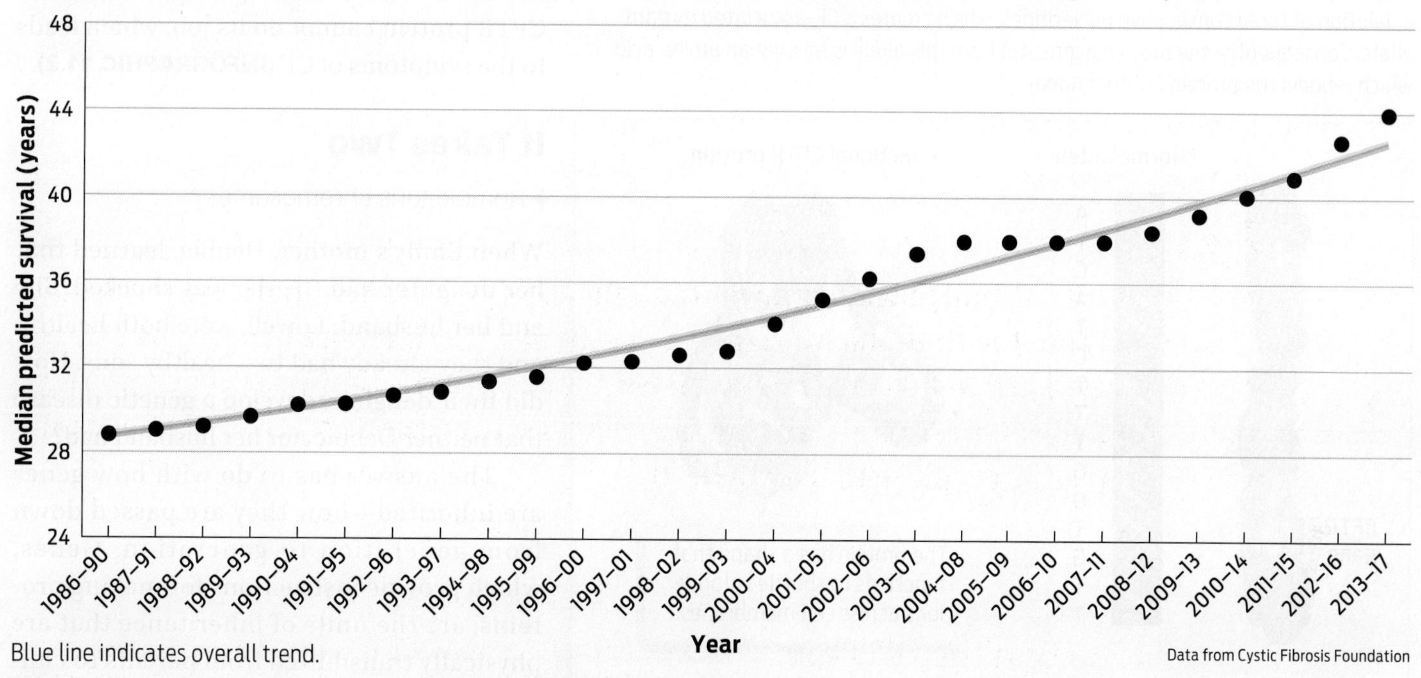

Blue line indicates overall trend.

Data from Cystic Fibrosis Foundation

? How does the median survival for people with CF born between 2013 and 2017 compare to the median survival for those born between 2000 and 2004?

a team of scientists led by Lap Chee Tsui, at Toronto's Hospital for Sick Children, and Francis Collins, then at the University of Michigan, homed in on the genetic basis for the condition: a mutation in a specific gene that sits on chromosome 7.

As we saw in Chapter 9, mutations are changes in the nucleotide sequence of DNA. They result when mistakes occur during the process of DNA replication, or when environmental insults cause DNA damage that isn't repaired properly.

Mutations create altered versions of genes called **alleles.** Because genes code for proteins, a change in the sequence of a gene can change the shape of the encoded protein, and therefore its function.

You can think of alleles as being like words with different spellings. Some of these variant spellings are harmless, and the variants retain the same meaning (for example, color, colour; theater, theatre). Others change the meaning of the word entirely (here, hear; read, red). Still other changes to a word's spelling can render it nonsensical (green, grken; song, sxng). A change in one letter—one nucleotide—is sometimes enough to alter the function of a protein, or make it nonfunctional.

Tsui and Collins (who would later head the Human Genome Project at the National Institutes of Health) found that the CF mutation occurs in a gene that provides the instructions to make a protein that shuttles chloride ions into and out of cells. They called the gene *CFTR*, for *cystic fibrosis transmembrane conductance regulator.*

The 1989 discovery was a milestone. Now that they knew which gene was responsible for CF, scientists could study how mutations in the gene make people sick. A variety of

ALLELE
Any of the alternative versions of the same gene that have different nucleotide sequences.

INFOGRAPHIC 11.2

CF Is Caused by Mutations in the *CFTR* Gene

Cystic fibrosis (CF) is caused by a variety of mutations in the cystic fibrosis transmembrane regulator (*CFTR*) gene that sits on chromosome 7. One such mutation consists of a deletion of three consecutive nucleotides, which creates a CF-associated mutant allele. Consequently, the protein expressed from this allele is missing an amino acid, which renders the protein nonfunctional.

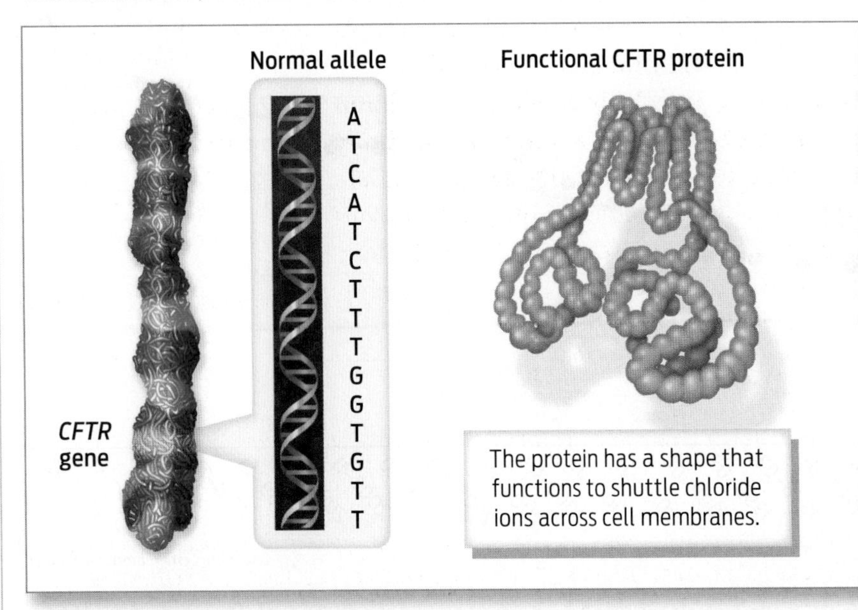

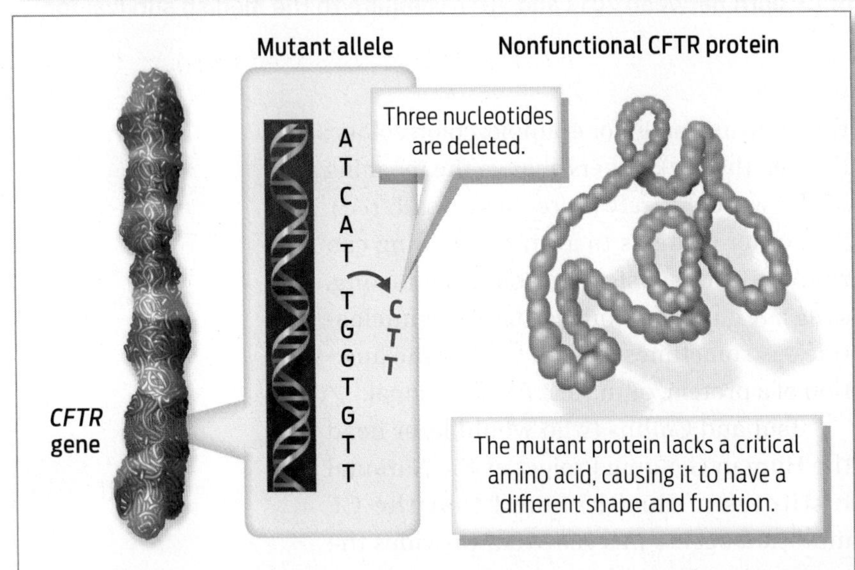

? How many nucleotides differ between the mutant and normal alleles of the *CFTR* gene illustrated? How many amino acids differ between the proteins encoded by the two alleles?

mutations in the *CFTR* gene can cause the disease. The most common is a deletion of three nucleotides within the *CFTR* gene. People who carry this mutation are missing a key amino acid in the CFTR protein. This faulty CFTR protein cannot do its job, which leads to the symptoms of CF **(INFOGRAPHIC 11.2)**.

It Takes Two

▸ Homologous chromosomes

When Emily's mother, Debbie, learned that her daughter had CF, she was shocked. She and her husband, Lowell, were both healthy, and they already had two healthy sons. How did their daughter develop a genetic disease that neither Debbie nor her husband had?

The answer has to do with how genes are inherited—how they are passed down from generation to generation. Genes, which provide instructions for making proteins, are the units of inheritance that are physically transmitted from parents to children. The particular alleles of genes you received from your parents are the reason you resemble your mother and father, and possibly also an uncle or a grandparent. But not every child of a couple receives exactly the same set of parental alleles, so children can and do differ from their parents and from one another.

Consider Emily's parents. Like all humans, they are **diploid** (from the Greek word for "double"), meaning that each of their body cells carries two copies of each chromosome. Of the 46 total chromosomes in our cells, 23 come from our mother and 23 come from our father. These 23 individual chromosome pairs—one from mom, one from dad—are called **homologous chromosomes.**

Because our chromosomes come in pairs, so too do our genes. Humans have two copies of nearly every gene in our body cells. These two gene copies can be either the same or slightly different in the specific nucleotide sequences making them up. In other words, each chromosome may carry a different allele of the same gene.

Because we have two alleles for a given gene, a nonfunctional allele doesn't always spell disaster. In the case of the *CFTR* gene, for example, a person can have one defective allele and remain healthy if his or her other *CFTR* allele is normal and produces a functional CFTR protein. That explains why Emily's parents, Debbie and Lowell, are healthy: each has one defective *CFTR* allele and one normal *CFTR* allele that makes up for the defective copy **(INFOGRAPHIC 11.3)**.

But as Debbie and Lowell illustrate, it's not always possible to know what genes a person has just from his or her outward appearance. For that reason, geneticists make a distinction between a person's observable or measurable traits, or **phenotype,** and his or her genes, or **genotype.** Debbie and Lowell have normal phenotypes, but both also carry a CF allele as part of their genotype, which they passed on to Emily.

But not all the Schaller children have the disease—Debbie and Lowell also have two healthy boys. Why didn't these children inherit CF? The short answer is that Emily inherited two copies of the defective allele and her bothers inherited only one copy of the defective allele. To better understand how this can happen, let's look more closely at how sexual reproduction works.

DIPLOID
Having two copies of every chromosome.

HOMOLOGOUS CHROMOSOMES
A pair of chromosomes that both contain the same genes. In a diploid cell, one chromosome in the pair is inherited from the mother, the other from the father.

PHENOTYPE
The visible or measurable features of an individual.

GENOTYPE
The particular genetic makeup of an individual.

INFOGRAPHIC 11.3

In Humans, Genes Come in Pairs

Humans are diploid organisms, meaning they have two complete sets of chromosomes—one set inherited from each parent. Because chromosomes come in pairs, so too do the genes located on them. The two gene copies, known as alleles, can be either the same or different. In the case of CF, having at least one normal allele is sufficient to remain healthy.

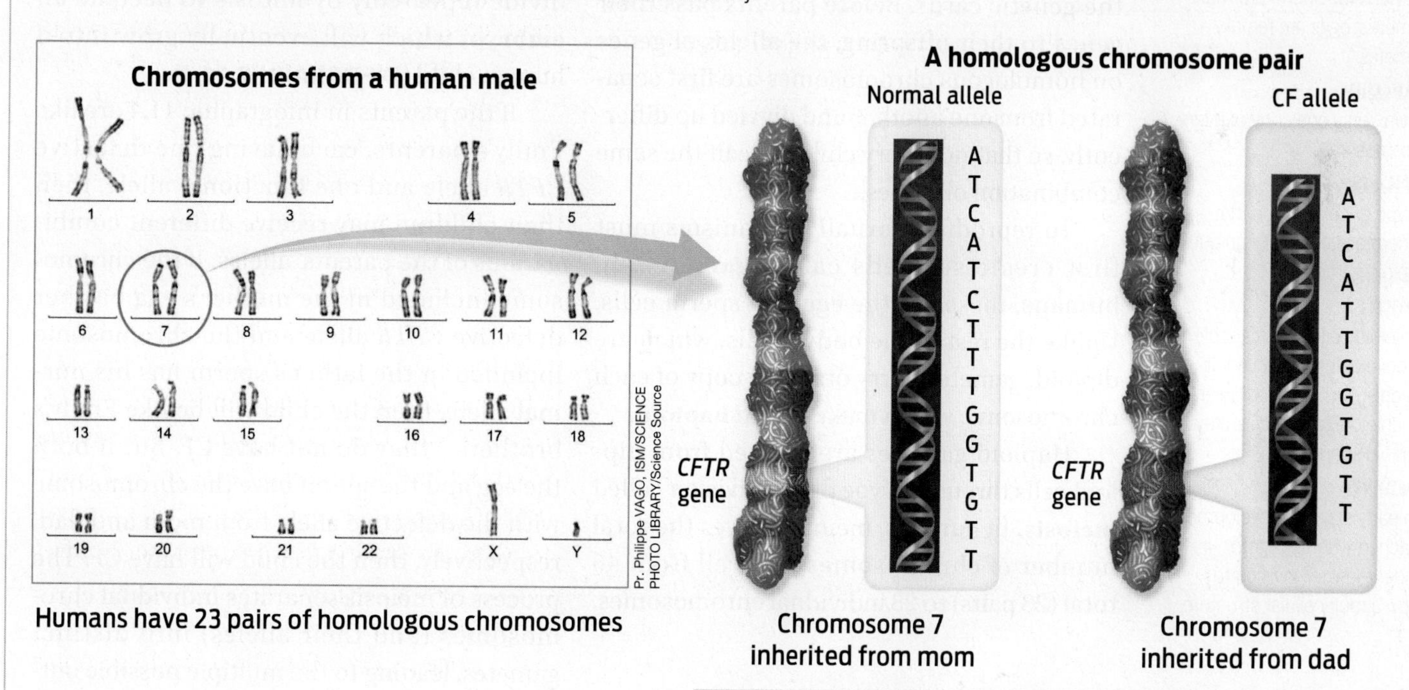

Emily's parents remain healthy because each has a normal allele, which produces enough of the normal protein for cells to function properly.

? **Explain why humans are diploid.**

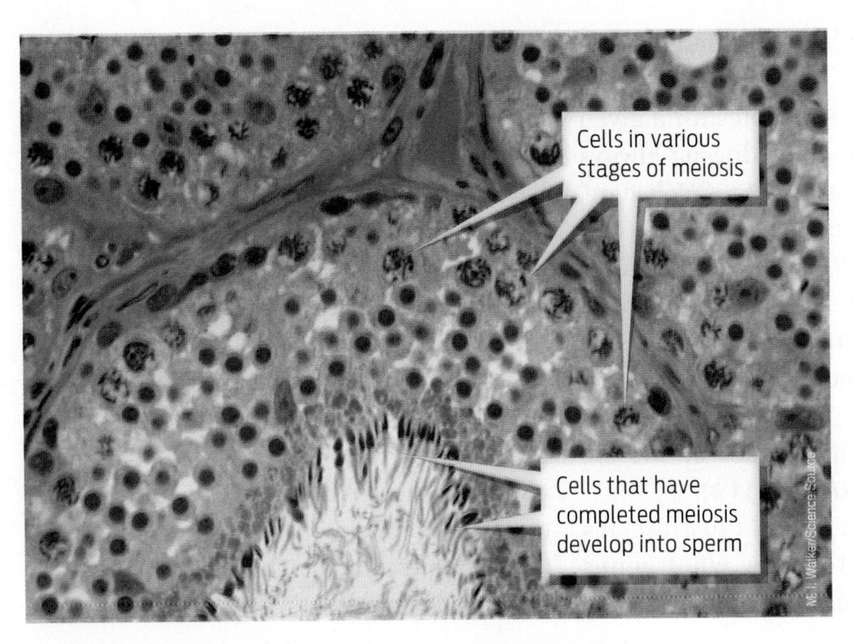

Cross section of a testis, the site of meiosis in males.

GAMETES
Specialized reproductive cells that carry one copy of each chromosome (that is, they are haploid). Sperm are male gametes; eggs are female gametes.

HAPLOID
Having only one copy of every chromosome.

MEIOSIS
A type of cell division that generates genetically unique haploid gametes.

ZYGOTE
A diploid cell that is capable of developing into an adult organism. It is formed when a haploid egg is fertilized by a haploid sperm.

EMBRYO
An early stage of development reached when a zygote undergoes cell division to form a multicellular structure.

Luck of the Draw

▶ Making gametes through meiosis

Sexual reproduction is a bit like shuffling the genetic cards. Before parents pass their genes to their offspring, the alleles of genes on homologous chromosomes are first separated from one another and divvied up differently, so that not every child is dealt the same combination of alleles.

To reproduce sexually, organisms must first create sex cells called **gametes.** In humans, these are the egg and sperm cells. Unlike the rest of the body's cells, which are diploid, gametes carry only one copy of each chromosome, which makes them **haploid.**

Haploid gametes are created from diploid cells through a type of cell division called **meiosis.** In humans, meiosis halves the total number of chromosomes in a cell from 46 total (23 pairs) to 23 individual chromosomes.

"They told us she would only live to be about 12 years old."

—Debbie Schaller

When a haploid sperm (with 23 chromosomes) fertilizes a haploid egg (with 23 chromosomes), the result is a diploid **zygote** (with 46 chromosomes). This zygote will then divide repeatedly by mitosis to become an **embryo,** which will eventually grow into a human child **(INFOGRAPHIC 11.4).**

If the parents in Infographic 11.4 are like Emily's parents, each having one defective *CFTR* allele and one functional allele, then their children may receive different combinations of the parents' alleles. If the chromosome included in the mother's egg has her defective *CFTR* allele and the chromosome included in the father's sperm has his normal allele, then the child will be like Emily's brothers—they do not have CF. But if both the egg and the sperm have the chromosome with the defective allele from mom and dad, respectively, then the child will have CF. The process of meiosis separates individual chromosomes (and their alleles) into distinct gametes, leading to the multiple possible outcomes when the parents' gametes combine during reproduction.

It's useful to explore how meiosis differs from mitosis, the type of cell division that reproduces body cells, which we

INFOGRAPHIC 11.4

Gametes Pass Genetic Information to the Next Generation

To reproduce sexually, diploid organisms produce specialized sex cells called gametes, which are haploid: they carry only one copy of each chromosome. When a sperm fertilizes an egg the resulting diploid zygote divides by mitotic cell division, eventually generating enough cells to form a baby. The baby is diploid.

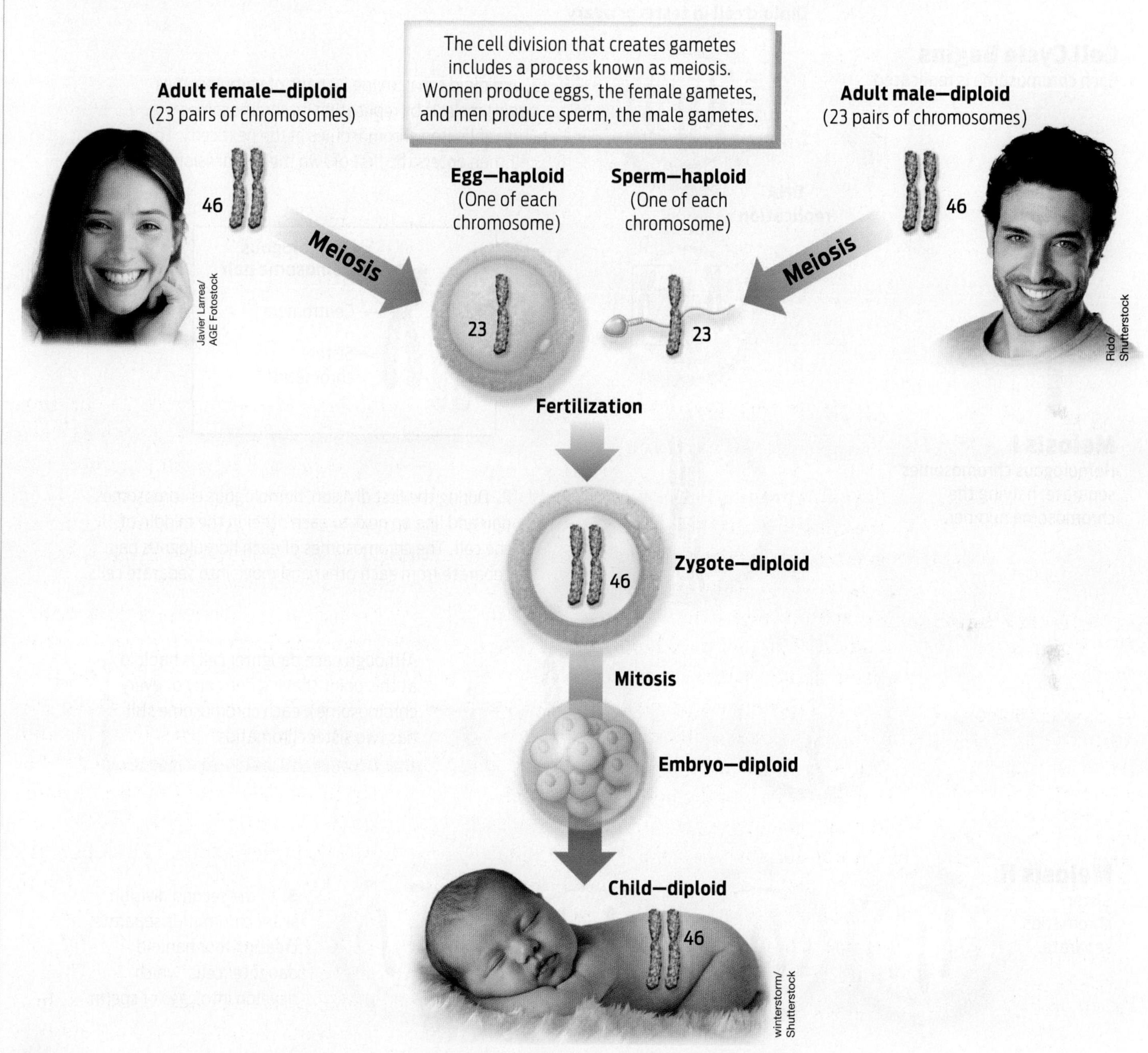

? Which of the following is/are haploid: a skin cell; a zygote; a sperm?

INFOGRAPHIC 11.5

Meiosis Produces Haploid Eggs and Sperm

Humans produce egg and sperm cells through meiosis, which takes place in the ovaries and the testes. Meiosis halves the chromosome number from 46 to 23, by placing one complete set of chromosomes in each egg or sperm cell.

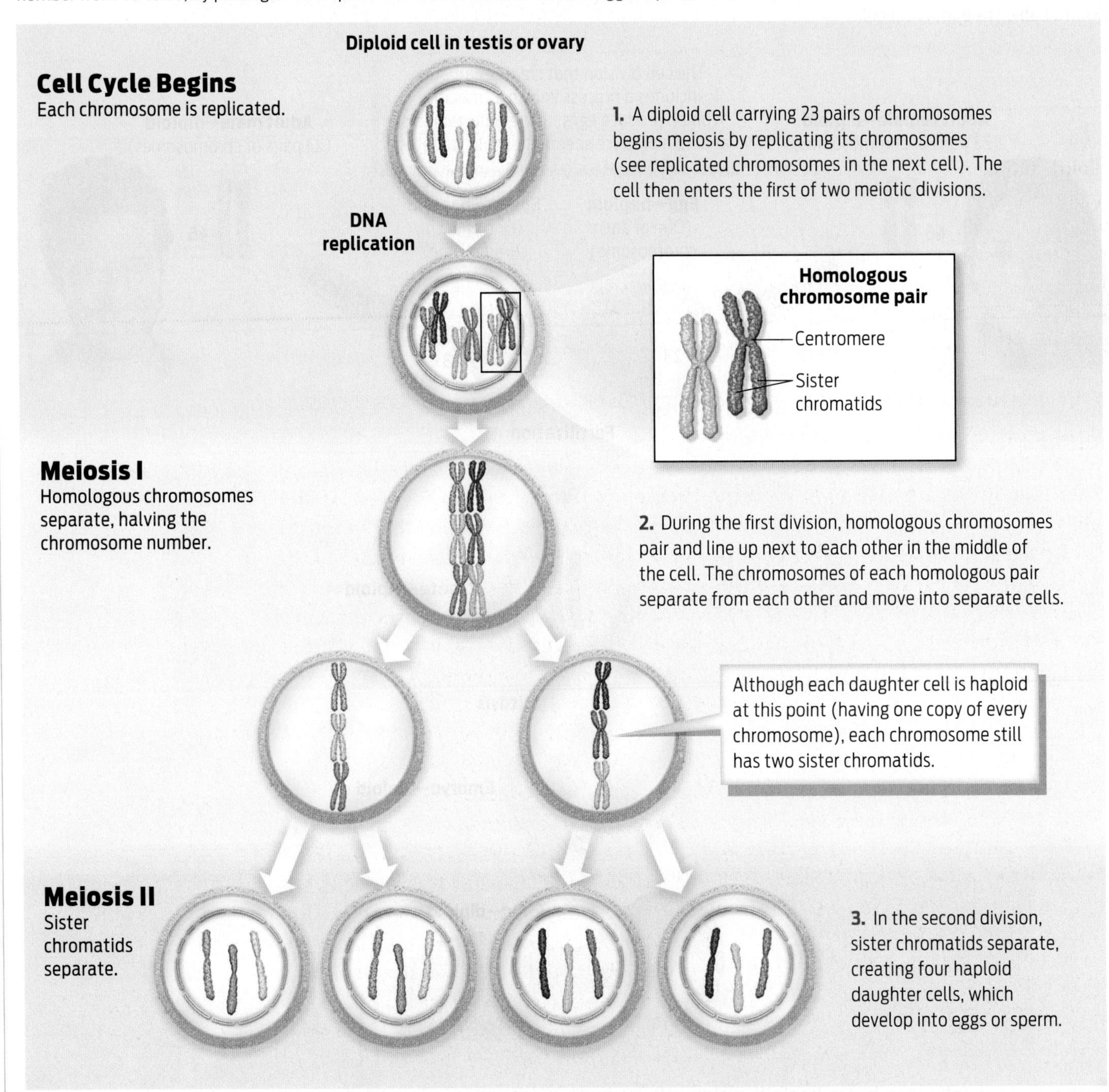

? Which meiotic division reduces the chromosome number from diploid to haploid?

encountered in Chapter 10. A key point is that the gametes produced by meiosis are not all identical to one another. This is one important difference between meiosis and mitosis. In mitosis, replicated chromosomes (consisting of a pair of sister chromatids) line up along the midline of the cell. The two sister chromatids of each chromosome then separate, becoming independent chromosomes. Each daughter cell ends up with an identical (diploid) complement of 46 chromosomes. In meiosis, two separate divisions occur, producing four daughter cells rather than two, and giving each cell only one copy of each chromosome. The first division separates homologous chromosomes from each other; the second division, like mitosis, then separates sister chromatids. At the end of meiosis, one (diploid) cell has divided into four (haploid) cells, which develop into egg or sperm **(INFOGRAPHIC 11.5)**.

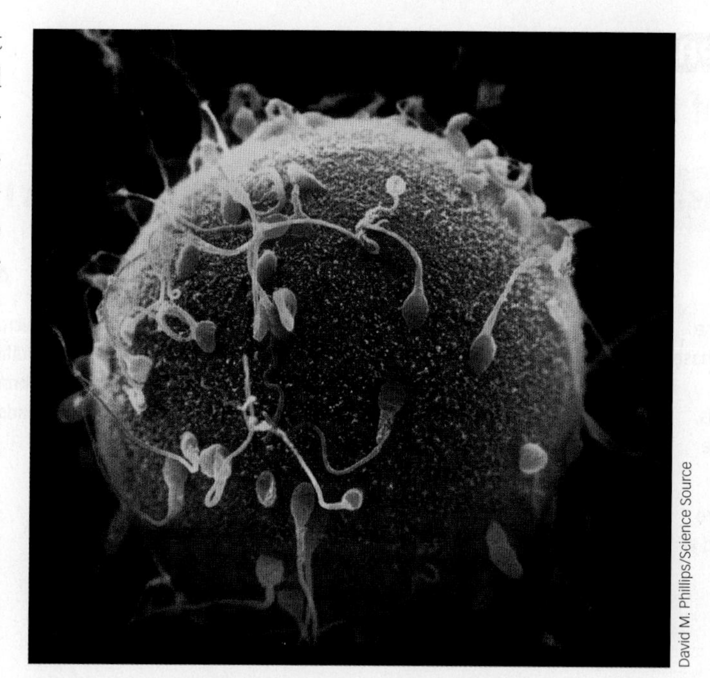

David M. Phillips/Science Source

Sperm attempting to fertilize a human egg.

Like mitosis, meiosis is a finely orchestrated process that can be divided into defined stages based on the movement of chromosomes. Replicated chromosomes condense early in the process. The nuclear envelope dissolves and a spindle apparatus forms. The movement of chromosomes is choreographed by spindle fibers that first attach to the centromeres of the chromosomes and then push and pull them to the proper location in preparation for the two separate divisions. The first division, which separates homologous chromosomes from one another, is termed meiosis I; the second division, which separates sister chromatids, is called meiosis II **(INFOGRAPHIC 11.6)**.

Generating Diversity

▶ Recombination and independent assortment

Importantly, no two sex cells produced by the same parent are genetically identical—they all have unique combinations of alleles from both parents. That is because of two major events that occur during meiosis. The first event is **recombination,** in which homologous maternal and paternal chromosomes pair up next to each other in the cell and physically exchange segments of DNA before they separate during meiosis I. As a result of recombination, which is also called crossing over, maternal chromosomes contain segments (and therefore alleles) from paternal chromosomes. Likewise, paternal chromosomes contain segments (and alleles) from maternal chromosomes. The process generates new combinations of alleles on individual chromosomes.

The second event is **independent assortment,** in which the two chromosomes of each homologous pair become distributed into daughter cells randomly with respect to all other chromosome pairs. As a consequence of independent assortment, the two alleles of a gene located on one chromosome are distributed independently of the two alleles of a gene located on another. This means, for example, that the specific alleles for hair color you inherit have no bearing on the specific alleles for dimples you inherit.

Let's see how independent assortment is achieved. During meiosis I, homologous

RECOMBINATION
An event in meiosis during which maternal and paternal chromosomes pair and physically exchange DNA segments.

INDEPENDENT ASSORTMENT
The principle that alleles of different genes are distributed independently of one another during meiosis.

INFOGRAPHIC 11.6 UP CLOSE

Phases of Meiosis

Interphase

- Each chromosome replicates in S phase, resulting in two sister chromatids, connected at the centromere.
- Chromosomes are loosely gathered in the nucleus.

Meiosis I

Prophase I

- Replicated chromosomes begin to coil up.
- The nuclear membrane begins to disassemble.
- Microtubule fibers begin to form the meiotic spindle.

Metaphase I

- Microtubule spindle fibers from opposite ends of the cell attach to each of the two homologous chromosomes.
- Paired homologous chromosomes become aligned along the middle of the cell.

Anaphase I

- Microtubules shorten, pulling the homologous chromosomes to opposite ends of the cell.

Telophase I

- Homologous chromosomes reach each pole.
- Microtubule spindle fibers disassemble.
- Nuclear membrane forms around each set of chromosomes, forming the haploid daughter cell nuclei.

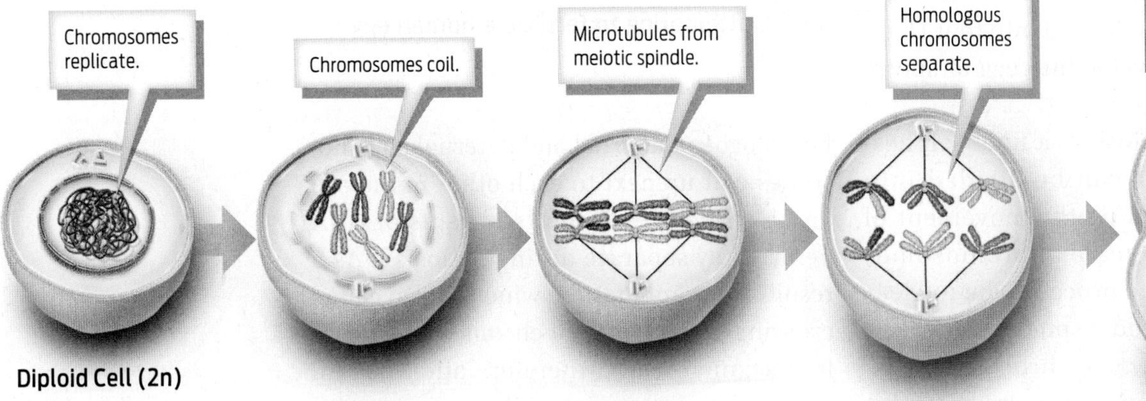

Plant Pollen Cells:

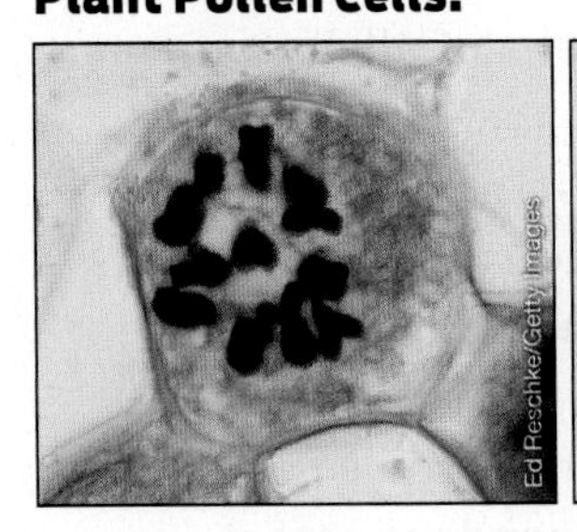

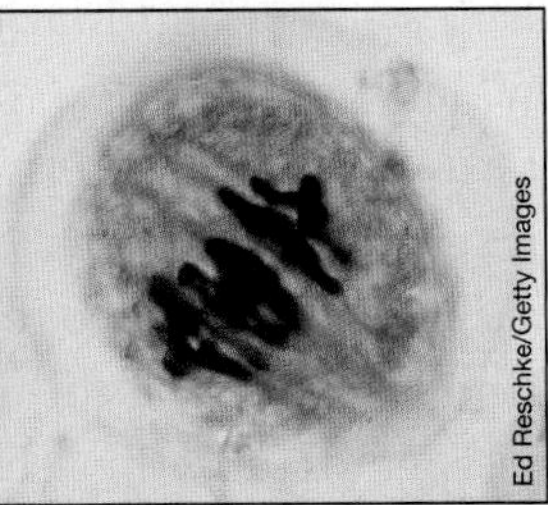

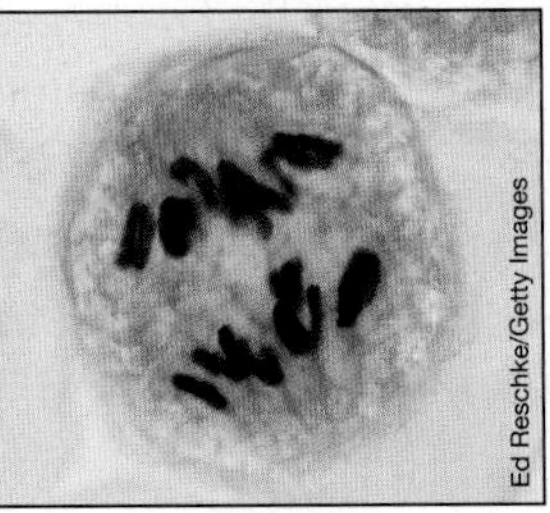

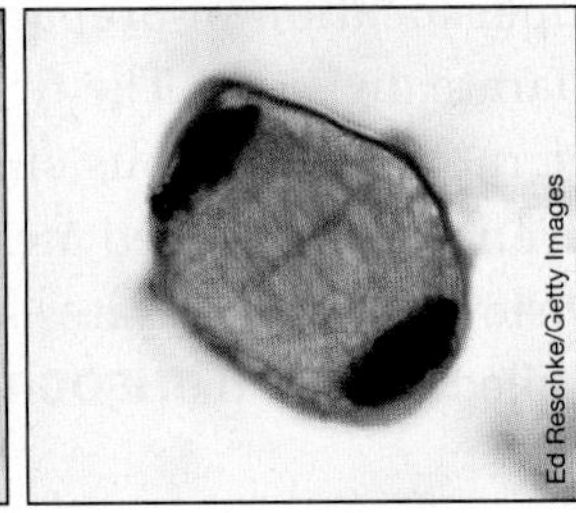

maternal and paternal chromosomes line up along the midline of the dividing cell. Each chromosome pair can line up in two different ways: sometimes the maternal chromosome is on the left, sometimes the paternal chromosome is. All the chromosomes on the left are pulled into one daughter cell, while all the chromosomes on the right are pulled into the other daughter cell. Because the identities of the chromosomes on the left and right vary, the exact combination of maternal and paternal chromosomes that each sperm or

Meiosis II

Prophase II
- The nuclear membrane disassembles.
- Microtubule fibers form the meiotic spindle.

Metaphase II
- Microtubule spindle fibers from opposite ends of the cell attach to the sister chromatids of each chromosome.
- Replicated chromosomes become aligned along the middle of the cell.

Anaphase II
- Microtubules shorten, pulling the sister chromatids to opposite ends of the cell.

Telophase II
- Identical sets of chromosomes reach each pole.
- Microtubule spindle fibers disassemble.
- Nuclear membrane forms around each set of chromosomes, forming the haploid daughter cell nuclei.

Daughter Cells
- Cells develop into eggs or sperm.
- Cells are haploid (n).

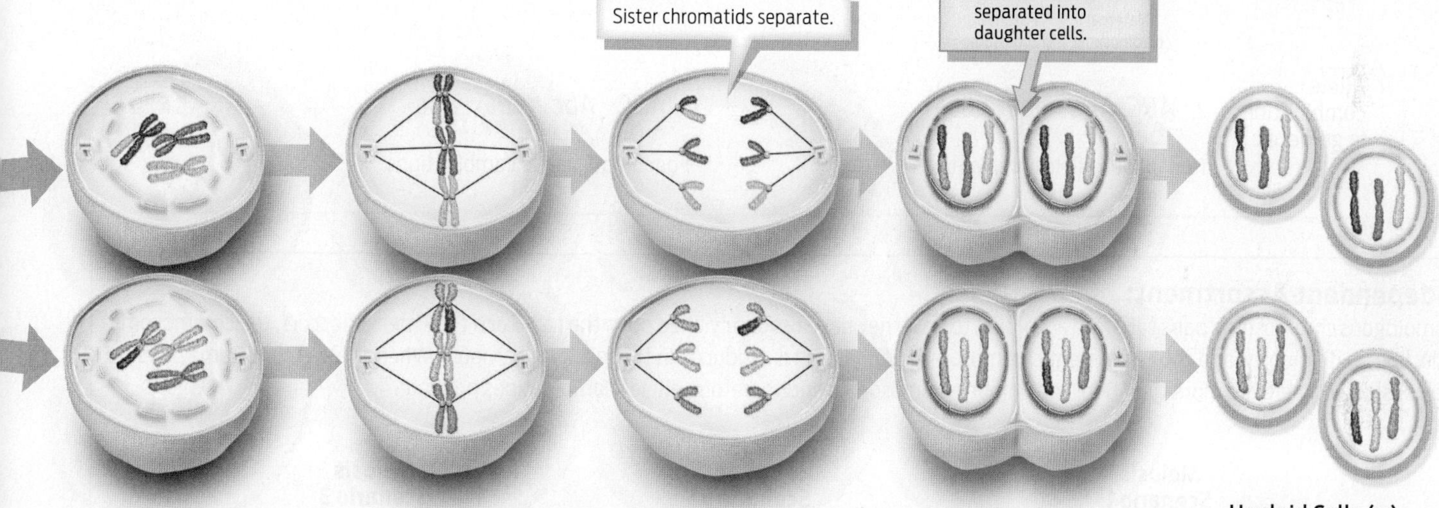

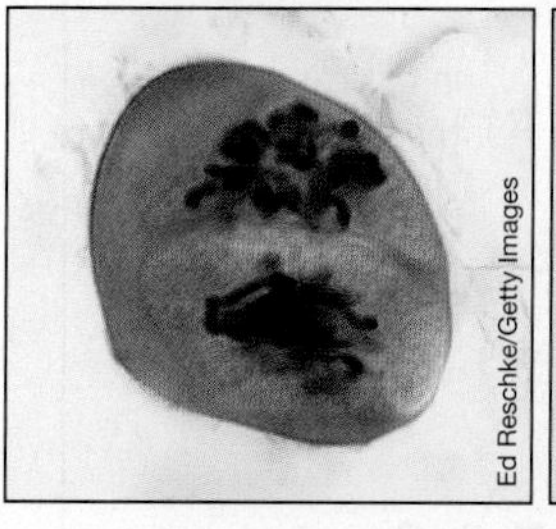

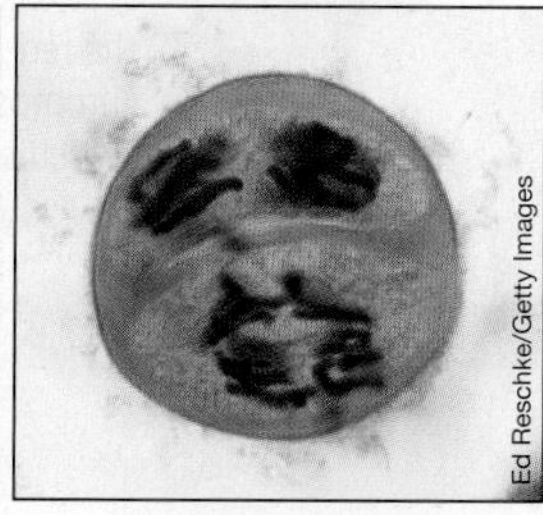

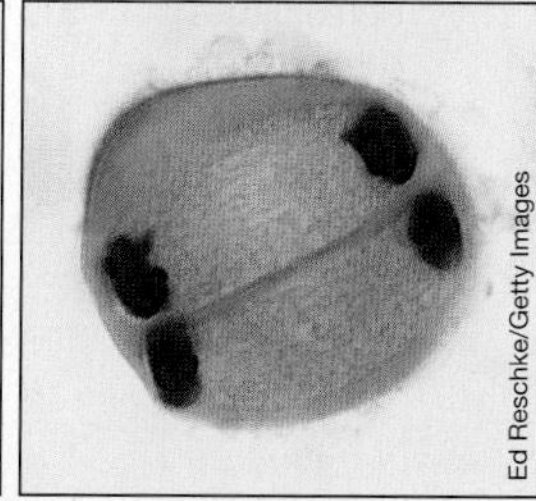

egg inherits is different every time meiosis occurs.

Because of recombination and independent assortment, no two gametes produced by an individual are exactly alike. Each gamete contains one copy of every gene, but the particular combination of alleles in each gamete is always unique. The incredible variety of allele combinations generated by the processes of recombination and independent assortment adds to the pool of genetic diversity in a population. As we'll see in

INFOGRAPHIC 11.7

Meiosis Produces Genetically Diverse Eggs and Sperm

Meiosis produces haploid gametes that are genetically unique. Each egg and each sperm has its own distinct combination of alleles. The two events that create this diversity are recombination and independent assortment.

Recombination:

Before separating at meiosis I, the maternal and paternal chromosomes line up next to each other and physically exchange segments of DNA. Consequently, maternal chromosomes contain segments (and therefore, alleles) from paternal chromosomes, and vice versa.

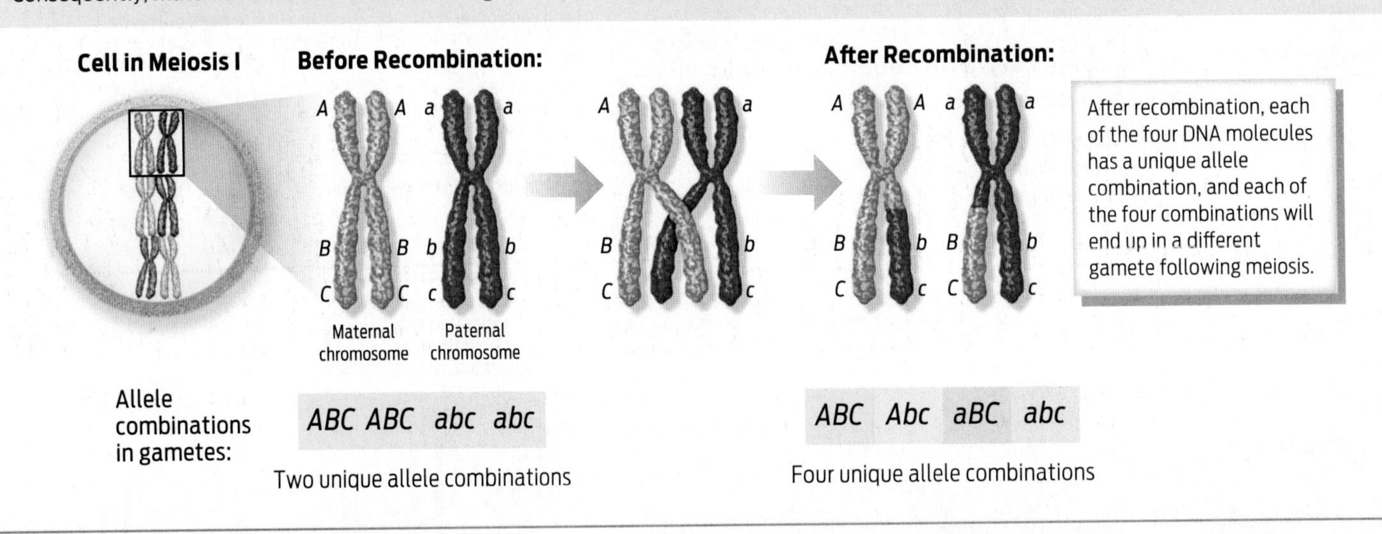

Independent Assortment:

Homologous chromosome pairs separate according to how they have randomly lined up in the cell. Each time meiosis occurs, the chromosome pairs line up differently, and thus a different chromosome combination is produced in the resulting gametes. When all 23 chromosome pairs are considered, there are more than 8 million unique chromosome (and therefore, allele) combinations possible.

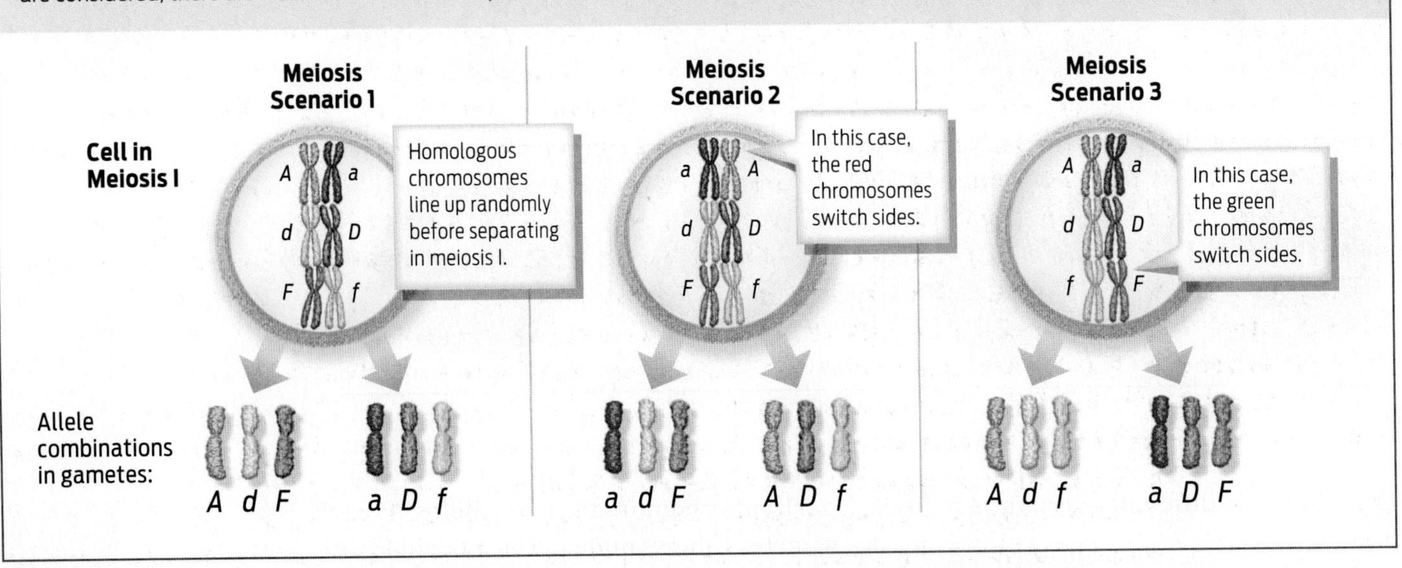

? **Which process (recombination or independent assortment) mixes up combinations of alleles of genes on a single chromosome? Explain your answer.**

later chapters, a population's stockpile of genetic variation forms the basis of evolution **(INFOGRAPHIC 11.7)**.

After doctors diagnosed Emily's cystic fibrosis, both Debbie and Lowell learned that their parents had relatives who had died at a very young age. At the time, the cause of death was thought to be a respiratory illness such as pneumonia. In reality, these relatives most likely had CF, Debbie now thinks; doctors at the time simply did not have the tools to diagnose the disease.

The Schallers now knew that the disease ran in both sides of the family. But they could still not help Emily. "They told us she would only live to be about 12 years old," Debbie recalls, adding, "We just put ourselves in the hands of medical professionals."

Living with the Disease

▶ The biology of cystic fibrosis

Growing up, Emily was scarcely aware of her own disability. The visits to doctors and periodic stays in the hospital were just a part of life. All her teachers and friends knew that she had CF. "My family and friends were all so supportive," she says. In high school Emily played volleyball, basketball, and soccer, and she participated in many walkathons to raise money for CF research.

It wasn't until the end of high school that the disease started to impede her level of activity. "My lung function started to decline, so I wasn't able to keep up with the other kids," Emily says. She began to be hospitalized more often.

In healthy people, the CFTR protein acts as a channel through a cell's membrane that allows chloride ions to move in and out of the cell, keeping the cell's chemistry in balance. But in people with CF, this channel is distorted or absent altogether, and the mechanism goes awry. The result is that mucus—a slippery substance that lubricates and protects the linings of the airways, digestive system, reproductive system, and other tissues—becomes abnormally thick and sticky.

This abnormal mucus blocks ducts throughout the body, including in the pancreas, where it causes digestive problems.

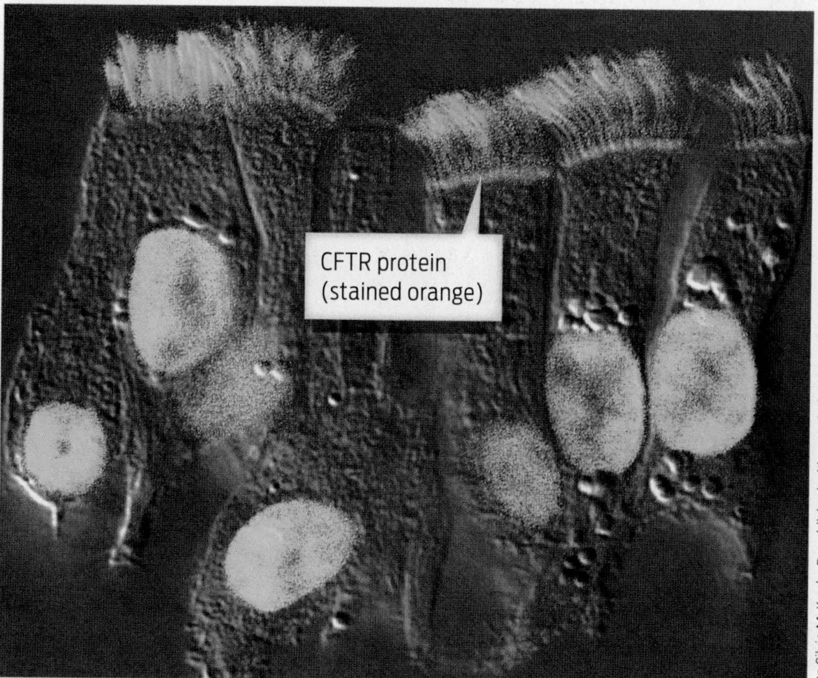

Dr. Silvia M. Kreda, Republished with permission of American Society for Cell Biology, from Kreda, S. M., et al. (2005). Characterization of wild-type and *ΔF508* cystic fibrosis transmembrane regulator in human respiratory epithelia. *Molec Biol Cell* 16(5); permission conveyed through Copyright Clearance Center, Inc.

The CFTR protein is expressed in the cell membrane of human lung cells.

(The name "cystic fibrosis" refers to the characteristic fibrous lumps—called cysts—that form within the pancreas of people with this condition.) The most problematic symptom, however, is that thick mucus builds up in the lungs. People with CF have trouble breathing, and the mucus provides fertile ground for bacteria and other organisms to grow. Over time, repeated infections permanently damage the lungs. As a result, people with CF may slowly lose their ability to breathe, eventually dying of suffocation **(INFOGRAPHIC 11.8)**.

To avoid lung damage, every morning Emily wears an inflatable vest that vibrates continuously, loosening mucus in her lungs. For 30 minutes she inhales a saltwater mist and another medication to thin her mucus, which she then coughs out periodically. To that regime she adds two other medications three times a week to kill infections and to

INFOGRAPHIC 11.8

The CFTR Protein and Cystic Fibrosis

The CFTR membrane protein facilitates the movement of chloride ions. When the CFTR protein is not working, chloride ions are trapped in the cell. The result is a thick, sticky mucus on the surface of cells that traps bacteria, leading to infections.

Normal Lung

In cells lining the lungs, the CFTR membrane protein allows passage of chloride ions. When ions can leave the cell, water is able to flow across the membrane freely, keeping the mucus on the outside of the cell thin and slippery. Bacteria and particles trapped in mucus are easily cleared from the lungs.

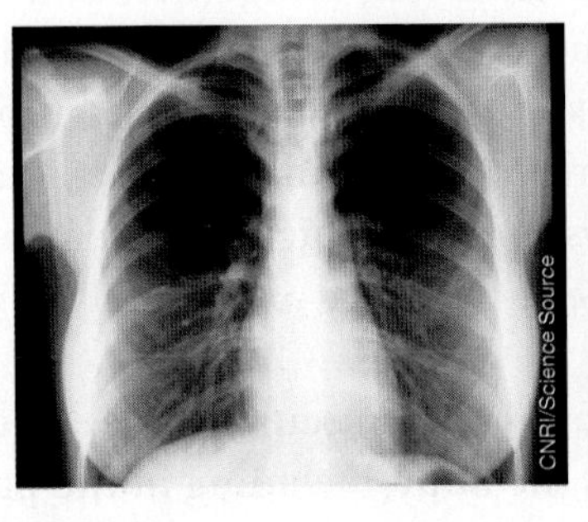

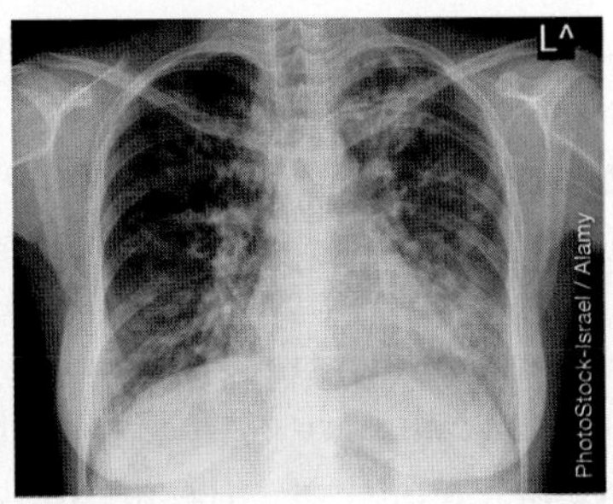

Cystic Fibrosis Lung

The CFTR protein is not working in these cells, disrupting chloride ion flow. These cells retain more water, causing the mucus outside the cell to become thicker and hard to clear out of the lungs. Bacteria trapped in this mucus remain in the lungs and cause infection.

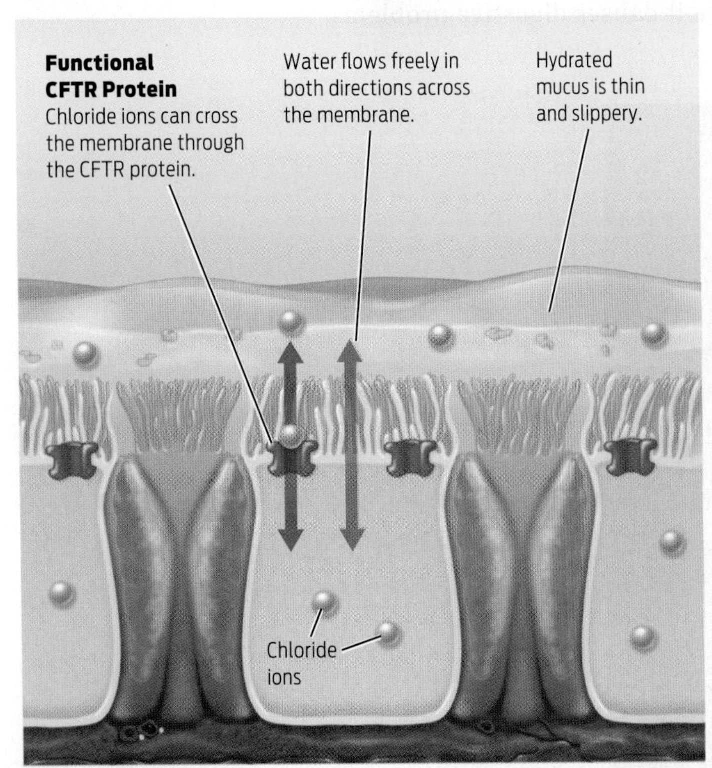

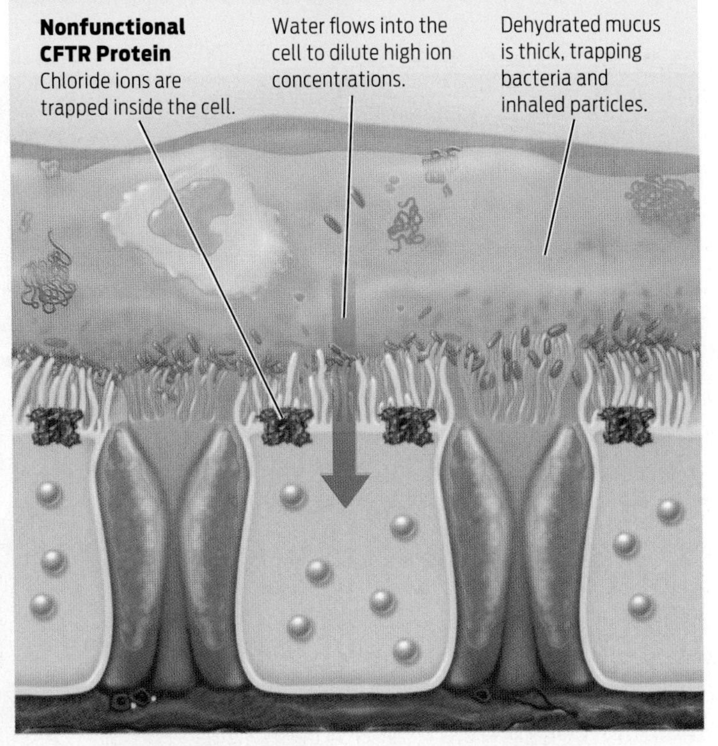

? Explain the relationship between the CFTR protein, chloride ions, water movement, and how thin or sticky the mucus is.

People with cystic fibrosis like Emily wear vibrating vests to loosen the mucus in their lungs while inhaling a saltwater mist to thin the mucus.

keep her lungs from becoming inflamed. And she runs every morning, which helps keep her lung capacity up.

All in the Family

▶ Dominant and recessive inheritance patterns

Emily hasn't ruled out having a family of her own one day. Even though she has CF, her children will not necessarily have the disease.

Why not? Remember that since Emily has CF, that means both of her parents must carry a CF allele. The CF allele produces a nonfunctioning CFTR protein, but Debbie and Lowell are healthy because each has a normal *CFTR* allele that encodes a normally functioning CFTR protein. Thanks to the normal allele, the effect of the CF allele in each of them is hidden. When one allele of a gene masks the effect of the other, the hidden allele is described as a **recessive allele** (designated by a lowercase letter, such as *a*). The other allele, which masks the effect of the recessive allele, is known as the **dominant allele** (designated by a capital letter, such as *A*). The inheritance of traits controlled by single genes with alleles that are either dominant or recessive is sometimes called Mendelian, after Gregor Mendel, the scientist who discovered this pattern (see **Milestone 3: Mendel's Garden**).

Geneticists call Lowell's and Debbie's genotypes, each with two different alleles, **heterozygous.** Their two healthy sons are either heterozygous like their parents, or else they have two normal alleles—that is, their genotype is **homozygous** for the normal allele. A genotype made up of two dominant alleles is known as homozygous dominant. Emily's genotype is homozygous recessive: she inherited one recessive CF allele from each parent, which is why she has the disease.

What were the chances that Debbie and Lowell would have a child with CF? To figure out the likelihood that parents will have a child with a particular trait, we can plot the possibilities on a **Punnett square** (named after the geneticist Reginald Punnett, who devised it). A Punnett square matches the possible parental gametes and shows the likelihood that particular parental alleles will combine. As heterozygous individuals, Debbie and Lowell each have a 50% chance of passing on their CF allele to a child, which means they have a 25% chance of having a child with CF and a 75% chance of having a healthy child. The chance that a child will

RECESSIVE ALLELE
An allele that reveals itself in the phenotype only if a masking dominant allele is not present.

DOMINANT ALLELE
An allele that can mask the presence of a recessive allele.

HETEROZYGOUS
Having two different alleles for a given gene.

HOMOZYGOUS
Having two identical alleles for a given gene.

PUNNETT SQUARE
A diagram used to determine probabilities of offspring having particular genotypes, given the genotypes of the parents.

be a heterozygous **carrier**—that is, that the child will carry the recessive allele for CF but will not have the disease because the allele's effect is masked by the dominant allele—is 50% **(INFOGRAPHIC 11.9)**.

Just as Emily's genotype is different from her parents' genotype, so Emily's children will have genotypes different from her own. Whether her children will develop CF depends on the father's genotype. Since Emily is homozygous, she can contribute only recessive CF alleles to her children. If Emily were to have children with a man who had two normal alleles, none of her children would have the disease—they would all have a heterozygous genotype but a normal phenotype. But as carriers, they could then pass on the recessive CF allele to their children. If Emily had children with a man who was heterozygous for the CF gene, then her children would have a 1 in 2, or 50%, chance of having CF.

Not all recessive alleles cause disease. Many physical traits are the result of inheriting two recessive alleles of a gene. For example, people with blue eyes or red hair have inherited recessive alleles that prevent the deposition of dark pigment. And not all genetic diseases are caused by recessive alleles: some, such as the neurodegenerative disorder Huntington disease, are determined by dominant alleles. Diseases caused by

CARRIER
An individual who is heterozygous for a recessive allele and can therefore pass it on to offspring without showing any of its effects.

INFOGRAPHIC 11.9

How Recessive Traits Are Inherited

Cystic fibrosis is a recessive trait, which means that the disease phenotype is caused by inheriting two recessive alleles, as in Emily's case. Emily's parents do not have CF because, even though each carries one recessive CF allele, they also each possess one dominant normal *CFTR* allele. In other words, they are both heterozygous carriers of the recessive CF allele. To calculate the probability that Debbie and Lowell will have a child with CF, we can create all possible alleles in their gametes and then join all possible combinations of these sperm and egg in a Punnett square.

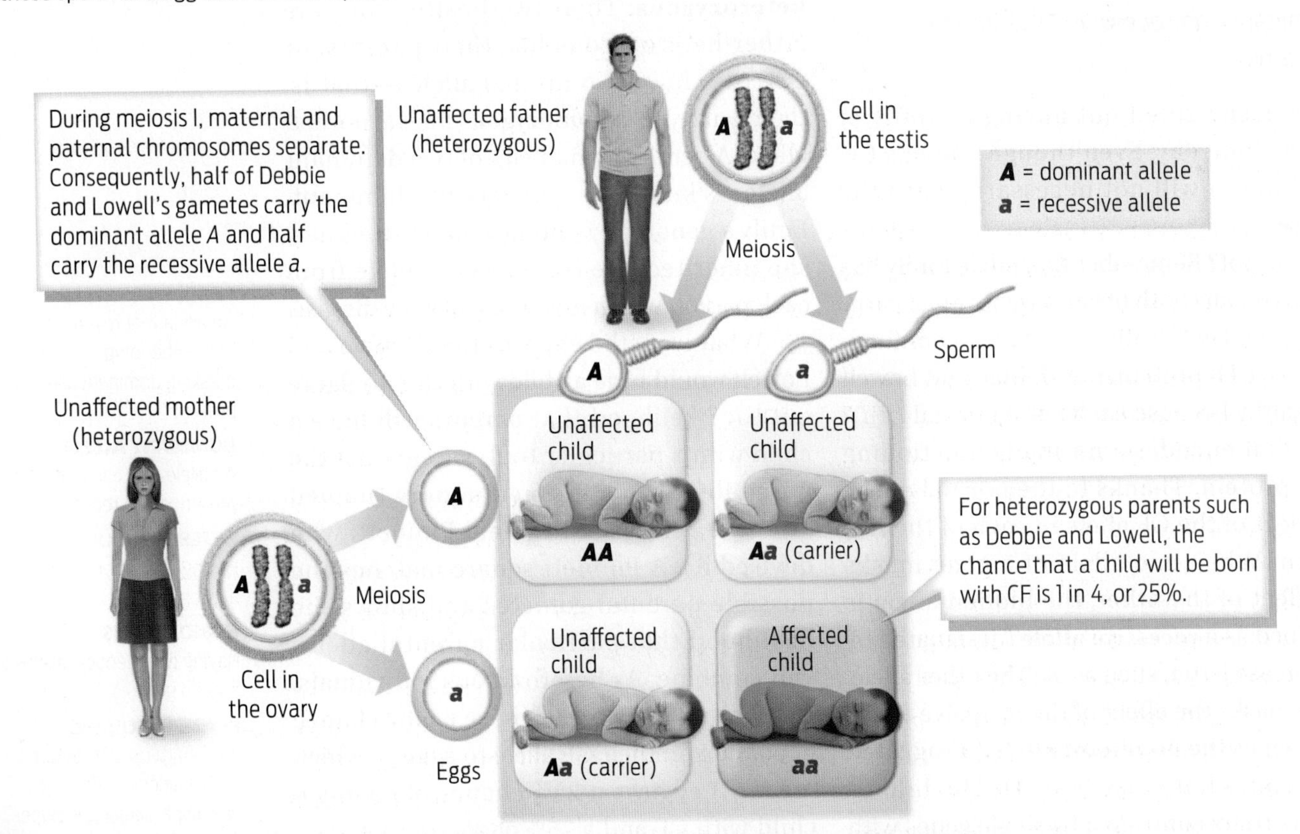

? What is the probability that these two parents will have a child who is a carrier for cystic fibrosis?

INFOGRAPHIC 11.10

How Dominant Traits Are Inherited

Some genetic conditions are caused by dominant alleles. Examples are Huntington disease, a degenerative neurological disease, and polydactyly, having more than five fingers or toes per limb. Many common traits such as dark eyes and dimples are also determined by dominant alleles. In these cases, having one copy of the dominant allele is sufficient to display the trait.

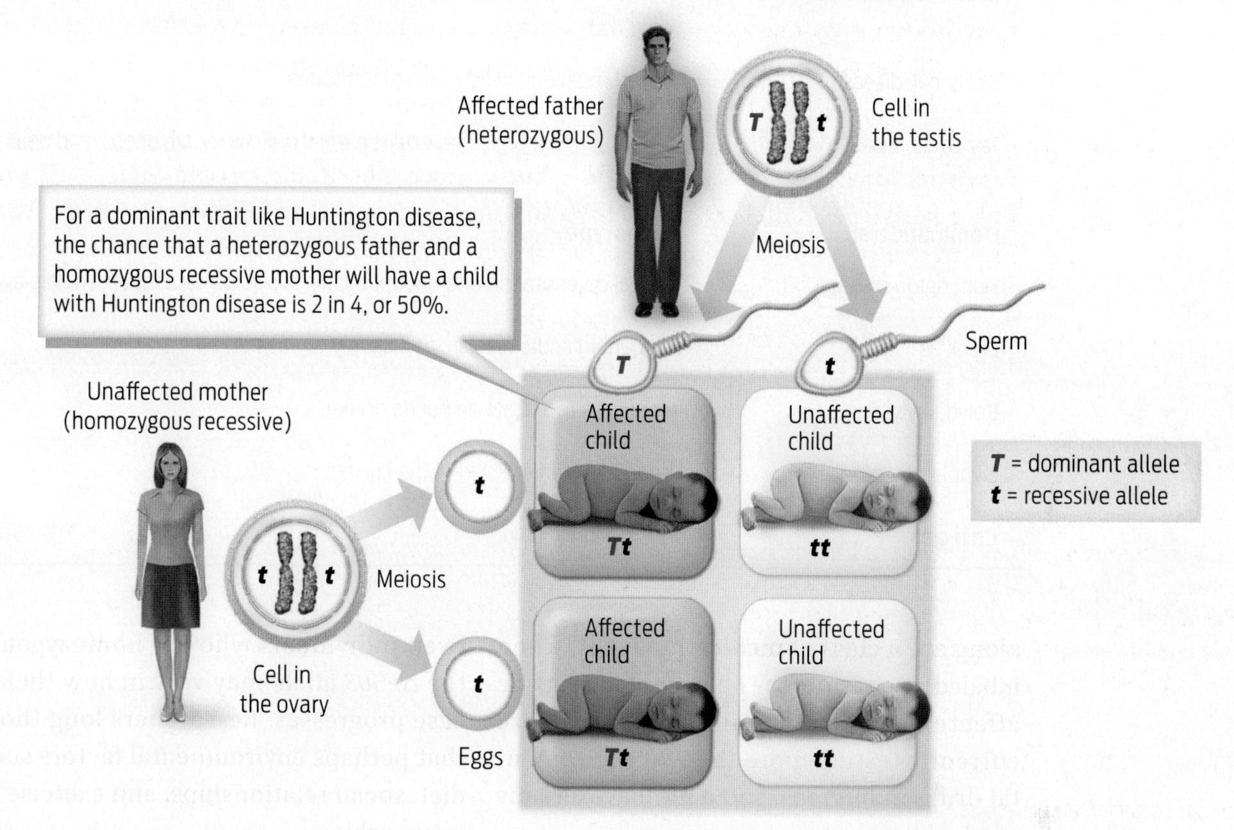

? **What is the probability that these parents will have a child with two copies of the dominant (*T*) allele?**

dominant alleles have a higher probability of showing up in the next generation because it takes only one disease allele to cause the trait **(INFOGRAPHIC 11.10)**.

In all cases, anyone with a genetic disease is at risk for passing it on to his or her children. The ultimate risk varies depending on whether the disease-related alleles are dominant or recessive, and on the genotype of the partner **(TABLE 11.1)**.

Couples who carry disease alleles needn't feel that having children is a roll of the dice, however. Science has provided some excellent ways to ensure that their children won't develop the diseases they could otherwise inherit. Many couples facing this situation use a technology called pre-implantation genetic diagnosis to detect and select embryos that do not carry defective alleles. Through in vitro fertilization, a man's sperm can fertilize a woman's eggs outside the body (see Chapter 31). The genes of each resulting embryo are then examined for specific alleles, and only embryos that don't contain defective alleles are subsequently implanted into the mother. Hundreds of thousands of babies have been born by this technique. Some couples, however, may choose not to undergo assisted reproduction because of religious or other personal reasons.

For children who do inherit a genetic disorder, health care providers can often offer ways to treat their symptoms or reduce their severity. In the case of CF, new treatments are coming online that could help Emily, and her children and grandchildren, too. Furthest

TABLE 11.1 Inherited Genetic Conditions in Humans

Recessive trait	Phenotype
Albinism	Lack of pigment in skin, hair, and eyes
Cystic fibrosis	Excess mucus in lungs, digestive tract, and liver; increased susceptibility to infections
Sickle cell disease	Sickled red blood cells; damage to tissues
Tay-Sachs disease	Lipid accumulation in brain cells; mental deficiency, blindness, and death in childhood
Dominant trait	**Phenotype**
Huntington disease	Mental deterioration and uncontrollable movements; onset at middle age
Freckles	Pigmented spots on skin, particularly on face and arms
Polydactyly	More than five digits on hands or feet
Dimples	Indentations in the skin of the cheeks
Chin cleft	Indentation in chin

along are a class of medications that, when inhaled, can restore the balance of ions inside affected cells in the lungs. Scientists are currently testing more than 20 experimental drugs in humans, and a few have already reached the market.

Other Genes That Influence CF

▶ Multi-gene crosses

Through basic research, scientists continue to learn more about cystic fibrosis. Over the past 25 years, scientists have discovered more than 1,000 different alleles of the *CFTR* gene. About one in 25 people in the general population is a carrier of a CF mutation. The most common mutation is *ΔF508:* approximately 70% of people with CF are homozygous for this mutation, and nearly 90% have at least one *ΔF508* allele. People with this particular CF allele have more severe disease, as the mutation causes the CFTR protein to be absent from the cell membrane altogether.

Researchers have long puzzled over why the severity of CF often varies significantly between any two people with the disease. Even individuals who are homozygous for the *ΔF508* allele may vary in how their disease progresses. Researchers long thought that perhaps environmental factors such as diet, social relationships, and exercise were responsible.

But in recent years, scientists have learned that there is more to the story. Other genes on different chromosomes contribute to the severity of CF symptoms. The genes so far discovered predominantly influence the immune system, which helps the body fight off infections.

For example, scientists have found that one allele of a gene called *TGFB1,* located on chromosome 19, is associated with more-severe lung disease in people with CF. This gene is involved in many processes, including influencing the immune response to infection. While the specific mechanism by which *TGFB1* influences the severity of CF symptoms remains unknown, one hypothesis is that people with this particular *TGFB1* allele mount a more vigorous response to infections than those with other *TGFB1* alleles. Such a heightened immune response can cause lung tissue to scar, making breathing more difficult. Thus the lungs of people with CF who inherit this specific allele of *TGFB1* are more likely to scar in response to infections, leading to greater breathing difficulties. The impact of such "modifier genes" on the CF phenotype makes it more complicated to predict how disabling any particular person's CF disease will be—but not impossible.

Parents who are heterozygous carriers of a CF allele, for example, have a 1 in 4, or 25%, chance of having a child who has CF. If these two parents are also heterozygous for *TGFB1*, then their chance of having a child who is homozygous recessive for *TGFB1* is also 1 in 4 (25%). The chance of two independent events occurring together is calculated by multiplying the two independent chances together. So the probability of these parents having a child who is homozygous recessive for both *CFTR* and *TGFB1* is $1/4 \times 1/4$, or 1 in 16. This probability can also be calculated by using a Punnett square **(INFOGRAPHIC 11.11)**.

Understanding how these modifier genes contribute to the disease may point the way toward even more therapies for CF. Drugs

INFOGRAPHIC 11.11

Tracking the Inheritance of Two Genes

People with CF differ in the severity of their disease. Some of this variability is influenced by alleles of other genes that sit on other chromosomes. One such gene, *TGFB1*, is located on chromosome 19. We can use a Punnett square to follow the inheritance of two genes, as in the example below. In this example, the *CFTR* gene is represented by the symbol *A* and the *TGFB1* gene by the symbol *D*.

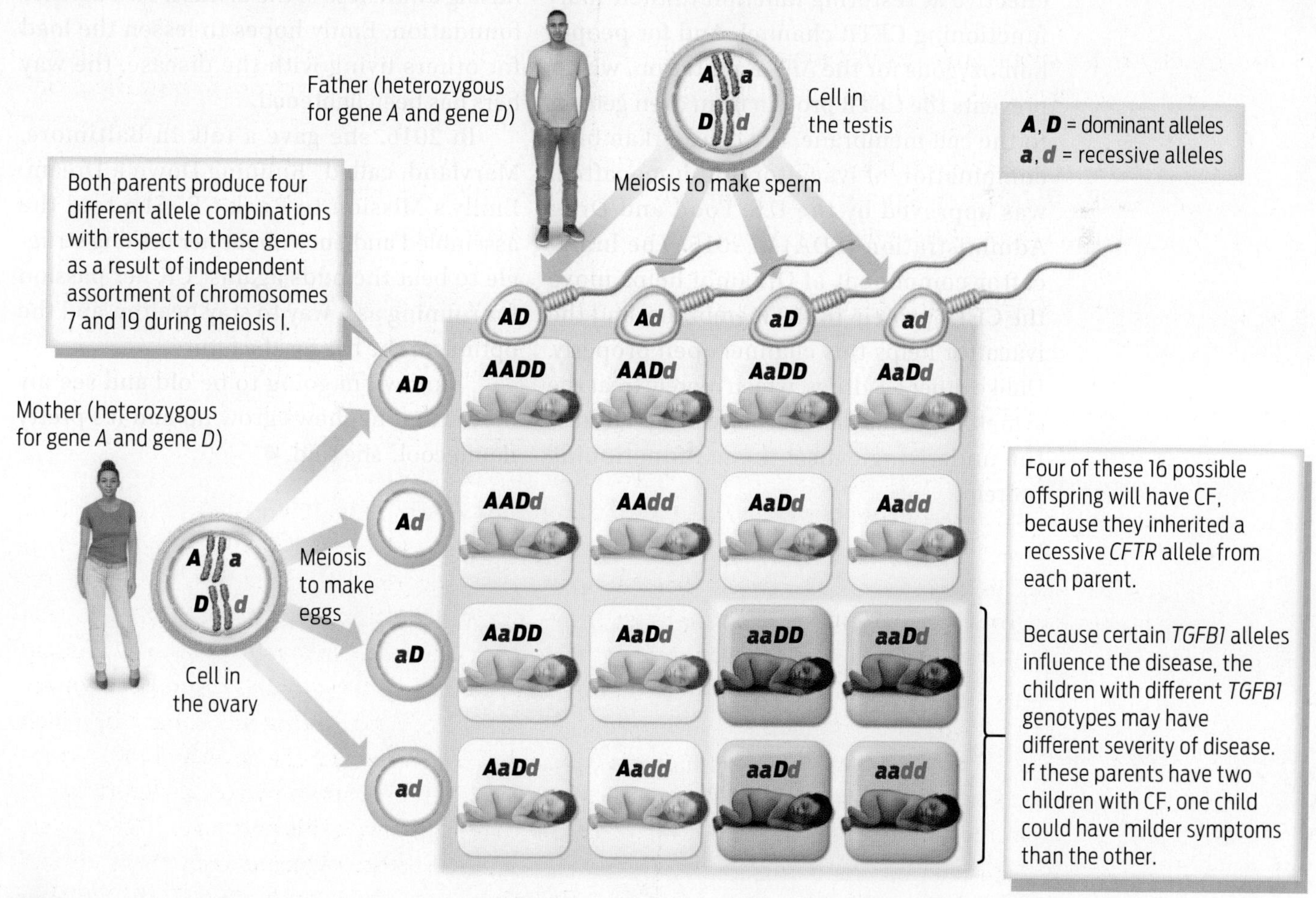

? What is the probability of having a child with two dominant *CFTR* and two dominant *TGFB1* alleles?

that reduce inflammation by targeting the TGFB1 protein, for example, may help reduce scarring in the lungs.

Emily has had her genotype tested. While she doesn't carry the *ΔF508 mutation*, she does have another allele associated with severe disease, *G551D*. This allele is much rarer than *ΔF508*—only about 4% of all people with CF have it. People with this mutation have a CFTR protein channel that doesn't open properly. There is good news for people with this mutation: a drug called Kalydeco (its generic name is ivacaftor) is remarkably effective at restoring function to their malfunctioning CFTR channel. And for people homozygous for the *ΔF508* mutation, which prevents the CFTR protein from even getting to the cell membrane, the drug Orkambi (a combination of ivacaftor and lumacaftor) was approved by the U.S. Food and Drug Administration (FDA) in 2015. The lumacaftor component of Orkambi helps move the CFTR protein to the membrane, and the ivacaftor helps this channel open properly. Unlike other CF drugs, which merely treat the symptoms of the disease, these drugs address the underlying cause: the malfunctioning protein.

"I just took the first deep breath I've ever taken in my life."

—Emily Schaller

Emily started taking Kalydeco in 2010. By day 4, she knew it was making a difference. "I just took the first deep breath I've ever taken in my life," she recalls telling her brother as they were walking down the street. She didn't cough, and there wasn't any mucus in her lungs.

Kalydeco and Orkambi are not cures for CF—the genetic mutation is still present in the cells of someone with CF. Even so, for those individuals with the appropriate alleles, they help immensely. Emily nicknamed Kalydeco "blue lightning" because of how quickly the small blue pill helped her breathing.

Keenly aware of how medical progress has extended her life, Emily conducts her own share of fund raising and education related to CF. In 2007, she started the Rock CF Foundation, a nonprofit organization devoted to improving the lives of people with CF and raising awareness of the disease. Through her foundation, Emily hopes to lessen the load for others living with the disease, the way hers has been lightened.

In 2016, she gave a talk in Baltimore, Maryland, called "Running Down a Dream: Emily's Mission to Rock CF." She told the assembled audience about her lifelong struggle to beat the odds against CF, her passion for running as a way to stay healthy, and the optimism she has for the future.

"I know I'm going to be old and see my nieces and nephews grow up and it's pretty damn cool," she said. ■

CHAPTER 11 SUMMARY

Driving Question 1 How does the organization of chromosomes, genes, and their alleles contribute to human traits?

- Genes, which code for proteins, are the units of inheritance; they are physically passed down from parents to offspring.
- An organism's physical traits constitute its phenotype; its genes constitute its genotype. You can't always determine an individual's genotype from his or her phenotype.
- Humans are diploid organisms, meaning they have two copies of each chromosome in their cells. Because chromosomes come in pairs, we have two copies of each gene in our body cells. These copies can be the same or different from each other.
- Different versions of the same gene are called alleles. Alleles arise from mutations that change the nucleotide sequence of a gene.

Driving Question 2 How does meiosis produce genetically diverse gametes?

- To reproduce sexually, diploid organisms produce haploid gametes—sperm and eggs.
- Meiosis is cell division that produces haploid gametes from diploid cells.
- During meiosis, homologous chromosomes recombine and assort independently to generate genetically diverse sperm and eggs. No two sperm or egg cells produced by the same person will be exactly alike.

Driving Question 3 Why do different traits have different patterns of inheritance?

- Alleles may be dominant or recessive. Dominant alleles can mask the effects of recessive alleles, which can be hidden.
- Many traits result from carrying two recessive alleles; others result from carrying one dominant allele.
- Cystic fibrosis (CF) is a recessively inherited genetic disease. Alterations in the CFTR protein cause disease by interfering with ion and water balance in cells, especially in the lungs.
- A Punnett square can help predict a child's genotype and phenotype when the pattern of inheritance—dominant or recessive—is known.

Driving Question 4 What are some practical applications of understanding the genetic basis of human disease?

- Alleles of other (non-*CFTR*) genes can influence the severity of the CF phenotype in people with cystic fibrosis.
- New drugs target specific CFTR proteins (encoded by specific *CTFR* alleles) to enhance their function.

More to Explore

- Cystic Fibrosis Foundation: www.cff.org/
- Rock CF Foundation: http://letsrockcf.org/rockcf.org/
- Genetics Home Reference, Cystic Fibrosis: https://ghr.nlm.nih.gov/condition/cystic-fibrosis
- Welsh, M. J., and Smith, A. E. (December 1995). Cystic fibrosis. *Sci Am* 52–59.
- Pearson, H. (2009). Human genetics: One gene, twenty years. *Nature* 460:164–169. http://www.nature.com/news/2009/080709/full/460164a.html
- Ramsey, B. W., et al. (2011). A CFTR potentiator in patients with cystic fibrosis and the *G551D* mutation. *N Engl J Med* 365:1663–1672.

CHAPTER 11 Test Your Knowledge

Driving Question 1 How does the organization of chromosomes, genes, and their alleles contribute to human traits?

By answering the questions below and studying Infographics 11.2, 11.3, and 11.4, you should be able to generate an answer to this broader Driving Question.

Know It

1. How do the two alleles of the *CFTR* gene in a lung cell differ?

a. They were inherited from different parents.
b. One is on chromosome 7 and one is on chromosome 3.
c. Only one is expressed.
d. all of the above
e. There is no difference because they are both the same gene.

2. Consider a liver cell.
  a. How many chromosomes are present?
  b. How many alleles of each gene are present?

3. Consider a gamete.
  a. How many chromosomes are present?
  b. How many alleles of each gene are present?

Use It

4. A diploid cell of baker's yeast has 32 chromosomes.
  a. How many pairs of homologous chromosomes are present in a diploid cell of baker's yeast?
  b. How many chromosomes are present in each of its haploid gametes?

5. In diploid organisms, having two homologues of each chromosome can be beneficial if one allele of a gene encodes a nonfunctional protein. Can haploid organisms survive if they have nonfunctional alleles? Explain your answer.

6. From which parent did Emily inherit cystic fibrosis? Explain your answer.

Driving Question 2 How does meiosis produce genetically diverse gametes?

By answering the questions below and studying Infographics 11.4, 11.5, 11.6, and 11.7, you should be able to generate an answer to this broader Driving Question.

Know It

7. A human female has ____________ chromosomes in each skin cell and ____________ chromosomes in each egg.
  a. 46; 46
  b. 23; 46
  c. 46; 23
  d. 23; 23
  e. 92; 46

8. A woman is heterozygous for the CF-associated gene (the alleles are represented here by the letters *A* and *a*). Assuming that meiosis occurs normally, which of the following represent eggs that she can produce?
  a. *A*
  b. *a*
  c. *Aa*
  d. *AA*
  e. *aa*
  f. *A* or *a*
  g. *A*, *a*, or *Aa*

9. Draw a maternal version of chromosome 7 in one color and a paternal version of chromosome 7 in another color. Maintaining this color distinction, now draw a possible version of chromosome 7 that could end up in a gamete following meiotic division.

10. **a.** During meiosis I, what separates? Is this the same or different from what happens during mitosis in human epithelial cells (e.g., skin cells)? Explain your answer.
  b. At the end of meiosis I, are the cells haploid or diploid? Is this the same or different from human epithelial cells that have completed mitosis? Explain your answer.

Use It

11. An alien has 82 total chromosomes in each of its body cells. The chromosomes are paired, making 41 pairs. If the alien's gametes are produced by meiosis, what are the number and arrangement (paired or not) of chromosomes in one of its gametes? Give the reason for your answer.

12. Describe at least two major differences between mitosis (discussed in Chapter 10) and meiosis.

13. If meiosis were to fail and a cell skipped meiosis I, so that meiosis II was the only meiotic division, how would you describe the resulting gametes?

Driving Question 3 Why do different traits have different patterns of inheritance?

By answering the questions below and studying Infographics 11.9, 11.10, and 11.11, you should be able to generate an answer to this broader Driving Question.

Know It

14. What is the genotype of a person with CF?
  a. homozygous dominant
  b. homozygous recessive
  c. heterozygous
  d. any of the above
  e. none of the above

15. Strictly on the basis of the following *CFTR* genotypes, what do you predict the phenotype of each person to be?
  a. heterozygous
  b. homozygous dominant
  c. homozygous recessive

16. How many copies of the CF-associated allele does a person with CF have in one of his or her lung cells? How does this compare to someone who is a carrier for CF? How does it compare to someone who is homozygous dominant for the *CFTR* gene?

17. Women can inherit alleles of a gene called *BRCA1* that puts them at higher risk for breast cancer. The alleles associated with elevated cancer risk are dominant. Of the genotypes listed here, which carries the lowest genetic risk of developing breast cancer?
  a. *BB*
  b. *Bb*
  c. *bb*
  d. *BB* and *Bb* carry less risk than *bb*.
  e. All carry equal risk.

Use It

18. A person has a heterozygous genotype for a gene associated with a particular inherited disease. However, this person does not have the disease phenotype. Does this disease have a dominant or a recessive inheritance pattern?

19. Assume that Emily (who has CF) decides to have children with a man who does not have CF and who has no family history of CF.

a. What combination of gametes can each of them produce?

b. Place these gametes on a Punnett square and fill in the results of the cross.

c. On the basis of the Punnett square results, what is the probability that the couple will have a child with CF?

d. On the basis of the Punnett square results, what is the probability that they will have a child who is a carrier for CF?

Mini Case

Apply Your Knowledge

20. You are a genetic counselor. A 21-year-old college student has scheduled an appointment because his 47-year-old mother has Huntington disease, and he is worried about developing this disease himself. You ask about other family members. The student's maternal grandmother (his mother's mother) does not have Huntington disease. The student's father is 62 years old and does not have Huntington disease. From this information, you draw a Punnett square to determine the probability that the student will develop Huntington disease.

a. What could you tell the student about his risk?

b. The student has a half-sister. She is 19 years old and has the same mother but a different father. Her father is 45 and does not have Huntington disease. However, the father's mother died of Huntington disease. How does the half-sister's risk compare to her brother's (the student's) risk? Could you give her a definitive answer about her risk? Why or why not?

21. According to the discussion in this chapter, why might one person with a homozygous recessive *CFTR* genotype have a somewhat different phenotype from another person who also has a homozygous recessive *CFTR* genotype?

22. Phenylketonuria is described as an inborn error of metabolism. In this recessive genetic condition, the enzyme that breaks down the amino acid phenylalanine is defective or missing. Testing of all newborns allows this condition to be detected at birth. A special diet that severely minimizes intake of phenylalanine (e.g., avoiding diet sodas and most usual sources of protein) can treat the condition. If two carriers of both cystic fibrosis and phenylketonuria were to have a child, what is the probability that the child will have

a. both cystic fibrosis and phenylketonuria?

b. cystic fibrosis and be a carrier for phenylketonuria?

c. neither condition?

d. neither condition and not be a carrier for either?

Driving Question 4 What are some practical applications of understanding the genetic basis of human disease?

By answering the questions below and studying Infographics 11.8, 11.9, 11.10, and 11.11, you should be able to generate an answer to this broader Driving Question.

Know It

23. Can cystic fibrosis be diagnosed prenatally by examining the fetal chromosomes (as shown in the left panel of Infographic 11.3)? Why or why not?

Interpreting Data

Apply Your Knowledge

24. Ivacaftor (trade name Kalydeco) is a drug designed to enhance the activity of the CFTR protein encoded by specific alleles, including the *G551D* allele. The protein encoded by this allele is present in the cell membrane, but is not very active. The drug has been shown in laboratory studies to enhance the activity of the CFTR protein. To determine if ivacaftor has an impact on disease symptoms in people with cystic fibrosis, the drug was tested in a randomized, double-blind clinical trial.

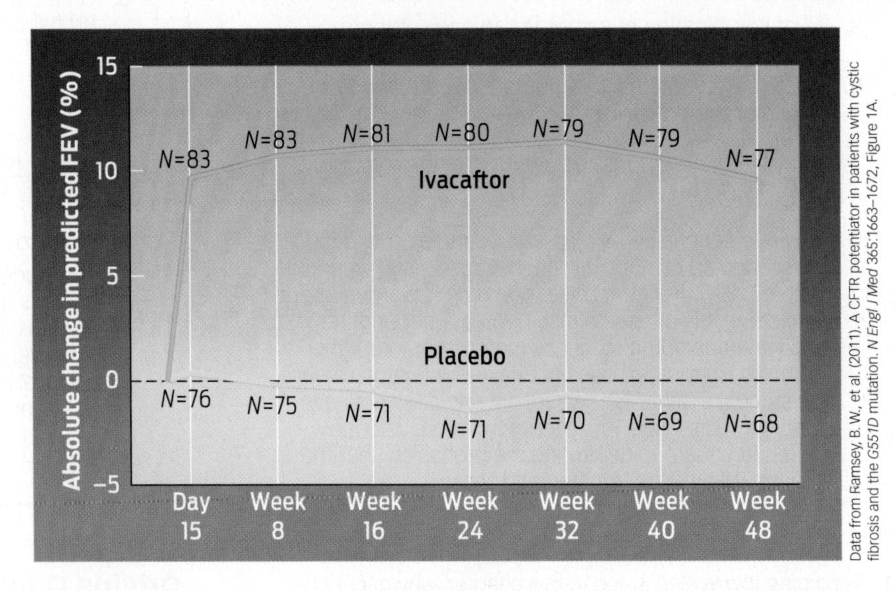

Data from Ramsey, B. W., et al. (2011). A CFTR potentiator in patients with cystic fibrosis and the *G551D* mutation. *N Engl J Med* 365:1663–1672, Figure 1A.

a. From the information provided here and in the graph at right, describe the experimental design of the trial. Include who the participants were, how participants were assigned to the experimental group and the control group, and what treatment(s) were given to participants in each group. Why was it important that this was a double-blind trial?

b. Before treatment started, the researchers obtained baseline measurements of the participants' lung function. One of these measurements was the amount of air that a participant could forcibly exhale from the lungs in 1 s (expressed as a percentage of total lung volume), known as the predicted forced expiratory volume (FEV). The drug was given to the experimental group every 12 hours; the predicted FEV was measured, and expressed as a change from the baseline measurement. The results are shown in the graph. The number of subjects (*N*) is shown for each group at various time points during the study.

- How soon did the experimental group experience an improvement in lung function as measured by predicted FEV?
- How long was the improvement sustained?
- By the end of the study, what was the absolute improvement of the experimental group relative to baseline?

c. The Δ*F508* allele causes the CFTR protein to be absent from the cell membrane. Is ivacaftor likely to be a viable treatment for patients whose CF is caused by this allele? Why or why not?

Bring It Home

Apply Your Knowledge

25. Emily took a genetic test to determine which CF alleles she inherited. The results revealed she has a *G551D* allele, making her a candidate for therapy with ivacaftor. Since taking this drug, Emily's breathing and lung function have improved. In this case, the genetic test opened up a treatment option for a patient. For some genetic diseases, such as Huntington disease, treatment is limited and there is no cure. If you were faced with the decision to take a genetic test, especially for a disease for which no cure was available, would you take the test? Why or why not?

The garden outside the Augustinian Abbey in Brno, where Gregor Mendel performed his experiments. ▼

MENDEL'S GARDEN

An Austrian monk lays the foundation for modern genetics

NLM/Science Source

Biophoto Associates/Science Source

DRIVING QUESTIONS

1 How was Mendel able to recognize the transmission of alleles before the discovery of DNA?

2 What do Mendel's two laws state about how offspring inherit alleles from their parents?

Many scientists of Mendel's day—the mid-19th century—believed that parental traits were blended together in offspring, like mixing paint colors.

Gregor Johann Mendel was an unlikely father of genetics. He was a depressed Austrian monk who by all accounts suffered from debilitating test-taking anxiety, failing his teaching exam twice. Mendel nevertheless collected the first clear evidence showing how traits are inherited: parents pass discrete "elements" for those traits to their children. These elements remain intact and can be passed on indefinitely to future generations without being diluted. Although he couldn't say at the time what these elements were, Mendel had actually discovered what came to be called genes. We now know that genes come in pairs, and that they exist in multiple discrete forms called alleles (see Chapter 11).

Many scientists of Mendel's day—the mid-19th century—believed that parental traits were blended together in offspring, like mixing paint colors. For example, a tall mother and a short father would have children of medium height, who would then pass on that trait—medium height—to their own children. Other scientists clung to an older idea, that a sperm or an egg contained a miniature adult waiting to be born. But through a series of simple yet elegant experiments conducted in a monastery garden, Mendel provided a new explanation for heredity, decades before the word "genetics" was coined **(INFOGRAPHIC M3.1)**.

In 1843, Mendel became a monk at the Augustinian Abbey of St. Thomas in Brünn (now Brno, in the Czech Republic). He studied theology and was ordained as a priest in 1847. When he failed his teaching exam (the Augustinians were a teaching order), the abbot at St. Thomas sent Mendel to the University of Vienna to brush up on his skills. For 2 years he studied math, physics, zoology, and botany, but once again he flunked the test. This depressing result encouraged him to turn from teaching to research.

Mendel returned to the monastery in 1853, and a year later began researching a topic that had sparked his interest in school: hybrids, the offspring of two different breeds or varieties. Mendel hoped to explain what he and many others had observed about hybrids—that physical traits (e.g., size, color) can sometimes skip a generation, disappearing in one generation only to show up again in the next. For example, hybrid crop plants would often produce offspring that looked much like themselves, but occasionally they produced offspring that resembled earlier generations.

Mendel began his research using mice, which he kept in cages in his two-room flat.

He intended to breed gray mice with albino (white) mice to see what color fur the offspring would have. He didn't get very far down this path because, as Robin Henig wrote in her 2001 book *The Monk in the Garden,* the local bishop thought that "toying with the reproduction of mammals was simply too vulgar an undertaking for a priest." So Mendel decided to work instead with pea plants, which proved a better model organism anyway. The plants grew quickly, he could better control their environment and breeding, and they smelled nicer than the mice, too.

Mendel began by choosing specific traits that he could see and study, among them seed texture, seed color, pod shape, pod color, flower color, and stem length. Each of the traits he chose to study existed in two forms. For example, seed texture was either round or wrinkled; seed color was either green or yellow. Mendel started his breeding experiments with plants that "breed true"—that is, they produce offspring that carry the same traits as the parents, generation after generation. Only then could he study what happened to particular traits when purebred, or true-breeding, plants of one variety were mated with purebred plants of another variety.

Pea plants can self-pollinate, which means that the pea flower contains both male and female sexual organs and a single plant can fertilize itself to produce offspring. To produce true-breeding plants, Mendel covered the pea flowers with a small bag so that he could control fertilization, manually fertilizing plants with their own pollen and preventing pollen from another plant from entering. Once he had established true-breeding plants, he could then cross-pollinate two different plants. What would happen if he crossed a true-breeding green-seeded plant with a true-breeding plant that produced yellow seeds? Or a purple-flowered plant with a white-flowered plant? For each cross, Mendel painstakingly pollinated individual flowers from the two plants by hand.

INFOGRAPHIC M3.1

Ideas of Inheritance before Mendel

Preformationist ideas, popular for centuries before Mendel, held that the next generation of life already existed fully formed in miniature inside the egg or sperm. More common in Mendel's time was the idea that substances from the mother and father blend together during conception to produce traits of the offspring.

Preformationist Ideas

Spermist theory maintained that sperm held a homunculus inside.

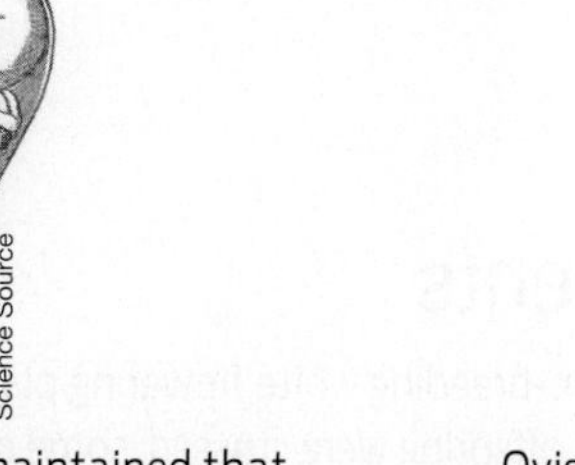

Ovist theory proposed that the preformed human was inside the egg.

Blending Ideas

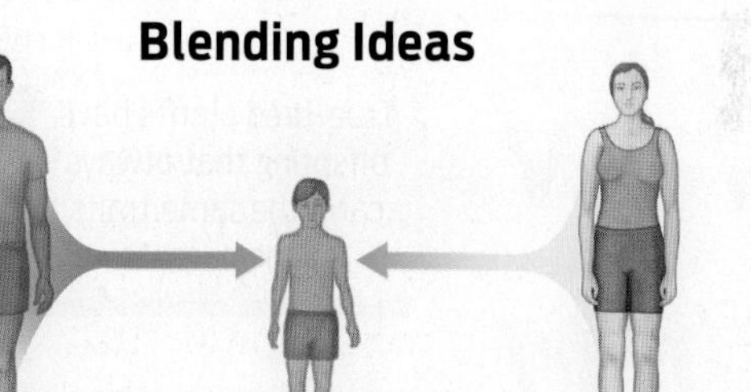

Fluid or particles from the father and mother blend together to make the traits of the offspring.

? What did the spermist and ovist theories have in common?

He also prevented self-pollination by removing the male reproductive parts from the plants to be fertilized.

Mendel noticed that when he crossed a true-breeding white-flowering plant with a true-breeding purple-flowering plant, the first generation of offspring (what we now call the F_1 generation) all had purple flowers. That the flowers were true purple rather than pale purple suggested that parental traits were not blended, as earlier hypotheses of inheritance would have predicted. But the trait for white flowers didn't disappear completely, either. When Mendel randomly selected two F_1 purple-flowering plants to breed, he found that on average 1 out of every 4 plants of the second generation of offspring (the F_2 generation) had white flowers. Mendel reasoned that a hidden white element must be present in the purple F_1 plants. So each F_1 plant must have two such elements, one representing purple (the trait that appeared) and the other representing white (the hidden trait) **(INFOGRAPHIC M3.2)**.

If these results sound familiar, there's a good reason for that: they reflect dominant and recessive patterns of inheritance, which we discussed in Chapter 11. Purple flower color is dominant over white, which is recessive. Mendel was the first to gather evidence showing that traits could be inherited in a dominant or recessive fashion; in fact, he coined these terms. While earlier scientists had noticed

INFOGRAPHIC M3.2

Mendel's Experiments

When Mendel cross-pollinated true-breeding white flowering plants with true-breeding purple flowering plants, all the F_1 offspring had purple flowers. When these F_1 offspring were crossed, some of F_2 had purple flowers, and some had white flowers, in a predictable ratio. This allowed Mendel to recognize that there are alternative "elements" for the flower color trait. These elements remain intact over generations and do not blend with one another.

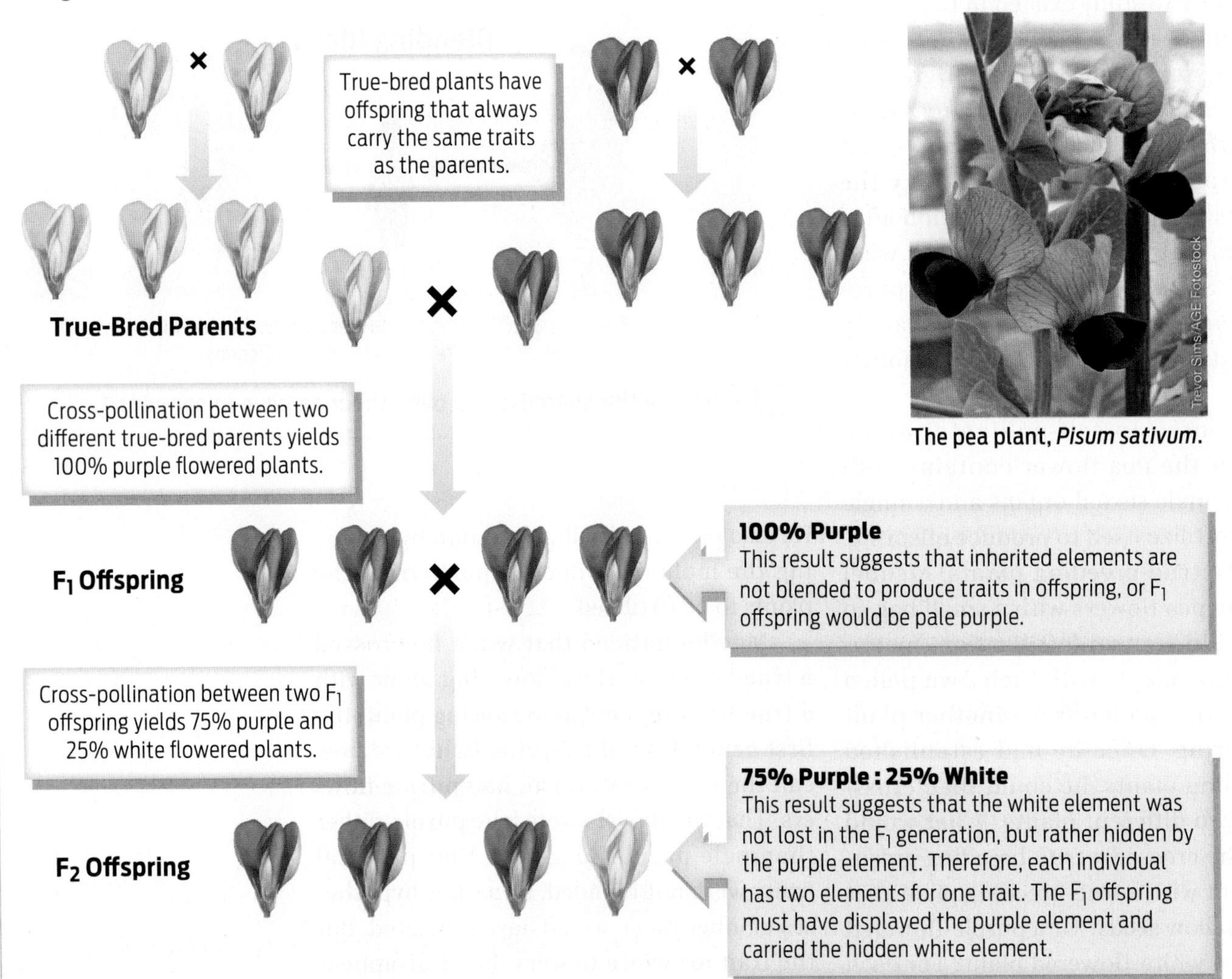

The pea plant, *Pisum sativum*.

? Which flower color (purple or white) is dominant?

TABLE M3.1 Mendel's Experimental Data

Mendel was meticulous in his data collection. He cultivated, crossed, and analyzed thousands of offspring to ensure accurate ratio calculations.

Trait	P_1 Cross	F_1 Generation	F_2 Generation	Actual Ratio	Probability Ratio
Seed texture	round × wrinkled	round	5,474 : 1,850 round wrinkled	2.96 : 1	3 : 1
Seed color	yellow × green	yellow	6,022 : 2,001 yellow green	3.01 : 1	3 : 1
Pod shape	inflated × constricted	inflated	882 : 299 inflated constricted	2.95 : 1	3 : 1
Pod color	green × yellow	green	428 : 152 green yellow	2.82 : 1	3 : 1
Plant height	tall × short	tall	787 : 277 tall short	2.84 : 1	3 : 1

that traits could disappear in one generation and reappear in later generations, Mendel was the first to offer a coherent explanation of *why* they did.

Over 7 years, Mendel cultivated and tested more than 28,000 pea plants, and in the process discovered the basic principles of inheritance. He published his results in 1866 **(TABLE M3.1)**.

Today we know that Mendel's "elements" are alleles of genes, and that genes are located on chromosomes. The principles he discovered have been formalized into two laws. The first is Mendel's law of segregation, which states that for any diploid organism, the two alleles of each gene segregate separately into gametes. That is, every gamete receives only one of the two alleles, with the specific allele that any one gamete receives being random. The physical basis of segregation is the separation of

Today we know that Mendel's "elements" are alleles of genes, and that genes are located on chromosomes.

INFOGRAPHIC M3.3

Mendel's Law of Segregation

Mendel's experiments enabled him to formulate the law of segregation. This law has held up over time, although today we know that Mendel's "elements" are alleles of genes. The law of segregation states that when an organism produces gametes, the two alleles for any given gene separate so that each gamete receives only one allele. When gametes come together in fertilization, the resulting offspring will have traits that reflect the particular combination of inherited alleles.

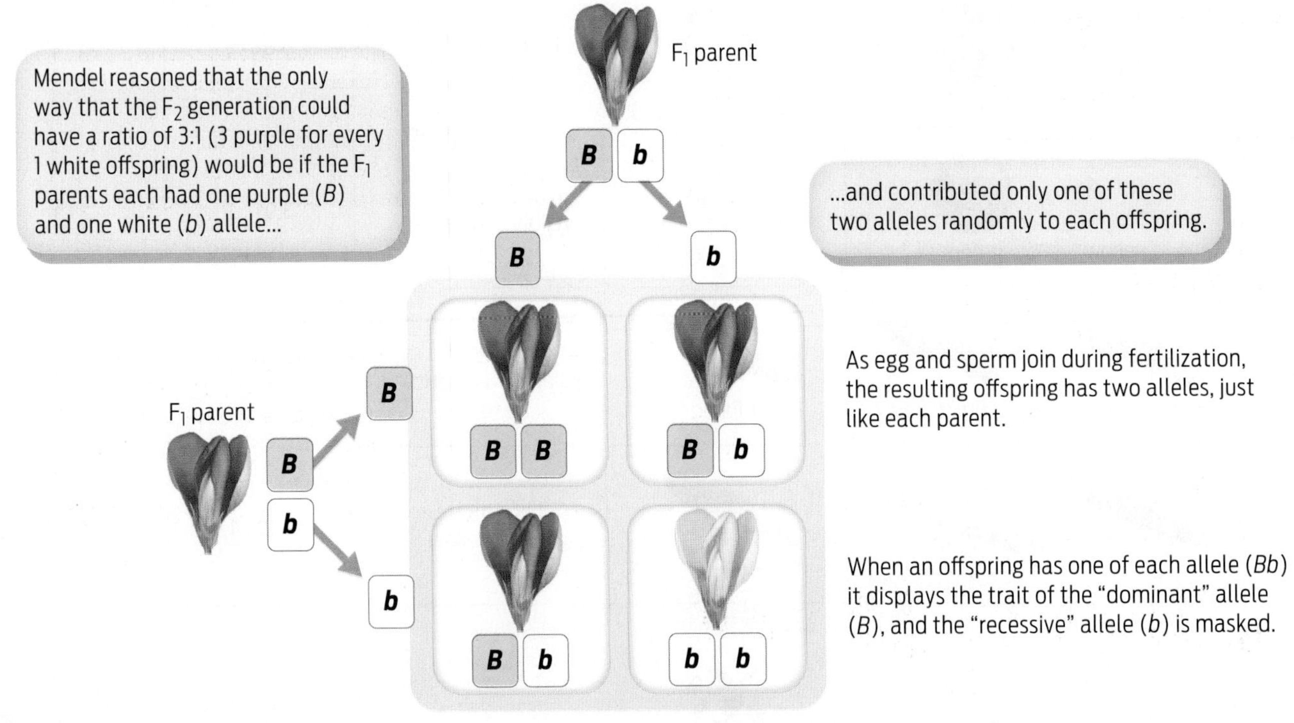

? How many alleles for any given gene are segregated into a single gamete?

maternal and paternal chromosomes during meiosis **(INFOGRAPHIC M3.3)**.

The second law, the law of independent assortment, states that the two alleles of any given gene segregate independently from the two alleles of any other gene. Because of independent assortment, offspring can display any combination of the different traits, rather than inheriting all of the traits together. We now know this law holds true only for genes that are located on different chromosomes, or far enough away from each other on a single chromosome to readily recombine. It was mere happenstance that Mendel chose traits for which the genes are located on different chromosomes and so assort independently **(INFOGRAPHIC M3.4)**.

Despite Mendel's groundbreaking research, no one realized the significance of his results at the time—not even Charles Darwin, whose *Origin of Species* was published in 1859.

In 1868, Mendel was elected abbot of St. Thomas and largely shifted his focus from science to monastic life and the administration of the abbey. Although Mendel's research was cited by other scientists, he didn't receive much notice until three botanists who were also studying how traits are inherited in plants rediscovered his work more than 30 years later, in 1900. While preparing to publish their ideas about inheritance, they looked through the research literature and found that Mendel had beaten them to the punch. Mendel was finally recognized as the researcher who had solved a crucial mystery of inheritance many years before. ■

INFOGRAPHIC M3.4

Mendel's Law of Independent Assortment

Mendel went on to study how multiple traits are inherited—for example, seed color and seed texture. Tracing two traits at a time helped him formulate the law of independent assortment. This law states that the two alleles of any given gene will segregate independently from the alleles of other genes when distributed into gametes. Consequently, each gamete may acquire any possible combination of alleles, and therefore traits.

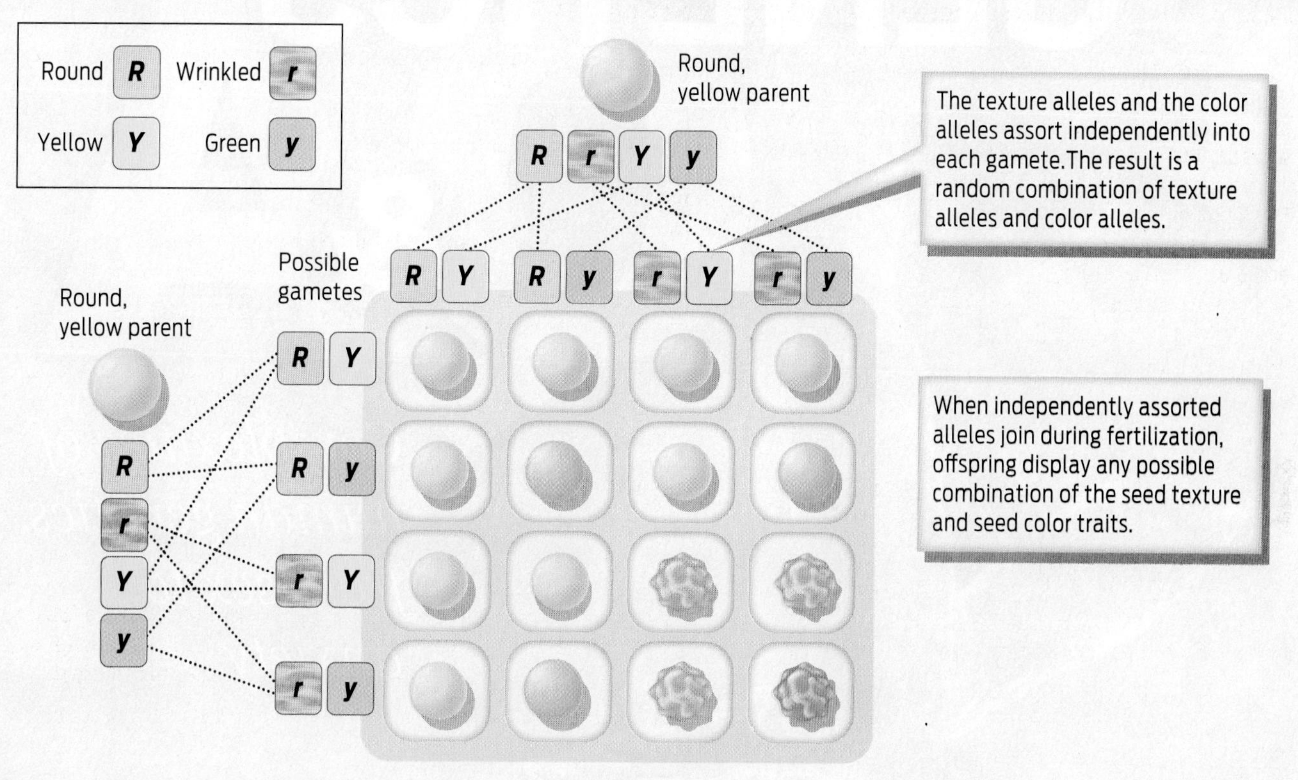

? Why are there not two alleles of the *R* gene or two alleles of the *Y* gene in any gamete?

More to Explore

- Abbott, S., and Fairbanks, D. J. (2016). Experiments on plant hybrids by Gregor Mendel. *Genetics 204*(2):407–422: www.genetics.org/content/204/2/407
- Henig, R. M. (2001). *The Monk in the Garden: The Lost and Found Genius of Gregor Mendel, the Father of Genetics.* New York: Mariner Books.
- DNA Learning Center, Cold Spring Harbor Laboratory, Gregor Mendel and pea plants: www.dnalc.org/view/16002-gregor-mendel-and-pea-plants.html
- Mukherjee, S. (2016). *The Gene: An Intimate History.* New York: Scribner. See also: https://www.pbs.org/show/gene/

MILESTONES IN BIOLOGY 3 Test Your Knowledge

1. You have a tall pea plant.

a. What are its possible genotypes?

b. What crosses could you do to determine whether it is true breeding? For each cross, give the expected phenotypes of the offspring.

2. In crossing pea plants:

a. If a true-breeding tall pea plant with purple flowers is crossed with another true-breeding tall pea plant with purple flowers, what will the offspring pea plants look like?

b. Will the offspring be true breeding? Explain your answer.

3. Half of the gametes of a heterozygous parent will carry the dominant allele, and half will carry the recessive allele. Which of Mendel's laws explains this division?

4. Consider Mendel's laws.

a. Which of Mendel's laws would be violated if the offspring of two heterozygous tall, purple-flowered pea plants (denoted as *TtBb*) were only tall, purple-flowered plants or short, white-flowered plants?

b. What would such a violation suggest about the two genes?

GENETICS Q&A

Complexities of human genetics, from sex to anxiety

Svisio/Getty Images

DRIVING QUESTIONS

1 How do chromosomes determine sex, and how does sex influence the inheritance of certain traits?

2 What inheritance patterns are observed when alleles are not simply dominant or recessive?

3 What inheritance patterns are observed when more than one gene and/or the environment influences a trait?

4 How does nondisjunction lead to numerical abnormalities of chromosomes, and what are the consequences of these abnormalities?

In Chapter 11, we discussed the basic principles of inheritance, including how gametes are formed and how simple dominant and recessive traits are passed from parents to offspring. In this chapter, we examine more-complex cases of inheritance, including examples of sex-linked traits, multi-gene traits, and the role of the environment in shaping phenotypes.

SEX DETERMINATION

Q What makes a man?

A A botched circumcision in 1966 on a little boy named Bruce Reimer in time became a landmark example of how biology shapes gender identity—the subjective experience of oneself as male or female. Doctors at the hospital where Reimer was circumcised used an experimental procedure that involved burning off the foreskin. The procedure went awry, and Bruce's penis was singed nearly completely off. For advice about what to do, Bruce's parents turned to John Money, a well-known psychologist at Johns Hopkins University in Baltimore, Maryland, who had written much about the importance of environment in determining a person's gender identity. Based on his recommendation, Bruce's parents decided to have their little boy surgically reassigned as a girl and to rear him as "Brenda." Bruce's testes were removed and he was given hormone treatments to promote female development.

But Brenda never behaved like a typical girl. She didn't like playing with girls' toys, didn't enjoy wearing dresses, and often got into fistfights at school. By the time Brenda reached puberty, her behavior had become so troublesome that her father broke down and told her what had happened to her. Brenda felt relieved rather than angry: "All of a sudden everything clicked. For the first time things made sense and I understood who and what I was."

Brenda eventually had reconstructive surgery to remove her breasts and create a penis, took hormones, and changed her name to David. David told his story in a book published in 2000, *As Nature Made Him: The Boy Who Was Raised as a Girl*, by John Colapinto.

> *"For the first time things made sense and I understood who and what I was."*
>
> **—David Reimer**

By that time, it had become increasingly clear that gender identity is significantly influenced by biology. Studies had shown that prenatal exposure to fetal sex hormones such as testosterone not only determine whether a fetus will develop female or male genitalia, but also influence a developing baby's brain, influencing behavior.

Sex hormones are produced by sex organs called **gonads**—ovaries in females, testes in males. Sex hormones include **androgens** (from the Greek *andros*, meaning "man"), such as testosterone, and **estrogens** (from the Greek *oistros*, meaning "frenzy" or "passion"), such as estradiol. Both males and females produce androgens and estrogens, but in most cases males produce higher levels

GONADS
Sex organs: ovaries in females, testes in males.

ANDROGENS
A class of sex hormones, including testosterone, that are present in higher levels in men and cause male-associated traits like deep voice, growth of facial hair, and defined musculature.

ESTROGENS
A class of sex hormones, including estradiol, that are present in higher levels in women than in men and that support female sexual development and function.

Brenda Reimer in elementary school, and later as David Reimer.

of androgens and females produce higher levels of estrogens. In a developing fetus, these hormones shape the development of both internal and external sexual anatomy.

Whether a fetus develops male or female gonads depends on the particular set of chromosomes it receives from its parents. Humans have 23 pairs of chromosomes; 22 pairs are **autosomes** and 1 pair are the **sex chromosomes,** X and Y. Females have two X chromosomes, while males have one X and one Y. The Y chromosome contains a gene called *SRY* that signals the testes to develop (*SRY* stands for "*s*ex-determining *r*egion on the *Y* chromosome"). The testes, in turn, produce masculinizing hormones such as testosterone that lead to the development of a male body. In the absence of a Y chromosome, a fetus will develop into a female. Sons inherit their Y chromosome from their father and their X chromosome from their mother. Daughters inherit one X chromosome each from their mother and father. Thus, fathers determine the sex of a baby by providing either an X or a Y sex chromosome in the sperm that fertilizes a mother's egg **(INFOGRAPHIC 12.1)**.

Although most individuals have internal and external genitalia that are either clearly male or clearly female, some exceptions do occur. Each year about 1 in every 1,500 babies born in the United States falls into an intermediate sex category termed "intersex." An intersex person is someone whose external genitalia do not match his or her internal sex organs or genetic sex—for example, a person with an XX chromosome pair who has

Nathan Devery/Science Source

X and Y chromosomes. The *SRY* gene on the Y chromosome signals the development of testes. In the absence of the *SRY* gene on the Y chromosome, the embryo will develop as a female.

AUTOSOMES
Paired chromosomes present in both males and females; all chromosomes except the X and Y chromosomes.

SEX CHROMOSOMES
Paired chromosomes that differ between males and females. Females have XX, and males have XY.

INFOGRAPHIC 12.1

X and Y Chromosomes Determine Human Sex

Human chromosomes can be isolated from cells, photographed under a microscope, and then arranged in their pairs for examination. As shown below, males and females differ by virtue of a pair of sex chromosomes. Females have two X chromosomes and males have a single X and a single Y chromosome. Every person must have at least one X chromosome, but it's the presence of a gene on the Y chromosome that initiates male development.

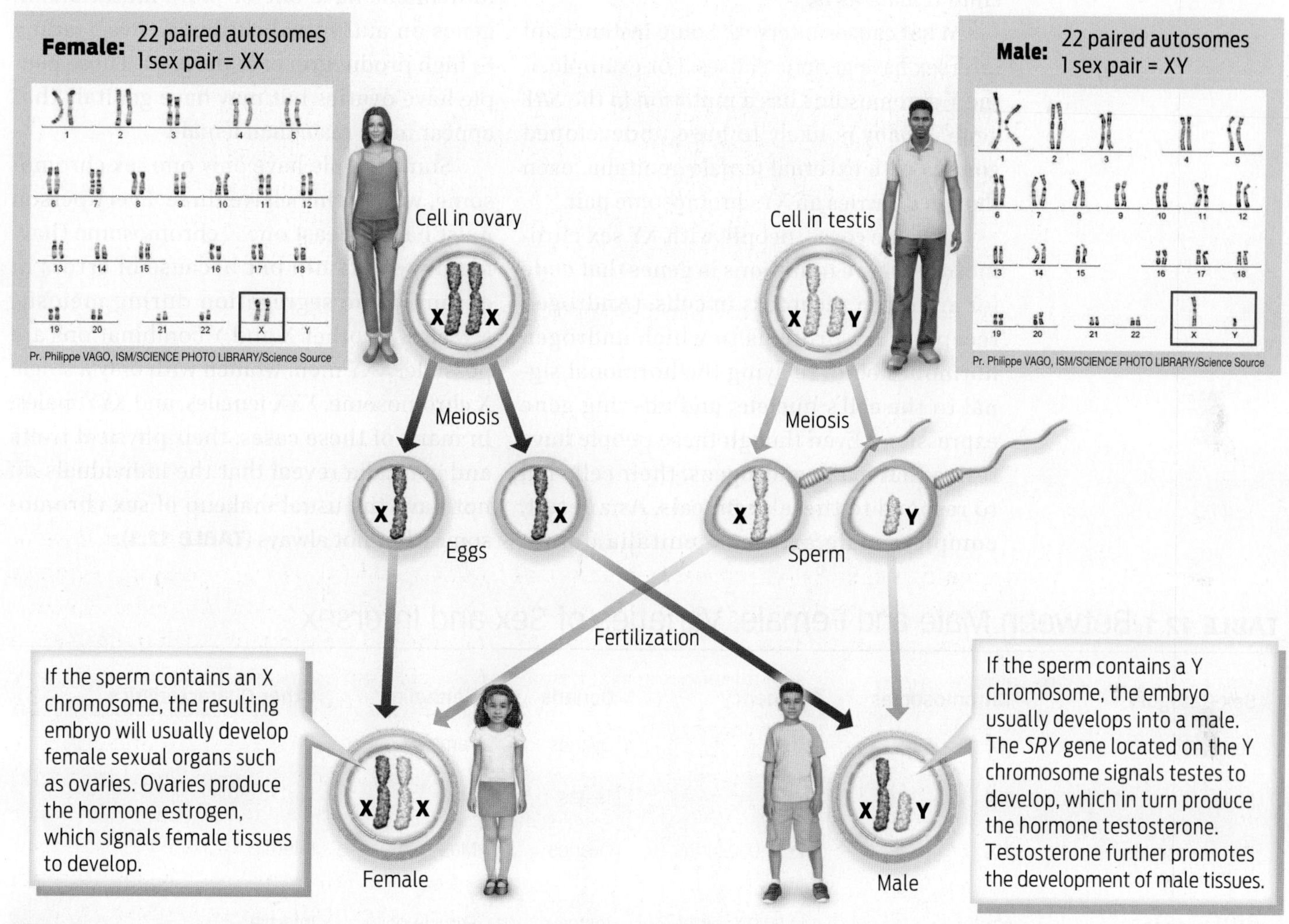

? What would you predict would be the biological sex of a person born with two X chromosomes and a Y chromosome?

internal ovaries but external genitalia that appear male. Also included in this category are babies born with ambiguous genitalia—for example, a penis that is very small or a clitoris that is exceptionally long.

Debate over the experience of David Reimer and other similar cases, as well as political activism from intersex advocates, has changed the care of intersex babies. In the past, doctors would typically operate on intersex babies immediately after birth to "fix" the genital anomaly. They might shorten a large clitoris in genetic females, for example. In genetic males with a very small penis, they might remove the penis and testes and assign the baby a female gender. Sometimes the surgery would be done even without consulting with parents.

Today, such babies are often assigned a sex by parents and doctors only after a period of observation to assess behavior patterns. Surgeons then perform surgery to create either male or female genitalia. Alternatively, parents may forgo surgery, preferring that their child remain as is.

What causes intersex? Some instances of intersex have genetic causes. For example, if the Y chromosome has a mutation in the *SRY* gene, a baby is likely to have undeveloped gonads with external female genitalia, even though it carries an XY chromosome pair.

In some cases, people with XY sex chromosomes have mutations in genes that code for androgen receptors in cells. (Androgen receptors are proteins to which androgen hormones bind, relaying the hormonal signal to the cell's nucleus and altering gene expression.) Even though these people have testes that make androgens, their cells fail to respond to these chemicals. As a result, complete male external genitalia do not develop, and these individuals appear to be female.

Similarly, occasionally people with XX sex chromosomes have male genitalia. In some cases, this anomaly is caused by a condition called congenital adrenal hyperplasia. These individuals have one or more mutations in genes on autosomal chromosomes, leading to high production of androgens. These people have ovaries but may have genitals that appear more male than female.

Some people have only one sex chromosome, while others have three. Every person must have at least one X chromosome (having none is fatal), but because of errors in chromosome segregation during meiosis, a variety of other X and Y combinations are possible: XXY men, women with only a single X chromosome, XXX females, and XYY males. In many of these cases, their physical traits and genitalia reveal that the individuals do not have the usual makeup of sex chromosomes, but not always **(TABLE 12.1)**.

TABLE 12.1 Between Male and Female: Varieties of Sex and Intersex

Sex Category	Chromosomes	Frequency	Gonads	Genitalia	Other Characteristics
Female	XX		Ovaries	Female	
Male	XY		Testes	Male	
Female pseudo-hermaphroditism	XX	1 In 13,000 births	Ovaries	Male	Infertile
Male pseudo-hermaphroditism	XY	1 in 80,000 births	Testes	Female or ambiguous	Infertile
True gonadal intersex	XX and/or XY	1 In 13,000 births	Ovaries and testes	Male, female, or ambiguous	Infertile; historically called true hermaphrodites
Triple X syndrome	XXX	1 in 3,000 births	Ovaries	Female	Fertile; taller than average; learning disabilities
Klinefelter syndrome	XXY	1 in 1,000 females	Testes	Male	Infertile; enlarged breast tissue
47, XYY syndrome	XYY	1 in 600 males	Testes	Male	Fertile; taller than average; elevated risk of learning and emotional disabilities
Turner syndrome	X	1 in 1,000 males	Ovaries	Female	Infertile; broad chest; webbed neck

As these varieties of intersex demonstrate, sex isn't always so clear cut. Sex encompasses numerous genetic, hormonal, anatomical, and psychological aspects, and these elements don't always align. Defining what counts as "masculine" and "feminine" is even more complicated. For example, some men have characteristics that we typically identify as female, such as a high voice and sparse body hair, yet they are genetically and anatomically male. Likewise, some women have what are considered to be more masculine features, such as angular faces and more muscle compared to body fat, yet they are genetically and anatomically female. In fact, very few physical or mental characteristics are entirely male or entirely female. In addition, some people—for example, transgender individuals—may mentally identify with one sex even though their genitalia and chromosomal makeup classify them as the other.

For biologists, the story of David Reimer—known in the medical literature as the "John/Joan case"—seriously undermined the idea that children are born psychosexually "neutral." Contrary to what some researchers had claimed, it was not possible to mold someone into whichever gender identity you wanted through surgery and child rearing—at least not without damaging psychological fallout.

David Reimer eventually married a woman and adopted her three children. Though he could not have children of his own since his testes had been removed, he took the responsibilities of fatherhood seriously. "From what I've been taught by my father," he told Colapinto, "what makes you a man is: You treat your wife well. You put a roof over your family's head. You're a good father. . . . That, to me, is a man."

David said he told his story so that others would be spared the nightmarish experience he went through. That experience may have led to the depression that cost him his life: David killed himself in 2004 at the age of 38.

SEX-LINKED INHERITANCE

Q Why do some genetic conditions affect sons more often than daughters?

A Some 10 million American men—about 7% of the male population—either cannot distinguish red from green, or else perceive these hues differently from the way other people do. Such red–green color blindness affects only 0.4% of women. Similarly, 1 in 5,000 boys worldwide is born with hemophilia, a blood-clotting disorder, yet hemophilia rarely afflicts girls.

Why do these disparities arise? These conditions are caused by genes found on the X chromosome. When a gene is located on either of the sex chromosomes, daughters and sons don't share the same probability of inheriting it.

As an example, consider the neuromuscular condition known as Duchenne muscular dystrophy (DMD). DMD is a disease in which muscles slowly degenerate, leading to paralysis. About 1 in 2,400 boys worldwide is born with the condition each year. Most affected boys need wheelchairs by the time they are teenagers, and they rarely live past 30.

Why does DMD primarily affect males? The gene for DMD is located on the X chromosome. Recall that a female has two X chromosomes. For a recessive trait like DMD, a normal allele on one X chromosome masks the effect of a recessive allele on the other X chromosome. A male, however, has a single X chromosome, so he will show the effects of any recessive alleles on that X chromosome. Statistically, it is much more likely for a male to inherit one rare DMD allele than it is for a female to inherit two such alleles.

Because females can carry an allele for DMD without having the disease, they may not even know they are carriers who can pass it on to their sons. Males with a DMD allele

INFOGRAPHIC 12.2

X-Linked Traits Are Inherited on X Chromosomes

Duchenne muscular dystrophy (DMD) is an example of an X-linked trait. Recessive mutations of the *dystrophin* gene on the X chromosome cause the disease. DMD primarily affects males because they inherit only one copy of the X chromosome (from their mothers). Therefore, the single DMD allele they inherit determines their phenotype. Since females have two X chromosomes, they may carry the DMD allele but have a healthy phenotype.

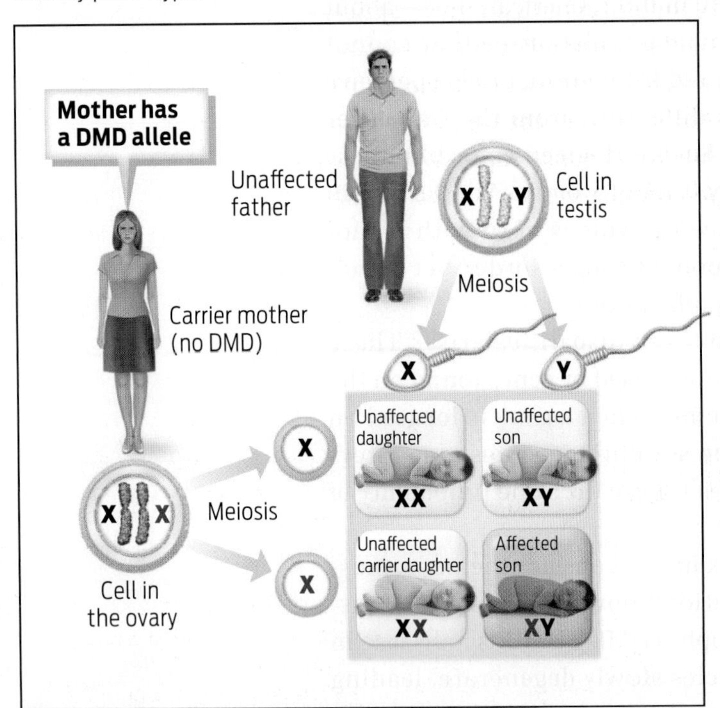

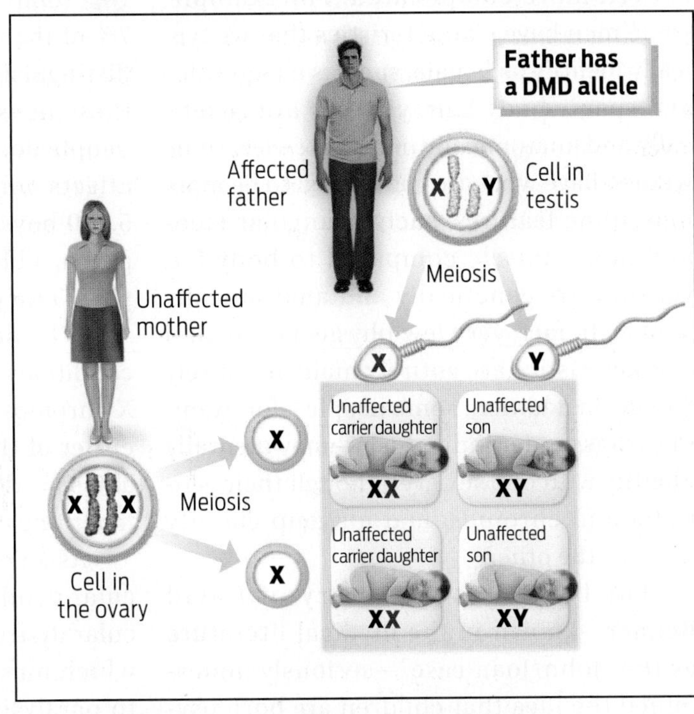

? Red–green color blindness is most commonly the result of inheriting recessive alleles of specific genes on the X chromosome. A woman and her husband both have the same type of red–green color blindness. Predict the frequencies of sons and daughters with red–green color blindness that this couple could have.

can pass it to their daughters, but not to their sons. Traits such as DMD, hemophilia, and red–green color blindness that are inherited on X chromosomes are called **X-linked traits** **(INFOGRAPHIC 12.2)**.

Sex-linked traits—those influenced by genes on either the X or Y chromosome—have distinct patterns of inheritance compared with traits influenced by genes on the autosomes. In autosomal inheritance, males and females inherit the trait with equal probability. With sex-linked inheritance, males and females are affected differently.

Determining the pattern of inheritance shown by a particular trait is easy to do with plants or laboratory animals, where you can perform a carefully controlled genetic cross—as Mendel did with his peas (see **Milestone 3: Mendel's Garden**). With humans, it's not possible to conduct such crosses. Instead, to determine a pattern of inheritance in humans, scientists rely on a tool called a **pedigree**, a diagram that shows the presence or absence of particular traits in family members across multiple generations. With the help of a pedigree that extends back several generations, scientists can determine whether a trait is inherited on autosomes or sex chromosomes.

One of the most famous pedigrees of all time is that of Queen Victoria of England (1819–1901) and her descendants, several of

X-LINKED TRAIT
A phenotype determined by an allele on an X chromosome.

PEDIGREE
A visual representation of the occurrence of phenotypes across generations.

whom were afflicted with the blood-clotting disorder hemophilia. In people with hemophilia, blood cannot clot in response to an injury. Before there were medicines for this condition, people sometimes bled to death as a result of a relatively minor cut.

Victoria and her husband, Prince Albert, had nine children—five girls and four boys. One of their sons, Leopold, had hemophilia; he died at age 30. Their daughter Alice had several children, including one son who had hemophilia. Although Alice's daughter Alexandra did not have hemophilia, Alexandra's son Alexei did. By the early 20th century, 10 of Victoria's descendants had hemophilia—all of them male. This pattern of a trait appearing more frequently in males, but inherited via mothers, is typical of rare, recessive X-linked traits **(INFOGRAPHIC 12.3)**.

Victoria's daughters Alice and Beatrice married into the royal families of Spain, Germany, and Russia. Because both were carriers, hemophilia passed into and became common in the royal houses across Europe. This is why hemophilia was once dubbed "the royal disease."

INFOGRAPHIC 12.3

A Pedigree Analysis Can Help Determine a Sex-Linked Pattern of Inheritance

This diagram, known as a pedigree, shows how an X-linked recessive trait (in this case hemophilia) passes through generations. Female carriers of the allele do not express the hemophilia trait, but their sons who inherit the allele have hemophilia.

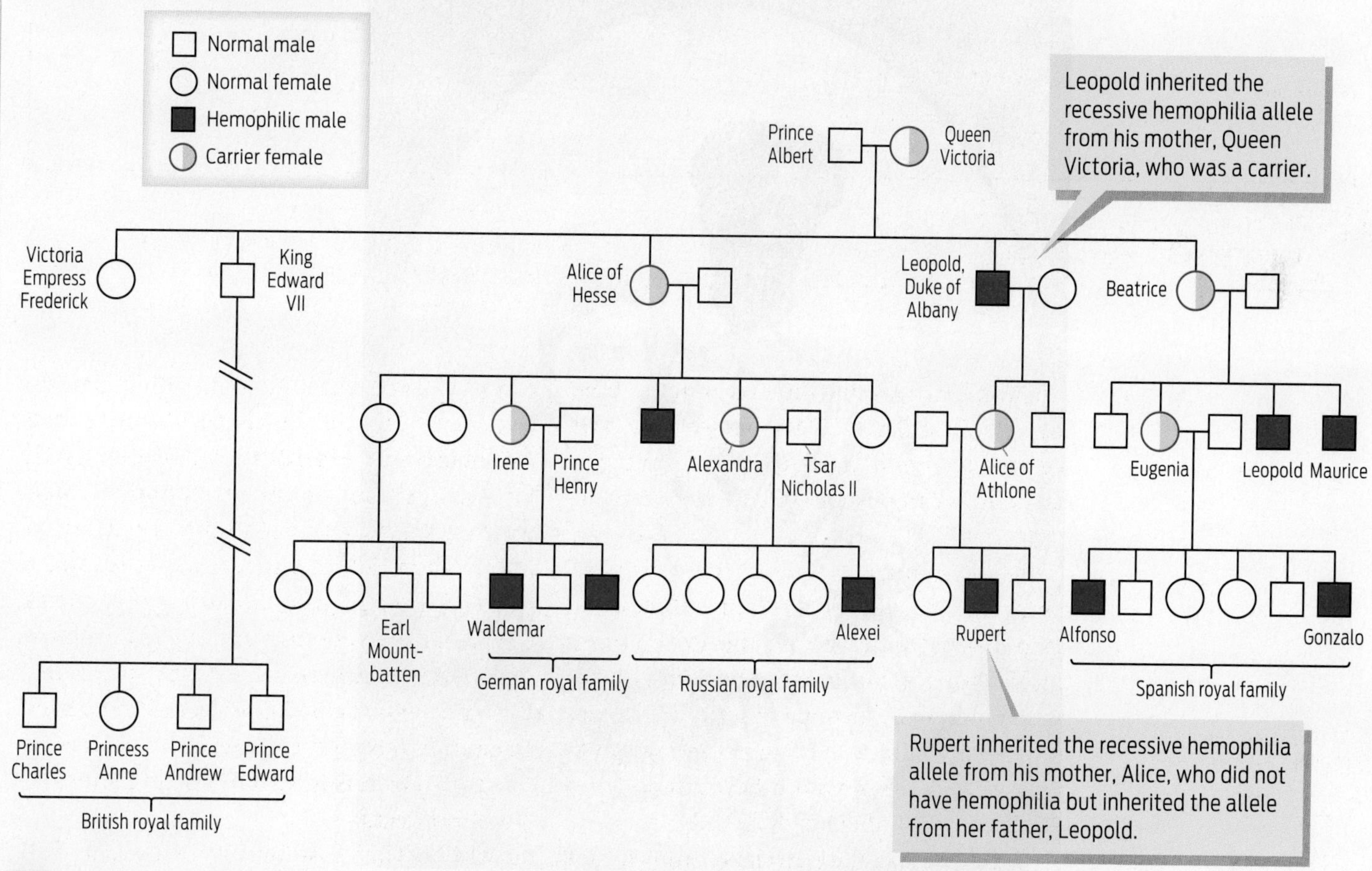

? Imagine that Irene married Maurice instead of Prince Henry. What would you predict about their sons and daughters?

Y-CHROMOSOME ANALYSIS

Q Did Thomas Jefferson father children with a woman who was enslaved?

A Thomas Jefferson was the third president of the United States, the principal architect of the Declaration of Independence, and the founder of the University of Virginia. He was also a slaveholder. Historians have long debated the meaning of this and other seeming contradictions in the Founding Father's life and politics. For example, although Jefferson's writings clearly show that he did not support the institution of slavery, he owned at least 200 slaves. He made disparaging comments about slaves, yet maintained close relationships with those living in his house—sometimes very close.

Y-CHROMOSOME ANALYSIS The comparison of sequences on the Y chromosomes of different individuals to examine paternity and paternal ancestry.

Jefferson was rumored to have fathered at least six children with Sally Hemings, an enslaved woman who tended to his family. For decades, historians dismissed the rumor as unreliable oral history. That situation changed in 1998, when scientists applied a genetic technique called **Y-chromosome analysis** to the case.

As the name implies, this technique examines the Y chromosome, which is very small and contains few genes. Sons inherit their Y chromosome from their fathers. These Y chromosomes are passed through generations, from fathers to sons, largely unchanged. That's because Y chromosomes have no homologous partner chromosome with which

Courtesy of Mason County Museum, Maysville, Kentucky

An enslaved woman named Lucy, born at Monticello in 1811. Slaves at Monticello often cared for the Jefferson children.

to pair and exchange DNA during meiosis (see Chapter 11). In other words, the Y chromosome rarely undergoes genetic recombination. Consequently, the Y chromosome that a son inherits from his father is almost identical to the Y chromosome that his father inherited from his father. Comparing DNA sequences on Y chromosomes, therefore, can reliably establish paternity **(INFOGRAPHIC 12.4)**.

During Jefferson's own time, people commented on the resemblance of Hemings's children to the president. But later historians either explained the resemblance away or proposed other explanations—for example, that one of Jefferson's nephews had fathered her children.

To set the record straight, in 1998 a team of geneticists led by Eugene A. Foster compared

INFOGRAPHIC 12.4

Y Chromosomes Pass Largely Unchanged from Fathers to Sons

Y-chromosome analysis for paternity testing relies upon the fact that the Y chromosome does not undergo recombination during meiosis and so passes unchanged from a father to his sons.

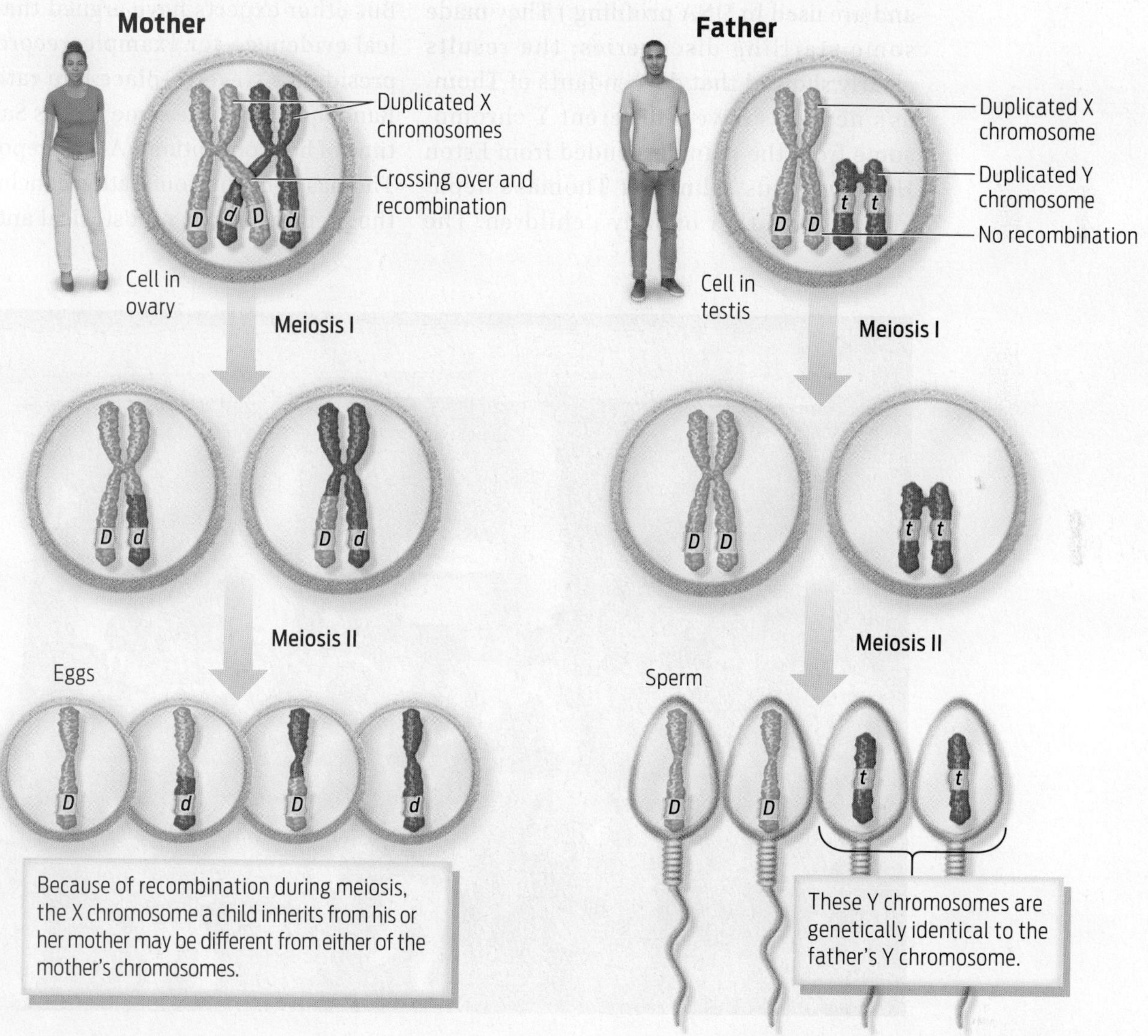

? A male has a particular dominant allele *F* on his Y chromosome. Which of his children (sons and/or daughters) will inherit this allele?

the Y chromosomes of three groups of men: descendants of Thomas Jefferson's paternal uncle Field Jefferson; one male descendant of Eston Hemings, Sally Hemings's son; and descendants of Jefferson's sister's sons (Jefferson's nephews). Since Jefferson's only surviving child from his wife was a daughter, he did not have any direct male descendants, which is why scientists tested descendants of Jefferson's uncle.

The team analyzed 11 short tandem repeats (STRs) on the Y chromosome. (Recall from Chapter 7 that STRs are short regions of DNA that are repeated a different number of times in different people and are used in DNA profiling.) They made some startling discoveries: the results clearly showed that descendants of Thomas's nephews have a different Y chromosome from the man descended from Eston Hemings, thus ruling out Thomas's nephews as the father of Sally's children. The results also clearly showed that Eston's descendant—John Weeks Jefferson—has the same Y chromosome as the descendants of Field Jefferson. Consequently, Thomas Jefferson *could* have fathered Eston Hemings. The study does not *prove* that he is the father, but it does show that the father was definitely a male Jefferson **(INFOGRAPHIC 12.5)**.

Some historians have argued that Thomas's younger brother Randolph Jefferson could have fathered Eston. (Or, indeed, that any of the other seven Jefferson males who periodically visited Monticello, where Thomas and Sally lived, could be the father.) But other experts have argued that historical evidence—for example, records of the president's travels—place him rather than Randolph under the same roof as Sally at the time of her conceptions. A 2000 report by the Thomas Jefferson Foundation concluded that the preponderance of historical and biolog-

LESLIE CLOSE/AP Images

Are all of these people descendants of Thomas Jefferson and Sally Hemings? The group gathered for a photo at Monticello in 1999.

ical evidence points to a "strong likelihood" of a sexual relationship between Thomas and Sally.

For the descendants of Eston Hemings, the DNA study was powerful vindication, even if debate continues. They had long argued that they were descended from Thomas Jefferson, but without hard evidence, most historians disregarded their claims. "I feel wonderful about it," Julia Jefferson Westerinen, a Staten Island artist and Eston's great-great-granddaughter, told *The New York Times* when the study results were published, "I feel honored."

INFOGRAPHIC 12.5

DNA Links Sally Hemings's Son to Jefferson

Scientists compared DNA sequences on the Y chromosome of Sally Hemings's and Thomas Jefferson's grandfather's descendants. The DNA sequences match at the 11 different STR locations analyzed.

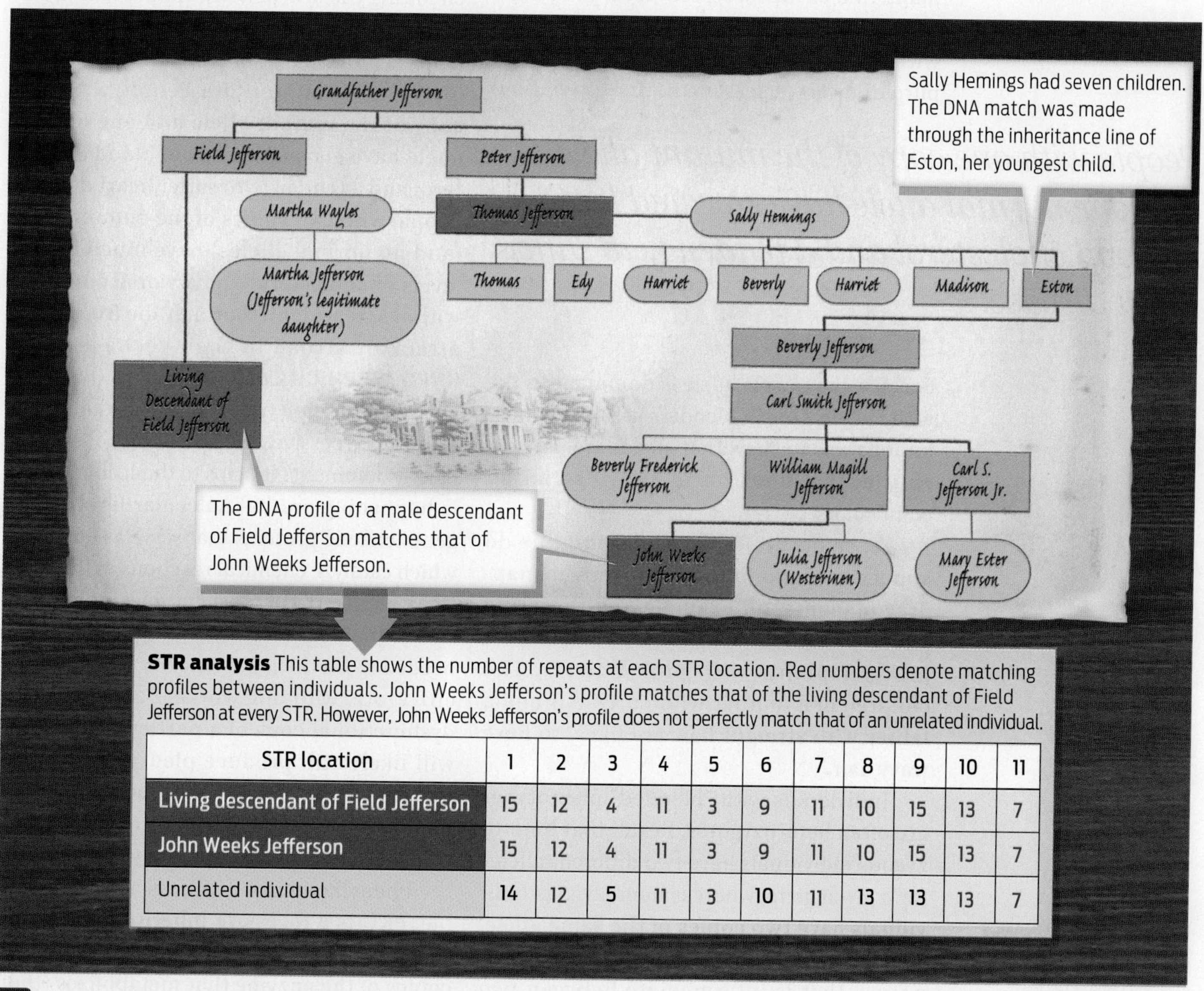

STR analysis This table shows the number of repeats at each STR location. Red numbers denote matching profiles between individuals. John Weeks Jefferson's profile matches that of the living descendant of Field Jefferson at every STR. However, John Weeks Jefferson's profile does not perfectly match that of an unrelated individual.

STR location	1	2	3	4	5	6	7	8	9	10	11
Living descendant of Field Jefferson	15	12	4	11	3	9	11	10	15	13	7
John Weeks Jefferson	15	12	4	11	3	9	11	10	15	13	7
Unrelated individual	14	12	5	11	3	10	11	13	13	13	7

? Of the proposed children of Thomas and Sally, who is predicted to have the same Y chromosome as Thomas? If Harriet had a son, would he share his Y chromosome with his grandfather Thomas?

INCOMPLETE DOMINANCE

Q Why do some traits appear to blend in offspring?

A We saw in Chapter 11 and Milestone 3: Mendel's Garden that many traits are inherited in a dominant or a recessive pattern. Emily Schaller has cystic fibrosis because she inherited two copies of the mutated *CFTR* allele, which is recessive. Her parents, however, are healthy because they each have one normal copy of the *CFTR* allele, which is dominant. And as Mendel discovered, pea plants with purple flowers crossed with pea plants with white flowers produce offspring with purple flowers (at least in the first generation) because the purple allele is dominant to the white allele. One of Mendel's greatest contributions was to show that alleles are discrete entities that don't blend away like mixed paint colors.

People with one copy of the mutant allele and one normal allele have elevated levels of blood cholesterol and a tendency to suffer heart disease.

Yet many traits we see around us do appear to blend in the offspring. In snapdragon plants, for example, red-flowering plants mated with white-flowering plants produce plants with pink flowers. The children of a mother with curly hair and a father with straight hair are likely to have wavy hair.

Individuals with "blended" phenotypes are often heterozygotes. Recall that heterozygous individuals have two different alleles for a given gene, whereas homozygous individuals have two copies of the same allele. When a heterozygous individual has a phenotype that is intermediate between two homozygous individuals, the inheritance pattern is called **incomplete dominance.** Incomplete dominance occurs when one allele does not completely mask the presence of the other allele, and there is a measurable effect on the phenotype of having one of each allele. (This is in contrast to complete dominance, in which one allele masks the other and the heterozygote has a phenotype that is indistinguishable from that of one of the homozygotes.)

INCOMPLETE DOMINANCE
A form of inheritance in which heterozygotes have a phenotype that is intermediate between the two homozygotes.

An example of incomplete dominance in humans is familial hypercholesterolemia (FH). This dangerous condition results from inherited mutations in the gene encoding the low-density lipoprotein (LDL) receptor, a protein on cell membranes that helps remove harmful cholesterol from the blood (see Chapter 28). People with one copy of the mutant allele and one normal allele have elevated levels of blood cholesterol and a tendency to suffer heart disease. People with two copies of the mutant allele (and no normal alleles) have much higher levels of cholesterol and a worse outcome; without treatment, they may die from heart attack or stroke in their teenage years **(INFOGRAPHIC 12.6)**.

Why do some traits behave as dominant, others as recessive, and still others as incompletely dominant? It helps to think about what is happening on the molecular level. Many genes encode proteins that serve as enzymes, which catalyze chemical reactions in the body (see Chapter 4). For most enzymes, only a very small quantity of the working molecule is needed to do the job. Someone who is a heterozygote, with one working copy and one dysfunctional copy of a particular enzyme, will likely still produce plenty of the normal enzyme to catalyze the reaction. So this mutant allele would behave as recessive, and you'd need two mutant versions to get sick.

Phenylketonuria (PKU) is a good example. PKU is a recessive inherited condition that results from inheriting two defective copies of the enzyme that metabolizes (that is, breaks down) the amino acid phenylalanine. People with two defective copies of

INFOGRAPHIC 12.6

The Inherited Trait of High Cholesterol Exhibits Incomplete Dominance

In incomplete dominance, heterozygotes display a phenotype intermediate between that of either homozygous genotype. In familial hypercholesterolemia (FH), blood levels of LDL cholesterol are determined by the genotype of the LDL receptor gene. The LDL receptor normally helps clear LDL cholesterol from the blood. People who are homozygous for the mutant allele have dangerously high levels of LDL cholesterol, heterozygotes have moderately elevated levels of LDL cholesterol, and people who are homozygous for the normal allele have normal cholesterol levels.

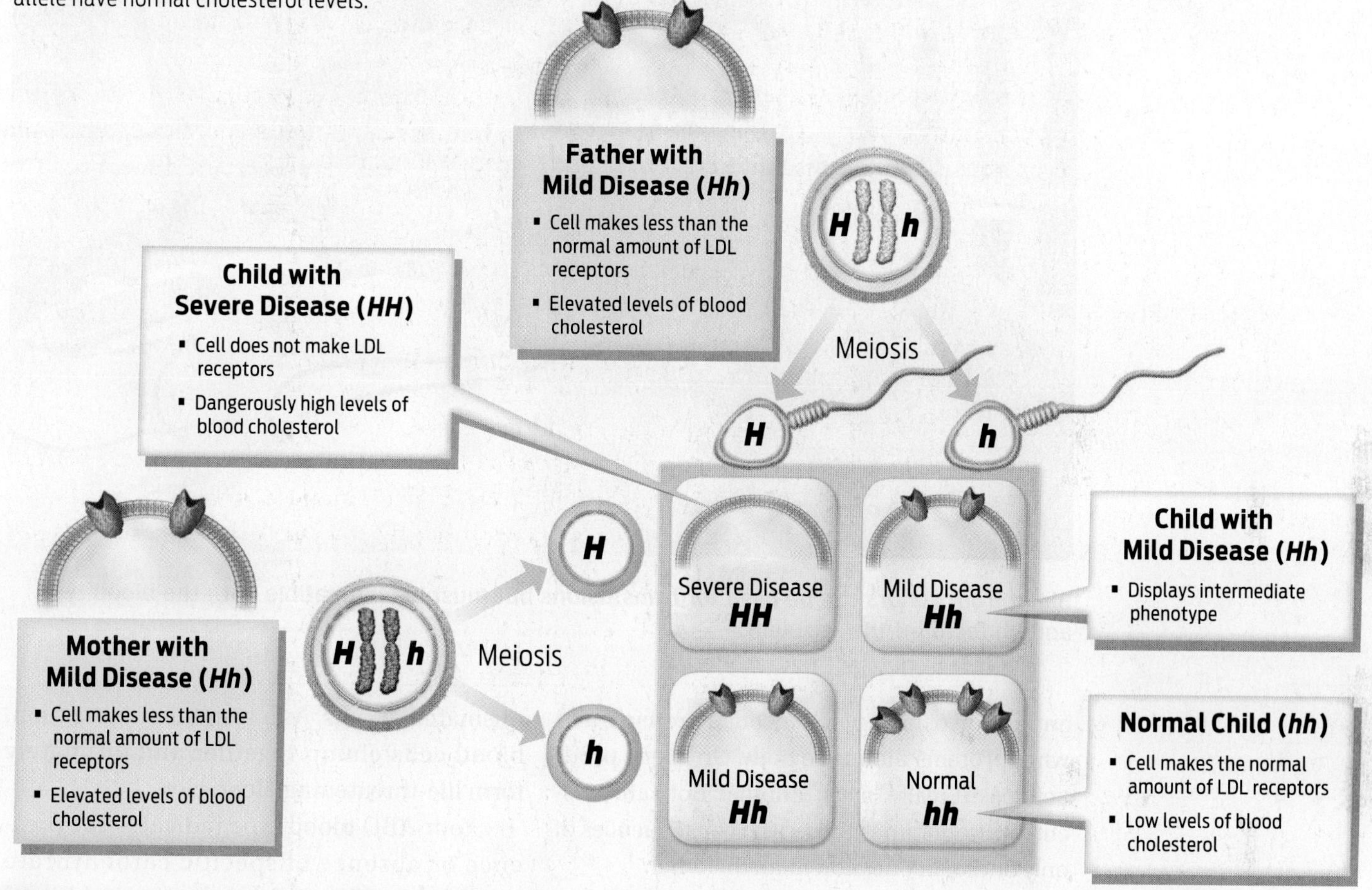

? A woman has the severe form of FH, and her husband does not have FH at all. What proportion of their children will have severe FH, mild FH, or no FH?

the gene are quite sick, but people with one working copy are just fine.

When it comes to processing cholesterol, however, being a heterozygote and having only half the number of normal LDL receptors is not sufficient to remain healthy. But it's also not as bad as having two mutant alleles and making no LDL receptors at all. In other words, the mutant allele in FH behaves as incompletely dominant: one copy makes you sick, but having two copies makes you much sicker.

For the neurodegenerative condition called Huntington disease, having one mutant version of the protein is enough to cause the illness. That's because the mutant version acts as a "poison" that interferes with the action of the normal protein. This condition is therefore inherited in a dominant fashion.

Though the concepts of dominant and recessive alleles are very useful in explaining inheritance patterns, it's helpful to recall that these terms were coined before scientists had

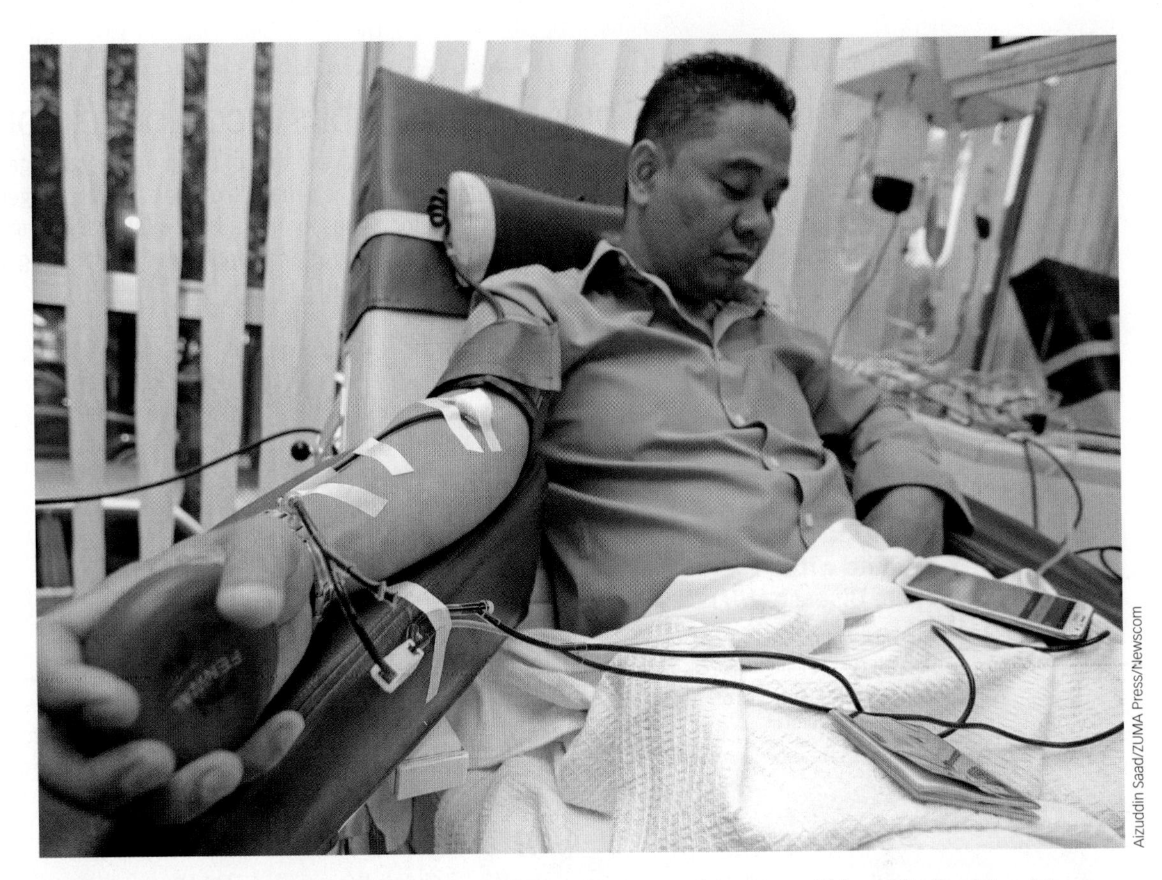

Aizuddin Saad/ZUMA Press/Newscom

Blood from donors can be used for transfusions but must be compatible with the blood type and Rh factor of the recipient.

any notion of how genes encoded proteins, and what proteins did in the body. One gene product doesn't really "mask" another. Both are present, but sometimes we see the consequences of only one, or of one more than the other.

CODOMINANCE

Q Who can be a universal blood donor?

A When someone needs a blood transfusion, the donated blood cannot come from just anyone. Instead, the transfused blood must be compatible in ways that are determined by genetics. The two most important genetic attributes are ABO blood type and Rhesus (Rh) factor, both of which must be compatible between donor and recipient. Mixing incompatible blood leads to an immune reaction against the genetically mismatched cells. As a result of this reaction, blood cells clump together and ultimately form life-threatening blood clots.

Your ABO blood type indicates the presence or absence of specific carbohydrate markers on the surface of your red blood cells. For example, if you have type A blood, your cells display A markers, whereas if you have type B blood, your cells have B markers. If you have type O blood, then you lack both A and B markers. Finally, if you have type AB blood, then you have both A and B markers.

Which blood type you have is determined by a single gene, *ABO,* found on chromosome 9. The *ABO* gene has three alleles: *A, B,* and *O.* The protein expressed by the *ABO* gene is an enzyme. The enzyme made from the *A* allele deposits A markers on the surface of red blood cells; the enzyme made from the *B* allele adds B markers; and the enzyme made from the *O* allele is nonfunctional and deposits neither. Since we inherit one allele from each parent, the possible combinations of the three alleles are *OO, AO, BO, AB, AA,* and *BB.*

Blood type AB is an example of **codominance,** meaning that the phenotype displays the contributions of both the maternal allele and the paternal allele in a heterozygous individual. Unlike incomplete dominance, in which heterozygotes have an intermediate phenotype, codominant alleles share the limelight: heterozygotes express both traits. Thus people with an *AB* genotype express both A and B markers on their cells and therefore have type AB blood.

While blood type alleles *A* and *B* are codominant, *O* is recessive to both *A* and *B.* Consequently, if you have blood type A, your genotype will be either *AA* homozygous or *AO* heterozygous. The same goes for blood type B: you will have a genotype of either *BB* or *BO* **(INFOGRAPHIC 12.7).**

Another important blood type marker is Rh factor, a protein found on the surface of red blood cells. The Rh factor is encoded by a gene found on chromosome 1. Your Rh status, (+) or (−), indicates the presence or absence of the Rh factor protein on the surface of your red blood cells. The inheritance pattern for Rh factor indicates that its alleles are either dominant or recessive: positive Rh factor alleles (*Rh*+) are dominant over negative Rh factor alleles (*Rh*−), which do not produce the protein. So if a person carries one positive allele and one negative allele, the positive allele will dominate and that person will have an Rh-positive phenotype.

Type O Rh-negative donors are known as universal donors because their blood can be transfused to people of any other blood type. Red blood cells from these donors lack the surface markers that the immune system recognizes as foreign and so will not trigger an immune response in a

CODOMINANCE
A form of inheritance in which the effects of both alleles are displayed in the phenotype of a heterozygote.

INFOGRAPHIC 12.7

ABO Blood Type Demonstrates Codominant Inheritance

In codominant inheritance, heterozygotes display the effects of both alleles in their phenotype. ABO blood type in humans is an example. Alleles for blood type code for the presence of different surface markers on red blood cells. A person with type AB blood, for example, displays both A and B markers, while type O blood displays neither A nor B markers. A person's ABO blood type must be considered when he or she gives or receives blood.

Blood Transfusions
The ability to donate or receive blood is based on immune rejection. If two people have the same surface markers, then their blood will be compatible. People with type O blood have no surface markers that provoke an immune response in a recipient (so O is the universal donor). People with type AB blood will not recognize either marker as foreign, so they can receive blood from any donor.

	Type A markers	Type B markers	Type A and B markers	Neither A nor B markers
Red blood cell type	A	B	AB	O
Genotype	*AA* or *AO*	*BB* or *BO*	*AB*	*OO*
Can donate to	Type A or AB recipient	Type B or AB recipient	Type AB recipient	Type A, B, AB, or O recipient
Can receive from	Type A or O donor	Type B or O donor	Type A, B, AB, or O donor	Type O donor

? Which blood type exemplifies codominance? What are the possible blood types of the parents of someone who has the codominant blood type?

recipient. Because any person can receive O Rh-negative blood, O-negative donors are always in demand. Blood banks can fall short of O-negative blood during such disasters as earthquakes or hurricanes in which many people are hurt and require blood. People with type AB Rh-positive blood can receive any type of blood and are known as universal recipients **(INFOGRAPHIC 12.8)**.

INFOGRAPHIC 12.8

A Mismatched Blood Transfusion Causes Immune Rejection

If donor and recipient are not compatible in ABO blood type and Rh factor, a recipient can have a life-threatening immune reaction to donated blood. A person with type B blood, for example, cannot donate blood to a person with type A blood.

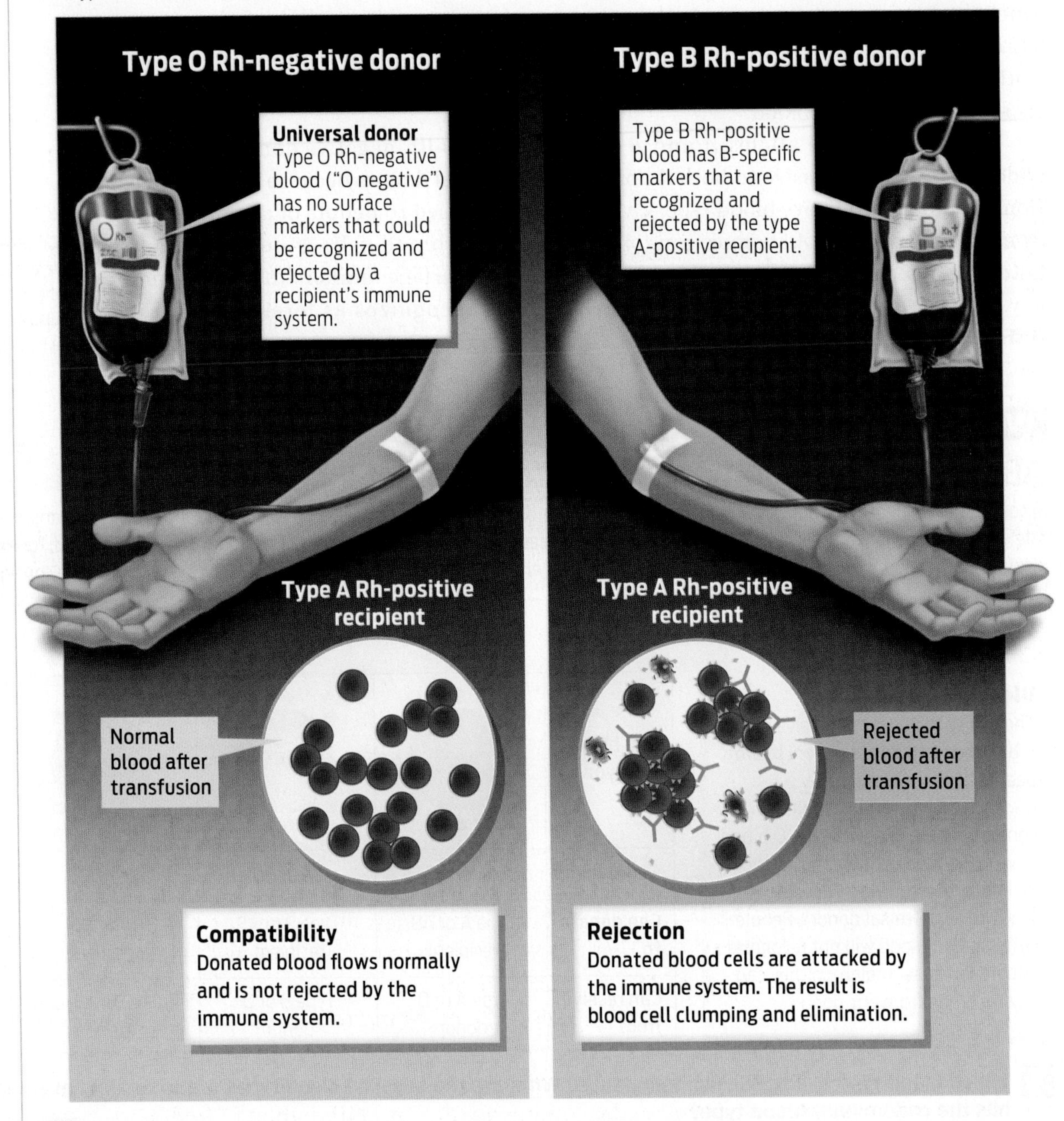

? Who can safely receive the blood of someone with AB-positive blood?

POLYGENIC INHERITANCE

Q How much of human height is genetic?

A The short answer is, a lot. Geneticists estimate that height is 60% to 80% genetic; in other words, genes account for 60% to 80% of the difference in height you see from person to person. But there isn't one single gene that determines height—instead, there are many. This is why we see such a range of heights among a population.

In the United States, most people fall somewhere between 5 feet and 6 feet 2 inches tall, and women tend to be shorter than men. If we plot height on a graph, the result resembles a bell curve: there are a range of heights, and the heights of most people fall near the middle of the curve under the bell. In other words, height is a trait that shows **continuous variation** in the population. In contrast, the other traits we've encountered so far have been discrete, also called discontinuous—individuals have one of only two or three possible phenotypes for a given trait. Examples are Mendel's round or wrinkled peas (see **Milestone 3: Mendel's Garden**) and ABO blood type.

Why does height show an unbroken range of phenotypes rather than discrete categories like tall or short? The main reason is that human height is a **polygenic trait**—one that is influenced by more than one gene. When multiple genes act together, their effects add up to produce a range of phenotypes.

Many common traits are polygenic and therefore exhibit continuous variation in the population. Another example is skin color: people exhibit a range of skin color shades, ranging from very light to very dark.

The fact that height is largely genetic means that, all other things being equal, two tall parents are very likely to have a child who is also tall; the same goes for two short parents. If you were to plot the height of students in your class against the average of their parents' height, most tall children would come from tall parents, and most short children would come from short parents. But genes aren't the whole story.

Even though 60% to 80% of the variation we see in height is due to genes, the other 20% to 40% is due to environmental factors such as nutrition. Why do these estimates of environmental influence vary so much? The answer is that the estimates depend on what environment you're talking about. In developed countries, most people have access to adequate nutrition, and height is more than 80% inherited, or heritable. In such settings, when scientists compare the height of a person to his or her relatives, they find that height varies less than 20% among close relatives. In contrast, in developing countries, many people are still malnourished, and

Courtesy Peter Morenus

Height is an example of a trait that shows continuous variation in any given population. In humans, it is determined by many alleles of many genes.

CONTINUOUS VARIATION
Variation in a population showing an unbroken range of phenotypes rather than discrete categories.

POLYGENIC TRAIT
A trait whose phenotype is determined by the interaction among alleles of more than one gene.

environment plays a larger role. Another way of looking at this difference between developed and developing countries is that more people in developed countries have reached their genetic potential compared to people in developing countries because most people in the developed world have access to adequate nutrition. In developing countries, access to nutrition varies much more, and this variation is reflected in larger variations in height between a given person and his or her relatives.

MULTIFACTORIAL INHERITANCE
An interaction between genes and the environment that contributes to a phenotype or trait.

When both genes and environment work together to influence a given trait, the trait is described as **multifactorial.** So height is both polygenic and multifactorial. Many complex human traits, such as our susceptibility to heart disease and depression and our intelligence, are multifactorial **(INFOGRAPHIC 12.9).**

INFOGRAPHIC 12.9

Human Height Is Both Polygenic and Multifactorial

Multiple genes as well as environmental factors such as diet, nutrition, and overall health act together to determine how tall we become.

Polygenic: multiple genes working together
Many genes act together to determine one's height. The combination of alleles a person inherits (*aabbcc*, *AabbCc*, etc.) predicts a distinct height phenotype. Within a population, the result of these polygenic interactions is a range of height phenotypes, from very short to very tall.

A Polygenic Punnett Square
A mating between two people of medium height (where three genes control height: *AaBbCc* × *AaBbCc*) produces seven distinct phenotypes, determined by the number of dominant alleles inherited.

	abc	*abC*	*aBc*	*Abc*	*aBC*	*AbC*	*ABc*	*ABC*
abc	*aabbcc*	*aabbCc*	*aaBbcc*	*Aabbcc*	*aaBbCc*	*AabbCc*	*AaBbcc*	*AaBbCc*
abC	*aabbCc*	*aabbCC*	*aaBbCc*	*AabbCc*	*aaBbCC*	*AabbCC*	*AaBbCc*	*AaBbCC*
aBc	*aaBbcc*	*aaBbCc*	*aaBBcc*	*AaBbcc*	*aaBBCc*	*AaBbCc*	*AaBBcc*	*AaBBCc*
Abc	*Aabbcc*	*AabbCc*	*AaBbcc*	*AAbbcc*	*AaBbCc*	*AAbbCc*	*AABbcc*	*AABbCc*
aBC	*aaBbCc*	*aaBbCC*	*aaBBCc*	*AaBbCc*	*aaBBCC*	*AaBbCC*	*AaBBCc*	*AaBBCC*
AbC	*AabbCc*	*AabbCC*	*AaBbCc*	*AAbbCc*	*AaBbCC*	*AAbbCC*	*AABbCc*	*AABbCC*
ABc	*AaBbcc*	*AaBbCc*	*AaBBcc*	*AABbcc*	*AaBBCc*	*AABbCc*	*AABBcc*	*AABBCc*
ABC	*AaBbCc*	*AaBbCC*	*AaBBCc*	*AABbCc*	*AaBBCC*	*AABbCC*	*AABBCc*	*AABBCC*

Multifactorial: genes and environment working together
Human populations show a continuous range of heights (red line), rather than a genetically predicted number of phenotypes (blue bars). Environmental influences blur the genetic boundaries, resulting in a seamless continuity across the phenotype range.

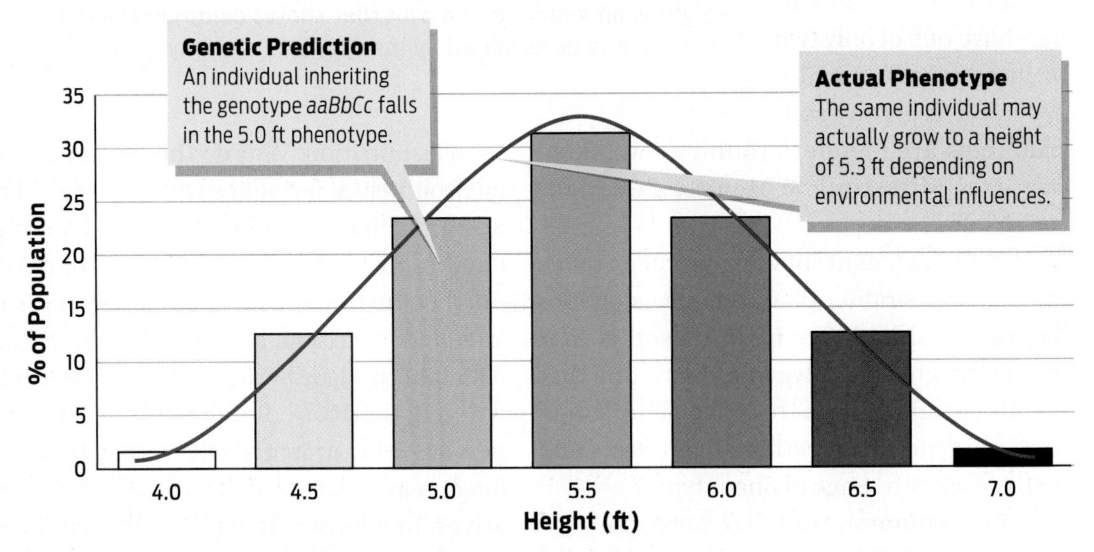

? A person is 5 feet 6 inches (5.5 feet) tall. Consider at least two ways that this person could reach this height (factor in the number of inherited dominant alleles and nutritional status).

EPIGENETICS

Q Do our genes explain everything about us?

A If you've ever known a pair of identical twins, you might have been struck by what they *didn't* have in common. Maybe one craved adventure, while the other preferred the safety of the couch. Maybe one was straight, the other gay. These anecdotal examples reveal a fundamental truth about genetics: even two people with the exact same genes can display different phenotypes.

How can this be, given what we've learned about DNA as the "blueprint of life"? For starters, it's important to understand that genes do not act in isolation from the environment. What we mean by "the environment" in this case is multidimensional: it includes the cells that contain the genes, the body that carries those cells, and the family and geographic region that a person grows up in, as well as their diet and other lifestyle factors. Environmental input can directly affect which genes are turned on or off in a cell. When we cut ourselves, for example, genes involved in healing damaged skin are turned on. When we practice a musical instrument, changes in gene expression lead to the formation of stronger connections between particular neurons in our brain. Carcinogens like cigarette smoke can damage DNA, leading to changes in gene expression.

Some traits are more likely to be influenced by environmental input than others. A trait like eye color does not vary much with a change in environment. At the other extreme, the particular language we speak is entirely dependent upon the country and family we grow up in and is not subject to genetic influence.

To estimate the contribution that genes make to a trait, scientists often compare traits in identical twins to those in fraternal twins. Identical twins are formed when one fertilized egg splits into two embryos. These individuals share essentially 100% of their DNA. Fraternal twins come from two separately fertilized eggs. They share a uterine environment but are otherwise as different from each other, genetically, as any other siblings are, which means they share about 50% of their DNA. When identical twins "match" for a particular trait—that is, are concordant—more often than fraternal twins, it is a sign that the trait has a genetic component. The larger the difference in concordance between identical and fraternal twins, the larger the genetic influence on the trait.

For example, consider the mental illness called schizophrenia, which causes auditory hallucinations and delusions. When one identical twin has been diagnosed with schizophrenia, the odds are about 50% that the other twin will be diagnosed as well. But for fraternal twins, the odds are only 17% that they will be concordant. This is strong evidence that schizophrenia has a genetic contribution.

When one identical twin has been diagnosed with schizophrenia, the odds are about 50% that the other twin will be diagnosed as well.

Or consider sexual orientation. Results vary, but one study found that identical twins were concordant for a gay sexual orientation about 65% of the time, whereas fraternal twins were concordant about 30% of the time. These results suggest a genetic influence, but the fact that identical twins are not 100% concordant for this trait suggests that environmental variables also play a role **(TABLE 12.2)**.

Researchers don't fully understand how the environment interacts with our genes to produce the wide variety of phenotypes we see around us, but they are actively exploring the issue. One mechanism that may help explain the mystery is **epigenetics.** The prefix *epi-* means "above or beyond," so the term *epigenetics* literally means "above the genes."

EPIGENETICS
Changes in gene expression that are not based on changes in the DNA sequence.

TABLE 12.2 Twin Concordance Rates Suggest the Extent of Genetic Influence on Traits

When identical twins raised in the same environment have a higher concordance rate than do fraternal twins raised in the same environment, the trait has a genetic component. The larger the difference in concordance between identical and fraternal twins, the larger the genetic influence on the trait.

	Percent Concordance	
Trait	Identical (Monozygotic) Twins	Fraternal (Dizygotic) Twins
Eye color	98%	49%
Baldness	92%	69%
Height at age 2 years	89%	58%
Autism spectrum disorder	88%	31%
Obesity	70%–80%	20%–40%
Homosexuality	66%	30%
Asthma	8%–66%	5%–45%
Stuttering	63%	19%
Schizophrenia	50%	17%
Type 1 diabetes	20%–50%	10%–15%
Cancer	38%	26%
Multiple sclerosis	28%	3%

Epigenetics refers to durable changes in gene expression that are not the result of changes in the DNA sequence itself.

In a cell, DNA doesn't exist as a naked molecule, but rather is wrapped up with proteins called histones. DNA wraps around histones like thread on a spool. DNA and histones can also be decorated with other molecules, such as methyl groups (CH_3), which alter how accessible genes are to being "read," or expressed. Changes to the environment can sometimes alter the placement of these epigenetic marks on the DNA and histones, allowing certain genes to be shut off and others to be turned on.

In 2004, researchers at McGill University in Quebec conducted a clever experiment demonstrating how the environment can alter gene expression through the placement of epigenetic marks. The experiment involved rat mothers and their pups. It tested whether maternal nurturing behavior toward pups affects the placement of epigenetic marks on a gene in a rat pup's brain that is important in handling stressful experiences.

Some rat mothers are highly nurturing of their pups, whereas others are less nurturing. Highly nurturing moms lick their pups a lot during their first week of life; less nurturing moms do not treat their pups to frequent licking. These variations in maternal nurturing are associated with the development of individual differences in how rat pups respond to stress. Pups with highly nurturing moms end up being calm rats as adults that are also highly nurturing to their own pups. Those with low-nurturing moms end up being anxious adults that are less nurturing towards their pups. This difference between calm and anxious rats may look like a simple inherited trait, passed from mothers to their offspring. However, if pups born to a highly nurturing mom are raised by a low-nurturing mom, those pups grow up to be anxious and low-nurturing themselves. Conversely, pups born to a low-nurturing mom that are switched at birth to be raised by a high-nurturing mom grow up to be calm and highly nurturing themselves.

What explains these long-lasting phenotypes that are not the result of inheriting different genes? The McGill investigators found that the explanation has to do with how environmental signals alter the epigenetic marks placed on the animals' DNA.

Maternal licking stimulates nerves in the skin that send signals to a rat pup's hypothalamus, a part of the brain. These signals alter the pattern of methylation on the gene for the glucocorticoid receptor (GR), which buffers an animal's response to cortisol, the stress hormone responsible for the fight-or-flight response. The release of cortisol (by the adrenal glands) puts the rats "on edge"—they are primed and ready to respond to danger. After a while, GRs in the brain's hippocampus sense this rise in cortisol and send a signal to shut off cortisol production. This tamps down the stress response, which calms the rats.

The investigators found that more licking leads to a reduced number of methyl groups attached to the *GR* gene and higher levels of the gene's expression. Rats with more GRs are better able to detect and respond to rising cortisol levels and return quickly to a calm state following the stressful situation. This produces less-anxious pups that become more nurturing moms. Pups that are licked less have more methyl groups attached to the *GR* gene and less *GR* gene expression. They don't respond as quickly to rising cortisol levels and remain stressed for longer. These rats become anxious, less-nurturing moms. In this way, a behavioral trait can be stably transmitted from parents to offspring, even though it's not being passed through the genes **(INFOGRAPHIC 12.10)**.

INFOGRAPHIC 12.10

The Environment Can Alter Gene Expression through Epigenetic Modification

Rats raised by low-nurturing mothers have more methyl groups on the glucocorticoid receptor (*GR*) gene. This methylation reduces the expression of this gene, resulting in lower levels of glucocorticoid receptors. These rats are less able to reduce the levels of cortisol (a stress hormone) produced in stressful situations. They become highly anxious adults that do not lick their pups. Rats that are highly nurtured as pups have fewer methyl groups on the *GR* gene, and produce sufficient glucocorticoid receptors to recover quickly from stress. They grow up to be calm adults that nurture their pups.

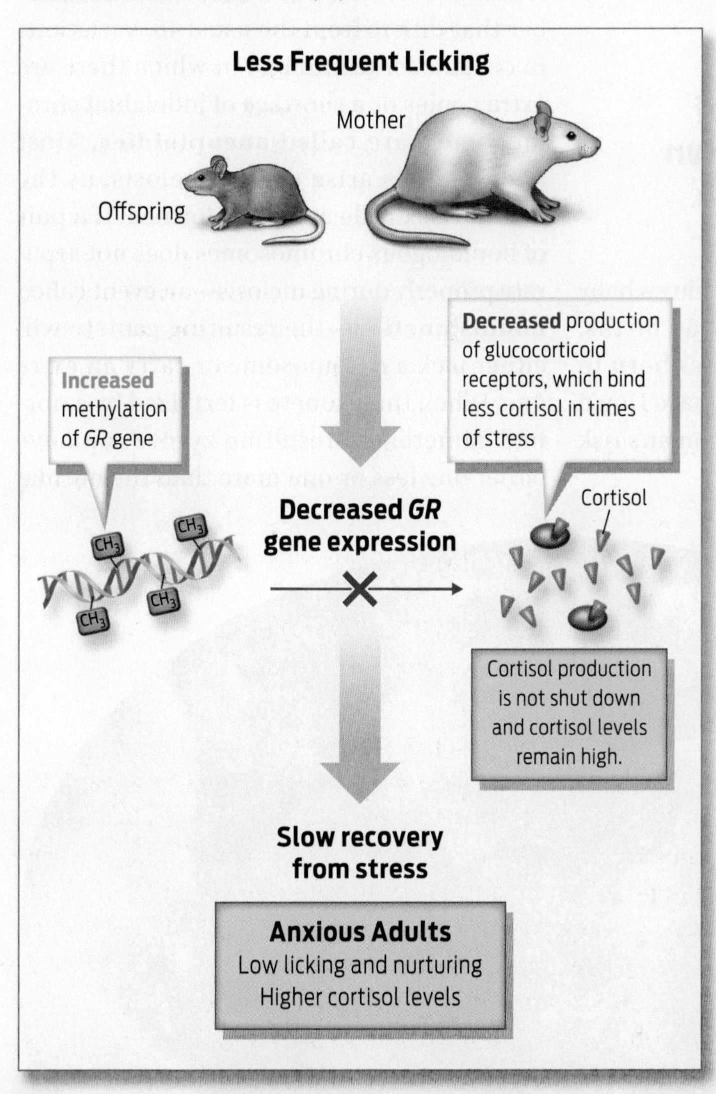

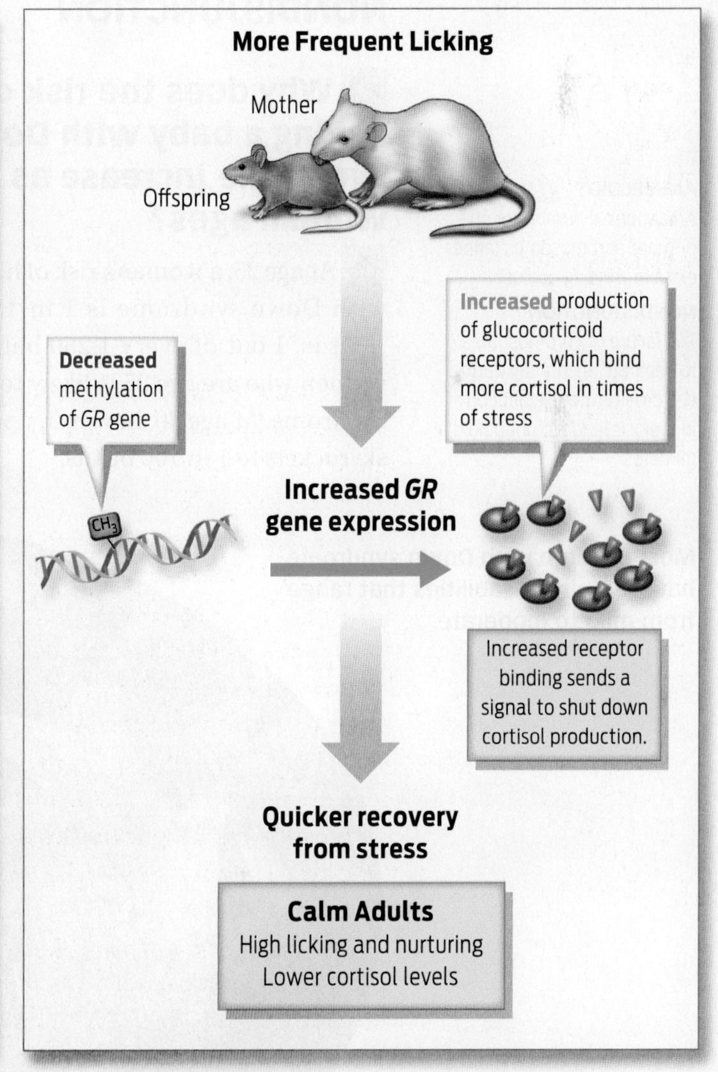

? **How can the nurturing environment of genetically identical rat pups lead to different nurturing behaviors as adults?**

This research has obvious potential implications for humans, who also require nurturance from parents to become healthy and happy adults. Increasing evidence suggests that methylation of the *GR* gene plays a role in how human children respond to mistreatment or abuse, similar to rat pups. Fortunately, unlike genetic changes, epigenetic changes may be reversible.

It can be reassuring to know that our fate isn't set in stone by our genes; experiences matter, too. Some researchers have suggested that epigenetic marks might be passed on to offspring through sperm and egg, just as genes are. This idea is very controversial and not supported by strong evidence at this time.

NONDISJUNCTION

Q Why does the risk of having a baby with Down syndrome increase as a woman ages?

ANEUPLOIDY
An abnormal number of one or more chromosomes (either extra or missing copies).

NONDISJUNCTION
The failure of chromosomes to separate accurately during cell division; nondisjunction in meiosis leads to aneuploid gametes.

A At age 25, a woman's risk of having a baby with Down syndrome is 1 in 1,250 births. That is, 1 out of every 1,250 babies born to women who are age 25 is likely to have Down syndrome. At age 40, however, a woman's risk skyrockets to 1 in 100 births.

As women age, the risk of giving birth to a baby with any chromosomal abnormality increases. That's because as a woman ages, so do her eggs. All the eggs that a woman will ever have were formed before she was born, during fetal development, and they have been aging like the rest of the cells in her body. Until puberty, a woman's developing eggs are paused in the middle of meiosis (at meiosis I); they haven't yet completed their cell division. During a menstrual cycle, one of these cells resumes meiosis and is ovulated. In older women, they are more likely to experience an error in chromosome segregation during this process, leading to an abnormal number of chromosomes in the egg.

A fetus with an abnormal number of chromosomes carries a chromosome number that differs from the usual 46. Variations in chromosome number, in which there are extra copies or a shortage of individual chromosomes, are called **aneuploidies.** Most aneuploidies arise during meiosis, as the parents' sex cells are being formed. If a pair of homologous chromosomes does not separate properly during meiosis—an event called **nondisjunction**—the resulting gamete will either lack a chromosome or carry an extra one. When that gamete is fertilized by a normal gamete, the resulting zygote can have either one less or one more than the normal

Most children with Down syndrome have learning disabilities that range from mild to moderate.

Markus Moellenberg/Getty Images

number of the affected chromosome. In most cases, the abnormality is so severe the zygote spontaneously aborts **(INFOGRAPHIC 12.11)**.

Sometimes the abnormality is not life threatening but does cause severe disability. The most common of these aneuploidies is **trisomy 21,** another name for Down syndrome. Trisomy 21 results when an embryo inherits an extra copy of chromosome 21. Anyone can conceive a child with the abnormality, but older women are at much higher risk.

Most children with Down syndrome have learning disabilities that range from mild to moderate, but some have profound mental disability. They are also at higher risk for other diseases and typically don't live beyond 50 years of age.

A variety of screening tests are used to detect chromosomal abnormalities early in pregnancy. Some measure the levels of specific proteins in the mother's blood, some analyze fetal DNA found in the mother's circulation, and some

TRISOMY 21
Having an extra copy of chromosome 21; also known as Down syndrome.

INFOGRAPHIC 12.11

Chromosomal Abnormalities: Aneuploidy

Birth defects can arise when chromosomes fail to separate normally during meiosis, a phenomenon called nondisjunction. The resulting gametes carry an abnormal number of chromosomes, a condition called aneuploidy.

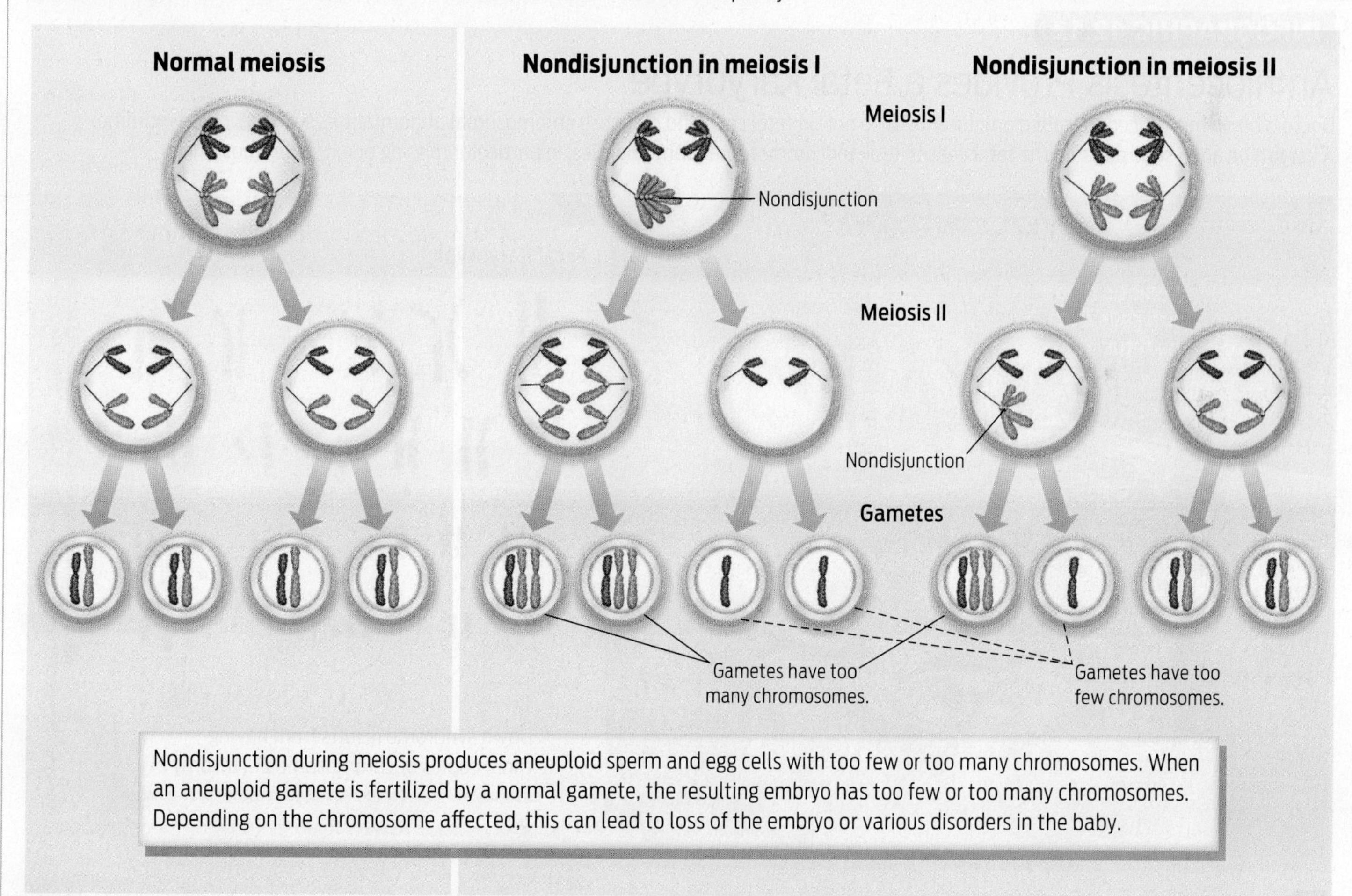

Nondisjunction during meiosis produces aneuploid sperm and egg cells with too few or too many chromosomes. When an aneuploid gamete is fertilized by a normal gamete, the resulting embryo has too few or too many chromosomes. Depending on the chromosome affected, this can lead to loss of the embryo or various disorders in the baby.

? An egg experienced nondisjunction during meiosis I. It has 22 autosomes and no sex chromosome. If this egg is fertilized by a sperm containing 22 autosomes and a Y chromosome, how many autosomes, sex chromosomes, and total chromosomes will the zygote have? Which of the egg, sperm, and zygote can be described as aneuploid?

AMNIOCENTESIS
A procedure that removes fluid surrounding the fetus so as to obtain and analyze the chromosomal makeup of fetal cells.

KARYOTYPE
The chromosomal makeup of cells. Karyotype analysis can be used to detect chromosomal disorders prenatally.

examine the chromosomes in fetal cells. The last type of analysis relies on **amniocentesis,** a procedure in which the clinician inserts a long, thin, hollow needle through the pregnant woman's abdominal wall and into her uterus. Through the needle, the health care practitioner removes the equivalent of 2 to 4 teaspoons of amniotic fluid, which surrounds the growing fetus. The fetal cells found in this fluid contain the fetus's DNA, and can therefore be used to determine a fetal **karyotype,** the chromosomal makeup in its cells **(INFOGRAPHIC 12.12).**

The reasons for undergoing amniocentesis vary from couple to couple. But if a test comes back positive for an abnormality, couples have options: they can begin to plan for a disabled child, or they can make the decision not to carry the child to term.

Although scientists have linked some of the most obvious birth defects to the age of a woman's eggs, recent research also shows that a man's age affects the quality of his sperm. Men who father children after age 45 are more likely to have children with cognitive disorders such as autism, for example. Male fertility declines over time, too, although much more gradually than does female fertility. Research shows that the older the man, the more likely he is to produce sperm with genetic defects. ■

INFOGRAPHIC 12.12

Amniocentesis Provides a Fetal Karyotype

Doctors perform a procedure called amniocentesis to obtain fetal cells and diagnose chromosomal abnormalities such as Down syndrome. A karyotype analysis is done on the fetal cells to look for chromosomal abnormalities, in particular missing or extra chromosomes.

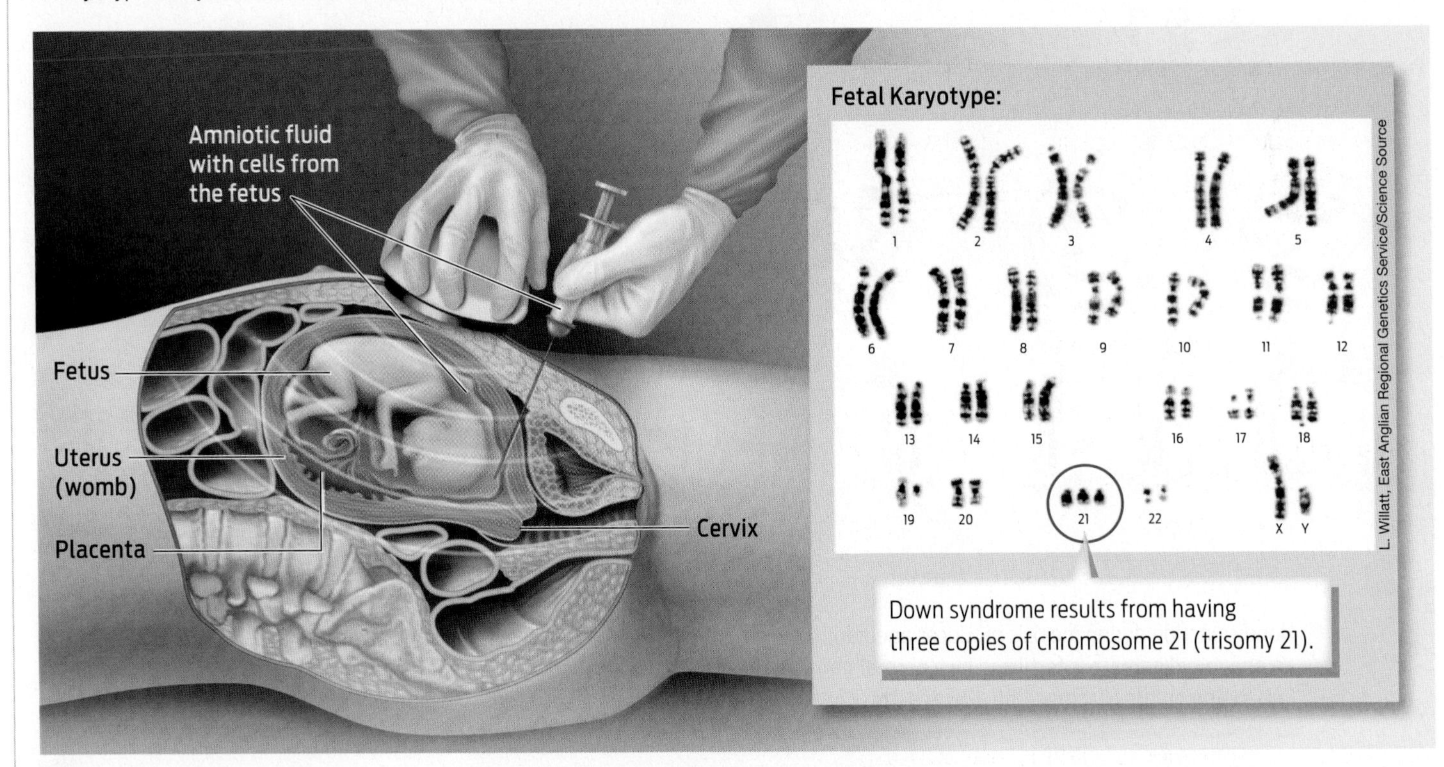

? **What would you observe in the karyotype of a fetus with Turner syndrome? (*Hint:* Refer to Table 12.1.)**

CHAPTER 12 SUMMARY

Driving Question 1 How do chromosomes determine sex, and how does sex influence the inheritance of certain traits?

- Humans have 23 pairs of chromosomes. One of these pairs is the sex chromosomes: XX in females and XY in males. Because the presence of the Y chromosome determines maleness, it is fathers who determine the sex of a baby.
- Sex as a body phenotype is determined by genetics working in combination with hormones; variations in either genes or hormones can lead to intersex or cases of ambiguous genitalia.
- Disorders and other traits inherited on X chromosomes are called X-linked traits, and are more common in males than in females.
- Because the Y chromosome in a male does not have a homologous partner, it does not recombine during meiosis. The Y chromosome a son inherits from his father is essentially identical to the Y chromosome his father inherited from his father (the grandfather), a fact that can be used to establish paternity.

Driving Question 2 What inheritance patterns are observed when alleles are not simply dominant or recessive?

- Familial hypercholesterolemia is a trait that exhibits incomplete dominance, a form of inheritance in which heterozygotes have a phenotype intermediate between the two homozygotes.
- ABO blood type is an example of a codominant trait—both maternal and paternal alleles contribute equally and separately to the phenotype.

Driving Question 3 What inheritance patterns are observed when more than one gene and/or the environment influences a trait?

- Many traits are polygenic—that is, they are influenced by the additive effects of multiple genes. Polygenic traits often show a continuous, bell-shaped distribution in the population. Human height is an example.
- In many cases, a person's phenotype is determined by both genes and environmental influences; this type of inheritance is described as multifactorial. Depression and cardiovascular disease are examples of multifactorial illnesses.
- The expression of genes can also be influenced by the environment, through epigenetic modifications.
- In rats, and potentially humans, maternal care early in life can establish epigenetic marks on a key gene involved in responding to stress during adulthood.

Driving Question 4 How does nondisjunction lead to numerical abnormalities of chromosomes, and what are the consequences of these abnormalities?

- Some genetic disorders result from having a chromosome number that differs from the usual 46. Down syndrome, or trisomy 21, is caused by an extra copy of chromosome 21.
- Aneuploidy, having one or more extra or missing chromosomes, is the result of nondisjunction—when chromosomes fail to separate properly during meiosis, generating aneuploid gametes.

More to Explore

- Zimmer, C. (2018). *She Has Her Mother's Laugh: The Powers, Perversions, and Potential of Heredity.* New York: Dutton.
- Colapinto, J. (2000). *As Nature Made Him: The Boy Who Was Raised as a Girl.* New York: HarperCollins.
- Foster, E. A., et al. (1998). Jefferson fathered slave's last child. *Nature* 396:27–28.
- Lick Your Rats: https://learn.genetics.utah.edu/content/epigenetics/rats/
- Allen, E. G., et al. (2009). Maternal age and risk for trisomy 21 assessed by the origin of chromosome nondisjunction: A report from the Atlanta and National Down Syndrome Projects. *Hum Genet* 125(1):41–52.
- Cowan, R. S. (2008). *Heredity and Hope: The Case for Genetic Screening.* Cambridge, MA: Harvard University Press.

CHAPTER 12 Test Your Knowledge

Driving Question 1 How do chromosomes determine sex, and how does sex influence the inheritance of certain traits?

By answering the questions below and studying Infographics 12.1–12.5 and Table 12.1, you should be able to generate an answer for this broader Driving Question.

Know It

1. Which of the following most influences the development of a female fetus?

a. the presence of any two sex chromosomes
b. the presence of two X chromosomes
c. the absence of a Y chromosome
d. the presence of a Y chromosome
e. either b or c

2. Why are more males than females affected by X-linked recessive genetic diseases?

3. If a man has an X-linked recessive disease, can his sons inherit that disease from him? Why or why not?

Use It

4. Which of the following couples could have a son with Duchenne muscular dystrophy (DMD)?

a. a male with Duchenne muscular dystrophy and a homozygous dominant female
b. a male without Duchenne muscular dystrophy and a homozygous dominant female
c. a male without Duchenne muscular dystrophy and a carrier female
d. a and c
e. none of the above

5. Predict the sex of a baby with each of the following chromosomal makeups. Use your answer to check your answer to Question 1.

a. XX
b. XXY
c. XY
d. X

6. Consider your brother and your son.

a. If you are female, will your brother and your son have essentially identical Y chromosomes? Explain your answer.
b. If you are male, will your brother and your son have essentially identical Y chromosomes? Explain your answer.

7. A wife is heterozygous for Duchenne muscular dystrophy alleles, and her husband does not have DMD. Neither has any other notable medical history. What percentage of their sons, and what percentage of their daughters, will have

a. Duchenne muscular dystrophy (which is determined by a recessive allele on the X chromosome)?
b. an X-linked dominant form of rickets (a bone disease)?

Driving Question 2 What inheritance patterns are observed when alleles are not simply dominant or recessive?

By answering the questions below and studying Infographics 12.6–12.8, you should be able to generate an answer for this broader Driving Question.

Know It

8. How does incomplete dominance differ from codominance?

9. If you are blood type A-positive, to whom can you safely donate blood? Who can safely donate blood to you? List all possible recipients and donors and explain your answer.

Use It

10. The MN blood group system is based on alleles of the glycophorin A (*GYPA*) gene.

a. A woman with type M blood marries a man with type N blood. Their first child has type MN blood. What does this tell you about the relationship between the M and N alleles? Explain your answer.
b. If that child with type MN blood marries a person with type MN blood, what frequencies of MN blood types do you expect in their children?

Apply Your Knowledge

Mini Case

11. A serious car crash on a freeway has resulted in multiple injuries causing substantial blood loss in three members of a family—a mother, a father, and their 2-year-old daughter. The local blood bank will be challenged to supply blood, as its supplies of all blood types were drained after the roof of a shopping plaza collapsed the week before and many transfusions were required to treat the injured people.

a. The EMTs must give blood immediately to all three members of the family. What blood type should they use? Consider both ABO blood type and Rh factor. Explain your answer.

b. Both parents carry a blood donor card. The mother is O-negative and the father is A-positive. From this information, what (if any) additional blood types (beyond your answer to part a) can be given to either parent? Explain your answer.

c. Does knowing the parents' blood types give you enough information about the daughter's possible blood type to use a different blood type for her transfusion? Why or why not? (*Hint:* Consider possible blood types for the daughter and the implications of using, for example, A-negative donor blood. Could you guarantee that using blood of that type would be safe?)

Driving Question 3 What inheritance patterns are observed when more than one gene and/or the environment influences a trait?

By answering the questions below and studying Infographics 12.9 and12.10, you should be able to generate an answer for this broader Driving Question.

Know It

12. What aspects of height make it a polygenic trait?

13. Which of the following inheritance patterns includes an environmental contribution?

a. polygenic
b. X-linked recessive
c. X-linked dominant
d. multifactorial
e. none of the above

14. What is the difference between polygenic inheritance and multifactorial inheritance?

Use It

15. If two women have identical alleles of the hundreds of suspected height-associated genes, why might one of those women be 5 feet 5 inches tall and the other be 5 feet 8 inches tall?

16. Two identical twin rat pups are separated at birth. One is raised by a low-nurturing mother and one is raised by a highly nurturing mother. Use the terms below to explain why one pup will grow up to be a highly nurturing parent and one pup will grow up to be a low nurturing parent.

Methyl groups	Cortisol
Environment	Glucocorticoid receptor
Nurturing phenotype	Epigenetics
GR gene	Genotype

17. From what you have read in this chapter, how can you account for two people with the same genotype for a predisposing disease allele having different phenotypes?

Driving Question 4 How does nondisjunction lead to numerical abnormalities of chromosomes, and what are the consequences of these abnormalities?

By answering the questions below and studying Infographics 12.11 and 12.12, you should be able to generate an answer for this broader Driving Question.

Know It

18. What is the normal chromosome number for each of the following?

a. a human egg
b. a human sperm
c. a human zygote

19. When looking at a karyotype—for example, to diagnose trisomy 21 in a fetus—is it possible to also use that analysis to tell if the fetus has inherited a cystic fibrosis allele from a carrier mother?

Use It

20. Which of the following can result in trisomy 21?

a. an egg with 23 chromosomes fertilized by a sperm with 23 chromosomes
b. an egg with 22 chromosomes fertilized by a sperm with 23 chromosomes
c. an egg with 24 chromosomes, two of which are chromosome 21, fertilized by a sperm with 23 chromosomes
d. an egg with 23 chromosomes fertilized by a sperm with 24 chromosomes, two of which are chromosome 21

21. From what you have read in this chapter, which of the possibilities in Question 20 is most likely to occur? Explain your answer.

Bring It Home

Apply Your Knowledge

22. What factors would lead you to consider prenatal genetic testing? In your opinion, what is the value of having this information? Use the table to the right to help organize your thoughts, and then make a conclusion about the use of prenatal genetic testing.

Factor	Value
Consider known risks (e.g., family history, mother's age).	*Consider how the information may be used.*

Interpreting Data

Apply Your Knowledge

23. The graph at right shows the average ("mean") age of women who had children with trisomy 21 ("Cases"), the average age of those who did not ("Controls"), and the average age of all women giving birth in the population studied. The data are presented for 15 years.

a. In general, how does the age (at time of birth) of women giving birth to a baby with Down syndrome compare to the age of women giving birth to a baby without Down syndrome?

b. During which year was the average age of the cases closest to the average age of the controls? How close were the average ages in this year?

c. During which year was the average age of the cases the most different from the average age of the controls? How different were the average ages in this year?

d. Using the data points for each year over this 15-year period, calculate the overall average age of women having babies with Down syndrome and the overall average age of women having babies who do not have Down syndrome.

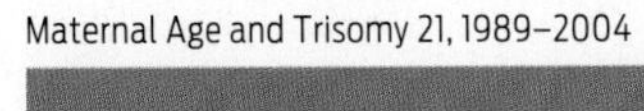

Maternal Age and Trisomy 21, 1989–2004

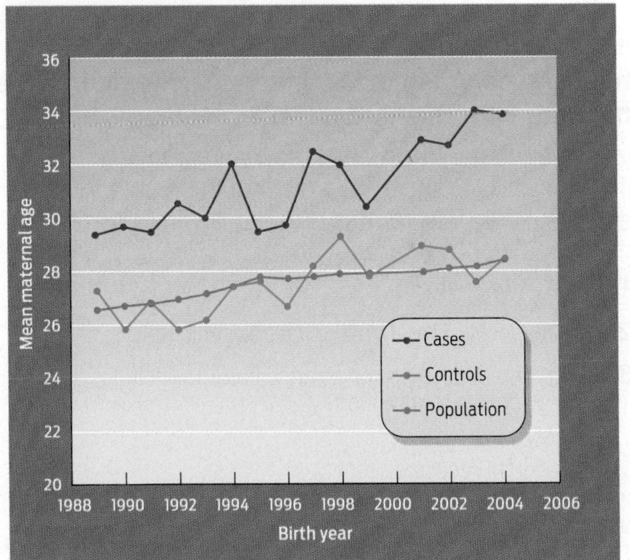

Data: Graves Allen, E., et al. (2009). Maternal age and risk for trisomy 21 assessed by the origin of chromosome nondisjunction: a report from the Atlanta and National Down Syndrome Projects. *Hum Genet* 125(1):41–52.

BUGS That Resist DRUGS

Drug-resistant bacteria are on the rise. Can we stop them?

CDC/Science Source

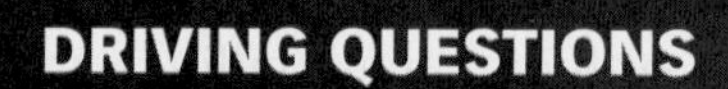

DRIVING QUESTIONS

1. What is MRSA, and how can bacteria resist the effects of antibiotics?
2. How can populations of bacteria become genetically diverse?
3. What is fitness, in an evolutionary sense?
4. How do populations evolve, and what is the role of antibiotics in the emergence of antibiotic resistance?

In May 2019, 15-year old Trey Sudderth was playing basketball with his friends when he injured his ankle. Three weeks later, he was dead.

In March 2018, Sammie Fallon, a 17-year-old college student, went to her doctor for a routine medical biopsy. Three weeks later, she was dead.

In February 2013, 17-year-old Tabby Meade attended a high school choir competition. Two days later, she was dead.

The list of sudden deaths like these goes on and on. But surprisingly, these young people weren't killed in accidents or by violence; they were all killed by an infectious bacterium known as methicillin-resistant *Staphylococcus aureus* (MRSA). This bacterium has become widespread in recent years and is difficult to treat with many existing antibiotics.

MRSA (pronounced "mer-sa") is estimated to sicken approximately 75,000 people each year and kill approximately 10,000 people in the United States. Formerly, most outbreaks of MRSA were confined to hospitals, where people are often very ill and have compromised immune systems. But since the mid-1990s, a growing number of healthy people have become infected outside hospitals. Day care facilities, military bases, and schools nationwide have reported outbreaks, and young, healthy people are getting sick.

For more than 70 years, we've successfully combated potentially deadly bacterial infections with antibiotics—those "wonder drugs" that Alexander Fleming first discovered in a moldy petri plate nearly a century ago (see Chapter 3). But many of our most trusted antibiotics are no longer effective at killing the bacteria they once defeated. Over time, these bacteria have changed genetically—evolved—to become resistant. As a result, scientists are now seeing bacterial infections that don't respond to any known antibiotics. Many fear the day when we run out of treatment options altogether.

"This is a major public health imperative," says Robert Daum, professor of microbiology at the University of Chicago, who studies antibiotic resistance. "We need a plan of attack now."

Staph the Microbe

▶ Bacterial colonization versus infection

Bacteria are everywhere: in the air, in food, on toothbrushes and computer keyboards, and even on and inside you. In fact, your body hosts a wide variety of bacteria. These range from helpful ones like *Lactobacilli* and *Bifidobacteria,* which live in the gut and aid digestion, to harmful ones like *Salmonella,* which causes food poisoning. The helpful bacteria on your skin, for example, produce acids that make your skin inhospitable to other, less helpful bacteria. Nevertheless, at any given moment, a fair number of us harbor potentially dangerous bacteria on our skin. Most of the time, this isn't a problem. But in certain circumstances, these bacteria can outgrow your helpful skin bacteria and cause serious infections.

MRSA infection is caused by the bacterium *Staphylococcus aureus*—often called simply "staph." Although all strains of staph can cause infection, the medical community is especially concerned about *S. aureus* strains that have developed resistance to the antibiotic drugs that once effectively killed them. "MRSA" is actually a misnomer because the antibiotic methicillin is no longer used to treat staph infections. Drug-resistant strains of staph are usually resistant to several different types of antibiotics, including penicillins and cephalosporins.

Remarkably, staph bacteria cause no harm to most of the people who carry them.

An estimated one-third of the U.S. population carries staph on their skin or in their nose, and about 2% of the population carries MRSA, according to the Centers for Disease Control and Prevention (CDC). If you carry staph of any strain but aren't sick, you are "colonized" but not infected. Staph spreads from person to person through skin-to-skin contact or through shared contaminated items such as towels and bars of soap.

Most healthy people can be colonized with any staph strain, including MRSA, and not become ill—for the most part, our skin and our immune systems protect us from these bacteria. But infections can occur if staph bacteria come into contact with a wound or otherwise enter the body. For example, athletes who have cuts and scrapes may acquire a staph infection in locker rooms or during contact sports. Staph infection usually causes only minor skin eruptions such as boils or pustules that can resemble spider bites. But if staph manages to enter the bloodstream or travel deeper into the body, then more serious complications can occur **(INFOGRAPHIC 13.1).**

INFOGRAPHIC 13.1

Staph Colonization and Infection

Staphylococcus aureus is a spherical bacterium that can cause pimples, boils, and wound infections. *S. aureus* can be passed from person to person by direct contact with contaminated skin or by transfer of the bacteria via contaminated objects or surfaces.

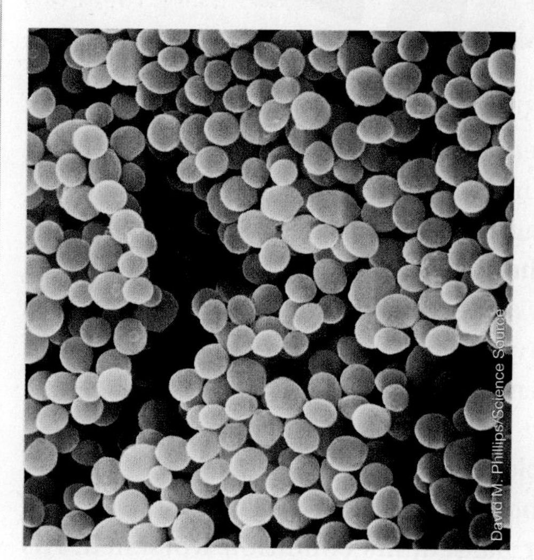

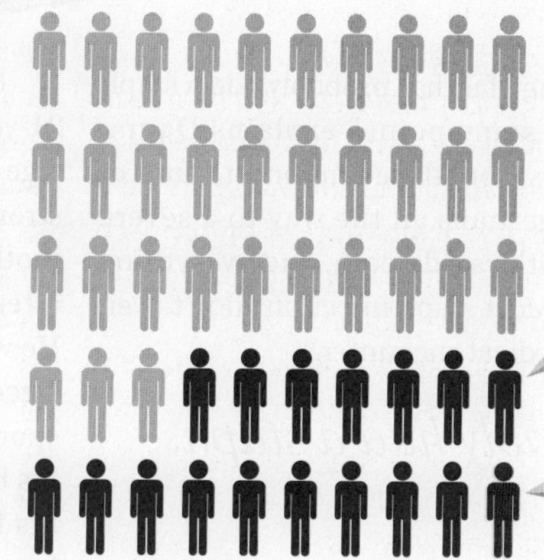

Experts estimate that the skin, nose, and throat of about one-third of the U.S. population (nearly 90 million people) is colonized by *S. aureus*. These bacteria do not usually cause illness unless they penetrate skin barriers through open hair follicles, cuts, and scrapes.

More than 2% of people in the United States (nearly 2 million people) carry MRSA strains of *S. aureus*. Most of these individuals are disease free.

Infections caused by *Staphylococcus aureus*

S. aureus can colonize normal skin surfaces without causing infection.

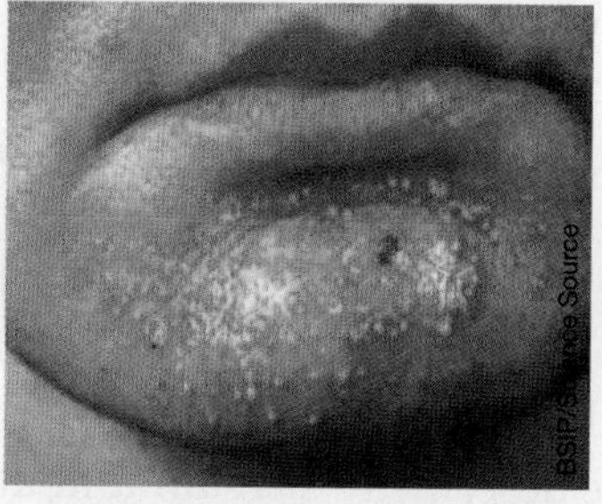

S. aureus can cause mild skin infections like pimples and boils.

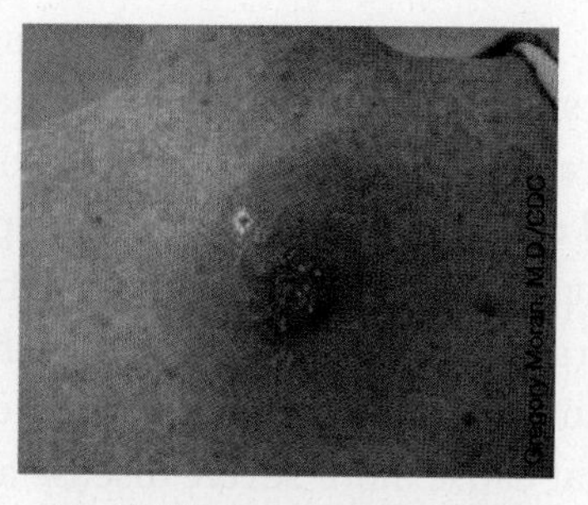

Methicillin-resistant *S. aureus* can cause more severe illness that is difficult to treat with existing medications.

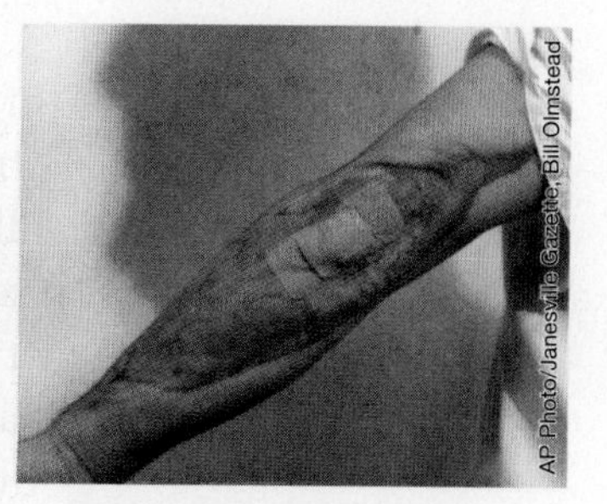

Some extremely virulent strains of MRSA can eat through skin and soft tissue.

? What is the difference between *S. aureus* colonization and a MRSA infection?

Ricky Lannetti as a healthy 21-year-old football player. Right: Ricky with his mother, Theresa Drew.

"Every one of us has probably had a staph infection at some point," explains Daum. "Staph ranges from the commonest cause of infected fingernails all the way to a severe syndrome with rapid death, and everything in between. Most staph infections don't even result in a medical encounter."

Every one of us has probably had a staph infection at some point.

—Robert Daum

Staph bacteria can cause such a range of symptoms in part because many different strains of staph exist. Each strain differs from all others in its genetic makeup. MRSA, for example, is composed of a number of unique strains of drug-resistant staph, and some cause more serious disease than others. Those that cause more-severe disease are described as "virulent." The increasing number of deadly MRSA infections among healthy people in the community likely reflects the emergence of a particularly virulent strain of the bacteria.

Ricky Lannetti, for example, was a healthy 21-year-old football player at Lycoming College in Williamsport, Pennsylvania. "He was strong as an ox and he ran like a deer," says his mother, Theresa Drew. Early one December after a game, Ricky came down with the flu. He wasn't recovering, and on the morning of December 6, Drew drove her son to Williamsport Hospital. By the time he was admitted, his blood pressure was dangerously low and his body temperature was erratic. As each hour passed, his condition worsened. His lungs began to fail. Doctors tried five different antibiotics, all without success. When his heart began to weaken, doctors prepared him to be flown to the cardiac center at a bigger hospital in Philadelphia. But it was too late: Ricky died that night.

It was only after an autopsy was performed that doctors discovered what had killed him: MRSA had infected Ricky's bloodstream and attacked his organs. Although doctors couldn't be sure how Ricky contracted MRSA, they suspect that it entered his body through a pimple on his buttocks. From there, it spread to his internal organs.

"Doctors tried every antibiotic imaginable, including vancomycin," says his father, Rick Lannetti. But the treatment came too late. Ricky's immune system was already weak because of the flu. When he contracted MRSA, his body was unable to fight back as well as it otherwise would have. "In the end," his father says, "MRSA had broken every one of his organs beyond repair" **(INFOGRAPHIC 13.2)**.

INFOGRAPHIC 13.2

Athletes Have an Elevated Risk of Skin Infections

Staphylococcus aureus and other harmful bacteria can transfer from person to person by direct or indirect contact. For this reason, athletes are at special risk of infection. To raise awareness, the CDC teamed up with the National Collegiate Athlete Association to create a series of educational posters.

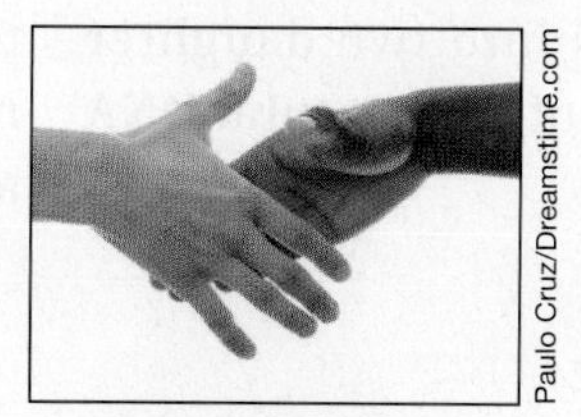

Paulo Cruz/Dreamstime.com

Skin-to-skin contact or contact sports

DNY59/iStockphoto

Offscreen/Dreamstime.com

Used soap, towels, and equipment

Genevieve Astrelli/iStockphoto

Contaminated surfaces

CDC

? List three ways that MRSA could spread among members of a college baseball team.

Acquiring Resistance

▸ Antibiotic resistance and genetic variation

Bacterial infections were a common cause of death before the 1940s, when antibiotics first became widely available. Since then, we've relied on antibiotics to treat most common bacterial infections, including ones that might otherwise have been killers. But almost immediately after antibiotics were introduced, bacteria that could survive antibiotics—drug-resistant bacteria—began to emerge. That trend continues today. For example, the drug ceftaroline was introduced in 2010, and ceftaroline-resistant strains of *S. aureus* were detected in 2011. Within the last decade or two, drug-resistant bacterial strains have become much more common.

Most people who are infected with drug-resistant bacterial strains are still treatable, but they have fewer treatment options. And sometimes—as in Ricky Lannetti's case—existing drugs are completely ineffective.

Drug-resistant strains of staph, for example, are typically resistant to an entire class of antibiotic drugs called the beta-lactams. Beta-lactams include penicillin and the cephalosporin antibiotics, such as methicillin and cephalexin. These drugs are the most commonly prescribed class of antibiotics, used to treat ear infections, bronchitis, and urinary tract infections, among others. They work by interfering with a bacterium's ability to synthesize cell walls **(INFOGRAPHIC 13.3)**.

Vancomycin, a non–beta-lactam drug, is the antibiotic of choice when a serious MRSA infection is confirmed. But even vancomycin isn't always effective; some staph strains are now resistant to vancomycin, too.

What's behind this rise in drug-resistant bugs? Like all organisms, bacteria can acquire genetic differences through mutations that occur when they replicate their DNA during cell reproduction (see Chapter 9). These random mutations create new alleles, including ones that may confer antibiotic resistance.

Bacteria reproduce asexually by **binary fission.** Unlike sexual reproduction, in which gametes from two parents fuse, binary fission does not require a partner. A single parental cell simply replicates its single chromosome, grows, and then splits into two daughter cells, each with a copy of the parental DNA **(INFOGRAPHIC 13.4)**.

BINARY FISSION
A type of asexual reproduction in which one parental cell divides into two.

Each time DNA is replicated during binary fission, there is a chance that mutations will occur. When they do, new alleles are produced, and these alleles will then be present in the daughter cells. Bacteria reproduce much more rapidly than other organisms—a generation of bacteria can double in as little as 20 minutes. Hence, they accumulate mutations at a relatively high rate. And once an individual cell has acquired a particular

INFOGRAPHIC 13.3

How Beta-Lactam Antibiotics Work

Antibiotics are grouped into classes, one of which is called the beta-lactams. Beta-lactam antibiotics interfere with a bacterium's ability to synthesize cell walls.

In the absence of antibiotic:

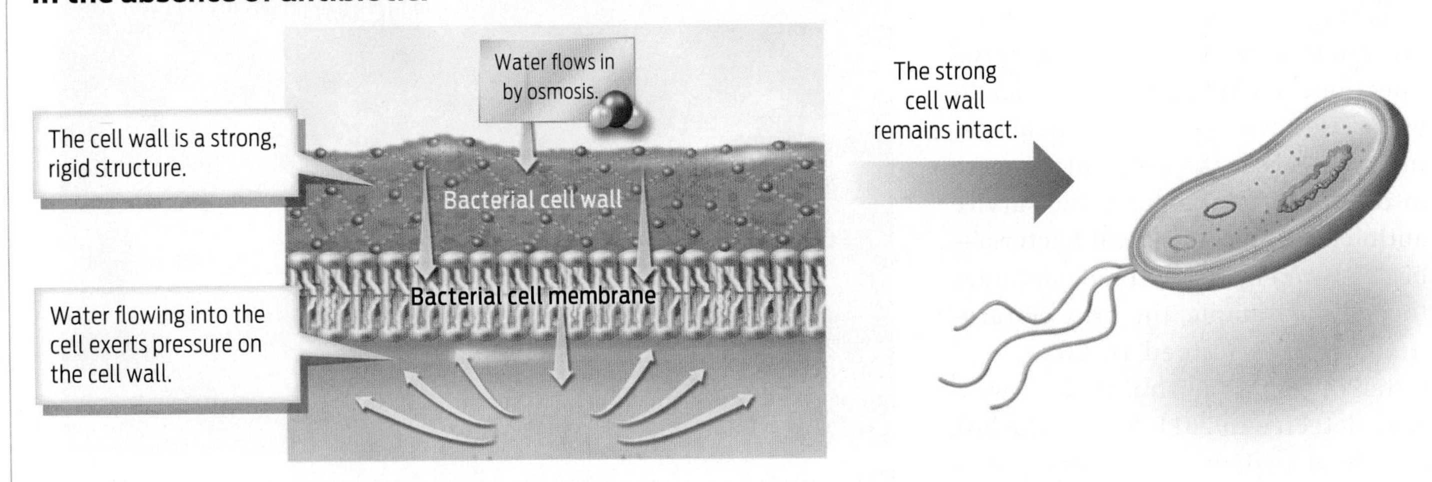

In the presence of antibiotic:

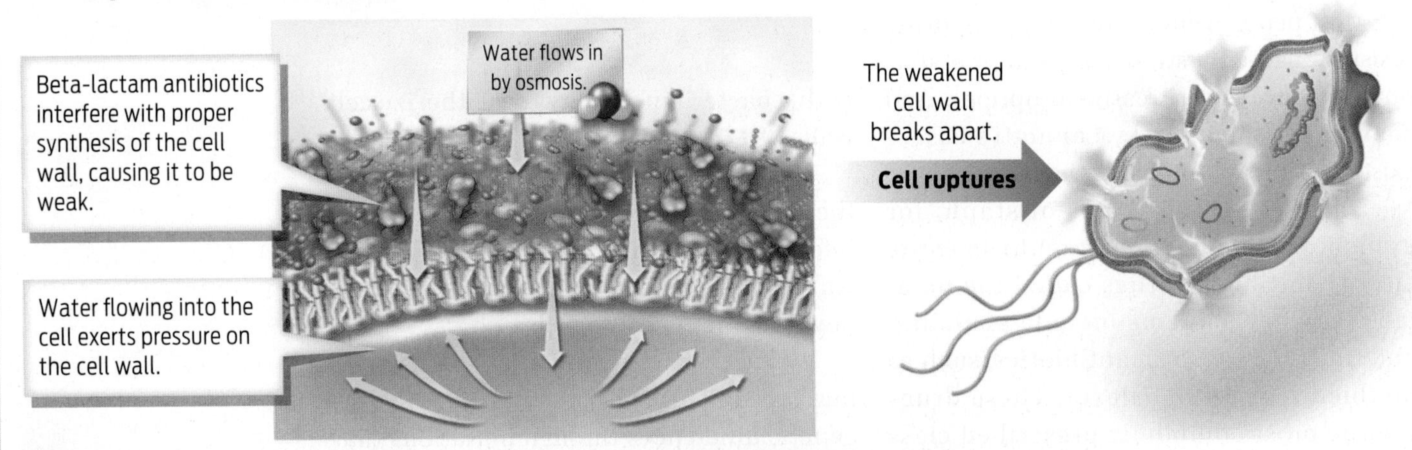

? If cells were grown in an isotonic solution that included a beta-lactam antibiotic, would they still rupture? Explain your answer.

INFOGRAPHIC 13.4

How Bacteria Reproduce

Bacteria reproduce through binary fission. Binary fission is a form of asexual reproduction in which a single parent cell replicates its contents and then divides into two daughter cells. Note that each daughter cell inherits all its DNA from the single parent cell.

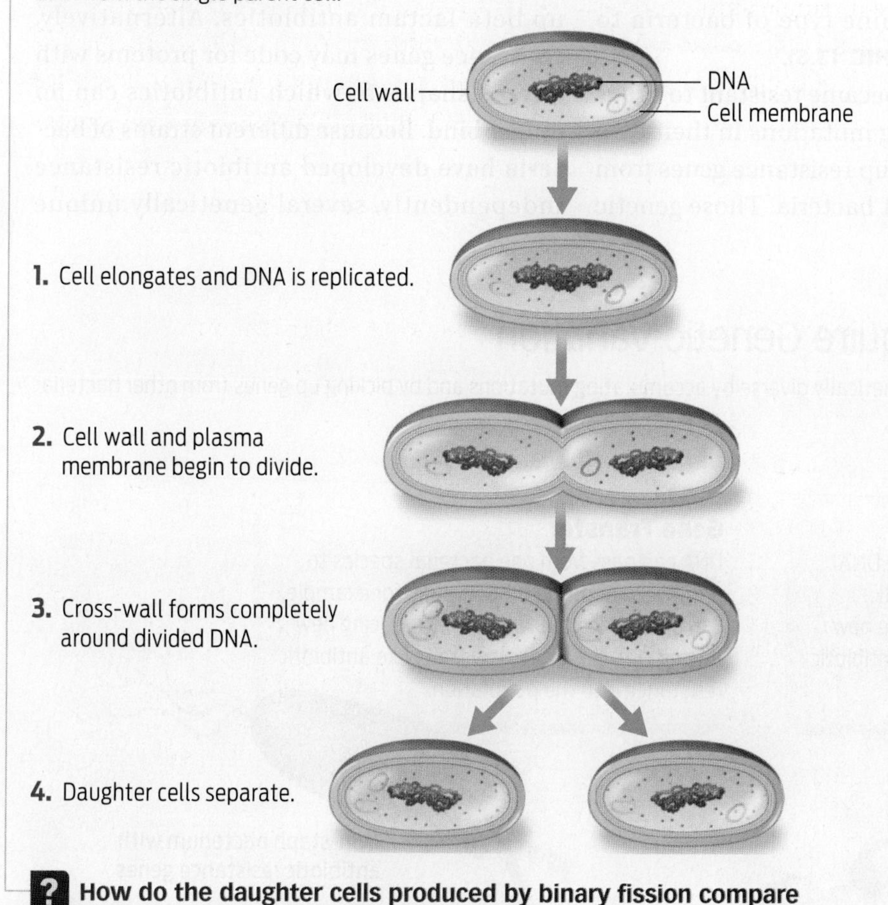

? How do the daughter cells produced by binary fission compare to each other and to the parent cell?

mutation, it will pass that mutation to its daughter cells, which in turn will pass it on to their daughters. At the same time, individual daughter cells can also acquire new mutations, in addition to the ones they have inherited. In this way, an entire **population** of bacteria that is genetically different from the original cell can arise very quickly. In biology, a population is a group of individuals of the same species (in this case *S. aureus*) living together in the same geographic area. That area could be a prairie, a pond, or a person's nose. New mutations can arise and become prevalent in a population of bacteria in a person's body over the course of a single illness.

Because of these mutations, there are genetic differences within this population of infecting bacteria. This population therefore has a high degree of genetic variation—the genetic differences between members of a population. If some of these specific genetic differences confer resistance to particular antibiotics, individual bacteria with those differences will no longer be killed by those antibiotics. The survivors will be able to reproduce and pass this resistance trait on to offspring.

POPULATION
A group of organisms of the same species living together in the same geographic area.

Mutations aren't the only way that populations of bacteria can acquire genetic variation and become more diverse, genetically speaking. Bacteria can also acquire new alleles, and even new genes, through a mechanism called **gene transfer,** in which pieces of DNA pass from one type of bacteria to another **(INFOGRAPHIC 13.5)**.

GENE TRANSFER
The process by which bacteria can exchange segments of DNA between them.

Staph bacteria became resistant to drugs either by developing mutations in their own genes or by picking up resistance genes from other drug-resistant bacteria. Those genetic changes ultimately altered staph proteins in ways that helped staph bacteria dodge antibiotic drugs. The altered or acquired genes may code for proteins that can disable antibiotics. For example, some bacteria produce enzymes called beta-lactamases that chew up beta-lactam antibiotics. Alternatively, resistance genes may code for proteins with altered shapes to which antibiotics can no longer bind. Because different strains of bacteria have developed antibiotic resistance independently, several genetically unique

INFOGRAPHIC 13.5

How Bacterial Populations Acquire Genetic Variation

Asexually reproducing bacterial populations become genetically diverse by accumulating mutations and by picking up genes from other bacteria of the same or different species.

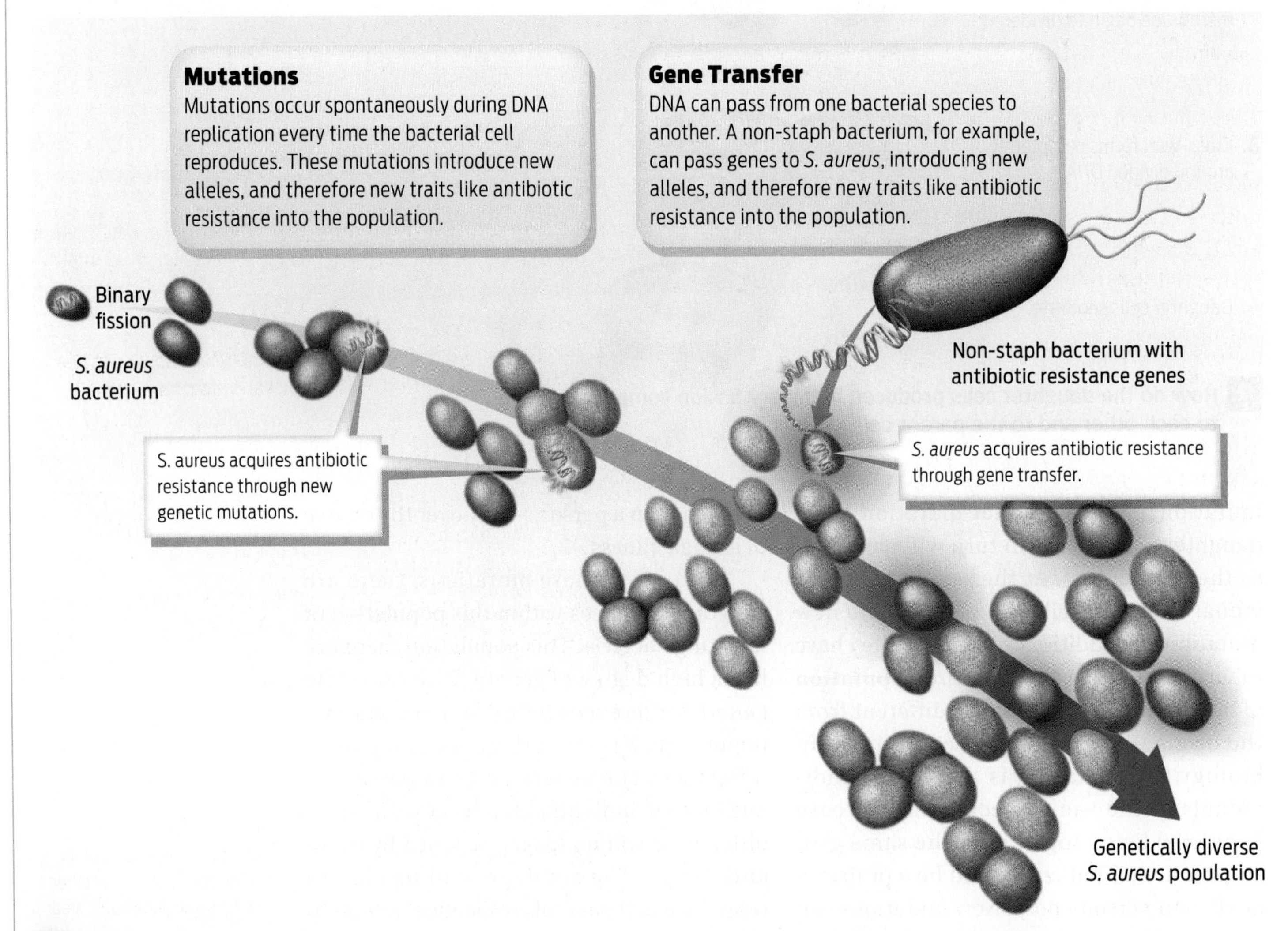

? Which mechanism (mutation or gene transfer) is more likely to introduce a new allele, rather than a new gene, into the population?

drug-resistant staph strains circulate through human communities at the same time.

An Ever-Changing Enemy

▶ Fitness and evolution

While an individual bacterium—or any individual organism—can undergo genetic changes that give it new traits, this doesn't entirely explain how entire populations of bacteria such as staph develop the trait of resistance to drugs. An entire population of organisms with a new trait can arise only when the environment favors that trait—that is, when the trait is advantageous to the organisms carrying it.

When a population's environment favors one trait over others, the frequency of alleles that code for that trait in the population will increase over time. Take the trait for drug resistance, for example. As a result of random mutations and gene transfers happening all the time in any bacterial population, some individual bacteria in a population may possess alleles that confer antibiotic resistance. In an environment free of antibiotics, any individual bacterium will have about the same chance of surviving and reproducing as any other, whether or not it carries a drug-resistance allele. In other words, the ability to resist antibiotics will confer neither an advantage nor a disadvantage, since there are no antibiotics around. But in the presence of an antibiotic, bacteria with an allele for drug resistance may survive, whereas other bacteria will die. The surviving bacteria then reproduce and pass their alleles for drug resistance on to future generations. Consequently, the frequency of the resistance trait increases. This is how a population evolves. **Evolution** is defined as a change in the frequency of alleles in a population over time.

An organism's ability to survive and reproduce in a particular environment is called its **fitness.** The higher an organism's fitness, the more likely that alleles carried by that organism will be passed on to future generations and increase in frequency. In an environment in which antibiotics are abundant, drug-resistant bacteria are fitter than nonresistant bacteria **(INFOGRAPHIC 13.6).**

EVOLUTION
Change in allele frequencies in a population over time.

FITNESS
The relative ability of an organism to survive and reproduce in a particular environment.

INFOGRAPHIC 13.6

An Organism's Fitness Depends on Its Environment

The term "fitness" describes the relative ability of an organism to reproduce in a particular environment. Fitness is determined by the interaction between phenotype and environment. Antibiotic-resistant bacteria, for example, have high fitness in the presence of antibiotics.

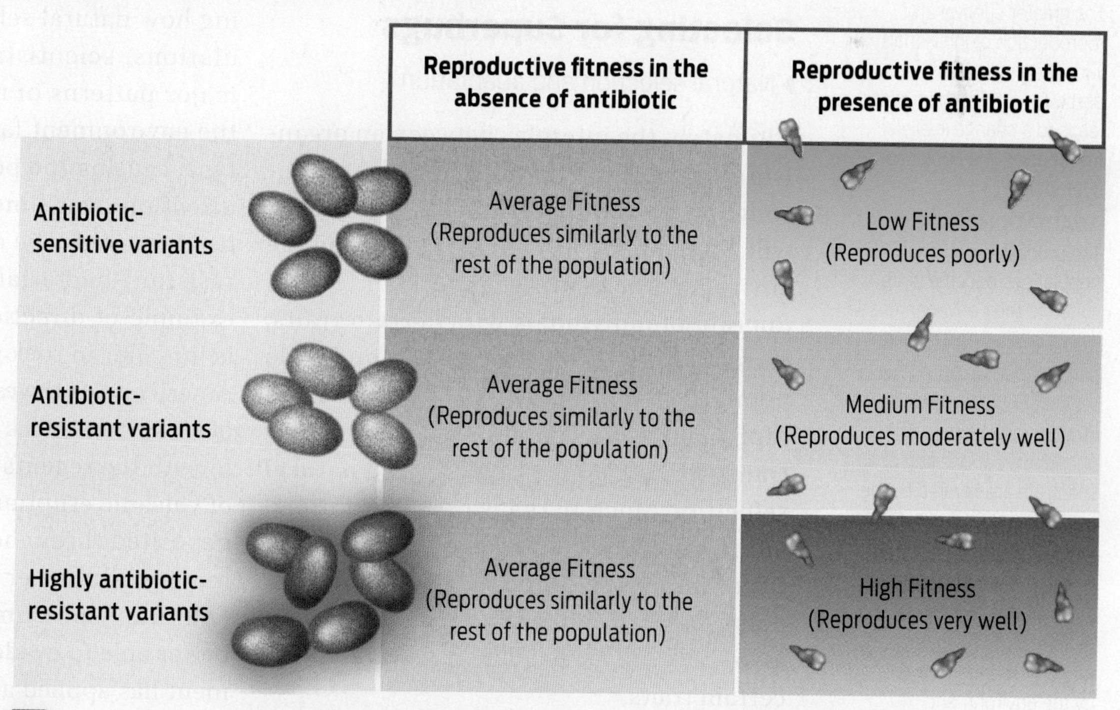

? In the absence of antibiotics, is there a difference between the fitness of antibiotic-sensitive and antibiotic-resistant variants?

Currently, in the United States, antibiotic use is widespread. In 2017, doctors prescribed more than 258 million courses of antibiotics—equivalent to 793 prescriptions for every 1,000 people in the country. The abundance of antibiotics in the environment has created the perfect breeding ground for antibiotic-resistant staph.

In 2017, doctors prescribed more than 258 million courses of antibiotics—equivalent to 793 prescriptions for every 1,000 people in the country.

In a different environment, in which antibiotics are less common, these same resistant bacteria will not necessarily have an edge over other bacteria. In other words, fitness is always relative to the environment; organisms can be fit in one environment and not in another.

Selecting for Superbugs

▸ Natural selection and adaptation

Ultimately, the interplay between an organism's traits—its phenotype—and its environment is what determines which traits will predominate in a population. When the environment favors the survival and reproduction of individuals with certain traits, those traits become more common in the population. The higher rate of survival and reproduction of individuals with certain traits within a population is called **natural selection.** Much in the way that plant and animal breeders have for centuries practiced artificial selection to produce individuals with desired traits, the environment also, in a sense, selects individuals with certain traits.

When natural selection acts on a population over time, advantageous traits become more common, and the population becomes better suited to its environment. In other words, evolution by natural selection leads to **adaptation.** This is what we see with antibiotic-resistant bacteria: the population has become better suited, or adapted, to an environment in which antibiotics are abundant. Charles Darwin was one of the first people to figure out how natural selection works and to study its results. He realized that natural selection will occur automatically anytime individuals in a population differ in ways that alter their fitness, as long as that variation is inherited (see **Milestone 4: Adventures in Evolution**).

Note that while natural selection depends on genetic variation existing between individuals in a population, it is the *population* as a whole that evolves, not the individuals. Individual organisms do not experience a change in allele frequencies over time. Therefore, individual organisms do not evolve **(INFOGRAPHIC 13.7)**.

Natural selection doesn't always affect populations in the same way. By studying how natural selection has shaped populations, scientists have identified three major patterns of natural selection. When the environment favors an extreme phenotype, causing the population to shift in one direction over time, **directional selection** has occurred. The emergence of antibiotic-resistant bacterial populations is a good example of directional selection. Another is fur color in rock pocket mice living in the American Southwest. Ancestral populations of rock pocket mice had light fur that blended in with the region's sandy soil. After a series of volcanic eruptions, black lava rocks were deposited throughout the area. Since then, mice with darker fur have become more common—presumably because they are better able to evade predators. The environment has applied a selective pressure on the

NATURAL SELECTION
The greater survival and reproduction of individuals with certain traits in a particular environment that leads to a change in allele frequencies in a population over time.

ADAPTATION
The process by which populations become better suited to their environment as a result of natural selection.

DIRECTIONAL SELECTION
A type of natural selection in which organisms with phenotypes at one end of a spectrum are favored by the environment.

STABILIZING SELECTION
A type of natural selection in which organisms near the middle of the phenotypic range of variation are favored by the environment.

INFOGRAPHIC 13.7

Evolution by Natural Selection

In a typical population of organisms, individuals will vary genetically. When the environment favors some genetic variants over others, those variants will have higher fitness—they will survive better and reproduce more successfully. Over generations, the frequency of alleles that confer higher fitness will increase, while those that confer lower fitness will decrease. This nonrandom change in allele frequencies over generations is called evolution by natural selection.

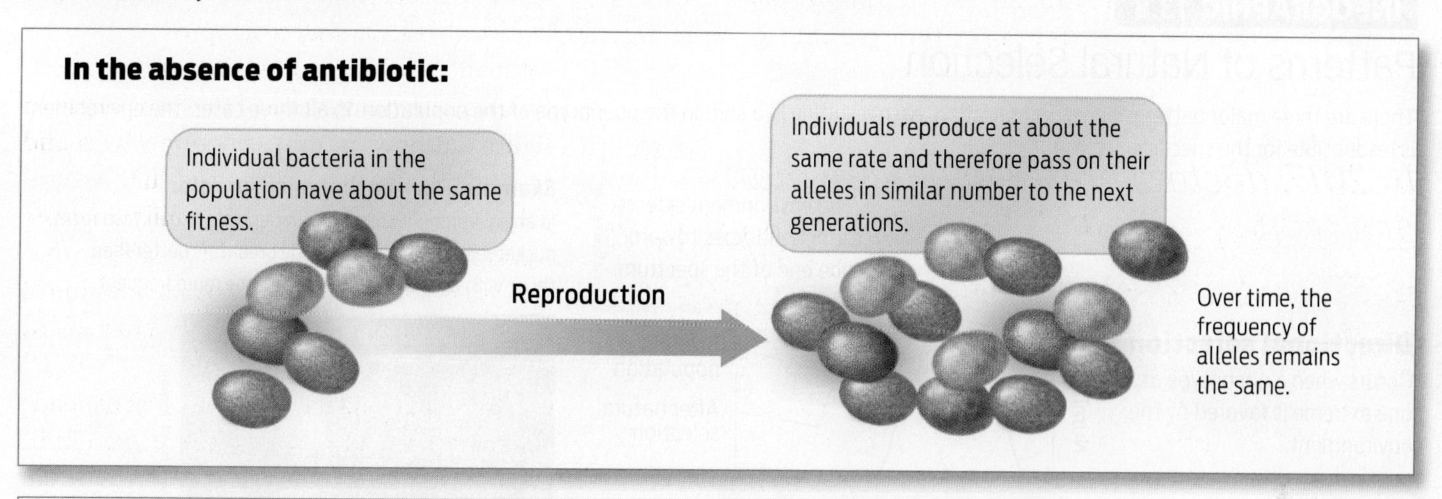

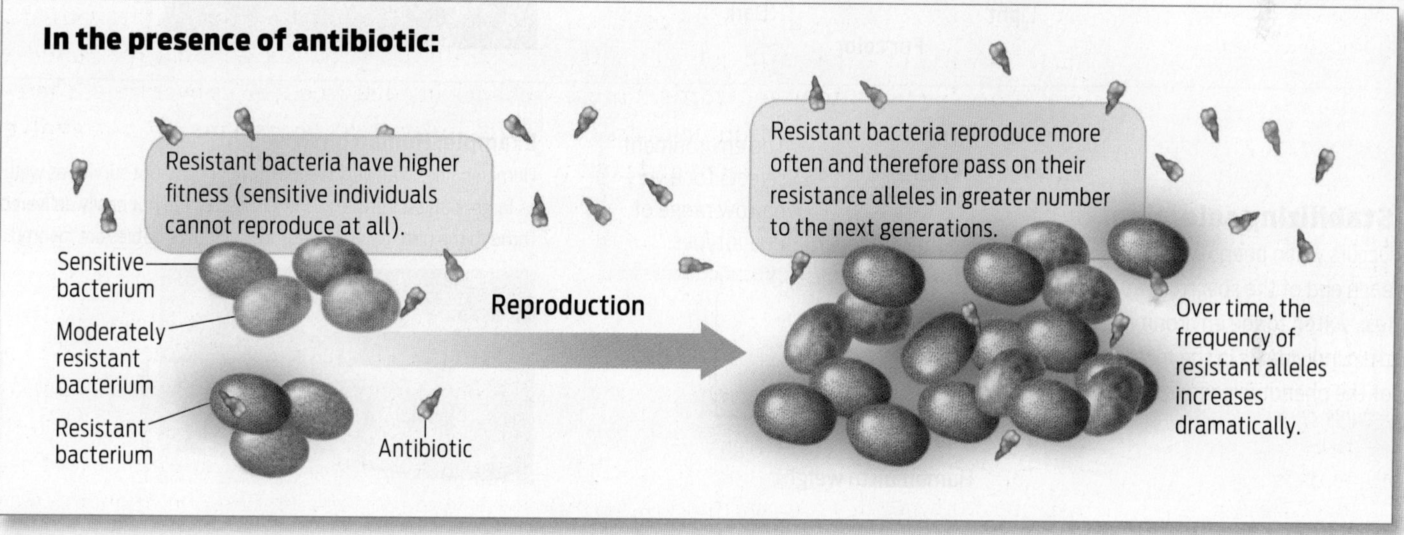

? Some blue mussels can develop thicker shells because of a specific allele that controls shell thickness. Blue mussels with this allele are more likely to survive attacks by crabs. If crabs are introduced into a genetically diverse population of mussels, what will happen to the mussel population over time?

population—in these cases, selecting for a trait at one end of the phenotypic spectrum.

When the environment favors the middle of the phenotypic spectrum, but selects against extremes, we call this **stabilizing selection.** Human birth weight is an example. Both very small and very large babies are less fit than medium-size babies, who are hearty but still small enough to fit through the birth canal.

Finally, when the environment favors both ends of a phenotypic spectrum, **diversifying selection** occurs. An example is finches in environments where only large seeds and

DIVERSIFYING SELECTION
A type of natural selection in which organisms with phenotypes at both extremes of the phenotypic range are favored by the environment.

small seeds are available; birds with medium-size beaks are not as successful at cracking either type of seed, so big-beaked birds and small-beaked birds have an advantage **(INFOGRAPHIC 13.8)**.

The particular pattern of natural selection that a population follows depends on the interaction of phenotypes with the environment. For example, in the absence of antibiotics, populations of staph bacteria might

INFOGRAPHIC 13.8

Patterns of Natural Selection

There are three major patterns of natural selection, each resulting in a shift in the phenotype of the population. In all three cases, the environment is responsible for the specific pattern.

Directional selection

Occurs when a phenotype at one extreme is favored by the environment.

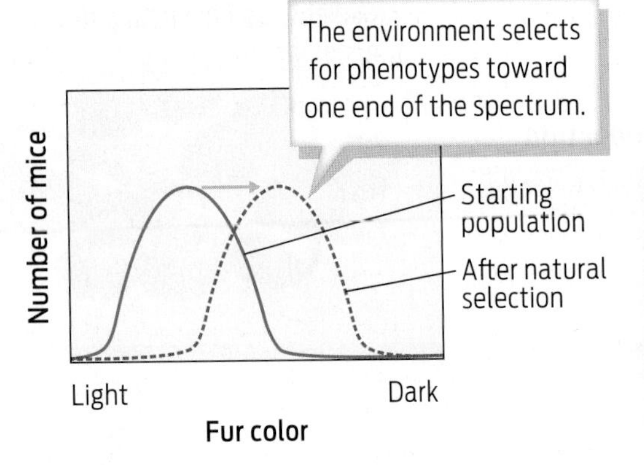

Example: Rock pocket mice fur color

In an environment covered with deposits of dark lava rocks, pocket mice with dark fur evade predators better than those with light fur. Dark mice become more frequent.

Stabilizing selection

Occurs when phenotypes at each end of the spectrum are less suited to the environment than individuals in the middle of the phenotypic range.

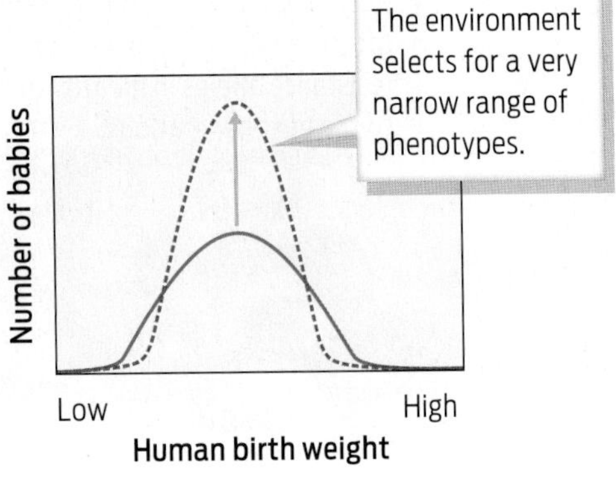

Example: Human birth weight

Human babies with very low birth weights do not survive as well as larger babies. Very large human babies are not easily delivered through the birth canal. Therefore, midrange babies are favored.

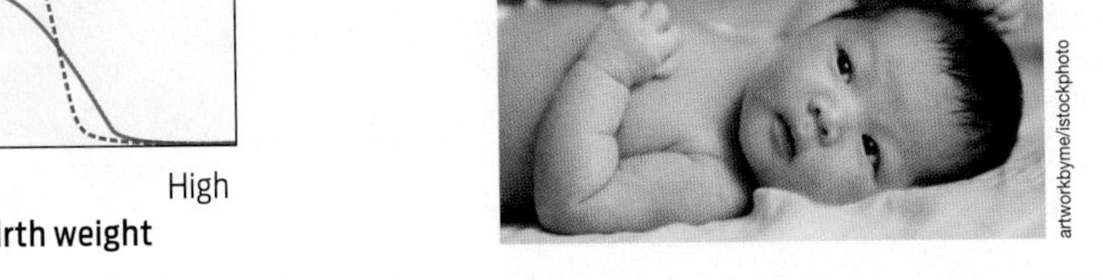

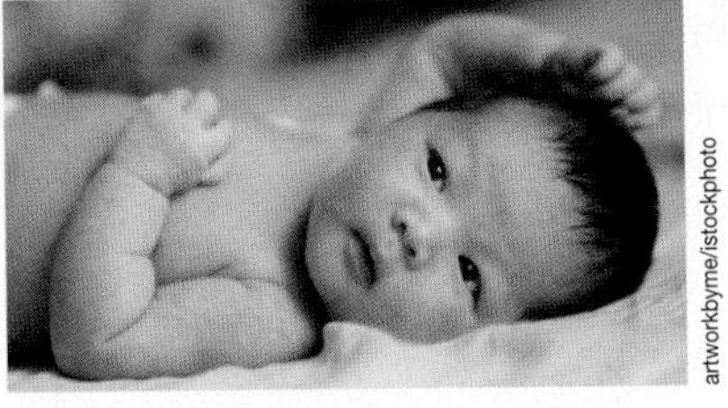

Diversifying selection

Occurs when the extremes of the phenotypic range are better suited to the environment than individuals in the middle of the phenotypic range.

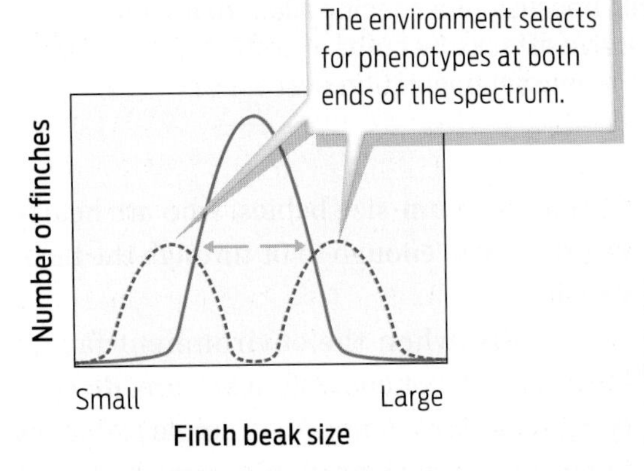

Example: Finch beak size

Finches living in an environment where only large, hard seeds and small, soft seeds are available are more successful if they have either large or small beak sizes. Medium beaks are not as successful at cracking either type of seed.

? An aquatic environment has light rocks and deep shadows. What pattern of selection will allow oysters to hide from predators in this environment?

have followed stabilizing or diversifying selection in response to some other pressure. Ultimately, though, the presence of antibiotics in the environment resulted in directional selection leading to the MRSA that killed Trey Sudderth, Sammie Fallon, Tabby Meade, and Ricky Lannetti.

MRSA in the Community

Drug-resistant staph strains first emerged in hospitals during the early 1960s, partly as a result of the selective pressure from antibiotics. Since then, hospitals have remained hot spots for staph infections. The combination of heavy antibiotic use, lots of sick patients, and close quarters makes hospitals a fertile breeding ground for resistant bugs.

In response, many hospitals have implemented measures to reduce infections. For example, studies have shown that simply requiring all health care workers to wash their hands before handling each patient can dramatically reduce the number of infections. "Washing hands well and often is absolutely critical," says Ruth Lynfield, state epidemiologist and medical director of the Minnesota Health Department, who studies infectious diseases in children.

More alarming than MRSA infections in hospitals are MRSA infections in the community at large. Though such illnesses were previously almost unheard of, in the mid-1990s the number of community-acquired infections in the United States began to spike, says Daum, the University of Chicago microbiology professor. In a study published in 1998, Daum noted that the prevalence of community-acquired MRSA infections requiring hospitalization increased from 10 cases per 100,000 hospital admissions in 1988–1990 to 259 cases per 100,000 admissions in 1993–1995: a more than 25-fold jump in only 5 years.

What happened in those years? New strains of *S. aureus* emerged and flourished in community environments where antibiotic use is rampant. Daum and his colleagues recently showed that a strain called USA300 is more virulent than other MRSA strains. "It appears to be juiced up," he says. Many USA300 genes are expressed at high levels. One gene in particular is turned on all the time—a gene that controls expression of a number of staph toxins, which can damage cells and tissues.

The result is that people infected with these strains have more-severe disease. In some cases, the bacteria literally eat through skin and soft tissues, an effect called necrotizing fasciitis. These "superbugs" can also kill more quickly. Symptoms can appear so suddenly that, according to Daum, "You could be healthy at 1:00 in the afternoon and be dead by 1:00 in the morning."

It was likely USA300 that killed Ricky Lannetti. "I've been an infectious disease guy for over 20 years now and we didn't talk about staph necrotizing pneumonia like we do now," says Daum.

Even more troubling, staph continues to evolve. Evidence shows that when the strains prevalent in the community mix with the strains prevalent in hospitals, the risk increases that a staph strain even more virulent than either original strain will emerge.

Stopping Superbugs

▸ Prevention and treatment strategies

Staph bacteria aren't the only ones that have grown resistant to antibiotics. It is getting harder to treat patients with severe *Salmonella* food poisoning caused by drug-resistant strains of this bug. *Neisseria gonorrhoeae,* the bacterium that causes gonorrhea, has become resistant to another important group of antibiotics, the fluoroquinolones. In addition, some cases of pneumonia are caused by strains of *Klebsiella* that are now resistant to every available antibiotic.

Because the very use of antibiotics drives bacterial populations to evolve resistance, antibiotic resistance is inevitable. But humans have hastened the emergence of drug-resistant strains of bacteria by their haphazard use and overuse of antibiotics. From the moment antibiotics were first introduced, physicians began prescribing them for colds, coughs, and earaches—even though most of these illnesses are caused by viruses, rather than bacteria, and viruses aren't killed by antibiotics.

Doctors aren't the only culprits. Approximately 80% of the antibiotics sold in the United States are used for livestock. Farmers give antibiotics in low doses to poultry, swine, and cattle to promote growth and to help prevent infections in crowded, unsanitary conditions. This practice can cause food-borne pathogens such as *Salmonella* or *Campylobacter* to develop antibiotic resistance. Eating improperly prepared foods from these animals can result in dangerous infections in humans, who then must be treated with even more antibiotics. What's more, undigested antibiotics in animal manure can leach into the environment through groundwater or when manure is used as fertilizer. In this environment, drug-resistant bacteria are more fit and will thus become more prevalent over time through natural selection.

INFOGRAPHIC 13.9

Preventing and Treating Infections by Antibiotic-Resistant Bacteria

Smart use of antibiotics can reduce the selective pressure that favors the emergence of antibiotic-resistant strains of bacteria. Keeping surfaces and hands clean can reduce the spread of all bacteria, including antibiotic-resistant strains. New research is exploring vaccination and treatment strategies for prevention and treatment of infections by antibiotic-resistant bacteria.

Gallo Images - Farming SA/Getty Images

Reduce antibiotics in livestock feed.
Excessive antibiotics in the environment create continuous selective pressure for all bacteria.

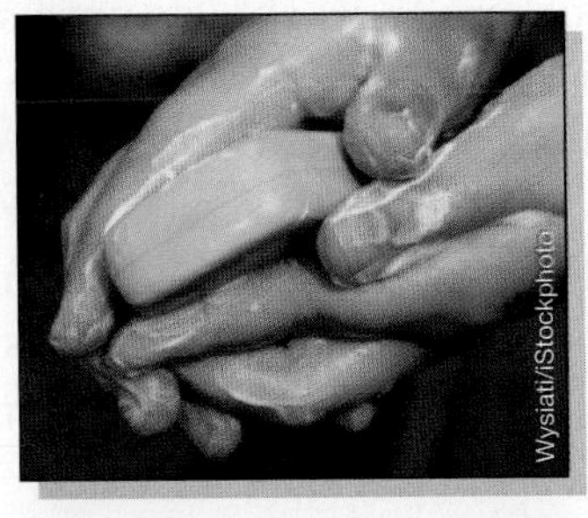

Wysiati/iStockphoto

Wash hands frequently.
Both regular and antibacterial soap, as well as hand sanitizer, are effective at preventing the spread of bacterial infection.

walik/iStockphoto

Keep locker rooms and sports equipment clean.
Protect athletes from contact with contaminated surfaces.

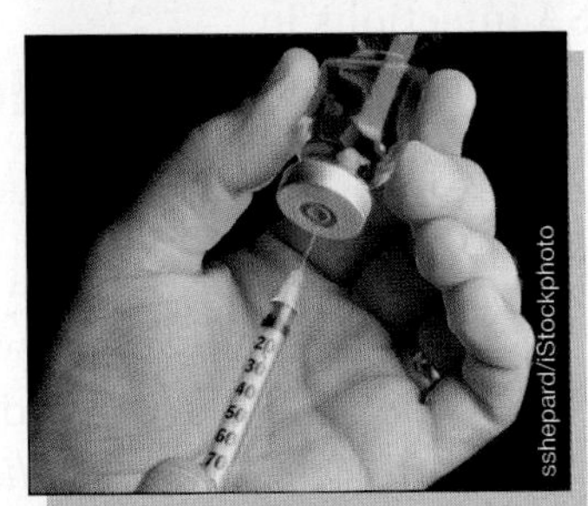

sshepard/iStockphoto

Research new vaccines.
Vaccines prevent resistant bacterial strains from making people ill.

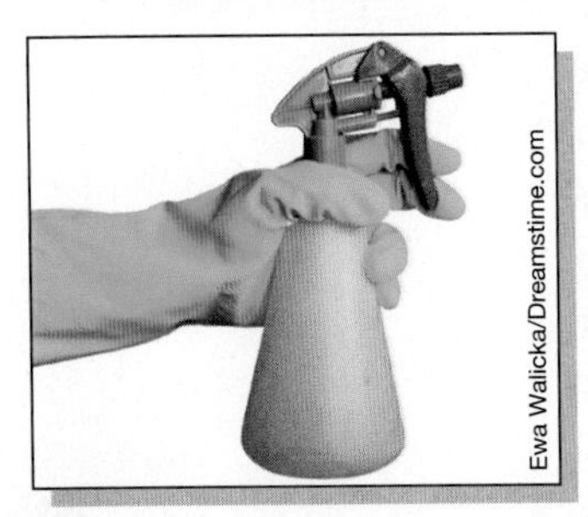

Ewa Walicka/Dreamstime.com

Disinfect common surfaces.
Disinfection reduces the transmission of infection by contact with contaminated surfaces, especially in facilities that serve a lot of people.

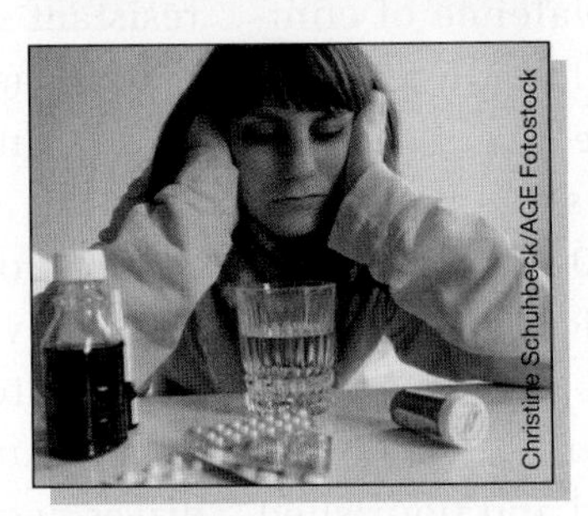

Christine Schuhbeck/AGE Fotostock

Do not take antibiotics for viral infections.
Viruses are not killed by antibiotics. Overuse of antibiotics causes resistant bacterial strains to become widespread.

? List three ways that you, in your everyday life, can contribute to reducing antibiotic-resistant infections.

"We really have to be careful about how we use antibiotics because antibiotic use is the biggest driver of antibiotic resistance," says Lynfield, of the Minnesota Health Department.

Clearly, developing stronger antibiotics isn't the only or the best solution to the problem of resistance, because bacteria will ultimately adapt to those new drugs, too. Perhaps the best way to control resistance, say experts, is to change those practices that enable resistant strains to thrive. It is critical, for example, that any antibiotic prescribed be taken precisely as directed, for the full course of treatment, no matter how much better the patient may be feeling. If bacteria are exposed to antibiotics at low levels or for short durations, the entire population may not be eradicated. The remaining bacteria may be resistant to the antibiotic and proliferate. And anyone taking antibiotics exposes all the bacteria in his or her body to the antibiotics, possibly enabling other drug-resistant bacteria to emerge. These drug-resistant bacteria might then be transmitted to other people.

At the community level, the more antibiotics are used, the more resistance will emerge. For this reason, doctors are heavily discouraged from prescribing antibiotics unnecessarily. Likewise, efforts are being made to crack down on the practice of feeding livestock low levels of antibiotics.

Of course, these measures won't fight the resistant strains that are already circulating. But there are ways to reduce and perhaps prevent infections in this case, too. Careful hygiene, especially hand washing, is one important tool. Another is making sure that frequently used common areas such as locker rooms are routinely disinfected. Perhaps the best way to stop the spread of MRSA would be to develop an effective vaccine that can prevent *S. aureus* infections, something researchers are actively exploring **(INFOGRAPHIC 13.9)**.

"We really have to be careful about how we use antibiotics because antibiotic use is the biggest driver of antibiotic resistance."

—Ruth Lynfield

A vaccine against *Streptococcus pneumoniae* introduced in 2000 caused the rate of infection—and especially the rate of drug-resistant infections—to drop dramatically. Not only did the rate of infection drop in vaccinated children, but other age groups benefited as well because the bacteria were not being transmitted as frequently.

As another example of the impact that vaccines have on infections, Daum points to the bacterium *Haemophilus influenzae,* which once frequently caused pneumonia, meningitis, and other serious diseases in children. Today, children are vaccinated against it. "When I was an intern we used to see 60 to 80 *Haemophilus* infections a month," he says. "Today we see none, it's gone. And MRSA needs to be gone, too." ■

CHAPTER 13 Summary

Driving Question 1 What is MRSA, and how can bacteria resist the effects of antibiotics?

- Populations of bacteria exist nearly everywhere, including on and in our bodies; most are harmless or even beneficial, but some can cause disease.
- Bacteria can resist the effects of antibiotics through mutations or new genes that allow them to degrade the antibiotic, or to block the binding of the antibiotic to its target in the cell.

Driving Question 2 How can populations of bacteria become genetically diverse?

- Populations are groups of individuals of the same species living together in the same geographic area.
- Genetic variation is introduced into a population by the process of mutation, which creates new alleles.
- Bacterial populations can acquire new alleles by both mutation and gene transfer—the exchange of DNA between bacterial cells.

Driving Question 3 What is fitness, in an evolutionary sense?

- Genetic variation in a population gives rise to a corresponding variation of phenotypes in the population.
- Individuals with different phenotypes have differing abilities to survive and reproduce in a particular environment; that is, they have different fitness.
- Individuals with higher fitness in a given environment reproduce and pass on their alleles more frequently than do individuals with lower fitness, resulting in a change in allele frequencies over time.

Driving Question 4 How do populations evolve, and what is the role of antibiotics in the emergence of antibiotic resistance?

- Evolution is defined as a change in allele frequencies in a population over time. Evolution affects *populations*. Individuals do not evolve.
- Natural selection is one cause of evolution. Evolution by natural selection occurs when individuals in a population vary genetically in ways that affect their ability to survive and reproduce in a particular environment. Some alleles become more common, while others become less common as a result.
- Over time, natural selection leads to adaptation: advantageous traits become more common in the population, which as a result becomes better suited, or adapted, to its environment.
- Natural selection can take several forms that shift phenotypes in a population in distinct ways: directional selection, diversifying selection, or stabilizing selection.
- Antibiotic-resistant populations of bacteria emerge by directional selection in the presence of antibiotics.

More to Explore

- The Evolution of Bacteria on a "Mega-Plate" Petri Dish (Harvard Medical School, 2016): www.youtube.com/watch?v=plVk4NVIUh8
- Centers for Disease Control and Prevention, Antibiotic Resistance: www.cdc.gov/drugresistance/
- Herold B. C., et al. (1998). Community-acquired methicillin-resistant *Staphylococcus aureus* in children with no identified predisposing risk. *JAMA* 279(8): 593–598.
- McKenna, M. (2011). *Superbug: The Fatal Menace of MRSA*. New York: Free Press.
- Rock Pocket Mice animation: https://www.biointeractive.org/classroom-resources/simulating-evolution-rock-pocket-mouse-population

CHAPTER 13 Test Your Knowledge

Driving Question 1 What is MRSA, and how can bacteria resist the effects of antibiotics?

By answering the questions below and studying Infographics 13.1–13.3, you should be able to generate an answer for this broader Driving Question.

Know It

1. Can *S. aureus* be present in or on a person who has no evidence of an infection?
- **a.** No; *S. aureus* is associated only with infections.
- **b.** Yes, but only non-MRSA strains are present in the absence of an infection.
- **c.** Yes, but only for very short periods of time (between touching a contaminated surface and washing the hands).
- **d.** Yes; *S. aureus* is a common skin bacterium.
- **e.** Yes; *S. aureus* is a common bacterium found in the bloodstream.

2. The term "MRSA" as it is used today refers to
- **a.** *S. aureus* bacteria that are resistant to many antibiotics.
- **b.** a collection of skin and other infections caused by a type of bacteria.
- **c.** *S. aureus* bacteria that are found only in humans with certain types of skin infections.
- **d.** *S. aureus* bacteria that are normal residents of human skin in the vast majority of the human population.
- **e.** all bacteria that are resistant to antibiotics.

3. What is the difference between *S. aureus* colonization and *S. aureus* infection?

4. MRSA is most likely to be problematic if found
- **a.** on the surface of the skin.
- **b.** in nasal passages.
- **c.** in the bloodstream.
- **d.** on the fingernails.
- **e.** The presence of MRSA in any of those locations indicates a serious infection.

5. In the presence of penicillin:
- **a.** What happens to a penicillin-sensitive strain of *S. aureus*?
- **b.** What happens to a penicillin-resistant strain of *S. aureus*?

6. How do beta-lactam antibiotics kill sensitive bacteria?
- **a.** by attracting water into cells
- **b.** by destabilizing the cell membrane
- **c.** by preventing DNA replication during bacterial reproduction
- **d.** by destabilizing the cell wall
- **e.** all of the above, depending on the specific strain of bacteria

Use It

7. An athlete has a nasty skin infection caused by MRSA. How might this infection have been contracted?

8. For the patient in Question 7, which general kinds of antibiotics would you choose (or avoid) in treating the infection? What other measures would you recommend to prevent spread of MRSA to the athlete's teammates and family? Explain your answer.

9. Why do beta-lactam antibiotics affect sensitive bacterial cells but not eukaryotic cells? (You may want to review cell structure, discussed in Chapter 3, to answer this question.)

10. A sensitive *S. aureus* bacterium acquires a new gene that allows it to resist the effects of beta-lactam antibiotics (that is, the bacterium is now resistant). What might the protein encoded by that gene do?
- **a.** synthesize beta-lactam antibiotics
- **b.** digest beta-lactam antibiotics
- **c.** produce a toxin
- **d.** enhance colonization of human skin
- **e.** enhance entry into the bloodstream

Driving Question 2 How can populations of bacteria become genetically diverse?

By answering the questions below and studying Infographics 13.4 and 13.5, you should be able to generate an answer for this broader Driving Question.

Know It

11. What are the two major mechanisms by which bacterial populations acquire genetic diversity?
- **a.** mutation and meiosis
- **b.** binary fission and evolution by natural selection
- **c.** gene transfer and mutation
- **d.** mutation and binary fission
- **e.** gene transfer and replication

Use It

12. Binary fission is asexual. What does this mean? How could two daughter cells end up with different genomes at the end of one round of binary fission?

Apply Your Knowledge

Interpreting Data

13. A single *S. aureus* cell gets into a wound on your foot. *S. aureus* divides by binary fission approximately once every 30 minutes.

a. Thirty minutes after the initial infection, how many *S. aureus* cells will be present?

b. In 1 hour, how many *S. aureus* cells will be present?

c. In 12 hours, how many *S. aureus* cells will be present? (*Hint:* The general formula is $2^{\text{number of generations}}$; you need to figure out how many generations occurred in 12 hours.)

d. Mutations occur at a rate of 1 per 10^{10} base pairs per generation. *S. aureus* has 2.8×10^6 base pairs in its genome. Therefore, approximately 0.0028 mutations will occur per cell in the population. At the end of 12 hours, how many mutations will be present in the population of *S. aureus* in the wound in your foot? What are the implications of this for treating a possible infection?

14. If a single bacterial cell that is sensitive to an antibiotic—for example, vancomycin—is placed in a growth medium that contains vancomycin, it will die. Now consider another single bacterial cell, also sensitive to vancomycin, that is allowed to divide for many generations to become a larger population. If this population is placed into a vancomycin-containing growth medium, some bacteria will grow. Why do you see growth in this case, but not with the transferred single cell?

Driving Question 3 What is fitness, in an evolutionary sense?

By answering the questions below and studying Infographic 13.6 you should be able to generate an answer for this broader Driving Question.

Know It

15. What is the evolutionary meaning of the term "fitness"?

16. A strain of MRSA is resistant to penicillin and vancomycin. For each antibiotic-containing environment below, predict whether its fitness will be high or low.

a. penicillin

b. ciprofloxacin

c. vancomycin

Use It

17. If we take the fittest bacterium from one environment—one in which the antibiotic amoxicillin is abundant, for example—and place it in an environment in which a different antibiotic is abundant, will it retain its high degree of fitness?

a. Yes; fitness is fitness, regardless of the environment.

b. Yes; once a bacterium is resistant to one antibiotic, it is resistant to all antibiotics.

c. Not necessarily; fitness depends on the ability of an organism to survive and reproduce, and it may not do this as well in a different environment.

d. No; what is fit in one environment will never be fit in another environment.

Driving Question 4 How do populations evolve, and what is the role of antibiotics in the emergence of antibiotic resistance?

By answering the questions below and studying Infographics 13.7–13.9 you should be able to generate an answer for this broader Driving Question.

Know It

18. What is the environmental pressure in the case of antibiotic resistance?

a. the growth rate of the bacteria

b. how strong or weak the bacterial cell walls are

c. the relative fitness of different bacteria

d. the presence or absence of antibiotics in the environment

e. the temperature of the environment

Nick D. Kim/CartoonStock

It was on a short-cut through the hospital kitchens that Albert was first approached by a member of the Antibiotic Resistance.

19. The evolution of antibiotic resistance is an example of

a. directional selection.
b. diversifying selection.
c. stabilizing selection.
d. random selection.
e. steady selection.

20. In humans, very-large-birth-weight babies and very tiny babies do not survive as well as midrange babies. What kind of selection is acting on human birth weight?

a. directional selection
b. diversifying selection
c. stabilizing selection
d. random selection
e. steady selection

Use It

21. In what sense do bacteria "evolve faster" than other species?

22. Imagine that a genetically diverse population of garden snails occupies your backyard, in which the vegetation is many shades of green with some brown patches of dry grass.

a. If birds like to eat snails, but they can see only the snails that stand out from their background and don't blend in, what do you think the population of snails in your backyard will look like after a period of time? Explain your answer.

b. Suppose you move the population of snails to a new environment, one with patches of dark brown pebbles and patches of yellow ground cover. Will individual snails mutate to change their color immediately? As the population evolves and adapts to the new environment, what do you predict will happen to the phenotypes in your population of snails after several generations in this new environment? How did this occur? Include the terms *gametes, mutation, fitness, phenotype,* and *environmental selective pressure* in your answer.

Mini Case

Apply Your Knowledge

23. Your friend has had a virus-caused cold for 3 days and is still so stuffy and hoarse that he is hard to understand. He seems to be telling you that his doctor called in a prescription for an antibiotic for him to pick up at his pharmacy. You hope that you misunderstood him, but you realize that you heard him perfectly well.

a. Will the antibiotic help your friend's cold?
b. What are the risks to your friend if he takes the antibiotic? (Think about what might happen if he should develop a wound infection in the future.)
c. Your friend is a wrestler. What are the risks to his teammates or competitors if he takes the antibiotic?

Bring It Home

Apply Your Knowledge

24. Your roommate has been prescribed an antibiotic for bacterial pneumonia. She is feeling better and stops taking her antibiotic before finishing the prescribed dose, telling you that she will save the remainder to take the next time she becomes sick. What can you tell your roommate to convince her that this is not a good plan?

MILESTONES IN BIOLOGY **4**

Adventures in Evolution

Charles Darwin and Alfred Russel Wallace on the trail of natural selection

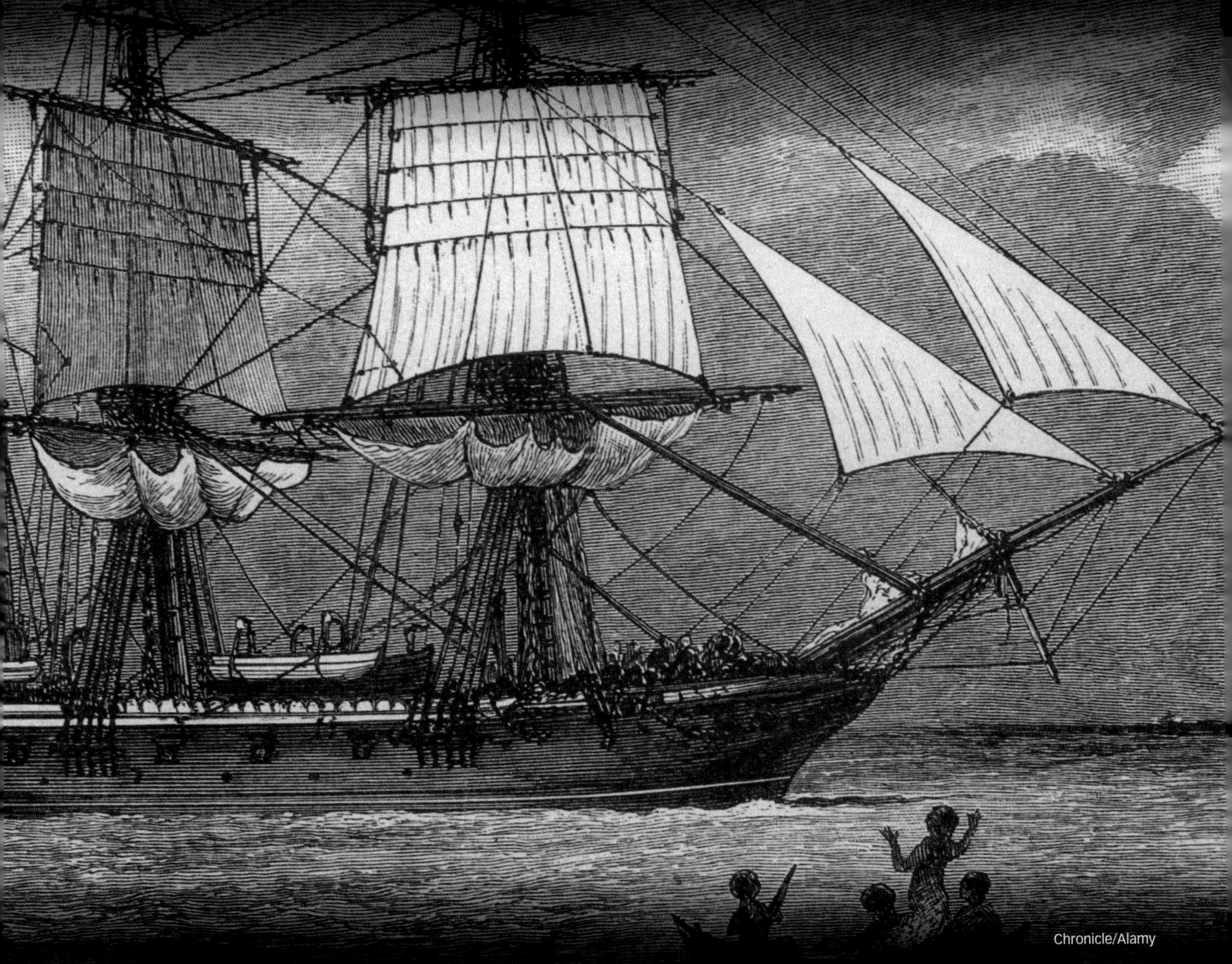

Chronicle/Alamy

DRIVING QUESTIONS

1. What observations did Darwin make about nature that helped shape his thinking about evolution?
2. What works by other scientists shaped Darwin's thoughts about evolution?

During his last term at Cambridge University, Charles Darwin faced a dilemma: what should he do with himself after graduation? He'd considered becoming a physician, like his father. But the sight of blood made him queasy and he hated rote memorization. He changed his focus to theology, intending to become a clergyman, but his real passion was bug collecting. Only that wasn't going to pay the bills.

Then a professor told him about the internship of a lifetime: a 5-year, around-the-world trip as a naturalist aboard a British surveying ship, the HMS *Beagle*. The ship's captain, Robert FitzRoy, wanted a travel companion who would also collect specimens along the way. Unsure what he wanted to do with his life but eager to see the world, the 22-year-old Darwin jumped at the chance. He later said of the trip, "The voyage of the *Beagle* has been by far the most important event in my life and has determined my whole career."

> *"The voyage of the* Beagle *has been by far the most important event in my life and has determined my whole career."*
>
> **—Charles Darwin**

Yet he almost didn't go. Darwin's father, Robert Darwin, thought his son should buckle down and prepare to enter the clergy. A trip around the world seemed to him a useless distraction—a "wild scheme," he called it—and he refused at first to let his son go. But eventually, at the cajoling of his family, he relented. Charles packed his bags, said goodbye to his girlfriend Emma, and set sail for South America. It was December 1831.

The passage aboard the 90-foot vessel was frequently harrowing, and Darwin suffered debilitating bouts of seasickness, but his journey aboard the *Beagle* set in motion one of the great revolutions in science. What he saw on that trip planted the seeds of ideas that have completely changed the way we view the world and our place in it. As the evolutionary biologist Stephen Jay Gould put it, "The world has been different ever since Darwin."

Journey to an Idea

▸ Evolution before Darwin

Though Darwin is the most famous figure associated with the theory of evolution, he did not invent the idea. Nor was he alone among his contemporaries in studying it. In fact, the notion that species change gradually over time had been around for generations. To be sure, most people in the 1830s—Darwin included—still assumed that species were fixed and unchanging, created perfectly by God. But evidence to the contrary had been accumulating for some time. Explorers and naturalists were traveling to faraway lands and finding unusual plants and animals they had never seen before. Fossils were being uncovered, providing evidence that some species no longer seen on Earth had lived in the past. And anatomists were noting uncanny physical resemblances between different species, including chimpanzees and humans. Evolution was already in the air when Darwin began thinking about it.

However, the ideas that people in Darwin's time had proposed to explain *how* species changed were flawed. One common misconception was Lamarckism, named after the French naturalist Jean-Baptiste Lamarck, who suggested that species could change through the inheritance of acquired characteristics. In the Lamarckian view, giraffes, for example, developed their long necks by continually stretching them to feed on tall trees. Once it acquired its long neck, a giraffe could then pass that advantageous trait on to its offspring **(INFOGRAPHIC M4.1)**.

"The world has been different ever since Darwin."
—Stephen Jay Gould

This idea of the inheritance of acquired characteristics, while incorrect, was a popular one in Darwin's time—one that even Darwin himself found it hard to fully shake off in his writings. (Mendel's work on pea plants, which helped establish our modern view of inheritance, was not rediscovered until 1900; see **Milestone 3: Mendel's Garden.**)

The *Beagle* and Beyond

▶ Darwin's intellectual journey

Though it would be years before Darwin proposed his own theory of evolution, his trip around the world provided him with an indispensable foundation. He kept a diary of his adventures, which was later published as *The Voyage of the Beagle* (1839).

While at sea, Darwin had plenty of time to read and think about the ideas then being discussed in scientific circles. He read, for instance, the work of Charles Lyell, whose *Principles of Geology* (1830–1833) argued that Earth was much older than the 6,000 years popularly accepted at the time (a figure

INFOGRAPHIC M4.1

Lamarckism: An Early Idea about Evolution

Lamarck believed that traits acquired in one's lifetime could be passed on to offspring. We now know this to be wrong.

? According to the Lamarck model, if you lift a lot of weights and develop a very muscular physique, would you expect your children to be equally "buff"?

INFOGRAPHIC M4.2

Darwin's Voyage on the *Beagle*

Darwin's 5-year-long trip around the globe aboard the HMS *Beagle* laid the foundation for his thinking about evolution. The plants, animals, fossils, and geological forces he encountered along the way led him to question the conventional wisdom of his day.

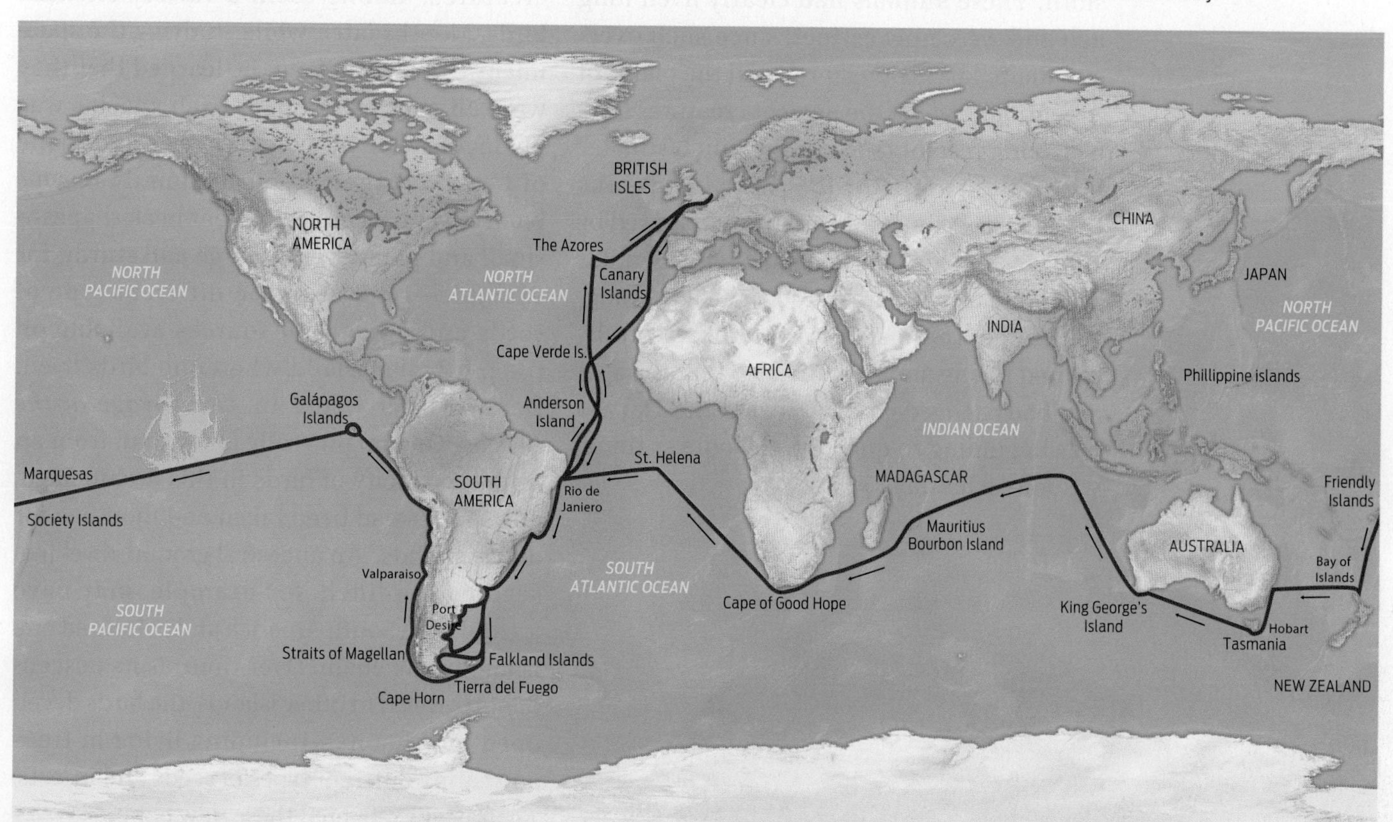

? **Which parts of the globe were not part of Darwin's voyage?**

based on a literal reading of the Bible), and that its geology had been shaped entirely by incremental forces operating over a vast expanse of time. Valleys, for example, were formed by the slow grinding forces of wind and water, not by catastrophic floods; mountains were pushed up gradually by the action of volcanoes and earthquakes. Lyell's view of incremental change producing dramatic results over great spans of time left an indelible impression on Darwin.

Not long into his trip, he saw something that seemed to confirm Lyell's view. While docked at the Cape Verde islands, 300 miles off the west coast of Africa, Darwin noticed a white layer of compressed seashells and corals embedded in a rock face 30 feet above sea level. This layer had clearly been formed in the sea, but had somehow been lifted up out of the water. Later on his trip, Darwin witnessed how this might happen. After an earthquake devastated the town of Concepción, Chile, he noticed that the coastline had been pushed up several feet, as shown by the rim of mollusks and barnacles that hovered out of the water around the bay. Lyell was right.

With such thoughts of an ancient, slowly changing Earth on his mind, Darwin studied the plants, animals, and geology at each stop on his trip, collecting fossils and specimens of local flora and fauna wherever he went **(INFOGRAPHIC M4.2)**.

While exploring the shore of Argentina in August 1833, Darwin unearthed a particularly prized find: the fossilized remains of several large mammals embedded in a sea cliff, including one that looked like a giant armadillo and another that resembled a giant sloth. These animals had clearly lived long ago and were now extinct, since such oversize mammals no longer roamed the plains of Argentina. And yet the ancient creatures bore a striking resemblance to the much smaller armadillos and sloths that were indigenous to the area. Why should animals separated by such vast epochs of time share such similar anatomical structures? To Darwin, this suggested that these creatures were ancestrally related to one another, and also that the species had changed gradually over time. Darwin was beginning to question the conventional wisdom of his day.

The Large Ground Finch (*Geospiza magnirostris*), one of several finch species that Darwin studied, has a sturdy bill for cracking large seeds.

In 1835, the young naturalist stepped ashore on the Galápagos Islands, off the coast of Ecuador. On this archipelago, or island chain, Darwin observed and collected many creatures, among them a variety of small birds. Months later, while studying the specimens back in England, he learned that they were all species of finch. Each species was distinguishable by a different size and shape of beak, but they all bore a family resemblance. Moreover, the different beak shapes—small and pointy versus large and sturdy, for example—related to the different type of seeds and other food sources available on the particular island where the birds lived. Darwin later wrote in *The Voyage of the Beagle,* "One might really fancy that, from an original paucity of birds in this archipelago, one species had been taken and modified for different ends." An ancestral ground-dwelling, seed-eating finch, for example, may have arrived from South America and colonized one island in the chain. Over time, as its descendants traveled to other islands, the birds developed new habits—including living in trees and eating different types of seeds and insects. Their beaks reflected these new habits.

This notion of one species giving rise over time to new species Darwin came to call "descent with modification," which he represented in his notebooks by a diagram that looked like a branching tree. (Darwin himself didn't like using the term "evolution" because he thought it gave a mistaken idea of progress or direction toward a goal, but "evolution" is the term that stuck.)

After 5 years traveling the globe, Darwin returned home to England in October 1836, but his intellectual journey was only just beginning. He began to think more seriously about how species might change over time. A key insight came to him in September 1838 while reading the work of the political economist Thomas Malthus, whose pessimistic book *An Essay on the Principle of Population* (1798) described how hunger, starvation, and disease would ultimately limit human population growth. The same must be true of plant and animal species, Darwin realized. If every individual in a population reproduced, even in a slowly reproducing population such as elephants, the world would be completely overrun with elephants in not that many generations. Since Earth is not overrun with elephants, factors must be limiting their population growth. Such limitations, Darwin reasoned, would lead to competition for resources that would put weaker individuals at a disadvantage. "It at once struck me," Darwin later wrote in his autobiography "that under these circumstances favourable variations would tend to be preserved, and unfavourable ones to be destroyed. The result of this would be the formation of new species."

These favorable variations needn't be very pronounced, he realized. All that was needed was for certain individuals to have a slight edge over others in the competition to survive and reproduce. These helpful variations would then be inherited by their offspring and become more common in the population. In effect, the environment was "selecting" for favorable traits, much as plant and animal breeders selected and perpetuated desirable varietals—a plant with especially large fruit, for instance, or the many breeds of dogs we see today.

This idea of "natural selection" (see Chapter 13) was Darwin's original contribution to the theory of evolution. Others had speculated at length about species change, but Darwin was the first to provide a clear mechanism of evolution. The philosopher of science Daniel Dennett has called natural selection "the single best idea anyone has ever had."

Among other things, natural selection provided a powerful explanation for the apparent design seen in nature. Before Darwin, the conventional view was that provided by William Paley, whose book *Natural Theology* (1802) Darwin had read in college. In that work, Paley made his famous "argument from design." That is, many creatures are so finely constructed and well adapted to their environment that their existence implies the existence of a designer—much in the way that the existence of a watch implies the existence of a watchmaker. To many people in the 19th century, that designer was God. With the idea of natural selection, Darwin did away with the need for a watchmaker: natural selection could accomplish the same thing without any input from a designer.

"It at once struck me that under these circumstances favourable variations would tend to be preserved, and unfavourable ones to be destroyed."

—Charles Darwin

By 1844, Darwin had developed his ideas into a 200-page manuscript that he hoped would be the definitive word on the subject. He did not rush his theory about natural selection into print, however. He knew that his ideas would be controversial, as they contradicted strongly held beliefs about God and the special creation of all animals, including humans. Other scientists with evolutionary ideas were causing quite a stir in England and being openly ridiculed. Even sharing his theory of evolution by natural selection with trusted colleagues, Darwin said, was "like confessing a murder." To withstand challenges, he knew he would need more detailed evidence.

And so, at age 37, Darwin began to investigate closely one large group of animals: barnacles, the small invertebrates that cling to ships or marine life. Darwin spent 8 years, from 1846 to 1854, carefully cataloguing the barnacles' tiny features, comparing and contrasting them with those of other known invertebrates. It was tedious work, leading Darwin to write, "I hate a Barnacle as no man ever did before." Yet the work proved valuable, for it reinforced his idea that a great deal of variation exists in nature—barnacles are nothing if not diverse—and it provided ample evidence of descent with modification since the creatures clearly shared adaptations with other invertebrates **(INFOGRAPHIC M4.3).**

In the summer of 1858, Darwin was hard at work on his "species" book when he received a letter from a young naturalist with whom he had a casual acquaintance. Alfred Russel Wallace was a collector who made a living selling rare butterflies and birds to other collectors and museums. Wallace's envelope was postmarked from an island in Indonesia. Inside was a 20-page manuscript describing the author's bold new idea about how species change over time, which he wanted Darwin to read and have published. Darwin had been scooped.

INFOGRAPHIC M4.3

The Evolution of Darwin's Thought

Darwin was influenced by the work of others, which informed the way he interpreted his own research and collections.

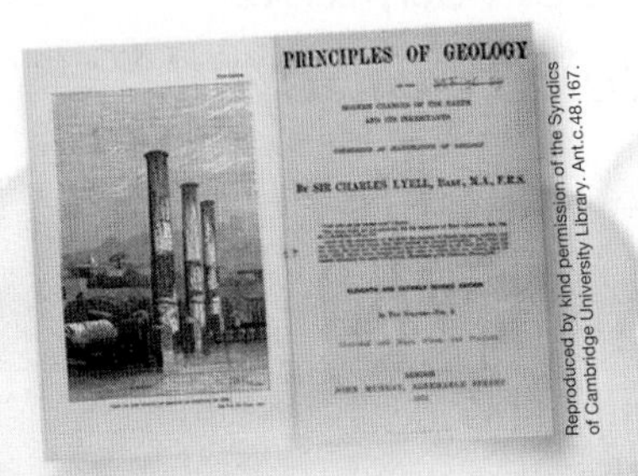

Lyell's work (1833)

- Earth's geology is formed by slow-moving forces.
- Earth is much older than thought at the time.

Fossil of giant land sloth (1833)

- It was discovered buried in a sediment layer below a deposit of shells.
- The large creature was clearly extinct, but it resembled smaller, living South American animals. Could they be related?

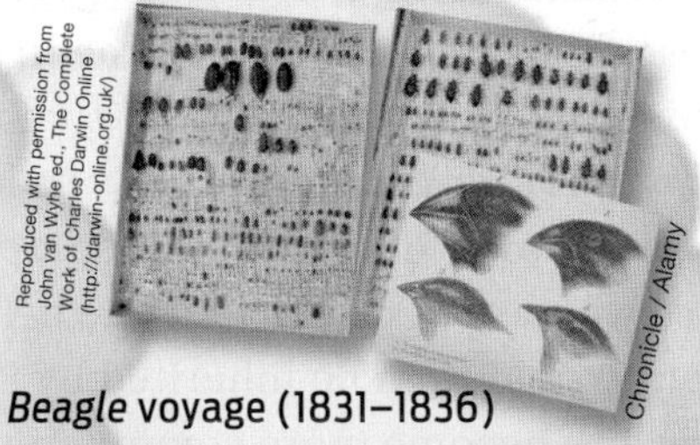

***Beagle* voyage (1831–1836)**

- Darwin collected plants, animals, and fossils from across the globe.
- Darwin observed similarities and differences and attempted to explain these characteristics.

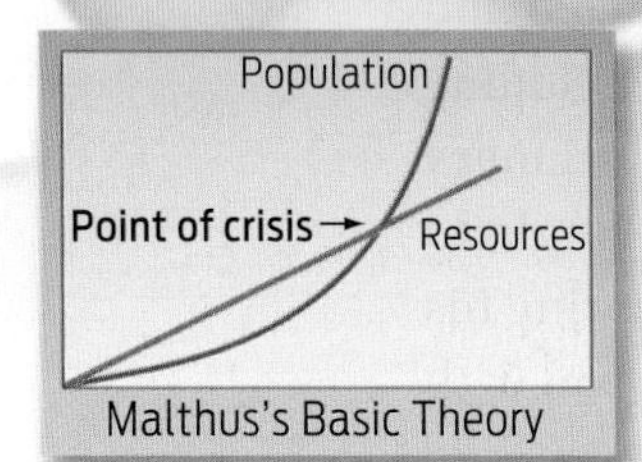

Malthus's work (1798)

- Populations are limited by a number of factors, including food, water, and disease.
- Some individuals die and some survive.

Barnacle research (1846–1854)

- Darwin observed the many differences among barnacles, reinforcing for him the idea that much variation exists in nature.

Charles Darwin (February 12, 1809–April 19, 1882)

? How many years of work influenced Darwin's thoughts on evolution?

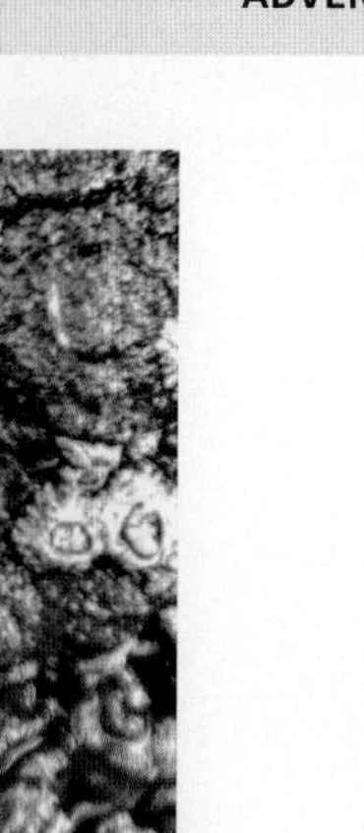

Danita Delimont/Getty Images

Barnacles are invertebrate animals that live attached to surfaces such as rocks near ocean shores. These plentiful and diverse creatures helped Darwin understand the importance of variation in nature.

In Darwin's Shadow

▶ Alfred Russel Wallace

Although we often credit Darwin with the discovery of natural selection, he was not alone in charting this intellectual territory. Another British naturalist was also hot on the trail. Like Darwin, Wallace was fascinated by natural history and had a thirst for adventure. In other ways, though, the two men couldn't have been more different. Darwin came from a wealthy family and had received a prestigious Cambridge education. He was greeted as a minor celebrity when he returned from his trip around the world and was accepted into the scientific establishment. Wallace, by contrast, was a man of more humble origins, for whom nothing in life had come easily.

The eighth of nine children, Wallace could not afford a university education. He attended night school and supported himself as a builder and railroad surveyor. His budding fascination with natural history, though, led him to read widely. Like Darwin, he read Lyell's work on geology and Malthus's work on human population. He also devoured Darwin's recently published travel account, *The Voyage of the Beagle*.

In 1848, having scrimped and saved, the 25-year-old Wallace set sail for Brazil, to the mouth of the Amazon River. There he hoped to earn his reputation as a respectable scientist by understanding the origin of species. Exploring the rain forest of the Amazon, Wallace was struck by the distribution of distinct yet similar-looking (what he called "closely allied") species, which were often separated by a geographic barrier such as a canyon or river. For example, he noted that different species of sloth monkey were found on different banks of the Amazon River. Over the course of his 4-year trip, Wallace scoured the Amazon and collected thousands of specimens.

Wallace was on his way home to London with his specimens in 1852 when disaster struck: his ship caught fire and sank. Wallace survived, but he lost everything—his notes, sketches, journals, and all his specimens. In spite of this catastrophe, Wallace was undeterred. Less than 2 years later, he was off on another collecting expedition, this time to the Malay archipelago (what is now Singapore, Malaysia, and Indonesia).

Wallace's first paper, "On the Law Which Has Regulated the Introduction of New Species," was published in September 1855. Based on his island work, it focused on the similar geographical distribution of "closely allied" species. For example, "the Galápagos Islands," he wrote, "contain little groups of plants and animals peculiar to themselves, but most nearly allied to those of South America." From these observations, Wallace deduced this law, as he called it: "Every species has

INFOGRAPHIC M4.4

The Evolution of Wallace's Thought

Like Darwin, Wallace was influenced by the writings and work of others, which shaped his interpretations of his own observations and research.

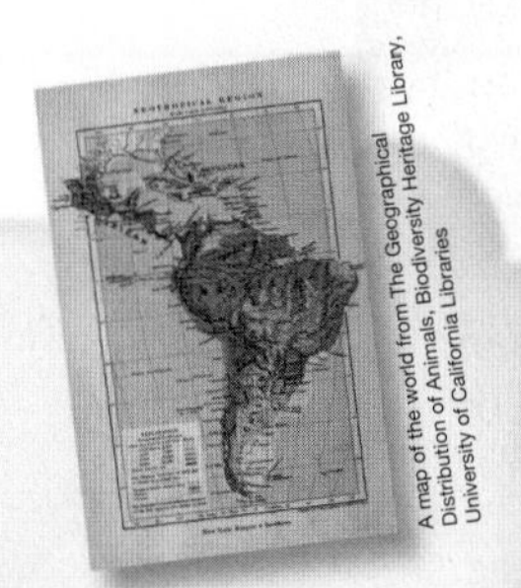

A map of the world from The Geographical Distribution of Animals, Biodiversity Heritage Library, University of California Libraries

Amazon trip (1848–1852)

- Wallace observed that related (or "closely allied") species occupied neighboring geographic areas.
- He noted the role of physical barriers (such as the Amazon River) in separating related species from one another.

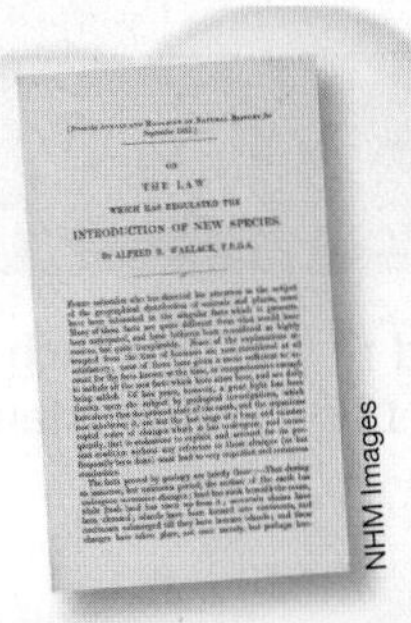

NHM Images

First publication (1855)

- Wallace's publication dealt with the physical distribution of species.
- He introduced the idea that new species are temporally and spatially connected to a related species.

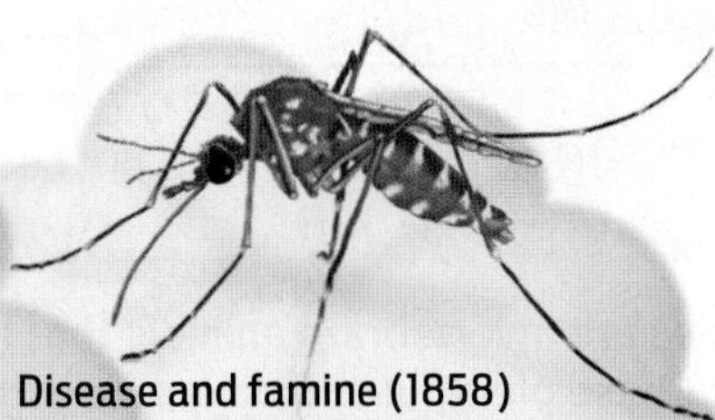

Disease and famine (1858)

- While suffering from malaria, Wallace pondered the role of disease and famine in keeping human populations in check.
- He wondered how these factors could apply to the evolution of animal species.

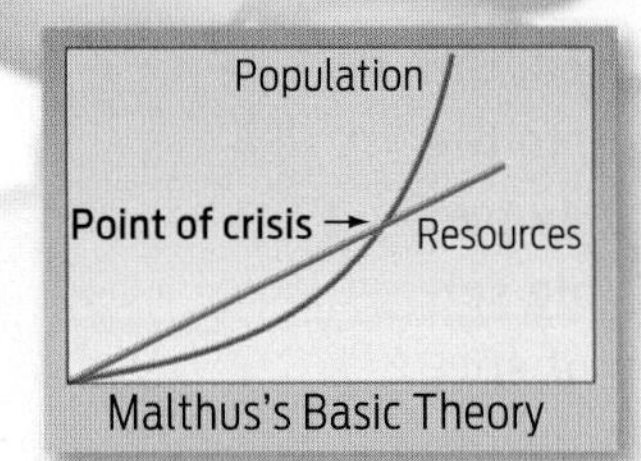

Malthus's work (1798)

- Populations are limited by a number of factors, including food, water, and disease.
- Some individuals die and some survive.

London Stereoscopic Company/Getty Images

Alfred Russel Wallace (January 8, 1823–November 7, 1913)

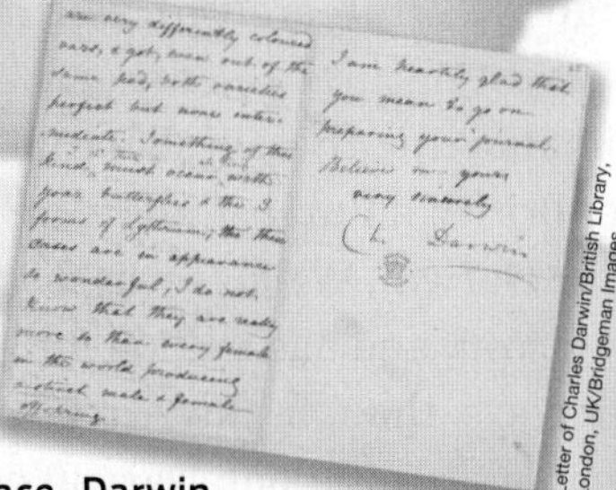

Letter of Charles Darwin/British Library, London, UK/Bridgeman Images

Wallace–Darwin correspondence (185?–1858)

- Wallace began corresponding with Darwin several years before sending Darwin his completed manuscript.
- The two men were clearly developing very similar ideas about the nature of evolutionary change.

? **Both Darwin and Wallace made long trips that influenced their thinking. When and where did each travel?**

come into existence coincident both in space and time with a pre-existing closely allied species." Wallace's article was groundbreaking, foreshadowing Darwin in a number of ways, but it lacked an explanation—a mechanism—of exactly how one species might have evolved from another.

Wallace continued his travels, but in early 1858 disaster struck again: he contracted malaria. Confined to bed, he let his mind wander. He thought about what Malthus had written about disease and how it kept human populations in check. How might these forces of disease and death, multiplied over time, influence animal populations, he wondered? Then came the flash of insight—like "friction upon the specially-prepared match," he recalled. In every generation, weaker individuals will die while those with the fittest variations will survive and reproduce; as a result, species will change and adapt to their surroundings, eventually forming new species. Wallace had worked out the mechanism for evolution that was missing from his earlier work. He quickly wrote down his idea and sent it to the one naturalist he thought might be able to appreciate it. This was the 20-page manuscript that arrived on Darwin's doorstep on June 18, 1858 **(INFOGRAPHIC M4.4)**.

How might these forces of disease and death, multiplied over time, influence animal populations, Wallace wondered?

Darwin was stunned. For 20 years he had been working diligently on the same idea and now it seemed someone else might get credit for it. "All my originality will be smashed," he wailed to his friend Lyell. Recognizing Darwin's predicament, Lyell and other colleagues devised a plan that would clearly establish Darwin's intellectual precedence. They would arrange to have papers by both men presented at a meeting of the Linnaean Society in London. The meeting took place on

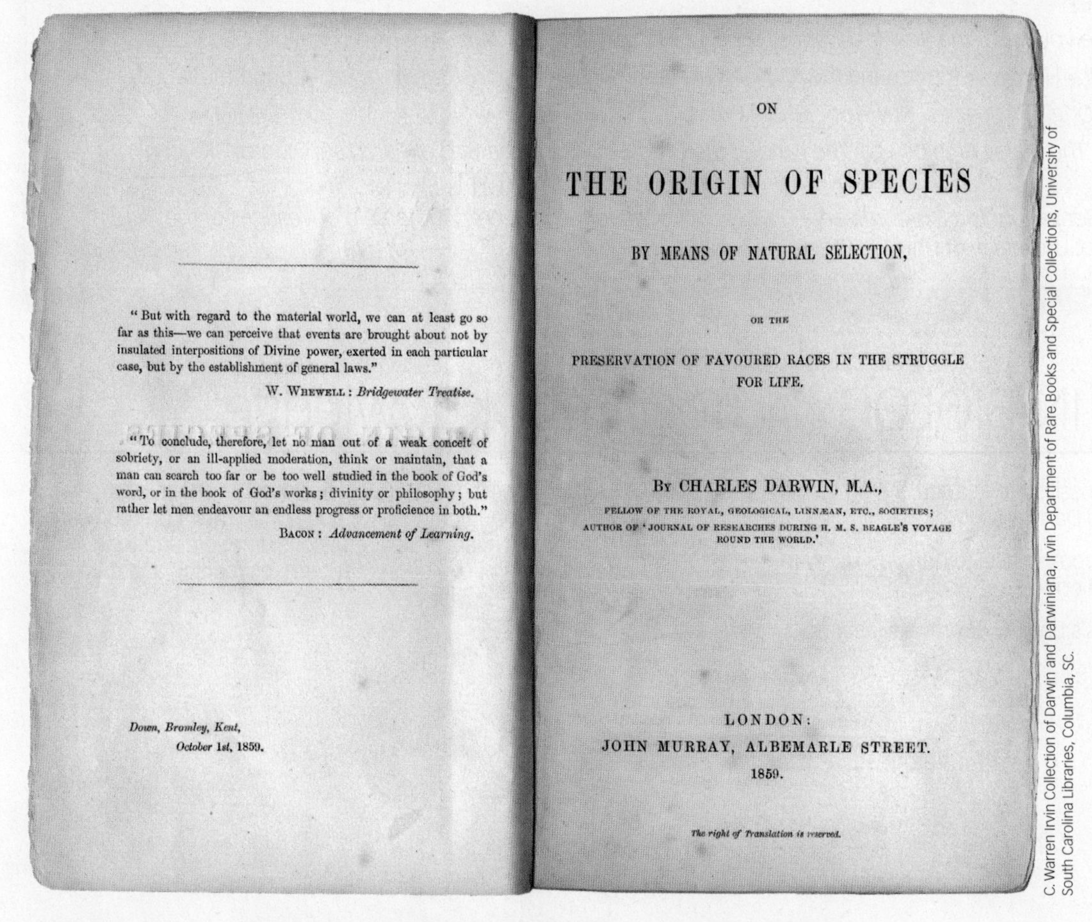

"But with regard to the material world, we can at least go so far as this—we can perceive that events are brought about not by insulated interpositions of Divine power, exerted in each particular case, but by the establishment of general laws."

W. WHEWELL: *Bridgewater Treatise.*

"To conclude, therefore, let no man out of a weak conceit of sobriety, or an ill-applied moderation, think or maintain, that a man can search too far or be too well studied in the book of God's word, or in the book of God's works; divinity or philosophy; but rather let men endeavour an endless progress or proficience in both."

BACON: *Advancement of Learning.*

Down, Bromley, Kent,
October 1st, 1859.

ON

THE ORIGIN OF SPECIES

BY MEANS OF NATURAL SELECTION,

OR THE

PRESERVATION OF FAVOURED RACES IN THE STRUGGLE FOR LIFE.

BY CHARLES DARWIN, M.A.,

FELLOW OF THE ROYAL, GEOLOGICAL, LINNÆAN, ETC., SOCIETIES;
AUTHOR OF 'JOURNAL OF RESEARCHES DURING H. M. S. BEAGLE'S VOYAGE ROUND THE WORLD.'

LONDON:
JOHN MURRAY, ALBEMARLE STREET.
1859.

The right of Translation is reserved.

C. Warren Irvin Collection of Darwin and Darwiniana, Irvin Department of Rare Books and Special Collections, University of South Carolina Libraries, Columbia, SC.

Charles Darwin's landmark publication on evolution by natural selection.

July 1, 1858. The papers were dutifully read, but there was no discussion or fanfare. In fact, neither of the authors was even present. Wallace was still traveling in Malaysia, and Darwin was mourning the recent death of his young son and too distraught to attend.

The scientific meeting secured Darwin's reputation, but still he was unsettled. Wallace's communication had lit a fire under his feet. He needed to finish his book. That work, *On the Origin of Species by Means of Natural Selection,* was published in November 1859. It would become one of the most famous books of all time, going through six editions by 1872.

Although it may seem that Wallace was cheated of his rightful recognition as a discoverer of evolution by natural selection, he was never bitter. On the contrary, he was delighted when he heard about his copublication with Darwin. He fully accepted that Darwin had formulated a more complete theory of natural selection before he did, and there is no trace of resentment in his later writings. In fact, Wallace titled his major work *Darwinism,* in recognition of the other man's intellectual influence.

After the presentation of 1858, Wallace stayed in the Malay archipelago for 4 more years, systematically recording its fauna and flora and securing his reputation as both the greatest living authority on the region and an expert on speciation. In fact, Wallace is responsible for our modern-day definition of "species." In his work on butterflies, he defined "species" as groups of individuals capable of interbreeding with other members of the group but not with individuals from outside the group. This idea—known today as the biological species concept (see Chapter 14)—remains one of the most important in evolutionary theory. ■

More to Explore

- Darwin, C. (1989 [1839]). *The Voyage of the Beagle: Charles Darwin's Journal of Researches.* New York: Penguin Classics.
- American Museum of Natural History, A Trip Around the World: www.amnh.org/exhibitions/darwin/a-trip-around-the-world/
- Browne, J. (2008). *Darwin's Origin of Species.* New York: Grove Press.
- Wallace, A. R. (2016 [1869]). *The Malay Archipelago: The Land of the Orang-Utan and the Bird of Paradise.* Oxford, UK: John Beaufoy Publishing.
- Secord, J. (2003). *Victorian Sensation: The Extraordinary Publication, Reception, and Secret Authorship of Vestiges of the Natural History of Creation.* Chicago: University of Chicago Press.

MILESTONES IN BIOLOGY 4 Test Your Knowledge

1. What did the discovery of a fossil sloth in a sea cliff on the coast of Argentina suggest to Darwin?
2. Why was Thomas Malthus's book critical to Darwin's thinking about descent with modification?
3. How did Wallace use Malthus's book to inform his ideas about species?
4. What did Darwin's and Wallace's field experiences in observing the natural world first hand, in addition to their communication with other scientists and careful consideration, add to their understanding of evolution that perhaps reading and thinking alone couldn't provide?

14 Nonadaptive Evolution and Speciation

Urban EVOLUTION

How cities are altering the fate of species

Bill Draker/imageBROKER/AGE Fotostock

Tarheelson/iStock/Getty Images

Tetra Images/Getty Images

DRIVING QUESTIONS

1. What is a gene pool?
2. How do different evolutionary mechanisms influence the composition of a gene pool?
3. How does the gene pool of an evolving population compare to the gene pool of a nonevolving population?
4. How do new species arise, and how can we recognize them?

Light rain is falling as biologists Jason Munshi-South and Stephen Harris traipse through the underbrush of Highbridge Park in the Washington Heights section of Manhattan, some 8 miles north of Times Square. They brush back a swatch of green to reveal a small, shoebox-size trap, with one unhappy camper inside: a tiny white-footed mouse. This diminutive rodent, only about 2 inches long, is one of a few urban species that scientists have begun to look at more closely for answers to questions about evolution.

Studying evolution in Manhattan—arguably the most unnatural place on the planet—might seem an odd choice. But Munshi-South and Harris belong to a new breed of biologist, one fascinated by the nature right under our noses. From rodents in parks to ants on median strips to the cockroaches and bedbugs inside buildings—even the bacteria brewing in our belly buttons—no location is too mundane for this scientific crew.

What's the advantage of staying local? Besides being quite convenient—Munshi-South works at Fordham University just north of

CUNY/PHOTO

Biologist Jason Munshi-South laying a trap for white-footed mice.

Manhattan—cities like New York offer some unique opportunities for the evolutionary biologist.

"There isn't really anywhere on the planet that isn't impacted by human activity," says Munshi-South. "If we want to understand how species are actually evolving now, we need to understand the inputs from human activity."

Nowhere is that input more evident than in the Big Apple. With a population of 8 million people packed into just 300 square miles, New York City is the largest and most densely populated city in the United States. All those restless urbanites have left a profound mark on the landscape, changing it in ways both dramatic and subtle. Just look at Manhattan, New York City's most densely populated borough: once an island of thick forest, Manhattan is now a sliver of concrete interspersed with oases of green. The skyscrapers, the bridges, the subway system, and the bars and coffee shops have all transformed a once wild place beyond recognition. In the process, says Munshi-South, the city's wildlife has been subject to "a grand evolutionary experiment."

Sex and the City

▶ Population genetics

To study evolution, Munshi-South has traveled to some pretty far-flung locations. For example, he's ventured to Southeast Asia, to study how the mating behavior of small mammals is affected by logging operations, and Africa, to research how the migration patterns and stress levels of forest elephants are affected by the petroleum operations there. Mice in Manhattan might seem a far cry from that earlier research, but it's really not, he says. The common thread is understanding how animals cope with a rapidly changing environment caused by human activity.

Urbanization isn't the only human-caused environmental change that animals face, of course—climate change is another big one. Nevertheless, urbanization has assumed a pressing urgency in recent years.

"Already 50% of people live in cities," says Munshi-South. "It's going to be 60% in 15 years, and it's just going to keep going up and up." In fact, demographers predict that

"There isn't really anywhere on the planet that isn't impacted by human activity."

—Jason Munshi-South

the human population will reach 9 billion in 2050, and that by then 70% of us—roughly 6 billion people—will live in cities. All those urban dwellers represent a significant evolutionary force to be reckoned with. So it's important to understand how our fellow animals are adapting—or not adapting—to city life.

If you're an evolutionary biologist who wants to understand how a group of organisms is coping with environmental changes, you need to know something about those organisms' underlying genetics—not just the genetics of individuals, but the genetics of the population as a whole. For that, you need the tools of **population genetics,** which allows scientists to understand the genetic makeup of a population and how that makeup changes over time. Essentially, population genetics is a way to take stock of who's reproducing and who isn't, and the consequences for the population as a whole.

From a population genetics perspective, each distinct population of organisms—whether mice in Manhattan or elephants in Africa—has its own particular collection of alleles, which together constitute its **gene pool.** Within the gene pool, each allele is present in a certain proportion, or

POPULATION GENETICS
The study of the genetic makeup of populations and how the genetic composition of a population changes.

GENE POOL
The total collection of alleles in a population.

ALLELE FREQUENCY
The relative proportion of an allele in a population.

allele frequency, relative to the total number of alleles for that gene in the population. For example, if a particular allele for a gene is present 50 times out of a total of 1,000 alleles, its allele frequency is 0.05. Over time, various forces can change the frequency of alleles—that is, how common they are in the population. When the frequency of alleles changes over time, a population evolves. Recall from Chapter 13 that the definition of evolution is a change in the allele frequencies of a population over time **(INFOGRAPHIC 14.1)**.

Evolutionary changes in a gene pool can have lasting consequences for a population. For example, they can result in the population becoming more adapted to its environment—think of the antibiotic-resistant bacteria we met in Chapter 13. The evolutionary mechanism that results in adaptation is natural selection.

But natural selection isn't the only mechanism of evolution. There are three others: mutation, genetic drift, and gene flow. We've already encountered mutation in previous chapters. Mutation is what introduces new alleles into a population and thus provides the genetic variation on which natural selection acts. Because it is random, however, mutation does not automatically lead to a

INFOGRAPHIC 14.1

Population Genetics: Studying How Allele Frequencies Change

Population geneticists study gene pools—the total collection of alleles in a population. If allele frequencies in a gene pool change over time, then evolution has occurred.

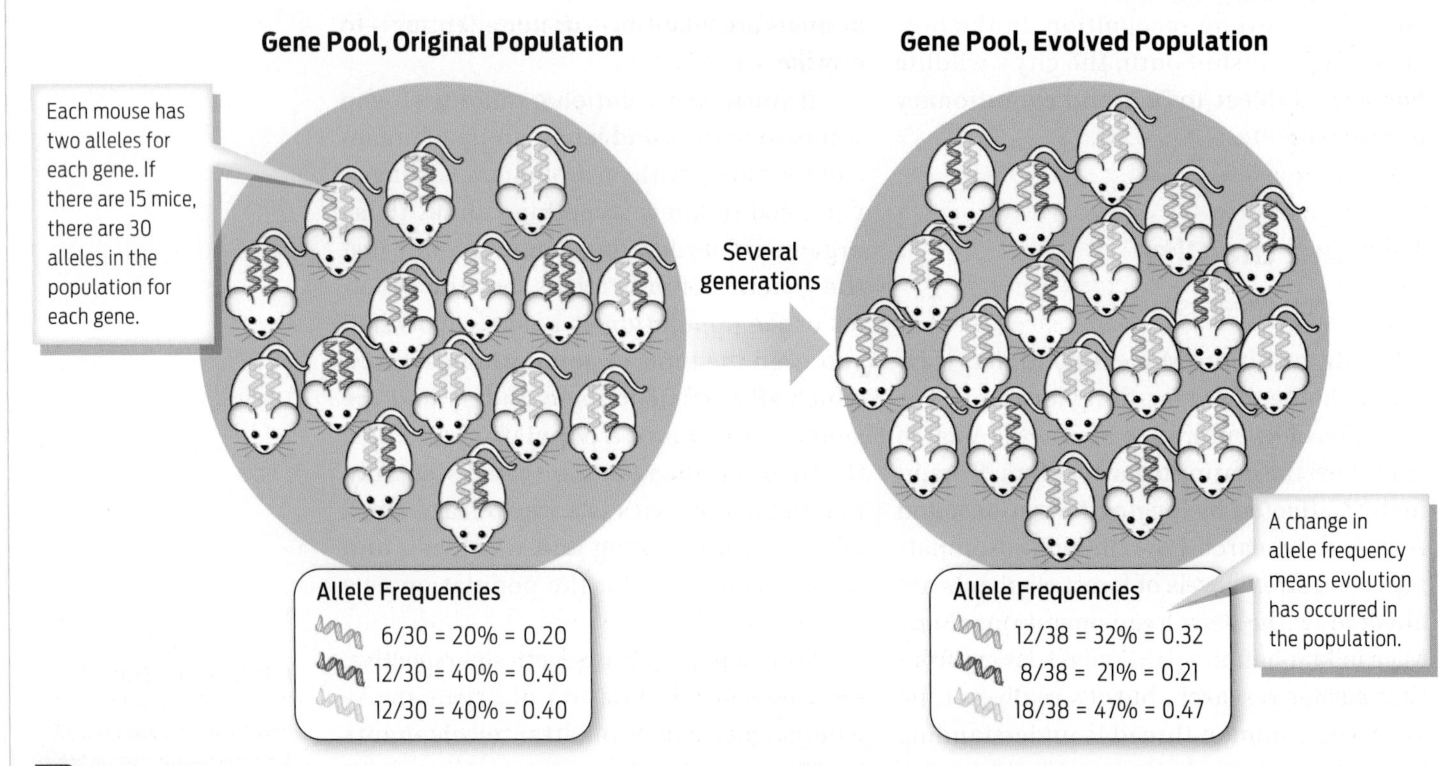

? The starting population (on the left) undergoes several generations of reproduction. In the resulting population, the frequency of the blue allele is 0.2, the frequency of the red allele is 0.4, and the frequency of the yellow allele is 0.4. Did evolution occur in this case? Why or why not?

population becoming more adapted to its environment. It is, rather, a type of **nonadaptive evolution:** a random change in the allele frequencies of a population.

Nonadaptive evolution isn't necessarily "bad," or maladaptive. If mutations didn't introduce genetic variation into a population, there would be no evolution at all. And many nonadaptive changes in allele frequency can be considered "neutral"—neither "good" nor "bad." But nonadaptive evolution can alter the genetic diversity of a population and, in turn, greatly influence the fate of a species. That's because genetic diversity—the total amount of genetic variation present in a population's gene pool—is important for the continued survival of populations in the face of a changing environment. For species that make cities their home, genetic diversity can mean the difference between making it big or losing it all.

Changing by Chance

▶ Genetic drift

Peromyscus leucopus—the white-footed mouse—is one of the oldest residents of Manhattan, long predating the arrival of the first European colonists in the 1600s. This rodent squeaks out a living in the green spaces of the city—in any park that has canopy cover. There, more than a dozen distinct populations of white-footed mice are living across the city.

When Munshi-South initially had the idea of studying mice in New York City, he thought there probably wouldn't be many genetic differences among the different populations living there. After all, New York is a fairly young city, and evolutionary change generally happens slowly. But the results of his study clearly showed that was not the case.

"It turned out that actually the populations are fairly distinct in the parks," says Munshi-South. And these differences have come about over a relatively short period of evolutionary time—a few hundred years at the most.

Over three centuries, as New York City has been transformed from a rich tapestry of

Nonadaptive evolution isn't necessarily "bad," or maladaptive. If mutations didn't introduce genetic variation into a population, there would be no evolution at all.

forest into a sparse patchwork of fragmented green spaces, populations of white-footed mice have become trapped in little islands of green, cut off from their distant cousins in the rest of the city. As a consequence, each population of mice has its own distinct gene pool, and evolution is occurring differently in each local population.

To peer into these gene pools, Munshi-South and his colleagues analyzed mice from 15 different populations around New York City. Using mousetraps baited with birdseed, they caught a total of 312 mice from these populations. After catching each mouse, the researchers cut off the tip of its tail and preserved the tissue sample in ethanol; DNA from the tail tissue could later be extracted and sequenced at specific chromosome locations (see Chapter 7). The mice, which were not seriously harmed by this procedure, were then released back into the wild.

When DNA sequences at 18 different chromosome locations from all 312 mice were assessed, the results showed distinct clustering of alleles, with mice within one population tending to share more alleles with one another than with mice from other populations. In fact, says Munshi-South, you can

NONADAPTIVE EVOLUTION
Any change in allele frequencies that does not by itself lead a population to become more adapted to its environment; the mechanisms of nonadaptive evolution are mutation, genetic drift, and gene flow.

INFOGRAPHIC 14.2

Gene Pools of New York City Mouse Populations

Researchers collected DNA from the tails of 312 mice at 15 locations in New York City. To determine how related the mouse populations were, researchers analyzed the DNA, assigning mice carrying similar genotypes the same color. When sorted by location, the data show that mice within a population shared more alleles with one another than they did with mice from other populations.

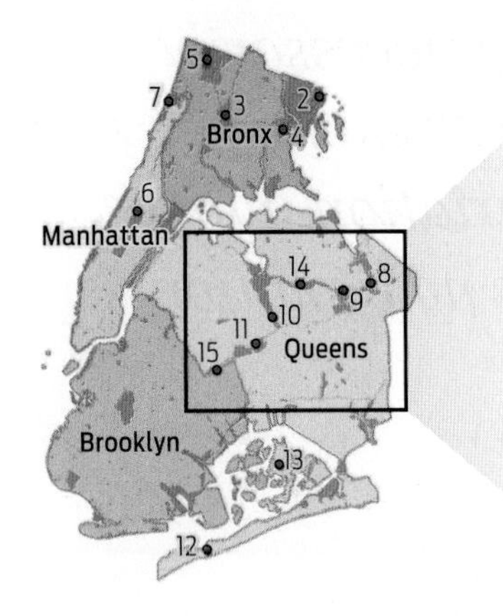

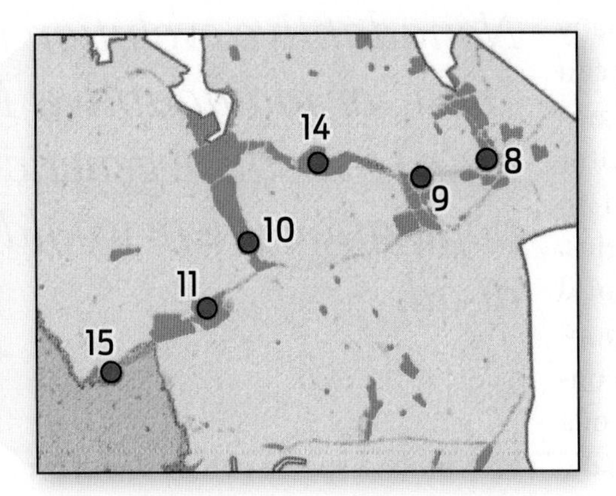

Habitat fragmentation has led to separated mouse populations in New York City.

The red dots show Munshi-South study sites. Shades of green represent park areas. Urbanization has fragmented green spaces into separate patches. Some patches are connected by "greenways" and some are not.

Mouse populations are genetically distinct from one another.

Each block represents a group of mice sampled from a particular location (population). The alleles of each sampled mouse are represented by a vertical bar within the block. When genetic sequences of these alleles match those of another mouse within the population, they are given the same color. When sequences don't match, they are shown in different colors.

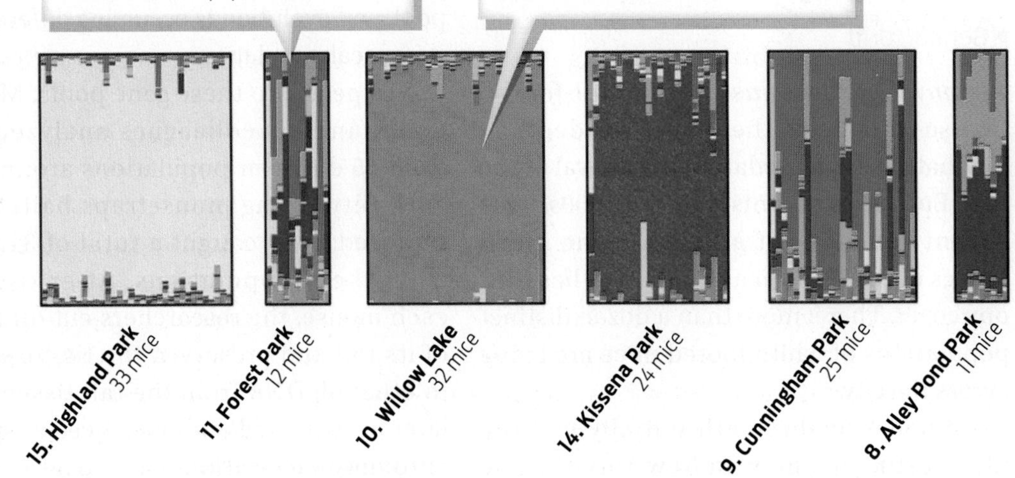

Overall, when you look at the blocks (populations), you can see that most mice in a given population have alleles (colors) that are similar to one another and distinct from other populations. For example, Highland Park mice have predominantly "yellow" alleles, while Willow Lake mice have predominantly "gray" alleles. Populations connected by green corridors to other populations may have alleles that are common to the populations to which they are connected.

Republished with permission of John Wiley and Sons, from Rapid, pervasive genetic differentiation of urban white-footed mouse (Peromyscus leucopus) populations in New York City. Munshi-South J., Kharchenko K., Molecular Ecology, 2010 Oct;19(19):4242–54 by BLACKWELL PUBLISHING LTD, permission conveyed through Copyright Clearance Center, Inc

? From the allele data, which populations are the most likely to be isolated (and not "trading" mice with other populations)?

accurately predict where a mouse is from just by looking at its DNA **(INFOGRAPHIC 14.2)**.

How did these genetic differences among populations come about? One possibility is that each population of mice evolved by natural selection as a result of local differences in the environment. Perhaps each different green space has different predators or food sources, for example, and these local differences could have selected for individuals with different alleles. This explanation is not the most likely one, however, given how close these green spaces are to one another—in some cases, less than a mile apart—and how similar the environments are. More likely, says Munshi-South, the cause of the genetic differences is **genetic drift.**

Genetic drift is a bit like rolling the evolutionary dice. By simple chance, some individuals survive and reproduce, and others do not. Those that pass on their genes aren't necessarily more fit or better adapted; they're just lucky. Perhaps their nest or burrow wasn't swept away in a flash flood, for example, unlike their unfortunate cousins.

Over time, genetic drift tends to decrease the genetic diversity of a population, as some alleles are lost completely and others sweep to 100% frequency. Genetic drift will have more dramatic effects in smaller populations than in larger ones. In a population with few individuals, any single individual that does not reproduce could spell the loss of alleles from the population. But all populations experience some measure of genetic drift, since chance is a fact of life.

Biologists refer to two general types of genetic drift: founder effects and bottlenecks. A **founder effect** occurs when a small group of settlers ("founders") splits off from a main population and establishes a new one. Because a founding population is by definition small, there is a good chance that the particular alleles the group carries will not be fully representative of the population it left. Thus, founder effects tend to reduce the genetic diversity of the new population.

If the founder population happens to contain a rare allele, then this allele may become much more common in the new population compared to the original group. Polydactyly (having extra fingers or toes) is an unusually common trait in the Amish population of eastern Pennsylvania, the result of Ellis-van Creveld syndrome, a recessive inherited condition. The Amish are a rural, isolated population who are descended from a small number of German immigrants who moved to the area in the 18th century. The allele for Ellis-van Creveld syndrome arrived with a single couple who immigrated to the area in 1744 and has since spread throughout the population.

Genetic drift is a bit like rolling the evolutionary dice. By simple chance, some individuals survive and reproduce, and others do not.

Polydactyly is also common in certain cat populations. The writer Ernest Hemingway adopted a six-toed cat when he lived in Key West, Florida. That cat mated with a local cat, and many of their offspring had extra toes. Today, descendants of these six-toed cats are permanent residents of Hemingway's estate.

Because white-footed mice are native to the New York region, it is unlikely that they have experienced founder effects; the current populations have all been in the city for a very long time. More applicable to these mice is the phenomenon known as the **bottleneck effect.** When a population is cut down sharply—forced through a "bottleneck"—there's a good chance that the remaining population will possess a less-diverse gene pool. Bottlenecks can occur from natural causes—say, a flood that sweeps through the city, killing many individuals—or from

GENETIC DRIFT
Random changes in the allele frequencies of a population between generations; genetic drift tends to have more dramatic effects in smaller populations than in larger ones.

FOUNDER EFFECT
A type of genetic drift in which a small number of individuals leaves one population and establishes a new population, resulting in lower genetic diversity than in the original population.

BOTTLENECK EFFECT
A type of genetic drift in which a population is suddenly reduced to a small number of individuals, and alleles are lost from the population.

INFOGRAPHIC 14.3

Genetic Drift Reduces Genetic Diversity in Populations

Allele frequencies can change from one generation to the next purely as a result of chance: this is genetic drift. Drift has more dramatic effects in smaller populations than in larger ones. There are two types of genetic drift, the founder effect and the bottleneck effect.

Founder Effect

The founder effect is a type of genetic drift that occurs when a small group of "founders" leaves a population and establishes a new one. If, by chance, alleles from the original population are absent from the founders, they will also be absent from the new population. This will result in the founding population being less diverse than the original.

A large, diverse original population

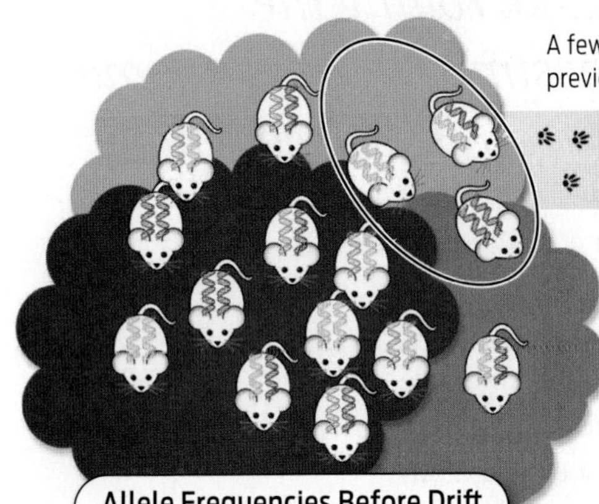

A few founders migrate to previously uninhabited territory.

By chance, these mice have a reduced frequency for some alleles compared to the original population.

A less-diverse founding population (after several generations)

Allele Frequencies Before Drift

6/30 = 20% = 0.20
12/30 = 40% = 0.40
12/30 = 40% = 0.40

Blue alleles are not represented in the founding population.

Allele Frequencies After Drift

10/22 = 45% = 0.45
12/22 = 55% = 0.55

Bottleneck Effect

Genetic bottlenecks occur when a population loses a large proportion of its members. If the original population is large, the reduced population may retain the same alleles present in the original population. But in a small starting population, bottlenecks are more likely to result in the loss of alleles from the population.

A more-diverse original population

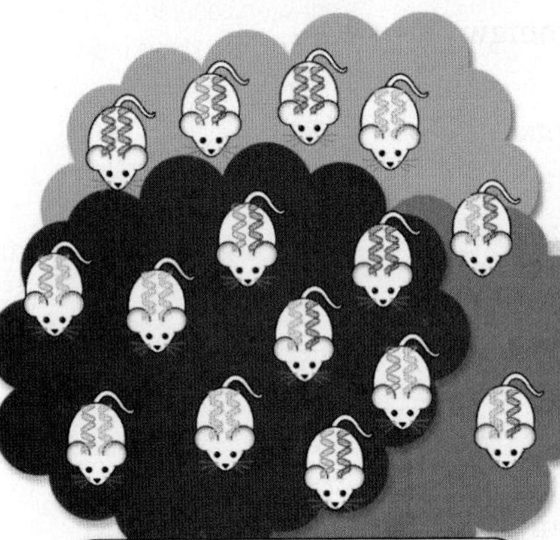

A bottleneck event, like rapid habitat loss, eliminates a large percentage of the population.

By chance, the surviving mice have a reduced frequency for some alleles compared to the original population.

A less-diverse bottleneck population (after several generations)

Allele Frequencies Before Drift

6/30 = 20% = 0.20
12/30 = 40% = 0.40
12/30 = 40% = 0.40

Red alleles are lost from the gene pool.

Allele Frequencies After Drift

6/18 = 33% = 0.33
12/18 = 67% = 0.67

? Compare and contrast the bottleneck effect and the founder effect.

human interference, such as the clearing of a forest. Either way, a population that is forced through a genetic bottleneck usually contains a fraction of the original diversity in the population **(INFOGRAPHIC 14.3)**.

As an example of a genetic bottleneck, consider the cheetah (*Acinonyx jubatus*), the fastest land animal. Cheetahs almost became extinct 10,000 years ago when the harsh conditions of the last ice age claimed the lives of many large vertebrates on several continents. Ultimately, a few cheetahs survived and reproduced, but the more than 12,000 individuals alive today are now so genetically similar that skin grafts between unrelated individuals do not cause immune rejection; nearly all genetic diversity has been eliminated from the population.

Why does genetic diversity matter? You can think of a gene pool as being like a population's portfolio of financial assets. Having a diverse array of investments is a better strategy for long-term success than having all your money tied up in one kind of stock—especially if that stock loses value in changed economic times.

Suppose a population of mice suddenly finds itself in an environment more crowded and polluted environment than it did before. If the population carries with it a rich variety of alleles, some of these alleles (those associated with stronger immune systems, for example) may help that population survive and reproduce in the altered conditions. Individuals with these alleles will be more fit in this environment, and the population will adapt by natural selection. With less diversity in the population, the gene pool may contain fewer alleles that can help the population survive and reproduce in the altered conditions. The opportunity for adaptation will be more limited, and the population may shrink. Preserving genetic diversity is thus a prime concern of conservation biologists interested in protecting natural populations from extinction, especially as humans encroach on more and more wild habitat.

The Daily Commute

▶ Gene flow

Once a population has lost genetic diversity because of genetic drift, genetic diversity can be reintroduced in only two ways: (1) by mutation, which as we saw in Chapter 9 continually introduces new alleles into the population, and (2) by **gene flow,** in which alleles move between populations as individuals leave and enter populations and breed with members of other populations. Like genetic drift, gene flow is a type of nonadaptive evolution that does not lead to a population becoming more adapted to its environment. Unlike genetic drift, gene flow tends to increase the genetic diversity of a population, not decrease it **(INFOGRAPHIC 14.4)**.

Munshi-South and his colleagues found that different New York City mouse populations had different levels of gene flow, meaning that some populations exchanged alleles with other populations more than did others. They wanted to understand why, so they built a statistical model and tested whether it could explain the data they collected at their study sites. Their idea was this: gene flow should be possible between populations where there is a corridor of tree canopy connecting them. To test this hypothesis, the researchers divided the city into categories based on the percentage of "green" landscape (tree cover) or "gray" landscape (concrete), and then measured how closely their actual data matched the hypothetical results. It was a simple idea, but one that worked surprisingly well. "That pretty much explained the variation we saw in gene flow and drift between the populations," Munshi-South says. In fact, the "green corridor" model explained the amount of gene flow even more than the absolute distance between the populations.

Putting it all together, the picture looks like this: urbanization has led to habitat fragmentation that has isolated and bottlenecked dozens of mouse populations, leaving each group with a distinct gene pool. Populations that were completely isolated (for example,

GENE FLOW
The movement of alleles from one population to another, which may increase the genetic diversity of a population.

INFOGRAPHIC 14.4

Gene Flow Increases Genetic Diversity in Populations

Migration and interbreeding of individuals move alleles between populations. Populations that can interbreed with other populations have more allele diversity than isolated populations.

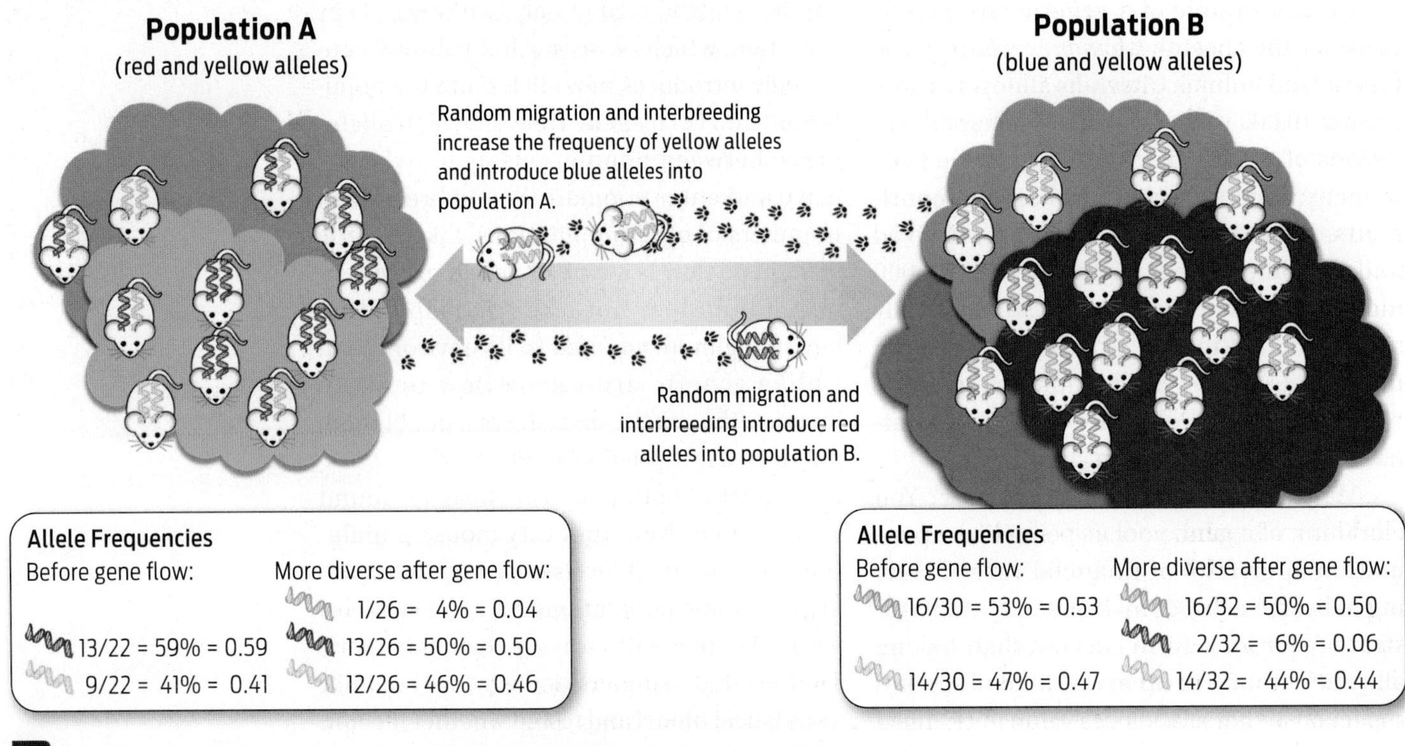

Population A — Allele Frequencies

Before gene flow:	More diverse after gene flow:
	1/26 = 4% = 0.04
13/22 = 59% = 0.59	13/26 = 50% = 0.50
9/22 = 41% = 0.41	12/26 = 46% = 0.46

Population B — Allele Frequencies

Before gene flow:	More diverse after gene flow:
16/30 = 53% = 0.53	16/32 = 50% = 0.50
	2/32 = 6% = 0.06
14/30 = 47% = 0.47	14/32 = 44% = 0.44

? Which new allele was introduced into population B by gene flow?

sites 10 and 15 in Infographic 14.2) have continued to diverge by genetic drift, while populations connected to other populations by a green corridor (for example, sites 8 and 9) have remained more similar to one another (and have more genetic diversity) because of gene flow between them. Furthermore, mutation randomly adds new alleles to these populations, contributing to diversity between them. "So it's basically this interplay between genetic drift, gene flow, and mutation," says Munshi-South, that is influencing the level of genetic diversity in the populations.

As we've seen, gene flow from one population into another may introduce new alleles and therefore enrich the diversity of that population's gene pool. This greater diversity allows the population to adapt to a changing environment. But that's not the only reason gene flow matters: small, isolated populations can be damaged by lack of genetic diversity even when the environment is not changing.

Consider the Florida panther (*Puma concolor*), for example. In the past, Florida panthers mated with puma populations from neighboring states where their ranges overlapped. This interbreeding—breeding among different populations of the same species—fostered an exchange of alleles that continually enriched the local populations' genetic diversity. The increased genetic diversity, in turn, allowed the populations to better respond to changing environments. By the mid-20th century, however, hunting and development had squeezed the Florida panther population into an isolated region at the state's southernmost tip. By 1967, only 30 panthers remained, and the U.S. Fish and Wildlife Service listed them as endangered.

By 1980, the panthers showed unmistakable signs of ill health—birth defects, low sperm count, missing testes, and bent tails—that resulted from **inbreeding,** mating between closely related members of a population.

Inbreeding can have dangerous consequences for a population. Because closely related individuals are more likely to share the same alleles, the chance of two recessive harmful alleles coming together during mating is high. When that happens, homozygous recessive genotypes are created in higher frequency in the population, and previously hidden recessive alleles start to affect phenotypes in negative ways. This effect is called **inbreeding depression.**

To counteract this dangerous trend, in 1995 the U.S. Fish and Wildlife Service brought in eight female pumas from Texas to mate with Florida's male panthers and thereby introduce genetic diversity. The program was successful: the hybrid kittens—30 in all—showed no signs of inbreeding depression. By 2007, more than 100 healthy panthers were roaming the swamps and grasslands of Florida, and their numbers continue to increase.

For the time being, says Munshi-South, mice in Manhattan seem to be doing just fine in maintaining adequate genetic diversity. While the populations are clearly distinct from one another, each contains within it a fair amount of genetic variation. This is probably because each population is still quite large, so the drift that is occurring has not dramatically reduced the number of alleles in each population. That factor, combined with mutation and occasional episodes of gene flow between some of the populations, has allowed these populations to maintain significant genetic diversity.

The same cannot be said of other species that Munshi-South and his colleagues are studying in New York City. The northern dusky salamander (*Desmognathus fuscus*), which makes its home in freshwater streams seeping out of the ground, was extremely common throughout much of the city just 60 years ago. Today, it clings to a single hillside in northern Manhattan and to parts of Staten Island, having suffered a severe bottleneck. Not only that, but its single hillside home has been bisected not once, but twice, by bridges connecting the boroughs of the Bronx and Manhattan. Together, these bridges host 14 lanes of traffic, which prevent salamanders from moving between sections of hillside. These human-made structures have thus divided this small, isolated population even further. Munshi-South and his colleagues have found that the populations living on either side of these bridges are genetically distinct—evidence that significant gene flow is not happening between them. Whether the dusky salamanders will be able to retain their tenuous hold along this hillside, only time will tell.

Spencer Platt/Getty Images

Human-made barriers, like New York City's High Bridge, interfere with gene flow between populations.

City Mouse, Country Mouse

▶ The Hardy–Weinberg principle

Fifteen miles north of Manhattan, on farms and apple orchards tucked away in the New York countryside, white-footed mice lead slower-paced lives, free from the stresses of urban living.

INBREEDING
Mating between closely related individuals. Inbreeding does not change the allele frequencies within a population, but does increase the ratio of homozygous individuals to heterozygous individuals.

INBREEDING DEPRESSION
The negative reproductive consequences for a population associated with having a high frequency of homozygous individuals possessing harmful recessive alleles.

Have city mice and country mice evolved differently as a result of living in different environments? To answer this question, Munshi-South and a graduate student, Stephen Harris, are collecting DNA samples from country mice in order to compare the gene pools of city and country mice. "The sort of low-hanging fruit that we hope to find are the genes that are consistently different between urban and rural populations," says Munshi-South.

They've only just begun this work, but have already uncovered some interesting results. The allele frequencies for a number of genes do seem to differ consistently between urban and rural populations. These alleles have to do with traits like immunity to pathogens, metabolizing different food sources, and breaking down toxic

INFOGRAPHIC 14.5

Urban Environments Select for Unique Traits in Mice

Country mice and city mice differ most in genes associated with immunity and metabolism of toxins and nutrients, with urban populations having less allele diversity for these genes. The reduction in allele diversity for these genes in city mice may reflect directional selective pressures in the urban environment, such as exposure to pathogens, pollution, and high-fat foods.

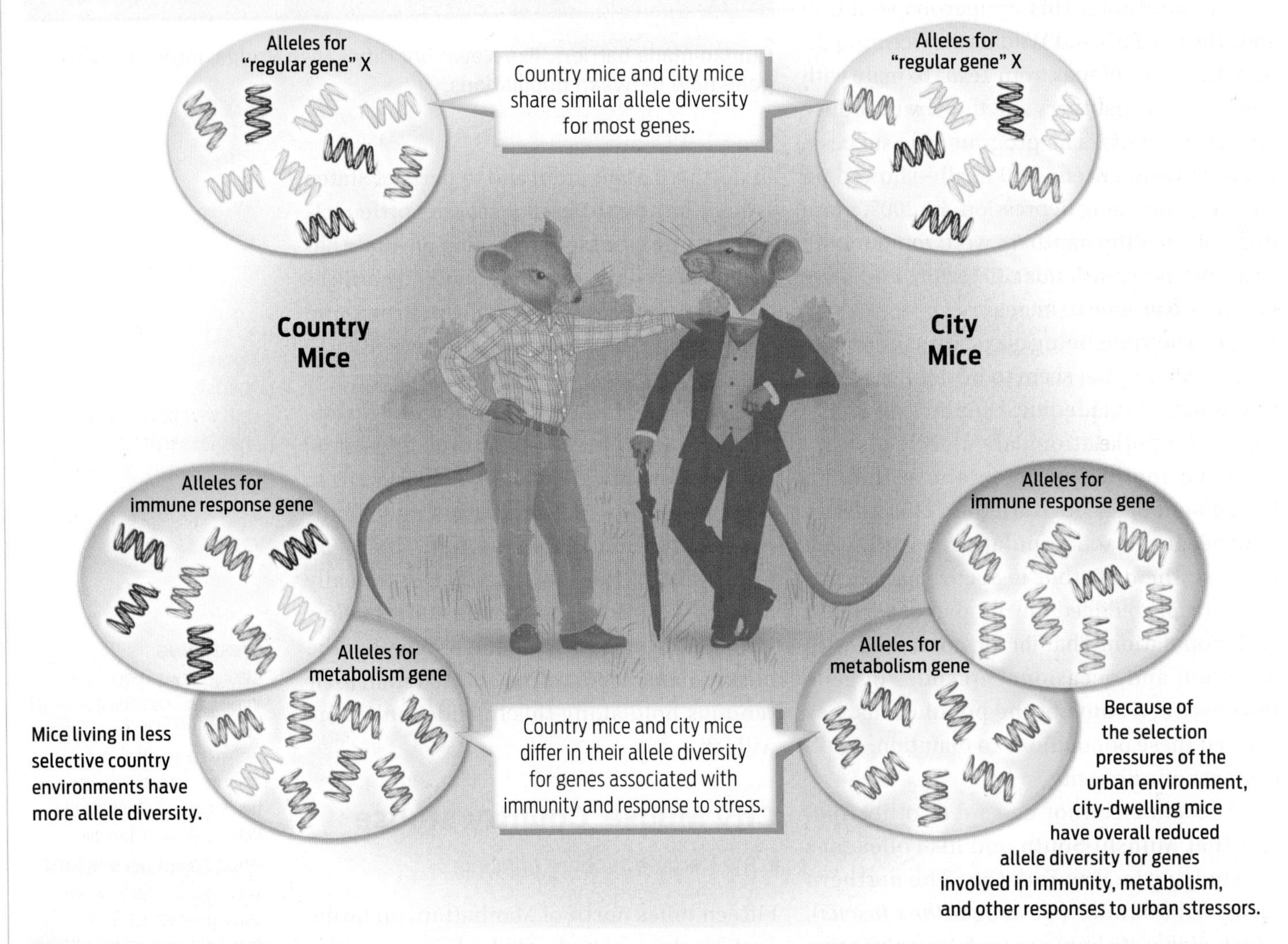

? Two populations of mice live in a similar warm climate with plentiful rainfall. One population lives near a large farm, and the other resides in undisturbed woods. Which of these genes would you predict to vary the most between the two populations: a gene involved in water balance, a gene involved in detoxification of chemicals (like pesticides), or a gene involved in temperature regulation?

chemicals. In urban environments, mice face different pathogens and have increased access to high-fat human food waste. They also must contend with higher levels of industrial toxins in the water and soil. City mice have higher frequencies of alleles that allow them to handle these urban stresses. This makes perfect sense, Munshi-South says: the city mice and the country mice have experienced different selective pressures, and these selective pressures have led to differences in the gene pools of these two large populations, with city mice having reduced allele diversity as a result **(INFOGRAPHIC 14.5)**.

While city mice have experienced selection for urban traits, this does not mean they are all the same. Many genetic differences still exist among the local populations of urbanized mice. That's because mutation, drift, and gene flow continue to occur even as natural selection is happening. In fact, for most natural populations, all four mechanisms of evolution—natural selection, mutation, gene flow, and genetic drift—are continually operating at the same time, fostering unique evolutionary outcomes **(TABLE 14.1)**.

How can scientists detect whether evolution is occurring in a population? The most direct way would be to determine the allele frequencies in a population, wait a few generations, and then determine the frequencies again. If the allele frequencies have changed, then the population has evolved (that's the definition of evolution).

But waiting for generations to go by isn't necessarily the most convenient approach. A shortcut that biologists sometimes use is to take a "snapshot" of the genotype frequencies in a population and compare these to the genotype combinations you would predict to find in a nonevolving population. If they differ, then you know the population is evolving.

How do genotypes behave in a nonevolving population? A common misconception is that dominant traits will tend to increase automatically in a population and recessive traits will be eliminated, thereby removing genetic variation from the population. Two mathematicians, G. H. Hardy and Wilhelm Weinberg, independently showed in the early 20th century that this was not the case. Through simple mathematical proofs, they showed that genotype frequencies (and therefore allele frequencies) in a population will remain constant, generation after generation, provided no evolutionary forces—such as genetic drift, natural selection, or gene flow—are bearing down on the population. In other words, in a nonevolving population, the genotype frequencies and allele frequencies will remain constant, and genetic variation will be maintained. Moreover, genotype frequencies in a nonevolving population can be predicted from the allele frequencies. This mathematical observation, which has come to be known as the **Hardy–Weinberg principle,**

HARDY–WEINBERG PRINCIPLE
The principle that, in a nonevolving population, both allele and genotype frequencies remain constant from one generation to the next.

TABLE 14.1 Adaptive and Nonadaptive Mechanisms of Evolution

Mechanism of evolution	How allele frequencies change	Adaptive or nonadaptive?	How genetic diversity is affected
Natural selection	Individuals with favorable alleles reproduce preferentially, increasing the frequency of these alleles.	Adaptive	Decreases—unfavorable alleles are eliminated from the population
Mutation	New alleles are created randomly.	Nonadaptive	Increases—new alleles are introduced into the population
Genetic drift (founder effect and bottlenecks)	Allele frequencies change due to chance events.	Nonadaptive	Decreases—alleles are eliminated from the population
Gene flow	Alleles move from one population to another.	Nonadaptive	Increases—new alleles are added to the population

INFOGRAPHIC 14.6 UP CLOSE

The Hardy–Weinberg Principle

The Hardy–Weinberg principle states that allele and genotype frequencies in a nonevolving population will remain constant generation after generation. These frequencies can be described mathematically. For a gene with two alleles, one dominant and one recessive, the allele and genotype frequencies can be calculated using two simple equations:

1. Allele frequencies equation:

   $p + q = 1$

   where p is the frequency of the dominant allele and q is the frequency of the recessive allele.

2. Genotype frequencies equation:

   $p^2 + 2pq + q^2 = 1$

   where p^2 is the frequency of homozygous dominants, $2pq$ is the frequency of heterozygotes, and q^2 is the frequency of homozygous recessives.

By describing how alleles and genotypes behave in a nonevolving population, the Hardy–Weinberg principle provides biologists with a baseline from which to evaluate whether or not evolution is occurring in a real population. For example, if biologists find that the actual genotype frequencies in a population differ significantly from the expected values, then they know that some evolutionary force is acting to shift them—such as genetic drift, selection, gene flow, or mutation.

By definition, a population is not evolving (and is therefore in Hardy–Weinberg equilibrium) when it has stable allele frequencies and stable genotype frequencies from generation to generation. This can be achieved only when all five of the following conditions are met:

1. No mutation introducing new alleles into the population
2. No natural selection favoring some alleles over others
3. An infinitely large population size (and therefore no genetic drift)
4. No gene flow between populations
5. Random mating of individuals

Example:

The Hardy–Weinberg principle is also useful for calculating allele frequencies in a population when you know the frequency of at least one genotype. For example, let's say you have a population of mice with two different alleles for a fur-color gene, *B* and *b*, where *B* is dominant and *b* is recessive. Let's further assume that there are two possible phenotypes associated with these genotypes, brown (*BB* or *Bb*) and white (*bb*).

Now let's say we find that white mice make up 9% of the mouse population. What, then, are the frequencies of the *B* and *b* allele in this population? We can use the Hardy–Weinberg equations to find out. Because white mice are *bb*, we know that the frequency of homozygote recessives, q^2, is 9%, or .09 (9/100). Taking the square root of this number, we get q, or the frequency of the *b* allele, which in this case is 0.3. Because there are only two alleles for fur color in the populations, their frequencies ($p + q$) must add up to 1, or 100%. Therefore, the frequency of the *B* allele is $1 - 0.3 = 0.7$, or 70%.

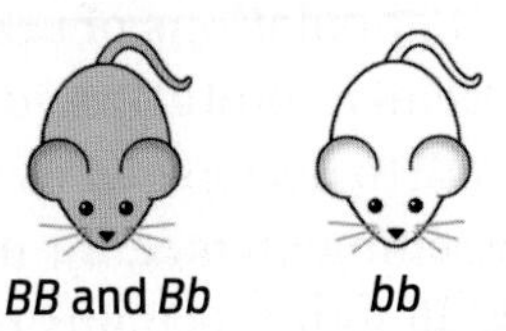

Genotype frequencies:

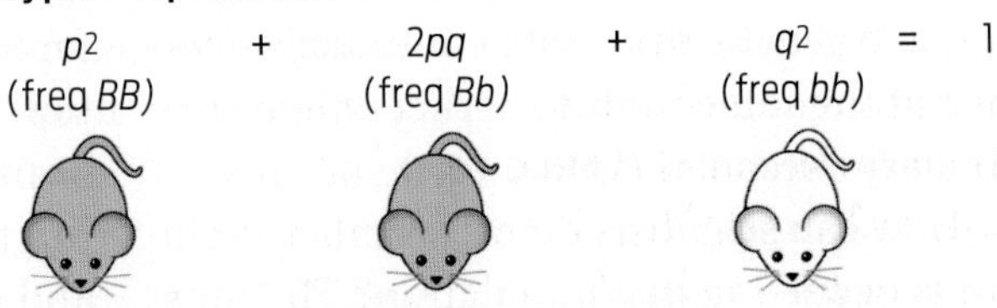

Allele frequencies:

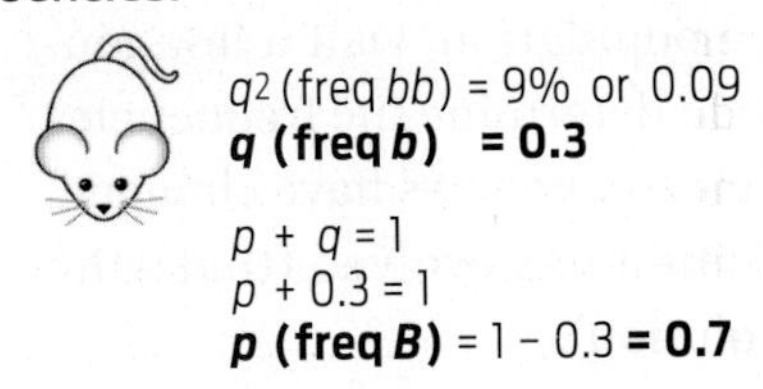

Now that we know the frequency of the two alleles, we know p and q, and we can calculate the genotype frequencies we would expect if this particular gene is in Hardy–Weinberg equilibrium in this population. For example, we would predict the frequency of heterozygotes to be $2pq$ (0.42). If we then sample mice in the population, analyze their genotypes, and find the frequency of heterozygotes to be 0.6, we would know that this gene is not in Hardy–Weinberg equilibrium in this population, and some evolutionary force is acting to change the allele or genotype frequencies.

? If the frequency of homozygote recessives is 10%, what are the frequencies of the recessive and dominant alleles in the population?

is a cornerstone of population genetics **(INFOGRAPHIC 14.6)**.

The Hardy–Weinberg principle can help researchers figure out whether some form of evolution, such as genetic drift or natural selection, is operating in a given population. To see how this works, let's say biologists obtain samples of DNA from a random sampling of mice in a population and look at the frequencies of genotypes at 10 different regions of DNA. Nine of those regions have genotype frequencies predicted by the Hardy–Weinberg principle, but one does not: it is far from Hardy–Weinberg equilibrium. Researchers then know that something interesting is happening at that one DNA location—some force of evolution is acting. In fact, this is how Munshi-South and his colleagues identified the candidate genes to compare between city and country mice.

"You can use certain deviations from Hardy–Weinberg equilibrium to find parts of the genome that are under selection," he says. "So, if they strongly deviate from Hardy–Weinberg, whereas the rest of the genome roughly fits it, those outliers are likely to have something interesting going on, like natural selection."

By understanding how city life has changed mice genetically, researchers will have a better understanding of how human activity is influencing mice evolution. That might not sound like a hugely important goal, especially if you're not a fan of mice. But there are larger lessons to take away. According to Munshi-South, "Manhattan offers a preview of what human activity will do to many other species in the coming years."

"Even just global warming alone is going to drive a lot of these processes in the future," he says. Indeed, some urban animals are already feeling the heat.

Biodiversity on Broadway

▶ Species and speciation

Just down the road from where Munshi-South works, researchers at Columbia University decided to look at what is perhaps the most urban of all green spaces: median strips. Their idea was to view these median strips as "islands" of wilderness within the city and to explore the diversity of ant species there. Were any species new to Manhattan, for example, or previously unidentified? And were native ant species able to compete successfully with introduced species?

The researchers collected 6,619 individual ants from 44 sites along three different avenues in New York—Broadway, Park Avenue, and the West Side Highway. Amid these crawling masses, they identified 13 different species of ant, including both native and introduced species. The most common ant species, found on nearly all medians, was an introduced species known as the pavement ant (*Tetramorium caespitum*), which hails originally from Europe, but ants from as far away as Japan were found. Somehow, despite the close quarters, all these different species coexist alongside one another while maintaining their distinct lifestyles. As the researchers noted in their study, "Manhattan

Neil Emmerson/Robert Harding World Imagery

New York City medians are small islands of vegetation surrounded by concrete.

Biologist Amy Savage, from Rutgers University, sampling ants on the Broadway median in New York City.

Lauren Nichols/Your Wild Life

INFOGRAPHIC 14.7

Diversity of Ant Species in New York City

Researchers collected ants from median strips in Manhattan. They have identified native and introduced ant species, and the characteristics that equip species for success in this environment.

Locations of the 44 street medians included in this study: 1, Broadway; 2, West Side Highway; 3, Park Avenue.

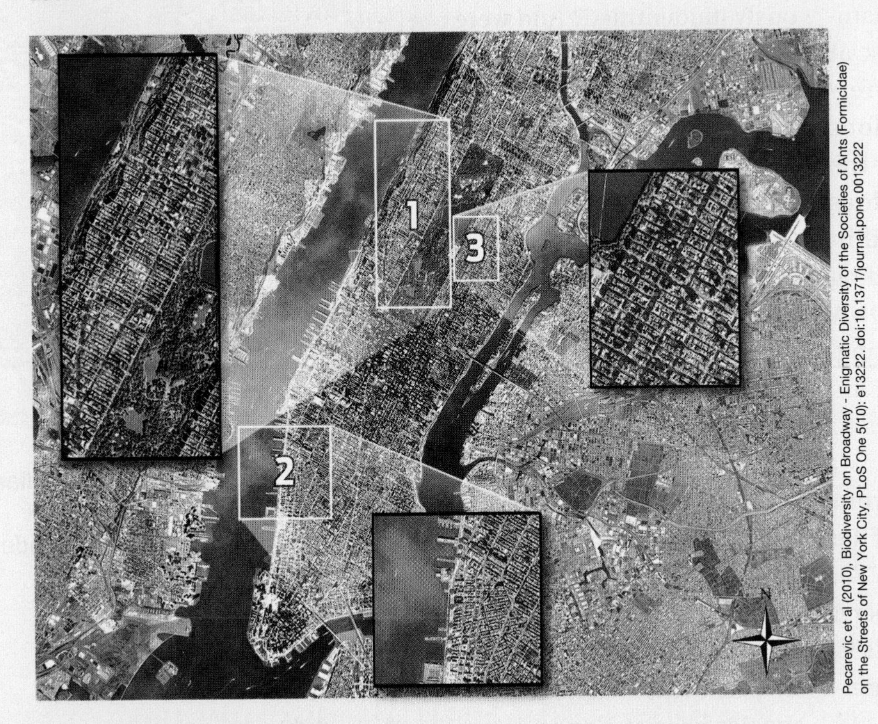

Pecarevic et al (2010), Biodiversity on Broadway - Enigmatic Diversity of the Societies of Ants (Formicidae) on the Streets of New York City. PLoS One 5(10): e13222. doi:10.1371/journal.pone.0013222

Alexander Wild/alexanderwild.com

The pavement ant, *Tetramorium caespitum*
Nonnative species; found in 93.2% of medians

Alexander Wild/alexanderwild.com

The eastern black carpenter ant, *Camponotus pennsylvanicus*
Native species; found in 15.9% of medians

Photo by April Nobile / From www.antweb.org. Accessed 7 August 2013

The yellow-footed ant, *Nylanderian flavipes*
Nonnative species; found in 52.3% of medians

Alexander Wild/alexanderwild.com

The thief ant, *Solenopsis molesta*
Native species; found in 63.6% of medians

Alexander Wild/alexanderwild.com

The cornfield ant, *Lasius neoniger*
Native species; found in 11.4% of medians

? Which species of ant is found on the highest proportion of medians in New York City?

is, if not quite a melting pot of ant species, at least a mixing bowl" **(INFOGRAPHIC 14.7)**.

From studying these and other urban ant populations, researchers have found evidence that the ant species that seem to do best in urban settings hail from warmer climates, like the Southwest. Because of all the heat-absorbing concrete, cities on average tend to be warmer than rural and suburban areas. If cities are any indication, global climate change will alter the ranges and diversity of ant species in an area, with unpredictable consequences.

The success of the median strip study gave the researchers the idea for an even more ambitious project: a nationwide study of ants, using community members who are not professional scientists as research assistants to collect specimens. The project is called, appropriately, School of Ants.

Anyone can enroll. All you need are index cards, ziplock bags, and a pecan sandy cookie—the bait of choice for seasoned ant collectors. Once the ants are collected, they're put in the freezer for a night and then shipped off to a lab at the North Carolina State University, where they will be studied and catalogued.

The aim of School of Ants is to create a map of ant species across the United States, especially in urban areas, where ants have been little studied. "Basically, we're mapping what's actually out there," says Andrea Lucky, an entomologist at the University of Florida who runs the project. "But instead of going off to the rain forest we're looking in people's backyards."

The most exciting part of the project, says Lucky, is discovering new ant species—as one School of Ants researcher did recently in Durham, North Carolina. The new species, which has extremely tiny eyes, behaves as a social parasite, taking over the worker ants of other colonies and making them into "slaves."

How did the researchers know the ant species was new? Basically, says Lucky, they have to know a lot about what is already known. They look in books, in museum collections, and in research papers to see if the ant has been previously described. Then they compare their specimen to the existing ones. "You end up looking at a lot of ants," Lucky says.

Mostly, researchers consider physical appearance—shape of the head, number of spines and segments of antennae, even how furry the creature is—but DNA evidence is important for recognizing more subtle differences. The idea is that these physical and genetic differences reflect adaptations to different environments that then prevent the species from interbreeding. In essence, that is what defines a species.

The term "species" comes from the Latin word meaning "kind" or "appearance." According to the **biological species concept**—the formal definition that biologists use—a species is a group of individuals whose members can interbreed and produce fertile offspring.

Members of different species cannot mate and produce fertile offspring with each other. In other words, they are reproductively isolated. Such **reproductive isolation** can be caused by a number of factors. For example, the two species may have different mating times, locations, or mating rituals—so, like ships passing in the night, they may never meet. Many ant species, for example, breed at different times of year. Or, two species may be able to mate—as zebras and horses can—but the hybrid offspring they produce is infertile **(INFOGRAPHIC 14.8)**.

Reproductive isolation explains why species remain separate, as do the variety of ant species that share a median strip. But how did the species form in the first place? New species emerge when a strong barrier prevents gene flow between populations. That barrier could be physical—like a road or river that divides a forest in two—or climatic, like the different temperatures that occur at different elevations on a mountainside. Once this barrier forms, the separated gene pools will evolve independently by the mechanisms we have already encountered: mutation, genetic drift, and natural selection. Eventually, if enough genetic changes accumulate between populations of the same species to make them reproductively isolated, the two populations may diverge into separate species, a process called **speciation.**

Speciation is happening all the time in nature, but it can be hard to see because it occurs so slowly. It generally takes many thousands of years for species to diverge. We see the results of speciation whenever we look at the diversity of nature—there are more than 12,000 known ant species, for example—but observing speciation as it happens is much harder.

The classic, and still the best, example of speciation comes from Charles Darwin himself: the finches he observed on the

BIOLOGICAL SPECIES CONCEPT
The definition of a species as a group whose members can interbreed to produce fertile offspring.

REPRODUCTIVE ISOLATION
Mechanisms that prevent mating (and therefore gene flow) between members of different species.

SPECIATION
The genetic divergence of populations, leading over time to reproductive isolation and the formation of new species.

INFOGRAPHIC 14.8

Species Are Reproductively Isolated

Species are reproductively isolated in a variety of ways. These include isolation on the basis of habitat, behavior, genetics, and anatomy as well as timing and outcomes of reproduction.

Ecological Isolation—species live in different environments. The Arctic fox and the desert fox live in such different places that they never encounter one another.

Temporal Isolation—species have different mating or fertility time frames. The leopard frog mates in early spring and the bullfrog mates in early summer.

Behavioral Isolation—species display different mating activities. The prairie chicken is not attracted to the mating display of the ring-necked pheasant.

Mechanical Isolation—species have incompatible mating organs. Plants pollinated by the hummingbird do not receive pollen from plants pollinated by the black bee.

Gametic Isolation—species have incompatible gametes. The egg and sperm from a dog and a cat cannot unite to form a zygote.

Hybrid Inviability—species' gametes unite but viable offspring cannot form. The goat and sheep can mate, but the zygote formed does not survive.

Hybrid Infertility—offspring are viable but cannot reproduce. Zebras and horses are different species because their hybrid offspring, zebroids, cannot produce offspring of their own.

? **Which forms of reproductive isolation could not be overcome by in vitro fertilization?**

Galápagos Islands, near Ecuador, while traveling aboard the *Beagle* (see **Milestone 4: Adventures in Evolution**). The original finch species came from the mainland of Ecuador. Descendants of this mainland population then island-hopped through the archipelago, creating founder populations on each island. Each island's founder population adapted, by natural selection, to the unique environment it encountered. From one ancestral species of finch have emerged 13 different species of finch, each with a distinctive beak size and shape adapted to a unique environment **(INFOGRAPHIC 14.9)**.

In some respects, ants on different median strips resemble Darwin's finches: they are geographically separated and do not typically interbreed. Given enough time, ants of one species located in different median strips might very well form different species, much like finches on different islands. The same goes for New York City's mouse populations:

INFOGRAPHIC 14.9

Speciation: How One Species Can Become Many

New species may form when a strong barrier to gene flow occurs between populations, as happened with Darwin's finches. The first species of finch originated on the mainland of South America and spread to the Galápagos archipelago, a series of islands off the coast of Ecuador. As the original finch species moved from island to island, individuals were physically separated from one another by the ocean between each island. These separated populations then evolved in isolation from one another. Ultimately, the separated populations evolved such that they could not interbreed, therefore becoming distinct finch species. At least 13 finch species have diverged from the original South American species.

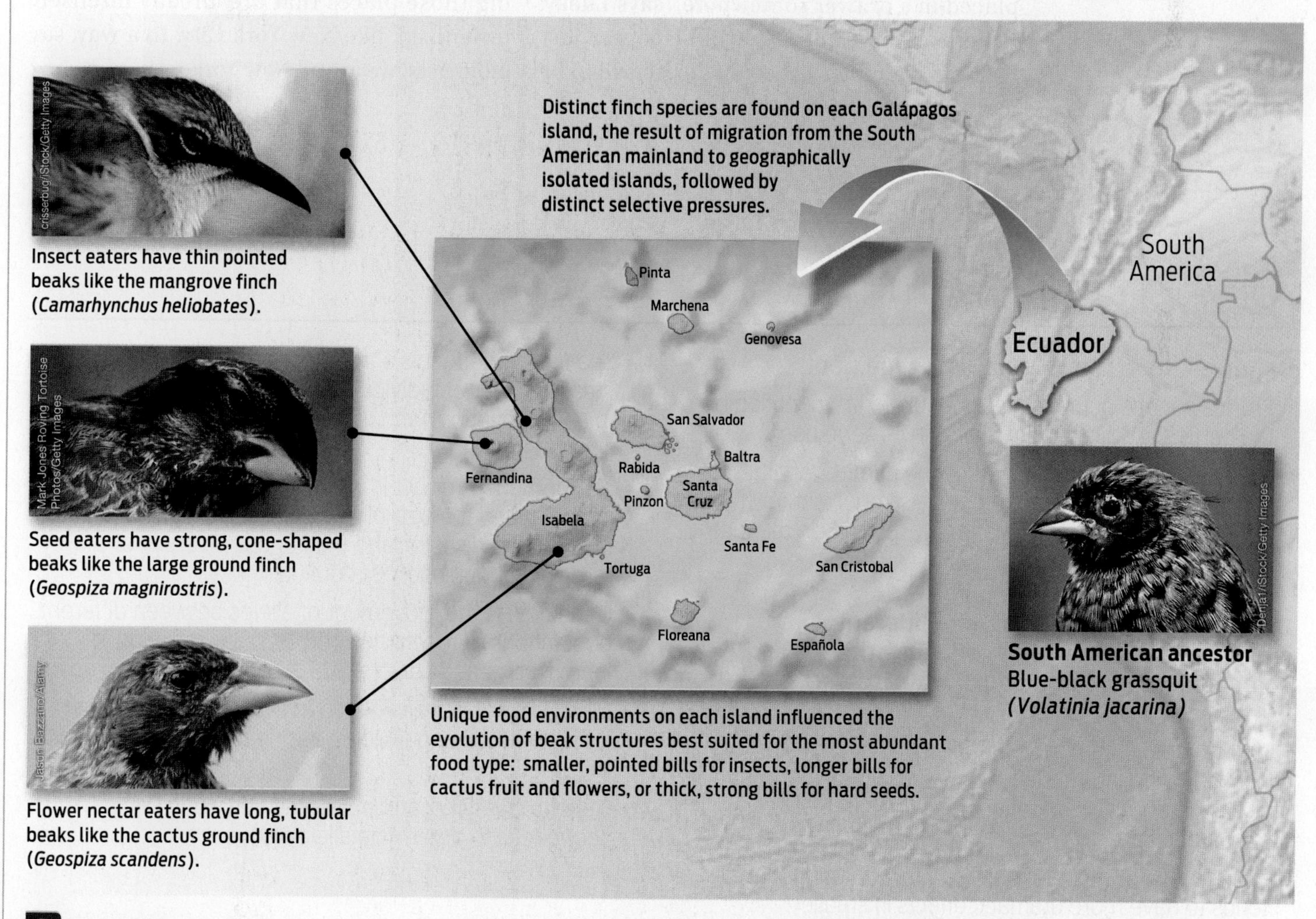

? The different species of finches have different beak shapes. Do these differences establish that the species are distinct?

they, too, might one day form new species, provided the populations remain isolated and continue to diverge genetically. But the timescale involved makes it hard to predict. It took thousands of years for Darwin's finches to evolve into different species; the median strips and parks themselves may be long gone before that happens for ants and mice.

Empire State of Mind

If cities are the new evolutionary laboratory, then researchers Munshi-South and Lucky are like modern-day versions of Charles Darwin—mapping uncharted biological territory, albeit in places found right under our noses.

"I think a lot of people are fairly shocked to find out that everything in North America isn't already named and catalogued and placed in a drawer somewhere," says Lucky. "There are multiple levels of discovery just waiting to happen." And you don't have to be a professional scientist to contribute to the process of discovery.

With more than half of all humans now living in urban environments, it makes sense for biologists to consider the effect that these many urban dwellers may have on populations of other organisms that share our living quarters. As Munshi-South points out, cities are active contributors to the evolutionary process, shaping the wildlife around us in profound ways. Along with climate change, urbanism is one of the main ways that humans are directly altering the face of the planet and thereby shaping the fate of species. Scientists are only just beginning to understand how cities drive evolutionary change, but they will have ample opportunity for continued research. The trend of urbanism shows no signs of stopping.

Those curious about the future can catch a glimpse of what lies in store for us by studying those places that are already intensely urbanized, like New York City. In a way, say biologists, we are all New Yorkers now. ■

CHAPTER 14 Summary

Driving Question 1 What is a gene pool?

- From a genetic perspective, a population is identified by the particular collection of alleles in its gene pool.
- Genetic diversity, as reflected by the number of different alleles in a population's gene pool, is important for the continued survival of populations, especially in the face of changing environments.

Driving Question 2 How do different evolutionary mechanisms influence the composition of a gene pool?

- Evolution is a change in allele frequencies in a population over time. Evolution can be adaptive or nonadaptive. Natural selection is an adaptive form of evolution, whereas mutation, genetic drift, and gene flow are nonadaptive forms of evolution.
- Genetic drift is a type of nonadaptive evolution that tends to have more dramatic effects in smaller populations.
- The founder effect is a type of genetic drift in which a small number of individuals establish a new population in a new location, with reduced genetic diversity as a likely result.
- The bottleneck effect is a type of genetic drift that occurs when the size of a population is reduced, often by a natural disaster, and the genetic diversity of the remaining population is reduced.
- Gene flow is the movement of alleles between different populations of the same species, often resulting in increased genetic diversity of a population.
- Inbreeding of closely related individuals may occur in small, isolated populations, posing a threat to the health of a species.
- Genetic diversity can be assessed by using DNA sequences to determine allele frequency.

Driving Question 3 How does the gene pool of an evolving population compare to the gene pool of a nonevolving population?

- The Hardy–Weinberg principle describes the frequency of alleles and genotypes in a nonevolving population. It can be used to detect evolutionary change in a population, and to calculate allele frequencies when at least one genotype frequency is known.

Driving Question 4 How do new species arise, and how can we recognize them?

- According to the biological species concept, a species is a group of individuals that can interbreed to produce fertile offspring.
- Speciation can occur when gene pools are separated and gene flow is restricted, so that populations diverge genetically over time.

More To Explore

- Munshi-South, J. (2012). TED Talk: Evolution in a big city: http://ed.ted.com/lessons/evolution-in-a-big-city
- School of Ants: http://schoolofants.org/
- Munshi-South, J. (2012). Urban landscape genetics: Canopy cover predicts gene flow between white-footed mouse (*Peromyscus leucopus*) populations in New York City. *Molec Ecol* 21:1360–1378.
- Pećarević, M., et al. (2010). Biodiversity on Broadway: Enigmatic diversity of the societies of ants (Formicidae) on the streets of New York City. *PLoS One* 5(10):e13222.
- Hardy, G. H. (1908). Mendelian proportions in a mixed population. *Science* 28(706):49–50.

CHAPTER 14 Test Your Knowledge

Driving Question 1 What is a gene pool?

By answering the questions below and studying Infographics 14.1 and 14.2, you should be able to generate an answer for this broader Driving Question.

Know It

1. Genetic diversity is measured in terms of allele frequencies (the relative proportions of specific alleles in a gene pool). A population of 3,200 mice has 4,200 dominant *G* alleles and 2,200 recessive *g* alleles. What is the frequency of *g* alleles in the population?

2. The allele frequencies for a particular gene are given for three populations. From this information, which population is most likely to be an isolated population? Note that this gene has seven possible alleles: *A1, A2, A3, a1, a2, a3*, and *a4*.

Population A: 30% *A1*, 30% *a1*, and 40% *a4*

Population B: 25% *A2*, 25% *A3*, 25% *a2*, 25% *a3*

Population C: 20% *A2*, 20% *A3*, 30% *a2*, 30% *a3*

Use It

3. A small population of 26 individuals has five alleles, *A* through *E*, for a particular gene. The *E* allele is present in only one homozygous individual.

Five individuals are *D/A* heterozygotes.

Five individuals are *A/A* homozygotes.

Five individuals are *A/B* heterozygotes.

Five individuals are *C/D* heterozygotes.

Five individuals are *C/C* homozygotes.

One individual is an *E/E* homozygote.

a. Calculate the allele frequency of each of the five alleles.

b. Five *A/E* heterozygotes migrate into the population. Now what are the allele frequencies of each of the five alleles in the population?

4. Some populations, such as cheetahs, have gene pools with very few different alleles. What approach(es) could be taken to try and introduce new alleles into these kinds of populations?

5. The global human population continues to grow, and more people than ever are living in crowded cities. Given this situation, what selective pressures might the human population currently face or be expected to face in the near future?

Driving Question 2 How do different evolutionary mechanisms influence the composition of a gene pool?

By answering the questions below and studying Infographics 14.3, 14.4, and 14.5 and Table 14.1, you should be able to generate an answer for this broader Driving Question.

Know It

6. Which of the following statements apply to the founder effect?
- **a.** It is adaptive.
- **b.** It decreases genetic diversity.
- **c.** It is a type of genetic drift.
- **d.** all of the above
- **e.** b and c

7. Which of the following are examples of genetic drift?
- **a.** founder effect
- **b.** bottleneck effect
- **c.** inbreeding
- **d.** a and b
- **e.** a, b, and c

8. A bottleneck is best described as
- **a.** an expansion of a population from a small group of founders.
- **b.** a small number of individuals leaving a population.
- **c.** a reduction in the size of an original population followed by an expansion in size as the surviving members reproduce.
- **d.** the mixing and mingling of alleles by mating between members of different populations.
- **e.** an example of natural selection.

9. A population of ants on a median strip has 12 different alleles, *A* through *L*, of a particular gene. A drunk driver plows across the median strip, destroying most of the median strip and 90% of the ants. The surviving ants are all homozygous for allele *H*.
- **a.** What is the impact of this event on the frequency of alleles *A* through *L*?
- **b.** What type of event is this?

Use It

10. From their gene pool and population size, which of the four populations in the accompanying table would concern you most from a conservation perspective? Why would you be concerned?

Population	No. of individuals	No. of alleles, gene 1	No. of alleles, gene 2	No. of alleles, gene 3
1	50	1	7	5
2	1,000	1	5	7
3	50	3	2	2
4	1,000	1	1	2

11. In humans, founder effects may occur when a small group of founders immigrates to a new country—for example, to establish a religious community. In this situation, why might the allele frequencies in succeeding generations remain similar to those of the founding population rather than gradually becoming more similar to the allele frequencies of the population of the country to which they immigrated?

12. Why is genetic drift considered a form of evolution? How does it differ from evolution by natural selection?

Apply Your Knowledge

Interpreting Data

13. The accompanying figure shows a bar plot of moles from different parks in New York City. As in Infographic 14.2, each vertical bar represents genotypes from 18 genomic locations in one animal. The bars are color coded, with similar genotypes represented by the same color. From the data presented in the figure:
- **a.** Are these three populations genetically isolated from one another? Explain your answer. What factors could explain their isolation, or lack thereof?
- **b.** Is one of the populations experiencing gene flow with another population? If so, which one, and how do you know?

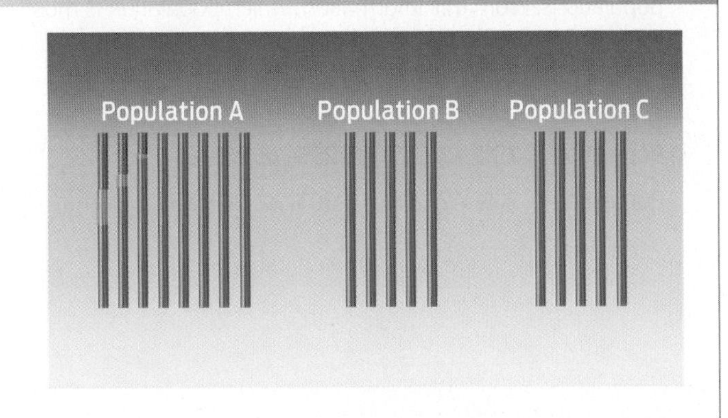

Driving Question 3 How does the gene pool of an evolving population compare to the gene pool of a nonevolving population?

By answering the questions below and studying Infographics 14.5 and 14.6, you should be able to generate an answer for this broader Driving Question.

Know It

14. Which of the following statements is/are true about a nonevolving population?

a. Allele frequencies do not change over generations.
b. Genotype frequencies do not change over time.
c. Organisms with the highest fitness are reproducing more frequently.
d. all of the above
e. a and b

15. A starting population of bacteria has two alleles of the *TUB* gene, *T* and *t*. The frequency of *T* is 0.8 and the frequency of *t* is 0.2. The local environment undergoes an elevated temperature for many generations of bacterial reproduction. After 50 generations of reproduction at the elevated temperature, the frequency of *T* is 0.4 and the frequency of *t* is 0.6. Has evolution occurred? Explain your answer.

16. Why is inbreeding detrimental to a population?

Use It

17. The Hardy–Weinberg principle has important applications in public health. For example, it can be used to estimate the frequency of carriers (heterozygotes) of rare recessive diseases in a population. Phenylketonuria (PKU) is a rare, recessive genetic condition that affects approximately 1 in 15,000 babies born in the United States. (You may have noticed the statement "Phenylketonurics: contains phenylalanine" on aspartame-containing products—a warning for people with PKU that they should avoid consuming that product.) For the purposes of this question, assume that the genotype and allele frequencies for this gene remain constant over generations.

a. If PKU is an autosomal recessive condition, what is the genotype of people with PKU?
b. Express their genotype in terms of Hardy–Weinberg (that is, whether they would be *pp, pq,* or *qq*).
c. Given the frequency of PKU in the population, what is the allele frequency of the recessive allele?
d. What is the frequency of the dominant allele?
e. Calculate the expected frequency of carriers in the U.S. population.

Driving Question 4 How do new species arise, and how can we recognize them?

By answering the questions below and studying Infographics 14.8 and 14.9, you should be able to generate an answer for this broader Driving Question.

Know It

18. The biological species concept defines a species

a. on the basis of similar physical appearance.
b. on the basis of close genetic relationships.
c. on the basis of similar levels of genetic diversity.
d. on the basis of the ability to mate and produce fertile offspring.
e. on the basis of recognizing one another's mating behaviors.

19. How does geographic isolation contribute to speciation?

Use It

20. Two populations of rodents have been physically separated by a large lake for many generations. The shore on one side of the lake is drier and has very different vegetation from that on the other side. The lake is drained by humans to irrigate crops, and now the rodent populations are reunited. How could you assess if they are still members of the same species?

21. If geographically dispersed groups of a given species all converge at a common location during breeding season, then return to their home sites to bear and rear their young, what might happen to the gene pools of the different groups over time?

Mini Case

Apply Your Knowledge

22. More than 50% of the global human population now lives in urban areas, and it is predicted that 70% will live in urban areas by 2050. Researchers have hypothesized that the emotional health of urbanites is influenced positively by interaction with nature. Given this information, and what you have read in this chapter, write a compelling paragraph on the need to conserve urban species and approaches to such conservation based on population genetics.

Bring It Home

Apply Your Knowledge

23. The School of Ants is a project carried out by community members to document the distribution and diversity of ants across the United States. Similarly, the Audubon Christmas Bird Count is a conservation-related project carried out by volunteer community members. The Audubon Society uses the data collected by these volunteers to evaluate the health of bird populations and make informed decisions about conservation. Many other projects are also carried out by volunteer community members—for example, Project Squirrel and the Gravestone Project. Carry out an Internet search to find a citizen-scientist or community-scientist project that you find interesting. Now take the next step: enroll, collect some data, and contribute to science.

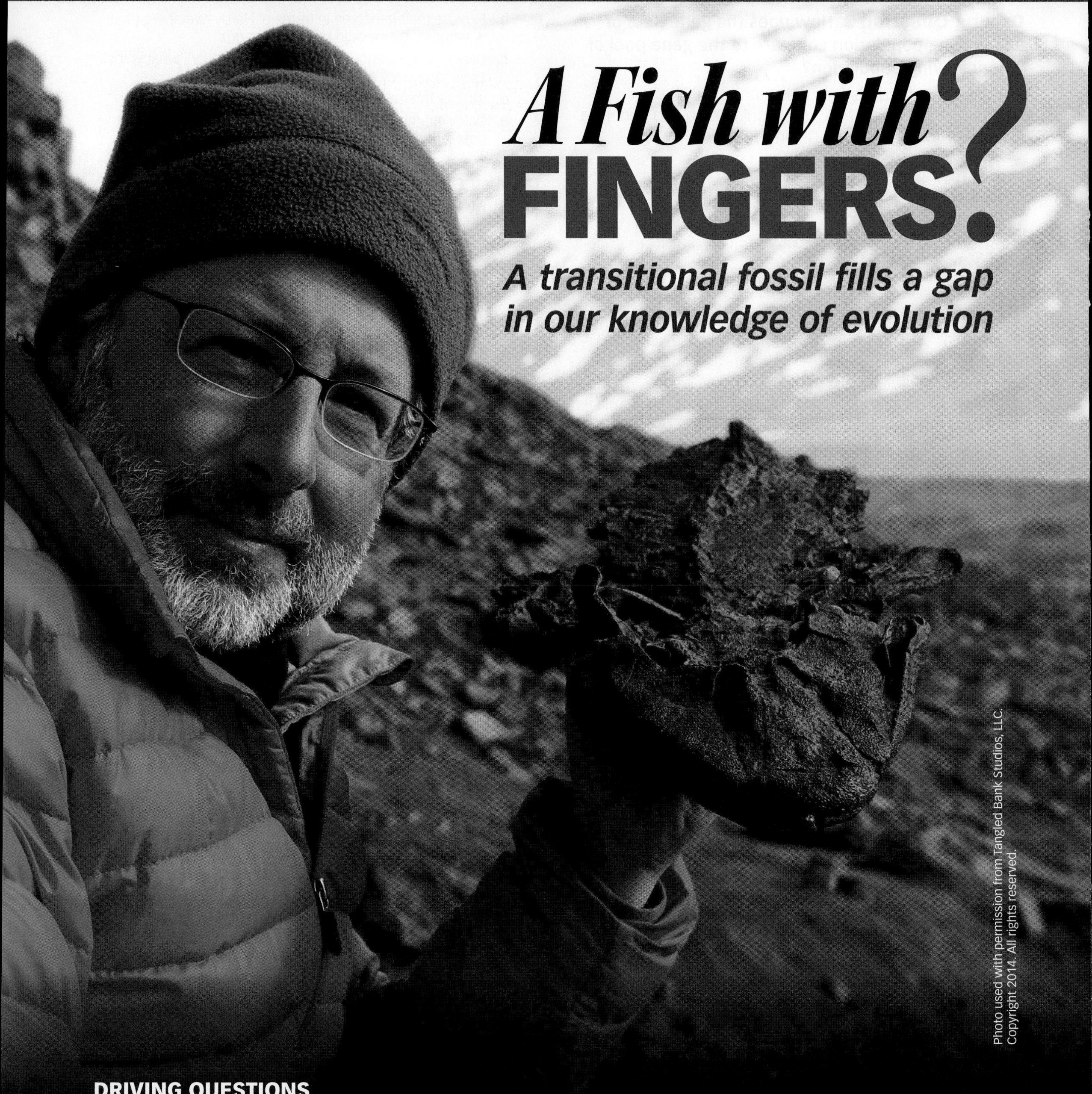

A Fish with FINGERS?

A transitional fossil fills a gap in our knowledge of evolution

Photo used with permission from Tangled Bank Studios, LLC. Copyright 2014. All rights reserved.

DRIVING QUESTIONS

1 How does the fossil record reveal information about evolutionary changes?

2 What features make *Tiktaalik* a transitional fossil, and what role do these types of fossils play in the fossil record?

3 What can anatomy and DNA reveal about evolution?

Fossil hunting in the vast wilderness of Ellesmere Island in the Canadian Arctic.

Ted Daeschler/Academy of Natural Sciences/VIREO

For 5 years, biologists Neil Shubin and Ted Daeschler spent their summers trekking through one of the most desolate regions on Earth. They were fossil hunting on remote Ellesmere Island, in the Canadian Arctic, about 600 miles from the north pole. Even in summer, Ellesmere is a forbidding place: a windswept, frozen desert where sparse vegetation grows no more than a few inches tall, where sleet and snow fall in the middle of July, and where the sun never sets. Only a handful of wild animals survive here, but those that do make for dangerous working conditions. Hungry polar bears and charging herds of muskoxen are known hazards of working in the Arctic, says Daeschler, who carried a shotgun for protection.

When not looking over their shoulders, the researchers drilled, chiseled, and hammered their way through rocks looking for fossils. Not just any rocks and fossils, but ones dating from 375 million years ago, when animals were taking their first tentative steps on land. For three summers, they scoured the site of what was once an active streambed but found little of interest. Then, in 2004, the team made a tantalizing discovery: the snout of a curious-looking creature protruding from a slab of pink rock. Further excavation revealed the well-preserved remains of several flat-headed animals between 4 and 9 feet long. In some ways, the animals resembled giant fish: they had fins and scales. But they also had traits that resembled those of land-dwelling amphibians—notably, a neck, wrists, and fingerlike bones. The researchers named the new species *Tiktaalik roseae; tiktaalik* (pronounced tic-TAH-lick) is a native word meaning "large freshwater fish." This ancient hybrid animal no longer exists, but it represents a critical phase in the evolution of four-legged, land-dwelling **vertebrates**—including humans.

VERTEBRATE
An animal with a bony or cartilaginous backbone.

A *Tiktaalik roseae* fossil.

Ted Daeschler/Academy of Natural Sciences/VIREO

Tiktaalik "splits the difference between something we think of as a fish and something we think of as a limbed animal," says Daeschler, a curator of vertebrate zoology at the Academy of Natural Sciences in Philadelphia. "In that sense, it is a wonderful transitional fossil between two major groups of vertebrates."

"It looks like a fish in that it has scales and fins, but when you look inside the skeleton you see how special it really is."

—Neil Shubin

DESCENT WITH MODIFICATION
Darwin's term for evolution, combining the ideas that all living things are related and that organisms have changed over time.

FOSSILS
The preserved remains or impressions of once-living organisms.

FOSSIL RECORD
An assemblage of fossils arranged in order of age, providing evidence of changes in species over time.

PALEONTOLOGIST
A scientist who studies ancient life by examining the fossil record.

Today, of course, four-legged animals roam far and wide over land. But 400 million years ago, it was a different story. Life was mostly aquatic then, restricted to oceans and freshwater streams. How life made the jump from water to land is a question that has long intrigued evolutionary biologists. In fact, scientists have been searching for evidence of this milestone ever since Charles Darwin first proposed that all life on the planet shares an ancient common ancestor and is therefore related by a tree of common descent. According to Shubin, a professor of biology at the University of Chicago and the Field Museum of Natural History, *Tiktaalik* is the most compelling example yet of an animal that lived at the cusp of this important transition. Not only does it fill a gap in our knowledge, but the discovery also provides persuasive evidence in support of Darwin's theory.

Ancient Snapshots

▸ Reading the fossil record

The theory of evolution—what Darwin called **descent with modification**—draws two main conclusions about life on Earth: that all living things are related, and that the different species we see today emerged over time as a result of natural selection operating over millions of years. Many lines of evidence support this theory (remember that in science a "theory" is an idea supported by a tremendous amount of evidence and which has never been disproved; see Chapter 1). One of the most compelling lines of evidence for evolution comes from **fossils,** the preserved remains or impressions of once-living organisms. Fossils serve as snapshots of past life, capturing what life was like at particular moments in time.

Fossils form in a number of ways: an animal or plant may be frozen in ice, trapped in amber (hardened tree sap), or buried in a thick layer of mud. The material that entombs the organism protects it from being eaten by scavengers or rapidly decomposed by bacteria. Over time, if conditions are right—for example, if the mud encasing the specimen remains undisturbed long enough for hardening to occur—the organism's shape is preserved. Not all organisms are equally likely to form fossils, however. Animals with bones or shells are more likely to be preserved than animals without such hard parts (think earthworms or jellyfish) that decay quickly. Likewise, the conditions permitting fossilization are rare. Put simply, to become a fossil, the organism has to be in just the right place at just the right time **(INFOGRAPHIC 15.1).**

Because not all organisms are preserved, the **fossil record** is not a complete record of past life. Nevertheless, the existing fossil record is remarkably rich and offers a revealing window into the past. **Paleontologists,**

INFOGRAPHIC 15.1

Fossils Form Only in Certain Circumstances

Not every organism that dies forms a fossil. Organisms are more likely to fossilize if they have bony skeletons or hard shells. In addition, the organism must be preserved quickly and kept undisturbed while mineralization or mud hardening occurs. Therefore, the fossil record is not a complete record of past life, but it has supplied an impressive body of evidence for evolution.

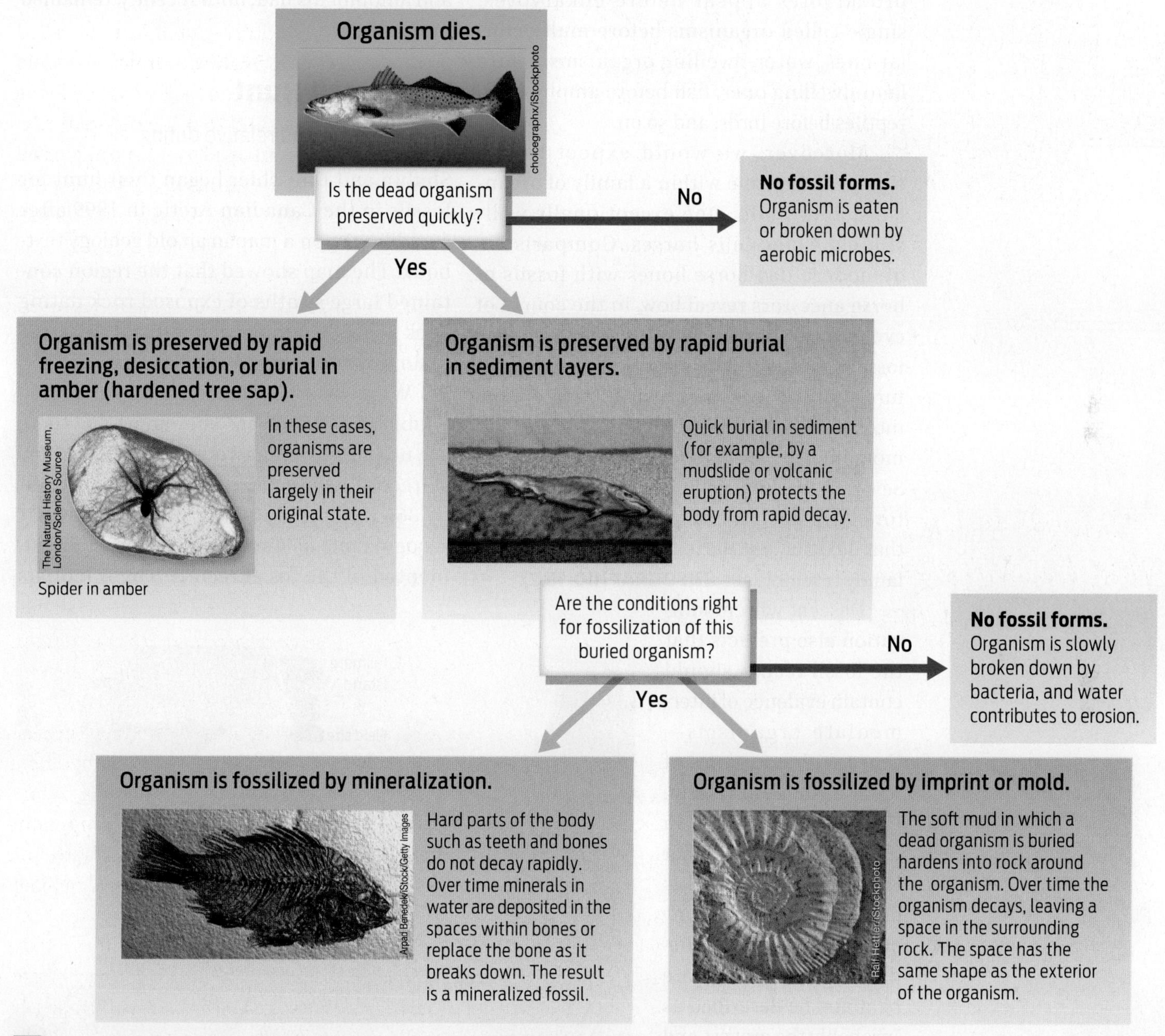

? Why are flies preserved in amber more common than fossilized flies?

scientists who study ancient life, have uncovered hundreds of thousands of fossils throughout the world, from many evolutionary time periods. When those fossils are arranged in order of age, they provide a tangible history of life on Earth. The fossil record also allows biologists to test certain tenets of Darwin's theory.

For example, the theory of evolution concludes that all organisms have descended from a single common ancestor that lived billions of years ago. If that is true, we would expect the fossil record to show an ordered succession of evolutionary stages as organisms evolved and diversified into new species. Indeed, that is exactly what we see: prokaryotes appear before eukaryotes, single-celled organisms before multicellular ones, water-dwelling organisms before land-dwelling ones, fish before amphibians, reptiles before birds, and so on.

Moreover, we would expect to see changes over time within a family of organisms, and we do. One exceptionally well-studied example is horses. Comparisons of modern-day horse bones with fossils of horse ancestors reveal how, in the course of evolution, horses lost most of their toes. The fossils show a series of changes in bone structure over time: the most recent fossils are the most similar to modern organisms, and the more ancient fossils are the most different. Several branches and lineages of horse ancestors have been identified, including many that died out, but the fossils all clearly share a family resemblance **(INFOGRAPHIC 15.2)**.

Descent with modification also predicts that the fossil record should contain evidence of intermediate organisms—those with a mixture of "old" and "new" traits. Darwin acknowledged in *The Origin of Species* that the fossil record of his day did not provide many examples of such intermediate organisms—a state of affairs he described as "probably the gravest and most obvious of all the many objections which may be urged against my views." Yet Darwin knew that if his hypothesis were correct, those intermediate fossils would eventually be found. He was right: many have been uncovered by paleontologists. Scientists have discovered animals with mixtures of reptile and bird characteristics, and animals with mixtures of reptile and mammal characteristics. But the transition between fish and amphibians had, until recently, remained more obscure.

The Fossil Hunt

▶ Radiometric and relative dating

Shubin and Daeschler began their hunt for fossils in the Canadian Arctic in 1999 after stumbling upon a map in an old geology textbook. The map showed that the region contained large swaths of exposed rock dating back 375–380 million years—just the period of time of interest to the researchers.

Why was this period so important to Shubin and Daeschler? They knew that there are no land-dwelling vertebrates in the fossil record before 385 million years ago. But by 365 million years ago, organisms easily recognizable as amphibians are well documented in the fossil record. The scientists

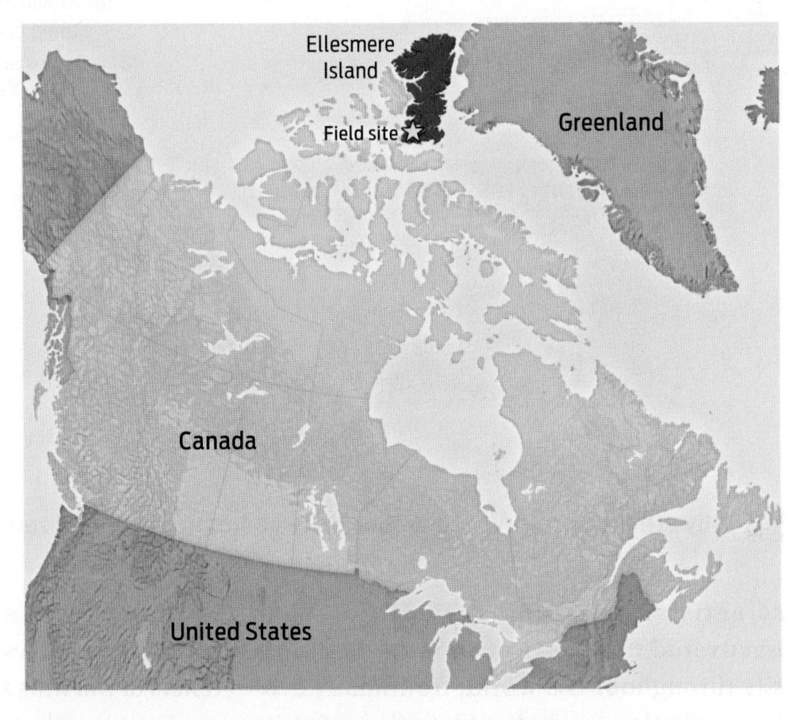

Ellesmere Island, which has an area nearly as large as Great Britain, contains Canada's most northern point.

INFOGRAPHIC 15.2

Fossils Reveal Changes in Species over Time

The fossil record of horses supports the theory of descent with modification. Forelimb fossils are similar to one another, but show changes over time from the earliest horse ancestors to modern-day horses as species diverged from a common ancestor. In the fossil record we can observe over time a reduction in toe number, as the central toe became dominant, allowing horses to move more rapidly in new prairielike environments.

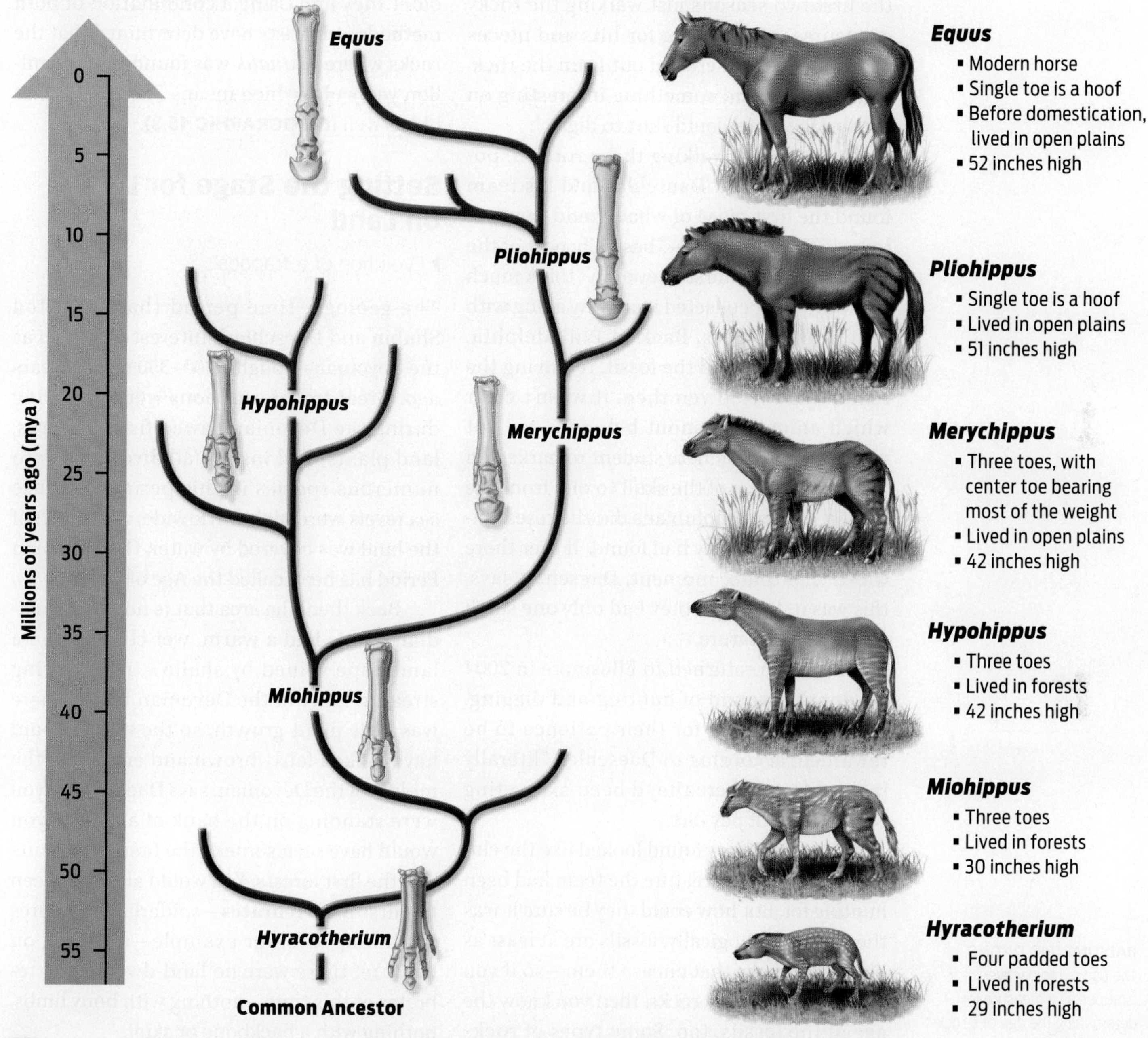

? **What changes can be observed between three-toed *Miohippus* and three-toed *Merychippus*?**

hypothesized that if they looked at rocks sandwiched in between these two time periods—those around 375 million years old—they might find one of Darwin's elusive "intermediates." Ellesmere Island is one of only three places on Earth where rocks of this time period are exposed. To Shubin and Daeschler's knowledge, no other paleontologists had explored the area, which meant it was a potential fossil gold mine.

Knowing exactly where to look for fossils was tricky, since Ellesmere Island covers 75,000 square miles. To locate the most promising dig site, the scientists first studied aerial photographs. Once on the ground, the scientists and their team split up and spent the first two seasons just walking the rocky exposures, prospecting for bits and pieces of fossils that had eroded out from the rock. When they found something interesting on the surface, they would start to dig.

It was while walking these rocky exposures in 2002 that Daeschler and his team found the first piece of what would turn out to be a *Tiktaalik* fossil—"basically part of the snout," he says. At first, they didn't think much of the find, but collected it anyway along with other fossil pieces. Back in Philadelphia, researchers cleaned the fossil, removing the remaining rock. Even then, it wasn't clear which animal the snout belonged to. Not until a visiting graduate student remarked on the resemblance of the skull to one from the earliest known amphibians did the researchers realize what they had found. If ever there was a "lightbulb" moment, Daeschler says, this was it. But, alas, they had only one small piece of the creature.

The team returned to Ellesmere in 2004 for another round of hunting and digging. It didn't take long for their patience to be rewarded: according to Daeschler, "literally inches" from where they'd been excavating before, they hit pay dirt.

The fossils they found looked like the elusive intermediate creature the team had been hunting for. But how could they be sure it was the right age? Logically, fossils are at least as old as the rocks that encase them—so if you know the age of the rocks, then you know the age of the fossils, too. Some types of rocks can be dated directly by **radiometric dating,** in which the proportion of certain radioactive isotopes in rock crystals serves as a geologic clock (isotopes and radiometric dating are described further in Chapter 16). Fossils found in or near these layers can be dated quite precisely. If fossils are found in rock layers that cannot be directly dated by radiometric dating, they can be dated indirectly by their position with respect to rocks or fossils of known age that are either deeper or shallower, a technique called **relative dating.** Generally speaking, the deeper the fossils, the older they are. Using a combination of both methods, scientists have determined that the rocks where *Tiktaalik* was found are 375 million years old, which means *Tiktaalik* is that old as well **(INFOGRAPHIC 15.3).**

RADIOMETRIC DATING
The use of radioactive isotopes as a measure for determining the age of a rock or fossil.

RELATIVE DATING
Determining the age of a fossil from its position relative to layers of rock or fossils of known age.

INVERTEBRATE
An animal without a backbone.

Setting the Stage for Life on Land

▸ Evolution of tetrapods

The geologic time period that attracted Shubin and Daeschler's interest is known as the Devonian—roughly 400–350 million years ago. Great transformations were occurring during the Devonian: jawed fishes, sharks, land plants, and insects all diversified into numerous species in this period. Because sea levels were high worldwide, and much of the land was covered by water, the Devonian Period has been called the Age of Fishes.

Back then, the area that is now the Canadian Arctic had a warm, wet climate and a landscape veined by shallow, meandering streams. Early in the Devonian Period there was little plant growth, so the world would have looked fairly brown and empty. By the middle of the Devonian, says Daeschler, if you were standing on the bank of a stream you would have seen some of the first land plants and the first forests. You would also have seen the first **invertebrates**—spiderlike creatures and millipedes, for example—crawling on land. Yet there were no land-dwelling vertebrates at this time—nothing with bony limbs, nothing with a backbone or skull.

By the late Devonian, things were changing quickly. By then, says Daeschler, "you had a green floodplain, a green world." This green world—a rich and productive ecosystem, with energy-rich leaf litter flowing into shallow streams—set the stage for the move of vertebrates onto land.

INFOGRAPHIC 15.3

How Fossils Are Dated

When an organism dies and is preserved, its remains may in time become fossilized and buried under layers of sediment, which accumulate on top of even older layers of sediment. When a fossilized organism is uncovered, its age can be determined by two main forms of dating: radiometric and relative.

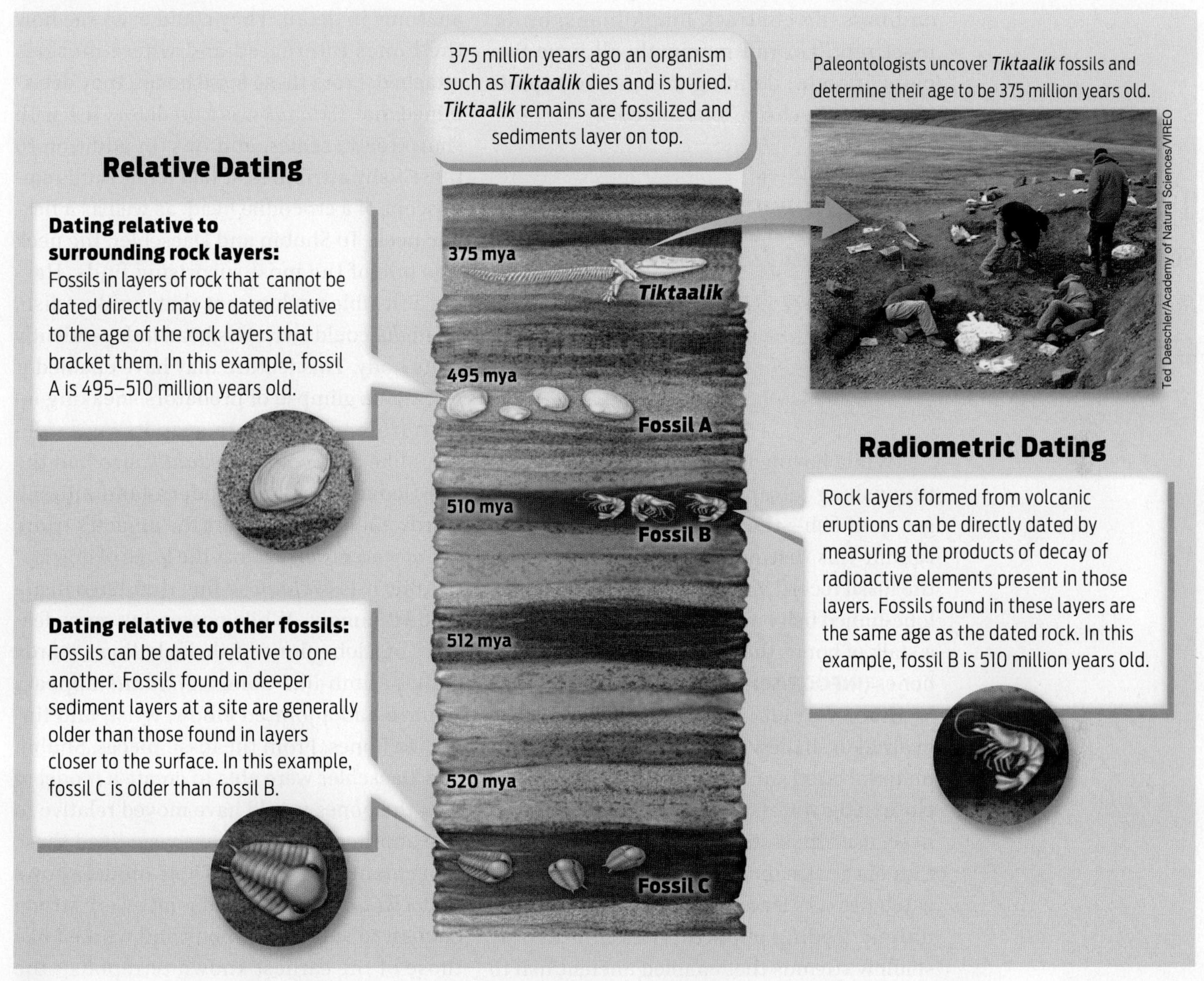

? **What is the approximate age of a fossil found in the blue-gray layer immediately below fossil B?**

The physical challenges of living on land are very different from those in water. Water provides buoyancy to aquatic animals, supporting their bodies and helping keep them afloat. By contrast, animals that walk on land have to cope with gravity. Air doesn't support land animals the way that water supports their aquatic counterparts, so their bodies need a sturdier structure. Animals on land can also dry out, which poses a danger to their survival because cells need water to function. And, of course, taking in oxygen is different on land and in water.

Of the many features that distinguish land animals from fish, biologists have singled out one as a key evolutionary milestone:

limbs. Fish do not have limbs, in the sense of jointed, bony appendages with fingers and toes. Instead, they have webbed fins. In most fishes, the fin bones are thin and fan out away from each other. These so-called ray-finned fishes include the modern-day perch, trout, and bass. By contrast, amphibians, birds, most reptiles, and mammals all have two pairs of limbs, defining them as **tetrapods** (from the Greek for "four-footed").

Tiktaalik *"splits the difference between something we think of as a fish and something we think of as a limbed animal."*

—Ted Daeschler

While having limbs is a key feature distinguishing tetrapods from fish, one small group of fish—the lobe-finned fish—seems to blur this distinction. First appearing in the fossil record about 400 million years ago, lobe-finned fish have fleshy fins supported by a stalk of bones that resemble primitive limb bones **(INFOGRAPHIC 15.4)**.

Lobe-finned fish are thought to have evolved in shallow streams, where rich plant material lured small fish and other creatures close to the water's edge. The lobe-finned fish likely used their sturdy fins to touch the bottom of the streambed while maneuvering to catch prey. As Daeschler explains, it was the unique feeding opportunities afforded by shallow streams that enabled ancient fish to start evolving features, like lobed fins, that were adaptive in shallow water. Through natural selection, these traits would have become more common in the fish population. But lobe-finned fish were still very far from being true tetrapods. *Tiktaalik* is a step closer: "It looks like a fish in that it has scales and fins," Shubin told reporters in 2006 after the discovery, "but when you look inside the skeleton you see how special it really is."

TETRAPOD
A vertebrate animal with four true limbs—that is, jointed, bony appendages with digits. Mammals, amphibians, birds, and reptiles are tetrapods.

The Fish That Did Push-Ups

▶ Analyzing intermediate fossils

Shubin and Daeschler were lucky: the fossils they found were so well preserved that they were able to study *Tiktaalik*'s skeletal anatomy in detail. They could even see how the bones interacted and where muscles attached. From these fossil bones, they determined that *Tiktaalik* was a predatory fish with sharp teeth, scales, and fins. In addition to these fishy attributes, it had a flat skull reminiscent of a crocodile head, as well as a flexible neck. To Shubin and Daeschler, the neck was one of the most surprising finds. Having a flexible neck meant that, unlike a fish, *Tiktaalik* could swivel its head independently of its body. This feature may have enabled it to catch a glimpse of predators sneaking up on it from behind, or to snap its jaws sideways like a crocodile. *Tiktaalik* also had the full-fledged ribs of a modern land animal, sturdy enough to support the animal's trunk out of water even against the force of gravity.

But it is *Tiktaalik*'s fins that have justly made it famous. While possessing many features of a lobe-finned fish, including a sturdy stalk of limb-like bones, *Tiktaalik* appears to have had a jointed elbow, wrist, and fingerlike bones. From the fossil pieces, Shubin and Daeschler were able to create a model of how the bones would have moved relative to one another, and they have visualized these movements on screen. Their model shows that *Tiktaalik*'s bones and joints were strong enough to support its body and worked like those of the earliest known tetrapods—the early amphibians. "This animal was able to hold its fin below its body, bend the fin out toward what we think of as a wrist, and bend the elbow," explains Daeschler. In other words, it was a fish that could do a push-up.

With this hybrid anatomy, *Tiktaalik* was not galloping across the land, of course. It probably lived most of the time in water, but Shubin and Daeschler suspect that *Tiktaalik* may have used its supportive fins to pull itself

INFOGRAPHIC 15.4

A Comparison of Fin Types

Tiktaalik's forelimbs were technically fins. However, they more closely resembled lobe fins than ray fins.

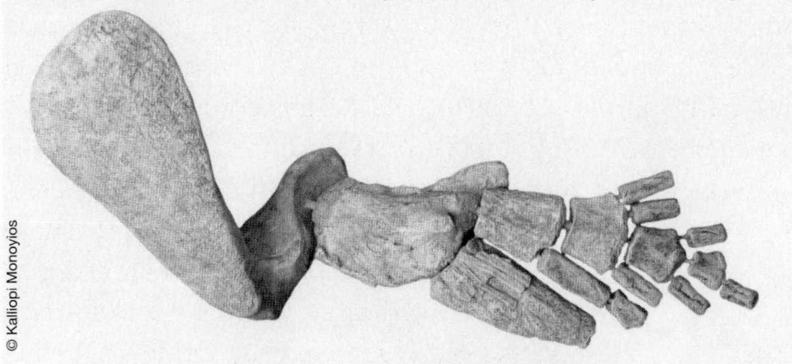

Tiktaalik Forelimb
Tiktaalik forelimbs had thick, weight-bearing bones covered in muscle and skin. The bones included a series of flexible wrist bones and digits ending in a small number of bony rays.

A physical model of *Tiktaalik's* forelimb fin, made from replicas of fossil bones fused with flexible wire

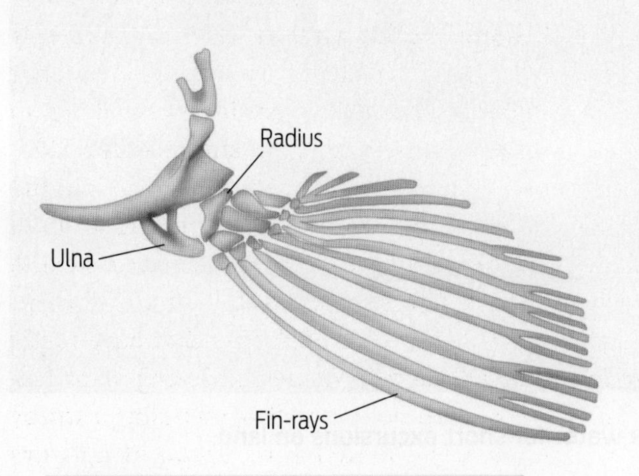

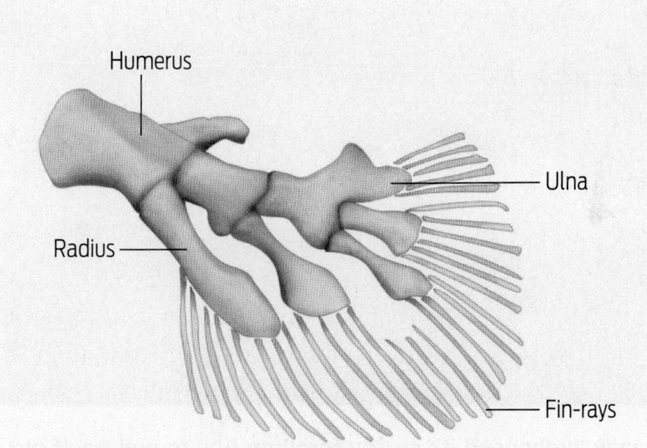

Ray Fin
Ray-finned fishes have fins that consist of webs of skin supported by a row of thin bony rays that branch from a series of bones connected to the body of the fish.

Lobe Fin
Lobe-finned fishes have sturdy fins that consist of rows of thick bones with short bony rays at the tips. These rows of bones all connect to a central bone, the humerus, which attaches to the body of the fish. The bony skeleton of the lobe fin is covered with muscle and skin.

? How does *Tiktaalik*'s forelimb differ from that of a lobe-finned fish?

out of the water for brief periods. "This is a fish that can live in the shallows and even make short excursions onto land," Shubin said. The ability to crawl onto land would certainly have been a useful trait in the Devonian, when open water was a brutal fish-eat-fish world but land was a predator-free paradise, full of nourishing bugs.

Like other fish living at the time, *Tiktaalik* is thought to have had both lungs and gills, which explains how it could breathe out of water for these short excursions. People sometimes assume that lungs were a late evolutionary adaptation, and that they came from modified gills, which modern-day fish use to breathe in water. But lungs—air-filled organs used for respiration—actually evolved very early in evolutionary history, more than 375 million years ago. They existed in ancient fish. We know this because scientists have

Claus Lunau/Science Source

Tiktaalik **likely used its sturdy forelimb fins to pull itself out of the water for short excursions on land.**

discovered fish fossils containing lung cavities, and calcified bones surrounding them, in former mudbanks. Most modern fish have retained their gills but lost their lungs over time (the lungs have evolved into a balloon-like structure called a swim bladder, which helps fish float). Some modern fish known as lungfish, however, have retained this ancestral trait. Lungfish are lobe-finned fish closely related to the lobe-finned fish from which *Tiktaalik* is believed to have descended.

Of course, no forethought was involved in the process of limb evolution or the emergence of other suitable land traits. Fish did not develop limbs for the purpose of walking on land. Rather, limbs first evolved in shallow water, where they proved adaptive and were thus retained in the descendants of the organisms that first developed them. Then, when an opportunity arose to take advantage of a tantalizing new habitat—land—the amphibious creatures already had the skeletal "toolkit."

For all its amphibian-like adaptations, *Tiktaalik* is still considered a fish because its limbs lack the true jointed fingers and toes that characterize tetrapod limbs (in other words, they're still fins). But it's by far the most tetrapod-like of all the ancient fishes discovered to date. Scientists have jokingly referred to it as a "fishapod" **(INFOGRAPHIC 15.5)**.

And that's what makes *Tiktaalik* such an important find: it occupies a midpoint between fish and tetrapods. "It very much fits in that gray area between things we typically call fish and things we typically call limbed animals," says Daeschler. Such intermediate, or transitional, fossils document important steps in the evolution of life on Earth. They help biologists understand how groups of organisms evolved, through natural selection, from one form into another. And they confirm that Darwin's theory of descent with modification—which predicts such intermediate forms—is correct.

Another famous transitional fossil is *Pakicetus,* an early whale. Unlike tetrapods, which evolved first in water and then spread to land, some land-dwelling creatures eventually

INFOGRAPHIC 15.5

Tiktaalik, an Intermediate Fossil

Tiktaalik possesses adaptations of both fish and tetrapods. The fish-like traits would have been useful when in water, while the tetrapod-like traits would have been useful when *Tiktaalik* ventured onto land. For all its amphibian-like characteristics, *Tiktaalik* is still technically a fish because its limbs lack true jointed fingers and toes, a defining feature of tetrapods.

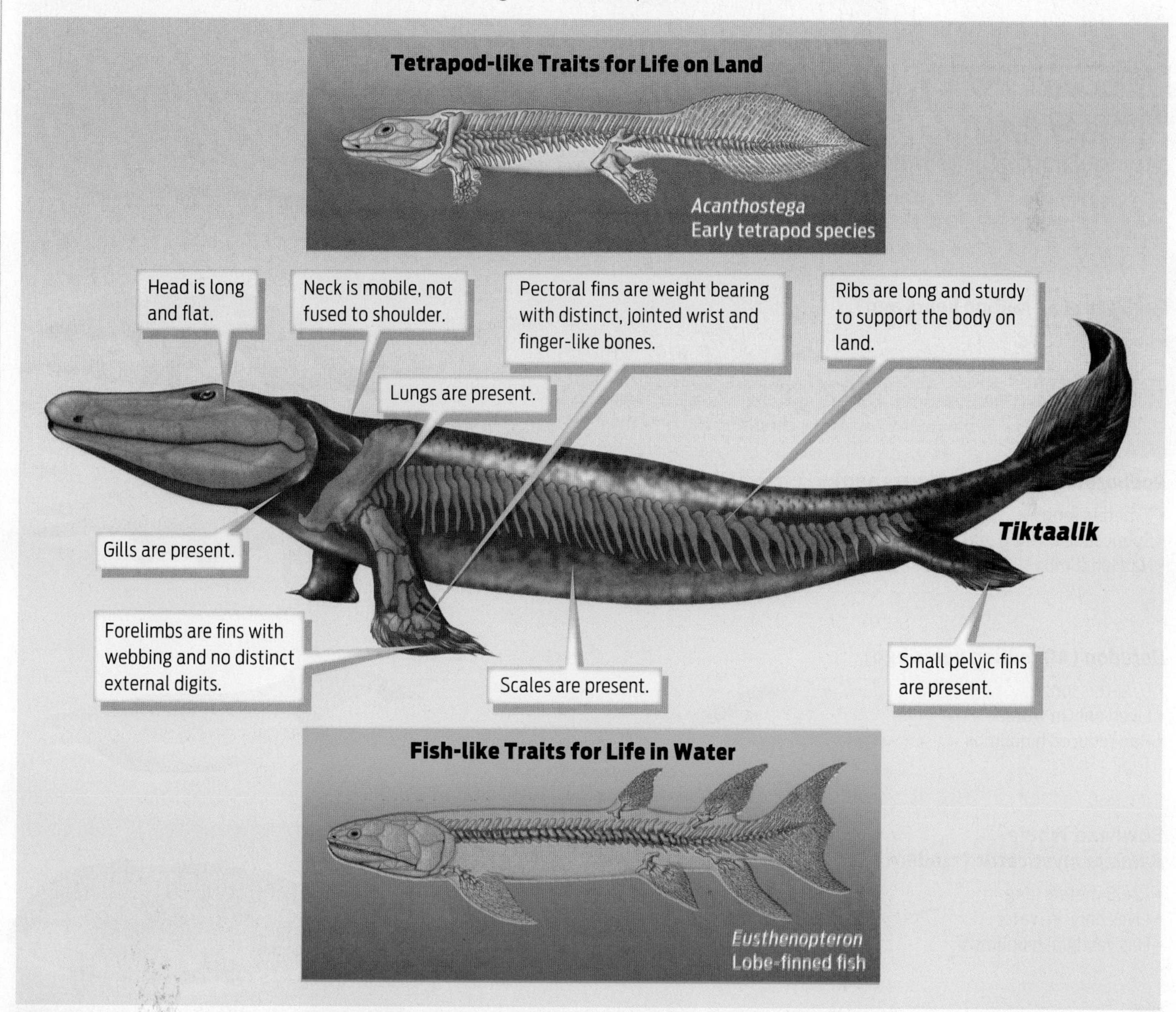

? Why would weight-bearing forelimbs be an advantage for *Tiktaalik*?

made their way back to the sea, adapting to an aquatic life once more. That group includes cetaceans—whales, porpoises, and dolphins. The ancestor of cetaceans was a wolf-size land-dwelling mammal that lived 50 million years ago. This animal, *Pakicetus,* had the body of a land animal, including four legs and paws. Yet its skull had features that are unique to whales, including a specific arrangement of ear bones. Fossils of early whales that came after *Pakicetus* show how whales became increasingly adapted to an aquatic existence **(INFOGRAPHIC 15.6)**.

A Fin Is a Paw Is an Arm Is a Wing

▸ Structural and developmental homology

In *The Origin of Species,* Darwin asked, "What can be more curious than that the hand of a man, formed for grasping, that of a mole for digging, the leg of the horse, the paddle of the porpoise, and the wing of the bat, should all be constructed on the same pattern, and should include similar bones, in the same relative positions?" To Darwin, this uncanny similarity was evidence that all these

INFOGRAPHIC 15.6

Another Evolutionary Transition: Land-Dwelling Mammals to Ocean-Dwelling Whales

Ancestors of whales were terrestrial mammals. A very early whale, *Pakicetus*, was also terrestrial. Transitional fossils illustrate adaptations for an increasingly aquatic existence, including specializations of the ear and neck and almost complete loss of the hindlimbs.

Pakicetus **(52 million years ago)**
- 1.8 meters long
- Spent time on land and in water
- Had hindlimbs

Rodhocetus **(46 million years ago)**
- 3 meters long
- Spent time on land and in water
- Had hindlimbs

Dorudon **(40 million years ago)**
- 6 meters long
- Lived only in water
- Had reduced hindlimbs

Bowhead Whale
Balaena mysticetus **(modern species)**
- 15–18 meters long
- Lives only in water
- Has vestigial hindlimbs

? What hindlimb differences do you observe between species that lived only in water compared to those that also lived on land?

INFOGRAPHIC 15.7

Forelimb Homology in Fish and Tetrapods

The number, order, and underlying structure of the forelimb bones are similar in all the groups illustrated below, evidence that they are homologous structures. The differences in the relative width, length, and strength of each bone contribute to the specialized function of each forelimb, and reflect evolutionary adaptations to different environments.

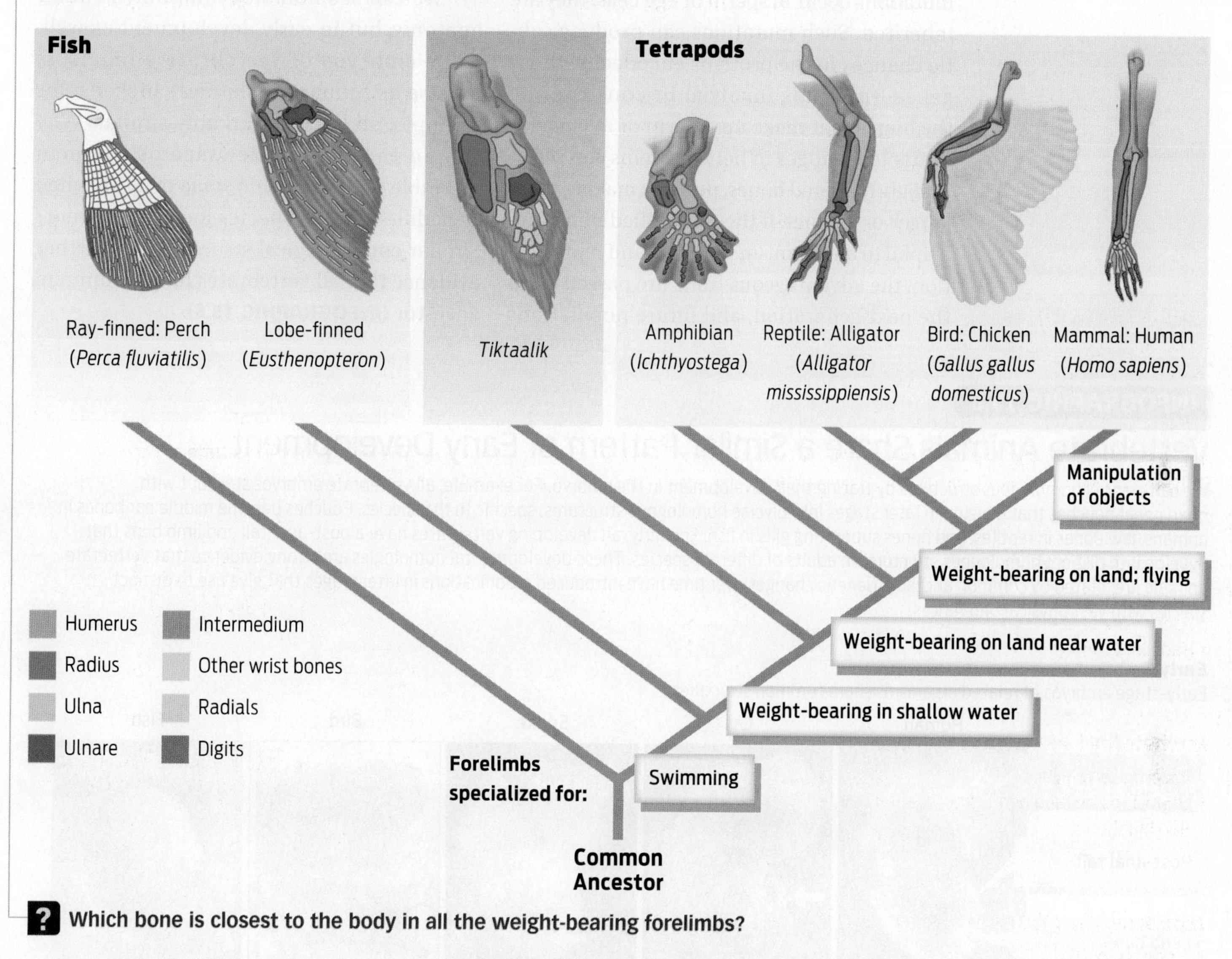

? Which bone is closest to the body in all the weight-bearing forelimbs?

organisms were related—that they share a common ancestor in the ancient past.

The fact that all tetrapods share the same forelimb bones, arranged in the same order, is an example of **homology**—a similarity due to common ancestry. Before Darwin, comparative anatomists had identified many such similarities in anatomy; what they lacked was a satisfactory explanation for why such similarity should exist. Darwin provided that explanation: homologous structures are similar because they are inherited from the same ancestor—in this case, an amphibious creature like *Tiktaalik*. Why is this significant? Think of it this way: every time you bend your wrist back and forth—to swipe a paint brush or hold a smartphone to your ear, for example—you are using structures that first evolved 375 million years ago in fish. As Shubin points out, "This is not just some archaic, weird branch of evolution; this is *our* branch of evolution" **(INFOGRAPHIC 15.7)**.

HOMOLOGY
Anatomical, genetic, or developmental similarity among organisms due to common ancestry.

If they have the same bones, why then do a human arm and a bird wing look so different? Remember that during the process of cell division, mutations are continually being introduced into the DNA of genes. When these mutations occur in sperm or egg cells, they are inherited. Such mutations can produce subtle changes in the proteins encoded by those genes—proteins involved in constructing the bones that make up an arm or a wing, for example. Changes in bone proteins can result in slightly altered bones, perhaps making them longer or thinner. If these modified bones are helpful to an organism's survival and reproduction, the advantageous traits are passed on to the next generation, and future populations possess the same adaptations. This "descent with modification" (Darwin's phrase again) results in diverse organisms sharing common—homologous—structures and putting them to different uses.

We can see homology not only in adult anatomy, but in early development as well. Early embryos of vertebrate animals as diverse as humans, fish, and chickens, for example, all look remarkably similar. Why should the embryonic stage of a human resemble the embryonic stage of a fish when the adults of each species look so different? Similar embryological structures are further evidence that all vertebrates have a common ancestor **(INFOGRAPHIC 15.8)**.

INFOGRAPHIC 15.8

Vertebrate Animals Share a Similar Pattern of Early Development

We can identify homologous structures by tracing their development in the embryo. For example, all vertebrate embryos start out with pharyngeal pouches that develop in later stages into diverse homologous structures, specific to the species. Pouches become middle ear bones in humans, jaw bones in reptiles, and bones supporting gills in fish. Similarly, all developing vertebrates have a post-anal tail and limb buds that develop into distinct homologous structures in adults of different species. These developmental homologies are strong evidence that vertebrate animals are related by common ancestry. Genetic changes over time have introduced modifications in later stages that give rise to distinct species with vast physical differences.

Early Embryos

Early-stage embryos of related organisms share common structures.

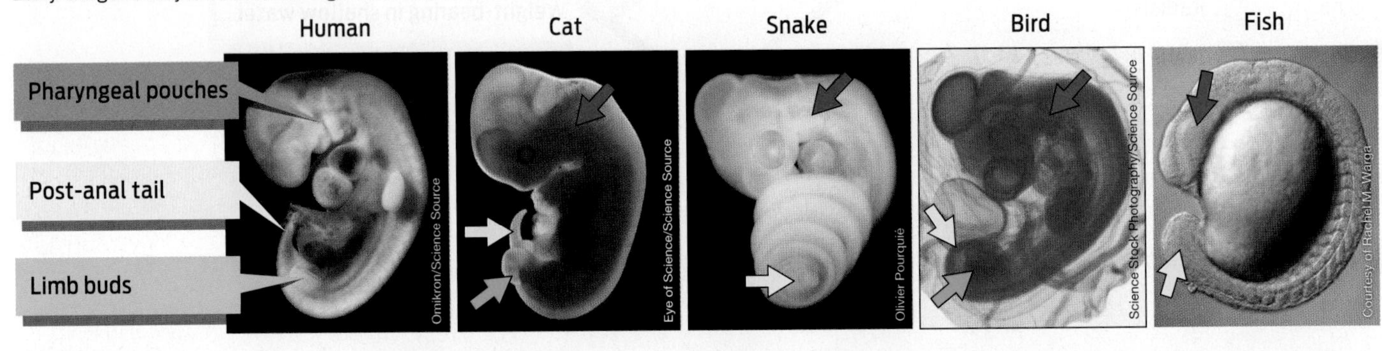

Adult Organisms

Later in development, these structures take on species-specific shape and function.

? Of the organisms shown, which have a post-anal tail as adults?

Development helps us solve other evolutionary conundrums as well, such as why snakes (which are reptiles) don't have limbs like other tetrapods. In fact, snake embryos *do* possess the beginnings of limbs, but these limb buds remain rudimentary and do not develop into full-fledged limbs (although you can still see stubby hindlimbs in some species of snake today). Such **vestigial structures,** which serve no apparent function in a modern organism, are strong evidence for evolution. These now apparently useless features are inherited from an ancestor in whom they *did* serve a function. Other examples of vestigial structures include the human tailbone and the phenomenon of "goose pimples" (technically called erector pili). In fur-covered animals, "goose pimples" help puff fur up to better maintain heat.

Zooming in even further, to the molecular level, we find still more examples of homology—and thus more evidence of common ancestry. Scientists have known since the 1950s that DNA is the molecule of heredity, and that it is shared by all living organisms on Earth. Every molecule of DNA—whether from a fish, maple tree, bacterium, or human—is made of the same four nucleotides (A, C, T, and G), and all organisms read the information encoded by those nucleotides to make proteins in the same basic way, using the universal genetic code (see Chapter 8). Why should all living things use the same system of encoding genetic information? The best explanation is that this system was the one used by the ancient ancestor of all living organisms, passed on to all of its descendants, and preserved throughout billions of years of evolution.

DNA and Descent

▶ Molecular homology

While all living organisms share DNA and the genetic code, the actual sequence of DNA nucleotides will differ between species in characteristic ways. That's because (as described in Chapter 9) errors in DNA replication and other mutations are continually introducing variation into gene sequences (and the proteins they encode). Over time, neutral and advantageous mutations will tend to be preserved in a species, while harmful mutations will tend to be selected against and eliminated. In addition, much of an organism's DNA consists of long stretches of noncoding sequences with no known function. Because mutations in these regions have no effect on an organism, they accumulate over time. As mutations are passed on to descendants, the number of sequence differences between the ancestor and its descendants grows. The accumulation of sequence differences is slow in the case of gene sequences coding for critical proteins whose structures are well adapted to their functions, but more rapid in the case of noncoding DNA, which is not involved in making proteins. The DNA sequences of closely related species will therefore be more similar than those of species that are more distantly related.

For example, when scientists looked at one specific region of DNA—the cystic fibrosis transmembrane regulator (*CFTR*) region—they discovered that human DNA in

VESTIGIAL STRUCTURE
A structure inherited from an ancestor that no longer serves a clear function in the organism that possesses it.

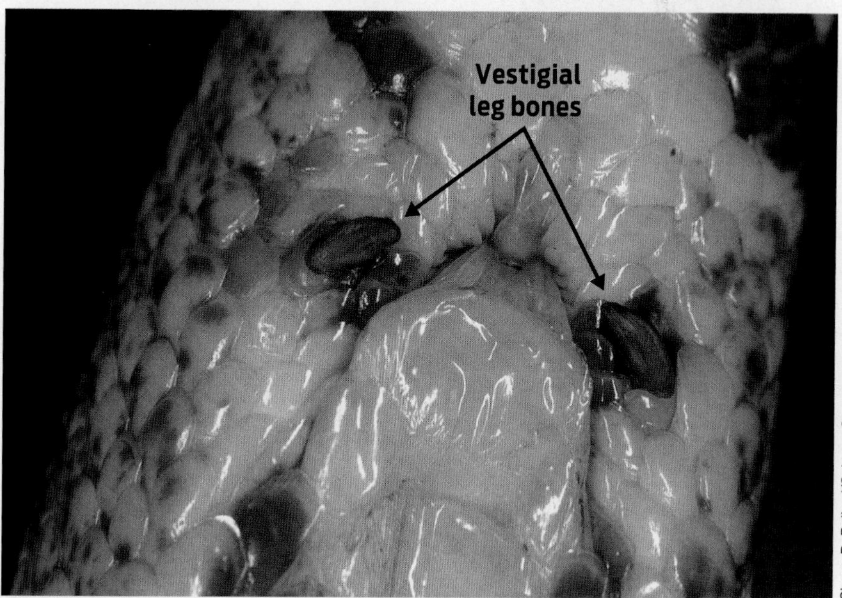

Some snakes have remnants of pelvis and leg bones left over from their four-legged ancestors.

this region is 99% identical to chimpanzee DNA. The near identity of this DNA sequence in these two species reflects the fact that humans and chimps share a common ancestor that lived relatively recently—just 5–7 million years ago. By contrast, human DNA is 85% identical to the DNA of a mouse at this same region, which makes sense given that humans and mice share a common ancestor that lived between 60 and 100 million years ago. Even less sequence identity would be seen between a human and a toad, whose common ancestor—a lobed-finned fish—lived roughly 375 million years ago. The more distantly related two species are, the more sequence differences you will see

INFOGRAPHIC 15.9

DNA Sequences Can Establish How Closely Species Are Related

Related organisms share DNA sequences inherited from a common ancestor. Over time, the sequence in each species acquires independent mutations. The more time that has passed, the greater the number of sequence differences that will be present. Thus, the percentage of nucleotides that differ between two species gives an indication of the evolutionary distance between them.

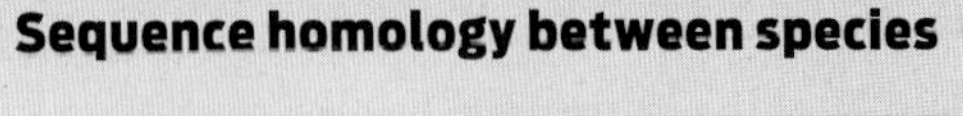

Species A GGTATCGAGGTTCTACATTGCAACTTCTAC

Close relative GGAAACGAGGTTCTACATTGCCACTTCTAC

Distant relative GGAAACGAGGTTCGACATAGCCACTTCTAC

3 differences in 30 nucleotides
3/30 = 10%, or 90% similarity

5 differences in 30 nucleotides
5/30 = 17%, or 83% similarity

Similarity to human DNA sequences for the *CFTR* region

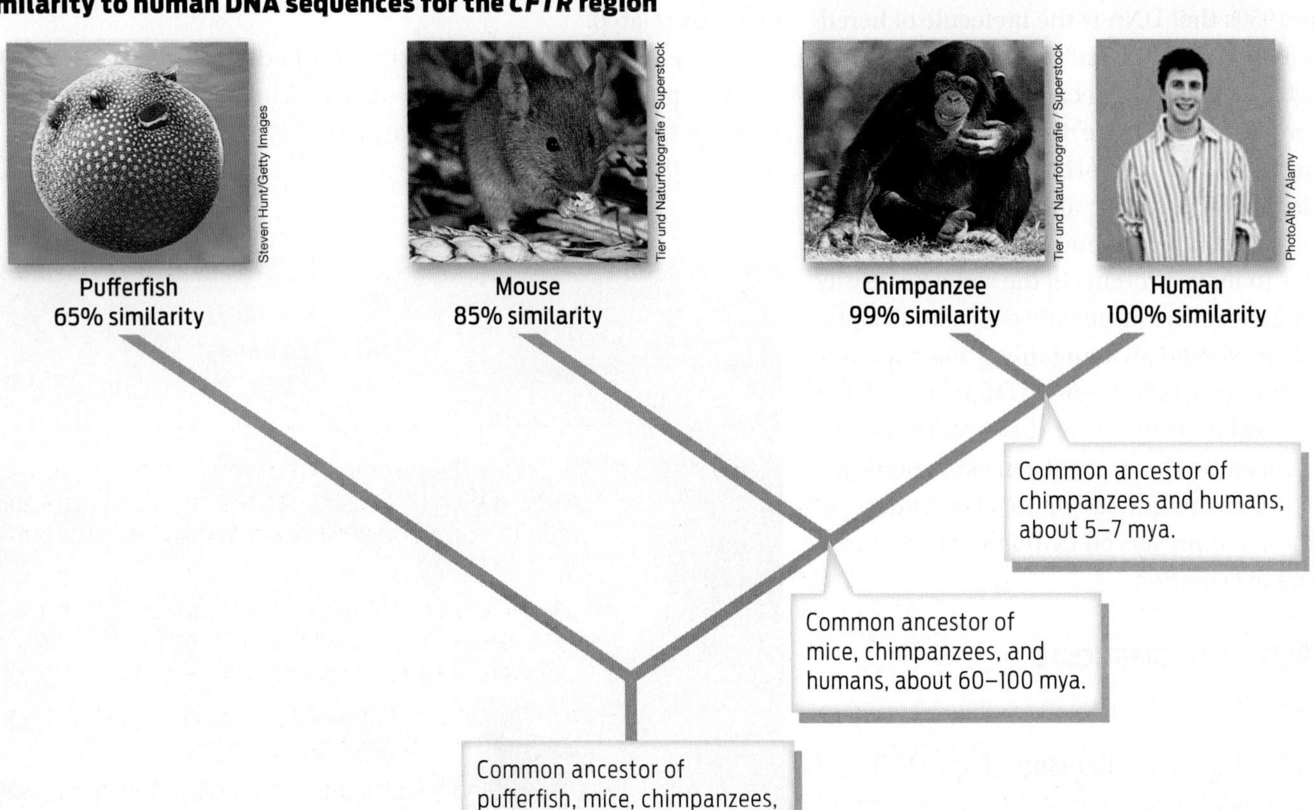

Data from Thomas, J.W. et al. (2003) *Nature* 424:788–793

? What is the percent similarity between the close relative and the distant relative shown in the top panel?

in DNA sequences, and the less homology. DNA serves as a kind of molecular clock: each additional sequence difference is like a tick of the clock, showing the amount of time that has elapsed since the two species shared a common ancestor **(INFOGRAPHIC 15.9)**.

When combined with evidence from the fossil record, anatomy, development, and biogeography (a fourth line of evidence, discussed in Chapter 16), molecular data become a powerful tool for understanding evolution. As we'll see in Chapter 16, DNA evidence is often a more reliable clue to common ancestry than is physical appearance, and can serve as a check on conclusions derived from the fossil record or anatomy. DNA is also deepening our knowledge of how limbs are constructed at the molecular level. Scientists working in Shubin's lab have shown that the same genes orchestrate limb development in animals as diverse as chickens, mice, and humans. In 2016, they discovered that these same genes coordinate fin development in fish—though with a different end result (fins rather than limbs). Learning how these genes work and how changes in their DNA sequences can produce large-scale changes in body plan or limb structure is an exciting area of biology right now, informally known as "evo-devo" (short for "evolutionary developmental biology").

Filling in the Gaps

Asked what he thinks is most interesting about the discovery of *Tiktaalik*, Daeschler homes in on what he says is a popular misconception about the fossil record—that it's "spotty" and "chaotic." That's simply not true, he says. Despite the fact that it does not record *all* past life, the fossil record is still "very good"—so good, in fact, that you can use it to make and test predictions. You can, for example, look at the fossil record of fish and tetrapods and—suspecting on the basis of anatomy that the two groups are related—hypothesize that an intermediate-looking animal must have existed at some point. Then you can go look for it. Daeschler refers to this process as "filling in the gaps," and it's exactly what he and Shubin did with *Tiktaalik*. They knew, from the existing fossil record, when such a creature was likely to have existed, so then it was just a question of where to look for it.

For Shubin and Daeschler, *Tiktaalik* is exciting because it shows that our understanding of evolution is correct: "It confirms that we have a very good understanding of the framework of the history of life," says Daeschler. "We predicted something like *Tiktaalik*, and sure enough, with a little time and effort, we found it." ■

Chapter 15 Summary

Driving Question 1 How does the fossil record reveal information about evolutionary changes?

- The theory of evolution—what Darwin called "descent with modification"—draws two main conclusions about life: that all living things are related, sharing a common ancestor in the distant past; and that the species we see today are the result of natural selection operating over millions of years.
- The theory of evolution is supported by a wealth of evidence, including fossil, anatomical, and DNA evidence.
- Fossils are the preserved remains or impressions of once-living organisms that provide a record of past life on Earth. Not all organisms are equally likely to form fossils.
- Fossils can be dated directly or indirectly: on the basis of the age of the rocks they are found in, or on their position relative to rocks or fossils of known ages.
- When fossils are dated and placed in sequence, they show how life on Earth has changed over time.
- As predicted by descent with modification, the fossil record shows the same overall pattern for all lines of descent: younger fossils are more similar to modern organisms than are older fossils.

Driving Question 2 What features make *Tiktaalik* a transitional fossil, and what role do these types of fossils play in the fossil record?

- Descent with modification predicts the existence of intermediate organisms that possess mixtures of "old" and "new" traits.
- Scientists have found fossils of many such intermediate organisms, including *Tiktaalik*, which has features of both fish and tetrapods (four-limbed vertebrates).

Driving Question 3 What can anatomy and DNA reveal about evolution?

- An organism's anatomy reflects adaptation to its environment. A changed environment provides opportunities for the selection of new adaptations.
- Homology—the anatomical, developmental, or genetic similarities shared among groups of related organisms—is strong evidence that those groups have descended from a common ancestor.
- Homology can be seen in the common bone structure of the forelimbs of tetrapods, the similar embryonic development of all vertebrate animals, and the universal genetic code.
- Many genes, including those controlling limb development, are shared among distantly related species—an example of molecular homology owing to common ancestry.
- DNA can be used as a molecular clock: more-closely related species show greater DNA sequence homology than do more-distantly related species.

More to Explore

- *Tiktaalik roseae* home page: http://tiktaalik.uchicago.edu
- PBS, *Your Inner Fish* (2014): http://www.pbs.org/your-inner-fish/home/
- Shubin, N. (2008). *Your Inner Fish: A Journey into the 3.5-Billion-Year History of the Human Body*. New York: Random House.
- Nakamura, T., et al. (2016). Digits and fin rays share common developmental histories. *Nature* 537:225–228.
- Daeschler, E. B., et al. (2006). A Devonian tetrapod-like fish and the evolution of the tetrapod body plan. *Nature* 440:757–763.
- PBS, *Nova* (2009). What Darwin Never Knew, based on Carroll, S. (2006) *Endless Forms Most Beautiful: The New Science of Evo Devo*. New York: Norton.

CHAPTER 15 Test Your Knowledge

Driving Question 1 How does the fossil record reveal information about evolutionary changes?

By answering the questions below and studying Infographics 15.1, 15.2, and 15.3, you should be able to generate an answer for this broader Driving Question.

Know It

1. Which of the following is most likely to leave a fossil that represents most of the organism?
   a. a jellyfish
   b. a worm
   c. a wolf
   d. an octopus (an organism that lacks a skeleton)
   e. All of the above are equally likely to leave a fossil.

2. Generally speaking, if you are looking at layers of rock, at what level would you expect to find the newest—that is, the youngest—fossils?

3. You are examining a column of soil that contains vertebrate fossils from deeper to shallower layers. Would you expect a fossil with four limbs with digits to occur higher or lower in the soil column relative to a "standard" fish? Explain your answer.

4. What can the fossil shown below tell us about the structure and lifestyle of the organism that left it? Describe your observations.

Mantonature/iStockphoto

Use It

5. You have molecular evidence that leads you to hypothesize that a particular group of soft-bodied sea cucumbers evolved at a certain time. You have found a fossil bed with many hard-shelled mollusks dating from the critical time, but no fossil evidence to support your hypothesis about the sea cucumbers. Does this cause you to reject your hypothesis? Why or why not?

6. A specific type of oyster is found in North American fossil beds dated from 100 million years ago. If similar oyster fossils are found in European rock, in layers along with a novel type of barnacle fossil, what can be concluded about the age of the barnacles? Explain your answer.

Bring It Home

Apply Your Knowledge

7. Do an Internet search to find out about fossils discovered in your home state. Determine what kinds of organisms they represent, how old they are, and where in your state you would need to go to have a chance of finding fossils in the field.

Driving Question 2 What features make *Tiktaalik* a transitional fossil, and what role do these types of fossils play in the fossil record?

By answering the questions below and studying Infographics 15.4, 15.5, and 15.6, you should be able to generate an answer for this broader Driving Question.

Know It

8. Which of the following features of *Tiktaalik* is not shared with other bony fishes?

a. scales
b. teeth
c. a mobile neck
d. fins
e. none of the above

9. *Tiktaalik* fossils have both fishlike and tetrapod-like characteristics. Which characteristics are related to supporting the body out of the water?

Use It

10. *Tiktaalik* fossils are described as "intermediate" or "transitional" fossils. What does this mean? Why are transitional organisms significant in the history of life?

11. *Tiktaalik* has been called a "fishapod"—part fish, part tetrapod. Speculate on the fossil appearance of its first true tetrapod descendant. What features would distinguish it from *Tiktaalik*? How old would you expect those fossils to be relative to *Tiktaalik*?

12. If some fish evolved modifications that allowed them to be successful on land, why didn't fish just disappear? In other words, why are there still plenty of fish in the sea if the land presented so many favorable opportunities?

Driving Question 3 What can anatomy and DNA reveal about evolution?

By answering the questions below and studying Infographics 15.7, 15.8, and 15.9, you should be able to generate an answer for this broader Driving Question.

Know It

13. Compare and contrast the structure and function of an eagle wing with the structure and function of a human arm.

14. Vertebrate embryos have structures called pharyngeal pouches. What do these structures develop into in an adult human? In an adult bony fish?

15. You have three sequences of a given gene from three different organisms. How could you determine how closely the three organisms are related?

Use It

16. What is the evolutionary explanation for the fact that both human hands and otter paws have five digits?

17. Could you use the presence of a tail to distinguish a human embryo from a chicken embryo? Why or why not?

18. If, in humans, the DNA sequence TTTCTAGGAATA encodes the amino acid sequence phenylalanine–leucine–glycine–isoleucine, what amino acid sequence will that same DNA sequence specify in bacteria?

19. Gene *X* is present in yeast and in sea urchins. Both produce protein X, but the yeast protein is slightly different from the sea urchin protein. What explains this difference? How might you use this information to judge whether humans are closer evolutionarily to yeast or to sea urchins?

Interpreting Data

Apply Your Knowledge

20. The gene responsible for hairlessness in Mexican hairless dogs is called corneodesmosin (*CDSN*). This gene is present in other organisms. Look at the sequence of a portion of the *CDSN* gene from pairs of different species, as shown in the table. For each pair, determine the number of differences. From the variations in this sequence, which organism appears to be most closely related to humans? Which organism appears to be least closely related to humans?

Species	Sequence
Homo sapiens (human)	ACTCCGGCCCCTACATCCCCAGCTCCCA
Canis lupus familiaris (dog)	ATTCTGGCTCCTACATTTCCAGCTCCCA
Homo sapiens (human)	ACTCCGGCCCCTACATCCCCAGCTCCCA
Pan troglodytes (chimpanzee)	ACTCCGGCCCCTACATCCCCAGCTCCCA
Homo sapiens (human)	ACTCCGGCCCCTACATCCCCAGCTCCCA
Sus scrofa (pig)	AGTCTGGCTCCTACATCTCCAGCTCCCA
Homo sapiens (human)	ACTCCGGCCCCTACATCCCCAGCTCCCA
Macaca mulatta (rhesus monkey)	ACTCTGGCCCCTACATCCCCAGCTCCCA

Mini Case

Apply Your Knowledge

21. Fossils allow us to understand the evolution of many lineages of plants and animals. They therefore represent a valuable scientific resource. What if *Tiktaalik* (or an equally important transitional fossil) had been found by amateur fossil hunters and sold to a private collector? Do you think there should be any regulation of fossil hunting to prevent the loss of valuable scientific information from the public domain?

EVOLUTION Q&A

From moon rocks to DNA, clues to the history of life on Earth

SINCLAIR STAMMERS/Science Source

DRIVING QUESTIONS

1 What do we know about the history of life on Earth, and how do we know it?

2 What factors help to explain the distribution of species on Earth?

3 What are the major groups of organisms, and how are organisms placed in groups?

Q&A

The history of life on Earth is a grand pageant, spanning eons, with a dizzying cast of characters—most of whom have long since exited the stage. Cataloguing all this diversity and understanding how it arose is a daunting challenge. To help make sense of this imposing sweep of history, scientists rely on a range of methods and disciplines, from geology and biochemistry to genetics and biogeography.

GEOLOGY

Q How old is Earth, and how do we know?

A When the *Apollo* 11 astronauts Neil Armstrong and Buzz Aldrin returned to Earth from their historic 1969 moon walk, they carried with them a cargo of lunar rock chipped from the moon's surface. Embedded within these hunks of shimmering anorthosite lay clues to the earliest history of our solar system, including the planet we call home.

RADIOMETRIC DATING
The use of radioactive isotopes as a measure for determining the age of a rock or fossil.

RADIOACTIVE ISOTOPE
An unstable form of an element that decays into another element by radiation—that is, by emitting energetic particles.

The Genesis Rock, a sample of lunar crust from about the time the moon was formed, was retrieved by *Apollo* 15 astronauts James Irwin and David Scott in 1971. The scale at the top is in centimeters; the cube is 1 cm on a side.

According to the most widely accepted explanation for the formation of our solar system, the planetary objects in our solar system are the result of a single event: the collapse of a swirling solar nebula. Both the sun and the planets formed out of the cosmic dust of this collapsing gas cloud. Since all the planets were formed at roughly the same time, we can date the age of the solar system by dating any planetary object within it. Of the many moon rocks obtained over the course of the six Apollo missions, the oldest have been calculated to be some 4.4 to 4.5 billion years old. That means that Earth is at least that old as well.

Why go to the moon to date Earth? With the exception of a few meteorite battle scars, the moon's surface has remained largely intact over the course of its existence. By comparison, Earth is a swirling ball of molten lava that continuously churns and digests its rocky outer crust. Because of this perpetual change in our planet, it is difficult to find original, undisturbed rocks from Earth's earliest period. (The oldest known rocks on Earth's surface, the Acasta Gneiss in a remote region of northern Canada, are 4.0 billion years old.)

From moon rocks as well as material from meteorites that have fallen to Earth, scientists estimate that the age of Earth—and of the solar system more generally—is 4.54 billion years, give or take a few million years.

How are such rocks—whether extraterrestrial or earthly—dated? The most important method is **radiometric dating,** in which the amount of radioactivity present in a rock is used as a geologic clock. When rocks form, the minerals in them contain a certain amount of **radioactive isotopes.** These isotopes are unstable atoms of elements such as uranium, potassium, and rubidium that decay into other atoms. They are identified by the name of the element and a mass number—protons plus neutrons—of the particular isotope. For example, the isotope uranium-238 has 92 protons and 146 neutrons, which are added together to get a mass number of 238.

Radioactive isotopes decay by releasing high-energy particles from the nucleus, a change that literally transforms one element into another. For example, an atom of the radioactive isotope uranium-238 eventually decays into a stable atom of lead-206. The time it takes for half the isotope in a sample to break down is called its **half-life.**

Different radioactive isotopes decay at different rates. Uranium-238 has a half-life of 4.5 billion years, whereas potassium-40 has a half-life of 1.3 billion years. Carbon-14 is useful for dating once-living, organic remains because its half-life is relatively short: it decays to nitrogen-14 in just 5,730 years. Because the isotopes decay at a known and constant rate, they can be used to determine the age of the materials in which they're found **(INFOGRAPHIC 16.1)**.

Scientists estimate that the age of Earth—and of the solar system more generally—is 4.54 billion years.

Rocks are typically characterized as either sedimentary or igneous. **Sedimentary rocks**

HALF-LIFE
The time it takes for one-half of a sample of a radioactive isotope to decay.

SEDIMENTARY ROCK
Rock formed from the compression of layers of particles eroded from other rocks.

INFOGRAPHIC 16.1

Unstable Elements Undergo Radioactive Decay

Radioactive isotopes are unstable versions of elements that undergo a process of radioactive decay, in which they emit energy and are converted to another element.

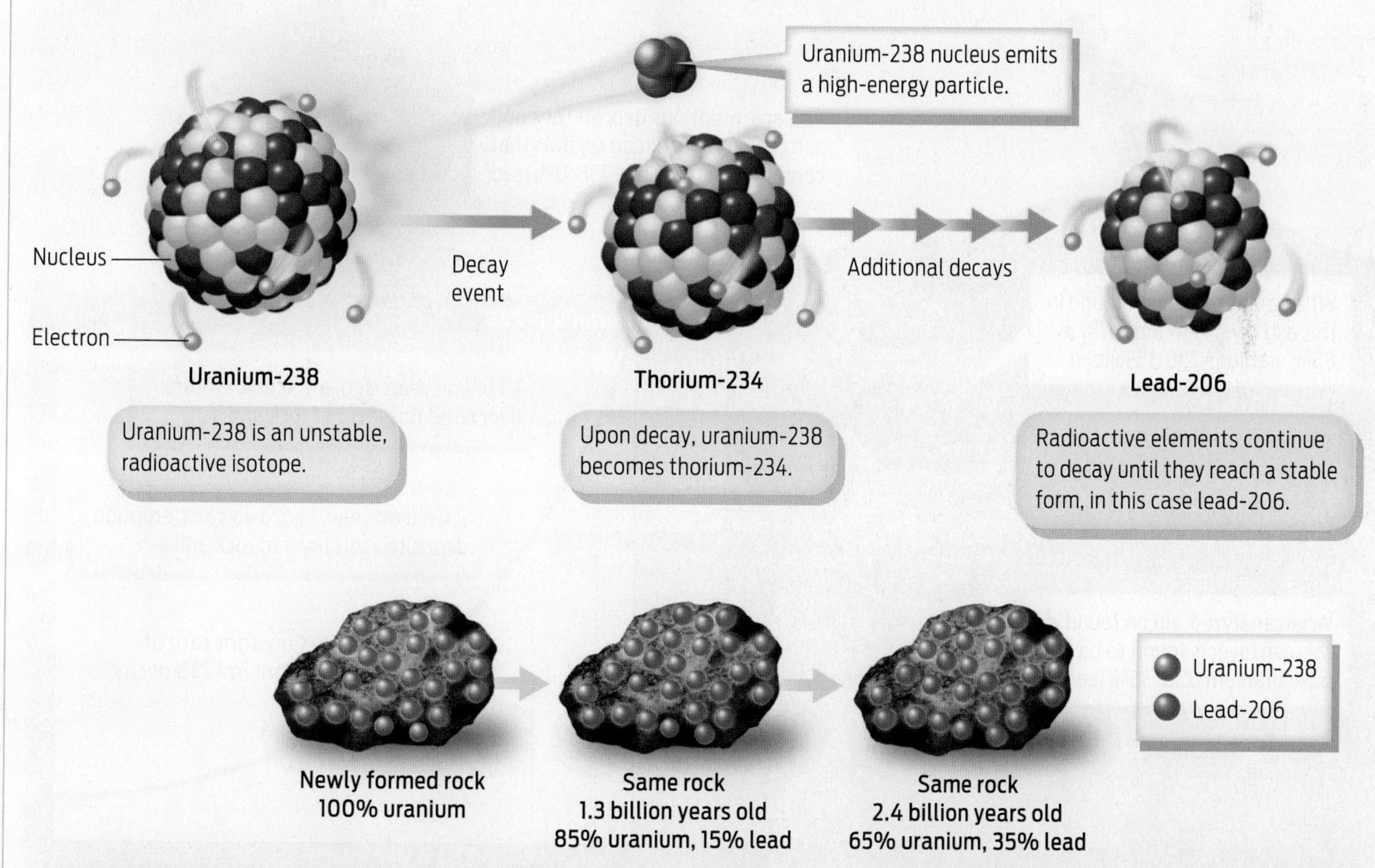

The half-life of uranium-238 is 4.5 billion years, meaning that it takes that long for half the amount of uranium in a sample to decay to lead-206. Over time, rock containing uranium-238 will have less radioactive uranium and more lead-206.

? **Define the half-life of a radioactive element.**

are formed gradually over time from the compaction of sand and dust, whereas **igneous rocks** are formed suddenly as a result of volcanic eruptions ("igneous" is Latin for "fire"). Radiometric dating is typically performed on igneous rocks. To date really old rocks, the uranium–lead method is used. Here's how it works: when igneous rocks form, crystals of the mineral zircon are produced within the rock. Zircon crystals have a highly ordered structure that incorporates uranium but excludes lead. Thus, when these crystals first form, there is no lead present and the radioactive clock is set to zero. Over time, the uranium decays at a constant rate into lead-206. By measuring the ratio of uranium to lead in these crystals, scientists can calculate the age of the rock **(INFOGRAPHIC 16.2)**.

By using radioactive dating, scientists have established that Earth is indeed quite old—old enough for evolution to have been acting for billions of years. They have also been able to establish precisely the dates of key events in the timeline of life on Earth as represented in the fossil record.

IGNEOUS ROCK
Rock formed from the cooling and hardening of molten lava.

INFOGRAPHIC 16.2

Radioactive Decay Is Used to Date Some Rock Types

Some rock types, like those produced during volcanic eruptions, contain radioactive minerals like zircon that can be used to determine the age of the rock. Because the isotope uranium-238 decays to lead at a constant rate, the age of rock layers containing these minerals can be calculated by measuring the ratio of uranium-238 to lead-206 in the mineral sample.

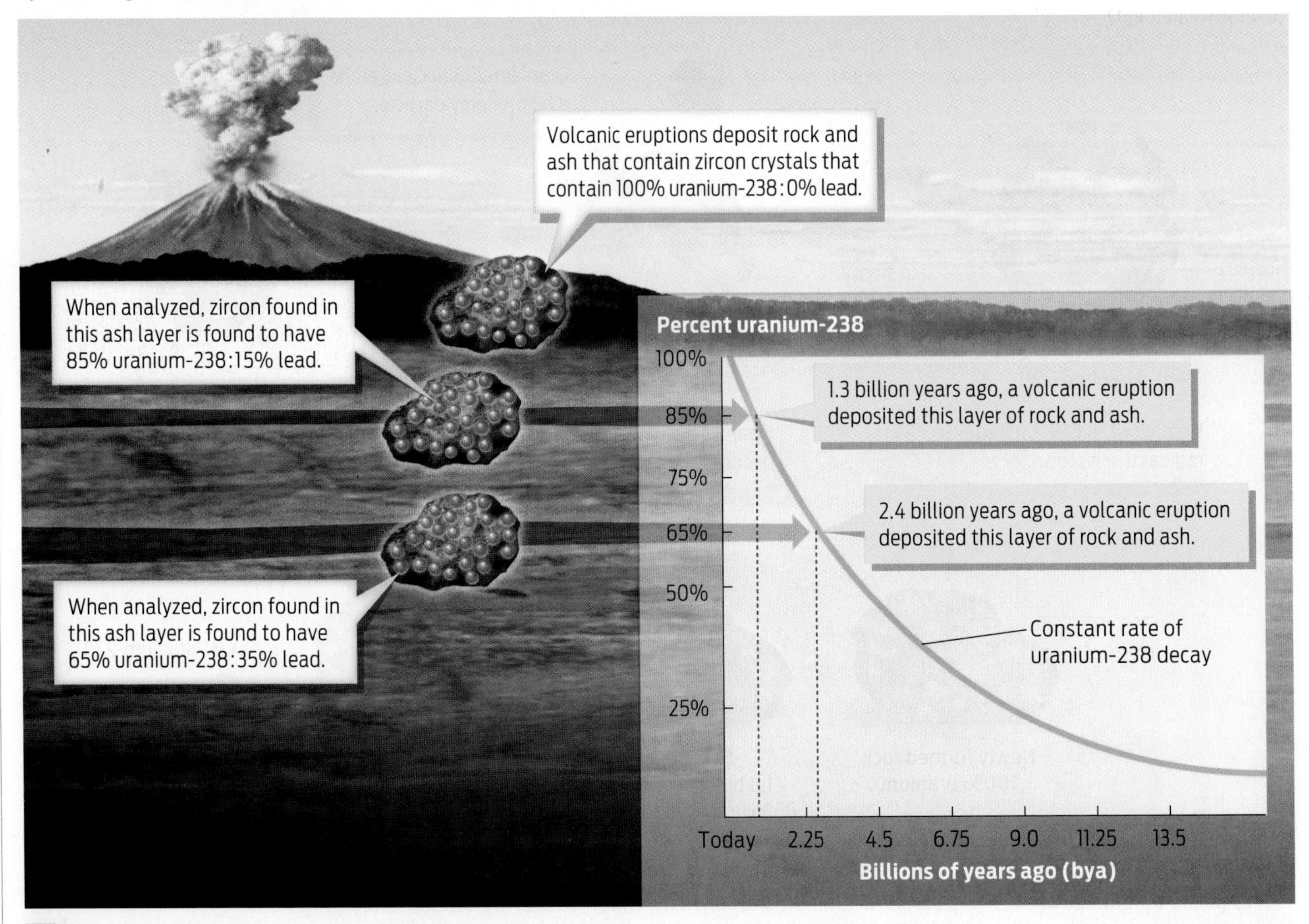

? **As rocks age, what happens to their amounts of uranium-238 and lead-206?**

BIOCHEMISTRY

Q When and how did life begin?

A At some point in Earth's distant past, life did not exist. Then, at a later point, it did. Where did this life come from? How did it start? The transition from nonliving to living occurred more than 3 billion years ago, a length of time so vast that reconstructing events from that period might seem hopeless. Even so, we can still formulate hypotheses and investigate them scientifically.

Scientists have offered a number of hypotheses to explain how life began on Earth, including the idea that it arrived here fully formed on an asteroid or meteorite from outer space. Others hypothesize that life emerged in stages over time, as inorganic chemicals combined into successively more complex molecules, including the ones making up living things. A landmark experiment lending support to this latter hypothesis was performed by University of Chicago chemist Harold Urey and his 23-year-old graduate student Stanley Miller in 1953 **(INFOGRAPHIC 16.3)**.

INFOGRAPHIC 16.3

Could Life Have Evolved in the Primordial Soup?

Urey and Miller replicated the chemical environment of the early Earth in a flask, then simulated lightning with electrical sparks. When the contents of the flask were analyzed, organic molecules, including amino acids, were present, formed from reactions between the inorganic precursors. The results suggest one way that life may have started.

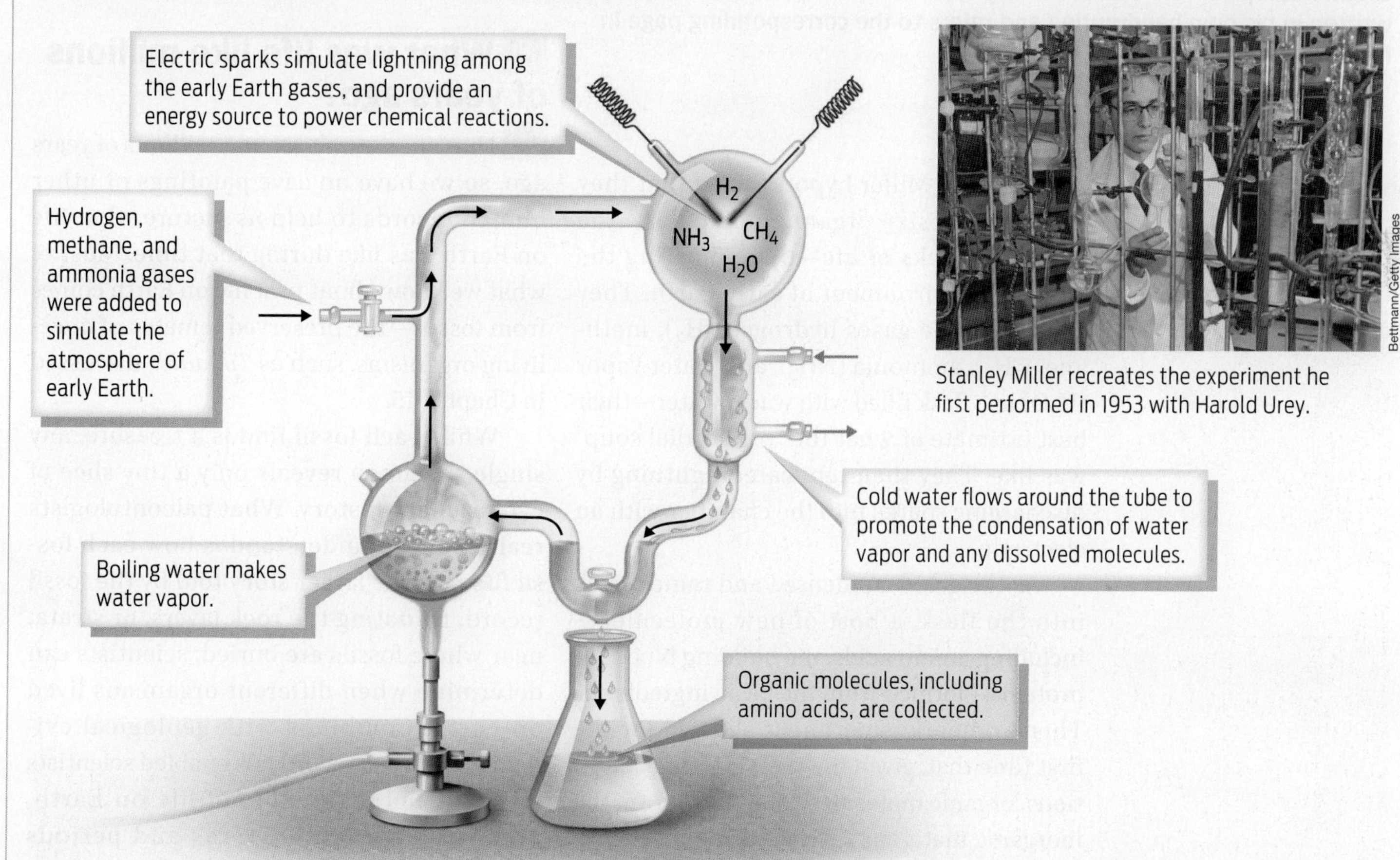

Stanley Miller recreates the experiment he first performed in 1953 with Harold Urey.

? **Why was this experiment important for developing and testing hypotheses about the origin of life on Earth?**

Jeffrey Bada and Robert Benson/Scripps Institution of Oceanography, University of California at San Diego

Original samples from one of Stanley Miller's experiments. The note is written in his own handwriting and refers to the corresponding page in his laboratory notebook.

Urey and Miller hypothesized that they could synthesize organic molecules—the building blocks of life—by replicating the chemical environment of early Earth. They combined the gases hydrogen (H_2), methane (CH_4), ammonia (NH_3), and water vapor (H_2O) in a flask filled with warm water—their best estimate of what the "primordial soup" was like. They then replicated lightning by discharging sparks into the chamber with an electrode.

As the gases condensed and rained back into the flask, a host of new molecules—including amino acids, the building blocks of proteins—formed from the basic ingredients. This landmark experiment showed for the first time that, given the right starting conditions, organic molecules could form from the inorganic materials believed to be present in the primordial soup.

Since Urey and Miller's experiment, other researchers have confirmed and extended their results. Experiments have shown that it is possible, by varying the composition of the starting materials, to produce from inorganic precursors essentially all the organic molecules used by living organisms. Organic molecules that have been produced in such experiments include all 20 amino acids, as well as sugars, lipids, nucleic acids, and even adenosine triphosphate (ATP)—the molecule that powers almost all life on Earth.

These results are significant because organic molecules are the building blocks of life, and without them life could not exist. Their synthesis in the primordial past would therefore have been a necessary first step in the emergence of life.

PALEONTOLOGY

Q What was life like millions of years ago?

A Humans weren't around millions of years ago, so we have no cave paintings or other human records to help us picture what life on Earth was like during that time. Most of what we know about past life on Earth comes from fossils—the preserved remains of once-living organisms, such as *Tiktaalik,* discussed in Chapter 15.

While each fossil find is a treasure, any single specimen reveals only a tiny slice of evolutionary history. What paleontologists really want to understand is how each fossil fits into the larger story told by the fossil record. By dating the rock layers, or strata, near where fossils are buried, scientists can determine when different organisms lived on Earth. Combined with geological evidence, the fossil record has enabled scientists to construct a timeline of life on Earth, organized by defining eras and periods **(INFOGRAPHIC 16.4)**.

INFOGRAPHIC 16.4

Geologic Timeline of Life on Earth

The fossil record reveals changes in organisms over time, including the appearance and diversification of multicellular organisms as well as mass extinctions, the rapid loss of a large proportion of the organisms present at a given time.

Era	Time Period		Prominent Life Events	Life in Water	Life on Land
Precambrian	Archean	3,500 mya	First evidence of prokaryotic cells.		
		2,500 mya	Prokaryotes dominate the oceans.		
	Proterozoic	545 mya	First eukaryotes appear in water. Soft-bodied invertebrates develop in water.		
Paleozoic Era	Cambrian	495 mya	Expansion of ocean animal diversity.		
	Ordovician	439 mya	First coral reef and primitive fish appear. Ocean plants diversify. First fungi appear in water.		
			Mass extinction		
	Silurian	408 mya	Seedless plants, primitive insects and soft-bodied animals appear on land.		
	Devonian	354 mya	Fish species diversify. First insects and seed-bearing plants appear on land.		
			Mass extinction		
	Carboniferous	290 mya	Amphibians appear and begin to diversify.		
	Permian	251 mya	Coral species abundant in oceans. Reptiles appear on land.		
			Mass extinction		
Mesozoic Era	Triassic	206 mya	Ocean life diversifies in recovery from Permian extinction. Dinosaurs and mammals appear on land.		
			Mass extinction		
	Jurassic	144 mya	First flowering plants and bird species appear. Large dinosaurs are plant-eaters.		
	Cretaceous	65 mya	Dinosaurs diversify. Cone-bearing and flowering plants dominate many habitats.		
			Mass extinction		
Cenozoic Era	Tertiary	1.8 mya	Mammals, birds and flowering plants diversify. Grasses appear. First primates appear.		
	Quaternary	Today	Many large mammals become extinct. Humans appear.		

? **During which period were prokaryotes the dominant organisms, and were they dominant on land or in water?**

During the Cambrian explosion, many new animal species appeared, including this five-eyed creature called *Opabinia*.

The timeline shows that during the 4.5 billion years that Earth has been around, its geography and climate have gone through dramatic changes. For the first few hundred million years or so of its history, Earth was a fiery inferno, coursing with seas of molten lava and bombarded by meteorites. Not until things simmered down and the surface cooled a bit, around 3.8 billion years ago, could it support life.

The oldest known fossils date from some 3.7 billion years ago, when Earth's climate was very different from what it is today. Notably, the atmosphere did not contain substantial amounts of oxygen (O_2). In this oxygen-deprived world, the only organisms that could thrive were unicellular prokaryotes that did not rely on oxygen for their metabolic reactions. Between 3.0 and 2.5 billion years ago, unicellular photosynthetic organisms that produce oxygen emerged and proliferated. Oxygen began to accumulate in the atmosphere, opening the door for more-complex eukaryotic organisms to evolve.

The first multicellular organisms to make use of this oxygen for cellular respiration were green algae, which appeared 1.2 billion years ago. Soft-bodied aquatic animals followed, about 600 million years ago. Nevertheless, only since about 545 million years ago, during the Cambrian Period, do we see fossil evidence of a truly diverse animal world. As part of an event termed the Cambrian explosion, ocean life swelled with a mind-boggling array of strange-looking creatures, including *Opabinia,* an organism with five eyes and a snout resembling a vacuum-cleaner hose, discovered in fossils from this period.

The first organisms to colonize land were primitive plants, which appeared roughly 450 million years ago. By 350 million years ago, forests of seedless plants covered the globe.

Then, 250 million years ago, the number of life forms was drastically cut down: roughly 95% of living species were extinguished in a mass die-off known as the Permian **extinction.** Scientists do not know what caused the Permian extinction, but some hypothesize that massive volcanic activity filled the atmosphere with heat-trapping gases that led to a rapidly changing climate. The Permian extinction wasn't bad for all organisms, though; some flourished as space and resources opened up for the survivors, who spread and diversified in a phenomenon known as **adaptive radiation.** Among these surviving organisms were reptiles, which thrived in the hot, dry climate of the Triassic Period. The most famous group of reptiles, the dinosaurs, dominated the land for nearly 200 million years, thanks to their drought-resistant skin and fast-moving legs. The dinosaurs died out in another **mass extinction** at the end of the Cretaceous Period, 65 million years ago.

Exactly why the dinosaurs went extinct remained a mystery for many years. Evidence now suggests that a massive 6-mile-wide asteroid killed off the dinosaurs (and 60% of the other species living at the time). The asteroid plowed into Earth with almost unimaginable force, sending a thick layer of soot and ash into the atmosphere and blocking out the sun for months. A crater 110 miles wide in Mexico's Yucatán peninsula, near the town of Chicxulub, is the accepted impact site.

With the extinction of the dinosaurs, mammals seized their chance to spread and diversify on land, giving rise to many of the species we see on the planet today. This pattern of irregular—rather than steady—change is a common feature of the fossil record.

BIOGEOGRAPHY

Q Why are there no penguins at the north pole, and no polar bears at the south pole?

A As habitats go, Earth's north and south poles are pretty similar: cold, lots of ice, plenty of fish to eat. Yet polar bears are found only in the north and penguins are found only in the south. Why? If environmental conditions were the only relevant factors, you might expect that polar bears and penguins would both live in the same location (as they sometimes do in TV commercials). But the distribution of organisms around the world reflects the details of their evolutionary history—their connection in time and space to the species from which they are descended and to which they are related. In fact, the distribution of organisms was one of the main pieces of evidence that Darwin and Wallace used to support their theory of evolution by natural selection (see **Milestone 4: Adventures in Evolution**). Their round-the-world voyages were among the first examples of **biogeography,** the study of the natural geographic distribution of species. Biogeography seeks to explain why particular organisms are found in some areas and not others.

Evidence now suggests that a massive 6-mile-wide asteroid killed off the dinosaurs (and 60% of the other species living at the time).

Penguins, for example, make their home in the southern hemisphere, especially in the coastal regions of Antarctica. According to fossil evidence, penguins first appeared about 65 million years ago near what is now southern New Zealand, when that region was much closer to Antarctica. They are believed

EXTINCTION
The elimination of all individuals in a species; extinction may occur over time or in a sudden mass die-off.

ADAPTIVE RADIATION
The spreading and diversification of organisms that occur when the organisms colonize a new habitat.

MASS EXTINCTION
An extinction of between 50% and 90% of all living species that occurs relatively rapidly.

BIOGEOGRAPHY
The study of the distribution of organisms in geographic space.

INFOGRAPHIC 16.5

Biogeography: Where Species Live Depends on Where Their Ancestors Evolved

Polar bears live at the north pole and penguins live at the south pole because that is where their respective ancestors evolved. Their adaptations for cold climates keep them from crossing the tropical equator.

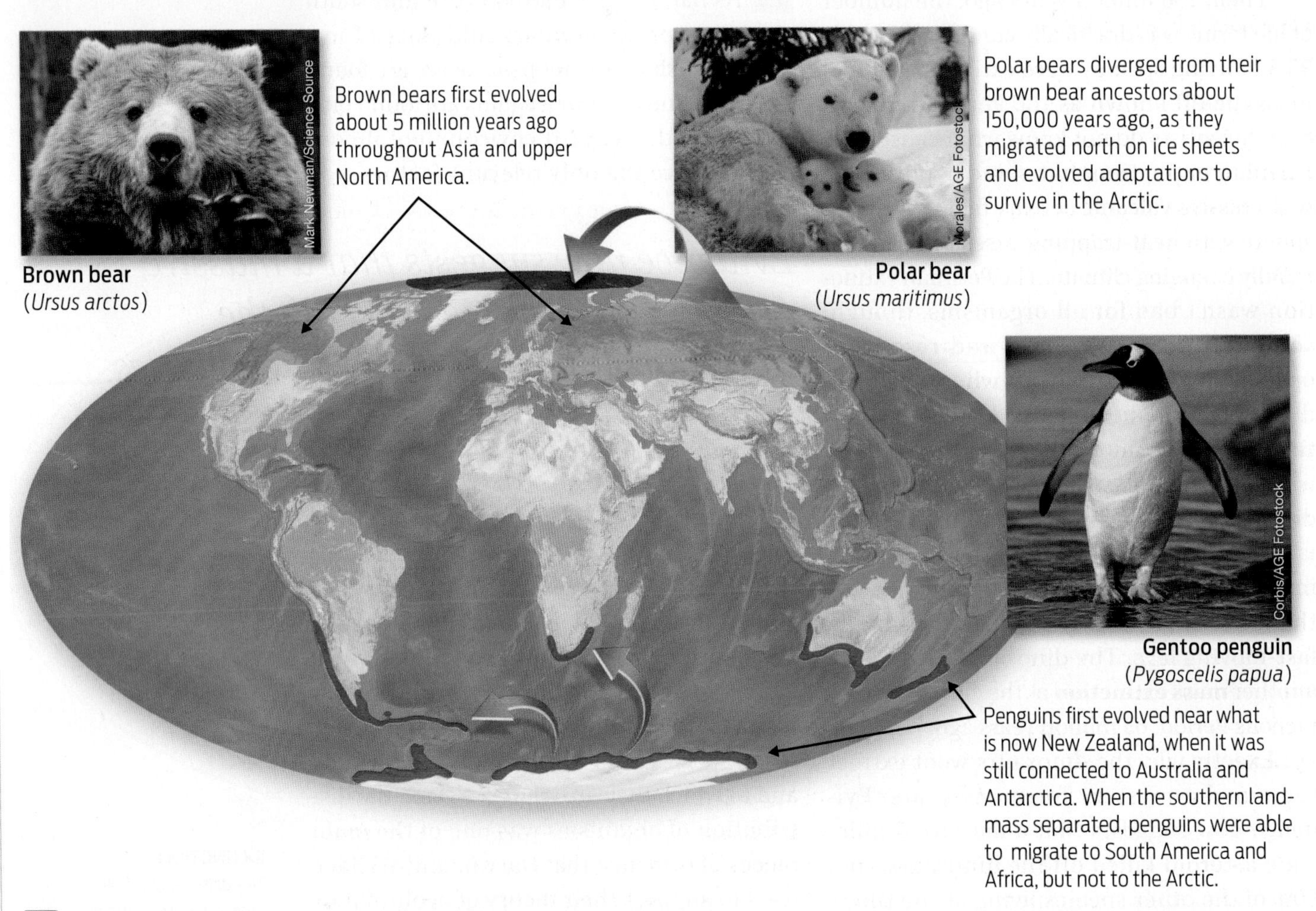

? If brown bears evolved in what is now New Zealand (instead of in Asia and upper North America), would polar bears and penguins coexist? If so, where?

to have evolved from bird ancestors that could fly. Over time, penguins lost their ability to fly but gained the ability to swim, with their wings essentially evolving into flippers. Penguins that migrated to Antarctica evolved insulation and other traits that helped them thrive in this cold, watery environment.

Polar bears, by contrast, live only in the Arctic. From fossil and DNA evidence, scientists conclude that polar bears evolved from brown bears roughly 150,000 years ago in Siberia, when that region became isolated by glaciers. Like penguins, polar bears have adapted to life in a cold, icy environment.

But though their habitats are similar climatically, and might conceivably support the coexistence of both species, there's no easy way for them to travel from one pole to the other: the equator marks a warm tropical barrier that prevents the migration of the cold-adapted species living on either side of it **(INFOGRAPHIC 16.5)**.

Q Why are marsupials found in Australia and the Americas but nowhere else?

A While traveling around the world, Darwin and Wallace were both struck by the presence on different landmasses of animals and plants that seemed to bear a family resemblance. Islands, for example, seemed to have species different from, but apparently related to, those on the mainland. This distribution of species seemed to support the theory of evolution by natural selection that each scientist had been pondering. Yet some facts about species distribution remained perplexing in Darwin's and Wallace's time. For example, why were Australia and South America the two places on the globe with a diversity of marsupials? (Marsupials are mammals that give birth to incompletely developed offspring, which they typically carry to term in a pouch on the outside of the mother's belly.) These continents are separated from each other by great distances, so it was unrealistic to suggest that a marsupial ancestor had swum between these regions.

It wasn't until the 20th century that scientists solved this conundrum. In 1915, Alfred Wegener, a German geologist, proposed that the continents are not fixed in place but rather move over time, a process termed **continental drift.** Wegener argued that at one time, the different continents we see today were all joined together into a single large landmass, called Pangaea, that broke apart slowly over time. His ideas weren't fully accepted until the 1960s, when scientists learned about **plate tectonics.** This theory states that the continents are part of large plates that sit atop Earth's mantle. As heat rises and falls in the mantle, it creates a kind of convection current that moves the mantle rock and, in turn, the plates sitting above them. Plates may drift apart or even sometimes collide, causing earthquakes and volcanoes.

Continental drift helps explain the distribution of marsupials we see today. The most recent common ancestor of marsupials and placental mammals (those animals, like humans, that bear young that have developed in a womb) appears to have lived about 120 million years ago in what is now China, when this region was part of a large landmass called Laurasia. Marsupials gradually diverged from placental mammals and began their march across Laurasia. About 65 million years ago, marsupials spread to what is now South America when it became connected to Laurasia through a land bridge.

Fossil evidence suggests that the most recent common ancestor of living marsupials lived in South America.

Fossil evidence suggests that the most recent common ancestor of living marsupials lived in South America. Marsupial fossils dating from 40 million years ago have been discovered there. The oldest marsupial fossils in Australia are 30 million years old, suggesting they arrived later. But how did they get there? The theory of continental drift proposes that Australia, South America, and Antarctica were all once part of a large landmass called Gondwana, which began to break up about 180 million years ago. When marsupials arrived in South America, Australia was still connected to South America by what is now Antarctica (which was then ice free). According to scientists, marsupials may have crossed over Antarctica and entered Australia, only for those two continents to then split apart. Evidence in support of this hypothesis comes from marsupial fossils discovered on

CONTINENTAL DRIFT
The movement of the continents relative to one another over time.

PLATE TECTONICS
The theory that the continents are part of large sections, or plates, that sit atop Earth's mantle and that move around and collide due to heat convection currents in the underlying mantle.

INFOGRAPHIC 16.6

Continental Drift Influences the Distribution of Species on Earth

Earth's landmasses are part of large plates that sit atop Earth's mantle. Over geologic time, these plates have moved with respect to one another, affecting the ability of organisms to move among the landmasses. The geographic distribution of organisms on Earth today reflects continental drift, as can be seen with the distribution of marsupials.

180 mya

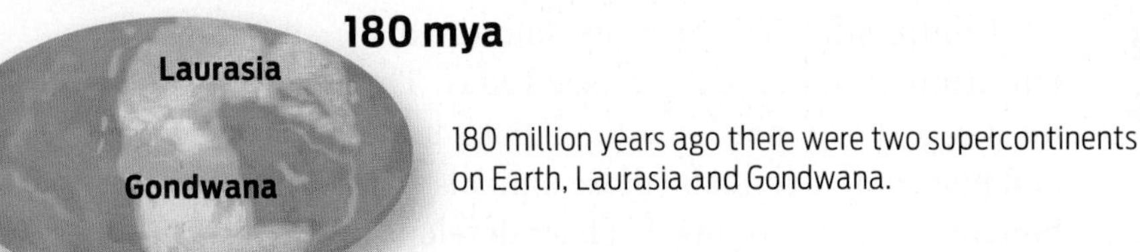

180 million years ago there were two supercontinents on Earth, Laurasia and Gondwana.

120 mya

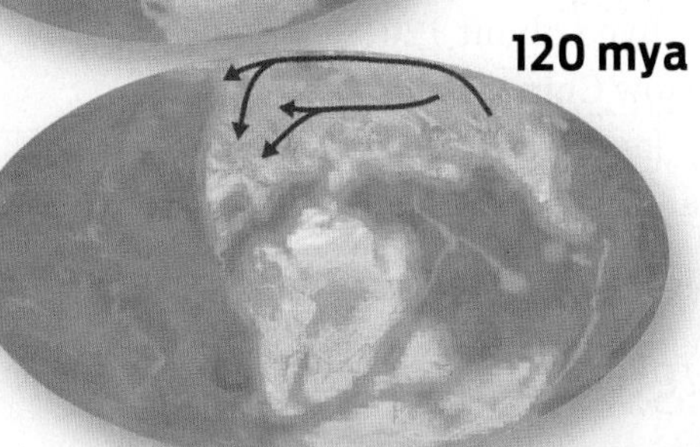

About 120 million years ago marsupial and placental mammals diverged from a common mammal ancestor in China. Marsupials migrated throughout Laurasia, which was not connected to Gondwana.

Michael Weber/imageBROKER/AGE Fotostock

Australia koala
(*Phascolarctos cinereus*)

65 mya

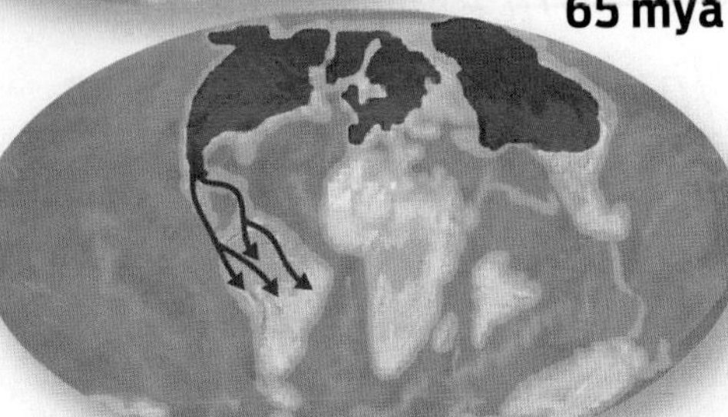

About 65 million years ago, the part of Gondwana we now call South America became connected with North America, allowing migration of marsupials into this region.

50 mya

About 50 million years ago, marsupials migrated from South America through Antarctica and into Australia until about 35 million years ago, when these pieces of Gondwana were no longer connected. The marsupials in Laurasia went through a mass extinction during this time and were almost completely eliminated from the northern land regions.

Today

Today, marsupials are found mainly in South America and Australia. The disconnection of Antarctica from South America and Australia halted migration between these continents. As Antarctica migrated south, its colder temperatures resulted in marsupial extinction there.

Mark Hamblin/Getty Images

South American opossum
(*Didelphis marsupialis*)

? When did marsupials first migrate into the southern hemisphere?

Seymour Island off the coast of Antarctica that date from 40–35 million years ago **(INFOGRAPHIC 16.6).**

Tectonic plates continue to move and collide, occasionally causing dramatic events such as earthquakes and volcanic eruptions. Global Positioning System (GPS) measurements, used to track the direction and velocity of plate movement, show that some plates are moving at about 2 to 5 cm per year—about as fast as your fingernails grow.

CONVERGENT EVOLUTION

Q Are creatures that look alike always closely related?

A Polar bears share many traits with their brown bear cousins—both species are recognizable as bears, despite obvious differences in color. The fact that polar bears resemble brown bears is suggestive evidence that the two species share a recent common ancestor. But common ancestry is not the only reason that two species might appear similar. Even species that are not closely related may share similar adaptations as a result of independent episodes of natural selection, a phenomenon called **convergent evolution.**

Take flying squirrels and sugar gliders. Flying squirrels are placental mammals that evolved in North America. Sugar gliders are marsupial mammals that evolved in Australia. Flying squirrels and sugar gliders are not closely related, yet both have independently evolved gliding wings to leap from tree to tree in the forested environments in which they live. Convergent evolution shows that sometimes a good solution to a common problem can evolve more than once **(INFOGRAPHIC 16.7).**

CONVERGENT EVOLUTION
The process by which organisms that are not closely related evolve similar adaptations as a result of independent episodes of natural selection.

INFOGRAPHIC 16.7

Convergent Evolution: Similar Solutions to Common Problems

Species that are not closely related can have very similar appearances because they independently evolved adaptations to similar environments and challenges.

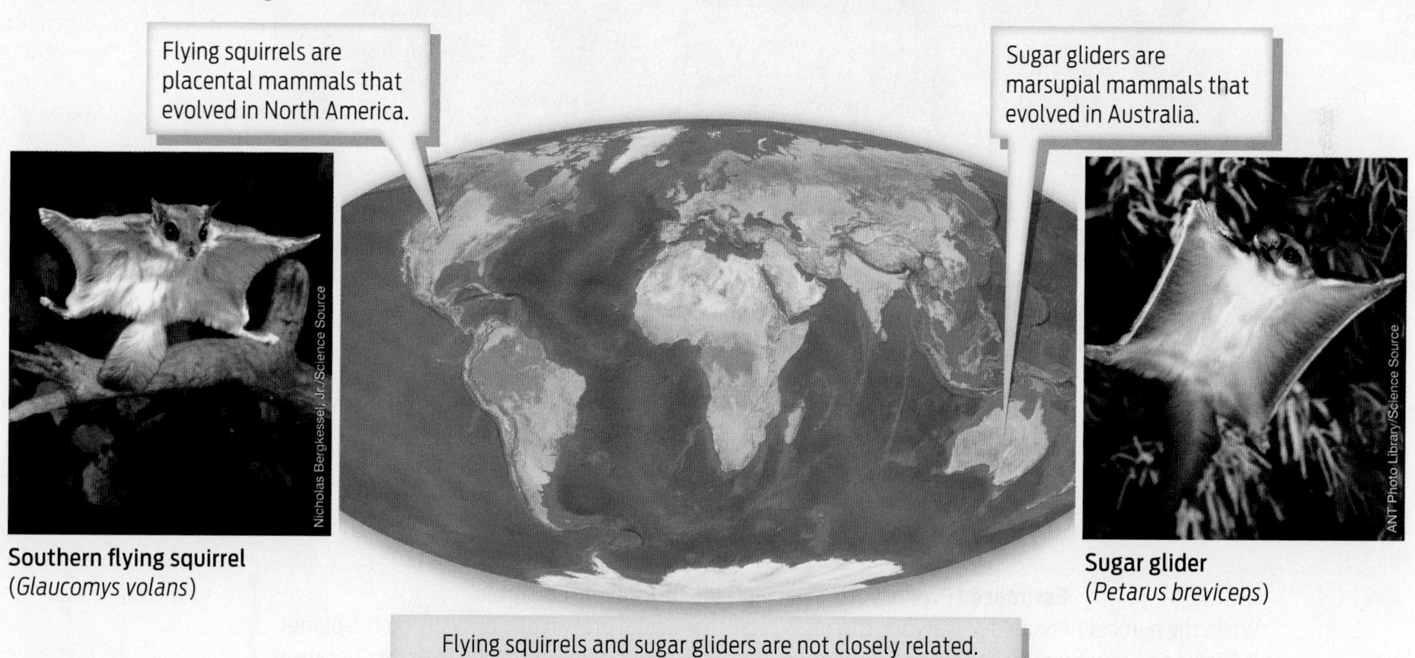

Southern flying squirrel (*Glaucomys volans*)

Sugar glider (*Petarus breviceps*)

? What are some key differences between the placental flying squirrel and the marsupial sugar glider?

DIVERSITY AND TAXONOMY

Q How many species are there on Earth, and how do scientists keep track of them?

A Current estimates of the total number of species on Earth range anywhere from 30 million to 2 billion. Of these, just 1.5 million or so species have been formally described. Many species are found in diversity hot spots like rain forests, and new species are constantly being discovered, on the order of 17,000 new species each year. Some noteworthy recent additions include the hog-nosed rat (*Hyorhinomys stuempkei*), a cartwheeling spider (*Cebrennus rechenbergi*), and the tiniest known snail (*Acmella nana*), which is smaller than the head of a matchstick **(INFOGRAPHIC 16.8)**.

TAXONOMY
The identification, naming, and classification of organisms on the basis of shared traits.

With so many species out there, how do scientists keep track of them all? The organizing principle by which scientists systematically identify, name, and classify organisms is called **taxonomy.** Taxonomy is

INFOGRAPHIC 16.8

How Many Species Are There?

The number of species in each group below represents only those that have been formally characterized and classified. The true number of species is likely to be much higher (estimated numbers are shown in parentheses).

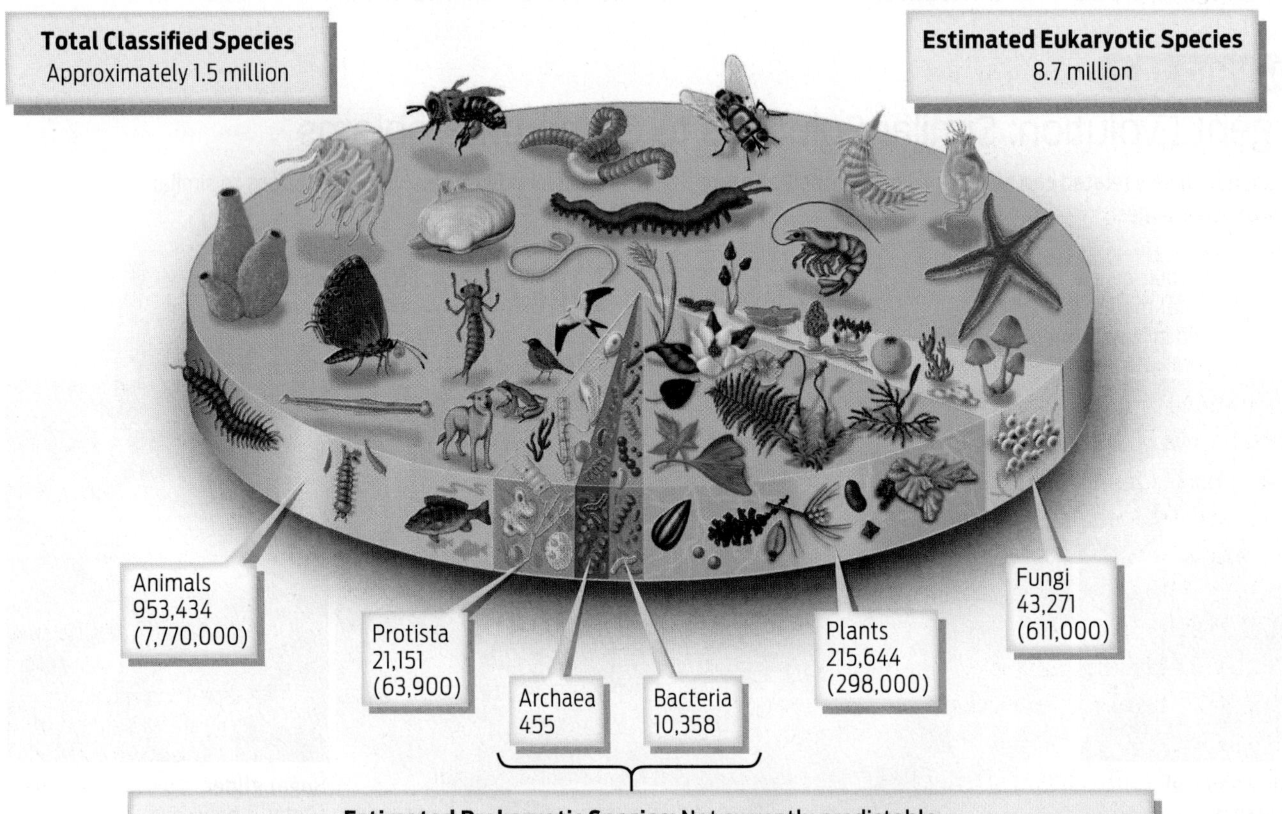

Estimated Prokaryotic Species: Not currently predictable
While the number of named prokaryotic species is small, the actual prokaryotic diversity on the planet is likely to be much larger than that of eukaryotes. True prokaryotic diversity is hard to assess because prokaryotes are so small and live just about everywhere, and many can't be grown in the lab.

? **Which eukaryotic group has the greatest proportion of estimated species already described?**

part of the broader study of systematics, the study of biological diversity of life on Earth.

In essence, taxonomy is an attempt to impose a human sense of order on the vast array of species found on our planet. Taxonomists categorize species on the basis of features they have in common, such as whether their cells are eukaryotic or prokaryotic, whether they photosynthesize, and whether they have four legs and fur.

To sort organisms, taxonomists use a system of nine progressively narrower categories: domain, supergroup, kingdom, phylum, class, order, family, genus, species. As you move down the list from domain to species, the categories become increasingly exclusive, until finally only one specific type of organism is included. The genus and species names provide a useful scientific identifier for every living organism. Because that scientific name is in Latin, it can be easily recognized in many languages.

Take humans, for example. Humans are members of the domain Eukarya, the supergroup Opisthokont, and the kingdom Animalia. Within the animal kingdom, they belong to the phylum Chordata, a group that includes the **vertebrates** (animals with a rigid backbone). Further, humans are **mammals,** members of the class Mammalia; along with all members of this class, they share features such as mammary glands and a body that is covered with hair. Humans belong to the Primate order, which also includes monkeys, apes, and lemurs. And humans are members of the Hominidae family, and so are closely related to their fellow hominids: chimpanzees, gorillas, and orangutans. Our scientific name—made up of our genus and species names—is *Homo sapiens* ("wise human") **(INFOGRAPHIC 16.9)**.

Classification would seem to be a simple matter—just observe, measure, and sort. But deciding which category an organism belongs in can sometimes be tricky, as the example of convergent evolution has shown. Sometimes, to properly classify organisms, scientists have to look a little deeper.

PHYLOGENY

Q Is a crocodile more closely related to a bird or to a lizard?

A The fact that all land vertebrates have four limbs and the same forelimb bones indicates that they all share a common ancestor (see Chapter 15). But how precisely are they related? In other words, who's more closely related to whom? Biologists want not only to categorize organisms, but also to have those categories reflect **phylogeny,** the actual evolutionary history of the organisms. This evolutionary history is represented visually by a diagram called a **phylogenetic tree,** which is similar in some respects to a family tree.

Current estimates of the total number of species on Earth range anywhere from 30 million to 2 billion.

Phylogenetic trees can be drawn in a number of ways, but most have certain features in common. At the base, or root, is the common ancestor shared by all organisms on the tree. Over time, and with different selective pressures, different groups of organisms diverge from that common ancestor and from one another, leading to separate branches on the tree. The points on the tree at which two branches diverge are called nodes. A node represents the common ancestor shared by all organisms on the branches above that node. At the very tips of the branches we find the most recent organisms in that lineage, including living organisms and organisms that became extinct. We can use this kind of phylogenetic tree to establish relationships between living organisms (at the tips of the branches) on the basis of the ancestors they share. The more recently two groups share a

VERTEBRATES
Animals with a rigid backbone.

MAMMALS
Members of the class Mammalia; all members of this class have mammary glands and a body covered with hair.

PHYLOGENY
The evolutionary history of a group of organisms.

PHYLOGENETIC TREE
A branching diagram of relationships showing common ancestry.

common ancestor, the more closely they are related **(INFOGRAPHIC 16.10)**.

A phylogenetic tree visually depicts the best hypothesis we currently have for how species are related, based on a shared evolutionary history. The evidence for the phylogenetic tree comes from many sources, including the fossil record, physical traits, and shared DNA sequences. For many years, biologists relied solely on observable physical or behavioral features to construct phylogenetic trees. But with the genetic revolution, it's become common practice to include DNA evidence as well. Typically, researchers compare sequence differences in a gene that is found in all living organisms, such as the gene encoding a particular component of the ribosome, the cell's protein synthesis machinery.

INFOGRAPHIC 16.9

Classification of Species

Scientists classify organisms into groups that are increasingly exclusive. The broadest category is the domain. Closely related organisms are grouped based on morphological, nutritional, and genetic characteristics. There are far fewer organisms in an order than in a phylum.

Taxonomic group	Example: humans	Number of living species
Domain	Eukarya	~4–10 million
Supergroup	Opisthokonta	> 1 million
Kingdom	Animalia	> 1 million
Phylum	Chordata	~50,000
Class	Mammalia	~5,000
Order	Primate	~250
Family	Hominidae	7
Genus	*Homo*	1
Species	*Homo sapiens*	1

? Which of the following groups has the greatest number of living species: phylum, class, family, kingdom, or supergroup?

INFOGRAPHIC 16.10

How to Read a Phylogenetic Tree

Evolutionary history, or phylogeny, is represented visually by a phylogenetic tree. Trees have a common structure, with a root, nodes, and branches. To determine evolutionary relationships among living or extinct organisms, consider the most recent common ancestors.

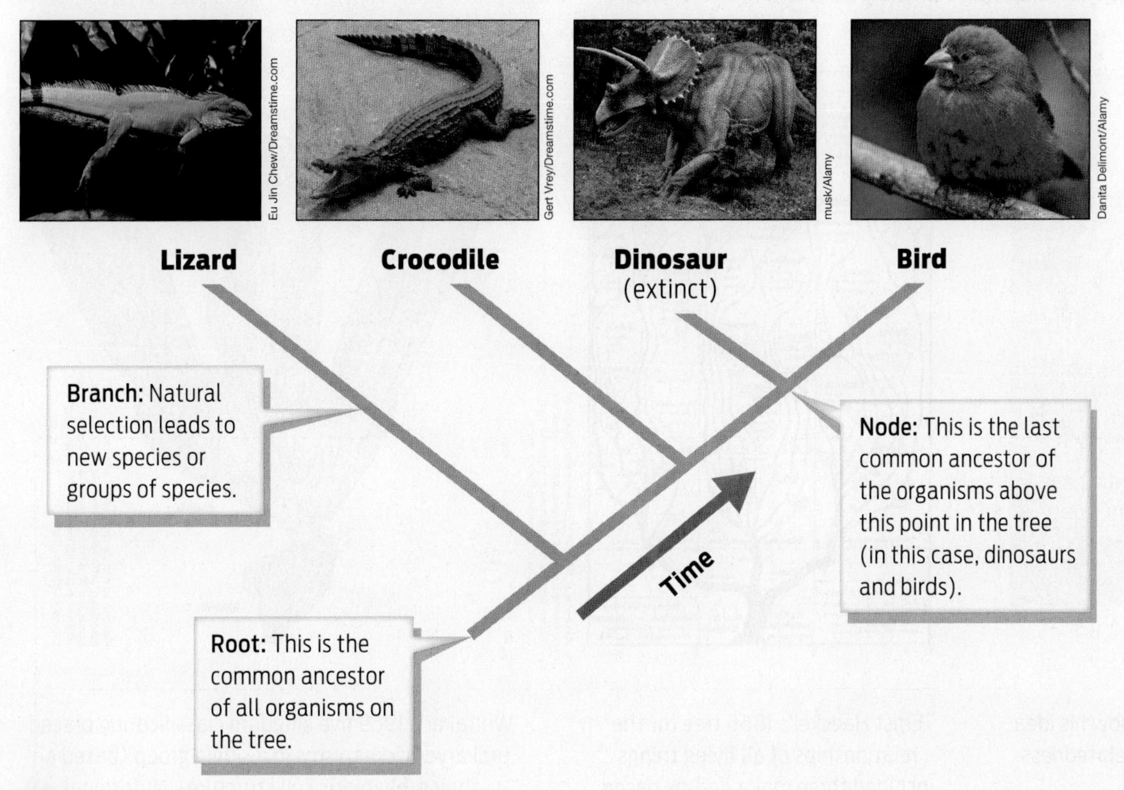

? **Is a crocodile more closely related to a bird or to a lizard?**

All cells have ribosomes, which is why this gene is so useful for establishing phylogeny.

Sometimes the new genetic information yields surprises. Modern genetic evidence shows, for example, that crocodiles are more closely related to birds than they are to lizards, appearances notwithstanding. Genetics, you might say, is shaking the evolutionary tree.

Q How many branches are on the tree of life?

A Since each living species sits on its own branch of a phylogenetic tree, the complete tree of life has as many branches as there are species in the world. Today's species are like the thinnest twigs in the upper branches of an enormous oak tree. Closer to the bottom of the tree, nearer to the ancient trunk, however, we find significant forks. Just how many forks appear at the bottom of the tree is a question that has been debated for decades.

Before the 18th century, biologists divided living things into just two main categories: animals and plants. This classification was based on whether an organism moved around and ate or did not move around and eat. By the mid-19th century, the microscope had revealed a whole new world of organisms, so a third branch was added to life's tree. The German biologist and artist Ernst Haeckel, in 1866, was the first to draw a phylogenetic tree that included these microscopic organisms, which he termed Protista, or protists.

INFOGRAPHIC 16.11

The Evolution of the Tree of Life

Phylogenetic trees represent our best understanding of the relationships between living organisms. Early trees were based on obvious physical similarities between organisms. More detailed structural and functional analyses generated new groups. The use of gene sequences revealed that organisms previously thought to be closely related were not. Gene and even entire genome sequences are continuing to shape our understanding of the relationships between organisms, and also the shape of the tree of life itself.

LETZ/SIPA/Newscom

Darwin's first drawings show how his idea of natural selection implied relatedness between species.

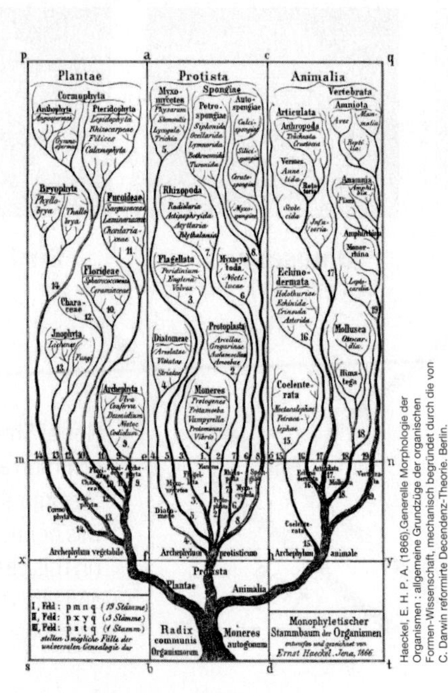

Haeckel, E. H. P. A. (1866).Generelle Morphologie der Organismen : allgemeine Grundzüge der organischen Formen-Wissenschaft, mechanisch begründet durch die von C. Darwin reformirte Decendenz-Theorie. Berlin.

Ernst Haeckel's 1866 tree for the relationships of all living things included three major groups based on similar physical characteristics.

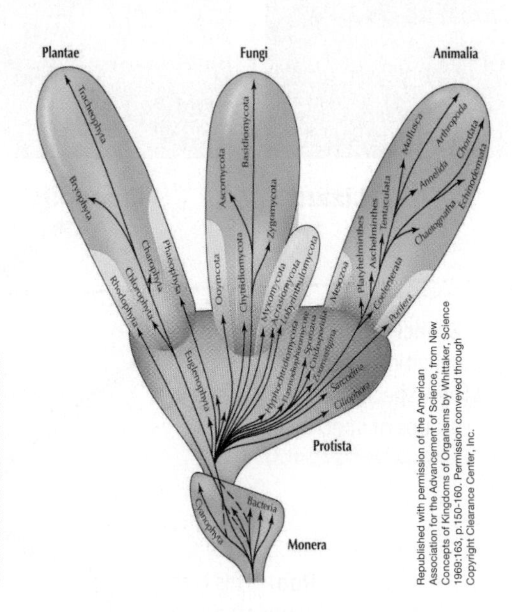

Republished with permission of the American Association for the Advancement of Science, from New Concepts of Kingdoms of Organisms by Whittaker, Science 1969:163, p.150-160. Permission conveyed through Copyright Clearance Center, Inc.

Whitaker's 1969 five kingdom classification placed prokaryotic organisms in a single group (based on their prokaryotic cell structure). Nutritional differences were used to place plants, fungi, and animals into different kingdoms.

Timeline of Phylogenetic Trees

? Which two of the three domains are most closely related to each other?

By the 1960s, taxonomists realized that even three branches did not fully capture the diversity of life. Many organisms, such as fungi, didn't fit neatly into any of these groups. So another classification scheme was proposed—one that grouped all living organisms into five large kingdoms on the basis of how they obtained their food (by eating, photosynthesizing, or decomposing) and whether they had eukaryotic or prokaryotic cells. These five kingdoms were Animalia, Plantae, Fungi, Protista, and Monera. Protista included mostly single-celled eukaryotic organisms (such as the amoeba), and Monera included all prokaryotic organisms (such as bacteria).

Yet even this revised classification scheme eventually had to be overhauled as more information became available. In the 1970s, genetic studies by Carl Woese revealed that, on the basis of genetic relatedness, not all prokaryotes could be lumped together. The original kingdom Monera, containing all prokaryotes, is now divided into two

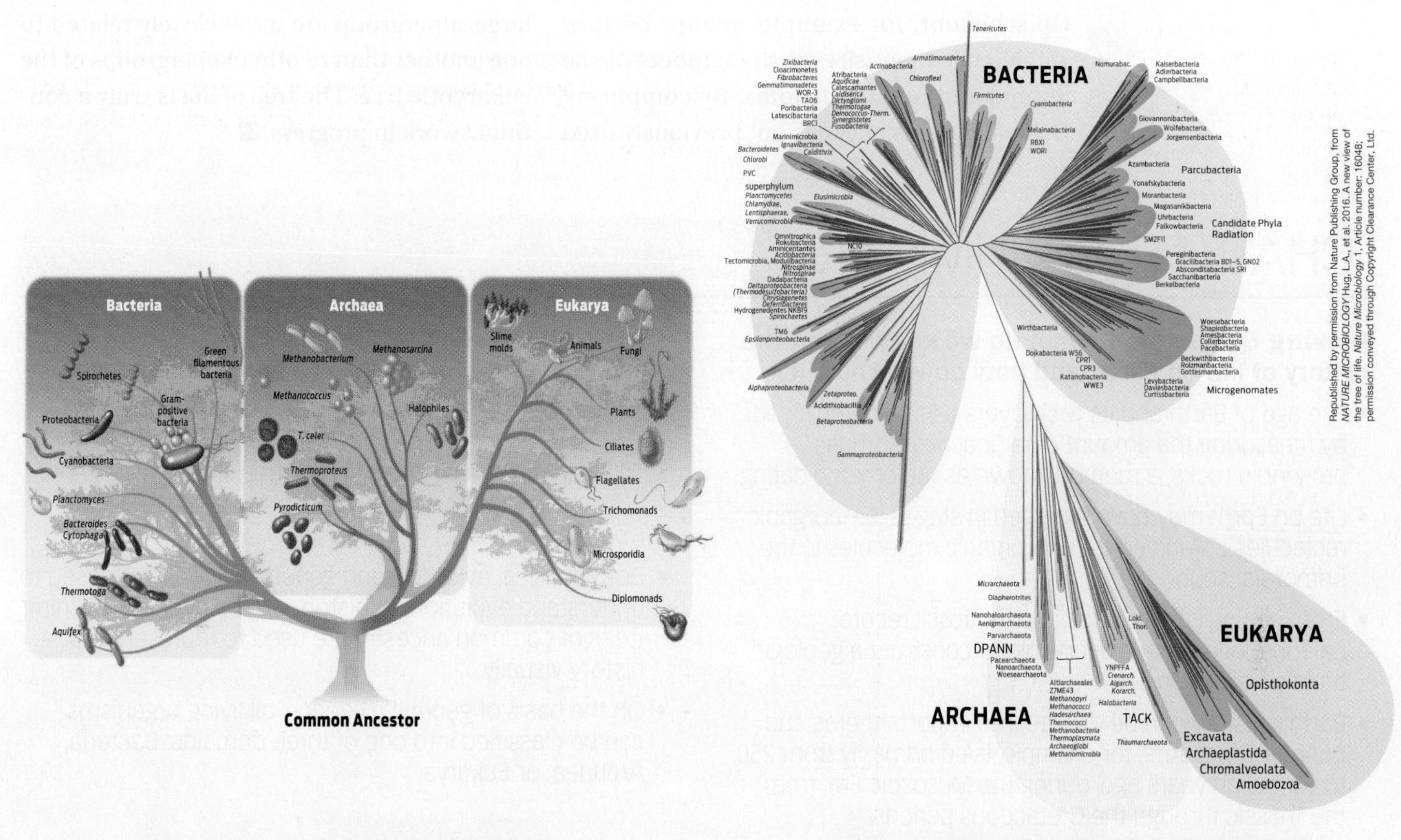

Woese's 1977 classification system split life into three domains, two of which are prokaryotic, based on analyzing DNA sequences.

A 2016 tree of life used whole genome sequences to confirm major eukaryotic supergroups and provide evidence for a wealth of previously uncharacterized bacterial diversity.

Republished by permission from Nature Publishing Group, from NATURE MICROBIOLOGY Hug, L.A., et al. 2016. A new view of the tree of life. Nature Microbiology 1, Article number: 16048; permission conveyed through Copyright Clearance Center, Ltd.

domains, Archaea and Bacteria, based on degree of genetic relatedness. Consequently, scientists now group organisms into one of three large **domains**—Bacteria, Archaea, and Eukarya—which represent three fundamental branching points in the trunk of the evolutionary tree. Genetic studies also revealed that protists are too genetically diverse to be put in one category. Within the domain Eukarya, the groups Animalia, Plantae, and Fungi remain recognized kingdoms, but the protists (members of the former kingdom Protista) are dispersed across the eukaryotic domain of life on the basis of genetic evidence **(INFOGRAPHIC 16.11)**.

Since about 2012, researchers have recognized a new scheme for categorizing eukaryotic life. The supergroup, a new classification, is meant to help clarify where protists should be placed on the eukaryotic tree. As mentioned earlier, grouping all single-celled eukaryotes together does not make sense from a genetic standpoint, since they are not all closely related. The new classification scheme attempts to classify protists by genetic similarity to other groups

DOMAIN
The highest (most inclusive) category in the modern system of classification. There are three domains: Bacteria, Archaea, and Eukarya.

on the tree and, as a result, has produced some unexpected pairings. The supergroup Opisthokont, for example, groups certain single-celled parasites with members of the animal and fungi kingdoms. By comparing DNA sequences of genes not previously used to assemble phylogenetic trees, scientists have discovered that the members of this large supergroup are more closely related to one another than to other supergroups of the eukaryotic tree. The tree of life is truly a continual work in progress. ■

CHAPTER 16 SUMMARY

Driving Question 1 What do we know about the history of life on Earth, and how do we know it?

- The age of Earth and its rock layers can be determined by measuring the amount of radioactive isotopes present in rocks, a method known as radiometric dating.
- Life on Earth may have emerged in stages, as inorganic molecules combined to form organic molecules in the primordial soup.
- From geological evidence and the fossil record, paleontologists have been able to construct a geologic timeline of life on Earth.
- Earth's history can be divided into important eras and periods. Dinosaurs, for example, lived primarily from 250 to 65 million years ago, during the Mesozoic Era, from the Triassic through the Cretaceous periods.
- The history of life on Earth has been marked by repeated extinctions and adaptive radiations, a phenomenon of irregular—rather than steady—change.

Driving Question 2 What factors help to explain the distribution of species on Earth?

- Ancient movement of Earth's landmasses due to continent drift affected the eventual distribution of species around the globe; the study of this phenomenon is known as biogeography.
- Convergent evolution is the evolution of similar adaptations in response to similar environmental challenges in groups of organisms that are not closely related.

Driving Question 3 What are the major groups of organisms, and how are organisms placed in groups?

- Life is astoundingly diverse. Current estimates of the total number of species on Earth range anywhere from 30 million to 2 billion. Of these, 1.5 million species have been formally described.
- Biologists sort organisms into a series of nested categories based on shared anatomical and genetic features: domain, supergroup, kingdom, phylum, class, order, family, genus, species.
- The scientific name of an organism is given by its genus and species names (the scientific name of humans is *Homo sapiens*).
- Both physical evidence and genetic evidence are used to understand evolutionary history, or phylogeny. Branching trees of common ancestry are used to represent that history visually.
- On the basis of genetic evidence, all living organisms can be classified into one of three domains: Bacteria, Archaea, or Eukarya.

More to Explore

- Witze, A. (2016, August 31). Claims of Earth's oldest fossils tantalize researchers. *Nature*. https://www.nature.com/news/claims-of-earth-s-oldest-fossils-tantalize-researchers-1.20506
- Gould, S. J. (1989). *Wonderful Life: Burgess Shale and the Nature of History.* New York: W. W. Norton.
- Mora, C., et al. (2011). How many species are there on Earth and in the ocean? *PLoS Biol* 9(8):e1001127.
- Woese, C. R., et al. (1990). Towards a natural system of organisms: Proposal for the domains Archaea, Bacteria, and Eucarya. *Proceedings of the National Academy of Sciences* 87:4576–4579.
- Milius, S. (2015). The tree of life gets a makeover: Schoolroom kingdoms are taking a backseat to life's supergroups. *Sci News* 188:22.

CHAPTER 16 Test Your Knowledge

Driving Question 1 What do we know about the history of life on Earth, and how do we know it?

By answering the questions below and studying Infographics 16.1–16.4, you should be able to generate an answer for this broader Driving Question.

Know It

1. What do uranium-238, carbon-14, and potassium-40 have in common?

2. To date a rock that contains what you suspect to be the very earliest life on Earth, which isotope would you use: uranium-238, carbon-14, or potassium-40? Explain your answer.

3. Place the following evolutionary milestones in order from earliest (1) to most recent (7), providing approximate dates to support your answer.

_____The first multicellular eukaryotes

_____The first prokaryotes

_____The Permian extinction

_____The Cambrian explosion

_____The first animals

_____The extinction of dinosaurs

_____An increase in oxygen in the atmosphere

Use It

4. Consider a rock formed at about the same time as Earth was formed.

a. How old is this rock?

b. How much of the original uranium-238 is likely to be left today in that rock?

5. Diverse animal fossils are found dating from the Cambrian Period but not earlier. Why might these organisms have made their first appearance in the fossil record only then, even though their ancestors may have been living, and evolving, for a long time before the Cambrian? (Think about what kinds of new structures might have evolved during the Cambrian Period that would have allowed these organisms to leave fossils.)

6. You are a paleontologist with a particular interest in early microbes. A microbiologist has brought you a fossilized cell that appears to be prokaryotic. The scientist claims to have used potassium-40 dating to date the fossil to 3.9 billion years ago. If the analysis is accurate, what percentage of potassium-40 should remain in the specimen? If you used uranium-238 dating to verify the microbiologist's claim, what proportion of uranium-238 should remain in the specimen?

Mini Case

Apply Your Knowledge

7. Along the banks of a river, some sedimentary rock strata have been revealed by erosion. By radiometric dating, the layer above these strata is determined to be about 290 million years old, and the layer beneath is dated to about 354 million years ago. A paleontologist starts to uncover fossils in the sedimentary rock strata. The fossils are clearly land-dwelling vertebrates. Are they more likely to be reptiles or amphibians? Explain your answer.

Interpreting Data

Apply Your Knowledge

8. You have carried out radiometric analysis on four igneous rocks uncovered at several sites you are exploring. From the percent lead you determine in each case, what is the approximate age of the rock?

Rock A: 75% lead _____

Rock B: 50% lead _____

Rock C: 30% lead _____

Rock D: 10% lead _____

Driving Question 2 What factors help to explain the distribution of species on Earth?

By answering the questions below and studying Infographics 16.5–16.8, you should be able to generate an answer for this broader Driving Question.

Know It

9. If two organisms strongly resemble each other in their physical traits, can you necessarily conclude that they are closely related? Explain your answer.

10. What did the arrangement of landmasses on Earth look like between 135 and 65 million years ago? What happened to these landmasses, and how does this change help explain the distribution of organisms found on the planet?

Use It

11. A cactus called ocotillo (*Fouquieria splendens*), which grows in New Mexico, looks very much like *Alluaudia procera*, a species of plant that grows in the deserts of Madagascar. These two plant species are not closely related—they are in different orders in the kingdom Plantae. Why, then, do they look so alike?

12. If penguins had evolved in northern regions of North America and not in New Zealand, what would you predict about the current distribution of polar bears and penguins?

13. Both bats and insects fly, but bat wings have bones and insect wings do not. Would you consider bat and insect wings to be a result of convergent evolution, or of homology—evolution based on inheritance of similar structures from a common ancestor? Explain your answer.

Driving Question 3 What are the major groups of organisms, and how are organisms placed in groups?

By answering the questions below and studying Infographics 16.8–16.11, you should be able to generate an answer for this broader Driving Question.

Know It

14. Which of the following is *not* a domain of life?

a. Animalia
b. Eukarya
c. Bacteria
d. Archaea
e. Plantae
f. Neither a nor e is a domain of life.

15. Put the following terms in order from most inclusive (1) to least inclusive (6).

_____ Domain
_____ Species
_____ Supergroup
_____ Kingdom
_____ Genus
_____ Phylum

16. A phylogenetic tree represents

a. a grouping of organisms on the basis of shared structural features.
b. a grouping of organisms on the basis of cell type.
c. a grouping of organisms on the basis of complexity.
d. a grouping of organisms on the basis of evolutionary history.
e. a grouping of organisms on the basis of where they are found.

Use It

17. Why was the kingdom Monera eventually split into two domains? What are these two domains?

18. On the tree below, which number represents the most recent common ancestor of humans and corn?

a. 1
b. 2
c. 3
d. 4
e. Humans and corn do not share any ancestors.

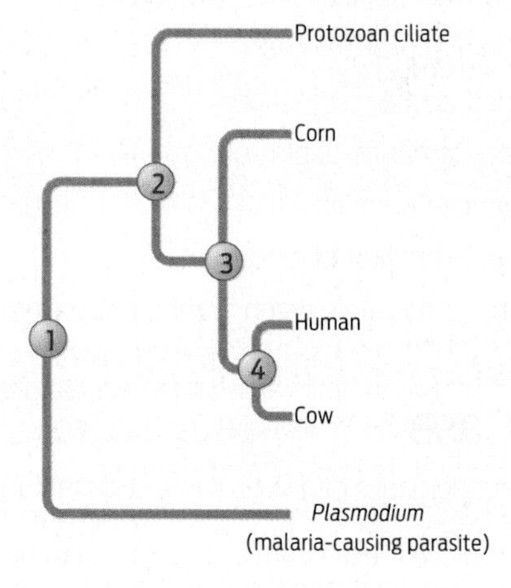

Apply Your Knowledge

Bring It Home

19. Carry out some online research on the fossils found in your home state. What groups of organisms are represented in the statewide fossil record? What is the oldest fossil found in your state? What do the fossils in your state suggest about the pattern(s) of evolution in your state?

Bill Abbott/CartoonStock

INVISIBLE YOU

A hidden world of microbes helps make you who you are

Amina Bouslimani et al., Molecular cartography of the human skin surface in 3D, PNAS April 28, 2015; 112(17) E2120–E2129; Figure 4 courtesy of Prof. Theodore Alexandrov

DRIVING QUESTIONS

1. What is the human microbiome, and what is its role in health, disease, and forensics?
2. What are prokaryotes, and why are they classified into two distinct domains of life?
3. What are features of bacteria and archaea?

"Because microbes are ubiquitous, they can be tiny witnesses to the events of our lives."

—Jessica Metcalf

Jessica Metcalf holds up an image of a human hand. It's a ghostly blue, with yellow and green splotches glowing against a black background. The hand looks as if it's been dipped in fluorescent paint, then pressed onto film, leaving an eerie afterglow (see the chapter-opening photo).

She explains that the different colors correspond to different molecules found on the hand. One splotch might represent caffeine from a heavy coffee drinker, another might be a bit of antidepressant left on a fingertip, and a third might be the DNA from a particular type of bacterium.

Metcalf is a pioneer in the emerging field of microbial forensics. She seeks to use information from microbes to aid forensic investigations. The glowing hand comes from a colleague's project aimed at mapping the skin's chemical and microbial landscape.

"Because microbes are ubiquitous, they can be tiny witnesses to the events of our lives," says Metcalf, an evolutionary biologist at Colorado State University in Fort Collins. "Skin microbes hold great potential for forensic science." We each have our own unique collection of microbes living on and in us, and scientists hope to use these traveling companions as a kind of microbial fingerprint. Like traditional fingerprints, our microbial signatures are left on surfaces and may provide clues to an identity. But microbial traces can potentially provide even more data—information about a person's lifestyle, what that individual recently ate, whether the person wears makeup or takes certain medications.

Peter Bolter/Alamy

Computer keyboards and computer mice retain microbial fingerprints that can identify users.

Studies like the glowing hand project would have been impossible to do just a few years ago. But thanks to advances in DNA sequencing, it is now possible to make visible to the naked eye what previously required high-powered microscopes. What these new techniques have revealed is an entire world of microbes living among our own cells. Scientists call the community of microbes living in or in our bodies the human **microbiome.**

Our skin, our gut, and our respiratory and reproductive tracts are all home to diverse microbial organisms—among them, bacteria, archaea, protists, and fungi. These microscopic companions influence our susceptibility to disease and contribute to our normal physiology. By some estimates, there are at least as many microbial cells living on and in us as there are human cells making up our body **(INFOGRAPHIC 17.1).**

MICROBIOME
A community of microbes at a particular location (e.g., on a person's skin or in the gut).

INFOGRAPHIC 17.1

The Human Microbiome

A huge diversity of microbes live on and in the human body. These resident microbes include prokaryotic bacteria and archaea, as well as eukaryotic fungi and protists. This community of microorganisms—the human microbiome—influences our health. While there are common organisms in the human microbiome, each person's individual microbiome is unique.

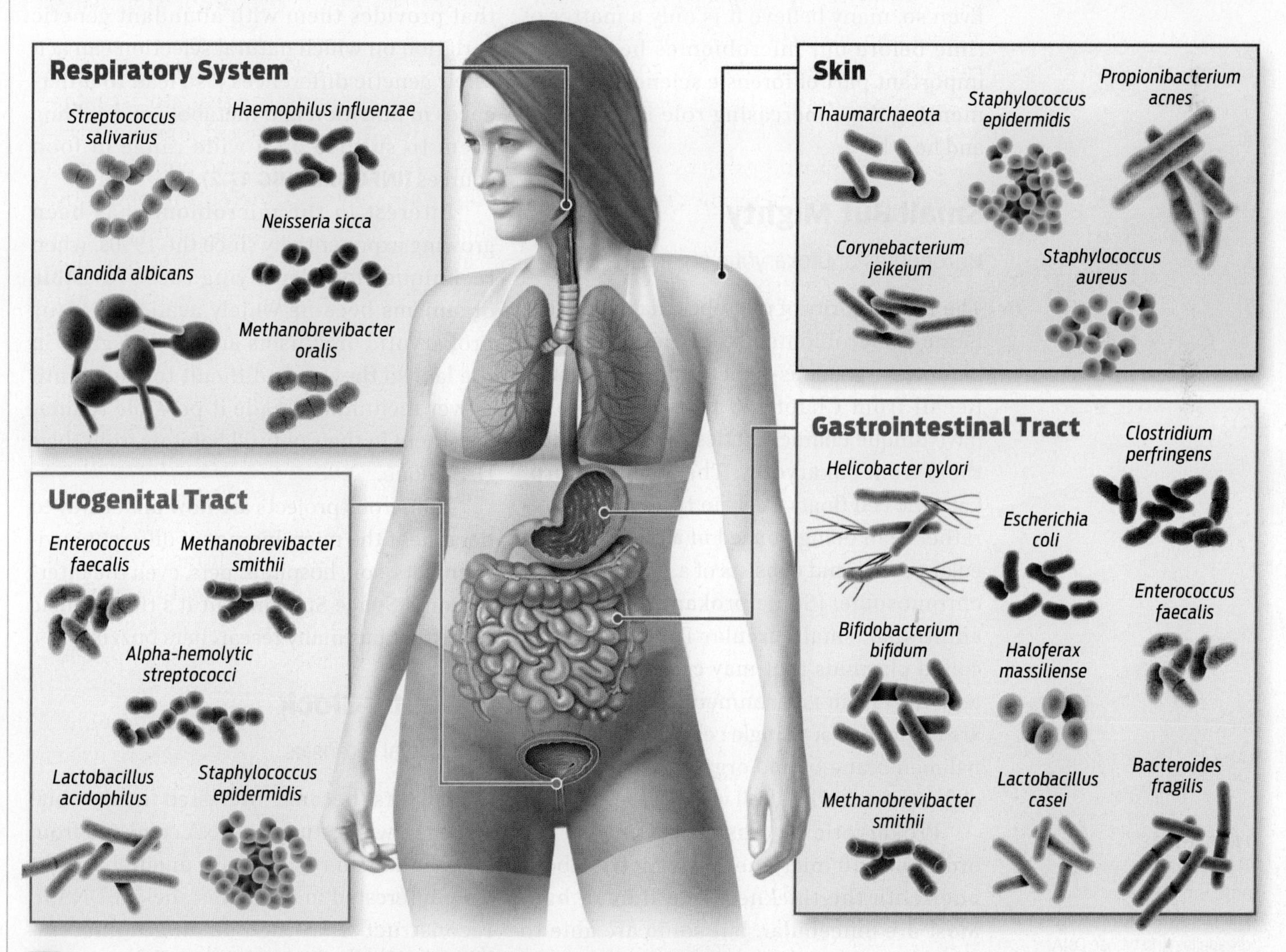

? List three microorganisms likely to be found in a microbial fingerprint.

The paper that many say launched the field of microbial forensics was published in 2010. Metcalf's former post-doctoral advisor, Rob Knight, showed that it was possible to identify individuals based on the bacterial profiles they left on computer keyboards and computer mice. Using this approach, he and his colleagues were able to correctly identify nine separate individuals based on the microbial fingerprints they left on these objects, even after the objects were left untouched for 2 weeks. Knight's lab is also responsible for the glowing human hand study.

The science of microbial forensics remains at the exploratory stage. So far, the only crimes microbes have helped solved have been on the television show *CSI: Miami.* Even so, many believe it is only a matter of time before our microbiomes become an important part of forensic science, complementing their increasing role in medicine and health.

Small But Mighty

▸ Properties of prokaryotic organisms

The vast majority of microbes making up the human microbiome are single-celled prokaryotic organisms—bacteria and archaea. Recall from Chapter 3 that **prokaryotes** have unique characteristics that distinguish them from eukaryotes. The DNA in a prokaryotic cell floats freely in the cytoplasm—rather than being housed in a nucleus, as in eukaryotes—and consists of a single, circular chromosome. (Some prokaryotes contain additional, small circular DNA molecules called plasmids that may confer an advantage in certain environments.) Prokaryotes are composed of a single cell that lacks internal membrane-bound organelles and is usually surrounded by a cell wall.

PROKARYOTE
A (typically) unicellular organism whose cell lacks internal membrane-bound organelles and whose DNA is not contained within a nucleus.

Prokaryotic organisms are tiny, on the order of 1–10 micrometers, which is about one-tenth the thickness of a human hair. Most are unicellular, but some are able to form multicellular colonies that can function as a single unit.

What prokaryotes lack in size, they make up for in numbers. Prokaryotes occupy virtually every niche on the planet, and most scientists agree that we have barely scratched the surface in cataloguing their numbers and diversity. There are more prokaryotic organisms in a handful of dirt than there are plants and animals in a rain forest. Each one of us carries trillions of these tiny organisms on and in our bodies.

There are several reasons why prokaryotes have been so phenomenally successful at the game of life. First, they reproduce quickly—sometimes dividing in less than an hour—and are thereby able to rapidly populate a given environment. They can shuffle and share DNA between their cells, an ability that provides them with abundant genetic variation on which natural selection can act. Their genetic differences also lead to differences in nutrition and metabolism, enabling them to survive on a wide range of food sources **(INFOGRAPHIC 17.2)**.

Interest in the microbiome has been growing exponentially since the 1990s, when techniques for identifying these invisible organisms became widely available. Many prokaryotic organisms are hard to grow in the lab, so they were difficult to study until newer techniques made it possible to analyze them in their natural habitats from their DNA alone.

Numerous projects are now under way to document the microbiomes of different environments: soil, hospitals, pets, even the International Space Station. But it's the forensic angle that has many researchers buzzing.

A Death Clock

▸ Microbial forensics

Metcalf first became interested in microbial forensics while studying DNA collected from ancient human remains. She and Rob Knight were interested in using this ancient DNA to reconstruct what the human gut microbiome was like before the modern invention of agriculture and antibiotics. Such information

INFOGRAPHIC 17.2

Prokaryotic Organisms Are Small but Highly Successful

Prokaryotic organisms are much smaller than eukaryotic organisms. Their cells lack organelles, and their DNA floats freely in the cytoplasm. Prokaryotic organisms divide very rapidly and can generate large amounts of genetic diversity, allowing them to adapt quickly to changing environments. Because of this, they can be found nearly everywhere on the planet.

Prokaryotic Organisms

- Typically single-celled
- No organelles
- DNA floats freely in cytoplasm
- Unique cell walls
- Divide and reproduce rapidly
- Adaptable to a variety of environments

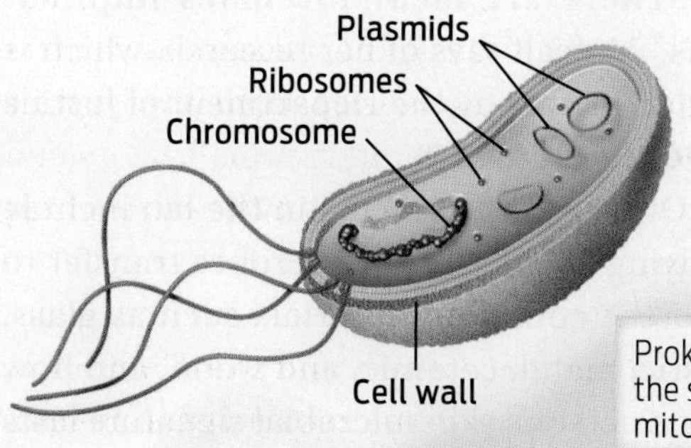

Prokaryotes are about the size of a eukaryotic mitochondrion.

Prokaryotes are about one-hundredth the size of a eukaryotic cell.

? List at least three features that distinguish prokaryotes from eukaryotes.

could be useful for understanding our modern susceptibility to particular diseases. But the researchers quickly encountered a stumbling block.

"I started thinking about how, in order to interpret the microbial DNA from ancient samples, I needed to know how the microbiome changes during death and decomposition," Metcalf says.

When animals die, microbes from the environment begin to decompose their flesh. These fellow travelers in ancient remains would make interpreting microbiome DNA from these samples difficult.

And so began one of Metcalf's current projects: the quest to understand the microbiology of death. How does the microbiome change after we die? Which species become more common, and which less common? Are the common species the ones we carried with us in life, or do they come from the environment in which we are decomposing?

"I started thinking about how, in order to interpret the microbial DNA from ancient samples, I needed to know how the microbiome changes during death and decomposition."

—Jessica Metcalf

Metcalf began studying the microbiology of death in mice. She let mice decompose atop tiny soil graves. Each day, she or a lab member would swab parts of the animal corpses—head, torso, abdominal cavity—and the surrounding soil. Then she analyzed the DNA extracted from these samples to identify which microbes were most abundant.

By studying these changes in microbial life over time, Metcalf has discovered that most of the microbes that grow on us after death come from the environment, with a reduced contribution from our own microbiome. What's more, she found that the sequence of microorganisms that appear is largely consistent and predictable. In fact, she and her team could predict the time of death to within 3 days, even after 48 days of decay. She realized that these microbial changes would be used as a kind of "microbial clock," a term she coined.

Metcalf wondered, would the clock run the same way in humans? It's obviously much harder to come by dead humans than it is dead mice. That's why in recent years she's partnered with anthropological research stations, sometimes called (somewhat crudely) "body farms." At these facilities, cadavers that have been donated to science are left to decompose in a protected field. The research stations are used by various scientists and forensic professionals to understand the changes that occur after death.

Metcalf worked with scientists at the Southeast Texas Applied Forensic Science Facility to conduct the experiment. Four fresh human corpses were left to decompose in the field. At regular intervals, project scientists would swab the skin of the corpse as well as the surrounding soil.

As in mice, the microbial clock of decomposition in humans allowed the researchers to accurately establish the time of death: in this case, with an error range of 2–4 days over 25 days.

"There are clear forensics implications," Metcalf says of her research, which is funded in part by the Department of Justice **(INFOGRAPHIC 17.3)**.

Other active projects in the lab include studying how well skin microbes transfer to different common materials such as glass, plastic, metal, ceramic, and wood, and how long a person's skin microbial signature lasts after death.

Unseen Roommates

Our microbiomes may identify us, but they aren't glued to us. We continually shed a cloud of microbes as we move through the world, creating a kind of microbial aura. Metcalf recently turned her microbial detective skills to understanding this microbial aura in the context of people's homes.

Called the Home Microbiome Project, the study followed seven families, including their pets—for a total of 18 people, three dogs, and one cat—for 6 weeks. Study participants swabbed their hands, feet, and noses daily. They also swabbed surfaces in the house, like countertops, floors, light switches, and doorknobs.

"One thing that's become clear with microbiome research is that people have a unique individual skin microbiome that

INFOGRAPHIC 17.3

The Microbiology of Death

Five stages of decay take place when an animal dies. Each stage has predictable characteristics, including a predictable series of microbes that aid in the decomposition process. The presence of these characteristic microbes can be used as a microbial death clock to determine an organism's time of death.

Fresh

Shortly after death, blood pools in the lower parts of the body, turning tissue blue. Muscles turn rigid. Cells begin to break down from their own enzyme activity.

Bloat

Microbes within the body feed on body tissues. Anaerobic bacteria produce gases that give the body a bloated appearance. Gas pressure can cause skin to rupture.

Microbes at work:

Proteobacteria: *Pseudomonas*; Enterobacteriaceae
Proteobacteria: *Wohlfahrtiimonas; Ignatzschineria*
Basidiomycota: *Lysurus*

Active Decay

Ruptured tissues are exposed to oxygen. Aerobic microbes and insect larvae enter from the environment and feed on body tissue. Decaying tissues release fluids and strong odors.

Microbes at work:

Bacteroidetes: *Myroides*
Ascomycota: *Yarrowia*
Nematoda: Rhabditidae
Proteobacteria: *Psudomonadaceae*; Chromatiaceae; *Proteus*

Advanced Decay

Insect larvae and microbes complete soft tissue decay and develop into beetles. Fluids from the body release carbon, nitrogen, and other nutrients to the surrounding soil.

Microbes at work:

Stramenopiles: *Plasmopara*
Zygomycota: *Mortierellaceae*
Firmicutes: *Planococcaceae*
Proteobacteria: *Acinetobacter*

Dry Remains

No soft tissue or fluid remains. Dry skin, cartilage, and bone are bleached white. Nutrients released to the surrounding soil result in increased plant growth.

Microbes at work:

Firmicutes: *Lactobacillales*
Firmicutes: *Sporosarcina*
Ascomycota: *Stromatonectria*

New Zealand Geographic issue 138, Mar-Apr 2016 "Game Over" by Dave Hansford

? A decaying corpse has an abundance of Bacteroidetes, Ascomycota, and Nematoda. At what stage of decay is this corpse?

is somewhat consistent over time," Metcalf says. "You leave it in spaces that you occupy and you transfer it to objects that you touch."

The home study showed just how quickly we deposit our microbiome in the areas that we occupy. Within 24 hours of people moving into a house, their skin microbiomes blanketed surfaces. When a person left on a trip, that person's skin microbiome disappeared from the house within just a few days.

Picking up a phone, drinking from a glass, or even just walking through a room will leave behind millions of individual microbes from hundreds of species. These microbial fingerprints will then disappear rapidly from surfaces that are infrequently touched, but can persist for longer on frequently touched objects.

"Our classic fingerprints contain no temporal information in them," Metcalf says. "They could have been there for 10 seconds or they could have been there for 10 years. Whereas the microbial fingerprint has a temporal aspect to it because it will disappear over time."

Microbial fingerprints also contain information about our lifestyles. In another study, based on a staged home break-in, researchers were able to predict that one of the "burglars" had a least 10 alcoholic drinks a week and that the other was taking medicines for migraines. These clues came from the types of microbes that were left at the crime scene.

Revising the Tree of Life

▸ Prokaryotic domains of life

Microbes are invisible to the naked eye, and sometimes they even look similar under a microscope. So Metcalf and her colleagues need a different sort of vision to tell different species of microbes apart. That's where DNA comes in handy. Because DNA is common to all life, yet its exact sequence changes over time as organisms evolve, it can be used to identify individual species. DNA sequences that are unique to each species serve as a kind of identifying barcode.

From a sample obtained via a swab, researchers can isolate DNA, amplify it using polymerase chain reaction (PCR) technology (Chapter 7), and then sequence it using automated sequencing techniques than can run thousands of samples at a time.

The methods that Metcalf and her colleagues use to study the microbiome are extensions of an approach pioneered in the 1970s by Carl Woese, an American biologist. Woese was interested in improving scientists' understanding of the evolutionary history of life. He wanted to base his tree of life not on differences in appearance or behavior among organisms, but on actual degrees of genetic relatedness.

At the time, no genomes had been sequenced, and large-scale sequencing of DNA and RNA was still in its infancy. Thus, Woese had to overcome technical as well as conceptual hurdles. His first conceptual hurdle was to identify a molecule that is common to all life, but whose sequence changes slowly over time. He settled on the 16S ribosomal RNA. This RNA molecule makes up one subunit of the prokaryotic ribosome, the molecular machine that manufactures proteins. One part of this molecule is almost identical across all prokaryotic organisms, while another part is more variable. These tiny variations in the sequence of the ribosomal RNA molecule, Woese realized, could be used to identify individual species of bacteria and measure their relatedness. Bacterial species with more similar 16S RNA sequences could be placed closer together on the tree, indicating their closer relationship.

All went well until Woese got to a group of microbes known as methanogens. These curious creatures don't use oxygen to live; in fact, oxygen is toxic to them, so great care must be used to grow them in the lab. Instead of carbon dioxide, these microbes release methane into the atmosphere (methanogen means "methane-generator"). They grow in low-oxygen environments like wetlands, as well as inside the guts of certain animals.

When Woese sequenced the methanogens' 16S RNA, he discovered that it looked nothing like the bacterial sequences he had collected. This finding bewildered Woese at first. But over time he came to what he thought was the inescapable conclusion: methanogens were not bacteria at all. They represented another domain of life, distinct from both bacteria and eukaryotes. He called this group archaea, from the Greek word for "ancient" **(INFOGRAPHIC 17.4)**.

Woese's proposal of a new domain of life did not immediately catch on with other biologists. In fact, many were skeptical of the idea at first. But over time, as more evidence accumulated, they were forced to admit that Woese was right. The familiar tree of life that many scientists had grown up with, featuring five kingdoms with a single prokaryotic kingdom called "Monera," needed to be overhauled (see Chapter 16).

In Woese's day, the process of sequencing a single 16 RNA molecule was painstaking and could take the better part of a year. Today, thanks to advances in technology,

INFOGRAPHIC 17.4

Bacteria and Archaea, Life's Prokaryotic Domains

All living organisms fall into one of the three domains of life. Within each domain, there are subgroups of organisms, grouped together based on their evolutionary relationships, as determined by 16S rRNA sequences. Two of the domains of life, Bacteria and Archaea, have organisms with prokaryotic cells, but each has a distinct evolutionary history. Archaea are more closely related to eukaryotes than to bacteria.

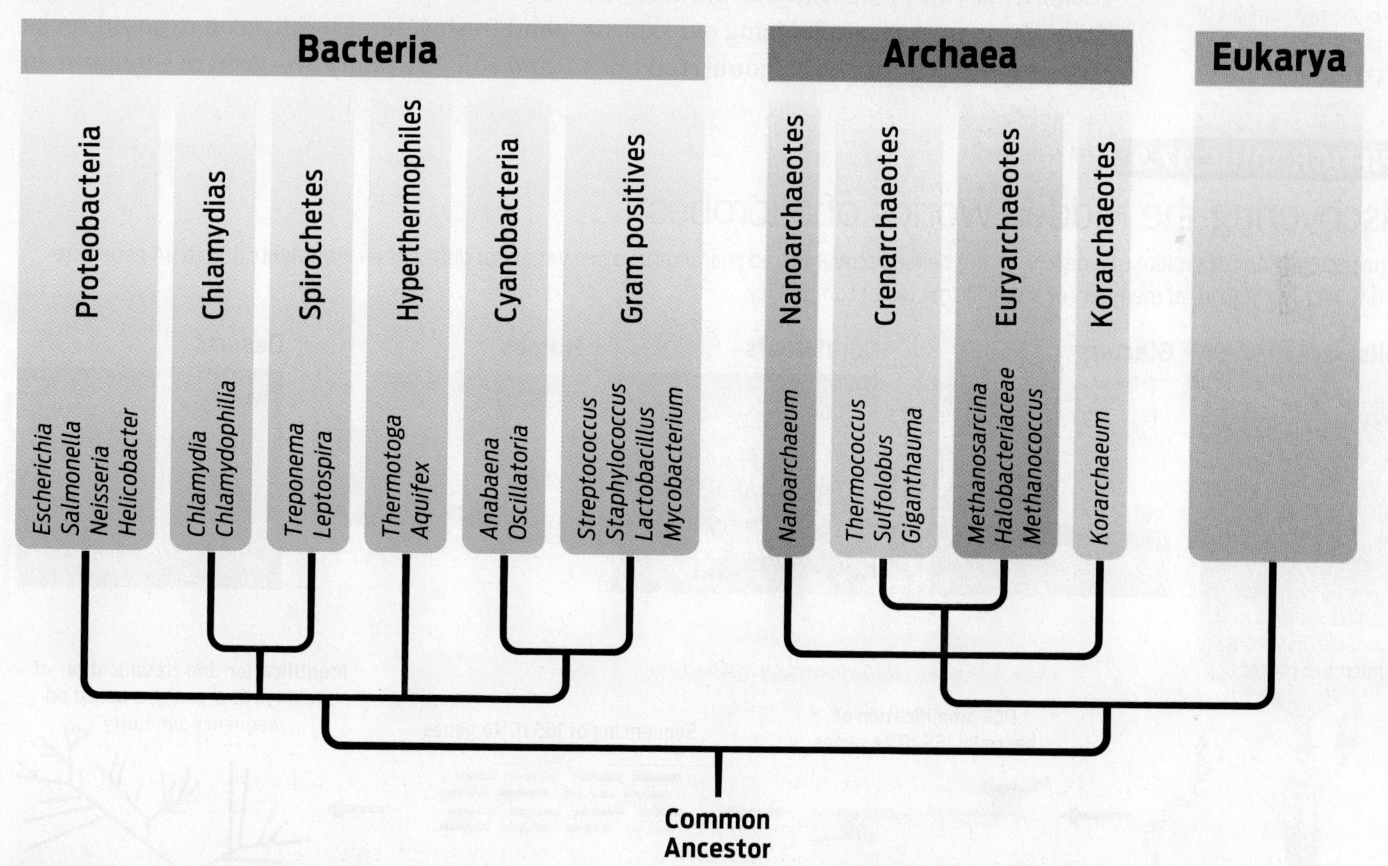

? What is the basis for the classification of bacteria and archaea into separate domains, despite their common prokaryote structure?

it's much easier and faster. But the premise of using genetic information to identify microbes is much the same, except that instead of using the 16S RNA molecule itself, scientists now use 16S RNA gene sequences to infer the presence and relative abundance of different microbial species. This technique has allowed scientists to learn more about the abundance and identities of prokaryotic organisms in different locations, from soil to glaciers to coral reefs **(INFOGRAPHIC 17.5)**.

Lifestyles of the Small and Infamous

▸ Bacterial structure and diversity

Much of the human microbiome is made up of **bacteria**, the unsung workhorses of the planet. These tiny organisms include commonly encountered microbes such as the *Escherichia coli* present in our gut and the *Staphylococcus aureus* colonizing our skin, as well as not so commonly encountered ones like the fluorescence-emitting bacteria living in partnership with a sea squid.

BACTERIA
One of the two domains of prokaryotic life; the other is Archaea.

While all bacteria are prokaryotic, and most possess a cell wall, their genetic diversity translates into a wide variety of differences in nutrition, metabolism, shape, and lifestyle. Bacteria live just about everywhere on the planet, from briny lakes submerged under miles of ice, to hazardous waste disposal sites, and everything in between.

Like all organisms, bacteria can be categorized by what they eat. Some bacteria are autotrophs (literally, "self-feeders"): they are able to make their own food directly, using material from the nonliving environment, ranging from carbon dioxide to rocks. Others are heterotrophs (literally, "other feeders"): they must consume material from other living organisms to obtain food.

One of the largest and most important groups of autotrophic bacteria is the cyanobacteria. These bacteria are found in oceans and freshwater, as well as on exposed rocks and soil—virtually everywhere sunlight can

INFOGRAPHIC 17.5

Discovering the Hidden Worlds of Microbes

Distinct collections of prokaryotic organisms are being discovered and characterized in a variety of different environments. 16S rRNA genes are used to identify individual members of a specific microbial community.

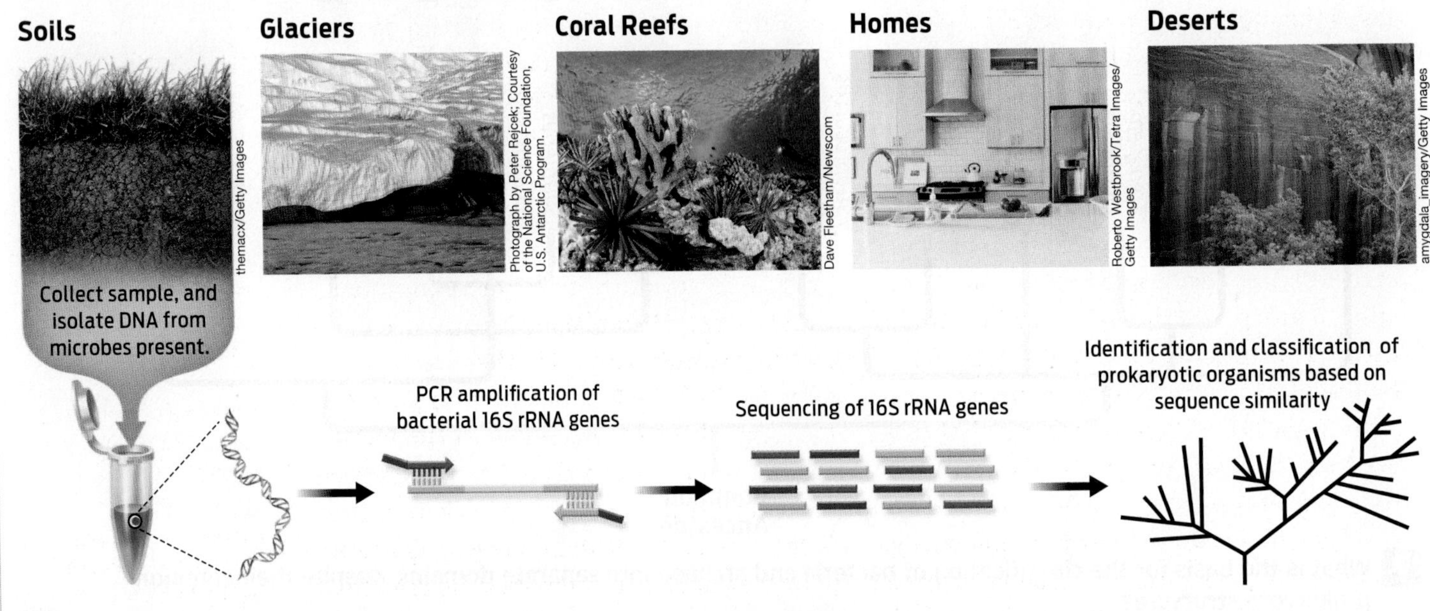

? Which sequence is used to identify members of a given microbiome?

reach. Cyanobacteria use the energy of sunlight to carry out photosynthesis in a manner similar to plants: they take in CO_2 to generate carbohydrates (food) and in the process produce much of the oxygen that other organisms, including humans, rely on.

Heterotrophic bacteria include those that obtain food by consuming material from living or dead organisms. Such bacteria play an important role in decomposition: they feed on dead organisms, allowing carbon and other elements—which would otherwise be trapped in dead organisms, sewage, or landfills—to be recycled. They are also useful in bioremediation projects. For example, some types of bacteria break down droplets of oil, much the same way that humans digest butter, so they help clean up oil spills naturally.

Bacteria break down their food molecules through a variety of metabolic pathways—some of which require oxygen, and some of which do not. For example, many bacteria employ the anaerobic process of fermentation (see Chapter 6) to get energy from food. The products of fermentation can be valuable (and tasty) to humans. You may have seen "L. bulgaricus" listed as an ingredient of yogurt. Live *Lactobacillus bulgaricus* bacteria are present and hard at work in the yogurt, fermenting sugars into lactic acid. This acid helps the milk solidify into yogurt and gives yogurt its tangy taste. Other bacteria use oxygen to break down organic molecules, like the aerobic bacteria that feast on an oil spill.

In addition to nutritional and metabolic differences, bacteria display a variety of structural adaptations that suit their various lifestyles. They come in different shapes: spherical (in which case they are known as cocci), rod-shaped (bacilli), and spiral (spirochetes). Many bacteria are equipped with **flagella,** tail like structures that project from the cell and rotate like propellers to help it move. For example, the bacterium *Helicobacter pylori,* the most common cause of stomach ulcers, uses its flagella to propel itself through the gastric mucus found in the

James Peake/Alamy

Hawaiian bobtail squid (*Euprymna scalopes*) live in symbiosis with fluorescent bacteria that reside in the squid's light organ. The illumination from the bacteria helps the squid avoid casting a shadow on a moonlit night, making it less noticeable to prey.

stomach. **Pili** are shorter, hairlike appendages that enable bacteria to adhere to a surface. *Neisseria gonorrhoeae,* the bacterium that causes the sexually transmitted disease (STD) known as gonorrhea, uses its pili to remain attached to the lining of the urinary tract. Without pili, the bacteria would be flushed out by the flow of urine.

Other bacteria are surrounded by a **capsule,** a sticky outer layer that helps the cell adhere to surfaces and to avoid the defenses of the host. *Streptococcus mutans,* for example, produces a capsule that allows it to adhere to teeth, where it forms the plaque that can lead to cavities.

The highly resourceful bacteria have forged a diverse range of living arrangements with other creatures. Many live in close association, or **symbiosis,** with other organisms—often to the benefit of one or both partners. Lactobacilli reside naturally in the female vaginal tract, for example. There they feed on naturally occurring sugars, which they ferment to lactic acid. The resulting acidity of the vaginal tract suppresses the growth of yeast, preventing yeast infections. Antibiotics

FLAGELLA (SINGULAR: FLAGELLUM)
In bacteria, long, slender appendages extending from some bacterial cells, used in movement of the cell.

PILI (SINGULAR: PILUS)
Short, hairlike appendages extending from the surface of some bacteria that enable them to adhere to surfaces.

CAPSULE
A sticky coating surrounding some bacterial cells that adheres to surfaces.

SYMBIOSIS
A relationship in which two different organisms live together, often interdependently.

Charles O'Rear/Getty Images

Yellowstone's Grand Prismatic Hot Spring. Heat-tolerant bacteria and archaea live in this 80°–100°C, low-pH, high-sulfur aquatic environment that bubbles up from below Earth's crust. The rainbow colors that ring this spring come from light reflecting off the pigments made by different microbe species.

taken for a bacterial infection are likely to kill the resident lactobacilli as well as the invaders, and a yeast infection may be an unhappy side effect.

Another example of beneficial bacterial symbiosis is provided by *Vibrio fischeri.* This bioluminescent bacterium lives and feeds inside the light organs of certain species of squid. The glow-in-the-dark *Vibrio* produces light beneath the squid. That light helps obscure the shadow that the squid might cast on a moonlit night, making it less noticeable to its prey as it hunts.

Unfortunately, not all bacteria are beneficial to the host. While the vast majority of bacteria do not cause human disease, some do. Organisms that cause disease are called **pathogens.** Many pathogenic bacteria cause disease by producing toxins that harm their hosts. For example, certain strains of *E. coli* secrete a potent toxin that causes bloody diarrhea and sometimes even kidney failure and death in its host. Washing hands before preparing or eating food will help prevent transfer of *E. coli* from dirty hands, and cooking meat to recommended temperatures will kill contaminating bacteria, reducing the risk of toxin-producing *E. coli* entering the body with a meal.

Not all pathogens produce toxins. Some cause disease by living and reproducing in the body and interfering with its normal processes. An example is the bacterium *Treponema pallidum,* which causes syphilis, an STD.

In some cases, bacteria grow as a biofilm, a collection of bacteria that adhere to a surface and to one another. Bacteria in biofilms secrete sticky substances that encase the biofilm, protecting the bacteria. Up to 65% of all bacterial infections may be associated with biofilms.

Sometimes the line between harmless and harmful bacteria can be blurred. Organisms that can—but don't always—cause disease are known as opportunistic pathogens. For example, most of us have *S. aureus* on our skin at many times during our lives. Most of the time, this bacterium does not cause any harm, but if it penetrates the skin—through a wound, for example—it can cause a serious infection and even death, as discussed in Chapter 13 **(INFOGRAPHIC 17.6).**

The bacterial members of the human microbiome are not distributed evenly across the body. Regions of the body such as the face, chest, and back that contain abundant oil-producing glands tend to harbor fat-loving microorganisms. In contrast, less exposed regions, such as the groin, armpit, and toes, tend to harbor species that prefer a warm and wet environment.

Some Like It Hot

▸ Archaeal structure and diversity

Archaea are similar to bacteria in that archaea are single-celled organisms that lack a nucleus, but genetically they are as different from bacteria as humans are. All those genetic differences add up to a number of unique features that distinguish archaea from bacteria. For example, whereas bacteria

PATHOGEN
A disease-causing agent or organism.

ARCHAEA
One of the two domains of prokaryotic life; the other is Bacteria.

have cell walls made of the molecule peptidoglycan, archaea have cell walls composed of other molecules. Archaeal cell membranes can also have a different chemical composition, and archaea rely on unique forms of metabolism.

Though archaea are found in many run-of-the-mill habitats such as rice paddies, forest soils, lake sediments, and our own respiratory tract, the most well-known species are the so-called extremophiles, found in extreme environments. Many of these extreme-loving archaea are hyperthermophiles—organisms that can survive only at extremely high temperatures. Many hyperthermophilic archaea are anaerobic and rely on sulfur instead of oxygen in capturing energy from food. Sulfur-rich hot springs like those in Yellowstone National Park are home to these archaea.

Some archaea are halophiles, or "salt lovers." These archaea prefer a home saturated in salt, which would shrivel most other living things. Their presence is detectable by the

INFOGRAPHIC 17.6

Exploring Bacterial Diversity

Bacteria live in every imaginable place on Earth and have a diverse array of lifestyles. Their unique structural and metabolic adaptations have enabled them to become a dominant force of life on the planet.

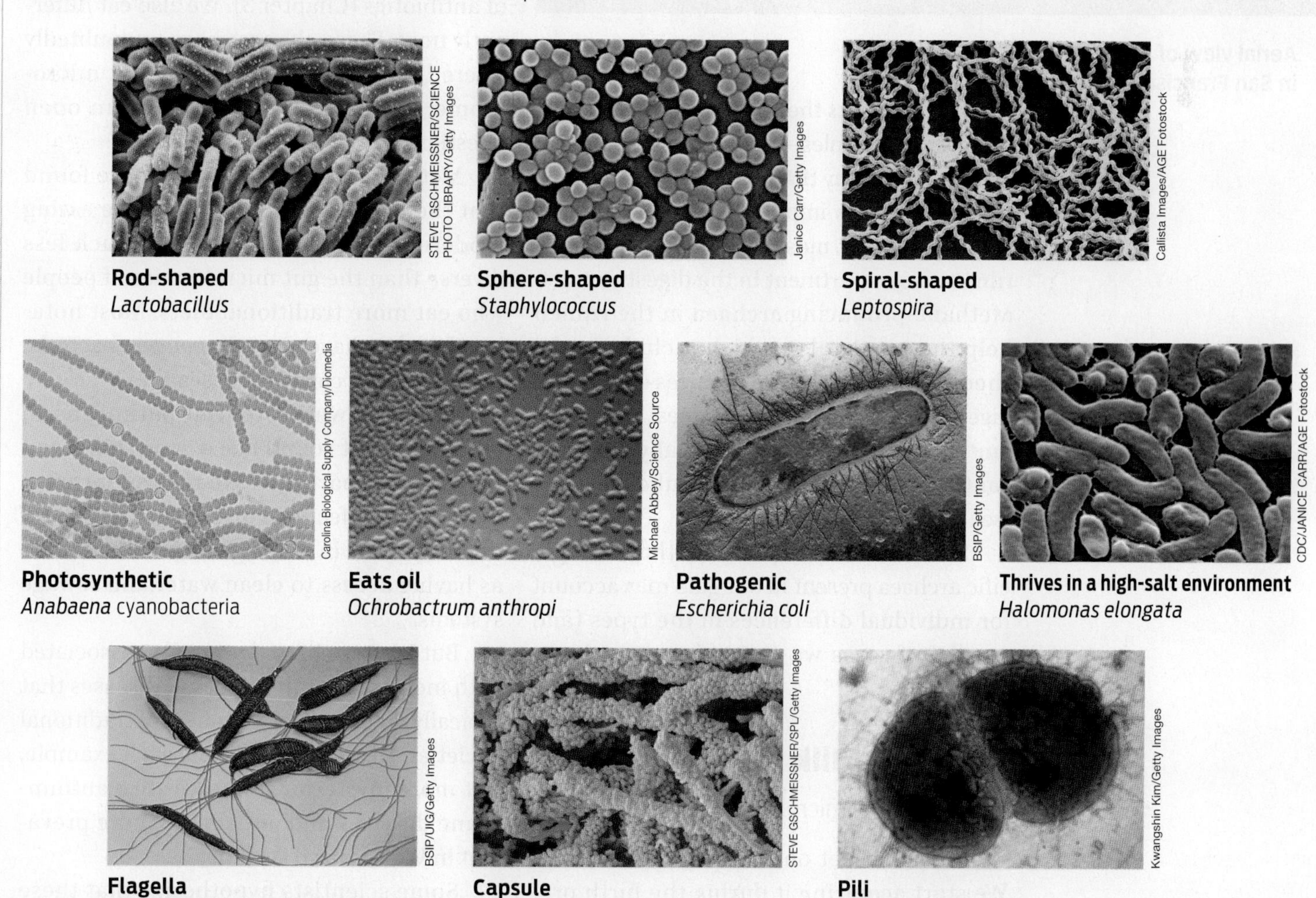

? **Flagella, capsules, and pili can all be found on pathogenic bacteria. Pick one of these structures, and describe how it might contribute to the development of disease.**

aerialarchives/Alamy/AGE Fotostock

Aerial view of salt ponds in San Francisco Bay.

colorful pigments they produce—bright reds, yellows, and purples—as seen in salt ponds in San Francisco Bay **(INFOGRAPHIC 17.7)**.

Archaea play important roles in the lives of many animals, most especially those with a rumen, a compartment in the digestive tract. Methane-producing archaea in the rumen help these animals—which include cattle, sheep, goats, deer, and giraffes—with the digestion of cellulose-rich grasses. The methane that cows produce is actually a significant contributor to rising greenhouse gases (see Chapter 23).

Some scientists speculate that the specific archaea present in our guts may account for individual differences in the types (and smells) of gases we emit—for example, as flatulence.

Mother's Milk

▶ Acquiring the microbiome

Where do we get our unique microbiome? We start acquiring it during the birth process. When a baby passes through the birth canal of its mother, it acquires a collection of microbes from her vaginal tract. [Babies born by cesarean section (C-section) do not pick up these bacteria in this way, but can be provided with them after birth.] Breastfeeding further contributes to the human microbiome. Children are essentially fully colonized by 1–2 years of age. Genetics, what we eat, and the medicines we take also contribute to differences in our microbiomes.

Metcalf is interested in these differences as well. She has studied the gut microbiome of people eating traditional diets, such as the Yanomami who live in remote regions of the Amazon rain forest. The goal of this research is to understand how the human gut microbiome has changed over time, from before industrialization until now.

One of the most significant changes in that time, of course, was the development of antibiotics (Chapter 3). We also eat differently now. These changes have undoubtedly altered the composition of our gut microbiome, but exactly how remains an open question.

Metcalf and her colleagues have found that the gut microbiomes of people eating modern, Westernized diets are much less diverse than the gut microbiomes of people who eat more traditional diets. Most notably, the latter have more *Treponema* bacteria, many of which can cause disease.

Having fewer potential pathogens in our guts might sound like a good thing, and for the most part it is. We now suffer from many fewer infectious diseases as a result of improvements in our hygiene practices, such as having access to clean water and sewage systems.

But Western lifestyles are also associated with increases in other types of diseases that typically do not affect members of traditional societies. Heart disease is one such example, but more mysterious illnesses like autoimmune diseases and autism are more prevalent in Westernized societies as well.

Some scientists hypothesize that these diseases may have a connection to the modern microbiome. The idea is this: as cultures around the world become more "Western"

INFOGRAPHIC 17.7

Exploring Archaeal Diversity

Archaea are sometimes known as "extremophiles." They live in diverse environments, often surviving and thriving in very harsh conditions.

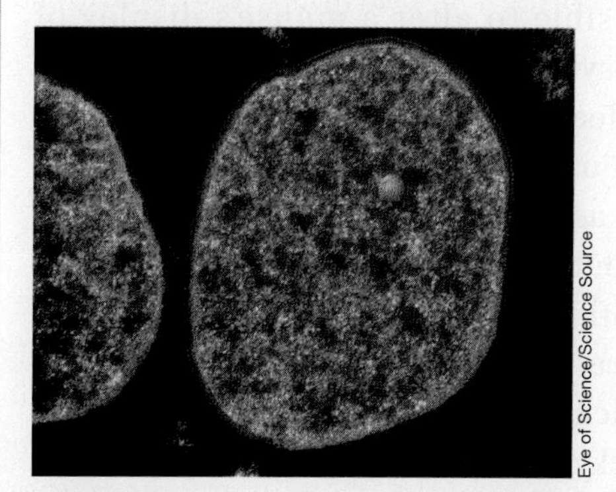

Eye of Science/Science Source

Sulfolobus
Considered extremophiles because they can grow at 70ºC and at pH 2.0

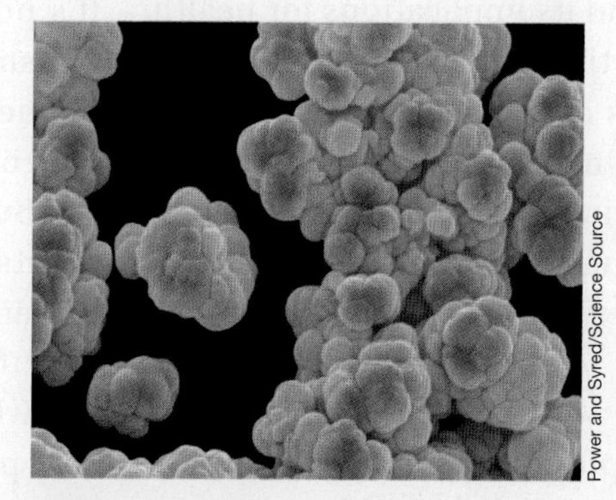

Power and Syred/Science Source

Methanosarcina
Has unusual cell walls and membranes, and produces methane

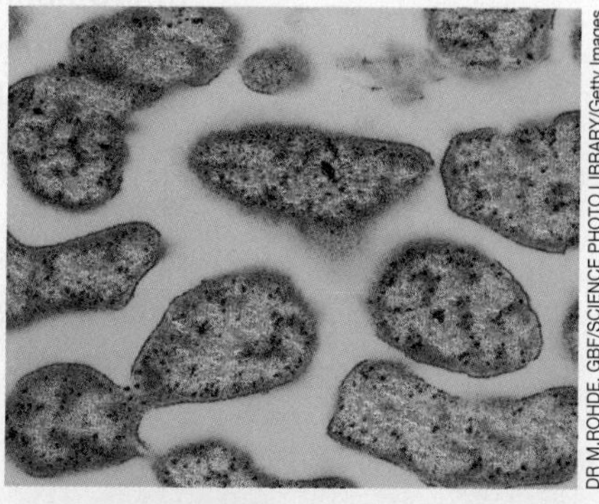

DR M.ROHDE, GBF/SCIENCE PHOTO LIBRARY/Getty Images

Methanococcoides
Grows in cold temperatures as low as −2.5ºC

Courtesy Mike Dyall-Smith

Haloquadratum walsbyi
Thrives in water temperatures well above boiling, at least 120ºC

Eye of Science/Science Source

Halobacteriales
Not only tolerates but requires a high-salt environment to live

? What are three extreme conditions in which some archaea can survive?

in terms of having access to antibiotics and safe drinking water, the diversity of their gut bacteria declines—as does the incidence of infectious disease. At the same time, we start having higher incidences of chronic illnesses connected to the immune system, such as allergies, Crohn's disease, autoimmune disorders, and multiple sclerosis. This correlation between low diversity of gut bacteria and higher risk of diseases related to the immune system has led to hypotheses about the role of a diverse microbiome in health and disease.

The possibility of a connection between a diverse microbiome and health has led to treatment approaches—none validated as of yet—that involve introducing these organisms back into the gut of affected individuals. The hope is that they will help reset the immune system and ease symptoms.

An additional remarkable finding from Metcalf's research is that genes for antibiotic resistance are found in microbes from indigenous peoples who have never encountered antibiotics. This evidence suggests that wild populations of bacteria already had these genes in their gene pool. The presence of these genes helps account for the rapid emergence of antibiotic resistance among many different bacterial groups today (see Chapter 13).

Microbiome Hope and Hype

▶ Health applications of the microbiome

People are understandably excited about the microbiome and its implications for health. The trillions of tiny organisms that share our bodies with us cannot help but shape our existence. An increasing body of evidence shows that having a less diverse microbiome can increase the risk of harmful conditions such as allergies, asthma, obesity, and type 1 diabetes.

Louise Murray/Science Source

Preparing a fecal transplant.

One of the clearest examples of the microbiome playing a role in health is the value of fecal transplants in curing people of dangerous gut infections. *Clostridium difficile,* or *C. diff* for short, causes a dangerous infection that produces intense diarrhea and inflammation of the colon; it can be fatal. People often acquire a *C. diff* infection after taking antibiotics while in the hospital. The antibiotics kill off pathogens and beneficial bacteria, but not *C. diff,* which can then gain a foothold in the body and grow out of control.

So far, the only treatment that has worked consistently to cure *C. diff* is a fecal transplant—transferring stool from a healthy person to the gut of the sick person. The procedure, while unappealing to some, is effective between 80% and 90% of the time.

Some evidence links the microbiome to the growing obesity crisis in industrialized countries. Scientists have shown that they can make a mouse obese by transferring the gut microbiome of an obese human to it using a fecal transplant. And numerous studies have shown that the gut microbiome of obese people is often different from that of non-obese people. Whether this is a cause or an effect of obesity is not yet clear.

Similar results have been found using mice engineered to display traits of autism. It's possible to alter a mouse's display of autism symptoms by altering its gut microbiome, leading some scientists to propose a kind of gut–brain axis.

But as microbiome expert Elisabeth Bik points out, it can very hard to rule out other explanations. According to Bik, "With autism specifically, there's a big confounding factor, which is that kids with autism are often very picky about food and have idiosyncratic diets. So that could be influencing their gut microbiome."

"There are clear hints that obesity might be in part due to the microbiome and might be transferable through the microbiome," she says. "But a lot of these experiments are done in mice that have a tendency to become obese. So these mice will already become obese very easily."

In many cases, people's excitement about the microbiome has outstripped the science, Bik suggests. Most of the relationships that have been observed are correlations, which of course do not always imply causation (Chapter 1). This has led, unfortunately, to an abundance of hype in the news.

Hype aside, can we alter our microbiome intentionally? Perhaps the best and easiest thing a person can do to ensure a healthy gut, Bik says, is to regularly consume fiber, which can be found in many fruits and vegetables, as well as in legumes, whole grains, and nuts. Many of our "good" bacteria thrive on fiber, which they help us digest, and in the process release helpful chemicals such as certain short-chain fatty acids. These molecules both strengthen the mucus layer of the gut lining and reduce gut inflammation. A thriving population of beneficial bacteria nourished on fiber can prevent the growth of pathogenic bacteria. If not kept in check, such pathogenic bacteria will explode in numbers and destroy the mucus layer that protects intestinal cells.

Bacteria in the gut can also use the nutrients from fiber to synthesize vitamins, such as vitamin K. These vitamins contribute to human health, possibly preventing vitamin deficiencies. For example, people on a diet deficient in vitamin K did not develop a vitamin K deficiency, but people who had been given antibiotics to reduce their gut microbes developed a vitamin K deficiency while on the same vitamin K–restricted diet. Fiber is an example of a **prebiotic,** a component in food that encourages the growth of beneficial gut microbes **(INFOGRAPHIC 17.8).**

Another common—if less scientifically supported—method of attempting to alter one's gut microbiome is to consume **probiotics,** which are edible concoctions of living microorganisms. Examples of

PREBIOTICS
Components of food (like fiber) that promote the growth of beneficial microbes.

PROBIOTICS
Living bacteria intended to be consumed and introduced into the microbiome (e.g., yogurt or capsules containing bacteria).

INFOGRAPHIC 17.8

A Healthy Gut Microbiome Defends Against Chronic Disease

Beneficial gut microbes flourish on a high-fiber diet and compete with pathogenic bacteria that can cause illness. They help maintain a thick mucus layer to protect epithelial cells from microbial damage. In addition, they use fiber as a source of nutrition and produce vitamins and chemicals that contribute to gut and overall health.

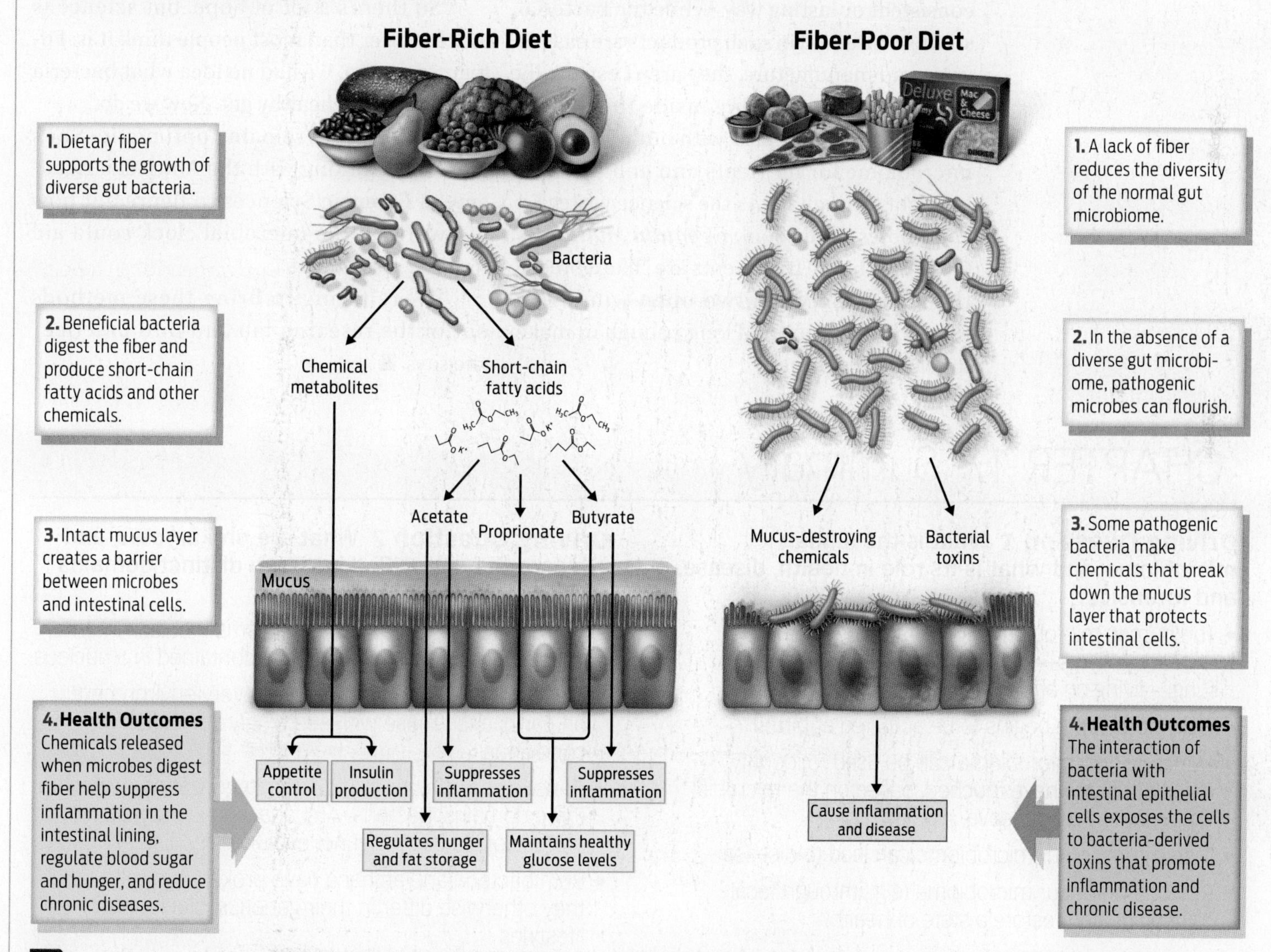

? From a microbiome perspective, why is a high-fiber diet important to health?

"There are clear hints that obesity might be in part due to the microbiome and might be transferable through the microbiome."

—Elisabeth Bik

commonly consumed probiotics include yogurt with "live cultures" and kefir, as well as capsules containing specific living microorganisms. Although the sale of such products is big business, very few probiotics have been shown to alter the human microbiome in any consistent or lasting way. While the bacterial strains contained in such products are easy to grow and manufacture, they aren't especially well adapted to growing inside the human gut. They can't compete with our normal gut microbiome for nutrients and end up moving right through us. As the science writer Ed Yong notes in his book *I Contain Multitudes* (2016), probiotic treatments are "like a breeze that blows between two open windows": they don't stick around long enough to make much of a difference.

So far, the strongest evidence in support of probiotics has come from their use in treating a very specific, life-threatening condition affecting the gut of preterm infants. In this case, the strains used are carefully chosen and grown for this particular purpose.

Bik isn't fazed by the many questions that remain about the microbiome and its role in health. She points to the Human Genome Project as a guidepost.

"Everybody thought, 'Okay, if we sequence the human genome, we'll be able to cure any disease.' That didn't happen. But it has led to a whole new field, personalized genomics. There's 23andMe and similar tests. There's a lot of new health information now.

"So there's a lot of hope, but science is much slower than most people think it is. Fifteen years ago, we had no idea what bacteria were living in a healthy gut. Now we do."

Metcalf, too, remains optimistic. She's currently working with the American Academy of Forensic Sciences to figure out how her work on the microbial clock could aid forensic researchers.

"We're trying to bring these methods out of the research lab and into practice," she says. ■

CHAPTER 17 Summary

Driving Question 1 What is the human microbiome, and what is its role in health, disease, and forensics?

- The human microbiome consists of the microorganisms—bacteria, archaea, protists, and fungi—living on and in us.
- Our microbiome begins to be acquired at birth.
- Our individual microbiome can be used to identify objects that we have touched, based on the microbial fingerprints that we leave.
- Disruptions to our microbiome can lead to disease.
- Restoration of our microbiome (e.g., through fecal transplant) can restore a state of health.

Driving Question 2 What are prokaryotes, and why are they classified into two distinct domains of life?

- Prokaryotes are unicellular organisms that lack internal organelles and whose DNA is not contained in a nucleus.
- Prokaryotes are found in virtually every environment on Earth, even those with seemingly inhospitable conditions.
- Genetic analysis has led to the categorization of life into three domains: Bacteria, Archaea, and Eukarya. Each domain of life has a distinct evolutionary history.
- Both bacteria and archaea have prokaryotic cells, but they otherwise differ in their genetics, biochemistry, and lifestyles.

Driving Question 3 What are features of bacteria and archaea?

- Bacteria are a diverse group of prokaryotic organisms with many unique adaptations, such as flagella and capsules, that allow them to live and thrive in many environments.
- Some bacteria are disease-causing pathogens, but most are harmless and many are even beneficial. Cyanobacteria, for example, are responsible for much of the photosynthesis that supports life on Earth.
- Often known as "extremophiles," archaea live in some of the most inhospitable conditions on Earth. Many archaea flourish in less extreme environments as well.

More to Explore

- Solving Crimes with the Necrobiome: https://www.youtube.com/watch?v=B_lHQSxz9GI
- Microbiomania: https://phylogenomics.blogspot.com/p/blog-page.html
- Microbiome Digest — Bik's picks https://microbiomedigest.com/
- Metcalf, J. L., Xu, Z. Z., Bouslimani, A., Dorrestein, P., Carter, D. O., and Knight, R. (2017). Microbiome tools for forensic science. *Trends Biotech* 35(9):814–823.
- Metcalf, J. L., Carter, D. O., and Knight, R. (2016). Microbiology of death. *Curr Biol* 26(13):R561–R563.
- Yong, E. (2016). *I Contain Multitudes: The Microbes Within Us and a Grander View of Life*. New York: Ecco.

CHAPTER 17 Test Your Knowledge

Driving Question 1 What is the human microbiome, and what is its role in health, disease and forensics?

By answering the questions below and studying Infographics 17.1, 17.3, and 17.8, you should be able to generate an answer for this broader Driving Question.

Know It

1. Where are members of the human microbiome found?
- **a.** skin
- **b.** gut
- **c.** feces
- **d.** respiratory tract
- **e.** all of the above

2. Most members of the human microbiome are prokaryotic. What does this mean?
- **a.** Their cells are very similar to human cells.
- **b.** They lack a nucleus, but contain other membrane-bound organelles.
- **c.** They lack all membrane-bound organelles.
- **d.** Their DNA is not contained in a nucleus.
- **e.** both c and d

3. How is the human microbiome acquired?

Use It

4. One baby is born by C-section and is formula-fed for its first year of life. Another baby is born by vaginal delivery and is breastfed for its first year of life. What differences, if any, might you expect in the microbiome of these two children at 1 year of age? Explain your answer.

5. People who are eating a "Western" diet have their gut microbiome sampled through a fecal sample. Half of the people then adopt a high-fiber, plant-based diet (similar to many traditional diets throughout the world). What do you predict about the gut microbiome in each group? If the "traditional" diet is followed for decades, what do you predict about the incidence of autoimmune diseases? How much confidence would you have in the results of such a study (in which you ask a group of people to follow a new diet for decades)?

Driving Question 2 What are prokaryotes, and why are they classified into two distinct domains of life?

By answering the questions below and studying Infographics 17.2, 17.4, and 17.5, you should be able to generate an answer for this broader Driving Question.

Know It

6. Organisms are placed into one or another of the three domains of life on the basis of
- **a.** cell type.
- **b.** physical appearance.
- **c.** evolutionary history as assessed by genetic relatedness.
- **d.** ability to cause disease.
- **e.** degree of sophistication—that is, how evolutionarily advanced they are.

7. Which domain(s) of life contain organisms with a prokaryotic cell structure?
- **a.** Archaea
- **b.** Bacteria
- **c.** Eukarya
- **d.** Archaea and Bacteria
- **e.** Archaea, Bacteria, and Eukarya

8. The absence of membrane-bound organelles in a cell tells you that the cell must be

a. from a member of the domain Bacteria.
b. from a member of the domain Archaea.
c. from a member of the domain Eukarya.
d. either a or b
e. either b or c

9. The term *prokaryotic* refers to

a. a type of cell structure.
b. a domain of life.
c. a group with a shared evolutionary history.
d. a type of bacterium.
e. a type of archaea.

10. What did Carl Woese use to group organisms into three distinct domains?

a. the presence or absence of a nucleus
b. the sequence of an rRNA molecule
c. the ability to survive and reproduce in extreme environments
d. the presence or absence of a cell wall

Use It

11. Why were bacteria and archaea originally grouped together?

12. When first discovered, archaea were called "archaebacteria." Why do you suppose this name was used? What are the strengths and weaknesses of this earlier term?

13. Can you use cell structure to classify a cell as either bacterial or archaeal? Explain your answer.

14. Many prokaryotic organisms can carry out photosynthesis. How is this beneficial to humans?

Driving Question 3 What are features of bacteria and archaea?

By answering the questions below and studying Infographics 17.6 and 17.7, you should be able to generate an answer for this broader Driving Question.

Know It

15. If you were looking for a bacterium, where would expect to find one?

a. on your skin
b. in soil
c. in the ocean
d. associated with plants
e. any of the above

16. When examining archaea, which of the following is *not* a trait that you could expect to find?

a. ability to grow at high temperatures
b. a nucleus
c. ability to grow at extremely acidic pH
d. a cell wall
e. ability to survive in a high-salt environment

Use It

17. If you are unable to culture archaea from an environmental sample, is it safe to conclude that there are no archaea present? Why or why not?

18. Halophilic (salt-loving) archaea are able to prevent osmotic water loss from their cells, even in high-salt environments. What is one mechanism by which they could prevent water loss and thrive in high-salt environments? Hint: Think about osmosis and the salt concentration in the environment versus inside their cells.

19. If *Neisseria gonorrhoeae* had no pili, would it still be a successful pathogen? Explain your answer.

Apply Your Knowledge

Interpreting Data

20. It has been suggested that increasing rates of immune disorders, such as asthma and some food allergies, are the result of a reduced exposure to a diversity of "friendly" bacteria in the human microbiome.

a. Based on the data shown in the bar chart, which population would you predict to have the lowest rates of asthma and allergies? Note that the Amerindians are from rural communities in Venezuela.
b. If your prediction is correct, is that evidence of correlation or causation?
c. Design an epidemiological study to test the hypothesis that reduced diversity of the human microbiome is responsible for increasing rates of immune disorders. (Hint: Refer back to Chapter 1 for observational and epidemiological studies.)

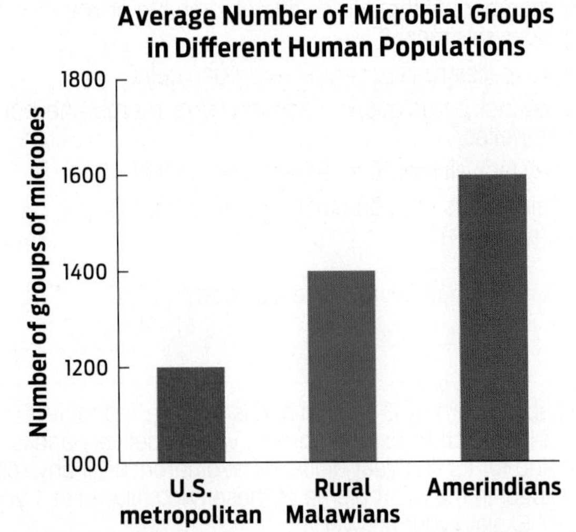

Data from Yatsunenko et al. 2012. Human gut microbiome viewed across age and geography. *Nature* 486: 222–228

Mini Case

Apply Your Knowledge

21. Your friend's mother has been hospitalized and has developed a *Clostridium difficile* (*C. diff*) infection of the colon. This is a common hospital-associated infection, in which the use of antibiotics to treat another infection reduces the normal microbiome of the colon, allowing *C. diff* to establish itself. Her doctor is recommending a fecal microbial transplant (FMT), in which feces from a healthy donor are transferred into the intestinal tract of a person with *C. diff* (either by enema or by a tube extending from the nose to the small intestine). Your friend and her mother find this idea to be disgusting, particularly when they realize that the donor fecal material is not sterilized before the transplant.

Data from a clinical trial comparing FMT to standard antibiotic therapy with vancomycin are shown in the two bar charts.

a. What do you observe about the microbial diversity in patients versus donors?

b. What do you observe about the microbial diversity in patients before and after treatment?

c. What can you infer about how the FMT works?

d. Based on your answer to part c, why can't the donor fecal material be sterilized?

e. Based on the data presented, what treatment do you encourage your friend's mother to choose? Explain your answer.

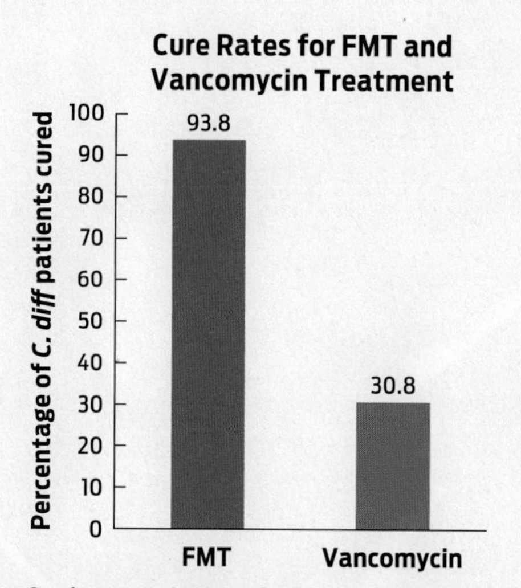

Data from van Nood et al. 2013. Duodenal Infusion of Donor Feces for Recurrent *Clostridium difficile*. *NEJM* 368(5):407–415.

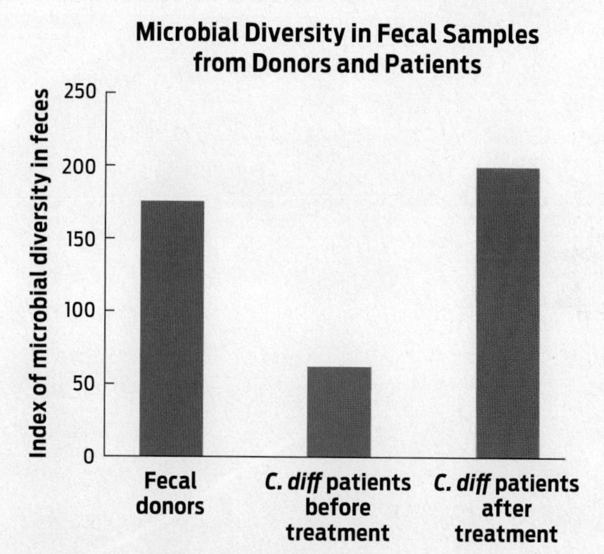

Data from van Nood et al. 2013. Duodenal Infusion of Donor Feces for Recurrent *Clostridium difficile*. *NEJM* 368(5):407–415

Bring It Home

Apply Your Knowledge

22. Your friend is thinking of starting an extreme, animal-based keto diet—a diet emphasizing meat, eggs, and cheese. Your friend has heard that keto diets can result in rapid weight loss and loves meat and cheese. You have heard that diet can alter the gut microbiome. You do a quick search and identify a study in which participants ate an animal- or plant-based diet, then the diversity of the gut microbiome was analyzed (both in terms of numbers of different groups and the specific microbes present). People following the animal-based diet experienced a large change in their gut microbiome, including a large increase in the number (and proportion) of bile-resistant bacteria. In mice, these bile-resistant bacteria can cause inflammatory bowel disease. Explain to your friend why this diet may have some side effects (beyond nutritional concerns and weight loss).

Source: L. A. David et al. (2014). Diet rapidly and reproducibly alters the human gut microbiome. *Nature* 505:559–563. doi:10.1038/nature12820

18 Eukaryotic Diversity

Can Rubber Save the Rain Forest?

A SMALL STATE IN BRAZIL AIMS TO FIND OUT

Chris Hellier/Science Source

DRIVING QUESTIONS

1. What are eukaryotic organisms, what is their evolutionary history, and what influences the number of eukaryotic species?
2. How are plants defined, and what adaptations have enabled their success?
3. How are fungi defined, and what adaptations have enabled their success?
4. How are animals defined, and what adaptations have enabled their success?
5. What are protists, and why are they hard to classify?

During the 2016 Summer Olympics, held in Rio de Janeiro, the Brazilian government distributed some 9 million condoms to athletes, local residents, and visitors. Amid rising concerns over spread of the Zika virus and ongoing concerns about human immunodeficiency virus (HIV) transmission, the effort was a reasonable public health measure aimed at reducing sexually transmitted infections. It was also advertising—for Brazil's locally made, all natural, rubber-based condom.

A factory in the small Brazilian state of Acre (pronounced "Ah-cray") manufactures the condoms from rubber harvested from the rubber tree (*Hevea brasiliensis*). The rubber tree is native to the Amazon rain forest, of which the largest part—about 60%—is located within Brazil's national borders.

Yasuyoshi Chiba/AFP/Getty Images

The Natex factory produces condoms made of rubber from native rubber trees.

Tapping trees for rubber is an ancient tradition in the Amazon and is still practiced the way it always has been, by rubber tappers skilled in the craft. The condom manufacturer, Natex, pays about 700 rubber-tapping families for the rubber that they collect. Natex is the world's only manufacturer of condoms made from rubber harvested from wild trees.

Acre's condom factory is just one of several efforts by the local government to create a market for sustainably derived forest products. The goal is to provide stable income for local people who would otherwise turn for subsistence to felling timber or raising cattle, both of which contribute to deforestation.

"Acre is kind of a laboratory for innovative forest policy," says Amy Duchelle, a senior scientist with the Center for International Forestry Research (CIFOR) in Bogor, Indonesia, who spent several years conducting research in the region. "The condom factory is a strategy to allow local people to profit from the rain forest without destroying it."

Through these efforts, Acre aims to halt deforestation at 18% of the state's surface area, which would make it the most protected region in Brazil. And environmentalists hope it will be a model for other states and countries to follow.

That's crucial, because rain forests around the world are being cut down at an

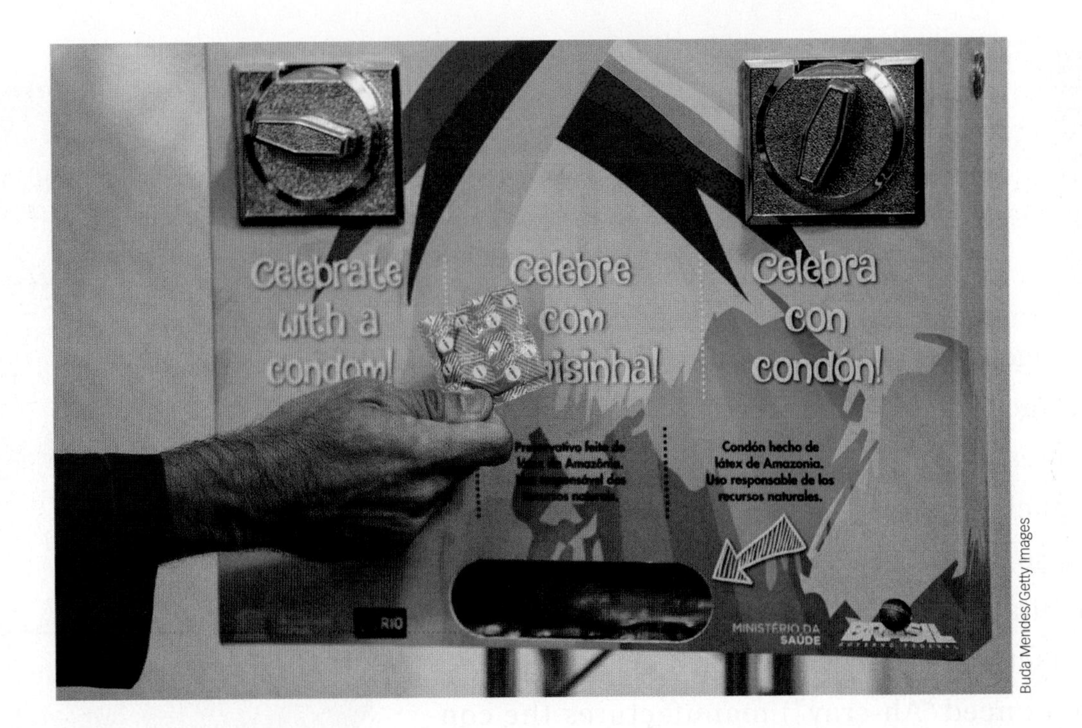

A condom dispenser at the 2016 Olympic Village.

Buda Mendes/Getty Images

alarming rate. In the Amazon alone, nearly 20% of the forest has already been destroyed. Scientists warn that if deforestation continues at the current pace, the Amazon could be gone by the turn of the next century **(INFOGRAPHIC 18.1)**.

> *"Acre is kind of a laboratory for innovative forest policy."*
>
> **—Amy Duchelle**

If that happens, Earth's **biodiversity** will suffer immeasurably. The Amazon is home to a staggering number of species, including many found there and nowhere else. Losing them from the Amazon means extinction from the planet. What's more, destruction of the Amazon would have far-reaching consequences for living things that don't live in the rain forest but rely on it for goods and services—both tangible ones like pharmaceuticals, foods, and latex, and less tangible ones like climate regulation.

BIODIVERSITY
The number of different species and their relative abundances in a specific region or on the planet as a whole.

Can rubber make a difference? Environmentalists have high hopes that it can. If the past is any guide, the fate of the rubber tree may be inextricably linked to the fate of the region as a whole.

Too Big to Fail

▸ Eukaryotic tree of life

At 2.7 million square miles, the Amazon is the largest tropical rain forest on Earth. It's roughly the size of the continental United States and covers about 40% of the South American continent. The Amazon is often referred to as the lungs of the planet, since the plants that grow here give off breathable oxygen as part of photosynthesis. But that's actually a bit of a misconception: the Amazon takes in about as much oxygen as it produces, so it's not a net producer of oxygen.

A more useful way to think about the rain forest is as the planet's air conditioner, helping to keep it from overheating. An enormous amount of water flows through the forest as it travels from snow in the Andes Mountains, down through the Amazon River, and out to the Atlantic Ocean. A significant portion of this water flow is absorbed by trees through their roots. It is then given off into

INFOGRAPHIC 18.1

Deforestation Is Highest in the Amazon Rain Forest

Globally, more than 361 million hectares of tree cover were lost between 2001 and 2018. Brazil, where much of the Amazon rain forest is located, had the greatest degree of deforestation during this time, and has lost the equivalent of 10% of its tree cover since 2000.

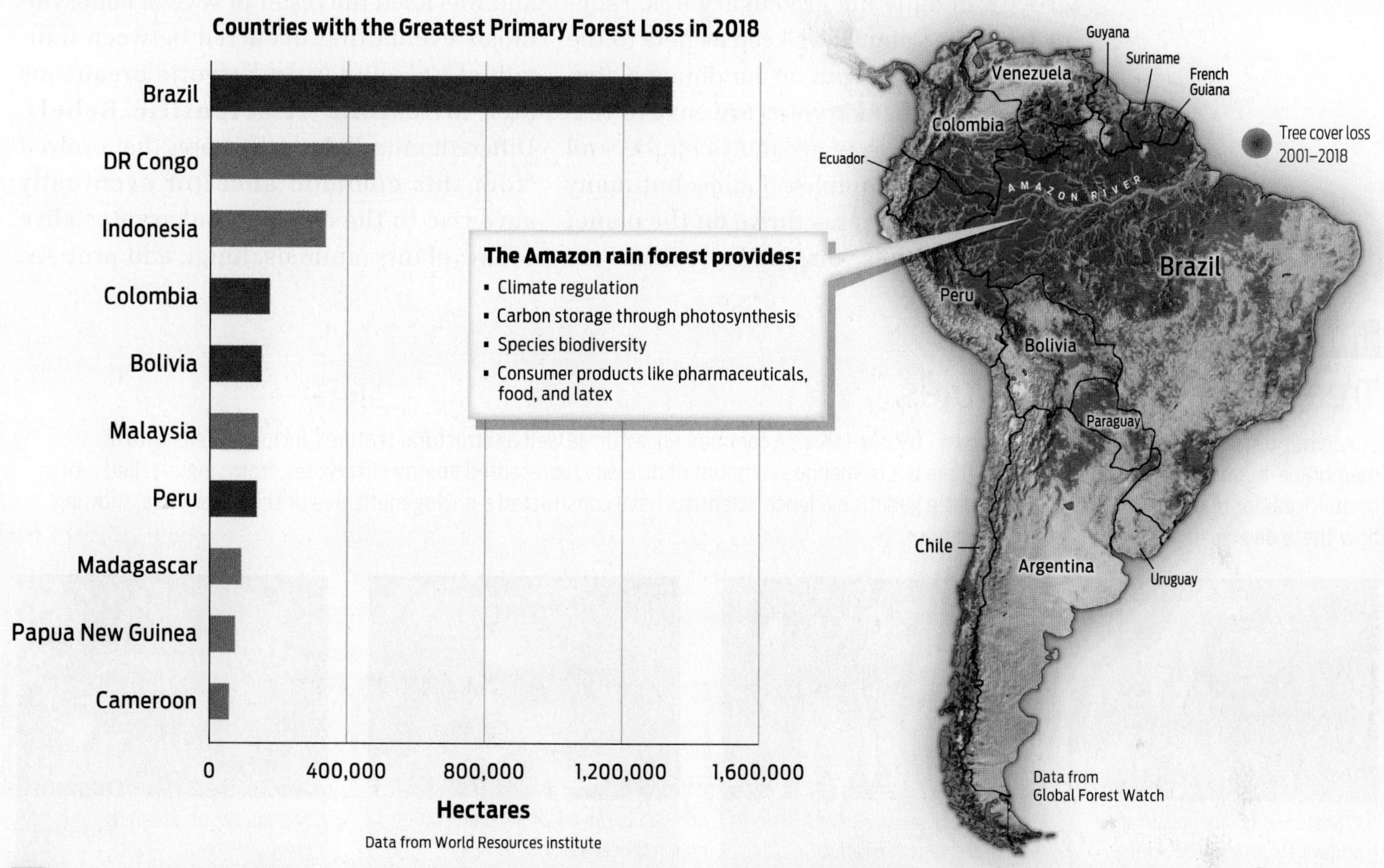

? Brazil has the highest rate of deforestation. What is the rate of deforestation in Indonesia and Colombia as a percentage of that in Brazil?

the atmosphere through the trees' leaves, creating invisible flowing rivers of water vapor in the sky that contribute to precipitation. All that water plays a significant role in shaping local and regional climate, with the Amazon creating as much as 80% of its own rainfall.

The trees that grow here also serve as a large reservoir of stored carbon—more than 100 billion metric tons. They capture carbon within their organic molecules through photosynthesis, thereby keeping this carbon out of the atmosphere. Burning or cutting down those trees releases that carbon into the atmosphere as carbon dioxide, a greenhouse gas that contributes to global climate change (see Chapter 23).

It's partly because of its vast size and warm, wet climate that the Amazon ranks as the most biodiverse region on Earth: more numbers and types of organisms live here than anywhere else. In addition to about 16,000 species of trees, the Amazon is home to many thousands of other plant, animal, fungi, and protist species—the group of organisms we collectively call **eukaryotes.** These organisms all have a similar type of cell—one with a membrane-enclosed nucleus and organelles—owing to their linked

EUKARYOTE
Any organism of the domain Eukarya; eukaryotic cells are characterized by the presence of a membrane-enclosed nucleus and organelles.

evolutionary history. Eukaryotes evolved more than 2 billion years ago and now make up one of life's three domains, alongside bacteria and archaea.

Most of the organisms we are familiar with in daily life are eukaryotes, ranging from the animals we keep as pets to the green vegetables we put on our dinner plates. These kinds of eukaryotes are easy to recognize because they are multicellular and grow large and complex bodies, but many unicellular eukaryotes thrive on the planet as well. All of these disparate creatures are related by ancestry, which can be visualized using a phylogenetic tree. At the root of the tree is the last common ancestor of all the various eukaryotes. This ancestor was unicellular, lived more than a billion years ago, and was itself the result of several endosymbiosis events that occurred between individual unicellular prokaryotic organisms (see **Milestone 1: Scientific Rebel**). Different unicellular eukaryotes that evolved from this common ancestor eventually gave rise to the different eukaryotes alive today: plants, animals, fungi, and protists.

INFOGRAPHIC 18.2

Tree of Life: The Eukaryotes

Eukaryotic organisms—members of the domain Eukarya—share a common ancestor, as well as structural features, including cells with a membrane-bound nucleus and organelles. There is a tremendous amount of diversity represented among eukaryotes, from single-celled protists to multicellular plants, animals, and fungi. Using genetic evidence, scientists have constructed a phylogenetic tree of the eukaryotes, showing how these diverse organisms relate to one another.

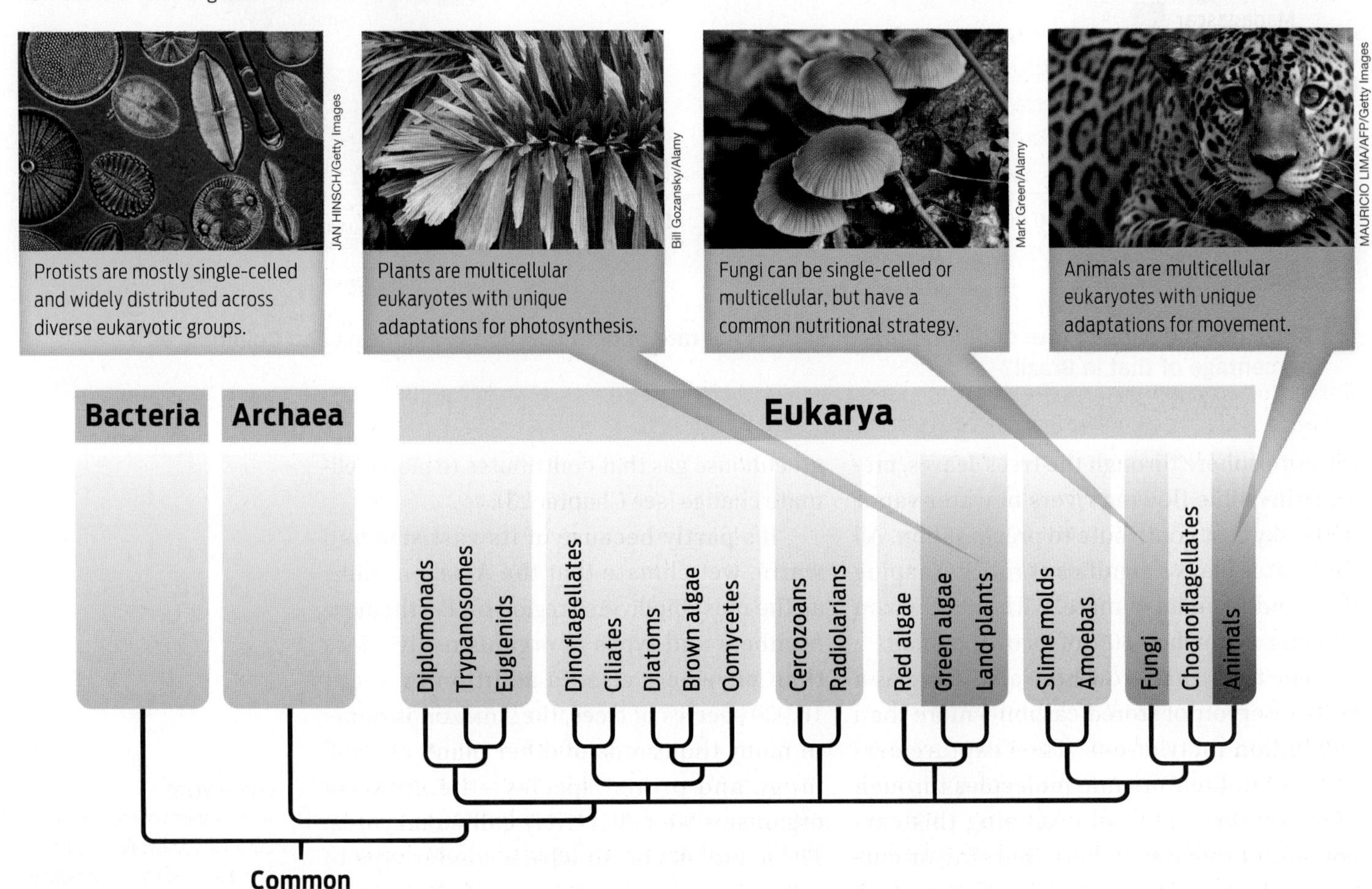

? **Are fungi more closely related to animals or to plants?**

Protists are single-celled eukaryotes that do not form a cohesive group of closely related organisms and instead are located throughout the eukaryotic tree, with some protists being more closely related to fungi and animals, and some more closely related to plants **(INFOGRAPHIC 18.2)**.

Eukaryotes live just about everywhere, but certain places on Earth are especially rich in eukaryotic diversity. In addition to rain forests like the Amazon, coral reefs are recognized as hot spots of eukaryotic diversity.

Despite their richness, these hot spots of eukaryotic diversity face serious threats. In the case of the Amazon, that threat comes in the form of deforestation. Much of this deforestation is carried out to provide land for agriculture. For most of human history, that meant small subsistence farmers clearing a few acres to grow food for their families. But by the late 20th century, most agriculture-related deforestation was aimed at opening up wide tracts of land for large-scale operations, primarily cattle ranching.

Cattle ranchers have a natural incentive to cut down the forest: they need pasture to graze their cattle and tree-free areas to grow soybeans and other feed crops. But their needs put them in direct conflict with rubber tappers, who require standing forest to ply their trade.

The conflict between ranchers and rubber tappers reached a fever pitch in 1988, when cattle ranchers murdered Chico Mendes, a rubber tapper and union organizer. "That really set off an international, national, and local outcry about human rights, and the rights of traditional people who were dependent on the forest for their livelihoods," says Duchelle, of CIFOR.

In reaction, Brazil moved to create protected forest areas called extractive reserves, where people can legally tap rubber, harvest Brazil nuts and other non-timber forest products, and even collect timber to a certain extent, provided it is done sustainably. The first extractive reserve was created in Acre, where Mendes was from, but they've now spread across the Brazilian Amazon.

Natex, the condom manufacturer, is located in the town of Xapuri, across the river from the Chico Mendes Extractive Reserve, where a lot of the rubber is harvested.

For products that truly must not fail—like airplane tires, surgical gloves, and condoms—natural rubber is the go-to choice.

"The goal with the condom factory is to both increase the price paid for rubber, and make sure that the value added benefits the collectors themselves, as opposed to going to middle men farther down the supply chain," Duchelle says.

White Gold

▸ Evolution of plant diversity

Natural rubber is a near-miracle substance. It is durable and waterproof, and, once treated or vulcanized, can maintain its elasticity in heat or cold. About 40% of the world's rubber market is for natural rubber harvested from trees; the remaining 60% is supplied by synthetic versions. But for products that truly must not fail—like airplane tires, surgical gloves, and condoms—natural rubber is the go-to choice.

Demand for rubber began to skyrocket in the late 19th century, with the rise of the automobile industry. Rubber was needed for hoses, belts, and—especially—tires. Many American and European "rubber barons" who sold rubber during this time became very wealthy.

But the rubber barons necessarily relied on the skill and know-how of rubber tappers who could identify rubber trees and coax rubber from them. Tapping rubber trees involves making a diagonal cut in the bark in line with the path of the spiraling vessels in the tree that carry the latex—a milky, white substance that coagulates in the air. Tappers then wind

PROTIST
A eukaryote that cannot be classified as a plant, animal, or fungus; usually unicellular.

INFOGRAPHIC 18.3

Plants: Getting Ahead while Standing Still

Plants are multicellular eukaryotes that carry out photosynthesis and are adapted to living on land. Because they are not mobile, they employ unique mechanisms of defense and reproduction to help them succeed despite being stuck in one place.

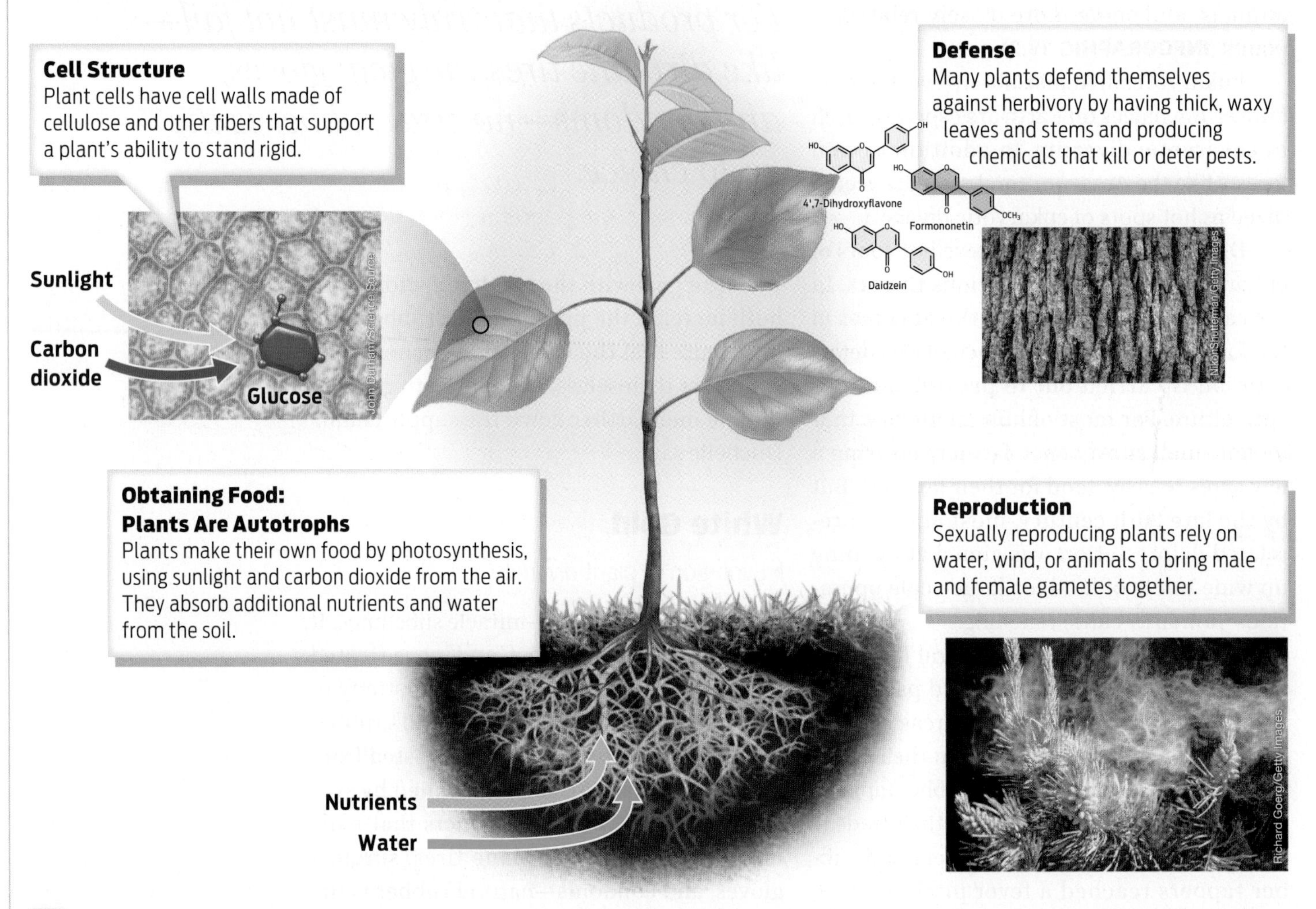

? List three characteristics of plants.

the congealing latex around a stick, forming heavy balls that they can sell to wholesalers and manufacturers.

Latex is obviously valuable to humans, but that's not why rubber trees make it. Latex is valuable to them, too. To understand just how valuable, it helps to know more about the rubber tree's evolution and its relationship to other eukaryotes.

The rubber tree is, of course, a **plant,** one of the main groups of multicellular eukaryotes. Every organism needs ways to obtain nourishment, reproduce, and defend itself, and plants are no exception. The particular ways in which plants have solved these fundamental life challenges is what defines them as plants and distinguishes them from other eukaryotes—animals and fungi, for example.

Plants' signature talent is being able to make their own food. Through photosynthesis (see Chapter 5), plants capture the energy of sunlight and use it to produce sugar. This sugar nourishes not only the plant but also the many nonphotosynthetic creatures on Earth that depend on plants for food—including humans.

PLANT
A multicellular eukaryote that has cell walls, carries out photosynthesis, and is adapted to living on land.

Plants are largely immobile—stuck in one place. So they must obtain the starting materials for photosynthesis right where they are. Those materials include carbon dioxide that they take in through their leaves, and water and a few other nutrients that they obtain from the soil.

Most plants reproduce sexually, but unlike animals they can't go in search of mates. Instead, they rely on water, wind, or animals to bring male and female gametes together. Often this means producing an overabundance of sex cells to increase the chance that at least one will find a partner.

Since they are immobile, plants also cannot run away from predators that seek to eat them. Many have evolved other solutions that help them to avoid becoming easy targets: thick, protective bark, for example, and defensive chemicals called **secondary metabolites.** Some plants taste bitter, which helps ward off leaf-munching insects and other herbivores. Latex is full of such secondary metabolites, which is the main reason the rubber tree makes this sticky substance **(INFOGRAPHIC 18.3).**

Plants in the rain forest have evolved an impressive variety of secondary metabolites that deter herbivores. Some of these secondary metabolites are quite familiar to us—for example, caffeine, nicotine, and cocaine. Many of our pharmaceuticals come from these plant secondary metabolites **(TABLE 18.1).**

One defensive plant chemical, known as quinine, plays a particularly important role in Brazil and surrounding countries of the Amazon basin. Found in the bark of the

SECONDARY METABOLITES
Chemicals produced by plants that are not directly involved in growth or reproduction but that help protect the plant by their impacts on other organisms.

TABLE 18.1 Useful Chemicals That Come from Plants

Drug/chemical	Action/clinical use	Plant source
Atropine	Blocks neurotransmitters	*Atropa belladonna*
Bromelain	Anti-inflammatory, proteolytic	*Ananas comosus*
Caffeine	Central nervous system stimulant	*Camellia sinensis*
Codeine	Analgesic, relieves cough	*Papaver somniferum*
Cyclosporine	Anti-transplant rejection drug	*Tolypocladium inflatum* fungus
Digitalin	Cardiac stimulant	*Digitalis purpurea*
Ephedrine	Antihistamine; stimulation of sympathetic nervous system	*Ephedra sinica*
Gossypol	Male contraceptive	*Gossypium* species
L-Dopa	Anti-tremor; treats Parkinson's disease	*Mucuna* species
Menthol	Dilates capillaries; treats muscle pain	*Mentha* species
Morphine	Analgesic; inhibits pain	*Papaver somniferum*
Novocaine	Local anesthetic	Coca plant (South America)
Quinine	Antimalarial, antifever	Cinchona tree
Salicin	Analgesic (aspirin)	*Salix alba* (willow bark tree)
Sennosides	Laxative	*Cassia* species
Stevioside	Sweetener	*Stevia rebaudiana*
Taxol	Antitumor agent	*Taxus brevifolia*

cinchona tree, which is native to Peru, quinine is a traditional treatment for malaria, a disease that is common in this region.

Though plants today cover every land habitat, they weren't always so well equipped to survive on land. Plants first evolved from water-dwelling green algae, a type of protist, about 450 million years ago. At that time, life on Earth was confined primarily to seas. The water in which plant ancestors floated served several purposes: it prevented their cells from drying out, and it allowed male and female gametes to drift easily to each other. Before plants could diversify on land, they needed to evolve the means to cope with drier surroundings.

The earliest plants to make the transition from water to land naturally relied on some of same solutions as their protist ancestors. These **nonvascular plants** lived in moist environments; they were small and low to the ground, so flowing water could diffuse in and out of their cells; and they released male gametes only when coated by a layer of moisture that allowed the gametes to swim to an egg. The modern-day plants most like these earliest terrestrial arrivals are the mosses you see growing like squat, spongy mats on the damp forest floor. One of the wettest places on Earth, the Amazon rain forest is a soggy paradise for these ancient plants.

In contrast to mosses, most of the plants we see around us are **vascular plants,** which have specialized tubular tissues adapted for transporting water and nutrients through the plant body. These plants take up water from the soil through roots; thus, as long as water is present in the soil, they can survive where the ground surface is dry. The first true vascular plants were **ferns.** Unlike mosses and other nonvascular plants, ferns can stand upright and grow tall, thanks to the vascular tissue that keeps their stems rigid and the plant well supplied with water. But for fertilization to occur, ferns still must release their male gametes into a layer of moisture, so they are typically found growing in moist environments. Both ferns and the nonvascular plants rely for dispersal on haploid reproductive cells called spores that can begin to grow in a moist environment.

At one time, ferns ruled the plant world, spreading their massive fronds across the entire landscape in the Carboniferous Period, 360 to 300 million years ago. But their reign came to an end when another group of plants evolved to challenge the ferns' dominance: those with seeds.

Seed plants first emerged about 360 million years ago, during the late Devonian Period. A seed, which envelops the plant's embryo, is an ideal package for withstanding harsh conditions and transporting the embryo to a location where it can grow into a new plant. The seed provides a protective shell for the plant embryo and comes complete with a built-in supply of food. Seeds fully solved the problem of how to disperse offspring on dry land and protect them until conditions are right for them to grow.

Along with this innovative embryo-protection method, seed plants evolved another key innovation necessary to succeed in dry environments: pollen. These tiny but tough moisture-retaining packets contain the male gametes—sperm—and serve as durable gamete-delivery vehicles. When sperm from the pollen meets an egg in the female ovule on another plant, the sperm fertilizes the egg to form an embryo. The ovule develops into a seed containing the embryo **(INFOGRAPHIC 18.4)**.

The first seed plants were the **gymnosperms,** a diverse group of plants that includes conifer, or cone-bearing, trees such as pine and spruce as well as ginkgo trees, the females of which produce foul-smelling fruits that besmirch many a city sidewalk. Gymnosperms rely primarily on wind to disperse pollen. Male cones release pollen grains. Pollen grains land on female cones and deliver sperm to an egg in the ovule. Fertilization occurs, and the zygote divides to form an embryo, which becomes enclosed in a seed. Seeds develop on the underside of scales on female cones and are released as

NONVASCULAR PLANT
A plant that lacks vascular tissue to transport water and nutrients through the plant body.

VASCULAR PLANT
A plant with tissues that transport water and nutrients through the plant body.

FERNS
The first true vascular plants; ferns do not produce seeds.

GYMNOSPERM
A seed-bearing plant with exposed seeds typically held in cones.

INFOGRAPHIC 18.4

Advantageous Adaptations: Vascular Tissue, Pollen, and Seeds

For plants to succeed and diversify on land, they needed to evolve adaptations for withstanding dry environments. Two key evolutionary milestones were the evolution of vascular tissue and of pollen and seeds.

Nonvascular Plants

Plants like mosses are restricted to moist environments because they do not have vascular tissue. They absorb water and nutrients into their cells like a sponge, directly from their immediate environment.

Vascular Plants

Plants like rubber trees use vascular tissue to move water and nutrients. These structures, apparent as veins in leaves and vessels in the stem, allow plants to take up water and nutrients via roots that penetrate deep into soil, and to transport water, nutrients, and food over long distances throughout their bodies.

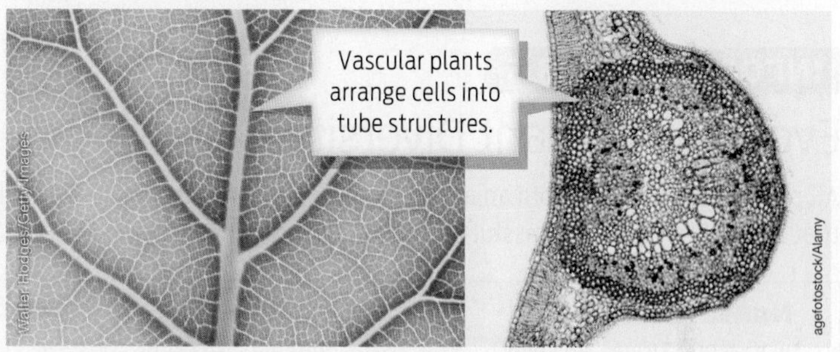

Cross section of a vascular stem

Sperm Survive Only in Water

Some plants have "swimming" sperm that must move through water to reach the female reproductive structures.

Pollen and Seeds Resist Drying

The male gametes of some plants are contained in pollen, which resists drying and can be transported in air. These plants also have seeds to protect the embryonic plant from drying.

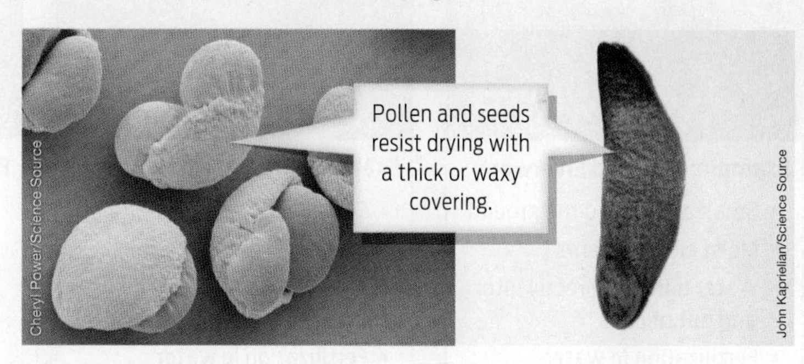

? **Give two reasons why mosses are short plants that live near water.**

the masses of "propellers" you've no doubt encountered on the ground in fall. Gymnosperm trees have colonized vast regions of the cold and dry northern regions of the planet.

Yet some environments posed a challenge to wind pollination. There is little wind in the understory of the Amazon rain forest, for example, and the dense vegetation would prevent pollen from spreading far through the air. This is one environment where the fourth major group of plants, the **angiosperms,** might have evolved. Angiosperms have flowers, a feature that makes pollination by animals possible. Bees, insects, bats, and birds visit the flowers in search of pollen or nectar to eat, and the pollen sticks to their bodies. Some of the pollen they pick up is then dropped off at the next flower they visit. Because each type of pollinator visits only certain types of plants, they act like postal workers who carry a package to a specific address—in this way, even a rare plant can receive pollen.

Seed plants were so successful that they quickly came to dominate forests by the time dinosaurs appeared in the Mesozoic Era,

ANGIOSPERM
A seed-bearing flowering plant with seeds typically contained within a fruit.

roughly 200 million years ago. Today, more than 90% of all living plants are seed plants. In angiosperms, seeds are contained in a fruit; *angio* is from the Greek word meaning "vessel" or "container." The seeds of gymnosperms are not protected by a fruit, which is how they got their name: *gymnos* is Greek for "naked," so the name literally means "naked seeds." The rubber tree is an angiosperm whose fruits look like green, three-lobed heads of garlic, each containing three brown seeds inside **(INFOGRAPHIC 18.5)**.

The Problem of Blight

▶ Evolution of fungal diversity

Although native peoples in the Amazon had been extracting latex from rubber trees for many generations, it was only in 1839 that rubber caught the attention of industrialists.

INFOGRAPHIC 18.5

Evolution of Plant Diversity

All plants have evolved from an ancient algal ancestor that lived in water. Different groups of plants have developed different specializations that allow them to be successful on land.

Nonvascular Plants
No Vascular Tissue, No Seeds

Example: Thalloid Liverwort
- Small and low to the ground
- Moist environments
- Water diffuses directly into and out of cells
- Fertilization in water

Ferns
Vascular, No Seeds

Example: Shuttlecock Fern
- Specialized tissues to transport nutrients and water
- Roots draw water from the soil
- Can grow tall and rigid
- Fertilization in water

Gymnosperms
Vascular, Seed-Producing in Cones

Example: Brazilian Pine
- Specialized tissues to transport nutrients and water
- Roots draw water from the soil
- Can grow tall and rigid
- Pollination by wind

Angiosperms
Vascular, Seed-Producing in Fruit

Example: Amazon Rubber Tree
- Specialized tissues to transport nutrients and water
- Roots draw water from the soil
- Can grow tall and rigid
- Pollination by wind and animal

Flowers and fruits

Pollen and seeds

Vascular tissue

Common Ancestor
450 million years ago

? **Which evolved first: flowers, seeds, or vascular tissue?**

That was the year Charles Goodyear discovered the process of vulcanization, a chemical treatment that makes latex more durable, and launched the rubber boom in the Amazon.

The boom was short lived. In 1876, a British botanist spending time in Brazil absconded with 70,000 rubber plant seeds, which the British government used to establish rubber tree plantations in its eastern colonies. By 1910, there were 50 million rubber trees growing in Asia, and Brazil's rubber market had collapsed. A decade later, more than 90% of the world's rubber came from plantations in Asia.

"This was the first really well-known instance of biopiracy, and Brazilians are still mad about it," says William Laurance, a conservation biologist at James Cook University in Australia, who lived and worked in Brazil for 20 years studying the rain forest.

The theft left many of the local tappers struggling to make a living. There was simply no way they could compete with the Asian colonies, which did not have to contend with certain disease problems faced by rubber tappers in the Amazon.

In Brazil, rubber trees are spread throughout the forest at relatively low density (about two trees per hectare). This pattern reflects the influence of a **fungus** called *Microcyclus ulei,* which causes a disease called leaf blight. *M. ulei,* which is native to the Amazon, spreads easily among rubber trees that grow close to one another. It kills them by feasting on and destroying the trees' leaves, so the tree cannot feed itself through photosynthesis. The presence of this fungus in the Amazon means that rubber trees can only grow safely when they are far enough apart. In contrast, in Asia, where blight is not a factor, rubber trees can be planted close together, making it easier to harvest more rubber.

Every attempt to establish rubber tree plantations in the Amazon has ended in failure. Most famously, the car maker Henry Ford tried to establish a plantation and a town to accompany it (called, appropriately, Fordlandia). But Ford neglected to consult any botanists. Within a few years, all the trees—some 2 million of them—were dead from leaf blight.

Fungi are a large and diverse group of eukaryotes. Like plants, fungi got their start in the water as single-celled organisms. An early adaptation, achieved about 750 million years ago, was the development of a cell wall containing the molecule chitin, which imparts strength and thus protection against the water pressure from osmosis. (Plants cells also have a cell wall, but it is made out of cellulose.)

Today, fungi come in a diverse array of forms. Unicellular varieties are collectively known as yeasts. Many species of yeast are useful to humans, including the ones that ferment sugar to alcohol to make wine and beer, and those that help make bread rise. Other varieties of fungus are multicellular. Multicellular fungi are known as molds and include parasites like the one that causes leaf blight and mushrooms like those you see growing on a tree trunk or sprouting from the ground.

Unlike plants, fungi cannot photosynthesize, so they must obtain nutrients by feeding off other organisms. Most fungi, including the one that causes leaf blight, obtain nutrients in a characteristic way: by secreting digestive enzymes onto their food source. These enzymes digest complex molecules into individual subunits such as amino acids and simple sugars. The fungi then absorb theses subunits into their cells. Because they digest their food externally, fungi have no need for a stomach or intestines.

Multicellular fungi have a body composed of threadlike structures called **hyphae.** Each individual hypha is a chain of many cells, capable of absorbing nutrients. Hyphae grow and interweave through nutrient-rich substrates like a dead tree or soil to form a spreading mass called a **mycelium.** These structures can vary in size and location. For example, the mold on a slice of bread has a small mycelium that's in plain view. By contrast, the mushrooms you see on the forest

FUNGUS
A unicellular or multicellular eukaryotic organism that obtains nutrients by secreting digestive enzymes onto organic matter and absorbing the digested product.

HYPHA (PLURAL: HYPHAE)
A long, threadlike structure through which fungi absorb nutrients.

MYCELIUM
A spreading mass of interwoven hyphae that forms the often subterranean body of multicellular fungi.

floor are merely one aboveground part of what can be a huge, underground mycelium.

DECOMPOSER
An organism such as a fungus or bacterium that digests and uses the organic molecules in dead organisms as sources of nutrients and energy.

SPORES (FUNGAL)
Fungal cells that are resistant to drying out and can be dispersed to new locations as part of sexual or asexual reproduction.

Many fungi feast on dead organisms. That makes them **decomposers,** organisms that break down organic molecules trapped in dead organisms. Their activity is necessary for the recycling of nutrients in the forest. Without fungi, dead trees and animal carcasses would pile up in the forest and smother everything in it. Thanks to the action of fungi, however, the organisms decompose and the elements they contained will nourish many organisms throughout the environment.

Many fungi can reproduce both sexually and asexually, while others are strictly asexual. In all cases, fungi produce **spores**—tiny, one-celled reproductive units that resist drying out and disperse from the parent. In many multicellular fungi, the structure that releases spores is called a **fruiting body.** In many cases, this fruiting body is what we typically think of as a "mushroom," with a recognizable stalk and cap.

INFOGRAPHIC 18.6

Fungi: The Decomposers

Fungi are unicellular or multicellular eukaryotes that obtain nutrients by feeding on other organisms by external digestion. Their approaches to nutrition, reproduction, and defense reflect the fact that they are generally stuck in one place. Because they feed on dead organisms, fungi are important decomposers, returning nutrients to the soil as they break down organic molecules.

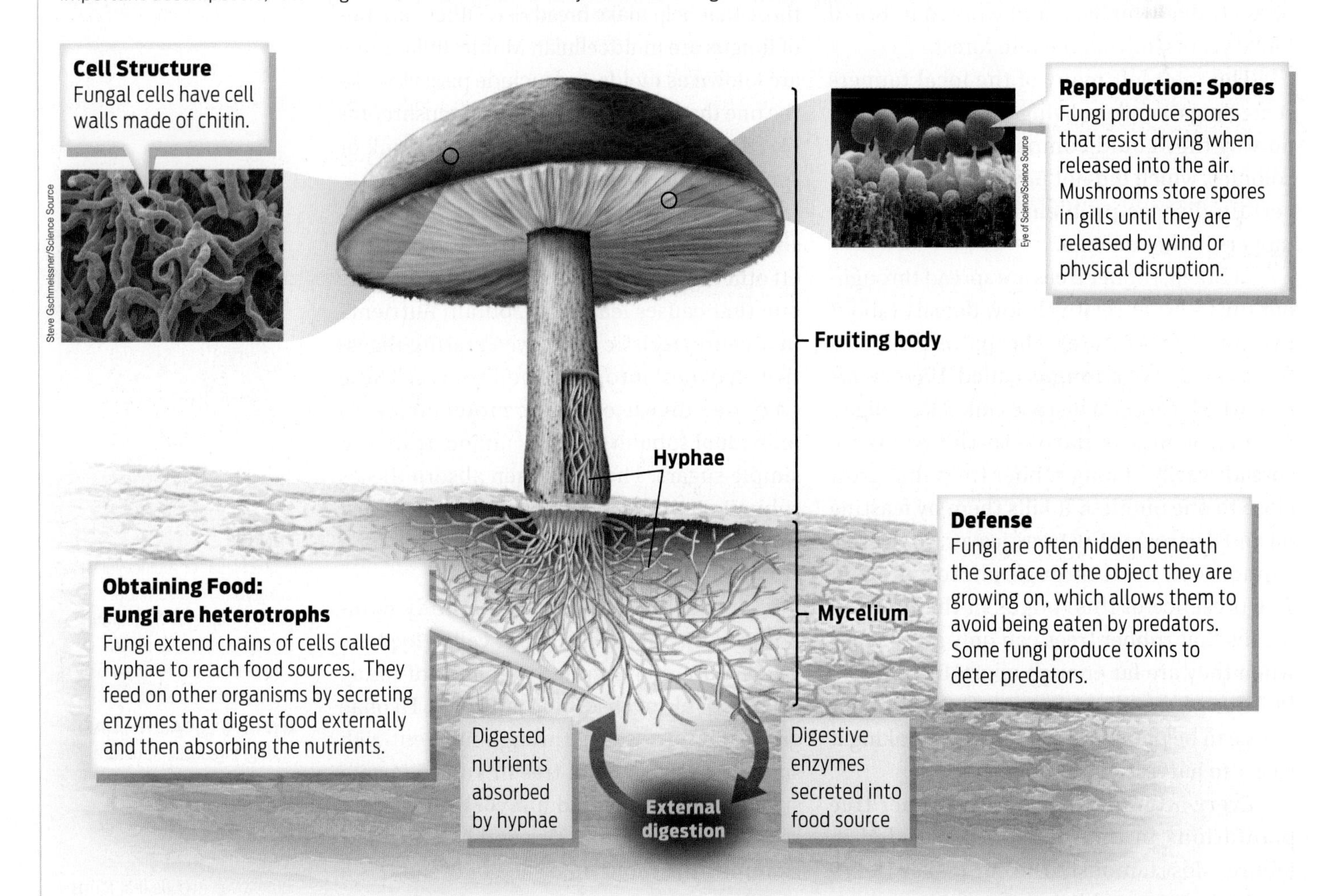

? List three characteristics of fungi.

Like plants, many fungi are immobile—a fact that influences their approaches to defense. To avoid being eaten by predators, they may live underground or produce poisonous toxins that dissuade would-be attackers, who may come to recognize their distinctive shapes, colors, and odors **(INFOGRAPHIC 18.6)**.

Fungi can be found in many different environments, including the Amazon, often growing on, in, and under vegetation. Some soil-dwelling species form a close relationship with the roots of many tree species. Their slender hyphae grow into microscopic spaces in the soil where the tree's roots can't fit, greatly enhancing a root's ability to absorb water and nutrients. In turn, trees supply nutrients to the fungi **(INFOGRAPHIC 18.7)**.

Perhaps as a result of having evolved together with the leaf blight fungus, rubber trees have developed seedpods that explode violently, sending seeds up to 120 feet away.

FRUITING BODY
A fungal structure that is specialized for the release of spores.

INFOGRAPHIC 18.7

Fungal Diversity

Fungi are a diverse group of organisms with a variety of reproductive and feeding strategies. Fungi may be referred to as yeasts or molds, depending on how they grow and reproduce.

Yeast

- Single-celled

Obtain Nutrients:

- Individual cells release digestive enzymes onto food source and absorb nutrients
- Some species contribute to tasty food and beverages when they ferment sugars
- Some species live symbiotically with plants or animals

Reproduction:

- Sexual and asexual

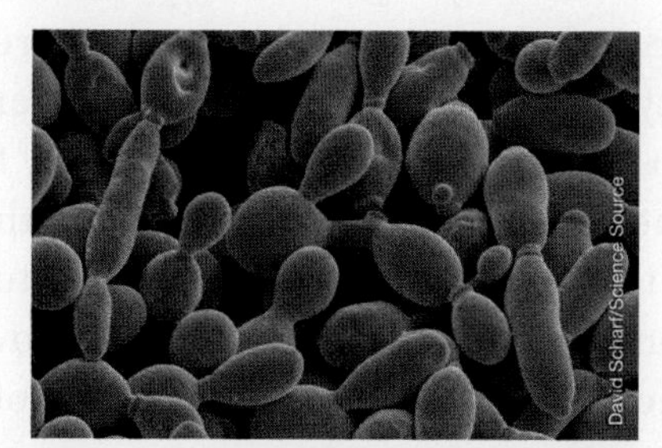

Saccharomyces cerevisiae yeast cells

Yeast on the surface of grapes

Molds

- Multicellular hyphae form mycelia in food source

Obtain Nutrients:

- Hyphal cells release digestive enzymes onto food source and absorb nutrients
- Many species decay leaf litter and dead organisms
- Some species associate with plant roots, supplying and receiving nutrients

Reproduction:

- Sexual and asexual
- Molds release spores that germinate to form new mycelia
- Some molds, like mushrooms, produce fruiting bodies that emerge from mycelia and release sexual spores that germinate to form new mycelia

Cordyceps fungus decomposing a moth

Aspergillus mold fruiting body

South American leaf blight (*Microcyclus ulei*)

Mushroom hyphae on a dead log

? Of the two types of fungi described, which are multicellular?

Dispersing seeds this way is an adaptation that keeps the rubber trees from growing too close to one another in the forest, and therefore prevents leaf blight spores from spreading easily between them.

Leaf blight, like the rubber trees themselves, is native to the Amazon.

Leaf blight, like the rubber trees themselves, is native to the Amazon. The fungus does not currently exist in Asian countries, but some scientists think that the arrival of blight is inevitable. If or when that does happen, it could mean a devastating blow to the worldwide rubber market.

Prime Real Estate

▸ Evolution of animal diversity

Rubber trees and rubber tappers wouldn't be the only victims of a global rubber tree blight. Anything that affects the ability of rubber trees to survive and grow in the Amazon will also affect the animals that depend on these trees for food or shelter. That includes a variety of insects, birds, fish, and monkeys.

ANIMAL
A eukaryotic multicellular organism that can move and that obtains nutrients by ingesting other organisms.

Animals are eukaryotes that cannot photosynthesize and so must obtain nutrients by feeding on other organisms. Unlike fungi, animals digest their food internally through a digestive system. Many animals are herbivores, eating only plants; others are carnivores, eating other animals. Animals reproduce sexually.

Like other eukaryotes, animals evolved from a unicellular protist. What this protist looked like is still a mystery, but some evidence points to an ancient protist that resembles modern-day choanoflagellates. These tiny unicellular organisms have a flagellum that beats and a "collar" of microvilli that trap bacterial prey. They can clump together to form colonies, which may have been a steppingstone to multicellularity.

The first true multicellular animals, which evolved 560 million years ago, were ocean-dwelling creatures, most likely resembling sponges. Sponges lack defined tissues or organs and have no distinct shape. They are stationary and obtain nutrients by filtering water through their collars. Interestingly, the cells that make up the collars of sponges are very similar in appearance to those of choanoflagellates—further evidence that choanoflagellates and animals share a common ancestor.

Animals as a group didn't stay stationary for long. Indeed, the trait that sets the great majority of animals apart from their eukaryotic kin is their nearly limitless capacity for movement. Whether it's snakes that slither quietly to sneak up on their prey, mosquitoes that home in on blood, birds that soar down from the sky to snatch rodents in their talons, or salmon that swim upstream to congregate and breed, movement is a tremendously useful adaptation for solving life's fundamental problems: finding food, seeking mates, and avoiding predators.

Because most animals can go in search of food, they don't have to make it for themselves (like plants) or grow until they find something to eat (like fungi). They also don't have to rely on water, wind, or pollinators to disperse their gametes; they can instead seek out and choose a particular mate. And

RGB Ventures/SuperStock/Alamy

Seeds and seed pod of the rubber tree.

INFOGRAPHIC 18.8

Animals: The Power of Movement

Animals are multicellular eukaryotes that obtain nutrients by eating other organisms. Animals have evolved the capacity for independent movement, which greatly aids the search for food and mates, as well as defense.

? **List three ways in which animals benefit from being able to move.**

instead of poisons to deter potential predators, many animals rely on speed to outrun them, although some of the slower ones, like slugs, resort to poisons and warning coloration.

Although not every animal can move so effortlessly, the evolution of movement was a key evolutionary milestone that permitted animals to spread and diversify around the globe. And even those animals that lead sedentary lives as adults tend to have larval stages that are motile **(INFOGRAPHIC 18.8)**.

Movement first evolved in the animal lineage along with a body plan that had symmetry. Animals such as jellyfish, starfish, sea urchins, and coral exhibit **radial symmetry:** they're circular, with no defined left and right sides. This is a useful shape for propelling oneself vertically through water, as jellyfish do. It also allows the animal to easily "swallow" its prey by essentially wrapping its body around it.

Most animals on the planet—everything from worms and insects to monkeys and humans—exhibit **bilateral symmetry:** if you draw a line down the middle, you produce left and right halves that are mirror images of each other. Bilateral symmetry has become the most common body form in the animal kingdom because it is a useful adaptation for seeking out food, stalking prey, and avoiding predators. For instance, most bilaterally symmetrical animals have an eye on each side of the face, enabling them to look straight

RADIAL SYMMETRY
The pattern exhibited by a body plan that is circular, with no defined left and right sides.

BILATERAL SYMMETRY
The pattern exhibited by a body plan with right and left halves that are mirror images of each other.

INFOGRAPHIC 18.9

Animals Have Evolved Different Body Symmetries

The earliest animals lacked distinct symmetry. The evolution of radial symmetry enabled animals like jellyfish to propel themselves up and down in the water to find food and new habitats. Animals with bilateral symmetry are able to see and move toward food and mates, and away from danger.

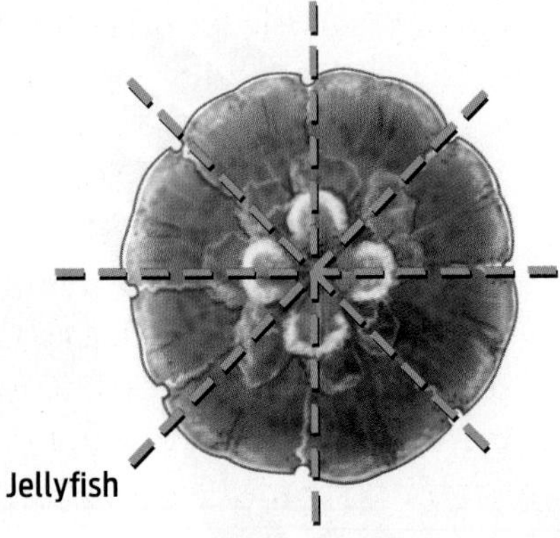

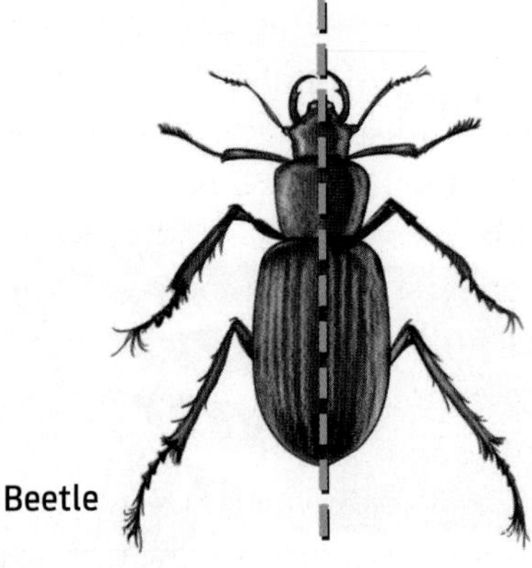

Asymmetrical

Asymmetrical animals have no defined shape.

Animals without symmetry are not mobile as adults and rely on whatever floats by for food. They often have physical protective mechanisms.

Radial Symmetry

Animals with radial symmetry are symmetrical in all directions, with no distinct right or left side.

Animals with radial symmetry either do not move in their adult life (corals and sea anemones) or depend on water currents and short-distance propulsion for mobility (jellyfish).

Bilateral Symmetry

Animals with bilateral symmetry have mirror-image left and right sides.

This type of symmetry enables purposeful, directed movement to find food and mates and to flee from danger.

? For each of the following animals, identify the symmetry of its body plan: sand dollar; parrot; jellyfish; earthworm; octopus.

ahead, as well as paired limbs to aid running or flying **(INFOGRAPHIC 18.9)**.

The early animals lived and evolved in the oceans for millions of years. The first animals to take tentative steps on land, roughly 420 million years ago, were invertebrates, animals that lack a backbone—most likely **arthropods.** This large group of animals includes crustaceans such as lobsters and crabs; millipedes and centipedes; arachnids such as spiders and scorpions; and insects like bees and flies. From the fossil record, we know that the first arthropods to be successful on land resembled millipedes.

These early land animals faced many of the same problems that plants faced as they left the oceans: how to keep from drying out, how to support themselves without the natural buoyancy of water, and how to reproduce in ways that didn't rely on water to transport gametes.

A key adaptation that helped animals to move onto land was the evolution of an **exoskeleton.** This nonbony skeleton forms on the outside of the body and helps to support the animal against the pull of gravity. In most land arthropods, the exoskeleton is also waxy and waterproof, so it helps these animals retain water in dry environments. Much like the internal **endoskeleton** of humans and other vertebrates, the exoskeleton of arthropods provides attachment sites for muscles, permitting rapid movement.

Arthropods are, by far, the most successful animals on the planet, at least according to their numbers. There are an estimated 2–4 million species of arthropods, although only 855,000 have been officially described. The number of individual arthropods on the planet is estimated to top 10^{18}—that's 1 with 18 zeros after it. The vast majority of all arthropods are **insects**—arthropods with

ARTHROPOD
An invertebrate having a segmented body, a hard exoskeleton, and jointed appendages.

EXOSKELETON
An external skeleton; in arthropods, the exoskeleton is made up of proteins and chitin.

ENDOSKELETON
An internal body skeleton, typically made of cartilage or bone.

INSECT
An arthropod with three pairs of jointed legs and a body with three segments.

three pairs of jointed legs and a body with three distinct sections.

Though arthropods colonized land first, and still dominate it today, many other invertebrate groups that originally evolved in the sea have joined the arthropods on land, attracted to plentiful food sources and hiding places. Sliding quietly amid leaf litter are many specimens of **mollusk,** including slugs and snails. By digesting dead plant material, mollusks help recycle nutrients. Thanks to their calcium-rich shells, snails provide this valuable mineral to the creatures that feast on them, such as rodents and birds. Some humans find mollusks a tasty treat as well: if you have enjoyed clams, oysters, or squid, then you have eaten some aquatic varieties of mollusk.

Lift any rock in a forest and you're likely to find many **annelids,** or segmented worms. Annelids such as earthworms create passageways in the soil as they move. These passageways allow air and water to enter the soil, making it fertile for other life. By eating and digesting leaf and other plant litter, earthworms also make nutrients available for plants.

Last to arrive on land were our own closest animal relatives, the vertebrates (members of the phylum Chordata). These organisms have a bony or cartilaginous backbone. Fish, birds, reptiles, amphibians, and mammals (including humans) are all vertebrates. Compared to the invertebrates, vertebrates make up a tiny fraction of the total animal diversity on the planet. We discussed the evolution of vertebrates in Chapter 15 **(INFOGRAPHIC 18.10).**

MOLLUSK
An invertebrate with a soft, unsegmented body enclosed in a hard shell.

ANNELID
An invertebrate with a soft, segmented body; annelids are commonly referred to as worms.

INFOGRAPHIC 18.10

Evolution of Animal Diversity

Over their 800-million-year history animals have adapted to many aquatic and terrestrial environments on Earth, enabled in part by their diverse body plans and varied adaptations for movement.

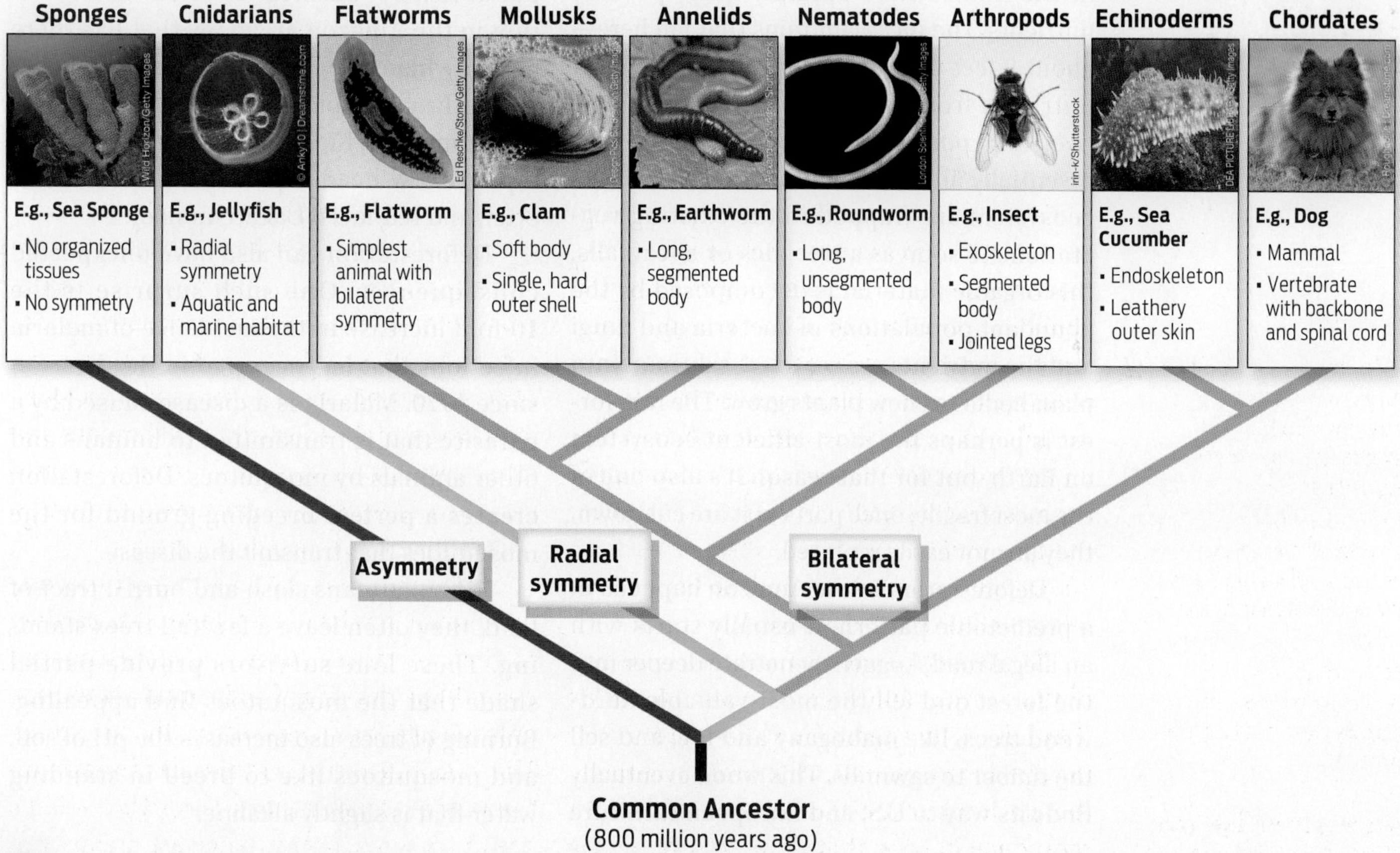

? List the three different animal body symmetries. For each type, name an animal group with that symmetry.

As impressive as animal diversity in the rain forest is, none of it would be possible without the plants that provide these animals with food and shelter. In fact, according to Laurance, there is a clear link between animal and plant diversity. "If you've got very high tree diversity, and then also a lot of these trees are growing very tall, you create a lot of different places for animals to live," he says. "So the animal diversity may be responding to the very high plant diversity."

The animals in the forest aren't simply freeloaders in this situation, however; they provide a valuable service to the many flowering plants in the forest. Insects, for example, play a crucial role in pollination of flowering plants. And larger animals help to disperse plants' seeds.

Deforestation Casualties

▶ Protist evolutionary diversity

What may seem surprising, given all the life supported in the rain forest, is that the soils in the Amazon are actually quite poor in nutrients. The incessant rains that fall here—about 9 feet of water every year—wash out nutrients from soil into the many smaller rivers that pour into the Amazon. As a result, essentially all the available nutrients in this ecosystem are trapped in the existing vegetation. As soon as a tree dies or a leaf falls, this organic material is decomposed by the abundant populations of bacteria and fungi and the nutrients are recycled, taken up into plant bodies as new plants grow. The rain forest is perhaps the most efficient ecosystem on Earth, but for that reason it's also one of the most fragile; once parts of it are cut down, they are not easily replaced.

Deforestation in the Amazon happens in a predictable pattern. It usually starts with an illegal road. Loggers penetrate deeper into the forest and fell the most valuable hardwood trees, like mahogany and ipe, and sell the timber to sawmills. This wood eventually finds its way to U.S. and European furniture and floor wholesalers, which may or may not be aware of its source.

Loggers then set fires, reducing the remaining trees to stubby ash skeletons. Next come the bulldozers, which push what's left into heaps and tear up the remaining roots from the soil. The land is now ready to be planted with soybeans or used as cattle pasture.

Burning trees deposits nutrients in the soil, which then support one or two seasons of crop growth. But because the underlying soil is naturally so nutrient-poor, these recently cleared lands quickly lose their fertility, leading to yet more slashing and burning.

Deforestation is threatening many of the eukaryotic species that live in the rain forest. According to a 2015 study by Laurance and his colleagues, roughly half of the 14,000 species of tree in the Amazon qualify as globally threatened under International Union for Conservation of Nature (IUCN) Red List criteria.

Among the animal species that could be put at risk by "business as usual" deforestation in the Amazon are many that live there and nowhere else, including the tree ocelot, hoary-throated spinetail bird, white-cheeked spider monkey, Rio Branco antbird, Brazilian tapir, yellow-headed poison dart frog, giant otter, and red-faced Uakari monkey.

Deforestation can also have unexpected consequences. One such surprise is the 10-fold increase in the incidence of malaria infections that has occurred in the Amazon since 1970. Malaria is a disease caused by a parasite that is transmitted to humans and other animals by mosquitoes. Deforestation creates a perfect breeding ground for the mosquitoes that transmit the disease.

When humans slash and burn a tract of land, they often leave a few tall trees standing. These lone survivors provide partial shade that the mosquitoes find appealing. Burning of trees also increases the pH of soil, and mosquitoes like to breed in standing water that is slightly alkaline.

Though transmitted by mosquitoes, malaria is actually caused by a tiny, single-celled eukaryotic organism called *Plasmodium*. These tiny parasites live inside the mosquito and are transferred to other animals through bites.

Plasmodium is an example of a protist, the fourth and final group of eukaryotes we will meet in this chapter. Unlike the other eukaryotes—plants, animals, and fungi—protists don't form a tightly related group. So in that sense, it's almost a misnomer to refer to them by a single term. Some protists are more closely related to plants, animals, or fungi than to other protists. Most share the trait of being unicellular, and until the DNA revolution it was hard to tell them apart, so biologists tended to lump them together (see **Milestone 5: Shaking the Tree**).

Though small in size, protists can claim a hugely important role in the evolution of eukaryotes as a whole. For about a billion years, protists were the only eukaryotic organisms on the planet; multicellular eukaryotes evolved much later, about 570 million years ago. Plants, animals, and fungi each evolved from unicellular protists, although not the same one.

Reflecting their distribution across the eukaryotic tree, protists have diverse lifestyles that in many cases resemble those of

Kitchin and Hurst/AGE Fotostock

luoman/E+/Getty Images

Yellow-headed poison dart frogs (*Dendrobates leucomelas*) make their home among the plants of the Amazon rain forest. They are just one of many species threatened by rain forest deforestation (pictured left).

other eukaryotes. For example, some protists carry out photosynthesis and in that way resemble plants. Others eat other organisms and in that way resemble animals. Despite this diversity, protists do share the common trait of being susceptible to drying out, so they are typically found in wet environments: lakes, oceans, ponds, moist soils, and living hosts **(INFOGRAPHIC 18.11)**.

Though protists are extremely resilient and unlikely to suffer in the same ways as other eukaryotes from destruction of rain forest habitats, it's clear that their impact on humans could shift dramatically as a result. This is especially true if, as is expected, climate change shifts the range of the insects that carry them, bringing the insects into regions where formerly they could not survive the winter.

A Tipping Point

Deforestation in the Amazon reached a tipping point in 2004, when some 11,000 square miles of forest was destroyed—an area about the size of Belgium. Since that time, the rate of deforestation has dropped dramatically—by about 80%.

What changed? Laurance credits several factors, including the high-profile murder of a Catholic nun, Dorothy Stang, who had been an outspoken critic of illegal logging and land theft. With the attention to her murder came a crackdown by the government on illegal logging and deforestation. Brazil also began

INFOGRAPHIC 18.11

The Challenge of Classifying Protists

Protists are a diverse group of organisms that are difficult to classify. They share features with animals, plants, and fungi, but are not classified as any one of these. Nor do they have a single unifying characteristic that places them within a single evolutionary group. Protists are currently found in multiple groups within the domain Eukarya.

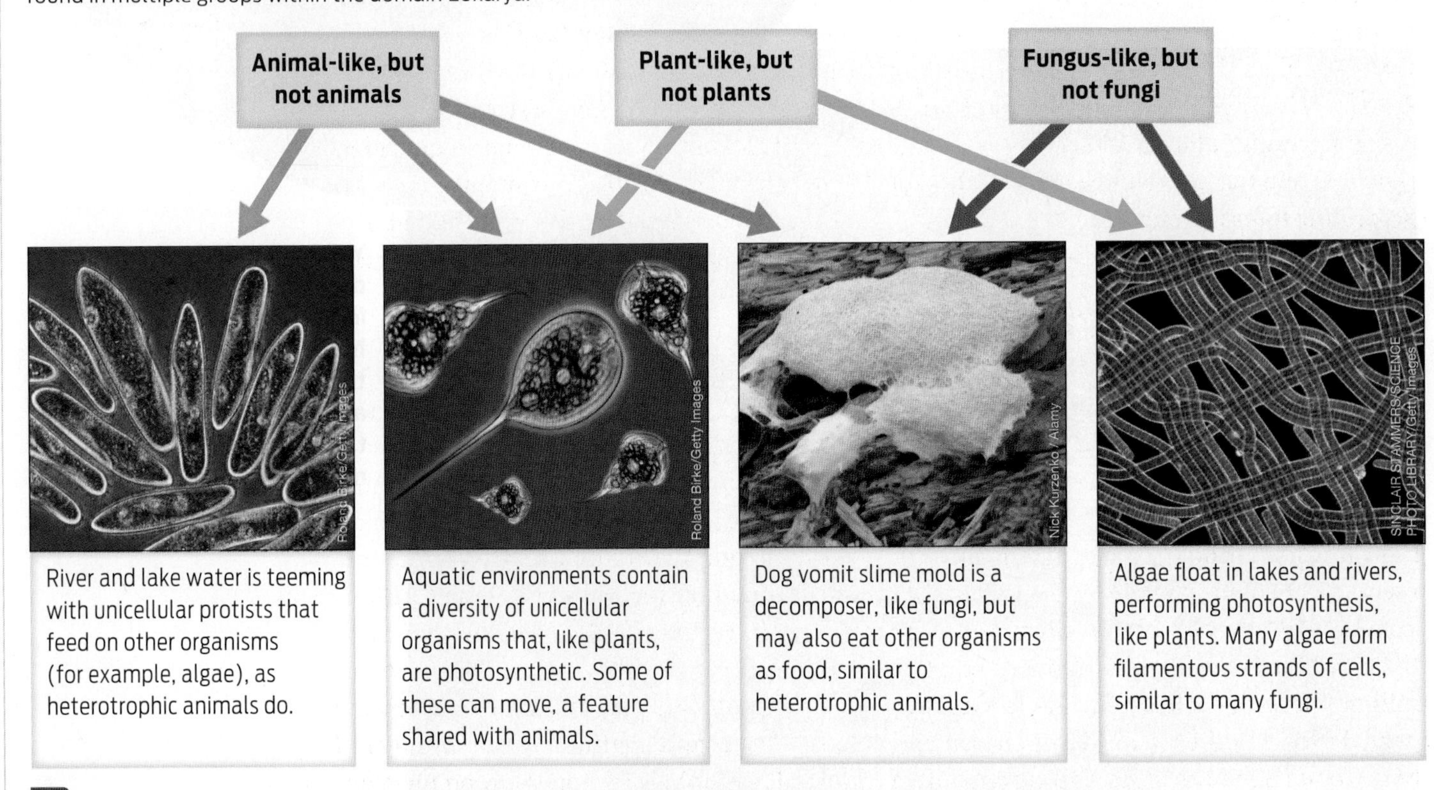

River and lake water is teeming with unicellular protists that feed on other organisms (for example, algae), as heterotrophic animals do.

Aquatic environments contain a diversity of unicellular organisms that, like plants, are photosynthetic. Some of these can move, a feature shared with animals.

Dog vomit slime mold is a decomposer, like fungi, but may also eat other organisms as food, similar to heterotrophic animals.

Algae float in lakes and rivers, performing photosynthesis, like plants. Many algae form filamentous strands of cells, similar to many fungi.

? Are all protists closely related to one another? Explain your answer.

using a satellite system to monitor deforestation from space.

Boycotts by U.S. companies have also likely played a role. In 2006, the nonprofit organization Greenpeace released a report showing that forests were being destroyed to grow soybeans to feed cattle, which were sold to companies like McDonald's and Wal-Mart. In response to international outrage, these companies agreed to stop buying meat from cattle fed soybeans grown on illegally deforested lands.

But deforestation has not stopped in Brazil. In 2014—a relatively "good" year, deforestation-wise—an area the size of Olympic National Park in Washington State was wiped out. And as of 2019, there were signs that deforestation might be ramping up again. That's why new ideas and new approaches, such as those being implemented in Acre, are badly needed **(INFOGRAPHIC 18.12)**.

INFOGRAPHIC 18.12

Brazilian Amazon Deforestation

The area of forest removed (deforestation) increased between 1991 and 2004, and then decreased steadily until 2014. Since then, deforestation has increased.

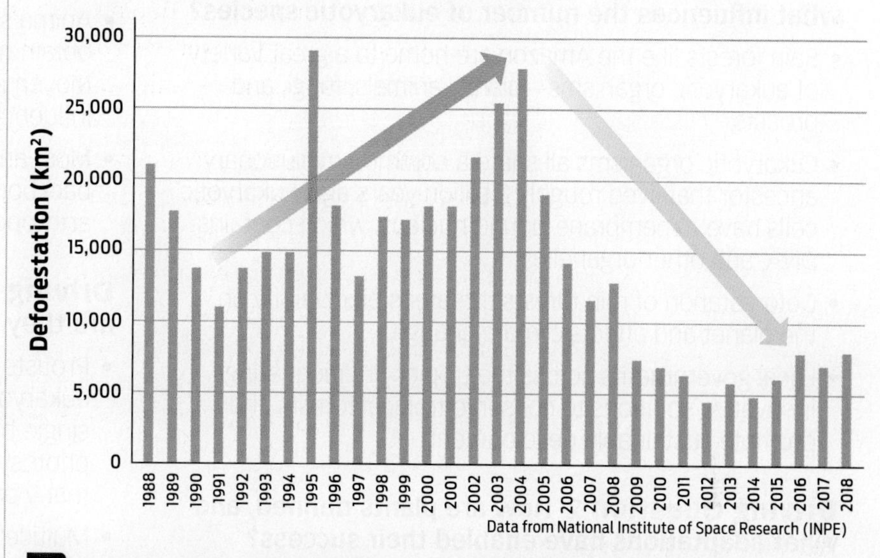

? In your own words, describe the pattern of Brazilian forest deforestation over time.

In 2010, the government in Acre passed a law called the State System of Incentives for Environmental Services, with the goal of making standing forest more valuable by linking it to markets for the services it provides—rubber, Brazil nuts, fish, even carbon storage. The incentives reward small farmers, indigenous communities, and even cattle ranchers and loggers with cash payments for protecting forest resources.

As part of the law, Acre established the first state-based REDD+ program. REDD+ stands for "reducing emissions from deforestation and forest degradation, and enhancing carbon stocks." This program, which was established by the United Nations, aims to reduce greenhouse gas emissions by encouraging tropical countries to better manage their forests. Through this program, governments can become eligible for payments after demonstrated results of keeping carbon in the trees. Germany has already invested $17.5 million into Acre, and California is currently considering whether to buy forest-based carbon credits from the state as part of its own cap-and-trade program.

How did Acre become such a leader in this area? "It really all started with Chico Mendes and the rubber tappers movement," says Duchelle, of the Center for International Forestry Research. Acre's self-proclaimed "Forest Government," which has held leadership roles in the state government for nearly 20 years, has worked on the premise that economic development need not be incompatible with forest sustainability.

In many ways, Acre is unique: its strong history of forest conservation movements makes it easier for sustainable practices to take hold here. But even much larger states in Brazil, with very different challenges—more soybean farmers, more cattle ranchers—have learned from Acre. Mato Grosso, a large state in central Brazil passed a REDD+ law in 2013 that was very much influenced by what Acre has done.

"It's a special environment," Duchelle says. "Again and again, Acre has pioneered innovative models for forest-based development." ■

CHAPTER 18 Summary

Driving Question 1 What are eukaryotic organisms, what is their evolutionary history, and what influences the number of eukaryotic species?

- Rain forests like the Amazon are home to a great variety of eukaryotic organisms—plants, animals, fungi, and protists.
- Eukaryotic organisms all share a common evolutionary ancestor that lived roughly 2 billion years ago. Eukaryotic cells have a membrane-bound nucleus, which contains DNA, and other organelles.
- Deforestation of rain forests threatens biodiversity on the planet and affects climate change.
- Local governments across the tropics are pioneering innovative solutions to conserve tropical forests and promote sustainable development.

Driving Question 2 How are plants defined, and what adaptations have enabled their success?

- Plants are multicellular eukaryotes that carry out photosynthesis and are adapted to living on land. Plant cells have a cell wall made of cellulose. Plants reproduce sexually.
- Some plants produce distasteful chemicals called secondary metabolites as a defense against herbivores; latex contains many such secondary metabolites.
- Plants can be divided into groups, such as the mosses, ferns, gymnosperms, and angiosperms, on the basis of their terrestrial adaptations.
- Two key evolutionary milestones in plants were the evolution of vascular tissues and reproduction using pollen and seeds.

Driving Question 3 How are fungi defined, and what adaptations have enabled their success?

- Fungi are decomposers, acquiring their nutrition by breaking down dead organic matter and absorbing the results. They can be either unicellular or multicellular. Fungal cells have a cell wall made of chitin. Fungi reproduce both sexually and asexually.

Driving Question 4 How are animals defined, and what adaptations have enabled their success?

- Animals are multicellular eukaryotic heterotrophs that obtain nutrients by ingestion. They reproduce sexually. Movement is a key evolutionary adaptation that has influenced the success of animals.
- Most animals are invertebrates—that is, they lack a backbone. The most abundant invertebrates by far are arthropods, especially insects.

Driving Question 5 What are protists, and why are they are hard to classify?

- Protists are a diverse group of mostly unicellular eukaryotic organisms that do not cluster on a single branch of the evolutionary tree. They include photosynthetic plantlike algae, animal-like parasites, and many other organisms.
- Multicellular eukaryotes (plants, animals, and fungi) evolved from single-celled, protist-like ancestors.

More to Explore

- NPR. (November 12, 2015). The rain forest was here: http://www.npr.org/series/455762230/the-rain-forest-was-here
- Duchelle, A. E., et al. (2014). Acre's State System of Incentives for Environmental Services (SISA), Brazil. In: Sills, E. O., et al., eds. *REDD+ on the Ground: A Case Book of Subnational Initiatives across the Globe*, pp. 68–85. Bogor, Indonesia: CIFOR.
- Ter Steege, H., et al. (2015). Estimating the global conservation status of more than 15,000 Amazonian tree species. *Sci Adv* 1(10): e1500936.
- Grandin, G. (2009). *Fordlandia: The Rise and Fall of Henry Ford's Forgotten Jungle City.* New York: Picador.

CHAPTER 18 Test Your Knowledge

Driving Question 1 What are eukaryotic organisms, what is their evolutionary history, and what influences the number of eukaryotic species?

By answering the questions below and studying Infographic 18.2, you should be able to generate an answer for this broader Driving Question.

Know It

1. What do all eukaryotes have in common?
- **a.** cell walls
- **b.** photosynthesis
- **c.** multicellularity
- **d.** cells containing a nucleus
- **e.** unicellularity

2. What are the defining features of eukaryotes, members of the domain Eukarya?

3. What do a rubber tree and a human latex collector have in common?

Use It

4. How do you think the diversity of eukaryotic organisms in each of the following areas would compare to the diversity in the Amazon rain forest? What factors might influence eukaryotic diversity in these areas?
- **a.** Lake Michigan
- **b.** the Sonoran Desert in Arizona
- **c.** a high alpine meadow in the Colorado Rockies

5. Which of the following would you predict to have the greatest impact on the number of different eukaryotic species in the Amazon rain forest? Explain your answer, and consider the impact of each on all eukaryotes.
- **a.** application of a herbicide
- **b.** application of an insecticide
- **c.** application of a fungicide

Interpreting Data

Apply Your Knowledge

6. A group of researchers designed an experiment to examine factors that influence the number of plant species in a rain forest in Belize. They began by analyzing seeds falling to the ground in a particular area, counting the number of seeds dropped by different species of plants. If all the seeds germinate, the seedlings present should match the diversity of species in the seeds. Then they analyzed seedlings that germinated in three small (1 m^2) areas in the same forest location: one that was sprayed weekly with water (control), one that was sprayed with an insecticide (a chemical that kills insects), and one that was sprayed with a fungicide (a chemical that kills fungi). They identified and counted seedlings to determine diversity (number of species) and abundance (number of seedlings) (*indicates a significant difference from control).
- **a.** Plot these data on two graphs (one for species number and one for seedling number).
- **b.** What can you conclude about the impact of insects on the number of seeds that germinate into seedlings? What can you conclude about the impact of fungi on the number of seeds that germinate into seedlings?
- **c.** Do insects or fungi have a greater impact on maintaining plant diversity (defined as number of plant species) in the rain forest?

	Control	Insecticide	Fungicide
Diversity (number of species)	3.7	3.4	3.2*
Abundance (number of seedlings)	30	58*	35

Data from Bagchi, R., et al. (2014). Pathogens and insect herbivores drive rainforest plant diversity and composition. *Nature* 506:85–88, doi:10.1038/nature12911

Driving Question 2 How are plants defined, and what adaptations have enabled their success?

By answering the questions below and studying Infographics 18.3, 18.4, and 18.5 and Table 18.1, you should be able to generate an answer for this broader Driving Question.

Know It

7. Which group of plants was the first to live on land? Why do we find these plants only in particular environments? (After all, if they were the first, shouldn't they have spread everywhere by now?)

8. A major difference between a fern and a moss is
- **a.** the presence of seeds.
- **b.** the presence of flowers.
- **c.** the presence of cones.
- **d.** the presence of a vascular system.
- **e.** the ability to carry out photosynthesis.

Use It

9. What is an advantage of having seeds?

10. What type of seed plant is likely to rely on hungry animals to spread its seeds? Explain your answer.

11. How did the evolution of vascular systems in plants change the landscape?

Driving Question 3 How are fungi defined, and what adaptations have enabled their success?

By answering the questions below and studying Infographics 18.6 and 18.7, you should be able to generate an answer for this broader Driving Question.

Know It

12. Consider how fungi obtain nutrients and energy.
- **a.** Can fungi carry out photosynthesis?
- **b.** Can fungi ingest their food?
- **c.** How do fungi obtain their nutrients and energy?

13. Which of the following meals include(s) a fungus?
- **a.** a bread and blue cheese platter with fruit
- **b.** mushroom soup
- **c.** both a and b
- **d.** a fruit salad
- **e.** yogurt

Use It

14. A very early classification scheme placed the fungi together with the plants. Why do you think fungi were grouped with plants? What features distinguish them from plants?

Driving Question 4 How are animals defined, and what adaptations have enabled their success?

By answering the questions below and studying Infographics 18.8, 18.9, and 18.10, you should be able to generate an answer for this broader Driving Question.

Know It

15. A sand dollar gets its name from its body shape—it resembles a large coin. What type of body symmetry does a sand dollar have?
- **a.** bilateral
- **b.** radial
- **c.** none (sand dollars are amorphous)
- **d.** hyphae
- **e.** mycelium

16. What do mosquitos, snails, and earthworms have in common?
- **a.** They are all insects.
- **b.** They are all mollusks.
- **c.** They are all arthropods.
- **d.** They are all invertebrates.
- **e.** They all have an exoskeleton.

17. Which of the following defense/predator avoidance features or strategies can be found in arthropods?
- **a.** an exoskeleton
- **b.** venom
- **c.** flight
- **d.** secondary metabolites
- **e.** all of the above
- **f.** a, b, and c

18. Which of the following is *not* a feature of all animals?
- **a.** defined body symmetry
- **b.** internal digestion
- **c.** motility
- **d.** cells that lack cell walls
- **e.** multicellularity

Use It

19. Many characteristics are used to classify animals. Why do we need to use so many different characteristics? Consider the following five animals: cockroach, earthworm, honeybees, parrot, and slug; and the following three characteristics: ability to fly, two-legged, exoskeleton.
- **a.** Which of the five animals could be grouped by each characteristic?
- **b.** Would this grouping reflect their real taxonomic relationships?
- **c.** By what feature(s) would you put earthworms and slugs together in their own group? What about parrots and honeybees? Are the organisms in each of these pairs in the same taxonomic group?

Driving Question 5 What are protists, and why are they hard to classify?

By answering the questions below and studying Infographic 18.11, you should be able to generate an answer for this broader Driving Question.

Know It

20. What do all members of the informal group known as protists have in common?
- **a.** nothing
- **b.** They are all eukaryotic.
- **c.** They all carry out photosynthesis.
- **d.** They are all human parasites.
- **e.** They are all decomposers.

Use It

21. Why do scientists no longer consider protists a separate kingdom? How are scientists placing protists into their new taxonomic "homes"?

22. Many protists have an organelle called the contractile vacuole that pumps out water that enters the cell by osmosis. Why is this a useful adaptation for a protist? What might happen to a protist if its contractile vacuole stopped working? (Think about where many protists live, and what happens to bacteria whose cell walls are disrupted by antibiotics.)

Mini Case

Apply Your Knowledge

23. Biopiracy has been described as the use of the knowledge of native peoples about nature and natural products by others for profit without obtaining permission of the indigenous people, and without sharing profits. Given the tremendous biodiversity in the Amazon and medicines that have already been developed from natural sources in the area, life-saving and lucrative medicines likely remain to be discovered by bioprospecting rain forest resources. (Bioprospecting is the search for natural resources that provide valuable products.) The Convention on Biological Diversity attempts to address many of the concerns arising from biopiracy and bioprospecting.

a. Do some research on the Convention on Biological Diversity and identify the issues that would be particularly relevant for bioprospecting in the Amazon.

b. Draft a plan setting out ways in which a drug company could fairly bioprospect in the Amazon.

c. How do you think that the various stakeholders would respond to your plan?

Bring It Home

Apply Your Knowledge

24. During his presidency, Barack Obama established over 30 new national monuments, protecting more than 550 million acres of land and water. Do some research to learn about one of these new national monuments.

a. Where is it?

b. What kind of eukaryotic biodiversity is present?

c. Are there any threats to the eukaryotic biodiversity present? Has the protected status of this national monument reduced any of these threats?

MILESTONES IN BIOLOGY **5**

Shaking the Tree

A revised view of eukaryotic diversity may be the key to tackling deadly diseases

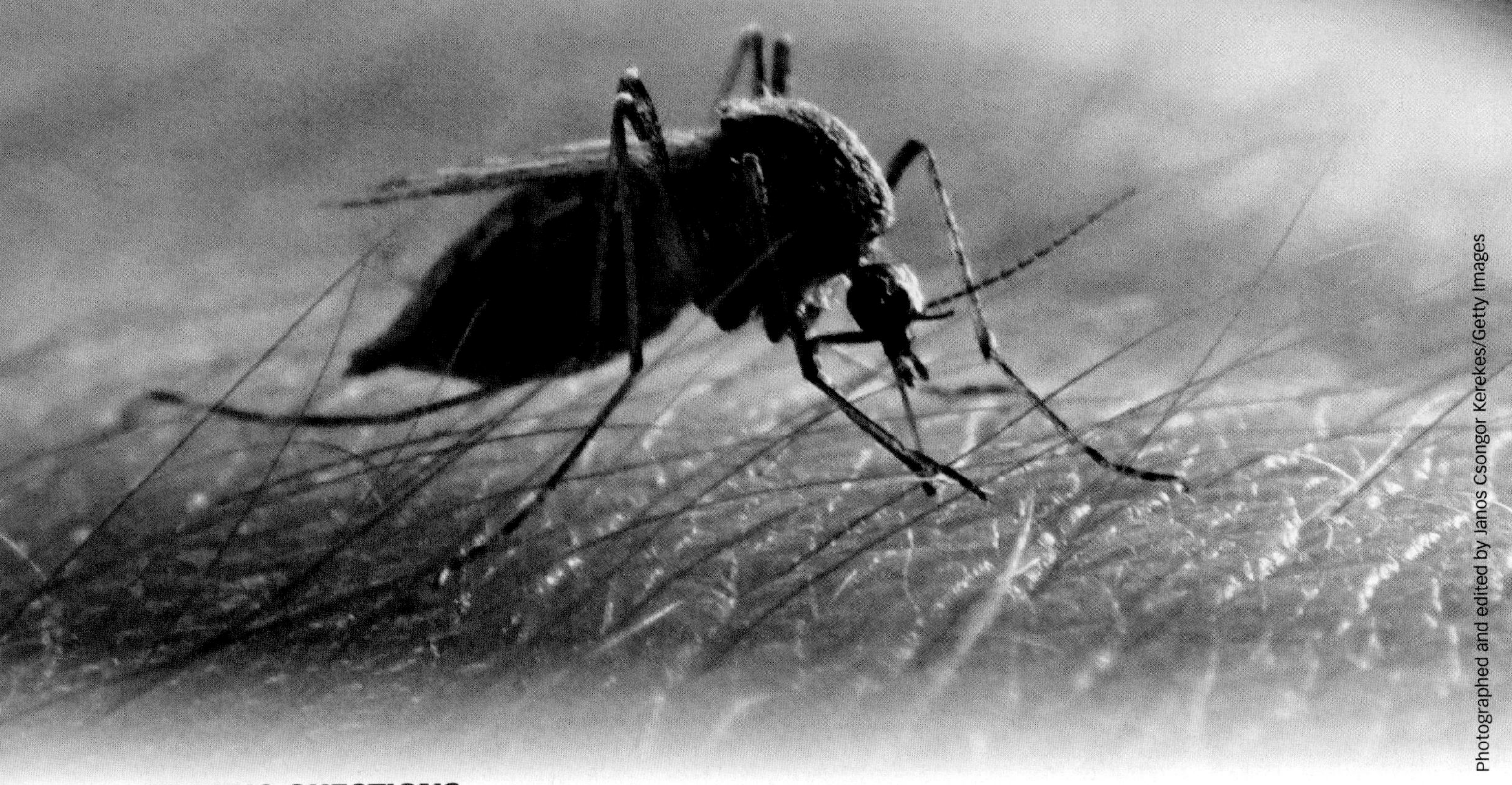

Photographed and edited by Janos Csongor Kerekes/Getty Images

DRIVING QUESTIONS

1 How has genetic evidence transformed the classification of protists?

2 Why is it important to classify protists accurately?

3 What are current challenges in preventing and treating malaria?

As the war in Vietnam raged, soldiers on both sides of the conflict faced an unrelenting enemy: malaria. This parasite-caused illness—transmitted through the bite of a mosquito—produces high fever, headache, chills, and vomiting in those infected. If not treated within 24 hours, the condition can be fatal.

Malaria had traditionally been treated with drugs such as quinine, derived from the bark of the cinchona tree, and its synthetic derivative chloroquine. But increasingly, these drugs were unable to stem the tide of infection; the parasites had begun to evolve resistance.

Worried that a malaria-weakened army would be unable to fight off U.S.-backed military forces, the prime minister of North Vietnam, Ho Chi Minh, turned to Communist China for help. Recognizing their common interest in defeating a shared set of enemies, China's leader, Mao Zedong, launched a secret mission to find a malaria cure.

Project 523 was launched on May 23, 1967—hence the name. It enlisted hundreds of Chinese scientists and traditional Chinese healers. They were charged with screening thousands of known plant compounds for antimalarial effects and scouring traditional sources of Chinese medicine for leads on promising new medicines.

Taking their cue from an ancient medical text, the scientists homed in on one

Xinhua News Agency/Getty Images

Xinhua News Agency/Getty Images

Youyou Tu at work and with the 2015 Nobel Prize in Physiology or Medicine she received for her research on the development of artemisinin for the treatment of malaria.

INFOGRAPHIC M5.1

The Discovery of Artemisinin

During Project 523, Chinese scientists discovered artemisinin, a compound found in the herb *Artemisia annua* that effectively kills the malaria parasite *Plasmodium falciparum*.

? In the ancient text, was it likely that the medicine was described as a hot tea? Why or why not?

particular herb, called qinghao (known outside of China as *Artemisia annua* or sweet wormword). Qinghao had been used for centuries in treating "intermittent fevers," a description that aptly describes a symptom of malaria. The scientists tested the herb against malaria-sickened mice but the results proved inconsistent.

Then, one scientist, Youyou Tu, got a flash of inspiration from another ancient text, which recommended soaking the herb in water and then drinking the liquid. Tu, a chemist, realized that the typical extraction process of boiling the herbs was likely destroying the chemical. When she changed the extraction process, the results were dramatic: nearly 100% of the mice were cured by the herbal remedy **(INFOGRAPHIC M5.1)**.

Tu's efforts would eventually result in the drug known as artemisinin, one of today's main pharmaceutical weapons against malaria. The discovery of artemisinin was truly a milestone in medicine, helping to save the lives of millions around the world. For her work, Tu shared the Nobel Prize for Physiology or Medicine in 2015.

But increasingly, scientists are seeing worrisome signs that malaria parasites are developing resistance to artemisinin, too. If that happens, the death toll due to malaria could rise once more. This frightening prospect has scientists racing to find out more about these tiny parasites and what makes them unique, as part of the quest to develop more effective weapons directed specifically against them.

What Are They?

▶ *Plasmodium*, protists, and supergroups

The malaria parasite is a single eukaryotic cell. It belongs to a genus of organisms called *Plasmodium*. The most common malaria-causing species are *P. falciparum* and *P. vivax*, with *P. falciparum* causing the more severe form of the illness.

The parasite lives inside several different species of mosquitoes of the genus *Anopheles*. It is transmitted to humans through bites from female mosquitoes in search of a blood meal to feed their developing eggs. Once inside a human host, the malaria parasites first infect liver cells and then move on to red blood cells. The parasites divide rapidly, eventually bursting out of these cells. The deaths of the red blood cells and the release of their contents produce the main symptoms of the disease. Eventually, the parasites are transmitted back to a mosquito as it feeds on the blood of an infected person, and the cycle continues **(INFOGRAPHIC M5.2)**.

In 2018, about 92 countries around the world, mostly in tropical regions of South

INFOGRAPHIC M5.2

Malaria Parasite's Life Cycle

Malaria is caused by organisms in the genus *Plasmodium*. *Plasmodium* has a complex life cycle, involving a mosquito and a human host. To complete its life cycle and reproduce, various stages of *Plasmodium* mature and replicate in human liver and red blood cells, as well as in the gut of mosquitoes. The stages present in human blood cause the symptoms of malaria.

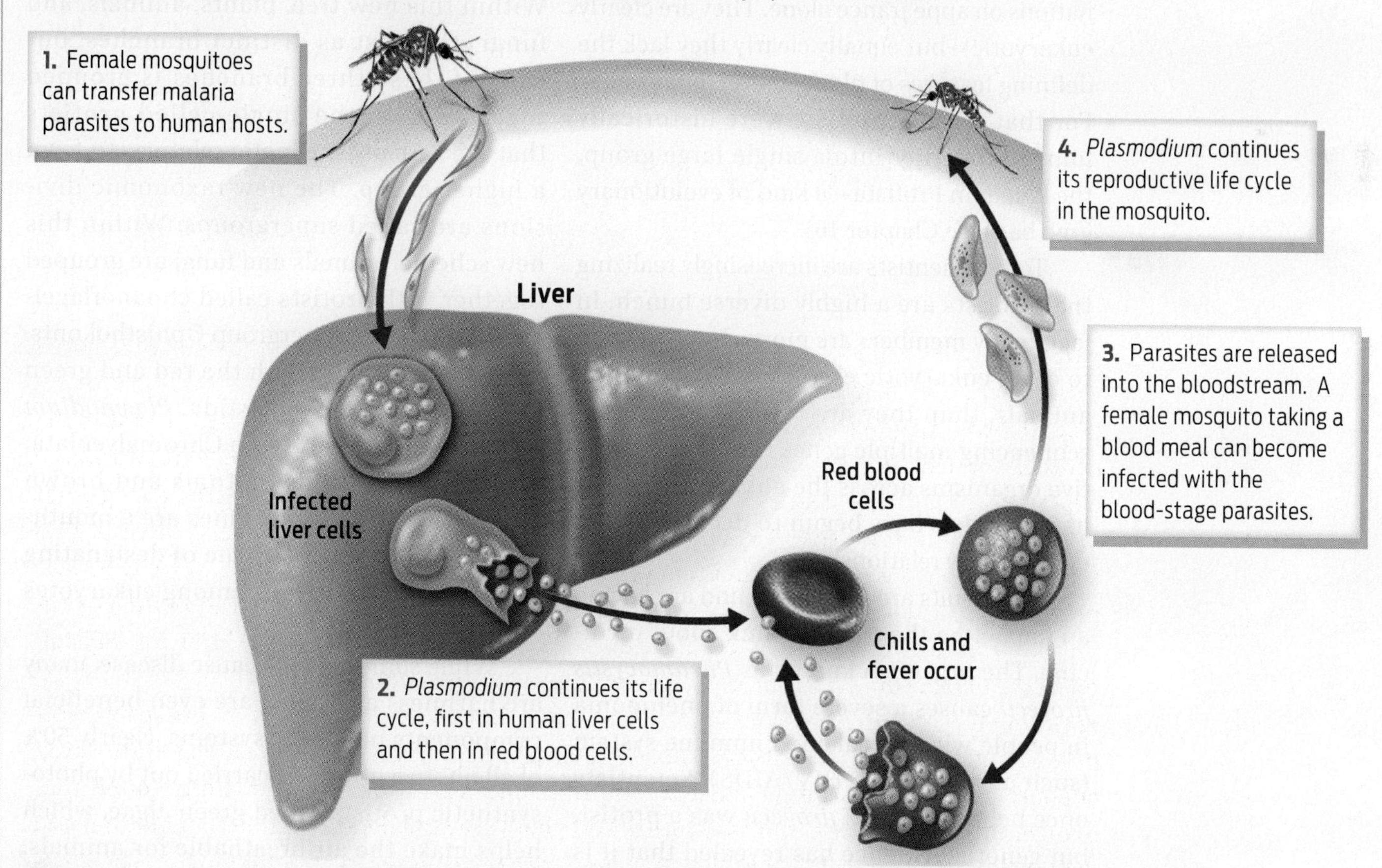

? **Does the mosquito just carry the parasites between human hosts? Explain your answer.**

America, Africa, and Asia, had ongoing malaria epidemics. About 3.4 billion people are at risk for the disease—nearly half the world's population. The malaria parasite infects roughly 250 million people every year and kills hundreds of thousands. The stakes of properly classifying and understanding this tiny parasite could not be higher.

About 3.4 billion people are at risk for the disease—nearly half the world's population.

Plasmodium is an example of a protist, a member of a diverse group of mostly single-celled eukaryotic organisms that we encountered in Chapter 18. Single-celled protists have long proved challenging for scientists to study and classify. Because of their small size, it's often hard to make useful discriminations on appearance alone. They are clearly eukaryotic—but equally clearly they lack the defining features of plants, fungi, or animals. For that reason, protists were historically lumped together into a single large group, the kingdom Protista—a kind of evolutionary grab bag (see Chapter 16).

Today scientists are increasingly realizing that protists are a highly diverse bunch. In fact, many members are more closely related to other eukaryotic organisms, like fungi or animals, than they are to one another. By sequencing multiple genes from representative organisms across the eukaryotic tree of life, scientists have begun to decipher these evolutionary relationships.

The results are surprising, and are having an impact on the way we think about medicine. The organism known as *Pneumocystis jirovecii* causes a severe form of pneumonia in people with a weakened immune system (such as those with HIV/AIDS). Scientists once believed that *P. jirovecii* was a protist, but genetic evidence has revealed that it is actually a fungus. This revised classification has allowed scientists to develop more effective treatments for the infection—in this case, antifungal medications.

Similarly, the single-celled organism *Phytophthora infestans*, the cause of potato blight, was long considered to be a type of fungus. This organism was responsible for about 1 million deaths from starvation during the potato famine in Ireland between 1845 and 1851. Newer molecular data have revealed it to be a protist belonging to a group of organisms called Chromalveolata, which are not closely related to fungi. Its revised classification explains why *Phytophthora* resists fungicides.

To make sense of these new data points, biologists have done away with the traditional eukaryotic tree that was divided into four kingdoms: plants, animals, fungi, and protists. In its place, they have constructed a new tree that groups eukaryotes based on their actual shared evolutionary history. Within this new tree, plants, animals, and fungi still exist as distinct branches, but each of these three branches is grouped together with the single-celled protists that are its closest genetic relatives to form a higher group. The new taxonomic divisions are called supergroups. Within this new scheme, animals and fungi are grouped together with protists called choanoflagellates to form the supergroup Ophisthokonts. Plants are grouped with the red and green algae to form Archaeplastida. *Plasmodium* belongs to the supergroup Chromalveolata, which also contains diatoms and brown algae. The supergroup names are a mouthful, but they have the virtue of designating the true family relations among eukaryotes **(INFOGRAPHIC M5.3)**.

While some protists cause disease, many are harmless and some are even beneficial components of our ecosystems. Nearly 50% of all photosynthesis is carried out by photosynthetic protists called green algae, which helps make the air breathable for animals. Other protists produce valuable compounds.

For example, red algae produce carrageenan, a polysaccharide that is used as a thickening agent for ice cream and yogurt.

Warm and Wet

▶ Climate and antimalaria efforts

Malaria is just one of many parasite-caused diseases that are common in regions with warm, wet climates. Others include Chagas disease, caused by a protist called *Trypanosoma cruzi* that is transmitted by blood-sucking assassin bugs, and African sleeping sickness, caused by a protist called *Trypanosoma brucei* that is transmitted by the tsetse fly.

What makes tropical regions uniquely susceptible to parasitic diseases? One main factor is that the insect vectors that carry the parasites are able to grow unchecked in the warm year-round temperatures. And the heavy rains provide lots of moist spots in

INFOGRAPHIC M5.3

Supergroups: A Revised Classification of Eukaryotes

Protists are typically single-celled eukaryotic organisms that lack distinguishing features of fungi, animals, and land plants. For this reason, they were historically lumped together in the kingdom Protista. An increasing amount of DNA sequence information has shown that protists are very diverse, sharing closer relationships with other eukaryotic groups than with one another. These revised groupings are called supergroups.

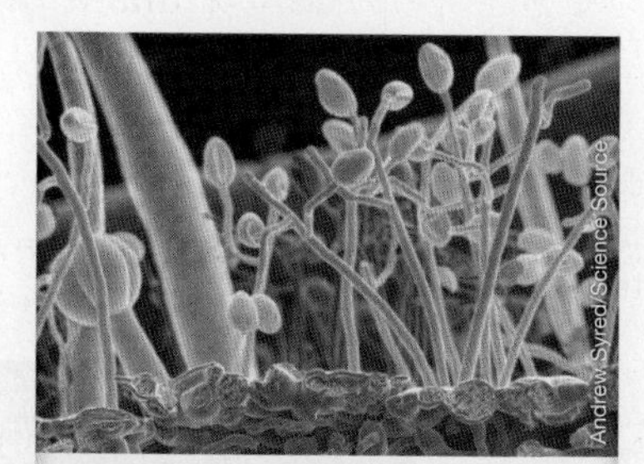

Phytophthora infestans, the cause of potato blight, was classified as a fungus until reclassified as a member of the Chromaveolata supergroup.

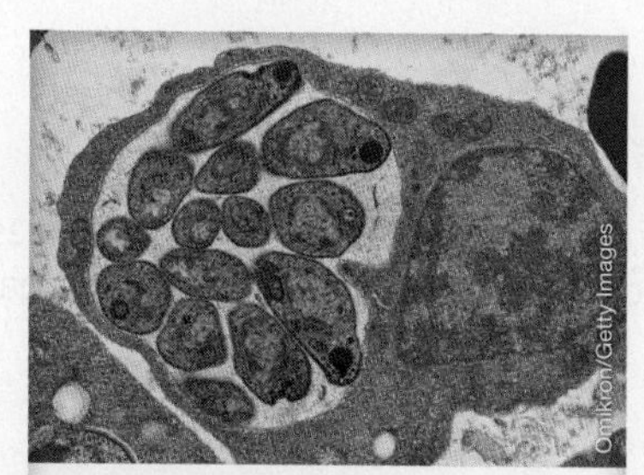

Plasmodium falciparum is a member of the Chromaveolata supergroup, which also contains brown algae.

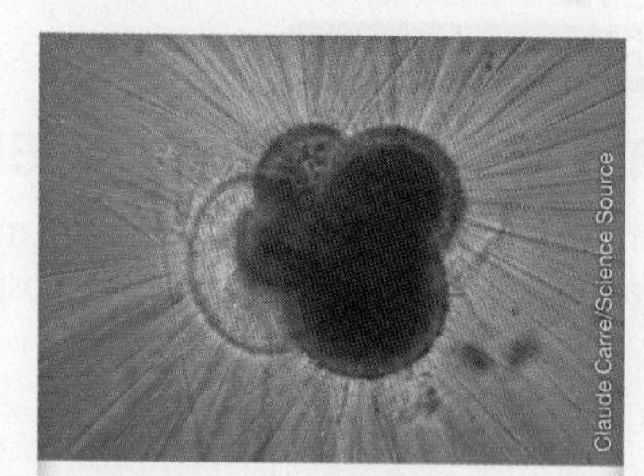

Photosynthetic *Globigerina bulloides* is a member of the supergroup Archaeplastida, which also includes land plants.

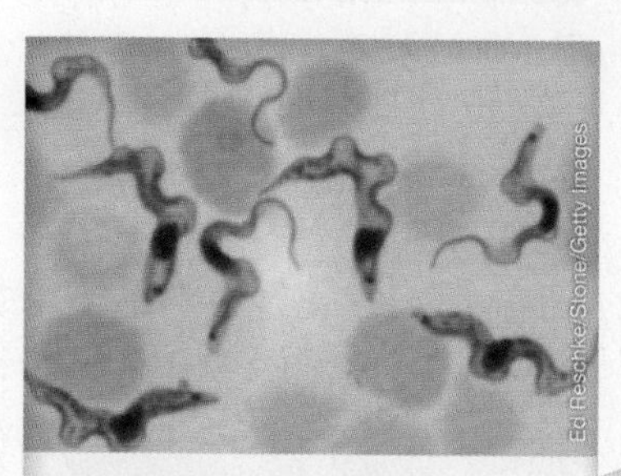

Trypanosoma brucei, the protist that causes African sleeping sickness, is classified in the Excavata supergroup. This group is named for a feeding structure that appears to be asymmetrically "excavated" from one side of the organism.

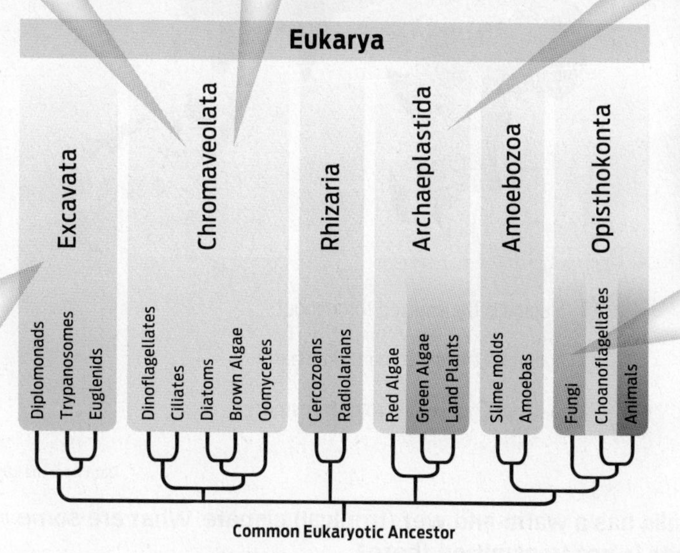

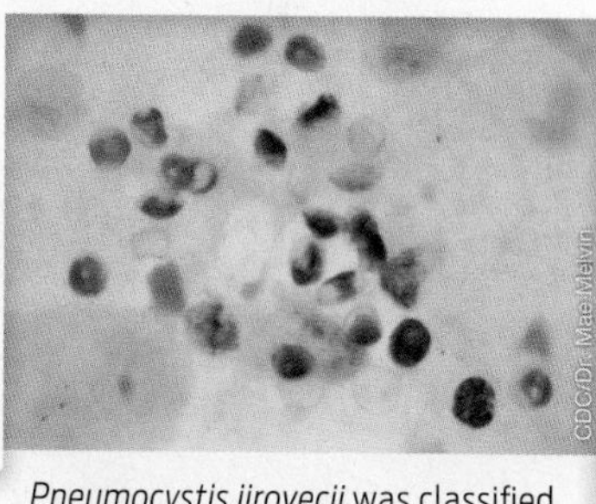

Pneumocystis jirovecii was classified as a protist until reclassified as a fungus in the Opisthokont supergroup that also includes animals.

? Which is more closely related to humans, *Phytophthora infestans* or *Pneumocystis jirovecii*?

which the larvae of these various insects can develop. *Anopheles* mosquitoes, for example, lay their eggs in pools of water. The eggs eventually hatch the larvae, which develop into adult mosquitoes. Many tropical regions are also relatively poor and often lack adequate public health protections, including mosquito control measures.

Sub-Saharan Africa shoulders the highest burden of malaria worldwide. In 2018, 93% of malaria cases and 94% of malaria deaths occurred in this region **(INFOGRAPHIC M5.4)**.

With the development of insecticides like DDT in the 1940s (see **Milestone 6: Progress or Poison**), public health workers began to hope that malaria could be eradicated completely by eliminating the mosquitoes that transmit the parasite. The World Health Organization (WHO) launched such an effort in 1955. As a result, malaria infections fell sharply in some countries, such as India and Sri Lanka. But the effort stalled when it became clear that mosquitoes were evolving resistance to the insecticides, including DDT. Some regions, including sub-Saharan Africa, were never part of the eradication program in the first place.

WHO currently endorses the use of DDT, sprayed indoors, as a way to control mosquitoes. But with resistance to the pesticide emerging as a large and growing problem, efforts that rely only on insecticides are likely to fail.

INFOGRAPHIC M5.4

Global Distribution of Malaria

Sub-Saharan Africa bears the largest burden of malaria. Malaria is also present in large areas of South America and southeast Asia. These areas all have climates that allow the mosquito vector to flourish and transmit the parasites.

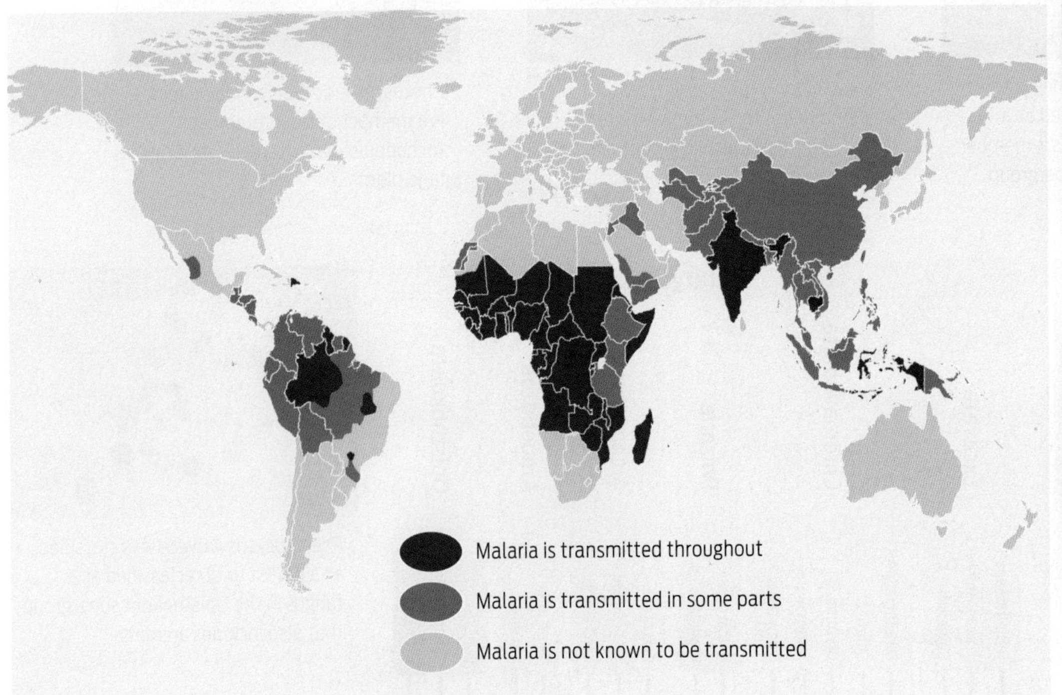

Data from the CDC.

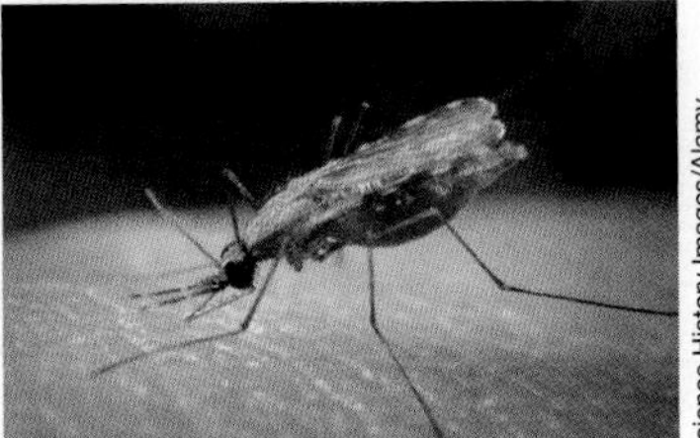

The *Anopheles* mosquito transmits the *Plasmodium* parasite while feeding.

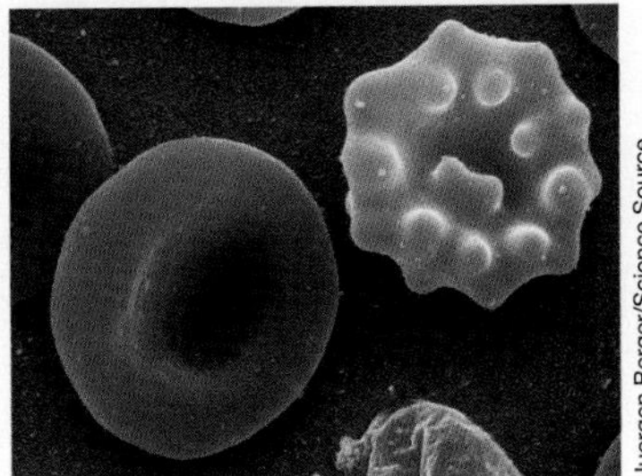

The *Plasmodium* parasite reproduces inside infected red blood cells, giving them this lumpy appearance.

? The northern part of Australia has a warm and wet (tropical) climate. What are some reasons that could explain the fact that the malaria parasite is not transmitted there?

Increasingly, scientists and governments are looking to forms of insect "birth control" as a way to combat malaria. One approach, called sterile insect technique, involves releasing large numbers of sterile male mosquitoes into an environment. The sterile males mate with females but cannot produce offspring, so the mosquito population declines over time.

A more radical solution—one that has yet to be tried in the wild—is the technique called a gene drive. This approach relies on the genome-editing tool CRISPR (Chapter 9) to engineer mosquitoes with specific traits—such as the inability to produce female offspring. Mosquitoes altered through this approach will reliably transmit the engineered traits ("drive it") through the population. In principle, gene drives could cause a population of malaria-transmitting mosquitoes to go extinct. Scientists and ethicists are currently debating the pros and cons of this method.

Climate change will also be a complicating factor. Climate change may shift the ranges of the disease-carrying insects, so regions of the world not currently plagued by malaria may find themselves facing this threat should they become warmer and wetter.

Revised Beginnings

▸ Evolution of eukaryotes

Protists have a unique claim to fame among eukaryotic organisms: they were the first eukaryotic organisms on the planet, and thrived here long before multicellular plants, animals, and fungi came along. Without protists, the other eukaryotes on the planet—including us—would not be here.

Scientists used to think that certain existing protists represented a kind of living relic—the closest thing to what ancient eukaryotes must have been like. *Giardia lamblia,* for example, is a unicellular parasite

Sub-Saharan Africa shoulders the highest burden of malaria worldwide.

that lives in the intestinal tract of animals such as beavers; unsuspecting campers who drink untreated pond water can contract giardiasis (also called beaver fever). *Giardia* appears to lack mitochondria—the eukaryotic cell's power plants—and their absence suggested to some researchers that a *Giardia*-like organism may have been the primordial organism that, 2 billion years ago, engulfed a free-living bacterium to form the first mitochondrion through endosymbiosis (see **Milestone 1: Scientific Rebel**).

More recent evidence indicates that *Giardia* do have remnants of mitochondria and mitochondrial genes; they've just shrunk over evolutionary time, leaving truncated versions in the present-day organisms.

This kind of swapping and rearranging of organelles among early protists has left some remarkable legacies among today's eukaryotic organisms. And some of these unique attributes are proving useful when it comes to treating parasite-caused infections in humans.

The malaria parasite, for example, bears within its cell the remnants of a chloroplast that it borrowed from a photosynthesizing protist that it had engulfed many millions of years ago. This lingering acquisition, called an apicoplast, no longer photosynthesizes (and does not need to, since the parasite lives in the dark inside animal hosts), but does perform a few essential functions for the malaria parasite, including making certain required lipids. This dependence makes the apicoplast a good target for new antimalarial drugs. Because of its chloroplast origins, the apicoplast is vulnerable to certain weed killers that target chloroplast enzymes. Scientists think that these or similar drugs might be the next potent weapons against the

INFOGRAPHIC M5.5

Unique Protist Traits Can Be Pharmaceutical Targets

Many protists have the "signatures" of ancient endosymbioses. These include nuclear genes transferred from various symbionts, and remnants of chloroplasts. While these chloroplast remnants do not function in photosynthesis in *Plasmodium*, they do provide targets for drugs that target chloroplast enzymes.

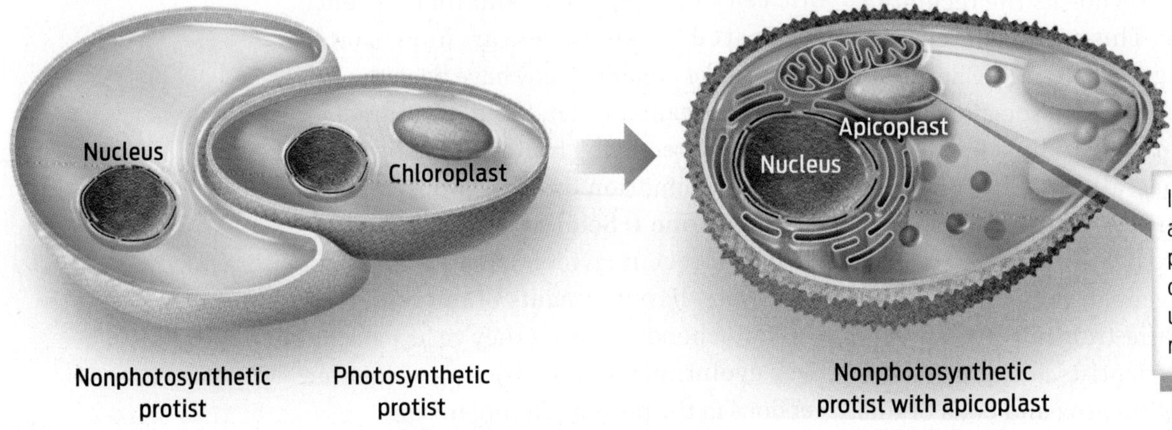

? Would the nonphotosynthetic protist host (far left) have been susceptible to the antimalarial herbicide before it engulfed the photosynthetic endosymbiont? Why or why not?

ever-evolving malaria parasite. Clinical trials of some of these drugs are now under way **(INFOGRAPHIC M5.5)**.

Battling Resistance

To help lower the risk of drug resistance, malaria is usually treated with a combination of drugs—typically, artemisinin and a quinine derivative such as piperaquine. Thanks to such combination therapy, WHO estimates that the incidence of malaria has fallen by 37% globally, and mortality rates have decreased by 60%. That's about 6 million deaths that have been avoided since 2001.

But some worrisome signs suggest that malaria is once again becoming resistant to our most effective drugs. In parts of Cambodia and Thailand, *P. falciparum* has developed resistance to both artemisinin and the drugs with which it's combined. If these resistant strains spread to other regions, particularly sub-Saharan Africa, the consequences could be devastating. For that reason, WHO has endorsed a plan to eradicate *P. falciparum* in this region of Asia by 2030.

Ultimately, the success of this effort may ultimately hinge on a better understanding of how these tiny protists evolved and where they fit in the overall eukaryotic tree of life. As the Chinese military general Sun Tzu once said, "the first rule of war is to know your enemy." ■

More to Explore

- Tu, Y. (2015). Nobel Prize lecture: Discovery of artemisinin: A gift from traditional Chinese medicine to the world.
- Milius, S. (July 29, 2015). The tree of life gets a makeover: Schoolroom kingdoms are taking a backseat to life's supergroups. *Sci News*. https://www.sciencenews.org/article/tree-life-gets-makeover
- Yong, E. (2014). How malaria defeats our drugs. *Mosaic Sci*. https://mosaicscience.com/story/how-malaria-defeats-our-drugs
- Matthews, D. (2018). A genetically modified organism could end malaria and save millions of lives—if we decide to use it. *Vox*. https://www.vox.com/science-and-health/2018/5/31/17344406/crispr-mosquito-malaria-gene-drive-editing-target-africa-regulation-gmo

MILESTONES IN BIOLOGY 5 Test Your Knowledge

1. About mosquitoes:

a. How is *Plasmodium falciparum* dependent on the *Anopheles* mosquito?

b. What is the evolutionary relationship between *Anopheles* mosquitoes and *Plasmodium falciparum*? (Hint: Consider their current classifications.)

2. Historically, *Phytophthora infestans* was classified as a fungus and *Pneumocystis jirovecii* and *Trypanosoma brucei* were classified as protists. Review and explain the basis for their current classifications.

3. What are the practical implications of the reclassification of *Pneumocystis jirovecii*? Be specific in your answer.

4. Insecticides and drugs are part of antimalarial efforts. For each of the substances listed below, state whether it is directed against the mosquito or the parasite, and describe any current challenges to its use.

a. artemisinin

b. DDT

c. quinine

5. Why is climate change an important factor to consider in the context of malaria?

6. Looking at Infographic M5.2, which depicts the malaria parasite's life cycle:

a. Which step of the life cycle is prevented by DDT or other insecticides?

b. If you wanted to design a drug to block the life cycle of the malaria parasite and minimize symptoms in humans, which steps would you block? (Explain your answer.)

7. Plants, animals, and fungi are all in separate kingdoms. How are they grouped (relative to one another) in supergroups?

Skin Deep

Science redefines the meaning of racial categories

Courtesy of American Anthropological Association

DRIVING QUESTIONS

1 What contributes to human skin color, and why is there so much variation in skin color among different populations?

2 Where did the earliest modern humans evolve, and how do we know?

3 What can genetics and the fossil record tell us about human evolution?

When Barack Obama was elected for his first term in 2008, he was hailed as America's first Black president. When Tiger Woods won the Masters Golf Tournament in 1997, he was lauded as the first Black man to win. When Halle Berry won an Oscar in 2002 for best actress, she was commended as the first Black woman to win in that category.

Why was the skin color of these individuals so remarkable? A 250-year history of slavery in the United States has left a bitter legacy. More than 150 years after slavery was legally abolished in this country, Black Americans still face systemic racism and discrimination. They are underrepresented in positions of power and prestige, so the recognition of the achievements of Obama, Woods, and Berry signaled a major change: barriers to social advancement were beginning to come down.

Yet, to shoehorn any of these three people into a simple racial category is misleading: Barack Obama was born to a White mother and a Black African father; Tiger Woods's background includes African, Chinese, Dutch, and Thai forebears; Halle Berry was born to a White mother and an African American father. So what does the term "Black" really mean?

Historically, racial categories like "White" or "Black" were employed by one group to maintain power over another and to justify forms of oppression, including slavery. In the United States, racial categories were reinforced by laws like the "one drop" rule adopted by several states in the 1910s and 1920s, which held that any American with even "one drop" of African blood was to be considered Black. People then continued to use these categories and their connotations to justify racial discrimination and, in some places, racial segregation.

Though social and political attitudes have changed, people continue to invoke racial categories like "Black" or "White" for various reasons, including simple physical description. From a biological perspective, however, it is increasingly clear that racial categories have little meaning. Biologically distinct human races do not exist. Groups of people can and do share similar physical characteristics, such as skin color and facial features, but these superficial differences obscure how fundamentally similar all human beings are.

INFOGRAPHIC 19.1

Humans Are Genetically Similar

Humans are 99.9% similar at the level of our DNA sequence. This similarity between humans includes all of the traits that define us as human. We are overwhelmingly more similar to one another than we are different and cannot be categorized into racial groups based on genetic analysis.

? Given variation between humans, how many nucleotide differences would you predict to find in a 1,000-nucleotide sequence in two unrelated people?

In fact, genetic studies have found that the DNA of any two people differs, on average, by just one DNA base pair out of every thousand. This means that humans are 99.9% identical, genetically speaking. Since the human genome contains roughly 3 billion DNA base pairs, that amounts to 3 million average DNA base pair differences per person. It is this tiny fraction—0.1% of our DNA—that accounts for the diversity of traits we see from person to person **(INFOGRAPHIC 19.1)**.

Moreover, the differences that do exist among human populations have all come about very recently—in the blink of an eye, evolutionarily speaking. Members of the human species—*Homo sapiens*—only began to migrate and disperse around the globe around 70,000 years ago. In that time, humans have evolved multiple traits that helped them survive the different environments they encountered along the way.

People with West African ancestry, for example, carry with high frequency an allele that helps them resist malaria, a disease that is common in West Africa. An allele common among Northern Europeans enables them to digest milk better than other populations do, an indication that at some point in history, dairy products provided an important source of nutrition to this population. Tibetans have a high frequency of an allele that helps their red blood cells compensate for the low oxygen level in their high-altitude environment.

Skin color is another example of a trait that differs among human groups. For many years, scientists had wondered why different skin colors evolved. But it wasn't until the work of Nina Jablonski, an anthropologist at Pennsylvania State University, that a convincing answer to this question emerged. Her research adds to scientific research that shows why racial categories are social rather than biological divisions.

The Biology of Skin Color

▶ Melanin, folate, and vitamin D

More than a decade ago, Jablonski and her husband, George Chaplin, a geographer, set out to understand why human populations evolved varying skin tones. They knew that skin tone largely reflects the amount of **melanin,** a pigment present in the skin. People naturally produce different levels of melanin. More melanin yields darker skin, while less melanin yields lighter skin. Skin also responds to sunlight by producing more melanin and temporarily becoming darker **(INFOGRAPHIC 19.2).**

Jablonski and Chaplin also knew that, in general, skin tone correlates with geography: people from regions closer to Earth's poles tend to be lighter-skinned, whereas those from areas closer to the equator tend to have

MELANIN
A pigment produced by a specific type of skin cell that gives skin its color.

INFOGRAPHIC 19.2

Melanin Influences Skin Color

Melanocytes are a type of cell located in the epidermis, the outermost layer of skin. Melanocytes make the pigment melanin and deposit it into other cells in the skin. A person's skin color depends largely on the amount and type of melanin that his or her skin melanocytes produce. Sunlight can also temporarily increase the amount of melanin in a person's skin.

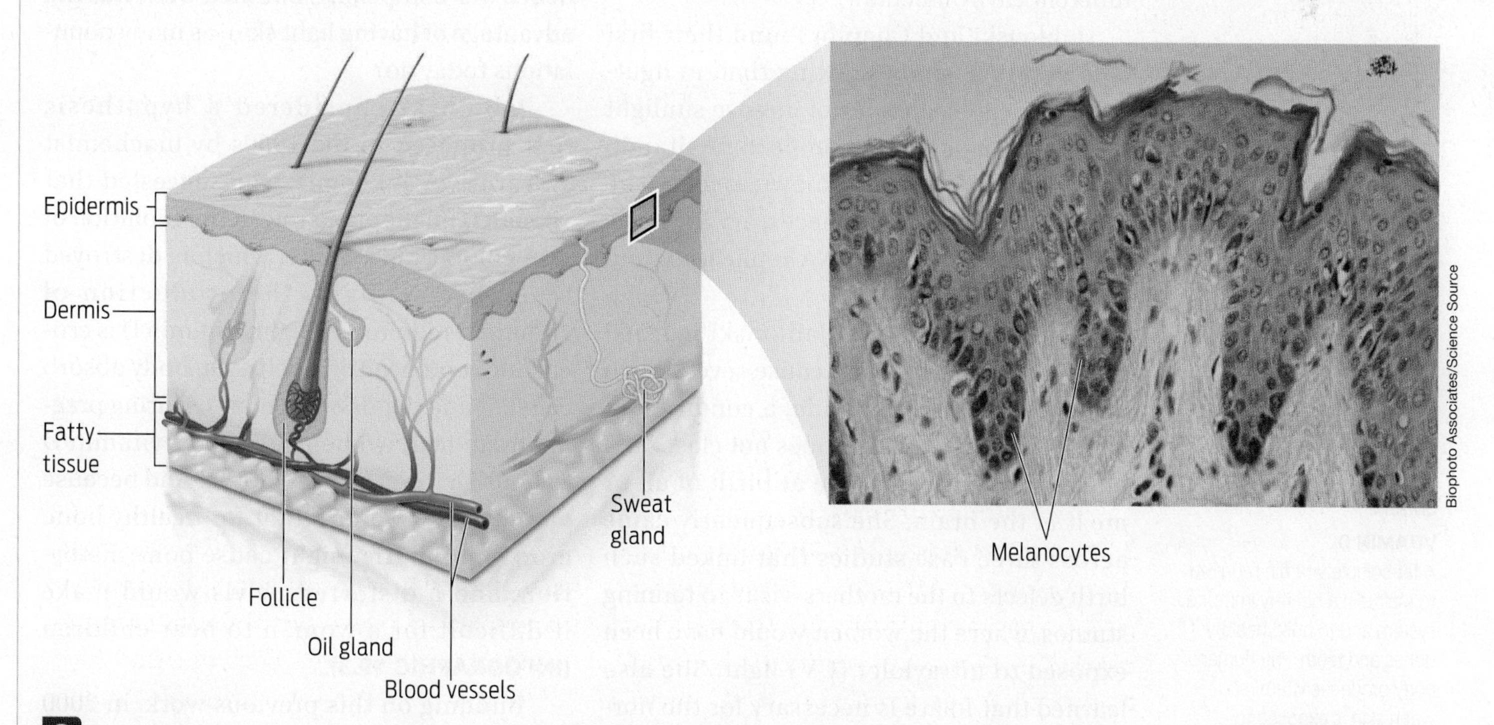

? **For sunlight to increase melanin production in skin, what cell type must be active following exposure to sunlight?**

George Chaplin and Nina Jablonski examine a map of predicted human skin color based on UV light intensity.

Mark Wilson/Getty Images

Nina Jablonski

In certain environments, is there an evolutionary advantage to light or to dark skin?

darker skin. Jablonski wanted to understand this relationship, so she searched the scientific literature. Might there be an evolutionary advantage to having light or dark skin in different environments?

Jablonski and Chaplin found their first clue in a 1978 study showing that, in light-skinned people, an hour of intense sunlight can halve the level of an important vitamin called **folate.** Folate, also known as folic acid, is an essential nutrient, necessary for basic cellular processes like DNA replication and cell division.

Then, at a seminar, Jablonski learned that low folate levels can cause severe birth defects such as spina bifida, a condition in which the spinal column does not close, and anencephaly, the absence at birth of all or most of the brain. She subsequently came across three case studies that linked such birth defects to the mothers' visits to tanning studios, where the women would have been exposed to ultraviolet (UV) light. She also learned that folate is necessary for the normal development of sperm.

Taken together, these observations suggested to Jablonski that people with light skin are more vulnerable to folate destruction than are darker-skinned people—presumably because melanin absorbs damaging UV light and dissipates it as heat. Could the need to protect the body's folate stores from UV light have favored the evolution of darker skin shades? The supporting evidence was compelling. But then what was the advantage of having light skin, as many populations today do?

Jablonski considered a hypothesis first proposed in the 1960s by biochemist W. Farnsworth Loomis, who suggested that **vitamin D** might play a role in the evolution of skin color. Unlike folate, which is destroyed by excess sunlight, the production of vitamin D requires UV light. Vitamin D is crucial for good health: it helps the body absorb calcium and deposit it in bones. During pregnancy, in fact, women need extra vitamin D to nourish the growing embryo. And because vitamin D is so important for healthy bone growth, too little might cause bone distortion, and a distorted pelvis would make it difficult for a woman to bear children **(INFOGRAPHIC 19.3).**

Building on this previous work, in 2000 Jablonski and Chaplin compared data on skin

FOLATE
A B vitamin also known as folic acid; folate is an essential nutrient, necessary for basic cellular processes such as DNA replication and cell division.

VITAMIN D
A fat-soluble vitamin required to maintain a healthy immune system and to build healthy bones and teeth. The human body produces vitamin D when skin is exposed to UV light.

color in indigenous populations from more than 50 countries to levels of global UV light as measured by NASA satellites. They found a clear correlation: the weaker the UV light, the fairer the skin—a compelling suggestion that both dark and light skin are linked to levels of global sunlight. The pair subsequently published their results in the *Journal of Human Evolution.*

The researchers now had a complete hypothesis: light skin provided an evolutionary advantage in less sunny parts of the world because it helped the body produce vitamin D, while dark skin was favored in sunnier regions because it helped protect the body's folate stores. The body's need to balance levels of these two important nutrients given varied levels of UV light explains why there

INFOGRAPHIC 19.3

Folate and Vitamin D Are Necessary for Reproductive Health

Folate, also known as folic acid, is especially critical during periods of rapid cell division, including embryonic and fetal development. Vitamin D is important for the absorption of calcium and phosphate from the small intestine, and also for bone mineralization. Folate is destroyed by UV, while vitamin D production is stimulated by UV. Skin color plays a role in folate protection and vitamin D production, with the balance between these depending on the amount of UV in the environment.

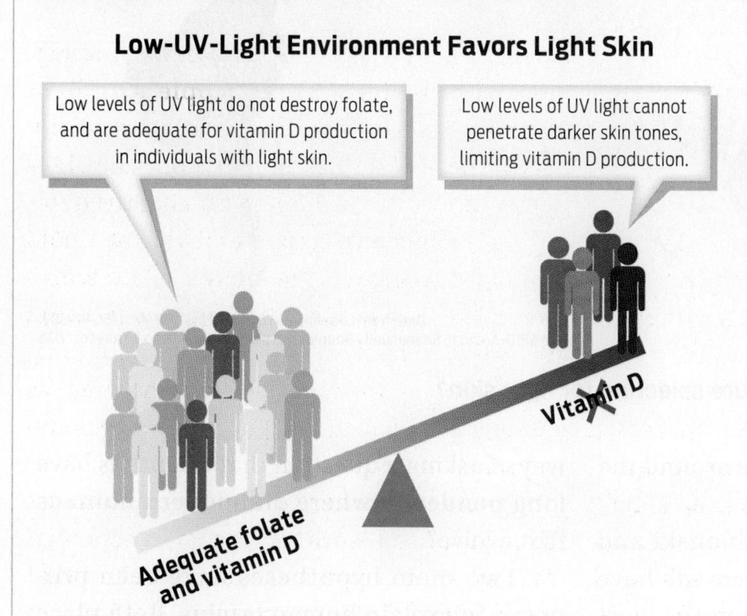

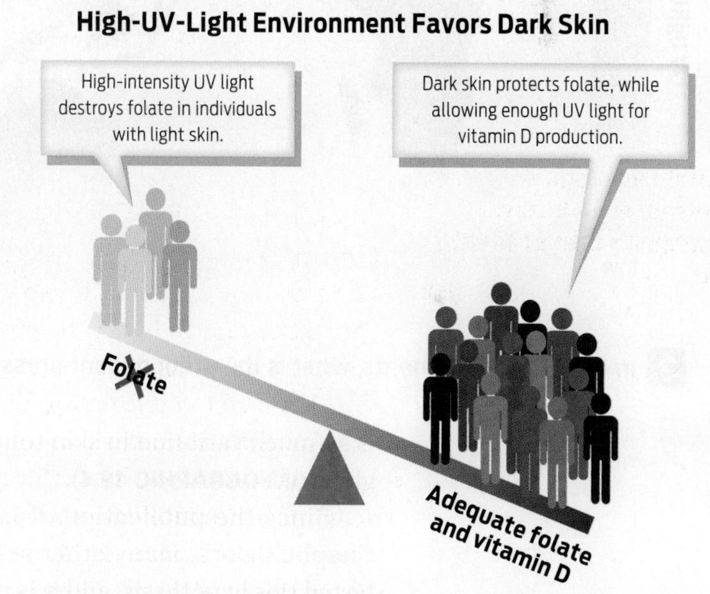

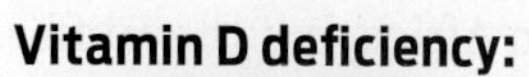

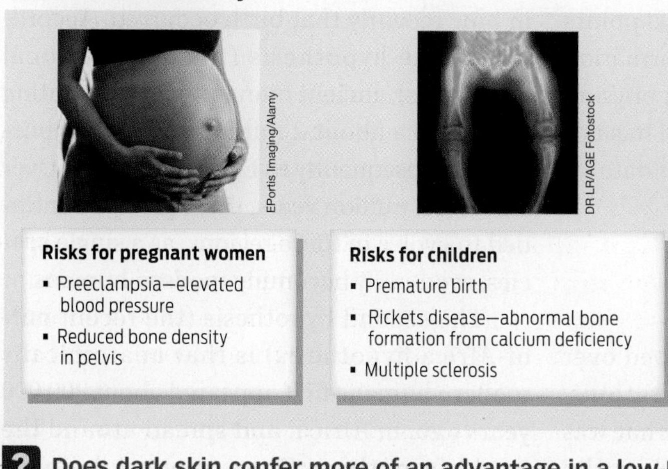

Folate deficiency:

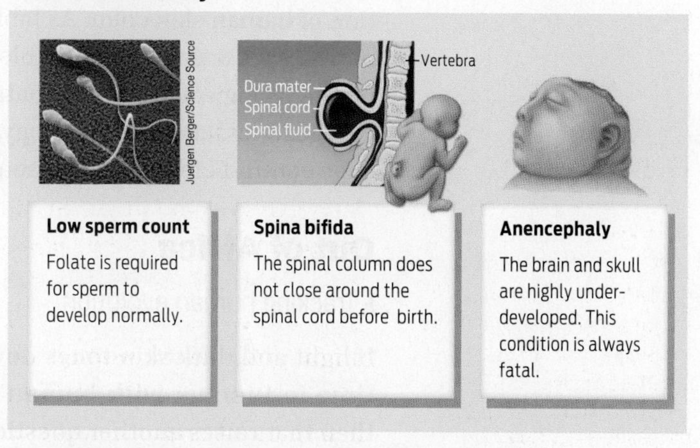

? **Does dark skin confer more of an advantage in a low-UV or a high-UV environment? Explain your answer.**

INFOGRAPHIC 19.4

Human Skin Color Correlates with UV Light Intensity

Nina Jablonski and George Chaplin used NASA satellite measurements of UV light intensity to predict the amount of skin pigment that would best block harmful UV rays yet still enable the body to produce sufficient vitamin D in populations around the globe. Their predictions closely match actual skin color variations around the world.

Predicted pigmentation of skin based on UV intensity

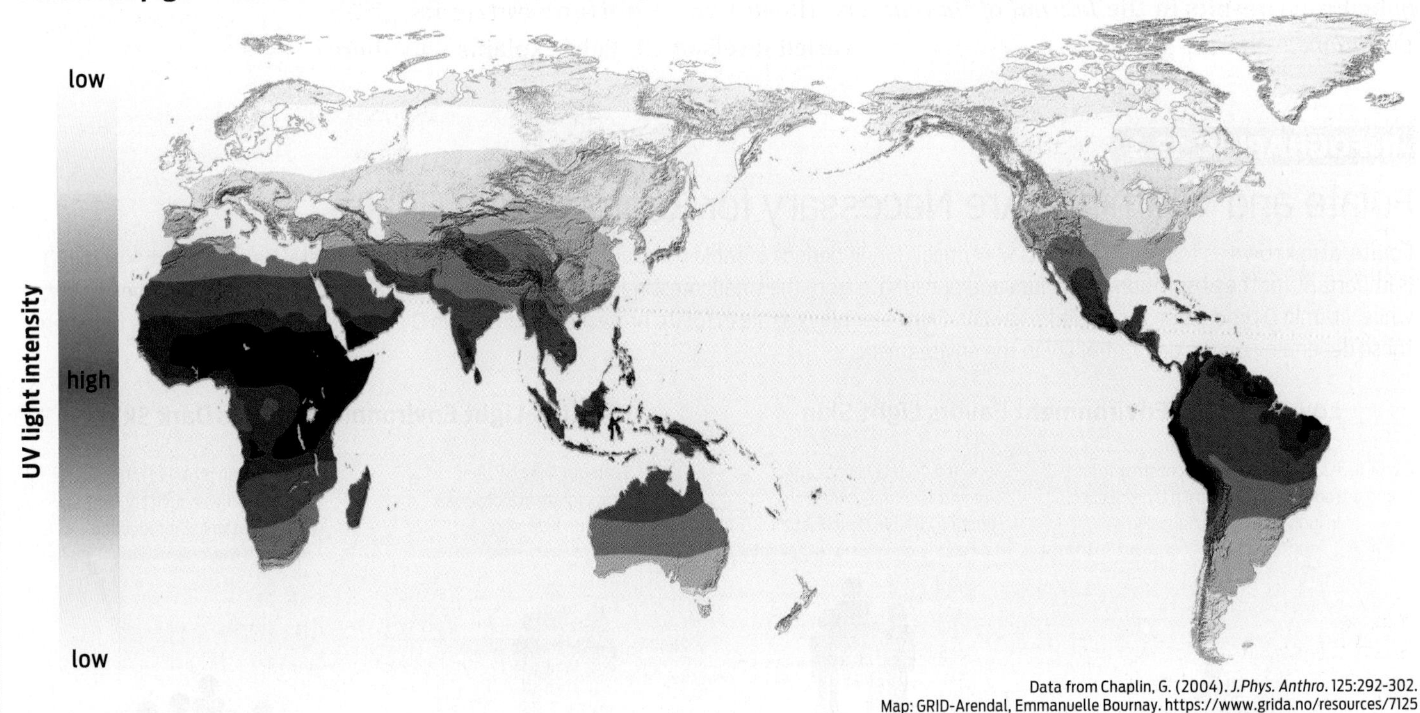

Data from Chaplin, G. (2004). *J.Phys. Anthro.* 125:292-302.
Map: GRID-Arendal, Emmanuelle Bournay. https://www.grida.no/resources/7125

? **In low-UV environments, what is the predominant pressure selecting for light skin?**

is so much variation in skin tone around the globe **(INFOGRAPHIC 19.4)**.

Since the publication of Jablonski and Chaplin's work, many other scientists have tested this hypothesis, and it is now the most widely accepted explanation for the evolution of human skin color. As Jablonski points out, "It synthesizes the available information on the biology of skin from anatomy, physiology, genetics, and epidemiology, and has not been contradicted by any subsequent data."

Out of Africa

▸ Tracking human evolution

If light and dark skin tones developed over time in tandem with human migrations, then that raises another question: what was the earliest human skin tone, and where did humans begin their journey? This is another way of asking a question that scientists have long pondered: where did modern humans first evolve?

Two main hypotheses have been proposed to explain human origins. Both place the human birthplace in Africa, but they differ in how recently that birth occurred. According to one hypothesis (the multiregional hypothesis), ancient humans began migrating from Africa about 2 million years ago, journeying subsequently to Europe and Asia. Over the next 2 million years, these groups continued to evolve in these regions, as a single species, eventually becoming modern humans.

The second hypothesis (the recent out-of-Africa hypothesis) is that anatomically modern humans first appeared about 200,000 years ago, in Africa, and spread around the world from there. These modern humans replaced earlier members of our genus, whose own species became extinct as a result. Deciding between these two hypotheses remained controversial until the 1980s, when genetic evidence started to make the picture clear.

In 1987, a team of geneticists led by Allan Wilson of the University of California at Berkeley used **mitochondrial DNA (mtDNA)**—genetic material we inherit solely from our mothers—to construct an evolutionary tree of humanity. Mitochondrial DNA is DNA located in the mitochondria of our cells. Unlike nuclear DNA, which is inherited from both parents and undergoes recombination during meiosis, mtDNA passes from mothers to offspring essentially unchanged. That's because sperm do not contribute their mitochondria to the newly formed zygote **(INFOGRAPHIC 19.5).**

MITOCHONDRIAL DNA (mtDNA)
The DNA within mitochondria; it is inherited solely from the mother.

INFOGRAPHIC 19.5

Mitochondrial DNA Is Inherited from Mothers

When egg and sperm fuse during fertilization, sperm contribute only nuclear DNA to the newly formed zygote. The egg provides all other organelles, including mitochondria. Consequently, only mothers contribute mitochondrial DNA (mtDNA) to their children.

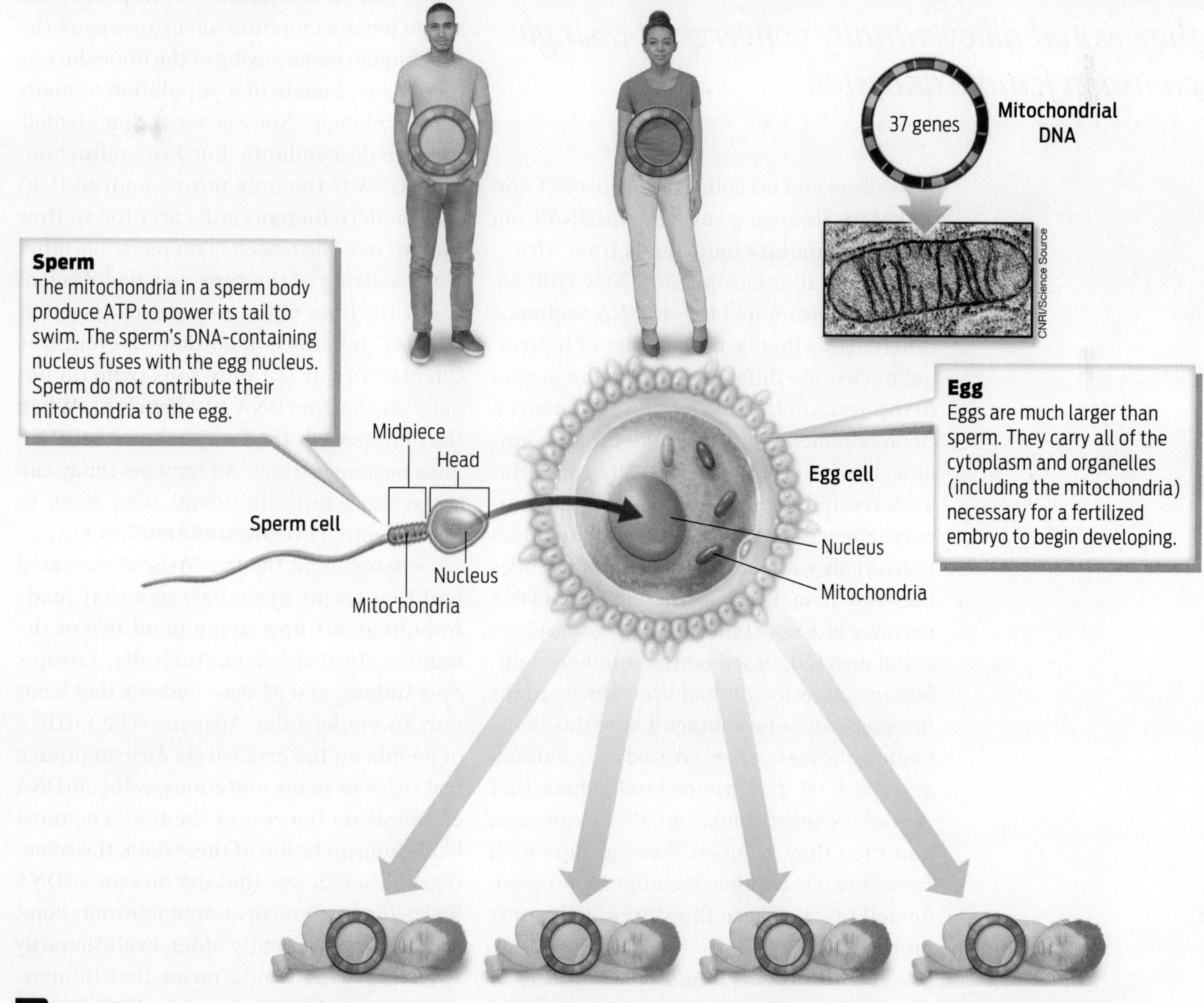

? If biologists used mtDNA to trace the evolutionary history of a group of people, would they be following mothers, fathers, or both?

Like nuclear DNA, mtDNA mutates at a fairly regular rate, but much faster. A mother with a mutation in her mtDNA will pass it to all her children, and her daughters will pass it to their children in turn. Because these mutations pass down without being combined and rearranged with paternal mitochondrial DNA, mtDNA is a powerful tool for tracking human ancestry back through thousands of generations.

If every person on the planet were to construct a family tree that listed every female ancestor for thousands of generations back in time, they would all eventually converge at a single common female ancestor.

Wilson and his colleagues Rebecca Cann and Mark Stoneking collected mtDNA from 147 contemporary individuals from Africa, Asia, Australia, Europe, and New Guinea. They then examined the mtDNA sequence differences—that is, the number of individual nucleotide differences from one person to the next. On the basis of these differences, the researchers were able to determine how closely or distantly related the individuals were: individuals sharing nearly identical sequences are more closely related than individuals with many sequence differences between them. Because mitochondrial DNA mutates at a constant rate, the researchers could also tell, based on the number of differences, how long it had been since groups diverged from one another. Using this information, the researchers created an evolutionary tree. First they grouped individuals that shared the most similar mtDNA sequences, and then they grouped these groups with respect to one another. A computer program helped them choose the statistically most probable tree.

The researchers found that branches of the tree from all five geographic areas could be traced back to a single female ancestor who lived in eastern Africa some 200,000 to 150,000 years ago. In other words, if every person on the planet were to construct a family tree that listed every female ancestor for thousands of generations back in time, they would all eventually converge at a single common female ancestor. As a newspaper reporter put it, the researchers had found "Mitochondrial Eve."

This was clear evidence in support of a recent African origin for our species, and it made a big splash in popular culture. (*Newsweek* featured an African Adam and Eve on its cover.) But the finding also led to confusion. While catchy, the nickname "Eve" is misleading in a number of respects. This single female common ancestor wasn't the only human female living at the time; she was merely one female in a population of many ancient humans. Nor was she the only female to leave descendants. But Eve's mitochondrial DNA is the only mitochondrial DNA that modern humans still carry today. How can this be? The reason is simple: while other females living at the time also had descendants, the lines of these descendants either died off, perhaps in a bottleneck event (see Chapter 14), or left only sons (who do not pass on their mtDNA to offspring). When that happened, their mitochondrial DNA lines became extinct. All humans today can trace their mitochondrial DNA back to Mitochondrial Eve **(INFOGRAPHIC 19.6)**.

Interestingly, the tree Wilson generated had two major branches: one that leads to individuals now living in all five of the regions studied—Asia, Australia, Europe, New Guinea, and Africa—and one that leads only to modern-day Africans. The mtDNA of people on the exclusively African branch had twice as many mutations as the mtDNA of people on the rest of the tree. The most likely interpretation of these data, the scientists reasoned, was that the African mtDNA had had more time to accumulate mutations, and was consequently older, evolutionarily speaking. This would mean that humans originated in Africa, where they likely formed several ancestral populations. After some

INFOGRAPHIC 19.6

Modern Human Populations Share a Common Female Ancestor

Many women of Eve's generation left descendants, but only Eve's mitochondrial DNA survives among humans today.

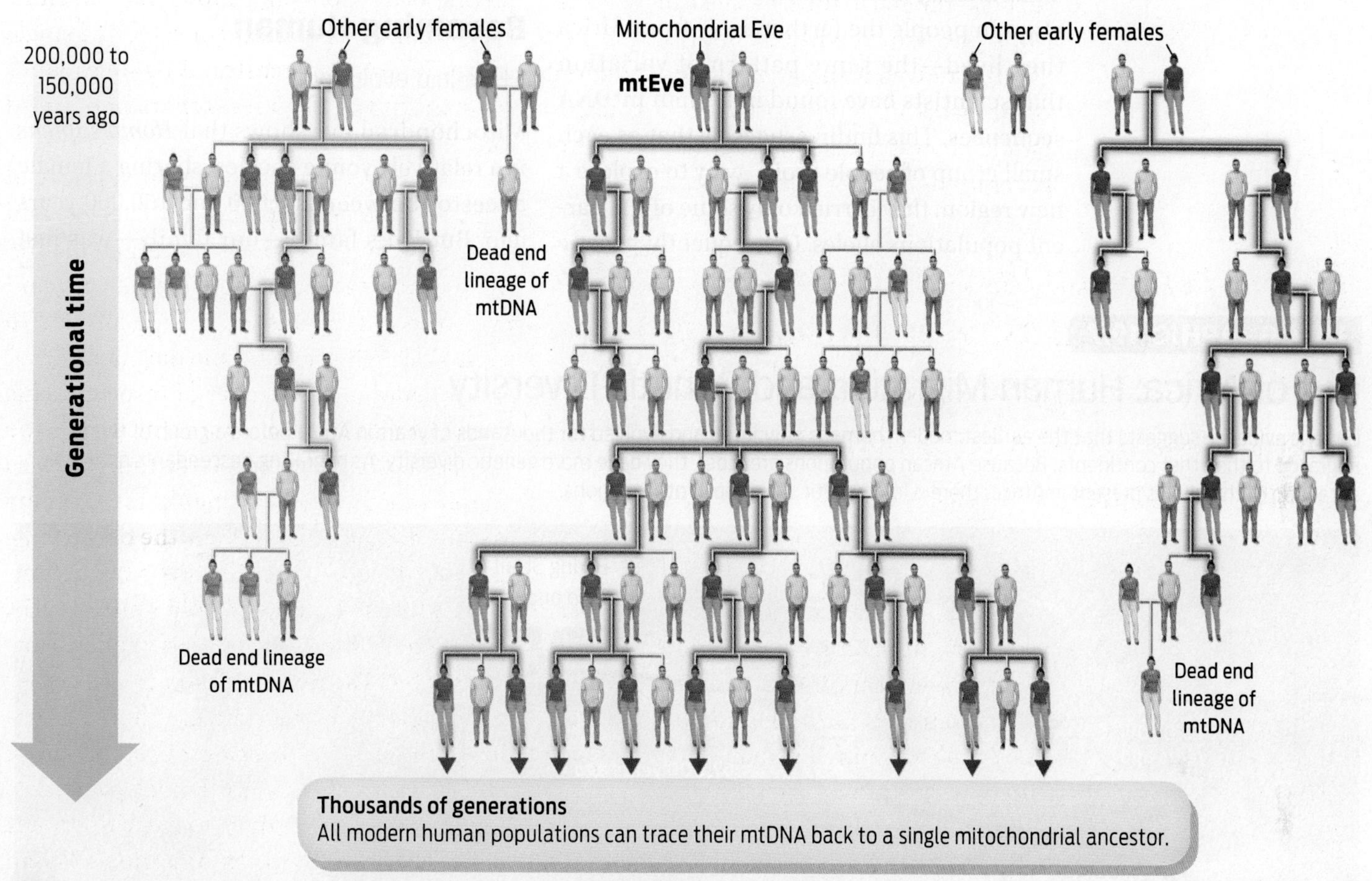

? If a male with the yellow mtDNA as shown here married a female with the purple mtDNA shown, which mtDNA would be passed on to their offspring?

period of time, one group of Africans left the continent, and their descendants continued to migrate to other continents. Eventually these migrants became the ancestors of modern-day Asians, Australians, and Europeans. Evidence suggests this migration began around 70,000 years ago.

Since Wilson's study, additional evidence has emerged that backs the recent out-of-Africa hypothesis. Fossils discovered in Ethiopia in 2003 and 2005 represent some of the oldest known fossils of modern humans, and they are 160,000 and 195,000 years old, respectively. Both sets of remains date from the time Wilson and his colleagues think that Mitochondrial Eve lived in eastern Africa. The fossil discoveries provide evidence that anatomically modern humans were living in that region around the same time that Mitochondrial Eve lived, and provide further evidence that modern humans originated in Africa.

In 2017, researchers discovered even older fossils of our species, this time in northern Africa. The fossils are approximately 300,000 years old, indicating that our ancestors were living and dispersing around Africa for thousands of years before Eve's descendants left the continent.

The recent out-of-Africa hypothesis is also supported by research that sampled genetic diversity from nuclear DNA. In a 2008 study,

Richard Myers, of the Stanford University School of Medicine, and his colleagues found that humans populations living in Africa have the most genetic variation of any human population. They found less and less genetic variation in people the farther away from Africa they lived—the same pattern of variation that scientists have found in human mtDNA sequences. This finding suggests that as each small group of people broke away to explore a new region, they carried only some of the parent population's alleles. Consequently, genetic diversity decreased in tandem with the distance people traveled away from Africa—a classic example of the founder effect described in Chapter 14 **(INFOGRAPHIC 19.7)**.

Becoming Human

▶ Hominid evolution

Mitochondrial Eve shows that *Homo sapiens* is a relatively young species, sharing a female ancestor between 200,000 and 150,000 years ago. But Eve's family—our family—was just

INFOGRAPHIC 19.7

Out of Africa: Human Migration and Genetic Diversity

Genetic evidence suggests that the earliest modern humans originated and evolved for thousands of years in Africa before a group of them migrated to the other continents. Because African populations are older, they have more genetic diversity. As migrating descendants only took a fraction of the alleles present in Africa, there is less genetic variation in other regions.

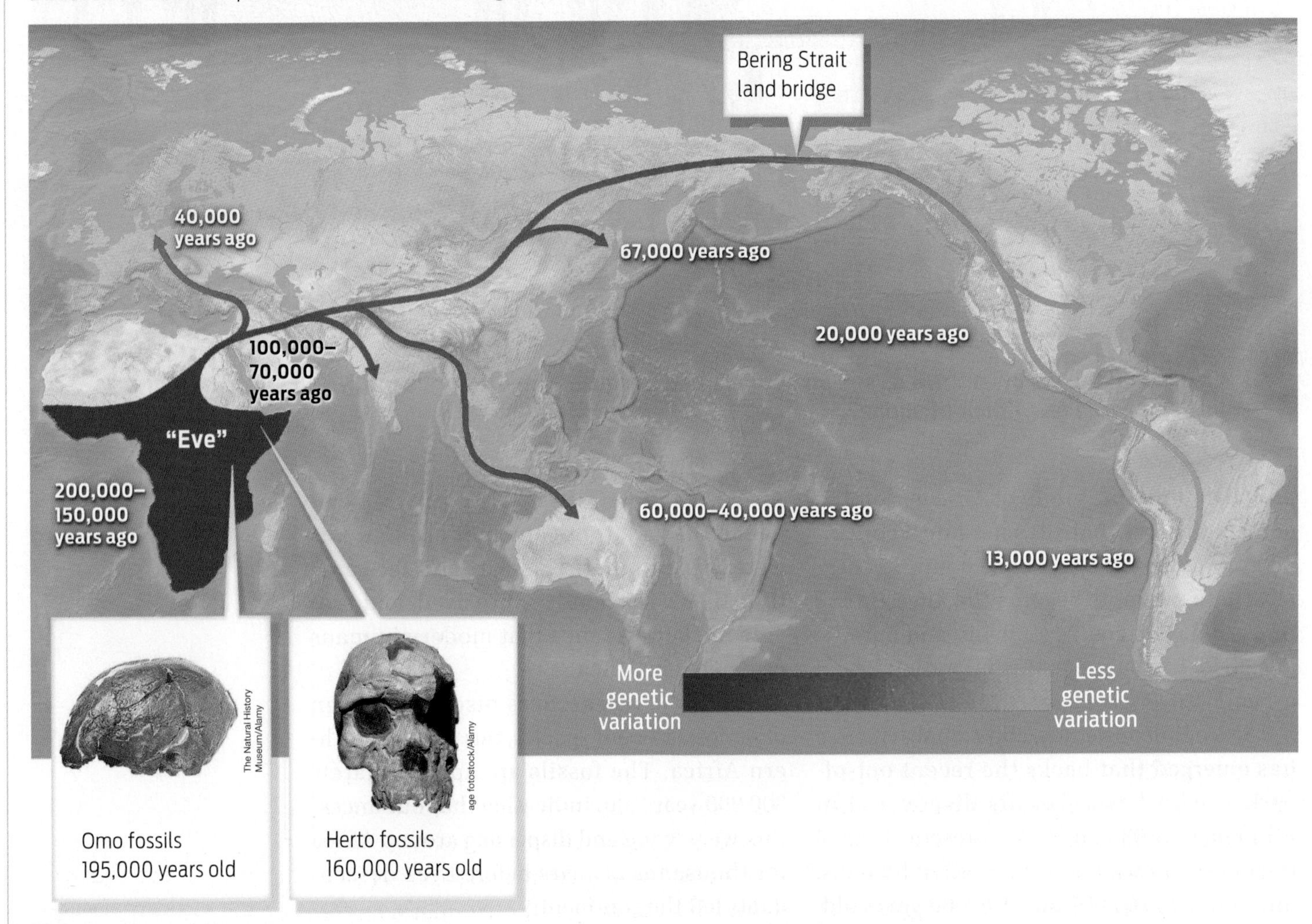

? If humans had originated and evolved in Australia and then, after thousands of years, migrated to Africa, what would you expect to be the level of genetic diversity in Australian and African populations?

one of many other human families that existed at the time, in Africa and in other places around the world. The term "human" refers to our genus name—*Homo.* Many other members of the genus *Homo*—many other ancient humans—once roamed the globe, among them *Homo habilis, Homo erectus,* and *Homo neanderthalensis.* Some even overlapped with us in time and space. Yet our single species—*H. sapiens*—is the only one that did not eventually die out.

Moreover, these other *Homo* species are just our immediate evolutionary relatives. Our extended family tree includes living and extinct primates belonging to the biological family Hominidae. These **hominids** include humans, orangutans, gorillas, chimpanzees, and bonobos.

Fossil evidence shows that humans and other hominids evolved from a common ancestor that lived 13 million years ago. Of the living members of this group, humans and chimpanzees are the most closely related, although their last shared ancestor lived about 7 million years ago. In the 7 million years since their lines separated, both humans and chimps have undergone a tremendous amount of evolutionary change, which is why living humans look and behave so differently from chimps—and from any other primate species living today.

Among the defining characteristics of *H. sapiens* are the ability to walk upright and the possession of a big brain. An upright gait leaves the hands free to make and use tools, and the big brain enabled *H. sapiens* to develop complex language. Ancient bones, tools, and other finds allow scientists to trace when these distinctly human traits emerged. Artifacts found at various archeological sites indicate that simple tool use began approximately 2.6 million years ago, most likely when our hominid ancestors began eating meat from large animals. The first hominid toolusers were members of the genus *Australopithecus.* This genus walked upright and appears to have lived on the ground, rather than in trees, as evidenced by the lack of an opposable big toe, which had helped the early hominids grip branches.

John Reader/Science Source

Fossil skulls of *Australopithecus africanus* (top), *Homo habilis* (center), and *Australopithecus robustus boisei* (bottom). All three hominid species were extracted from the same excavation site in East Turkana, Kenya, indicating that these species all lived at the same time, around 1.5 million years ago.

Another milestone was the ability to use and control fire, which appeared about 800,000 years ago. Evidence such as ashes of plant matter and fragments of burned bone found inside South African caves show that *Homo erectus,* a member of our human genus, was likely the first species able to control fire. Using fire enabled *H. erectus* to cook meat, to stay warm, and probably to fight off predators.

At some point between 800,000 and 200,000 years ago, hominid brain size began to expand rapidly. Studies of fossilized organisms from ocean sediments show that this was also a time of climate instability. The temperature dramatically shifted from high to low and back several times during this period. Scientists hypothesize that a larger brain would have enabled better communication and problem solving, which would

HOMINIDS
Any living or extinct member of the family Hominidae, the great apes—humans, orangutans, gorillas, chimpanzees, and bonobos.

have been very useful to our hominid ancestors as they coped with a changing climate. This was also around the time that anatomically modern humans, *H. sapiens,* appeared **(INFOGRAPHIC 19.8)**.

Evolutionary Trade-offs

▶ Natural selection in human evolution

The fact that anatomically modern humans originated in Africa suggests that the first humans likely had dark skin. Dark skin probably evolved early in our history, in tandem with a reduction in body hair. Before dark skin evolved, our ancestors would have had light skin, just as chimpanzees do today.

Fossil and genetic evidence suggests that about 2 million years ago hominids became "bipedal striders, long-distance walkers, and possibly even runners," according to Jablonski. To sustain such activities, hominids needed an effective cooling system—a feature they could have developed only by losing excessive body hair and gaining more sweat glands. In contrast, hairy chimpanzees, our closest living animal relatives, can sustain only short bouts of activity without getting overheated. "It's like sweating in a wool blanket," Jablonski

INFOGRAPHIC 19.8

Humans: An Extended Family Tree

Homo sapiens is the only surviving lineage in the evolutionary history of humans. In other words, several hominids have existed or coexisted as related but distinct species in the past. The physical traits of modern-day humans, such as upright walking and big brains, reflect adaptations to evolutionary pressures.

Present
1
2
3
4
5
6
7 million years ago
Walking on two feet
Divergence of humans from chimps
Making fire
Using stone tools
Sahelanthropus tchadensis
Ardipithecus ramidus
Australopithecus anamensis
Australopithecus afarensis
Kenyanthropus platyops
Australopithecus africanus
Paranthropus boisei
Paranthropus robustus
Australopithecus garhi
Homo habilis
Homo rudolfensis
Homo erectus
Homo heidelbergensis
Homo naledi
Homo neanderthalensis
Homo sapiens

? **Which human species were living at the same time as *Homo habilis*?**

INFOGRAPHIC 19.9

Natural Selection Influences Human Evolution

The environment selects for specific genetically determined traits. Different environments select for different traits, and therefore different alleles.

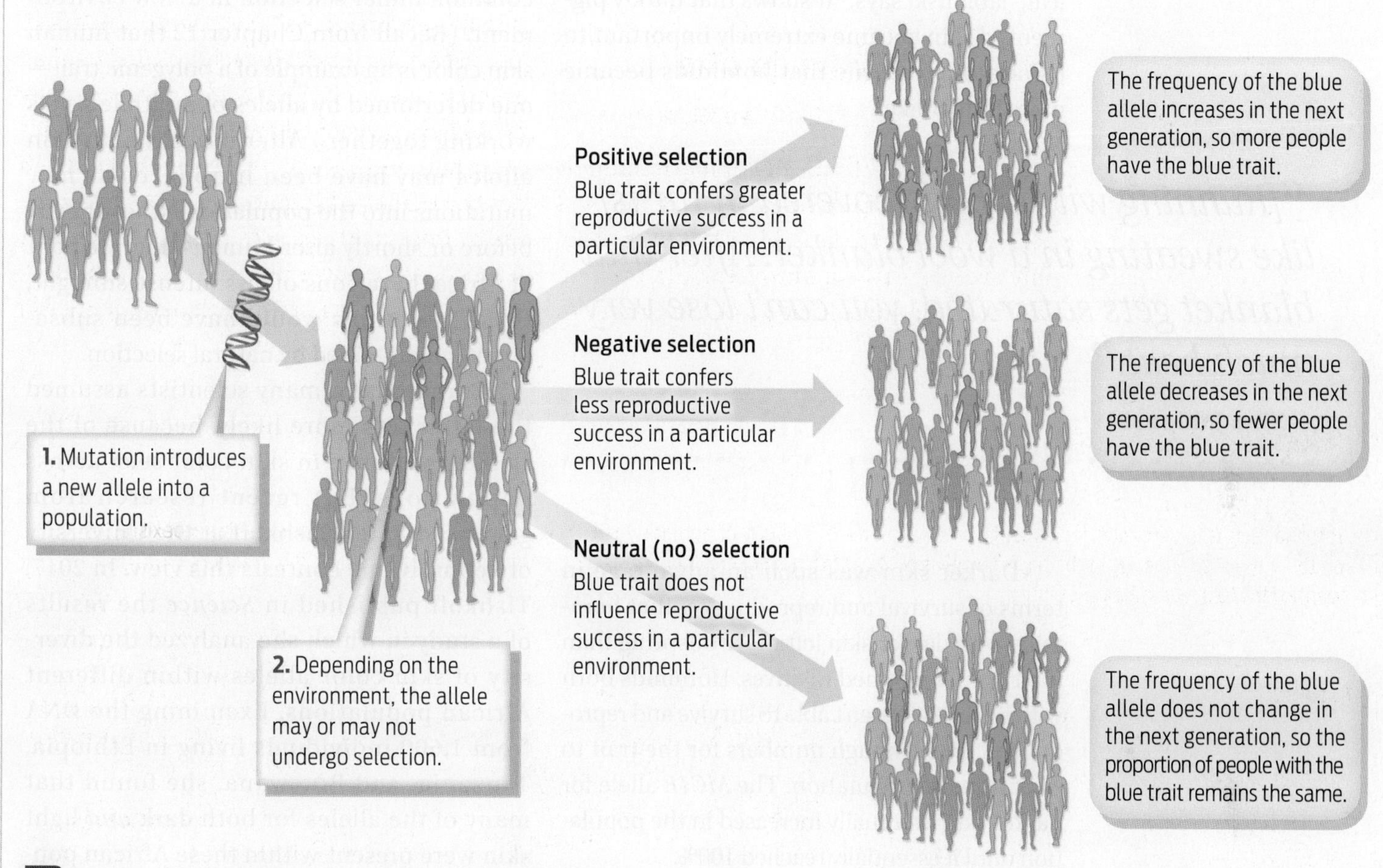

? **What kind of selection is predicted for alleles that contribute to dark skin in a low-UV environment?**

explains. "After that blanket gets saturated, you can't lose very much heat."

Eventually, some factor—food scarcity, perhaps—forced ancient hominids out of the forests and into the open savannahs to hunt for food. Hominids with less hair and more sweat glands were likely better hunters because they could sustain long bouts of activity without getting overheated. Like modern-day chimpanzees, these hominids likely had fair skin under their hair. Without hair to protect their light skin, they were exposed to the intense African sun. And, according to Jablonski's hypothesis, exposure to the sun would have reduced their folate levels and thus their fitness in the sun-drenched environment.

Any of these ancient hominids that carried alleles that increased their ability to produce more melanin would have been able to spend more time in the sun without detrimental effects. Darker coloration would have protected their skin, and consequently their folate levels, from the sun, enabling them to hunt and travel in the open fields. As a consequence, alleles for dark skin would have increased in frequency in the population **(INFOGRAPHIC 19.9)**.

Genetic research has provided evidence that supports this hypothesis about the evolution of dark skin color. In 2004, Alan Rogers, of the University of Utah, and his colleagues studied a gene called *MC1R* that influences skin color. They discovered that more than a million years ago, one allele of *MC1R* that contributes to dark skin became fixed—that is, its frequency approached 100%—in the African population of hominids. "This is critical," Jablonski says. "It shows that darkly pigmented skin became extremely important to us" around the time that hominids became more humanlike.

"[Running with a body covered in fur is] like sweating in a wool blanket. After that blanket gets saturated, you can't lose very much heat."

—Nina Jablonski

Darker skin was such an advantage in terms of survival and reproduction that hominids with darker skin left more offspring than their lighter-skinned relatives. Hominids born with light skin weren't able to survive and reproduce in great enough numbers for the trait to persist in the population. The *MC1R* allele for darker skin eventually increased in the population until it essentially reached 100%.

But populations that migrated north, away from the African sun, faced different environmental circumstances. Folate was not as easily destroyed in this lower-UV-light environment. In fact, the high levels of melanin present in dark skin became a disadvantage in this setting: they prevented the body from producing enough vitamin D. In a low-UV-light environment, fair skin allowed the body to soak up more UV light and produce essential vitamin D. In this environment, fair-skinned people thus were more fit and left more descendants than did dark-skinned people. Consequently, the frequency of alleles for light skin in northern climates increased with each generation.

If the earliest humans migrating from Africa to northern regions all had dark skin, where, then, did alleles for light skin come from? There are two possibilities. One is that the alleles may have still been present in the African population and then became more common under selection in a new environment. (Recall from Chapter 12 that human skin color is an example of a polygenic trait—one determined by alleles of multiple genes working together.) Alternatively, light skin alleles may have been introduced as new mutations into the population either shortly before or shortly after humans migrated out of Africa. In regions of less intense sunlight, these mutations would have been subsequently maintained by natural selection.

Traditionally, many scientists assumed the latter was more likely, because of the striking variance in skin color seen across populations. But recent research from geneticist Sarah Tishkoff at the University of Pennsylvania contests this view. In 2017, Tishkoff published in *Science* the results of a study in which she analyzed the diversity of skin color alleles within different African populations. Examining the DNA from 1,600 individuals living in Ethiopia, Tanzania, and Botswana, she found that many of the alleles for both dark *and* light skin were present within these African populations. Her results show that alleles for light skin are an ancient—rather than a recent—evolutionary development.

Consider, for example, an allele of a gene called *SLC24A5* that confers light skin. This allele is commonly found in Western European populations and has therefore been considered "European." But Tishkoff's team showed that the allele is common in people from Ethiopia and Tanzania as well.

"When people think of skin color in Africa, most would think of darker skin," Tishkoff says, "But we show that within Africa there is a huge amount of variation, ranging from skin as light as some Asians to the darkest skin on a global level and everything in between." Nilotic peoples in eastern Africa have very dark skin, for example, while the

San of southern Africa have light skin. Even among those with dark skin, there is a lot of hidden genetic variation.

The mutations that created different skin color alleles were genetically minor, yet phenotypically significant: in most cases, they consist of just single nucleotide changes in a handful of genes. That's a handful of nucleotides among the 3 billion total in our DNA. Skin color really is only skin deep **(INFOGRAPHIC 19.10).**

The Illusion of Race

▶ Patterns of human genetic variation

A thousand years or so ago (before widespread immigration), if you were to walk from equatorial Africa to the tip of Finland, stopping to visit with locals along the way, you'd never hit a point where all the people to your south had dark skin and all the people to your north had light skin. All you would find are gradations of skin tone blending imperceptibly into one

INFOGRAPHIC 19.10

The Evolution of Skin Color

Human skin color is an example of a trait that has undergone natural selection. Varying levels of UV light have selected for a range of skin tones around the globe. In each case, the amount of melanin represents a compromise between the need to protect folate and the need to make vitamin D.

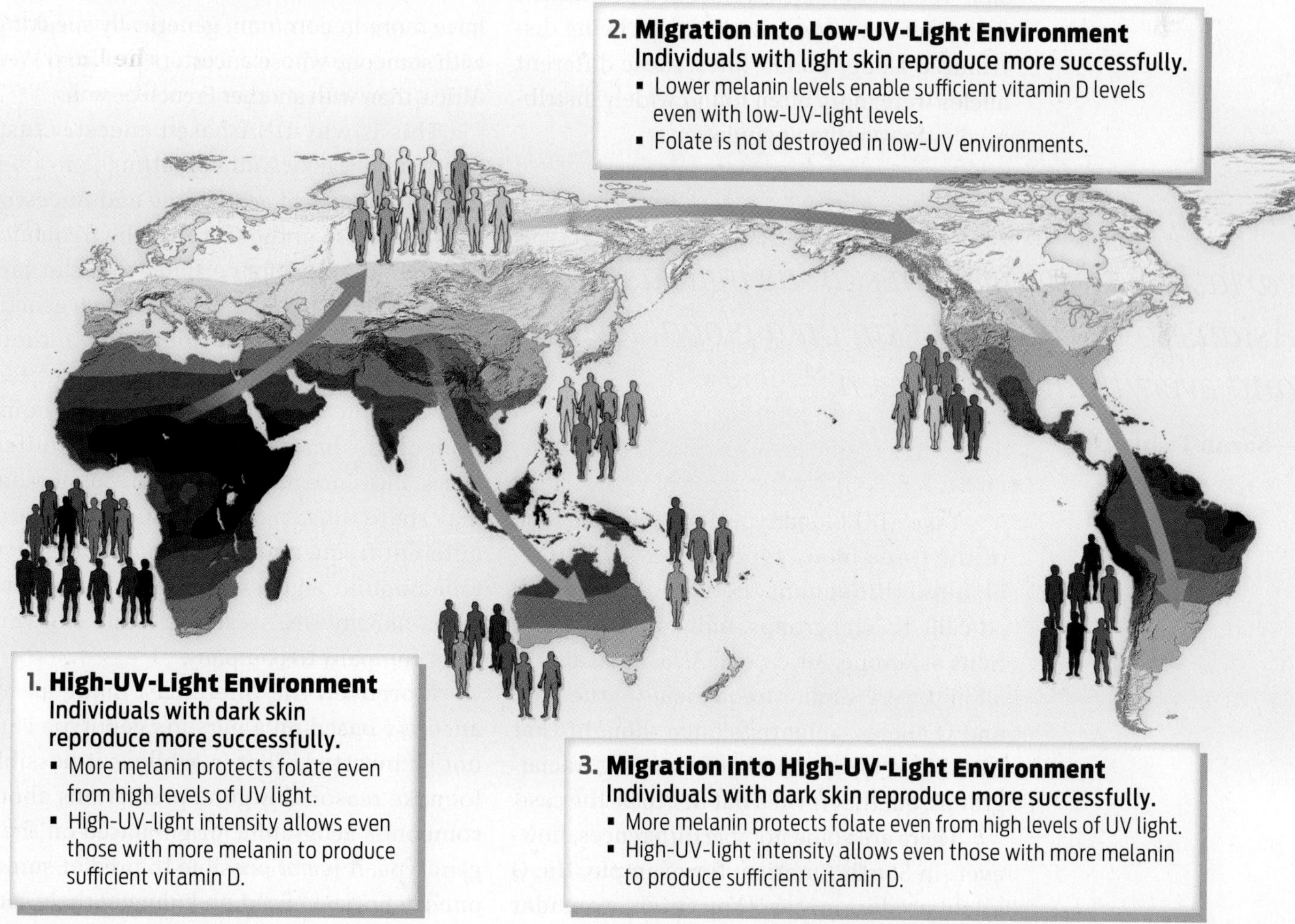

Data from Chaplin, G. (2004). *J.Phys. Anthro.* 125:292-302.
Map: GRID-Arendal, Emmanuelle Bournay. https://www.grida.no/resources/7125

? Why is light skin a disadvantage in high-UV environments?

another. In other words, people don't sort into neat boxes on the basis of skin color. That's one reason why racial categories based on this trait are misguided.

Another is the fact that skin color is actually somewhat unusual in being a trait that varies so widely between populations. Much of the genetic diversity that early humans brought with them when they migrated out of Africa has not been selected in the same way as skin color.

Scientists came to this understanding when they began collecting DNA samples from thousands of individuals living around the world. Among the 0.1% of DNA that differs from person to person, most of this DNA is polymorphic—meaning that multiple possible versions, or alleles, exist in a population for each DNA region. Rather than being distributed along "racial" lines, these different alleles were more often found widely distributed among human populations.

"Within Africa there is a huge amount of variation, ranging from skin as light as some Asians to the darkest skin on a global level and everything in between."

—Sarah Tishkoff

Take ABO blood type, for example. Each of the three blood type alleles—A, B, O—is found throughout the world, among all so-called racial groups. Indigenous populations of Europe, Africa, and Asia, for example, all have very similar frequencies of the A, B, and O alleles. Scientists once thought that ABO blood types would differ among "racial" groups, but that turned out not to be the case.

There are some notable differences, however: in South America, for example, the O allele predominates. (You might consider why this is the case, given what you know about genetic drift and the distance of South America from Africa.) But nothing about this pattern of geographic distribution resembles the pattern of skin color distribution. As a result, it's not possible to predict what blood type a person has, just by knowing their skin color. The same is true for the vast majority of genetic traits **(INFOGRAPHIC 19.11)**.

This observation shows that another aspect of the traditional conception of "race"—the idea that the color of a person's skin says a lot about them underneath—is fundamentally flawed. Genetic studies have found over and over again that there is much more genetic variation *within* so-called "racial" groups than there is *between* them. In fact, for any given trait (with the notable exception of skin color), roughly 90% of all variation in that trait is found within any one group. As a consequence, for any number of genetically determined traits, someone whose ancestors hail from, say, France might have more in common, genetically speaking, with someone whose ancestors hail from West Africa, than with another French person.

This is why DNA-based ancestry testing can be tricky, and sometimes misleading. Companies like 23andme and Ancestry.com make use of the fact that the frequency of different alleles for certain genes can vary among human populations. If enough genetic regions are considered (hundreds to thousands), geneticists can make sound inferences about where one's ancestors come from on the basis of these frequency differences. But since multiple human populations may share those particular alleles, albeit at different frequencies, there is always a certain amount of guesswork involved—which explains why DNA testing results can vary from company to company.

Moreover, the inferences made about ancestry based on a person's genotype cannot be inverted. That is, while is it possible to make reasonably good predictions about someone's geographic origin based on their genotype, it is *not* possible to predict someone's genotype based on knowing their geographic origin. To take an example that we have encountered elsewhere in this book, knowing that someone has the allele for sickle cell anemia (*HbS*) tells you that this

INFOGRAPHIC 19.11

Human Genetic Variation Does Not Sort into Racial Categories

The genetic differences that exist among people do not sort into groups defined by skin color. A population of people with a similar skin color will have a huge amount of genetic variation in other traits and alleles. This means that a person with a certain skin color may share the same alleles for a trait like blood type with someone with a very different skin color.

Grouped by Skin Color

While skin color has historically been used to sort people into racial groups, the underlying genetic diversity within these groups is greater than it is between them. Individuals who share a similar skin color may differ in many other genetic traits, like ABO blood type.

Skin color is a trait that shows a great deal of variation across the human population, meaning there are no sharp dividing lines between different skin-color groups.

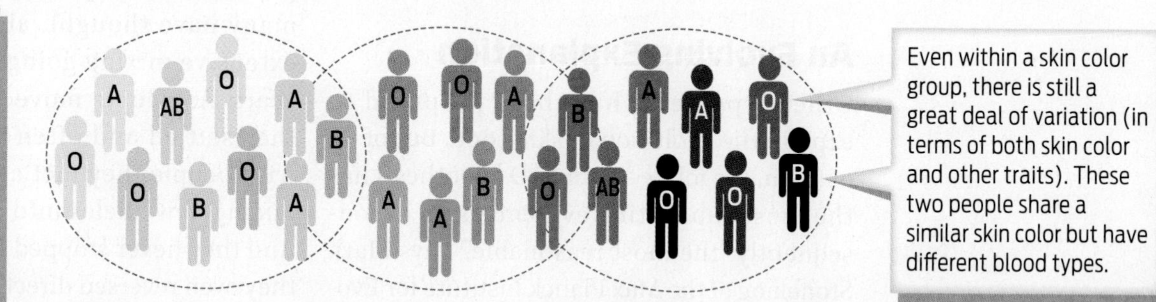

Even within a skin color group, there is still a great deal of variation (in terms of both skin color and other traits). These two people share a similar skin color but have different blood types.

Grouped by ABO Blood Type

Because most genetic variation in the human species is found everywhere, regardless of skin color, individuals who differ in skin color may share many other traits, like ABO blood type. For this trait, two individuals with very different skin color may be more genetically similar.

People of every skin color are distributed among all blood type groups.

Type A Blood

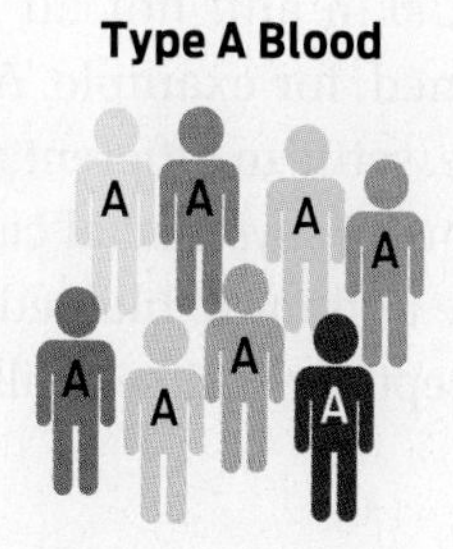

Type B Blood

Type AB Blood

Type O Blood

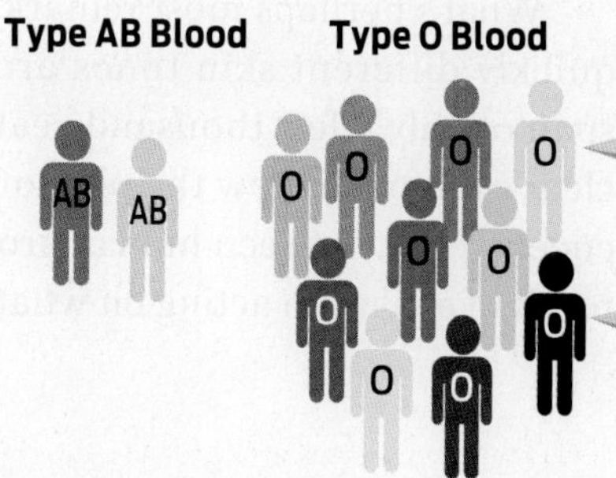

This light-skinned individual with O blood type is more genetically similar to this dark-skinned individual, than to a light-skinned individual with A blood type, when considering the blood type gene.

ABO Blood Type Allele Frequencies Do Not Correlate with Skin Color

As with most genes, the different alleles for the ABO blood type gene are present across the world, but in slightly different frequencies. Knowing a person's skin color does not allow you to make assumptions about their blood type or any of the thousands of other traits that vary among humans.

Skin Color and ABO Allele Frequencies in Indigenous Human Populations

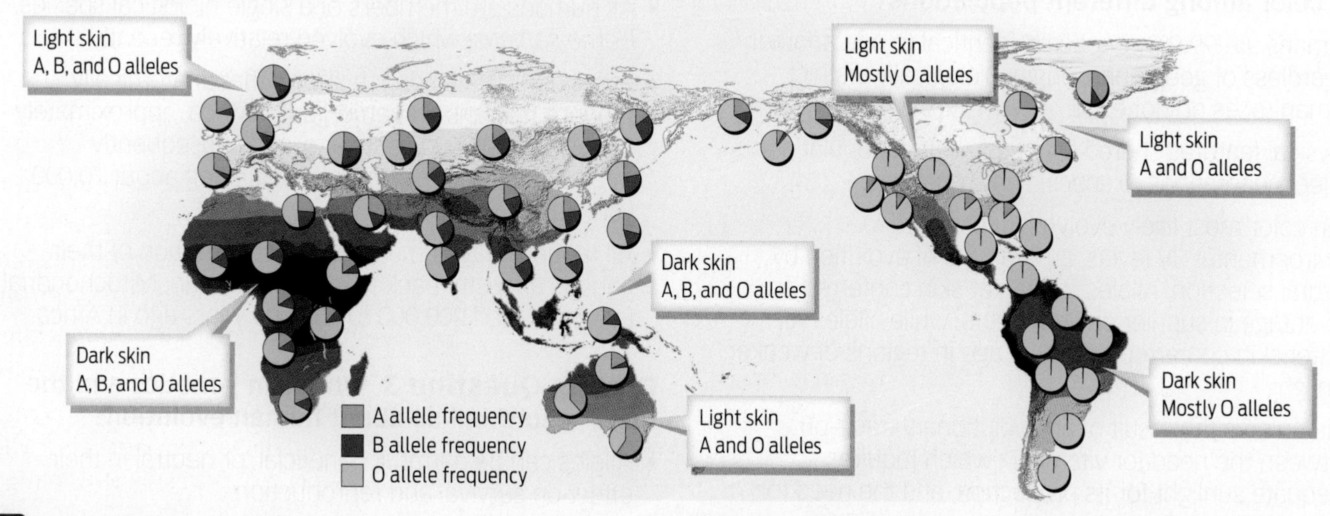

? Skin color was selected for by UV intensity. Does it look like blood type was also selected for by UV intensity? Explain your answer.

person likely had ancestors who lived in a region of the world where malaria was common (for example, West Africa, Saudi Arabia, or central India). Conversely, knowing that someone has ancestors from these regions does not tell you that they have the *HbS* allele.

An Evolving Explanation

Other hypotheses have been proposed to explain the evolution of skin tone. But of all of them, the folate–vitamin D hypothesis has the most supporting evidence and is consequently "the most reasonable," says Mark Stoneking of the Max Planck Institute for Evolutionary Anthropology in Leipzig, Germany, who collaborated on the Mitochondrial Eve study. In fact, Stoneking says, "Skin tone is one of the best examples of human evolution."

What's perhaps most remarkable is how quickly different skin tones arose—in the span of only a few thousand years. This is a clear example of how the phenotypic differences we see between human groups are the result of evolution acting on what amount to very small genetic differences. Rather than reinforcing old notions of race, the science of skin color actually tears them down.

Throughout human history, the lines between supposed "races" have been fluid. Genetic studies show that hardly any human population today is "pure" in the way that many have thought; all show evidence of extensive mixing going back thousands of years. As people moved around the globe, they settled and often conceived children with people they met along the way, introducing new alleles into the local gene pool. And they never stopped moving. Sometimes, they even reversed direction.

Though people tend to create racial groupings based on obvious physical characteristics, such features can be shared with other groups, says Jablonski. Not all Africans have equally dark skin and not all Europeans are light skinned, for example. And as humans travel more, settle in different areas, and intermarry, Jablonski says, racial categories will become more and more blurred. Perhaps in time the concept of race itself will disappear. ■

CHAPTER 19 Summary

Driving Question 1 What contributes to human skin color, and why is there so much variation in skin color among different populations?

- Humans are 99.9% genetically identical to one another regardless of geographic origin. Biologically distinct human races do not exist.
- Physical features shared by people within populations reflect adaptations to specific environments.
- Skin color most likely evolved in response to environmental UV levels, an example of evolution by natural selection. Alleles for darker skin conferred an advantage in sunnier environments, while alleles for lighter skin conferred an advantage in regions of weaker sunlight.
- Skin color is the result of an evolutionary trade-off between the need for vitamin D, which requires adequate sunlight for its production, and the need for folate, which is destroyed by too much sunlight.

Driving Question 2 Where did the earliest modern humans evolve, and how do we know?

- All humans are members of a single biological species, *Homo sapiens*, which evolved relatively recently.
- Fossil and DNA evidence shows that anatomically modern humans first emerged in Africa, approximately 200,000 to 300,000 years ago, and subsequently migrated to other continents, beginning about 70,000 years ago.
- All modern-day humans can trace a portion of their genetic ancestry back to a single woman, Mitochondrial Eve, who lived 200,000 to 150,000 years ago in Africa.

Driving Question 3 What can genetics and the fossil record tell us about human evolution?

- Alleles can be harmful, beneficial, or neutral in their effect on survival and reproduction.

- Fossil evidence shows that humans and apes descended from a common ancestor and that walking upright preceded development of a big brain. There were many species that could walk upright before *Homo sapiens* appeared.
- Humans evolved from apelike primate ancestors who likely had light skin. Darker skin emerged in tandem with loss of body hair as our hominid ancestors ventured into the hot savannah, while lighter skin emerged as humans migrated farther north.

More to Explore

- Jablonski, N. G., and Chaplin, G. (2010). Human skin pigmentation as an adaptation to UV radiation. *Proc Natl Acad Sci* 107:8962–8968.
- Crawford, N. G., Kelly, D. E., Hansen, M. E., et al. (2017). Loci associated with skin pigmentation identified in African populations. *Science* 358(6365):p.eaan8433.
- Goodman, A, Moses, Y., and Jones, J. (2020). *Race: Are We So Different?* 2nd ed. Hoboken, NJ: Wiley-Blackwell.
- Marks, J. (2010). Ten facts about human variation. In: Michael P. Muehlenbein, ed., *Human Evolutionary Biology*. Cambridge, UK: Cambridge University Press. https://webpages.uncc.edu/~jmarks/pubs/tenfacts.pdf
- Feldman, M., and Lewontin, R. (2008). Race, ancestry, and medicine. In: *Revisiting Race in the Genomic Age*. New Brunswick: Rutgers University Press. http://beck2.med.harvard.edu/week4/lewontin_feldman_sm.pdf
- Roberts, D. (2012). *Fatal Invention: How Science, Politics, and Big Business Re-create Race in the Twenty-First Century.* New York: New Press. See also: Roberts, D. (2015). www.ted.com/talks/dorothy_roberts_the_problem_with_race_based_medicine

CHAPTER 19 Test Your Knowledge

Driving Question 1 What contributes to human skin color, and why is there so much variation in skin color between different populations?

By answering the questions below and studying Infographics 19.1, 19.2, 19.3, 19.4, and 19.10, you should be able to generate an answer for this broader Driving Question.

Know It

1. The ancestors of modern humans evolved in a high-UV environment. What does this suggest about their skin color?

2. In the course of human evolution, which of the following environmental factors likely influenced whether a population had mostly light-skinned individuals or mostly dark-skinned individuals?

a. average annual temperature
b. average annual rainfall
c. levels of UV light
d. the vitamin D content of the typical diet
e. mitochondrial DNA inheritance

3. Jablonski and Chaplin hypothesized that darker skin is advantageous in ____________ UV environments because darker skin ____________.

a. high-; reduces vitamin D production
b. high-; protects folate from degradation
c. high-; increases the rate of folate synthesis
d. low-; allows more vitamin D to be produced
e. low-; allows more folate to be produced

Use It

4. If vitamin D production did not require exposure to UV, predict the skin color you might find in populations living at the equator. What about in populations living in Greenland? Explain your answers.

5. Which of the following would help darker-skinned people who live in low-UV environments remain healthy?

a. folate supplementation
b. sunscreen
c. increased production of melanin
d. vitamin D supplementation
e. calcium supplementation

6. What can you infer about the skin-color alleles and the geographic origins of the ancestors of a light-skinned person and the ancestors of a dark-skinned person?

7. Our closest primate relatives, chimpanzees, have light-colored skin yet live in tropical (high-UV) environments. How would the Jablonski–Chaplin hypothesis explain this observation?

a. Chimpanzees don't need folate for successful reproduction.
b. Chimpanzees are not susceptible to skin cancer.
c. The hair of chimpanzees protects their light skin from UV light.
d. Chimpanzees require much higher levels of vitamin D than humans do.
e. In chimpanzees, a light-colored pigment offers UV protection.

8. Vitiligo is a disease in which melanocytes are destroyed, with resulting loss of pigmentation.

a. If a dark-skinned person develops vitiligo and therefore lighter-colored skin, will his or her race change?

b. Given your answer to part a, can you think of one or more factors that might have led people to classify (or misclassify) themselves or others as members of one race or another?

Driving Question 2 Where did the earliest modern humans evolve, and how do we know?

By answering the questions below and studying Infographics 19.5, 19.6, 19.7, and 19.8, you should be able to generate an answer for this broader Driving Question.

Know It

9. Who is Mitochondrial Eve, and when did she live?

10. According to the out-of-Africa hypothesis of human origins and migration, which group of people should show the highest level of genetic diversity?

a. Africans
b. Europeans
c. Asians
d. South Americans
e. Australians

11. The following three types of DNA can be used to trace evolutionary history. Match each DNA type with the best description of how it is transmitted in a population.

Type of DNA	Description
____ mtDNA	a. inherited by all children only from their mother
____ nuclear DNA	b. inherited by all children only from their father
____ Y-chromosome DNA	c. inherited only by sons from their father
	d. inherited by all children from both their mother and their father
	e. inherited only by daughters from their mother

Use It

12. Rank the levels of genetic diversity you would expect to find within the five populations listed in Question 10 from highest to lowest. Justify your ranking.

13. If there were many human females living approximately 200,000 years ago, why do we find that the mitochondrial DNA in all living humans is all related to a single woman from that time? What kind of evidence could you look for to test your explanation? (*Hint:* Think about all the human fossils that have been uncovered, and consider that it is possible to extract DNA from fossils.)

Apply Your Knowledge

Mini Case

14. A mother with medium skin tone gives birth to a baby with darker skin than she has. Her lighter-skinned husband accuses her of infidelity.

a. How reasonable is this accusation, given the genetics of skin color? (*Hint:* The genetics of skin color are similar to the genetics of height, shown in Infographic 12.9.)

b. Could an mtDNA analysis be used in a paternity test? Why or why not?

Driving Question 3 What can genetics and the fossil record tell us about human evolution?

By answering the questions below and studying Infographics 19.1, 19.9, 19.10, and 19.11, you should be able to generate an answer for this broader Driving Question.

Know It

15. What percentage of DNA sequences do all humans share?

a. 0%
b. 25%
c. 50%
d. 75%
e. >99%

16. Place the following ancestors in order of earliest (1) to most recent (5).

____ *Homo sapiens*
____ Last common ancestor of chimpanzees and humans
____ *Australopithecus*
____ *Ardipithecus ramidus*
____ *Homo erectus*

17. A person has dark skin. Based on that knowledge, can you predict what blood type this individual is likely to have? Conversely, if you know that a person has type A blood, can you predict what skin color this individual is likely to have? Explain your answer.

Use It

18. Where would the last common ancestor of gorillas and humans fit into the ordering in your answer to Question 16? Explain.

19. Why would individual australopithicines who could make and use tools have had a selective advantage (that is, higher fitness) over individuals who could not make or use tools?

20. Members of the genus *Australopithecus* walked upright, and their fossilized footprints show no evidence of an opposable big toe.

a. What foot structure and lifestyle might have been selected for if early hominid evolution occurred in a forested environment? In a grasslands environment? Would you predict any differences because of the selective pressures in each environment? Why or why not?

b. What other traits would you expect to be favored in a forested environment? In open grasslands?

Interpreting Data

Apply Your Knowledge

21. An extensive study of a hominid fossil dating from approximately 2 million years ago was published in 2013. For each of the features described below, consider whether they are closer to an ancestral state or to modern humans. On the basis of the features described, where would you place this fossil on the lineage between the chimpanzee–human ancestor and modern humans—what genus is it likely a member of?

- The shoulder structure and very long arms suggest the ability to climb and perhaps hang or swing.
- The spine and other skeletal features suggest an upright stance.
- There is no opposable toe.
- The heel is very narrow and pointed (not flat and wide, which would better accommodate body weight when walking).
- The skeleton suggests that when walking on two feet, the feet would roll inward with each step.
- The skull is very small.
- The chest is not cylindrical but wider at the base and narrow at the shoulders, much like a triangle.

Bring It Home

Apply Your Knowledge

22. The U.S. Census Bureau provides information on classifying race (https://www.census.gov/topics/population/race/about.html). What races does the U.S. Census Bureau recognize? What about people of mixed race? What about people who identify themselves as Hispanic? How easy is it for you to identify yourself with respect to race given the racial categories on the U.S. Census?

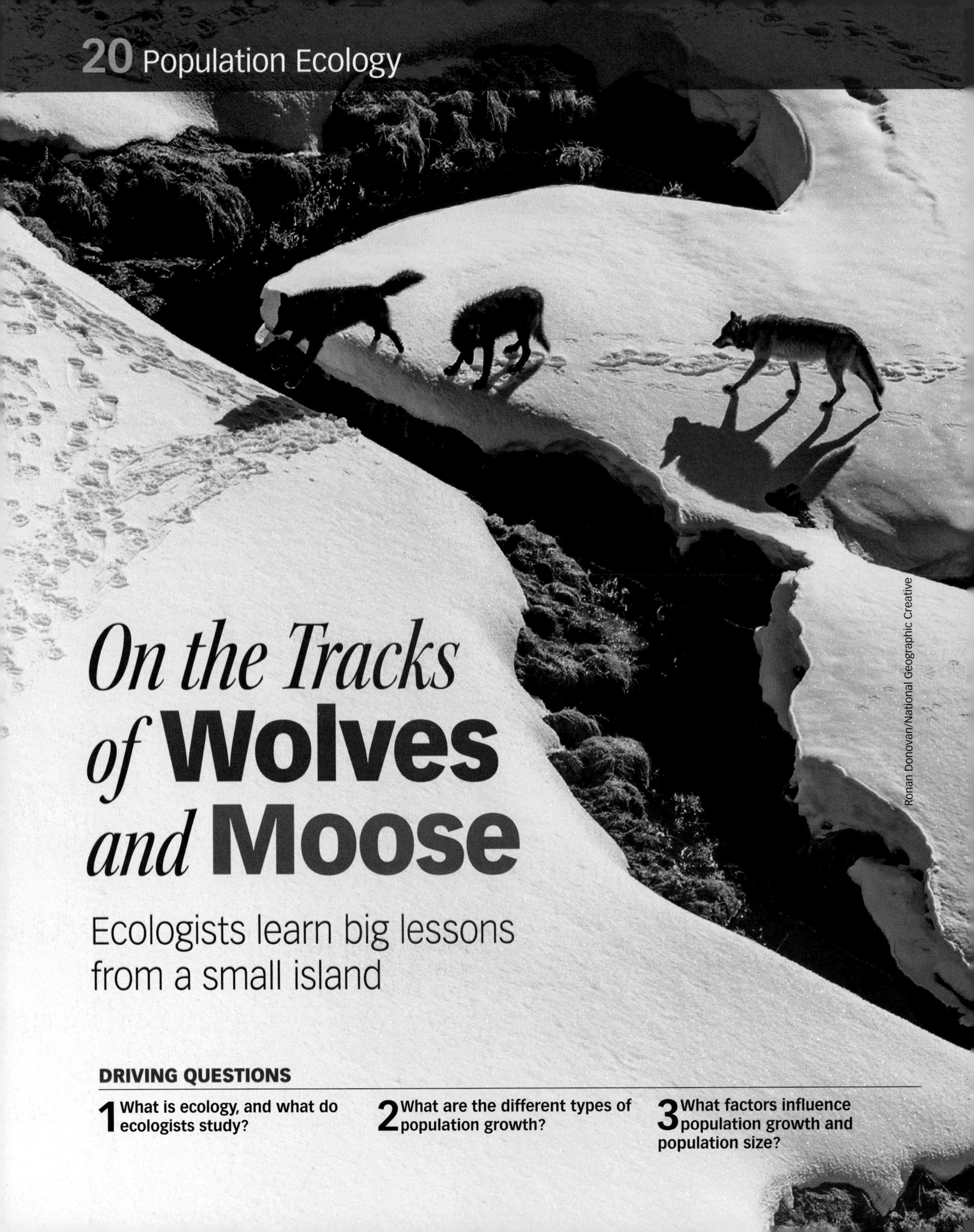

Ronan Donovan/National Geographic Creative

On the Tracks of **Wolves** *and* **Moose**

Ecologists learn big lessons from a small island

DRIVING QUESTIONS

1 What is ecology, and what do ecologists study?

2 What are the different types of population growth?

3 What factors influence population growth and population size?

John Vucetich spent a freezing February day trudging through knee-high snow on an island in Lake Superior. He was tracking a young gray wolf he called Romeo. The tracks led to a site in the forest where cracked branches and crimson-stained snow were evidence that a violent struggle had taken place hours earlier. Later, thawing in his cabin beside a wood-burning stove, Vucetich wrote in his field journal:

Teeth, hooves, blood, bruises, adrenaline, exhaustion. Romeo killed a moose. Very likely, this is the first moose he'd ever killed. He'd seen his parents, the alpha pair of Chippewa Harbor Pack, do it many times. . . . He'd wounded moose a couple of times this winter, but never killed one.

For nearly 30 years, Vucetich has been shadowing wolves like Romeo and his kin on Isle Royale, a remote island about 15 miles off the Canadian shore in the northwest corner of Lake Superior. A 200-square-mile slice of roadless wilderness that is accessible only by boat and seaplane, Isle Royale may seem an unlikely place for a scientific laboratory, but that's exactly what it is for Vucetich and his colleagues. Every summer, and for a few weeks every winter, they investigate the island's packs of gray wolves (*Canis lupis*) and the herd of moose (*Alces alces*) that are their prey.

Begun in 1958, the Isle Royale wolf and moose study is the longest-running predator–prey study in the world. For six decades, researchers have studied how these two groups of island inhabitants interact and coexist in this wild place. They are motivated by a simple yet increasingly pressing goal: to observe and understand the dynamic fluctuations of Isle Royale's wolves and moose in the hope that such knowledge will inspire a new, flourishing relationship with nature.

Islands have long provided biologists with important lessons about nature—think

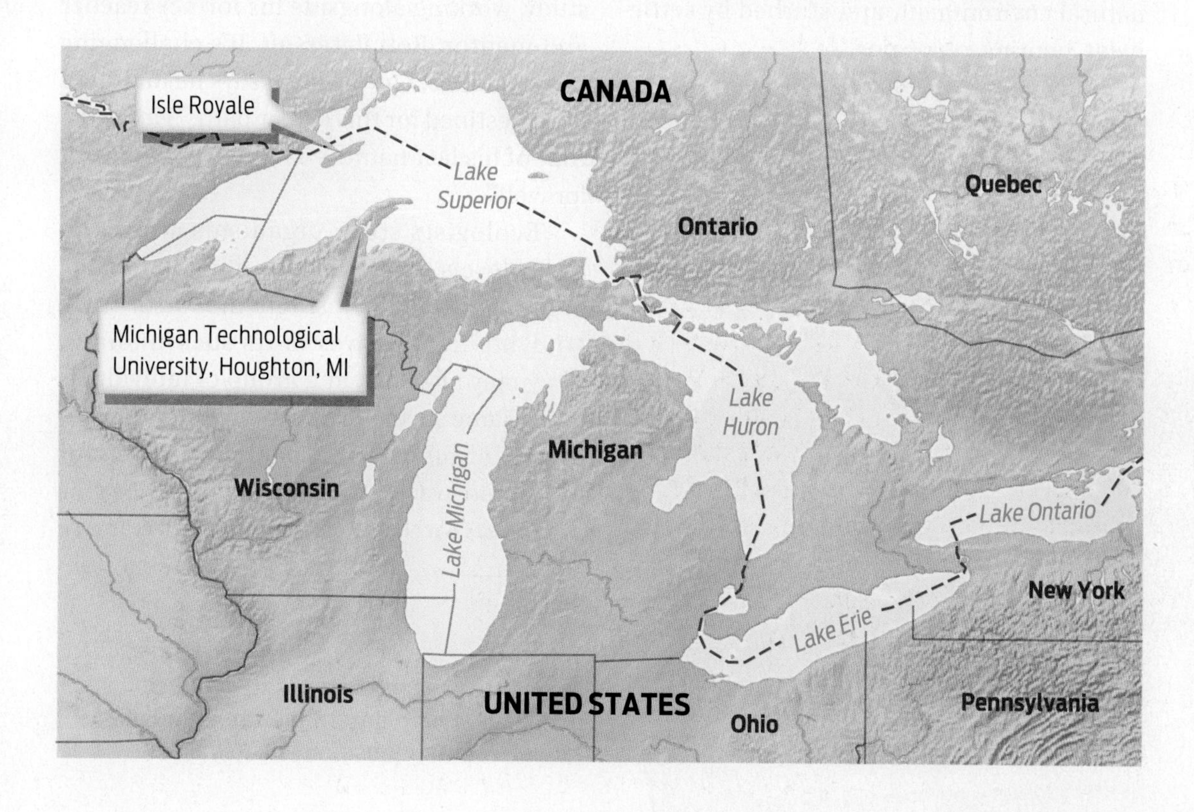

"Isle Royale is not too big and it's not too small and it's not too close and not too far."

—John Vucetich

of Darwin and Wallace and their island adventures (see **Milestone 4: Adventures in Evolution**). But no island has ever produced such a mountain of data for researchers interested in **ecology**—the study of the interactions among organisms and between organisms and their nonliving environment—as Isle Royale. At a moment when wild populations are threatened all over the world, the lessons of Isle Royale couldn't have come at a better time.

In Nature's Laboratory

▶ Ecology as a science

ECOLOGY
The study of the interactions among organisms and between organisms and their nonliving environment.

POPULATION
A group of organisms of the same species living and interacting in a particular area.

A number of features make Isle Royale an ideal place in which to study ecology. Because the island is uninhabited by humans and is protected as a national park, scientists can study moose–wolf interactions in a nearly natural environment, undisturbed by settlement, hunting, or logging.

Isle Royale is also an ideal distance from shore—close enough to the mainland for moose and wolves to have reached it, but far enough away that other animals do not easily migrate to it. Because there are no other predators or prey on the island, the only things eating moose are wolves, and moose are just about the only thing the Isle Royale wolves eat. These simplified conditions allow scientists to get a good look at the two types of residents' behavior and ecological impact.

Another feature that makes Isle Royale good for research is its size. The island is not so big as to have an unmanageably large population of moose, and not so small as to be unsupportive of a wolf population. "It's a little bit of the Goldilocks thing," says Vucetich, a professor in the School of Forest Resources and Environmental Science at Michigan Technological University. "Isle Royale is not too big and it's not too small and it's not too close and not too far. It's just the right size to have a population of wolves and moose that we can study."

Vucetich began studying wolves as a student at Michigan Tech in the early 1990s. In 2001, he became coleader of the Isle Royale study, working alongside his former teacher and mentor, Rolf Peterson. It's challenging work at times, but Vucetich says he may have been destined for this career path: "*Vuk*"—the root of his last name—"is the Croatian word for wolf."

Ecologists study organisms at a number of levels. They can look at an individual organism, such as a single moose or wolf, studying how it fares in its surroundings. They may also look at a group of individuals of the same species living in the same place—a herd of moose or a pack of wolves, for example—watching what happens to this **population** over time. Two or more

Scientists use small planes to conduct aerial surveys to determine the numbers of wolves and moose on Isle Royale.

interacting populations of different species constitute a **community.** Isle Royale, for example, is home to a community of wolves, moose, and the plants the moose feed on.

Finally, ecologists may want to understand the functioning of an entire **ecosystem,** all the living organisms in an area and the nonliving components of the environment—such as temperature, rain, and sunlight—with which they interact. When moose eat trees, for example, they reduce the available habitat for other animals, such as birds. However, the heat of summer can reduce the ability of moose to feed, which in turn improves tree growth **(INFOGRAPHIC 20.1)**.

Numbers in Nature

▶ Population sampling, dispersion patterns

Vucetich was initially drawn to ecology as a way to experience the outdoors, his first love. Only later did he realize he was actually quite good at something ecology has a lot of: math. "As a high school student, I didn't like math at all," he says. But when he saw that math allowed him to spend more time outdoors, he became "interested and inspired to learn a great deal about math."

Vucetich is a population ecologist, and population ecology is all about numbers. On Isle Royale, the main numbers that interest the researchers year after year are the numbers of wolves and moose. "In any given season there are more or less of those species and we want to understand why," says Vucetich. Answering the "why" involves a lot of time, patience, and—of course—counting.

Much of the counting is done from the air. Sitting one in front of the other inside a tiny two-seat plane, pilot and observer circle the island scanning for evidence of wolves and moose. Wolves are relatively easy to find and count, especially in the snow: "You follow the wolf tracks until you find the wolves," says Vucetich. The other thing that makes counting wolves easy is that they live in packs: if you find one wolf, you've generally found the others. And since there are usually no more than a dozen or so wolves on the island at any time, researchers can count every one.

INFOGRAPHIC 20.1

Ecology of Isle Royale

Ecologists study organisms at a number of levels, each of which involves different types of interactions.

U.S. Fish and Wildlife Service

Individual

A single organism of a particular species

- One wolf

John Vucetich

Population

A group of individuals of the same species living and interacting in the same region

- A pack of wolves whose individuals live and reproduce with one another

Accent Alaska.com/Alamy

Community

Interacting populations of different species

- Wolves prey on moose
- Ticks infest moose
- Moose feed on trees

sherwoodimagery/iStock/Getty Images

Ecosystem

Species interacting with other species and the environment

- Moose eat trees. Less vegetation alters the landscape for other species.
- Hot summers reduce food for moose, affecting their winter survival

? Which level of ecology involves the interactions of organisms with their environment?

It's a different story with moose. The island can be home to more than a thousand moose—too many to count all at once.

COMMUNITY
Interacting populations of different species in a defined habitat.

ECOSYSTEM
All the living organisms in an area and the nonliving components of the environment with which they interact.

INFOGRAPHIC 20.2

Distribution Patterns Influence Population Sampling Methods

Wolves and moose have different lifestyles and distribution patterns on Isle Royale. Determining the size of each population requires a distinct counting strategy.

Jupiterimages/Getty Images

Individual wolves cluster together in packs, making them easier to spot, track, and count.

2013 Wolf Territories and Kill Locations

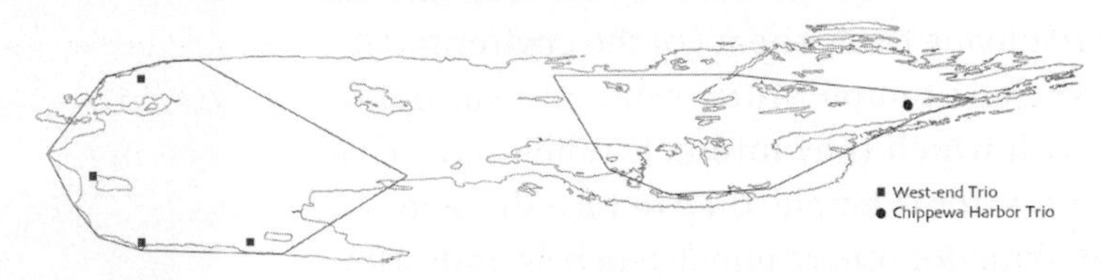

Tom Hansch/Dreamstime.com

Moose are largely solitary creatures and are distributed more randomly on the island. Counting them requires a different strategy.

2013 Moose Distribution

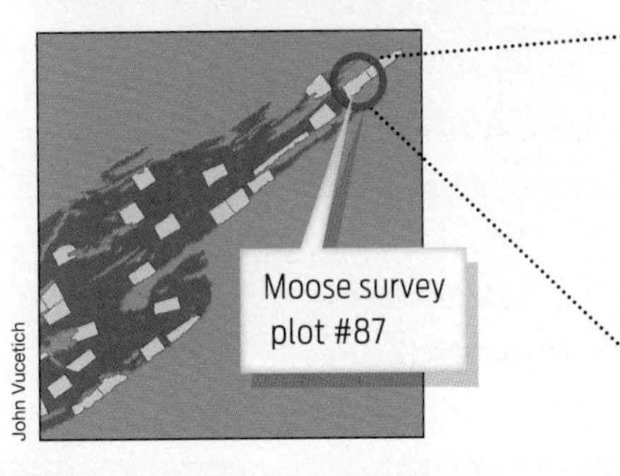

John Vucetich

This is an aerial view of plot #87, outlined in orange. The green dotted line defines the 9 overlapping circles flown as moose are counted. This sampling is done for 91 plots across 20% of the island to gather enough data for an accurate estimate of moose population size and density on the island.

Data from Vucetich, J. A., and Peterson, R. O. (2013). *Ecological Studies of Wolves on Isle Royale: Annual Report* 2012–2013.

? To establish the population size for the moose on Isle Royale, do the researchers try to count all the moose?

Besides, moose are relatively solitary creatures, and their brown coloring makes them harder to spot against the backdrop of dark evergreen trees. When moose are feeding in the forest—which is much of the time—counting them is, according to Vucetich, "like trying to count fleas on a dog from across the room." It's simply not possible to count them all.

Instead, the team uses a shortcut: they count all the moose in a series of defined plots of land representing about 20% of the island, average the number of moose per plot, and then extrapolate to the rest of the island. But even this shortcut requires many careful hours of study in the plane, straining to see the moose through the trees. To help himself concentrate, Vucetich recites a sort of mantra: "Think moose, think moose, look for the moose" **(INFOGRAPHIC 20.2)**.

The somewhat random distribution of individually roaming moose represents one type of **dispersion pattern** found in nature. Dispersion patterns generally reflect the way resources are distributed across an area, as well as characteristic interactions among other members of the population. Moose are found wherever there are edible trees. Being

DISPERSION PATTERN
The way organisms are distributed in geographic space, which depends on resources and interactions with other members of the population.

solitary and randomly dispersed may help protect moose from predation, since a single moose is harder to spot in the forest than a large group would be. For some plants, random dispersion may result from the way their seeds are spread. For example, gusty winds transport the air-blown seeds of pine trees far and wide, resulting in a random dispersion of trees in the forests on the island.

A truly random dispersion is rare in nature; even wind-blown seeds must fall on fertile soil to grow, and this does not always happen. More common is a clustered, or clumped, dispersion pattern. A clumped dispersion pattern occurs when resources are unevenly distributed across the landscape, or when social behavior dictates grouping, as it does with the highly social wolves. Clumping has its advantages: for wolves, clumping helps them to gang up on moose; they can circle their prey and close in for the kill. Clumping can also be a defense against predation, as it is for a school of fish.

A third dispersion pattern found in nature is uniform dispersion. In this case, individuals keep apart from one another at regular distances, usually because of some kind of territorial behavior. Birds such as penguins that nest in defined spaces a few feet away from one another are a good example **(INFOGRAPHIC 20.3)**.

To help himself concentrate, Vucetich recites a sort of mantra: "Think moose, think moose, look for the moose."

Population Boom and Bust

▶ Logistic and exponential population growth

Moose have not always roamed Isle Royale. The first antlered settlers likely arrived around 1900, when a few especially hardy individuals swam across the 15-mile-wide channel from Canada. With an abundant food supply and no natural predators on the island, the moose population exploded, growing from a handful of individuals around the turn of the century to more than a thousand by 1920.

This rapid increase in numbers reflected the population's high **growth rate,** defined as

GROWTH RATE
The difference between the birth rate and the death rate of a given population; also known as the rate of natural increase.

INFOGRAPHIC 20.3

Population Dispersion Patterns

Different organisms have different distribution patterns. There are three main types, but few organisms in nature fall into strictly one category.

Individuals are equally likely to be anywhere within the area.

High-density clumps are separated by areas of low abundance.

Individuals maximize space between them by being uniformly spaced.

? If plant seeds are carried by animals that congregate at watering holes, what type of dispersion would you predict for the plants?

IMMIGRATION
The movement of individuals into a population.

EMIGRATION
The movement of individuals out of a population.

EXPONENTIAL GROWTH
The unrestricted growth of a population increasing at a constant growth rate.

the birth rate minus the death rate. Because it denotes the simple balance between birth and death, the growth rate is also known as the rate of natural increase. When the birth rate of a population is greater than the death rate, the population grows; when the death rate is greater than the birth rate, the population declines; and when the two rates are equal, the result is zero population growth.

In many populations, **immigration,** the movement of individuals into a population, and **emigration,** the movement of individuals out of a population, make substantial contributions to population growth. But because the moose and wolves of Isle Royale are isolated, and individuals neither come to nor go from the island on a regular basis, their population growth rates are due only to births and deaths.

Ecologists describe two general types of population growth. The rapid and unrestricted increase of a population growing at a constant rate is called **exponential growth.** When a population is growing exponentially, it increases by a certain fixed percentage every generation. Thus, instead of a constant number of individuals being added at each generation—say, the population going from

INFOGRAPHIC 20.4

Population Growth and Carrying Capacity

A population growing at a constant rate without checks will increase exponentially. However, for most populations found in nature, as the population reaches its carrying capacity, the growth rate slows and eventually remains at or near zero.

? What can cause an exponentially growing population to switch to logistic growth?

100 to 120 to 140 to 160—the increase is more like credit card interest, with each fixed percentage increase added to the principal (the population) before the percentage is added again in the next cycle. And so, with an exponential growth rate of 20%, a population of 100 would increase at each generation from 100 to 120 to 144 to 173 to 207. If the population continued to grow exponentially, it would quickly get out of control, not unlike a credit card bill you don't pay on time.

Such unrestricted growth is rarely, if ever, found unchecked in nature. Instead, as populations increase, various environmental factors tend to limit an organism's ability to reproduce. Maybe not enough food is available for the growing population or there is limited access to **habitat,** the physical environment where an organism lives and to which it is adapted. When population-limiting factors slow the growth rate, the result is **logistic growth,** a pattern of growth that starts rapidly and then slows.

Eventually, after a period of rapid growth, the size of the population may level off and stop growing as the population reaches the environment's **carrying capacity**—the maximum number of individuals that an environment can support given its space and resources. Carrying capacity places an upper limit on the size of any population; no natural population can grow exponentially forever without eventually reaching a point at which resource scarcity and other factors limit population growth. This is true even of the human population (see Chapter 24).

The size of a population may fluctuate around the environment's carrying capacity, briefly exceeding it and then dropping back. After an initial overshoot of carrying capacity, factors like disease or food shortage will cause the population to shrink. This drop, in turn, may allow the environment time to recover its food supply. At that point the population may begin to grow again, briefly exceeding carrying capacity, and so on, in a cycle of "boom and bust" **(INFOGRAPHIC 20.4).**

When moose first arrived on Isle Royale, their population grew exponentially. This unchecked proliferation of hungry mouths took a severe toll on the island; by 1929, the moose had munched their way through

HABITAT
The physical environment where an organism lives and to which it is adapted.

LOGISTIC GROWTH
A pattern of growth that starts off fast and then levels off as the population reaches the carrying capacity of the environment.

CARRYING CAPACITY
The maximum population size that a given environment or habitat can support given its food supply and other natural resources.

John Vucetich

A pack of wolves moving in for the kill. Ecologist John Vucetich examining a kill site.

most of its vegetation. Once the island's food supply was gone, the moose population crashed. The moose population had exceeded the island's carrying capacity, and by 1935 it had dwindled to a few hundred starving individuals.

Even for wolves, there's no such thing as a free lunch: they pay a price for predation in the form of a declining food supply.

The herd got lucky, though. The next summer, fire consumed 20% of the island, and the scorched areas provided space for new trees to grow. But as soon as the forest recovered, moose numbers again began to explode, ravaging the forests once more.

Then, around 1950, everything changed. One especially cold winter, a pair of gray wolves crossed an ice bridge connecting Canada to Isle Royale, forever altering the ecology of the island. Since then, the fates of the wolves and moose have been tightly linked.

At the beginning of the Isle Royale study, in 1959, there were about 550 moose and 20 wolves on the island. Moose numbers climbed for about 15 years, reaching a peak of approximately 1,200 animals in 1972, and then declined rapidly, to a low of approximately 700 moose in 1980. As moose numbers fell, wolf numbers rose—from a low of 17 wolves in 1969 to a high of 50 animals in 1980. These two trends were linked: the wolves were feeding well enough to increase their own population, but by hunting and killing so many moose they caused the moose death rate to exceed the birth rate. With a negative growth rate, the moose population shrank.

What would happen next? Would the wolf predators simply drive their moose prey to extinction? No one knew. The only thing to do was watch and wait. Eventually, it became clear that the two populations were rising and falling together in a specific pattern, with the size of the wolf population peaking several years after the size of the moose herd and then dropping.

Why does the wolf population decrease? Because even for wolves, there's no such thing as a free lunch: they pay a price for predation in the form of a declining food supply. The result is a repeating cycle for the numbers of predator and prey. Rather than growing exponentially and leveling off, the populations cycle through repeated rounds of boom and bust **(INFOGRAPHIC 20.5)**.

Ecological Detectives

▸ Ecological methods, population interactions

The size of the wolf population affects the ecology of Isle Royale in other ways than just by limiting the size of the moose population. Another pattern has emerged in the decades

INFOGRAPHIC 20.5

Population Cycles of Predator and Prey

The wolf and moose populations are intimately linked and inversely proportional. When wolf populations are high, moose populations are low. When wolf populations are low, moose populations are high.

Moose and Wolf Populations 1959–2018

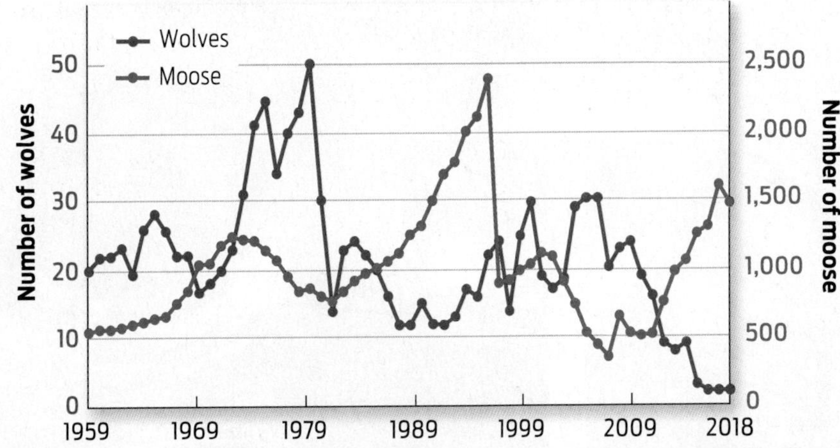

Data from Vucetich, J. A., and Peterson, R. O. (2016) *Ecological studies of Wolves on Isle Royale, Annual Report 2015–2016.*
Data from Hoy, S. R., et. al. (2019). *Ecological Studies of Wolves on Isle Royale: Annual Report 2018–2019.*

? Why does the population of moose (prey) decline as the population of wolves (predator) increases?

of data collected on Isle Royale: there is a correlation between a large wolf population and vigorous tree growth. When wolves are plentiful, they keep the moose population in check. Because trees are the primary food source for moose, they grow larger when fewer moose are eating them. It's therefore possible to follow the rise and fall of the wolf population by monitoring the state of the forest.

One way ecologists can determine forest growth is by counting and measuring the width of tree rings, which reflect how much trees have grown season by season. They also measure the height of the trees. Taller and bigger trees mean that fewer moose have been foraging on them, and the presence of fewer moose in turn indicates that more wolves have been keeping the moose population in check **(INFOGRAPHIC 20.6)**.

INFOGRAPHIC 20.6

Patterns of Population Growth

Wolf, moose, and tree populations are all interconnected. Trees provide food for moose, and moose provide food for wolves. Anything that affects the size of one population will affect the size of the others.

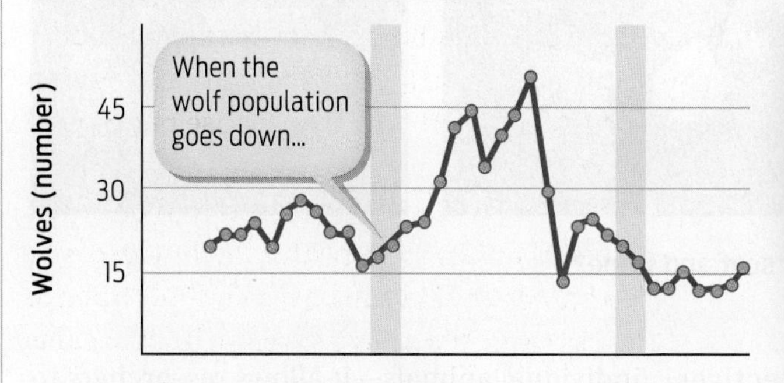

The main diet of Isle Royale wolves is moose. The wolf population grows and diminishes in response to the availability of this food resource.

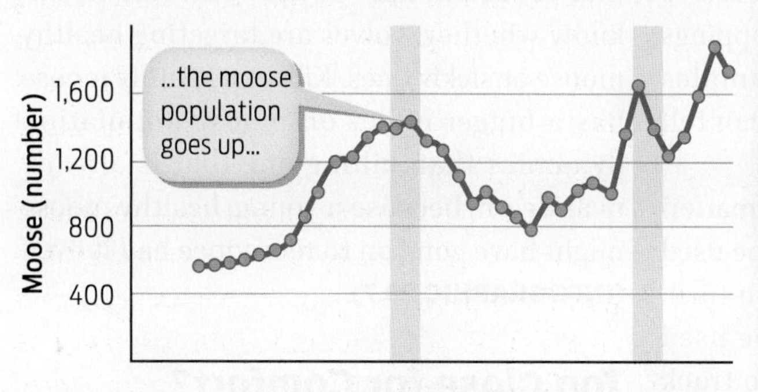

The main diet of moose is trees. A larger moose population means that more tree material is eaten, so tree growth slows.

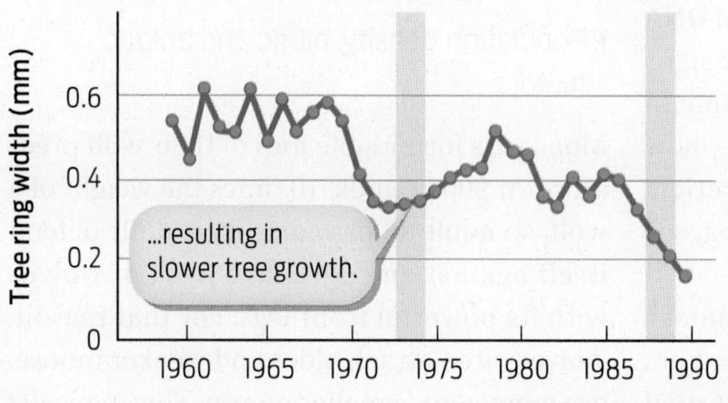

Tree growth can be measured by the width of each tree ring. One ring represents the amount of growth in 1 year: the wider the ring, the more the growth in that year.

Data from McLaren, B. E., and Peterson, R. O. (1994). Wolves, moose, and tree rings on Isle Royale. *Science* 266(5190):1555–1558.

? What does an abundance of large trees suggest about moose and wolf populations?

INFOGRAPHIC 20.7

A Variety of Methods Can Be Used to Monitor Populations

In addition to information about population size, researchers collect other data that are essential for monitoring the physical health of moose and wolf populations.

Moose Scat

Reveals the vegetation preferences of the moose populations

Moose Bones

Provide information on the presence of arthritis and osteoporosis, as well as bone marrow fat content and thus nutritional health

Wolf Scat

A source of DNA, providing a genetic profile for each wolf on the island

Urine-Soaked Snow

Tested to determine the ratio of urea and creatinine in the urine, an indication of a moose's nutritional health

? **What can researchers learn about moose from studying their scat and urine?**

Other clues the ecological detectives look at are urine-soaked snow and droppings (scat). By analyzing moose scat samples under the microscope, researchers can tell exactly what moose have been eating.

Scat also provides important information about an animal's genetics. Scat can be used as a source of DNA for that animal. In turn, DNA profiles of scat samples can be used to confirm population counts and to track which wolves helped kill which moose. DNA can also be used to look for diseases or signs of inbreeding. "Through the DNA we can get a good sense of individual wolves—how they live and how they die," says Vucetich. (For more information on DNA profiling, see Chapter 7.)

At a kill site, researchers gather moose bones. From these bones, the researchers can tell how old a moose was when it died and what its health was like, such as whether it had arthritis or osteoporosis. The value of this information goes beyond understanding individual animals—it allows researchers to know whether wolves are targeting healthy moose or sickly ones. Killing a healthy moose has a bigger effect on moose population dynamics than killing one that is already near death, because a young, healthy moose might have gone on to reproduce had it lived **(INFOGRAPHIC 20.7)**.

Too Close for Comfort?

▸ Population density, biotic and abiotic factors

Moose are formidable foes of their wolf predators. At 900 pounds, 10 times the weight of a wolf, an adult moose can successfully defend itself against an aggressive pack of wolves with its powerful front legs. For that reason, wolves often attack older and weaker moose, or younger and smaller moose. They typically target the nose and hindquarters, where they bite and latch onto the flesh like a steel trap. When enough wolves are attached to the prey, their collective weight brings down the moose, and the feeding begins.

A number of factors can influence the likelihood that wolves will kill moose. One of the simplest is **population density,** the number of organisms per unit area. Because the total area of Isle Royale stays the same, as the size of the moose population increases, so does its density. When present at high population density, moose are easier for wolves to locate and kill. Further, food scarcity can be a problem under this condition, leaving moose hungry and weak and therefore more vulnerable to attack.

Because wolf predation and plant abundance have a greater effect on moose when the moose population is large, these are examples of **density-dependent factors**—factors that exert different degrees of influence depending on the density of the population. As living parts of the environment, they are also examples of **biotic** factors influencing growth.

Some environmental pressures take a toll on a population no matter how large or how small it is. In an exceptionally cold winter with deep snow, for example, moose can die of cold or starvation, or the weather can weaken them so the wolves find it easier to hunt and kill them. Since cold weather affects moose regardless of population size, it is considered a **density-independent factor:** whether there are 10 moose or 1,000, a harsh winter affects them all.

The wolf population is also affected by density-independent factors such as weather. "A mild winter is always tough on the wolves," says Peterson, who notes that moose can more easily escape wolves when snow cover is light. Density-independent factors like snowfall can be considered nature's form of bad luck. Most density-independent factors

"Through the DNA we can get a good sense of individual wolves—how they live and how they die."

—John Vucetich

POPULATION DENSITY
The number of organisms per unit area.

DENSITY-DEPENDENT FACTOR
A factor whose influence on population size and growth depends on the number and crowding of individuals in the population (for example, predation).

BIOTIC
Refers to the living components of an environment.

DENSITY-INDEPENDENT FACTOR
A factor that can influence population size and growth regardless of the numbers and crowding within a population (for example, weather).

AP Images/National Park Service

One of the first wolves to be released into Isle Royale in late 2018 as part of a plan to restore a healthy wolf population.

INFOGRAPHIC 20.8

Abiotic and Biotic Influences on Population Growth

Both nonliving (abiotic) and living (biotic) environmental factors influence the size and growth of populations. Some factors influence the population regardless of population density, whereas others are density-dependent.

Precipitation
Snowfall stresses moose, making it hard for them to find food and avoid predators.

Photobac/Dreamstime.com

Fire
Fire is an immediate danger to moose; it can destroy their habitat and make food scarce.

Temperature
High summer temperatures can cause heat stress in moose and increase insect parasitism.

terrymorris/iStockphoto.com

Abiotic Factors
(Density-independent)

James Mattil/AGE Fotostock

Biotic Factors
(Density-dependent)

Jim Kruger/iStock/Getty Images

Predators
The presence of a wolf predator limits moose population growth, especially when moose density is high.

Oksana Churakova/Dreamstime.com

Food
The availability of trees as a food source influences moose population growth, especially when moose density is high.

Disease
Infectious organisms and parasites such as ticks can reproduce more and spread more easily between moose when moose density is higher. Tick infestations weaken moose, limiting their reproduction.

? **Would widespread flooding be characterized as biotic or abiotic, and as a density-dependent or density-independent factor?**

are **abiotic,** or nonliving, and include things like temperature, precipitation, and fire **(INFOGRAPHIC 20.8).**

Watching and Waiting versus Intervening

For the scientists on Isle Royale, population ecology is full of unexpected twists and turns. There is often no sure way to know how various environmental factors will influence the growth of a population. Even on an isolated island with only one large predator and one large prey, population dynamics are never simple. Scientists gather data, look for patterns, and form hypotheses, but predicting what will happen next is much more difficult. "What Isle Royale has shown us—and has shown us convincingly for the past 50 years—is that we're lousy at predicting the future," says Vucetich. "What we're a fair bit better at is explaining the past."

For example, beginning around 1980, a disease called canine parvovirus (CPV) infected Isle Royale's wolves. This disease typically strikes domestic dogs and was likely brought unintentionally to the island on the boots of hikers. The disease killed all but 14 of the island's wolves. Over the next 10 years the moose population skyrocketed, providing further support to the idea that wolves exert a strong influence on the abundance of their prey. The event was useful from a scientific standpoint—but entirely unexpected. "There's no way that anyone could have predicted that. Not in a million years," says Vucetich.

That wasn't the end of the surprises. In the last 15 years, it's become apparent that a warming climate—not just predation by wolves—is influencing moose population size. The first decade of the 21st century was one of the hottest on record. Sweltering summer temperatures hit moose especially hard. The large herbivores get hot easily, and they don't perspire; they escape the heat by resting in the shade. A lot of time spent resting means less time for eating, and a moose that eats less all summer has less insulation for the harsh winter cold.

J. A. Vucetich and R. O. Peterson, www.isleroyalewolf.org

A young moose on Isle Royale (note the patchy fur, a sign of tick infestation).

Warmer temperatures have affected moose in a more insidious way as well. About 10 years ago, Vucetich and his colleagues began to notice that a tick parasite was bothering the moose, and that warm weather seemed to favor ticks. Ticks suck the moose's blood and cause them to itch. The moose scratch themselves against trees and chew their hair out trying to rid themselves of the itchy freeloaders. Since a single moose may host many thousands of ticks, the combination of tick-related blood loss and heat-induced weight loss can be deadly. In 2004, the average moose had lost more than 70% of its body hair, the result of carrying more than 70,000 ticks.

By 2007, the deadly combination of blood-sucking ticks, hot summers, and relentless predation from wolves had driven the moose population to its lowest point in at least 50 years—385, down from 1,100 in 2002. Predictably, the wolf population followed suit, declining from 30 individuals in 2005 to 21 in 2007. In recent years, winters have been colder and more long-lasting, which has

ABIOTIC
Refers to the nonliving components of an environment, such as temperature and precipitation.

led to fewer tick infestations and healthier moose **(INFOGRAPHIC 20.9)**.

Hunted by wolves, preyed on by ticks, dogged by oppressive heat—moose certainly do not have it easy. They can live to be 17 years old, but most moose on Isle Royale die before reaching their 10th birthday.

Life is no picnic for Isle Royale's wolves, either. Although they can live to be 12 years old, most die by age 4. The most common cause of death is starvation. With few available food sources, a wolf may go 10 days without eating. Obtaining a meal on the 11th day may mean having to wrestle a 900-pound moose on an empty stomach.

The difficulty of finding food is just one obstacle for wolves. They also have a high incidence of bone deformities, which cause back pain and partial paralysis of the hind legs. In the early years of the Isle Royale study, such deformities were rare, but they've become more common in recent years, almost certainly a result of inbreeding (see Chapter 14). For the last 15 years, every dead

INFOGRAPHIC 20.9

A Warming Climate Influences Wolf and Moose Population Size

In recent years, climate change has become a significant influence on moose and wolf populations on Isle Royale. Warmer temperatures lead to increased tick infestations of moose, resulting in loss of hair and a weakened and depleted population. In cooler years, moose experience less hair loss and their populations grow.

One moose may be home to tens of thousands of ticks at a time.

Ticks cause moose to lose their hair, their appetite, and a good deal of blood.

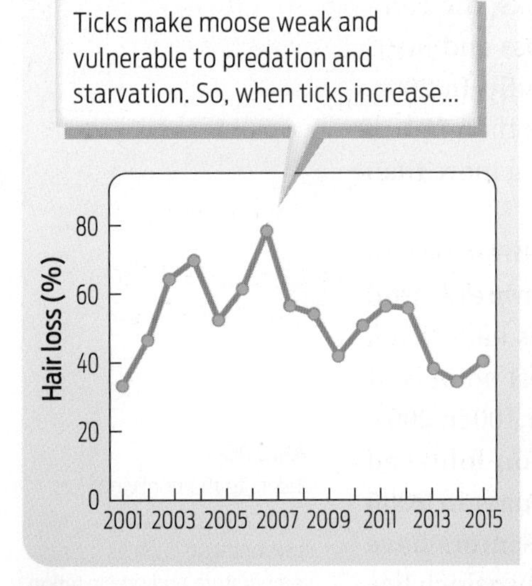

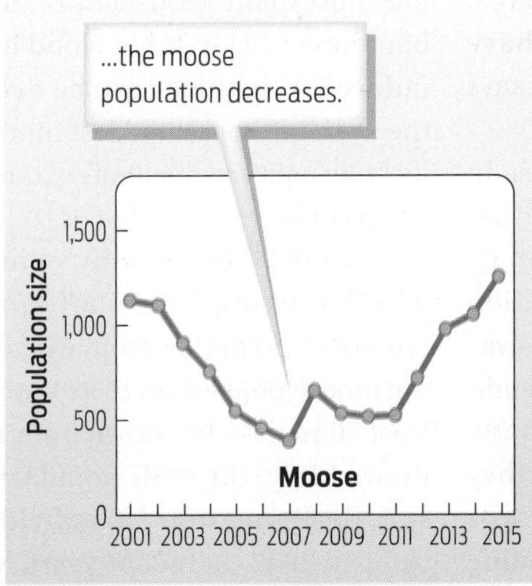

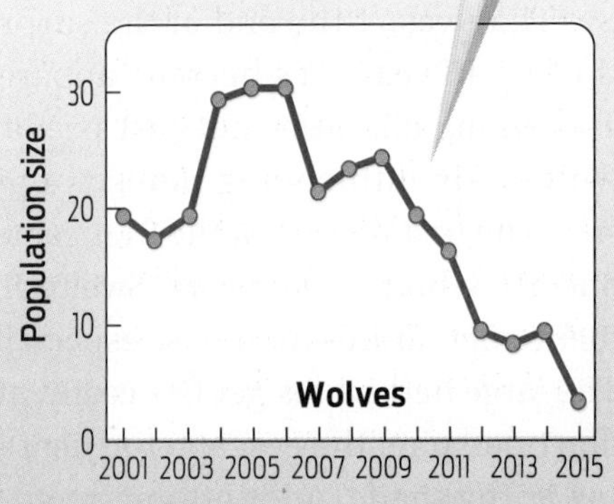

? How do particularly hot summers impact the moose population? Be as specific as you can.

wolf on Isle Royale has had such deformities. Vucetich thinks that climate change may be an indirect cause of the inbreeding: it lessens the frequency with which ice bridges form between Canada and Isle Royale—bridges over which new wolves could reach the island and diversify the wolf gene pool.

It's not the first time wolf populations have been in trouble. When European settlers first arrived in North America, the gray wolf roamed throughout all of the future 48 contiguous U.S. states. By 1914, hunting and trapping had greatly reduced the population, and survivors were limited to remote wooded regions of Michigan, Wisconsin, and Minnesota. The federal government officially listed the species as endangered in the early 1970s, when it seemed on the verge of extinction. Since then, as a result of conservation efforts, wolves in other areas have started to bounce back: there are an estimated 7,000 to 11,000 gray wolves in Alaska, and approximately 5,500 in the rest of the country. But on Isle Royale, wolves are at their lowest numbers since the study began: just two wolves were spotted in 2016. And these individuals—one male, one female—are severely inbred and unlikely to produce viable offspring.

The Isle Royale wolves' latest plight poses an ethical dilemma: should scientists intervene on their behalf—say, by importing wolves from another population to reintroduce genetic diversity—or should they just let nature take its course? It's a question that Vucetich thinks about a lot. The answer, he says, will require balancing a number of competing values—not just the value of individual populations, but the values of ecosystem health, scientific knowledge, and wilderness. Without wolves, for example, would the moose population once again explode and decimate the island's forest? Would healthier wolves be able to completely overwhelm moose, and drive them to extinction on the island? Does the fact that human-caused climate change has worsened these problems mean that we have a moral obligation to alleviate them? These are some of the difficult questions that wildlife managers will need to consider when debating whether and how to intervene.

"Every citizen has a stake in this question of how we relate to nature."

—John Vucetich

The dilemma is a familiar one to conservation biologists. According to Vucetich, these competing values show up in varying degrees in almost any management question that we have in any part of the world. They represent, he says, "this grand question of how should humans relate to nature." There are no easy or obvious answers. Nevertheless, he believes it is important for people to debate and discuss these issues—not just scientists and experts, but lay people, too, because "every citizen has a stake in this question of how we relate to nature."

In December 2016, the National Park Service proposed introducing up to 30 wolves over the next several years into Isle Royale—a recommendation based on results of an environmental impact report that states this is the only way to save the wolf population on the island and to preserve the natural laboratory. After a period of public engagement, the plan was approved, and in September 2018, the National Parks Service began to reintroduce wolves.

Will the new arrivals survive and form stable packs that can help keep the moose population in check? Only time will tell, but scientists and the public are watching closely. ■

Ken Canning/Getty Images

Wolves on Isle Royale are facing extinction.

CHAPTER 20 Summary

Driving Question 1 What is ecology, and what do ecologists study?

- Ecology is the study of the interactions among organisms and between organisms and their nonliving environment.
- Ecologists study these interactions at a number of levels, including individual, population, community, and ecosystem.
- Living organisms may have a clumped, random, or uniform dispersion pattern, depending on ecological and behavioral adaptations. Few organisms fall into strictly one category.

Driving Question 2 What are the different types of population growth?

- Population growth is an increase in the number of individuals in a population. The growth rate of a population, also known as the rate of natural increase, is defined as the birth rate minus the death rate.
- Exponential growth is the unrestricted growth experienced by a population growing at a constant rate. Logistic growth is the slowing of the growth of a population because of environmental factors such as crowding and lack of food.
- Carrying capacity is the maximum population size that an area can support, given its food supply and other life-sustaining resources. Populations cannot grow exponentially forever; eventually, they hit the carrying capacity for the region and stop growing.

Driving Question 3 What factors influence population growth and population size?

- Population growth can be limited by a variety of factors, including biotic (living) and abiotic (nonliving) parts of the environment.
- Density-independent factors, such as a severely cold winter, can affect a population of any size.
- Density-dependent factors, such as the presence of predators, have different impacts on the population depending on the population size and crowding of individuals.
- Populations in a community are interconnected, with the fate of one often influencing the fate of the others.

More to Explore

- Wolves and Moose on Isle Royale: http://isleroyalewolf.org
- Hoy, S. R., Peterson, R. O., and Vucetich, J. A. (2019). *Ecological Studies of Wolves on Isle Royale: Annual Report 2018–2019.*
- PBS. (2007). *Nature*: *In the Valley of the Wolves.*
- Mlot, C. (2019). Relocated island wolves outlasting mainland wolves in new Isle Royale home: www.sciencemag.org/news/2019/12/relocated-island-wolves-outlasting-mainland-wolves-new-isle-royale-home
- Seabird Monitoring on the Isle of May, Scotland: www.ceh.ac.uk/sci_programmes/IsleofMayLong-TermStudy.html

CHAPTER 20 Test Your Knowledge

Driving Question 1 What is ecology, and what do ecologists study?

By answering the questions below and studying Infographics 20.1, 20.2, 20.3, and 20.7, you should be able to generate an answer for this broader Driving Question.

Know It

1. What is the difference between a community and a population?

2. An ecosystem ecologist might study
   a. plant populations.
   b. herbivores that eat the plants.
   c. predators in the population.
   d. the impact of precipitation patterns on the plant populations.
   e. all of the above

3. Palm trees can be found at oases in deserts. How would the dispersion pattern of desert palm trees be described?

Use It

4. How would you explain to a 10-year-old child what ecologists do?

5. Your local environmental group wants to determine the population size for squirrels in a nearby nature preserve. What are some methods you could use to estimate the size of the squirrel population? Would the same approaches be as useful in determining the population size of maple trees in the same area? Why or why not?

6. An invasive plant species has been introduced into a particular region. It is displacing native plants in the area. What kinds of impact might you predict on other species in the community? Explain your answer.

Driving Question 2 What are the different types of population growth?

By answering the questions below and studying Infographic 20.4, you should be able to generate an answer for this broader Driving Question.

Know It

7. Which of the following would cause a population to grow?
   a. identical increases in both the birth rate and the death rate of a population
   b. a decrease in the birth rate and an increase in the death rate of a population
   c. an increase in the birth rate and a decrease in the death rate of a population
   d. an increase in the birth rate and a larger increase in the death rate of a population
   e. identical decreases in the birth rate and the death rate of the population

8. Which of the following statements describes an example of population growth?
   a. The average weight of Americans has increased substantially in the past decade.
   b. Tropical fish have been found in waters more northerly than their usual habitat.
   c. The number of people in a town has increased by 25% in the past 5 years.
   d. The number of butterflies in a region stayed the same from 1950 to 2010.
   e. all of the above

9. When a population reaches its carrying capacity, what happens to its growth rate?

Use It

10. The size of the wolf population on Isle Royale was nine in 2014, three in 2015, and two in 2016, 2017, and 2018.
    a. What must have happened to wolf birth and/or death rates between 2013 and 2016?
    b. What must have happened to wolf birth and/or death rates between 2016 and 2018?
    c. In late 2018 and early 2019, 15 wolves were introduced onto Isle Royale. One of these died, and one left the island on an ice bridge, leaving a total of 13 wolves on the island in March 2019. What was the primary mechanism affecting wolf population growth in this interval?
    d. What do you predict will happen to the moose population in the next year? Explain your answer.

Interpreting Data

Apply Your Knowledge

11. Population Q has 100 members. Population R has 10,000 members. Both are growing exponentially at a 5% annual growth rate.
    a. Which population will add more individuals in 1 year? Explain your answer.
    b. After 5 years, what will be the size of each population?
    c. If the larger population reaches its carrying capacity at the end of the third year, what will its size be after 5 years?

Driving Question 3 What factors influence population growth and population size?

By answering the questions below and studying Infographics 20.5, 20.6, 20.7, 20.8, and 20.9, you should be able to generate an answer for this broader Driving Question.

Know It

12. The Mexican gray wolf has been reintroduced into parts of New Mexico and Arizona. There have been several wolf deaths due to shootings and traffic kills. Are these influences on the wolf population biotic or abiotic factors?

13. Drought causes a pond to dry up almost completely. The pond's frog population drops to 10% of its initial size. What kind of factor—biotic or abiotic—has influenced the frog population?

14. Which of the following is a density-dependent factor influencing population growth?
    a. elevated temperature
    b. prolonged winters
    c. a viral disease
    d. a devastating forest fire
    e. all of the above

Use It

15. Why is it important for researchers to determine the cause of death of moose on Isle Royale? Can this information be used to help make predictions about moose and wolf populations? Explain your answer.

16. A group of predatory fish lives in a school in a large lake. If a parasite were introduced to the lake—for example, by a vacationing fisherman—would you expect it to have a greater impact on the population if the fish were at high density or at low density? (Assume the parasite is passed from one fish to another through the water but can remain alive in the water only for a very short period of time.) What would happen to this same population if there were a severe drought and very hot summer?

17. Classify each of the following as a biotic or an abiotic factor in an ecosystem. Then predict the impact of each factor on the moose population of Isle Royale. Explain your answers, keeping in mind possible interactions between the various factors and between the moose and wolf populations.

a. hot summer temperatures
b. ticks that parasitize moose
c. declining numbers of balsam fir trees
d. a parvovirus in wolves
e. deep winter snowfall

18. Assume that a new herbivore is added to Isle Royale that is not a prey for wolves. Predict the effect of this introduction on

a. the populations of trees.
b. the moose population.
c. the wolf population.

19. Using the estimates provided in the table, graph the moose population size on Isle Royale since 2009. Given the trend in the moose population, how can you explain the reduction in wolves from 24 in 2009 to 2 in 2016–2018?

Year	Estimated moose population
2009	530
2010	510
2011	515
2012	750
2013	975
2014	1,050
2015	1,250
2016	1,300
2017	1,600
2018	1,600
2019	2,060

Mini Case

Apply Your Knowledge

20. The wolves of Isle Royale are suffering from bone deformities, probably as a result of inbreeding in their small population.

a. Do you agree with the decision to allow humans to intervene in the wolf population on Isle Royale? Explain your answer.

b. Do you think that introducing new wolves to Isle Royale will be a long-term solution to wolf inbreeding on Isle Royale? Why or why not? *Hint:* Think about the likelihood of ice bridges to the mainland forming as the climate continues to warm.

Bring It Home

Apply Your Knowledge

21. Use your knowledge of ecology to plan a home saltwater aquarium. Think about the factors you need to consider to have a healthy community of fish and plants in your aquarium. Consider both biotic and abiotic factors and the population densities of the organisms in your aquarium.

What's Happening to Honey Bees?

The mysterious loss of bees and other pollinators has scientists worried

bo1982/Getty Images

DRIVING QUESTIONS

1. What are ecological communities, and what role do pollinators play in communities?
2. What types of feeding relationships exist within communities?
3. How do symbiotic relationships influence members of a community?
4. How do species interact in a space with shared resources?

"One in every three bites of food we eat is pollinated directly or indirectly by honey bees."

—Dennis vanEngelsdorp

Dave Hackenberg has been keeping bees for more than 40 years. Every spring, as flowering plants start to bloom, he trucks bees from his home in central Pennsylvania to farms around the country, where they help farmers pollinate local crops—everything from California almonds to Florida melons. In November 2006, as he had done for years, Hackenberg brought his buzzing cargo to his winter base in central Florida. When he dropped them off, his 400 healthy hives were "boiling over" with bees, he says. Three weeks later, when he returned to check on them, the bees had essentially vanished; only 40 healthy hives remained.

Mysteriously, there were no dead bees lying in or near the hives. Nor were there any signs of intruders who might have destroyed the hives in search of honey. The bees were simply gone. It was, as Hackenberg said, a bee ghost town.

"I literally got down on my hands and knees and looked between the stones for dead bees," says Hackenberg, but the beekeeper found none. "I was kind of speechless. And people know I'm not speechless."

Hackenberg's was the first reported case of what has since become known as colony collapse disorder (CCD). Hackenberg was not alone. Surveys conducted in 2007 and 2008 by the U.S. Department of Agriculture (USDA) and the Apiary Inspectors of America found that beekeepers all across the United States had suffered similar unexplained devastation, losing anywhere from 30% to 90% of their colonies during the winter season.

COMMUNITY
A group of interacting populations of different species living in the same area.

Since those first losses, more than 10 million honey bee colonies across the United States have been wiped out, with American beekeepers losing an average of 30% of their colonies every winter since 2006. While some annual loss of bee colonies is considered normal, these numbers are more than double historical averages. In 2019, average winter losses surpassed 37%, the highest level reported since the survey began. These levels of colony loss are unsustainable and pose a real threat to the beekeeping industry **(INFOGRAPHIC 21.1)**.

What's happening to honey bees? No one knows for sure, but it's a predicament that has beekeepers, farmers, and scientists racing to understand and combat the plight of the precious pollinator. At stake are not only billions of dollars' worth of agricultural crops, but also the health and diversity of natural communities that rely on the valuable services of bees.

Community Partners

▶ Pollination and pollinators

To an ecologist, a **community** is a group of populations of different species that live and interact together in a particular area.

INFOGRAPHIC 21.1

Annual Losses of Managed Honey Bee Colonies

Beekeepers complete annual surveys to report bee colony losses, allowing the problem to be tracked by region and over time. Losses have been increasing over time. In 2019, colony losses of between 12% and 66% were recorded across the United States. Researchers are working to identify causes of colony loss.

Few bees still remain in this hive displaying colony collapse disorder.

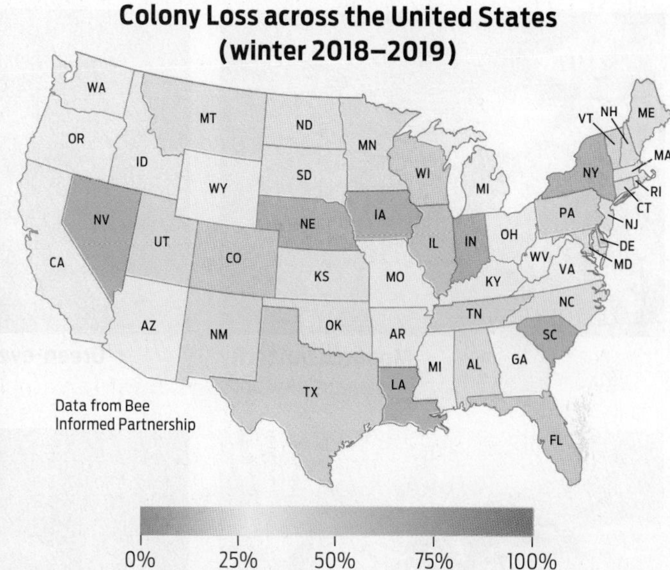

? Name one state with particularly high honey bee colony losses, and one state with particularly low colony losses.

Members of a community can depend on each other in complex ways as a result of coevolving together in the same space. Nowhere is that more apparent than in the relationship between pollinators, like bees, and the flowers they pollinate. This relationship took time to evolve, and the consequences for both groups of organisms were profound.

A hundred million years ago, when dinosaurs roamed Earth, the plant world was dominated by cone-bearing conifers such as pine and redwood trees. These trees spread their **pollen** by the wind; when the pollen reaches another plant, the sperm in the pollen can fertilize an egg. But a new type of **pollination** was evolving at this time, one that would forever change the landscape: pollination by insects.

With the arrival on the evolutionary scene of pollinating insects like bees and butterflies, flowering plants multiplied, diversified into new species, and spread around the globe. Pollination is how flowering plants have sex. A flower is the reproductive hub of the plant—where its reproductive organs are located. For a flowering plant to reproduce, sperm from pollen must find its way to the eggs of a female plant of the same species. Pollinators make this possible for a huge number of species. The great success of flowering plants owes everything to the reproductive advantage of relying on insects and other animals, rather than wind, to deliver pollen **(INFOGRAPHIC 21.2)**.

Wind pollination is like junk mail: you need to send thousands of letters (pollen grains) to hook just one receptive customer—or in this case, to fertilize just one egg. Pollinating insects, on the other hand, are like FedEx: they deliver a pollen package directly to the appropriate recipient, contributing to the reproductive success of the plants. This focused delivery method allows animal-pollinated plants to manufacture many fewer pollen grains and thus waste less energy.

POLLEN
Small, thick-walled plant structures that contain cells that develop into sperm.

POLLINATION
The transfer of pollen from male to female plant structures so that fertilization can occur.

INFOGRAPHIC 21.2

Most Flowering Plants Rely on Animal Pollinators for Their Reproduction

Flowering plants attract pollinators with their colorful flowers and fragrant nectar. When a pollinator such as a bee visits a flower to collect energy-rich nectar, it also picks up pollen. When it visits another flower, the pollen is transferred to that flower, helping the plant to reproduce. A variety of animals pollinate plants that are important to agriculture and to natural communities.

Silrshoot/Getty Images

Monarch butterfly
Danaus plexippus

Alexander Gabrysch/Getty Images

Green-eyed hook-tailed dragonfly
Onychogomphus forcipatus

Biosphoto/Claudius Thiriet

Bumble bee
Bombus

Glenn Bartley/AGE Fotostock

Fork-tailed woodnymph hummingbird
Thalurania furcata

Dr. Merlin D. Tuttle/Science Source

Long-nosed bat
Leptonycteris curasoae

? Bees can transfer pollen between plants. What is the benefit to the bees?

Of the 250,000 species of flowering plants that exist worldwide, 90% depend on animal pollinators to reproduce. The vast majority of these pollinators are insects—mostly beetles, bees, butterflies, and moths.

The majority of beekeepers in the United States and Europe cultivate the bee species *Apis mellifera,* the Western honey bee. It is hard to overestimate the importance of this tiny pollinator to modern agriculture. In the United States alone, at least 90 different commercial crops—worth an estimated $15 billion annually—depend on honey bee pollination, including apples, oranges, blueberries, melons, pears, pumpkins, cucumbers, cherries, raspberries, broccoli, avocados, asparagus, clover, alfalfa, and almonds. Globally, 87 of the 115 leading food crops depend on pollinators, the most important of which are honey bees. Pollinator-dependent crops represent about 35% of global crop production.

"One in every three bites of food we eat is pollinated directly or indirectly by honey bees," says Dennis vanEngelsdorp, an apiarist at the University of Maryland and project director for the Bee Informed Partnership, which has been monitoring bee losses. Without honey bees, he says, we wouldn't starve: we would still have wheat, rice, corn, and other crops that are either wind- or self-pollinated. Nevertheless, many of our

INFOGRAPHIC 21.3

Commercial Crops Require Bees

Many of the crops that we rely on for food, fuel, and fiber rely on bees for their pollination and reproduction.

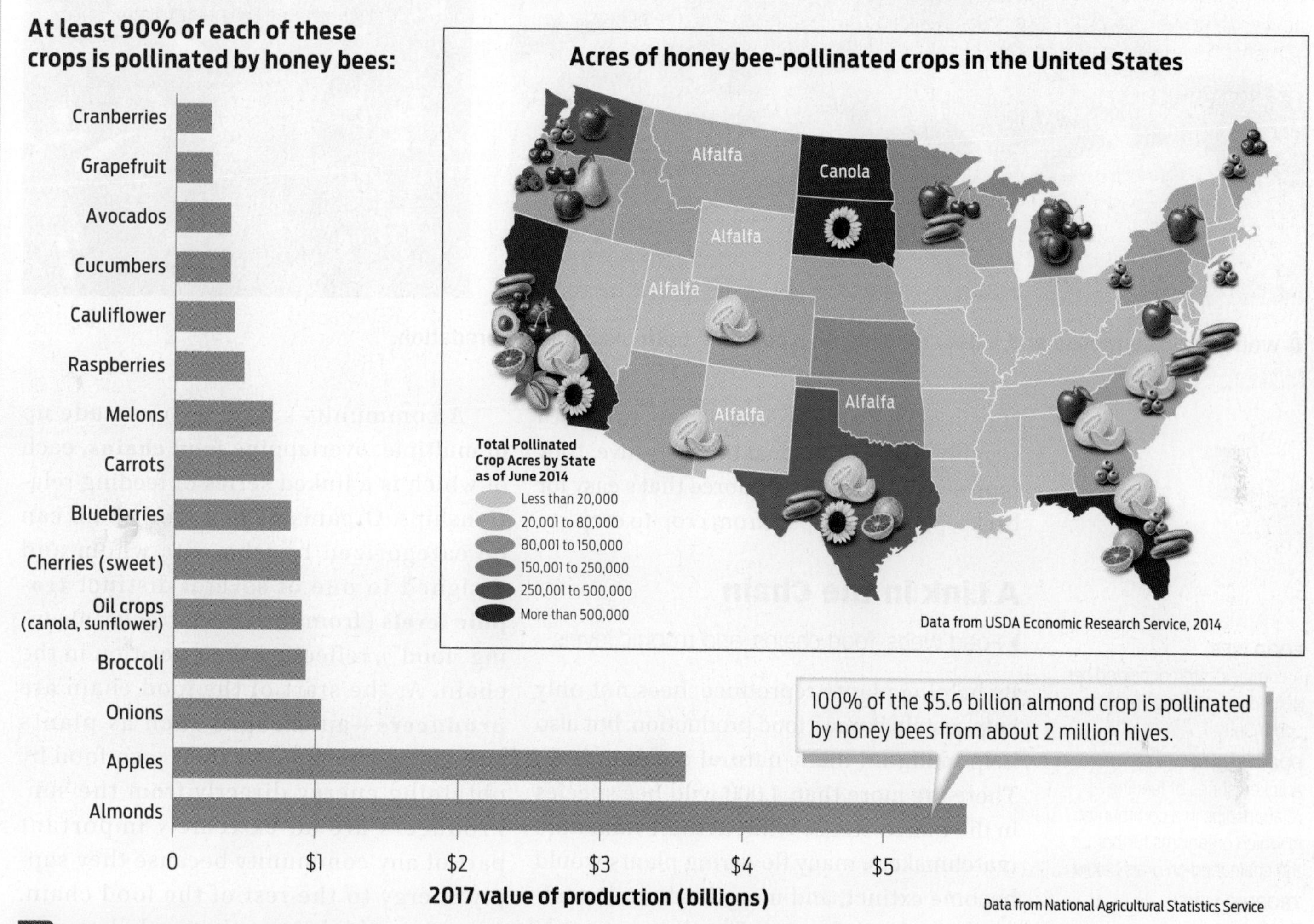

? **From the data presented, what is the total market value of honey bee pollination of raspberries, avocados, and apples added together?**

favorite foods might no longer grace our tables **(INFOGRAPHIC 21.3)**.

For bees, flowers provide food: they contain the protein-rich pollen and sugary nectar that bees need to nourish themselves and their hives. By using their long tonguelike structure called a proboscis, bees are able to reach deep into a flower to draw out the nectar. Because they are fuzzy and have a slight electrical charge, bees attract pollen as they snuggle up to a flower, much the same way that warm socks attract other clothes as they come out of the dryer. The bees then transfer this pollen to other plants as they continue their hunt for food.

Honey bees are more efficient at pollination than many other types of pollinators, which is why farmers have come to depend on them. The average honey bee will make 12 or more foraging trips a day, visiting several thousand flowers. On each trip, she (the foragers are all female) will confine herself to flowers from a single plant species, thus ensuring delivery of the proper pollen. Because they come and go from a central home base—their hive—honey bees can be counted on to

Carnivory

Gail Shumway/Getty Images

Herbivory

John Serrao/Science Source

A wolf hunting a mouse and a deer nibbling on a bush are both examples of predation.

stay in a fixed area around a crop. And with roughly 40,000 individual bees per hive, they represent a versatile workforce that's easy for beekeepers to transport from crop to crop.

A Link in the Chain

▶ Food webs, food chains, and trophic levels

By helping plants reproduce, bees not only help sustain human food production, but also help maintain many natural communities. There are more than 4,000 wild bee species in the United States. Without these miniature matchmakers, many flowering plants would become extinct, and many birds and mammals would go hungry. That's because, as pollinators, bees help maintain the integrity of the **food web,** the complex, interconnected set of feeding relationships found within a community.

Take the tiny blueberry bee (*Osmia ribifloris*), which feeds on the flowers of blueberry plants. Just one of these speedy pollinators can visit 50,000 blueberry flowers over the course of a few weeks in spring, helping to produce more than 6,000 blueberries. All those blueberries are an important food source for many wild animals, including bluebirds, robins, foxes, mice, rabbits, chipmunks, and deer. In turn, the small animals that eat the blueberries are fed upon by larger carnivorous animals, such as hawks and coyotes. Without the pollinating bees, the web begins to unravel.

A community's food web is made up of multiple, overlapping **food chains,** each of which is a linked series of feeding relationships. Organisms in a food chain can be categorized by who eats whom and assigned to one of several distinct **trophic levels** (from the Greek *trophé*, meaning "food"), reflecting their position in the chain. At the start of the food chain are **producers**—autotrophs such as plants and algae, which make their own food by obtaining energy directly from the sun. Producers are an extremely important part of any community because they supply energy to the rest of the food chain. Organisms higher up the food chain are **consumers**—heterotrophic organisms that eat the producers or eat other organisms lower on the chain to obtain energy.

A food chain may include several levels of consumers. The first level of consumers (primary consumers) are **herbivores,** animals that eat plants. At the second level (secondary consumers) are meat-eating **carnivores** that eat the herbivores. Both herbivory and carnivory are forms of **predation,** an interaction in which one organism feeds on another. At the top of the food chain are the top consumers—animals such as coyotes, hawks, and wolves (as well as meat-eating humans), which have few natural predators and are not generally eaten by anything else in the community **(INFOGRAPHIC 21.4).**

FOOD WEB
A complex, interconnected set of feeding relationships in a community.

FOOD CHAIN
A linked series of feeding relationships in a community in which organisms further up the chain feed on ones below.

TROPHIC LEVEL
The feeding level of an organism, reflecting its position in a food chain.

PRODUCERS
Autotrophs (photosynthetic organisms) that obtain energy directly from the sun and form the base of every food chain.

CONSUMERS
Heterotrophs that eat other organisms lower on the food chain to obtain energy.

HERBIVORE
An organism that eats plants.

CARNIVORE
An organism (typically an animal) that eats animals.

PREDATION
An interaction between two organisms in which one organism (the predator) feeds on the other (the prey).

INFOGRAPHIC 21.4

Who Eats Whom: Food Webs and Food Chains

Based on the way they obtain food, organisms in a community can be arranged into a food web, which contains multiple intersecting food chains. The energy in a food web is captured from the sun by producers. Energy and nutrients are then transferred to other organisms in other trophic levels as they feed on producers (herbivores) and other consumers (carnivores).

Food Web

A food web shows the complex network of feeding relationships in a community. Arrows point to the organism doing the eating.

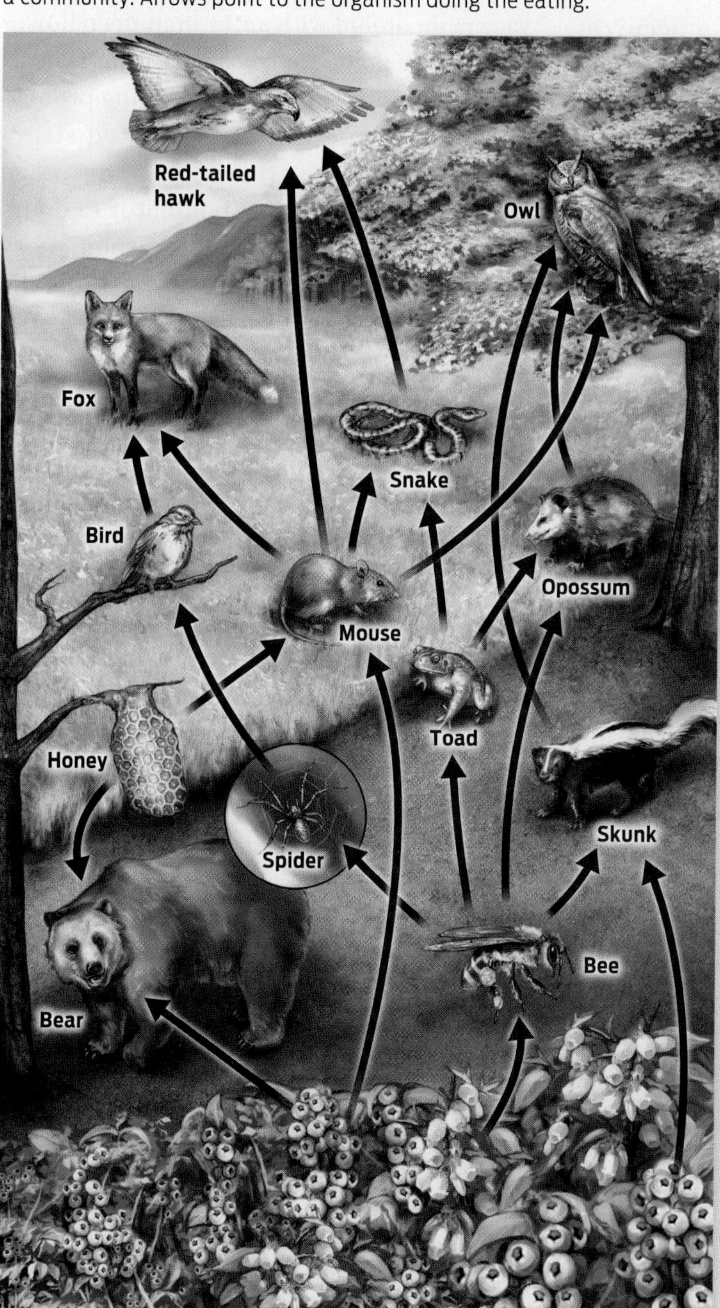

Food Chain

A food chain is a linked sequence of feeding relationships within a food web. Individuals in a food chain are classified into trophic levels based on their position within the chain.

? **Bears eat plants and berries, and also eat other animals (from insects to fish). How can bears be characterized: as producers? as consumers? as herbivores? as carnivores?**

A Swarm of Problems

▶ Parasitism, mutualism, commensalism

The United States is home to approximately 1,000 commercial beekeepers, who together cultivate about 2.5 million bee colonies. To a beekeeper—a small business owner—losing 40% of their colonies is an unsustainable financial loss. The sudden bee disappearances that have been occurring since 2006 are a serious worry for beekeepers.

While recent losses have indeed been considerable, this is actually not the first time that beekeepers' livelihoods have been hit hard. Since 1987, beekeepers have had to battle significant annual losses from an aggressive pest: the blood-sucking varroa mite.

"All the evidence so far has really supported the idea that it's likely a combination of factors that are stressing the bees beyond their ability to cope."

—Maryann Frazier

A non-native species, the varroa mite was likely introduced into the United States on the backs of imported bees. The sesame seed–sized freeloader is a parasite that feeds on bees' blood, weakening the bees' immune systems and spreading viruses. This feeding relationship is a type of **symbiosis**—a close relationship between two species. The form of symbiosis observed between a parasite and its host is called **parasitism.** In parasitism, one species (in this case, the mite) clearly benefits, and one species (in this case, the honey bee) clearly loses. Because it involves one species feeding on another, parasitism is also a form of predation.

Not all symbiotic relationships are harmful to one of the partners; indeed, these partnerships can sometimes be mutually beneficial. Bees and flowering plants are a perfect example of one such **mutualism.** Bees can't survive without the flowers, which provide food, and plants depend on the bees to help them reproduce. Honey bees have other mutualistic symbioses as well, including with bacteria that live safely inside the bees and benefit their hosts by helping them combat disease. Mutualisms can sometimes be very specific. Cacao trees, from which we get chocolate, have tiny flowers that can only be pollinated by the tiny midge fly. In Madagascar, an orchid with an 11-inch-deep flower can only be pollinated by a particular species of sphinx moth with an equally long proboscis. These sometimes-exaggerated structures testify to the evolutionary advantage for a plant of securing a pollinator to do its reproduction business, and the evolutionary advantage for a pollinator of having a source of food that only it can access.

A third type of symbiotic relationship is **commensalism,** a relationship in which one species benefits while the other is unaffected or unharmed. Bees living in a hollowed-out oak tree is one example. Another is the way certain bird species, such as egrets, travel on the backs of larger animals, such as water buffalo, to capture insects that are stirred up from the ground by the grazing animal's movements **(INFOGRAPHIC 21.5).**

As devastating as the parasitic varroa mite infestation has been and continues to be, it is unlikely to be the sole or even the primary factor responsible for the honey bee colony collapses. Research by apiarist vanEngelsdorp and others has shown that levels of mite infections in collapsing colonies are no higher than they had been in previous years. Moreover, a mite infestation does not explain the most curious aspect of the condition: the sudden disappearance of entire hives.

Not surprisingly, the sudden disappearances have fueled intense speculation among beekeepers and laypeople alike about what's going on. Hypotheses have included everything from pesticides, viruses, and genetically modified crops to cell phone radiation, global warming, and even alien abduction. It's a baffling who-done-it with many suspects but no smoking gun.

SYMBIOSIS
A relationship in which two different organisms live together, often interdependently.

PARASITISM
A type of symbiotic relationship in which one member benefits at the expense of the other.

MUTUALISM
A type of symbiotic relationship in which both members benefit; a "win–win" relationship.

COMMENSALISM
A type of symbiotic relationship in which one member benefits and the other is unharmed.

INFOGRAPHIC 21.5

Organisms May Live Together in Symbioses

Symbioses are relationships in which different species live together in close association. These associations can provide benefits, harm, or have no effect on the partners involved.

Mutualism: Both species benefit from the relationship

Bees and flowers
Bees get nectar and pollen for food. Flowers benefit from successful reproduction.

Anand Varma/National Geographic Image Collection

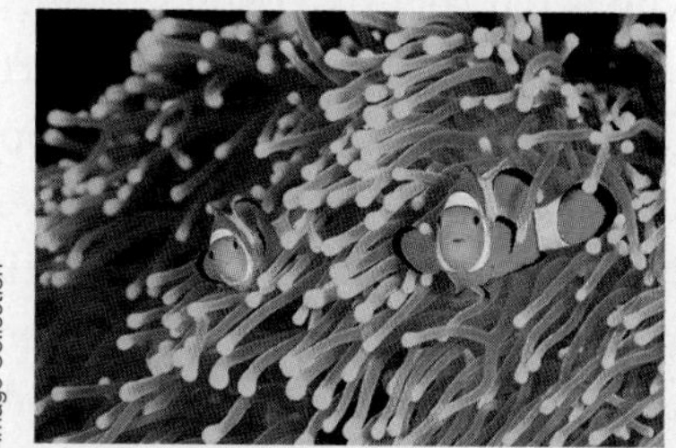

ullstein bild/Getty Images

Clownfish and anemone
Clownfish nestle in the tentacles of a poisonous anemone. The clownfish is protected from predators, and the anemone feeds on other fish attracted by the clownfish.

Parasitism: One species benefits and the other is harmed

Varroa mites and bees
The varroa mite feeds on bee adults and larvae. The bee's immune system is compromised and it becomes susceptible to disease.

Mirko Graul/Shutterstock

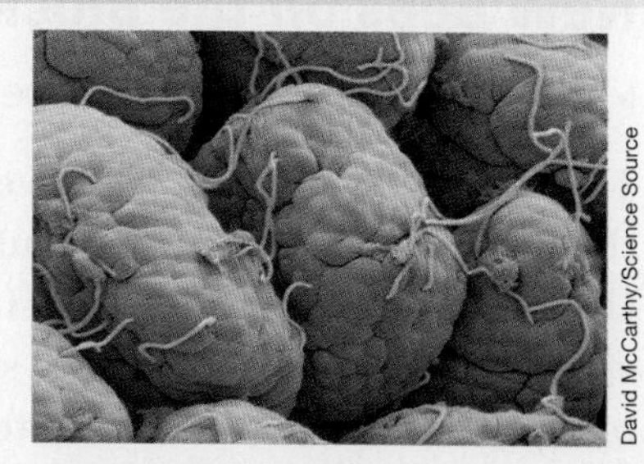

David McCarthy/Science Source

Pinworms and humans
The pinworm infects humans and feeds on food in the human gut. The human suffers discomfort.

Commensalism: One species benefits and the other is unharmed

Bees and trees
Natural bee swarms make safe hives on trees. The tree is not harmed.

Tim Graham/Getty Images

John Elk III/Getty Images

Egrets and water buffalos
The egret feeds on bugs that are stirred up by the African water buffalo. The water buffalo neither benefits nor is bothered.

? **What is the difference between commensalism and mutualism?**

Honey Bee Forensics

Among the first to investigate the die-offs was a team of Pennsylvania State University biologists headed by vanEngelsdorp and Diana Cox-Foster. It was Cox-Foster whom beekeeper Hackenberg called the day in 2006 he saw that his bees had gone missing.

The team started their investigation by performing autopsies on the few remaining bees in Hackenberg's colonies. When vanEngelsdorp looked through his microscope, he was shocked: "I found a lot of different scar tissue, and [what] looked like foreign organs," he says. There were also signs of multiple infections, including those caused by a parasitic fungus called *Nosema ceranae*. The bees' insides were overrun with pathogens.

Though the bees were clearly sick, each collapsed colony seemed to suffer from a different spectrum of ailments. The researchers hypothesized that something had damaged the bees' immune system, leaving them vulnerable to infections that a healthy colony could normally fend off. Some observers have even likened the condition to "bee AIDS."

In fact, there may be no single cause of honey bee losses, but rather a complex combination of causes. "All the evidence so far has really supported the idea that it's likely a combination of factors that are stressing the bees beyond their ability to cope," says Maryann Frazier, a bee researcher at Penn State who is part of Cox-Foster's team.

One factor that almost certainly plays a role in honey bee losses is poor nutrition. Just like humans, bees need a well-balanced diet that contains all the essential nutrients to remain healthy. For a number of reasons, honey bees are finding it harder and harder to obtain a nutritious diet.

New Bees on the Block

▸ Niche, competition, invasive species

The Western honey bee is native to Europe, the Middle East, and Africa, but not to the United States, making it an **introduced species** here. When European settlers first brought the Western honey bee to the United States in the 1600s, the bees quickly spread from managed colonies into the wild, in some cases displacing native bee species. Native Americans even noted a correlation between honey bees and European settlers, with the presence of one signaling the arrival of the other.

Honey bees are excellent at colonizing new habitats because they are largely **generalists** when it comes to flower choice. Though they tend to visit a single species of flower on each foraging trip, honey bees may visit more than 100 species of flowers within a single geographic region over the course of a season. In warm climates, they are active year round. They tend to feed throughout the day and start foraging earlier in the morning than many native bee species. In other words, honey bees have a broad ecological **niche**—the space, environmental conditions, and resources (including other living species) that a species needs to survive and reproduce.

Many other pollinators are **specialists** when it comes to foraging, meaning they prefer to visit only one or a few species of flower. For example, blueberry bees are partial to flowers of the blueberry bush. They forage on this one type of flower and don't bother visiting others. You might think it would be a great disadvantage to be specialist, since it limits your feeding options. But being a specialist has its advantages, especially if other pollinators are not as good as obtaining pollen from that flower. That's exactly the case with blueberry bees. While honey bees may visit blueberry flowers, they are not as effective at releasing pollen from them. Because of the flowers' unique shape, they require a bee species that practices "buzz pollination," beating their wings to shake the pollen loose. Blueberry bees excel at this maneuver, while honey bees do not, and they get more pollen as a result.

When two or more species must rely on the same limited resources—that is, when their niches overlap—the result is **competition.** Competition tends to limit the size of competing populations and may even drive one out. An important idea in ecology called the **competitive exclusion principle** holds that no two species can successfully coexist in identical niches in a community because one will eventually out-compete the other. That's because even a small advantage in one species' ability to find and secure food will, over time, lead to that species out-reproducing the other and driving it to extinction in that niche.

One of the fiercest competitors of the Western honey bee is the African honey bee (*Apis mellifera scutellata*). Bees of this sub-species were brought from Africa to Brazil in 1956 and quickly expanded their range to include Central America and the southern United States. African honey bees are better foragers than Western honey bees and they are less vulnerable to varoa mite infections. They are also much more aggressive than Western honey bees. They will chase away other pollinators from food sources, and even swarm and sting animals that get near their hives—behaviors earning them the colloquial name "killer bees" **(INFOGRAPHIC 21.6).**

INTRODUCED SPECIES
Species that are not native to a particular environment and that have arrived as a result of human activity.

GENERALIST
An organism that can be successful in a variety of environments, often because of a varied diet.

NICHE
The space, environmental conditions, and resources that a species needs to survive and reproduce.

SPECIALIST
An organism that requires a specific resource or habitat to be successful.

COMPETITION
An interaction between two or more organisms that rely on a common resource that is not available in sufficient quantities.

COMPETITIVE EXCLUSION PRINCIPLE
The concept that when two species compete for resources in an identical niche, one is inevitably eliminated from that niche.

INFOGRAPHIC 21.6

Organisms Compete for Resources

Species living in the same area may compete for resources such as food. Species may out-compete one another for the resource, or find a balance, depending on their abilities and behaviors.

Resource Competition between Western Honey Bees and African Honey Bees

Western honey bee (*Apis mellifera*)

VS

African honey bee (*Apis mellifera scutellata*)

Outcome: Competitive Exclusion

African honey bees use aggressive behavior to outcompete Western honey bees for flower resources.

Aggressive Behavior

- E.g., Invasive African "killer" honey bee
- Feeds on a variety of flowers also used by other bee species
- Aggressively chases other pollinators away from available food

Resource Competition between Western Honey Bees and Blueberry Bees

Western honey bee (*Apis mellifera*)

VS

Blueberry bee (*Osmia ribifloris*)

Outcome: Resource Partitioning

Blueberry bees have a variety of adaptations to outcompete honey bees for a single flower type. Honey bees survive on a variety of other flower types.

Specialist

- E.g., native blueberry bee
- Feeds on one food source (blueberry flowers)
- Successful in competition with other bee species, like the honey bee, that are not specialized to feed on this flower

Generalist

- E.g., Introduced honey bee
- Forages over great distances
- Feeds on a wide variety of flowers, rather than relying on a single flower type

? **Are killer bees relying on resource partitioning or on competition for success in their niche?**

Competition puts selection pressure on species to evolve adaptations to carve out a new niche—for example, by taking advantage of a different, less coveted food source. This selection pressure helps explain why many pollinators have distinct flower preferences, especially in terms of color. Many bees, for instance, are attracted to brightly colored blossoms with yellow, blue, or purple petals—but not red ones. Most butterflies, by contrast, are commonly attracted to red flowers that are large and easy to land on. Moth-pollinated flowers tend to have pale or white petals with no distinctive color pattern but with strong fragrance. These different adaptations allow different pollinators to occupy the same physical area but have different niches, based on how they are using resources. The use of different resources by different species in a given area enabling them to divide a niche is known as **resource partitioning** (or niche partitioning) **(INFOGRAPHIC 21.7)**.

RESOURCE PARTITIONING
The use of different resources by different species in a given area, enabling them to divide a niche.

INVASIVE SPECIES
Introduced species that do harm in their new environment.

African honey bees are an invasive species. **Invasive species** are introduced species that compete with native species for resources and may even drive some native species to extinction. Because an introduced species by definition didn't evolve in this new environment, it may not have natural predators, so it can breed and spread quickly. Native

INFOGRAPHIC 21.7

Organisms Exist in the Same Environment by Resource Partitioning

Pollinators have different ecological niches, even within the same general space. They may require the same seasonal temperatures, structures for shelter, yearly rainfall, and food from the nectar and pollen of flowers, but prefer flowers of different size, color, and shape. Such specialization helps species avoid competition for shared resources.

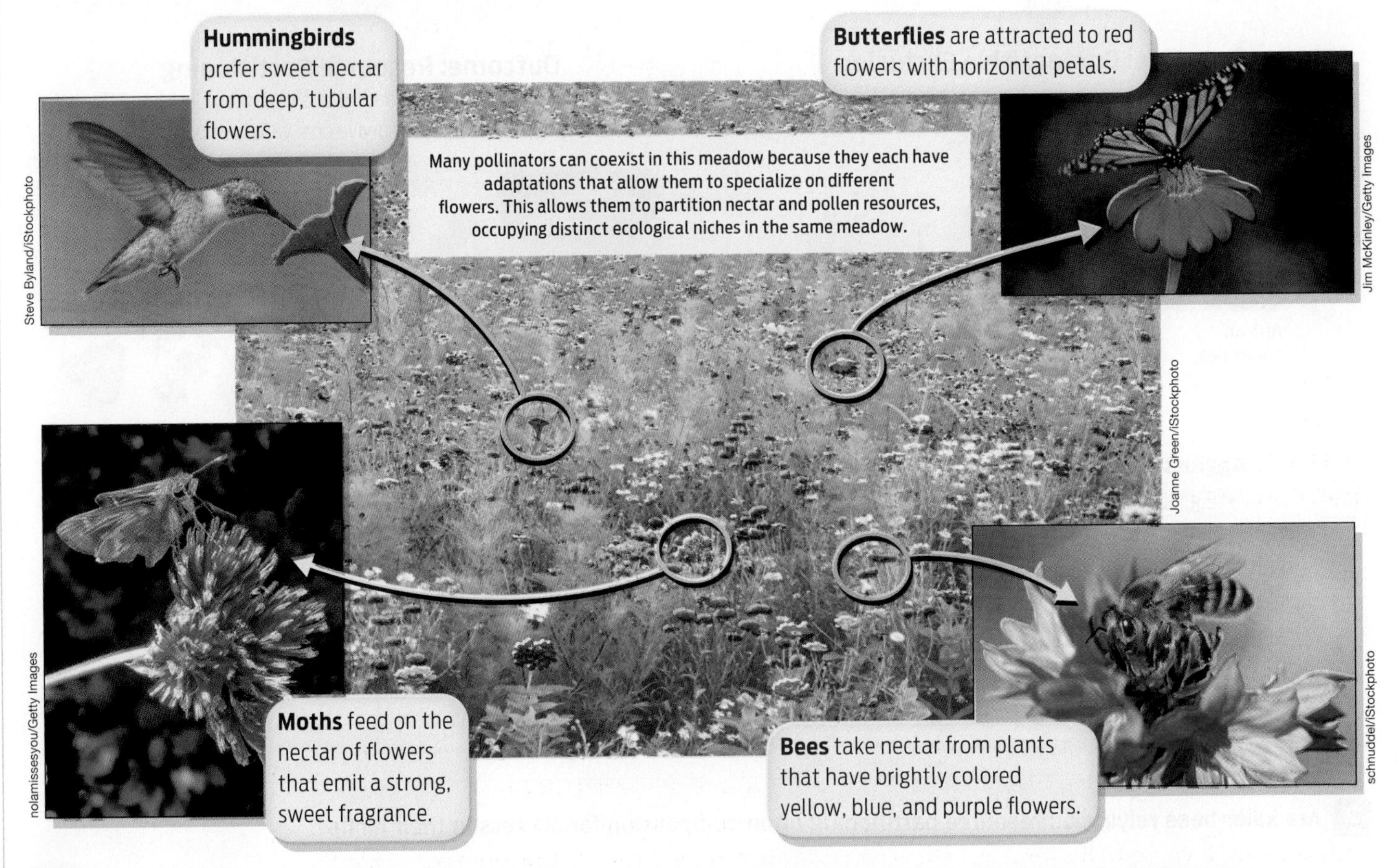

? **If bees, butterflies, and hummingbirds all rely on nectar, how can they coexist in a meadow of wildflowers?**

Curt Pickens/iStockphoto

mrolands/Featurepics.com

clu/iStock/Getty Images

Well-manicured lawns (top left) and farms planted with monocrops (top right) are "a desert to pollinators." Yards with wildflowers and diverse plants (bottom) are more welcoming to pollinating creatures.

wildlife may not be able to compete with a species that has no predators. Some scientists are worried that killer bees from Central America may displace or interbreed with U.S. populations of Western honey bees, with potentially disastrous results for beekeeping and agriculture. This hasn't happened—yet.

A much more serious problem for the Western honey bee, it seems, is competition from an even more dangerous species: humans. Humans and their activities have limited the resources used by bees and other pollinators. Agriculture, suburban sprawl, and development, for example, have all decreased bees' natural foraging areas and fragmented their habitat into nonoverlapping zones. Unable to access as many resources in a single foraging trip, bees must compete with each other in the patches that remain.

The familiar well-manicured lawn is particularly problematic. An immense stretch of green grass and no flowers, a lawn is "basically a desert to pollinators," says Frazier. There is literally nothing for them to eat. Likewise, many agricultural areas are planted with monocultures (that is, single crops), which all bloom at the same time, leaving no flowers for the rest of the year. This is why farmers have to bring in honey bees to fertilize certain crops, and hives are moved from crop to crop around the country; native bees can't easily live in these agricultural deserts, eating just once a year.

Worse yet, certain genetically engineered crops are pollen free. Bees visit these crops searching for food, only to leave them hungry. And certain non-native plants have floral structures that are inaccessible to native pollinating insects.

"It's not only the honey bees that are in trouble. All the beneficial insects are in a bad situation."

—Dave Hackenberg

As a consequence of these and other human actions, in some geographic regions bees must compete both with one another and with other pollinators for a dwindling supply of food-providing flowers. Many are going hungry, and some are in danger of being forced out completely.

Honey Bee in the Coal Mine?

Honey bees aren't the only pollinator in peril. According to a 2016 report issued by the Intergovernmental Science-Policy Platform on Biodiversity and Ecosystem Services, the number and abundance of pollinator species have declined greatly over the last several years. In fact, several bumble bee species are becoming or have become extinct in North America. The rusty patched bumble bee (*Bombus affinis*), for example, was once found throughout the eastern United States and southeastern Canada. Now, it is found only in small populations in a few Midwestern states. Over the course of just 20 years, its population has shrunk by approximately 90%. In 2017, the U.S. Fish and Wildlife Service officially listed it as an endangered species. This decline is especially disconcerting because the rusty patched bumble bee is what ecologists call a "keystone species"—it plays an outsized role in supporting other species in the community (see Chapter 22).

The concern among researchers and beekeepers is that honey bees may be the "canary in the coal mine," forecasting what's in store for other pollinators. "It's not only the honey bees that are in trouble," says Hackenberg. "All the beneficial insects are in a bad situation."

What's ailing these insects? In addition to a shrinking and fragmented habitat, a disquieting possibility is that they are being poisoned by pesticides. Penn State researcher Frazier and her colleagues have looked at pollen and wax from beehives and found large amounts of many different kinds of pesticides. Some are approaching toxic levels for the bees. "Pesticides are definitely in the mix, and we think they are definitely a player in the stresses that bees are experiencing," says Frazier.

Of particular concern to beekeepers is a class of pesticides known as neonicotinoids, or "neonics" for short. Neonics are an artificial form of nicotine heavily used in commercial agriculture all over the world. (Nicotine, produced by tobacco plants, is a natural deterrent to plant-eating insects.) Virtually all corn and most soybean seeds in the United States are treated with neonics. Research has shown that neonics can impair honey bees' ability to find and return to their hives—a possible explanation for why collapsing colonies are often missing most of their bees. The U.S. Environmental Protection Agency (EPA) acknowledges that neonics are "highly toxic to honey bees" and has begun a detailed review of neonic toxicity. In May 2019, the EPA announced it was banning the sale of 12 neonic-containing pesticides. A recent report produced by the USDA acknowledged that neonics and other pesticides may be a contributing factor to CCD but stopped short of singling them out as a primary cause. "The bottom line is we have not been able to put together, here in the United States, the evidence that neonicotinoids are causing the decline in honey bees," says Frazier. "But we are not convinced that they are *not* playing a role."

The situation is a bit different in Europe. A report by the European Food Safety Authority identified a number of risks posed to bees from the pesticides. In response, in 2014, the European Union imposed a temporary continent-wide ban on applying neonicotinoids to flowering crops such as corn, rapeseed, and sunflowers that are attractive to bees. The ban allowed researchers the chance to study the effects of neonics more thoroughly—especially at the sublethal doses bees are likely to encounter. As a result, three neonics were banned for outdoor use in 2018.

Although researchers have not yet been able to prove definitely that neonics are playing a role in bee losses—"the jury is still out," says Frazier—beekeepers like Hackenberg are understandably cautious about the pesticides to which they expose their colonies **(INFOGRAPHIC 21.8)**.

INFOGRAPHIC 21.8

What Is Causing Colony Losses?

The collapse and loss of colonies all over the world is of great concern. The causes of these losses are likely to be complex and to involve an interplay of several factors.

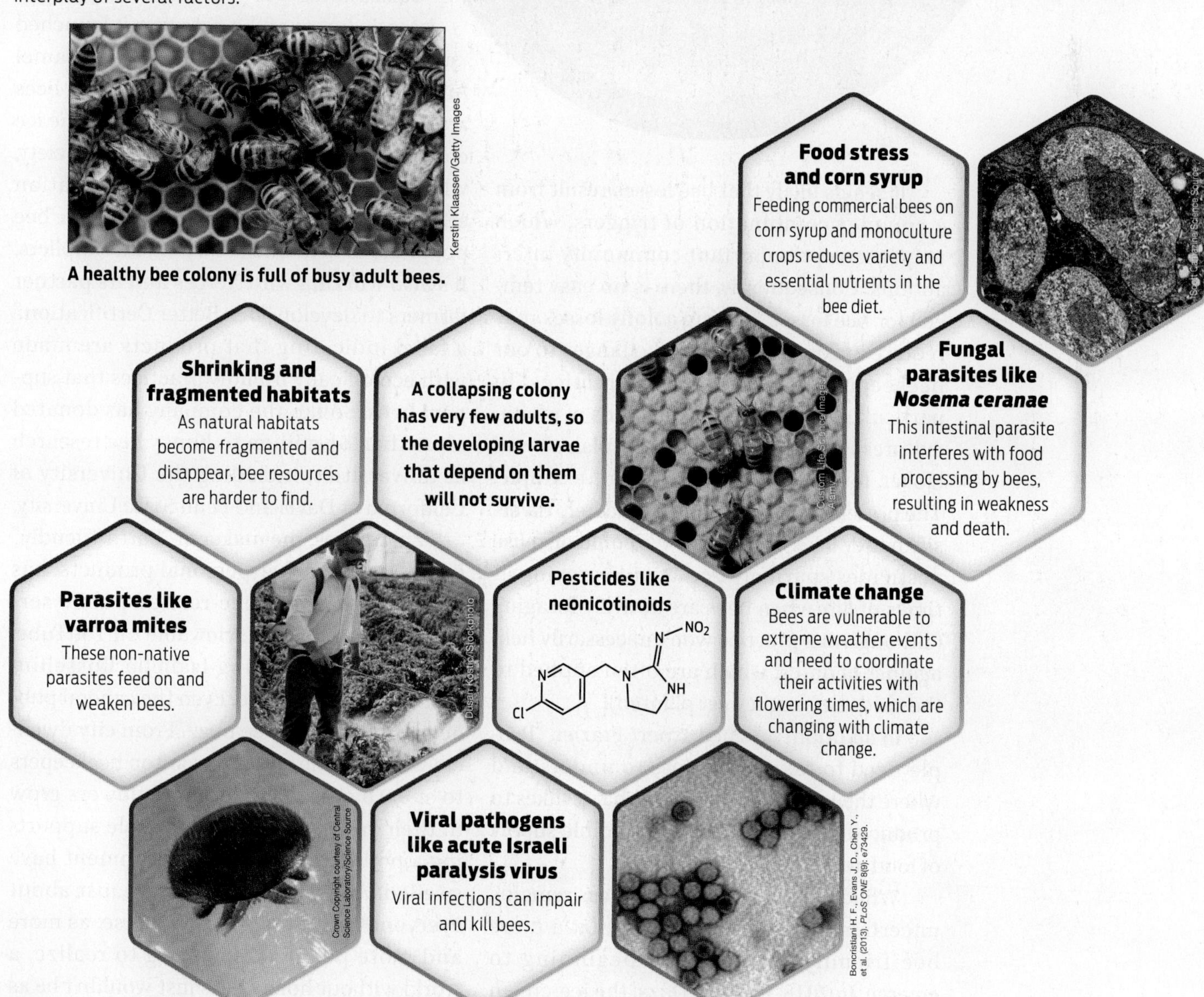

? **Why is the fate of honey bees important to humans?**

Häagen-Dazs donates funds to support honey bee research, acknowledging the essential role that bees play in 26 of its 60 ice cream flavors.

Valentyn75/Dreamstime.com

© HDIP, Inc.

It seems likely that bee losses result from a complex combination of triggers, which may interrupt important community interactions. Consequently, there is no easy remedy for bee losses. Halting colony losses may require making fundamental changes to our beekeeping and agricultural practices. In particular, we could break up fields of monocultures with varied bee-friendly plants—red clover, foxglove, and bee balm, for example. The presence of such plants would help diversify honey bees' food webs. We could also use pesticides sparingly and avoid spraying at times of day when bees are actively foraging (although this practice won't necessarily help against neonics, which are often applied to the seeds before they get planted).

In addition, says bee expert Frazier, "People need to take more time to understand where their food comes from, what it takes to produce food and have this incredible supply of food available to us."

While the fate of the honey bees remains uncertain, some signs suggest that a more bee-friendly awareness is beginning to emerge. In 2018, Häagen-Dazs, the ice-cream maker, celebrated the 10th anniversary of its "Help the Honey Bee" campaign, which raises funds and awareness in support of bees. In honor of the occasion, it launched a new flavor called Honey Salted Caramel Almond. Products that depend on honey bees are used in 25 of its 60 flavors. In 2017, the ice cream maker worked with the Xerces Society, which supports invertebrate conservation efforts, to plant a large habitat for native bee populations near one of its almond suppliers. It's also working with Xerces and its partner farmers to develop "Bee Better Certification," a label indicating that products are made with ecologically friendly practices that support bees. So far, the company has donated more than $1 million to honey bee research at universities, including the University of California at Davis and Penn State University.

Burt's Bees, the maker of "Earth-friendly" lip balms and other personal products, has created a series of bee-related public service announcements, viewable on YouTube, including one starring Isabella Rossellini dressed as a honey bee. Even the general public is catching the bee buzz. From city dwellers who become amateur rooftop beekeepers to suburbanites who let more flowers grow in their yards, the ranks of people supporting a pollinator-friendly environment have swelled. And that's a cause that just about everyone can get behind—because, as more and more people are coming to realize, a world without honey bees just wouldn't be as sweet. ■

CHAPTER 21 Summary

Driving Question 1 What are ecological communities, and what role do pollinators play in communities?

- An ecological community is made up of populations of different species living and interacting in a particular area.
- Flowering plants depend on pollinators such as bees, butterflies, bats, and moths to transfer sperm-containing pollen between plants of the same species.
- Honey bees are important pollinators of many crop plants used in human agriculture.
- In recent years, honey bees and other pollinators have suffered many losses, which may result from many interacting factors, including poor nutrition, pathogens, and pesticides.

Driving Question 2 What types of feeding relationships exist within communities?

- The complex set of interconnected feeding relationships in a community makes up a food web. The food web itself consists of linked series of feeding relationships called food chains.
- Organisms in a food chain can be placed into different trophic levels, based on their position in the chain. Organisms at the base of a food chain are producers—they obtain energy directly from the sun and supply it to the rest of the food chain; organisms higher up a food chain are consumers—they obtain energy by eating organisms lower on the chain.
- Food chains can have several types of consumers, including primary and secondary consumers. Primary consumers eat producers (plants), and secondary consumers eat primary consumers.
- Herbivores are animals that eat plants; carnivores are animals that eat other animals.
- Both herbivory and carnivory are forms of predation, in which one organism feeds on another.

Driving Question 3 How do symbiotic relationships influence members of a community?

- Symbiotic relationships are partnerships in which organisms live in close association.
- In mutualistic symbioses, both members benefit; in parasitism, one member benefits while the other suffers; in commensalism, one member benefits while the other is unharmed.
- The relationship between flowering plants and their pollinators is an example of a mutualism.

Driving Question 4 How do species interact in a space with shared resources?

- The space and resources that a species uses to survive and reproduce define its ecological niche.
- Some species have overlapping niches, leading to competition for resources.
- If competing species are unable to partition resources, one species may out-compete another.
- Because they are native to Europe, honey bees are an introduced species in the United States. Introduced species that compete with and drive out native species are called invasive species.

More to Explore

- Bee Informed: http://beeinformed.org
- Graham, K. (2018). Beyond honey bees: Wild bees are also key pollinators, and some species are disappearing: https://theconversation.com/beyond-honey-bees-wild-bees-are-also-key-pollinators-and-some-species-are-disappearing-89214
- Woodcock, B. A., et al. (2016). Impacts of neonicotinoid use on long-term population changes in wild bees in England. *Nat Comm* 7:12459.
- vanEngelsdorp, Dennis. (2009). Colony collapse disorder: A descriptive study. *PLoS One* 4(8):e6481.
- Cox-Foster, D. L. (2007). A metagenomic survey of microbes in honey bee colony collapse disorder. *Science* 318(12):283–286.

CHAPTER 21 Test Your Knowledge

Driving Question 1 What are ecological communities, and what role do pollinators play in communities?

By answering the questions below and studying Infographics 21.1, 21.2, 21.3, and 21.8, you should be able to generate an answer for this broader Driving Question.

Know It

1. How does a community differ from a population?

2. What service are bees providing when they pollinate a plant?
- **a.** nutrition
- **b.** reproduction
- **c.** defense against disease
- **d.** protection against herbivores (animals that eat plants)

3. What benefit do bees gain through pollinating plants?
- **a.** nutrition
- **b.** reproduction
- **c.** defense against disease
- **d.** protection from predators (animals that eat other animals)

Use It

4. If you have pollen allergies, are you more likely to be suffering from the effects of bee-carried pollen or wind-carried pollen? Explain your answer.

5. Sugar maple trees occur in forests in the northeastern United States and southeastern parts of Canada. Maple trees are wind pollinated. Do you predict bees to be abundant in a forest community dominated by maple trees? Explain your answer.

6. If a meadow of wildflowers were converted to a field of corn, would you predict that the number and diversity of bees in the community would increase or decrease? Explain your answer.

Driving Question 2 What types of feeding relationships exist within communities?

By answering the questions below and studying Infographic 21.4, you should be able to generate an answer for this broader Driving Question.

Know It

7. In relation to their position in a food chain, what do plants and photosynthetic algae have in common?
- **a.** nothing
- **b.** They are both producers.
- **c.** They are both first-level consumers.
- **d.** They are both top-level consumers.

8. A bear that eats both blueberries and fish from a river can be characterized as
- **a.** a primary consumer.
- **b.** a heterotroph.
- **c.** a secondary consumer.
- **d.** a producer.
- **e.** all of the above
- **f.** a, b, and c

9. In a land-based food chain, producers are
- **a.** herbivores.
- **b.** photosynthetic carnivores.
- **c.** carnivores.
- **d.** plants.

10. What is the difference between a primary and secondary consumer?
- **a.** Primary consumers eat plants; secondary consumers eat other consumers.
- **b.** Primary consumers eat other animals; secondary consumers eat plants.
- **c.** Primary consumers are plants; secondary consumers eat plants and other consumers.
- **d.** Primary consumers are plants; secondary consumers eat other consumers.

Use It

11. Describe a natural food web that includes both a bee-containing, land-based food chain (one that contains bees) and also at least one aquatic organism (that is, an organism from a lake, river, or ocean).

12. Predict what could happen to the community illustrated in Infographic 21.4 if the bees were to disappear. Make at least three specific predictions, and explain your rationale for each.

Driving Question 3 How do symbiotic relationships influence members of a community?

By answering the questions below and studying Infographic 21.5, you should be able to generate an answer for this broader Driving Question.

Know It

13. Which of the following characterizations best describes a symbiotic relationship?
- **a.** Both organisms benefit.
- **b.** The organisms live in close association.
- **c.** Only one organism benefits.
- **d.** The relationship is mutually harmful.
- **e.** Neither organism benefits.

14. Which of the following characterizations best describes mutualism?

- **a.** Both organisms benefit.
- **b.** Only one organism benefits.
- **c.** The organisms are neither hurt nor harmed.
- **d.** The relationship is mutually harmful.

Use It

15. Would you characterize the relationship between the disease-combating bacteria that live symbiotically within bees and their bee hosts as competition, parasitism, mutualism, or commensalism? Explain your answer.

16. We all have *Escherichia coli* bacteria living in our intestinal tracts. Occasionally these *E. coli* can cause urinary tract infections. From this information, which of the following terms would you say describe(s) the relationship between us and our intestinal *E. coli*? Why did you choose those term(s)?

- **a.** competition
- **b.** mutualism
- **c.** parasitism
- **d.** symbiosis
- **e.** predator–prey

Driving Question 4 How do species interact in a space with shared resources?

By answering the questions below and studying Infographics 21.6 and 21.7, you should be able to generate an answer for this broader Driving Question.

Know It

17. What are some important features of a honey bee niche? How is it that other nectar-feeding organisms can coexist with bees as part of a community?

18. Competition is most likely to occur

- **a.** when one species eats another.
- **b.** when two species occupy different niches.
- **c.** when one species helps another.
- **d.** when two species occupy overlapping niches.
- **e.** when two species help each other.

Use It

19. On a rocky intertidal shoreline (the area between the highest and lowest tidelines, so the intertidal zone is alternately exposed and covered by seawater), mussels and barnacles live together attached to rocks, where they obtain food by filtering it from ocean water. Because these two species coexist in the same habitat, we predict that they do not have identical niches. What might be separating their niches enough to allow them to occupy the same rocky intertidal zone?

Bring It Home

Apply Your Knowledge

20. Many people consider bees a stinging nuisance. What could you say to such people to dissuade them from killing all the bees in their backyards?

Mini Case

Apply Your Knowledge

21. Farmers often plant large acreage of a single crop to maximize yield and simplify harvesting. This is true of almonds in the Central Valley of California.

- **a.** From what you have read in this chapter, what are some of the pros and cons of monoculture?
- **b.** Do some online research to develop a specific model for an alternative to monoculture that addresses at least one of the issues you have identified.

Apply Your Knowledge

Interpreting Data

22. Scientists carried out an experiment to test the hypothesis that a neonicotinoid pesticide called imidacloprid could cause colony collapse disorder. They had a total of 20 hives (colonies) that were broken into five groups (with four hives per group). Four groups received imidacloprid at different dosages (400 μg/kg, 200 μg/kg, 40 μg/kg, and 20 μg/kg). One group did not receive imidacloprid. Hives were monitored for 23 weeks after the initial dose. The data are summarized in the table below.

a. Graph these data.

b. What patterns do you observe?

c. Do you think that the data support the hypothesis? Why or why not?

Dose of imidacloprid (μg/kg)	Number of dead hives at 12 weeks	Number of dead hives at 14 weeks	Number of dead hives at 16 weeks	Number of dead hives at 18 weeks	Number of dead hives at 21 weeks	Number of dead hives at 23 weeks
400	0	2	2	4	4	4
200	0	0	2	2	3	4
40	0	1	1	2	3	3
20	0	0	0	0	2	4
0	0	0	0	0	1	1

Data from Lu, C., et al. (2012). In situ replication of honey bee colony collapse disorder. *Bull Insectol* 65:99–106.

Bringing Bison Back

A controversial plan to rewild the American prairie

Avalon/Getty Images

DRIVING QUESTIONS

1 What are ecosystems, and how do keystone species affect ecosystems?

2 How do nutrients, water, and energy move through ecosystems?

3 How are biomes characterized, and how are human activities influencing prairie ecosystems within the temperate grassland biome?

Picture a prairie: tall grasses swaying in the breeze, a smattering of wildflowers, maybe a cornfield. This relatively flat and open landscape, typical of the American midwestern plains, isn't exactly known for its exotic wildlife.

But that's a relatively recent development, says Matthew Moran, a professor of biology at Hendrix College in Arkansas. "A hundred fifty years ago, on the American prairie, you would have seen more large mammals than you do today in the African Serengeti," he says.

As many as 100 million large mammals once lived on the prairies of the North American Great Plains, an area stretching from Canada to Mexico and covering a million and a half square miles. The region was chock full of antelope, elk, moose, wolves, prairie dogs, and bears.

Most iconic of all was the American bison (*Bison bison*). Some 30 million of these large, hairy herbivores once roamed the Great Plains. Explorers Meriwether Lewis and William Clark, who crossed the continent in 1905, described them as a "moving multitude" that "darkened the whole plain."

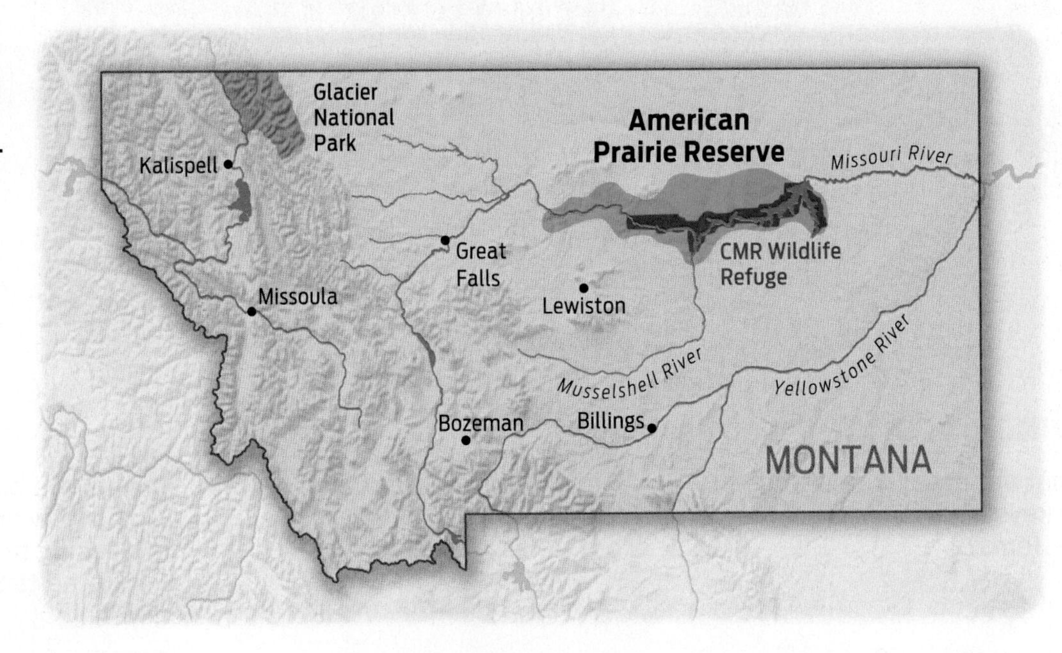

The American Prairie Reserve is working to create 3.2 million acres of uninterrupted prairie.

TCallahan/iStock/Getty Images

A group of bison on a sprawling grass-covered North American prairie.

But by the 1890s, due to widespread hunting, fewer than 500 bison remained. In fact, bison might have gone completely extinct except for the efforts of early hunters-turned-conservationists, including future president Theodore Roosevelt. Roosevelt and his friends saved a few herds and spirited them away to private lands and zoos.

Today, thanks to these and other efforts, approximately 500,000 bison live in North America. About 90% of these are raised as livestock on private ranches. But in the past 30 years, a movement to restore free-roaming bison has gained momentum, inspired by both environmental and ethical considerations.

Most ambitious—and controversial—of all these efforts is the American Prairie Reserve (APR) in northeastern Montana. Founded in 2001 with help from the World Wildlife Fund, this public–private partnership seeks to restore native wildlife on an area of approximately 3.2 million acres of prairie. That includes a herd of more than 10,000 bison, the largest single population in the world.

Funded by donations from individuals, including some very wealthy supporters, the APR has been buying up parcels of land from Montana residents—many of them cattle ranchers. Along with these private lands comes access to grazing rights on adjacent publicly owned lands. APR's goal is to purchase approximately 500,000 acres of private land, which will connect roughly three million acres of existing public land, including the 1-million-acre Charles M. Russell National Wildlife Refuge. So far, the reserve has cobbled together a patchwork of more than 100,000 acres of private land. When it's complete, the APR will be the largest wildlife refuge in the lower 48 states—nearly the size of Connecticut.

The APR's goal is nothing short of restoring the health of an entire prairie ecosystem. As the name suggests, an **ecosystem** is a complex, interwoven network of interacting components. In addition to the community of living things in an area, it includes physical conditions such as temperature, moisture, light, and the chemical resources found in soil, water, and air. Because the living and nonliving parts of an ecosystem can and do change, ecosystems are not static entities, but rather dynamic systems. And because the parts of an ecosystem are so interconnected,

ECOSYSTEM
All the living organisms in an area and the nonliving components of the environment with which they interact.

INFOGRAPHIC 22.1

An American Prairie Ecosystem

An ecosystem includes the community of living things in a particular location and the nonliving environment with which they interact in that location. Features of the nonliving environment that influence ecosystems include physical conditions such as temperature, moisture, light, and the chemical resources found in soil, water, and air. The North American prairie has an abundance of grasses, and a variety of animals. Bison once thrived in this ecosystem, and there are efforts to restore bison to the American prairie.

Living, Biotic:

- Grasses with robust root systems
- Shrubs
- Flowering plants
- Non-burrowing mammals: bison, elk, pronghorn antelope, red fox, jackrabbit
- Burrowing mammals: pocket gopher, prairie dog
- Bird species: prairie chicken, golden eagle, burrowing owl
- Reptiles: rattlesnake
- Amphibians: great plains toad
- Insects: grasshopper, pollinators

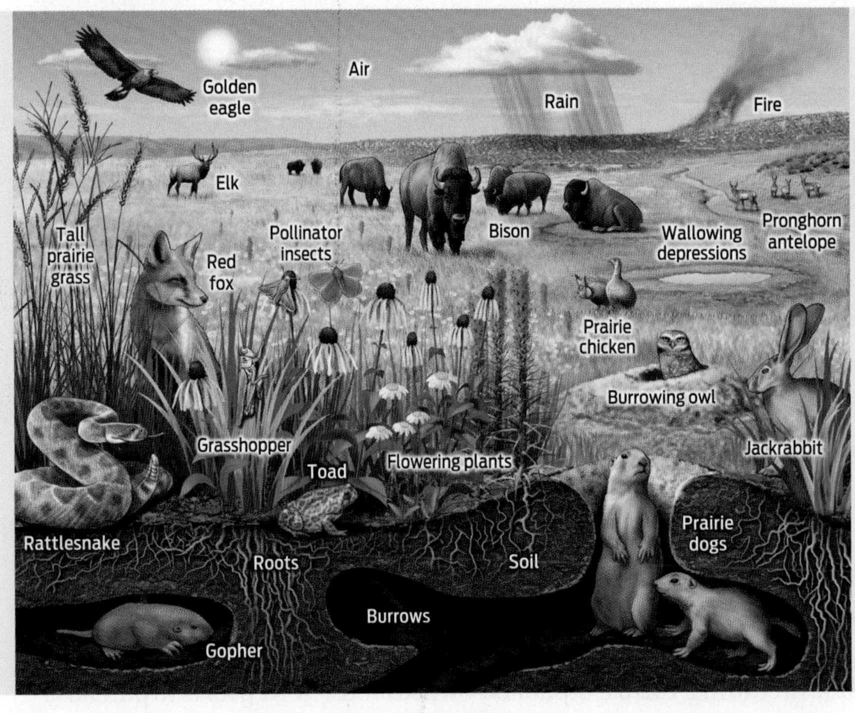

Nonliving, Abiotic:

- Moderate temperature: −30°C to 30°C (−22°F to 86°F)
- Moderate rainfall: 10–30 inches per year
- Soil
- Standing and flowing water
- Soil nutrients
- Air: CO_2, weather
- Light
- Fire
- Grazing
- Burrows
- Wallowing depressions

? If the temperature were to increase and rainfall were to decrease, cacti may replace shrubs and flowering plants in this ecosystem. In this scenario, which abiotic factors are influencing which biotic factors?

even a small change in one part of an ecosystem can have a domino effect that leads to changes in the rest of the ecosystem **(INFOGRAPHIC 22.1)**.

That was certainly the case with bison. Losing these animals triggered a chain of other species losses. By bringing bison back, the APR hopes to bring back other community members as well. It's an approach to wildlife conservation called rewilding.

Environmentalists and nature lovers are thrilled at the prospect of renewing the wild prairie, but the plan has detractors, too—mainly cattle ranchers who see the wildlife reserve as a direct threat to their way of life. The conflict between these groups reveals some fundamental rifts in American society, and helps to explain why conservation efforts in the United States have been controversial from the beginning.

More Than a Symbol

▸ Keystone species

It can be hard to fathom the scale of bison slaughter that took place over the span of just 40 years, from about 1850 to 1890. Hunters killed nearly 30 million of the animals in that period. That averages out to two or more bison killed every minute, of every day, of every month, of every year for 40 years. A famous photograph taken in 1870 shows a mountain of bison skulls stretching several stories up, waiting to be ground into fertilizer.

Before their disappearance, bison had roamed the Great Plains for at least 300,000 years. They likely arrived in North America from Siberia via a land bridge connecting the continents during the Ice Age.

For at least 12,000 years, Native American peoples—including the Apache, Blackfoot, Cree, Kiowa, and Sioux—had traveled the plains along with the bison, hunting them for food and materials for shelter.

Things began to change when European settlers began moving westward in the early 1800s. They went in search of land and gold, and hunted bison for food along the way. William "Buffalo Bill" Cody made a name for himself by hunting bison to feed laborers building the Kansas Pacific Railroad. (Bison were incorrectly identified as buffalo by European settlers and the name stuck.)

A mountain of bison skulls, waiting to be ground into fertilizer, circa 1870.

By the 1870s, a different form of bison hunting became popular—hunting for sport and for profit, enabled by the completion of the transcontinental railroad. Wealthy easterners came west for a taste of the "Wild West." Commercial bison hunters killed the animals for their valuable hides, which made excellent industrial-strength leather belts for steam engines.

This unregulated hunting had the tacit support of the U.S. government, which saw bison removal as a way to push Native Americans off of valuable lands and onto reservations. "Every buffalo dead is an Indian gone," said Richard Dodge, a colonel in the U.S. Army, in 1867.

By 1900, as a result of this slaughter, there were fewer than 500 bison in North America. Ultimately, Native Americans were forced into the reservation system.

Partly because they represent a bygone era in American history, bison have become a kind of mythic national symbol. In 2016, in a rare instance of bipartisanship, Democratic and Republican members of Congress voted to make the bison the first national mammal of the United States. The bison joined the iconic bald eagle as one of only two officially recognized national animals.

But bison are more than just an icon of the American West. They play a crucial role in shaping the prairie ecosystem of which they are a part. Through their behaviors, they alter the environment in ways that permit other organisms to thrive in this ecosystem. Their removal has had a significant impact, the full extent of which scientists are still trying to understand.

To ecologists, bison are a **keystone species**—a member of an ecosystem that plays a crucial role in holding the ecosystem together. A keystone species is not necessarily the most abundant member of a community, but it has an outsized impact. You can think of a keystone species as functioning like the keystone in an arch. It doesn't support the most weight in the structure, but if you remove it, the whole arch collapses. Similarly, if you remove bison, the prairie ecosystem begins to unravel.

Bison are able to play this important role in part because they act as ecosystem "engineers"—they change the natural environment in ways that create (or destroy) habitats

KEYSTONE SPECIES
Species on which other species depend, and whose removal has a dramatic impact on the ecosystem.

INFOGRAPHIC 22.2

Keystone Species Hold Ecosystems Together

As keystone species, bison play a fundamental role in supporting the entire prairie ecosystem. Keystone species are not necessarily the most abundant species in an ecosystem, but if they are lost, they have a huge impact.

The presence of a keystone species supports a diversity of species in an ecosystem, just as a keystone supports the weight of other stones in an arch.

The loss of a keystone species devastates an ecosystem, just as the removal of the keystone from an arch causes it to collapse.

With Bison

Without Bison

Bison are ecosystem engineers. They contribute to the health of a prairie ecosystem in the following ways:

- Grazing on grasses promotes growth of other plant species that support a diverse insect population
- Grazing patterns create ideal habitat for prairie chicken mating and offspring rearing
- Heavy grazing provides ideal habitat for prairie dogs that provide food and burrow holes for other animals
- Grazing promotes nutrient cycling through the ecosystem
- Wallowing provides depressions used by other species for water collection
- Bison are food for predators and scavengers

Without bison, the prairie ecosystem changes in the following ways:

- Without grazing, grass species will overgrow and outcompete other native plant species
- Without grazing, prairie chickens are a threatened species
- Without grazing, prairie dogs are less abundant
- Without grazing, nutrients remain locked up in ungrazed plant litter
- Without wallowing, the prairie is dryer, endangering the health of other species
- Without bison, the prairie ecosystem is much less diverse

? How can researchers determine whether a given species is a keystone species in an ecosystem?

for other organisms. The most famous example of ecosystem engineers are beavers. By gnawing and collecting tree logs, they build dams in creeks and help form ponds and wetlands across the landscape. These alterations create a diversity of habits that allow more species to flourish there than would otherwise be possible. Likewise, bison engineer the prairie environment in ways that support a diverse array other living creatures, including many species of plants, insects, birds, and mammals **(INFOGRAPHIC 22.2)**.

Beavers engineer ecosystems by blocking water flow with wood dams, creating still ponds and wetlands that support a diversity of organisms.

Through their grazing and wallowing, bison engineer habitat for other organisms like the prairie chicken.

One of the biggest ways that bison engineer the grassland ecosystem is through grazing. Bison prefer to eat grass rather than other flowering plants or trees. In fact, about 95% of their diet is made up of grasses.

Because grasses grow rapidly, they tend to out-compete other plants. Bison's selective grazing on grasses helps promote plant diversity by allowing other plants to coexist with these grasses. Often left ungrazed are non-grass flowering plants such as clover, milkweed, and sunflower. Bison grazing gives these ungrazed plants a fighting chance against the grasses. In turn, the ungrazed flowering plants help support a more diverse community of insects.

Bison also engineer their environment through a behavior called wallowing: they roll around in the grass to ward off biting insects and to shed loose fur. In the process, they create depressions in the ground. These depressions have unique properties—they're wetter than surrounding regions, for example—that other animals exploit for their own purposes. For example, amphibians like frogs may use them as breeding grounds.

The patchiness in the landscape that bison create through grazing and wallowing supports the lifestyles of several members of this ecosystem. Take prairie chickens, for example. Male prairie chickens attract females by performing an elaborate mating dance in a group called a lek. Males prefer to lek on flat or heavily grazed ground. The animals mate there, and then the females build their nests in places populated with tall grass, which gives them cover from predators. To reproduce successfully, prairie chickens need access to both heavily grazed (or wallowed) and ungrazed areas. Bison are thus key engineers of the prairie chicken habitat.

The World Wildlife Fund estimates that grasslands are disappearing at a rate of 1 million acres per year. That's faster than the deforestation of the Amazon.

Once a common fixture of the Great Plains, these dramatic birds are now restricted to a few small regions. Prairie chickens are considered threatened or vulnerable under the Endangered Species Act.

And they're not alone. "Now that the Plains are less variable, a lot of animals are having difficulty," Moran says.

NUTRIENT CYCLING
The movement of the atoms of nutrients as they cycle between different molecules in living organisms and the environment.

NITROGEN CYCLE
The movement of nitrogen atoms as they cycle between different molecules in living organisms and the environment.

NITROGEN-FIXING BACTERIA
Bacteria that convert gaseous nitrogen to ammonia, a form of nitrogen usable by plants.

Where the Buffalo (Don't) Roam

▶ Nutrient cycling

Although bison no longer roam the Great Plains the way they used to, they can still be found in managed herds. Moran has been studying bison on a site owned by the Nature Conservancy, the Tallgrass Prairie Preserve in Oklahoma. "We're doing science there to try to understand how the original prairie functioned with bison present," he says. He and his colleagues hope that these insights will help provide policymakers with sound science to create smart conservation policies.

From studying bison in managed colonies like these, Moran has found that bison have a profound effect on **nutrient cycling** in the ecosystem. Living things rely on several key nutrients that are continually replenished as their atoms cycle between living organisms and the environment. The most important nutrients for living organisms are nitrogen, phosphorus, and carbon. Without nutrient cycling, the availability of these nutrients in the environment would be extremely limited.

The movement of nitrogen through the environment is called the **nitrogen cycle.** Although nitrogen is plentiful in the atmosphere as nitrogen gas (N_2), this nitrogen is not in a form that most organisms can use. To obtain usable nitrogen, plants rely on **nitrogen-fixing bacteria** that live in the soil. These organisms—among the most important on Earth—convert gaseous nitrogen into ammonia (NH_3), a form that plants can take up and use. Only certain types of bacteria can perform nitrogen fixation, and no nonmicrobial organisms have this ability. Without these bacteria, the planet's eukaryotic life simply wouldn't exist.

From ammonia in soil, plants take up nitrogen into their bodies. When animals eat and digest these plants, some of that nitrogen is returned to the soil through feces and urine, which are also rich in ammonia. Some is stored in animal bodies in the form of amino acids and nucleotides. When animals die, bacteria and fungi decompose these nitrogen-rich molecules into ammonia, returning usable nitrogen to the soil. Nitrogen gas is returned to the air from the soil through the action of other bacteria.

Phosphorus is another important nutrient that cycles through ecosystems. It is a crucial component of nucleotides and cell membranes. Phosphorus exists naturally in

INFOGRAPHIC 22.3

Nutrients Cycle through Ecosystems

Nutrients such as nitrogen and phosphorus move into and out of the bodies of living organisms and through the environment. Nutrients are taken up and incorporated into molecules of living organisms. When organisms and their waste products decompose, the nutrients are returned to the environment. These movements make up a cycle for each nutrient.

Nitrogen Cycle

Nitrogen is a critical component of the amino acids that make up proteins and the nucleotides of DNA and RNA. Nitrogen atoms cycle between different chemical and biochemical compounds as they move from organisms to the soil, water, and air and back to organisms.

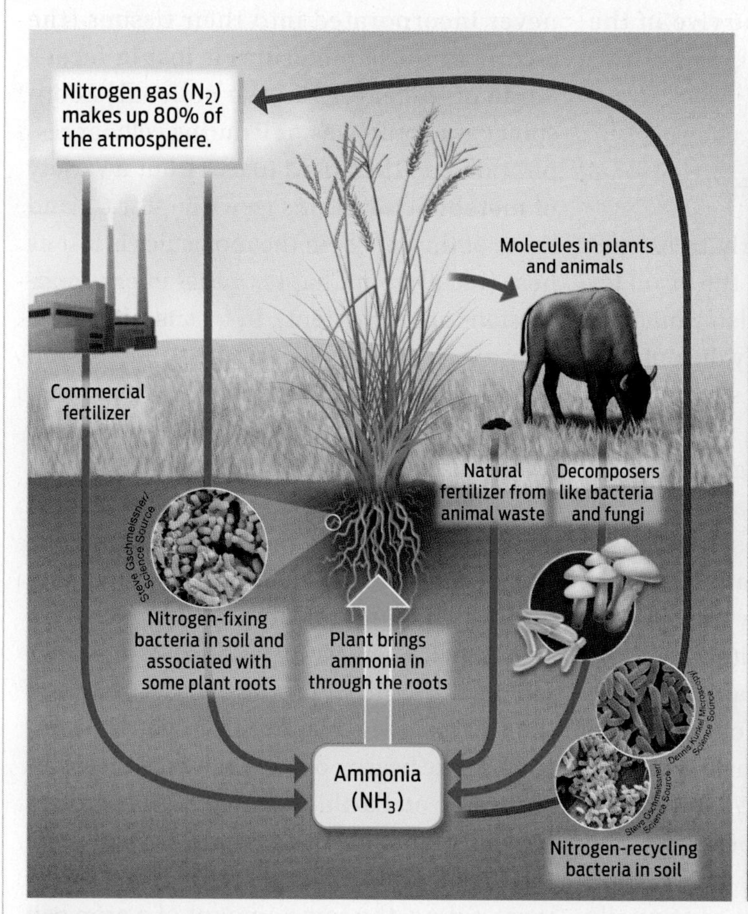

Phosphorus Cycle

Phosphorus is critical for the structure of nucleic acids and phospholipids. In animals, it is critical for bones and teeth. Phosphorus is added to an ecosystem by the weathering of rocks, or through human activities.

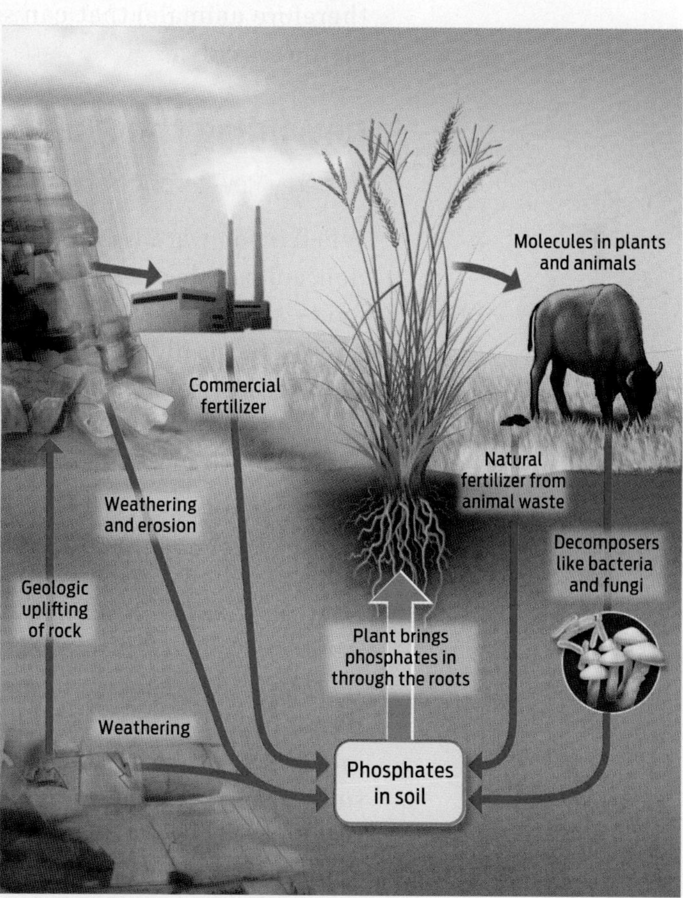

When looking at nitrogen and phosphorus, which is more abundant in the atmosphere? Is that atmospheric form usable to plants?

rocks and enters soil through the weathering of these rocks. Plants take up phosphorus through their roots. When animals eat these plants, they, in turn, obtain their phosphorus. When plants and animals die, phosphorus is returned to the soil. This is the **phosphorus cycle.**

Humans can increase the amount of nitrogen, phosphorus, and other nutrients that are present in the soil by applying commercial or natural fertilizer, both of which contain significant quantities of these nutrients **(INFOGRAPHIC 22.3).**

Bison assist nutrient cycling through their grazing. As plants grow, a lot of nutrients get taken up into the plant body. And as plants mature, they actually become less palatable to various animals, so those

PHOSPHORUS CYCLE
The movement of phosphorus atoms as they cycle between different molecules in living organisms and the environment.

nutrients remain locked up in those plants until they die and decompose. If usable nutrients become scarce in an area, only the hardiest plants will be able to survive. As bison graze on grasses, their urine and feces fertilize the soil with these nutrients. In turn, these released nutrients promote new grass growth, which can provide nourishment to other animals. In this way, bison keep vital nutrients cycling at a faster rate and help increase the number and types of plants (and therefore animals) that can survive in the prairie ecosystem.

Rewilding the Plains

▶ Energy flow through ecosystems

When bison were hunted nearly to extinction, animals that depended on the bison for food—such as grizzly bears—also vanished. These animals didn't completely die out, but they became restricted to the surrounding mountains, Moran explains.

"People don't realize that grizzly bears were very common on the Plains," Moran says. "They were probably denser on grasslands than they are in the mountains today. They've retreated to where they can survive, but it wasn't their preferred habitat."

The loss of species in a habitat partly comes down to the availability of food. Food provides two things to the animals that consume it: nutrients and energy. We saw earlier that nutrients cycle through ecosystems. The atoms of nitrogen, phosphorus, and carbon that are currently on Earth are continually recycled as animals eat, poop, pee, and die. Energy, on the other hand, follows a one-way path through an ecosystem: it flows.

The molecules in the body of an animal store energy. When one animal eats another, the energy stored in those molecules becomes available for the consumer. Sometimes, a consumer becomes the consumed, and the energy is transferred to the second consumer. In this way, energy continues on its one-way path through the ecosystem.

As we saw in Chapter 21, the linked series of feeding relationships in a community is called a food chain. It is essentially a description of who eats whom in a particular ecosystem. As consumers prey on organisms below them in the chain, energy is transferred up the chain through each trophic level.

Not all the energy stored in a lower level ends up being stored in the tissues of the organisms in the next trophic level. Some of the molecules consumed by organisms are never incorporated into their tissues (the energy in those molecules is lost in feces). Some of the energy in the molecules consumed is captured as ATP during cellular respiration and then used to carry out a variety of metabolic reactions (see Chapter 6), and some of the energy in the molecules is lost as heat (recall from Chapter 5 that energy conversions are inefficient). In fact, at each step, only approximately 10% of the energy stored in molecules in one trophic level ends up being stored in the molecules of organisms at the next trophic level **(INFOGRAPHIC 22.4)**.

This pattern of energy flow has several consequences. One consequence is that tertiary consumers (consumers that eat secondary consumers) tend to be rare in an ecosystem. There simply isn't enough energy left in the food chain to sustain many of them. This is the main reason that top consumers like bears, coyotes, hawks, and wolves are scarce on Earth, and why these creatures have no predators. (It's also why human vegetarianism is more energetically efficient than meat-eating: the same amount of a crop can feed many more vegetarians than meat-eaters who eat the animals that eat the crop.)

Another consequence is that removing one link in this chain can have cascading effects for the rest of the chain. Where they are present, bison provide food and nourishment not only to the animals that directly hunt and kill them, like grizzly bears and wolves, but also to those that scavenge on dead bison carcasses, like coyotes and eagles.

INFOGRAPHIC 22.4

Energy Is Lost as It Flows through a Food Chain

In a food chain, energy flows in one direction: from producers to consumers. The passage of energy is not efficient, however, as only 10% of energy from one trophic level ends up stored in the bodies of the next. The result is an energy pyramid, which shows that fewer top consumers can be supported in an ecosystem compared to organisms lower down the food chain.

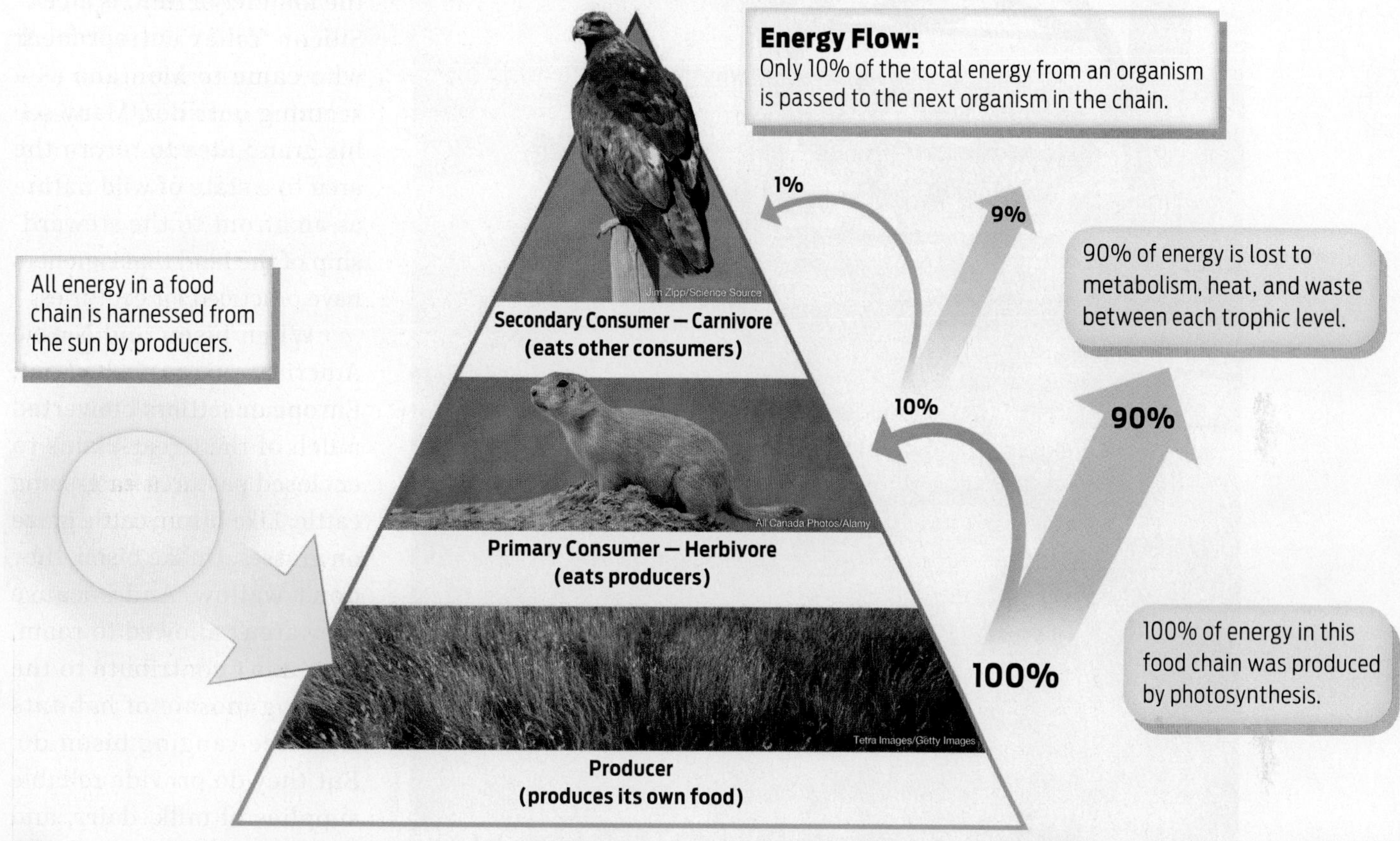

? When a cow eats grass, how much of the energy stored in the grass is actually stored in the cow? If a human eats a beef burger, how much of the energy stored in the grass would end up stored in the human?

Bison also indirectly affect food availability in the ecosystem. For example, they help increase the size of prairie dog populations, because prairie dogs forage more successfully in grasses that have been grazed. In turn, prairie dogs provide food for foxes, hawks, and eagles. And prairie dog burrows provide habitats for small mammals, amphibians, and reptiles, which are themselves food for snakes and birds.

Through these many influences, you can see how loss of the bison—the keystone species in this community—could lead to dramatic changes in the prairie ecosystem. This keystone role in prairie ecosystems explains why conservationists want to bring bison back. But a fair question can be asked: can you bring bison back to an area without infringing on the way of life of the humans who live there?

Home, Home on the Range

▶ Grasslands and other biomes

On lawns and storefronts in small towns around northeast Montana, it's common to see signs reading, "Save The Cowboy, Stop The American Prairie Reserve." The signs

Cody Oxarart/United Property Owners of Montana

A campaign sign opposing the American Prairie Reserve produced by the United Property Owners of Montana (UPOM).

are part of a campaign run by a group called United Property Owners of Montana (UPOM). This group, formed mostly of cattle ranchers, opposes the creation of the American Prairie Reserve.

UPOM believes that the reserve threatens the cattle ranching way of life and will ultimately disrupt their economic livelihood. "To me it means numerous family ranches, including mine, either selling out or relocating from that area," says Deanna Robbins, co-founder of UPOM. "If APR gets their way, those farms and ranches will be turned into playgrounds for the world's elite and curious."

Opposition to APR among some ranchers in northern Montana is fierce. It probably doesn't help that Sean Gerrity, the founder of APR, is an ex–Silicon Valley entrepreneur who came to Montana as a seeming outsider. Many see his grand idea to return the area to a state of wild nature as an afront to the stewardship of the land that ranchers have practiced for centuries.

When bison and Native Americans were pushed out, European settlers converted much of the Great Plains to enclosed pastures for grazing cattle. Like bison, cattle graze on grasses. Unlike bison, they don't wallow. And because they aren't allowed to roam, they don't contribute to the shifting mosaic of habitats that free-ranging bison do. But they do provide reliable supplies of milk, dairy, and meat.

Cattle ranchers in Montana generally don't look fondly on the return of native wildlife if that means predators like wolves and grizzlies preying on their livestock. Nor does the idea of "free-roaming" bison sound as romantic to ranchers as it does to environmentalists.

"The issues that concern property owners are the risk of disease transmission from bison to cattle, and the potential for massive damage to private property that would be caused by free-roaming bison," says Charles Denowh, UPOM's Policy Director. Ranchers fear that APR will be absolved of liability for the animals and ranchers will be left holding the bill.

Distrust of state and federal environmental regulation runs deep in this part of the country. Another popular sign seen on property in the area reads, "Don't Buffalo Me. No Federal Land Grab." Many existing wildlife conservation areas, including some national parks, have been established through federal designation of certain areas as national monuments. The APR isn't going that route. Instead, they are obtaining land through the free-market system, buying from willing sellers. Nevertheless, cattle ranchers know that every plot of land that APR buys means less buffer for their own ranches, fewer ranching families living in the area, fewer children in local schools, and fewer sales of tractors and other agricultural equipment. The population in this area was already declining; now, the APR is providing added incentive for ranchers to leave.

It's not hard to understand why these lands have become such contested territory. The same traits that make prairies prime cattle-raising territory also make them a good home for bison: they're flat, there are few obstacles to moving herds, and—most important—there's lots of grass.

Why do grasses—as opposed to some other form of vegetation—dominate prairie ecosystems? The reasons have to do with how hot prairies are and the low amount of precipitation they receive. Grasses are well adapted to thrive in these hot, relatively dry environments. These regions might be challenging for humans to settle in, but grasses, cows, and bison do just fine.

In fact, in every region of the globe, the type of vegetation that grows in an ecosystem determines the type of animals that can live and thrive there. Scientists have identified ten or so large geographic regions that can be distinguished on the basis of their characteristic vegetation, which in turn is dependent upon climatic factors like temperature and rainfall. These large regions are called **biomes.**

The American prairie ecosystem is part of a biome called temperate grassland. Grasses are plentiful in this biome, while trees are largely absent except right next to rivers. The annual average rainfall is about 25 to 89 cm (10–35 inches)—about one-third as much rain as temperate forests receive and one-fifth as much as tropical rainforests receive.

Temperatures in temperate grasslands vary over the course of the year. Summer temperatures can top 38° C (100° F), while winter temperatures can fall below −18° C (0° F).

Fires regularly sweep through temperate grasslands. Because the growth-generating part of their stems is located underground, grasses are not permanently harmed by the fires. But any trees and shrubs present are burned beyond regeneration, keeping the biome consistent in its plant growth.

> *"If APR gets their way, those farms and ranches will be turned into playgrounds for the world's elite and curious."*
>
> **—Deanna Robbins**

The climatic factors—temperature and rainfall—that shape life in each biome correlate with a region's position on the globe with respect to the equator. Biomes closer to the equator are warmer and wetter, while those closer to the poles are colder and drier. These differences reflect the fact that the sun shines more directly on the equatorial regions of the planet, whereas less direct sunlight hits the poles. Thus, temperature falls as you move from the equator to the poles. And because warm air can hold more moisture than cold air, rainfall also decreases as you move from the equator to the poles. The result is a distinct pattern of terrestrial and aquatic biomes located around the globe. Each of these biomes is characterized by a different combination of temperature and precipitation, which together promotes the growth of a different type of vegetation, which in turn supports a different variety of life **(INFOGRAPHIC 22.5)**.

BIOME
A large geographic area defined by its characteristic plant life, which in turn is determined by temperature and levels of moisture.

INFOGRAPHIC 22.5 UP CLOSE

Earth's Biomes

Biomes are large geographic areas defined by their characteristic plant life. The characteristic vegetation is, in turn, determined by patterns of temperature and rainfall, which vary according to the intensity of sunlight that hits different parts of Earth. For terrestrial biomes, average temperature and rainfall as well as seasonal variations in both are critical to determining the types of vegetation. For aquatic biomes, the temperature and salinity (saltiness) of the water, its depth, and whether it is still or moving are the critical factors.

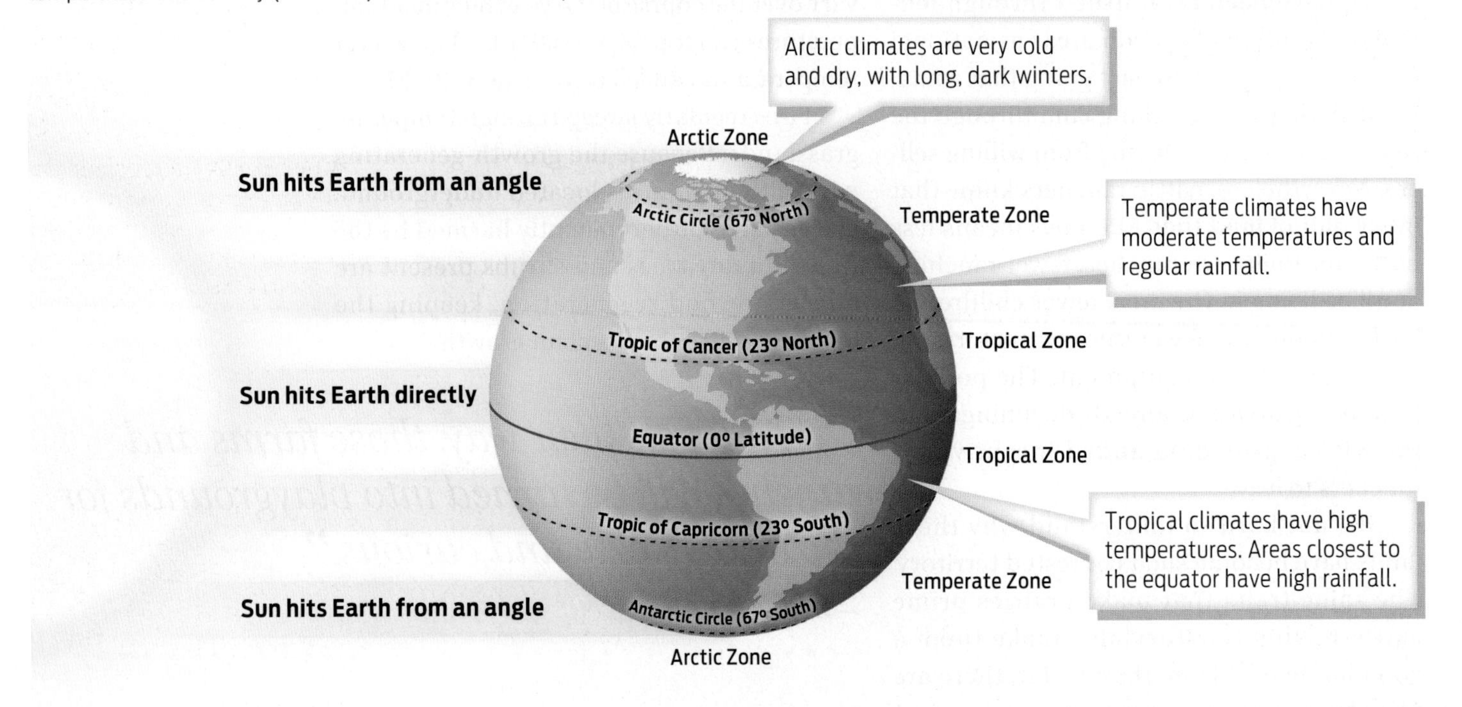

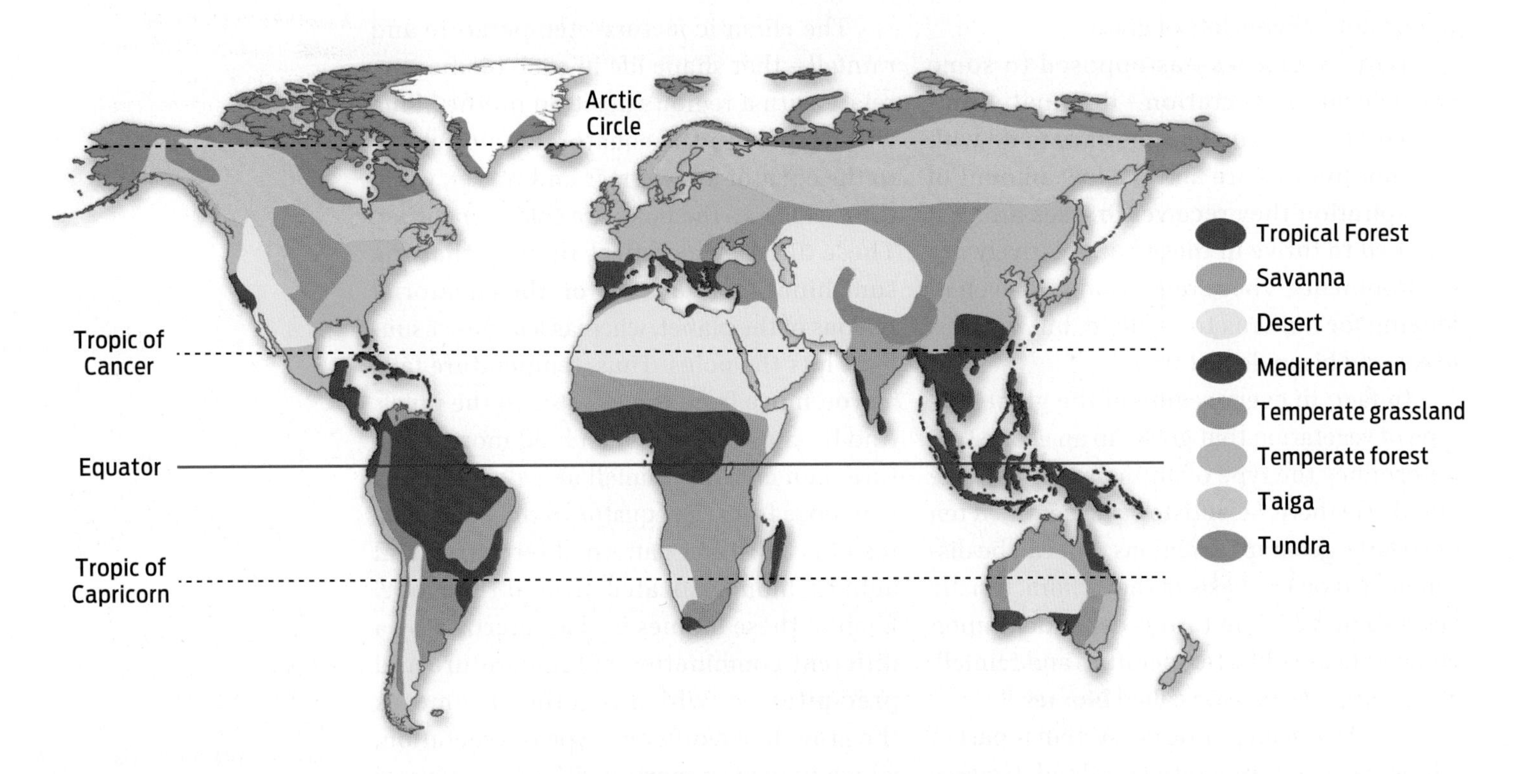

Taiga
A biome characterized by evergreen trees, with long and cold winters and only short summers.

Tundra
A biome that occurs in the Arctic and mountain regions. Tundra is characterized by low-growing vegetation and a layer of permafrost soil (frozen all year long) very close to the surface of the soil.

Tropical Forest
Tropical forests are biomes characterized by warm temperatures and sufficient rainfall to support the growth of trees. Tropical forests may be deciduous or evergreen, depending on the presence or absence of a dry season.

Temperate Deciduous Forest
Temperate forests are characterized by moderate winters and rainfall. Trees are mostly evergreen or deciduous, dropping their leaves in winter.

Savanna
A biome characterized by warm temperatures and two seasons (dry and rainy). The primary vegetation is grasses, with some shrubs and rare trees.

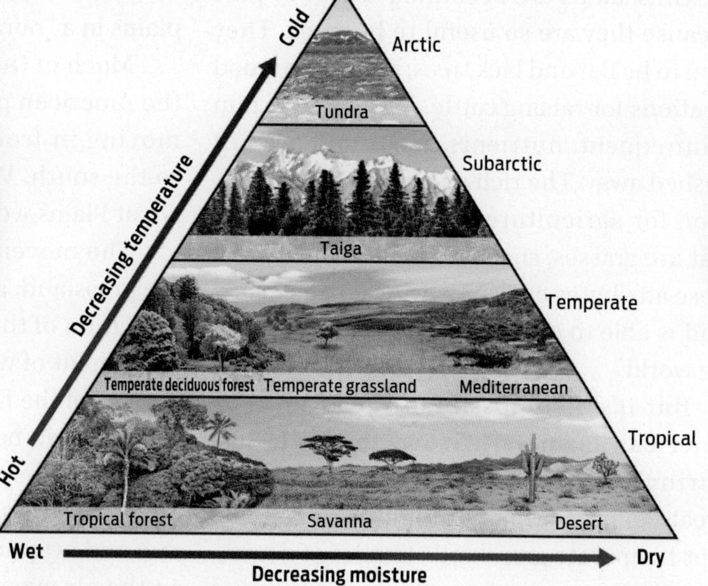

Mediterranean
A biome characterized by long, hot, and dry summers and cool, damp winters. The plant life includes characteristic short evergreen trees and shrubs with leathery leaves.

Temperate Grassland
A biome characterized by perennial grasses and other nonwoody plants. In North America, the prairies are examples of grasslands.

Desert
A biome characterized by extreme dryness. Cold deserts experience cold winters and hot summers, while hot deserts are uniformly warm throughout the year.

Aquatic: Marine
This biome covers about three-fourths of Earth and includes the oceans, coral reefs, and estuaries (where rivers meet the sea).

Aquatic: Freshwater
A Earth's Biomes characterized by a low salt concentration. Freshwater biomes include ponds and lakes, rivers and streams, and wetlands.

According to Ellen Anderson, a project manager with APR, temperate grasslands are "one of the least protected ecosystems in the world." In fact, she says, this region of Montana is one of only four places left across the world where large-scale conservation of grasslands is still possible. That's because large portions of the native sod have not been plowed under.

Today, less than 10% of temperate grasslands in the world remain intact, and that number is rapidly decreasing. The World Wildlife Fund estimates that grasslands are disappearing at a rate of 1 million acres per year. That's faster than the deforestation of the Amazon.

Grasslands are becoming scarce in part because they are so useful to humans. They tend to be flat and lack trees, so they are good locations for raising cattle. And because rain is infrequent, nutrients in the soil are not washed away. The rich soil makes grasslands good for agriculture—especially for crops that are grasses, such as wheat and corn. All these attributes explain why America's heartland is able to serve as the "breadbasket for the world."

But it's also possible to overdo it, as American farmers learned in the 1930s. During this period, drought struck the Great Plains. Normally, droughts do not disrupt temperate grasslands to a large degree, because the roots of grasses help to secure soil in place. But because so much of the plains had been converted to agriculture, the soil was no longer protected and essentially blew away in large dust clouds—a phenomenon known as the Dust Bowl.

Today, parts of the Great Plains region are lush and green, with acres and acres of crops, most abundantly corn. But that's only because modern feats of engineering allow farmers to access water stored beneath the ground in the Ogallala Aquifer, a vast underground reservoir of freshwater that formed over thousands of years. Without this supplement, America's breadbasket would have trouble feeding Kansas, let alone the world.

WATER CYCLE
The continuous movement of water on, above, and below Earth's surface.

AQUIFER
An underground layer of porous rock that contains water.

Thirsty Ground

▶ Water cycle and aquifers

Rainfall in the Great Plains, as elsewhere, reflects the amount of moisture in the air as it passes over an area. Air that flows from west to east across the plains tends to be dry, limiting the amount of rainfall. That's because any water vapor it did contain falls on the western side of the Rocky Mountains. Warm, wet air from the Pacific Ocean rises as it moves toward the mountains. As it rises, it cools, expands, and releases the water as rain before ever reaching the plains, placing the plains in a "rain shadow."

Much of the water that falls as rain over the American prairies begins as water vapor moving in from the warm Gulf of Mexico to the south. Without this water vapor, the Great Plains would essentially be a desert.

The movement of water in the temperate grassland, as in every biome, reflects the operation of the **water cycle,** the continuous movement of water on, above, and below the surface of the Earth. Like all cycles, the water cycle has no beginning or end, but it can be divided into stages. Water on the ground or in lakes, rivers, and oceans is heated by the sun and evaporates into the air. Water vapor in the air eventually cools down and forms liquid water again, which falls to Earth as precipitation in the form of rain or snow. On the ground, water collects in lakes and rivers. Some seeps into soil and shallow rock layers, forming groundwater. Groundwater can also seep deeper into porous rock layers called **aquifers,** which can store immense quantities of water. The energy of the sun sets evaporation in motion and is therefore what powers the water cycle.

The water cycle is a global phenomenon: water that originated in a lake can find its way, as water vapor, halfway around the world before being deposited back on Earth. But not all water on Earth is cycling at any given moment. Water at the bottom of the ocean or trapped in a glacier might take thousands of years to cycle.

Though water is considered a renewable resource, only a tiny portion of the total water on the planet is in the form of freshwater that is available for people to drink—less than 1% **(INFOGRAPHIC 22.6)**.

The water cycle follows the same general pattern of evaporation and condensation in every biome. Nonetheless, the amount of precipitation that occurs in each biome varies greatly. The amount of precipitation in the temperate grasslands of the American prairie is much lower than, say, in temperate forests. This difference in precipitation is largely why the American prairie lacks trees: they can't survive on the amount of rain the area gets.

But the hardy grasses do fine here. These thriving plants are an abundant food source for the many animals that make their home in this ecosystem, including, for the first time in more than 120 years, bison.

INFOGRAPHIC 22.6

The Water Cycle

Freshwater is a valuable resource. In addition to its role in keeping us hydrated, it irrigates crops, sustains fisheries, and provides recreational opportunities. Although water is "used," it is not "used up." It is ultimately returned to the global ecosystem as it evaporates to the atmosphere, flows into rivers and streams, or enters underground aquifers.

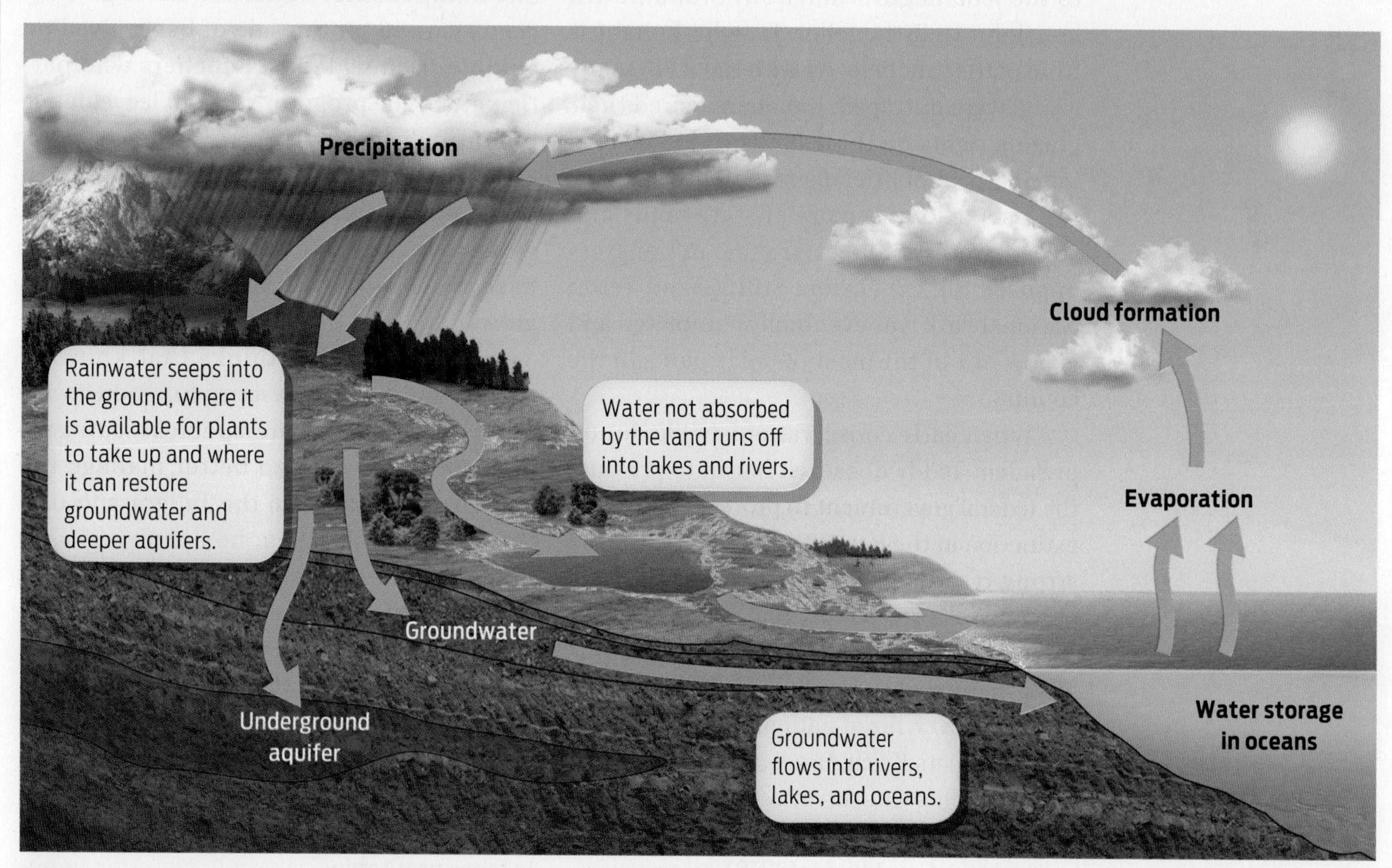

? **Describe the roles of oceans, groundwater, and aquifers in the water cycle.**

APR introduced the first 16 bison, including several pregnant females, into the area in October 2005. Since then, the herd has grown to about 850 animals. To accommodate the growing herd, APR has acquired more plots of land and expanded its reach into northern Montana. And that has meant fewer plots owned by cattle ranchers.

A Battleground of Competing Ideologies

The disagreements between ranchers and the APR have a familiar ring to them. Early conservation efforts directed at establishing national parks in the United States sparked equally strong opposition. Grand Teton National Park in Wyoming, for example, was created in 1950 only after decades of controversy. The park was to include land donated to the federal government by Standard Oil heir John D. Rockefeller, Jr., who bought it from cattle ranchers. As with the APR, many locals were outraged. Ranchers feared their grazing rights on these lands would be curtailed. Local politicians resented what they saw as a federal seizure of state lands and staged protests, and even some in Congress opposed the expansion. Still, Grand Teton National Park was eventually completed and is now one of the most popular parks in the country.

When early conservationists like future president Teddy Roosevelt began to pressure the federal government to protect bison from extinction in the late 19th century, they faced strong resistance as well. "The mindset of most people was simply that nature was for the taking, and for the country to become great, nature must be exploited and utilized to develop a civilized nation," write conservation biologists Keith Aune and Glenn Plumb, of the Wildlife Conservation Society, in their book *Theodore Roosevelt and Bison Restoration in the Great Plains* (2019).

It was governmental failure to protect bison that motivated private citizens to take matters into their own hands. Several conservation groups formed around the turn of the 20th century, including the Audubon Society (1904) and the American Bison Society (1905). Nongovernmental organizations like these remain crucial players in conservation efforts today.

The recent conflict between ranchers and the APR in northeastern Montana illustrates the challenges that often accompany efforts to set aside valuable lands for wildlife conservation. When someone's livelihood depends on using those lands for a different purpose, there's bound to be resistance.

The organizers of the American Prairie Reserve understand that their project is controversial to many locals and will unavoidably impact the local economy. That's why they have taken steps to mitigate that impact. For example, a program called Wild Sky compensates ranchers through proceeds earned on the sale of beef raised on wildlife-friendly lands. Ranchers who participate in the program agree to leave native prairie undisturbed.

But the conflict between wildlife preservation and agricultural development isn't going away anytime soon. In fact, it is only likely to get worse as the human population grows.

Biologist Moran thinks that knowing more about how ecosystems function—with and without keystone species—can give us ideas about how to better manage the resources we have. On the Tallgrass Prairie Preserve in Oklahoma, he's studying ways that bison alter ecosystems in the hopes that the lessons will decrease conflict between conservationists and ranchers.

"There's some evidence that if you mimic the bison grazing patterns, you actually increase productivity of the land," Moran says. That insight could help ranchers ranch in a way that both preserves biodiversity and is more profitable.

"It's might be a win–win situation for both wildlife and the ranchers," he says. ■

CHAPTER 22 Summary

Driving Question 1 What are ecosystems, and how do keystone species affect ecosystems?

- Ecosystems are made up of the living and nonliving components of an environment, including the communities of organisms present and the physical and chemical environment with which they interact.
- Bison are keystone species because they play a fundamental role in supporting an entire ecosystem, much like the keystone in an arch.
- Loss of a keystone species from an ecosystem (for example, by hunting) leads to a cascading loss of many other species.

Driving Question 2 How do nutrients, water, and energy move through ecosystems?

- Elements like nitrogen, phosphorus, and carbon cycle through ecosystems as animals eat other organisms, and then poop, pee, and die.
- Energy flows in one direction through a food chain, from producers to consumers.
- Only 10% of the energy from one trophic level ends up being stored in the molecules in the next trophic level. The rest is lost in feces, used to generate ATP to fuel chemical reactions, and lost as heat.
- The water cycle keeps freshwater moving between the atmosphere, rivers, lakes, and oceans, as well as groundwater and underground aquifers.

Driving Question 3 How are biomes characterized, and how are human activities influencing prairie ecosystems within the temperate grassland biome?

- Biomes are large regions defined by their characteristic plant life, which in turn is determined by temperature and precipitation.
- The American prairie is an ecosystem found in the temperate grassland biome.
- The amount and type of vegetation in a biome provide the base of the food chain and determine the variety of life that can thrive in that biome.
- Humans have removed keystone species, are changing how water is drawn from aquifers for agriculture, and are changing how prairie lands are used.
- Rewilding is an approach to conservation that involves restoring the full complement of creatures in an ecosystem and letting them shape its development.

More to Explore

- Moran, Matthew. (2019). Bison are back, and that benefits many other species on the Great Plains. *The Conversation*: https://theconversation.com/bison-are-back-and-that-benefits-many-other-species-on-the-great-plains-107588
- Knapp, A. (1999). The keystone role of bison in North America tallgrass prairie. *BioScience* 29(1): 39–50.
- Hegyi, Nate. (2019). The next Yellowstone: How big money is building a new kind of national park." Wyoming Public Media: https://www.wyomingpublicmedia.org/post/next-yellowstone-how-big-money-building-new-kind-national-park
- Aune, Keith, and Glenn Plumb. (2019). *Theodore Roosevelt and Bison Restoration on the Great Plains.* Charleston, SC: History Press.
- CBS Sunday Morning (2018). Prairie preservation: How to protect America's grasslands: https://www.youtube.com/watch?v=Ws5sJP7lTD0

CHAPTER 22 Test Your Knowledge

Driving Question 1 What are ecosystems, and how do keystone species contribute to ecosystem stability?

By answering the questions below and studying Infographics 22.1 and 22.2, you should be able to generate an answer for this broader Driving Question.

Know It

1. Which of the following are parts of an ecosystem?

a. the plant life present in a given area
b. the animals living in a given area
c. the amount of annual rainfall in a given area
d. the soil chemistry in a given area
e. none of the above
f. all of the above

2. What are keystone species?

Use It

3. A rocky shoreline is covered at high tide but exposed at low tide. This shoreline supports a community of mussels, algae, barnacles, and starfish. An ecologist systematically removes species from different areas of the beach. Removing the mussels or the barnacles doesn't substantially change the community, but removing the starfish dramatically changes the mix of species in the area. Which is the keystone species?

a. mussels
b. barnacles
c. algae
d. starfish
e. all of the above

Bring It Home

Apply Your Knowledge

4. Consider an ecosystem that you are familiar with. From what you know about the communities of organisms in this ecosystem, choose what you think might be a keystone species, and defend your choice.

Interpreting Data

Apply Your Knowledge

5. In trying to restore the bison to the American prairie, conservationists use bison from different herds as a source of new animals to increase the size of different populations (much as we saw with the introduction of new wolves onto Isle Royale in Chapter 20). It is important to ensure that the "source" population of bison is "genetically healthy." A genetically healthy population of bison has not interbred with cattle (that is, does not have evidence of cattle-specific DNA sequences in their genomes) and contains sufficient genetic variation to avoid the negative effects of inbreeding (Chapter 14) and to adapt to future environmental changes. Higher numbers of different alleles and a higher proportion of heterozygosity (having two different alleles for given genes) are considered to be positive indicators of genetic variation.

Different bison populations were genetically tested, to determine their suitability as source populations for bison recovery projects. The genetic testing involved analyzing mitochondrial DNA (inherited from mothers; see Chapter 19) for a cattle-specific sequence and a bison-specific sequence, nuclear DNA sequences for cattle-specific and bison-specific alleles, as well as the extent of heterozygosity at each DNA sequence. The results of the genetic analysis are shown in the accompanying table.

a. Which columns in the table show the extent to which the bison have interbred with cattle?
b. Which columns in the table show data relevant to genetic diversity of the different bison herds?
c. Based on the data presented in the table, which population(s) would be good sources of bison for restoration of other herds? Provide evidence supporting your choice,

Genetic Analysis of Bison from Different Herds

Herd	Number of bison	Cattle-specific sequence in mitochondrial DNA (+/–)*	Bison-specific sequence in mitochondrial DNA (+/–)*	Number of sequences with cattle-specific alleles (out of 14)	Average number of alleles at each sequence	Average heterozygosity (0–1)
Henry Mountains	129	–	+	0	3.88	0.554
National Bison Range	179	+	+	1	4.92	0.632
Yellowstone National Park	505	–	+	0	4.62	0.603
Badlands National Park	328	–	+	2	4.46	0.592
Cattle (control)	NA	+	–	14	NA	NA

* + = detected in population; – = not detected in population.

Based on Ranglack, D. H., L. K. Dobson, J. T. Du Toit, and J. Derr. (2015). Genetic analysis of the Henry Mountains bison herd. *PLoS One* 10(12): e0144239. doi:10.1371/journal.pone.0144239

Driving Question 2 How do nutrients, water, and energy move through ecosystems?

By answering the questions below and studying Infographics 22.3, 22.4, and 22.6, you should be able to generate an answer for this broader Driving Question.

Know It

6. Which form of nitrogen is abundant in the atmosphere? Which form of nitrogen can be easily taken up and used by plants?

7. Match each nitrogen conversion process on the left with the organism that can carry out that process.

__ $N_2 \rightarrow NH_3$	**a.** nitrogen-recycling bacteria in soil
__ proteins $\rightarrow NH_3$	**b.** nitrogen-fixing bacteria
__ $NH_3 \rightarrow N_2$	**c.** decomposers

8. Which of the following is *not* a reservoir of phosphorus?

a. rocks
b. bodies of plants and animals
c. atmospheric gases
d. the soil
e. bodies of water

9. If phosphorus is important for bones and teeth, why do plants need phosphorus?

10. When rain falls from the air on the American prairie, where did the air pick up that water?

a. evaporation from an underground aquifer
b. evaporation from a distant ocean
c. evaporation from nearby lakes
d. evaporation from groundwater

Use It

11. Explain why the kilograms of grass a cow eats do not produce the equivalent amount of energy in the form of meat. What happens to the energy stored in the grass once it is ingested by the cow?

12. Compare the diet of a human who is a herbivore with that of a human who is a top consumer. Consider what each might actually eat; how much energy from a producer is captured in the herbivore human; and how much energy from a producer is captured in the top consumer human.

Driving Question 3 How are biomes characterized, and how are human activities influencing prairie ecosystems within the temperate grassland biome?

By answering the questions below and studying Infographics 22.5 and 22.6, you should be able to generate an answer for this broader Driving Question.

Know It

13. What is the major feature that defines a biome?

a. its plant life
b. its animal life
c. its soil microorganisms
d. all of the above

14. The American prairie is part of a ____________ biome.

a. tundra
b. savanna
c. temperate grassland
d. desert

Use It

15. If the American prairie were to become both hotter and drier, what type of vegetation would you expect to find?

16. Look at the photo montage provided.

a. Describe what you observe over time. How would you characterize the biome in the earliest photo? How would you characterize the biome in the most recent photo?

b. Explain (in your own words) how a combination of overgrazing by cattle and reduced rainfall could have contributed to what you observe.

Photomontage by Rob Wu, photos The University of Arizona

Mini Case

Apply Your Knowledge

17. Agriculture on the American prairie depends on irrigation water pumped from the Ogallala Aquifer. The aquifer is being depleted far faster than natural processes can replenish it, and there are concerns about how long it can continue to provide freshwater.

Do some research to investigate the following three strategies. What are some advantages and disadvantages of each? Then choose and defend one as the strategy most likely to conserve the aquifer as well as to support agriculture to feed a growing population and contribute to the health of the planet.

a. develop genetically modified drought-resistant crops
b. let acres of cultivated fields return to grasslands
c. adopt dryland farming techniques

PROGRESS OR POISON?

Rachel Carson, pesticides, and the birth of the environmental movement

Bettmann/Getty Images

DRIVING QUESTIONS

1. Why can carefully considered environmental policies still have unintended consequences?
2. What is biomagnification, and how does it occur?
3. Why was DDT widely adopted? What properties of DDT permit it to negatively affect organisms at a variety of levels in a food chain?

Silent Spring is universally credited as the spark that ignited the modern environmental movement.

In 1958, the biologist and science writer Rachel Carson received a disturbing letter from a friend in Massachusetts. The letter described an event that had recently taken place on the friend's property near Cape Cod: an airplane had sprayed a thick cloud of a pesticide called DDT as part of a coordinated campaign to eradicate mosquitoes. In the days following the spraying, the friend noticed many dead songbirds in the area. "Their bills were gaping open, and their splayed claws were drawn up to their breasts in agony," she wrote. Could anything be done to stop these aerial sprayings, the friend wanted to know?

At the time she received the letter, Carson was a well-known science writer who had written some widely popular books about the sea. She cared deeply about nature and was horrified by her friend's report—especially since it wasn't the first time she'd heard about DDT toxicity. Carson had been concerned for a number of years about the effects DDT might be having on beneficial insects, birds, and fish, ever since she had worked as a marine biologist with the U.S. Fish and Wildlife Service, which had conducted studies on the pesticide in 1945. But the letter from her friend was a tipping point.

Carson decided that someone should research and write an article about the dangers of pesticides. At first, she tried to persuade other writers to take on the topic. When no one would, she realized she had to tackle it herself. Four years later, the outcome was *Silent Spring*, a book that sparked a national conversation about pesticides and ushered in a new way of thinking about human impacts on the environment.

In the book's famous opening chapter, "A Fable for Tomorrow," Carson asked readers to imagine a town "in the heart of America" where "a strange blight" had crept over the land. The birds that had once greeted the coming of spring with a chorus of song now lay sick, dying, and silent. "It was a spring without voices," she wrote. What had caused this strange blight? "No witchcraft, no enemy action had silenced the rebirth of new life in this stricken world," Carson wrote. "The people had done it themselves."

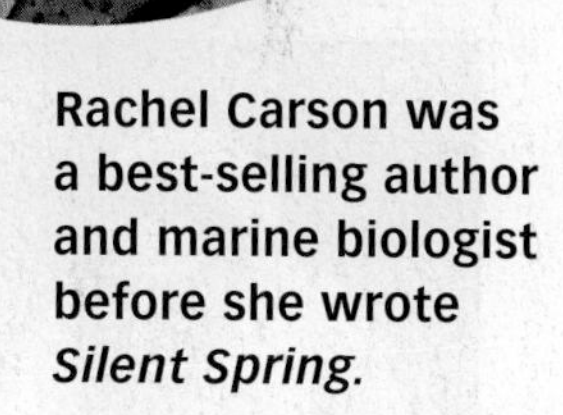

Alfred Eisenstaedt/The LIFE Picture Collection/Getty Images

Rachel Carson was a best-selling author and marine biologist before she wrote *Silent Spring*.

I Want My DDT

▸ Pesticides and biomagnification

DDT (dichlorodiphenyltrichloroethane), a synthetic chlorinated hydrocarbon, was discovered to be a potent insecticide in 1939 by the Swiss chemist Paul Müller. DDT poisons the nervous system of insects and other animals. It was first widely used during World

War II to combat insect-borne diseases, such as typhus and malaria, among U.S. soldiers. Typhus is spread by lice, which infested many soldiers living in cramped quarters. Malaria is transmitted by mosquitoes, which live in tropical regions of the globe. The disease plagued soldiers fighting in the islands of the South Pacific, and was also a problem in the American South, where many U.S. military training bases were located. DDT was mixed with powder and dusted directly on clothing; it was also sprayed on walls in barracks. As a public health measure, DDT succeeded marvelously, saving countless lives during the war. For his work, Müller was awarded a Nobel prize in 1948.

After the war, DDT became available to the public as a commercial pesticide and was widely used by farmers to protect their crops and by public health officials to contain insect-borne disease. The chemical companies that manufactured the potent bug-killer advertised it as an aid to domestic comfort and tranquility. Photos and newsreels from the era document mass spraying of suburban neighborhoods as children played happily in the chemical clouds **(INFOGRAPHIC M6.1)**.

But by the late 1950s, a number of scientists and citizens had become concerned that indiscriminate spraying of pesticides like DDT was doing more harm than good. Efforts by the U.S. Department of Agriculture (USDA) to eliminate gypsy moths and fire ants in a number of states had reportedly wiped out scores of other insects, fish, and birds. By 1958, when Carson received the

INFOGRAPHIC M6.1

Widespread Use of DDT to Kill Mosquitos and Lice

As an insect neurotoxin, DDT was used to combat a number of diseases associated with insects, including those borne by lice and mosquitoes.

DDT was applied directly to human hair, skin, and clothing to eliminate lice that spread typhus, a microbial disease.

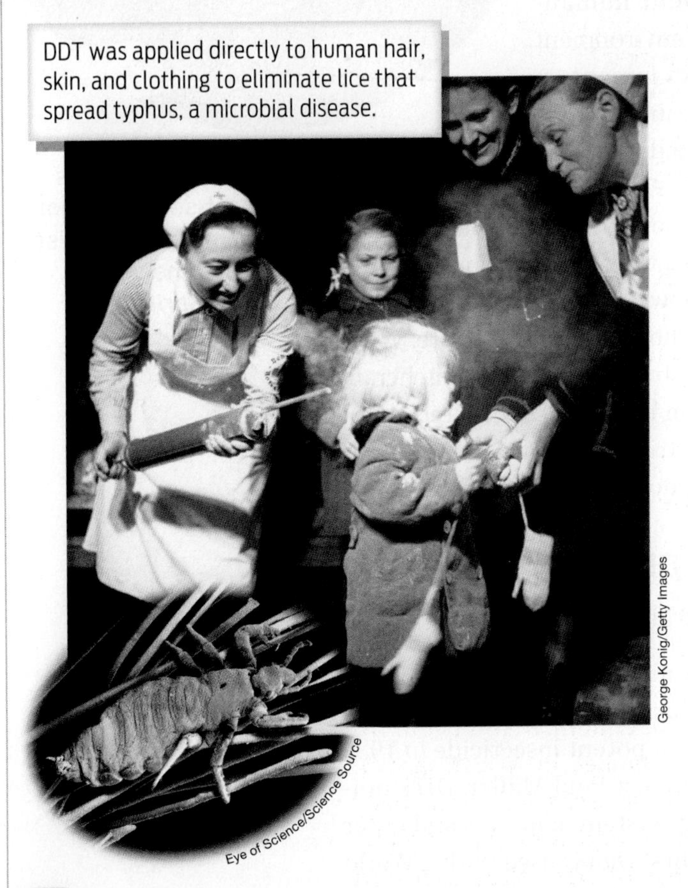

DDT was sprayed freely throughout neighborhoods to eradicate mosquitoes and other insects that carry human pathogens like the one that causes malaria.

? **Name at least one disease that is transmitted by insects.**

INFOGRAPHIC M6.2

Unintended Consequences of Using DDT

While DDT was effective against its insect targets, it also made its way into the food chain of many organisms, with unintended consequences.

? **Were the robins harmed by being sprayed by DDT, by contacting DDT-coated leaves, or by eating earthworms?**

letter from her friend, several court cases had been filed against the USDA in an effort to stop the indiscriminate spraying.

Carson began approaching scientists around the country, including many of her former colleagues in the Fish and Wildlife Service, for information about pesticide use and toxicity. "The more I learned about the use of pesticides, the more appalled I became," she later said. Over the next four years, she interviewed more scientists, scoured the research literature, and culled through newspaper reports to produce *Silent Spring,* a comprehensive treatise with 50 pages of footnotes.

Carson's book was filled with example after depressing example of how synthetic pesticides were wreaking destruction on natural populations of animals, most famously birds. For example, in the mid-1950s, government officials in states across the Midwest attempted to deal with the growing problem of Dutch elm disease, which was killing off the stately trees, by using DDT. Dutch elm disease is caused by a fungus, but it is spread from tree to tree by beetles that feed on the trees' leaves. To combat the beetles, scientists sprayed the trees with DDT. In the autumn, leaves coated in DDT fell to the ground, where earthworms feeding on the leaves took up the chemical and accumulated it inside their bodies. In the spring, robins migrated to these areas, fed voraciously on the earthworms, and were poisoned by the high amounts of DDT they ingested **(INFOGRAPHIC M6.2).**

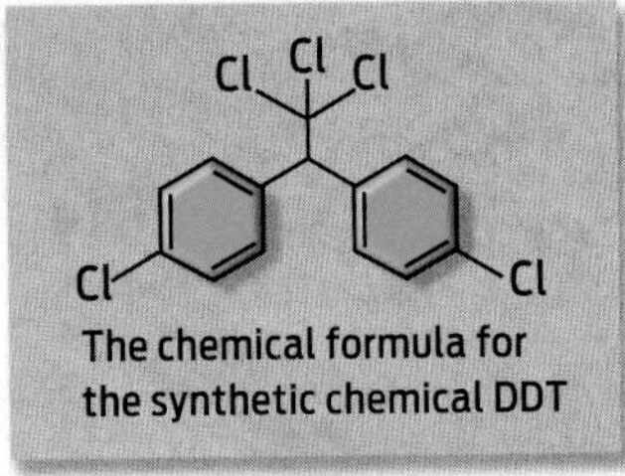

The chemical formula for the synthetic chemical DDT

Basemen: Stock/Alamy

Bettmann/Getty Images

Paul Müller, a Swiss chemist, showed that DDT was an effective insecticide. As a result of his work, DDT became a widely used pesticide.

Carson presented evidence that aquatic ecosystems are also affected by DDT. Since 1945, coastal waters around the United States had been sprayed with DDT to combat the salt-marsh mosquito. DDT entered the water supply, where it and its breakdown product, DDE, were taken up into the bodies of small fish and crabs. Larger fish would eat the smaller fish, thus taking in higher quantities of stored DDT and DDE, until finally the larger fish were eaten by predatory birds such as eagles, pelicans, and falcons, which ingested the highest quantities of all. Because of the interactions of organisms at different trophic levels in a food chain (see Chapter 21), organisms at the highest trophic levels can have high concentrations of DDT in their tissues, even if they were not directly exposed to this chemical. DDT is fat soluble (it can dissolve in fat) and not easily excreted in urine; therefore it accumulates in animal bodies. When other animals eat these animals, they eat their stored DDT, too. The process by which environmental toxins accumulate as they move up the food chain is called biomagnification **(INFOGRAPHIC M6.3)**.

The accumulated DDE impaired reproduction in birds of prey, especially pelicans, eagles, and falcons: their eggshells became so thin that mothers would crush their eggs when they sat on their nests to incubate them. This problem, combined with overhunting and habitat destruction, led to a steep drop in predatory bird populations in the United States. From numbers in the hundreds, for example, the bald eagle population in the United States plummeted in the 1950s and appeared on the verge of extinction—an ominous fate for our national emblem.

Besides the threat to wild animals, Carson pointed to possible dangers to humans. Though DDT was deemed safe for use as an insecticide on humans, there were concerns about long-term effects. "We have to remember that children born today are exposed to these chemicals from birth, perhaps even before birth," Carson said in a 1962 interview on CBS News. "Now what is going to happen to them in adult life as a result of that exposure? We simply don't know." These uncertainties were made more unnerving when DDT was shown to persist in the environment for many years, long after its application. We now know that DDT stays in soil for at least 15 years, and much longer in aquatic environments, possibly for hundreds of years.

The U.S. Environmental Protection Agency (EPA) currently classifies DDT as a "probable carcinogen"; because it is fat soluble, DDT accumulates in human tissues and in breast milk, and it has been associated with neurological disorders in places where it has been used heavily to prevent malaria. A 2014 study carried out by researchers at Rutgers and Emory Universities found an association between levels of DDT in the blood and a person's risk of developing Alzheimer's disease.

To Carson, the most galling and stupefying aspect of all the examples she found was the blind faith people put in pesticides—using them even when they had not been tested and when other methods had proved effective. As a means of controlling Dutch elm disease in the Midwest, for example, DDT spraying was a colossal failure; it did not save the trees, and may have actually made the problem worse because it had the unintended effect of killing the birds that served as a natural form of pest control. Yet effective means of controlling the disease—through the removal and burning of contaminated wood—had been used successfully for decades in other parts of the country, such as in New York.

How to account for these lapses of scientific judgment? A big part of the problem, Carson argued, was an unquestioned faith in the power of experts to control nature, to bend it to their whim with technology. In other words, it was as much a mindset as a specific technology that was the source of the trouble.

"Who has made the decision that sets in motion these chains of poisonings, this ever-widening wave of death that spreads out, like ripples when a pebble is dropped into a still pond?" she asked in her book. Carson minced no words with her answer: it was shortsighted individuals in positions of power whose capacity for destruction was not matched by an informed awareness of the associated costs.

From Silent Spring to Noisy Summer

Silent Spring was serialized in *The New Yorker* in the spring of 1962 and published as a book by Houghton Mifflin that fall. It caused an immediate sensation, as people around the country began to question the safety of these omnipresent and unseen chemicals. "'Silent Spring' Is Now Noisy Summer," *The New York Times* headlined a story documenting the furor that erupted in the wake of the book's publication. Though Carson found many supporters and admirers, among them President John F. Kennedy, she also found herself attacked by angry representatives of the chemical industry.

Velsicol Chemical, a maker of DDT and other pesticides, threatened to sue both Houghton Mifflin and *The New Yorker*.

INFOGRAPHIC M6.3

Biomagnification

Chemicals such as DDT are retained in the bodies of organisms that take them up. When those organisms are eaten by others, the concentration of chemicals increases at each level of the food chain.

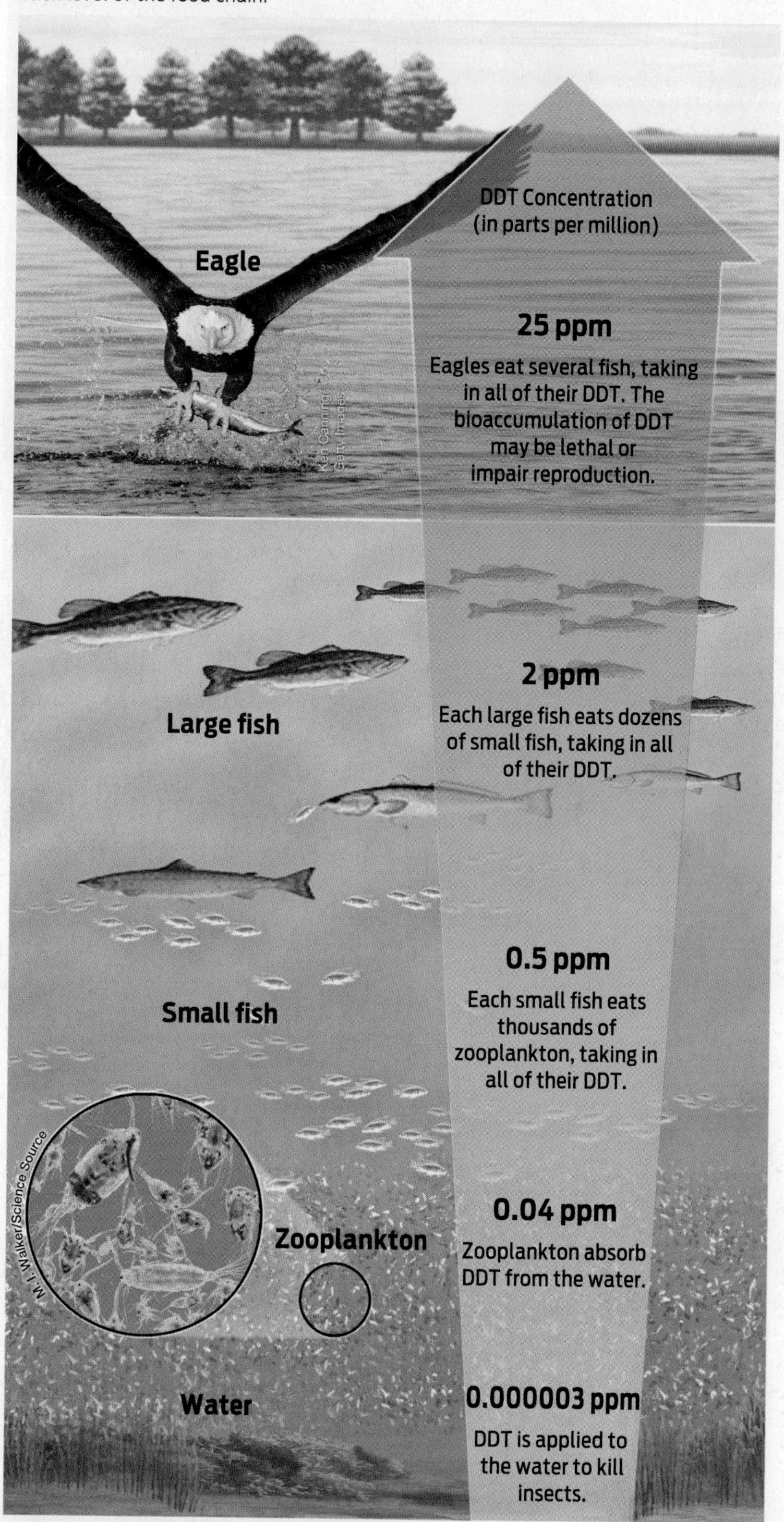

? Explain how a chemical like DDT when used to treat water can eventually be detected at high concentrations in the tissues of birds like eagles.

AP Photo

Rachel Carson testifying before Congress in 1963 on the dangers of pesticides.

A lawyer for the company accused Carson of being a Communist sympathizer who wanted to shrink the American food supply. Monsanto, another pesticide producer, published and distributed 5,000 copies of a brochure parodying Carson's book titled "The Desolate Year," which argued that without pesticides to help agriculture, food supplies would plummet and millions around the world would suffer from hunger and starvation. "If man were to faithfully follow the teachings of Miss Carson," scoffed an executive of American Cyanamid, "we would return to the Dark Ages, and the insects and diseases and vermin would once again inherit the earth."

Some of the criticism Carson faced was clearly because she was a woman. Her credentials as a scientist were routinely attacked (although she had a master's degree in zoology from Johns Hopkins), and she was often referred to as "hysterical." A former U.S. Secretary of Agriculture wondered aloud to the press "why a spinster with no children was so interested in genetics." A physician, writing in a medical journal, stated that reading *Silent Spring* "kept reminding me of trying to win an argument with a woman. It can't be done."

But Carson had done her homework, and her view ultimately carried the day. A commission convened by President Kennedy to investigate the claims made by Carson in her book found them to be sound. Soon thereafter, in 1963, congressional hearings were held at which Carson testified. She argued that a commission should be established to review pesticide issues and make decisions on the basis of broad public interest rather than the profit motives of a few.

Silent Spring is universally credited as the spark that ignited the modern environmental movement. In less than 10 years, what began as one woman's fight grew to enlist an army of activists, including many college students,

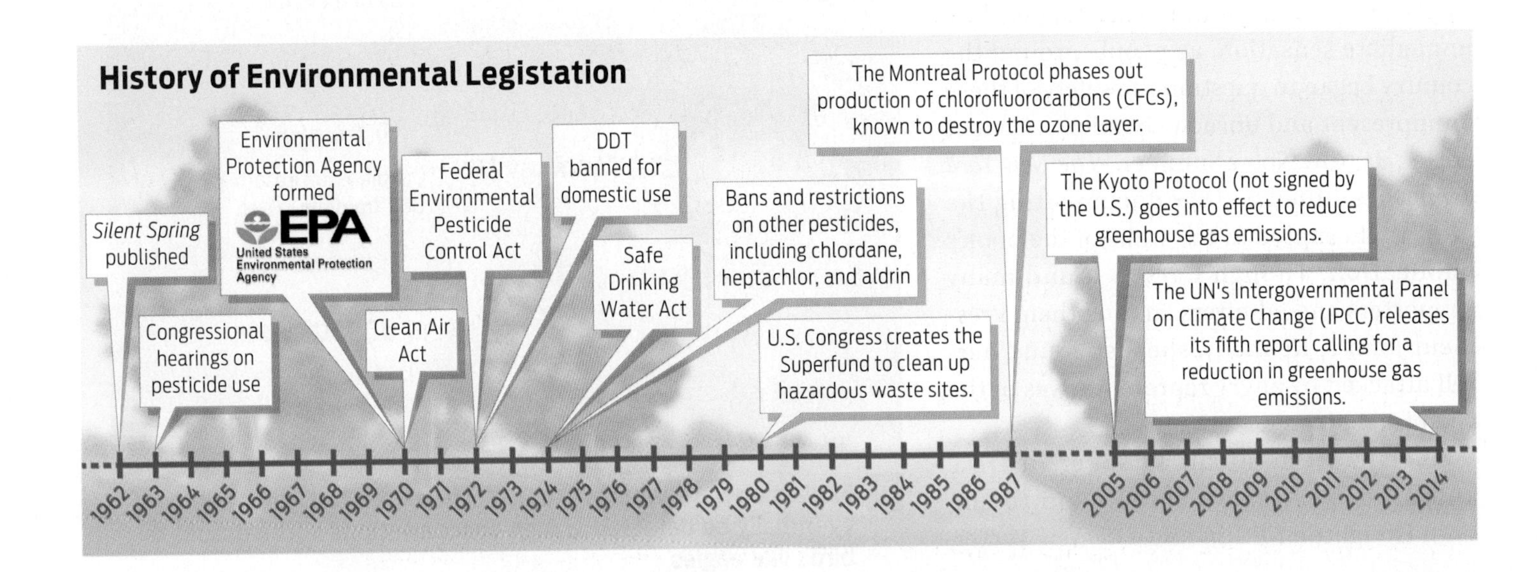

who together changed the national conversation. The first national Earth Day celebration was held on April 22, 1970. Later that year, President Richard Nixon created the Environmental Protection Agency and gave it the authority to set safe levels for chemicals. Shortly thereafter came several pieces of environment-related legislation, including the Clean Air Act, the Federal Environmental Pesticide Control Act, and the Safe Drinking Water Act. The EPA banned DDT for domestic use in 1972, except for cases of public health (for example, to prevent malaria). In 1974, it further banned or severely restricted the other pesticides that Carson had written about in her book: chlordane, heptachlor, dieldrin, aldrin, and endrin.

Carson would surely have been gratified to see the enactment of these milestones, but she didn't live to witness them. She died of breast cancer in 1964 at the age of 56, less than two years after *Silent Spring* was published. Though she had been sick for much of the period when she wrote her book, she never mentioned her illness publicly, fearing that her analysis of the human dangers of pesticides—cancer among them—might be labeled as biased. But her legacy lives on in the many people who credit her with changing their worldview and relationship to nature.

"For me personally," former vice president Al Gore wrote in an introduction to the 1992 edition of her book, "*Silent Spring* had a profound impact. . . . Indeed, Rachel Carson was one of the reasons that I became so conscious of the environment and so involved with environmental issues."

But not everyone praises Carson as a hero. For those who would like to reduce environmental regulation (including the scope of the EPA itself) on the grounds that it hinders free enterprise, she is a frequent target of criticism. Some of her detractors have argued that the ban on DDT was ultimately responsible for a rise in malaria rates and many thousands of deaths from this disease in Africa and Asia. But other researchers who have investigated these claims have found them dubious, noting that the World Health Organization (WHO) had stopped using DDT in its international efforts to control malaria even before it was banned in the United States, principally because mosquitoes had grown resistant to it. (Today, WHO permits indoor spraying of DDT as one way to combat malaria, but given the problems of resistance as well as ecological and human dangers, it intended to phase it out by 2020.)

George Silk/The LIFE Picture Collection/Getty Images

The beliefs of the time: An ad from *Life* magazine in 1948 featured model Kay Heffernon at Jones Beach, New York, apparently demonstrating that DDT was safe.

"Can anyone believe it is possible to lay down such a barrage of poisons on the surface of the earth without making it unfit for all life?"

—Rachel Carson

Carson herself never advocated the complete elimination of chemical pesticides. Rather, she opposed their indiscriminate use without conducting studies of their ecological effects. "It is not my contention that chemical insecticides must never be used. I do contend that we have put poisonous and biologically potent chemicals indiscriminately into the hands of persons largely or wholly ignorant of their potentials for harm," she wrote.

It was the height of human arrogance and stupidity, she argued, to think that we could inflict great damage on one part of the environment without harming others. The issues she raised are as applicable today as they were then, foreshadowing our current debates over greenhouse gases and global warming, as well as a new crop of synthetic pesticides (see Chapters 21 and 23). "Can anyone believe it is possible to lay down such a barrage of poisons on the surface of the earth without making it unfit for all life?" she asked. ■

More to Explore

- PBS, American Experience. (2017). *Rachel Carson* [documentary]: www.pbs.org/wgbh/americanexperience/films/rachel-carson/
- Haberman, C. (January 22, 2017). Retro report: Rachel Carson, DDT and the fight against malaria. *The New York Times* (includes video).
- Conniff, R. (2015). Rachel Carson's critics keep on, but she told truth about DDT. *YaleEnvironment360.* https://e360.yale.edu/features/rachel_carsons_critics_keep_on_but_she_told_truth_about_ddt
- Griswold, E. (September 23, 2012). How *'Silent Spring'* ignited the environmental movement. *The New York Times Magazine.*
- Oreskes, N., and E. Conway. (2010). *Merchants of Doubt: How a Handful of Scientists Obscured the Truth on Issues from Tobacco Smoke to Global Warming.* New York: Bloomsbury.

MILESTONES IN BIOLOGY 6 Test Your Knowledge

1. On the origins of DDT:

a. When was the chemical DDT first widely used, and what was its intended purpose?

b. How effective was it for that purpose?

2. On the mechanism of DDT:

a. How does DDT kill insects?

b. How does DDT harm top predator birds?

3. What is biomagnification?

4. PCBs (polychlorinated biphenyls), a type of chlorinated hydrocarbon, were used for a variety of purposes (including electrical insulation) until their use was banned in 1979. A 2000 survey of top predator fish in the Great Lakes showed that the concentration of PCBs in these fish ranged from 0.8 to 1.6 ppm. The wildlife protection value (the concentration that should not be exceeded to protect the safety of wildlife) is 0.16 ppm. How could there be such high levels of PCBs in top predator fish 21 years after PCBs were banned?

5. In 2013, a group of beekeepers launched a lawsuit against the EPA concerning the use of neonicotinoid pesticides and possible unintended impacts on honey bees (see Chapter 21). From what you have read here, what kinds of testing would you want the EPA to require before approving a pesticide applied widely to crops such as corn or soybeans?

Vital Signs

From melting ice caps to unquenchable fires, ominous signs of a changing planet

David Merron/Getty Images

DRIVING QUESTIONS

1 What are some ways that changing temperatures and changing climate impact ecosystems?

2 What is the greenhouse effect, and what does it have to do with global warming?

3 How does carbon cycle through ecosystems, and how are humans impacting the carbon cycle?

Beneath a deep layer of snow and ice, in a remote region of Siberia, a mysterious bacterium lies dormant. Sealed inside the frozen flesh of the horned animal it killed, the microbe hasn't stirred for nearly a hundred years. One summer, an unusual warm snap melts the ice, exposing the long-dead animal. The bacterium reawakens and spreads to humans, sickening and killing villagers.

This sounds like the plot of a B-grade horror movie. But it actually happened, in 2016, in the Siberian town of Salekhard, located 1,200 miles northeast of Moscow in the Arctic Circle. The bacterium was *Bacillus anthracis,* the organism that causes anthrax. The animals were reindeer that were infected with the bacteria around the turn of the 20th century. They lay there undisturbed until unusually warm weather in the region melted their frozen tombs, releasing the deadly microbe upon the unsuspecting town.

The anthrax outbreak in Salekhard was a vivid example of how warming temperatures around the globe threaten public health. But it's far from the only one. Cold winters serve as checks on disease-carrying pests like ticks, for example. In parts of the United States, tick larvae that would normally be killed by lengthy frosts are now surviving until spring, leading to more adult ticks. The ticks are also expanding their ranges northward, bringing with them diseases such as Lyme disease.

Pathogenic organisms like certain bacteria and parasites are themselves becoming more widespread due to global warming. The past few years have seen a spike in "flesh-eating" infections caused by the bacterium *Vibrio vulnificus,* which is found in shellfish such as oysters. Historically, *Vibrio* illnesses in the United States have been linked to shellfish from the Gulf of Mexico, where water temperatures are regularly warmer than 20°C (69°F). But in the past 20 years, outbreaks have occurred in the Pacific Northwest, Alaska, and parts of the Northeast, now that waters in these regions are regularly exceeding that temperature. Because of these changes in distribution, *Vibrio* is becoming known as a poster child for global warming **(INFOGRAPHIC 23.1)**.

What's behind this warming trend that is unleashing new threats upon the planet and making old ones worse? In short, humans. Through their activities, humans are increasing the concentration of gases in the atmosphere that trap the sun's warmth, just as a greenhouse traps warmth inside its clear walls. The result is a changing **climate**—the long-term average of atmospheric conditions in a region.

Note that climate is not the same as weather. **Weather** describes local atmospheric conditions over a short interval—the sun or clouds, and wind or rain, predicted in your weekly forecast. If the weather demonstrates a consistent change over a long period of time—consistently warmer winters, for example—that can indicate a change in the

CLIMATE
The long-term average of atmospheric conditions.

WEATHER
Local atmospheric conditions over a short period of time.

INFOGRAPHIC 23.1

Climate Change Poses Public Health Risks

Infectious diseases are occurring across broader ranges in response to environmental changes associated with climate change. Air and water temperatures have increased, and warm seasons are longer. With this change in climate, insect populations that spread infectious disease have increased, and infectious microbes have expanded into newly warm territories. These changes have brought a new level of concern about the health impacts of climate change.

Increased Air Temperatures

Lyme disease is caused by the bacterium ***Borrelia burgdorferi***, and is transmitted through the bite of the deer tick.

Ixodes scapularis

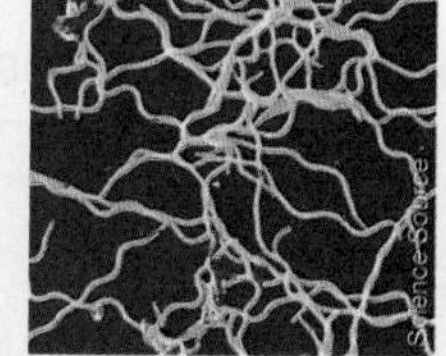
Borrelia burgdorferi

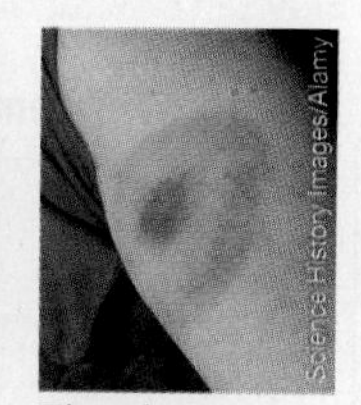
The classic bull's-eye rash at the site of infection

Due to the increase in global air temperatures, deer tick populations have increased and expanded into new territories. Incidence of Lyme disease has increased accordingly.

Reported distribution of deer tick (*Ixodes scapularis*):

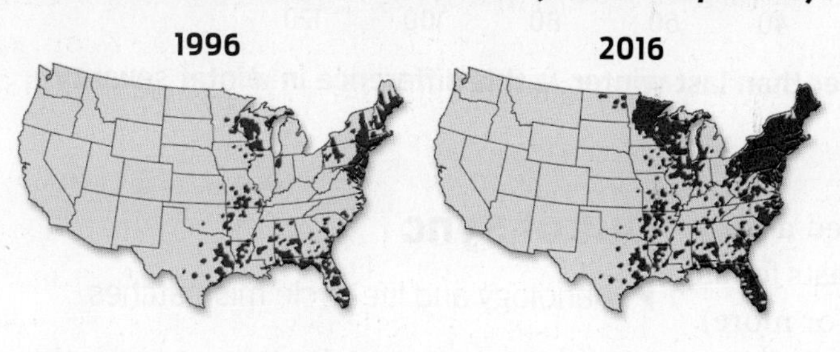

Reported cases of Lyme disease:

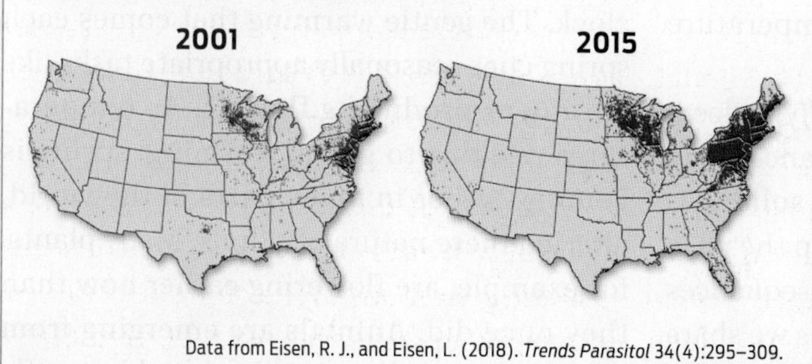

Data from Eisen, R. J., and Eisen, L. (2018). *Trends Parasitol* 34(4):295–309.

Increased Water Temperatures

Infection by the **flesh-eating bacterium, *Vibrio vulnificus***, is transmitted by contact with contaminated shellfish, like oysters.

Oysters

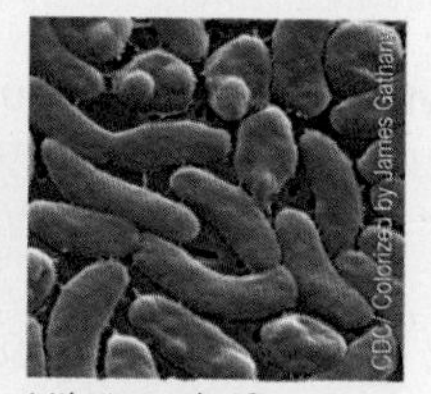
Vibrio vulnificus

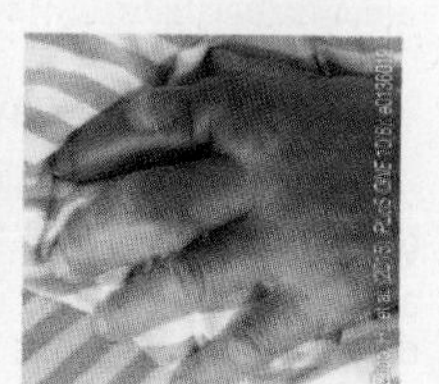
The "flesh-eating" symptoms of *Vibrio* infection

Due to the increase in global water temperatures, *Vibrio* infections are occurring more frequently in northern locations. As the oceans continue to warm, *Vibrio* bacteria are predicted to continue their expansion, along with the shellfish that harbor them.

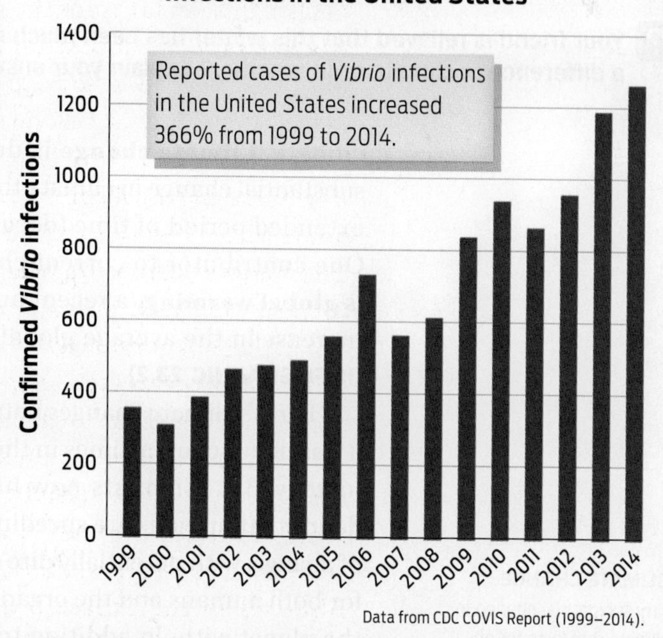

Data from CDC COVIS Report (1999–2014).

? Dengue virus is transmitted by *Aedes aegypti* mosquitoes. Currently, dengue virus infections have only been reported in Florida. What do you predict about the risk of dengue virus infections if a warming climate allows *Aedes aegypti* mosquitos to expand their range?

INFOGRAPHIC 23.2

Weather versus Climate

Weather refers to short-term and local atmospheric conditions, including temperature and precipitation. Climate is the long-term regional average of atmospheric conditions.

Weather

Short-term local atmospheric conditions

Weather data include daily or hourly descriptions of local temperature, precipitation, solar radiation, cloud cover, humidity, air pressure, wind speed, storm variables, etc.

Climate

Long-term regional averages

Climate data include yearly or longer averages of temperature, precipitation, solar radiation, air pressure, wind speed, air and water quality, snow cover, soil moisture, ocean currents, etc.

Daytime high temperatures on June 20, 2016

Map by Dan Pisut, NOAA Environmental Visualization Lab, based on NOAA RTMA data.

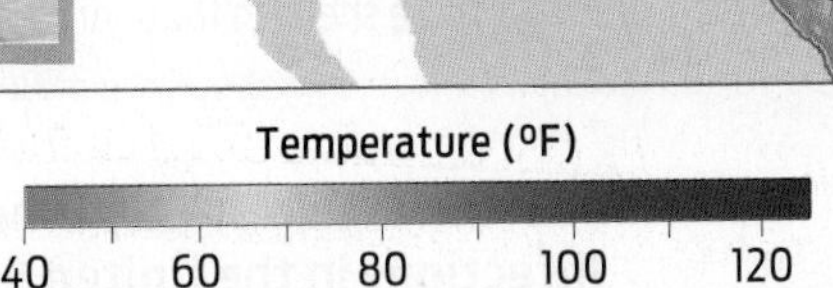

Average high temperatures in June (1981–2010)

Climate data reveal patterns that can be used to predict future weather. For example, based on these 30-year averages, when you travel to Las Vegas next June, temperatures are likely to be warm enough for shorts and T-shirts.

WY NV UT CO CA Las Vegas AZ NM TX

NOAA, data: NCEI

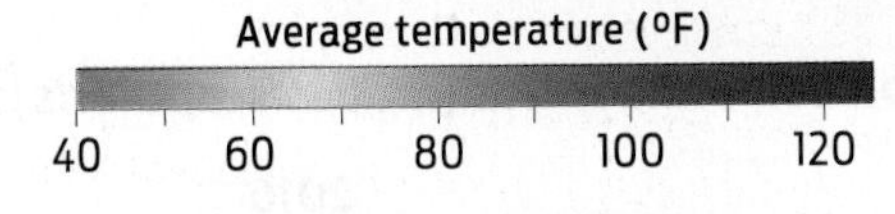

? Your friend is relieved that this winter has been much milder than last winter. Is this difference in winter severity a difference in climate or in weather? Explain your answer.

climate. **Climate change** is defined as any substantial change in climate that lasts for an extended period of time (decades or more). One contributor to current climate change is **global warming,** a recent and continuing increase in the average global temperature **(INFOGRAPHIC 23.2).**

Earth's climate changes naturally; indeed, it has done so many times in the planet's long history. But scientists now have solid evidence that humans are speeding up the pace of change, with potentially dire consequences for both humans and the organisms we share the planet with. In addition to threats from infectious disease, climate change is throwing off natural rhythms that have been stable for thousands of years, with cascading consequences for entire ecosystems.

CLIMATE CHANGE
Any substantial change in climate that lasts for an extended period of time (decades or more).

GLOBAL WARMING
An increase in Earth's average temperature.

Out of Sync

▸ Phenology and life-cycle mismatches

In nature, as in comedy, timing is everything. And for many species, temperature is nature's clock. The gentle warming that comes each spring cues seasonally appropriate tasks like mating or producing flowers. As temperatures rise due to global warming, spring is coming earlier in many parts of the world, altering these natural rhythms. Many plants, for example, are flowering earlier now than they once did. Animals are emerging from hibernation earlier. And bird and butterfly species are migrating north and breeding earlier in the spring than they did a few decades ago. It's a pattern of change that scientists are seeing around the globe and that you may have noticed in your own backyard **(INFOGRAPHIC 23.3).**

So what if flowers bloom earlier or animals come out of hibernation sooner? Why should we care? By themselves, these changes wouldn't necessarily be a big deal. But because living things are exquisitely adapted to their environments and are interdependent, a change in one part of an ecosystem may upset others.

The study of nature's interlinked rhythms is called **phenology.** Some of these rhythms are cued by rising temperature or snowmelt, while others are triggered by the lengthening of daylight hours that occurs during springtime. Because global warming has caused spring warming to come earlier while daylight has stayed the same, the result can be a mismatch between events that have been linked for millennia.

Consider the relationship between roe deer and their food source, the shoots of young plants that emerge in springtime. Roe deer give birth during one month in spring, based on the length of day. Vegetation in the area of eastern France where roe deer live begins to grow when temperatures are warm enough to melt snow and support growth. Because roe deer give birth based on daylength rather than springtime temperatures, a mismatch occurs between the time that roe deer are born and the time that plentiful young shoots are available. Researchers in France have tracked the timing of plant growth and the survival of roe deer offspring over the course of nearly 30 years. They have found that fewer roe deer

PHENOLOGY
The study of cyclic life events such as plant flowering and animal migration and how these are influenced by climate and seasonal changes.

INFOGRAPHIC 23.3

Changing Temperatures Affect Plant Behavior

Changes in average temperatures are changing the seasonal behavior of plants and animals. Changes in average spring temperatures result in earlier flowering dates in some locations, and later flowering dates in other locations.

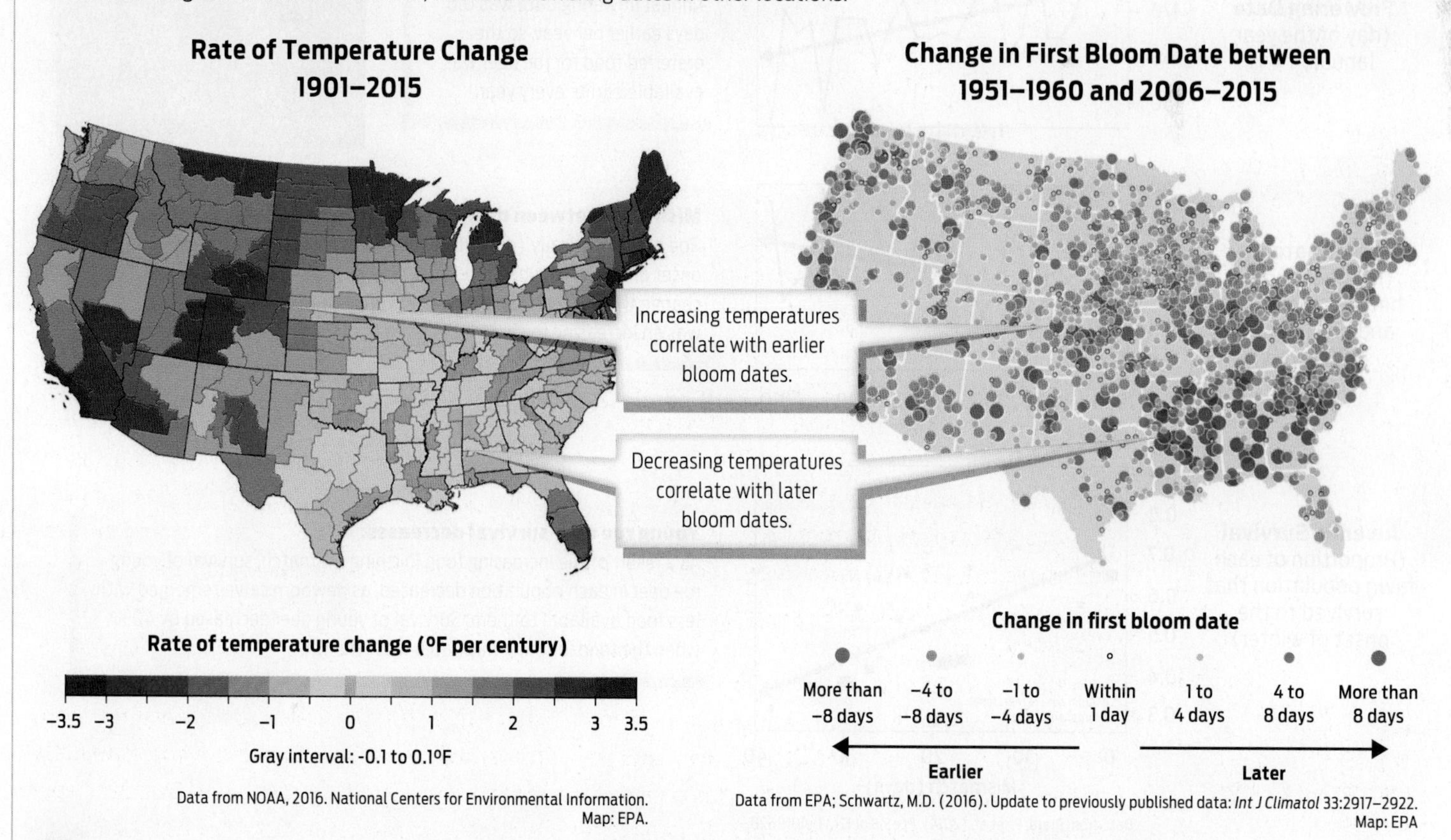

? Why are bloom dates in the southeastern United States occurring later and bloom dates in New England occurring earlier?

offspring are surviving due to this mismatch **(INFOGRAPHIC 23.4)**.

What will be the outcome of these mismatches multiplied across an ecosystem? Scientists don't really know. "We're seeing changes to systems that have been relatively stable for thousands of years," says Hector Galbraith, director of the Climate Change Initiative at the Manomet Center for Conservation Sciences, in Plymouth, Massachusetts. "The really scary thing about climate change is it's very difficult to predict the ecosystem effects of these changes."

Nevertheless, some disturbing scenarios are already coming to pass. Take the relationship between drought, trees, insect pests, and

INFOGRAPHIC 23.4

Warming Temperatures Can Disrupt the Timing of Seasonal Events

Scientists studied the roe deer and their ecosystem for nearly 30 years. Warmer springs have made the preferred food for deer available earlier, but the deer have not shifted their birthing dates to match the food availability. As a result, the deer population is being negatively affected.

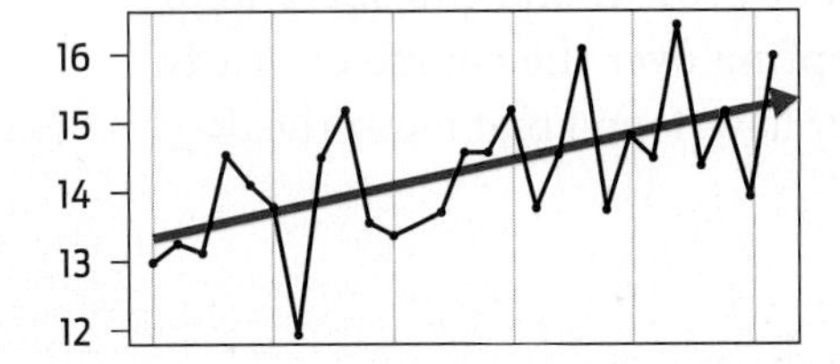

Temperature goes up: Annual spring (April to June) temperature increased by 0.07°C per year.

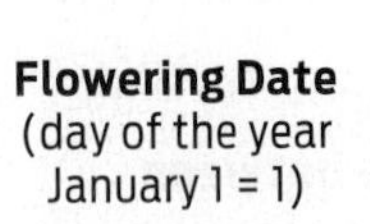

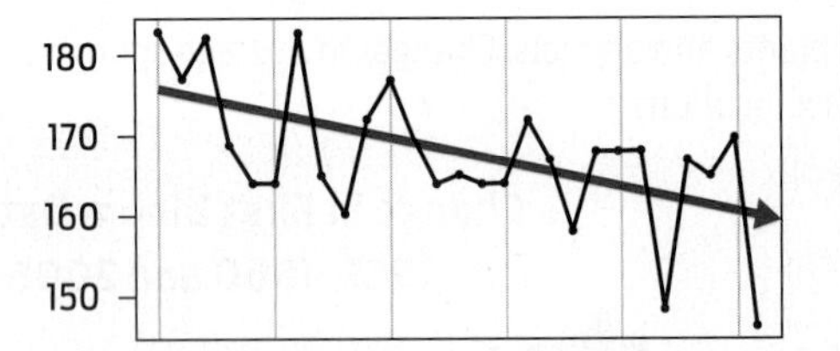

Flowering date comes early: Annual flowering date was 0.6 days earlier per year, so the preferred food for roe deer was available earlier every year.

Mismatch (Number of days between flowering and birth beyond 1985 levels)

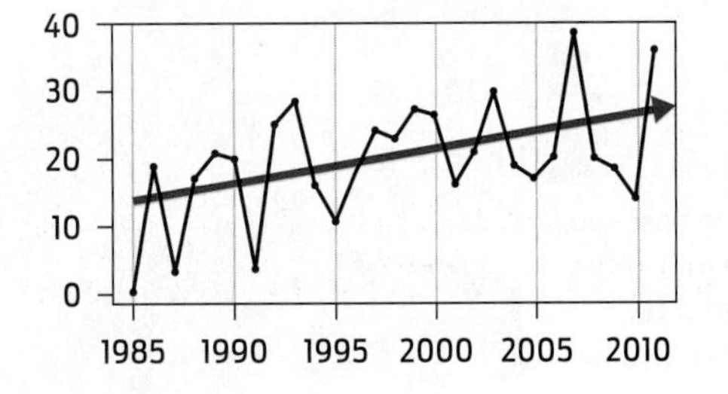

Mismatch between deer birth and food availability increases: Roe deer historically (pre-1985) gave birth about 1 month before the onset of flowering when food was most available. Since roe deer did not change their birthing date as the flowering date became earlier, there was an increasing mismatch between the two.

Juvenile Survival (Proportion of each fawn population that survived to the onset of winter)

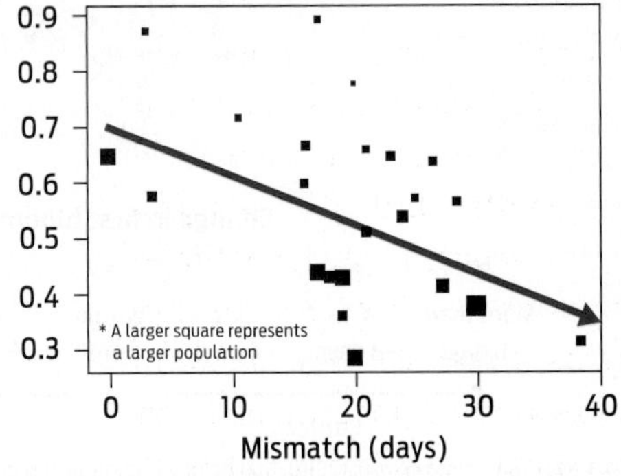

Young roe deer survival decreases: As a result of the increasing food/birthing mismatch, survival of young roe deer in each population decreased, as newborn calves emerged with less food available to them. Survival of young deer decreased by 40% when the food/birthing mismatch was 1 month.

Data from Plard, F., et al. (2014). *PLoS Biol* 12(4):e1001828.

? **Why is increasing spring temperature having a negative impact on deer populations in this ecosystem?**

An unprecedented number of Australian fires burned in 2019, due to higher temperatures during the fire season and extended periods of drought.

fire. Drought, an extended period of dryness, is becoming more common with climate change. In warmer temperatures, moisture evaporates more quickly from soils and is sucked more quickly from trees. Trees that are confronted by dry soils and loss of water caused by extended drought are more vulnerable to insect pests that can kill them.

Bark beetles are a common pest in the conifer forests that cover large swaths of the Pacific Northwest and Canada. They burrow into the trees' bark and lay eggs. As a result of living with these pests for thousands of years, trees have evolved an effective defense mechanism against them: they release sap that smothers the insects and prevents them from recruiting other beetles. But trees that are stressed by lengthy droughts are less able to defend themselves. Once a few beetles penetrate a tree's defenses, they send out chemical signaling molecules called pheromones that attract reinforcements, and the result is an increase in the number of dead trees, which provide kindling for forest fires.

INFOGRAPHIC 23.5

Rising Temperatures Mean Widespread Ecosystem Change

Climate change has cascading consequences for entire ecosystems. Organisms such as bark beetles have expanded their range and populations, resulting in widespread death of trees. Trees are also threatened by the changing temperatures and precipitation associated with climate change. Dead trees, drought, and high temperatures contribute to severe wildfires.

? Explain how bark beetles contribute to increases in wildfires in the Mountain West region.

Historically, beetles have been kept in check by freezing winter temperatures that kill their larvae. But rising winter temperatures have made for milder winters that lack deep freezes. According to some reports, the coldest winter night in many areas of the United States is now 4°C (7°F) warmer than it was 50 years ago. As a result, more beetle larvae are surviving through the winter. In forests that are already weakened by drought, trees may not contain enough water to make sufficient sap to defend themselves against the growing population of invaders. When the beetles reproduce unchecked, trees begin to die. In parts of California, Colorado, Montana, and other western states, the combination of extended drought and widespread bark beetle infestations has led to devastating wildfires that have proved hard to contain.

Bark beetles are also increasing their northern range. Scientists fear they may eventually hit conifers in the boreal forests of Canada that lack the defenses of more southern-dwelling trees. A mass die-off of these trees could spell disaster for the animals that rely on these forests for habitat **(INFOGRAPHIC 23.5)**.

Longer droughts, hotter temperatures, and more dead trees are creating a "perfect storm" of conditions for devastating fires in regions not historically swept by fire. And in places like California and Australia, where fires are historically common, they are getting worse and more frequent. In 2019, Australia experienced one of its hottest, driest summers on record, fueling intense fires that raged across the country. These fires caused catastrophic loss of wildlife, harmed or killed humans, and destroyed billions of dollars' worth of property.

While some species may be able to adapt to a changing climate by shifting their ranges, future climate change will likely exceed the ability of many species to adapt, as these species can no longer reach hospitable habitats. If the world continues to warm at the current rate, Earth's average temperature is predicted to increase by 4.3°C (7.8°F). In that case, according to a 2015 study published in *Science*, an estimated 1 in 6 (16%) species will be driven to extinction. The natural residents of mountaintops are especially vulnerable: as temperatures rise, species may move up to higher, colder elevations, but eventually they will have nowhere left to go.

This potential catastrophic loss of biodiversity could have unimaginable effects on ecosystems. As we saw in Chapter 22, loss of even one species from an ecosystem can have a cascading negative effect on the ecosystem as a whole, reducing the number of other species that can survive there.

Warming Up

▸ Greenhouse gases and the greenhouse effect

Although temperature swings and shifts in the ranges of organisms have been natural phenomena in the past, the amount of warming in recent years goes beyond any increase in temperature observed in the past, and evidence indicates that this change is not part of a natural cycle. From 1880 until 2018, Earth's surface warmed, on average, by about 1.1°C (2.0°F), according to a 2019 report by NASA's Goddard Institute for Space Studies. That may not sound like a lot. But consider this: the difference in average global temperatures between today and the last ice age—10,000 years ago, when much of North America was buried under ice—is only about 5°C (9°F). Where global temperatures are concerned, even a 1° change is significant.

The difference in average global temperatures between today and the last ice age—10,000 years ago—is only about 5°C (9°F).

Moreover, that's the average for the whole planet. Some parts of the globe, like the poles, are nearly 4°C (7°F) warmer than they were in the 19th century.

The rate of warming has increased as well. Eighteen of the 19 warmest years on record have occurred since 2001. And each year seems to bring a new heat record: 2015 was hotter than 2014 globally, and 2016 was hotter still. The year 2019 was the second warmest year on record. Much of this warming is attributable to the **greenhouse effect,** the trapping of heat in Earth's atmosphere.

As sunlight shines on our planet, it warms Earth's surface. This heat radiates back to the atmosphere. Some of this heat escapes into space, but the rest is absorbed by **greenhouse gases** in the atmosphere

GREENHOUSE EFFECT
The natural process by which heat from sunlight is radiated from Earth's surface and trapped by gases in the atmosphere, helping to maintain Earth at a temperature that can support life.

GREENHOUSE GAS
Any of the gases in Earth's atmosphere that absorb heat radiated from Earth's surface and contribute to the greenhouse effect—for example, carbon dioxide and methane.

INFOGRAPHIC 23.6

The Greenhouse Effect

The greenhouse effect is a natural process that helps maintain steady and life-sustaining surface temperatures on Earth. Sunlight heats Earth's surface and that heat radiates back to the atmosphere. While some of the heat escapes to space, certain gases in Earth's atmosphere, known as greenhouse gases, trap heat within the atmosphere. This trapped heat warms the atmosphere and Earth's surface.

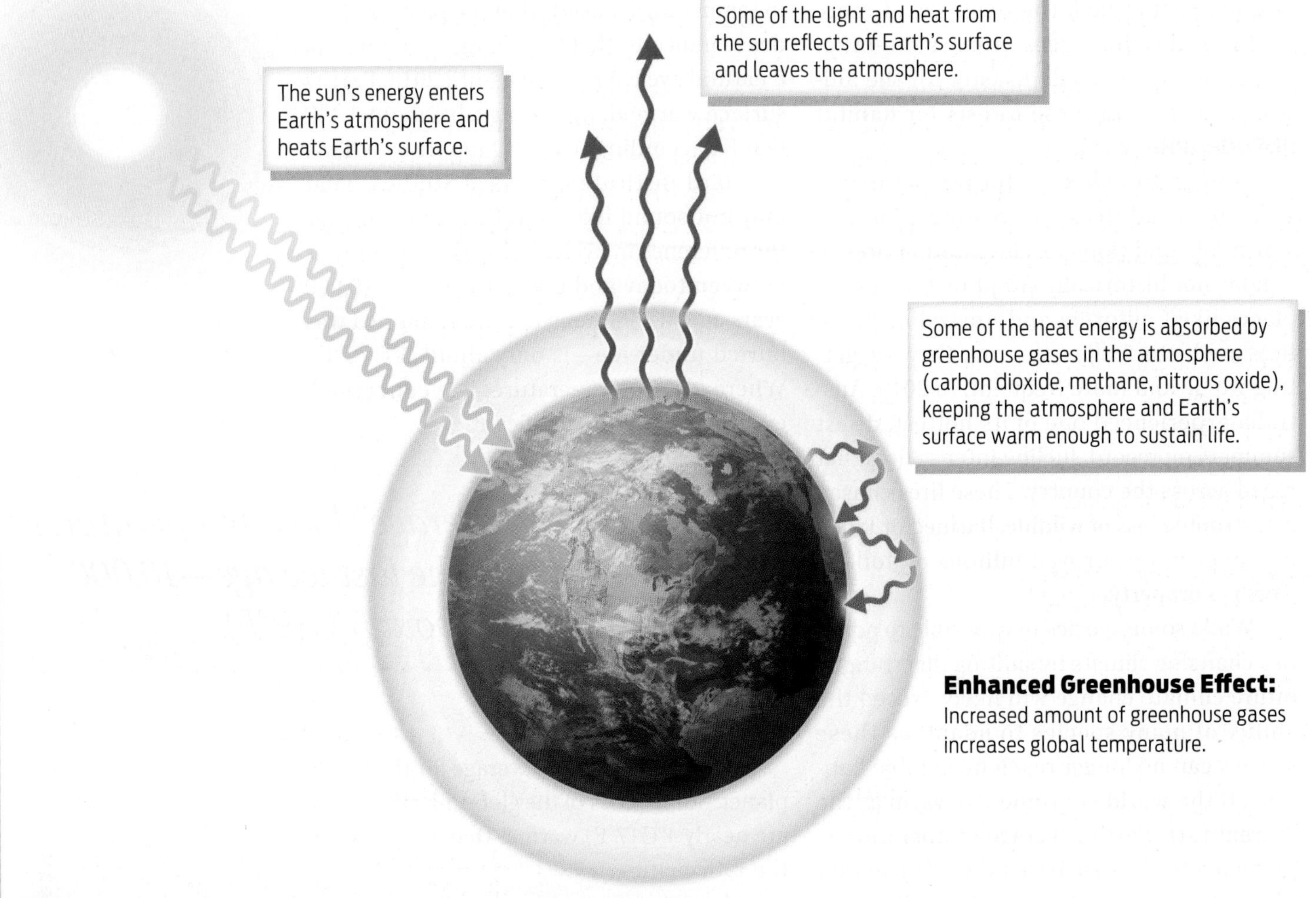

? **How do greenhouse gases contribute to keeping Earth warm?**

such as carbon dioxide. The heat trapped by greenhouse gases raises the temperature of the atmosphere and, in turn, of Earth's surface **(INFOGRAPHIC 23.6)**.

If you've ever stepped into a greenhouse, then you already have an intuitive sense of how the greenhouse effect works. As the sunlight shines through the glass, it warms the floor and other objects inside the greenhouse. The glass walls of the greenhouse trap that heat, keeping the greenhouse warm and cozy for the plants. Similarly, gases in Earth's atmosphere act like the wall of a greenhouse. As the sun shines on Earth, it warms the surface. The gases in atmosphere seal this heat in, warming the planet.

The greenhouse effect is a natural process that helps maintain life-supporting temperatures on Earth. Without this greenhouse effect, the average surface temperature of the planet would be a frigid −18°C (0°F). In recent years, however, levels of greenhouse gases have been rising. These rising levels have increased the strength of the greenhouse effect, a phenomenon known as the enhanced greenhouse effect. As the amount of greenhouse gases in the atmosphere has increased, so have temperatures. The result is global warming, an overall increase in Earth's average temperature **(INFOGRAPHIC 23.7)**.

Knowledge of the greenhouse effect is not new. In fact, the Swedish physicist Svante Arrhenius predicted in 1896 that changes in the levels of carbon dioxide in the atmosphere could substantially alter the surface temperature through the greenhouse effect. More than a hundred years later, the evidence is overwhelming that rising levels of greenhouse gases are making Earth's surface warmer. As NASA explains on its website, "There is no question that increased levels of greenhouse gases must cause the Earth to warm in response." And with that warming comes consequences for the many ecosystems around the globe.

Arctic Meltdown

▸ Sea ice and land ice changes

Predictably, snow- and ice-covered regions such as the Arctic stand to suffer most immediately from a warming climate: as temperatures rise, frozen habitats start to melt. But the situation is even worse than one might imagine. As Mark Serreze, director of the National Snow and Ice Data Center at the University of Colorado, Boulder, notes, the Arctic has warmed, on average, twice as much as the rest of the planet. This phenomenon, which climate scientists call Arctic amplification, has to do with the way sea ice affects temperature. As Serreze explains, sea ice reflects solar radiation (because it is white) and also insulates the ocean. As global temperatures rise, ice begins to melt. With less sea ice, more solar radiation is absorbed by the (dark) ocean, raising ocean temperature, and more of the relatively warm ocean is exposed to air, raising the air temperature

INFOGRAPHIC 23.7

Earth's Surface Temperature Is Rising with Increases in Greenhouse Gases Like Carbon Dioxide

Long-term measurements of global temperature and atmospheric CO_2 concentration show a clear, positive correlation.

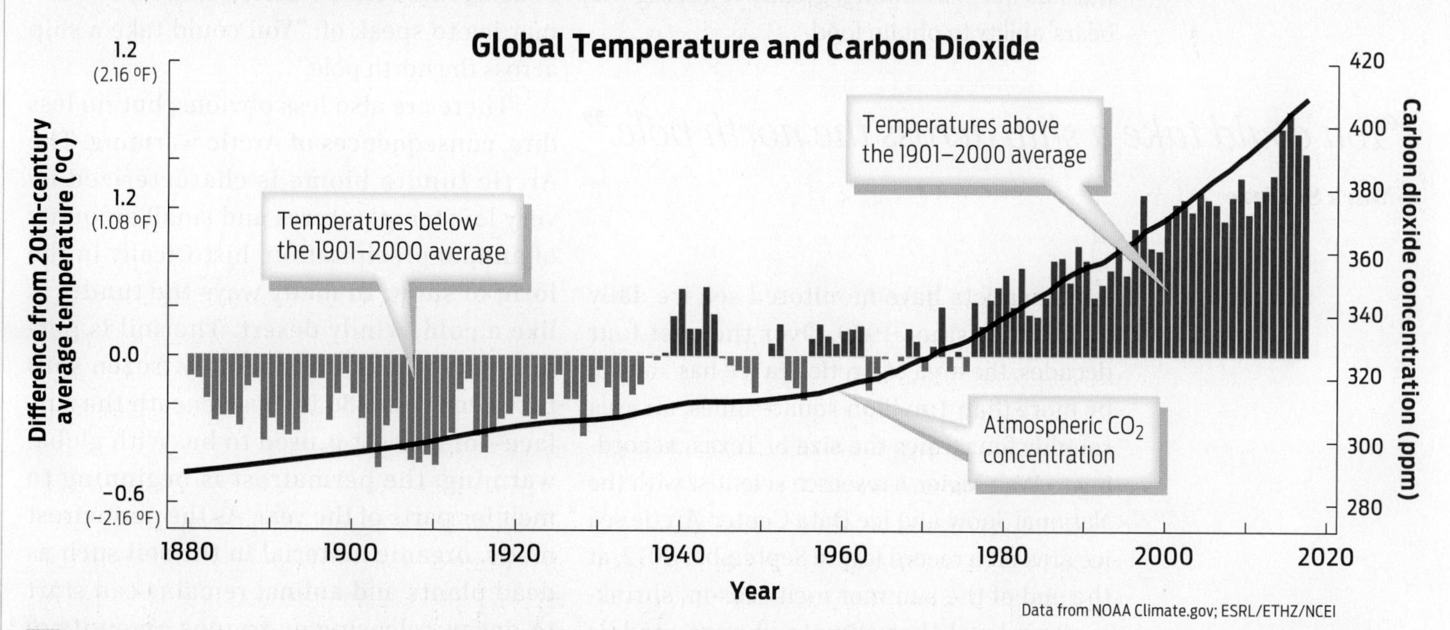

? From the most current data shown in the graph, how many degrees warmer is Earth than the 20th-century average? When did Earth's temperature start to be consistently higher than the 20th-century average?

Paul Souders/Getty Images

Polar bear on melting sea ice.

even more. It's a positive feedback loop: as additional ice is lost, temperatures rise at a faster pace.

Warming temperatures could spell disaster for species that call the Arctic their home. Polar bears, for example, spend most of the year roaming the Arctic on large swaths of floating sea ice that blanket a good portion of the Arctic Ocean from September through March. These massive mammals travel on sea ice to hunt for seals, which they nab as the seals periodically pop up through "whack-a-mole"–like breathing holes in the ice. The size of this frozen habitat has been shrinking, greatly reducing the bears' ability to obtain food.

"You could take a ship across the north pole."

—Mark Serreze

Scientists have monitored sea ice daily by satellite since 1979. Over the past four decades, the area of Arctic sea ice has shrunk by more than 1 million square miles, an area roughly four times the size of Texas, according to Walt Meier, a research scientist with the National Snow and Ice Data Center. Arctic sea ice area hit a record low in September 2012, at the end of the summer melt season, shrinking to a level that climate change models had predicted wouldn't happen until at least 2050. September 2019 had the second lowest area of Arctic sea ice on record, a three-way tie between that year, 2007, and 2016. Scientists now fear that nearly all of the polar bears' summer sea ice could vanish by 2040—and possibly sooner **(INFOGRAPHIC 23.8)**.

Warming temperatures are also melting glaciers and ice caps on land. Like an ice cube melting in a glass of water, the melting of sea ice doesn't raise the water level. But the melting of glaciers and ice caps does. And since water expands slightly when heated, water expansion adds to the problem of rising sea levels. Sea levels have risen 3.4 inches, on average, since 1993. How much will they rise in the future? "By 2100, you're looking at probably about a meter," says Serreze. "Here in Boulder we're at 5,400 feet—we're not worried about that. But if you're living in Miami, this is something that should concern you."

It's important to note that much of the data we have on climate change indicate global, long-term trends. From year to year, there may be slight variations—slightly warmer summers and less sea ice one year, slightly cooler summers and more sea ice the next. And indeed, from a low in 2007, sea ice area did indeed bounce back a bit in 2008 and 2009. But the overall trend is still unmistakably downward, toward less sea ice. By 2030 or 2040, says Serreze, there could be no summer ice to speak of: "You could take a ship across the north pole."

There are also less obvious, but no less dire, consequences of Arctic warming. The Arctic tundra biome is characterized by very low temperatures and small amounts of annual precipitation, historically in the form of snow. In many ways the tundra is like a cold, windy desert. The soil is permafrost (that is, permanently frozen soil) between 10 and 36 inches beneath the surface—or at least it used to be. With global warming, the permafrost is beginning to melt for parts of the year. As the permafrost melts, organic material in the soil such as dead plants and animal remains can start to decay, releasing enormous amounts of carbon dioxide as a result. In other words, the Arctic is an important storage reservoir

INFOGRAPHIC 23.8

Arctic Temperatures Are Rising and Sea Ice Is Melting

Current measurements suggest that the Arctic is warming faster than other parts of the Earth. Rising temperatures have caused Arctic sea ice to melt and break apart earlier in the season. Reduction in the extent of summer sea ice results in Arctic amplification by exposing more heat-absorbing ocean to solar radiation.

Global Mean Surface Temperature, 2015–2019

Earth's average global surface temperature in 2019 was the second warmest since modern record-keeping began in 1880, 1.8°F (0.98°C) warmer than the 1951–1980 mean. The past five years have been the warmest of the last 140 years.

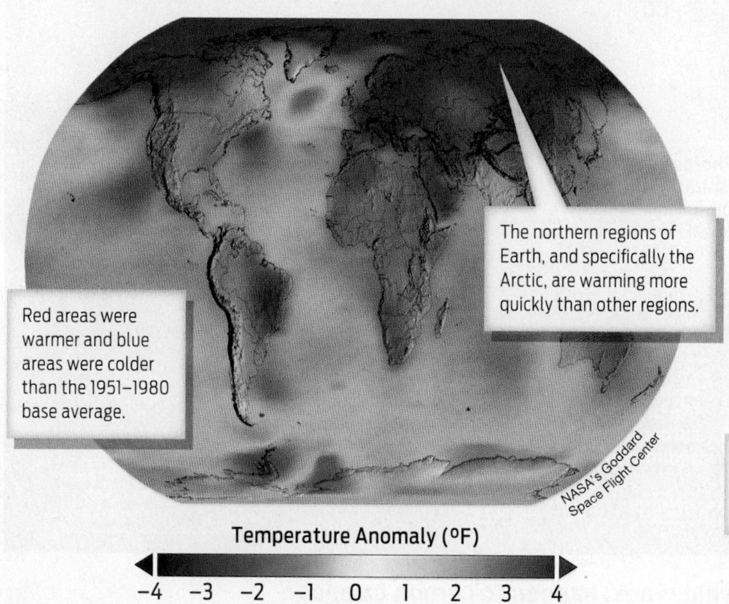

Arctic Sea Ice Extent, September 18, 2019

Arctic sea ice extent in 2019 was the second lowest on record, a three-way tie with 2016 and 2007.

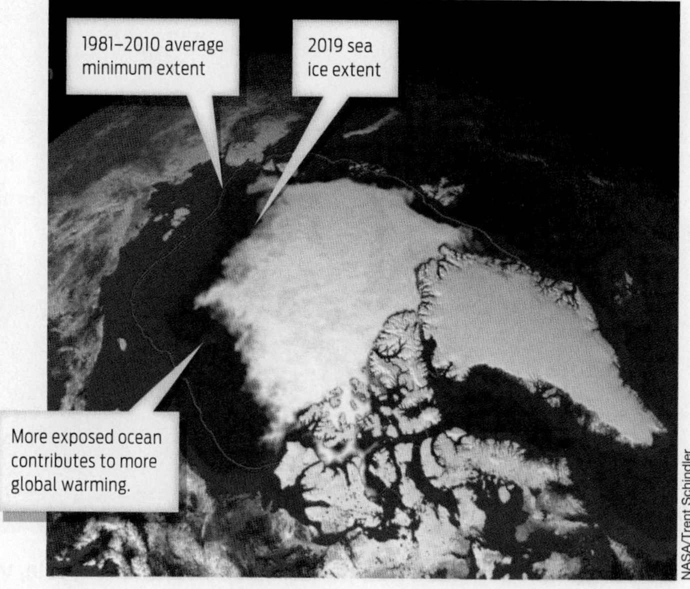

? Based on the map of global surface temperatures, which areas of Earth have higher than average temperatures, and which have lower than average temperatures?

of carbon, and its disturbance can further enhance global warming.

Follow the Carbon

▶ The carbon cycle

Carbon dioxide is the most notorious player in the greenhouse effect, and scientists believe it is responsible for most of the warming that has occurred. Atmospheric carbon dioxide levels have risen dramatically since the late 19th century. In fact, they are higher now than they have been in more than 800,000 years. So what has changed in that time? Through our activities, humans have begun to alter the **carbon cycle**, the movement of carbon atoms through the ecosystem, from living and nonliving things on Earth to the atmosphere and back again.

Carbon is a natural ingredient in every living organism, part of the backbone of all organic molecules (Chapter 2). Carbon also exists in inorganic forms: as carbon dioxide in the atmosphere, as carbonic acid dissolved in water, as calcium carbonate in limestone rocks. If dead organisms are fossilized before they can be digested by decomposers, the organic molecules contained within their bodies become trapped below Earth's surface or under the seas. Compressed over millions of years, these organic molecules become **fossil fuels**—coal, oil, and natural gas.

Like other chemical elements, the total amount of carbon on Earth remains

CARBON CYCLE
The movement of carbon atoms through the ecosystem as they cycle between organic molecules and inorganic CO_2.

FOSSIL FUEL
A carbon-rich energy source, such as coal, petroleum, or natural gas, formed from the compressed, fossilized remains of once-living organisms.

INFOGRAPHIC 23.9

The Carbon Cycle

The carbon cycle involves the movement of carbon atoms through the ecosystem as they cycle between organic molecules and inorganic CO_2. Natural processes, such as photosynthesis and respiration, are responsible for most carbon cycling. Since the late 1700s, human activities, including burning fossil fuels and deforestation, have made significant contributions to the carbon cycle, primarily by increasing the amount of carbon in the cycle in the form of CO_2. There is currently more CO_2 released into the atmosphere each year than is removed.

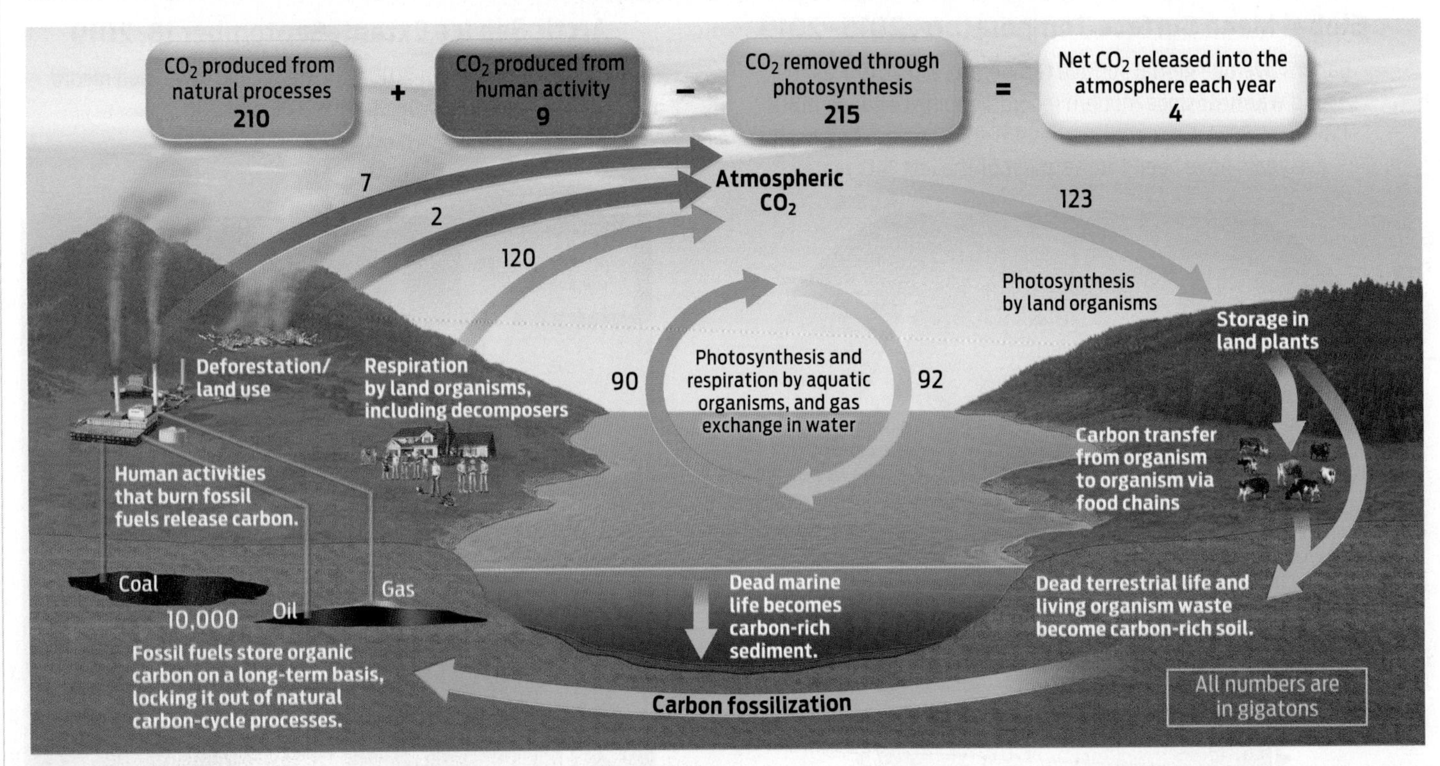

? **If human activities had no effect on the carbon cycle, what would happen to carbon balance?**

essentially constant. As it cycles through the environment, carbon moves between inorganic and organic forms. Plants, photosynthetic bacteria, and algae take up (inorganic) CO_2 from the atmosphere during photosynthesis and convert it into (organic) sugar molecules, thereby reducing atmospheric CO_2 levels. Animals take in organic carbon when they eat plants and other organisms and release inorganic CO_2 gas into the atmosphere as a by-product of cellular respiration. Similarly, when organisms die, decomposers in the soil use the dead organic material for food and energy, releasing some of the carbon during respiration as CO_2.

Photosynthesis and respiration form a cycle that keeps carbon dioxide at a relatively stable level in the atmosphere. But human actions, such as deforestation and burning fossil fuels, inject carbon dioxide that was not otherwise moving into the cycle (**INFOGRAPHIC 23.9**).

For most of human history, the amount of carbon present in the atmosphere as carbon dioxide has remained fairly constant. But since the industrial revolution, beginning in the late 1700s, humans have added increasing amounts of CO_2 to the atmosphere. In this way we have begun to alter the carbon cycle.

Before the industrial revolution, fossil fuels, buried deep underground, couldn't be accessed easily, so the carbon trapped in these fossil fuels wasn't part of the carbon cycle. But modern drilling and mining methods have unlocked the deep reserves of this ancient planetary energy. When humans burn fossil fuels, the carbon those fuels contain is released into the atmosphere as CO_2. That CO_2 is added to the carbon cycle and is a major contributor to the enhanced greenhouse effect.

How do scientists know that carbon dioxide levels are much higher now than in the past? There are two main sources of evidence. First, air bubbles trapped in glacial ice from Greenland and Antarctica provide a measure of carbon dioxide in the atmosphere at times in the distant past. Ice cores drilled at these sites provide data on very long-term changes in CO_2 levels (and temperature), going back hundreds of thousands of years. These data show, for example, that levels of CO_2 have cycled in patterns that correlate with major ice ages. Second, since 1958, scientists have directly measured CO_2 in the atmosphere—for example, at the Mauna Loa Research Station, which sits atop an inactive volcano in Hawaii. When combined, these data show that atmospheric CO_2 has been rising steadily since the industrial revolution—increasing from about 280 parts per million (ppm) in 1800 to 411 ppm in 2019—or more than 45% **(INFOGRAPHIC 23.10)**.

INFOGRAPHIC 23.10

Measuring Atmospheric Carbon Dioxide Levels

Examining data from ancient air bubbles and present-day air measurements, scientists have recorded an approximately 45% increase in atmospheric CO_2 levels since 1800.

Historical carbon dioxide levels are measured in glacial ice cores

As snow compacts into ice cores, bubbles of atmospheric gas remain trapped in the ice. The deepest ice is the oldest, and the air bubbles in those old layers contain levels of CO_2 that were present in the atmosphere at that time.

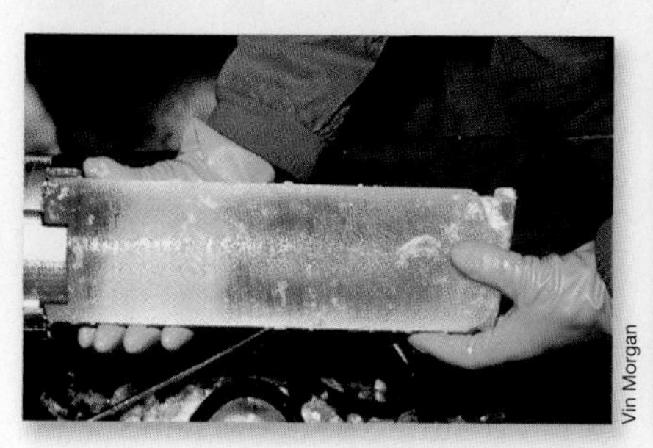

Vin Morgan

SPL/Science Source

Present-day carbon dioxide levels are measured directly from the air

Direct measurements of carbon dioxide are currently taken from the Mauna Loa Research Station in Hawaii.

U.S. Dept. of Commerce, NOAA, Earth System Research Laboratory

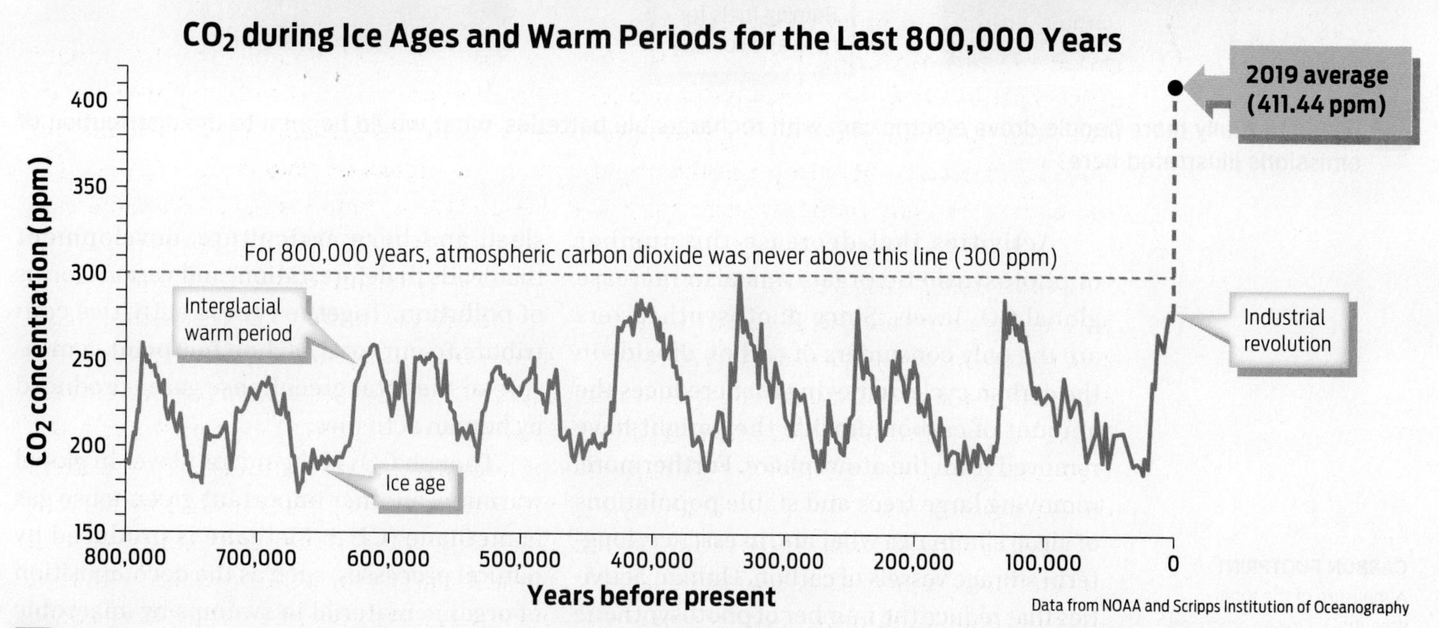

Data from NOAA and Scripps Institution of Oceanography

? **What is the value (in ppm) that CO_2 levels had not exceeded for 800,000 years? What is the most recent CO_2 concentration in the atmosphere? What percent increase over the 800,000-year maximum does this represent?**

INFOGRAPHIC 23.11

Human Activities Produce Greenhouse Gases

A variety of human activities are increasing the levels of greenhouse gases in the atmosphere. Power generation and transportation account for the largest proportion of greenhouse gas emission. In 2017, U.S. greenhouse gas emissions totaled 6,456.7 million metric tons.

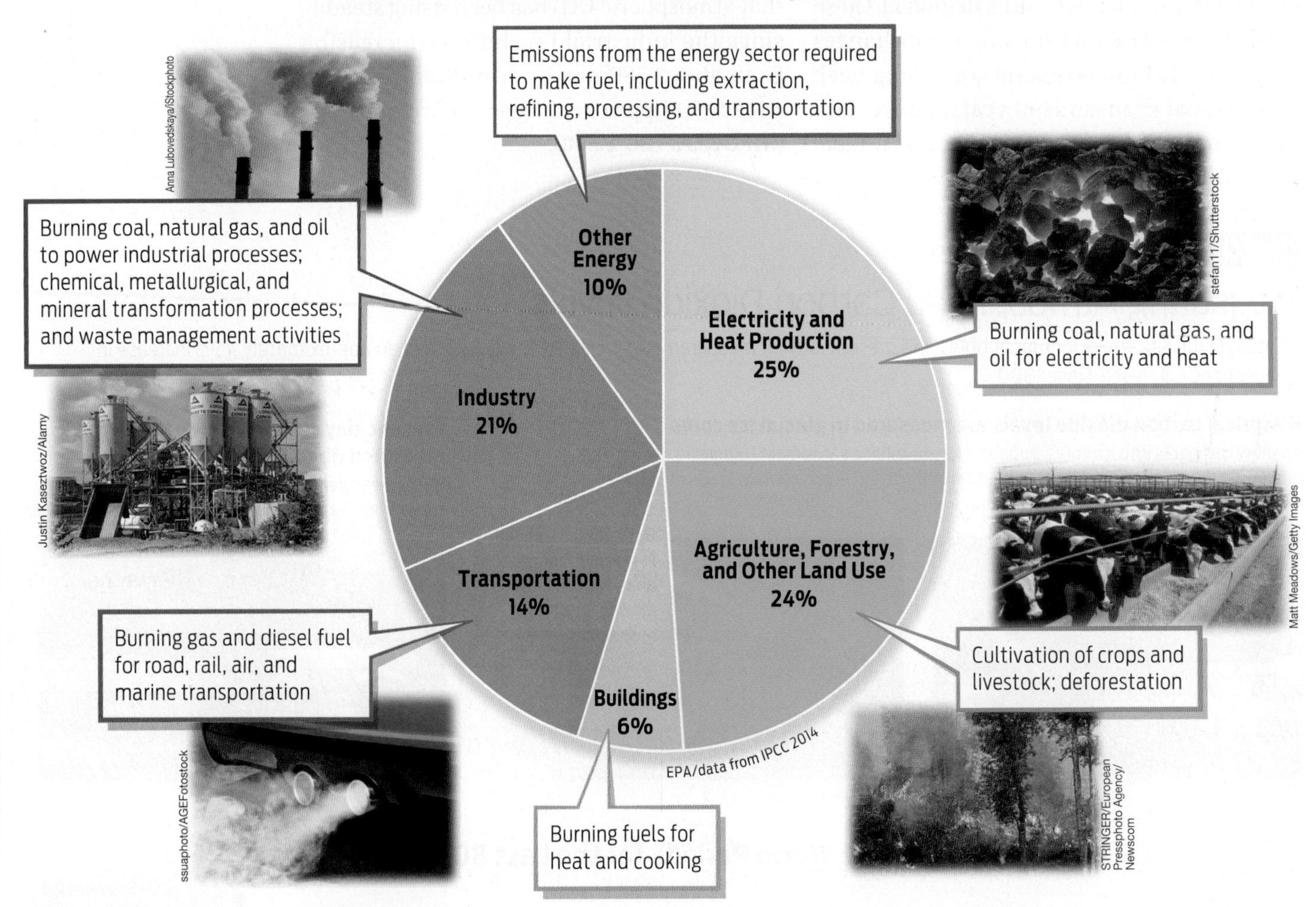

? If substantially more people drove electric cars with rechargeable batteries, what would happen to the distribution of emissions illustrated here?

Activities that decrease the number of photosynthetic organisms also increase global CO_2 levels. Since photosynthesizers are the only consumers of carbon dioxide in the carbon cycle, removing them reduces the amount of carbon dioxide they might have removed from the atmosphere. Furthermore, removing large trees and stable populations of algae eliminates what are, in essence, long-term storage vessels of carbon. Human activities that reduce the number of photosynthetic organisms on the planet include large-scale slash-and-burn agriculture, development that leads to deforestation, and various forms of pollution. Together, these activities contribute to our total **carbon footprint,** a measure of the total greenhouse gases produced by human activities.

CARBON FOOTPRINT
A measure of the total greenhouse gases produced by human activities.

Though CO_2 is the major player in global warming, another important greenhouse gas is methane (CH_4). Methane is produced by natural processes, such as the decomposition of organic material in swamps by anaerobic microbes. However, agriculture, including raising cattle and growing rice in submerged and swamp-like rice paddies, now accounts for more than half the total methane being pumped into the atmosphere. One of the main sources of methane is the gas produced by archaea that live in the digestive systems of cattle. Emitted as flatulence, it adds an estimated 100 million tons of methane a year to the atmosphere. Although the atmospheric concentration of methane is far less than the level of CO_2, methane is particularly worrisome because it is 30 times more potent than CO_2 as a greenhouse gas.

With this steep rise in greenhouse gases have come steadily rising temperatures around the globe, with most of that warming occurring since the 1970s. Virtually all climate scientists agree that greenhouse gases emitted by human activities—primarily driving gasoline-powered cars and burning coal to generate electricity—have caused most of the global rise in temperature observed since the mid-20th century. Thus, this global warming is clearly caused by humans **(INFOGRAPHIC 23.11)**.

The increase in carbon dioxide in the atmosphere is also changing the chemical composition of the oceans. Carbon dioxide absorbed by the oceans becomes carbonic acid. Thus, the oceans are becoming more acidic, which in turn harms marine life. Many have called this change the "evil twin of global warming."

Some scientists think that humans are changing Earth so dramatically that we've entered a new geological and evolutionary period called the **Anthropocene** (*anthro* is the Greek word for "human").

Gone for Good

▶ Renewable versus nonrenewable resources

Because they take millions of years to form naturally, fossil fuels are considered **nonrenewable resources**—natural resources that are not replenished at the same speed at which they are consumed. Once depleted, nonrenewable resources are essentially gone for good. In contrast, **renewable resources** do not run out or can be replenished easily; they include sunlight, wind, and sustainably harvested timber. There's a good reason we have historically relied on fossil fuels so heavily to meet our energy demands: they are plentiful and relatively cheap to obtain. But their supply is not endless, and their use contributes to climate change.

Mischa Keijser/Getty Images

Wind farms are often placed in the ocean, as is this offshore wind farm in the Netherlands.

From an ecological standpoint, the appeal of renewable sources of energy such as wind and sunlight is undeniable: they are plentiful and powerful, and do not add to carbon emissions. Solar power alone could, in theory, provide more than enough carbon-neutral energy to supply the needs of everyone on the planet many times over—assuming we could harvest enough of it cheaply.

The major problem with wind and solar power is that the technologies to harness them are currently more expensive to build and operate than ones based on oil, coal, and natural gas. What makes fossil fuels such convenient and inexpensive sources of energy is that the difficult work of harvesting the energy of sunlight has already been done by the photosynthetic organisms that were

ANTHROPOCENE
The geological period of the present day, dominated by human activities.

NONRENEWABLE RESOURCES
Natural resources that are not replenished at the same speed at which they are consumed and are therefore considered finite; fossil fuels are an example.

RENEWABLE RESOURCES
Natural resources that do not run out or can be replenished easily; solar power, wind power, and sustainably harvested timber are examples.

INFOGRAPHIC 23.12

Nonrenewable and Renewable Resources

Most of the natural resources we use to supply our energy needs are nonrenewable. Coal, petroleum, and natural gas are fossil fuels that take millions of years to form as organic material is compressed by layers of sedimentary rock. Our use of renewable resources, such as solar, geothermal, and wind energy, is growing as we consider ways to curb greenhouse gas emissions. The use of both nonrenewable and renewable resources comes with a variety of economic and environmental considerations.

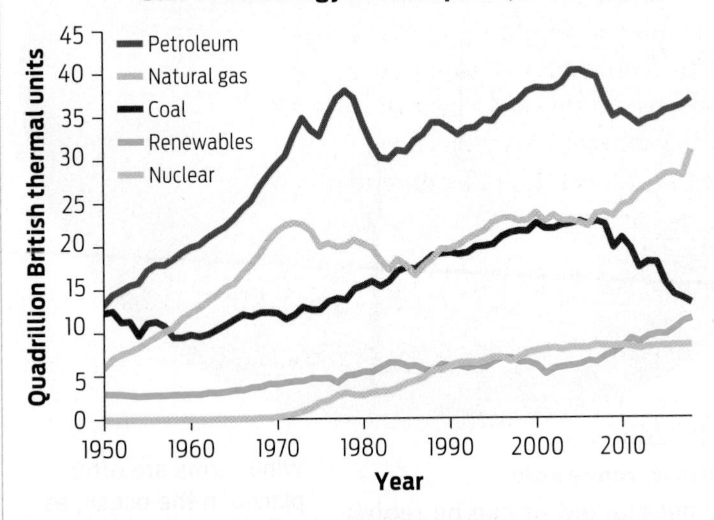

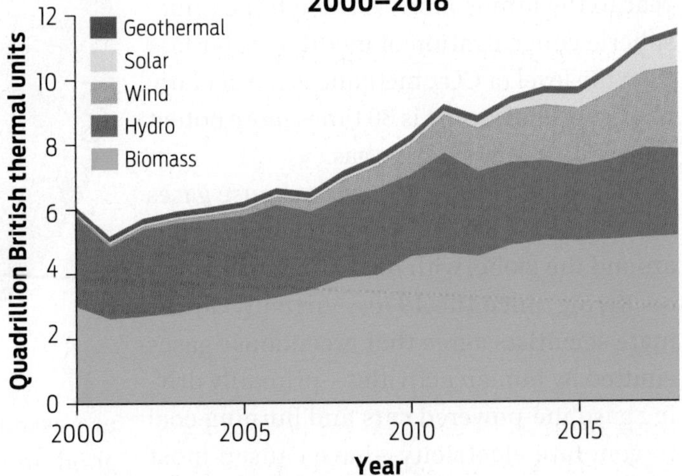

Data from U.S. Energy Information Administration, Monthly Energy Review

Nonrenewable Energy (Petroleum, Natural Gas, Coal)

Pros	Cons
• Relatively cheap technology • Because we have used these methods for so long, our energy infrastructure is set up for them • Currently an abundant energy source	• Extracting fossil fuels is damaging to local habitats and human water sources • Burning fossil fuels produces greenhouse gases and pollutes the environment • Burning fossil fuels can cause serious health problems, including asthma and cardiovascular disease • Fossil fuel deposits are not equally accessible, making their acquisition politically and economically complex

Renewable Energy (Geothermal, Solar, Wind, Hydro, Biomass)

Pros	Cons
• Uses an energy source that is not depleted • Produces much less or no carbon dioxide • The carbon stored in fossil fuels remains stored and is not released into the atmosphere	• More expensive until more widely used • Can modify the physical environment in ways that can damage local habitat and human environments • Some of the technology uses metals and chemicals that produce hazardous waste

? **Which nonrenewable resource has experienced the greatest increase in consumption since 2010?**

compressed over millions of years into oil, coal, and gas. So in a way, we are already, indirectly, using the energy of sunlight to power our lifestyles **(INFOGRAPHIC 23.12)**.

Currently we know of no way of dealing with the threat of climate change without shifting away from fossil fuels and toward renewable sources of energy. Even using technology for **carbon capture**—removing carbon dioxide from the atmosphere—will be no match for the problem unless we also stop adding this gas to atmosphere.

CARBON CAPTURE Using technology to remove CO_2 from the atmosphere.

Dangerous Denial

The evidence is overwhelming that humans are causing climate change, and there is a clear scientific consensus on this issue. As the U.S. National Academy of Sciences puts it, "There is compelling, comprehensive, and consistent objective evidence that humans are changing the climate in ways that threaten our societies and the ecosystems on which we depend." Yet despite this evidence, large sectors of the U.S. population remain skeptical that humans are really to blame for recent planetary warming. How can this be? Part of the answer, according to science historians Naomi Oreskes and Eric Conway, authors of the book *Merchants of Doubt* (2010), is that a small group of industry-funded scientists and public relations professionals have deliberately fostered the notion that climate science is "unsettled" and therefore not strong enough to support changes in energy policy. This is a tactic, Oreskes and Conway argue, borrowed from the tobacco industry's playbook, which similarly sought to cast doubt on the strength of the evidence linking smoking to lung cancer. In both cases, the peddling of "doubt" is a deliberate strategy that is more about politics than science. It is a way to prevent policy actions that would compromise the economic interests of those who benefit from the status quo.

This doubt-peddling strategy has been successful and helps explain why the United States, although among the world's biggest emitters of greenhouse gases, has so far been reluctant to make significant reductions in these emissions. It was one of the few countries that refused to ratify the Kyoto Protocol, a United Nations agreement adopted in 1997 that obligates endorsing countries to reduce carbon dioxide emissions. In 2015, under the Obama administration, the United States joined 194 other nations in signing the Paris Agreement, which commits nearly every country on Earth to reducing its greenhouse emissions through concrete plans. But the commitment lacked the support of the U.S. Congress, and the Paris Agreement is itself nonbinding. In 2019, under the Trump administration, the United States officially withdrew from the Paris Agreement.

The extent of climate change denialism in the United States is alarming given that the window for taking action is shrinking. Even if all of the world's greenhouse gas emissions were turned off today like closing a faucet, we would still face decades of warming because of past emissions—what climate scientists refer to as "heat in the pipeline." Basically, the climate has not yet caught up with the effects of burning fossil fuels in past decades because the oceans are slower to heat than the land (as any beachgoer knows). But eventually, the oceans will catch up, leading to further warming of the atmosphere.

There is compelling, comprehensive, and consistent objective evidence that humans are changing the climate.

This grim reality could easily lead some to take a fatalistic attitude, but that would be a dangerous mistake, says Hector Galbraith: "We've got to get beyond the deer in the headlights stage and begin to think as conservation biologists about what we're going to do about this to help to mitigate the impact." These preventive measures are what he and other climate experts call "adaptation."

Adaptation will not be easy. And it will not prevent painful events from coming to pass. But doing nothing, say scientists, risks turning a bad problem into a catastrophic one.

In concrete terms, adaptation means planning for the inevitable: more frequent droughts, heat waves, and severe storms, as well as rising sea levels. Increasingly, cities and towns all across the country are taking steps to incorporate adaptation into urban and land-use policies—restricting new construction in flood zones, for example, and working to conserve water in drought-prone areas, as well as protecting wetlands.

But these efforts won't do much to stem the tide unless we deal with the underlying cause of rapid climate change—runaway

INFOGRAPHIC 23.13

What You Can Do to Live More Sustainably

While climate change is a huge problem that will require national and international efforts to address, individuals can make choices every day to minimize their impact on the environment.

Drive Less and Invest in Fuel Efficiency

Bike more, take public transit, and use fuel-efficient vehicles such as scooters and hybrid or electric cars to reduce carbon emissions and dependence on nonrenewable fossil fuels.

Reduce Reuse Recycle

An estimated 75% of our trash can be recycled. Reducing your consumption of nonrenewable resources saves energy, and lowers air and water pollution. Recycle what you use, and reuse what you can, including reusable grocery bags, to lower demand for petroleum-based products.

Educate

Learn what you can about the impact of climate change, and help others understand the importance of conserving natural resources. Write your government representatives to promote broad-scale change in your community.

Conserve Water

U.S. households use 22,000 gallons of heated water per year for showers and baths. Reducing water usage by using efficient faucet aerators and shower heads saves electricity and reduces runoff and wastewater that eventually end up in the ocean.

Reduce Wasted Home Electricity

Electronics use energy even when they are turned off. This "vampire energy" accounts for 5% to 8% of a single family's home electricity use. Plugging electronics into a power cord that is turned off at night and during vacations can save the equivalent of one month's electric bill each year.

Eat Less Feedlot Beef

Producing 1 pound of beef can require 5 pounds of grain and more than 2,400 gallons of irrigation water. Eating lower on the food chain (plant-based foods) requires less water and energy.

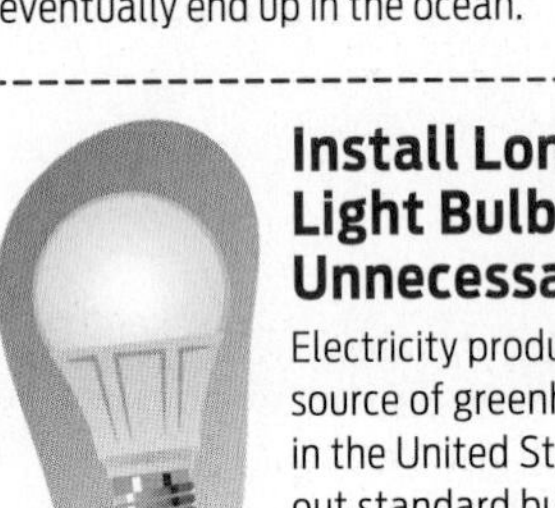

Install Long-Lasting Light Bulbs and Turn Off Unnecessary Lights

Electricity production is the largest source of greenhouse gas emissions in the United States. By swapping out standard bulbs for compact fluorescent bulbs, you can save significant money on your electricity bill and reduce carbon emissions.

Protect Natural Ecosystems

Natural ecosystems support global health. Plants provide food and oxygen, influence weather, and moderate the impact of climate change. Biodiverse systems protect water supplies and sustain commercial interests. Support strategic development plans that minimize human impact.

Choose Sustainable Seafood

Overfishing has put stress on reduced fishing populations. Eat seafood caught by fishers who operate under sustainable fishery management systems, using methods that are safe for coral reefs and nontarget marine organisms.

Volunteer to Help

Volunteer to help in cleanup operations in your community. Remove trash from water sources, pull invasive plant species, and start recycling programs in your neighborhood.

Vote with Your Dollars

Production, packaging, and shipping contribute to energy use and carbon emissions. Purchasing less, making environmentally conscious purchases, and shopping locally and at secondhand stores influence business decision making toward more sustainable practices.

? What are three specific changes you could make in your lifestyle that would improve sustainability and/or reduce emissions?

greenhouse gas emissions. And that means modifying the practices that collectively produce more than 90% of greenhouse gases in the United States today: burning fossil fuels for electricity, heat, transportation, and industry **(INFOGRAPHIC 23.13)**.

"Fossil fuels are incredibly efficient sources of energy," says Serreze, of the University of Colorado. "We've built our whole infrastructure around that. But what we didn't realize is that it's a trap, and that's what we're coming to grips with now." ■

CHAPTER 23 Summary

Driving Question 1 What are some ways that changing temperatures and changing climate impact ecosystems?

- Climate is not the same as weather. Climate refers to the long-term average of atmospheric conditions. Weather describes local atmospheric conditions over a short interval.
- Temperature is an important physical feature of any ecosystem and serves as a clock that cues many biological events, such as breeding, blooming, and hibernation.
- Climate change, especially global warming, is having widespread effects on plant and animal life on the planet—altering seasonal life cycles, shifting ranges, and contributing to species loss by extinction.

Driving Question 2 What is the greenhouse effect, and what does it have to do with global warming?

- The greenhouse effect is a natural process by which heat from sunlight is radiated from Earth's surface and absorbed by heat-trapping gases in the atmosphere, maintaining a global temperature that can support life. Rising levels of greenhouse gases have led to the enhanced greenhouse effect.
- Global warming is caused by an increase in the amount of carbon dioxide and other greenhouse gases in the atmosphere.
- Global warming is leading to the melting of sea ice in the Arctic, which is diminishing the available habitat for the organisms that rely on sea ice and creating a positive feedback loop for increased warming. Melting glaciers and ice caps on land are leading to rising sea levels.

Driving Question 3 How does carbon cycle through ecosystems, and how are humans impacting the carbon cycle?

- The carbon cycle is the movement of carbon atoms through living and nonliving components of an ecosystem by the biotic processes of photosynthesis and cellular respiration, as well as by long-term compression of fossilized organisms.
- Scientists agree that the warming that has occurred since the mid-20th century is due largely to human activities such as burning fossil fuels and deforestation that inject carbon dioxide and other greenhouses gases into the atmosphere.
- Fossil fuels (oil, coal, and gas) are considered nonrenewable resources, since they take millions of years to form and are not replenished on human timescales. Renewable resources are those that do not run out or can be replenished easily, including sunlight, wind, and sustainably harvested timber.
- Many in the United States deny the reality of human-caused climate change, despite the clear scientific consensus on the matter.

More to Explore

- NASA Global Climate Change: https://climate.nasa.gov/
- National Snow and Ice Data Center: https://nsidc.org/ http://nsidc.org/
- NOAA, Earth System Research Laboratory, Trends in Atmospheric Carbon Dioxide (animation): www.esrl.noaa.gov/gmd/ccgg/trends/history.html
- Cook, J., et al. (2016). Consensus on consensus: A synthesis of consensus estimates on human-caused global warming. *Environ Res Lett* 11(4), 048002.
- Urban, Mark. (2015). Accelerating extinction risk from climate change. *Science* 348(6234):571–573.
- Oreskes, N., and Conway, E. M. (2010). *Merchants of Doubt: How a Handful of Scientists Obscured the Truth on Issues from Tobacco Smoke to Global Warming.* New York: Bloomsbury Press. See also the 2014 documentary film of same name.
- Katz, J., and Daniel, J. (December 2, 2015). What you can do about climate change: Seven simple guidelines for thinking about carbon emissions. *The New York Times*.
- Quiz: How much do you know about climate change?: https://www.washingtonpost.com/climate-solutions/2019/11/22/quiz-how-much-do-you-know-about-climate-change/

CHAPTER 23 Test Your Knowledge

DRIVING QUESTION 1 What are some ways that changing temperatures and changing climate impact ecosystems?

By answering the questions below and studying Infographics 23.1, 23.2, 23.3, 23.4, and 23.5, you should be able to generate an answer for this broader Driving Question.

Know It

1. Distinguish between climate and weather.
2. List several examples of species discussed in this chapter that have changed their geographic distributions or the timing of events in their life cycles as a result of global climate change.
3. Describe three ways in which temperature changes and drought contribute to an increase in wild fires in Mountain West conifer forests.

Use It

4. Although trees may not be able to walk away from increasingly warm regions, evolutionary adaptations may allow trees to survive in warmer regions. Discuss each of the adaptations listed below and decide if it is likely to be helpful or harmful in a warming environment. (Think about water—water is taken up by the roots of plants and lost through pores in the leaves; CO_2 levels—CO_2 is taken up by plants through pores in leaves, then used by leaves for photosynthesis; and the movement of other species, such as insects, in response to global warming.)
   a. having smaller leaves
   b. having a larger number of pores on each leaf
   c. having thicker and waxier bark
5. What is a possible risk for humans if insects that carry pathogenic bacteria or viruses expand their ranges northward?
6. How can a change in spring flowering time in a particular region affect the animals of that particular ecosystem?

DRIVING QUESTION 2 What is the greenhouse effect, and what does it have to do with global warming?

By answering the questions below and studying Infographics 23.6, 23.7, and 23.8, you should be able to generate an answer for this broader Driving Question.

Know It

7. Which greenhouse gas is emitted every time you breathe out?
   a. oxygen
   b. carbon dioxide
   c. methane
   d. nitrogen
   e. water vapor
8. Which of the following organisms contributes to reducing atmospheric CO_2 levels?
   a. maple trees
   b. most algae
   c. polar bears
   d. Arctic foxes
   e. a and b
   f. a, b, and d
9. Could we live in the absence of the greenhouse effect? Explain your answer.
10. If global warming causes Arctic sea ice to melt, what will be the effect on sea levels in a low-lying region like Miami? If large parts of the Antarctic polar ice cap should melt, what would be the effect on sea level?

Use It

11. Explain how each of the following contributes to an elevation of levels of greenhouse gases.
   a. engaging in large-scale slash-and-burn agriculture
   b. driving gasoline-fueled cars
   c. producing cattle for beef and dairy products
   d. producing rice
12. Look at Infographic 23.8. From the data presented, how much warmer than the 1951–1980 base average was most of the continental United States in the period 2015–2019? Which part of the United States had the greatest degree of warming? How much warmer was this region than the 1951–1980 base average?

DRIVING QUESTION 3 How does carbon cycle through ecosystems, and how are humans impacting the carbon cycle?

By answering the questions below and studying Infographics 23.9, 23.10, 23.11, 23.12, and 23.13, you should be able to generate an answer for this broader Driving Question.

Know It

13. Plants like maple trees ____________ CO_2 by ____________.
   a. emit; photosynthesis
   b. take up; photosynthesis
   c. emit; cellular respiration
   d. take up; cellular respiration
   e. store; cellular respiration
14. Fossil fuels are most immediately derived from
   a. organic molecules.
   b. CO_2.
   c. methane.
   d. melting ice caps.
   e. photosynthesis.

15. Fill in the blanks in the diagram below.

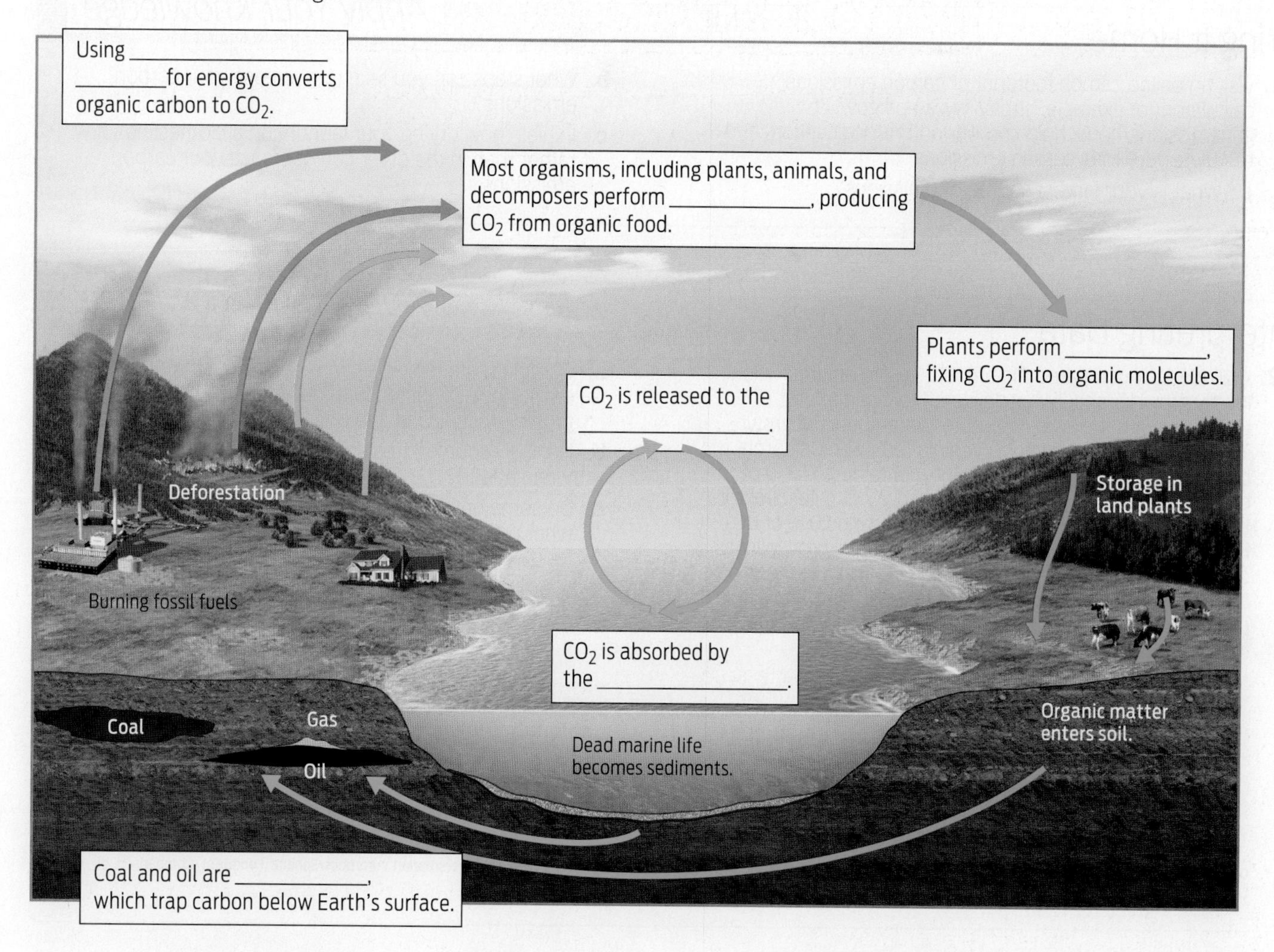

16. Name at least two human activities that increase CO_2 levels in the atmosphere and two natural processes that contribute CO_2 to the atmosphere.

Use It

17. How is ice useful in the measurement of atmospheric levels of CO_2?

18. Describe the evidence that increasing levels of greenhouse gases are responsible for global climate change. What if someone suggested to you that global climate change is due to increased intensity of solar radiation (that is, the amount of sunlight reaching Earth)? What evidence would you ask for in support of this hypothesis?

19. Which of the following data would you use to determine the levels of atmospheric CO_2 in 1750? Justify your choice, including an explanation of why the other choices would not be as effective.

a. historical weather records of daily temperatures

b. records for 1750 from the archives of the Mauna Loa Research Station

c. tree-ring analysis (to look for evidence of extreme fires)

d. ice cores from ice formed in 1750

Mini Case

Apply Your Knowledge

20. Annie is considering buying a new car. She is contemplating several options, and has decided that environmental impact (particularly greenhouse gas emissions) is her top priority. After all, she tries to eat primarily grains, fruits, and vegetables, and she regularly embraces "meatless Mondays." Annie is looking at a fuel-efficient gasoline-powered car, a traditional hybrid (which uses gasoline, but can also run on a battery that recharges while the car is being driven), and a fully electric plug-in car. Discuss the pros and cons of her three vehicle options. What other factors (which may be beyond her control) could affect the emissions associated with owning and driving some of the vehicles she is contemplating?

Apply Your Knowledge

Bring It Home

21. Visit an online carbon footprint or carbon emissions calculator (for example, https://www.epa.gov/ghgemissions/household-carbon-footprint-calculator), and calculate your total carbon emissions.

a. What is your largest source of emissions?

b. What steps can you take to decrease your carbon emissions?

c. Explain how drying your laundry on a clothesline rather than in the dryer can decrease your carbon emissions.

Apply Your Knowledge

Interpreting Data

22. A 2010 study compared the amount of CO_2 emitted when locally grown broccoli was delivered to Virginia Tech University with the amount emitted when broccoli grown in California was delivered to Virginia Tech. The California broccoli was delivered in shipments containing 768 lb of broccoli in a tractor-trailer that traveled 2,786 miles. Tractor-trailer fuel efficiency is 5 miles per gallon, and 20 lb of CO_2 is released per gallon of fuel burned. The local broccoli was delivered in shipments of 587 lb of broccoli in a cargo van that traveled 19.1 miles. Cargo van fuel efficiency is 16 miles per gallon, and 20 lb of CO_2 is released per gallon of fuel burned.

a. Complete the table below to determine the CO_2 emissions associated with delivering 1 lb of locally grown broccoli and 1 lb of nonlocally grown broccoli.

b. Is locally sourced fresh broccoli a year-round option at Virginia Tech?

c. Do some online research to determine approximately what proportion of CO_2 emissions are associated with food delivery versus food production.

Source of broccoli	Miles per shipment	Gallons of fuel burned per shipment	CO_2 released per shipment (lb)	CO_2 released per pound of broccoli delivered (lb)
Nonlocal	2,786			
Local	19.1			

Data from Schultz, J., and Clark, S. (2010). Foodprint comparison of local vs. nonlocal produce: http://www.blacksburgfarmersmarket.com/docs/Schultz_Foodprint_Comparison_of_Local_vs_Nonlocal.pdf.

Plants 2.0

Is genetic engineering the solution to world hunger?

Sandralise/iStock/Getty Images

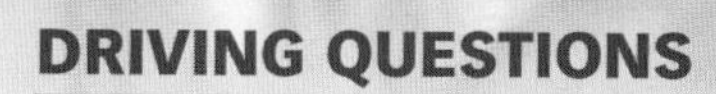

DRIVING QUESTIONS

1 What nutrients do plants need, and how do agricultural practices supplement these nutrients?

2 How do important crops and other angiosperms reproduce?

3 How can the use of genetically modified plants make agriculture more efficient?

Tomas Munita/The New York Times/Redux Pictures

Workers collecting guano on Isla Guañape off the coast of Peru.

In 1856, the U.S. government tried to corner the market on bird poop. That was the year Congress passed the Guano Islands Act, which led to the country's acquiring several small islands in the South Pacific—along with their stockpile of bird droppings.

Not just any bird droppings, but those of seabirds such as boobies, pelicans, and cormorants. Thanks to the fish these birds eat, their droppings—that is, guano—are rich in nutrients, making them an excellent fertilizer.

In the mid-19th century, farmers around the world were desperate for ways to boost crop production. Cities were growing and populations were rising fast. But manure from cattle was pricey and hard to transport. Guano seemed a perfect solution, leading to what contemporaries called "guano mania." The United States imported nearly 200,000 tons of guano in 1856 for use on cropland.

Today, farmers rely on industrially produced fertilizer rather than guano as a source of nutrients for their crops. But the need to improve crop yields continues unabated. If anything, the need is greater now than ever.

By 2050, there will be an estimated 2 billion more people on the planet, bringing the total population up to 9.7 billion. That's 2 billion more people who will need food, clothing, and fuel—not to mention jobs and health care. The United Nations predicts that to accommodate them, by 2050 we will need to produce 70% more food globally than we currently do. Most of the population growth will take place in developing countries, and in these countries food production will need to double. How to produce so much food sustainably is a challenge with no clear answer **(INFOGRAPHIC 24.1)**.

"You only have a few options for how to deal with food production," says Joseph Jez, a plant biologist at Washington University in St. Louis, Missouri, whose lab focuses on engineering plants to help solve ecological problems. "You either boost productivity or you use more land to grow more crops."

But agricultural lands in many parts of the world are already stressed and at near-maximum capacity. Converting more land to farms means sacrificing other irreplaceable resources, like rain forest. For this reason, many scientists are increasingly

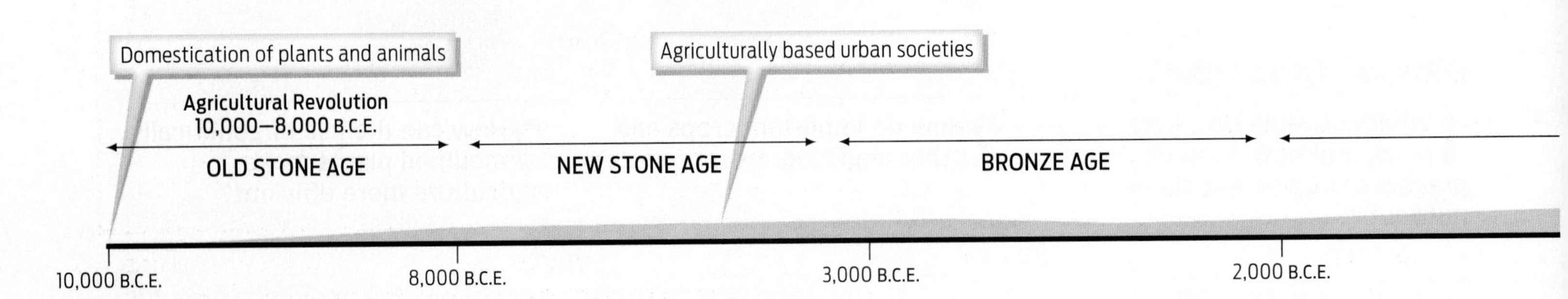

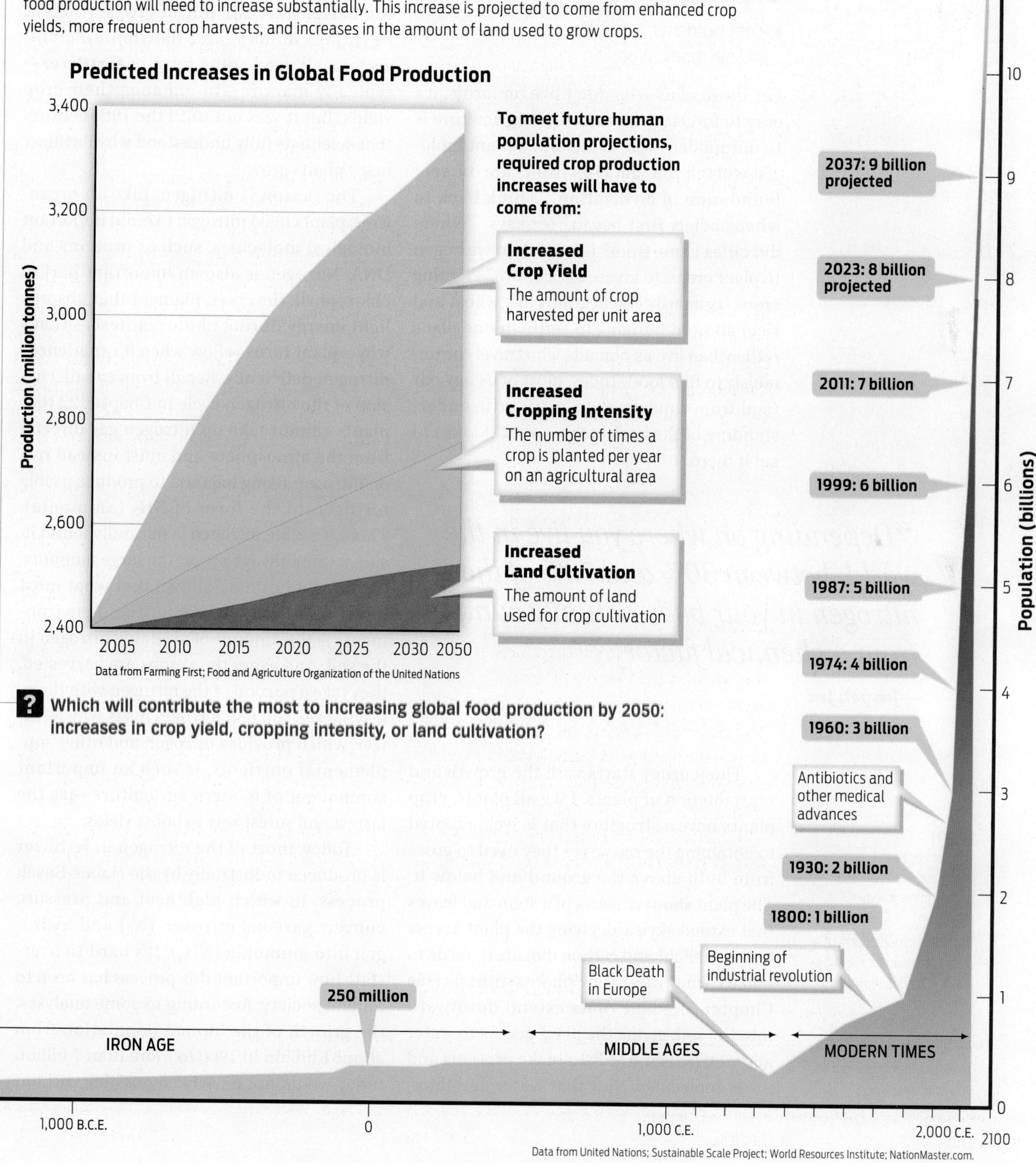

INFOGRAPHIC 24.1
The Challenge of Feeding a Growing Population
The human population has been growing exponentially, and is expected to reach 9.7 billion by 2050, before leveling off at nearly 11 billion people by 2100. To keep up with a growing human population, global food production will need to increase substantially. This increase is projected to come from enhanced crop yields, more frequent crop harvests, and increases in the amount of land used to grow crops.
Predicted Increases in Global Food Production
Production (million tonnes)
3,400
3,200
3,000
2,800
2,600
2,400
2005
2010
2015
2020
2025
2030
2050
Data from Farming First; Food and Agriculture Organization of the United Nations
To meet future human population projections, required crop production increases will have to come from:
Increased Crop Yield
The amount of crop harvested per unit area
Increased Cropping Intensity
The number of times a crop is planted per year on an agricultural area
Increased Land Cultivation
The amount of land used for crop cultivation
? Which will contribute the most to increasing global food production by 2050: increases in crop yield, cropping intensity, or land cultivation?
Population (billions)
11
10
9
8
7
6
5
4
3
2
1
0
2037: 9 billion projected
2023: 8 billion projected
2011: 7 billion
1999: 6 billion
1987: 5 billion
1974: 4 billion
1960: 3 billion
Antibiotics and other medical advances
1930: 2 billion
1800: 1 billion
Beginning of industrial revolution
Black Death in Europe
250 million
IRON AGE
MIDDLE AGES
MODERN TIMES
1,000 B.C.E.
0
1,000 C.E.
2,000 C.E.
2100
Data from United Nations; Sustainable Scale Project; World Resources Institute; NationMaster.com.

looking to genetic engineering to help improve yields. Think of it as building better plants.

Growing Pains

▸ Plant nutrients, fertilizer, and eutrophication

For those of us who don't live on farms, it's easy to forget just how central agriculture is to our modern way of life. But a plant biologist will tell you that crop plants are the very foundation of civilization. "Think back to when society first began," Jez says. "Where did cities come from? It was when we began to plant crops, to create agriculture." Planting crops (primarily cereal grains like wheat and rice) allowed humans to settle in one place rather than live as nomads who travel continuously to find food. Today, most of us buy our food from supermarkets, with little understanding of the back story—what it takes to get it there.

> *"Depending on where you live in the world, between 30% and 70% of the nitrogen in your body actually came from a chemical factory."*
>
> **—Jospeh Jez**

The journey starts with the growth and reproduction of plants. Like all plants, crop plants have a structure that is well adapted to obtaining the resources they need to grow, from both above the ground and below it. The plant **shoot** consists of a stem and leaves that extend skyward, giving the plant access to the sunlight and carbon dioxide it needs to make carbohydrates by photosynthesis (see Chapter 5). Plant **roots** extend downward into the soil, giving the plant access to water and nutrients it needs to make proteins and other molecules. How that water and those nutrients move around the plant is a longer story, which we'll describe in Chapter 25. For now, this basic plant body plan is all you need to know **(INFOGRAPHIC 24.2)**.

SHOOT
The aboveground parts of a plant: the stem and photosynthetic leaves.

ROOT
The belowground parts of a plant, which anchor it in the soil and absorb water and nutrients.

FERTILIZER
A substance applied to soil so as to provide one or more nutrients to plants.

Farming is one long experiment in coaxing better harvests out of crops. Ancient Egyptian, Roman, and Babylonian civilizations all used some form of **fertilizer**—typically manure—to enhance their crop yields. But it was not until the 19th century that scientists fully understood why fertilizer helps plants grow.

The reason is nitrogen. Like all organisms, plants need nitrogen to build important biological molecules, such as proteins and DNA. Nitrogen is also an important part of chlorophyll, the green pigment that absorbs light energy during photosynthesis—that's why a plant turns yellow when it experiences nitrogen deficiency. Recall from the discussion of the nitrogen cycle in Chapter 22 that plants cannot take up nitrogen gas directly from the atmosphere and must instead rely on nitrogen-fixing bacteria to produce usable nitrogen in the form of NH_3 (ammonia). Though usable nitrogen is naturally found in soil, it's not always present in large amounts. In fact, studies have shown that what most limits plant growth in many land environments is the amount of available nitrogen in the soil. And every time crops are harvested, they take a portion of the nitrogen with them, leaving the soil less fertile. This is why fertilizer, which provides nitrogen and other supplemental nutrients, is such an important component of modern agriculture—it's the fastest and surest way to boost yields.

Today, most of the nitrogen in fertilizer is produced industrially by the Haber-Bosch process, in which high heat and pressure convert gaseous nitrogen (N_2) and hydrogen into ammonia (NH_3). It's hard to overstate how important this process has been to human society. According to some analyses, the growth of the human population from about 1 billion in 1900 to more than 7 billion today would not have been possible without

INFOGRAPHIC 24.2

What a Plant Needs to Grow

Plants access sunlight (energy) and carbon (CO_2) through their aboveground shoot system. The root system extends into the soil to take up water and nutrients such as nitrogen and phosphorus, which are transported through the stem to the leaves.

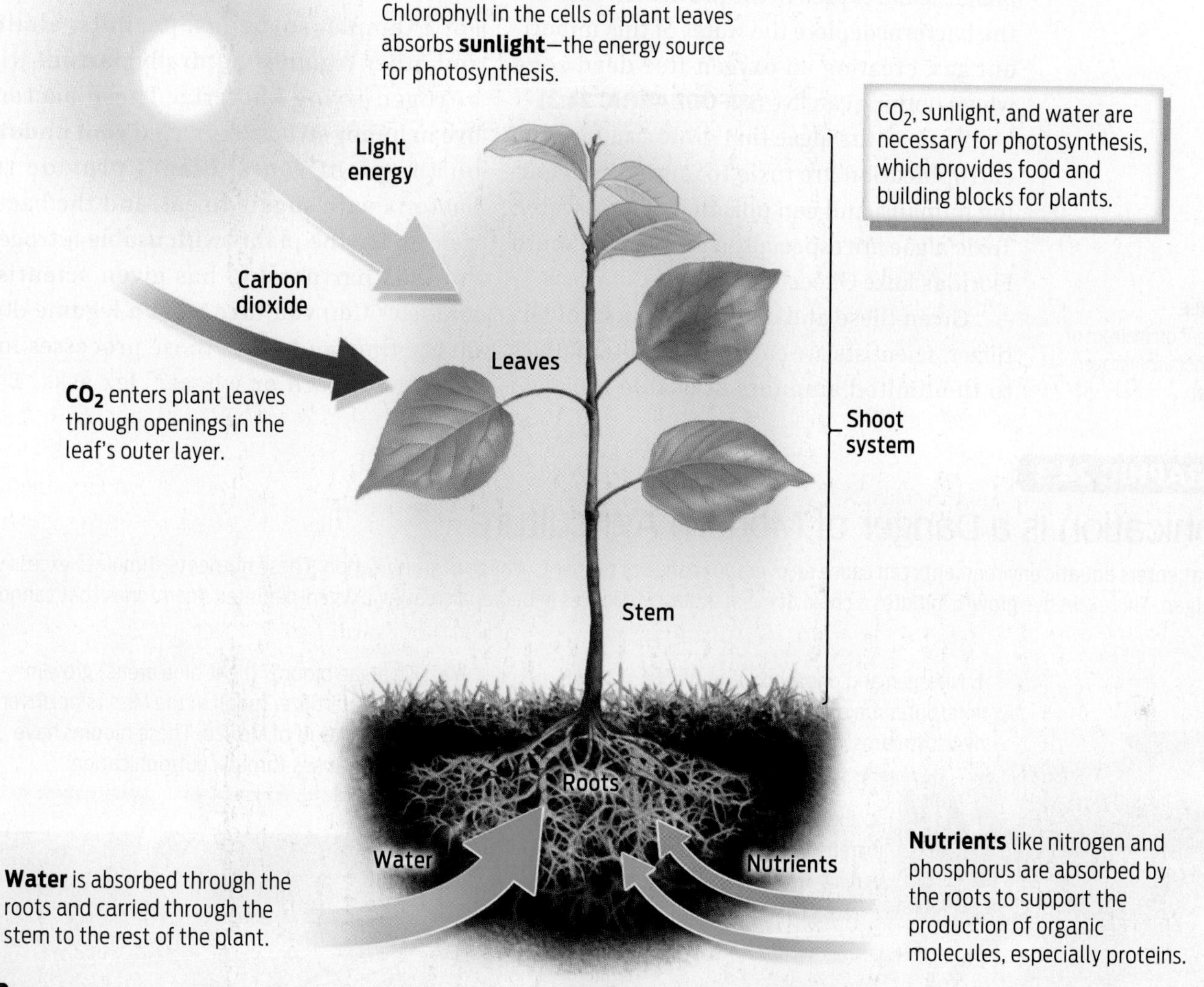

? How do plants acquire water, CO_2, and nitrogen—through leaves or through roots?

the ammonia produced through the Haber-Bosch process.

"There's the old saying 'you are what you eat,'" Jez says. "Well, depending on where you live in the world, between 30% and 70% of the nitrogen in your body actually came from a chemical factory."

In addition to nitrogen, industrial fertilizer often contains the elements phosphorus and potassium. Phosphorus is a key component of cell membranes, and potassium is an important ion, influencing many aspects of physiology.

Though we cannot feed the world without it, industrial fertilizer has unwanted side effects on the environment. When too much fertilizer is used, the excess washes out of fields and into lakes and rivers, leading to a damaging process called **eutrophication.** This enrichment of nutrients permits algae in

EUTROPHICATION
An overabundance of nutrients in bodies of water that can cause excessive growth of algae.

these bodies of water to bloom out of control. The rapidly growing algae eventually block sunlight from reaching lower regions of the lake or river, causing aquatic plants along the bottom to die. Bacteria decompose the dead plants, using oxygen in the process. Eventually, the bacteria deplete the water of this important gas, creating an oxygen-free dead zone where nothing can live **(INFOGRAPHIC 24.3)**.

Some of the algae that grow as a result of eutrophication are toxic to animals, including humans, and can poison drinking water. Toxic algae are especially a problem in South Florida's Lake Okeechobee.

Given these and other drawbacks of fertilizer, scientists are eager to find alternatives to the limited amounts of usable nitrogen naturally found in soil. What if there were a way to engineer plants to make their own usable nitrogen so that farmers didn't need to provide it? This is indeed an approach that scientists are actively pursuing.

Some plants are already halfway there. For example, soybeans, peanuts, alfalfa, and other legumes naturally partner with nitrogen-fixing bacteria. These bacteria live in lumpy structures called **root nodules** on the plant roots. Plants provide the bacteria with sugars to eat, and the bacteria provide the plants with usable nitrogen.

This partnership has given scientists an idea. "Can you take what a legume does all the time and move those processes into plants like corn or wheat?" Jez asks. This

ROOT NODULE
An enlargement on the root of a plant that contains nitrogen-fixing bacteria.

INFOGRAPHIC 24.3

Eutrophication Is a Danger of Modern Agriculture

Fertilizer that enters aquatic environments can cause an overabundance of nutrients, called eutrophication. These nutrients stimulate excessive growth of algae. This excessive growth initiates a chain of events that culminates in bacteria creating oxygen-depleted dead zones that cannot support life.

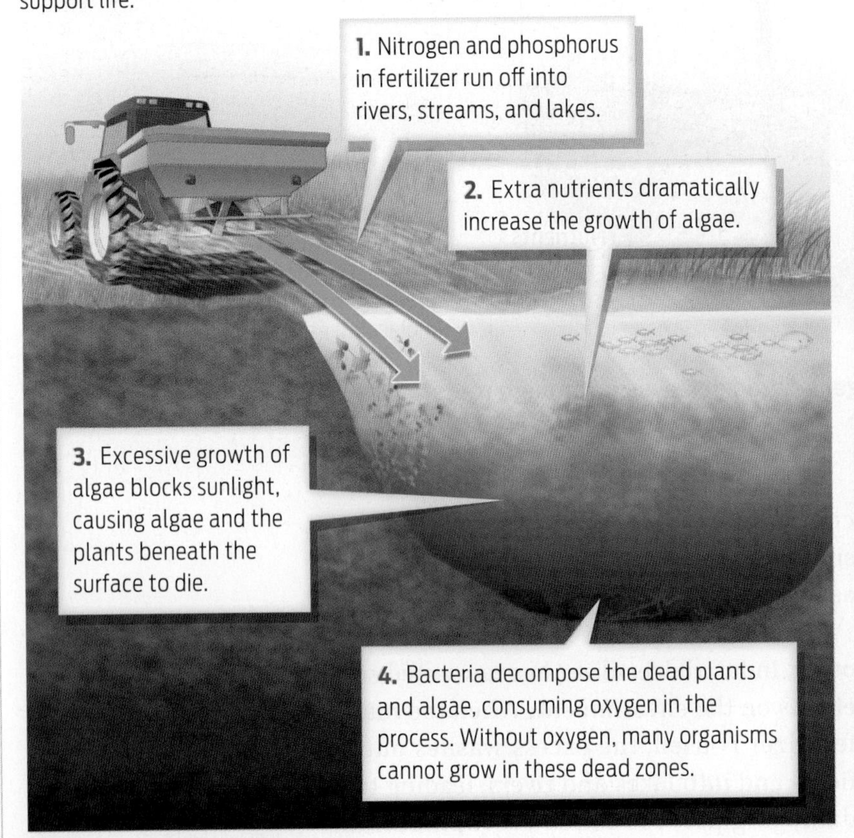

? Why are organisms, like fish, that require nitrogen and phosphorus harmed by the addition of these nutrients to a body of water?

INFOGRAPHIC 24.4

Nitrogen-Fixing Bacteria Produce Usable Nitrogen for Plants

Nitrogen-fixing bacteria live in the soil and in the root nodules of legumes like soybeans. These bacteria convert nitrogen gas to ammonia, which the legume can use for growth. Scientists are exploring the possibility of engineering other plants to form these nodules, or engineering other bacteria to fix nitrogen in association with a wider variety of plants.

Bacteria in Root Nodules

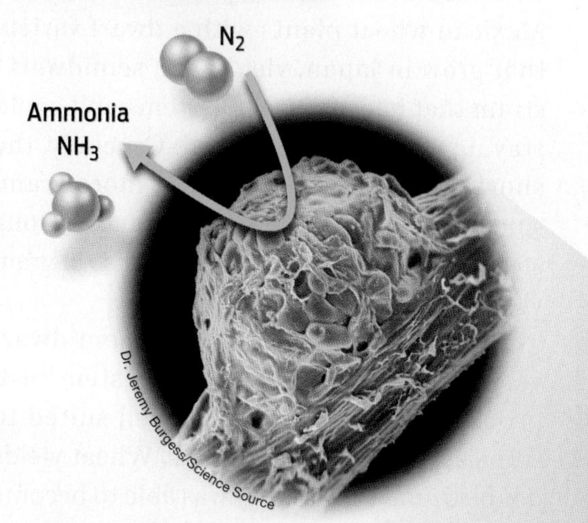

Bacteria in Soil

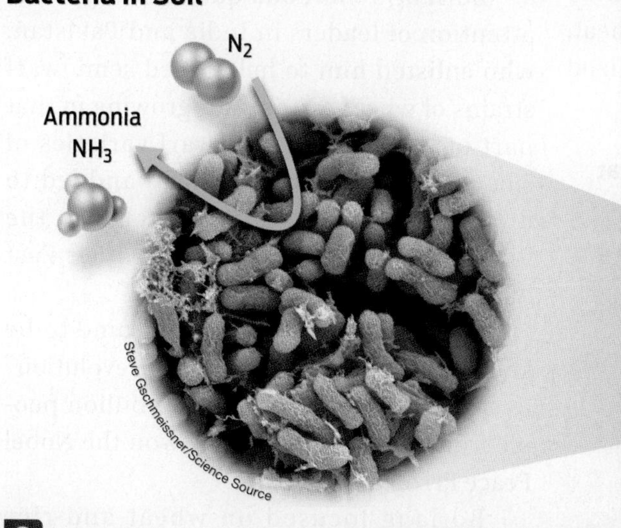

? **Why might legumes be more successful than other plants in nutrient-poor soil?**

would mean engineering corn and wheat to partner with nitrogen-fixing bacteria to produce root nodules.

Another approach might be to transfer the nitrogen-fixing abilities of a type of bacteria called *Rhizobium* into other bacteria and fungi that already associate with plants—"to provide an analogous nitrogen source that doesn't require the *Rhizobium*," Jez explains **(INFOGRAPHIC 24.4)**.

These are still pie-in-the-sky solutions at the moment, he says, but they're important to pursue. Limited nitrogen in the soil is the primary reason for low crop yields in sub-

For his work, Borlaug has come to be known as the father of the "green revolution." By some estimates, he saved a billion people from starvation.

Saharan Africa, for example. "Any way you can make nitrogen use more efficient is critical," Jez says.

Seeds of Plenty

▸ Plant reproductive anatomy and pollination

This isn't the first time that concerns over feeding a growing human population have fueled innovations in plant science. In the 1950s and 1960s, advances in public health spurred rapid population growth in developing countries around the globe. Yet agricultural gains lagged behind, threatening food shortages and mass hunger.

STAPLE CROP
A crop that is eaten in large quantities and provides most of the energy and nutrients in the human diet.

Fertilizer by itself only went so far to improve yields of common crops like wheat. On farms in Mexico, for example, fertilized wheat grew rapidly and produced large seed heads filled with grain. But beyond a certain point, the top-heavy plants would fall over easily in wind (called lodging), ruining crops and decreasing yields. A fungal disease called stem rust also plagued wheat crops in the region.

Norman Borlaug, an American plant biologist working in Mexico, eventually came up with a solution to both problems. He bred Mexican wheat plants with a dwarf variety that grew in Japan, yielding a "semidwarf" strain that had short, thick stems and could stay upright when fertilized. Crucially, the shorter plants produced even more grain, since less energy was going into making long stems. The result was a huge increase of grain yield per acre.

Borlaug then crossed this semidwarf variety with strains resistant to stem rust, producing plants that were well suited to tropical climates like Mexico's. Wheat yields doubled, and the country was able to become self-sufficient at feeding its people.

Borlaug's methods quickly caught the attention of leaders in India and Pakistan, who enlisted him to help breed semidwarf strains of wheat suitable for growing in that part of the world. Semidwarf varieties of rice (*Oryza sativa*) came next, and led to dramatic increases in rice yields in the Philippines, China, Taiwan, and other East Asian countries.

For his work, Borlaug has come to be known as the father of the "green revolution." By some estimates, he saved a billion people from starvation. Borlaug won the Nobel Peace Prize in 1970.

Borlaug focused on wheat and rice because they are two of the world's most important **staple crops,** providing nourishment to billions of people every day. Wheat and rice—as well as corn, barley, oats, and rye—are cereals (the name comes from Ceres, the Roman goddess of harvest and agriculture). Cereals are members of the grass family. They make good staples because their fruit—what we call grain—contains a seed that is full of starch, which is a good source of

Dr. Norman Borlaug with a high-yield, disease-resistant strain of wheat that he developed.

Art Rickerby/The LIFE Picture Collection/Getty Images

energy. Cereal grains can also be stored easily and do not rot since they are dry.

In addition to the edible and nutritious seeds of cereal grains like wheat and rice, humans eat many other plant parts. Apples, tomatoes, eggplants, and peaches are fruits. Fruit serves as a delicious incentive for animals to disperse the plant's seeds. Carrots, radishes, beets, and sweet potatoes are plant roots. Broccoli is a flower, and asparagus is a stem **(INFOGRAPHIC 24.5)**.

Today's crop plants are generally easy to harvest, eat, and digest. But that wasn't always the case. Most modern crop plants look almost nothing like their wild counterparts. That's because, over thousands of years, humans have been selectively breeding crop plants to possess traits that make

INFOGRAPHIC 24.5

Edible Parts of a Flowering Plant

Many parts of flowering plants are edible and make up important parts of the human diet.

? Many salads have tomatoes, beans, and cucumber as ingredients. Which part of a plant do all these represent?

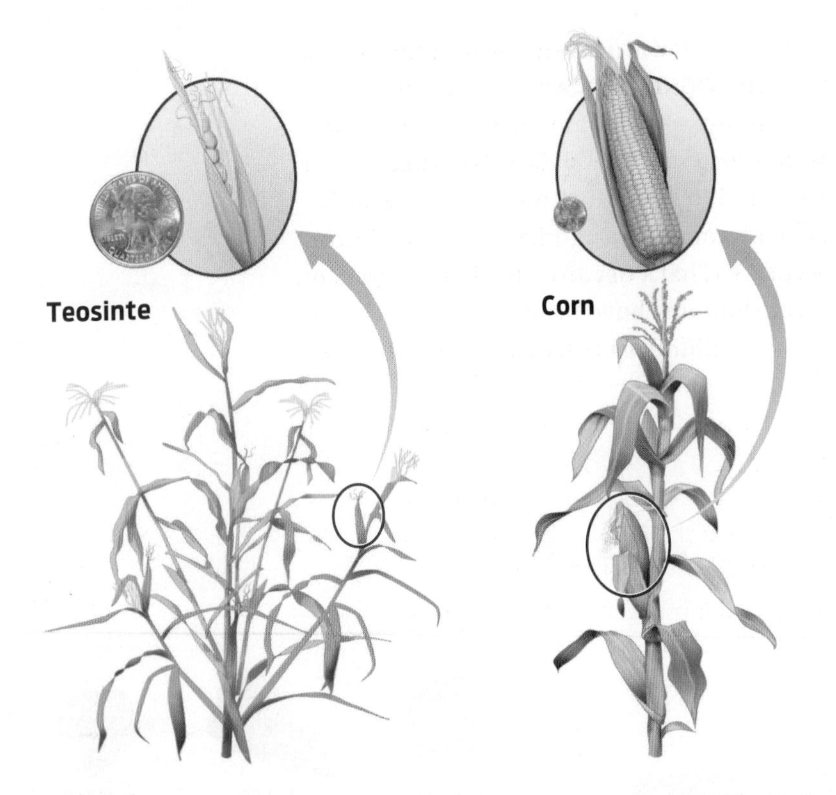

The ancestor of corn, called teosinte, had much smaller cobs and bushier stalks. Humans bred teosinte into the corn we know today.

them useful to humans, such as having larger fruits or seeds, sweeter taste, and softer texture.

Take corn, for example. The ancestor of corn (or maize, as it's known outside the United States) is a Mexican grass called teosinte. It is a much bushier plant than corn, with many small (4- to 5-inch) cobs on its branches. Each teosinte cob contains just a handful of small kernels arranged in two jagged rows, and each kernel is encased in a hard shell (called a fruitcase) that would chip your tooth if you tried to eat one.

Modern genetic evidence shows that changes (mutations) in just four or five genes account for the dramatic differences between teosinte and corn. Obviously, early farmers didn't know about genes or mutations, but they paid close attention to differences among teosinte plants. When mutations occurred that produced changes in the plants, like softer kernels, these plants were bred and cultivated. The result was a dramatic transformation in the plant: over thousands of years, early farmers bred teosinte plants into the corn plants we know today. Archeological evidence suggests a modern-looking corn plant had already been bred by 9,000 years ago.

STAMEN
The male reproductive structure of a flower, made up of a filament and an anther.

ANTHER
The part of the stamen that produces pollen.

CARPEL
The female reproductive structure of a flower, made up of a stigma, style, and ovary.

STIGMA
The sticky "landing pad" for pollen on the pistil.

OVARY (PLANT)
The structure at the base of the pistil that contains the ovules.

OVULE
The part of a flower that develops into a seed after fertilization.

The development of crop plants shows that farmers had at least an intuitive understanding of plant reproduction, even if they didn't have biologists' fine-grained understanding of plant anatomy. That's because you can't breed better plants without getting plants to have sex.

Like other sexually reproducing organisms, plants have distinct male and female reproductive organs that produce haploid gametes: sperm and eggs. In angiosperms, or flowering plants, male and female sex organs are housed within a plant's flowers. Though they might not look like it, crop plants like wheat, corn, and rice are types of flowering plants. Their flowers tend to be small and dwarfed by their larger leaves, but these small structures are still key to their reproduction.

The male part of the flower is called the **stamen.** It consists of a stemlike filament topped with a pollen-producing **anther.** Pollen grains contain the male gametes. The female part of the flower is called the **carpel.** The top of the carpel is a sticky "landing pad" for pollen called the **stigma.** At the base of the carpel is the **ovary,** a protective structure that holds the egg-producing **ovules.** The tube connecting the stigma to the ovary is the **style.**

For flowering plants to reproduce, sperm from pollen produced by the male anther must join with the eggs in a plant ovule. How that happens takes different forms, depending on the plant. Many flowering plants have "perfect" flowers, meaning that male and female reproductive parts are found within the same flower. In other flowering plants, male and female organs are found in separate flowers, and in still others, on separate male and female plants.

In each case, the first step of angiosperm reproduction is **pollination**—the transfer of pollen from male to female plant structures that occurs when a pollen grain from an anther lands on a stigma. This encounter triggers the growth of a **pollen tube** from the pollen grain down the style to the ovule, where fertilization occurs. The end result of fertilization is the formation of a plant embryo. In seed plants, the embryo is contained within a seed. And in angiosperms, seeds are encased in a fruit **(INFOGRAPHIC 24.6).**

To see how all this relates to crop plants, consider corn. The male organ of a corn plant, called the tassel, emerges from the top of the plant stalk. It contains many small flowers that produce pollen grains. The tassel is the corn stamen. The female reproductive organ is the ear, which develops on the plant stalk about halfway between the ground and the tassel. (Note that in corn, male and female reproductive organs are housed in different flowers on the same plant, making the flowers "imperfect.") The ear consists of a cob studded with egg-containing ovules that, when fertilized, will develop into seed-containing kernels (the corn's fruit). Each ovule produces a hairlike structure called a silk that grows along the ear and eventually emerges from the tip of the husk, the leaves that enclose the ear. The silk is the corn style.

Pollination in corn occurs when wind blows the pollen to an exposed silk at the top of the husk. The encounter between a pollen grain and a silk triggers the growth of the pollen tube, which extends from the pollen grain down the inside of the silk. Male sperm cells from the pollen grain travel through the pollen tube to the egg cell on the cob, where fertilization occurs.

STYLE
The tubelike structure that leads from the stigma to the ovary.

POLLINATION
The transfer of pollen from male to female plant structures so that fertilization can occur.

POLLEN TUBE
A hollow tube that grows from a pollen grain after pollination and transports the male gametes to the egg.

INFOGRAPHIC 24.6

Sexual Anatomy of a Flower

Flowers are the reproductive structures in angiosperms. The anthers at the top of stamens produce pollen, which lands on the stigma during pollination. Fertilization occurs when a male gamete reaches an egg within an ovule within the ovary. Following fertilization, the ovary becomes fruit, which contains the seeds.

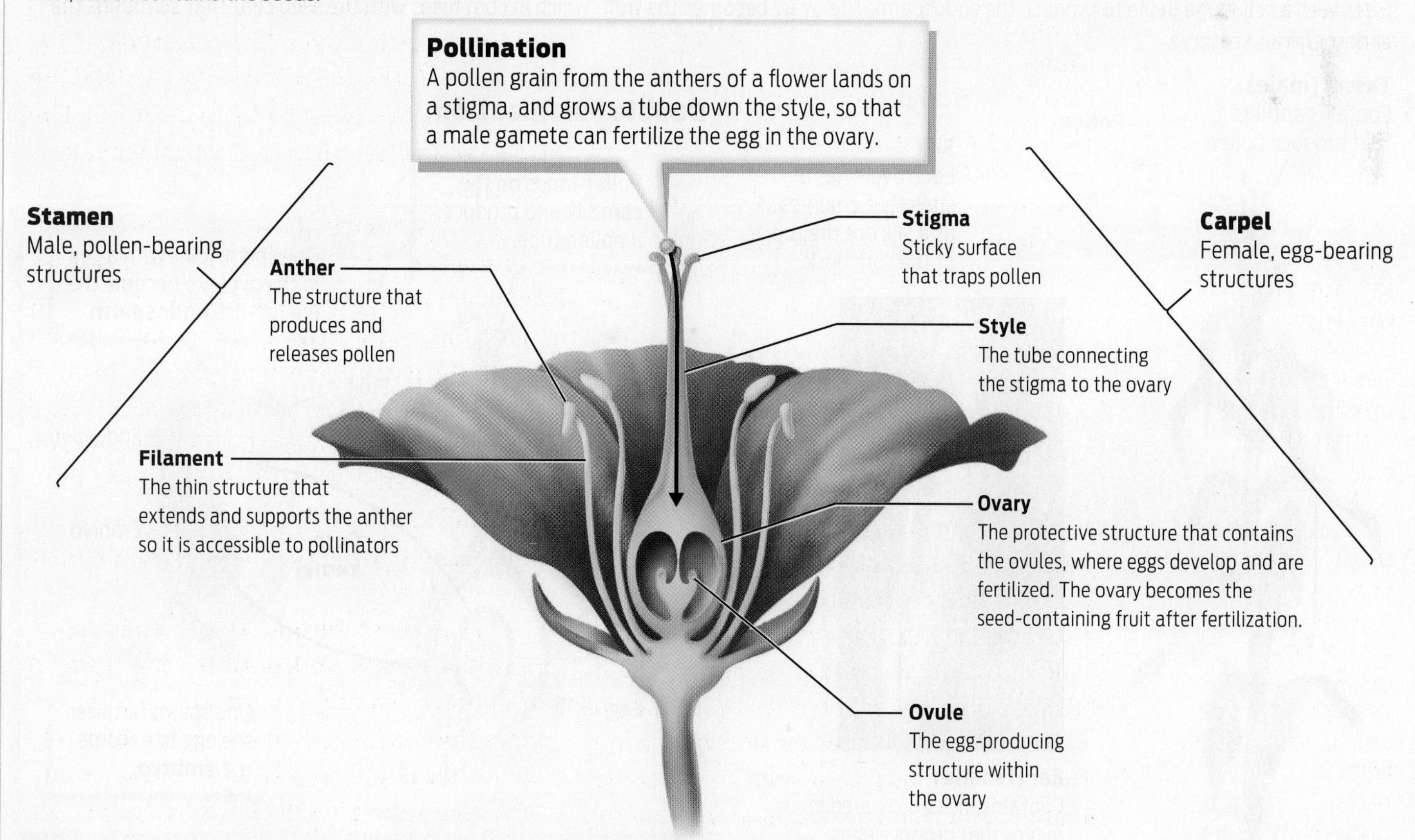

? **Where in the plant does pollination occur, and where does fertilization occur?**

DOUBLE FERTILIZATION
The process in angiosperms in which one sperm fertilizes the egg and one sperm fuses with another cell in the ovule.

ENDOSPERM
A part of the seed containing nutrients that is the seed's source of energy; produced when a sperm fuses with a cell in the ovule during double fertilization.

SEED COAT
The hardy outer covering of a seed that protects the developing embryo.

Like all angiosperms, corn plants carry out **double fertilization,** in which two sperm cells travel down the pollen tube to the egg-containing ovule. One sperm fuses with an egg to form the embryo. The other sperm cell joins with two nuclei in a central cell to form the **endosperm** of the seed. The endosperm is the source of stored nutrients and energy for the growing embryo; it is also what makes cereals an energy-rich food source for humans. The rest of the ovule becomes the **seed coat,** a hardy covering that encases the embryo and endosperm. Finally, the tissues of the ovary become the fruit, which in the case of corn is intimately fused with the seed coat. Every kernel on an ear of corn is a tiny fruit **(INFOGRAPHIC 24.7).**

Some flowering plants, like corn and wheat, rely on wind to transfer pollen. For many other flowering plants, pollinators such as bees and hummingbirds transfer pollen grains from the anther of one flower to the stigma of another. To these animal pollinators, a flower's scent is a potent draw, irresistibly attracting them to the blossom as they search for nectar to eat. Some flowers practice sexual deception, mimicking the shape or smell of a female insect to lure a pollen-carrying male into the flower to copulate.

Not all plants are so sexually adventurous. Some are asexual, or have asexual phases, and can produce new individuals by extending runners or producing bulbs

INFOGRAPHIC 24.7

Angiosperms Rely on Double Fertilization

Corn is an angiosperm, and like all angiosperms relies on double fertilization to reproduce. When a pollen grain lands on a female stigma at the top of a silk, it produces a pollen tube through which sperm travel to the ovule. One sperm fertilizes the egg to produce the embryo, and the other fuses with a cell in the ovule to produce the endosperm. The ovary becomes the fruit, which in corn fuses with the seed coat that surrounds the endosperm and embryo.

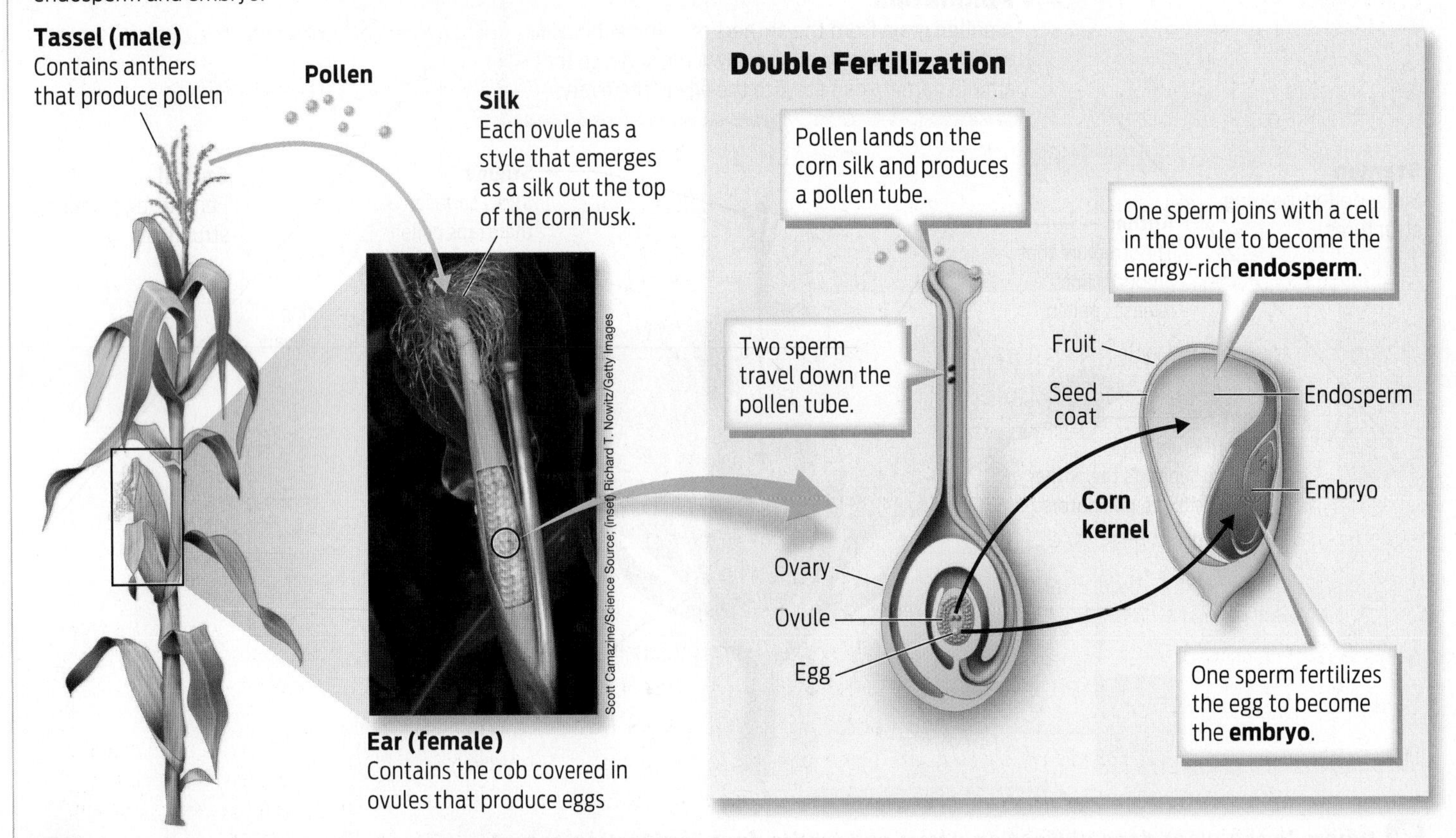

? **What are the two events of double fertilization?**

INFOGRAPHIC 24.8

Plants Reproduce Sexually and Asexually

In sexual reproduction, pollen must make its way to female sexual organs. Flowering plants use their flowers to attract pollinators. Other plants rely on wind to transfer pollen. Plants can also reproduce asexually. Some plants extend runners that can generate new plants. Grafting can attach shoots to different root systems, allowing asexual propagation of the stems on new plants.

Animal Pollination
Sexual

Flowers of the fly orchid (*Ophrys insectifera*) imitate the shape, color, and smell of a female insect. Male bees and wasps transfer pollen from plant to plant as they attempt to mate with the flowers.

Wind Pollination
Sexual

Windborne pollen is released in clouds from male pinecones to neighboring female plant structures. The plant must release huge amounts of pollen into the wind, to increase the chance of at least one pollen grain finding its target. These clouds of pollen plague people with allergies during pollen and allergy season.

Root Runners
Asexual

Plants like this wild strawberry can extend long runners away from the parent plant. These runners can generate new plant clones. The plants reproduce without the need for male and female gametes.

Grafting
Asexual

In grafting, a stem cutting from one plant is surgically attached to another plant, which will provide the root system. Grape vines used for winemaking are often grafted to roots that are adapted to the local environment.

? **A plant has flowers with a distinctive fragrance, and insects are often buzzing around it. Which reproductive strategy is the plant using?**

that develop into whole new plants. Asexual reproduction, in the form of taking cuttings and grafting, is especially important in agriculture. A cutting taken from one plant can be planted in soil to form a new plant; sugar cane and pineapples are often reproduced this way. A cutting taken from a plant can also be surgically grafted to the root system of a different plant—a procedure commonly used to perpetuate vineyard grapes **(INFOGRAPHIC 24.8)**.

Limiting Losses

▸ Pests and pesticides

Historically, one of the biggest threats to crop yields has been pests—insects, fungi, and bacteria that eat plant parts, sometimes sickening or killing the plant. Almost as soon as humans started planting large fields of a single crop, so-called **monocrops,** they had to confront pests attracted to this easy-to-access bounty. And pests continue to be a problem in worldwide agriculture: the Food and Agriculture Organization of the United Nations estimates that 20% to 40% of global crop yields are lost each year because of pest damage.

To combat pests, farmers rely on a variety of natural and synthetic **pesticides.** Some of the earliest pesticides were toxic elements such as arsenic, mercury, and lead. Unfortunately, these were toxic not only to the intended pest targets but also to people, creating health risks for farmers who applied these elements to their crops. Today, farmers use a wide variety of chemicals as

MONOCROP
A single crop species grown in the same field over many seasons.

PESTICIDE
A substance that is toxic to pests, organisms that can damage crops or farm animals.

pesticides, including ones made by other living organisms. One particular chemical, made by a bacterium, has had far-reaching effects.

The bacterium *Bacillus thuringiensis* (*Bt*) produces a chemical that is toxic to certain insect larvae, but harmless to humans and other mammals. Farmers first used *Bt* in 1928 as a natural biopesticide to combat the European corn borer, a caterpillar that eats corn leaves and stems, and it continues to be a commonly used biopesticide today.

INFOGRAPHIC 24.9

Genetic Modification Has Lowered Pesticide Use

Genetically modified crops have been engineered to have a variety of traits, including resistance to pests. *Bt* corn is widely planted, reducing the need to apply pesticides to millions of acres of corn.

Genetic Modification Process

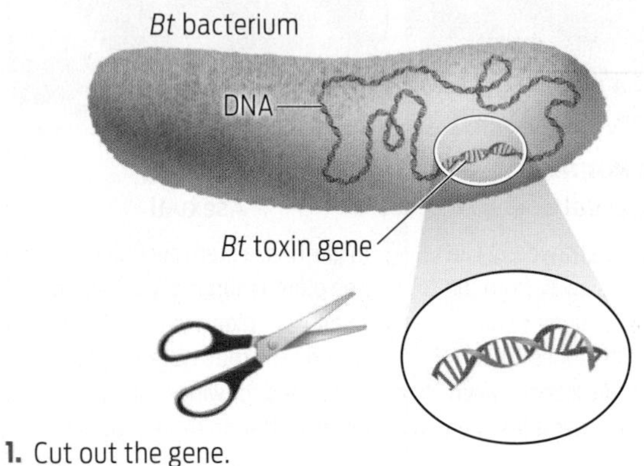

1. Cut out the gene.

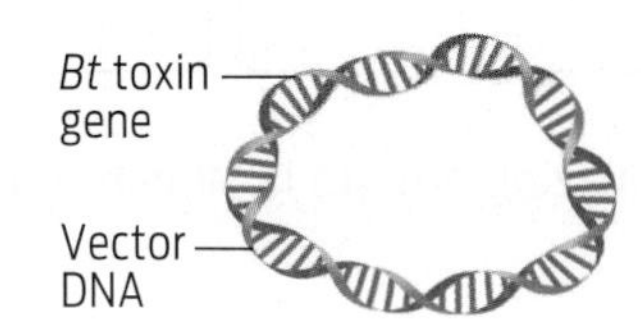

2. Insert the gene into a vector.

3. Coat particles ("bullets" for the gene gun) with the recombinant vector.

4. Shoot the coated particles at plant cells, allowing the recombinant vector to enter the nucleus.

5. Grow the engineered cells into small plants that can then be grown to maturity in soil.

Purpose:
Bt corn prevents attack by pests like the corn borer without the need for pesticides.

Corn without the toxin gene from the bacterium *Bacillus thuringiensis* (*Bt*) is eaten by the corn borer insect.

Results:
Insecticide use on U.S. farms has dropped dramatically with increased use of genetically modified *Bt* corn varieties. However, insecticide use may rise in the future due to the emergence of *Bt*-resistant insects.

U.S. *Bt* Corn Usage and Insecticide Use on *Bt* and non-*Bt* Crops

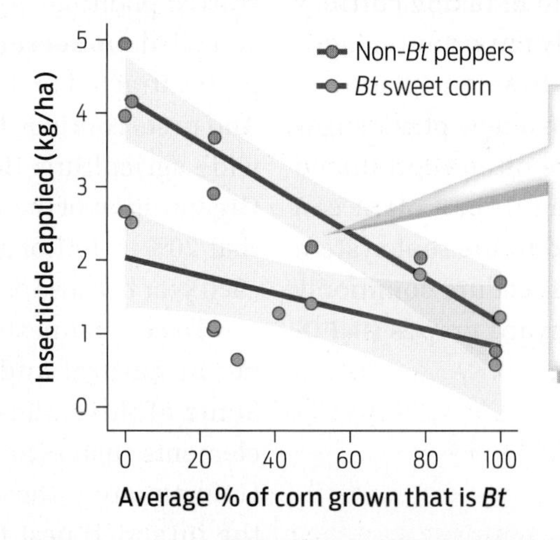

? In your own words, explain the reduction in pesticide use with the increased use of *Bt* corn.

A major innovation came in 1995 when scientists unveiled a variety of corn that had been genetically modified to contain genes from *Bt*. In effect, this corn produces its own pesticide, so there is no need to apply *Bt* pesticide to the crop. This approach has been wildly successful—in fact, more than 80% of corn and cotton planted in the United States today is the *Bt*-engineered variety.

Genetically modified (GM) crops are created by variations of the genetic engineering techniques we encountered in Chapters 8 and 9. First, the desired gene—in this case, the one for the *Bt* pesticide—is cut out of the source organism, the bacterium. It is then incorporated into a circular piece of DNA called a vector, which will deliver the bacterial gene to the plant nucleus. Because plant cells are surrounded by a sturdy cell wall, it's harder to introduce the DNA vector into their cells than into animal cells. To get around this problem, scientists often use a "gene gun" to bombard plant cells with the DNA vector and help it penetrate the cell walls. The vector enters the cell and the transgene is incorporated at random into the plant genome **(INFOGRAPHIC 24.9)**.

The effort to build better plants through genetic engineering doesn't stop with pest control. Scientists have used GM technologies to develop varieties of corn that are resistant to drought and tomatoes that ripen more slowly. They've designed plants that can grow on land contaminated with toxic metals to help bioremediation projects. Recently, scientists have applied GM technologies to improve the efficiency of photosynthesis itself, with the goal of producing higher-yielding crops.

One new technique in particular has transformed the genetic engineering landscape. Called CRISPR, it allows scientists to quickly and easily edit DNA at very specific regions of the genome (Chapter 9).

"CRISPR is one of those things I feel like I've been waiting for my entire career," says Joyce Van Eck, a plant biologist at the Boyce Thompson Institute, affiliated with Cornell University in Ithaca, New York. "It's really transformed the way that we work in my lab."

"CRISPR is one of those things I feel like I've been waiting for my entire career."

—Joyce Van Eck

Van Eck and her colleagues are using the CRISPR technique to study and modify genes that affect plant growth and the size of flowers and fruit in different plants, with the goal of increasing productivity. Recently, her lab published a study showing that it was possible to increase the size and hardiness of groundcherries, a finicky fruit related to the tomato, by cutting out certain sequences of the fruit's DNA with CRISPR.

CRISPR will not replace traditional breeding methods. The fastest and surest way to improve a crop is still through cross-breeding two different varieties, one of which contains a particular trait that farmers want to introduce into the crop—say, resistance to a particular disease. But sometimes no existing variety has natural disease resistance, in which case traditional breeding is powerless to provide protection against the disease.

Zachary Lippman/Cold Spring Harbor Laboratory/Howard Hughes Medical Institute

Groundcherries before (left) and after (right) genetic modification with CRISPR.

That's what's happening right now with oranges and grapefruits in Florida, Van Eck points out. "A disease called citrus greening is wiping out the citrus industry," she notes. "A lot of citrus growers in Florida are now growing peaches instead."

Citrus greening is caused by a bacterium that is spread by an insect. The bacteria choke the plant's vascular system, causing fruit to shrivel and, eventually, the plant to die. Similar problems confront certain varieties of bananas, coffee, and chocolate—the plants lack natural resistance to diseases that are wreaking havoc on the populations. But new varieties of CRISPR-modified plants may offer hope. "Genetic engineering might be the only solution at this point," Van Eck says.

Changing Nature

▶ Pros and cons of genetic engineering

Using genetic engineering technologies to "improve" on nature is not without controversy. Could GM plants have unforeseen negative consequences for health or the environment? Groups like the Non-GMO Project and Greenpeace have expressed concerns. For example, they have raised the question of whether GM foods containing new proteins pose a risk of generating allergies in unsuspecting people. No solid evidence supports this claim. They have also expressed concern that "super-weeds" could emerge that are harder to control. And indeed, available evidence does suggest that overuse of GM crops has led to the emergence of pesticide resistance in insects and herbicide resistance in plants.

There are also economic concerns associated with GM crops. The Monsanto Company, for example, sells genetically engineered seeds that are resistant to a commonly used herbicide called Roundup (glyphosate), which the company also produces. Farmers planting these seeds can spray Roundup on their fields to control weeds without killing their crops. But because Monsanto holds a patent on the seeds, farmers are not legally allowed to replant seeds from their harvest. Instead, they must buy new seeds each year from the company. Though many farmers don't mind this obligation, given the benefits of the technology, others find the increasing reliance on GM products to be a threat to traditional ways of farming.

While these are all understandable concerns, some opposition to GMOs seems misplaced. For example, some critics object to scientists' fiddling with the genetic blueprints of plants and other organisms that have evolved naturally. Yet, as we have seen, there is nothing "natural" about the genetic makeup of the crops we eat. Essentially every crop—from corn to wheat to kale to kiwi fruit—is the product of deliberate breeding designed to produce or enhance desired traits.

Many common crops that we have enjoyed for a long time have even been deliberately subjected to DNA-altering treatments. After World War II, under the auspices of the International Atomic Energy Agency, scientists began exposing crop plants to atomic radiation. Their hope was that the radiation would increase the frequency of useful mutations that could be artificially selected. This "radiation breeding" produced many of the most common varieties of a large fraction of the world's crops, including rice, wheat, barley, pears, peas, cotton, peppermint, sunflowers, peanuts, grapefruit, sesame, and bananas. These crops are sold in grocery stores without any special labeling—and include some **organic** crops raised without the use of pesticides. It doesn't make logical sense to oppose genetic engineering but not other technological approaches that change plant genomes even more radically.

To be fair, there is an important difference between these historical methods of plant breeding and modern genetic engineering approaches. Traditional plant breeders work with the natural genetic variation that exists in a plant population at any one time. Even with radiation breeding, scientists are really just speeding up a process (mutation) that happens naturally. But with the advent of genetic engineering, it's possible for

ORGANIC
Describes a way of growing crops that conforms to several regulations, among them that synthetic pesticides must not be used.

scientists to create new genetic traits that previously did not exist in a species. They can also transfer traits from one species into other to create transgenic organisms. These truly are revolutionary capabilities (**INFOGRAPHIC 24.10**).

This leads to a question: are GM crops safe to eat? The scientific consensus is that

INFOGRAPHIC 24.10

Crop Modification Techniques

Food crops have been modified by humans for thousands of years. Selective breeding relies on variation in a population, and selecting for desirable traits. Radiation can be used to increase variation through the introduction of new mutations, and potentially new and desirable traits. Modern techniques introduce or modify specific gene sequences to introduce or alter a particular trait.

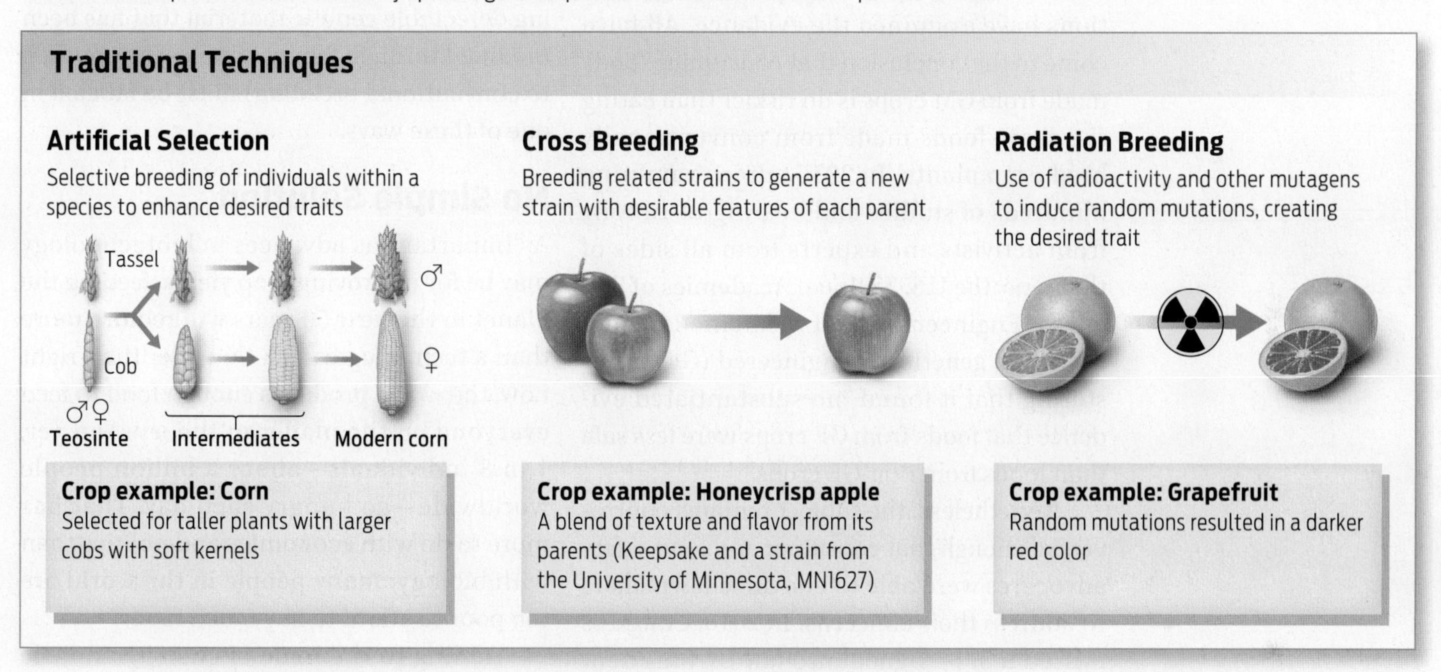

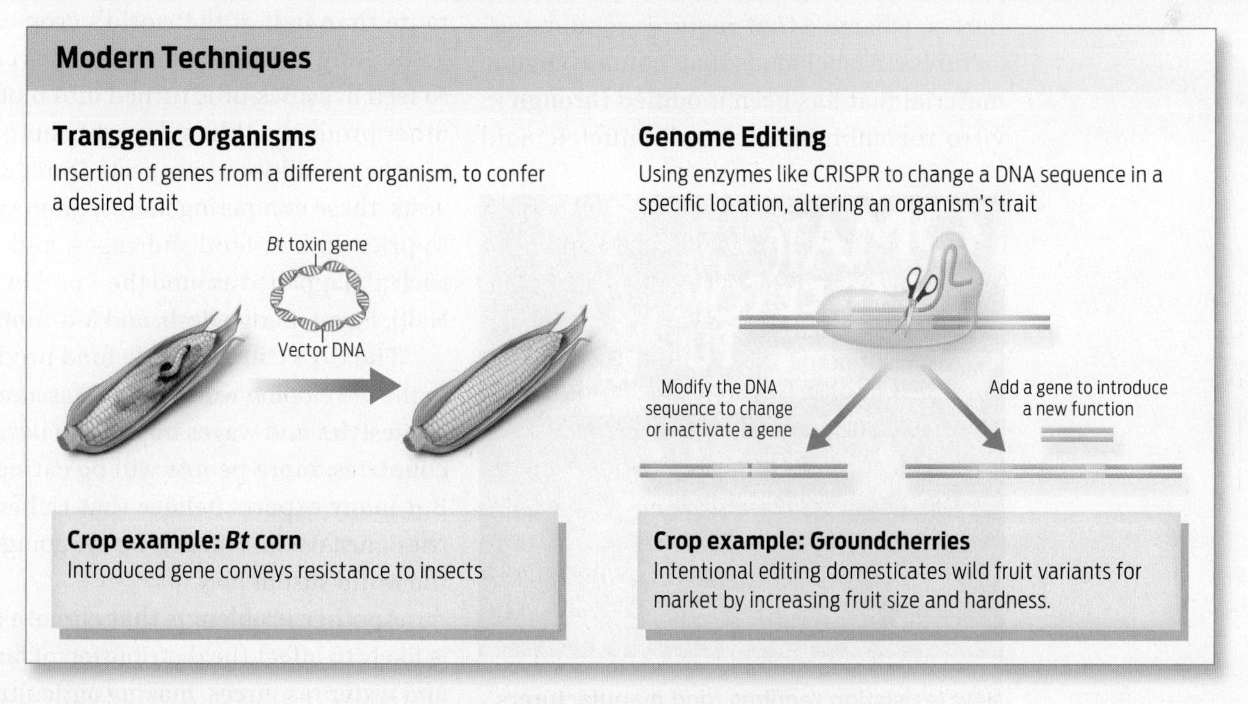

? Which of the five techniques can introduce new genes from an unrelated species?

Many common crops that we have enjoyed for a long time have even been deliberately subjected to DNA-altering treatments.

they are. The World Health Organization, the American Medical Association, the U.S. National Academy of Sciences, the Food and Drug Administration (FDA), the British Royal Society, and many other scientific organizations have examined the evidence. All have come to the conclusion that consuming foods made from GM crops is no riskier than eating the same foods made from conventionally bred crop plants. In 2016, after examining hundreds of studies and hearing testimony from activists and experts from all sides of the issue, the U.S. National Academies of Sciences, Engineering, and Medicine issued a report on genetically engineered (GE) crops stating that it found "no substantiated evidence that foods from GE crops were less safe than foods from non-GE crops."

Nonetheless, the subject remains controversial enough that consumer and food safety advocates were able to convince lawmakers to address their concerns. In 2016, Congress passed a bill—signed into law by President Barack Obama—that requires food manufacturers to label foods that "contain genetic material that has been modified through in vitro recombinant deoxyribonucleic acid (DNA) techniques; and for which the modification could not otherwise be obtained through conventional breeding or found in nature." Manufacturers have three ways to comply with the new law: they can provide a label that includes a symbol indicating the presence of genetically modified organisms (GMOs); print a label with GMO information in words; or add a Quick Response (QR) code, readable by a smartphone, that links to ingredient information. By 2022, all foods containing detectable genetic material that has been modified through lab techniques (as opposed to conventional breeding) must be labeled in one of these ways.

Robert Brook/Science Source

New legislation requires food manufacturers to label GMOs.

No Simple Solution

As important as advances in biotechnology may be for improving crop yields, feeding the planet in the next 50 years will require more than a technological fix. Consider that, right now, the world produces enough food to feed everyone on the planet in theory. And yet, 1 in 8 individuals—about a billion people worldwide—go hungry each day. That has more to do with economics and politics than with biology: many people in the world are too poor to afford to buy much food.

Adding to the issue of cost, just a little more than half of the world's crop calories go directly to feeding people. The rest goes to feed livestock or is turned into biofuels or other products. When droughts hit or other weather conditions unexpectedly reduce harvests, these competing uses of grain can lead to price spikes, food shortages, and riots—such as happened around the world in 2008 in Haiti, Egypt, Bangladesh, and Mozambique.

The UN's call to double food production in the developing world by 2050 assumes that as lifestyles and wages improve in developing countries, more people will be eating meat. But many experts believe that richer countries must eat less meat if we are going to feed the world sustainably.

Another problem is that climate change is likely to affect the distribution of farmland and water resources, making agriculture as a whole less predictable.

"It's basically a swirling mass of all sorts of issues," says Jez. "I don't think any one person actually has their mind wrapped around all of the issues completely."

That's why, he says, it's important to for people from different areas of policy and science and government and economics to talk to one another. The next generation will have to weigh all these different issues to solve this problem.

But given their intimate connection to human livelihoods, plants are a good place to start. "Plants are cool," Jez says. "When you actually stop and look at what they're able to do, you realize they're worth paying attention to." ■

CHAPTER 24 Summary

Driving Question 1 What nutrients do plants need, and how do agricultural practices supplement these nutrients?

- Plants have a structure that is well suited to obtaining the materials they need to grow. Through their aboveground shoot system, they absorb sunlight and take in carbon dioxide. Through their belowground root system, they absorb water and nutrients like nitrogen and phosphorus.
- Fertilizer helps plants grow by providing nitrogen and other supplemental nutrients that may be present in limited amounts in soils.
- A danger of fertilizer use is eutrophication: the enrichment of nutrients in aquatic environments causes overgrowth of algae, leading ultimately to oxygen depletion and the creation of a dead zone where nothing can grow.

Driving Question 2 How do important crops and other angiosperms reproduce?

- Staple crops are those that are eaten in large quantities and provide most of the energy and nutrients in the human diet. Many staple crops are grasses with edible seeds. Seeds are a product of plant reproduction.
- Many crop plants are angiosperms, or flowering plants, which reproduce sexually.
- Fertilization of female eggs by male sperm occurs when pollen (which contains sperm) travels from a male sex organ (the stamen) to a female sex organ (the carpel).
- Pollen may be transferred by animals or wind.
- Angiosperms carry out double fertilization: one sperm fertilizes the egg to form the plant embryo, and the other joins with two nuclei in a central cell in the ovule to produce the endosperm, a food source in the seed for the developing embryo.
- Some plants reproduce asexually—for example, through runners.

Driving Question 3 How can the use of genetically modified plants make agriculture more efficient?

- Through genetic engineering techniques, scientists can modify plants to give them unique properties, such as larger fruits or resistance to pests.
- A new genetic engineering technique called CRISPR allows scientists to easily add or edit specific genes in cells.
- Genetically engineering better crops may be an important way to address the nutritional needs of a growing human population.
- The scientific consensus is that genetically modified crops are safe to eat, but there are potential environmental and economic concerns.

More to Explore

- Jez, J. (2016) TEDx Talks: The once and future green factory: www.youtube.com/watch?v=7NCOKLqdEHE
- Jez, J. M., et al. (2016). The next green movement: Plant biology for the environment and sustainability. *Science* 353(6305):1241–1244.
- Lemmon, Z. H., Reem, N. T., Dalrymple, J., et al. (2018). Rapid improvement of domestication traits in an orphan crop by genome editing. *Nat Plants* 4:766–770.
- National Academy of Sciences. (2016). *Genetically engineered crops: Experiences and Prospects:* www.nas-sites.org/ge-crops/
- Saletan, William. (2015). Unhealthy fixation. *Slate*: http://www.slate.com/articles/health_and_science/science/2015/07/are_gmos_safe_yes_the_case_against_them_is_full_of_fraud_lies_and_errors.html
- HHMI BioInteractive. (2015). *Popped Secret: The Mysterious Origin of Corn* (film): https://www.youtube.com/watch?v=mBuYUb_mFXA

CHAPTER 24 Test Your Knowledge

Driving Question 1 What nutrients do plants need, and how do agricultural practices supplement these nutrients?

By answering the questions below and studying Infographics 24.1, 24.2, 24.3, and 24.4, you should be able to generate an answer for this broader Driving Question.

Know It

1. Which form of nitrogen is abundant in the atmosphere? Which form of nitrogen can be easily taken up and used by plants?

2. Plants take up nitrogen through their ____________ and carbon through their ____________.

a. roots; roots
b. shoots; roots
c. roots; shoots
d. shoots; shoots

3. What nutrients are provided by commercial fertilizers?

a. oxygen
b. carbon
c. nitrogen
d. phosphorus
e. both nitrogen and phosphorus

Use It

4. Why is carbon not included in industrial fertilizers?

5. If you applied to the soil around your plants a chemical that kills bacteria (but not plants), why might your plants die?

6. You have five containers with specific soils, and for each container you can control the composition of the surrounding air. Note that sterile soil contains no living organisms (plants, bacteria or fungi). The setup of each container is as follows:

A: sterile soil with no ammonia; limited N_2 in the air
B: sterile soil with no ammonia; abundant N_2 in the air
C: soil with no ammonia but with *Rhizobium* bacteria; limited N_2 in the air
D: soil with no ammonia but with *Rhizobium* bacteria; abundant N_2 in the air
E: sterile soil with ammonia added; abundant N_2 in the air

Which container do you predict will support the most robust growth of plants? Explain your answer.

Driving Question 2 How do important crops and other angiosperms reproduce?

By answering the questions below and studying Infographics 24.5, 24.6, 24.7, and 25.8, you should be able to generate an answer for this broader Driving Question.

Know It

7. Match the parts of a flower with the corresponding letter from the diagram below:

___ Anther	___ Carpel
___ Filament	___ Stamen
___ Ovary	___ Style
___ Ovule	___ Stigma

8. Mark each of the flower parts listed below as a male reproductive structure (M) or a female reproductive structure (F).

___ Stamen	___ Filament
___ Carpel	___ Ovule
___ Ovary	___ Stigma
___ Anther	___ Style

9. What is the role of the endosperm in a seed?

10. Complete this sentence using the terms *ovary, egg,* and *ovule*.

The ____________ is the female gamete. In angiosperms, it is found in an ____________, which in turn is found in the ____________.

Use It

Bring It Home

Apply Your Knowledge

11. Many everyday products contain compounds derived from plants. A few of these are:

- Caffeinated sodas or energy drinks
- Camphor- and menthol-containing vapor rubs and steams (for congestion and muscle aches)
- Nicotine
- Tea tree oil for minor skin irritations and infections
- Quinine-containing tonic water

You, members of your family, or your friends may have used some or all of these products. For each item listed, do some Internet research to identify the plant source (find both the scientific and common names) and something else about it that interests you (for example, where the plant is found, whether it has historically been used for medicinal purposes, or other sources of the compound).

Driving Question 3 How can the use of genetically modified plants make agriculture more efficient?

By answering the questions below and studying Infographics 24.9 and 24.10, you should be able to generate an answer for this broader Driving Question.

Know It

12. List three ways in which genetic engineering is being used in agriculture. What is the intended advantage or benefit of each?

13. Compare and contrast *Bt* and Roundup Ready GM plants. Discuss their purpose, benefits, and potential risks or unintended consequences.

14. Which of the following is consistent with organic farming practices?

a. using a synthetic chemical fertilizer
b. using *Bt* toxin proteins as an insecticide
c. using genetically modified *Bt* plants
d. using a synthetic pesticide

15. Of the practices listed in Question 14, which is most likely to contribute to eutrophication? Explain your answer.

Use It

16. Some of your friends recently read a news article on radiation breeding of plants. The friends were already concerned about genetically modified organisms, fearing that they were going to be "eating DNA" in GM foods. Now they are concerned that they may also end up eating radioactive food. What can you say to your friends to reassure them that these two particular fears are unfounded?

Mini Case

Apply Your Knowledge

17. Many strategies can be used to help increase food availability for a growing global population. If you were the head of a major funding agency, which of the following strategies would you prioritize for funding? Justify your decision.

a. investing in research to make photosynthesis more efficient, thereby enhancing plant growth
b. designing corn and wheat to form root nodules
c. public campaigns to promote reduced amounts of meat consumption

Apply Your Knowledge

Interpreting Data

18. In Turkey, the growth of GM crops has been banned, but it is legal to import approved animal feeds that contain GM products. The use of GM products for human food is not approved in Turkey. A 2015 study used polymerase chain reaction (PCR) amplification (see Chapter 7) to analyze a variety of corn-containing animal feeds and human foods for the presence of a foreign promoter sequence used to drive the expression of transgenes in GM corn. The scientists isolated DNA from 33 different animal feeds and a variety of human foods. They used PCR to detect the presence of a corn-specific gene, as well as the transgenic promoter sequence. The results are shown in the table below.

a. What was the purpose of testing for the corn-specific gene?

b. Complete the far right column by calculating the percentage of samples for each product that was positive for the foreign transgene.

c. Which sample(s) is/are problematic, given Turkish policies governing GM crops?

d. What recommendations would you make to the Turkish government, based on these findings?

Product	Number of samples tested	Number of samples positive for corn gene	Number of samples positive for transgenic promoter	Percentage of samples positive for transgenic promoter
Animal feeds	33	33	4	
Human food products				
Corn flour	6	6	0	
Corn starch	4	4	1	
Canned corn	10	10	0	
Popcorn	6	6	0	
Corn chips	9	9	1	
Corn biscuits	4	4	0	

Data from Turkec, A., Lucas, S. J., and Karlik, E. (2016). Monitoring the prevalence of genetically modified maize in commercial animal feeds and food products in Turkey. *J Sci Food Agric* 96:3173–3179. doi:10.1002/jsfa.7496.

A mosquito caught on a carnivorous sundew plant. ▼

PLANTS Q&A

Exploding seeds, carnivorous flowers, and other colorful adaptations of the plant world

Julie DeRoche/All Canada Photos/Alamy

DRIVING QUESTIONS

1 What are key structural features of plants?

2 What adaptations do plants have to cope with limiting nutrients and harsh conditions?

3 How do plants respond to stimuli and protect themselves?

In Chapter 24, we learned how plants grow and reproduce. In this chapter, we discuss features of plant anatomy and physiology that help plants survive in a variety of conditions.

PLANT STRUCTURE

Q Plants lack bones, so how do they stand up?

A From tiny 3-inch-tall crocus flowers to massive redwood trees standing 350 feet high, plants share the same basic design: a belowground root system for absorbing water and nutrients and an aboveground shoot system made up of stems and leaves. This body arrangement works well for plants: they are able to simultaneously reach for the sky and anchor themselves safely in the soil—all without the help of bones, tendons, or muscles. In the absence of such a musculoskeletal system, what keeps a plant from flopping over?

Like animals and other eukaryotes, plants are made of cells packed with organelles, including a nucleus, endoplasmic reticulum, and mitochondria. Plant cells contain a few plant-specific parts as well. All plants contain chloroplasts, the sites of photosynthesis. And all plant cells have a supportive cell wall made of a complex carbohydrate called cellulose and a **central vacuole,** essentially a large water balloon occupying the center of the cell. When filled with water, vacuoles exert **turgor pressure** against the cell wall, keeping a plant body rigid and upright. When the vacuoles are less than full, turgor pressure is reduced, and the plant wilts.

CENTRAL VACUOLE
A fluid-filled compartment in a plant cell that contributes to cell rigidity by exerting turgor pressure against the cell wall.

TURGOR PRESSURE
The pressure exerted by the water-filled central vacuole against the plant cell wall, giving a stem its rigidity.

LIGNIN
A stiff strengthening agent found in the inner cell wall of plant stems.

VASCULAR SYSTEM
A system of tube-shaped vessels that transports water and nutrients throughout an organism's body.

This decrease in turgor pressure explains why plants that dry out will wilt. Water flows into and out of the vacuole by osmosis, the movement of water across a cell membrane from an area of lower solute concentration to an area of higher concentration (see Chapter 3). Vacuoles have a relatively high solute concentration and tend to draw water in. When a plant dries out, the relative concentration of solutes outside the cell increases, and water moves out of the cell, decreasing turgor pressure against the cell wall.

Cell walls contribute to a plant's stiffness in another way. Plant cells are packed tightly together, much like bricks in a wall; carbohydrate molecules associated with the cell walls act like glue, helping adjacent cells stick together. With many cells held together in this way, plant tissues are exceptionally strong, which is why they make such durable ropes and fabrics.

Some plant tissues, such as those that make up the stems in woody plants, are made of cells with two cell walls: an outer cell wall containing cellulose, and a second, inner cell wall containing cellulose and **lignin.** Lignin is a hard, durable material that lends added support and strength to plant tissues; it is what makes them "woody." A plant with a thick, woody stem (or trunk, in the case of a tree) can grow tall and still not topple over. The tallest plants on Earth are the California redwoods (*Sequoia sempervirens*), which can reach heights of more than 350 feet **(INFOGRAPHIC 25.1).**

The tallest plants on Earth are the California redwoods (Sequoia sempervirens), *which can reach heights of more than 350 feet.*

Q Why don't plants bleed?

A Like animals, plants have a **vascular system** of tube-shaped vessels for transporting fluids. Instead of blood, however, a plant's vascular system transports water and nutrients throughout the plant's body. Plants need water to live and grow (you may have learned this the hard way if you've ever killed a plant through neglect). Among other things, water is crucial for photosynthesis in leaves.

But you don't water a plant's leaves—you water the roots by pouring water in the soil. So how does water get from the roots to the rest of the plant?

A plant's water-carrying tissue is called **xylem** (pronounced ZYE-lum). Xylem tissue is made of cells arranged into long, stiff tubes; the cells have holes in each end and are

XYLEM
Plant vascular tissue that transports water from the roots to the shoots.

INFOGRAPHIC 25.1

How Plants Stand Up

Plants have water-filled vacuoles that prevent wilting. The pressure in these vacuoles keeps cells rigid and the plant standing upright. Woody stems have a second layer of lignin, in addition to the fibrous cellulose in their cell walls. Lignin provides additional strength to stems.

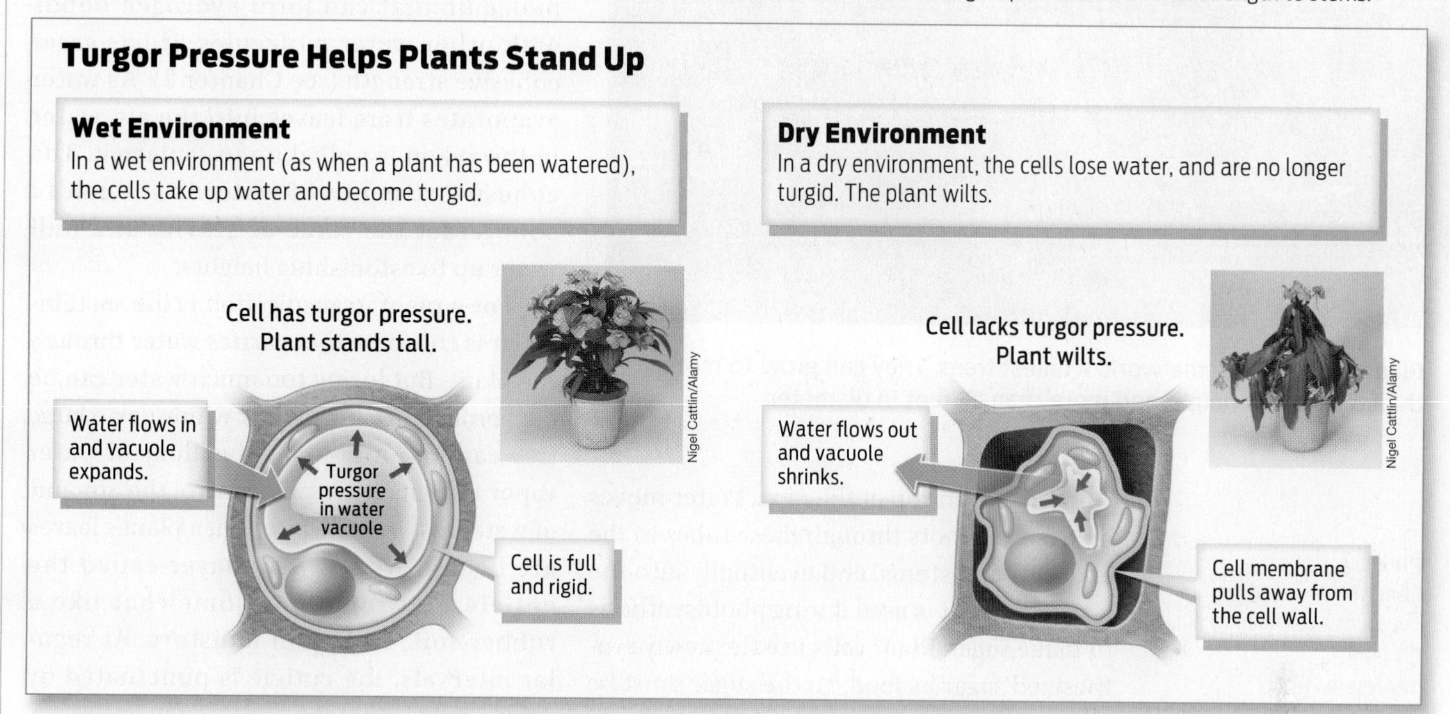

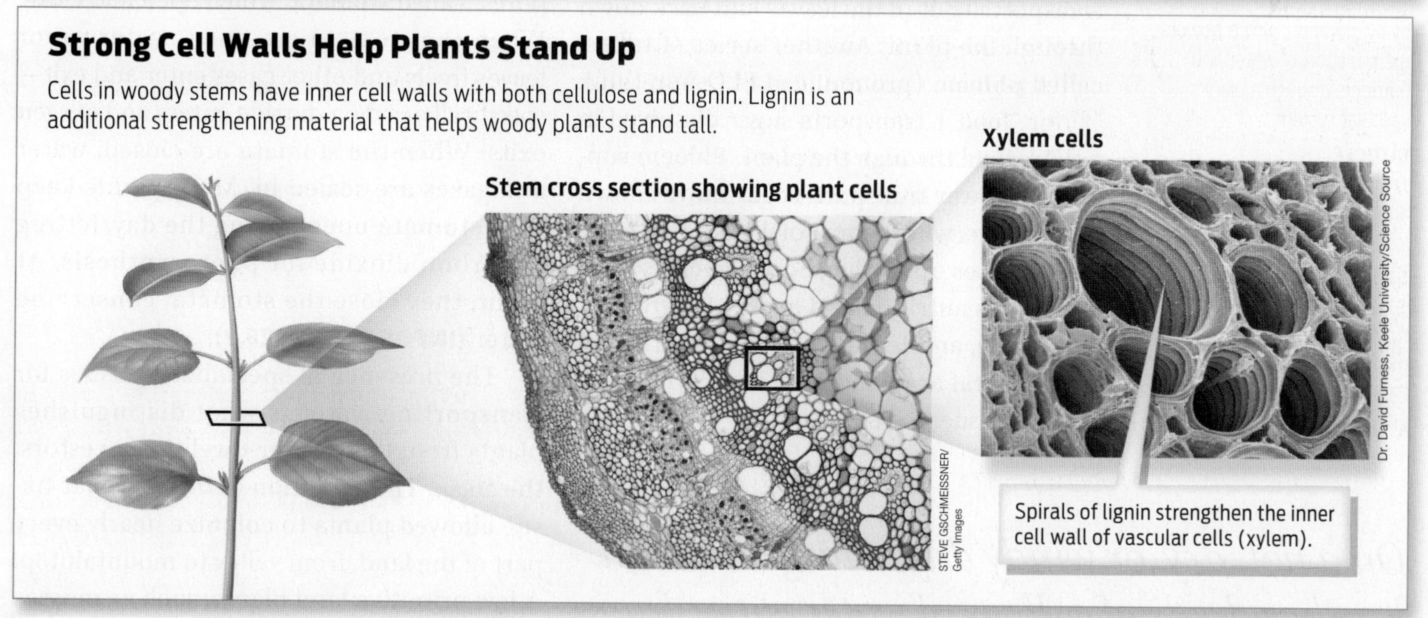

? **Why does watering plants with an excessive amount of fertilizer (which contains a high concentration of solutes) cause the plant to wilt?**

Giant sequoias are the world's tallest trees. They can grow to more than 350 feet in height and more than 25 feet in diameter.

stacked one on top of the next. Water moves up from the roots through these tubes to the aboveground stems, and eventually into the leaves, where it is used during photosynthesis to make sugar. Plant cells use the newly synthesized sugar as food, so the sugar must be transported out of the leaves and back down through the plant. Another series of tubes, called **phloem** (pronounced FLO-um; think "f" for "food"), transports sugar dissolved in a thick fluid through the plant. Phloem supports two-way transport: sugar moves down to the roots, where some of it is stored; later, sugar moves up to the shoot system, where it provides nutrition and energy for growing fruit, buds, and leaves. The veins you see in a plant's leaf are bundles of both xylem and phloem tissue.

PHLOEM
Plant vascular tissue that transports sugars throughout the plant.

TRANSPIRATION
The loss of water from plants by evaporation, which powers the transport of water and nutrients through a plant's vascular system.

CUTICLE
The waxy coating on leaves and stems that prevents water loss.

STOMATA (SINGULAR: STOMA)
Pores on leaves that permit the exchange of oxygen and carbon dioxide with the air and allow water loss.

On a hot, dry, or windy day, a large tree can lose hundreds of gallons of water vapor from its leaves.

Moving water and other nutrients up to the top of a 300-foot tree against the force of gravity is no easy task—hundreds of pounds of water must be lifted through what is essentially a long water pipe. Adding to the challenge, unlike animals, plants have no heart to pump fluid along. Instead, plants rely on evaporation of water from the leaves to draw water up the plant, a process called **transpiration.** Because water is a polar molecule that can form hydrogen bonds with other water molecules, it has great cohesive strength (see Chapter 2). As water evaporates from leaves into the air, water in the xylem is pulled up to replace it. The cohesive strength of water is enough to counteract the force of gravity and pull water up to astonishing heights.

For a plant, transpiration is life sustaining: it is the force that carries water through the plant. But losing too much water can be dangerous. On a hot, dry, or windy day, a large tree can lose hundreds of gallons of water vapor from its leaves. To control the amount of water lost by transpiration, a plant's leaves are coated with a waxy layer called the **cuticle** that functions somewhat like a rubber suit, sealing in moisture. At regular intervals, the cuticle is punctuated by pores, called **stomata,** which open and close. When the stomata are open, water vapor leaves freely and other gases enter and exit—specifically, carbon dioxide enters and oxygen exits. When the stomata are closed, water and gases are sealed in. Many plants keep their stomata open during the day, letting in carbon dioxide for photosynthesis. At night, they close the stomata, conserving water **(INFOGRAPHIC 25.2).**

The presence of specialized tissues for transporting water is what distinguishes plants from their water-dwelling ancestors, the algae. The evolution of this vascular tissue allowed plants to colonize nearly every part of the land, from valley to mountaintop. A few primitive land plants, such as mosses and liverworts, lack true roots and shoots containing vascular tissue (see Chapter 18). Without specialized tubes for transporting

INFOGRAPHIC 25.2

A Plant's Vascular System Transports Water and Sugar

Plants move water and dissolved nutrients from the soil throughout the plant body in vascular tissue called xylem. The transport of water relies on a type of evaporation from leaves called transpiration. Sugars generated by photosynthesis in the leaves are transported throughout the plant body in phloem. Some sugars may be used immediately as a source of nutrition and energy, and some sugars may be stored in the roots for future use.

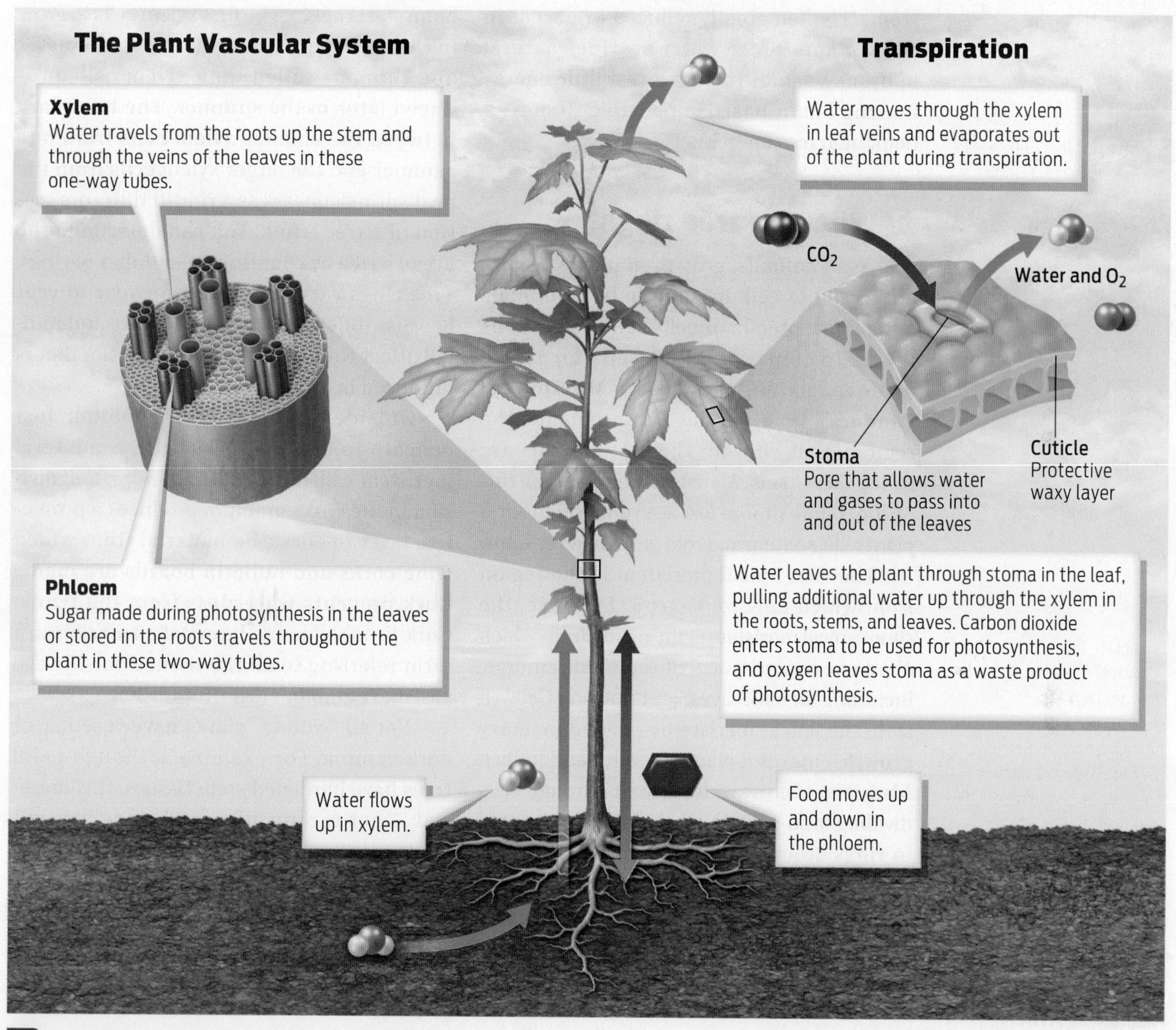

? Compare and contrast xylem and phloem.

water, most of these nonvascular plants are limited to environments that are saturated with water, where they grow close to the ground in squat, spongy mats.

Some vascular plants have made the evolutionary journey in the other direction, to life in the water. These aquatic plants include water lilies, cattails, and the *Elodea* popular in fish tanks. All have unique adaptations that permit them to thrive in wetlands and other environments where they are partially or completely submerged. For example, many aquatic plants lack the external protective cuticle layer that prevents water loss; carbon dioxide and oxygen can therefore pass in and out of any cell on the surface of the aquatic plant body. In addition, aquatic plants have no trouble obtaining water directly from their surroundings—so the xylem tissue, which normally transports water from the roots to stems, is often greatly reduced or absent in these plants. Since water provides natural buoyancy, aquatic plants invest little energy making lignin-based supportive tissues to help them resist gravity.

Q What are tree rings?

A As in animals, growth in plants occurs as a result of cell division. In both animals and plants, precursor cells called stem cells divide to contribute new cells for tissue maintenance and growth (see **Milestone 8: Pandora's Dish**). In developing plants, the regions that contain these stem cells are called **meristems.** Meristems found at the tips of roots and shoots are active throughout a plant's life, so stem and root growth never stops.

The root apical meristem is the region from which new roots grow. Likewise, the shoot apical meristem is the region from which all aboveground parts of the plant emerge, including stems, leaves, and flowers. Growth from the apical meristems—called **primary growth**—makes a plant get taller and longer. Many plants grow wider as well, through cells dividing in so-called lateral meristems located in rings around the stem or trunk—this is **secondary growth.**

In woody plants and trees, division of cells in a lateral meristem called the vascular cambium produces a thickened, woody trunk. Stem cells in the ring of vascular cambium divide to produce xylem cells on the ring's inside and phloem cells on its outside. As new xylem tissue (called secondary xylem) is added, the vascular cambium layer is pushed outward and the diameter of the trunk expands. The inner layer of secondary xylem, which eventually dies and becomes hardened with lignin, is termed **wood.** Wood provides strength that allows trees to grow tall; in turn, their greater height gives them an evolutionary advantage over nonwoody plants in the competition for sunlight.

In temperate regions—like those found in most of the United States—xylem growth is dormant, or inactive, in the winter. In the spring, xylem growth from the vascular cambium starts again. The first xylem cells to grow in the spring are usually larger in diameter and thinner-walled than xylem cells produced later in the summer. The boundary between the smaller xylem cells from one summer and the larger xylem cells from the next spring appears as a ring in the cross section of a tree trunk. You can determine the age of a tree by counting the number of rings. Tree rings vary in width from year to year because differences in temperature and rainfall affect the amount of xylem tissue that is produced in any one season.

Outside of the vascular cambium, just beneath to the trunk's surface, is a lateral meristem called the cork cambium. Cell division in the cork cambium produces a protective layer of cork, the material from which wine corks and bulletin boards are made. Cork prevents water loss from the trunk. Cork is one of many layers of a tree's bark, a term referring to all the tissues outside the vascular cambium **(INFOGRAPHIC 25.3).**

Not all "woody" plants have vascular or cork cambia. For example, although palm trees have hardened stem tissues, this material is not technically wood since it is not made of xylem. Nor do palm trees experience secondary growth resulting from the addition of new xylem every year—which is why palm trees may be quite tall, but are usually very thin. It's also why, if you cut a palm tree in half, you won't find any tree rings.

MERISTEM
A plant tissue consisting of stem cells that can divide and contribute to the growth of the plant.

PRIMARY GROWTH
Plants growing taller and roots growing longer as a result of cell division in apical meristems.

SECONDARY GROWTH
Plants growing wider as a result of cell division in lateral meristems.

WOOD
Hard, secondary xylem tissue found in the stem of a woody plant.

PLANT ADAPTATIONS

Q Why are some plants carnivorous?

A Plants are autotrophs: they make their own food through photosynthesis (see Chapter 5). Give a plant some sunshine, carbon dioxide, and water—plus a few soil

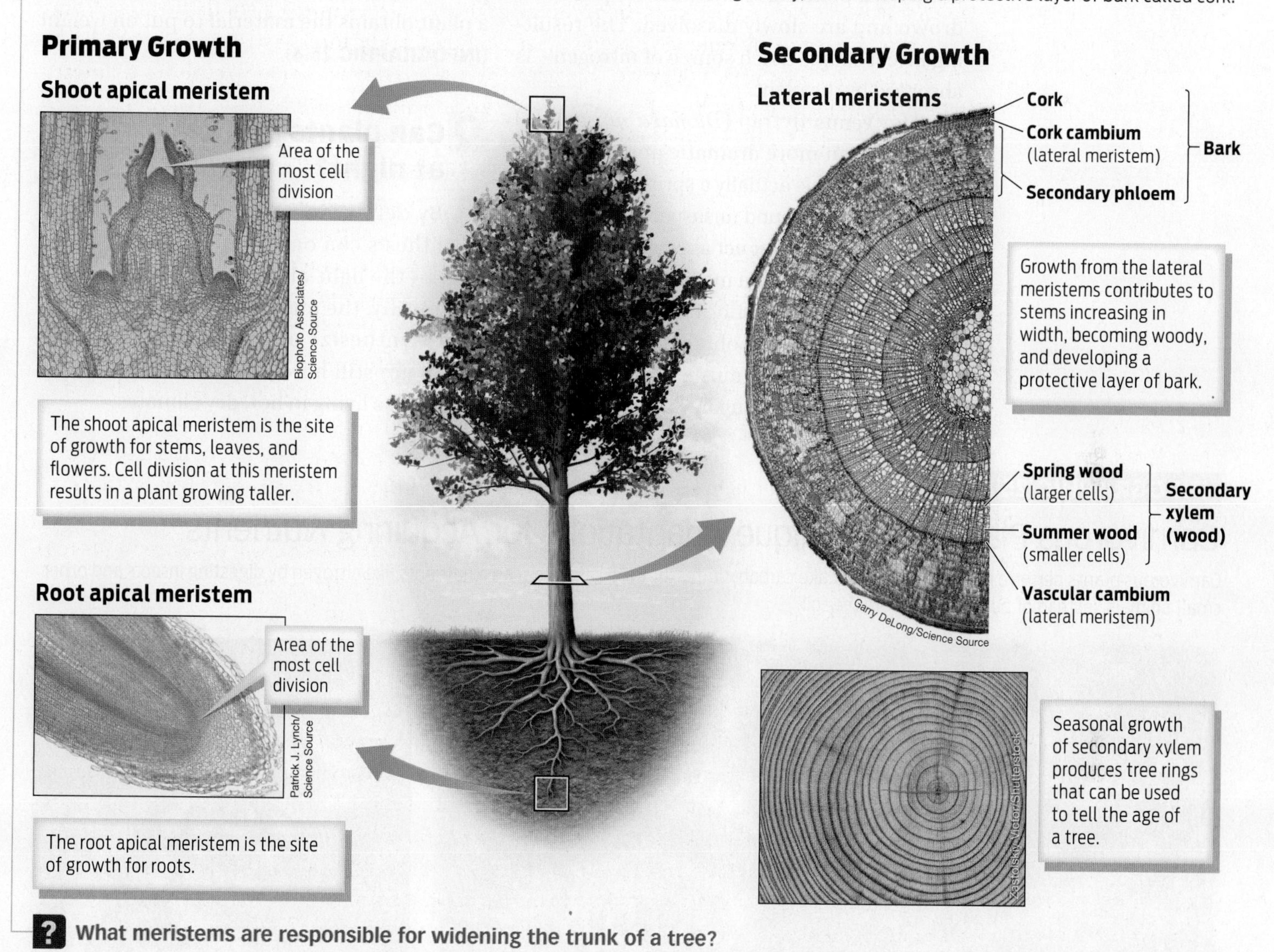

nutrients—and you will have a happy plant, capable of feeding and nourishing itself. So why, then, do some plants prey on animals, earning them the title "carnivorous"?

Plants rely on soil bacteria to fix nitrogen into a form that is usable by the plant (see Chapter 24). Fertile soil naturally contains these nitrogen-fixing bacteria, so it provides adequate supplies of usable nitrogen that plants can absorb and use to grow. In addition, soil can be supplemented with fertilizer, giving plants a boost of artificial nitrogen and other nutrients. But in certain natural environments, such as bogs or rock outcroppings, it's hard for plants to obtain the nitrogen they need. The acidity of a bog, for example, prevents organic matter from breaking down, so nutrients are recycled more slowly. In these environments, plants have evolved novel ways to obtain scarce nitrogen—some of which would put animal carnivores to shame.

Trumpet pitchers (*Sarracenia*), for example, lure insects with brightly colored flowers and nectar "bribes." But the rim of the plant's trumpet-shaped flower is slippery. Unsuspecting trespassers climb onto the rim, lose their grip, and tumble into a deep cavity filled with digestive juices. Prevented from escape by downward-pointing spikes, the tiny prisoners drown and are slowly dissolved. The resulting insect soup—a rich source of nitrogen—is absorbed by the plant.

The Venus flytrap (*Dionaea muscipula*) takes an even more dramatic approach. The plant's "flower" is actually a spring-loaded trap that snaps shut around unsuspecting prey. Tiny hairs inside the flower act as sensors; when the sensors are tripped by a moving insect, the trap slams shut and the feasting begins.

Even when plants obtain nitrogen in this carnivorous way, they must still perform photosynthesis to make sugar. The plant body is composed of complex carbohydrates, such as cellulose, which the plant makes by stringing sugar molecules together. And the starting material for photosynthesis is carbon dioxide gas. Thus the air, rather than the soil, is where a plant obtains the material to put on weight **(INFOGRAPHIC 25.4)**.

Q Can plants photosynthesize at night?

A By definition, one crucial part of photosynthesis can occur only during daylight hours—the light-absorbing "photo" part. So technically, the answer is, no—plants can't photosynthesize at night. However, some plants are still hard at work at night, especially those living in hot, dry climates.

INFOGRAPHIC 25.4

Carnivorous Plants Have Unique Adaptations for Acquiring Nutrients

Carnivorous plants perform photosynthesis to make carbohydrates, but they acquire scarce nutrients like nitrogen by digesting insects and other small animals, instead of pulling them from the soil.

The soil in acidic bog habitats recycles organic matter slowly, making the soils low in nutrients like nitrogen. Instead of pulling them from the soil, carnivorous plants acquire scarce nutrients by digesting insects and other small animals.

demaimiel62/Shutterstock

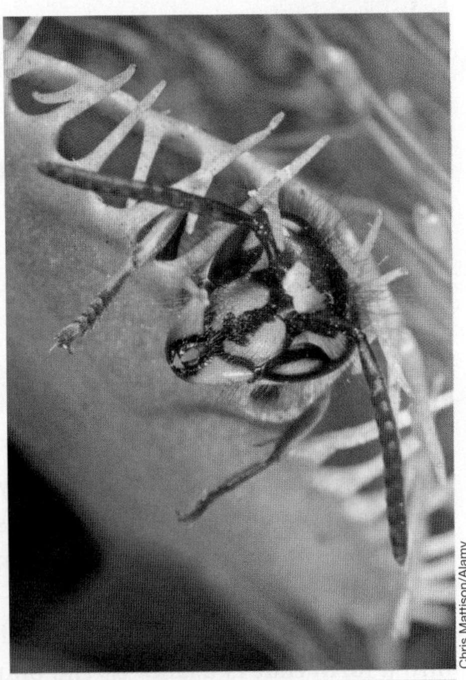

Chris Mattison/Alamy

The Venus flytrap (*Dionaea muscipula*) attracts insects into its spring-loaded trap. Projections interlock at the opening, preventing the insect's escape.

? Are insects primarily a source of carbon or of nitrogen for carnivorous plants?

INFOGRAPHIC 25.5

Beating the Heat: Photosynthesizing without Drying Out

All plants must perform the "photo" reactions of photosynthesis during the day. But many plants experience reduced levels of photosynthesis in hot and dry conditions. Other plants are adapted to live in hot and dry climates, and have different strategies to take in CO_2 while minimizing water loss.

Stomata open all day: too much water loss

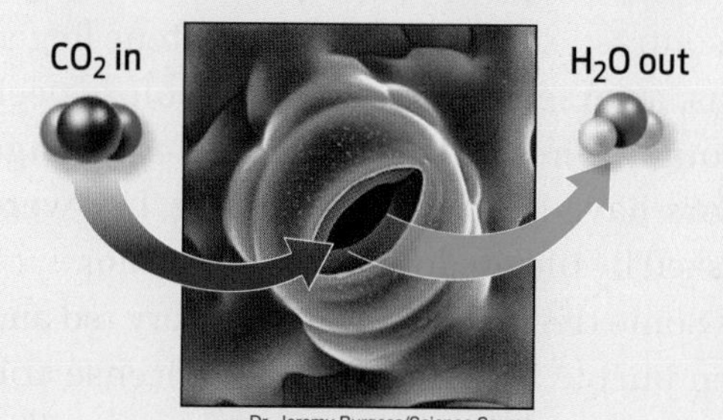

Dr. Jeremy Burgess/Science Source

When stomata are open, carbon dioxide can enter the leaves for photosynthesis. However, this increases transpiration and therefore water loss.

Stomata closed all day: too little sugar produced

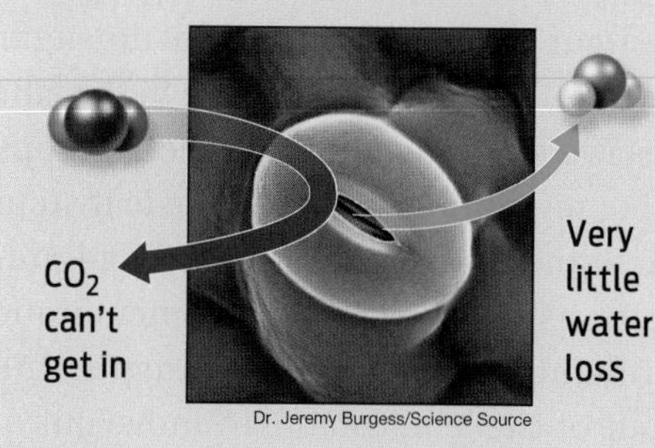

Dr. Jeremy Burgess/Science Source

In dry climates, some plants keep stomata closed during the hot hours of the day. This conserves water, but inhibits the uptake of carbon dioxide for photosynthesis. These plants make less sugar food.

EggImages/Alamy

Aloe ferox has succulent leaves.

Strategies to combat water loss in dry climates:

- Thick, waxy succulent leaves
- Extract CO_2 from airspaces even when stomata are mostly closed
- Open stomata at night to capture CO_2 in cooler temperatures

? Why do many plants make less sugar on very hot, sunny, dry days?

Because daylight hours tend to be the time when the environment is at its hottest and driest, plants lose a lot of water during the day. To conserve water, plants can close their stomata. But closing stomata also prevents carbon dioxide—the raw material of photosynthesis—from entering the leaf. In many plants, such as wheat and rice, the result of this trade-off is reduced output of plant food—that is, less sugar.

Some plants have adapted to—and flourish in—their sun-scorched surroundings. For example, corn and sugar cane can thrive in hot, sunny climates by keeping their stomata partially closed at most times of day. An enzyme in these plants can capture CO_2 even when it is present only at low concentrations in air pockets in the leaf.

Still other plants have adapted to hot, dry conditions by splitting up two parts of photosynthesis that usually occur together: taking in CO_2 and making sugar. Pineapples as well as cacti and other succulent plants conserve water by keeping their stomata closed during the day when it's hot. But at night, when it's cooler, they open their stomata, allowing carbon dioxide in. The CO_2 that is captured at night isn't used right away, but instead is stored. During the day, while their stomata are closed, these plants use the stored CO_2 to complete photosynthesis and make sugar. By separating in time the steps of photosynthesis, these well-adapted plants can thrive in conditions that would wither their less physiologically adapted cousins **(INFOGRAPHIC 25.5)**.

Unsuspecting trespassers climb onto the rim, lose their grip, and tumble into a deep cavity filled with digestive juices.

Like other eukaryotic organisms, plants use sugar to perform aerobic respiration to make ATP (see Chapter 6). All plants respire both during the day and at night. Because the total amount of carbon dioxide given off by plants during cellular respiration is less than the total amount taken in for photosynthesis, plants are carbon dioxide sinks that ultimately lower the amount of carbon dioxide in the atmosphere.

Q Why do leaves change color in the fall?

A For most of the year, leaves are photosynthesis factories, churning out sugar. A key component of their photosynthetic machinery is the pigment **chlorophyll,** which absorbs red and blue wavelengths of light and reflects green wavelengths. Chlorophyll is the reason plants appear green. To keep photosynthesis running, plants make abundant chlorophyll in spring and summer. But they also produce smaller amounts of other pigments—yellow-reflecting xanthophyll and orange-reflecting carotene. In the leaves, these pigments capture additional wavelengths of light, thereby expanding the range of light that is useful for photosynthesis. You can't see these other pigments in leaves during spring and summer because leaves are chock full of green chlorophyll, camouflaging the other hues (although you can see them elsewhere in some plants: in the flesh of a pineapple, for example, or the taproot that is a carrot).

After a summer of intense sugar stockpiling, many trees, bushes, and other deciduous plants drop their leaves in the autumn. As these plants start to settle in for the winter, they begin to shut down their photosynthesis machinery. During the winter months in temperate regions, there isn't enough water to drive photosynthesis; water in the ground is frozen and can't be absorbed. As temperatures cool, plants turn off the production of their light-absorbing pigments. Of all the pigments, chlorophyll is the most unstable chemically and therefore the shortest lived: its levels fall quickly once production stops. By contrast, xanthophyll and carotene linger. Thus, as green fades, the other colors peek through. These mostly yellow and orange colors have been there all along, but were masked by the predominant green color.

Some trees and bushes turn fiery red and deep purple in autumn. These intense colors are the result of a fourth pigment, called anthocyanin—the same pigment that gives apples and acai berries their color. Anthocyanin is produced in leaves in response to a high sugar concentration. Why do sugar concentrations in these plants rise in the fall? To conserve water in winter, a tree develops a corklike membrane between the leaf stem and the branch, sealing off the leaf and preventing transpiration. This corky membrane also prevents phloem from transporting sugar out of the leaf. Unable to move out of the leaf, sugar begins to collect. More sugar means more anthocyanin is produced, yielding the bright colors we associate with fall.

Trees such as maple and oak produce brilliant colors because they naturally produce lots of anthocyanin. Others, such as aspen and poplar, produce less anthocyanin and turn yellow or orange **(INFOGRAPHIC 25.6).**

Eventually the corky membrane that seals off the leaf from the tree causes the leaf to dry out. With a slight gust of wind, the leaf flutters to the ground. The tree is now prepared for winter.

Q Plants can't move, so how do they find good places to live?

A Unlike animals, plants cannot move to seek out more hospitable living conditions for their offspring when resources are scarce or

CHLOROPHYLL
The dominant pigment in photosynthesis, which makes plants appear green.

INFOGRAPHIC 25.6

Plants Produce Multiple Light-Capturing Pigments

Leaves contain chlorophyll and other pigments. These pigments help capture a wide range of wavelengths of light to maximize the efficiency of photosynthesis.

More Chlorophyll in Summer

Green leaves produce an abundance of the green pigment chlorophyll during the warm, sunny months, and smaller amounts of red and yellow pigments.

Less Chlorophyll in Fall

When the hours of daylight are shorter and the temperature is cooler, the trees perform less photosynthesis. Chlorophyll is no longer made and levels quickly decrease. The red and yellow pigments produced by the leaves are revealed.

? What is the function of the red and yellow leaf pigments that become visible during the fall?

the neighborhood gets too crowded. Forest or desert, valley or mountaintop—plants are stuck where nature put them. Their solution to this enforced sedentary existence is to disperse their offspring far and wide through seeds. A seed is a small embryonic plant contained within a sac of stored nutrients—essentially, a plant starter kit (see Chapter 24). It develops from a fertilized egg and is a perfect package for delivering an immature plant to its new home and protecting it from harsh conditions.

Seeds come in many shapes and sizes, and are dispersed in many different ways. Some are small and hitch a ride on fur or clothing by means of tiny hooks or burrs. Some, like coconuts, are large and buoyant and can float across oceans to reach distant beaches. The delicate parachutes of dandelions, the spinning helicopters of maples and pine trees, and other lightweight seeds sail on the wind. Cottonseeds are little more than hairy specks of dirt, but on a steady breeze they can windsurf for miles. Other seeds are packaged inside fruit. Tempted by the fruit's bright color and sugary content, animals eat the fruit and then deposit the seeds in feces some distance away from the original plant.

Some seeds are dispersed through a ballistic mechanism. The seedpod of the squirting cucumber (*Ecballium elaterium*), for example, fills with slimy juice as it ripens. Eventually, the mounting pressure of the increased volume of juice causes the cucumber to shoot off the plant like a rocket, trailing a plume of seeds and slime in its wake.

At the slightest touch, the bulging seedpods of the aptly named touch-me-not plant explode, spraying seeds like bullets. When a seed lands in favorable conditions, with enough water, it will germinate and grow into a young plant, or seedling **(INFOGRAPHIC 25.7)**.

Eventually, the mounting pressure of the increased volume of juice causes the cucumber to shoot off the plant like a rocket, trailing a plume of seeds and slime in its wake.

PLANT HORMONES

Q Plants can't see, so how do they know which way is up?

A Observe an old building and you'll likely see ivy scaling up its walls. Keep houseplants next to a window and you'll find them bending toward the sunlight. How does a plant know where it's going? While plants don't have eyes and therefore can't see, they are quite adept at sensing and responding to their environment. They are able to respond to the environment through various kinds of tropism (from the Greek *tropos,* meaning "turn"). A tropism is a turning in a specific

INFOGRAPHIC 25.7

Seeds Carry a Young Plant to a New Destination

Seeds are important for the dispersal of plants. As seeds are dispersed, offspring reach new locations, often far from their parents.

Seeds Disperse in a Variety of Ways

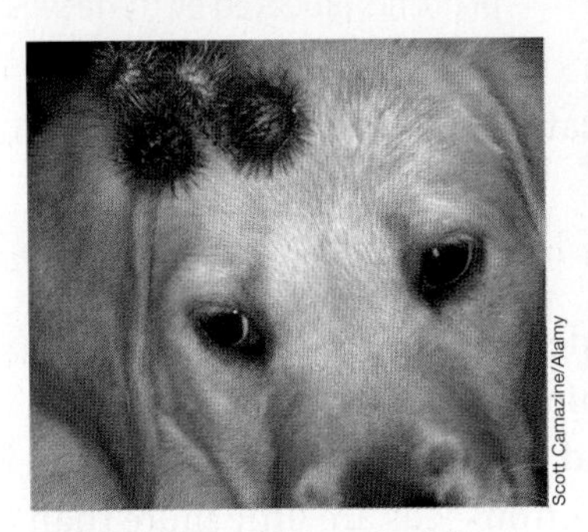
Scott Camazine/Alamy

Burrs hitch a ride on animal fur.

Suijo/Dreamstime.com

Small dandelion seeds fly for miles in the wind.

AlessandrZocc/Shutterstock

Maple seeds spin like the blades of a helicopter to disperse in the wind.

Mark A. Johnson/Alamy

Coconuts float, relocating seeds from one shore to another.

Juniors Bildarchiv GmbH/Alamy

Animals eat fruits, dispersing the seeds in their feces.

Danilo Donadoni/AGE Fotostock

The squirting cucumber shoots its seeds out with water.

? Do burrs need to have a tasty interior to make possible their dispersal? Why or why not?

direction in response to an external stimulus. This type of response is triggered by plant hormones, which are signaling molecules that regulate growth and development.

The growth of a plant shoot toward light is called **phototropism.** Through phototropism, leaves get the sunlight they need for photosynthesis. This quest for light is why ivy climbs up walls and houseplants bend toward the window. Shoots' growth toward the light is enabled by **auxin,** a plant hormone that promotes cell elongation as one of its effects. Auxin is produced in the tip and travels continuously down the stem. When light hits one side of a plant shoot, auxin moves to the shaded side, creating a gradient of the hormone in the stem. The side receiving the most direct sunlight contains the least auxin, while the shaded side contains the most. Auxin thus promotes elongation of cells on the shady side of the stem. Because the shaded side elongates faster than the sunny side, it pushes the stem toward the sun. The whole stem doesn't sense light, though—just the tip. Cover the tip of a young plant and it won't turn toward the light. Instead, the plant will grow straight up because, in the absence of sunlight, auxin does not accumulate more on one side of the stem than the other.

Other mechanisms also help a plant sense where it is and orient itself in space. **Gravitropism** is the growth of plants in response to gravity—roots grow downward, with the force of gravity, and shoots grow upward, against it. Auxin is again the main player in this mechanism. When a plant is placed on its side, more auxin is sent to the down side of the stem, in the direction of gravity. The presence of more auxin causes the cells on the down side to elongate more. Stems begin to curve away from gravity. Root cells, however, respond the opposite way to auxin: more auxin on the gravity side of roots inhibits root cell elongation on the down side, so roots bend downward in the direction of gravity. Gravitropism allows a planted seed to send its shoots toward the light and its roots toward the soil.

A plant's response to touch is called **thigmotropism.** This response enables vines to wind their way around poles or trellises and carnivorous plants to sense their prey. Like the other tropisms, touch-sensitive growth is controlled by auxin. Vine cells touching a pole, for example, get less

PHOTOTROPISM
The growth of the stem of a plant toward light.

AUXIN
A plant hormone that causes elongation of cells as one of its effects.

GRAVITROPISM
The growth of plants in response to gravity. Roots grow downward, with gravity; shoots grow upward, against gravity.

THIGMOTROPISM
The response of plants to touch and wind.

An example of gravitropism: leaf-bearing stems bend upward, away from gravity.

© INRA/R. Bastien, S. Douady, B. Moulia, (2013). Unifying model of shoot gravitropism reveals proprioception as a central feature of posture control in plants. PNAS 110(2):755–760.

auxin and consequently elongate less, while cells on the other side elongate more. The result is another kind of lopsided growth in the shoot, which eventually causes the vine to coil around whatever it's touching. A plant's sense of touch can be exquisitely sensitive—more sensitive than a human's. A human can detect the presence of a thread weighing 0.002 mg laid across the arm. By contrast, the feeding tentacle of the insectivorous sundew plant can sense a thread of less than half that weight. The legs of a single gnat are enough to trigger the tentacle into swift action **(INFOGRAPHIC 25.8)**.

ETHYLENE
A gaseous plant hormone that promotes fruit ripening as one of its effects.

Q Does one bad apple really spoil the whole bunch?

A This old adage is true, and can be shown empirically: put a ripe apple in a bowl of unripe ones, and the unripe neighbors will quickly ripen. That's because ripe fruit—bananas and apples, especially—emit **ethylene,** a gaseous plant hormone that promotes ripening. In a confined space, the ethylene gas collects and causes nearby fruit to ripen through the loss of chlorophyll and the breakdown of cell walls. The result is the conversion of a hard, green fruit to a soft, ripe one.

INFOGRAPHIC 25.8

Plants Sense and Respond to Their Environment

Plants can respond to a variety of stimuli, including light, touch, and gravity. Many of these responses are triggered by a plant hormone called auxin. Unequal distribution of auxin leads to a corresponding unequal pattern of cell elongation in stems and roots, and thus growth in a particular direction.

Phototropism: Plants respond to light.

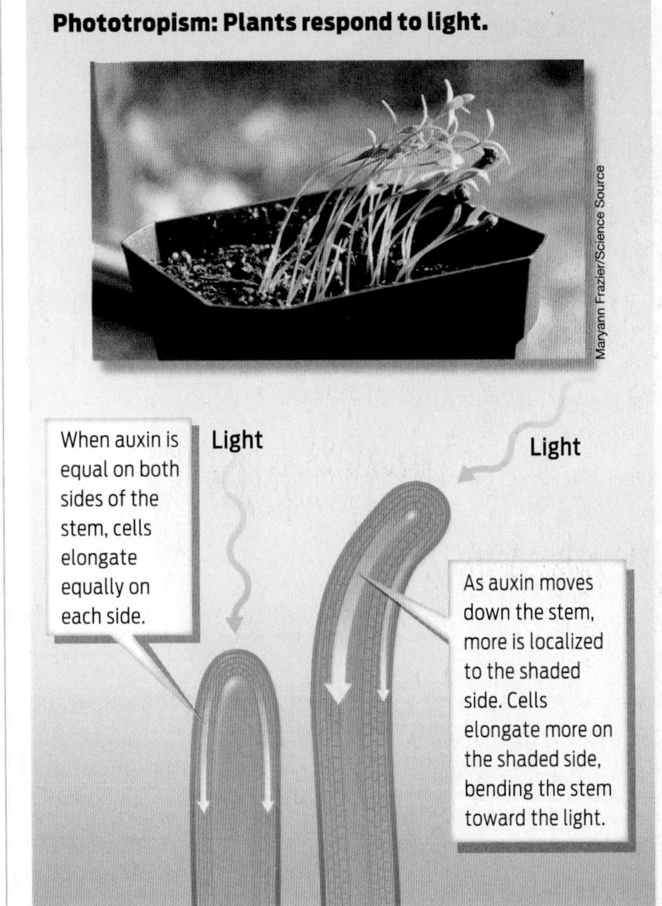

Gravitropism: Plants respond to gravity.

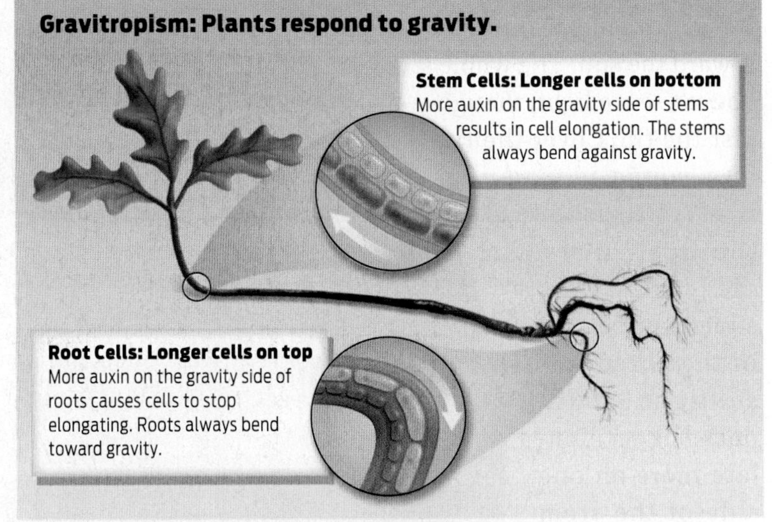

Thigmotropism: Plants respond to touch.

Vine cells touching this stem get less auxin and therefore elongate less, and cells on the other side of the vine elongate more. This causes the vine to bend toward the stem, allowing it to coil as it grows.

? What happens to cells of stems and roots as a result of auxin action?

National Geographic Image Collection/Alamy

Ethylene is used commercially to ripen fruit that is picked and transported while still green.

Commercial fruit growers take advantage of the action of ethylene when they ship fruit to distributors. Often, fruit is picked while still green and then exposed to natural or synthetic ethylene just before arrival in grocery stores to induce ripening at the preferred time.

Q Can plants take hormones to improve their performance?

A In a manner of speaking, yes. **Gibberellins** are plant hormones that promote growth. Scientists have identified more than 100 different types of gibberellin. One effect of gibberellins is to provide the chemical cue for seeds to germinate and grow. When conditions are right for a seed to germinate—when rising temperatures begin to melt frost, for example—these growth-promoting hormones give the green light for a seedling to grow.

Applying gibberellins to a young plant can increase the length of its stem, which also indirectly increases the size of its fruits—a fact that makes them very useful in agriculture. Gibberellins are commonly used to increase the size of seedless grapes, for example. On an untreated seedless grape plant, the stem remains relatively short, so the bunches of grapes growing on the stem are clustered densely together, resulting in small grapes. When sprayed with gibberellins, the stems grow longer, giving the grapes more room to grow. Because seeds are the natural source of gibberellins, seedless grapes have no source of gibberellins to help them grow naturally, which is why farmers need to spray them in the first place.

In many species of plants, seed germination is controlled by the balance between the growth-promoting properties of gibberellins and the growth-inhibiting effects of **abscisic acid (ABA),** another hormone. ABA keeps seeds from growing—that is, keeps them dormant. For example, seeds are dormant over the cold winter months, when conditions for growth are not ideal. With appropriate environmental cues—the right combination of temperature, moisture, and light—ABA is degraded and gibberellins are synthesized, leading to the end to dormancy and the start of germination. You can think of ABA as the brakes and gibberellins as the gas pedal: as the brakes on growth are lifted, the gas pedal is pressed, ushering in a new season's blooms **(INFOGRAPHIC 25.9)**.

GIBBERELLINS
Plant hormones that cause cell division and stem elongation.

ABSCISIC ACID (ABA)
A plant hormone that helps seeds remain dormant.

INFOGRAPHIC 25.9

Hormones Trigger Plant Growth and Development

Certain plant hormones regulate growth and development of plants. When humans apply these hormones to crops, plant growth can be dramatically enhanced.

Gibberellin hormones enhance plant growth

Spraying plants with gibberellins enhances the number and size of the fruit produced.

Treated with gibberellins

Not treated with gibberellins

Photo by Stephen Vasquez, used with the permission of UC Regents.

The relative amounts of ABA and gibberellins trigger germination

Environmental cues cause inhibitory ABA to be degraded and growth-promoting gibberellins to be produced. This results in germination and plant growth.

? **If you wanted to stimulate a seed to germinate, which hormone would be more helpful, ABA or gibberellins?**

PLANT DEFENSES

Q Why are some plants poisonous?

A From juicy peaches and succulent strawberries to zesty basil and peppery arugula, many plants are incontestably delicious. But humans aren't the only ones who think so: many herbivores, such as insects, birds, rodents, and other small mammals, find plant parts tasty and irresistible. This is both helpful and harmful to a plant. On the one hand, plants rely on animals to eat their fruits and disperse their seeds. On the other hand, plants must ensure that other organisms eat only noncrucial parts of the plants. Eating a plant's fruit is one thing; eating all of its leaves is quite another.

Plants have evolved many defenses to protect their important parts from herbivores' chomping. Some defenses are mechanical: the stems of a raspberry plant are covered in prickly spines to prevent unwanted chewing; holly leaves are waxy and difficult for insect jaws to grasp; and cacti have sharp needles that deter all but the most fearless animals. Other defenses are chemical: leaves of the tobacco plant produce nicotine, which is toxic to insects; and the bark of the South American evergreen cinchona tree produces quinine, an extremely bitter substance that many animals find distasteful (except certain humans, who use it in their gin and tonics). Such antiherbivory chemicals are a highly effective way to deter pests from eating a plant's leaves.

While a plant's fruits are often tasty and meant to be digested, seeds generally are not. The seed contains a new plant, so it must be protected. Many seeds are encased within an indigestible shell that keeps them from being destroyed by an animal's stomach juices. The unlucky animal that succeeds in breaking open the shell and eating the seed is in for an unwelcome surprise. Seeds are sources of some of the most potent poisons on Earth, including ricin, cyanide, and strychnine. Ricin, found in castor beans, can be lethal to animals in quantities as little as two beans. Cyanide, which is found in small doses in the seeds of peaches, apricots, and apples, kills by interrupting cellular respiration in mitochondria; unable to make ample amounts of the short-term energy-storage molecule ATP, nerve and muscle tissues quickly shut down **(INFOGRAPHIC 25.10)**.

Although antiherbivory chemicals can complicate herbivores' lives, they are often quite useful to humans. Some of our most important medicines are extracts of plant chemical defenses. These medicines include aspirin, morphine, digitalis, quinine, and the anticancer drug Taxol, which was originally obtained from the bark of the Pacific yew tree. ■

INFOGRAPHIC 25.10

How Plants Defend Themselves

Plants use both physical and chemical defenses to protect themselves against a variety of herbivores.

Physical Defenses

Cathleen Abers-Kimball/iStockphoto

Raspberry thorns can impale hungry insects.

hisforhome/iStockphoto

The jaws of insects that try to feed on holly leaves may not be strong enough to penetrate their waxy coat.

Chemical Defenses

Serghei Piletchi/Dreamstime.com

Peach pits have a tough exterior that protects the embryo inside. If an animal is successful in cracking the seed open, it is met with cyanide, a lethal toxin.

timsa/iStockphoto

When a hornworm feeds on a wild tobacco plant, the plant releases chemicals that both repel the worm and attract the worm's predators.

? Which defenses, chemical and/or physical, protect a peach seed?

CHAPTER 25 Summary

Driving Question 1 What are key structural features of plants?

- All plant cells have chloroplasts and are surrounded by a cell wall made of the complex carbohydrate cellulose. A central water-filled vacuole creates turgor pressure against the cell wall and helps a plant stand up.
- Some plant tissues have cells with an additional inner cell wall made of cellulose and lignin. Lignin, an extremely durable material, makes plant tissues woody.
- Plants have a vascular system. Xylem transports water and nutrients from the roots to the shoots. Phloem transports sugars throughout the plant body.
- Water transport is powered by evaporation through stomata in leaves, a process called transpiration. Plants have mechanisms that control the amount of water lost by transpiration.
- Plants undergo primary (vertical) and secondary (horizontal) growth as a result of cell division in meristems. Secondary growth occurs from cell division in rings of lateral meristems. Tree rings reflect the secondary growth of xylem tissue from season to season.

Driving Question 2 What adaptations do plants have to cope with limiting nutrients and harsh conditions?

- Plants are autotrophs: they make their food through photosynthesis by using sunlight, carbon dioxide, and water. They also require nutrients from the soil, including nitrogen, phosphorus, and sulfur. In nutrient-poor soils, some plants obtain nitrogen by trapping and digesting insects.
- Some plants have adapted to hot, dry climates, keeping their stomata closed during the day and collecting carbon dioxide for photosynthesis primarily at night. Other plants have adapted to hot, dry climates by keeping their stomata partially closed during the day; these plants can capture carbon dioxide inside the leaf, even when its concentration is low.
- A variety of pigments assist with photosynthesis. The dominant pigment, chlorophyll, is responsible for the green color of plant leaves. Other contributing pigments become visible in the fall, after chlorophyll has broken down.
- Seeds, which contain an embryonic plant, are dispersed in a variety of ways.

Driving Question 3 How do plants respond to stimuli and protect themselves?

- Plants respond to their environment by various tropisms. Phototropism is growth toward light; gravitropism is growth in response to gravity; and thigmotropism is growth in response to touch.
- Plants produce hormones that promote growth and development: ethylene stimulates fruit ripening; gibberellins and ABA regulate germination and stem growth; and auxin controls cell elongation.
- Plants have physical defenses (for example, thorns) and chemical defenses (for example, toxins) that fend off herbivores. Humans can often make use of these chemical defenses in developing pharmaceuticals.

More to Explore

- Chamovitz, D. (2013). *What a Plant Knows: A Field Guide to the Senses.* New York: Scientific American/Farrar, Straus, Giroux.
- PBS. (1999). *Nature*: The Seedy Side of Plants: www.pbs.org/wnet/nature/episodes/the-seedy-side-of-plants/introduction/1268/
- PBS. (2013). *Nature*: What Plants Talk About: www.pbs.org/wnet/nature/episodes/what-plants-talk-about/video-full-episode/8243/
- U.S. Forest Service, Ethnobotany: www.fs.fed.us/wildflowers/ethnobotany/
- Walters, D. (2017). *Fortress Plant: How to Survive When Everything Wants to Eat You.* Oxford, UK: Oxford University Press.

CHAPTER 25 Test Your Knowledge

DRIVING QUESTION 1 What are key structural features of plants?

By answering the questions below and studying Infographics 25.1, 25.2, and 25.3, you should be able to generate an answer for this broader Driving Question.

Know It

1. Which organelle generates turgor pressure?

a. the chloroplast
b. the central vacuole
c. the nucleus
d. the mitochondrion
e. the endoplasmic reticulum

2. Which of the following statements represents a true distinction between xylem and phloem?

a. Xylem provides support only; phloem provides transport.
b. Xylem provides water and nutrient transport; phloem provides sugar transport.
c. Xylem transports materials from shoots to roots; phloem transports materials in either direction.
d. Xylem transports sugars in either direction; phloem transports water from roots to shoots.
e. all of the above

3. What is the function of the cuticle?

a. It enables neighboring cells to stick together.
b. It provides rigidity to the cell wall.
c. It is toxic to many herbivorous insects.
d. It prevents water loss.
e. It is sticky and helps pollen adhere to a plant during pollination.

4. The cork cambium and the vascular cambium form rings. How do they contribute to plant growth?

a. They are responsible for roots growing longer.
b. They are responsible for shoots growing taller.
c. They are responsible for trunks growing wider.
d. They are responsible for increasing the branches from roots and shoots.

5. Transpiration occurs because of ________ and ________.

a. evaporation; gravity
b. turgor pressure; cohesion
c. turgor pressure; gravity
d. evaporation; cohesion

Use It

6. Paper is made from wood that is broken down to pulp. Why are lignin-digesting enzymes included in the pulping process? Would these enzymes have to be included in the pulping process if paper were made from green leaves? Explain your answer.

7. To crisp up some wilted celery, what should you soak it in?

a. pure water
b. a solution with the same solute concentration as the celery
c. a solution with a higher solute concentration than the celery
d. you should let the celery air dry for a few minutes
e. any of the three solutions (a, b, or c) will be equally effective

8. If someone strips a ring of bark from around the trunk of a birch tree, what will be the impact on that tree?

DRIVING QUESTION 2 What adaptations do plants have to cope with limiting nutrients and harsh conditions?

By answering the questions below and studying Infographics 25.4, 25.5, 25.6, and 25.7, you should be able to generate an answer for this broader Driving Question.

Know It

9. Plants are autotrophs and can make sugar from CO_2. How do they obtain CO_2?

a. through stomata
b. by absorption through the root system
c. by digesting insects
d. by breaking down carbon-rich carbohydrates stored in roots
e. a and b

10. When stomata are open, what is happening?

a. O_2 is entering the plant for photosynthesis.
b. CO_2 is entering the plant for photosynthesis.
c. H_2O is entering the plant for photosynthesis.
d. H_2O is leaving the plant.
e. a, b, and c
f. b and d

11. Which pigments are present in green leaves in midsummer?

a. chlorophyll
b. xanthophyll
c. carotene
d. anthocyanin
e. all of the above
f. a, b, and c

12. Why might the bright coloration of a trumpet pitcher (a carnivorous plant) have a different function from that of bright yellow or orange squash blossoms? (Hint: Squash are pollinated by bees).

Use It

13. Describe the "conflict" that plants face with respect to opening and closing their stomata.

14. If a plant could not make chlorophyll, would you expect it to survive? Why or why not?

Apply Your Knowledge

Interpreting Data

15. Scientists carried out an experiment to examine the effect of CO_2 concentrations on plant growth in a semiarid (that is, a dry) grassland environment in Colorado. They set up several plots in the field, consisting of chambers that allowed them to control the concentration of CO_2. One set of plots (A) was kept at ambient CO_2 concentration, and one set (B) was kept at elevated CO_2 concentration (two times the ambient concentration). In mid-July, the total plant mass in each plot set was recorded. The data for three consecutive years are shown in the table. (To establish a baseline, the CO_2 levels in 1996 were not manipulated.)

a. Graph these data.

b. Are there any differences between different plot sets in any given year? If so, describe the differences observed.

c. Are there any differences in the same plot set between years? If so, describe the differences and propose an explanation.

d. What are the implications of this study for grassland productivity (at least in Colorado) with rising CO_2 levels?

Plot set	Average plant mass (g/m^2), 1996	Average plant mass (g/m^2), 1997	Average plant mass (g/m^2), 1998
A (ambient CO_2)	110	108	145
B (elevated CO_2)	112	145	205

Data from Morgan, J. A., et al. (2001). Elevated CO_2 enhances water relations and productivity and affects gas exchange in C_3 and C_4 grasses of the Colorado shortgrass steppe. *Glob Change Biol* 7:451–466.

DRIVING QUESTION 3 How do plants respond to stimuli and protect themselves?

By answering the questions below and studying Infographics 25.8, 25.9, and 25.10, you should be able to generate an answer for this broader Driving Question.

Know It

16. If you wanted a plant to grow very tall, which hormone should you apply?

a. auxin
b. ethylene
c. gibberellins
d. anthocyanin
e. ABA

17. Which plant hormone is responsible for a plant's bending toward light?

a. auxin
b. ethylene
c. gibberellins
d. anthocyanin
e. ABA

Use It

18. Why do seedless grapes need hormone treatment to develop big clusters of big grapes, whereas seeded varieties can develop large fruits without the application of hormones?

19. Nopales are cactus pads (the large, thick, and spiny "paddles" of the prickly pear cactus) and make a delicious salad. What antiherbivory mechanism fails when we succeed in making ensalada de nopales—prickly pear salad?

Mini Case

Apply Your Knowledge

20. The Natural Products Branch of the National Cancer Institute looks for defensive compounds produced by plants, microbes, and marine organisms that may have anticancer activity. Once compounds have been isolated, they can be chemically modified to enhance their activity. Several drugs have come out of this program, including eribulin mesylate, a chemically modified compound originally purified from a sea sponge. This drug has been approved for women with metastatic breast cancer whose disease has not responded to previous treatment. In a clinical trial, patients taking eribulin mesylate had a significantly longer survival time (13.1 months) than did patients in chemotherapy regimens prescribed by their oncologists (10.6 months). (Data are from www.cancer.gov/ncicancerbulletin/041911/page5.)

The first step in developing drugs such as these is to prepare extracts by grinding up the natural product. Design a procedure to test such extracts for anticancer activity. Remember that preclinical trials often use cell lines grown in the laboratory or animal models. Consider what tests you will use to determine if an extract has an anticancer activity, and what variables you will measure and manipulate to find promising candidates to advance along the drug discovery pipeline.

Bring It Home

Apply Your Knowledge

21. Many household and garden plants are toxic to household pets. For example, rhododendrons are toxic to dogs and cats, and oleanders are toxic to cats, dogs, horses, birds, and cows. Do some Internet research to identify at least three other plants that are toxic to household pets or farm animals. For each plant, find both the scientific and common names and which part(s) of the plant is/are toxic to animals. Do any of the plants you identified have medicinal uses for humans?

26 Overview of Animal Physiology

MAN VS MOUNTAIN

Physiology explains a 1996 disaster on Everest

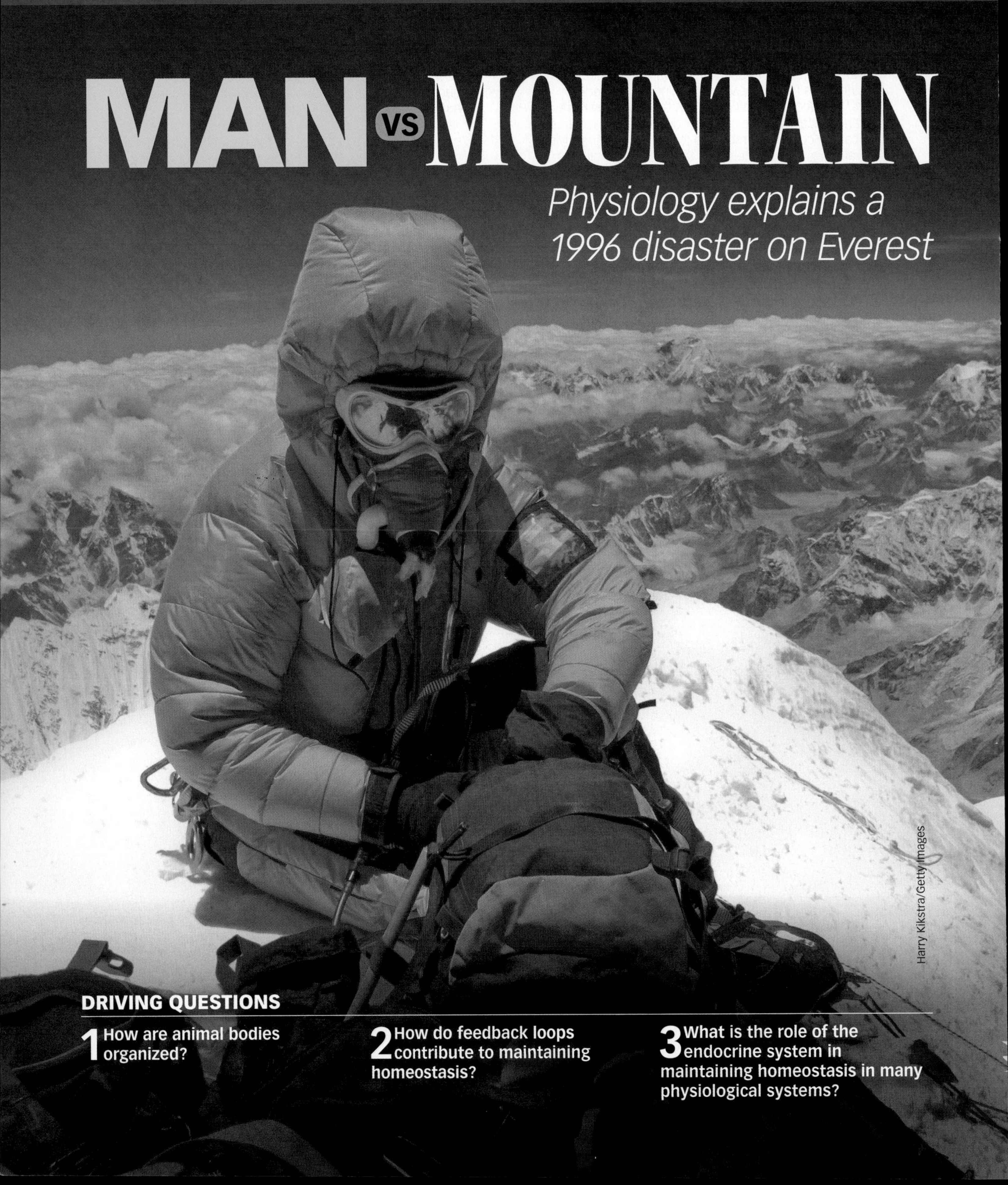

Harry Kikstra/Getty Images

DRIVING QUESTIONS

1. How are animal bodies organized?
2. How do feedback loops contribute to maintaining homeostasis?
3. What is the role of the endocrine system in maintaining homeostasis in many physiological systems?

At 1:17 P.M. on May 10, 1996, Jon Krakauer planted one foot in China, the other in Nepal, and stood on the roof of the world. He was at the highest point above sea level that any human has ever reached—short of standing on the moon. Yet he didn't feel like celebrating. It had taken him 6 long weeks to climb to the top of Mount Everest, and now that he was here his toes ached in the subzero cold, his breath came in short, painful bursts, and his head pounded from the altitude. It was a struggle just to stay upright. "I cleared the ice from my oxygen mask, hunched a shoulder against the wind, and stared absently at the vast sweep of earth below," wrote Krakauer in an account of his climb for *Outside* magazine later that year.

Krakauer's journey to Everest was a lifetime in the making. While other kids were idolizing astronaut John Glenn and baseball pitcher Sandy Koufax, Krakauer's childhood heroes were Tom Hornbein and Willi Unsoeld—two men from his hometown in Oregon who, in 1963, became the first climbers to scale the daunting western ridge of Everest. As a teenager, Krakauer became a skilled climber, vanquishing many of the world's most difficult peaks, and he dreamed of one day climbing Everest himself. By his mid-twenties, though, he had largely abandoned the idea as a boyhood fantasy. But old dreams die hard.

In 1995, Krakauer was working as a journalist when the opportunity came to shadow an Everest climb and report on it for *Outside* magazine. The 42-year-old writer-adventurer jumped at the chance. He would join a team headed by the celebrated climbing guide Rob Hall, whose company, Adventure Consultants, had successfully put 39 amateur climbers on top of Everest. Reaching the summit himself would mean enduring a weeks-long ascent from Base Camp, giving his body time to adjust to the high altitude.

It would also mean risking his life on a daily basis.

The icy tip of Mount Everest sits at 29,035 feet above sea level. For perspective, consider that the cruising altitude of most commercial jetliners is 30,000 feet. A human plucked from sea level and deposited at this altitude would quickly lose consciousness and die from a shortage of oxygen. A climber who has spent weeks adjusting to the altitude can function better at the summit, but not very well, and not for very long. Moreover, Everest is not only the highest place on Earth, but also one of the coldest. At the summit, where wind-chill temperatures average −53°C (−63°F), freezing to death is a real possibility.

Climbers use aluminum ladders to bridge dangerous crevasses in the Khumbu Icefall region during their ascent of Everest.

Andrew Bardon/National Geographic Creative

Despite these dangers—or perhaps because of them—about 150 fearless men and women try to climb Everest every season. And every season, some of them don't come back. At least 19 people died on Everest in 2015 alone. There are many reasons for these disasters—poor training, unforeseen accidents, raw egotism—but among the most important is basic biology: the human body is not equipped to survive at such extreme altitude, and in such extreme temperatures, for long.

The Body as Machine

▶ Anatomical organization of animal bodies

Like a car or a computer, a human body is made up of many parts working together in a coordinated fashion. The parts are organized hierarchically, so that smaller components are organized into increasingly larger units, which are themselves organized into more complex systems. The study of all this intricate hardware is called **anatomy.**

The product of millions of years of evolution, human bodies have an anatomical structure that is impressively well adapted to living in certain environments and performing certain functions. Our species evolved in the hot, flat savannahs of Africa, where environmental conditions favored big brains, opposable thumbs, and bipedal posture—as well as the ability to keep cool (see Chapter 19). As a result, modern humans excel at grasping a pencil or looking through a microscope; we do less well swimming at the bottom of the ocean or living on the highest mountaintops. Fundamentally, that's because of the way we're put together.

For all living things, the smallest anatomical unit is the cell. Human bodies are made up of trillions of cells, each of which can be classified as one of a few hundred different types. Cells, in turn, are organized into **tissues**—groups of specialized cells working together to execute a particular function. Humans and other animals have four different types of tissues—epithelial, connective, muscle, and nervous—that carry out specific tasks in the **organs** of which they are a part. The stomach, for example, is an organ composed of the four types of tissues organized into a compartment for churning and digesting food. At the highest level of organization, organs interact chemically and physically as part of **organ systems.** For instance, as part of the digestive system, the stomach works with the esophagus, small intestine, and liver to digest and absorb food **(INFOGRAPHIC 26.1).**

If the body is like a machine, then physiologists—scientists who study **physiology**—are interested in how this machine keeps running smoothly. Physiologists want to understand how organ systems cooperate to accomplish basic tasks, such as obtaining energy from food, taking in nutrients to build new molecules during growth and repair, and ridding the body of wastes. To the physiologist, the body is an integrated system for processing inputs and outputs and maintaining **homeostasis**—an internal environment that remains relatively stable even when the external environment changes.

ANATOMY
The study of the physical structures that make up an organism.

TISSUE
An organized collection of cells working to carry out a specific function.

ORGAN
A structure made up of different tissue types working together to carry out a common function.

ORGAN SYSTEM
A set of cooperating organs within the body.

PHYSIOLOGY
The study of the way a living organism's physical parts function.

HOMEOSTASIS
The maintenance of a relatively stable internal environment even when the external environment changes.

THERMOREGULATION
The maintenance of a relatively stable internal body temperature.

Balancing Act

▶ Thermoregulation and homeostasis

Like many other animals, humans have an optimal operating temperature and are exquisitely sensitive to temperature changes. Although we can tolerate a wide range of external temperatures in our daily lives, we cannot tolerate large fluctuations in our internal temperature. That's because the enzymes that catalyze the chemical reactions in our body function only within a very narrow temperature range. The body works hard to maintain a relatively constant internal temperature compatible with life, with this temperature fluctuating only about 0.5°C. Through this process of **thermoregulation,** our body temperature is kept at a consistent—and toasty—37°C (98.6°F).

INFOGRAPHIC 26.1

How the Human Body Is Organized

In the human body, the smallest anatomical units, cells, are organized into increasingly complex units, including tissues, organs, and organ systems. These organ systems then work together to allow the organism to function.

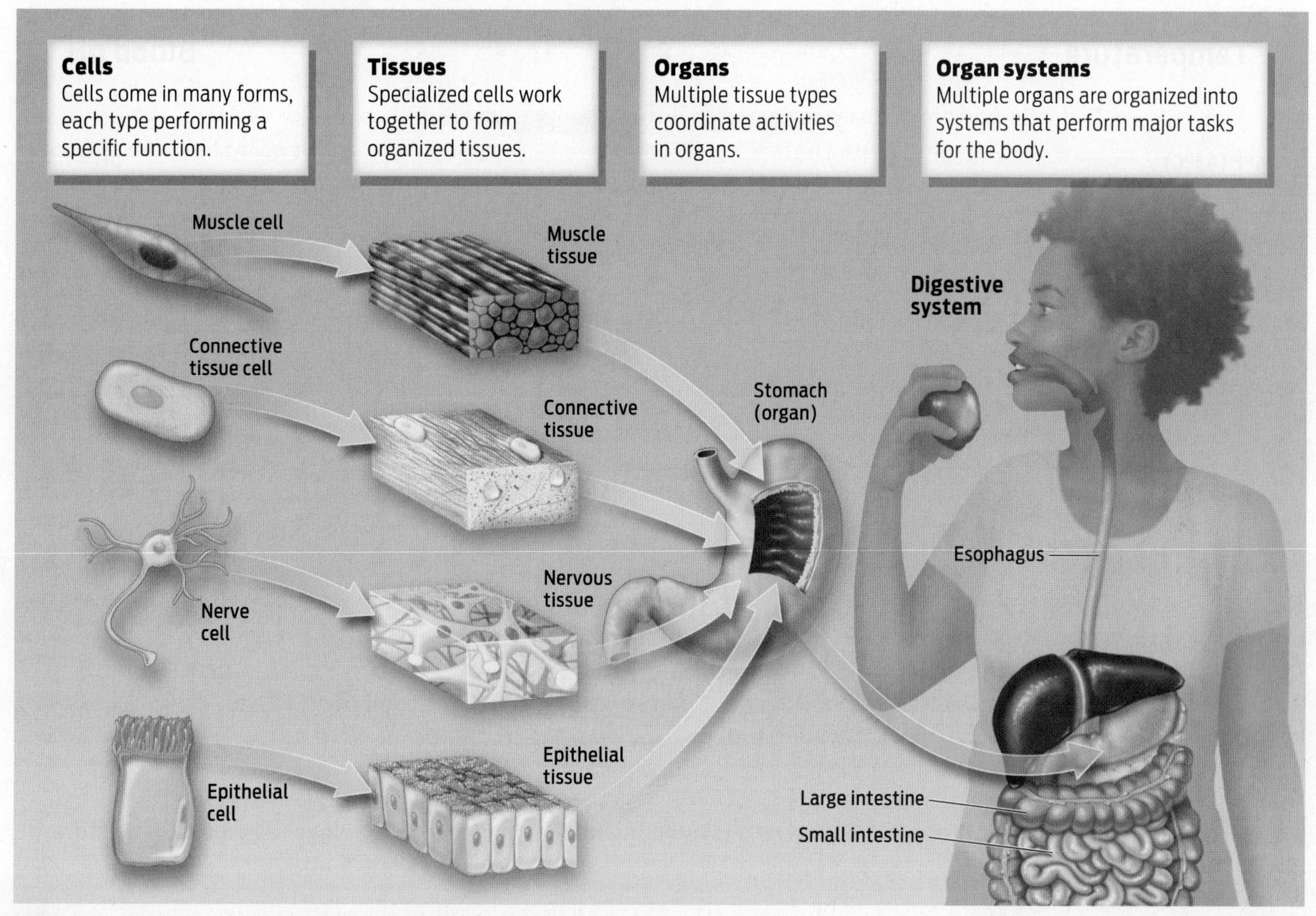

? **What is a primary difference between a tissue and an organ?**

Keeping a consistent body temperature is just one example of how the body maintains homeostasis. "What we're really thinking about with homeostasis are certain set points that your body needs to maintain," explains Robert Kenefick, an exercise physiologist with the U.S. Army Research Institute of Environmental Medicine in Natick, Massachusetts. He notes that the body has a number of such set points for things like temperature, blood pH, and blood pressure, and it works hard to keep these factors balanced within a very narrow range, even in the face of a changing external environment. The consequences of not maintaining this balance can be deadly:

> *"What we're really thinking about with homeostasis are certain set points that your body needs to maintain."*
>
> **—Robert Kenefick**

INFOGRAPHIC 26.2

The Body Works to Maintain Homeostasis

The body expends a great deal of energy to maintain a constant internal environment. Only small fluctuations are tolerated, even in extreme external conditions.

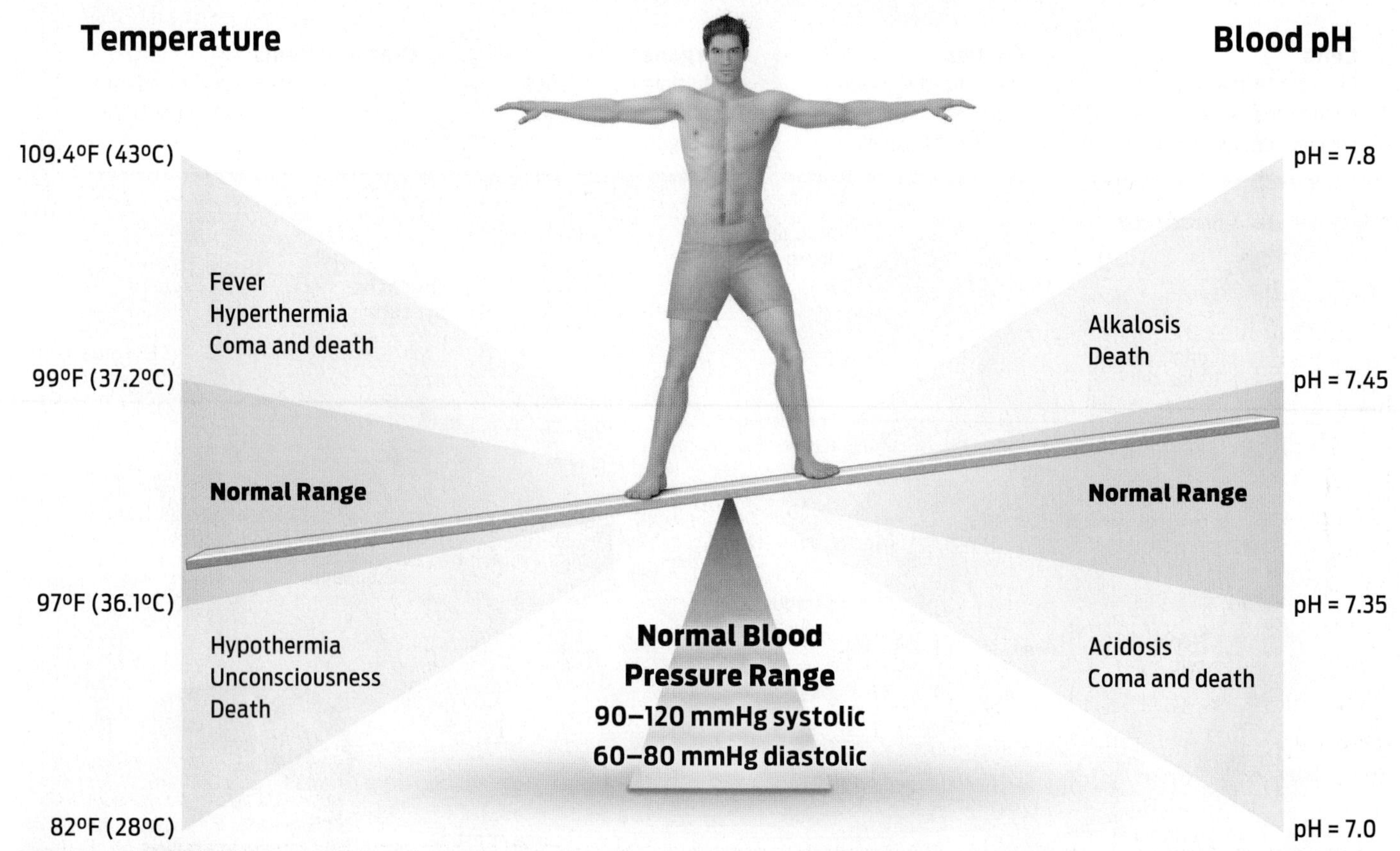

? **Is a body temperature that drops to 36.5°C (97.7°F) within the normal range, or will it trigger a response to restore homeostasis?**

a body temperature increase or decrease of just a few degrees can be lethal **(INFOGRAPHIC 26.2)**.

As an exercise physiologist, Kenefick has spent his career trying to understand how the human body maintains homeostasis during strenuous activities like hiking and running marathons. He works in the Army Research Institute's Thermal and Mountain Medicine Division, where a main focus of his research is understanding how the body performs in extreme cold.

Staying warm is hard to do when ambient temperatures drop below −50°C (−58°F), as they routinely do on Everest. To seal in heat, mountain climbers wear layers of protective gear designed to trap heat, wick away moisture, and insulate their bodies from the wind and cold. When not hiking, climbers consume copious amounts of hot tea or coffee to warm their insides. But insulated clothes and hot beverages would be of little help without the body's natural way of keeping warm.

Human bodies respond to cold in two main ways: by conserving the heat they already have, and by generating more. To conserve heat, the body performs peripheral **vasoconstriction**—a decrease in the diameter of blood vessels just below the surface of the skin. By constricting blood vessels in the skin, peripheral vasoconstriction pushes

VASOCONSTRICTION
The reduction in diameter of blood vessels, which helps to retain heat.

blood from the skin to the body core, where the internal organs are.

"A lot of people believe this is done to increase the amount of blood that goes to your core to help protect those organs," says Kenefick. There's some truth in that, but a more important reason for peripheral vasoconstriction is, as he explains, "to decrease the amount of heat loss from your skin to the environment." Like most things in the universe, heat moves along a gradient from higher to lower. Notes Kenefick, "If the temperature is higher in your skin and lower in the air, then you're going to lose heat to the air. By bringing [these temperatures] closer together, you lose much less heat to the environment." The clamping down of blood flow near the skin surface is the reason hands, feet, and noses are the first to feel cold on a cold day, and it's a sign that your body is trying to retain heat.

Climbers spend weeks moving from camp to camp, acclimatizing to increasingly higher elevations before attempting to reach Everest's summit from Camp IV (C4) on the South Col.

The second way the body responds to cold is by trying to generate more heat. It does this by shivering, the involuntary contraction of normally voluntary muscles. "We know that the by-products of cellular respiration—anytime cells work, and that includes your muscle cells—are CO_2, heat, and water," says Kenefick. "So when your muscles contract through shivering, they [generate] heat, and that heat helps to warm up the core of your body." (While the main product of cellular respiration is ATP, the process is not 100% efficient. Some of the energy stored in food is released as heat, rather than being captured as ATP.)

Of course, to maintain a constant temperature, our bodies must not only keep from getting too cold, but also keep from getting too hot. Two main physiological responses help prevent overheating: peripheral **vasodilation,** the expansion of the diameter of blood vessels, which increases blood flow to the skin; and evaporative cooling, or sweating, which cools the body by releasing heat to the air.

In other words, you have a set point for body temperature: if you get too cold, you vasoconstrict and shiver; if you get too hot, you vasodilate and sweat. A precise balance between the two responses must be maintained to keep tissues healthy. If peripheral

VASODILATION
The expansion in diameter of blood vessels, which helps to release heat.

vasoconstriction goes on for too long, for example, the result is frostbite—the death of tissues caused by lack of blood flow **(INFOGRAPHIC 26.3)**.

Krakauer and his Everest teammates were no strangers to the cold. After weeks of slowly ascending from camp to camp, they reached the launching pad for the summit, the South Col, at 1 P.M. on May 9. "It is one of the coldest, most inhospitable places I have ever been," Krakauer wrote. A windswept saddle of rock and ice that sits between the peaks of Everest and neighboring Lhotse, the Col rests at 26,000 feet above sea level. Climbers pitch their tents on the relatively flat terrain and try not to think about the fact that they have entered what's known as the death zone.

Conditions were particularly bad on the Col that day. Gale-force winds blew through the camp, limiting visibility, and a climb seemed unlikely. But by 11 P.M., the wind had died down and above their heads, the stars appeared, while a gibbous moon reflected off the mountain snow. It was the perfect night for a climb.

INFOGRAPHIC 26.3

Thermoregulation in Response to Cold and Heat

When the outside temperature is cold, the body works to maintain a constant internal temperature by conserving the heat it has and generating additional heat as well. When the outside temperature is hot, the body works to bring heat to the surface and release it into the environment.

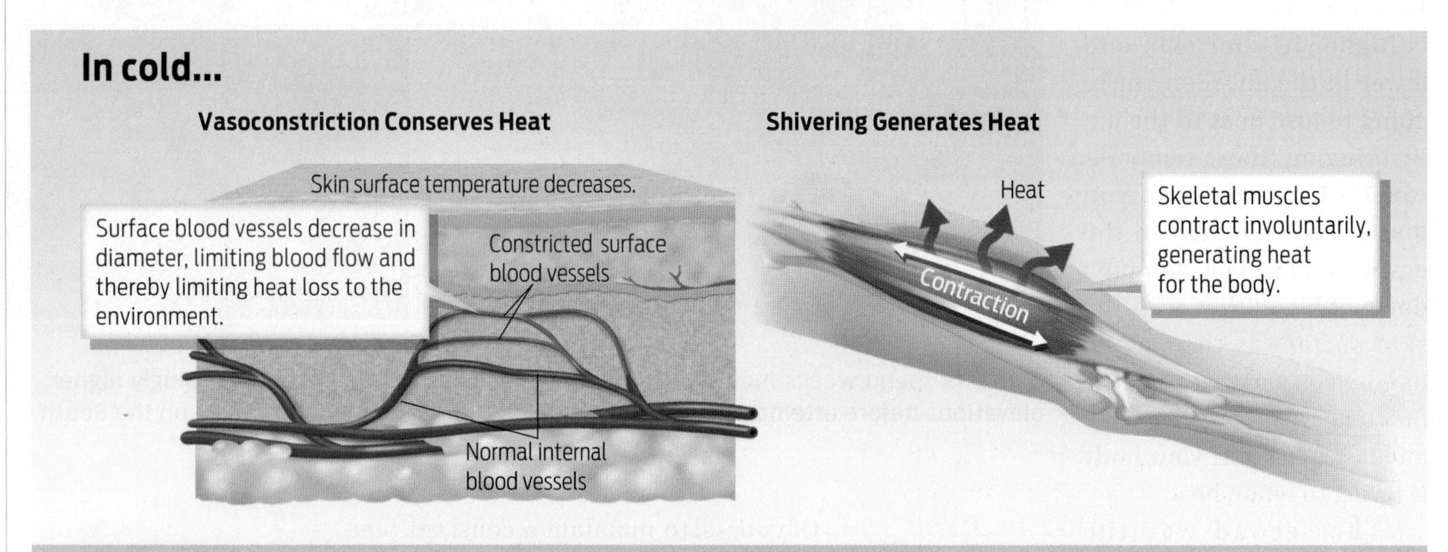

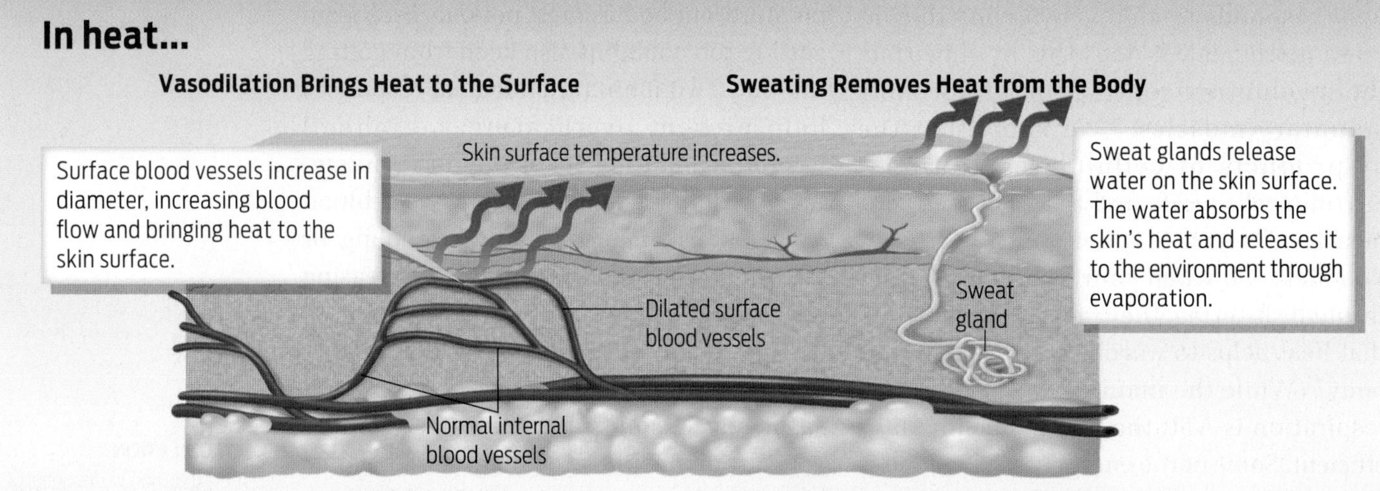

? How does the body generate heat? Is there a way to generate "cool," or do we rely on dissipating heat to reduce body temperature?

"Into Thin Air"

▶ Oxygen and hypoxia

The 15-member team left camp shortly after 11 P.M. Night climbing is necessary on Everest so that the team members will arrive at the most difficult parts of the climb during daylight hours and still have enough time to get back down to camp before nightfall. Krakauer led the pack that night, along with the team's head Sherpa, Ang Dorjee.

The pair reached the Southeast Ridge, the second-to-last stop before the summit, at 5:30 A.M., just as the sun was peering over the eastern peaks. By this time, Krakauer's hands and feet felt like unwieldy blocks of ice, nearly useless in performing the careful work of laying ropes and scaling ice. But it wasn't just the cold he had to deal with. His brain and body were also showing the effects of altitude: "Plodding slowly up the last few steps to the summit," wrote Krakauer in *Into Thin Air,* his 1997 book about the expedition, "I had the sensation of being underwater, of life moving at quarter speed."

At high altitudes, there is lower barometric pressure, which means there are fewer air molecules banging around in the atmosphere—including fewer molecules of oxygen. The percentage of oxygen in the air is the same as at sea level (about 20%), but since there are many fewer air molecules overall, the pressure of oxygen is much less, and therefore fewer oxygen molecules are taken in per breath of air (see Chapter 29). The lower pressure means that fewer oxygen molecules bind to the hemoglobin in blood, which means that blood is less saturated with oxygen, a condition called **hypoxia.** Since all cells require oxygen to function, hypoxia has many bodily consequences. The most serious and immediate occur in the brain. "I've been at 19,000 and 20,000 feet climbing myself," says physiologist Kenefick, "and I can tell you that doing simple tasks like tying your shoes—even though you've tied your shoes many times—is much more difficult." For the climbers on Everest, he says, each step would have been a struggle.

To help cope with conditions of low oxygen, climbers spend about 6 weeks **acclimatizing** their bodies to the conditions, spending a few nights at progressively higher elevations. Their bodies respond by increasing the production of red blood cells, the cells that contain hemoglobin and carry oxygen (see Chapter 9).

The physiological adjustment of acclimatization allows a climber to carry more oxygen than someone coming straight from sea level could. But even well-acclimatized climbers usually need bottled oxygen to reach the summit successfully.

At 1:17 P.M., after more than 12 hours of climbing, Krakauer finally reached the summit of Mount Everest. It was smaller than he expected—a patch of ice the size of a picnic table. Beyond the peaks of the surrounding Himalayas below, endless miles of continent stretched to the horizon.

"Plodding slowly up the last few steps to the summit, I had the sensation of being underwater, of life moving at quarter speed."

—Jon Krakauer

Standing on top of the world, Krakauer was surprised by his lack of elation. He had just cleared a huge personal hurdle, yet the victory felt hollow. Partly, he was too exhausted to truly care: he hadn't slept soundly in more than 50 hours, and the only food he had been able to choke down in 3 days was a bowl of ramen soup and some peanut M&Ms (sleep disturbances and digestive difficulties are additional side effects of high elevation). But another thought lurked in his brain: the oxygen tank he had slung on his back to help him breathe was running low, and he still had to get down the mountain.

"With enough determination, any bloody idiot can get up this hill," guide Rob Hall had famously said. "The trick is to get back down alive." Keenly aware of the clock, Krakauer

HYPOXIA
A state of low oxygen concentration in the blood.

ACCLIMATIZATION
The process of physiologically adjusting to an environmental change over a period of time. Acclimatization is generally reversible.

STR/Getty Images

High traffic on the Hillary Step.

snapped a few perfunctory photos, and within 5 minutes was headed back down the mountain toward Camp IV on the Col.

Fifteen minutes of descent later, he caught a glimpse of an alarming sight: a queue of 20 climbers, from three separate expeditions, waiting to come up a 40-foot wall of rock and ice called the Hillary Step. Getting up the Step requires ropes, so climbers must go up one by one, and on this day there was a traffic jam.

While waiting for his turn to get down the Step, Krakauer peered into the distance and saw something he hadn't noticed before: on the horizon, dark clouds were sweeping in from the south, filling up a corner of what had been a clear blue sky. A storm was brewing.

By this point, it was well past the agreed-upon turnaround time of 1 P.M. set by Hall. The climbers who were still headed up the mountain at this hour were willfully flouting safety rules. Not only that, but weather conditions were getting worse. Snow had started to fall, and it had become hard to see where mountain ended and sky began. The lower Krakauer got on the mountain, the worse the weather became.

Krakauer made it back to Camp IV on the Col just before 6 P.M. The bedraggled climber fell into his tent and quickly passed out. He was delirious, shivering uncontrollably, and exhausted. But he was alive.

HYPOTHALAMUS
A master coordinator region of the brain responsible for a variety of physiological functions.

SENSOR
A specialized cell that detects specific sensory input like temperature, pressure, or solute concentration.

EFFECTOR
A cell or tissue that responds to information relayed from a sensor.

Sensors Working Overtime

▶ Feedback loops and the endocrine system

Even as he slept, Krakauer's body was working hard to thermoregulate. Like many physiological processes, thermoregulation is not something that requires conscious thought. It is more like the automated response of a home heating system, triggered when the thermostat is tripped.

The body's thermostat is the **hypothalamus,** a grape-size structure that sits at the base of the brain, right above the brain stem. The hypothalamus receives signals from many different **sensors,** specialized cells in the body that detect changes in both the internal and external environment. For cold, the major sensors are thermoreceptors in the skin and in the hypothalamus itself. Information from the various sensors is fed to the hypothalamus, which integrates this information and directs an appropriate response.

Acting as a thermostat, the hypothalamus has a specific temperature set point below which a warning message is triggered that body temperature is dropping. When that happens, the hypothalamus "tells" the body to take corrective action. For example, it can send a signal to blood vessels in the skin, causing them to constrict in peripheral vasoconstriction. It can also send a signal to muscles to start shivering. Both signals are sent from the hypothalamus to their target tissues by nerve fibers running from the brain to the rest of the body. The cells, tissues, or organs that respond to such signals are known as **effectors:** they act to cause a change in the internal environment. Once the effectors have raised the body temperature, the sensors detect the changed conditions and the signals are turned off.

INFOGRAPHIC 26.4

Homeostasis Feedback Loops Require Sensors and Effectors

By means of sensors, the body constantly monitors factors like body temperature. The sensors relay temperature information to the hypothalamus. If the temperature is too hot or too cold, the hypothalamus sends signals to effector tissues and organs that work to return the temperature to homeostasis levels.

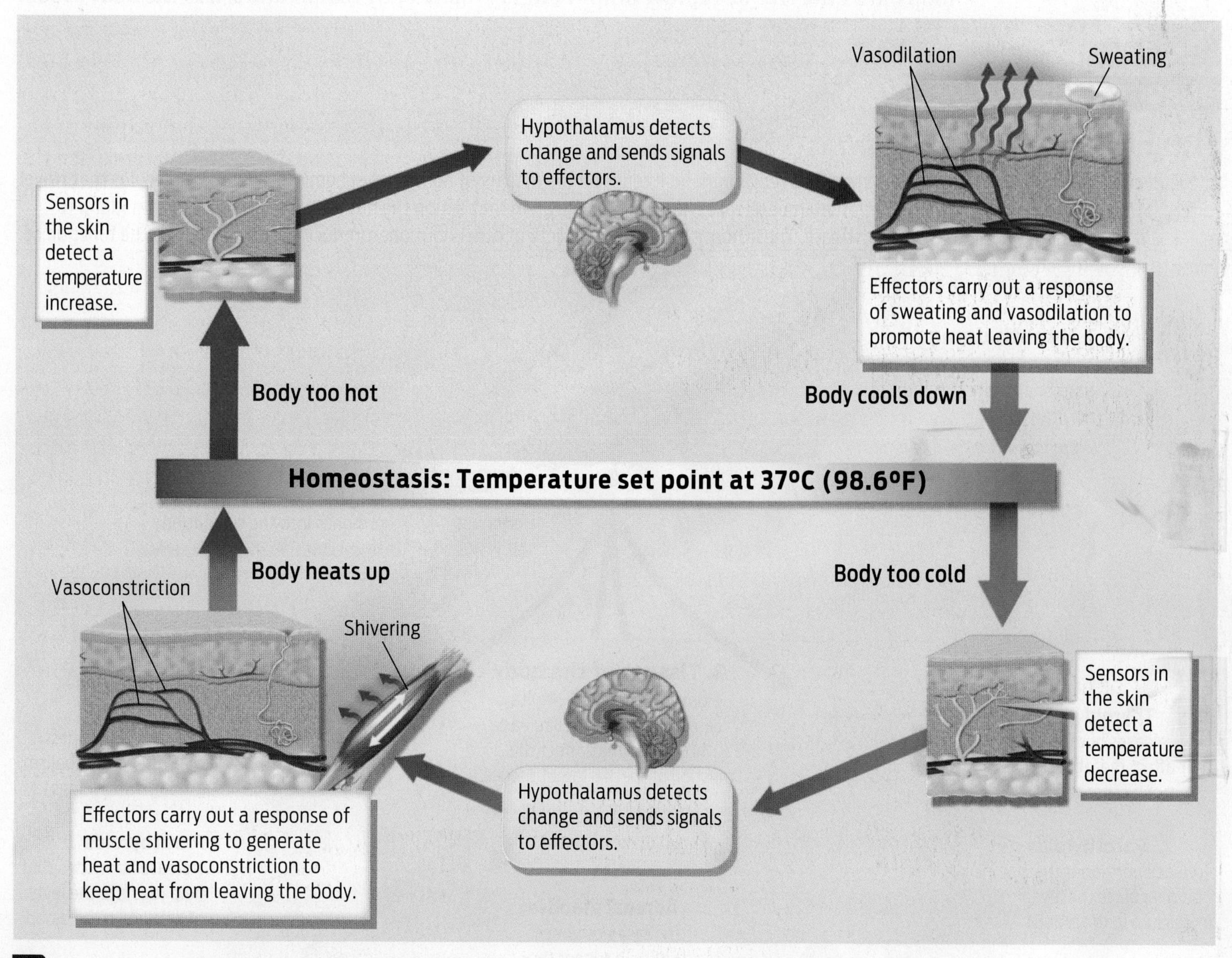

? Identify at least two effectors involved in homeostasis for body temperature.

This circuit of sensing, processing, and responding is an example of a homeostatic **feedback loop.** In this case, the loop is a negative feedback loop because the output of the loop—an increase in body temperature—feeds back to the sensors to *decrease* the response. As body temperature rises, the responses that would further increase body temperature—shivering and vasoconstriction—are no longer needed, so they are turned off. This loop helps keep the system at the set point **(INFOGRAPHIC 26.4).**

Not all feedback loops act in a negative fashion. Positive feedback loops occur when the output of a system acts to *increase* the response of the system. An example is the formation of a blood clot when you cut yourself—a response critical to preventing

FEEDBACK LOOP
A pathway in which the output from an effector feeds back to a sensor and changes further output.

blood loss. Blood platelets stick to damaged blood vessels and release molecules that attract even more platelets to the area, which in turn attract even more platelets, and eventually a blood clot forms. Positive feedback loops are effective at rapidly amplifying a response, but negative feedback loops tend to be more common in physiology because they aim to return the body to its set point, thereby helping to maintain homeostasis.

The hypothalamus does more than regulate body temperature: it is the body's main

INFOGRAPHIC 26.5

The Endocrine System

The endocrine system helps us respond to changes in environment. It relies on the activity of hormones—chemical signaling molecules that travel to target tissues via the circulation. The hypothalamus releases hormones in the brain that act on the pituitary gland, signaling that gland to release additional hormones into the bloodstream. These hormones signal a variety of other hormone-producing glands to affect the function of many tissues in the body.

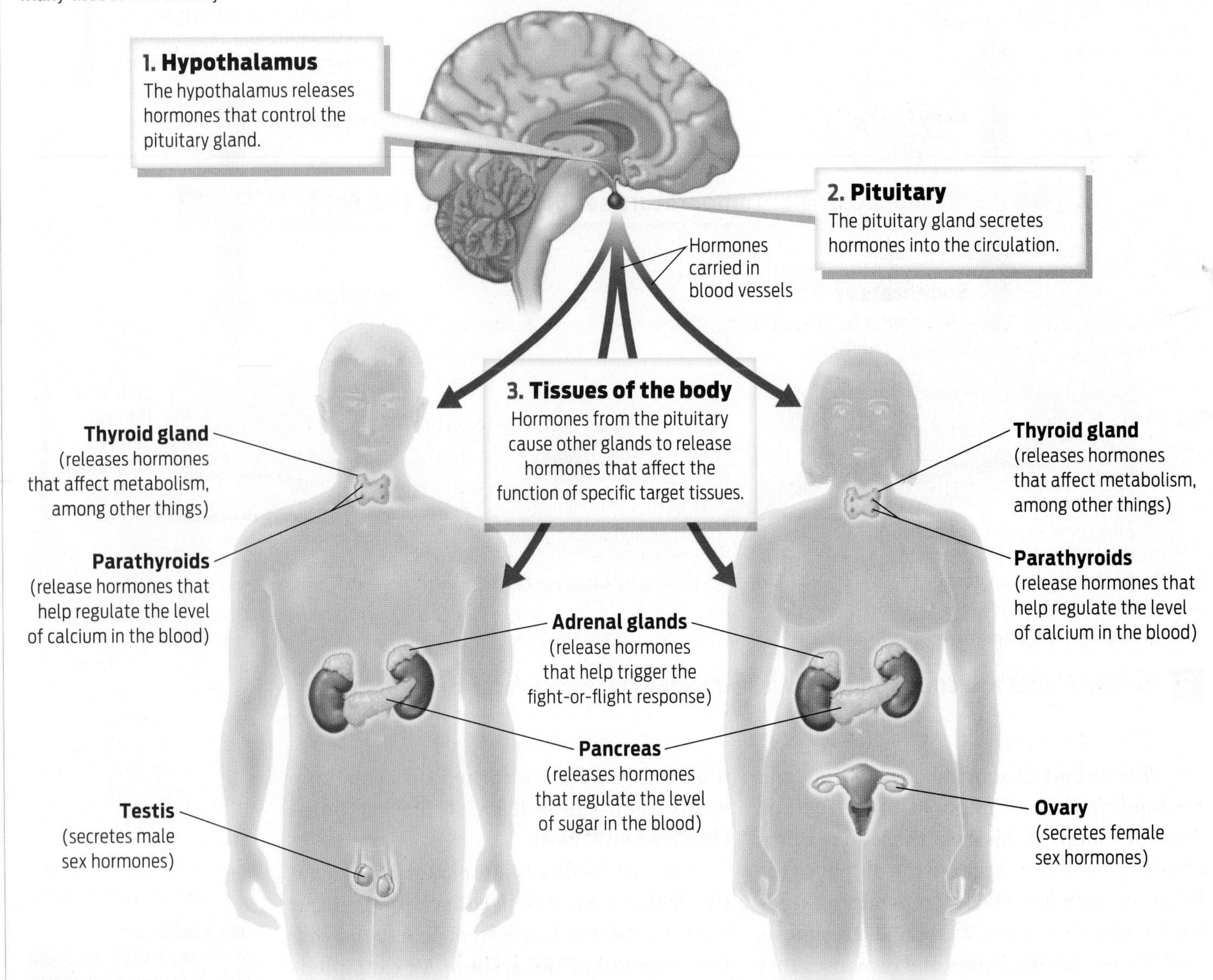

? When hormones are released from the hypothalamus, what organ to do they tend to act on?

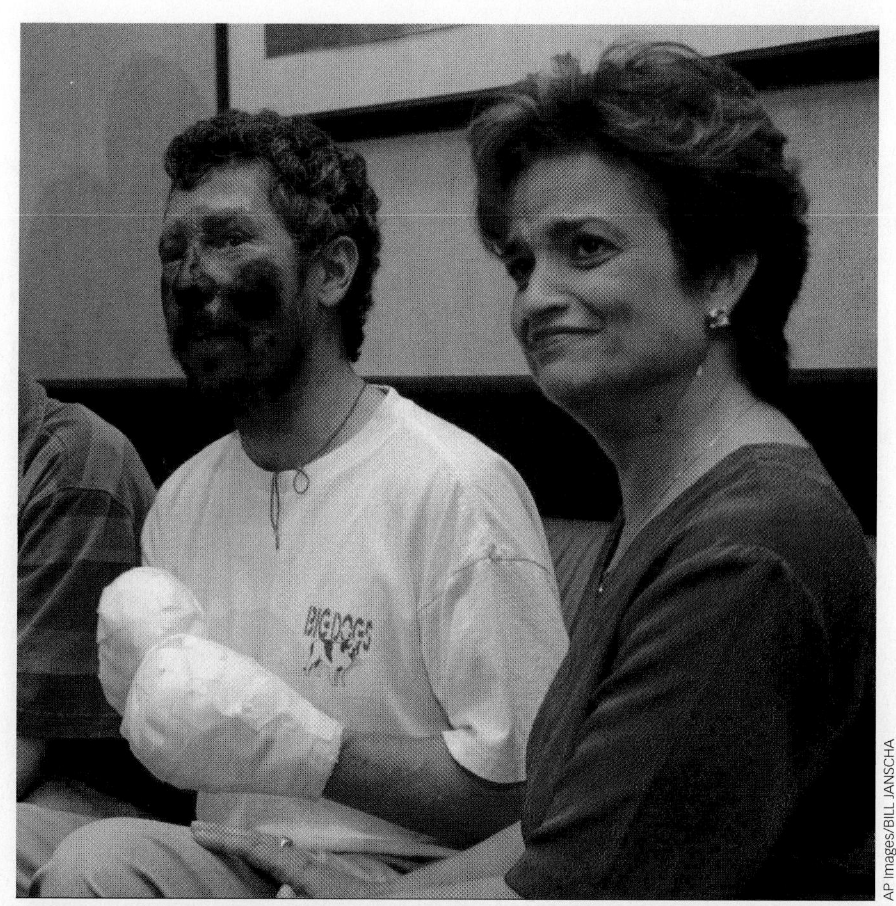

AP Images/BILL JANSCHA

Beck Weathers and his wife, Peach, in 1996, on his return from Everest. He lost his nose and both hands to frostbite.

homeostasis control center, regulating many bodily states, including hunger, thirst, sexual arousal, and sleep. The hypothalamus is part of what Kenefick calls our "lizard brain"—the evolutionarily ancient parts of the brain, which control our most basic physiological responses through unconscious reflexes.

The hypothalamus is able to fill its important role in homeostasis because it is so well connected to sensors and effectors. The hypothalamus is a key part of the **nervous system,** connected to parts of the body through nerves, and it is intimately associated with the **endocrine system,** which produces changes in the body through the action of chemical signaling molecules called **hormones,** which travel through the blood. (Scientists sometimes refer to the nervous and endocrine systems as a collective entity, the neuroendocrine system, since the two systems work so closely together.) The hypothalamus connects to the endocrine system via a direct circulatory connection to the **pituitary gland,** a pea-size structure that sits right below the hypothalamus and is sometimes referred to as the body's "master gland." Hormones released by the hypothalamus travel directly to the pituitary gland, signaling it to release more hormones. These hormones, in turn, travel through the bloodstream and act on many tissues in the body, including other hormone-secreting glands, such as the thyroid gland, adrenal glands, and testes and ovaries **(INFOGRAPHIC 26.5).**

The endocrine system, with its many hormone-secreting glands, is just one of the 11 organ systems in the human body that work together to perform the tasks necessary for survival—ranging from food intake and waste removal to self-defense and

NERVOUS SYSTEM
The collection of organs that sense and respond to information, including the brain, spinal cord, and nerves.

ENDOCRINE SYSTEM
The collection of hormone-secreting glands and organs with hormone-secreting cells.

HORMONE
A chemical signaling molecule that is released by a cell or gland and travels through the bloodstream to exert an effect on target cells.

PITUITARY GLAND
An endocrine gland in the brain that secretes many important hormones.

INFOGRAPHIC 26.6 UP CLOSE

Organ Systems

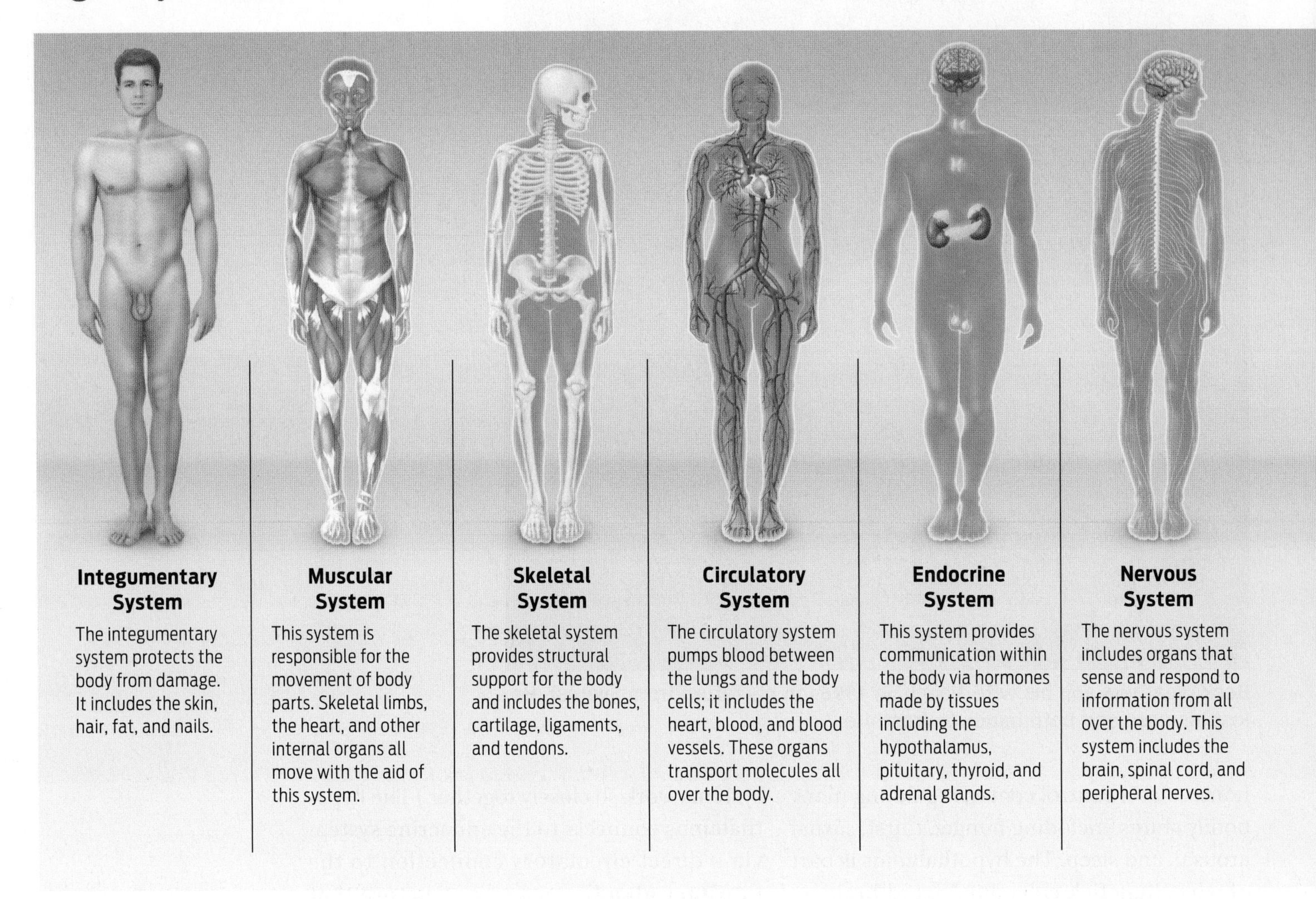

reproduction (see **INFOGRAPHIC 26.6** and subsequent chapters in this unit).

During the night, Krakauer was awakened by a teammate who gave him grave news: a number of his teammates, including Rob Hall, had not yet returned to Camp IV. They were still out in the biting subzero cold somewhere above 26,000 feet. Krakauer's heart sank. He knew the chances of surviving in the cold for that long were slim. By 5 P.M., everyone's oxygen tank would have been empty. It was now midnight. Krakauer feared for the others' lives. But he was also dumbfounded: Hall and the rest of his team had not been far behind him on the mountain. What had gone wrong?

The storm he had spotted on the horizon had gained in ferocity and strength as it climbed up the mountain. "One minute, we could look down and we could see the camp below. And the next minute, you couldn't see it," recalled Lou Kasischke, a member of Krakauer's team, who was one of 11 people trapped on the Col when the storm hit and who recounted his experience in the PBS documentary *Storm over Everest*. "Within the space of 5 minutes, it changed from really a good day with a little bit of wind to desperate conditions, something I'd never experienced the ferocity of before," said John Taske, another member of Krakauer's team, on the same PBS program.

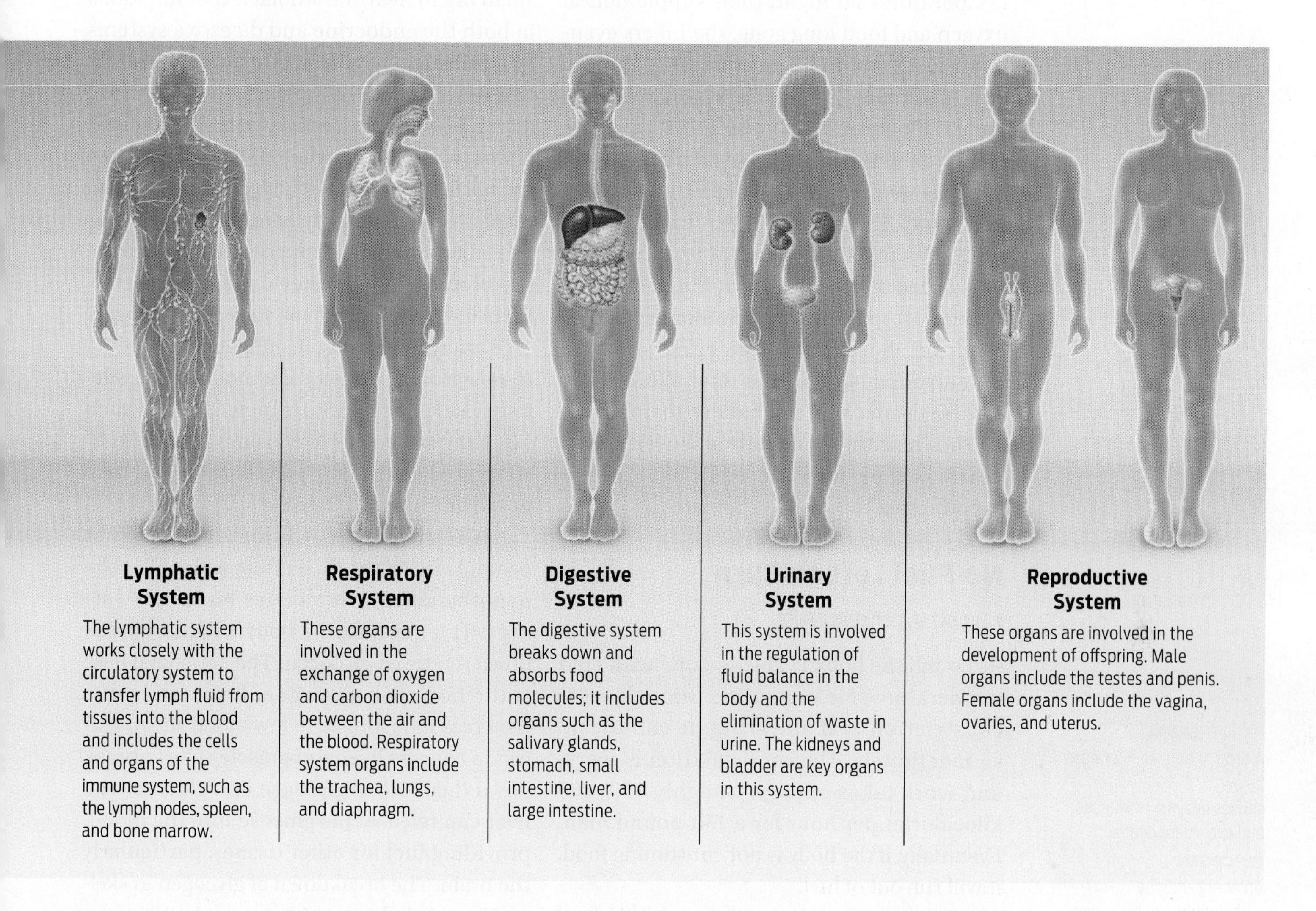

According to Kent Moore, a physics professor at the University of Toronto who has studied the disaster, the storm that hit Everest that day also caused a particularly severe drop in barometric pressure, greatly reducing the availability of oxygen. The sudden drop in pressure may have caused severe physiological distress in the climbers. In particular, they would have experienced the mental side effects of extremely low oxygen levels, which include confusion and disorientation.

Unable to tell in which direction they were going, and not wanting to take a wrong turn and step off a cliff, the climbers were forced to hunker down in the hurricane-force winds and wait for the storm to abate. Eventually, after 4 long hours, the clouds parted long enough for them to see where they were. Six climbers who were able to walk made it back to camp during a brief lull. An additional three were brought back safely through the efforts of Anatoli Boukreev, a Russian guide who, having descended to Camp IV, went back to search for them.

But others were not so lucky. Two climbers, too weak to make it back to camp, suffered severe frostbite before being rescued. One, Krakauer's teammate Beck Weathers, had to have both hands and his nose amputated. Those climbers stuck higher on the mountain—including Hall—could not be rescued. Trapped without shelter in the subzero temperatures all night, their supplemental oxygen and food long gone, the hikers eventually lost their ability to cope with the cold and succumbed to **hypothermia,** a drop in body temperature below 35°C (95°F). In all, eight climbers died on Everest that day.

This was not the first time that disaster had struck the mountain's summit. A 2008 study of all reported Everest deaths from 1921 to 2006 led by researchers at Massachusetts General Hospital found that more than 80% occurred above 26,000 feet, either during a summit attempt or the day after. While many of these deaths were attributable to traumatic injuries resulting from falls and avalanches, nearly as many were caused by hypoxia and hypothermia.

No Fuel Left to Burn

▸ Blood sugar regulation

Although the body is able to cope with cold temperatures for some time through vasoconstriction and shivering, it cannot do so indefinitely. Thermoregulation is work, and work takes energy—roughly 150–300 kilocalories per hour for a 150-pound man. Eventually, if the body is not consuming food, it will run out of fuel.

The main fuel the body uses in times of intense activity is the sugar glucose, a breakdown product of carbohydrate digestion (see Chapter 4). When we eat carbohydrates, sugars are released and absorbed into the circulation, and blood sugar (the concentration of glucose in the blood) increases. Some of this sugar may be used immediately as fuel for aerobic respiration in cells of the body (see Chapter 6). Whatever is not needed right away is converted into **glycogen,** a carbohydrate made up of linked glucose molecules that is stored in muscles and the liver.

Blood sugar is maintained within a relatively narrow range—too low and there is insufficient fuel for cells; too high and complications can ensue. Blood sugar is monitored and controlled by the **pancreas,** a small organ near the stomach that functions in both the endocrine and digestive systems (see **Milestone 7: Stumbling on a Cure**). In response to high blood sugar (for example, after a person eats a carbohydrate-rich meal), endocrine tissue in the pancreas produces the hormone **insulin,** which acts on liver and muscle cells, signaling them to remove sugar from the blood. The glucose taken up from the blood will be either used immediately for cellular respiration or stored as glycogen. Like many hormones, insulin acts by binding to **receptors** on target cells, much as a key fits into a lock. Receptors are molecules to which signaling molecules like insulin bind to "tell" a targeted cell what to do (in this case, take up sugar from the blood).

When blood sugar is low, the body first prompts us to eat by sending a signal to the hypothalamus, which cues hunger. If eating isn't an option, the body begins to break down its stored glycogen. The key signal here is the hormone **glucagon.** Released by the pancreas in response to low blood sugar, glucagon triggers liver and muscle cells to break down their stored glycogen into glucose. The liver can release this glucose into the blood, providing fuel for other tissues, particularly the brain. The breakdown of glycogen in skeletal muscle helps provide energy for the muscle itself.

The amount of glucose in the blood is therefore tightly regulated: when blood-sugar levels are high, excess glucose is stored in cells as glycogen (in response to insulin). The stored glycogen represents a source of glucose to be released into the blood during periods of starvation (in response to glucagon). The opposing effects of insulin (reducing blood glucose) and glucagon (increasing blood glucose) illustrate another example of homeostasis, in this case maintaining a relatively stable blood-sugar level **(INFOGRAPHIC 26.7).**

The trapped climbers hadn't eaten in hours, which meant they were operating on stored energy. Glycogen is the main stored

HYPOTHERMIA
A drop of body temperature below 35°C (95°F), which causes enzyme malfunction and eventually death.

GLYCOGEN
An energy-storing carbohydrate found in liver and muscle.

PANCREAS
An organ that secretes the hormones insulin and glucagon as well as digestive enzymes.

INSULIN
A hormone secreted by the pancreas that causes a decrease in blood sugar.

RECEPTOR
A molecule on or in a cell that binds to a specific signaling molecule, allowing the signaling molecule to exert an effect on that cell.

GLUCAGON
A hormone secreted by the pancreas that causes an increase in blood sugar.

INFOGRAPHIC 26.7

The Pancreas Regulates Blood-Glucose Levels

The pancreas responds to variation in blood-glucose levels by secreting insulin, inducing cells to take in excess glucose, or by secreting glucagon, inducing liver and muscle cells to release stored glucose.

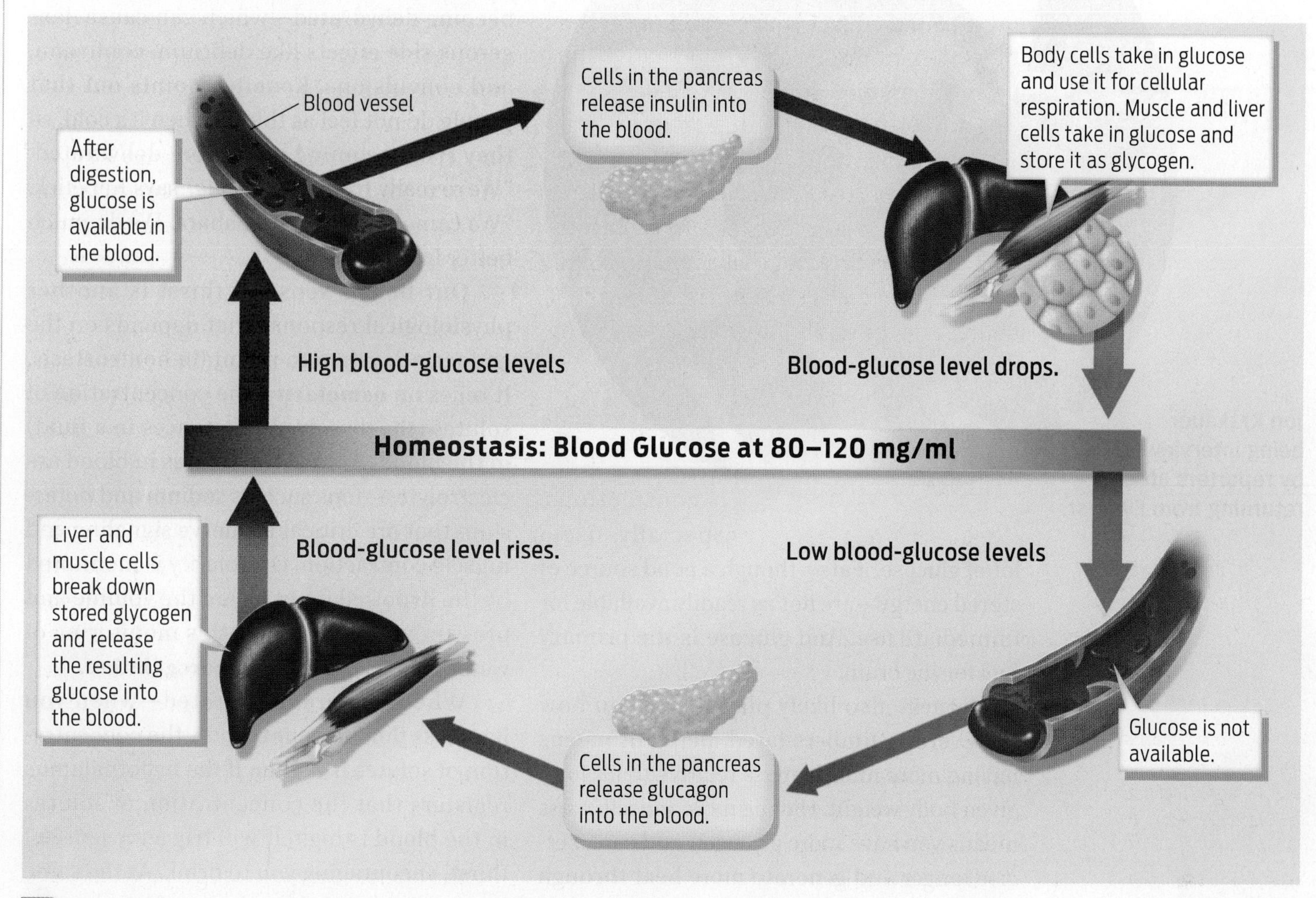

? What effects do insulin and glucagon have on blood sugar?

fuel that is tapped during vigorous exercise. But the human body can store only so much glycogen. Eventually, after hiking and shivering for many hours, you will exhaust this fuel supply. Without this fuel, your body will struggle to remain active and shiver. And if you can't remain active and shiver, then you can't generate heat and your body temperature will fall. That's when hypothermia can set in.

The average adult has enough stored glycogen to power about 12 to 14 hours of routine activity. When a person is exercising strenuously—say, running or hiking—glycogen stores can be depleted in as little as 2 hours. Marathon runners often refer to this point, which occurs at about mile 20, as "hitting the wall." To continue exercising beyond this point, you must eat something—preferably something with carbohydrates.

"A lot of mountaineering communities think you need fat," says Kenefick. "And that's true—fat has more calories per gram—9 kcals per gram compared to 4 kcals per gram of protein or carbohydrate. But when you're doing things like shivering, those types of

Devendra M. Singh/Getty Images

Jon Krakauer being interviewed by reporters after returning from Everest.

contractions, especially, use a lot of glucose." Fats—though a good source of stored energy—are not as readily available for immediate use. And glucose is the primary fuel for the brain.

Fitness also likely played a role in how the Everest climbers fared. Being fit means having more muscle mass relative to fat for a given body weight. Having more muscle mass means you have more glycogen and can exercise longer and generate more heat through cellular respiration. Someone who is less fit, or who simply has less muscle mass, will tire sooner, need to sit down and rest, and continue to lose heat to the environment. This is likely what happened to the climbers who were too weak to hike back to camp: their glycogen stores ran out sooner than those of other climbers.

OSMOLARITY
The concentration of solutes in blood and other bodily fluids.

OSMOREGULATION
The maintenance of relatively stable volume, pressure, and solute concentration of bodily fluids, especially blood.

KIDNEY
An organ involved in osmoregulation, filtration of blood to remove wastes, and production of several important hormones.

Triggering Thirst

▸ Water balance

Another exacerbating factor would have been dehydration—a little-known cold-weather risk. In cold conditions, our bodies must work harder under the extra weight of heavy clothing they carry, and sweat evaporates quickly in cold, dry air. We also lose a significant amount of water as water vapor when we exhale. The human body is about 65% water by weight, and when the total amount of water drops by only a few percent, we become dehydrated—which can cause dangerous side effects like delirium, confusion, and convulsions. Kenefick points out that people do not feel as thirsty when it's cold, so they risk becoming even more dehydrated. "We're really tropical animals," says Kenefick. "We came from the sub-Sahara. We do much better in the heat."

Our body's sense of thirst is another physiological response that depends on the endocrine system to maintain homeostasis. It relies on **osmolarity,** the concentration of solutes (the dissolved substances in a fluid) in the blood. Among the solutes in blood are electrolytes—ions such as sodium and potassium that are critical for nerve signaling and muscle contraction. Osmolarity is monitored by the hypothalamus, as are the volume and pressure of bodily fluids; this monitoring of water balance is called **osmoregulation.**

When you are dehydrated—when you have less fluid in your blood—the concentration of solutes is higher. If the hypothalamus registers that the concentration of solutes in the blood is high, it will trigger a sense of thirst, encouraging you to drink. At the same time, it triggers the release of a hormone called antidiuretic hormone (ADH) from the pituitary, which travels through the bloodstream and acts on the **kidneys.** ADH signals the kidneys to excrete less water in the urine. By reducing the amount of water lost in urine, ADH causes more water to be reabsorbed by the kidneys back into the bloodstream. Water in the bloodstream dilutes solutes and lowers osmolarity. That's why people who are dehydrated have darker urine—it contains less water and so is more highly concentrated.

Osmoregulation also depends on sensors that detect changes in blood volume and pressure. Sensors in the heart, for example, sense how full the heart's chambers are; sensors in blood vessels sense how stretched the vessels are. When low blood volume and pressure are detected, the hypothalamus again responds by triggering the release of ADH from the pituitary into the blood; this ADH then acts on the kidneys to help them return water to the bloodstream **(INFOGRAPHIC 26.8)**.

With these multiple sensors for detecting dehydration, why do we feel less thirsty in the cold? The reason, says Kenefick, is that peripheral vasoconstriction pushes blood toward the core. All that blood pushed centrally is sensed by the body as a normal amount of hydration. As a result, the sensation of thirst is reduced, despite the fact that you're dehydrated. This is why it's very important to drink adequate amounts of water in winter, even when you aren't thirsty.

"Because water plays such a large role in cellular function," says Kenefick, "being dehydrated is going to put a greater stress on your body." Dehydration can alter the concentrations of electrolytes in the blood, which then alter nerve function and muscle contraction. Dehydration also lowers blood pressure, and thus makes the heart work harder. Together, these effects can have dangerous consequences, impairing thinking and

INFOGRAPHIC 26.8

The Kidneys Respond to Changes in Water Balance

The amount of water in the circulation controls the concentration of dissolved molecules in the blood and also determines blood volume and blood pressure. The kidneys control water availability by responding to a variety of signals.

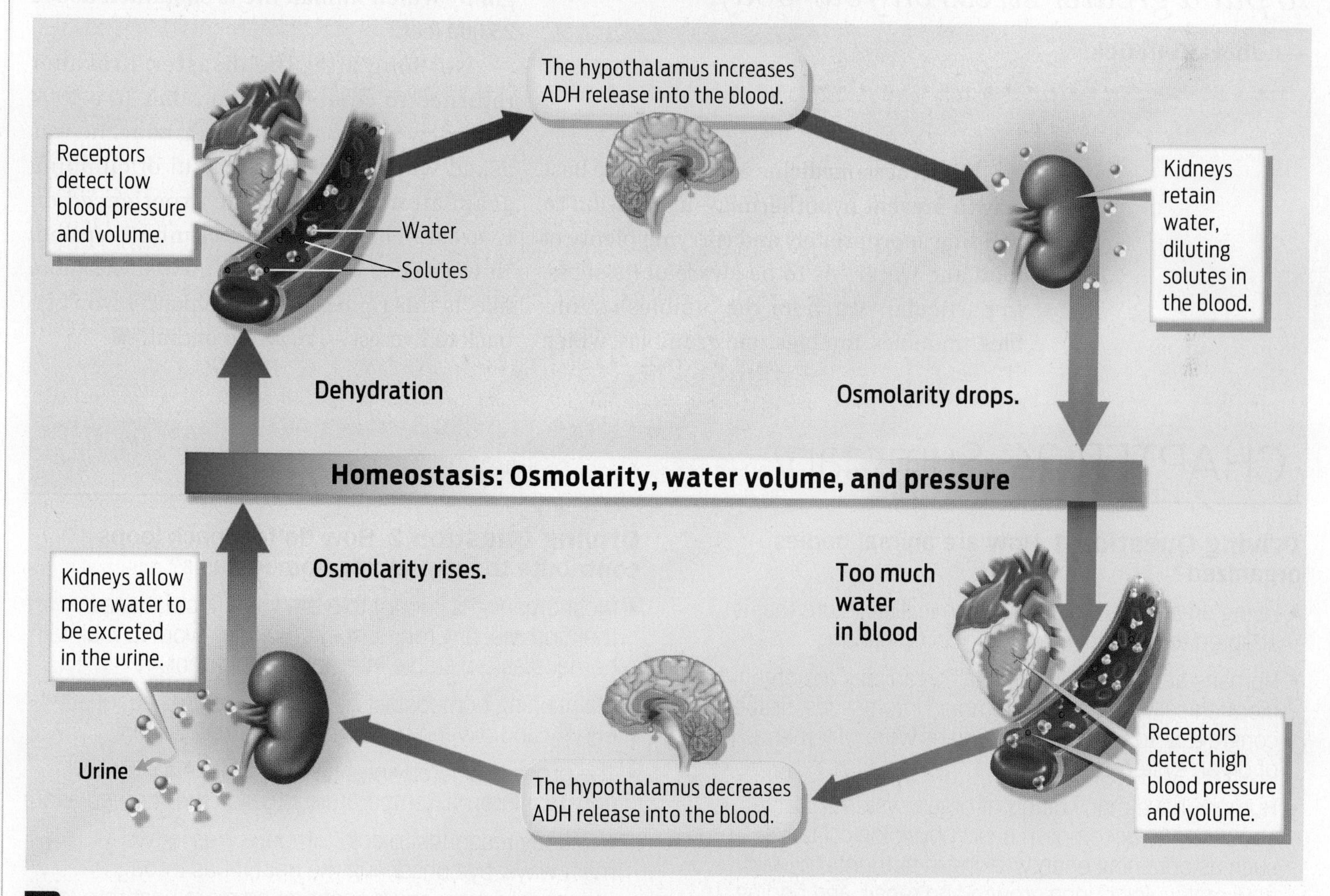

? **Where is ADH made, what organ does it act on, and what impact does it have on its target organ?**

coordination—two things that matter a great deal when you're navigating the treacherous terrain of the world's tallest mountain during a blizzard.

Warning Signs

Hypothermia isn't only a danger for death-defying mountain climbers: it's a leading cause of death during outdoor recreation like rafting and skiing, and is the number one way to lose your life while outdoors in cold weather. The Centers for Disease Control and Prevention estimates that hypothermia causes more than 1,000 deaths each year in the United States.

"Because water plays such a large role in cellular function, being dehydrated is going to put a greater stress on your body."

—Robert Kenefick

Wilderness-medicine experts say the best way to prevent hypothermia—in addition to dressing appropriately and carrying plenty of food and water—is to be aware of its signs. In particular, watch for the "umbles": stumbles, mumbles, fumbles, and grumbles, which show changes in motor coordination and altered brain function. If you experience any of these signs in cold conditions, it's time to seek shelter.

None of the climbers who died on Everest in 1996 was an inexperienced climber—three of them, in fact, were professional guides. Why didn't they heed these physiological warning signs? Part of the reason is that there was simply no time. The swift-moving storm made the decision for them. But the climbers had also earlier made questionable choices that affected their fate. Whether from over-confidence or brain-addled thinking, they continued climbing toward the summit even when the hour was late. In the end, the climbers made a fatal wager with biology: in their race to the summit, they pushed themselves beyond the breaking point, overestimating, in Krakauer's words, "the thinness of the margin by which human life is sustained above 25,000 feet."

Not long after the disaster, Krakauer returned to climbing mountains. In a 1997 interview with *Bold Type* magazine, he was asked whether he was fearful of climbing again after the trauma he experienced on Everest. The chastened climber replied: "It wasn't like 'Am I afraid of this?' It was more like 'Is this right? Is it too selfish?' I won't go back to Everest—I'm afraid of that." ■

CHAPTER 26 Summary

Driving Question 1 How are animal bodies organized?

- Living organisms have an anatomical structure that is adapted to suit their physiological functions.
- Humans and other multicellular organisms are organized hierarchically: cells are assembled into tissues; tissues congregate to form organs; organs work together as part of organ systems.
- Humans have many different organ systems that cooperate to accomplish basic physiological tasks, such as obtaining energy, taking in nutrients to build new molecules during growth and repair, and ridding themselves of wastes.

Driving Question 2 How do feedback loops contribute to maintaining homeostasis?

- Most organisms cannot tolerate wide fluctuations in their internal environment; their bodies work to maintain homeostasis, a stable internal environment.
- Maintaining homeostasis is work; it requires adequate energy and oxygen to power cellular respiration.
- The process whereby organisms maintain a relatively constant internal temperature is called thermoregulation.
- The body responds to cold temperatures in two main ways: by conserving the heat it has through vasoconstriction, and by generating more heat through

shivering. When overheated, the body releases heat by vasodilation and sweating.

- Maintaining homeostasis requires both sensors and effectors. Sensors include nervous system receptors that detect changes in a variety of internal states (for example, temperature and blood pressure). Effectors include the endocrine glands and muscles that respond to an abnormal state in an effort to correct it.
- Sensors and effectors work together as part of a circuit or feedback loop. Negative feedback loops are important in homeostasis.
- Osmoregulation is the control of water balance in the body. Sensors detect blood pressure, blood volume, and solute concentration.

Driving Question 3 What is the role of the endocrine system in maintaining homeostasis in many physiological systems?

- The endocrine system produces hormones—chemical messengers that travel through the bloodstream, bind to receptors on a target cell, and effect a change in that cell. Insulin and glucagon are hormones that regulate blood-glucose levels.
- The hormone ADH contributes to the maintenance of water balance by the kidneys.

More to Explore

- Krakauer, J. (1997). *Into Thin Air: A Personal Account of the Mt. Everest Disaster*. New York: Random House.
- *Everest* [Film]. (2015). Universal Pictures. Based on Weathers, B. (2002). *Left for Dead: My Journey Home from Everest*. New York: Villard Books.
- PBS. (2008). *Frontline: Storm over Everest*: www.pbs.org/wgbh/pages/frontline/everest/.
- Firth, P. G., et al. (2008). Mortality on Mount Everest, 1921–2006: descriptive study. *British Medical Journal* 337:a2654.
- Moore, G. W. K., et al. (2006). Weather and death on Mount Everest: an analysis of the *Into Thin Air* storm. *Bull Am Meteorol Soc* 87:465–480.

CHAPTER 26 Test Your Knowledge

Driving Question 1 How are animal bodies organized?

By answering the questions below and studying Infographics 26.1 and 26.6, you should be able to generate an answer for this broader Driving Question.

Know It

1. Compare and contrast anatomy and physiology.

2. Organize the following terms on the basis of level of structure, from the simplest (1) to the most complex (4).

_____Small intestine
_____Mucus-secreting cell of the small intestine
_____Digestive system
_____Layer of muscle that contributes to the function of the small intestine

3. Which of the following groups is in the correct order of organization from most inclusive level to lowest level?

a. tissues, cells, organ systems, organs
b. organ systems, organs, tissues, cells
c. cells, organ systems, tissues, organs
d. cells, tissues, organs, organ systems
e. cells, organs, organ systems, tissues

Use It

4. An emergency room doctor setting a complex bone fracture is relying primarily on knowledge of

a. anatomy.
b. physiology.
c. thermoregulation.
d. homeostasis.
e. osmoregulation.

5. Is a personal trainer who works with clients to help them lose weight through a combination of diet and exercise focusing primarily on anatomy or physiology? Explain your answer.

6. Why is the heart considered an organ and not a tissue?

Driving Question 2 How do feedback loops contribute to maintaining homeostasis?

By answering the questions below and studying Infographics 26.2, 26.3, 26.4, 26.7, and 26.8, you should be able to generate an answer for this broader Driving Question.

Know It

7. What is homeostasis?

8. Describe the feedback loop involved in thermoregulation in cold conditions. Use the following terms in your answer: hypothalamus, sensor, muscle, effector, low body temperature, normal body temperature.

9. What internal signals are associated with dehydration?
   a. high osmolarity
   b. low blood volume
   c. low blood pressure
   d. all of the above
   e. a and b

10. People who are severely dehydrated produce ____________ of urine that is ____________.
    a. a high volume; highly concentrated and dark in color
    b. a high volume; dilute and light in color
    c. a low volume; highly concentrated and dark in color
    d. a low volume; dilute and light in color
    e. a normal volume; a normal color (neither very light nor very dark)

11. Tibetan Sherpas, many of whom serve as guides and rescuers on Everest, often do not require bottled oxygen to reach the summit. Why might Tibetans, who have lived at high elevations for many generations, have an easier time than others with hypoxia? (Think about both short-term and long-term changes.)

Use It

12. How could damage to the hypothalamus prevent shivering even if the core body temperature drops dramatically?

13. Name two or three physiological responses that could help the body dissipate heat during exertion on a hot day. For each mechanism that you propose, explain how it would dissipate heat.

Driving Question 3 What is the role of the endocrine system in maintaining homeostasis in many physiological systems?

By answering the questions below and studying Infographics 26.5, 26.7, and 26.8, you should be able to generate an answer for this broader Driving Question.

Know It

14. What is the target of hormones released by the hypothalamus?
    a. the ovaries
    b. the pancreas
    c. the pituitary
    d. the thyroid gland

15. Which hormone regulates water conservation by the kidneys?
    a. insulin
    b. ADH
    c. glucagon
    d. ADH and insulin work together

16. Insulin is released from the ____________ in response to ____________.
    a. pancreas; elevated blood sugar
    b. pancreas; low levels of blood sugar
    c. hypothalamus; elevated blood sugar
    d. hypothalamus; low levels of blood sugar
    e. pituitary; signals from the hypothalamus

17. Describe the feedback loop active in maintaining water balance in a person who has become dehydrated. Use the following terms in your answer: blood pressure, blood volume, hypothalamus, ADH, effector, kidney.

Use It

18. What conditions might cause high levels of insulin in the bloodstream? What events would follow?

19. Glucagon is released as part of the response to a drop in body temperature. Why do you think this happens?

Apply Your Knowledge

Mini Case

20. A 65-year-old woman has been diagnosed with hypertension (high blood pressure). She needs medication that will return her blood pressure to normal (and safe) levels. Her doctor tells her that there are two main categories of drugs for hypertension: thiazides, a type of diuretic, and angiotensin-converting enzyme (ACE) inhibitors, which help relax blood vessels and prevent their constriction. Explain how both of these could reduce blood pressure.

Interpreting Data

Apply Your Knowledge

21. Jonas and Jennifer have abnormal fasting glucose levels in their blood. They each had a blood test to measure the levels of insulin in their blood (also measured after fasting). Their blood values for both glucose and insulin are shown in the table. Based on these data, is Jonas or Jennifer more likely to have type 2 diabetes (which is characterized by an inability of cells to respond to insulin)? Which of them is more likely to have type 1 diabetes (which results from a failure of insulin production)? Explain all their blood test results.

	Fasting glucose level (mg/dL)	**Fasting insulin level (international units, μIU/mL)**
Jonas	115	1
Jennifer	119	35
Normal range	70–100	5–25

Bring It Home

Apply Your Knowledge

22. The U.S. National Park Service frequently has to rescue stranded hikers, often at great expense. Do you think that hikers' level of preparation should be a factor in determining whether they should bear the cost of their rescue? What factors would you consider to determine whether a hiker was adequately prepared? Give a physiological reason for each factor that you propose.

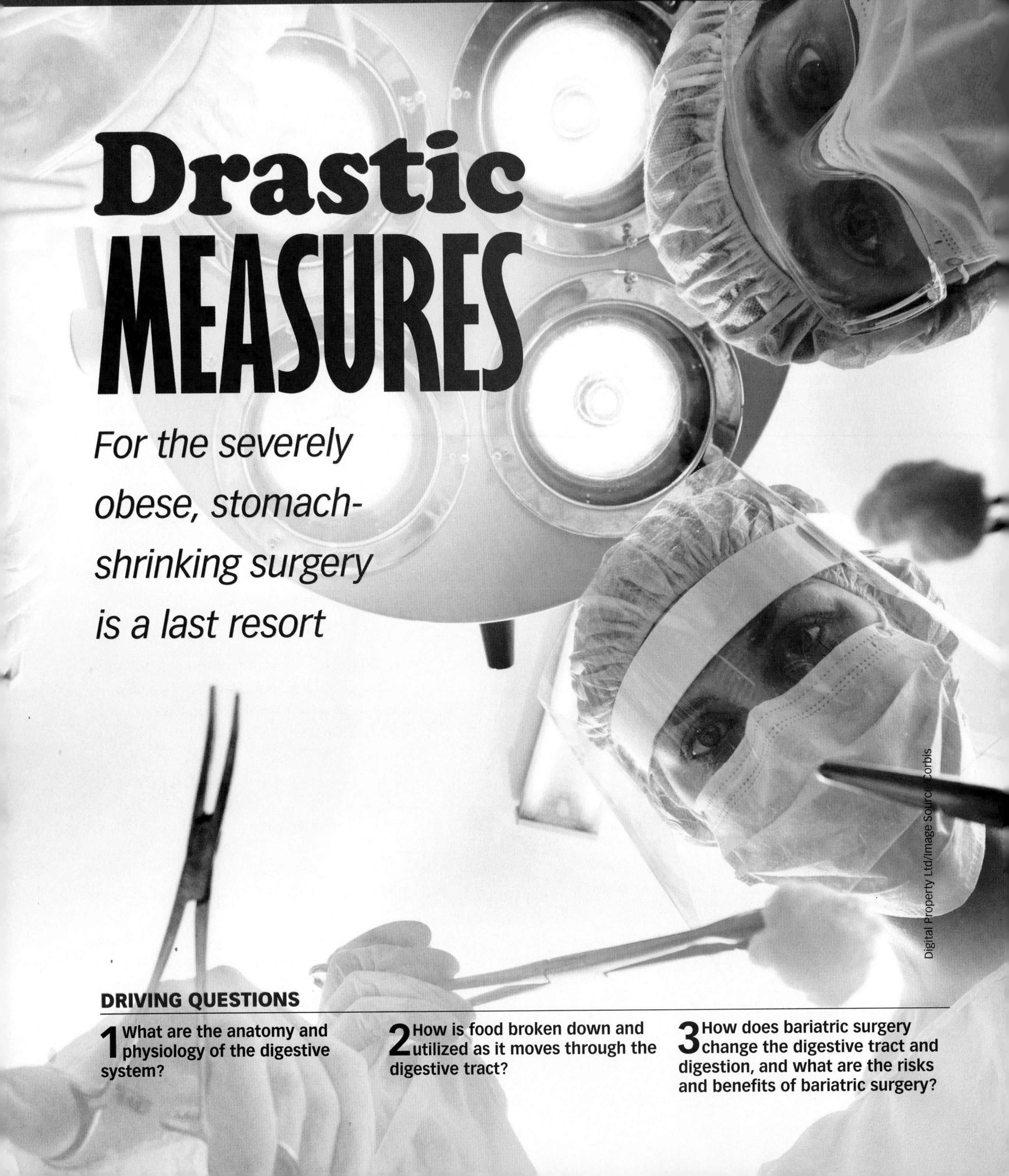

Drastic MEASURES

For the severely obese, stomach-shrinking surgery is a last resort

Digital Property Ltd/Image Source/Corbis

DRIVING QUESTIONS

1 What are the anatomy and physiology of the digestive system?

2 How is food broken down and utilized as it moves through the digestive tract?

3 How does bariatric surgery change the digestive tract and digestion, and what are the risks and benefits of bariatric surgery?

Amy Jo Smith hardly recalls a time growing up when her family wasn't dieting. Her parents were both obese, and they were always trying to lose weight.

Smith herself was relatively slender until her senior year in high school, when her weight began to creep up. She grew up on a horse farm in northeast Maryland, and as a teenager spent much of her spare time on the road, taking her horses to shows. She attributes her weight gain to a diet that consisted primarily of fast food. "I was always eating on the run," says Smith, now 38 years old and a computer literacy teacher. But, she says, her growing girth "never stopped me from doing the things I wanted to do." The extra weight did bother her, though, and she tried several diets and diet pills, only to see her weight yo-yo up and down.

Courtesy Amy Jo Smith

Amy Jo Smith in 2004.

In 2004, at a routine checkup, Smith's doctor noticed that Amy Jo was suffering a number of problems that were likely caused by Smith's 264-pound weight. For one thing, she had been experiencing migraines. She also had stress incontinence: her bladder would leak when she coughed or laughed. "I thought that was just normal," she says. And she went months at a time without having a period—a telltale sign of a hormonal imbalance often associated with obesity. Smith's physician suggested that, to lose weight, Smith consider having a surgical procedure that would shrink her stomach to the size of a golf ball—a drastic measure, but one that her doctor believed was warranted given the health

INFOGRAPHIC 27.1

Medical Complications of Severe Obesity

Severe obesity is a risk factor for the development of medical conditions, many of which have serious health consequences.

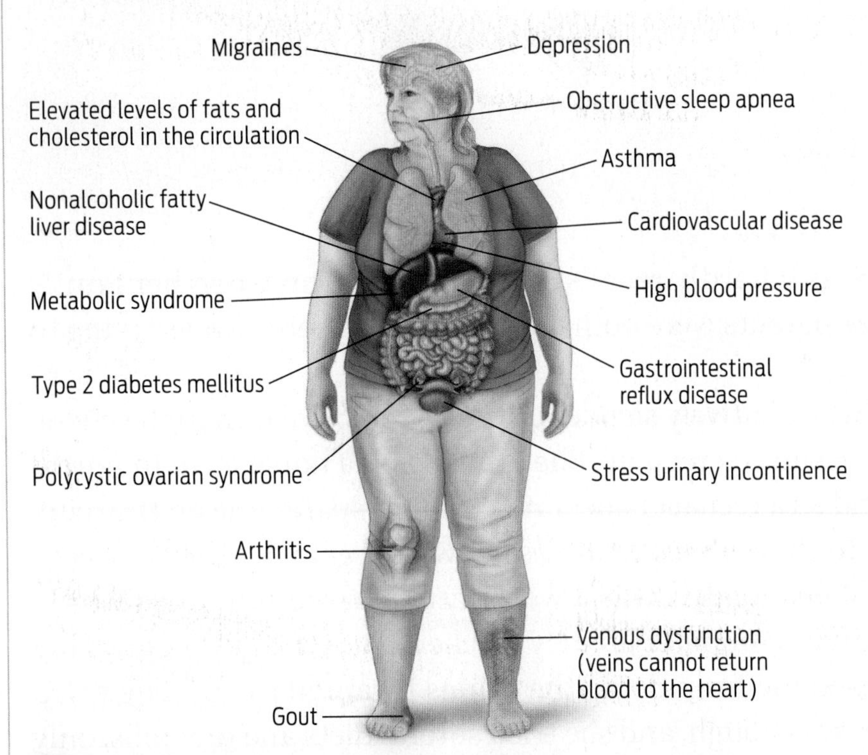

? List one respiratory, one cardiovascular, and one skeletal complication of severe obesity.

complications caused by Smith's weight **(INFOGRAPHIC 27.1)**.

Obesity is a medical condition defined as weighing 20% or more than is recommended for one's height or having a body mass index (BMI) of 30 or higher (see Chapter 6). Severe (or morbid) obesity is defined as being 100 pounds or more overweight or having a BMI of 40 or higher. Obesity becomes severe when it significantly increases the risk of one or more obesity-related health conditions or serious diseases, such as heart disease and diabetes. At 5′ 2″ and with a BMI of 48, Smith had become severely obese.

Even then, Smith had a hard time accepting that she needed such a drastic method to lose weight. "At first I thought he was a quack," she says of her doctor. But she began to think more about the surgery after a friend underwent the stomach-shrinking surgery and experienced dramatic results.

There are several types of bariatric, or weight-loss, surgery. (The word "bariatric" comes from the Greek *báros,* meaning "heaviness.") All involve surgically reducing the size of the stomach, either temporarily or permanently. The specific type of surgery recommended depends on the individual patient's medical history and weight-loss goal. Because these procedures carry many associated risks, a National Institutes of Health panel of experts has recommended surgery only for people considered severely obese—people whose risk of death from diabetes or heart disease because of excess weight is five to seven times greater than for those of average weight.

But bariatric surgery is no miracle cure. "It's sort of barbaric," says Monica Skarulis, director of the Metabolic Clinical Research Unit at the National Institutes of Health. Because the surgery so drastically reduces the size of the stomach and restricts how much a person can eat, it amounts to "forced behavior control," since patients must live on a strict diet, she says. If they overeat, they suffer nasty side effects such as vomiting and diarrhea. On top of that, some bariatric-surgery patients suffer mineral and vitamin deficiencies over the long term that cause bone loss and potentially other health impairments. And the surgery itself is risky: as many as 20% of patients suffer complications a year after the surgery that are severe enough to put them back in the hospital.

For some severely obese people, however, the risk of dying from obesity-related diseases is higher than the risk of surgical complications. And for weight reduction, the surgery is more effective than lifestyle changes alone. Almost all patients lose 30% to 50% of their excess weight in the first 6 months and 77% of their excess weight after about a year. Studies show that even 10 years after surgery, most patients still weigh 25% to 30% less than they did before the surgery. Consequently, demand for the surgery has soared in the United States since the early 2000s, and today the number of procedures performed hovers around 230,000 per year.

STOMACH
An expandable muscular organ that stores and mechanically breaks down food. Specific enzymes in the stomach digest proteins.

DIGESTIVE SYSTEM
The organ system that breaks down food molecules into smaller subunits, absorbs nutrients, and eliminates waste; it consists of the digestive tract and accessory organs.

DIGESTION
The mechanical and chemical breakdown of food into subunits, enabling the absorption of nutrients.

Gut Reactions

▸ Digestive system anatomy and physiology

The rationale behind bariatric surgery is simple: by reducing the amount of food the **stomach** can hold, the surgeries prevent patients from overeating. Reducing the amount of food taken in means less food is digested, fewer calories are absorbed into the body, and eventually weight is lost.

The stomach is an easy target for a surgical fix of overeating because it's relatively simple to reduce its size surgically. But care must be taken to make sure that the rest of the **digestive system** still works as it's supposed to—breaking down food molecules into smaller subunits, absorbing nutrients, and eliminating waste. The first of these actions—breaking down food molecules—is called **digestion.**

The digestive system can be thought of as having two main components: a central digestive tract—essentially a long tube lined with muscles that extends from the mouth to the anus—and accessory organs that flank the tract and assist in digestion. (The stomach is one part of the central digestive tract.) As the muscles of the digestive tract alternately relax and contract, food is pushed along. The accessory organs located along the length of the tract secrete enzymes and other chemicals into the tract to help break down food molecules. Through these coordinated actions, the digestive system transforms the food we eat into a form our bodies can use and rids the body of the waste left over once usable nutrients are removed from food we have taken in **(INFOGRAPHIC 27.2)**.

Digestion relies on both mechanical and chemical processes. These begin as soon as we put food into our mouths—that is, as soon as we ingest it. The act of chewing mechanically breaks food down into smaller pieces, while **salivary glands** secrete enzymes into saliva that chemically dismantle macromolecules into their subunits. The enzyme salivary amylase, for example, breaks down carbohydrates into simpler sugars. The tongue compresses the food into a ball and works it to the back of the mouth.

INFOGRAPHIC 27.2

The Digestive System

The digestive system consists of a long tube with specialized sections and accessory organs that secrete enzymes and other chemicals into the digestive tract. As food travels down the digestive tract, macromolecules are broken into subunits, nutrients are absorbed, and waste is eliminated.

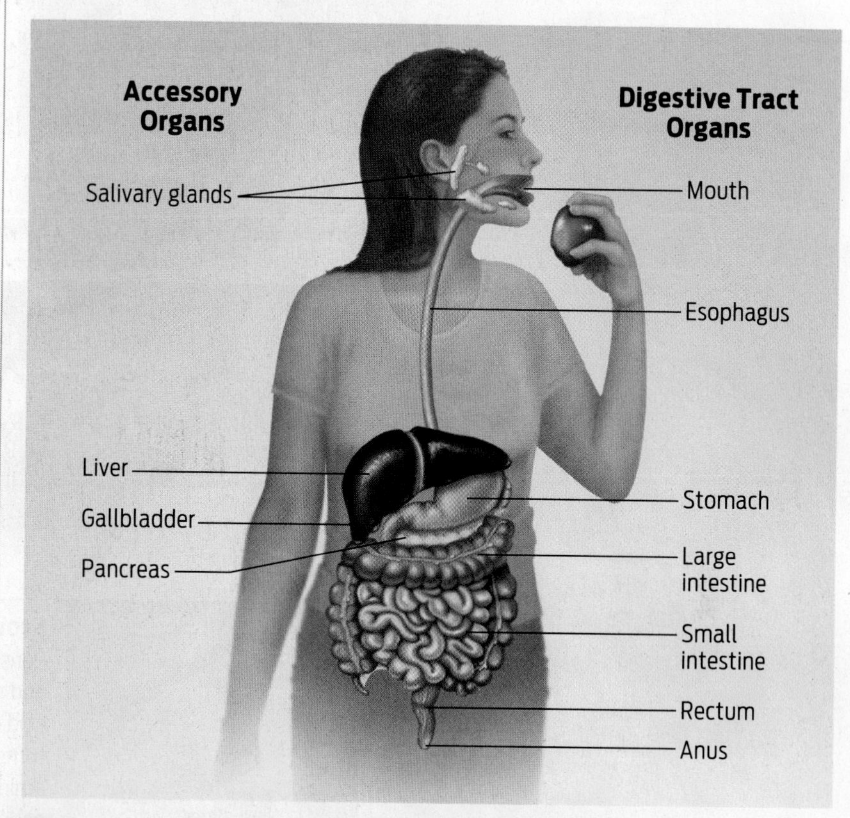

? List the accessory organs in the digestive tract.

When we swallow, food is propelled along the **esophagus** by rhythmic waves of muscle contractions called **peristalsis.** Food then enters the stomach, where stomach acid destroys harmful bacteria to protect us against food-borne diseases. Stomach acid also causes proteins in food to lose their three-dimensional shapes, turning them into linear chains of amino acids. This makes it easier for the enzyme **pepsin,** which is produced in the stomach, to chemically break proteins apart into individual amino acids.

Like the esophagus, the stomach is muscular, expanding and contracting as it accepts food and churns it. Each time it contracts, stomach acid mixes with food,

SALIVARY GLANDS
Glands that secrete enzymes into the mouth to break down macromolecules in food. One such enzyme is salivary amylase, which digests carbohydrates.

ESOPHAGUS
The section of the digestive tract between the mouth and the stomach.

PERISTALSIS
Coordinated muscular contractions that force food down the digestive tract.

PEPSIN
A protein-digesting enzyme that is active in the stomach.

INFOGRAPHIC 27.3

The Upper Digestive System

The upper digestive tract includes the mouth, esophagus, and stomach, as well as enzymes and other chemicals secreted by the salivary glands and the stomach.

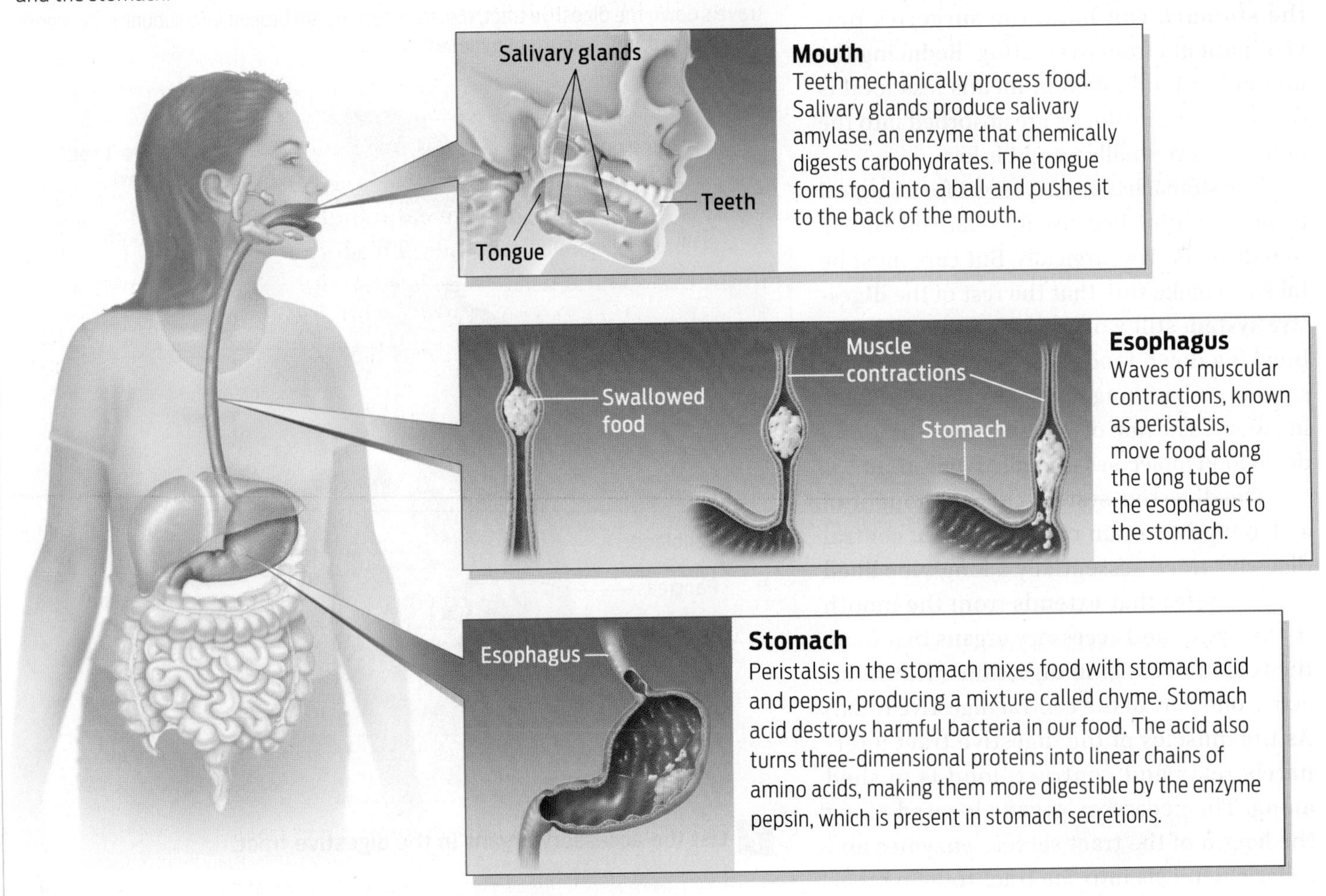

? **Name two key events that occur in the stomach.**

producing a soupy mixture called **chyme** (**INFOGRAPHIC 27.3**).

Despite the powerful acid churning inside it, the stomach remains intact. This is possible because the stomach is lined with a thick layer of protective mucus. Occasionally, this mucus layer is damaged—by a bacterial infection, for example—and the stomach lining becomes more vulnerable to gastric juices; the result is a painful sore called an ulcer.

While the stomach can absorb some substances directly into the bloodstream—water, ethanol, and certain drugs, for example—most of the chyme is pushed farther down the digestive tract, where it is processed even more.

Although the stomach is only a small part of the upper digestive tract, it plays a large part in weight gain. Evolutionarily speaking, the reason we have a stomach in the first place is to enable us to temporarily store the food we eat. Without a stomach, we would have to eat constantly to fuel our activities. When we eat a large meal, the stomach expands greatly to accommodate and store all that food. It's partly because of this elasticity that we can eat enough to sustain us for hours. But this elasticity also means that we can eat more than our bodies need at any particular moment.

The stomach leads into the **small intestine.** Most chemical digestion of food occurs there, assisted by the secretions of several

CHYME
The acidic "soup" of partially digested food that leaves the stomach and enters the small intestine.

SMALL INTESTINE
The organ in which the bulk of chemical digestion and absorption of food occurs.

PANCREAS
An organ that helps digestion by producing enzymes (such as lipase) that act in the small intestine, and by secreting a juice that neutralizes acidic chyme.

accessory organs. The **pancreas,** for example, secretes specific enzymes into the small intestine that help break down organic macromolecules such as carbohydrates, proteins, and fats into smaller molecules. It also secretes alkaline pancreatic juice into the small intestine to neutralize the acidic chyme, which would otherwise damage the small intestine. The first part of the small intestine, where these various enzymes and juices mix, is called the **duodenum.** Other enzymes, produced by the small intestine itself, further break down the smaller molecules into individual building blocks such as amino acids, nucleotides, and sugars.

Whereas proteins and carbohydrates are easily digested by this powerful mixture of digestive enzymes, fats pose a special challenge. Because they are hydrophobic (see Chapter 2), fats don't mix well with the watery solutions in the small intestine. This makes it difficult for fat-digesting enzymes to break them down. Helping the process along, the **liver** secretes **bile salts,** which divide large hydrophobic fat globules into smaller droplets—that is, emulsify them. These bile salts pass from the liver into the **gallbladder,** which stores them for future use. When we eat a high-fat meal, bile salts pass from the gallbladder into the duodenum, where they help emulsify the fats. Once the fats are emulsified, the enzyme **lipase,** secreted by the pancreas, chemically breaks them down to release their constituent fatty acids and glycerol **(INFOGRAPHIC 27.4).**

DUODENUM
The first portion of the small intestine, where mixing of chyme and digestive enzymes occurs.

LIVER
An organ that aids digestion by producing bile salts that emulsify fats.

BILE SALTS
Chemicals produced by the liver and stored by the gallbladder that emulsify fats so that they can be chemically digested by enzymes.

GALLBLADDER
An organ that stores bile salts and releases them as needed into the small intestine.

LIPASE
A fat-digesting enzyme active in the small intestine.

INFOGRAPHIC 27.4

Accessory Organs and the Small Intestine Work Together to Digest Food

The small intestine is the major organ that digests food. Accessory organs secrete enzymes and other substances into the small intestine, which itself also produces digestive enzymes.

Liver
The liver makes bile salts and secretes them into the gallbladder. When released into the small intestine, bile salts emulsify fats, breaking them up into smaller droplets. This allows the enzyme lipase to more efficiently break down fat molecules. Lipase is made by the pancreas and secreted into the small intestine.

Pancreas
Pancreatic juice secreted into the small intestine neutralizes acids in chyme. Enzymes in the pancreatic juice also break down carbohydrates, proteins, fats, and nucleic acids into their smallest subunits.

Duodenum
Food from the stomach is mixed with digestive secretions in the duodenum, the first portion of the small intestine.

Small intestine
The small intestine produces some digestive enzymes and is the site of most chemical digestion of food.

Gallbladder
The gallbladder stores bile salts and passes them into the small intestine when food enters the duodenum.

? **What do the liver, gallbladder, and pancreas have in common?**

LARGE INTESTINE
The last organ of the digestive tract, in which remaining water is absorbed and solid stool is formed.

Once digested into their smallest subunits, food molecules are taken up—absorbed—by cells lining the small intestine. The lining of the small intestine is folded into fingerlike projections called villi that greatly increase the surface area through which the intestine can absorb nutrients. The food molecules then pass into blood vessels of the circulatory system, and the bloodstream transports them throughout the body, where they serve as a source of the nutrients needed to build and maintain cells **(INFOGRAPHIC 27.5)**.

The final stop on this journey is the **large intestine,** or colon, which functions like a trash compactor—holding and compressing material that the body can't use or digest, such as plant fiber. Within the large intestine, fiber, small amounts of water, vitamins, and

INFOGRAPHIC 27.5

The Small Intestine Absorbs Nutrients

The small intestine is the primary organ that absorbs nutrients from food. Nutrients enter the circulatory system via blood vessels connected to the small intestine.

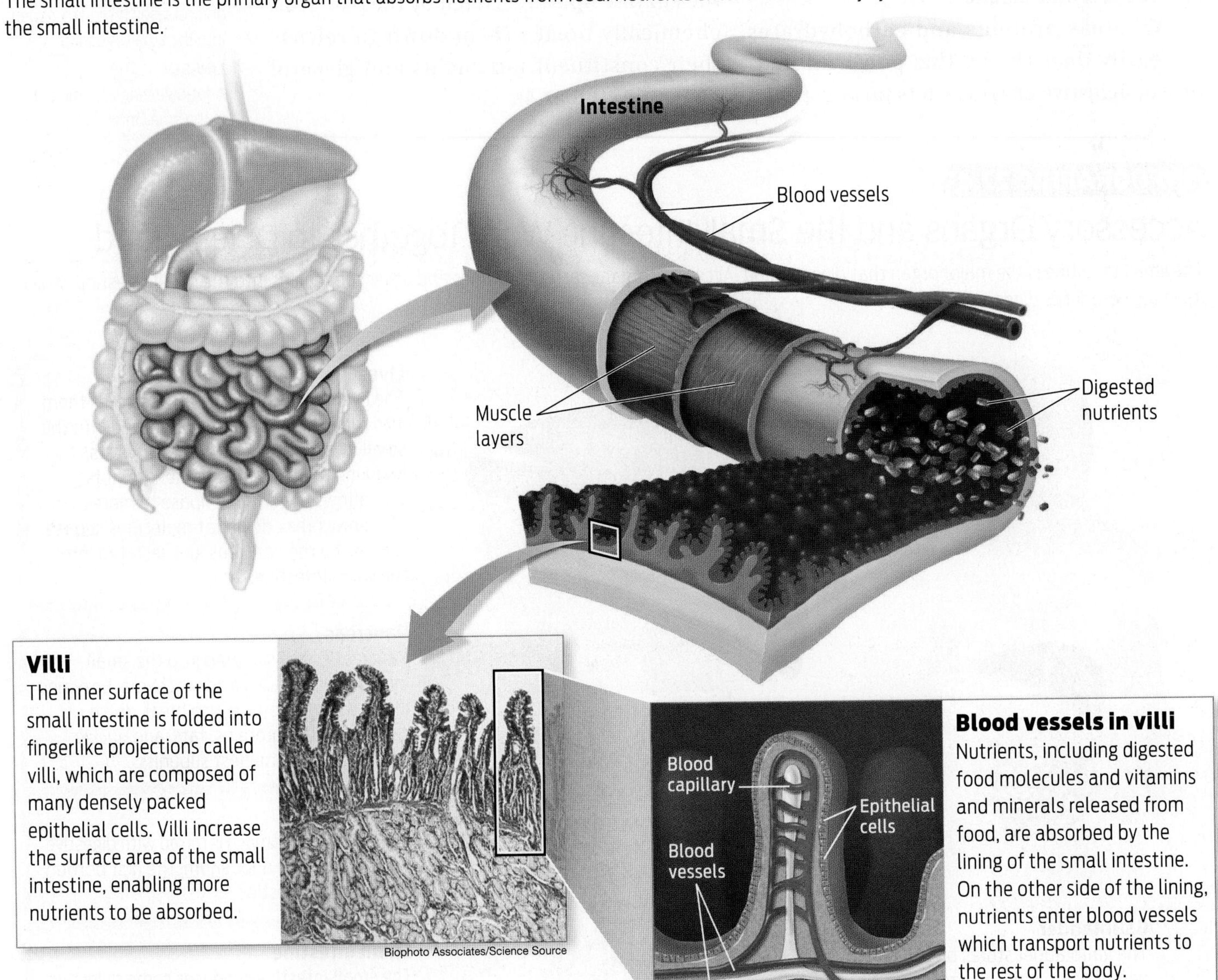

? What would you predict about the ability to absorb nutrients in someone who lacked normal villi?

INFOGRAPHIC 27.6

The Large Intestine

The large intestine absorbs water and some nutrients. It also packages waste material into stool.

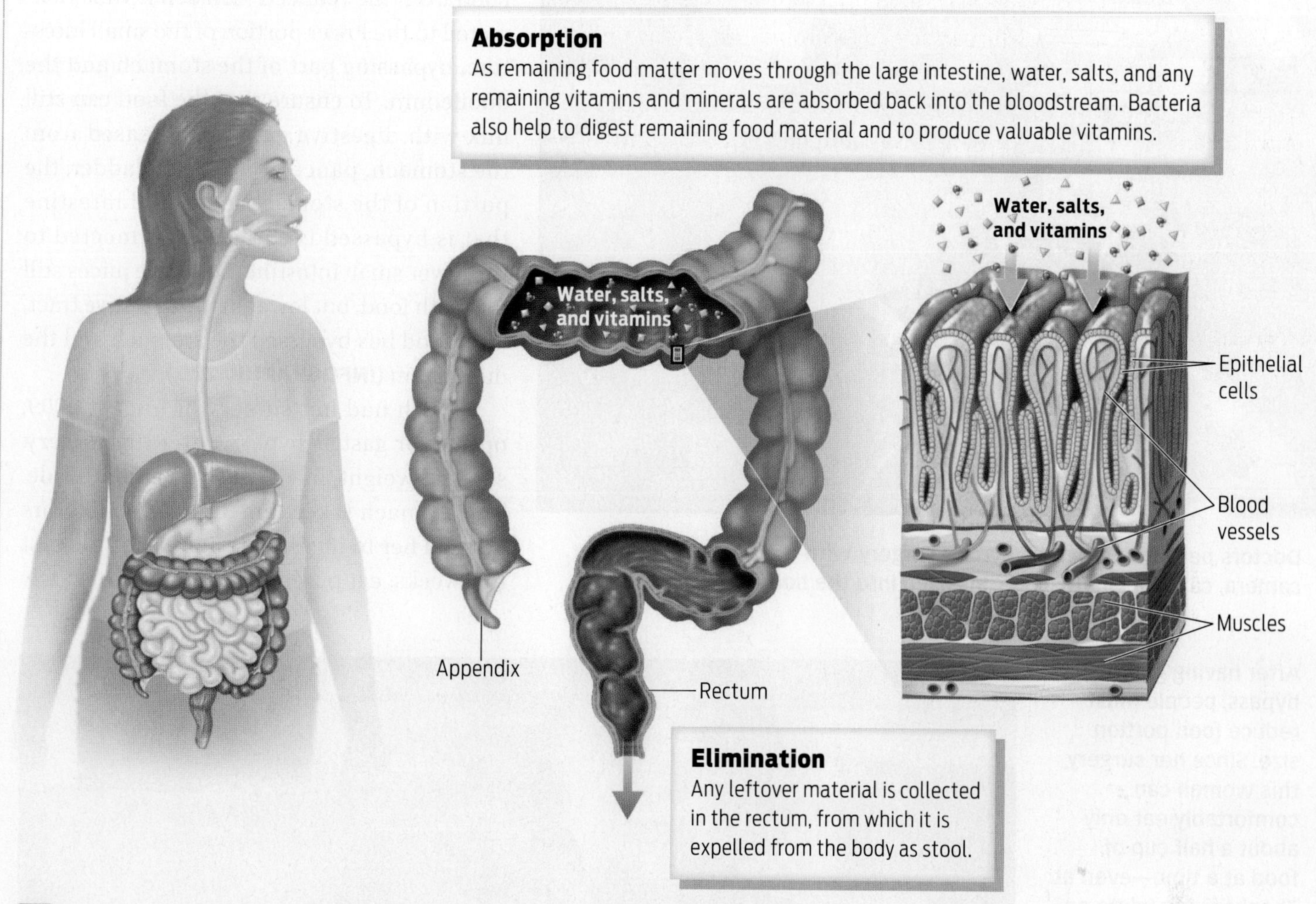

? **Should undigested fats normally be found in stool? Why or why not?**

other substances mix with mucus and bacteria that normally live in the large intestine. As this waste travels through the large intestine, most of the water and some vitamins and minerals are reabsorbed into the body through the intestinal lining. Bacteria chemically break down some of the fiber both to produce nutrients for their own survival and to provide valuable vitamins, which is one reason fiber is an important dietary nutrient. As the large intestine expands and contracts, it creates **stool,** which is pushed into the rectum and eliminated from the body through the anus **(INFOGRAPHIC 27.6)**.

Weighing the Options

▶ Types of bariatric surgery

Because of Smith's weight-related illnesses, her doctor recommended that she have bariatric surgery. The doctor explained that bariatric surgery changes the anatomy of the digestive system to limit the amount of food a person can eat and digest before feeling full.

Of the different types of bariatric surgery, sleeve gastrectomy is currently the most common, followed by gastric bypass. Both of these dramatically reduce the size of the stomach. In the gastric sleeve, the reduced stomach still

STOOL
Solid waste material eliminated from the digestive tract.

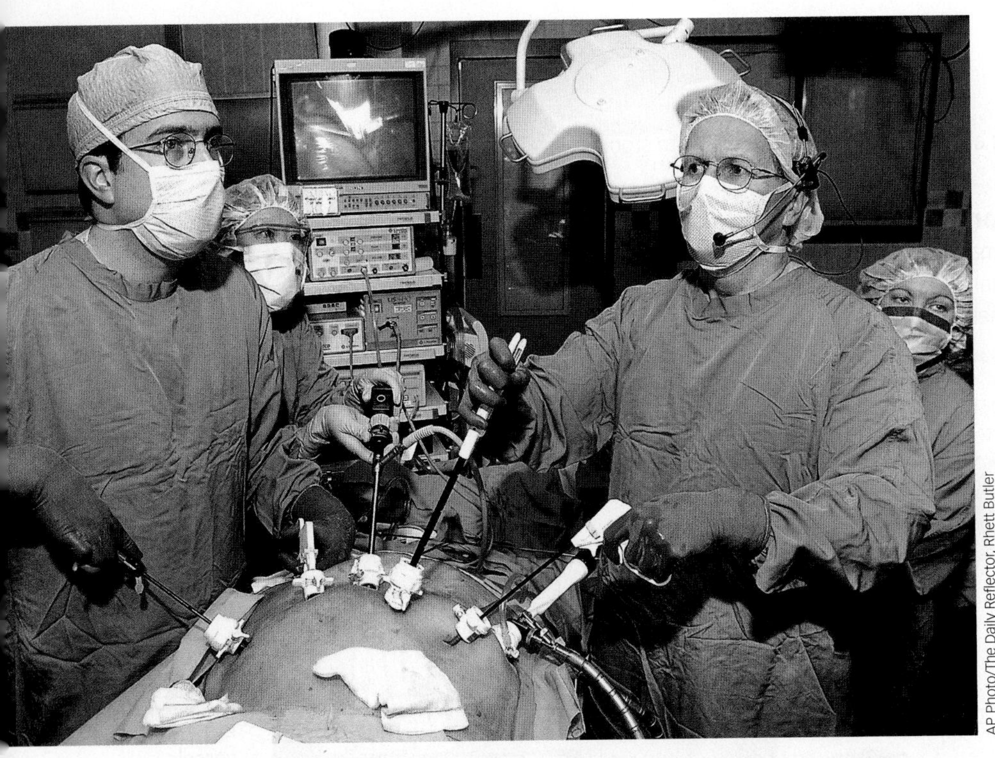
AP Photo/The Daily Reflector, Rhett Butler

Doctors performing gastric bypass surgery with the help of a tiny camera, called a laparoscope, inserted into the body.

feeds into the duodenum. There are no other changes to the digestive tract or digestion. In gastric bypass, the small intestine is cut and surgically attached to the reduced stomach. Food from the reduced stomach is thus redirected to the lower portion of the small intestine, bypassing part of the stomach and the duodenum. To ensure that the food can still mix with digestive enzymes released from the stomach, pancreas, and gallbladder, the portion of the stomach and small intestine that is bypassed is surgically connected to the lower small intestine. Digestive juices still mix with food, but lower in the digestive tract, after food has bypassed the stomach and the duodenum **(INFOGRAPHIC 27.7)**.

Smith had her surgery in August 2009, opting for gastric bypass. After the surgery, she lost weight, but it wasn't a smooth ride. The stomach takes time to heal, so doctors advised her to ingest only liquids for the first few weeks, eat puréed foods for the next few

The Post Standard/Syracuse.com—Peter Chen

After having gastric bypass, people must reduce food portion size. Since her surgery, this woman can comfortably eat only about a half cup of food at a time—even at Thanksgiving (plate on the left).

INFOGRAPHIC 27.7

Types of Bariatric Surgery

Sleeve gastrectomy is currently the most common form of bariatric surgery. It dramatically reduces the size of the stomach but does not redirect the flow of food and chyme through the digestive tract. Gastric bypass reduces the size of the stomach and redirects food from the upper stomach pouch to the lower small intestine. Additional anatomical rearrangements are necessary to expose food to digestive secretions in gastric bypass.

Before Surgery

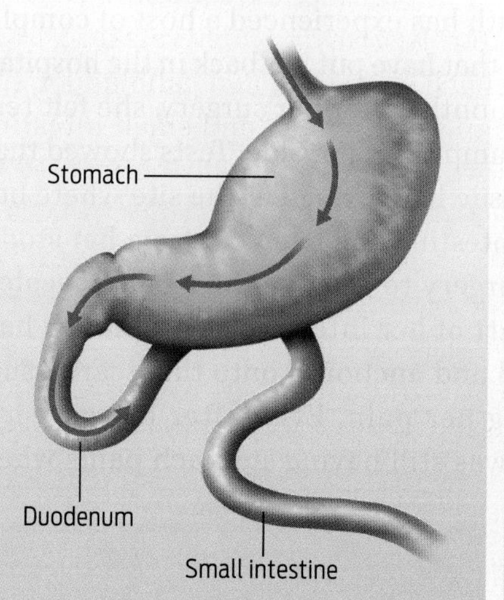

Food passes into the stomach, which can stretch from the size of a large sausage when empty to hold 3–4 liters of food. Chyme made in the stomach enters the small intestine, where digestion releases food nutrients.

Sleeve Gastrectomy

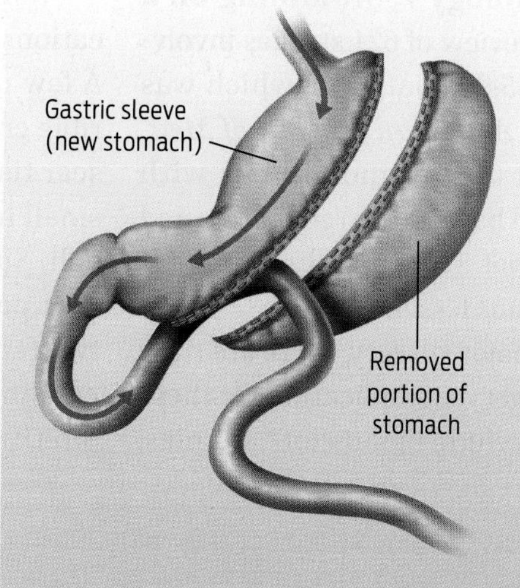

The stomach is dramatically reduced, to the size of a banana. Its small size limits the amount of food (and therefore Calories) that can be consumed. Stomach contents are still released into the upper small intestine for further processing.

Gastric Bypass

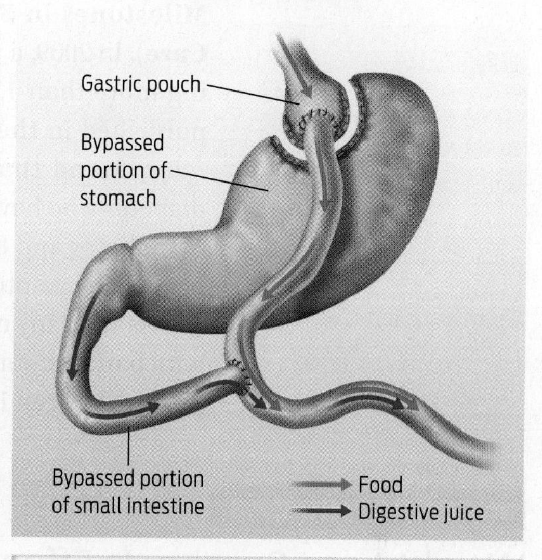

The stomach is reduced to the size of a golf ball, diminishing drastically the amount of food it can hold. Also, food bypasses the upper part of the small intestine, reducing the amount of nutrients (and therefore Calories) that can be absorbed. Digestive juices from the stomach and accessory organs still flow into the small intestine, only lower down.

? Compare and contrast sleeve gastrectomy and gastric bypass.

weeks, and then gradually progress to solid foods. Because the stomach is made so small in gastric bypass, it can hold only about an ounce of food at a time—a handful of crackers or a few broccoli florets. Eating too much at once can cause vomiting or intense stomach pain.

There are financial costs, too: the surgery runs anywhere from $11,000 to $26,000, and it's not always covered by insurance.

But the surgery does lead to weight loss. A 2016 study published in *JAMA Surgery* looked at the weight-loss patterns of nearly 1,800 American veterans who had gastric bypass surgery and compared them with those of a group of more than 5,000 obese individuals who did not have weight-loss surgery. The results were striking: at 1 year post-surgery, patients undergoing gastric bypass had lost 31% of their baseline weight; 10 years out, they still weighed 21% less than the control group. Another study, published in 2007 in the *New England Journal of Medicine,* found that bariatric surgery also saves lives: among 2,000 obese patients who underwent weight-loss surgery, over the course of 16 years of follow-up, there were 129 deaths in the diet-only group, mostly from weight-related heart disease and cancer, and 101 deaths in the surgery group—a large difference statistically **(INFOGRAPHIC 27.8)**.

Also encouraging to some doctors is the finding that weight-loss surgery can reverse type 2 diabetes in some people (see **Milestones in Biology 7: Stumbling on a Cure**). In 2009, a review of 621 studies involving more than 135,000 patients, which was published in the *American Journal of Medicine,* found that 78% of individuals with diabetes who have bariatric surgery are cured of diabetes and that 87% are either cured or have their symptoms lessen.

As with any major surgery, there are risks with bariatric surgery. Complications of these procedures can include blood clots, hernias, bowel obstructions, and—rarely—death. Patients can also end up back in the hospital to repair intestinal leaks that can lead to serious infection. For these and other reasons, some obesity and diabetes experts think that patients should consider less radical options first, such as a diet that sharply restricts carbohydrates.

Smith has experienced a host of complications that have put her back in the hospital. A few months after her surgery, she felt terrible cramping in her side. Tests showed that scar tissue had formed at the site where her small intestine had been cut from her stomach. Surgery to remove the tissue revealed that part of her intestine and stomach had twisted and anchored onto this scar tissue, causing her pain. Even after the surgery, Smith was still having stomach pains when

INFOGRAPHIC 27.8

Weight-Loss Surgery Is Effective and Saves Lives

A large study of weight-loss surgery showed that people who had gastric bypass lost substantially more weight than people who attempted to lose weight using nonsurgical methods. People who had gastric bypass also experienced fewer deaths in the 16 years following surgery compared with people who relied on nonsurgical methods for weight loss.

Difference in Weight Loss among Patients Undergoing Gastric Bypass and Nonsurgical Methods

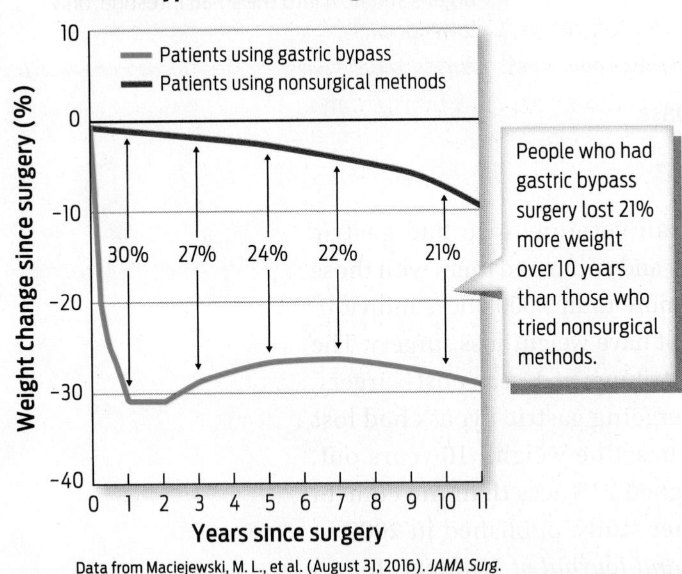

Data from Maciejewski, M. L., et al. (August 31, 2016). *JAMA Surg.*

Difference in Mortality among Patients Undergoing Gastric Bypass and Nonsurgical Methods

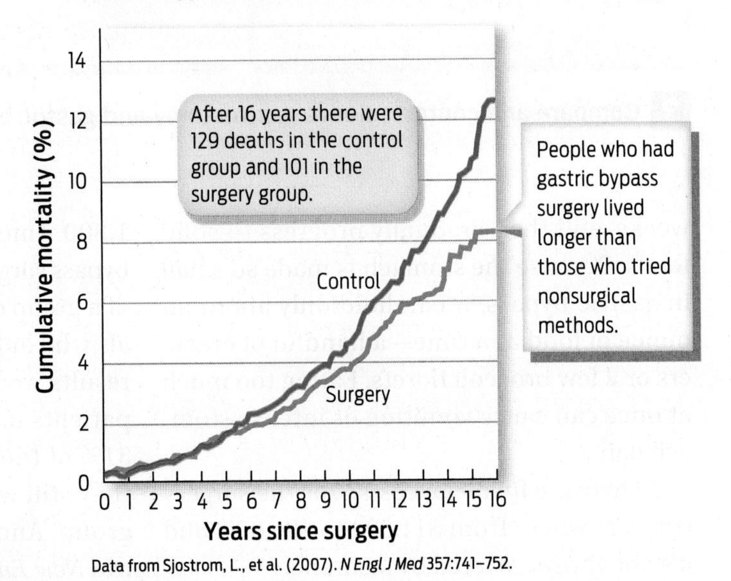

Data from Sjostrom, L., et al. (2007). *N Engl J Med* 357:741–752.

? A patient weighs 300 pounds at the time of surgery. What will the patient's weight be in 10 years if 28% of body weight is lost (as happened with the people in the 2016 study plotted above)?

she ate, so doctors temporarily put a feeding tube into her stomach and a catheter into a vein in her arm through which she could take nutrients directly into her bloodstream. Smith spent weeks in and out of the hospital between January and April of 2010.

Smith's health has improved since then, but she is still at risk. Since people who have gastric bypass surgery end up with part of the small intestine bypassed, they absorb fewer of the micronutrients they eat. Patients must take vitamin supplements for the rest of their lives. To monitor her micronutrient levels, Smith has a blood test every 3 months.

What's more, the surgery may not be a permanent cure for obesity. Most people who have bariatric surgery regain some weight over time. The stomach and intestines can sometimes expand to allow greater ingestion of food. And hormonal changes may occur after surgery that alter body metabolism. While scientists are still studying the hormonal mechanisms of weight gain, they know that if people do not exercise control over their diet and lifestyle, even those who have had surgery can regain significant amounts of weight.

Hunger Pangs

▶ Hormones, digestion, and appetite

The original rationale for bariatric surgery was to produce weight loss solely by making anatomical changes: reducing the size of the stomach and small intestine, it was thought, would prevent patients from putting too many Calories into their body, and would reduce the number of Calories that are absorbed by the small intestine.

In recent years, scientists have realized that there is more to bariatric surgery than just Calorie reduction. Bariatric surgery also profoundly alters the complex system of digestive hormones that regulate hunger and satiety (the feeling of fullness).

A variety of hormones act to ensure that the digestive tract secretes its enzymes and juices only when food is present. For example, the stomach produces the hormone gastrin in response to the stretching of the stomach that occurs when food enters. Like all hormones, gastrin is released into the circulation, which enables it to travel to its target. Gastrin travels back to act on the stomach itself, stimulating the secretion of gastric acid into the chamber of the stomach. This gastric acid then activates a protein-digesting enzyme in the stomach.

Because the stomach is made so small in gastric bypass, it can hold only about an ounce of food at a time—a handful of crackers or a few broccoli florets.

Hormones also regulate our appetite, in both the short and long terms. For example, in the short term, specific cells lining the stomach produce ghrelin, a hormone that lets the body know when the stomach is empty and it's time to eat. Levels of ghrelin rise before a meal and fall when the stomach is full. Ghrelin travels to the brain, where it signals a feeding center in the hypothalamus. In response, the hypothalamus releases two feeding hormones that stimulate eating. If people are injected with ghrelin, they become extremely hungry, even if they just ate a meal. For that reason, ghrelin is sometimes called the "hunger hormone."

Other hormones act as satiety hormones, causing the hypothalamus to release hormones that inhibit eating. For example, the hormone PYY is produced by both the small intestine and the colon after meals. When it reaches the hypothalamus, it has a two-fold effect: it blocks the release of one of the feeding hormones, and it stimulates the release of an "anti-eating" hormone.

In the long term, appetite is regulated by leptin, a hormone produced by adipose (fat) tissue. Leptin is a satiety hormone: it keeps appetite low by acting on the hypothalamus both to inhibit the release of feeding hormones and to stimulate the release of the "anti-eating" hormone. People who are unable to produce leptin feel chronically hungry, and overeat to an extreme extent.

All types of bariatric surgery alter levels of gastric hormones, and researchers now believe that these hormonal changes may account for a significant amount of weight loss that occurs following surgery. In particular, the levels of the hormone PYY become elevated following all types of bariatric surgery and have been shown to reduce food intake by promoting satiety. Sleeve gastrectomy, which removes a large portion of the stomach, also removes a large portion of the cells that produce ghrelin, the hunger hormone. Lower levels of ghrelin may help reduce food cravings in individuals who have had this surgery. The question of how surgery modifies the hormonal control of appetite is still open, and of great interest to physicians and researchers **(INFOGRAPHIC 27.9)**.

INFOGRAPHIC 27.9

Hormones Coordinate Digestion and Regulate Hunger

A variety of hormones coordinate digestive processes and regulate hunger. Coordination of digestion ensures that digestive enzymes and juices are secreted only when food is present in the digestive system. Hormones such as ghrelin and PYY control appetite in the short term, while leptin controls appetite in the long term.

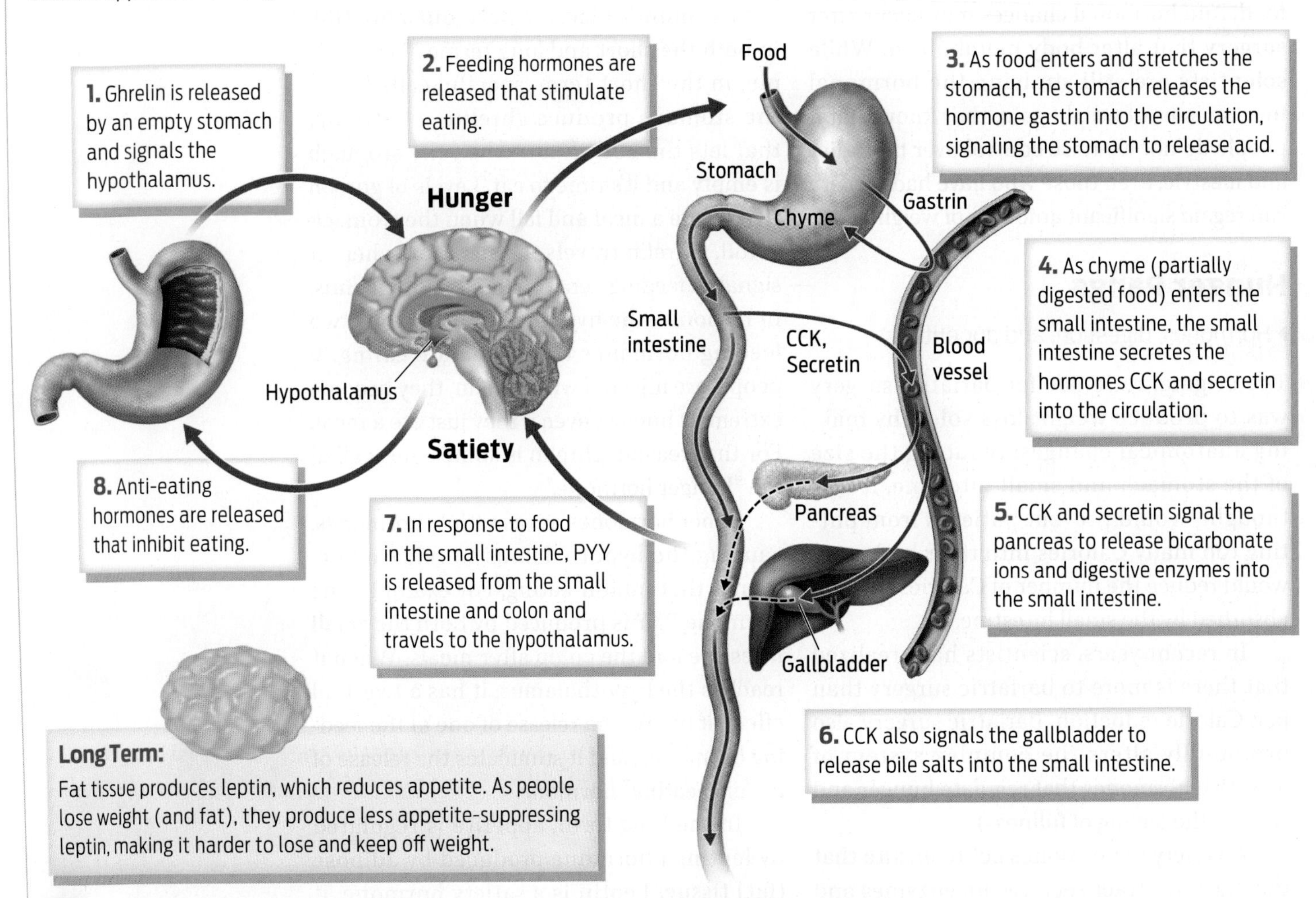

? For each of the following hormones, state whether they signal hunger or satiety: leptin, PYY, ghrelin.

Hormones also likely explain why many people have difficulty losing weight and keeping it off once it's lost. The main culprit is likely leptin, the hormone produced by fat cells. When people lose weight suddenly through diet and exercise, their leptin levels plummet. With less leptin, these people do not experience a sense of satiety, so they may feel hungry. Leptin levels may recover somewhat over time, but never fully return to their level before weight loss. Keeping weight off, therefore, means fighting a nearly constant battle with hunger.

Why would our bodies fight so hard to regain weight that we've intentionally lost? It probably has to do with how our ancestors evolved. Periodic starvation was likely the norm for our ancestors. In that context, being able to hold on to fat stores in the body would have been an advantage. Only in an environment of food abundance does this thrifty trait become a liability.

Evidence suggests that bariatric surgery can interrupt this cycle by changing the body's set point for leptin and other hormones. Even the gut microbiome (see Chapter 17) may be affected by bariatric surgery in ways that produce weight loss.

If people are injected with ghrelin, they become extremely hungry, even if they just ate a meal. For that reason, ghrelin is sometimes called the "hunger hormone."

Courtesy Amy Jo Smith

Amy Jo Smith in 2011 after her weight-loss surgery.

Not an Easy Road

Smith's own experience with bariatric surgery has been mixed. She still struggles with nausea every day; strong smells can cause her to vomit. She also feels pain in her left side, for which she takes medication.

Despite these complications, Smith has had not "one day of regret," she says. On her first "surgiversary"—her surgical anniversary—she wrote a letter to her surgical team in which she said:

> I have been blessed with 35 birthdays but none can compare to my surgiversary. I never imagined in a year that I would lose over 100 lbs, run a 5K the day of my surgiversary . . . sit sideways in a student desk, wear a size 12 pants from a 24–26 . . . be able to sit comfortably in a restaurant booth, and be able to stand on a table or chair without thinking, "My gosh, am I going to break this?"

Within a year and half of her surgery, Smith had dropped down to 146 pounds. "I don't recognize myself anymore," she says. ■

CHAPTER 27 Summary

Driving Question 1 What are the anatomy and physiology of the digestive system?

- The digestive system consists of a central digestive tract and accessory organs. Its function is to break down food molecules into smaller subunits, absorb nutrients, and eliminate waste.
- Digestion is coordinated by a variety of hormones, including the hunger hormone ghrelin and the satiety hormone leptin.

Driving Question 2 How is food broken down and utilized as it moves through the digestive tract?

- Digestion begins in the mouth, where teeth chew food and the tongue compresses food and pushes it to the back of the mouth (mechanical digestion), and where salivary enzymes begin breaking down carbohydrates (chemical digestion).
- Food passes from the mouth into the stomach through the esophagus, propelled by waves of muscular contractions called peristalsis.
- The stomach is muscular and acidic and contains pepsin, a protein-digesting enzyme. It is elastic and can expand after a large meal to store food for a few hours.
- Food processed in the stomach is called chyme. Chyme passes into the small intestine, where enzymes further digest it.
- Enzymes from the pancreas help to digest organic molecules in the small intestine.
- Bile salts, produced in the liver and stored in the gallbladder, emulsify fats and help the body digest them.
- The small intestine absorbs the broken-down products of food; once absorbed, food molecules enter the bloodstream and are transported throughout the body.
- The large intestine absorbs water and forms solid stool from indigestible matter in food such as fiber.

Driving Question 3 How does bariatric surgery change the digestive tract and digestion, and what are the risks and benefits of bariatric surgery?

- Bariatric surgery changes the anatomy of the digestive tract by decreasing the size of the stomach alone and/or by rerouting food to bypass portions of the digestive tract.
- Bariatric surgery may also change hormonal regulation of appetite and hunger.
- Bariatric surgery can result in weight loss and improvements in overall health, but is not without risks, including surgical complications and micronutrient deficiencies.

More to Explore

- National Institutes of Health, bariatric surgery: https://www.niddk.nih.gov/health-information/weight-management/bariatric-surgery
- Maciejewski, M. L., et al. (2016). Bariatric surgery and long-term durability of weight loss. *JAMA Surg* 151(11):1046–1055.
- Sjöström L., et al. (2007). Effects of bariatric surgery on mortality in Swedish obese subjects. *N Engl J Med* 357:741–752.
- Taubes, G. (2011). *Why We Get Fat: And What to Do about It.* New York: Anchor Books.
- Hallberg, S., and Hamdy, O. (September 2, 2016). Before you spend $26,000 on weight-loss surgery, do this. *New York Times.* https://www.nytimes.com/2016/09/11/opinion/sunday/before-you-spend-26000-on-weight-loss-surgery-do-this.html

CHAPTER 27 Test Your Knowledge

Driving Question 1 What are the anatomy and physiology of the digestive system?

By answering the questions below and studying Infographics 27.2, 27.3, 27.4, 27.5, 27.6, and 27.9, you should be able to generate an answer for this broader Driving Question.

Know It

1. Place the following structures of the digestive system in order from the entry of food (1) to the exit of waste (6).

_____ Esophagus
_____ Mouth
_____ Large intestine
_____ Small intestine
_____ Stomach
_____ Anus

2. Which part of the digestive tract has the most acidic pH?
   a. esophagus
   b. colon
   c. small intestine
   d. stomach
   e. mouth

3. Why is it helpful to have an expandable stomach?

4. What hormone is responsible for hunger pangs if you eat lunch later than usual?
   a. leptin
   b. gastrin
   c. ghrelin
   d. CCK

Use It

5. What do the gallbladder, liver, and pancreas have in common with respect to the digestive system? How do they differ from the mouth, stomach, and small intestine?

6. Muscle paralysis in the digestive tract would compromise which digestive function?
   a. chemical digestion in the stomach
   b. chemical digestion in the small intestine
   c. absorption in the small intestine
   d. chemical digestion in the mouth
   e. movement of food from the mouth to the stomach

7. Pepsin is most effective at a pH of about 2. Digestive enzymes in the small intestine are most effective at a pH of or near 7. If the pancreas were unable to secrete bicarbonate (the basic component of pancreatic juice that neutralizes acid), what would you predict about the waste eliminated from the large intestine?

8. A person has a mutated form of leptin that cannot properly transmit a satiety signal. What do you predict about the BMI and the circulating leptin levels in this person? Explain your answer.

Driving Question 2 How is food broken down and utilized as it moves through the digestive tract?

By answering the questions below and studying Infographics 27.5 and 27.6, you should be able to generate an answer for this broader Driving Question.

Know It

9. Where does the majority of chemical digestion take place?
   a. small intestine
   b. esophagus
   c. mouth
   d. stomach
   e. colon

10. What do pepsin and salivary amylase have in common? How do their activities differ?

11. Which organ produces lipase?

Use It

12. Someone whose gallbladder has been surgically removed will have trouble processing
    a. fats.
    b. carbohydrates.
    c. minerals.
    d. vitamins.
    e. proteins.

13. Compare and contrast the functions of bile salts and lipase.

14. Why would someone with a blocked duct between the pancreas and the small intestine experience pancreatic inflammation (pancreatitis)? Note that in this case inflammation is a response to tissue damage.

15. If you stand on your head, can processed food still pass from your small intestine into your large intestine? Explain your answer.

Mini Case

Apply Your Knowledge

16. Alicia has had her gallbladder removed. She must be careful not to eat high-fat meals, or else she is likely to experience greasy diarrhea. Her friend Tammy is taking Alli, a weight-loss drug that inhibits the fat-digesting enzyme lipase. Tammy must also avoid high-fat meals if she wants to avoid oily stools, a possible side effect of the drug. Why are both Alicia and Tammy at risk for similar digestive upsets when their situations are different (gallbladder removal, taking Alli)?

Driving Question 3 How does bariatric surgery change the digestive tract and digestion, and what are the risks and benefits of bariatric surgery?

By answering the questions below and studying Infographics 27.7 and 27.8, you should be able to generate an answer for this broader Driving Question.

Know It

17. Gastric bypass surgery causes the ___________ to become ___________.

- **a.** stomach; smaller
- **b.** small intestine; larger
- **c.** stomach; less acidic
- **d.** small intestine; less acidic
- **e.** stomach; larger

18. What is one similarity and one key difference between sleeve gastrectomy and gastric bypass?

19. From what you've read in this chapter, are there any other benefits to gastric bypass besides weight loss? Explain your answer.

Use It

20. Consider this patient:

- **a.** A 5′ 11″ man weighs 320 pounds. What is his BMI? (See Infographic 6.1.)
- **b.** If he has gastric bypass, and if his weight-loss trajectory is exactly that shown in the left panel of Infographic 27.8, how much will he weigh in 10 years? What will be his BMI at that point? Will he still be considered obese?
- **c.** What health outcomes might he experience as a result of his weight-loss surgery?

Bring It Home

Apply Your Knowledge

21. If a morbidly obese person who is considering gastric bypass surgery asked for your opinion on the procedure, what would you say about its known risks, benefits, and any unknowns? Would you say the same thing to someone considering the surgery who is simply overweight, not morbidly obese? Explain your answer.

Interpreting Data

Apply Your Knowledge

22. A 2012 study compared the impacts of medical therapy and bariatric surgery in obese people with uncontrolled type 2 diabetes. Patients were randomly assigned to receive aggressive medical therapy for their diabetes (including medications and diet and lifestyle modifications) or bariatric surgery. Several dependent variables were measured for 1 year. Two of these variables—average BMI and average number of diabetes medications—are shown in the table below.

- **a.** Draw two graphs, one plotting the average diabetes medications over time for the group receiving medical therapy and for the group receiving surgery, and one plotting the change in BMI from baseline for the two groups (set the baseline values at 0 on the time axis because by definition no change can have taken place yet).
- **b.** From these graphs, how does gastric bypass compare to medical therapy for diabetes management in obese patients with type 2 diabetes?
- **c.** The data shown in the table are for 41 patients who had medical therapy and 50 patients who had gastric bypass. From this information and any other limitations you can identify, are these data sufficient to make a recommendation of surgery for diabetes management? Why or why not?

			Time after treatment			
Treatment	**Variable**	**Baseline**	**3 Months**	**6 Months**	**9 Months**	**12 Months**
Medical therapy	Average BMI	36.3	35.4	34.8	34.5	34.4
	Average number of diabetes medications	2.8	3.1	3.1	3.0	3.0
Gastric bypass	Average BMI	37.0	31.8	28.2	26.9	26.8
	Average number of diabetes medications	2.6	1.1	0.6	0.4	0.3

Data from Schauer, P. R., et al. (2012). Bariatric surgery versus intensive medical therapy in obese patients with diabetes. *N Engl J Med* 366(17):1567–1576.

Stumbling on a cure

Banting, Best, and the discovery of insulin

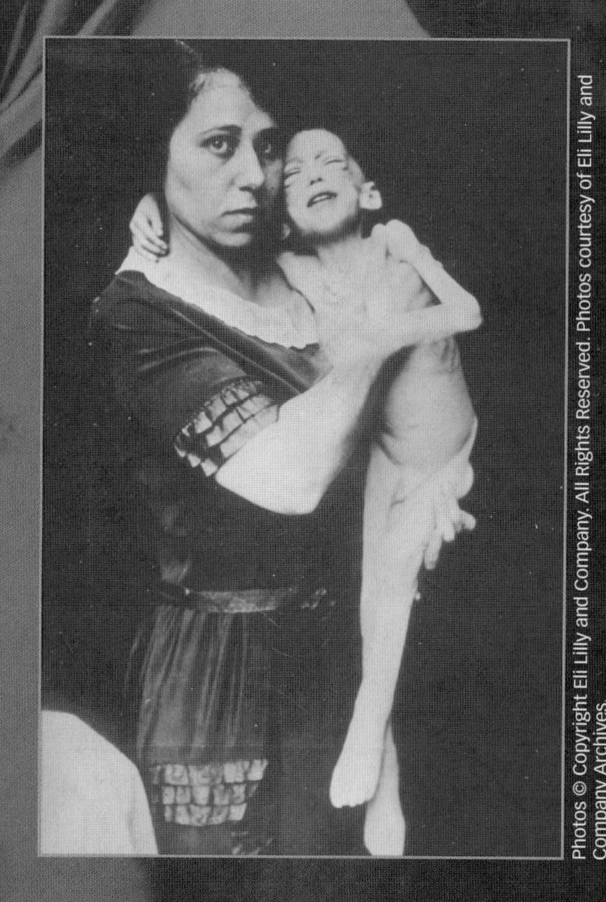

Photos © Copyright Eli Lilly and Company. All Rights Reserved. Photos courtesy of Eli Lilly and Company Archives.

DRIVING QUESTIONS

1 What is the role of insulin in blood-sugar regulation and diabetes?

2 In general terms, what is a hormone?

3 What features are shared by type 1 diabetes and type 2 diabetes, and what features are unique to each type?

Until 1921, there was no effective treatment for diabetes, and a diagnosis was in essence a death sentence.

In December 1921, a 14-year-old boy named Leonard Thompson lay sick and dying in a Canadian hospital bed. He weighed just 65 pounds and was lapsing in and out of consciousness. Elevated levels of sugar in his blood, caused by the disease known as diabetes mellitus, were wreaking havoc on his internal organs, and doctors told his parents he would likely not survive more than a month. With nothing to lose, Leonard's parents agreed to let the doctors try something unusual: they injected the boy with a chemical that scientists had recently isolated from the pancreas of a dog. That chemical radically altered Leonard's fate—and the lives of countless others since he received the treatment.

For most of human history, diabetes was a dreaded and deadly disease. People with diabetes have unusually high blood-sugar levels—a sign that the body's cells are not taking up sugar from the blood. Since cells require sugar (principally glucose) as fuel to power their activities, this lack of uptake eventually starves the body of nourishment, while the high blood-sugar levels cause a host of problems of their own.

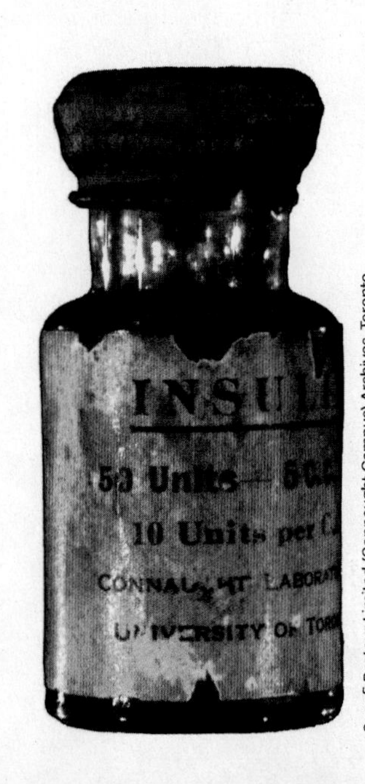

Sanofi Pasteur Limited (Connaught Campus) Archives, Toronto

The word "diabetes" comes from a Greek word meaning "to pass through," and refers to the fact that people with this disease tend to urinate excessively. By eliminating excess sugar, urinating restores normal blood-sugar levels. Before blood tests were available, doctors diagnosed diabetes by testing for the presence of sugar in a patient's urine, which would often attract flies because it was so sweet (the word "mellitus" means "like honey"). Until 1921, there was no effective treatment for diabetes, and a diagnosis was in essence a death sentence.

The breakthrough came in a laboratory located just footsteps from where Leonard was being treated at the University of Toronto. A team of ragtag researchers, working on a shoestring budget, succeeded where many other scientists had failed. With nothing more than a pack of stray dogs and some borrowed laboratory equipment, they discovered an effective treatment for millions of people.

Hunting Down a Mystery Chemical

▶ The search for insulin

In the fall of 1920, a young Canadian surgeon named Frederick Banting was preparing a lecture on the physiology of the pancreas, the carrot-shaped organ that sits next to the stomach. At the time, the pancreas was known to be important for the digestion of food molecules. Digestive juices secreted from the pancreas into the small intestine contain enzymes that help break down carbohydrates, proteins, and fats. But scientists were learning that the pancreas has other functions as well. Experiments in 1890 had shown that surgical removal of the pancreas in dogs led quickly and inevitably to all the symptoms of severe diabetes: high blood-sugar levels, sugar in the urine, coma, and death. This outcome suggested that the pancreas plays a critical role in the regulation of blood sugar.

Scientists were beginning to suspect that the pancreas was essentially two organs in one. One part of the pancreas released digestive enzymes into the intestines through ducts. Another part, made up of discrete islands of cells, released an unidentified substance into the blood that regulated blood-sugar levels. A number of researchers had tried, unsuccessfully, to isolate this mystery chemical from these groups of cells (called islets of Langerhans after Paul Langerhans, who first noticed them under a microscope in 1869). So convinced were researchers that the substance existed that they gave it a name, "insuline"—from the Latin *insula*, "island" **(INFOGRAPHIC M7.1)**.

The notion that certain "internal secretions" produced by glands in the body and released into the bloodstream could influence physiology was a relatively new idea at the beginning of the 20th century—the term "hormone" was first used in 1905. Before that time, conventional wisdom held that physiological processes, like the regulation of blood sugar, were controlled primarily by the brain through nerves. But evidence was accumulating that blood-borne chemicals—hormones—controlled a number of different physiological processes, including blood-sugar regulation. The study of hormones and

INFOGRAPHIC M7.1

The Pancreas Produces Digestive Enzymes and Insulin

The pancreas plays two distinct roles. Some pancreatic cells secrete digestive enzymes into the small intestine, where those enzymes help break down organic molecules, including carbohydrates, into smaller components. Pancreatic cells in the islets of Langerhans secrete hormones—including insulin—into the circulatory system.

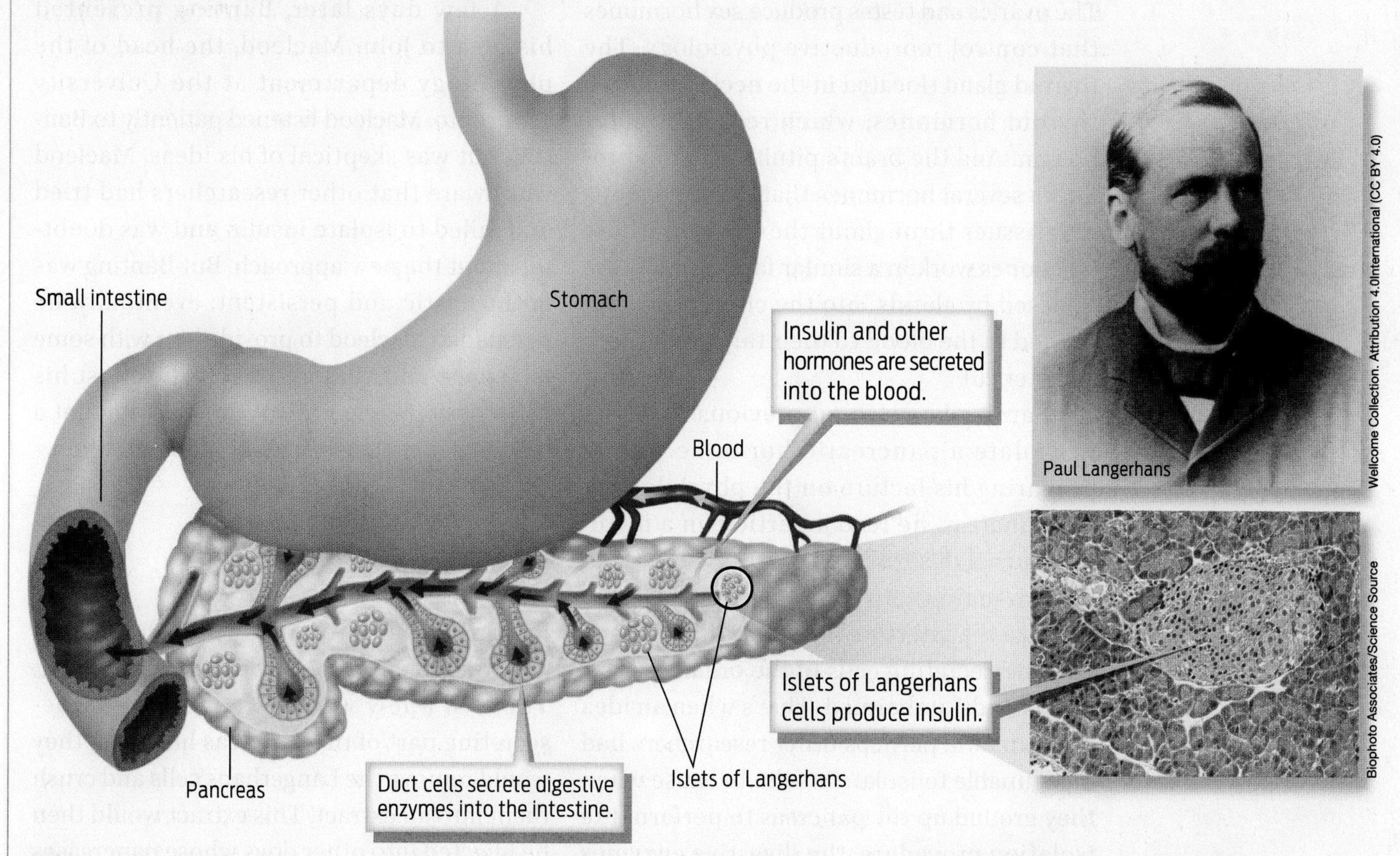

? **Which pancreatic cells (duct cells or islet of Langerhans cells) produce insulin? Which produce digestive enzymes?**

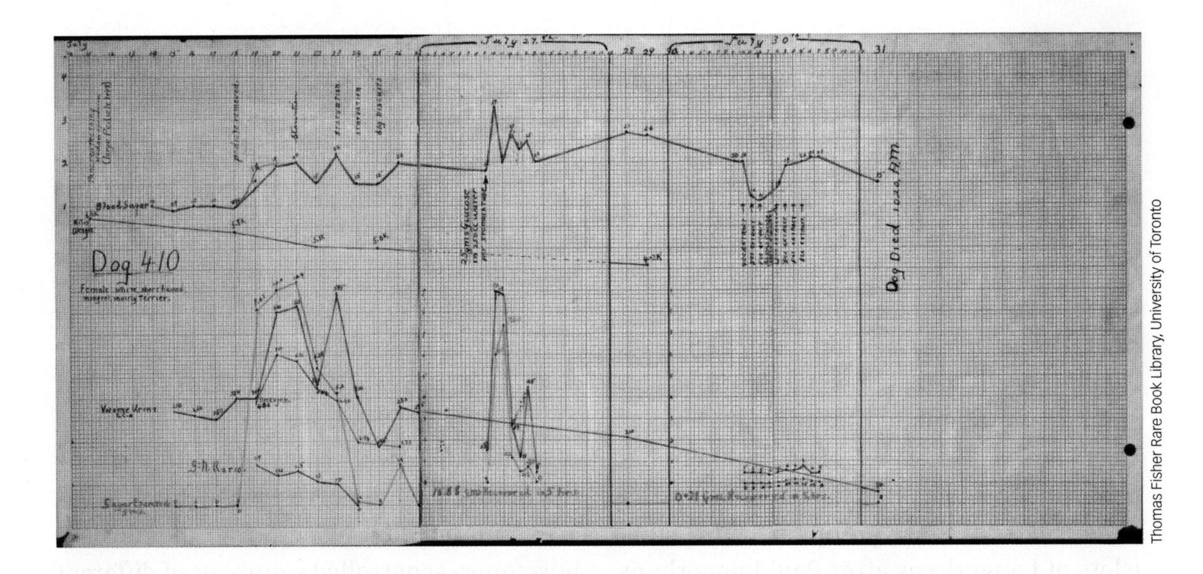

A page from Banting's laboratory notebook documenting glucose levels in dogs.

Thomas Fisher Rare Book Library, University of Toronto

the glands that produce them would come to be known as endocrinology.

Today we know that the body has many hormone-secreting glands, which together coordinate many physiological processes. For example, the adrenal glands (located above the kidneys) secrete the hormone adrenaline, which controls the "fight or flight" response. The ovaries and testes produce sex hormones that control reproductive physiology. The thyroid gland (located in the neck) produces thyroid hormones, which regulate metabolism. And the brain's pituitary gland produces several hormones that control glands and tissues throughout the body. All these hormones work in a similar fashion: they are released by glands into the circulation and carried in the blood to their target cells (see Chapter 26).

Banting knew about previous attempts to isolate a pancreatic hormone. While preparing his lecture on the physiology of the pancreas, he read an article in a medical journal describing how surgically tying off the main duct leading from the pancreas to the intestine led to the death of the enzyme-secreting cells of the organ but left the "islands" unharmed. That's when an idea came to him: perhaps other researchers had been unable to isolate insulin because when they ground up the pancreas to perform the isolation procedure, the digestive enzymes that the pancreas produces destroyed the insulin. Perhaps he could avoid that problem by first performing surgery on an animal to tie off the pancreatic duct and then waiting for the enzyme-secreting part of the organ to die. He jotted down some hasty notes, then went to bed, but was so excited he was unable to sleep.

A few days later, Banting presented his idea to John Macleod, the head of the physiology department at the University of Toronto. Macleod listened patiently to Banting but was skeptical of his ideas. Macleod was aware that other researchers had tried and failed to isolate insulin and was doubtful about the new approach. But Banting was enthusiastic and persistent; eventually, he persuaded Macleod to provide him with some lab space and 10 dogs on which to test his hypothesis. Macleod also agreed to appoint a graduate student, Charles Best, to assist him.

A Rocky Start

▶ Isolating and testing insulin

Banting and Best began by performing surgery on dogs to tie off the pancreatic duct. Then, in a few weeks, when the enzyme-secreting part of the pancreas had died, they would remove the Langerhans cells and crush them into an extract. This extract would then be injected into other dogs whose pancreases had been surgically removed, to see if the extract would prevent them from getting diabetes. At least, that was the idea—it proved to be easier said than done.

The initial results were disappointing: 7 of the 10 dogs died of complications from the surgery even before the researchers could perform the rest of the experiment. To obtain more research animals, they resorted to buying dogs off the street for $1, no questions asked. Banting practiced the surgery on these new dogs. Finally, he managed to keep the dogs alive long enough to extract the "island" cells. Banting and Best were now ready to inject the extract into another dog made diabetic by removing its pancreas. The procedure worked beautifully. The extract significantly lowered the dog's blood-sugar levels and also reduced the amount of sugar in the dog's urine. A pancreatic hormone did indeed seem to regulate blood sugar. Banting and Best reported these results to Macleod, who realized a major breakthrough had been made **(INFOGRAPHIC M7.2)**.

INFOGRAPHIC M7.2

Banting and Best Isolated Insulin from Dog Pancreases

Banting and Best developed a method of extracting insulin from dog pancreases. The extracted insulin reversed the symptoms of diabetes in diabetic dogs.

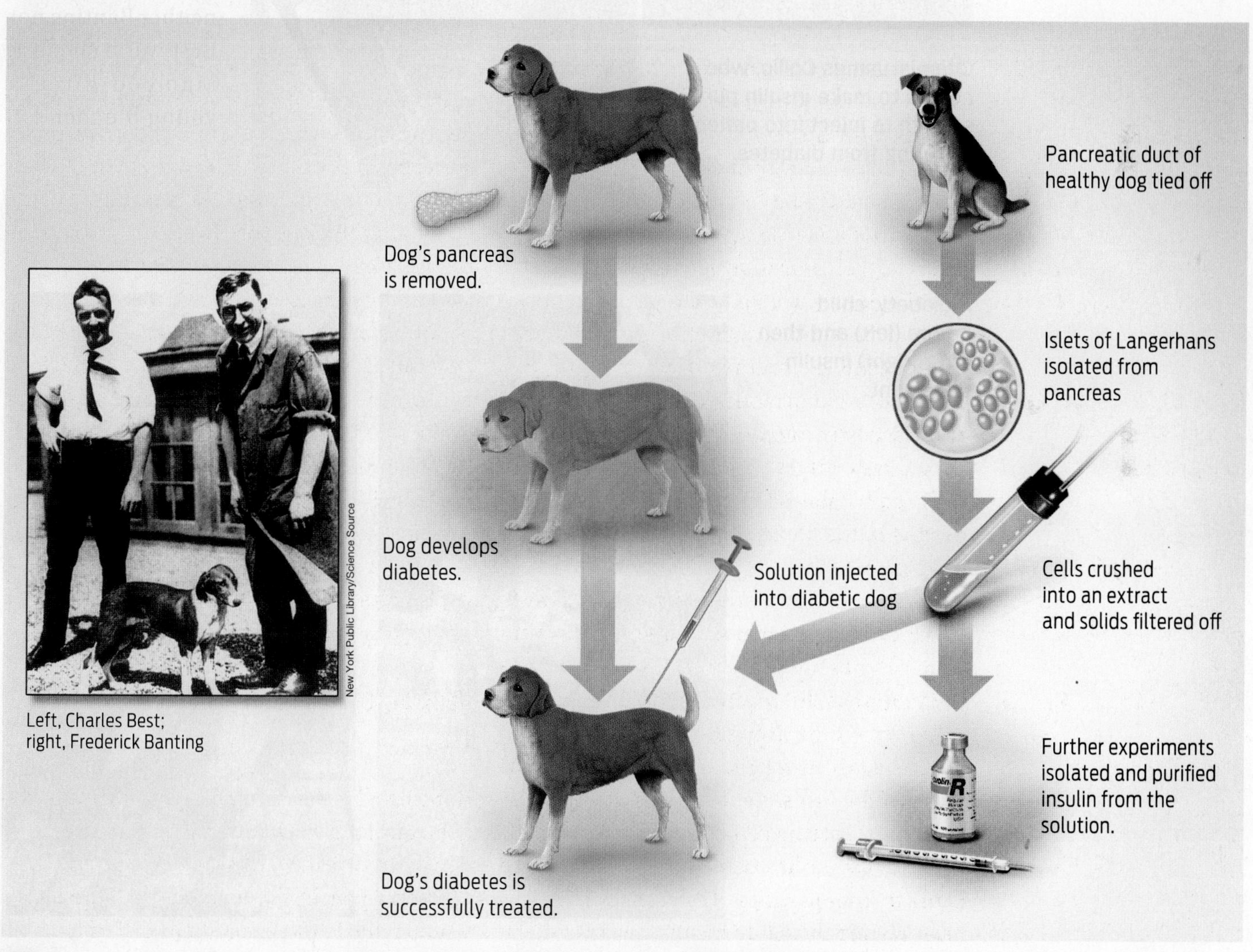

Left, Charles Best; right, Frederick Banting

? Why did removal of the pancreas cause the dog on the left to develop diabetes?

Banting and Best were excited to have isolated the elusive hormone insulin, but it quickly became clear that producing insulin from stray dogs was not a viable way to create a treatment for thousands of human patients, so they kept looking for other sources. They discovered that fetal calf pancreas was also an abundant source of insulin. Unfortunately, the insulin prepared from ground-up calf pancreas often produced a severe allergic reaction in the animals who received it. Macleod believed that these reactions occurred because the insulin was impure—it was mixed with other chemicals of the pancreas. Since neither Banting nor Best was a chemist, Macleod hired a young biochemist,

The Thomas Fisher Rare Book Library, University of Toronto

Chemist James Collip, who helped to make insulin pure enough to inject into patients suffering from diabetes.

Science & Society Picture Library/Getty Images

A diabetic child before (left) and then after (right) insulin treatment.

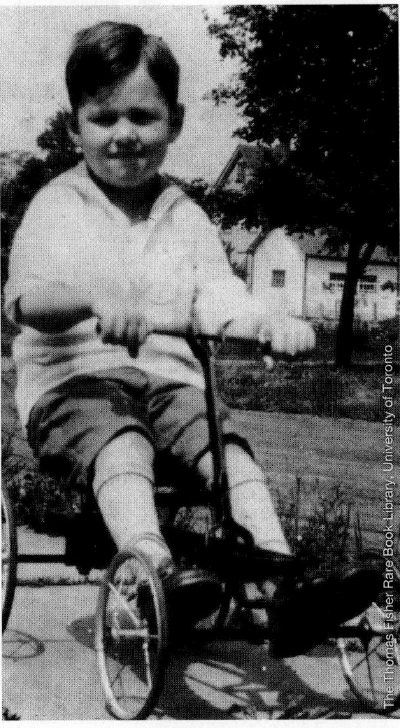

James Collip, to help prepare a chemically purer version of the drug. Such a cleaned-up version would be needed if the drug were to be used in people.

Eventually, using a method that included alcohol and evaporation, Collip was able to produce a pure form of the drug, and the team was ready to test insulin on humans. They talked to doctors at the hospital across the street at the University of Toronto, who were desperate for a new way to treat patients with diabetes. Leonard Thompson, the 14-year-old patient with diabetes who was falling in and out of consciousness, received this purified version on January 23, 1922. Within moments, he was alert and smiling, to the astonishment of his parents. Encouraged by these results, the doctors next went from bed to bed in the diabetes ward, injecting one patient after another. Before they reached the last patient, the first patients to be injected were regaining consciousness. Insulin as a treatment for diabetes was born.

Frederick Banting.

Balancing Blood Sugar

▶ Insulin and diabetes

Today we know that diabetes is caused by the body's inability to produce or properly respond to insulin. Normally, insulin binds to receptors on target cells, triggering those cells to take up glucose from blood. This glucose is then used by the cells as a source of energy. Without insulin or properly functioning insulin receptors, cells cannot take up sugar from the blood—and so they starve.

There are two main types of diabetes. Type 1 diabetes, which represents about 5% of cases, is caused by the death of the insulin-producing cells in the islets of Langerhans. (The cells die when they are attacked by the immune system, in a form of autoimmunity; see Chapter 32.) People with type 1 diabetes do not produce any insulin, so they require daily injections of insulin to survive. Type 2 diabetes, which accounts for about 95% of cases, begins when cells in the body don't respond to insulin. Scientists don't fully understand why this happens, but they think that inflammation resulting from excess body fat may play an important role. Having cells that are unresponsive to insulin is as bad as not making insulin at all. Type 2 diabetes is not treated with insulin, but instead is mostly managed through dietary changes, which aim to control blood

Ironically, Banting's original hypothesis about how to isolate insulin turned out to be wrong.

INFOGRAPHIC M7.3

Insulin Controls Blood Sugar

Insulin normally signals cells to take up sugar from the blood, thereby reducing blood-sugar levels. In diabetes, insulin is either not produced or cells don't respond to it, leading to elevated blood-sugar levels.

Normal Insulin Response

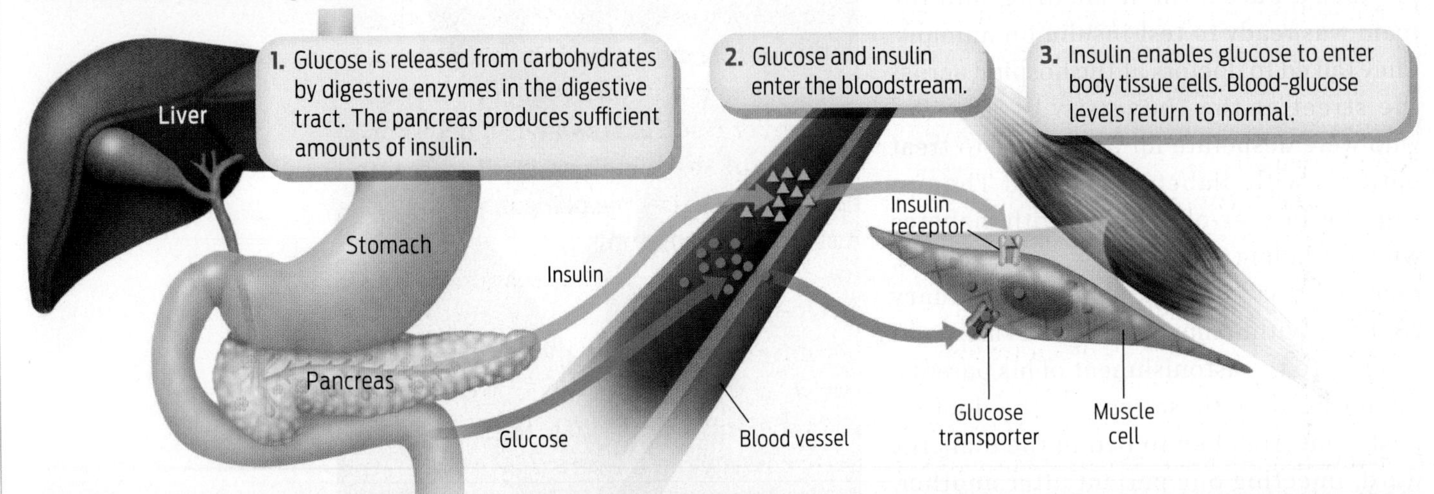

	Digestive System	Insulin	Body Tissue Cells	Bloodstream Glucose
Normal Insulin Response	Digestive enzymes release high levels of glucose into the blood. ↑	Pancreas produces sufficient amount of insulin. ↑	Insulin enables glucose to enter cells.	Blood-glucose levels return to normal.
Type 1 Diabetes	Digestive enzymes release high levels of glucose into the blood. ↑	Pancreas produces little or no insulin. ↑	Very little insulin allows only a small amount of glucose to enter cells.	Blood-glucose levels remain high. ↑
Type 2 Diabetes	Digestive enzymes release high levels of glucose into the blood. ↑	Pancreas produces sufficient amount of insulin. ↑	Cells are unresponsive to insulin.	Blood-glucose levels remain high. ↑

? What is the primary difference between type 1 and type 2 diabetes?

sugar spikes, and through weight loss **(INFOGRAPHIC M7.3)**.

The pancreas produces other hormones as well, including glucagon, which raises blood sugar. Glucagon and insulin work together to regulate blood sugar through a classic negative feedback loop. (For more on glucose metabolism, including the role of glucagon, see Chapter 26.)

Left untreated or unmanaged, chronic diabetes (both type 1 and type 2) can cause a host of health problems, including increased risk of heart disease and stroke, loss of vision, kidney disease, nerve damage, and even death. An estimated 34 million Americans (roughly 10.5% of the total population) suffer from diabetes, making it the most common endocrine disorder in the United States.

Reception and Controversy

Overnight, Banting went from being an unknown surgeon to a world-famous scientist, receiving many accolades. Because of his efforts, Leonard Thompson would live

13 more years, to age 27, when he died of pneumonia.

Ironically, Banting's original hypothesis about how to isolate insulin turned out to be wrong. Contrary to what Banting believed, it was not necessary to remove the digestive enzymes from the pancreas to isolate insulin. With Collip's purification method, the researchers were able to extract insulin from a whole pancreas, without first performing Banting's surgery.

Some scientists thought Banting's fame and status were undeserved because his initial premise was wrong. But these detractors, others said, misunderstood the nature of the scientific method, which often involves a good deal of luck.

"Nobody can deny that a discovery of first-rate importance has been made, and, if it proves to have resulted from a stumble into the right road, where it crossed the course laid down by a faulty conception, surely the case is not unique in the history of science," wrote Henry Hallet Dale, a British physiologist, in a letter to the *British Medical Journal* in 1922.

For their work on developing insulin, Banting and Macleod shared the 1923 Nobel Prize in Physiology or Medicine. What about Best and Collip? The Nobel committee deemed their contributions less critical to the discovery, a view that didn't sit well with the participants in the research. When Banting heard that Best had not been recognized, he was furious and refused at first to accept the award. Eventually, he accepted, but shared half his prize money with Best. Macleod, in turn, shared half his prize money with Collip. The earlier researchers who had determined the role of the pancreas in diabetes and deduced the existence of insulin—even gave it a name—got nothing.

Since Banting and his team made their important discoveries, a number of other scientific milestones have occurred. Researchers learned that insulin was a protein and, in 1952, the British biochemist Frederick Sanger determined its amino acid sequence—the first such protein sequence to be described. Today, insulin is no longer obtained from pancreas harvested from animal tissues. Instead, it is synthesized by genetically engineered bacterial cells modified to contain the human insulin gene (see Chapters 8 and 9 for a discussion of genetic engineering). Insulin remains one of the most important drugs of modern medicine. ■

More to Explore

- American Diabetes Association: www.diabetes.org
- Rosenfeld, L. (2002), Insulin: discovery and controversy. *Clin Chem* 48:2270–2288.
- Henderson, J. (2005). Ernest Starling and "hormones": an historical commentary. *J Endocrinol* 184:5–10.
- Taubes, G. (2016). *The Case against Sugar*. New York: Knopf.

MILESTONES IN BIOLOGY 7 Test Your Knowledge

1. What organ produces insulin?
   a. the liver
   b. the stomach
   c. the pancreas
   d. blood
   e. skeletal muscle

2. Someone who cannot produce insulin will likely have blood-sugar levels that are
   a. normal.
   b. lower than normal.
   c. higher than normal.

3. In general terms, what is a hormone?
   a. a signaling molecule that is transported by the circulatory system and acts on target cells
   b. a signaling molecule that regulates blood sugar
   c. a signaling molecule that is produced by the pancreas
   d. a signaling molecule that increases the concentration of blood sugar
   e. a protein that is released by one cell and acts on an adjacent cell

4. Knowing that someone has chronically elevated blood-sugar levels, are you able to say if this person has type 1 diabetes or type 2 diabetes? Why or why not?

5. Insulin is a
   a. monosaccharide.
   b. protein.
   c. triglyceride (fat).
   d. phospholipid.
   e. any of the above, depending on the diet

6. What are the key differences in glucose levels and insulin levels between type 1 and type 2 diabetes? Consider levels during periods of fasting and levels after a carbohydrate-rich meal.

DEATH *in* Bogalusa

From tragic deaths in a southern town, insight into heart disease

Kevin Curtis/Science Source

DRIVING QUESTIONS

1 What structures make up the cardiovascular system, and how does blood flow through the system?

2 What is the structure of the heart and of the different types of blood vessels?

3 What is the composition of blood, and what does blood do?

4 What is cardiovascular disease, and what are some of the risk factors for developing cardiovascular disease?

Some scientific discoveries begin in a laboratory. This one begins in a funeral home, in 1978. Under glaring lights, in the back room of the Cook-Richmond Funeral home in Bogalusa, Louisiana, two pathologists hover over the lifeless body of a young African American male. There is no morgue in Bogalusa, so an autopsy is being conducted here, with newspapers spread beneath the body. The pathologists slice through skin and muscle with scalpels, looking for the usual suspects—internal bleeding, broken bones—and then write up their report for the coroner.

The pathologists aren't quite done, however. Before closing up, they carefully remove the victim's heart, pack it in saline, and prepare it for a trip to a medical school in New Orleans. It is an unusual step, not standard for an autopsy. But the pathologists are heeding the instructions of a prominent local physician, who has the blessing of the family.

Greg Miles

Gerald Berenson, 1922–2018.

Gerald Berenson was a cardiologist with a passionate interest in heart disease. Since 1972, as a professor at the Louisiana State University School of Medicine, he had spearheaded what was then a novel study: an epidemiological study of heart disease in Bogalusa, a rural town of about 16,000 people located 60 miles north of New Orleans.

The study began with a pretty simple idea: follow a large group of children over a period of time and correlate their physical and lifestyle characteristics with the development of heart disease in adulthood. The biggest hurdle was a logistical one—how to enlist the thousands of children necessary to produce a robust data set and keep them coming back for evaluation year after year.

But Berenson wanted to do more than make statistical correlations. He also wanted to document the progression of heart disease directly. And for that he needed a different type of evidence.

The heart that Berenson obtained in 1978 was one of more than 200 such organs collected from young people in Bogalusa over the next 20 years. In fact, nearly every young person who died, of whatever cause, was autopsied and had the heart removed. From this medical detective work has emerged a detailed understanding of heart disease in young people and the evidence needed to clinch the case against a cold-blooded killer.

CARDIOVASCULAR DISEASE (CVD)
A disease of the heart or blood vessels, or both.

CARDIOVASCULAR SYSTEM
The system that transports nutrients, gases, and other critical molecules throughout the body. It consists of the heart, blood vessels, and blood.

One in three deaths in the United States is caused by cardiovascular disease—roughly 2,300 deaths per day.

A Silent Epidemic

▶ Cardiovascular disease and the cardiovascular system

Cardiovascular disease (CVD), which includes heart disease and stroke, is the number one killer of men and women in the United States and the developed world. According to the American Heart Association, one in three deaths in the United States is caused by cardiovascular disease—roughly 2,300 deaths per day **(INFOGRAPHIC 28.1)**.

CVD claims so many lives each year for two main reasons. First, the disease is largely "silent"—people don't know they're sick until it's too late. Second, the system it affects is one of the most important to our survival, so when things go wrong, the result tends to be fatal.

The **cardiovascular system** transports nutrients, gases, hormones, and other critical molecules throughout the body. It consists of the heart, the blood, and the blood vessels. The **heart** is essentially a pump. It is about the size of a fist and very muscular. By repeatedly contracting and relaxing, the heart pumps blood throughout the body.

Blood is a fluid that contains cells, molecules, and dissolved gases. One of its main roles is to deliver oxygen to tissues of the body and carry away carbon dioxide waste. Oxygen is carried in red blood cells, while carbon dioxide is dissolved in blood.

Blood travels through the body in blood vessels. **Arteries** carry blood away from the heart. **Veins** return blood to the heart. These vessels form circuits that carry oxygen-rich blood in one direction and oxygen-depleted blood in the reverse direction. For example, the **aorta** is a major artery carrying oxygen-rich blood from the heart to body tissues. Oxygen-depleted blood from body tissues returns to the heart in veins. The **coronary arteries** split off from the aorta and curve back to carry oxygen-rich blood

HEART
The muscular pump that generates force to move blood throughout the body.

BLOOD
A circulating fluid that contains several types of cells and transports substances, including nutrients, gases, and hormones.

ARTERIES
Blood vessels that carry blood away from the heart.

VEINS
Blood vessels that carry blood toward the heart.

AORTA
The large artery that receives oxygenated blood from the left ventricle and delivers it to the body.

CORONARY ARTERIES
The blood vessels that deliver oxygen-rich blood to the heart muscle.

INFOGRAPHIC 28.1

Cardiovascular Disease Is the Leading Killer in the United States

Of the top ten causes of death, cardiovascular disease (including heart disease and stroke) continues to be the leading cause of death in both men and women in the United States.

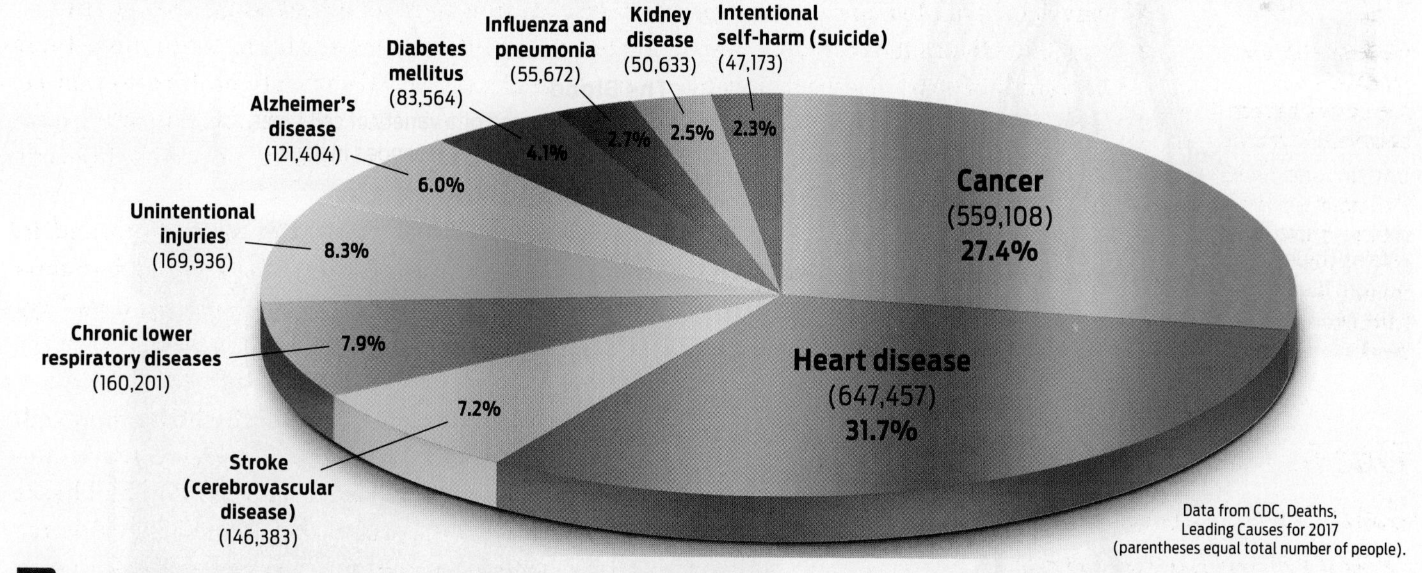

? Of the top ten causes of death in the United States, what percentage is attributable to heart disease and cerebrovascular disease (stroke) combined?

to the tissues of the heart itself. Not all arteries carry oxygenated blood: the pulmonary artery carries oxygen-depleted blood from the heart to the lungs. The pulmonary vein carries oxygen-rich blood from the lungs to the heart **(INFOGRAPHIC 28.2)**.

For decades, doctors have known that CVD is essentially a matter of "bad plumbing." In this disease, fatty deposits develop in the arteries, cutting off or reducing blood flow to the body or to the heart muscle itself. This is dangerous because tissues then do not receive enough of the oxygen that they need to carry out aerobic respiration (see Chapter 6). When the coronary arteries that supply blood directly to the heart muscle are blocked, the cells of the heart begin to die—an event called a **heart attack.**

HEART ATTACK
Damage to the heart muscle resulting from the restriction of blood flow to heart tissue.

INFOGRAPHIC 28.2

The Cardiovascular System

The cardiovascular system consists of the heart, the blood vessels, and the blood, which transport nutrients and gases throughout the body. The cardiovascular system delivers oxygen to body tissues for aerobic respiration and removes carbon dioxide waste.

The Blood Vessels
Blood vessels are the tubes that carry blood throughout the body.

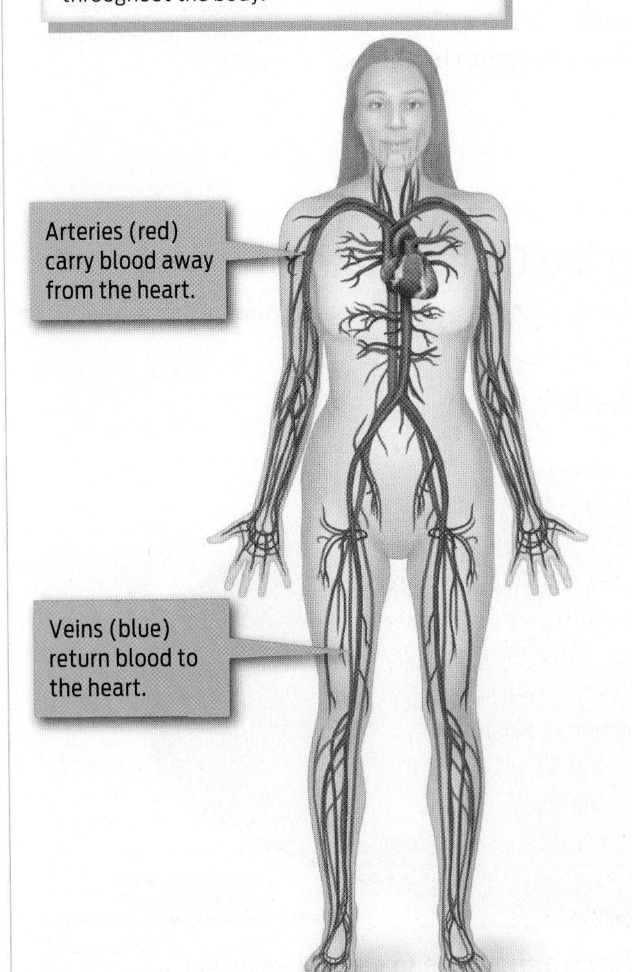

The Heart
The heart is a muscular pump that contracts to move blood through the blood vessels. It has four chambers that coordinate to receive and direct blood to the lungs and body.

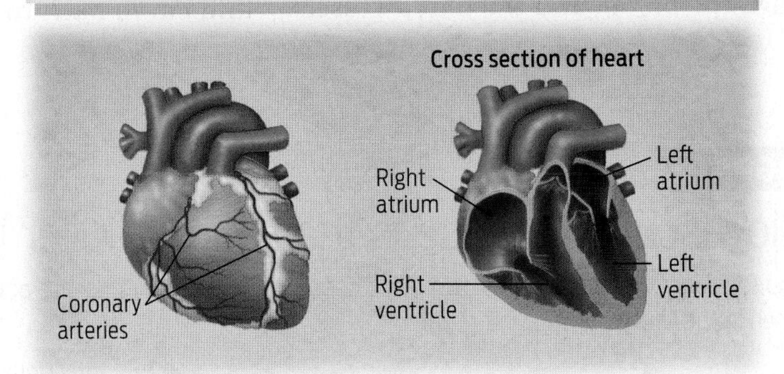

The Blood
Blood is composed of a variety of cell types, molecules, and gases suspended in liquid.

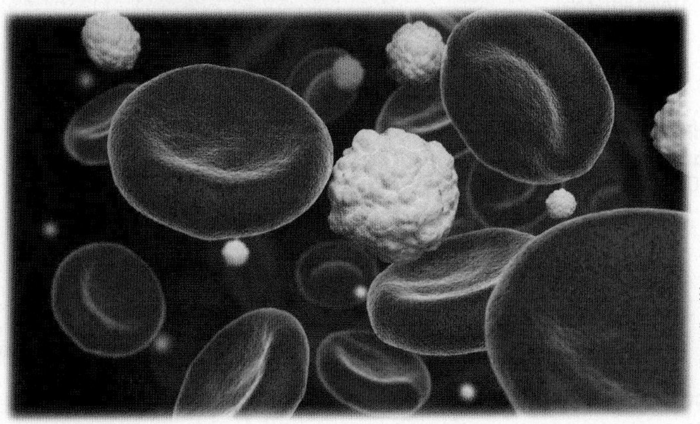
Alexey Kotelnikov/Alamy

? **Is blood traveling from your heart toward your big toe traveling in an artery or a vein?**

For the most part, young people do not have heart attacks. But just because teenagers do not die from heart disease, that does not mean they can ignore heart health. That's because heart disease can be insidiously gaining a foothold long before we have any obvious symptoms.

This is common knowledge now, but when Gerald Berenson started his study in 1972, it was far from accepted wisdom. Cardiologists—influenced by the prevailing beliefs of the day—focused more on treatment than on prevention. But Berenson had been trained in pediatrics as well as cardiology, and he knew that it was important to understand the beginnings of heart disease as well as its endings.

Under the Knife

▸ Atherosclerosis and plaque

William Newman was one of three pathologists on Berenson's team who analyzed the tissues collected from the autopsies. He was also directly involved in harvesting the specimens. Whenever a young person in Bogalusa died, Newman and a colleague would drive the 66 miles up from Louisiana State University School of Medicine in New Orleans to Bogalusa to perform the autopsy for the coroner.

What Newman and his colleagues saw when they opened the hearts of the young people shocked them: fatty streaks lined the coronary arteries, which supply the heart muscle with blood, as well as the aorta, the major artery leaving the heart that ultimately supplies blood to the rest of the body. Virtually all the individuals autopsied had these fatty streaks, and the extent of the streaking increased with age.

As Newman explains, a fatty streak is an early form of **atherosclerosis,** a condition in which fatty deposits and other substances build up in the lining of blood vessels and restrict blood flow in those vessels. Atherosclerosis, sometimes called hardening of the arteries, is a common cause of heart disease. It's a complex process, and scientists aren't entirely sure what triggers it. One hypothesis is that it begins when blood vessels become damaged in some way. Once a vessel is damaged, a waxy lipid called **cholesterol** begins to accumulate in the lining of the vessel. White blood cells that normally help to heal injuries collect at the site of cholesterol deposition, but instead of helping the situation, they make it worse, causing more cholesterol to accumulate. A fatty streak is the result. Newman says you can identify these streaks when you cut open a vessel because they "look a bit yellow." They will also take up a lipophilic ("fat-loving") dye, which makes them more apparent and easily quantified.

What Newman and his colleagues saw when they opened the hearts of the young people shocked them.

To quantify the extent of atherosclerosis, each pathologist would look at the vessels in every autopsy specimen and assign the stained fatty streaks a score, and the scores were then averaged. The study was blind, so the pathologists didn't know the source of the tissues beforehand. Once they had the data, statisticians correlated these anatomical measurements with known **risk factors** for heart disease—such as high blood pressure, smoking, and high cholesterol—exhibited by the young people when they were alive. The result? "We found that those individuals who had higher levels of known risk factors on the average had more fatty streaks in their coronary arteries and aorta than individuals who had lower levels," Newman says.

In other words, though many of them were not even old enough to vote, these young people already had telltale signs of heart disease. Some even had evidence of more severe atherosclerosis: thick and hardened deposits of cholesterol, fat, calcium, and

ATHEROSCLEROSIS
A condition in which fatty deposits build up in the lining of arteries, restricting blood flow; also known as hardening of the arteries.

CHOLESTEROL
A lipid that is an important component of cell structures; it is used to make important molecules and also plays a role in heart disease.

RISK FACTOR
A behavior, exposure, or other factor that increases the probability of developing a disease.

other materials collectively called **plaque** had begun to obstruct blood flow. Had these young people lived, those with the higher levels of atherosclerosis would have been at risk for a heart attack or other complications of heart disease **(INFOGRAPHIC 28.3)**.

PLAQUE
Deposits of cholesterol, other fatty substances, calcium, blood clotting proteins, and cellular waste that accumulate inside arteries, limiting the flow of blood.

CAPILLARIES
The smallest blood vessels, which are the sites of gas, nutrient, and waste exchange between the blood and tissues.

A Series of Tubes

▶ Blood vessels and gas exchange

The evidence that Newman and his colleagues found of atherosclerosis in the heart's coronary arteries in young people was particularly troubling. The coronary arteries branch off the aorta and supply the heart muscle with blood. It is these small arteries that are the usual sites of blockages that lead to a heart attack.

If the heart has blood continually pumping through it, why does it need its own blood supply from arteries leading from the aorta back into the heart muscle? The answer has to do with how gases like oxygen and carbon dioxide, and nutrients like sugar, are exchanged between blood vessels and tissues.

Arteries and veins are like major highways for blood transport—their main responsibility is bringing blood to and from the heart. Situated between these two large types of vessels, however, are capillaries. **Capillaries** are tiny blood vessels, located in tissues, where gas and nutrient exchange occurs.

INFOGRAPHIC 28.3

Atherosclerosis: A Common Cause of Cardiovascular Disease

Atherosclerosis begins with fatty streaks that thicken the wall of a blood vessel. A fatty streak is composed of lipids, such as cholesterol, and white blood cells. These fatty streaks may, over time, increase in size and harden, developing into plaques. Plaques obstruct blood flow. When blood flow in the coronary arteries is blocked, the result is a heart attack.

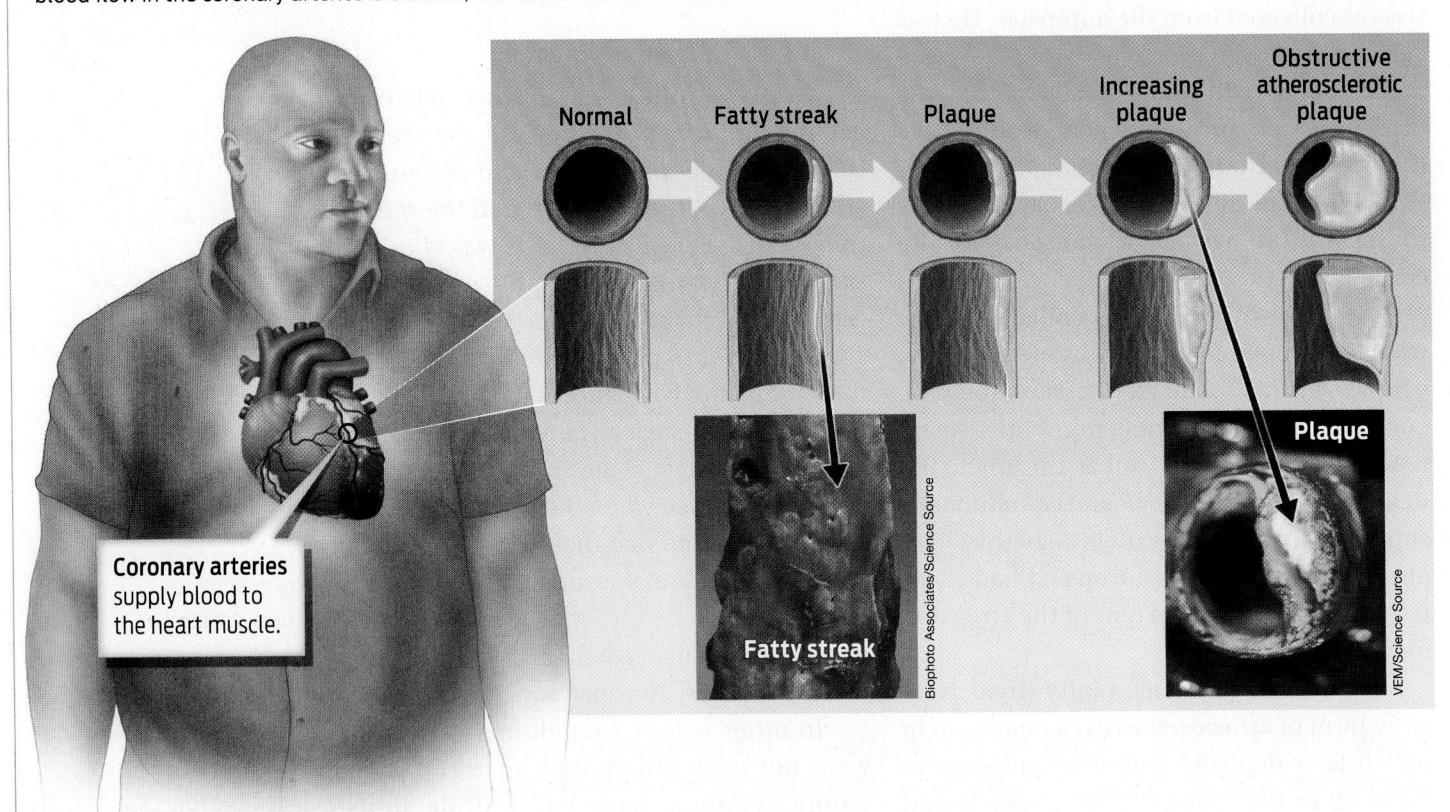

? **If fatty streaks don't obstruct blood flow, are they still a cause for concern? Why?**

INFOGRAPHIC 28.4

Capillaries Are Sites of Nutrient and Gas Exchange

Blood is carried by three different types of blood vessels. Arteries are large vessels that carry blood away from the heart. Arteries branch into smaller vessels in body tissues. The smallest blood vessels are the capillaries. The slow flow of blood in capillaries allows plenty of time for the exchange of nutrients and gases between the blood and tissue cells. Veins collect blood from the capillaries and deliver it back to the heart.

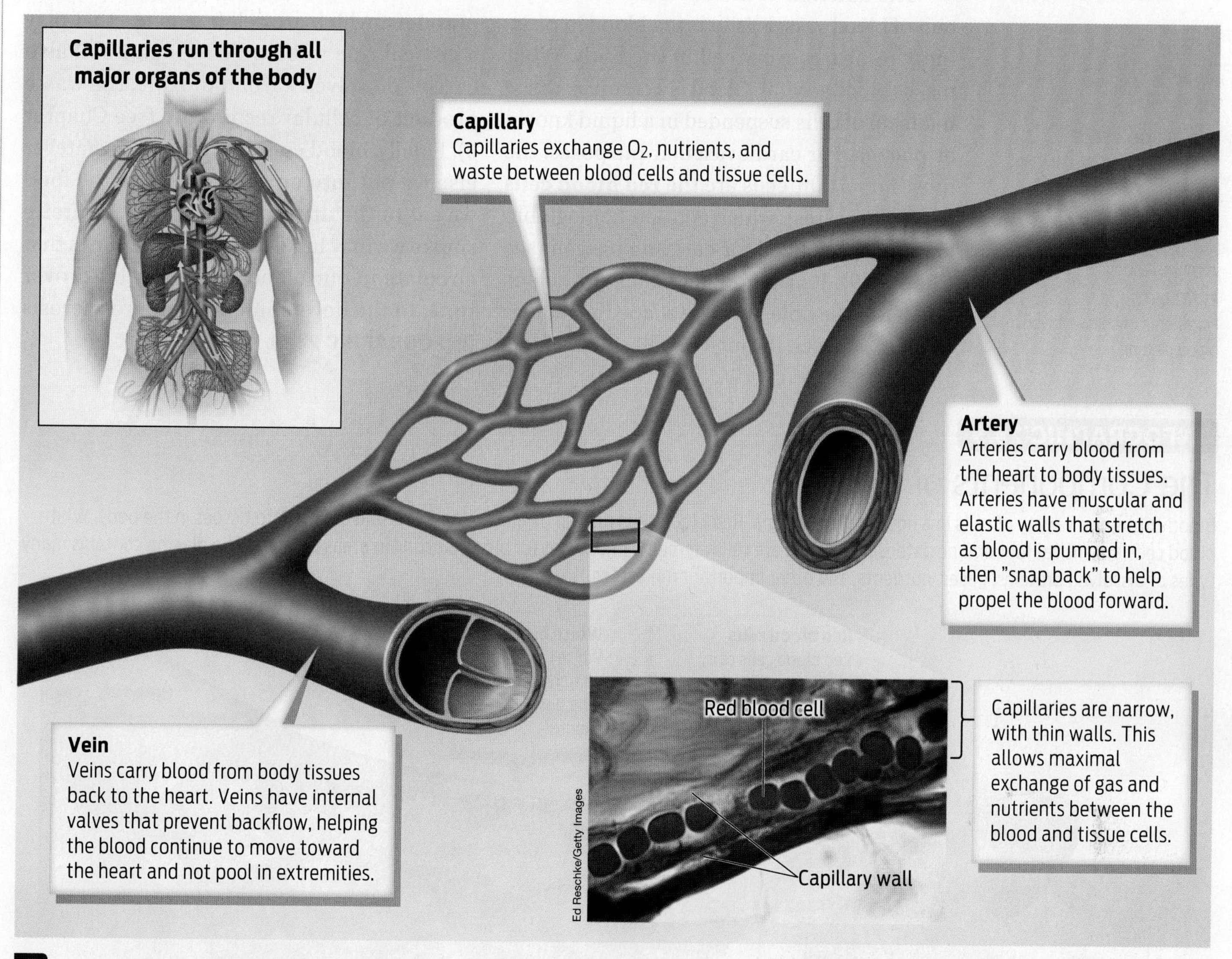

? When a capillary enters a tissue (like the muscle tissue in your leg), what happens to the oxygen and carbon dioxide carried by the capillary as it passes through the tissue?

They are where the action is. Oxygen and nutrients diffuse *out* of capillaries into tissues, and carbon dioxide and wastes diffuse from tissues *into* capillaries. Capillaries are ideally suited to their job because of their thin walls and their slow rate of blood flow, which allows ample opportunities for gas, nutrient, and waste exchange between blood and tissues. Every tissue in the body, including the heart, is infiltrated with capillaries **(INFOGRAPHIC 28.4)**.

Arteries are high-pressure vessels; the blood that flows through them is propelled by the force exerted by the heart muscle. After blood slows down and loses pressure in the capillaries, it enters the veins at low pressure.

As low-pressure vessels, veins rely on contractions of the skeletal muscles in which they are found to return the blood to the heart, and on valves located in these vessels to prevent the blood from flowing backward.

The function of this interconnected system of blood vessels is to bring blood in proximity to nearly every cell in the body. What makes blood so vital? Blood is a complex tissue made up of cells suspended in a liquid known as plasma. For cardiovascular purposes, the most important cells are the **red blood cells** (or erythrocytes), which contain hemoglobin and are specialized for carrying oxygen (see Chapter 9). Red blood cells lack a nucleus and have a flexible, concave shape that allows them to squeeze through capillaries in single file. Oxygen diffuses directly across the membranes of these cells into tissues.

RED BLOOD CELLS
Cells specialized for carrying oxygen.

WHITE BLOOD CELLS
Cells involved in the body's immune response.

PLATELETS
Fragments of cells involved in blood clotting.

Also present in blood are **white blood cells** (or leukocytes), which protect us from infection and respond to injuries as part of the immune response (see Chapter 32), and **platelets,** which are cell fragments that play a critical role in blood clotting. Blood also carries dissolved carbon dioxide, the waste product of cellular respiration (see Chapter 6). Finally, blood contains traces of whatever else we put into our bodies—from the food we eat to the drugs we take to the cigarette smoke we may inhale. Some of the things that circulate in our blood may contribute, over time, to the development of atherosclerosis **(INFOGRAPHIC 28.5).**

INFOGRAPHIC 28.5

The Components of Blood

Blood contains a variety of cells in a liquid called plasma. Red blood cells pick up oxygen in the lungs and deliver it to tissues in the body. White blood cells are critical for immune defense. Platelets are cell fragments important for blood clotting in areas of injury. The plasma contains many types of dissolved proteins, gases, nutrients, and other important molecules.

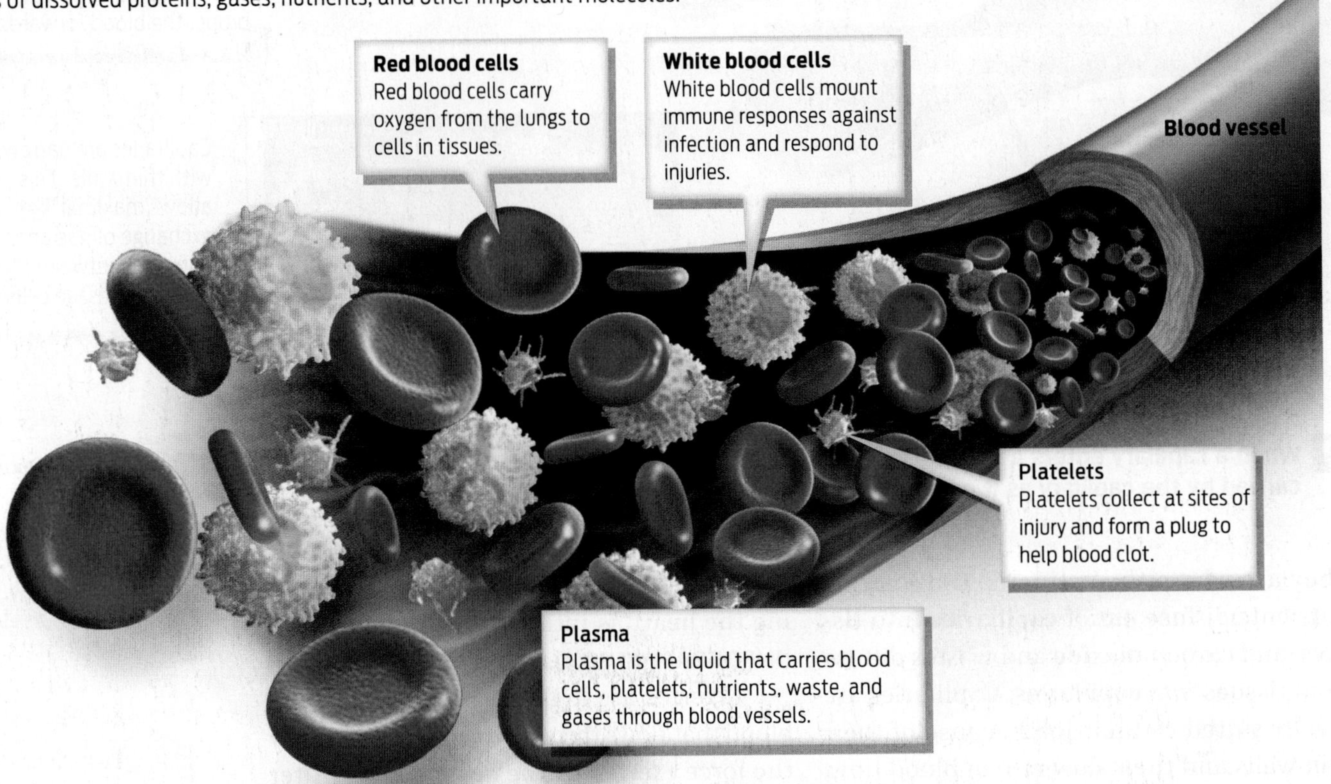

? **What are the functions of red blood cells, white blood cells, and platelets?**

Matters of the Heart

▸ Heart structure and double circulation

Before his death in 2018, Gerald Berenson was Director of the Tulane Center of Cardiovascular Health and Principal Investigator of the Bogalusa Heart Study. He was a seasoned researcher, yet still retained an air of the old-fashioned family doctor. Over the years, he had even been known to make house calls to many of the participants in the study.

"If there was any kind of medical problems I went and examined them myself and took care of them. I had 4,500 patients to look at initially," Berenson said in an interview.

It was this personalized approach to medicine, combined with Berenson's brand of southern tenacity, that ensured success of the study over the years. "I'm often asked 'Why Bogalusa?'" The answer is simple, he said: "It's where I'm from."

Because he grew up there, Berenson knew Bogalusa like the back of his hand. And that firsthand knowledge of this small Louisiana town proved crucial in solidifying support for the study. In fact, nearly everyone in the town participated in the study in some way. Teachers and nurses at local schools served as study liaisons. The pathologists who conducted the autopsies had Berenson as an instructor in medical school. Even Berenson and the coroner were old friends, and it was because of that relationship that Berenson was able to work out the arrangement that enabled the heart autopsies to be performed.

"Eighty percent of the known deaths in the area we were able to autopsy," Berenson said proudly. "Nobody gets that kind of rate."

The heart autopsies allowed doctors to see inside to the heart's unique plumbing. The heart muscle is divided into four chambers that contract and relax in a coordinated way: two paired **atria** sit above two paired **ventricles.** The atria receive blood into the heart, and the ventricles pump blood out of it.

Key to understanding heart function is the fact that blood passes through the heart twice for every trip through the body—it follows a double circuit. One circuit consists of the trip blood makes between the heart and the lungs, known as the **pulmonary circuit;** the other encompasses the trip of blood makes between the heart and the rest of the body, known as the **systemic circuit.**

"I'm often asked 'Why Bogalusa?' The answer is simple. It's where I'm from."

—Gerald Berenson

The two circuits can best be understood by tracing the path of blood as it makes one complete trip through the body. Oxygen-depleted blood from the body returns to the heart through veins and enters the right atrium. This oxygen-poor blood is then pumped into the right ventricle. From there, the pulmonary artery carries the blood to the lungs, where it picks up oxygen and drops off carbon dioxide. The now oxygen-rich blood returns to the left atrium of the heart through the pulmonary vein, completing the pulmonary circuit. Oxygen-rich blood is then pumped into the left ventricle, and from there into the aorta, the major artery leading out of the heart to the rest of the body. The systemic circuit is completed when oxygen-depleted blood from the body returns to the right atrium through the veins.

The heart is equipped with four valves—two atrioventricular valves (located between the atrium and the ventricle on the left and right sides of the heart) and two semilunar valves (located between the ventricles and the arteries leaving the heart). The coordinated opening and closing of these valves ensures a one-way flow of blood through the heart, with no backflow. The "lub dub" sound the heart makes as it beats is actually the

ATRIA
The upper chambers of the heart that receive blood. In humans, the right atrium receives oxygen-poor blood from the body, and the left atrium receives oxygen-rich blood from the lungs.

VENTRICLES
The lower chambers of the heart that pump blood away from the heart. In humans, the right ventricle pumps oxygen-poor blood to the lungs, and the left ventricle pumps oxygen-rich blood to the body.

PULMONARY CIRCUIT
The circulation of blood between the heart and the lungs.

SYSTEMIC CIRCUIT
The circulation of blood between the heart and the rest of the body.

sound of these valves closing in pairs; first the two atrioventricular valves snap shut ("lub"), then the two semilunar valves close ("dub"). On average, it takes about 1 minute for blood to make a complete trip around the body **(INFOGRAPHIC 28.6)**.

Obviously, for the heart to do its job properly, it has to keep pumping. A group of cells located in the right atrium is the heart's natural pacemaker, setting the heart's tempo—on average, 60–90 beats per minute when we are at rest. In fact, for as long as you are

INFOGRAPHIC 28.6

Blood Flow Follows a Double Circuit

For every trip through the entire body, blood makes two passes through the heart. In one of these passes, the pulmonary circuit, blood is pumped from the right ventricle of the heart to the lungs, where it picks up oxygen. Blood then returns to the left atrium of the heart. The systemic circuit then pumps blood from the left ventricle to the rest of the body's tissues, where it delivers oxygen. Blood then returns to the right atrium of the heart.

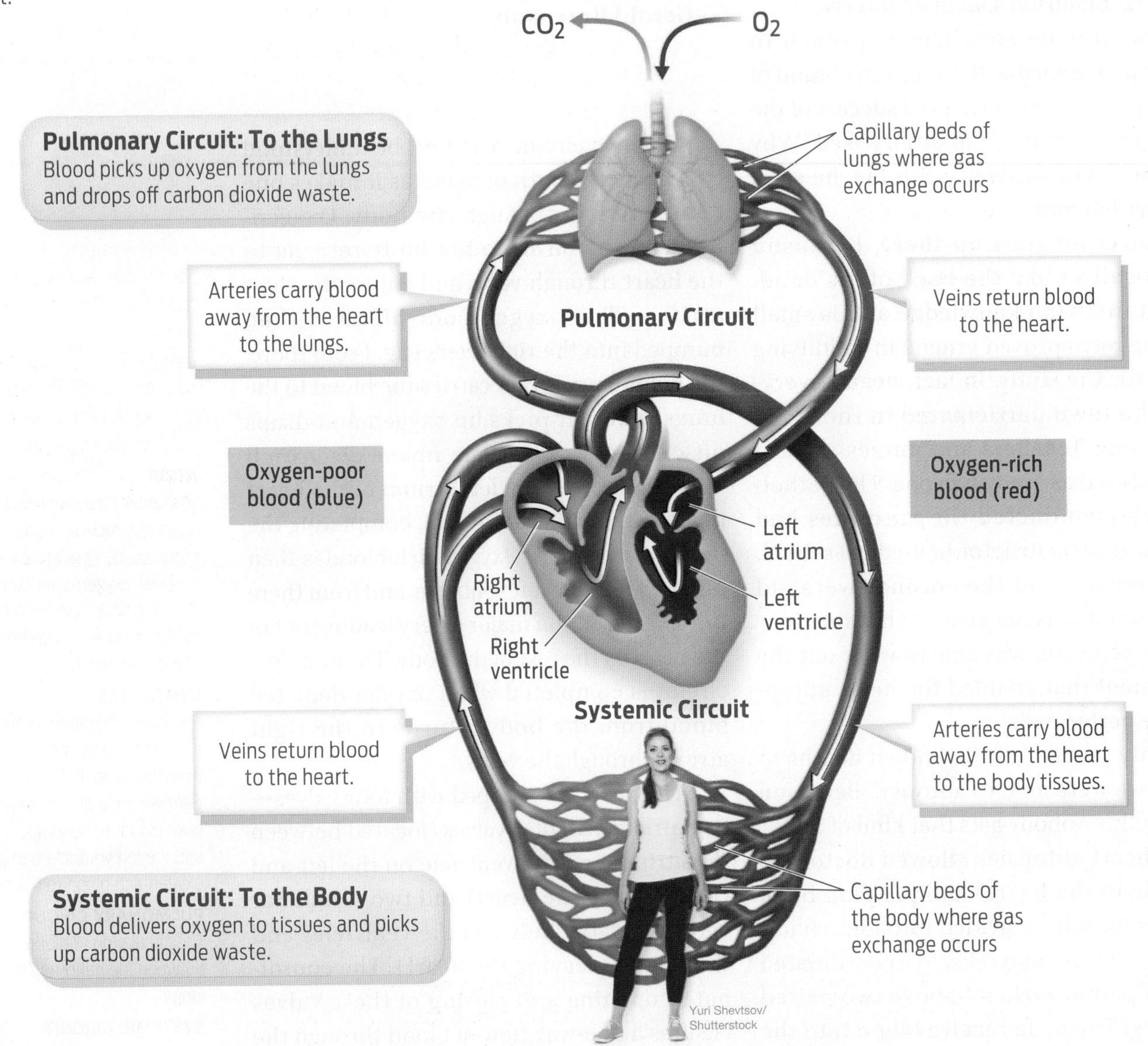

? **Is blood in a vein in the pulmonary circuit traveling to the heart or to the lungs? Is it oxygen rich or oxygen poor?**

alive, the heart muscle never stops beating. In that sense, it is the only muscle in the body that never gets a rest. The human heart beats more than 2.5 billion times in an average lifetime.

Like any muscle, the heart requires oxygen-rich blood to power the work of contraction. If a part of the heart loses its blood supply, that portion of the muscle begins to die, and the heart as a whole can begin to beat irregularly, or stop beating altogether. At that point, blood circulation stops—and very quickly, so do you. Understanding how blood flow becomes obstructed, then, is key to combating heart disease.

Under Pressure

▶ Blood pressure and hypertension

When the Bogalusa Heart Study began in 1972, Berenson and his colleagues were interested in identifying risk factors that contribute to heart disease, specifically in children. Other studies had identified risk factors in adulthood, but not in children.

One of the main risk factors the researchers were interested in was elevated **blood pressure.** Blood pressure is expressed as two numbers that relate to the action of the heart muscle as it pumps. With every heartbeat, the ventricles contract and blood is forced out of the heart and into the arteries. This relatively high pressure is the **systolic pressure.** The force of the blood entering the arteries can be felt as the **pulse**—for example, in your neck or your wrist. When the ventricles relax, the pressure in the arteries drops; this lower pressure during ventricular relaxation is the **diastolic pressure.** Blood pressure is expressed as systolic pressure over diastolic pressure. The normal measure for blood pressure is systolic pressure less than 120 mmHg (millimeters of mercury, a unit of pressure) and diastolic pressure less than 80 mmHg; the combination of these two pressures is expressed as 120/80 mmHg. Chronic elevation of either number increases the risk of CVD, including atherosclerosis, heart attack, and stroke.

High blood pressure, or **hypertension,** is dangerous because it can put stress on the walls of arteries, causing microscopic tears. These tears provide sites for the buildup of cholesterol, fats, and other substances, ultimately forming plaques.

When a plaque in a vessel grows large enough, it can rupture, exposing its contents to the bloodstream. This can cause blood clots to form on the plaque, obstructing blood flow. Blood clots in the coronary arteries are a common cause of heart attack. A piece of a plaque or a clot can also break off, travel to another spot in the body, and clog other vessels. If

BLOOD PRESSURE
The overall pressure in blood vessels, expressed as the systolic pressure over the diastolic pressure.

SYSTOLIC PRESSURE
The pressure in arteries at the time the ventricles contract.

PULSE
The detectable force of blood entering arteries, which can be felt in the neck or wrist.

DIASTOLIC PRESSURE
The pressure in arteries when the ventricles are relaxed.

HYPERTENSION
Elevated (high) blood pressure.

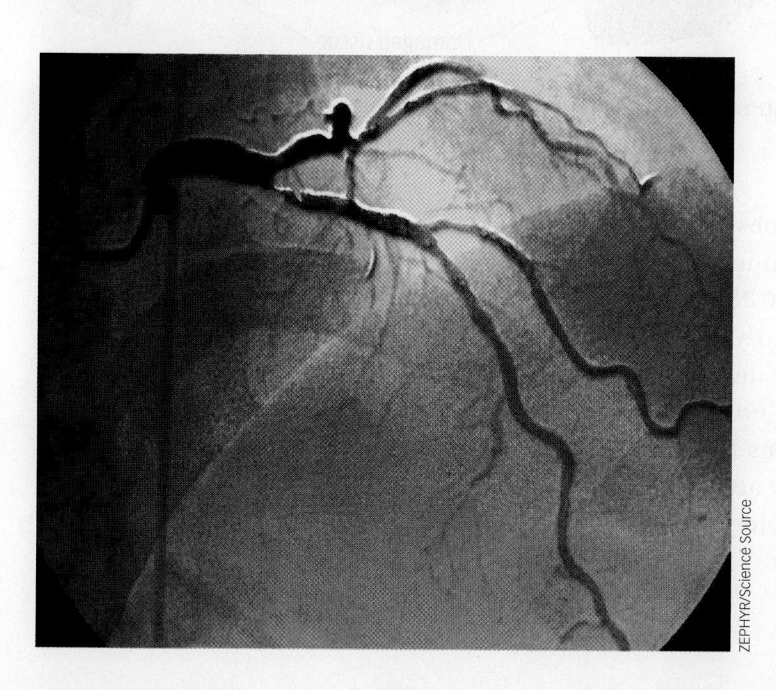

ZEPHYR/Science Source

Narrowing of a coronary artery from atherosclerosis (upper left, at the start of the lower branch) prevents adequate blood supply to the heart. The result can be a heart attack.

INFOGRAPHIC 28.7

High Blood Pressure Can Result in Atherosclerosis

When the heart ventricles contract, they force blood out of the heart. The force of the blood leaving the heart puts pressure on the walls of the arteries—the systolic pressure. When the ventricles relax, there is less pressure on the arteries—the diastolic pressure. Chronically elevated blood pressure can cause small tears in the artery walls. These are sites where plaque can build up.

Blood Pressure

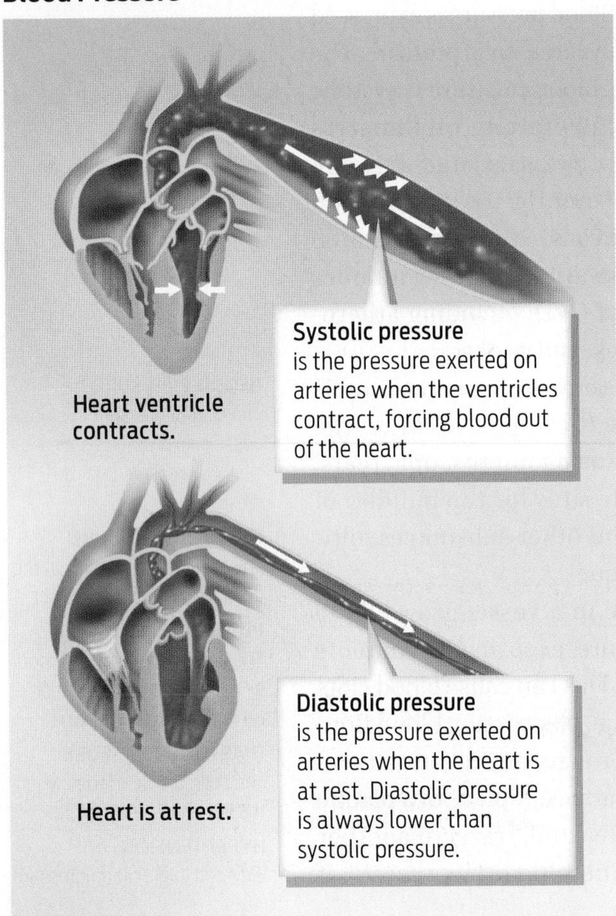

Lengthwise Section of an Artery and Its Wall

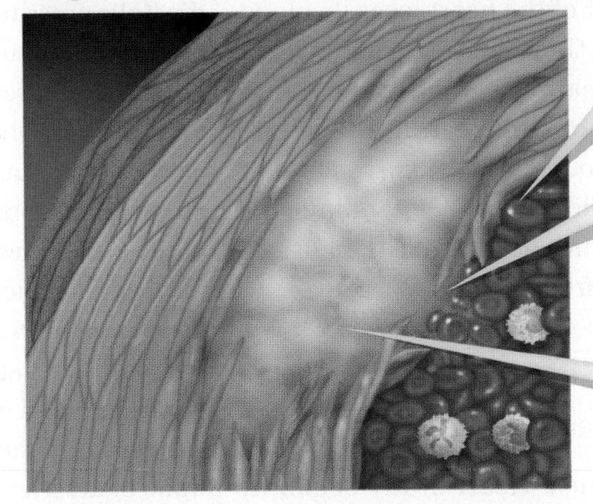

Blood clots can form on the surface of a plaque and obstruct blood flow. High blood pressure can also cause plaques to rupture, sending small fragments to other vessels where they may form clots. In coronary arteries, blood clots cause a heart attack. In the brain, they cause a stroke.

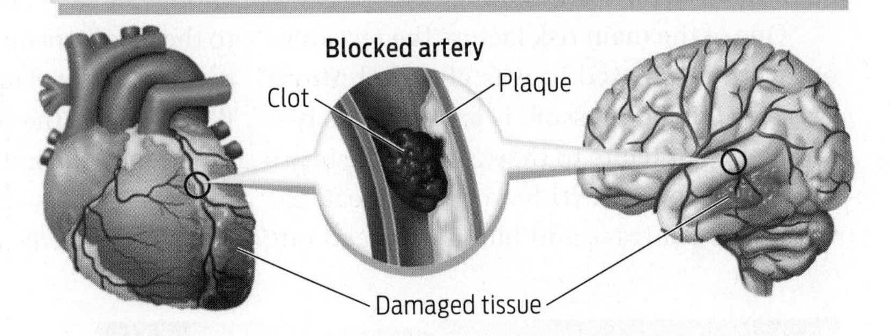

? If an artery can sustain high blood pressure without bursting, why is high blood pressure dangerous?

blood vessels in the brain are obstructed, the supply of oxygen to the brain is disrupted, causing a **stroke (INFOGRAPHIC 28.7).**

Hypertension is particularly dangerous because although it can have deadly consequences, there are typically no symptoms associated with it. This means that people have no way of knowing they are hypertensive unless they have their blood pressure checked and monitored.

STROKE
A disruption in blood supply to the brain.

What causes hypertension? Researchers are not entirely sure. However, they do know which factors predispose people to this condition. Some of these factors are controllable, some aren't. Uncontrollable factors include genetics, family history, older age, being African American, and being male; controllable ones include low physical activity, eating a high-salt diet, being overweight or obese, drinking too much alcohol, and stress.

Obviously, when the bodies of the young people arrived at the funeral home in Bogalusa, they no longer had any vital signs, so it was not possible to take their blood pressure. However, because nearly every child in the town had been followed as part of the heart study, researchers were able to compare the victims' blood pressure when alive with the autopsy results. They found that having had high blood pressure correlated closely with the extent of atherosclerosis found in the arteries of the victims.

High blood pressure wasn't the only risk factor that researchers were able to link with the autopsy results. Strong relationships were also found with high body mass index (BMI), high blood-cholesterol levels, and smoking cigarettes. When all the known risk factors were considered, a clear pattern emerged: the amount and extent of atherosclerosis increased with age and was strongly correlated with the known risk factors for heart disease. The young people with the most risk factors when alive were the ones with the most evidence of heart disease at the time of death **(INFOGRAPHIC 28.8)**.

"The autopsy studies were really landmark studies," says Peter Katzmarzyk, an epidemiologist at the Pennington Biomedical Research Center in Baton Rouge. Before them, he says, scientists didn't really understand that health problems like heart attack and stroke have their genesis in childhood. "The Bogalusa Heart Study really put that on the map," he says. And it made a strong case that averting severe consequences of heart

INFOGRAPHIC 28.8

Risk Factors and Heart Disease

Researchers measured the percentage of the inside lining of the aorta and coronary arteries covered in fatty streaks and fibrous plaques. These percentages were compared with the number of risk factors for heart disease a person had. The study found a direct correlation between the two. On average, young people with more risk factors had more evidence of heart disease.

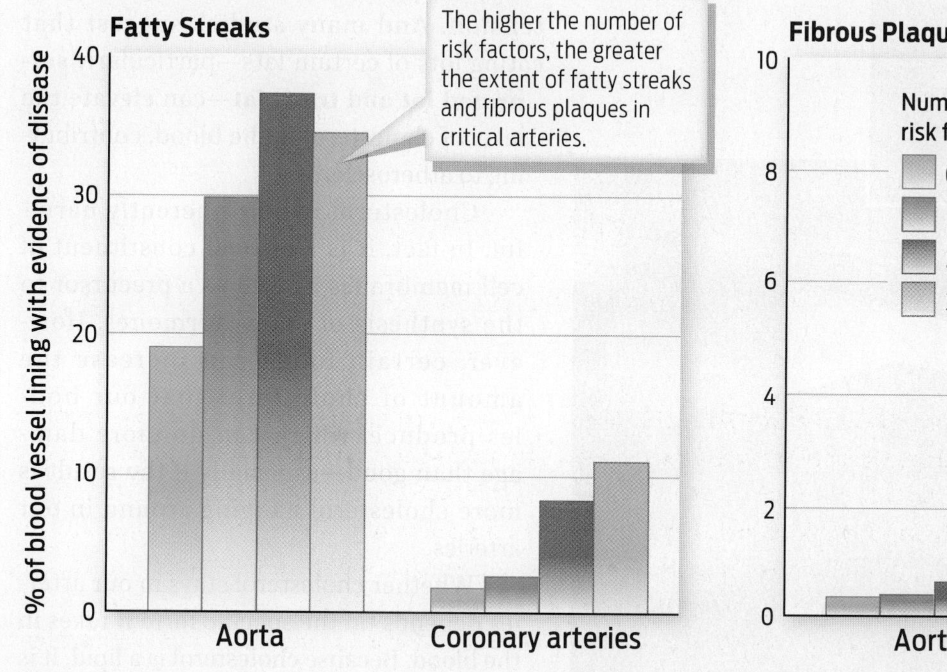

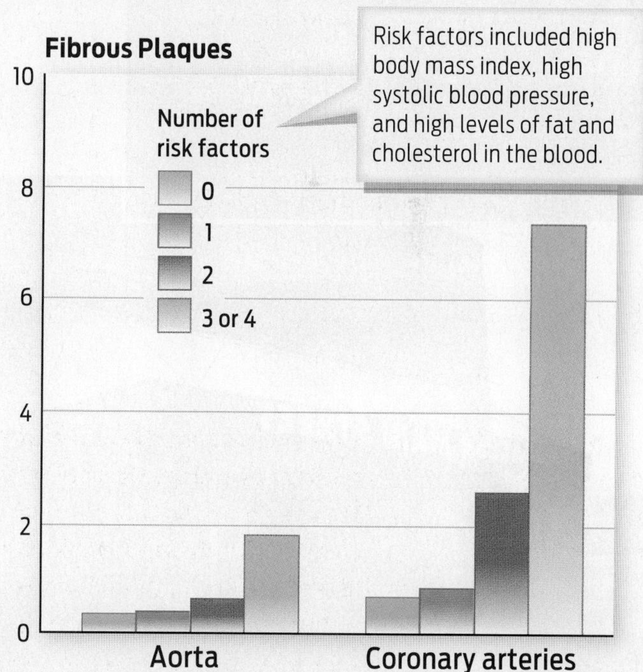

Data from Berenson, G.S., et al. (1998). *N Engl J Med* 338:1650–1656

? What would you predict about the presence and extent of fibrous plaques on the lining of the aorta and coronary arteries in someone with two risk factors for heart disease?

disease later in life would mean addressing the lifestyle choices we make when we are young.

Southern Discomfort

▶ Dietary fat, cholesterol, and lipoproteins

SATURATED FAT
A fat, such as butter, that is solid at room temperature; often found in animal products.

TRANS FAT
An unhealthful fat made from vegetable oil that has been chemically altered to make it solid at room temperature; often found in snack foods.

If Louisiana is shaped like a boot, then the town of Bogalusa sits in ideal kicking position—right at the easternmost toe. It's a fitting location for a town that has been beaten down over the years, its bruises visible in the form of crumbling playgrounds, dilapidated storefronts, and damage from Hurricane Katrina in 2005. Once known as the Magic City because of its rapid growth around the burgeoning timber industry, Bogalusa has since lost some of its luster, with the acrid smell of exhaust from the local paper mill serving as the sole reminder of its former glory days.

Once known as the Magic City, Bogalusa has since lost some of its luster.

Bogalusa is typical of many rural southern towns in which poverty and poor education have exacerbated problems of public health. More than 40% of individuals in Bogalusa live below the poverty line (contrasting with 23% for Louisiana as a whole and 15% nationally). And CVD disproportionately affects people of lower socioeconomic status.

Partly, the issue is access to health care. "One of the number one things related to health is socioeconomic status," says epidemiologist Katzmarzyk. "If you're wealthy and you have access to health care, you will be healthier." But there are likely cultural and behavioral factors at work, too.

Take diet, for example. Southern cuisine is famous for its fried foods, including fried chicken, fried fish, and fried potatoes. Also popular are bacon, ham, and sugar-sweetened beverages such as sweet tea and soda. In short, there's no shortage of salt, fat, and sugar in the southern diet.

Unhealthful diet choices can directly contribute to heart disease. Consuming large amounts of salt, for example, is known to raise blood pressure, contributing to hypertension. And many studies suggest that eating lots of certain fats—particularly **saturated fat** and **trans fat**—can elevate the level of cholesterol in the blood, contributing to atherosclerosis.

Cholesterol is not inherently harmful. In fact, it is a critical constituent of cell membranes as well as a precursor to the synthesis of many hormones. However, certain foods can increase the amount of cholesterol that our bodies produce, which can do more damage than good—especially if the result is more cholesterol hanging around in our arteries.

Whether cholesterol stays in our arteries depends on the specific form it takes in the blood. Because cholesterol is a lipid, it is hydrophobic (see Chapter 2)—which means it is not soluble in water-based solutions such as blood. To be carried in the blood, cholesterol must therefore be packaged along with other lipids in specialized particles called lipoproteins. Lipoproteins are spherical particles made of protein and lipids that contain a cargo of cholesterol (and triglycerides) in their hydrophobic center. They come in two main varieties. **Low-density lipoprotein (LDL)** carries cholesterol to body cells and can accumulate in the walls of blood vessels, contributing to atherosclerosis. In contrast, **high-density lipoprotein (HDL)** carries cholesterol out of blood vessels and delivers it to the liver, where it can be processed into a form that permits its elimination from the body—for example, as bile salts (see Chapter 27).

In general, HDL is "good" and LDL is "bad" (think "H" for "healthy" and "L" for "lethal"). Too much LDL relative to the amount of HDL in the blood can lead to cholesterol becoming deposited in a blood vessel wall. This is the start of atherosclerosis. Over time, as atherosclerosis progresses, the affected area can form into a thickened plaque that may eventually obstruct blood flow **(INFOGRAPHIC 28.9)**.

Eating saturated fats (found in meats, butter, and cheeses) tends to raise LDL, whereas **unsaturated fats** (from olive oil, fish, and nuts) tend to raise HDL. For this

LOW-DENSITY LIPOPROTEIN (LDL)
A form of cholesterol and protein in the blood that contributes to CVD.

HIGH-DENSITY LIPOPROTEIN (HDL)
A form of cholesterol and protein that is protective against CVD.

UNSATURATED FAT
A fat, such as olive oil, that is liquid at room temperature; often found in plants and fish.

INFOGRAPHIC 28.9

Cholesterol: Good and Bad

Cholesterol is an important lipid that the body makes to provide building materials for cell membranes, hormones, and vitamins. It is carried in the blood in different kinds of lipoprotein particles—HDL and LDL. Cholesterol in HDL ("good cholesterol") is easily removed from the circulation and transported to the liver for processing. Cholesterol in LDL ("bad cholesterol") tends to build up in the walls of arteries, contributing to atherosclerosis.

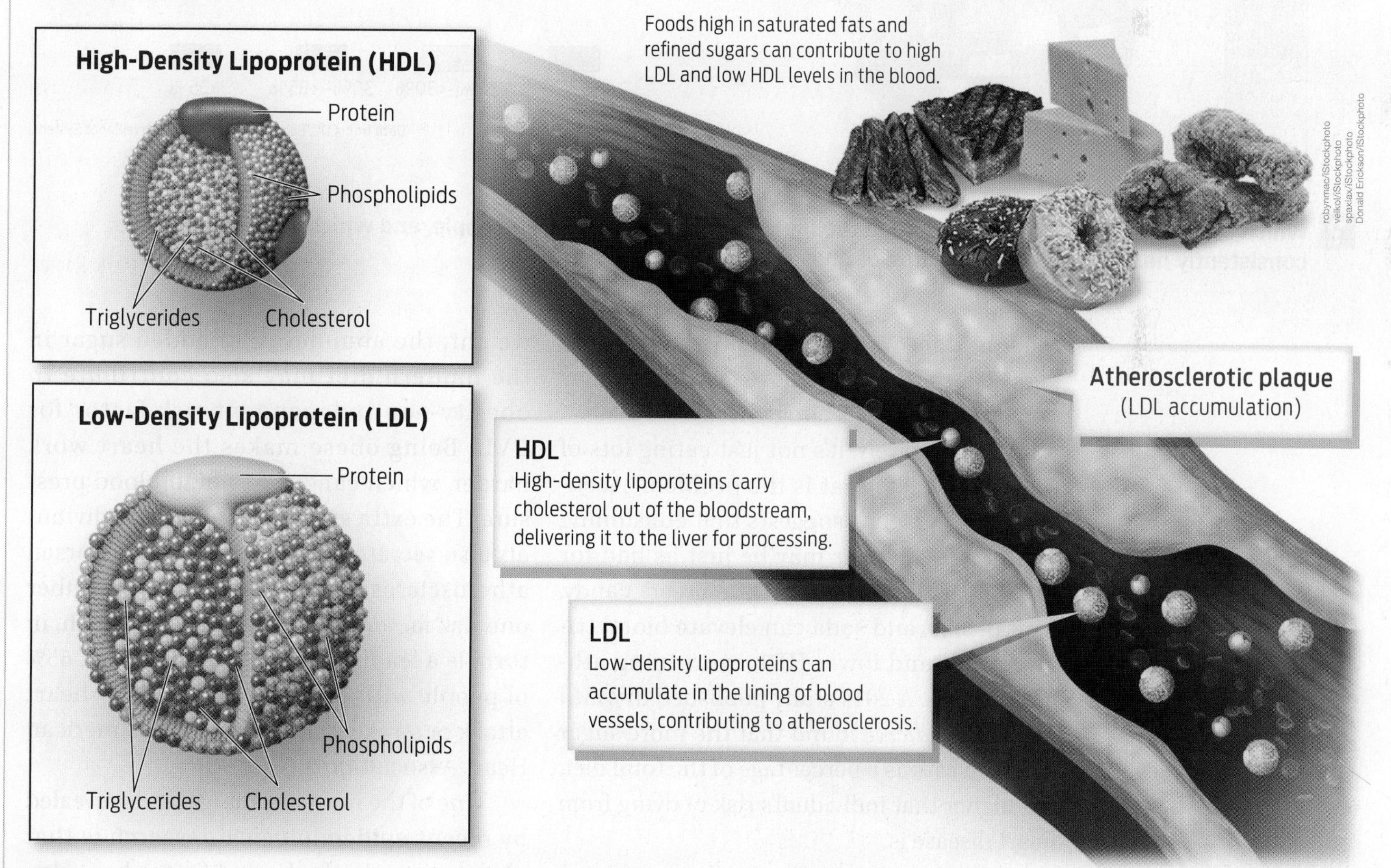

? Why is cholesterol in the circulation found packaged in HDL and LDL particles and not present as free cholesterol?

INFOGRAPHIC 28.10

Obesity in the United States Continues to Rise

The percentage of individuals in the United States who are overweight (BMI≥25) and obese (BMI≥30) has been steadily increasing. Currently, there are no states that report less than 20% obesity, with the highest levels of obesity found in the southern and midwestern states.

Historical Record of Obesity (BMI≥30) Among U.S. Adults

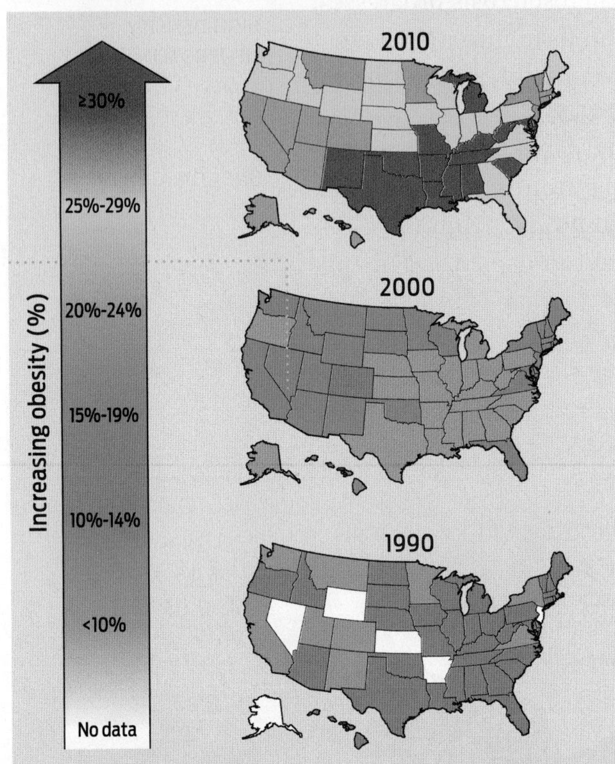

Percentage of Self-Reported Obesity (BMI≥30) Among U.S. Adults, 2018

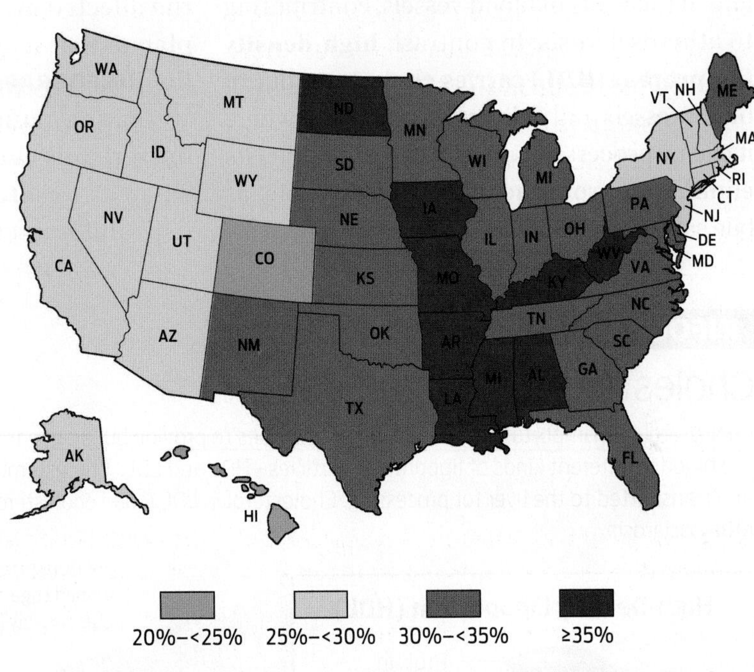

Data from CDC, Behavioral Risk Factor Surveillance System.

? **Which state(s) has/have consistently had the highest proportion of obese people, and which state(s) has/have consistently had the lowest proportion of obese people?**

reason, doctors recommend that people choose unsaturated fats over saturated ones (and avoid trans fats altogether).

Unfortunately, it's not just eating lots of unhealthful fats that is the problem. Accumulating evidence suggests that consuming lots of added sugar may be just as bad for you. Sugars in foods like white bread, candy, fruit drinks, and soda can elevate blood triglycerides and lower HDL, worsening atherosclerosis. A 2014 study published in *JAMA Internal Medicine* found that the more sugar a person eats as a percentage of the total diet, the higher that individual's risk of dying from heart disease is.

By making it easy for people to eat more calories than they need to maintain their weight, the abundance of added sugar in the modern diet may also contribute to obesity—an independent risk factor for CVD. Being obese makes the heart work harder, which can lead to high blood pressure. The extra stored fat in obese individuals also secretes hormones that can worsen atherosclerosis. And obesity is the number one risk factor for type 2 **diabetes,** which, in turn, is a leading cause of CVD. In fact, 65% of people with diabetes die of CVD—heart attack or stroke—according to the American Heart Association.

One of the most alarming trends revealed by recent epidemiological research is that obesity rates in the United States have skyrocketed since the 1980s. Nearly 36% of adults

DIABETES
A disease characterized by chronically elevated blood-sugar levels.

and 17% of children in the United States are obese (BMI ≥ 30), according to the Centers for Disease Control and Prevention **(INFOGRAPHIC 28.10)**.

In some parts of the country, such as in Bogalusa, the problem is even worse. Since 1973, when the Bogalusa study began, the proportion of children and young people ages 5 to 17 who are obese has increased more than fivefold, from 6% in 1973 to 31% today.

Louisiana is one of a series of states—those that lie between Texas and Florida—that epidemiologists sometimes refer to as the "diabetes belt," "stroke belt," or "obesity belt," because CVD cases seem to cluster in this region. What's behind this regional concentration? No one knows for sure. In addition to diet and other lifestyle factors, such as smoking, it may have something to do with climate. "It's so hot for large parts of the year, people tend not to exercise outside as they would in the north," says Katzmarzyk. It could be that a confluence of factors has collectively created "a perfect storm" for CVD in this part of the country **(INFOGRAPHIC 28.11)**.

> *"One of the number one things related to health is socioeconomic status."*
>
> **—Peter Katzmarzyk**

INFOGRAPHIC 28.11

Southern States Have the Highest Average Incidence of Cardiovascular Disease

States in the Southeast and the Southwest have the highest levels of heart disease in the United States. Societal factors such as diet, activity level, and smoking can influence the risk for developing cardiovascular disease.

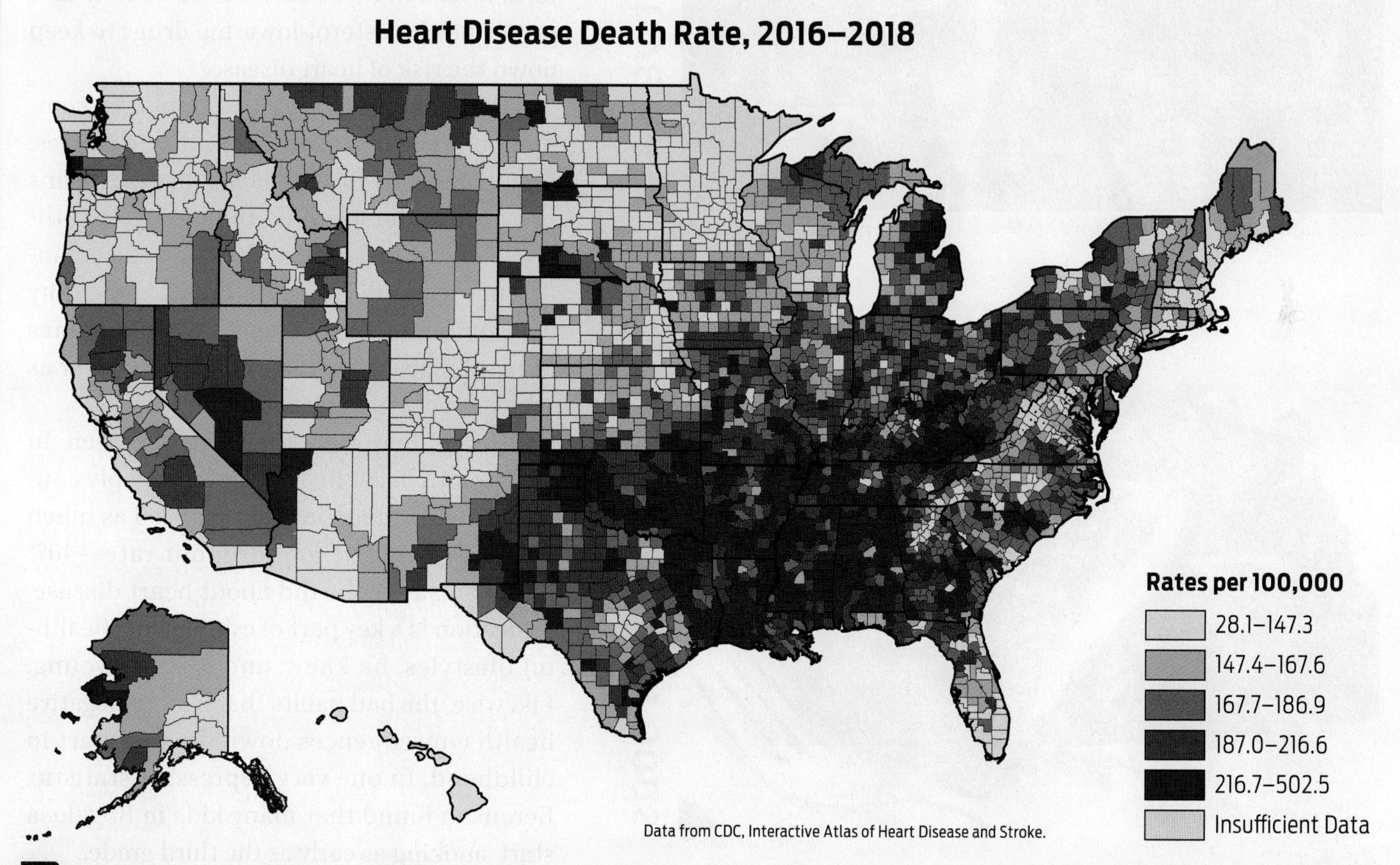

? **When comparing states that have high rates of obesity (see Infographic 28.10) with states that have high rates of death due to heart disease, is there any overlap?**

To reduce your risk of heart disease, follow the ABCs of heart health.

Avoid Tobacco

alptraum/iStockphoto

Be More Active

FatCamera/iStockphoto

Choose Healthful Foods

NightAndDayImages/iStockphoto

An Ounce of Prevention

Given the observed link between risk factors for heart disease and the extent of atherosclerosis in young people, it's obvious that reducing one's risk factors will reduce one's risk for heart disease. Eating better, quitting smoking, and starting or continuing to exercise will almost certainly reduce your chances of developing heart disease. In fact, a 2012 study found that eating more fruits and increasing one's activity level can positively affect blood cholesterol levels and reduce BMI. So that is the good news: you can take positive steps to reduce your risk of heart disease, including avoiding tobacco, getting more physical activity, and choosing healthful foods.

Unfortunately, for some people, especially those with a family history of heart disease or a genetic predisposition to it, lifestyle changes may not be enough. Some people, for example, naturally have higher "bad" cholesterol levels. In these cases, doctors may also prescribe cholesterol-lowering drugs to keep down the risk of heart disease.

Heart disease is a good example of a complex modern disease with multiple causes. Dealing effectively with heart disease means not only understanding the biology of atherosclerosis but also addressing the socioeconomic conditions that create unhealthy environments. In that sense, when it comes to heart disease, your zip code can matter as much as your genetic code.

Gerald Berenson knew this very well. In addition to heart health, he was deeply concerned with education and worried as much about the high school dropout rate—40% in Bogalusa—as he did about heart disease. Education is a key part of establishing healthful lifestyles, he knew, and it starts young. Likewise, the bad habits that lead to negative health consequences down the line start in childhood. In one very depressing statistic, Berenson found that many kids in Bogalusa start smoking as early as the third grade.

Keith Beaty/Toronto Star/Zuma Press/Newscom

Heart health starts young.

That's why Berenson was also passionate about getting heart disease prevention taught in elementary schools alongside standard subjects like reading and math. He and his colleagues developed a curriculum called Health Ahead/Heart Smart that builds on the lessons of the Bogalusa Heart Study and attempts to apply them in practical ways to help *prevent* heart disease.

More than 40 years later, the heart study that Berenson created continues to yield important insights. "There's no other study in the world like Bogalusa," he said. ■

CHAPTER 28 Summary

Driving Question 1 What structures make up the cardiovascular system, and how does blood flow through the system?

- The cardiovascular system transports nutrients, oxygen, hormones, and other substances throughout the body. It consists of the heart, blood vessels, and blood.
- The heart pumps blood in a double circulation pattern: a pulmonary circuit moves blood between the heart and the lungs, and a systemic circuit moves blood between the heart and the rest of the body.

Driving Question 2 What is the structure of the heart and of the different types of blood vessels?

- The heart has four chambers—two (upper) atria that receive blood and two (lower) ventricles that pump blood out of the heart.
- Different types of blood vessels have different structures and functions. Arteries transport blood away from the heart, and veins transport blood to the heart. Capillaries are the sites of nutrient and gas exchange in tissues.

Driving Question 3 What is the composition of blood, and what does blood do?

- Blood is composed of cells, water, gases, hormones, nutrients, and traces of whatever else we put in our body.
- Red blood cells carry oxygen, white blood cells participate in immune defenses, and platelets help the blood clot.

Driving Question 4 What is cardiovascular disease, and what are some of the risk factors for developing cardiovascular disease?

- Atherosclerosis, or hardening of the arteries, is a common cause of CVD. Cholesterol, fats, and other substances build up in the wall of a blood vessel, forming plaques that can obstruct the flow of blood.
- A heart attack results when the blood supply to the heart muscle is cut off and the muscle begins to die; atherosclerosis in coronary arteries is a common cause of a heart attack.
- High blood pressure (hypertension), obesity, high cholesterol, and smoking are known risk factors for atherosclerosis.
- Blood pressure is expressed as systolic pressure over diastolic pressure. Normal blood pressure is approximately 120/80 mmHg.
- Hypertension is a dangerous condition if left untreated. Some causes of hypertension are controllable, but others are not.
- Cholesterol is a normal constituent of cells and is transported through the body in the form of two lipoproteins, LDL or HDL. Too much LDL cholesterol relative to HDL in the blood can lead to the accumulation of plaque in arteries.
- Unhealthful diet choices can raise LDL, lower HDL, and contribute to atherosclerosis.
- Cardiovascular disease begins early in life, long before symptoms occur.

More to Explore

- American Heart Association: www.heart.org
- Berenson, G. S., et al., for the Bogalusa Heart Study. (1998). Association between multiple cardiovascular risk factors and atherosclerosis in children and young adults. *N Engl J Med* 338:1650–1656.
- Broyles, S., et al. (2010). The pediatric obesity epidemic continues unabated in Bogalusa, Louisiana. *Pediatrics* 125:900–905.
- HBO. (2012). *Weight of the Nation* (film).
- Yang, Q., et al. (2014). Added sugar intake and cardiovascular diseases mortality among US adults. *JAMA Intern Med* 174(4):516–524.

CHAPTER 28 Test Your Knowledge

Driving Question 1 What structures make up the cardiovascular system, and how does blood flow through the system?

By answering the questions below and studying Infographics 28.2, 28.4, 28.5, and 28.6, you should be able to generate an answer for this broader Driving Question.

Know It

1. List the three components of the cardiovascular system.
2. Blood enters the systemic circuit from
   a. the lungs.
   b. the left side of the heart.
   c. the right side of the heart.
   d. the body.
   e. both b and c
3. Which of the following is the correct order of structures through which blood flows as it passes through the cardiovascular system?
   a. right side of heart → body → left side of heart → lungs
   b. left side of heart → body → right side of heart → lungs
   c. left side of heart → right side of heart → body → lungs
   d. right side of heart → left side of heart → lungs → body

Use It

4. In which circuit (pulmonary or systemic) does blood pick up nutrients such as sugar?
5. Are the coronary arteries part of the pulmonary circuit or the systemic circuit? Explain your answer.
6. During heart surgery, patients are often placed on a heart–lung machine (known as a cardiopulmonary bypass). This machine circulates and oxygenates blood during the operation, taking over this work from the patient's heart and lungs. At what points would the patient's blood have to be diverted into the machine and back into the body to properly bypass the flow of blood through the heart and lungs?

Driving Question 2 What is the structure of the heart and of the different types of blood vessels?

By answering the questions below and studying Infographics 28.4 and 28.6, you should be able to generate an answer for this broader Driving Question.

Know It

7. Blood returning to the heart from the lungs enters the
- **a.** right atrium.
- **b.** right ventricle.
- **c.** left atrium.
- **d.** left ventricle.
- **e.** coronary arteries.

8. Which chamber of the heart pumps oxygen-rich blood to the body?
- **a.** the right atrium
- **b.** the right ventricle
- **c.** the left atrium
- **d.** the left ventricle
- **e.** both the right atrium and the left atrium (to the right and left sides of the body, respectively)

9. The heart obtains oxygen from
- **a.** the blood in its chambers.
- **b.** the coronary arteries.
- **c.** the systemic arteries.
- **d.** the coronary veins.
- **e.** the systemic veins.

10. What is the defining feature of an artery?
- **a.** It carries blood away from the heart.
- **b.** It has very thin walls.
- **c.** It has valves.
- **d.** It carries oxygenated blood.
- **e.** both a and d

Use It

11. Atrial fibrillation is a condition in which the atria contract very rapidly and irregularly. The ventricles also contract faster than normally, but their contraction is not coordinated with the atrial contraction. Explain how this condition can disrupt blood supply to the body.

12. Which of the following is most likely to be associated with left-side heart failure?
- **a.** fluid backing up in the feet and legs, causing swelling
- **b.** increased systolic pressure in the aorta
- **c.** reduced oxygenation of the blood leaving the left ventricle
- **d.** fluid backing up in the lungs, causing shortness of breath
- **e.** increased oxygenation of the blood returning to the right atrium

13. Some babies are born with a congenital heart defect in which the septum, or wall, between the left and right ventricles has an opening in it, and the aorta therefore can accept blood from both the left and right ventricles. These babies are known as "blue babies." Why?

Driving Question 3 What is the composition of blood, and what does blood do?

By answering the questions below and studying Infographic 28.5, you should be able to generate an answer for this broader Driving Question.

Know It

14. Describe the function of each of the following cellular components of blood: white blood cells; red blood cells; platelets.

15. Which of the following statements is *not* true of capillaries?
- **a.** Blood moves through them slowly.
- **b.** They have a very small diameter.
- **c.** They have very thin walls.
- **d.** They occur between arteries and veins.
- **e.** They have valves.

16. What is happening in a capillary in your big toe?
- **a.** Oxygen is diffusing from the blood into your toe tissue.
- **b.** Carbon dioxide is diffusing from your toe tissue into the blood.
- **c.** Nutrients are diffusing from the blood into your toe tissue.
- **d.** all of the above
- **e.** a and c

Use It

17. TV crime dramas often analyze arterial spray patterns of blood at crime scenes. Why do you never hear about venous spray patterns?

18. Why are soldiers standing at attention for long periods of time advised to contract their calf muscles to avoid becoming light-headed?

Driving Question 4 What is cardiovascular disease, and what are some of the risk factors for developing cardiovascular disease?

By answering the questions below and studying Infographics 28.1, 28.3, 28.7, 28.8, 28.9, 28.10, and 28.11, you should be able to generate an answer for this broader Driving Question.

Know It

19. Match each disease in the left column with the letter corresponding to its underlying cause:

___ heart attack	**a.** elevated systolic and diastolic pressure
___ hypertension	**b.** fatty streaks in arteries
___ stroke	**c.** blockage of a coronary artery
___ atherosclerosis	**d.** blockage of blood vessels in the brain

20. List at least four risk factors for cardiovascular disease. For each, state whether it is modifiable.

21. How does atherosclerosis increase the risk of stroke?

22. Which type of cholesterol is dangerous with respect to cardiovascular disease?

a. HDL
b. LDL
c. both
d. neither; cholesterol is a healthy lipid
e. HDL is a risk for hypertension and LDL is a risk for atherosclerosis.

Use It

23. The data generated from the Bogalusa study show a correlation between atherosclerosis and risk factors for cardiovascular disease. Does this mean that everyone with a risk factor (for example, high BMI) will develop atherosclerosis? Explain your answer.

Apply Your Knowledge

Mini Case

24. Steven is 14 years old. He is an ace goalie for his soccer team, which practices twice during the week and plays a game every weekend. He is also an ace online gamer, holding the highest player skill level in World of Warcraft. After many long afternoons of playing, Steven has become a very good graphic designer and his mother pays him to design computer-generated fliers for her business. She prepares healthful meals with lots of whole grains, fruits, and dairy (milk and cheese), and she gets regular checkups to monitor her elevated blood pressure. Steven spends some of his money on chocolate bars and potato chips. He is a big guy, and in the heaviest 5% of his age group. From what you've read in this chapter, what cardiovascular risk factors do you identify in Steven? What could you say to Steven and his mother about Steven's potential risk and reducing that risk?

Apply Your Knowledge

Interpreting Data

25. Refer to Infographic 28.10 and record your answers to the following questions in the table.

a. Look at the 2018 data for Texas, Arizona, California, South Dakota, and your home state. For each state, record the average percentage of adults with BMI ≥ 30.

b. Determine the total population of each state as of July 1, 2018 (you can find census data at https://www.census.gov/quickfacts/table/PST045216/00).

c. Determine the total number of people in each of those states with a BMI of at least 30.

State	Adults with BMI ≥ 30 (%)	Total state population	Number of adults with BMI ≥ 30
Texas			
Arizona			
California			
South Dakota			
Your home state			

Apply Your Knowledge

Bring It Home

26. Evaluate your own health and lifestyle with respect to risk for cardiovascular disease. Consider what kinds of changes you might want to talk about with your physician.

PEAK *performance*

An inside look at altitude training among elite athletes

Clive Rose/Getty Images

DRIVING QUESTIONS

1. What structures make up the respiratory system?
2. How do the respiratory and cardiovascular systems cooperate to deliver oxygen to body cells and remove carbon dioxide from tissues?
3. What factors influence the oxygen-carrying capacity of blood and breathing rate?
4. How can scientific knowledge of the respiratory system be used to design training regimens for elite athletes?

With a wingspan of 6 feet 7 inches, feet the size of flippers, and just 4% body fat at his fittest, swimmer Michael Phelps was built for speed. He is a 23-time Olympic gold medalist, holds four world record titles, and is widely recognized as the greatest swimmer of all time. With stats like these, it's hard to imagine how he could possibly improve his performance—short of growing gills. Yet the elite athlete proved he still had a few tricks up his swimsuit when he announced before the London Summer Olympics in 2012 that he sleeps in a "contraption" designed to give him an athletic boost.

"Once I'm already in my room, I still have to open a door to get into my bed," Phelps told Anderson Cooper on *60 Minutes.* "It's like the boy in the bubble."

The hypoxic chamber in Michael Phelps's bedroom.

Phelps wouldn't show the device on TV, but he later tweeted a picture of the glass-walled structure to his followers. Known as a hypoxic chamber, the device creates an artificial low-oxygen environment, equivalent to that at 8,500 feet above sea level. The swimmer claimed that sleeping in the chamber every night while he was training helped his body heal faster. In fact, it was a central part of his training regimen as he prepared for the London Games, at which he broke the world record at the time for most medals won by an Olympian—22 in all; his current record is 28 Olympic medals.

Phelps isn't the only elite athlete to resort to a technical fix to improve performance. Manufacturers of hypoxic chambers have seen an increased interest in their product from athletes such as tennis pro Novak Djokovic and triathlete Jonathan Brownlee. Even entire Olympic sports teams, like the U.S. men's rowing team, are getting in on the action.

But what does sleeping in a low-oxygen chamber have to do with athletic performance? Does it work? And is it ethical?

According to Randall Wilber, a sports physiologist with the U.S. Olympic Training Center in Colorado Springs, Colorado, the use of hypoxic chambers is a variation on the theme of natural altitude training. Athletes often spend several weeks living at high altitude, typically defined as 6,000 feet or more above sea level, and then return to sea level to train or compete. The widely used practice is based on the idea that the body will acclimatize to higher elevation by becoming better able to take up and transport oxygen, ultimately improving performance.

"If you look at the number of athletes who win medals in winter sport, summer sport, across many endurance-based sports, I would say in my estimation 90% to 95% of those athletes at minimum are using altitude training in some way or another," Wilber says.

But with more and more athletes using technical means to enhance performance, many in the sports field are taking a closer look at the practice, and some are asking questions about fairness.

Live High, Train Low

▶ Respiratory system anatomy and gas exchange

Altitude training isn't new. According to Wilber, altitude training first became popular after the 1968 Olympic Games, held in Mexico City, which sits at a relatively high elevation of 7,350 feet above sea level. A number of strange things happened at those Games that drew the attention of scientists and athletes alike. World records were set in a number of short-distance track and field events—like the 100-m and 200-m dash—but athletes performing in endurance events, like the 1,500-m run, fared worse than usual **(TABLE 29.1)**.

TABLE 29.1 Comparison of Performances in Short Races in the 1964 and 1968 Olympic Games

Olympic Games	Short Races: Men				Short Races: Women			
	100 m	***200 m***	***400 m***	***800 m***	***100 m***	***200 m***	***400 m***	***800 m***
1964 (Tokyo)	10.0 s	20.3 s	45.1 s	1 m 45.1 s	11.4 s	23.0 s	52.0 s	2 m 1.1 s
1968 (Mexico City)	9.9 s	19.8 s	43.8 s	1 m 44.3 s	11.0 s	22.5 s	52.0 s	2 m 0.9 s
% change*	+1.0	+2.5	+2.9	+0.8	+3.5	+2.2	0	+0.2

* + sign indicates improvement over 1964 performance.

Data from Howley, E. T. (1980). Effect of altitude on physical performance. In G. A. Stull and T. K. Cureton. *Encyclopedia of Physical Education, Fitness and Sports: Training, Environment, Nutrition, and Fitness*. Copyright © 1980 American Alliance for Health, Physical Education, Recreation and Dance, Reston, VA.

Agence France Presse/Getty Images

Bettmann/Getty Images

American long sprinter Tommie Smith (left) setting a World and Olympic record, and American long-jumper Bob Beamon (right) setting a World record, both at the 1968 Olympics in Mexico City.

"From 400 meters on down, including the long jump—think of Bob Beamon—all those people were just smashing world records left and right," says Wilber. "But from 800 meters on up, it went the other way: people were running slower than the world records." These results make perfect sense, he says. At altitude, the air is thinner, its molecules less densely packed, so there is less air resistance—which explains why records were set in sprinting events. But at high altitude, there is also less oxygen, so endurance events requiring prolonged exertion, such as distance running, predictably suffered.

The question that emerged was, what would happen if you did the opposite—lived at high elevation and then competed at lower altitude? Would the body respond differently?

There was tantalizing evidence that it would. In Mexico City, athletes coming from higher-altitude countries like Kenya and Ethiopia in general performed better. They did not experience a dip in performance as other athletes did, and thus had a competitive advantage. The race was on to understand these results and incorporate the lessons into training regimens.

As a sports physiologist with the U.S. Olympic Committee, Wilber serves as a consultant to Olympic athletes, helping them to design appropriate training regimes using the latest science. He has worked with an impressive roster of elite athletes, including cyclist Kristin Armstrong and speed skaters Apolo Ohno and Christine Witty, as well as swimmer Michael Phelps. One of Wilber's main areas of research is the effect of altitude on the **respiratory system** of athletes.

RESPIRATORY SYSTEM
The organ system that allows us to take in oxygen and unload carbon dioxide.

The respiratory system is what allows us to take in oxygen from the air and release

INFOGRAPHIC 29.1

The Human Respiratory System

The respiratory system allows us to take in oxygen from the air and eliminate carbon dioxide waste. Air travels through a series of tubes to small sacs in the lungs called alveoli. The alveoli have a close association with capillaries, permitting the uptake of oxygen into blood and the release of carbon dioxide from blood.

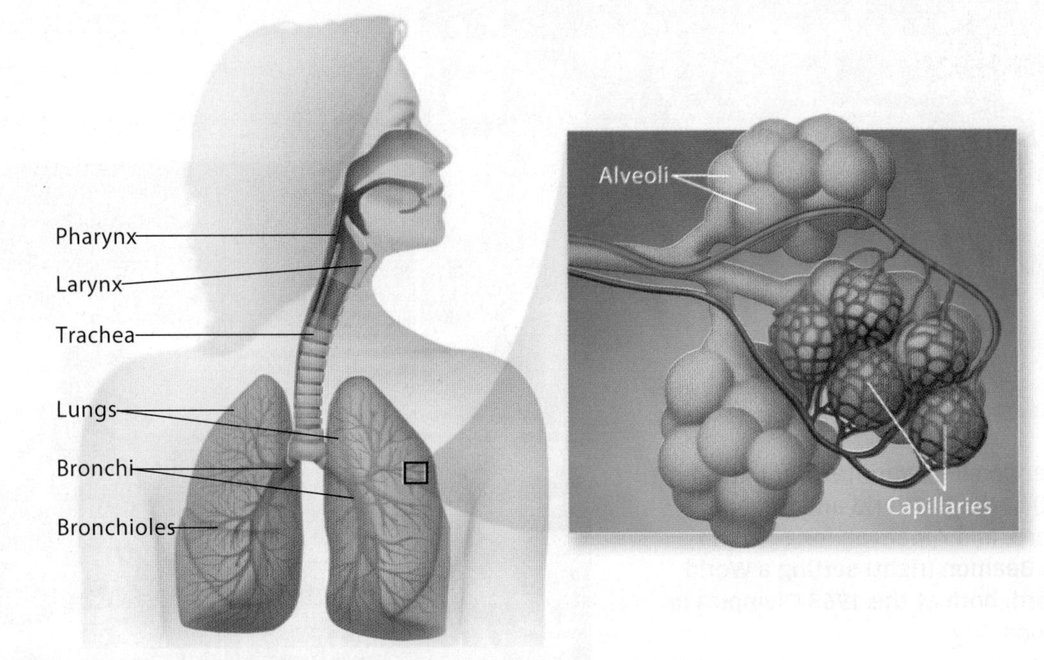

? Does oxygen move from the blood to alveoli, or from alveoli to the blood?

carbon dioxide. When we breathe, air first enters the nose and mouth and then moves through a series of branched tubes on its way to the **lungs.** Inhaled air first passes through a single tube having several distinct sections: initially through the throat, or **pharynx;** then through the **larynx,** which houses our vocal cords; and then through the **trachea,** or windpipe. At the end of the trachea, air is diverted into two major **bronchi,** one traveling to each lung. Much like an inverted tree, the tubes continue to branch and split into smaller and smaller **bronchioles** until finally air hits the many tiny sacs called **alveoli,** located deep within the lungs.

Alveoli, which are in close contact with capillaries (see Chapter 28), are the sites of **gas exchange** between the air and blood in the lungs. That is, oxygen diffuses from the air into the blood within the capillaries, and carbon dioxide diffuses from the blood into the airspaces in the lungs **(INFOGRAPHIC 29.1).**

"If you wanted to, you could run a 100-meter dash holding your breath."

—Randall Wilber

Each lung's **respiratory surface** is made up of millions of gas-exchanging alveoli. In an adult human, that respiratory surface is enormous: if all the alveoli were splayed out flat, they would occupy an area about the size of a tennis court (although the one-cell-thick layer would be so thin you wouldn't be able to see it). This large respiratory surface ensures that our bodies are able to take in enough oxygen to power the work of the trillions of cells in our body. Blood vessels

ATS-Altitude Training Systems

Australian swimmer James Magnussen using the Simulated Altitude Training Pool at NSW Institute of Sport.

LUNG
The major respiratory organ, the site of gas exchange between air and the blood.

PHARYNX
The throat.

LARYNX
The part of the airway between the pharynx and the trachea; also known as the voice box.

TRACHEA
A large airway between the larynx and the lungs.

BRONCHI
Two airways that branch from the trachea; one bronchus leads into each lung.

BRONCHIOLES
Small airways that branch from the bronchi.

ALVEOLI
Air sacs in the lung across which gases diffuse between air and blood.

GAS EXCHANGE
In the lungs, the process of taking up and releasing oxygen and carbon dioxide across capillaries.

RESPIRATORY SURFACE
A surface across which oxygen enters the blood and carbon dioxide leaves the blood.

carry the oxygen-rich blood from the lungs to the heart, which then pumps it around the body to various tissues. The oxygen supplied in this way enables cells in these tissues to perform aerobic respiration that generates ATP (see Chapter 6) **(INFOGRAPHIC 29.2).**

The respiratory and cardiovascular systems thus work together in close coordination. In fact, it's probably more useful to think of a single cardiorespiratory system rather than two separate systems acting individually.

If oxygen is so critical to athletic performance, why would athletes willingly deprive themselves of the needed gas by traveling to a place where oxygen is scarce? It may seem counterintuitive, but the lower oxygen levels actually help an athlete in the long run, because of the way the body responds to the altered conditions.

INFOGRAPHIC 29.2

Gas Exchange and Transport

Oxygen-rich air enters alveoli and oxygen enters the bloodstream. Carbon dioxide from the blood enters the alveoli and is breathed out. The oxygen-rich blood eventually reaches tissues, where oxygen is delivered, and carbon dioxide produced by aerobic respiration enters the blood to be carried back to the lungs.

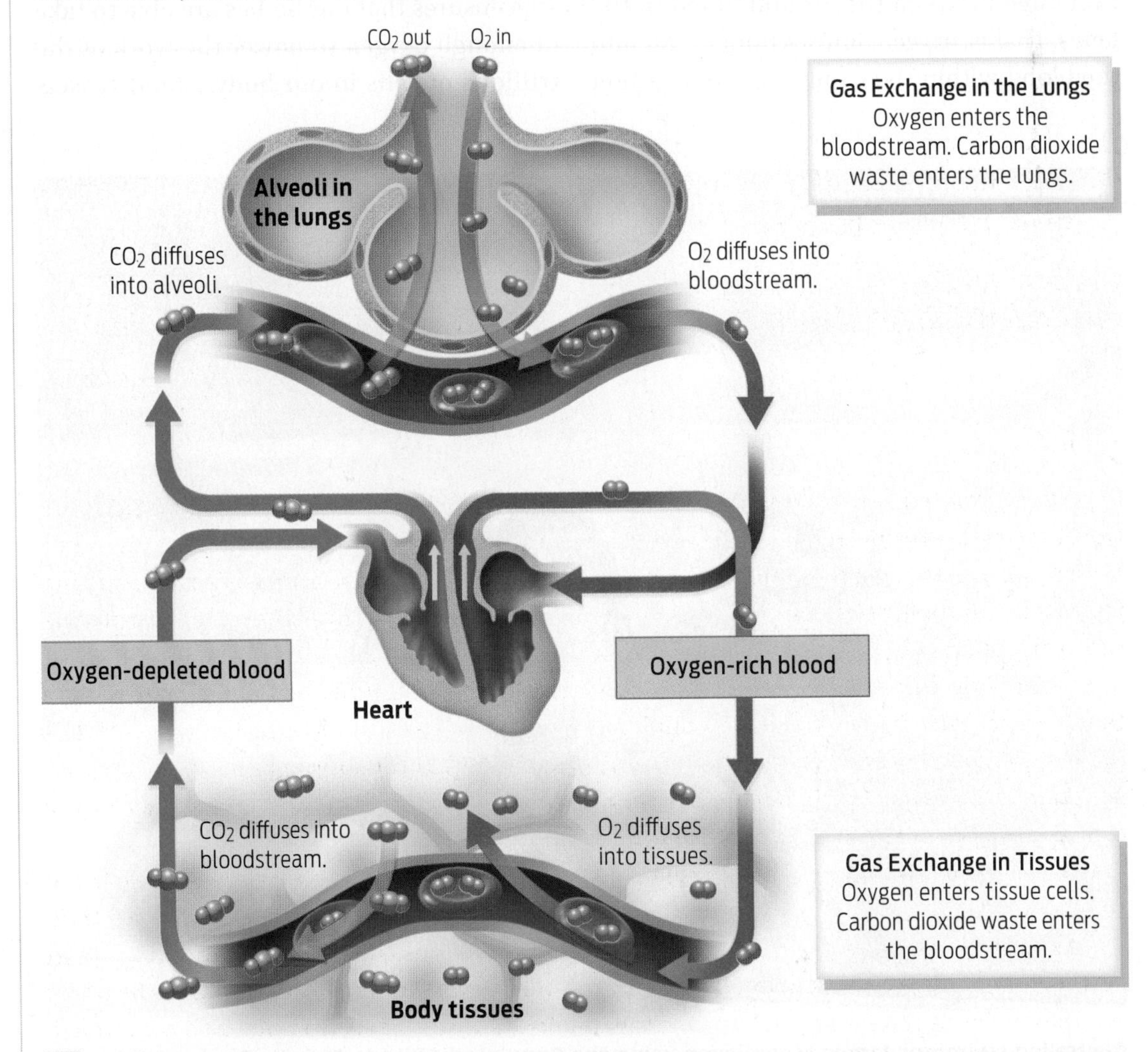

? Compare and contrast the movement of the gases O_2 and CO_2 in the lungs and in a tissue (such as muscle).

INFOGRAPHIC 29.3

The Effect of Altitude on Blood

The amount of available oxygen decreases with increasing altitude. Under these conditions, RBC production is stepped up to increase the oxygen-carrying capacity of the blood.

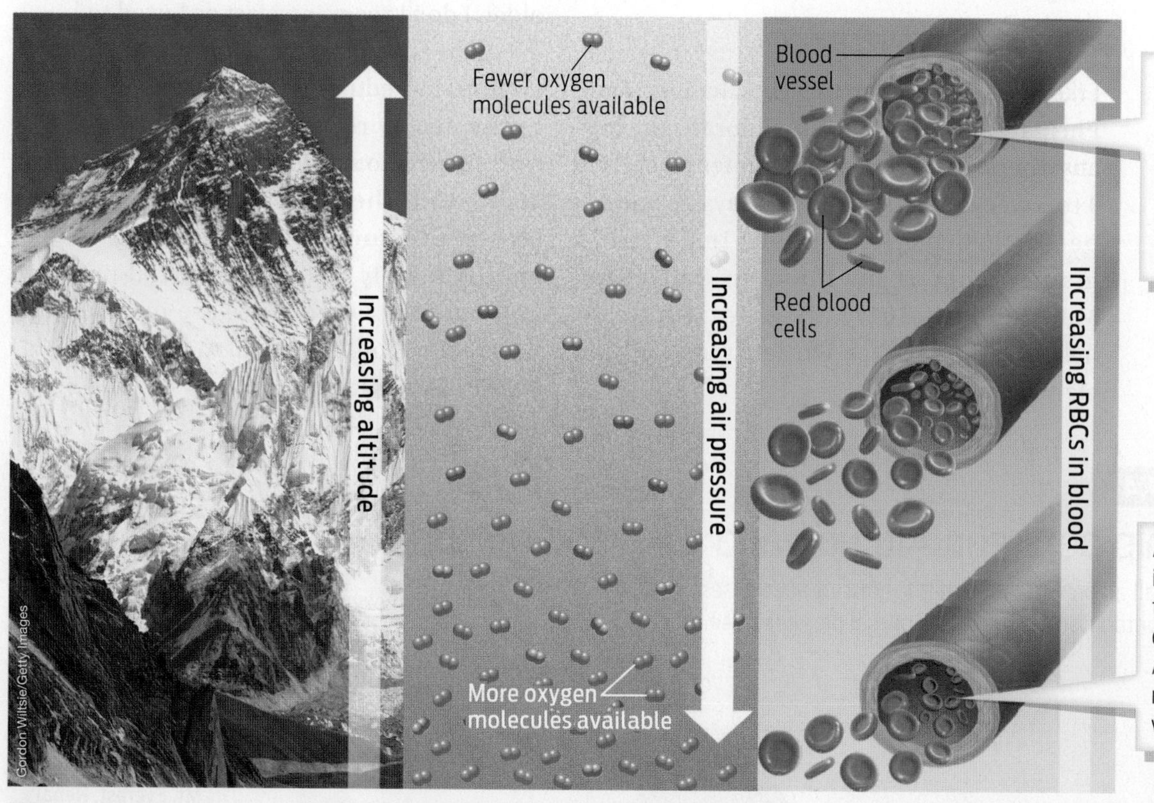

? Would you expect someone using supplemental oxygen while at altitude to have an increased RBC count? Explain your answer.

As Wilber explains, an athlete who travels to high altitude encounters less barometric air pressure, which means there are fewer total air molecules bouncing around in the atmosphere—including fewer molecules of oxygen per volume of air. "Thus you have a tougher time running or walking or climbing or swimming," he says. "The good side of the situation is that when you come to [high] altitude your body senses, *Hey, there's not enough oxygen here, I need to do something about it to compensate!*"

As scientists have learned, the body indeed compensates for the reduction of oxygen molecules in the atmosphere by increasing the number of oxygen-carrying **red blood cells (RBCs)** in the blood. With more RBCs, the body is able to take in roughly the same amount of needed oxygen, despite the altered conditions, thus maintaining homeostasis. When the athlete returns to sea level, after a period of acclimatization to altitude, he or she will have a competitive advantage because of the enhanced ability of the blood to carry oxygen (because of the higher number of red blood cells) **(INFOGRAPHIC 29.3)**.

The most popular form of altitude training in use today is known as "live high, train low": athletes live for a period of time at higher elevations and then return to lower altitudes to train or compete. This approach is better than living and training full time at high elevation, say some experts, since at

RED BLOOD CELLS (RBCS)
Blood cells specialized for transporting oxygen throughout the body.

high elevations one can't exercise or train as rigorously as at sea level—thus the benefits of acclimatization are canceled out.

Boy in a Bubble

▶ Oxygen and altitude training

Like many athletes on the U.S. Olympic team, Phelps typically traveled to Colorado Springs anywhere from three to five times a year, for 4 to 5 weeks at a time. Phelps says he had his personal hypoxic chamber installed in his bedroom at his home in Baltimore, Maryland, in 2011, after noticing that he recovered faster while living at high elevation.

PARTIAL PRESSURE The proportion of total air pressure contributed by a given gas.

"We've been able to realize after going to Colorado Springs so many times that it is something that helps me recover," Phelps told the Associated Press in 2012. "That's something that is so important to me now being older. I don't recover as fast as I used to."

Hypoxic chambers simulate altitude by artificially reducing the amount of oxygen in the air. By reducing oxygen concentration from about 21% (in untreated air) to about 15%, the chambers reduce oxygen's **partial pressure,** simulating an altitude of approximately 8,000 feet. Partial pressure and barometric pressure are measured in units called millimeters of mercury (mmHg) **(INFOGRAPHIC 29.4).**

INFOGRAPHIC 29.4

The Relationship between Altitude and Oxygen Pressure

As altitude increases, overall air pressure (barometric pressure) decreases. This means the pressure contributed by oxygen (its partial pressure) also decreases.

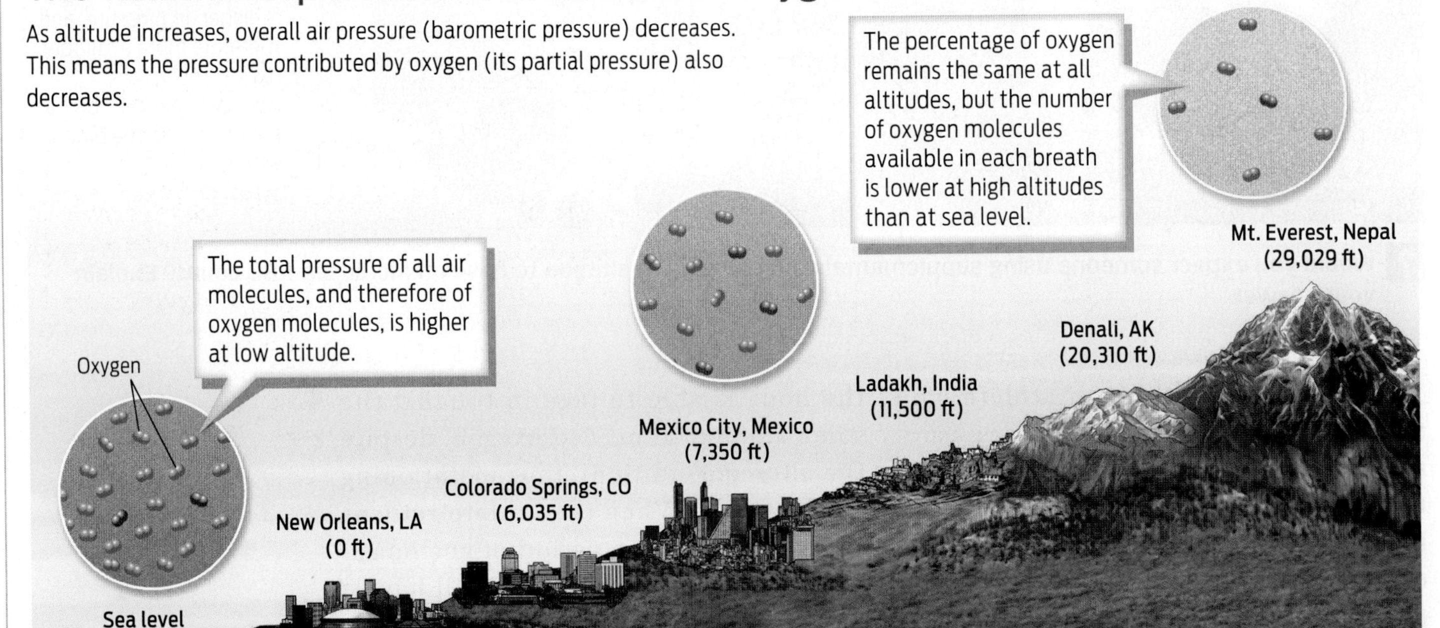

	New Orleans, LA	Colorado Springs, CO	Mexico City, Mexico	Ladakh, India	Denali, AK	Mt. Everest, Nepal
Barometric pressure (total air molecules, mmHg)	760	616	585	505	360	253
Partial pressure of oxygen (oxygen molecules, mmHg)	159	129	125	105	75	53
% O_2 in the air	20.93	20.93	20.93	20.93	20.93	20.93

Data from Altitude.org

? What is the partial pressure of oxygen at the peak of Mount Everest; in Colorado Springs, Colorado; and at sea level?

INFOGRAPHIC 29.5

Altitude Training Hypothesis

Both hypoxic chambers and natural altitude training are based on the same principle. By living or sleeping at lower partial pressures of oxygen, athletes will elevate their RBC count. This enhances the oxygen-carrying capacity of the blood when competing at sea level.

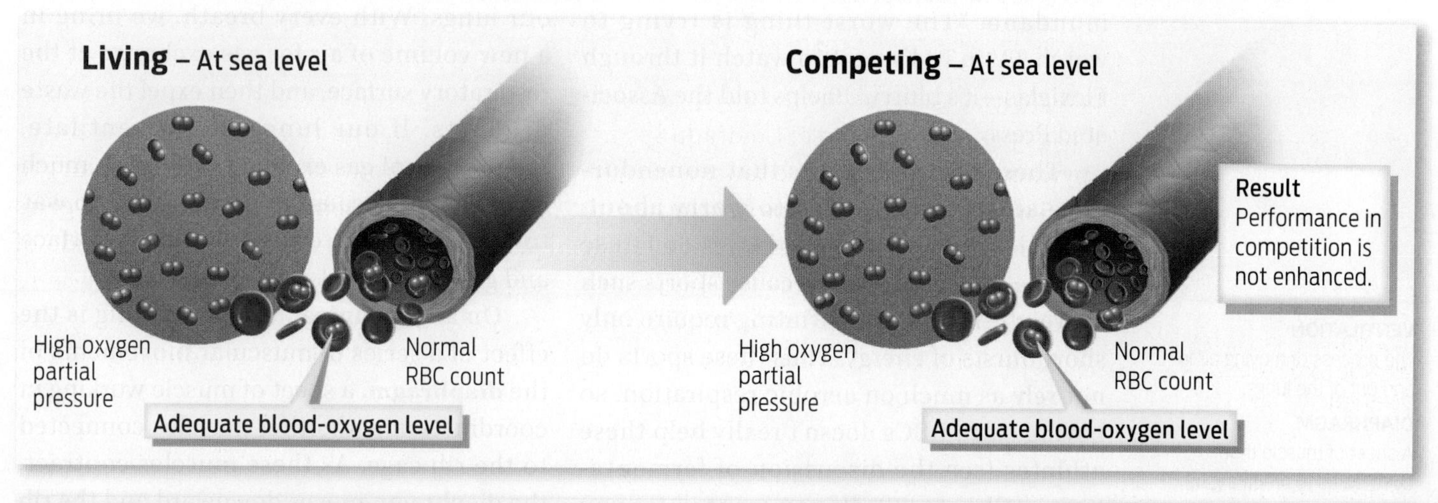

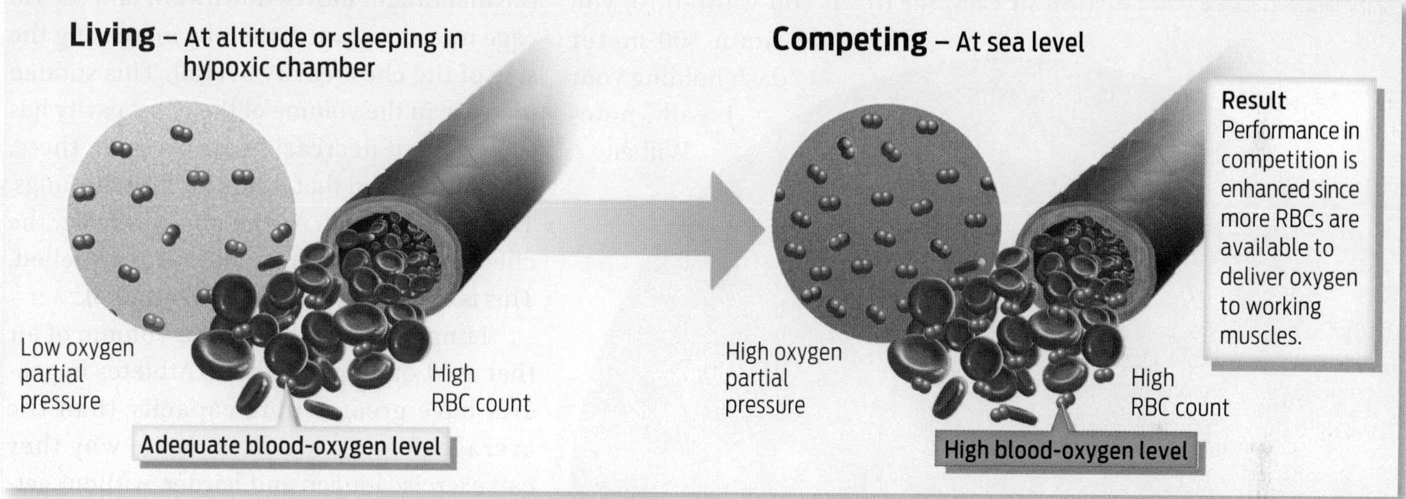

? What happens to the RBC count of an athlete living and sleeping in hypoxic conditions, and how is this advantageous when competing at sea level?

The lower oxygen pressure Phelps experienced while he slept stimulated his body to produce RBCs, while the regular oxygen concentration during the day helped him train at his normal, vigorous levels. That pattern is a simulated version of the "live high, train low" scenario **(INFOGRAPHIC 29.5)**.

Are there downsides to sleeping in hypoxic conditions? At least over the short term, there don't appear to be. In fact, many humans, including some Tibetan and Andean peoples, live easily at much higher elevations (above 10,000 feet) with even less oxygen. However, evidence suggests that these populations have adapted over many years to the hypoxic conditions by becoming better able to absorb and transport oxygen. Someone coming from sea level, who lacked these adaptations, would have a hard time thriving at these elevations. In fact, such a person may succumb to **altitude sickness,** a condition whose symptoms resemble those of an alcohol-induced hangover: headache, tiredness, and nausea. People can experience milder forms of altitude sickness when they ascend rapidly to altitudes above 8,000

ALTITUDE SICKNESS
An illness that can occur as a result of an abrupt move to an altitude with a reduced partial pressure of oxygen.

feet—for example, on a ski trip in Colorado. Wilber notes that some athletes find that the hypoxic chambers give them headaches or that they have trouble sleeping.

Phelps's biggest complaint was more mundane. "The worst thing is trying to watch TV in it. I've got to watch it through Plexiglas—it's blurry," Phelps told the Associated Press.

These are complaints that nonendurance athletes don't have to worry about. Typically, nonendurance athletes don't use altitude training. That's because sports such as weight-lifting and sprinting require only short bursts of energy. Thus these sports do not rely as much on aerobic respiration, so having extra RBCs doesn't really help these athletes (see the discussion of fermentation in Chapter 6). "If you wanted to, you could run a 100-meter dash holding your breath," notes Wilber.

VENTILATION
The process of moving air in and out of the lungs.

DIAPHRAGM
A sheet of muscle that contributes to breathing by contracting and relaxing.

Craig Lovell/Eagle Visions Photography/Alamy

The Ladakhi people live at an elevation of more than 11,500 feet with no adverse health effects.

Every Breath You Take

▸ Ventilation and breathing

Obviously, to take in oxygen, we need to breathe. Breathing is the way we **ventilate** our lungs. With every breath, we bring in a new volume of air for gas exchange at the respiratory surface, and then expel the waste products. If our lungs didn't ventilate, the process of gas exchange would be much less efficient because there would be no way to move fresh air to the respiratory surface and get stale air out of the way.

On a mechanical level, breathing is the effect of a series of muscular movements of the **diaphragm,** a sheet of muscle working in coordination with other muscles connected to the rib cage. As these muscles contract, the diaphragm moves downward and the rib cage moves outward, thereby increasing the size of the chest cavity overall. This sudden increase in the volume of the chest cavity has the effect of decreasing air pressure there, creating suction that draws air into the lungs. This is inhalation. As the muscles relax, the chest cavity contracts, and air is expelled. This is exhalation **(INFOGRAPHIC 29.6).**

Lung capacity is the total volume of air that a set of lungs can hold. Athletes in general have greater lung capacity than the average person, which is partly why they can exercise longer, and harder, without getting as winded as someone who is not as fit. And while you can't increase the size of your lungs, you can train the muscles of the chest to work more effectively, as many musicians and singers, as well as athletes, do. Exercising in a pool, for example, makes your chest muscles work harder, which strengthens them, ultimately enhancing lung capacity. Some populations of humans who live at high elevations (like the Bods, whose home, Ladakh, is at 11,500 feet in the Himalaya mountains) have wider chests with stronger chest muscles, permitting them to draw in more air so they can get sufficient oxygen.

Although getting oxygen is an essential part of breathing, the rate of our breathing

INFOGRAPHIC 29.6

Ventilation: How We Take a Breath

The contraction and relaxation of the diaphragm and rib cage muscles ventilate the lungs. Contraction decreases pressure in the lungs and air flows in. Relaxation of the muscles forces air out of the lungs.

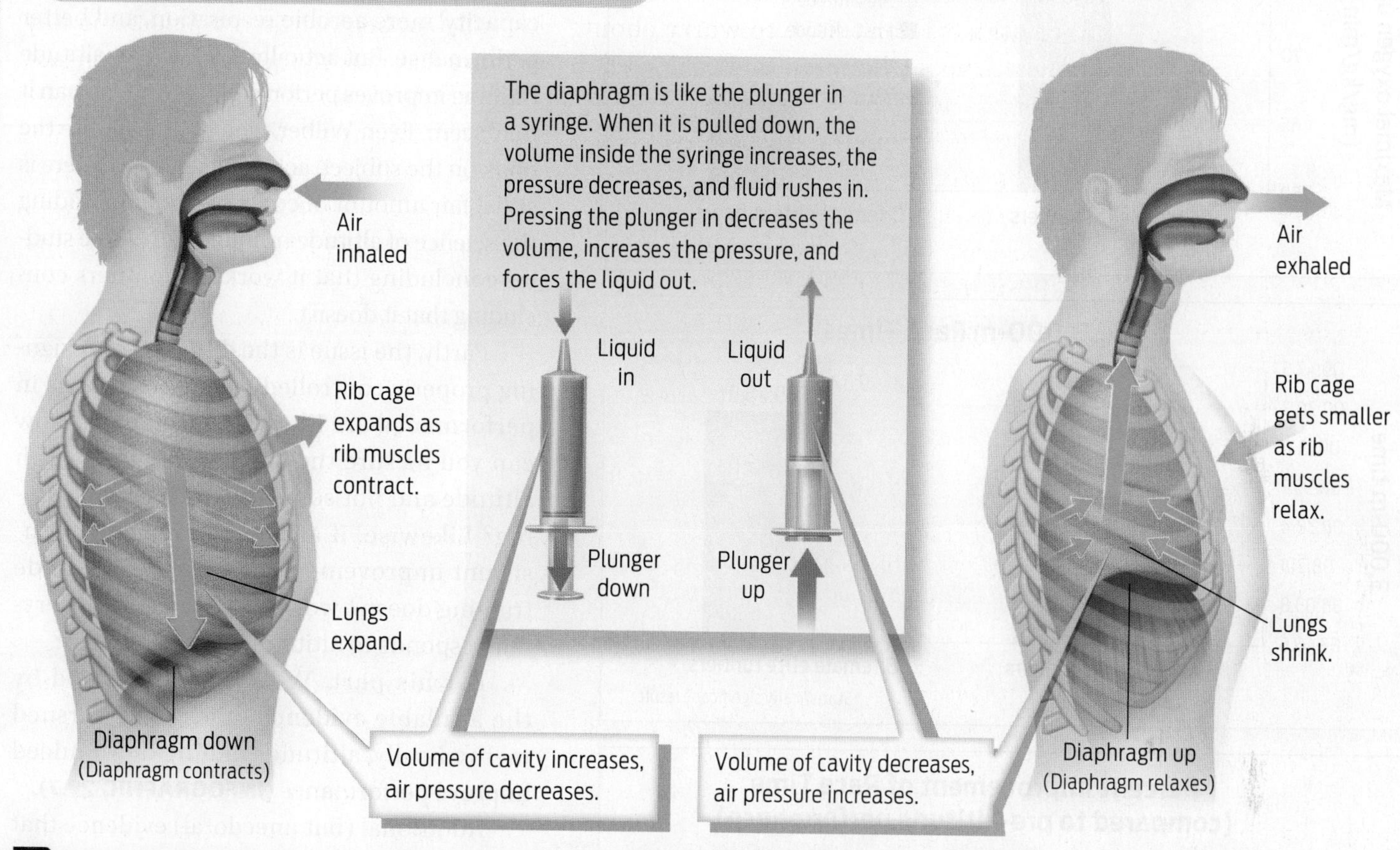

? **When the diaphragm and rib muscles contract, does air flow into or out of the lungs?**

is not actually determined by the demand for oxygen in our body. It's actually carbon dioxide that is more important. When we exercise, we are using more oxygen and producing more carbon dioxide. The increased concentration of carbon dioxide in the blood causes a decrease in blood pH; when CO_2 dissolves in the water in the blood—the plasma—it produces carbonic acid, which has a low pH (see Chapter 2). The brain senses this drop in pH and responds by sending a signal to the lungs to increase the breathing rate. The increased breathing rate increases the rate of gas exchange, bringing more oxygen into the body while unloading excess carbon dioxide, and thus raising blood pH back to normal.

Sometimes a person is unable to unload carbon dioxide from the lungs—as can happen when breathing is hindered by asthma or obesity, for example. When the lungs do not unload enough carbon dioxide, the concentration of CO_2 in the blood can rise to dangerous levels, creating a potentially fatal condition known as **acidosis.** The opposite

ACIDOSIS
A dangerous condition in which blood is too acidic.

INFOGRAPHIC 29.7

Altitude Exposure Improves Performance

In a study of 22 elite distance runners, maximal oxygen uptake and 3,000-m race performance were measured before and after a 4-week "live high, train low" altitude-training regimen (living at 2,500 m and training at 1,250 m). Performance and oxygen uptake were improved after living at altitude. Similar performance results were seen for collegiate athletes.

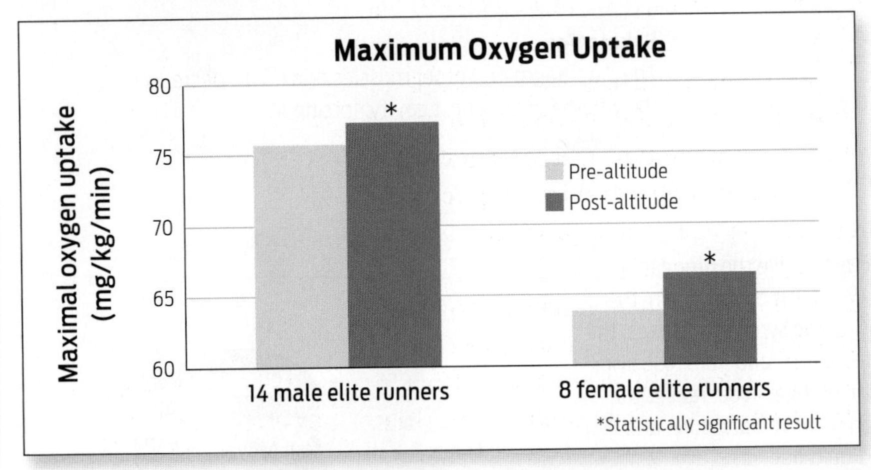

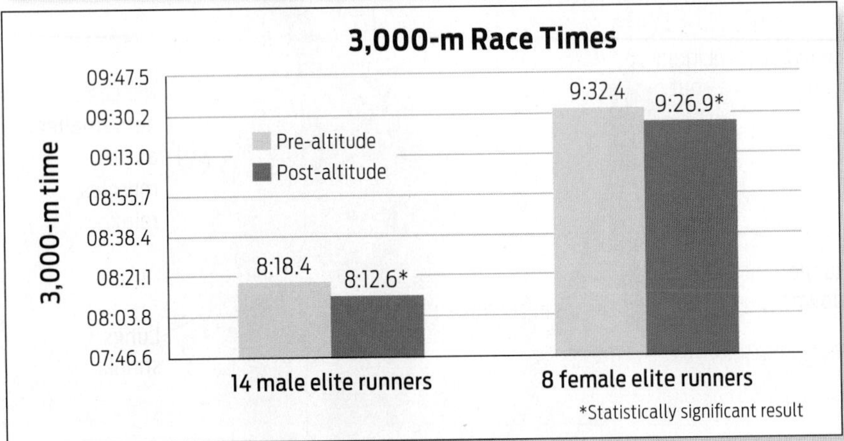

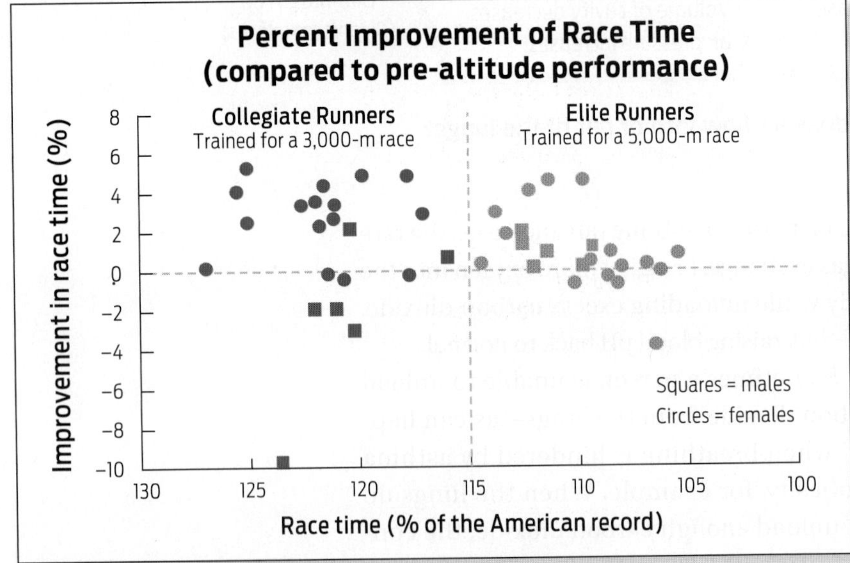

Data from Stray-Gunderson, *J Appl Physiol* 91: 1113-1120,2001.

? **Look at the absolute change in maximal oxygen uptake in males and females, and in the 3,000-m race times for males and females. Who had the larger increase in oxygen uptake and who had the larger change in race time?**

problem occurs when someone hyperventilates and releases too much carbon dioxide, which can cause dizziness and numbness; an easy fix is to breathe into a paper bag to re-inhale the exhaled carbon dioxide.

Does It Work?

The logic of altitude training seems airtight: increase the concentration of RBCs in the blood and you will have greater oxygen-carrying capacity, more aerobic respiration, and better performance. But actually proving that altitude training improves performance is harder than it may seem. Even Wilber, who literally wrote the book on the subject, acknowledges that there is still a fair amount of controversy surrounding the science of altitude training, with some studies concluding that it works, and others concluding that it doesn't.

Partly, the issue is the difficulty of designing properly controlled studies. Any gains in performance are likely to be small, so how can you be sure the gains were due to high altitude and not some other aspect of training? Likewise, if athletes don't show consistent improvement, is it because altitude training doesn't work, or because not everyone responds to altitude in the same way?

For his part, Wilber is persuaded by the available evidence that, when pursued methodically, altitude training does indeed improve performance **(INFOGRAPHIC 29.7)**.

Additional (but anecdotal) evidence that it works, he says, comes from the loyalty of athletes and coaches themselves. "When you see a Michael Phelps and his coach and other athletes coming back to altitude year after year after year after year, you know that it's working for them."

At the same time, Wilber cautions that altitude training requires consistent and long-term practice. For that reason, ordinary gymgoers are unlikely to see any significant results from periodic workouts in the numerous "hypoxic rooms" that have sprung up in commercial gyms all over the world.

Picking Up, Dropping Off

▸ Hemoglobin and oxygen binding

If an athlete experiences hypoxia regularly, for 12 to 20 hours a day, for at least 3 weeks, according to the research he or she will start to see benefits upon returning to sea level. The biggest benefit will be the ability to carry more oxygen in the blood. But unlike carbon dioxide, which travels easily in the blood plasma as carbonic acid, oxygen does not readily dissolve in the watery plasma and so is not transported this way. Instead, it's carried inside RBCs, bound to a protein called **hemoglobin.** RBCs are essentially bags of hemoglobin, with each red blood cell having between 250 and 300 million molecules of hemoglobin. RBCs lack a nucleus, and are streamlined for doing one thing very well: carrying oxygen.

Hemoglobin is a complex molecule, composed of multiple interworking parts. Each molecule of hemoglobin is made up of four interacting protein chains called "globins" (see Chapter 9) and four iron-containing **heme groups.** The heme groups are what bind oxygen. Since each hemoglobin molecule has four heme groups, each hemoglobin molecule can carry up to four molecules of O_2. Because the heme groups of hemoglobin contain iron, a dietary deficiency of iron can limit the production of hemoglobin, and in turn reduce the overall oxygen-carrying capacity of blood, leading to a condition known as iron-deficiency anemia **(INFOGRAPHIC 29.8).**

Key to understanding how hemoglobin functions in oxygen transport is the fact that it binds oxygen reversibly. It can pick up (that is, bind) oxygen in the lungs and drop off (that is, release) oxygen in the tissues. Whether hemoglobin is picking up or dropping off oxygen depends mainly on the partial pressure of oxygen (P_{O2}). Fresh air entering the

HEMOGLOBIN
A protein found in red blood cells that is specialized for transporting oxygen.

HEME GROUPS
Iron-containing structures on hemoglobin, the sites of oxygen binding.

INFOGRAPHIC 29.8

Hemoglobin Carries Oxygen in Red Blood Cells

Hemoglobin has four globin chains and four heme groups. Each heme group contains an iron atom and has the capacity to bind to one oxygen molecule.

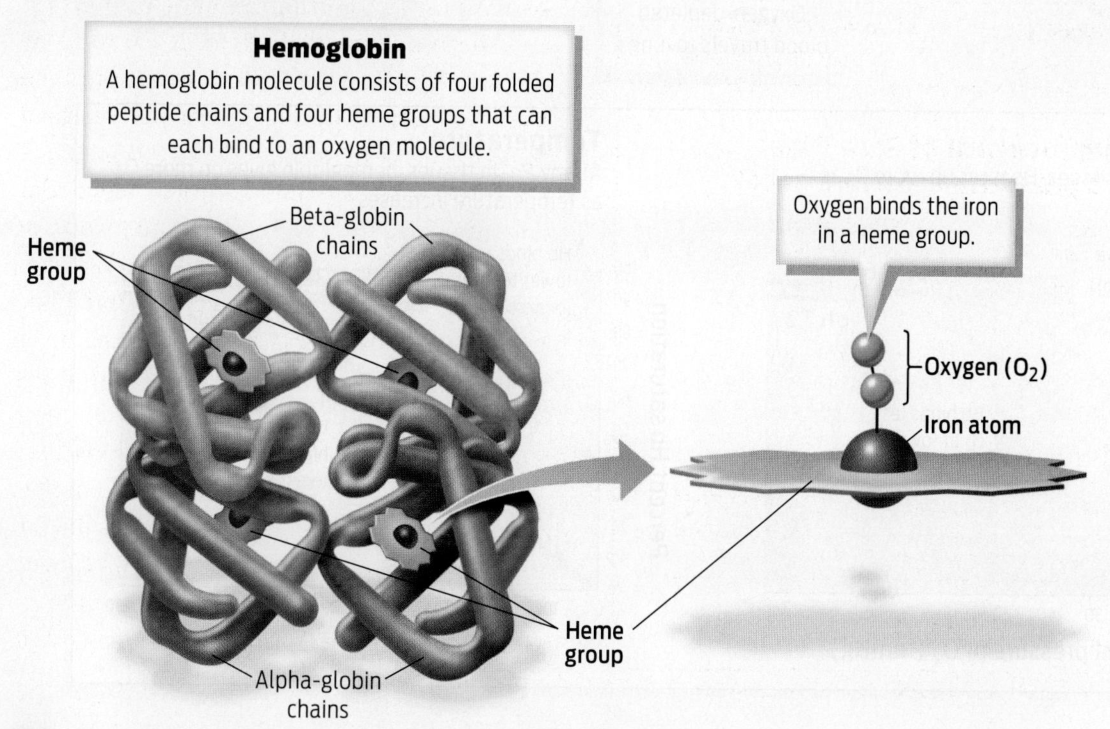

? **How would a dietary deficiency of iron affect the transport of oxygen in red blood cells?**

lungs has a relatively high partial pressure of oxygen, so the hemoglobin will become saturated (that is, fully loaded) with oxygen. In tissues, the partial pressure of oxygen tends to be lower, because cells in tissues are consuming oxygen as they carry out aerobic respiration, so hemoglobin tends to give up some of its oxygen to tissues.

Another important factor affecting oxygen binding is temperature. Hemoglobin is more likely to release oxygen as temperature increases—as it does, for example, in an

INFOGRAPHIC 29.9

Oxygen Binding to Hemoglobin Is Reversible

Hemoglobin reversibly binds oxygen. When the partial pressure of oxygen (P_{O_2}) is high, hemoglobin will bind oxygen. At low P_{O_2}, hemoglobin will release oxygen. The binding is also influenced by temperature and pH.

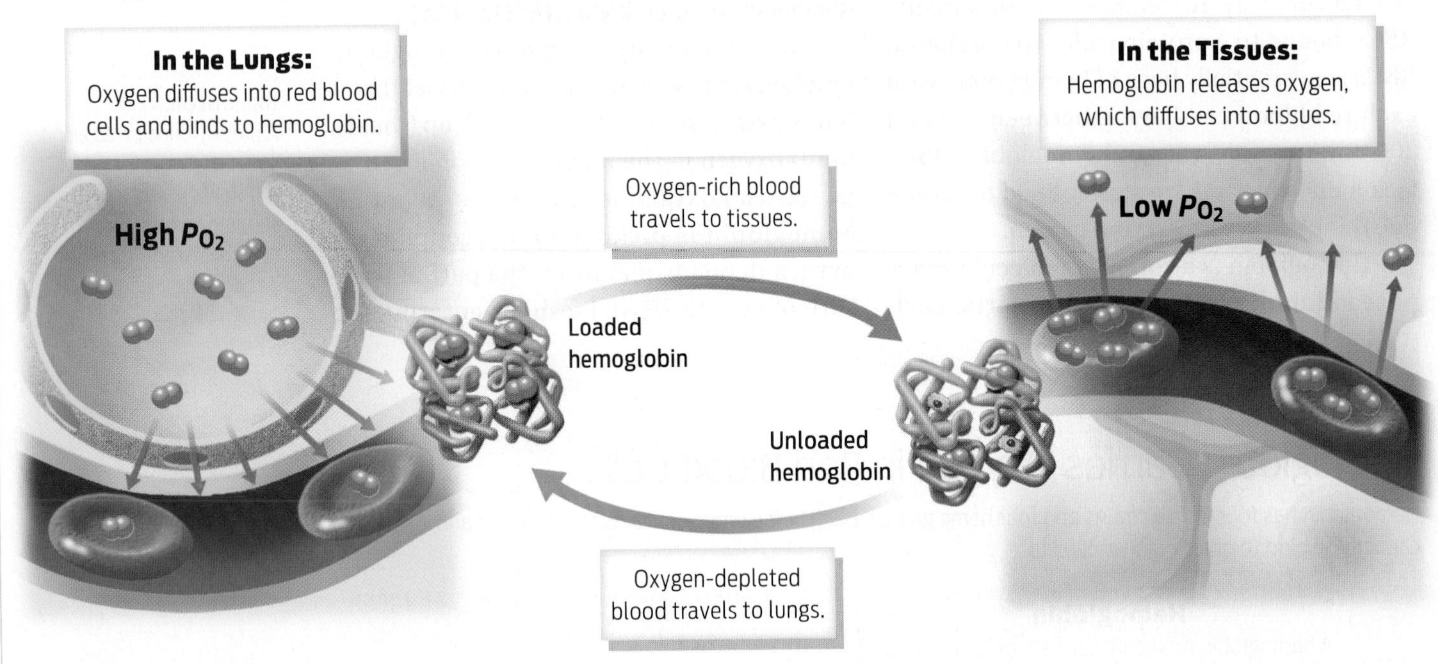

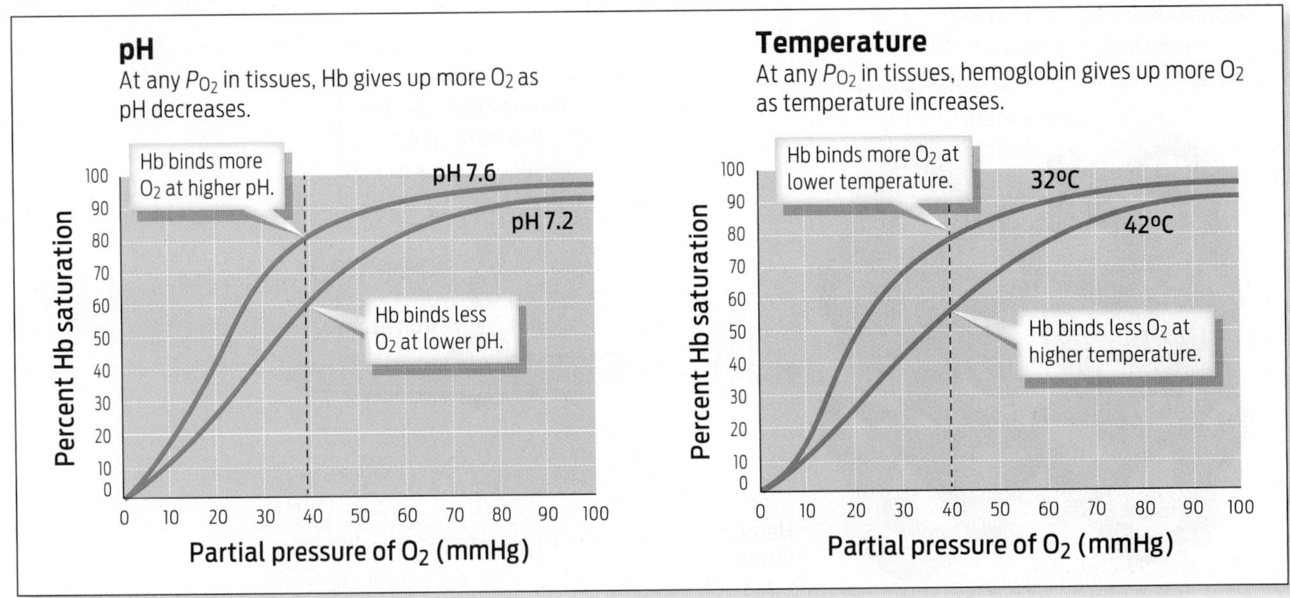

? At pH 7.2 and P_{O_2} of 30 mmHg, what is the percent saturation of hemoglobin?

Adam Pretty/Getty Images

Michael Phelps celebrates after winning one of his 23 Olympic gold medals.

actively contracting muscle that is burning energy.

Lastly, pH plays a role. At lower pH, hemoglobin releases more oxygen. As we've seen, one of the main factors that can lower pH is the concentration of CO_2. Since CO_2 is produced during aerobic respiration, the pH in muscle tissue goes down, becoming more acidic during exercise. In turn, the lower pH causes the hemoglobin to give up more of its oxygen, thus ensuring a continuous supply of this critical gas for the active muscle **(INFOGRAPHIC 29.9)**.

So hemoglobin has an extremely important role to play in respiration. If there were some easy way to increase the amount of hemoglobin in blood, without sleeping in a hypoxic chamber or living at altitude, then an athlete would be at a competitive advantage in terms of conducting aerobic respiration.

High-Tech Doping?

When an athlete (or anyone) spends time at altitude, his or her kidneys respond by secreting a hormone called **erythropoietin (EPO),** which stimulates RBC production in the bone marrow. By producing more RBCs—each of which contains millions of hemoglobin molecules—the body increases its ability to absorb and transport oxygen. This is a natural process, which goes on anytime the body needs to make more RBCs—for example, after blood loss from a wound.

Is it fair that some athletes can afford to buy a hypoxic tent that costs $7,000–$10,000, while other athletes cannot?

Because there is such a clear relationship between the concentration of RBCs in the blood and the ability to perform aerobic exercise, many athletes in elite sports in the past legally resorted to blood doping—artificially increasing the concentration of RBCs in their

ERYTHROPOIETIN (EPO)
A hormone produced by the kidneys that stimulates red blood cell production.

blood by undergoing blood transfusions or by injecting themselves with synthetic EPO **(INFOGRAPHIC 29.10)**.

But blood doping with EPO can have dangerous side effects, including abnormal blot clots and stroke. Because EPO doping literally thickens the blood by increasing the concentration of red blood cells, the heart has to work harder to pump it, and the thickened blood is more likely to cause a blockage in a vessel. A few athlete deaths have been attributed to the practice.

The International Olympic Committee officially banned blood doping by transfusion in 1986 and by EPO injection in 1990. However, the difficulty and expense of testing has made enforcement difficult.

Blood doping became very common in cycling and caused a major scandal in 1998, when it was discovered that many cyclists were using EPO. That scandal led to the formation of the World Anti-Doping Agency (WADA) in 1999, which has worked to develop tighter and more effective testing programs, including urine and blood tests that can detect synthetic EPO.

Of course, for testing to be effective, it has to be done systematically and rigorously. Cyclist and seven-time Tour de France winner Lance Armstrong, who admitted to doping and

INFOGRAPHIC 29.10

EPO and Blood Doping

Erythropoietin (EPO) stimulates the production of red blood cells in the bone marrow. After EPO treatment, more immature red blood cells are detectable, and maximum oxygen uptake is enhanced.

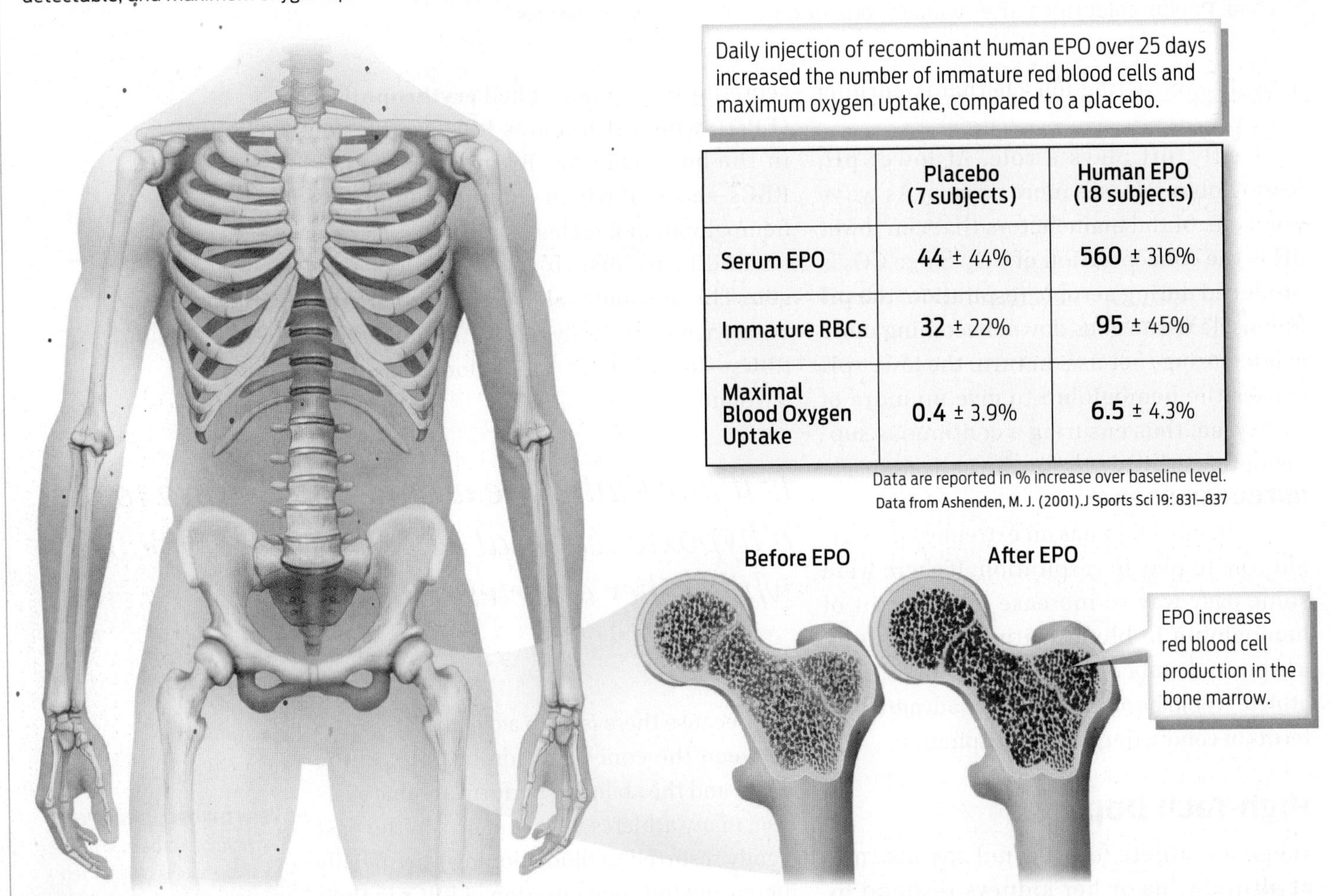

	Placebo (7 subjects)	Human EPO (18 subjects)
Serum EPO	44 ± 44%	560 ± 316%
Immature RBCs	32 ± 22%	95 ± 45%
Maximal Blood Oxygen Uptake	0.4 ± 3.9%	6.5 ± 4.3%

Data are reported in % increase over baseline level.
Data from Ashenden, M. J. (2001).J Sports Sci 19: 831–837

? What is the natural source of EPO?

was stripped of his titles in 2012, managed to avoid being tested numerous times by simply not answering the door when testers arrived, according to the testimony of his fellow cyclists.

There was speculation during the 2008 Olympics in Beijing that Phelps was doping—how else to account for his superhuman performance, eight gold medals in eight events? However, repeated tests proved that he was clean.

Unlike doping with EPO, the use of hypoxic chambers is not currently banned by WADA. Even though the underlying goals of blood doping and the use of hypoxic chambers are the same—increasing the production of RBCs—leaders in the sporting world have so far deemed the use of such chambers a variation on natural training regimens. So for now, it is not illegal to use them. However, it does raise the question of whether it is ethically proper or goes against the spirit of sport. Is it fair that some athletes—especially those with lucrative sponsorships, such as Phelps—can afford to buy a hypoxic tent that costs $7,000–$10,000, while other athletes, especially those from poorer countries, cannot?

As Wilber notes, "You're not going to find altitude houses or chambers in South Sudan, or a majority of the countries in the world, but yet in the U.S. or Finland or China, yeah, you're going to find them there. And you're going to find elite athletes using them."

Ethical or not, it is easy to see the appeal of such practices among elite athletes. When fractions of a second can mean the difference between winning gold and coming in fourth, any extra boost can help. This is a dilemma that Phelps knows all too well, having lost to South African swimmer Chad le Clos by less than a quarter of a second in the 200-m butterfly event at the 2012 Summer Games in London. Even with the potential added advantage of a few more RBCs, he was still not able to capture the gold in that race, having to settle instead for silver.

Phelps repaid the debt in 2016 in Rio, winning gold in the same event. That was his last Olympic Games, he said—so maybe now he can retire his bubble as well. ■

CHAPTER 29 Summary

Driving Question 1 What structures make up the respiratory system?

- The respiratory system takes in oxygen and eliminates carbon dioxide waste.
- The respiratory system consists of branched tubes that deliver air to the alveoli in the lungs.

Driving Question 2 How do the respiratory and cardiovascular systems cooperate to deliver oxygen to body cells and remove carbon dioxide from tissues?

- The respiratory system works in close conjunction with the cardiovascular system. Capillaries in the alveoli of the lungs are the sites of gas exchange between air and blood.
- Breathing is the way we ventilate our lungs and bring fresh air to the site of gas exchange. Ventilation requires a coordinated set of muscle movements that create negative pressure in the lungs to draw air in and positive pressure to expel air out.

Driving Question 3 What factors influence the oxygen-carrying capacity of blood and breathing rate?

- Breathing rate is controlled by the brain and relies on sensors that detect pH, which is directly related to the concentration of carbon dioxide in the blood.
- Oxygen is carried by hemoglobin in red blood cells. Hemoglobin binds reversibly to oxygen, picking it up from the lungs and dropping it off in tissues.
- Extended exposure to low oxygen pressure, such as occurs at high altitude, stimulates the production of red blood cells in the bone marrow. The growth signal comes from the hormone erythropoietin (EPO), which is secreted by the kidneys.

Driving Question 4 How can scientific knowledge of the respiratory system be used to design training regimens for elite athletes?

- Altitude training programs such as "live high, train low" and using a hypoxic chamber can increase the number of red blood cells in the blood and improve athletic performance.

More to Explore

- World Anti-Doping Agency: www.wada-ama.org/
- Wilber, R. (2004). *Altitude Training and Athletic Performance*. Champaign, IL: Human Kinetics Publishers.
- Stray-Gundersen, J., et al. (2001). "Living high-training low" altitude training improves sea level performance in male and female elite runners. *J Appl Physiol* 91:1113–1120.
- Beall, C. M., et al. (2010). Natural selection on EPAS1 (HIF2a) associated with low hemoglobin concentration in Tibetan highlanders. *Proc Natl Acad Sci* 107:11459–11464.
- Matson, J. (July 30, 2012). Rope a dope: drug testing in sports enters a more aggressive era. *Sci Am*. https://www.scientificamerican.com/article/olympic-drug-testing/

CHAPTER 29 Test Your Knowledge

Driving Question 1 What structures make up the respiratory system?

By answering the questions below and studying Infographics 29.1 and 29.6, you should be able to generate an answer for this broader Driving Question.

Know It

1. Add the names of the structures indicated by the labels A–F in the diagram below.

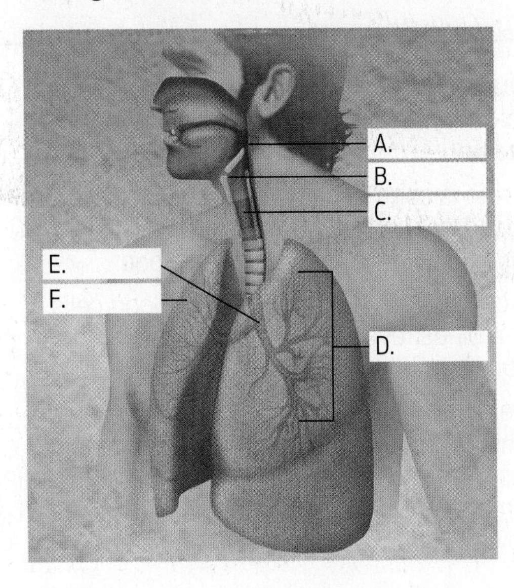

2. Which part of the respiratory system is the site of exchange of gases between blood and air?
   a. alveoli
   b. bronchioles
   c. trachea
   d. pharynx
   e. bronchi

3. Inhalation is accompanied by
   a. muscular relaxation and a decrease in lung volume.
   b. muscular relaxation and an increase in lung volume.
   c. muscular contraction and a decrease in lung volume.
   d. muscular contraction and an increase in lung volume.
   e. muscular contraction and no change in lung volume.

Use It

4. Pneumonia tends to cause an accumulation of fluid in the lungs. This makes it harder for oxygen to diffuse between the air in alveoli and red blood cells in capillaries in the lungs.
   a. From this description, explain why people with pneumonia experience shortness of breath.
   b. Explain why supplemental oxygen is often a treatment for severe cases of pneumonia.

5. Asthma is a disease that causes swelling and constriction of the airways in the lungs. Compare the predicted symptoms of asthma with the symptoms of pneumonia. Can you think of any treatments for asthma that might be different from treatments for pneumonia?

Driving Question 2 How do the respiratory and cardiovascular systems cooperate to deliver oxygen to body cells and remove carbon dioxide from tissues?

By answering the questions below and studying Infographics 29.2, 29.6, 29.8, and 29.9, you should be able to generate an answer for this broader Driving Question.

Know It

6. How is O_2 transported throughout the body?

a. dissolved in the plasma of blood
b. bound to hemoglobin in plasma
c. bound to hemoglobin in white blood cells
d. bound to hemoglobin in red blood cells
e. dissolved in the cytoplasm of red blood cells

7. What can cause a drop in blood pH?

a. a decrease in O_2
b. an increase in O_2
c. a decrease in CO_2
d. an increase in CO_2
e. b or d

Use It

8. If blood pH drops, what happens to the breathing rate? Explain your answer.

9. Oxygen diffuses from the air in alveoli to the blood in lung capillaries. Diffusion occurs rapidly over short distances, but decreases dramatically with increases in distance. Pneumonia is an accumulation of fluid in the alveolar air spaces. Why does pneumonia cause shortness of breath and give a bluish tint to the skin and nails?

10. Breathing in and out of a paper bag will ____________ pH and therefore ____________ ventilation.

a. not change; not change
b. increase; increase
c. increase; decrease
d. decrease; decrease
e. decrease; increase

Driving Question 3 What factors influence the oxygen-carrying capacity of blood and breathing rate?

By answering the questions below and studying Infographics 29.3, 29.4, 29.6, 29.8, and 29.9, you should be able to generate an answer for this broader Driving Question.

Know It

11. Relative to a tissue at rest, actively exercising tissues have

a. higher temperature, higher P_{O2}, and higher pH.
b. higher temperature, lower P_{O2}, and lower pH.
c. higher temperature, higher P_{O2}, and lower pH.
d. lower temperature, higher P_{O2}, and higher pH.
e. lower temperature, lower P_{O2}, and lower pH.

12. What is the particular feature of altitude that increases the oxygen-carrying capacity of the blood?

a. the actual height (elevation)
b. the reduced atmospheric (barometric) pressure
c. the reduced partial pressure of oxygen
d. the increased atmospheric (barometric) pressure
e. the decreased relative humidity

Use It

13. Hemoglobin releases O_2 at low pH. Give two reasons why a tissue might have a low pH.

14. Carbon monoxide (CO) binds to hemoglobin more tightly than does oxygen. In fact, CO can displace oxygen from hemoglobin. Predict the symptoms of CO poisoning, and provide an explanation.

15. Why do people "suck wind" (that is, breathe very heavily) with vigorous exercise?

Driving Question 4 How can scientific knowledge of the respiratory system be used to design training regimens for elite athletes?

By answering the questions below and studying Infographics 29.5, 29.7, and 29.10, you should be able to generate an answer for this broader Driving Question.

Know It

16. Which of the following mimics high altitude?

a. sleeping in a high-O_2 chamber
b. sleeping in a low-O_2 chamber
c. transfusing RBCs into the circulation
d. a and c
e. b and c

17. What does EPO do?

a. stimulates RBCs to release stored O_2
b. stimulates RBC production
c. increases the number of heme groups per molecule of hemoglobin
d. increases ventilation rate
e. b and c

18. Design an experiment to determine whether hypoxic chambers confer an advantage relative to altitude training. Consider how you will set up your experiment, including appropriate controls and the variables that you will consider and measure.

Use It

19. WADA must be able to test for a variety of banned substances. How could WADA test for each of the following? Rank them from easiest to detect (1) to hardest to detect (4).

_____ EPO doping
_____ Transfusion of whole blood
_____ Transfusion of RBCs
_____ Use of a hypoxic chamber

Mini Case

Apply Your Knowledge

20. EPO doping has been a persistent problem in sports, particularly endurance sports. Some have questioned how much EPO doping benefits those athletes who are already living and training at altitude. A 2019 study looked at the impact of EPO doping on two groups of well-trained male athletes: 29 Kenyans (KEN) living and training at altitude (2,150 m/~7,000 feet) and 19 Scots (SCOT) living and training at sea level. Researchers measured the participants' hematocrit (percentage of the blood made up by red blood cells), hemoglobin, maximum oxygen-carrying capacity (VO_{2max}), and 3,000-m race time. Measurements were taken 2 weeks before EPO (baseline) was administered and at the end of the EPO doping period.

a. Complete the table below by calculating the percent change in each variable from baseline.

Answer the following questions after considering the data:

b. Were the Kenyans and Scots equivalent at baseline?

c. Did the Kenyans experience any improvements as a result of EPO? If so, in which parameters?

d. Did the Scots experience any improvements as a result of EPO? If so, in which parameters?

e. Can EPO doping improve performance in athletes who are already trained at altitude? Explain your answer.

	Baseline	End of EPO	Percent change from baseline
KEN hematocrit (%)	45	49	
SCOT hematocrit (%)	42	49	
KEN hemoglobin (g/dL)	15.3	16.8	
SCOT hemoglobin (g/dL)	14.4	16.5	
KEN VO_{2max} (mL/kg/min)	66.2	70.7	
SCOT VO_{2max} (mL/kg/min)	57.7	62.1	
KEN 3,000 m	9:24 min (564 s)	8:57 min (537 s)	
SCOT 3,000 m	11:08 min (668 s)	10:30 min (630 s)	

Data from Haile, D. W., et al. (2019). Effects of EPO on blood parameters and running performance in Kenyan athletes. *Med Sci Sports Exerc* 51(2):299–307. doi: 10.1249/mss.0000000000001777. https://strathprints.strath.ac.uk/68962/1/Haile_etal_MSSE_2019_Effects_of_EPO_on_blood_parameters_and_running_performance.pdf

Interpreting Data

Apply Your Knowledge

21. Look at Infographic 29.7.

a. Start with the 3,000-m performance data for the elite athletes. How many seconds did the men improve by? How many seconds did the women improve by? Now determine that improvement as a percentage of the pre-altitude race time. What percent improvement did altitude training confer? In your opinion, is this method of training worth it for that percent improvement?

b. Now look at the graph with both the collegiate and the elite athletes. Even if the graph were not labeled, how could you know which group was which? (*Hint:* Look at their race times.) Did every athlete experience an improvement in race time? Was the change in race time identical for every athlete? Given the variability in results in this small sample size of conditioned athletes, what do you think you can extrapolate about altitude training in a larger population of active people?

Bring It Home

Apply Your Knowledge

22. A friend wants to join a new gym. The gym has many amenities, including personal trainers (for an additional fee) and a hypoxic chamber for "performance training." There is a steep sign-up fee and high monthly dues at this gym. Another gym is offering a no-fee sign-up and lower monthly dues. This gym has the same cardiovascular equipment and access to personal trainers (at an additional fee comparable to that at the other gym). Your friend is fit. She runs local 10K races and often places in the top 10 in her age group in your small town. Given the costs of the two gyms and your friend's fitness level and goals, which gym would you advise her to join? Explain your reasoning.

SMOKE on the Brain

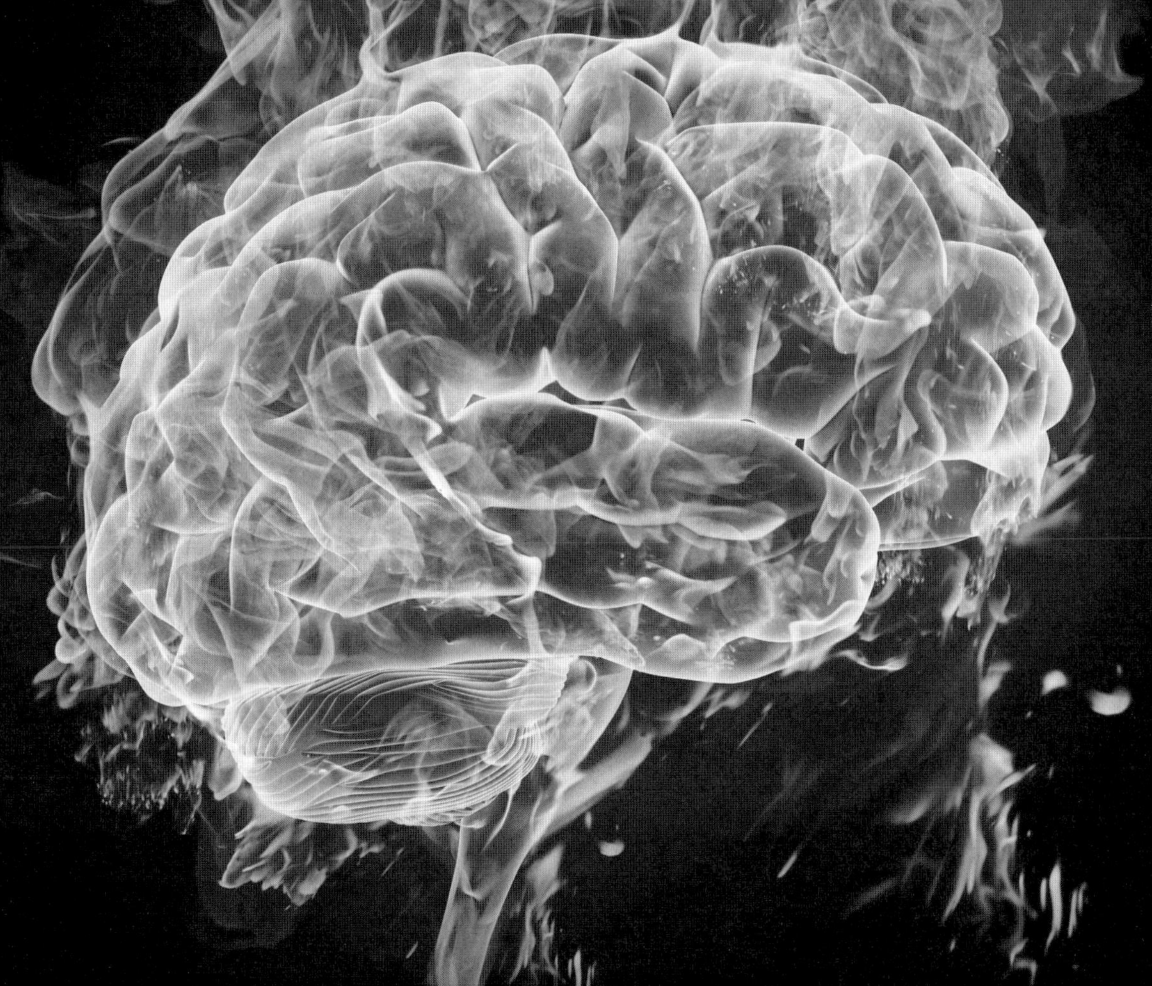

Neuroscience explains why nicotine and other drugs are hard to kick

Vladimir/Getty Images

DRIVING QUESTIONS

1. How is the nervous system organized?
2. How do cells in the nervous system transmit signals?
3. Why are some drugs (and some behaviors) addictive?

Jack Ward thought he could resist picking up a cigarette, but the temptation was too strong. He had successfully quit smoking in 2006. But when he walked into a smoke-filled poker room 3 years later at a friend's house in Brooklyn, New York, the scene before him seemed to run in slow motion. He watched as smokers took deep satisfying drags on their cigarettes and exhaled wispy ringlets of contentment. He resisted that night. But poker night became a weekly event, and finally he gave in: he began smoking again.

Ward knew that smoking is risky. He'd had countless arguments with his wife, who urged him to stop smoking to protect his own health as well as hers, which is partly why he quit in the first place. But during poker night, none of that seemed to matter.

Ward lit his first cigarette in high school. By the time he went to graduate school, he was smoking a pack and a half a day. "I was surrounded by people who smoked," Ward recalls. "It never seemed unusual to smoke so much."

Cigarette smoking is highly addictive: smokers can develop a physical and psychological need to smoke. And while anyone might be able to smoke one cigarette or even several and not become addicted, most people find it extremely difficult to stop if they have smoked for an extended length of time.

For many years, addiction was seen as a personal or moral failing—the result of a weak will or a set of bad choices. Increasingly, however, researchers are looking to biology to explain addiction's awesome power. Brain scientists now know that cigarettes, alcohol, cocaine, and other drugs of abuse stimulate the brain's reward system. This complex circuit of brain cells evolved to make us feel good after eating or having sex—activities we must engage in if we are to survive and pass along our genes. Without a feeling of pleasure from these activities that ensure our survival, we might never seek out food or sex, and our species would die out. Drugs of abuse stimulate this same reward system: that's why they make people feel so good, and why they can be so hard to resist.

Chemical substances like drugs and alcohol aren't the only things that can become addictive. Certain behaviors—gambling, shopping, sex, exercise—can start out as pleasurable habits but slide into addictions. Even eating can be addictive. Studies of obese people, for example, have shown that the brains of compulsive eaters are hyperactive in areas that respond to food. For these people, the mere thought of eating floods the brain with pleasure. In other words, almost anything deeply enjoyable can become an addiction.

Despite public health campaigns, smoking remains common among adolescents.

David Perry/AP Images

INFOGRAPHIC 30.1

Tobacco Use among U.S. Adults and Adolescents

Tobacco use is prevalent among adults and adolescents in the United States. Nearly 40 million adults smoke cigarettes, and more than 4.5 million adolescents use some kind of tobacco product. E-cigarette use (vaping) is particularly common among middle and high school students.

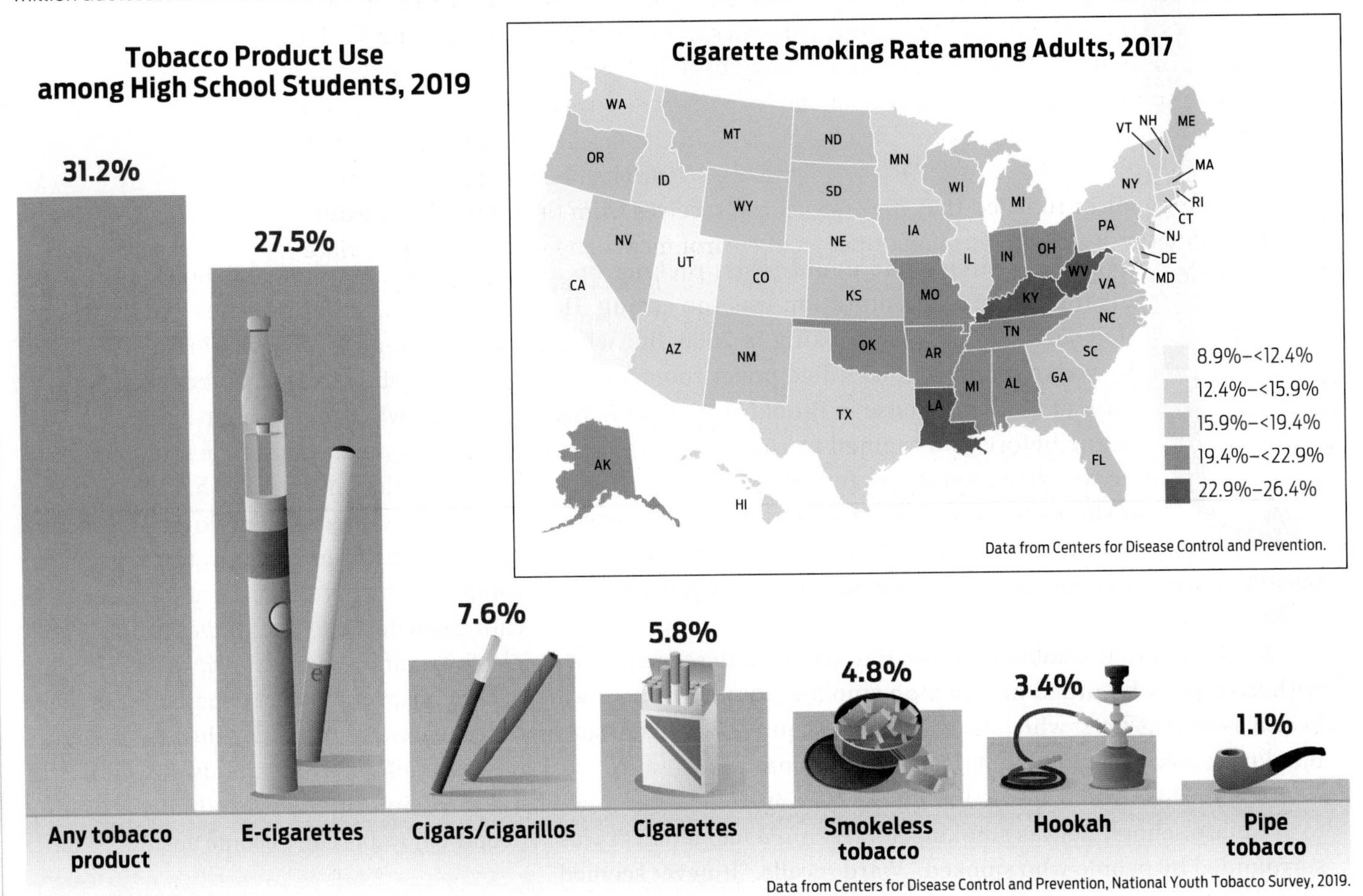

? **If a high school has a class of 500 graduating seniors, how many are likely to vape? How many are likely to smoke cigarettes?**

But the pleasure comes at a massive cost. After prolonged use, drugs of abuse change the structure and function of the brain in ways that can wreak havoc on users' lives. Severely addicted people may stop eating, stop working—in fact, stop all activities because nothing matters except the drug. And they will do anything to get the drug: without it, they have intense cravings, obsessive thoughts, and are deeply depressed.

Once established, addiction is difficult to treat. The stories of celebrities who have been in and out of rehab and the deaths of superstars such as Michael Jackson, Heath Ledger, and Prince from drug overdoses are testament to just how difficult overcoming addiction to drugs—whether prescription or illicit substances—can be. But scientists today are gaining a better understanding of how physiological changes in the brain cause addiction, and that knowledge is leading to better treatments, which will help addicts of all kinds reclaim their brains and their lives.

Addiction and the Brain

▸ Organization of the nervous system

Despite a massive public health campaign aimed at curbing cigarette smoking, it remains a common practice in the United States.

In 2018, about 14% of American adults smoked cigarettes regularly, according to the U.S. Centers for Disease Control and Prevention (CDC), down from about 45% in 1955. A 2019 survey conducted by the CDC found that more than 12% of middle school students and more than 30% of high school students use some form of tobacco. These numbers testify to tobacco's addictive potential **(INFOGRAPHIC 30.1)**.

The component in tobacco that makes cigarettes addictive is the chemical nicotine. When a person smokes, nicotine floods the lungs, where it is absorbed into the bloodstream. Once in the bloodstream, nicotine travels quickly throughout the body, reaching the brain within 8 seconds after tobacco smoke is inhaled. Chewing tobacco also contains nicotine, although the body absorbs it into the bloodstream through the mucous membranes that line the mouth rather than through the lungs.

Nicotine is addictive because it stimulates feelings of pleasure in the brain. The **brain** is the master organ of the body; it controls virtually all of the body's activities, including voluntary actions like moving and thinking, and involuntary actions like heart rate, breathing, and digestion. The brain integrates information it receives from the internal and external environments and produces appropriate actions. The brain is also the seat of memory. Pleasurable experiences recorded in our memories serve as motivation; we tend to seek out the same pleasurable experiences over and over. Ultimately, the decision to smoke a cigarette, or to engage in any other behavior, is made in the brain **(INFOGRAPHIC 30.2)**.

BRAIN
The organ of the central nervous system that integrates information and coordinates virtually all functions of the body.

INFOGRAPHIC 30.2

Nicotine Stimulates Pleasure in the Brain

The brain coordinates the vast majority of all the body's activities. When nicotine is absorbed, it stimulates feelings of pleasure in the brain. The brain ultimately makes the decision to smoke a cigarette or not.

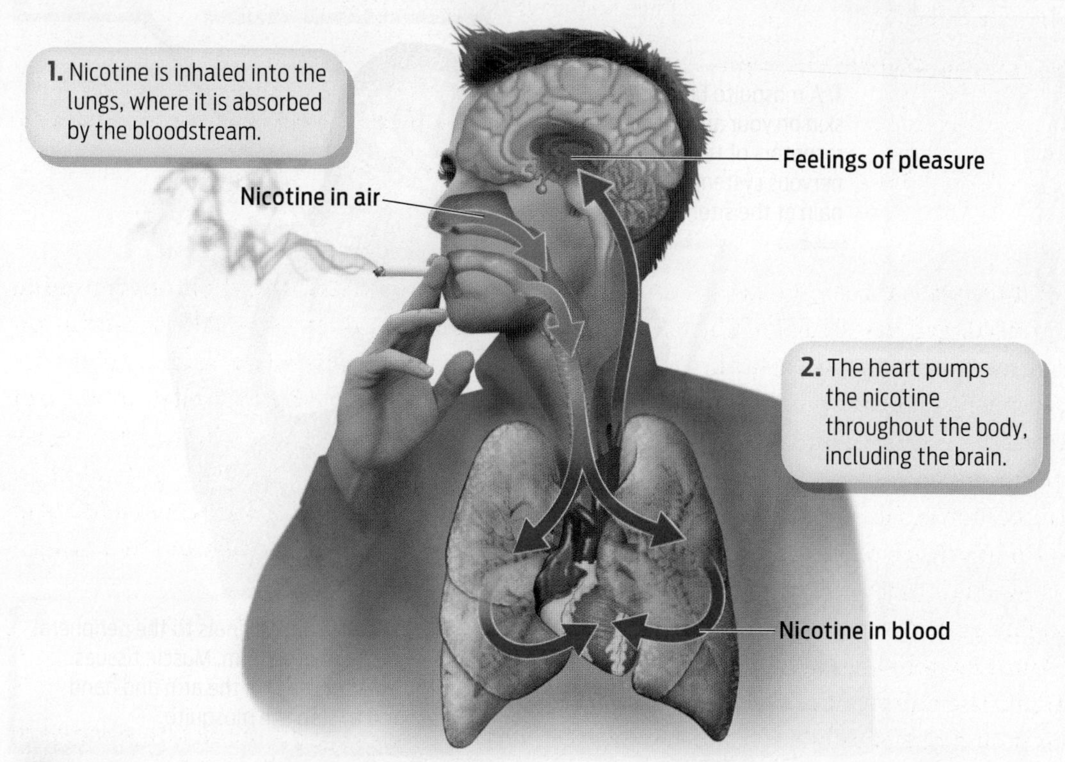

? Why can nicotine stimulate feelings of pleasure in the brain when it is chewed (that is, used in the form of chewing tobacco), rather than smoked?

NERVE
A bundle of specialized cells that transmits information.

SPINAL CORD
A bundle of nerve fibers, contained within the bony spinal column, that transmits information between the brain and the rest of the body.

CENTRAL NERVOUS SYSTEM (CNS)
The brain and the spinal cord.

PERIPHERAL NERVOUS SYSTEM (PNS)
All the nervous tissue outside the central nervous system that transmits signals from the CNS to the rest of the body.

The brain doesn't act in isolation: it communicates with the rest of the body, sending out and receiving signals to coordinate our many activities. When we decide to reach for a cigarette, for example, our brain sends a signal to muscles in our arm. These signals travel along **nerves,** long fibers made up of specialized cells and supportive tissue that transmit signals through the nervous system. A major collection of nerves is the **spinal cord,** which extends from the base of the brain down to the lower back and is contained in and protected by our bony spinal column. Together, the brain and the spinal cord make up the **central nervous system (CNS).** Nerves that travel from the spinal cord to distant body sites, like the fingers, toes, and heart muscle, compose the **peripheral nervous system (PNS).** The peripheral nervous system includes all nervous tissue outside the brain and spinal cord leading to and from our limbs and organs.

The CNS and the PNS work together to sense and respond to stimuli in the environment. Sensory cells in the PNS, such as those in our ears, our eyes, and our skin, enable us to detect changes in our environment—a sudden loud noise, for instance, or the sting

INFOGRAPHIC 30.3

The Central and Peripheral Nervous Systems Work Together

The nervous system has two main branches: the central nervous system (yellow) and the peripheral nervous system (orange). The central nervous system consists of the brain and spinal cord. The rest of the nervous system is the peripheral nervous system, which connects the CNS to the rest of the body.

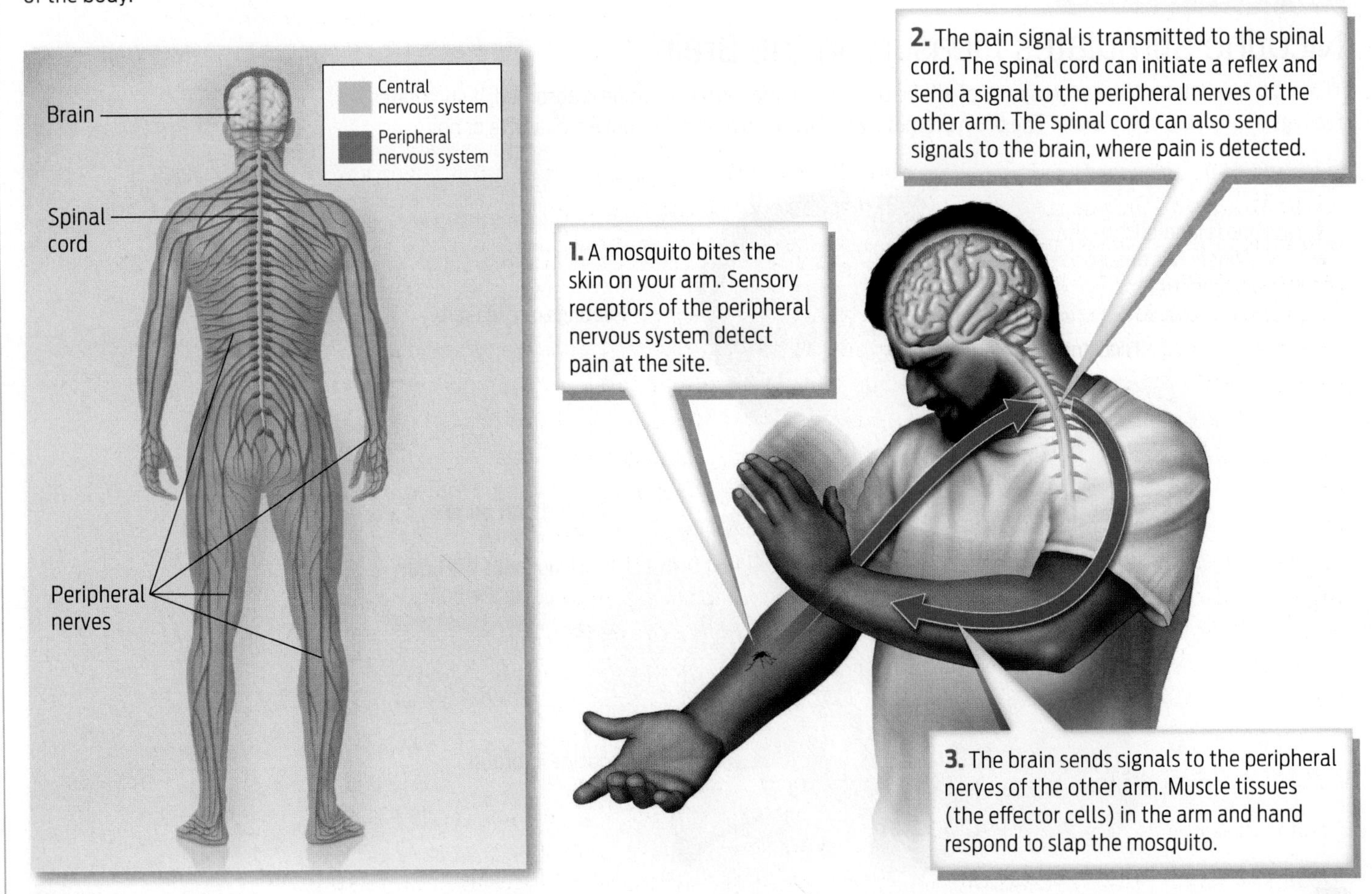

? Which functions—that is, sensing, integrating signals, and responding to the mosquito—are carried out by the peripheral nervous system?

INFOGRAPHIC 30.4

Neurons Are Highly Specialized

Neurons, the highly specialized cells of the nervous system, have a structure that enables them to send and receive electrical signals quickly.

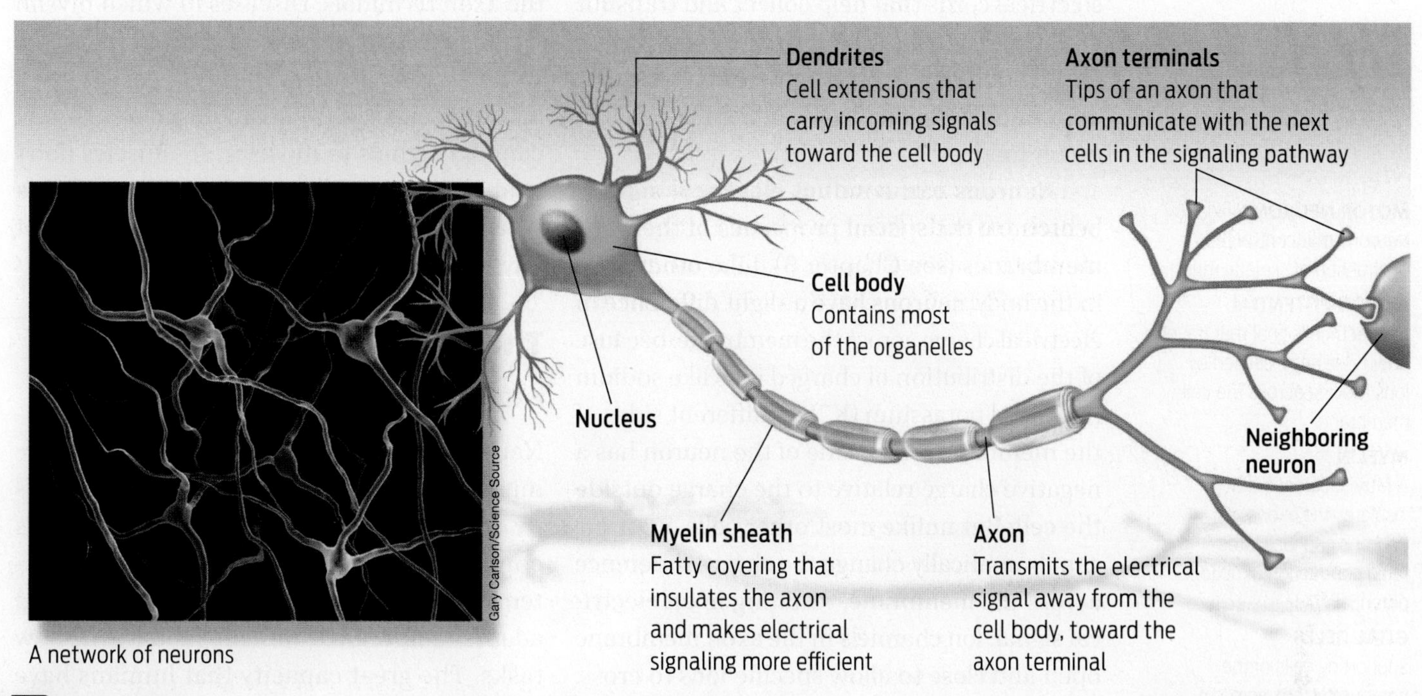

A network of neurons

? Which part of a neuron receives signals, and which part transmits those signals to the next cell in a signaling pathway?

of a mosquito. These stimuli are transmitted to the brain and spinal cord, which receive and integrate the incoming signals and trigger an appropriate response, such as a reflex. The reflex is carried out by an effector—cells or tissues that respond to information relayed from a sensor. In this case, the effector tissues are muscles that contract to allow us to jump on hearing the noise, or swat the mosquito on our arm. Peripheral nervous tissue in other organs, including the lungs, heart, and digestive organs, keeps our bodies operating without conscious thought. The nervous system allows us to perceive and understand the world around us and to translate thought into action **(INFOGRAPHIC 30.3)**.

Excitable Cells

▶ Neurons and action potentials

Addiction is a complex psychological and physiological process that affects many parts of the nervous system, but it begins at the level of the cell. Fundamentally, the brain—like any organ—is a collection of cells. The cells of the brain are organized into networks that receive, process, and send information. Highly specialized cells called **neurons** are the individual units of this elaborate network.

Neurons have a unique structure that enables them to send and receive signals. A neuron consists of a large **cell body** with branched extensions called **dendrites** that receive signals and a single large extension called the **axon** that carries signals away from the neuronal cell body. The **axon terminals,** at the end of the axon, transmit the signals to the next cell or cells in the network; the next cell could be another neuron, a muscle cell, or an endocrine cell **(INFOGRAPHIC 30.4)**.

The nervous system includes different types of neurons, including sensory neurons and motor neurons. **Sensory neurons,** such as the receptor cells in the eyes and skin, receive information from the external world and transmit it to the CNS. **Motor neurons**

NEURONS
Specialized cells of the nervous system that generate electrical signals in the form of action potentials.

CELL BODY
The part of a neuron that contains the nucleus and most of the cell's other organelles.

DENDRITES
Branched extensions from the cell body of a neuron, which receive incoming information.

AXON
The long extension of a neuron that conducts electrical signals away from the cell body toward the axon terminal.

AXON TERMINALS
The tips of an axon, which communicate with the next cell or cells in the pathway.

SENSORY NEURONS
Cells that convey information from both inside and outside the body to the CNS.

transmit information from the CNS to muscle cells, signaling them to contract or relax.

The signals sent along neurons are electric currents. Dendrites and axons are like electrical cords that help collect and transmit information in the nervous system. A nerve is just a bundle of axons of neurons. The axon of a single motor neuron can be more than a meter long.

Neurons can conduct electrical signals because of the special properties of their cell membranes (see Chapter 3). Like other cells in the body, neurons have a slight difference in electrical charge across the membrane: because of the distribution of charged ions like sodium (Na^+) and potassium (K^+) on different sides of the membrane, the inside of the neuron has a negative charge relative to the charge outside the cell. But unlike most other cells, neurons can dramatically change the charge difference across the membrane, resulting in an electrical signal. Ion channels in the axon membrane open and close to allow specific ions to cross. When the axon of a neuron allows positive ions to cross the cell membrane, the charge across the membrane changes. This sudden change of charge initiates an **action potential,** a coordinated pattern of ion flow across the membrane that is conducted down the axon like a wave **(INFOGRAPHIC 30.5).**

Action potentials travel down the length of a neuron very quickly. Just think how quickly your arm moves in response to touching a hot stove—the movement is almost instantaneous. An action potential can travel down the length of an axon at a rate of 120 m/s—about 270 mph.

Electrical signals can travel so quickly down neurons because, much like electrical wires, neurons are insulated: they are coated with a sheath of **myelin** that wraps around the axon at repeating intervals and prevents the electrical charge from leaking across the membrane. Because ion channels open only at unmyelinated regions, the action potential appears to "jump" from one unmyelinated region of the axon to another and so travels faster down the length of the axon. Myelin is critical to nerve transmission: without it, action potentials weaken, losing strength as they pass along the axon, and can peter out before reaching the axon terminals. Diseases in which myelin degenerates over time, such as multiple sclerosis, can cause progressive paralysis because in the absence of myelin, motor neurons can't conduct signals to muscles, so muscles don't contract. The myelin sheath is produced by a type of supportive cell called a **glial cell** that physically wraps around the axon.

The Anatomy of Addiction

▶ The brain's reward system

Neurons communicate not only to move muscles that control limbs, but also to solidify thoughts and lay down memories. This constant chatter within the nervous system is the way in which the brain grows and adapts to new environments and learns new tasks. The great capacity that humans have for conscious thought is a function of the sheer number of neurons found in the human brain—approximately 100 billion.

But the brain isn't just a uniform mass of neurons; like any organ, it has specialized regions that perform distinct functions. For example, one part of the brain processes visual information, while another specializes in hearing.

Scientists typically divide the brain into four major regions that orchestrate different functions. The **cerebellum,** located in the rearmost portion of the brain, controls movement, coordination, and balance. The **brain stem,** at the base of the brain, coordinates involuntary (automatic) actions like reflexes, heart rate, digestion, and breathing. The **diencephalon,** located above the brain stem, regulates homeostatic functions like body temperature, hunger, thirst, and sex drive. The largest part of the brain, the **cerebrum,** sits right on top. Its outer layer, the **cerebral cortex,** is the seat of our more advanced brain functions, including perception and thinking,

MOTOR NEURONS
Neurons that control the contraction of skeletal muscle.

ACTION POTENTIAL
An electrical signal that travels down a neuron, caused by ions moving across the cell membrane.

MYELIN
A fatty substance that insulates the axons of neurons and facilitates rapid conduction of action potentials.

GLIAL CELLS
Supporting cells of the nervous system, some of which produce myelin.

CEREBELLUM
The part of the brain that processes sensory information and is involved in movement, coordination, and balance.

BRAIN STEM
The part of the brain that is closest to the spinal cord and which controls vital functions such as heart rate, breathing, and blood pressure.

DIENCEPHALON
A brain region located between the brain stem and the cerebrum that regulates homeostatic functions like body temperature, hunger, thirst, and sex drive.

CEREBRUM
The region of the brain that controls intellect, learning, perception, and emotion.

CEREBRAL CORTEX
The outer layer of the cerebrum, which is involved in many advanced brain functions.

INFOGRAPHIC 30.5

Electrical Signals Are Transmitted along Axons

Neurons conduct electrical signals called action potentials, which are generated as ions flow across the cell membrane. When a neuron "fires," sodium (Na^+) and potassium (K^+) ions enter and leave the axon in a characteristic pattern, creating an action potential. The action potential in one patch of membrane generates a flow of ions that triggers an action potential in a neighboring patch of membrane. In this way, the electrical signal is conducted sequentially down the axon—much like a falling domino causes the next domino in the series to fall.

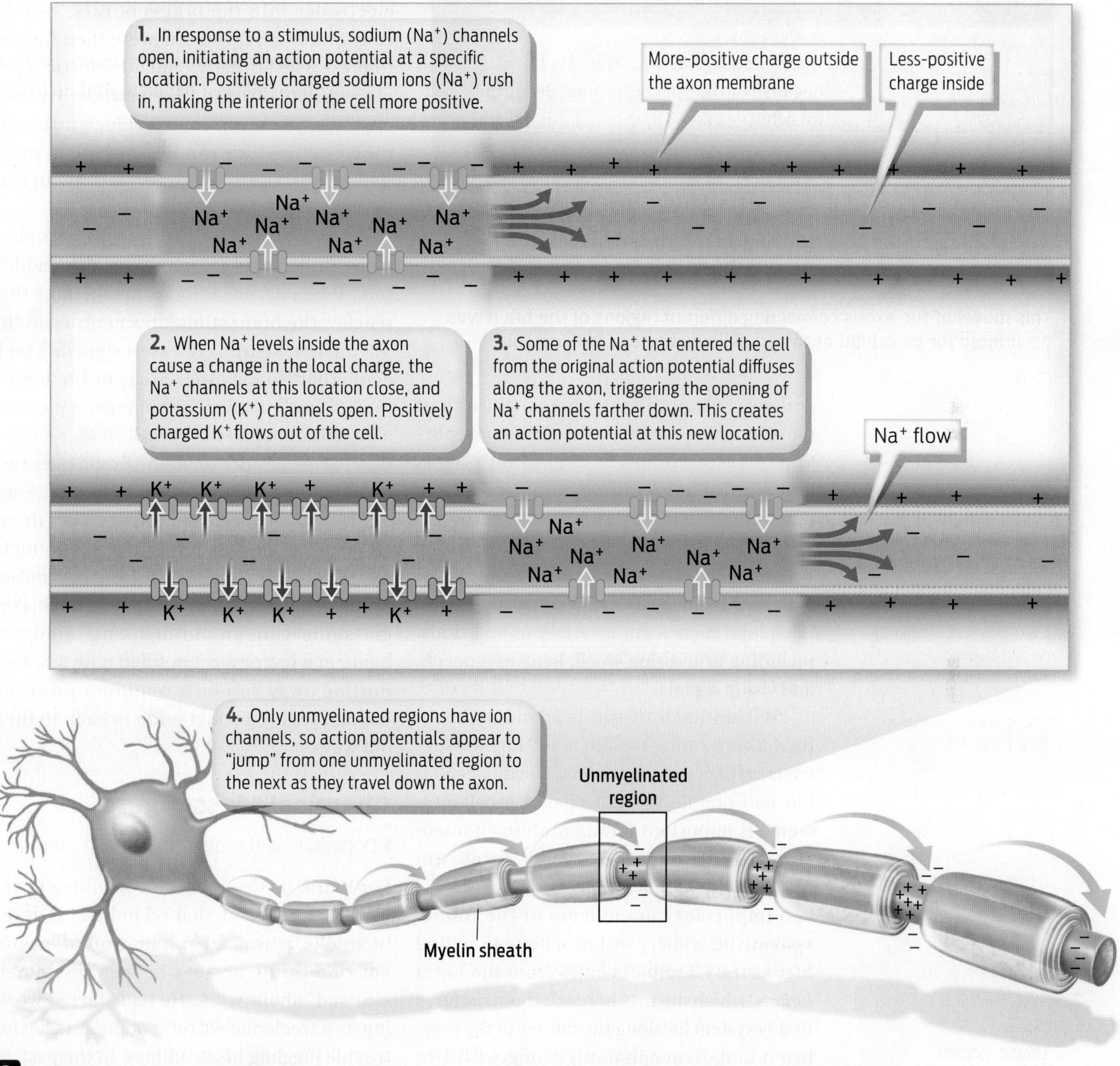

? **Which ion, flowing in which direction, triggers the initiation of an action potential at a given spot along a neuron?**

The Franklin Institute

This model of the axons connecting different regions of the brain was 3D-printed for an exhibit at the Franklin Institute in Philadelphia.

and gives us our distinct personalities and most human characteristics. The inner portion of the cerebrum contains brain regions that control more primitive functions, like responding to fearful situations. The cerebrum is made up of about 10 billion neurons and is divided into left and right hemispheres, with each hemisphere divided into four lobes. Each lobe carries out a variety of functions, including processing smell, hearing, speech, and vision signals.

Within each of the four major brain regions are subregions that are further specialized for certain functions. The diencephalon, for example, contains the hypothalamus, which is important for maintaining homeostasis (see Chapter 26). The cerebrum includes the hippocampus and amygdala—two important components of the **limbic system,** the primary seat of our emotions and memories. ("Limbic" comes from the Latin *limbus,* which means "border"; the structures in this system lie along the border of the cerebrum and diencephalon.) Along with two nearby regions, the nucleus accumbens and the ventral tegumental area (VTA), the limbic system is stimulated during pleasurable activities. It is this system that researchers sometimes refer to as the "pleasure center" or "reward system" of the brain.

LIMBIC SYSTEM
A set of brain structures that is stimulated during pleasurable activities and which is involved in addiction.

In 1954, psychologists James Olds and Peter Milner of McGill University provided the first evidence for the existence of such a pleasure center in the brain. They implanted electrodes into the brains of rats, specifically in the VTA. The rats were then trained to press a lever that would administer a jolt of electricity through the electrode. The rats apparently found the experience intensely pleasurable—they would continue to press the lever without rest for up to 24 hours, as many as 5,000 times per hour!

Scientists now know that the brain's pleasure center plays an important role in addiction. When a smoker lights up, nicotine that reaches the brain stimulates neurons in the VTA. These neurons release a signaling molecule that floods other areas of the brain's reward system, giving the smoker a pleasurable sensation **(INFOGRAPHIC 30.6).**

Because drugs affect the brain's pleasure centers, they can be very hard to kick. Even addicts who successfully stop taking drugs for decades can relapse simply by being in an environment that conjures up memories of the drug. The tinkling of ice in a whiskey glass, meeting an old drinking buddy, or being in a bar or a room filled with smokers puffing away can be a reminder powerful enough to lure former addicts back to their old ways.

Chemical Messengers

▸ Dopamine and other neurotransmitters

For Ward, it wasn't just the smoking at the weekly poker game that rekindled his desire to smoke, it was a combination of events. For one thing, he says he was "extremely stressed" about work. He had started working as a freelance writer, and he was having trouble juggling his deadlines. In the past, he would often rely on smoking to calm himself. "I'm a nervous worker," he says. "Some people shake their legs, or pace, or exercise to

INFOGRAPHIC 30.6

Specialized Regions of the Brain

The brain has specialized regions that carry out distinct functions. Several areas are stimulated by pleasurable activities and are important in addiction.

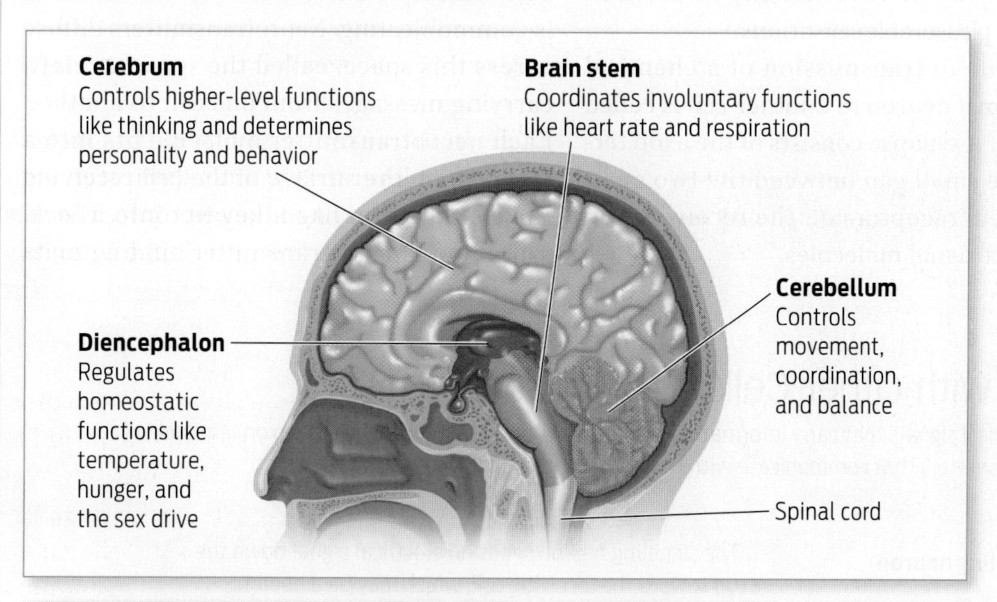

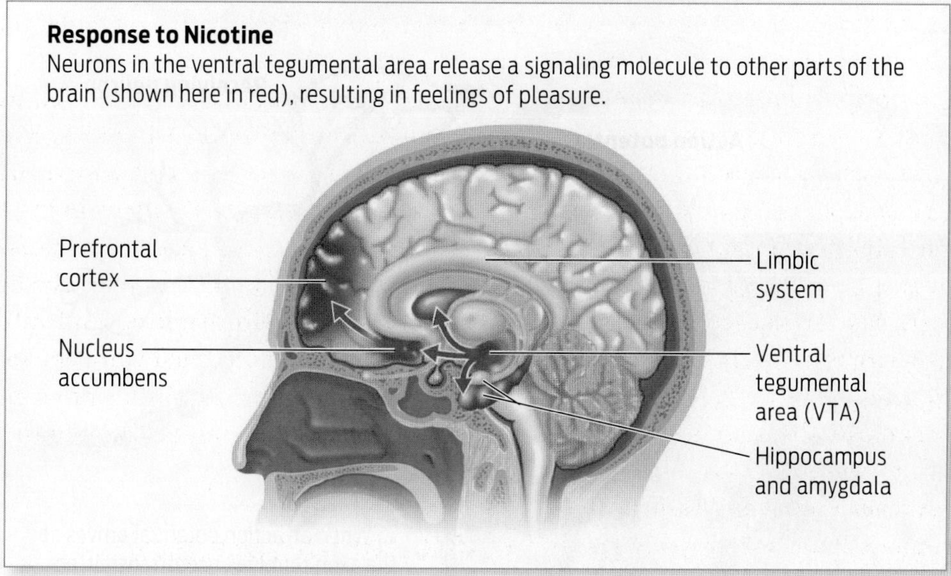

? What could be the consequences of an injury to the cerebellum?

work off stress. For me it was smoking." And then a difficult family situation arose. Up to that point, he had managed to avoid smoking, even in the presence of smokers, because, he says, he had learned to think of himself as a nonsmoker: holding to that self-concept was a way to persuade himself not to pick up a cigarette. But in the end, the combination of stressful events and easy access to cigarettes pushed him to smoke again.

As Ward took a long-awaited drag, neurons in his VTA released a chemical called **dopamine** that excited neurons in the pleasure center of his brain. Dopamine is one type of chemical messenger, or **neurotransmitter,** used by nerve cells to communicate. These

DOPAMINE
A chemical messenger that is involved in conveying a sense of pleasure in the brain.

NEUROTRANSMITTER
A chemical signaling molecule released by a neuron to transmit a signal to a neighboring cell.

SYNAPSE
The site of transmission of a signal between a neuron and another cell; includes the axon terminal of the signaling neuron, the space between the cells, and receptors on the receiving cell.

SYNAPTIC CLEFT
The physical space between a neuron and the cell with which it is communicating.

chemical signals transmit information from one cell to another cell, which could be a neuron or another type of cell, such as a muscle cell or a cell of an endocrine gland. As a neurotransmitter, dopamine is commonly involved in producing pleasurable sensations.

The site of transmission of a chemical signal from a neuron to another cell is called a **synapse.** A synapse consists of the axon terminal, the small gap between the two cells, and protein receptors on the receiving cell that detect signal molecules.

When the electrical signal of an action potential reaches a neuron's axon terminal, it stimulates the neuron to release neurotransmitters from its axon terminal into the space between the neuron and the cell with which it is communicating. Neurotransmitters diffuse across this space, called the **synaptic cleft,** carrying messages from one cell to another. Each neurotransmitter molecule fits into a receptor on the surface of the cell receiving the signal much like a key fits into a lock. The act of a neurotransmitter binding to its

INFOGRAPHIC 30.7

Neurons Communicate with Other Cells by Chemical Signals

Within a neuron, action potentials are electrical signals that carry information. When an action potential reaches the axon terminal, a neuron releases molecules called neurotransmitters that communicate with the next cell in the pathway at the synapse.

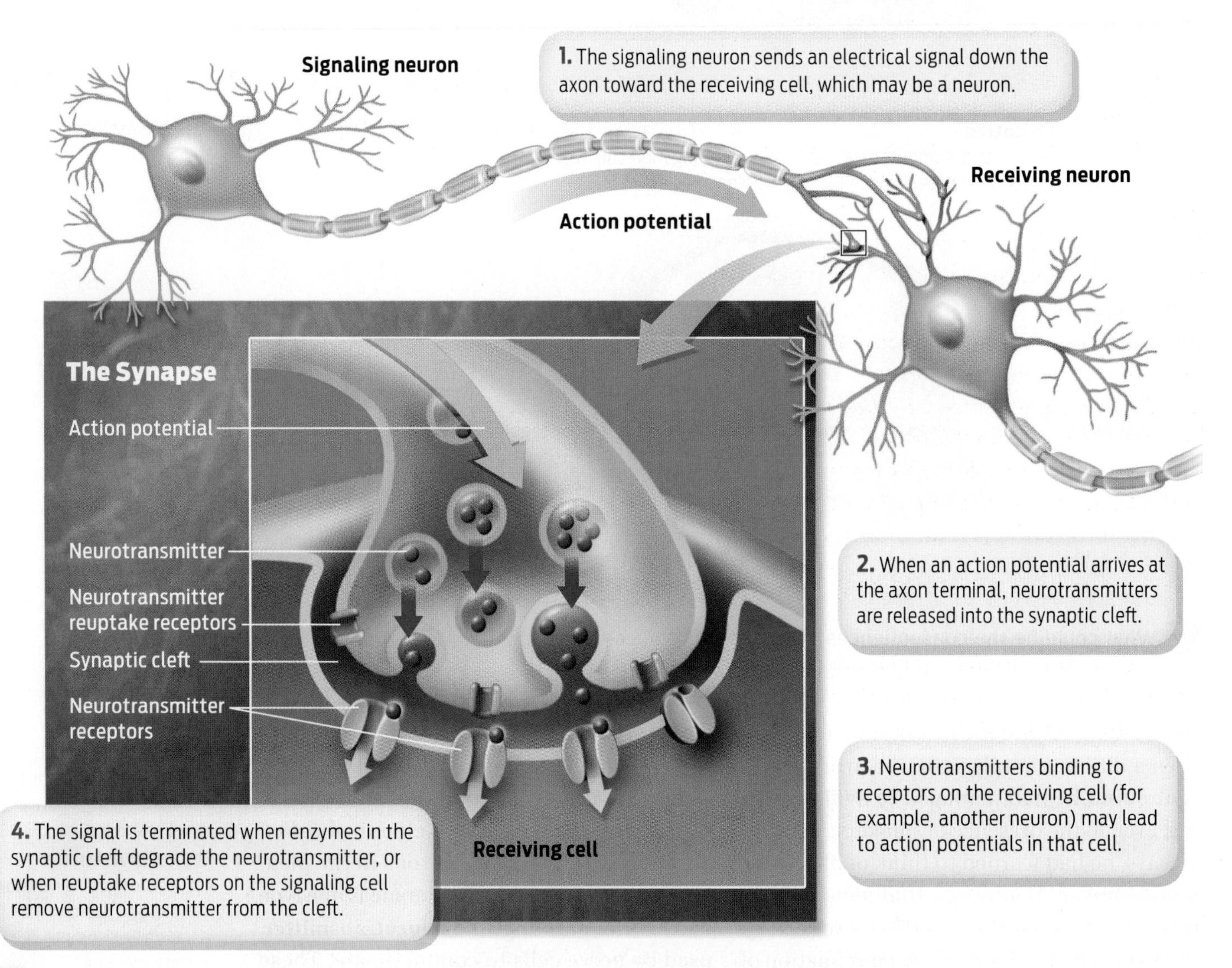

? **Which part of a neuron releases neurotransmitter molecules into the synaptic cleft?**

receptor initiates a response in the receiving cell. If the receiving cell is another neuron, the binding of neurotransmitter may spark another action potential, thereby perpetuating the signal. Signaling between the two cells is terminated when the neurotransmitter is removed from the synaptic cleft. The neurotransmitter may be degraded by enzymes in the cleft, or it may be taken back up into the signaling neuron by reuptake receptors **(INFOGRAPHIC 30.7)**.

Cells in the nervous system use several different neurotransmitters to communicate different messages. For example, the neurotransmitter serotonin, which is active in the CNS and the gastrointestinal tract, regulates anxiety, appetite, and sleep. The neurotransmitter acetylcholine is involved in learning, memory, and muscle contraction.

Many drugs—including both drugs of abuse and therapeutic medicines—influence the levels of neurotransmitters. Antidepressant drugs known as selective serotonin reuptake inhibitors (SSRIs), which include Prozac and Zoloft, influence the levels of the neurotransmitter serotonin in synapses. SSRIs help treat depression by preventing the reuptake of serotonin from the synaptic cleft, thus prolonging the activity of this mood-elevating neurotransmitter. Nicotine binds to receptors on neurons and triggers the release of multiple neurotransmitters in the brain. Nicotine's impact on multiple neurotransmitters accounts for the range of sensations that smokers experience: a reduction in anxiety, for instance, as well as an increased ability to concentrate. And as we saw, nicotine dramatically increases the release of dopamine from certain neurons, such as those in the VTA.

Too Much of a Good Thing

▶ Dopamine signaling and addiction

In normal circumstances, dopamine's main job is to convey information related to elation and pain. The joy we get from a meal, sex, a winning poker hand, or indeed anything that gives us pleasure is conveyed in part by dopamine. Drugs of abuse stimulate dopamine production, which is why they can become so addictive.

But these drugs cause such a high that over time they can alter the dopamine signaling system so that the pleasures of everyday life pale in comparison to the pleasure evoked by the drug. Normally the brain produces dopamine at a relatively constant rate, and dopamine occupies only a portion of dopamine receptors at any given time. But when a person smokes, snorts cocaine, or takes heroin, for example, dopamine levels in the synaptic cleft increase dramatically. With so much dopamine available, practically all of the brain's dopamine receptors become activated simultaneously.

> *"Some people shake their legs, or pace, or exercise to work off stress. For me it was smoking."*
>
> **—Jack Ward**

The immediate effect is euphoria. But there is a downside. Because so much dopamine is produced, the brain becomes overwhelmed and tries to dampen the drug's effect by switching off some of its dopamine receptors. When the drug wears off, fewer receptors are functioning—bringing down mood.

The resulting low can be so low that normal pleasures such as eating or socializing become dull and listless affairs. In fact, the user's mood may be even lower than it was before taking the drug. As dopamine receptors shut down, ever-larger quantities of the drug are required to produce a high—and the high may never be as high as the user experienced the first time.

As they come down from a high, addicts will likely feel even more unhappy and depressed as the dopamine response system is dampened. Eventually, many addicts need to take drugs simply to feel normal. Without drugs, they suffer the physical symptoms of withdrawal, which may include depression, anxiety, and intense cravings for a dopamine fix. The specific symptoms and their intensity vary depending on the drug **(INFOGRAPHIC 30.8)**.

Altered dopamine signaling isn't the only brain change that scientists have observed in drug addicts. Researchers have also shown that drugs such as cocaine can change the shape of neurons in specific parts of the brain

INFOGRAPHIC 30.8

Addictive Drugs Alter Dopamine Signaling

Addictive drugs increase dopamine release, causing initial feelings of pleasure. Continued use of the drug alters dopamine signaling, requiring more drug to achieve the same high. Attempts to quit can be difficult because of symptoms of withdrawal, the result of diminished dopamine signaling.

a. Normal Dopamine Signaling

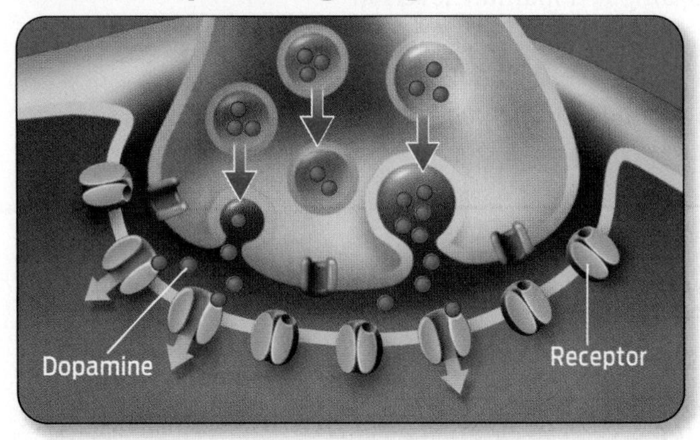

Cells release moderate amounts of dopamine into the synapse and not all dopamine receptors are occupied.

b. After Drug Use

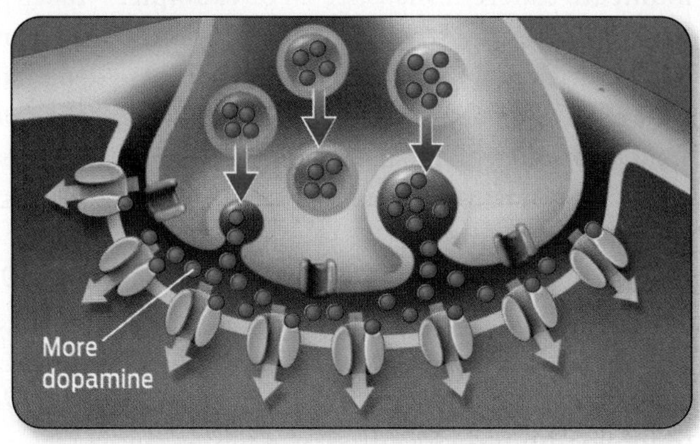

Certain drugs cause massive dopamine release into the synapse. Many more dopamine receptors become occupied. The result is an intense feeling of pleasure.

c. After Repeated Drug Use

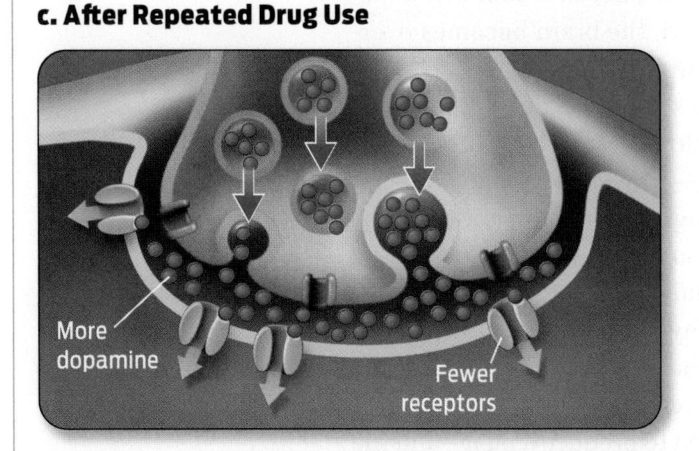

Dopamine overstimulation causes the receiving cell to down-regulate (that is, shut down) some dopamine receptors. Because the user now has fewer receptors available, more of the drug is required to feel high.

d. After Drug Withdrawal

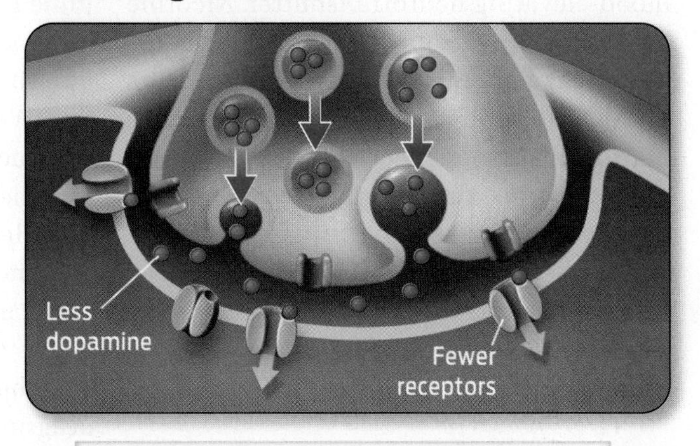

Removing the drug reduces the amount of dopamine released. In combination with fewer dopamine receptors, the user may feel sick and depressed.

? Why do people who are addicted to drugs need to take increasing amounts of the drug to experience the same high?

and consequently may impair their ability to transmit signals. Brain-imaging studies have also shown that addicts consistently have lower than normal levels of blood flow in the frontal regions of the cerebrum during withdrawal from cocaine, and higher than normal levels of this blood flow while they are on the drug. This region of the cerebrum is involved in decision making. In a variety of tests, drug addicts seem to make poorer decisions, with harmful consequences.

Even more troubling, adolescents who take drugs may be preventing their brains from developing normally. A 2004 study by researchers at the University of California, Los Angeles, and the National Institute of Mental Health (NIMH) that imaged the brains of 13 children and young people age 4 to 21 over 10 years showed that that some parts of the brain, such as the prefrontal cortex, are not fully developed until the mid-twenties or so. Taking drugs at an early age may hinder normal development of this region.

And that's not the end of the addiction story. In addition to dopamine, several other neurotransmitters are involved in addiction, says Joe Frascella, director of the division of Clinical Neuroscience and Behavioral Research at the National Institute on Drug Abuse (NIDA). Scientists are just starting to study how drugs of abuse affect these other neurotransmitters. While scientists have shown that almost every drug of abuse affects the dopamine system in varying degrees, Frascella says, "It's certainly more complex than just dopamine." Scientists have only just scratched the surface when it comes to learning exactly how long-term drug use affects the brain **(TABLE 30.1)**.

Born Addicts?

▶ Causes and consequences of addiction

Not everyone who takes drugs becomes addicted. Exactly why some people seem to be more at risk than others isn't clear. But researchers have a few hypotheses based on existing evidence. For some of us, it may be a matter of biology: a predisposition to addiction may be caused by a shortage of dopamine receptors or of other types of receptors. Some people may have been born with fewer receptors, or their brains may have lost receptors over time as a result of difficult life experiences. Consequently, drugs provide these people with a high that they can't get from any other stimulus. And the drug feels too good to stop.

And just as some of us may be biologically predisposed to addiction, others may be predisposed to avoid it. In particular, some people's brains may simply be better at overriding the pleasure-seeking impulse.

A 2011 study by researchers at the Scripps Research Institute Florida, for example, identified a brain pathway involved in nicotine addiction. The researchers found that a nicotine receptor called alpha-5 influences how susceptible mice are to nicotine addiction, with the number of receptors making an important difference. When given the opportunity to self-administer nicotine, mice with a normal amount of alpha-5 receptors will stop after reaching a certain dose. Mice with no alpha-5 receptors, however, won't stop until they've taken a much higher dose. Humans also have alpha-5 receptors, and scientists hypothesize that people with fewer receptors are less sensitive to nicotine and may become more easily addicted.

Regardless of the reason people become addicted, addiction is a serious public health problem. Tobacco, for example, is responsible in some way for one out of every five deaths in the United States, according to NIDA. Smokers have a higher incidence of both heart and lung disease. In addition, smoking causes cancer. Putting these effects together, tobacco use is the leading preventable cause of death and disease in the United States, killing some 480,000 people a year.

Drug use also affects health in ways that diminish the quality of life. For example, deficits in the dopamine system caused by drug use weaken memory and motor skills.

TABLE 30.1 Potentially Addictive Drugs and Their Effects

	Mode of action	Effect
COCAINE	• Causes a large release of dopamine into the synapse • Inhibits reuptake receptors • Causes an amplified signal between neurons	• Highly addictive • Causes powerful feelings of well-being and confidence • Users lose interest in life activities • High doses lead to paranoia, anxiety, and increased blood pressure at high doses
HEROIN	• Mimics natural endorphins • Binds opiate receptors in specific regions • Affects mood, respiration, and pain response	• Rush of euphoria followed by a foggy feeling • Powerful withdrawal symptoms make this drug extremely addictive • Overdose slows breathing to dangerous levels
CAFFEINE	• Mimics adenosine; blocks the natural sleep response • Causes release of adrenaline and dopamine; result is alertness and sense of pleasure	• Inhibits sleepiness and increases alertness • Withdrawal symptoms include headaches, jittery feelings, and increased anxiety • Interferes with deep-sleep cycles, causing exhaustion and depression • The adrenaline produced constricts blood vessels, affecting heart rate
NICOTINE	• Increases levels of dopamine	• Enhanced short-term feelings of pleasure, relaxation • Increased concentration • Very addictive
ECSTASY	• Causes excessive release of serotonin • Destroys nerve cells that produce serotonin	• Intense euphoria, followed by a depressive "crash" • Side effects include paranoia, anxiety, confusion, and difficulty concentrating • Long-term users can't distinguish between reality and fantasy
ALCOHOL	• A general depressant of the central nervous system • Changes communication patterns between neurons in specific brain regions	• Acts as an anesthetic • Influences breathing, motor, and behavior control • Diminishes senses • Destroys cells in the brain and other organs
MARIJUANA	• Mimics the chemical anandamide • Stimulates anandamide receptors in areas of the brain that affect memory, emotion, and sensory perception	• Relaxation, mild euphoria, and appetite stimulation • Can cause paranoia • Weakens short-term memory and can block the production of long-term memory • Diminishes problem-solving ability and coordination
INHALANTS (GLUE, HAIR SPRAY, PAINT THINNER, ETC.)	• Vapors destroy the myelin sheath of axons • Damages cells in the brain, lungs, heart, liver, kidneys, and bones	• Causes headaches, nausea, and disorientation • Can diminish the ability to learn, remember, and solve problems • May cause a rapid and irregular heartbeat
METHAMPHETAMINE	• Causes release of dopamine and norepinephrine into the synapse • Dopamine leads to feelings of pleasure • Norepinephrine increases blood pressure and heart rate	• Creates feelings of pleasure and euphoria, paranoia, and hallucinations • May alter dopamine-producing neurons connected with Parkinson's disease
RITALIN	• Causes increased levels of dopamine	• At lower doses, increases the ability to focus and concentration • At higher doses, can inhibit formation of new nerve pathways, interfering with cognition and brain development • Prescribed for attention-deficit disorder, but has become a common street drug

Nora Volkow, head of NIDA, has shown that methamphetamine (crystal meth) users have poorer short-term memory and score much lower on tests of motor skills, such as quickly walking a straight line **(INFOGRAPHIC 30.9)**.

Kicking the Habit

Research on the neurobiology of drug addiction is informing new and better ways to treat addiction. Volkow and other researchers have shown that after a period of abstinence from drugs, the dopamine system can repair itself—but it generally takes longer than a year. And while studies have shown that skills such as short-term memory and motor control do come back, it's not clear whether they are fully restored.

Given the many drug-induced brain changes, experts now see addiction as a chronic disease. Like people with heart disease or diabetes, people with addictions require long-term treatment plans. And the occasional relapse is only a predictable setback, not a failure of the treatment, says Volkow.

Experts also now know that the most effective treatments will likely target addictive behaviors in several ways. They can decrease the reward value of the drug, for example, by counseling addicts to seek out other pleasurable experiences and to repeat those experiences to reinforce their value in the brain. Avoiding the drug and focusing on other pleasurable experiences will, over time, weaken conditioned memories of the drug and drug-related stimuli.

A better understanding of the effects of substance abuse on the brain is also informing efforts to develop medications that can help an addicted person kick the habit. Two popular antismoking drugs, Chantix and

Tobacco is responsible in some way for one out of every five deaths in the United States.

INFOGRAPHIC 30.9

Addictive Drugs Can Diminish Memory and Motor Skills

Researchers used a series of measurements to estimate the number of dopamine reuptake receptors in the brains of methamphetamine users. Meth users had lower numbers of dopamine reuptake receptors than nonusers, and this reduction was correlated with a reduction in motor skills and memory.

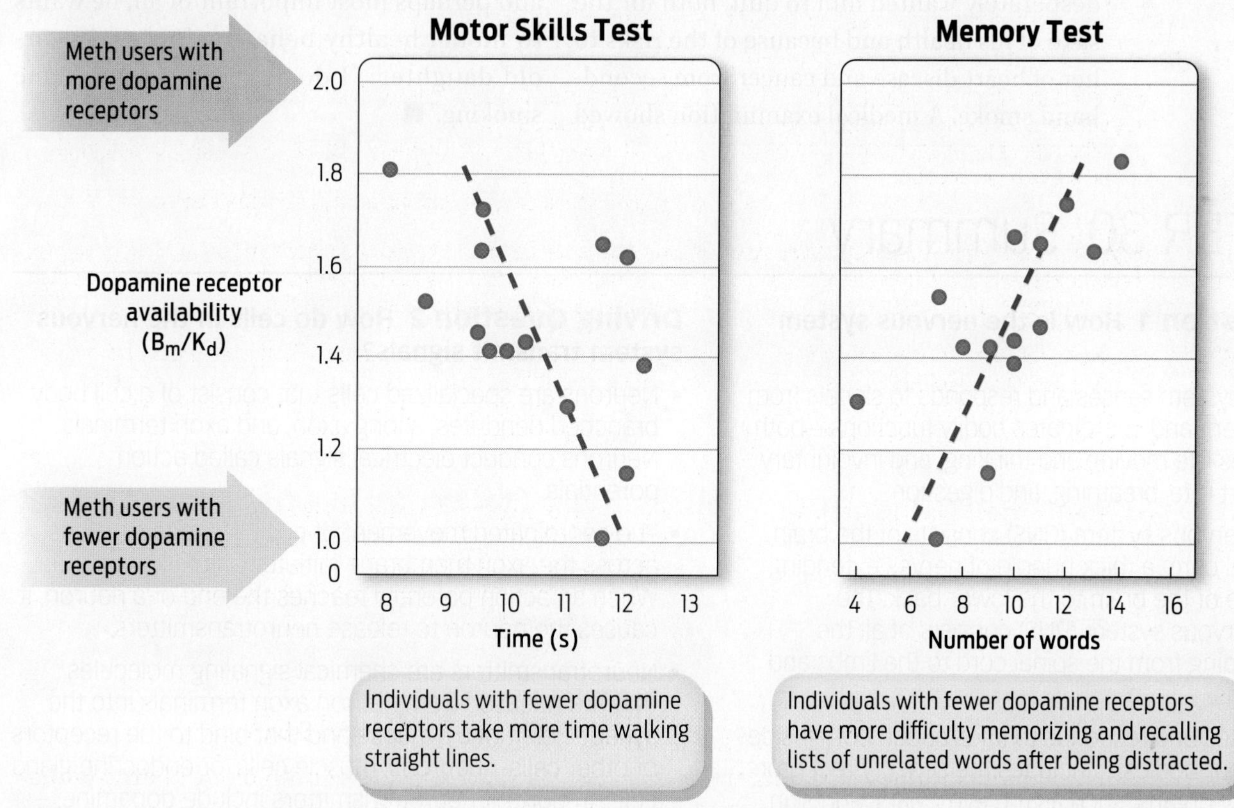

Data from Volkow, N., et al. (2001). *Am J Psychiatry* 158:377–382.

? How does the number of dopamine reuptake receptors for the participant who recalled the greatest number of words compare to that of the participant who recalled the fewest number of words?

Zyban, work by competing with nicotine for binding sites on neurons. By partially activating dopamine release, the drugs help reduce nicotine cravings in someone trying to quit smoking. They also make smoking less pleasurable because nicotine from cigarettes cannot bind to nicotine receptors while the drugs are present. As a result, smoking becomes less enjoyable, and therefore easier to stop. Research has shown these medications to be two to three times more effective than a placebo at aiding smoking cessation However, reports also cite serious side effects of these drugs, such as suicidal thoughts, so use of these medications should be considered carefully. A combination of behavioral strategies and medications that target specific neurotransmitters or brain circuits will likely work the best to help addicts kick the habit, says NIDA's Frascella.

Before Ward quit smoking in 2006, he was up to a pack and a half a day. Smoking had increasingly become a point of tension between him and his nonsmoking wife. She desperately wanted him to quit, both for the sake of his health and because of the risks to her of heart disease and cancer from secondhand smoke. A medical examination showed that Ward already had risk factors for heart disease, including high blood pressure. His doctor told him to stop smoking. "She didn't ask me to stop," Ward says. "She simply said, 'You are going to stop smoking in two weeks.' " The doctor prescribed a nicotine patch—a skin patch that delivers nicotine to the bloodstream and eases nicotine withdrawal symptom—and other medication, which helped him quit.

But after that fateful poker night in 2009, Ward caved into his cravings: he went back to smoking. This time, instead of his old pack-and-a-half-per-day habit, he managed to cut down to only a few cigarettes a week, all of them smoked on his weekly poker night.

Two years after Ward started smoking again, he moved. The weekly poker nights ended, and so did the social cues that had tempted him to restart smoking in the first place. There are more antismoking social influences in his life now than ever. Most of his friends do not smoke, he exercises more, and perhaps most important of all, he wants to model healthy behavior for his 4-year-old daughter: "I don't want her to see me smoking." ■

CHAPTER 30 Summary

Driving Question 1 How is the nervous system organized?

- The nervous system senses and responds to signals from the environment and coordinates bodily functions—both voluntary ones like moving and thinking, and involuntary ones like heart rate, breathing, and digestion.
- The central nervous system (CNS) consists of the brain and the spinal cord, a thick bundle of nerves extending from the base of the brain to the lower back. The peripheral nervous system (PNS) consists of all the nerves extending from the spinal cord to the limbs and internal organs.
- The PNS senses and responds to information both inside and outside our bodies. It includes the sensory receptors of our sensory organs, such as the eyes, ears, and skin, as well as the effectors that respond to signals sent from the CNS, such as muscles and endocrine glands.

Driving Question 2 How do cells in the nervous system transmit signals?

- Neurons are specialized cells that consist of a cell body, branched dendrites, a long axon, and axon terminals. Neurons conduct electrical signals called action potentials.
- The coordinated movement of positively charged ions across the axon membrane initiates an action potential. When an action potential reaches the end of a neuron, it causes the neuron to release neurotransmitters.
- Neurotransmitters are chemical signaling molecules that are released from neuron axon terminals into the synaptic cleft of a synapse and that bind to the receptors of other cells—neurons, muscle cells, or endocrine gland cells. Important neurotransmitters include dopamine, serotonin, and acetylcholine.

- Neurotransmitters bind to receptors on their target cells. The number of receptors can be reduced in response to persistently high levels of a neurotransmitter.
- Different parts of the brain coordinate different functions. The four main regions of the brain are the cerebellum, brain stem, diencephalon, and cerebrum.

Driving Question 3 Why are some drugs (and some behaviors) addictive?

- Dopamine is a neurotransmitter that produces feelings of pleasure. It is one of the primary neurotransmitters involved in addiction.
- A subregion of the brain called the limbic system makes up the brain's "pleasure center" and has been implicated in addiction.

More to Explore

- National Institute of Drug Abuse: www.drugabuse.gov/
- Volkow, N. D., et al. (2011). Addiction: Beyond dopamine reward circuitry. *Proc Natl Acad Sci* 108(37):15037–15042.
- Kringelbach, M. L., and Berridge, K. C. (2010). The functional neuroanatomy of pleasure and happiness. *Discovery Med* 9(49):579–587.
- Proctor, R. N. (2011). *Golden Holocaust: Origins of the Cigarette Catastrophe and the Case for Abolition.* Berkeley: University of California Press.
- Olds J. (1956). Pleasure centers in the brain. *Sci Am* 195:105–116.

CHAPTER 30 Test Your Knowledge

DRIVING QUESTION 1 How is the nervous system organized?

By answering the questions below and studying Infographics 30.3 and 30.6, you should be able to generate an answer for this broader Driving Question.

Know It

1. Mark each of the following structures as a part of the central nervous system (CNS) or of the peripheral nervous system (PNS).

_________ Light-detecting receptor in the eye
_________ Amygdala
_________ Pain receptor in the skin
_________ Spinal cord
_________ Thalamus

2. Which part of the brain coordinates movement? Which part of the brain maintains body temperature?

Use It

3. Is information flow in the spinal cord one way or two way? Explain your answer.

4. How does multiple sclerosis cause muscle weakness? Does multiple sclerosis directly affect muscles? Which part of the nervous system is affected?

5. A brain injury (caused by a blow to the head, for example) that results in the loss of the ability to speak most likely affected the
- **a.** cerebellum.
- **b.** cerebrum.
- **c.** diencephalon.
- **d.** brain stem.
- **e.** hypothalamus.

DRIVING QUESTION 2 How do cells in the nervous system transmit signals?

By answering the questions below and studying Infographics 30.4, 30.5, and 30.7, you should be able to generate an answer for this broader Driving Question.

Know It

6. Neurons receive information through their
- **a.** axons.
- **b.** axon terminals.
- **c.** cell bodies.
- **d.** dendrites.
- **e.** nuclei.

7. Action potentials are a type of ____________ signaling that relies on ____________.
- **a.** electrical; neurotransmitters
- **b.** electrical; charged ions
- **c.** electrical; electrons
- **d.** chemical; neurotransmitters
- **e.** chemical; charged ions

8. Neurons release neurotransmitters from their
- **a.** cell bodies.
- **b.** dendrites.
- **c.** axon terminals.
- **d.** all of the above
- **e.** b and c

9. What happens when a neurotransmitter is released into a synaptic cleft?

10. Compare and contrast electrical and chemical signaling by neurons.

Use It

11. Gatorade and other sports drinks contain replacement electrolytes (ions necessary to enable muscles to continue to contract, especially the ions lost during sweating). Gatorade contains sodium and potassium ions. Other than in the muscle, where else might these ions be crucial during sustained exercise?

12. Botox is a chemical treatment injected into skin to prevent wrinkling. It is a bacterial toxin that prevents certain neurons from releasing the neurotransmitter acetylcholine. Acetylcholine is normally released by motor neurons to signal muscles to contract. Does Botox paralyze muscles in a relaxed state or a contracted state?

13. Is more or less of the neurotransmitter acetylcholine released by the axon terminals of neurons in individuals with multiple sclerosis compared to those in people who do not have multiple sclerosis? Explain your answer.

DRIVING QUESTION 3 Why are some drugs (and some behaviors) addictive?

By answering the questions below and studying Infographics 30.2, 30.6, 30.8, and 30.9, you should be able to generate an answer for this broader Driving Question.

Know It

14. Addictive substances confer a sense of pleasure because they

- **a.** decrease the amount of dopamine in synaptic clefts.
- **b.** increase the amount of dopamine in synaptic clefts.
- **c.** increase the number of dopamine receptors on the axon terminals of cells that release dopamine.
- **d.** increase the number of dopamine receptors on dendrites of cells that release dopamine.
- **e.** c and d

15. Why do drug users need to take ever-increasing amounts of drugs to get the same high?

Use It

16. Cocaine prevents dopamine from being removed from the synapse. Why does this cause feelings of pleasure?

17. Would you expect a person born with a relatively low number of dopamine receptors to be happier or sadder than the average? Explain your answer.

Mini Case

Apply Your Knowledge

18. Parkinson's disease is caused primarily by a gradual loss of dopamine-producing neurons in the brain. Why is depression often among the debilitating symptoms of Parkinson's disease? A variety of medications are available to treat people with Parkinson's disease. Do some Internet research to match each medication listed in the left column with its probable mechanism of action.

_____ Mirapex — **a.** can be used by neurons to make dopamine

_____ Eldepryl — **b.** binds to and activates the dopamine receptor

_____ Levodopa — **c.** inhibits an enzyme that breaks down dopamine

How can drugs with different mechanisms of action all help treat Parkinson's disease? (*Hint:* What do all the underlying mechanisms have in common in terms of their effect?)

Interpreting Data

Apply Your Knowledge

19. Study Infographic 30.9.

- **a.** What are the independent and dependent variables in each experiment?
- **b.** The performances of 14 participants are plotted in each graph. Calculate the average time (in seconds) these participants took to walk the straight line, and the average number of words recalled by these participants.
- **c.** In the right-hand graph in Infographic 30.9, you will note a few outliers (two are circled in red in the copy of the graph shown here). For the outlier in the lower left part of the graph, is this participant recalling more or fewer words than would be predicted based on his or her availability of dopamine receptors? Explain your answer.
- **d.** For the outlier on the top right part of the graph, is this participant recalling more or fewer words than would be predicted based on his or her availability of dopamine receptors? Explain your answer.

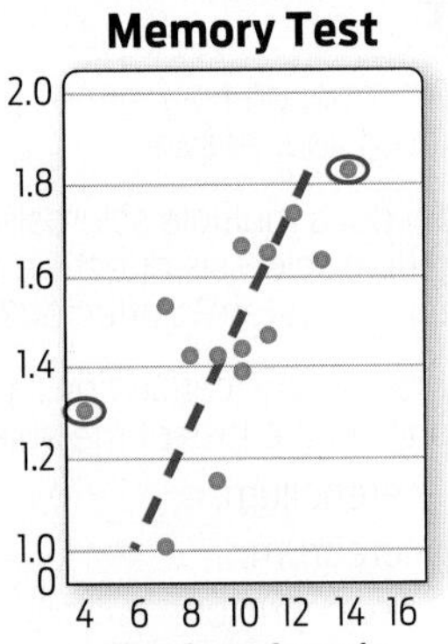

Apply Your Knowledge

Bring It Home

20. Replicate the motor skills experiments shown in Infographic 30.9 with some students. On a flat surface, set up a start line and, 10 yards away, a finish line. Instruct each subject to walk in a straight line from the start to the finish and back as fast as they can without running. Start timing when they cross the start line, and stop timing when they cross the start line again at the end of the trial. Have each participant do the trial three times, and record the average of the three trials as the participant's final time. Now calculate the average time in your set of participants. How does this compare to the 14 methamphetamine users whose performances are plotted in Infographic 30.9?

Method from Robertson, K. R., et al. (2006). Timed gait test: Normative data for the assessment of the AIDS dementia complex. *J Clin Exper Neuropsychol* 28:1053–1064.

Too Many Multiples?

The birth of octuplets raises questions about the fertility business

Nancy Pastor/Polaris Images

DRIVING QUESTIONS

1 What is the anatomy of the male and female reproductive systems, and how does the anatomy contribute to the function of the reproductive system?

2 What hormones are involved in reproduction, and how do they work in the reproductive systems of males and females?

3 What are the different types of assisted reproduction, and how do they work?

The live birth of octuplets is an extremely rare event, having occurred only once in recorded U.S. history before the year 2000. So the arrival of a second set in a California hospital in January 2009 was greeted with fanfare. Headlines screamed "Octuplets Stun Doctors" and "Eight Babies!"

But days after news of the miracle multiple birth spread worldwide, the public reception turned sour when it came to light that the 33-year-old mother, Nadya Suleman, already had six children all younger than the age of 7. Like the octuplets, those children were conceived using a form of assisted reproductive technology called **in vitro fertilization (IVF),** in which sperm and egg are brought together outside the body to form an embryo that is then implanted into a woman's uterus.

Even more disturbing to some, Suleman was an unemployed single mother on welfare. The public outcry was fierce: How could she support her children? Was she psychologically disturbed? And why had her doctor agreed to give her fertility treatments when she already had six children?

The case also cast a spotlight on the business of fertility treatment: in the United States, fertility clinics are largely unregulated. Although the American Society of Reproductive Medicine (ASRM) issues guidelines on how doctors should administer fertility services, most states do not have laws regulating what doctors can and cannot do in this regard. Though critics called Suleman's doctor irresponsible, he had not violated any laws.

IN VITRO FERTILIZATION (IVF)
A form of assisted reproduction in which eggs and sperm are brought together outside the body and the resulting embryos are inserted into a woman's uterus.

Infertility treatment wouldn't be nearly as controversial if it didn't increase the odds that a woman will conceive more than one child during the treatment. "Multiples," as these babies are called, are often born prematurely. Consequently, they are underweight and have underdeveloped organs, which puts them at risk for birth defects such as cerebral palsy and infant respiratory distress syndrome. Carrying multiples also increases the risk that the mother will develop dangerously high blood pressure, diabetes, vitamin deficiencies, or other medical conditions during her pregnancy that may affect her health or that of her unborn children. Even women who have twins have a higher risk of medical complications during pregnancy, and their babies are at a higher risk of premature birth than are singletons. Suleman's eight children are the only surviving set of octuplets ever.

Multiple births without fertility treatment are extremely rare. Scientists estimate that the incidence of natural triplets is 1 in 6,000 to 8,000 births—and the chance of having quadruplets is dramatically lower, at 1 in 500,000 births. The incidence of higher-order (five or more) multiple births is significantly rarer. But as the use of assisted reproductive technologies has skyrocketed,

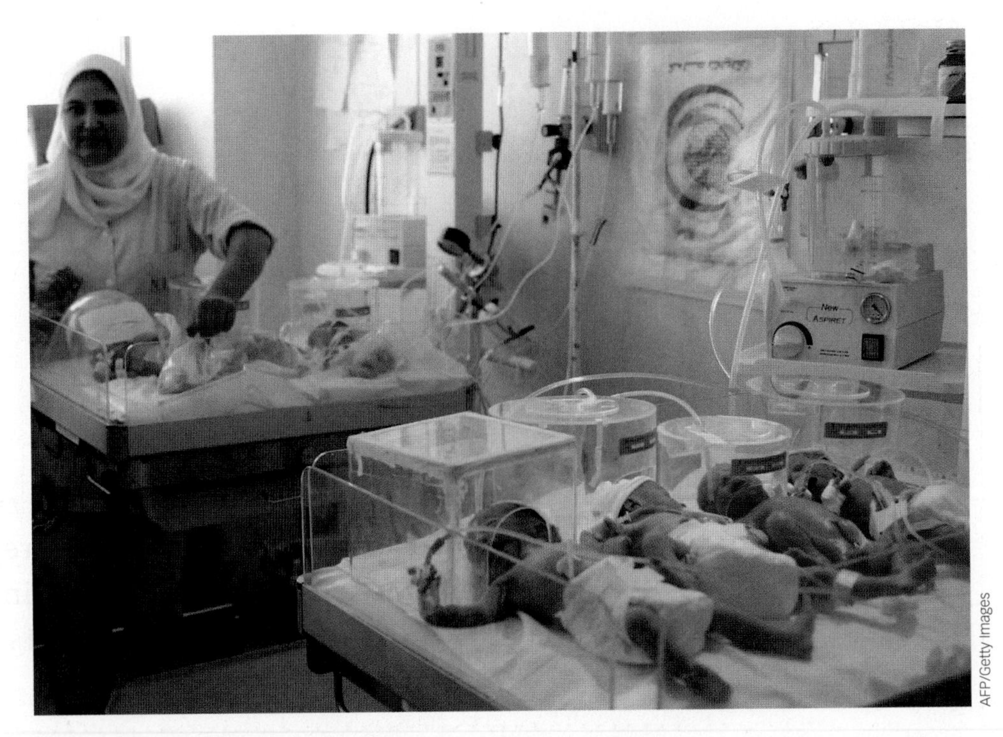

AFP/Getty Images

Multiples are often born prematurely and underweight, as were these septuplets born in Alexandria, Egypt, in 2008.

so, too, have the number of multiple births. The number of triplet births rose 400% between 1980 and 1998, according to statistics compiled by the Centers for Disease Control and Prevention. Since then, the rate of triplets and higher multiple births has been slowly declining, largely because doctors performing IVF are implanting fewer embryos, while the number of births of twins remains high **(INFOGRAPHIC 31.1)**.

Although there have been efforts to improve assisted reproductive technology and therefore reduce the likelihood of multiple births, extreme cases such as Suleman's octuplets reveal the holes in the regulation of assisted reproduction. No single governmental agency is empowered to oversee the roughly 500 fertility clinics that currently exist in the United States. While the ASRM issues recommendations about how many embryos should be implanted at any one time, it does not punish those practitioners who violate its guidelines. Also left unregulated: the number of children who may be conceived from one donor, leading some to worry that unsuspecting donor offspring could meet and have children of their own—dubbed "accidental incest." Inconsistent health insurance coverage for fertility treatment is a further complication. Because many patients pay out of pocket for such services, patients with limited financial resources often pressure doctors to be aggressive with treatment, despite the health risks associated with carrying multiples. Nadya Suleman's doctor claimed he was pressured in this very way.

Fertility Troubles

▸ Male and female reproductive anatomy

Suleman, who grew up as an only child, said she had always wanted a large family to help make up for her lonely childhood. She told Ann Curry on NBC's *Today* show in 2009 that, before the birth of her first child, she had tried to become pregnant on her own for several years but had been unable to. That's when she turned to a fertility doctor for help.

In the same interview, Suleman said she suffered from a medical condition that

INFOGRAPHIC 31.1

Multiple Births Have Become More Common

The rate of pregnancies resulting in multiple births has increased substantially since 1980. The increase is most dramatic in women older than the age of 35. Since 1998, the rate of triplet (and more) births has decreased.

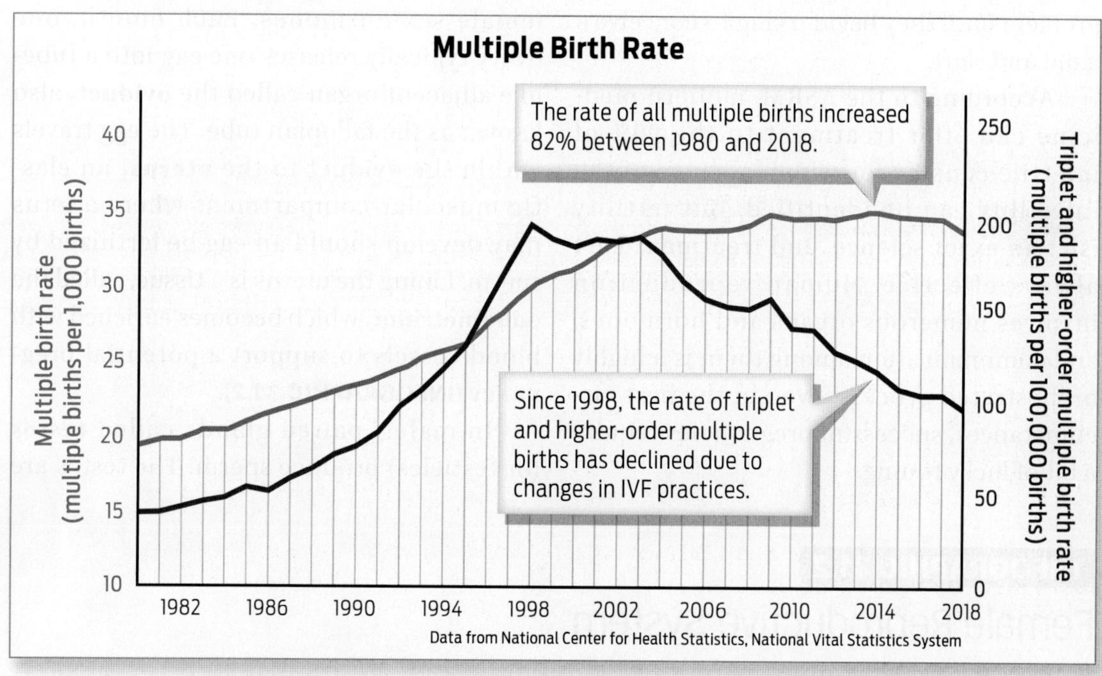

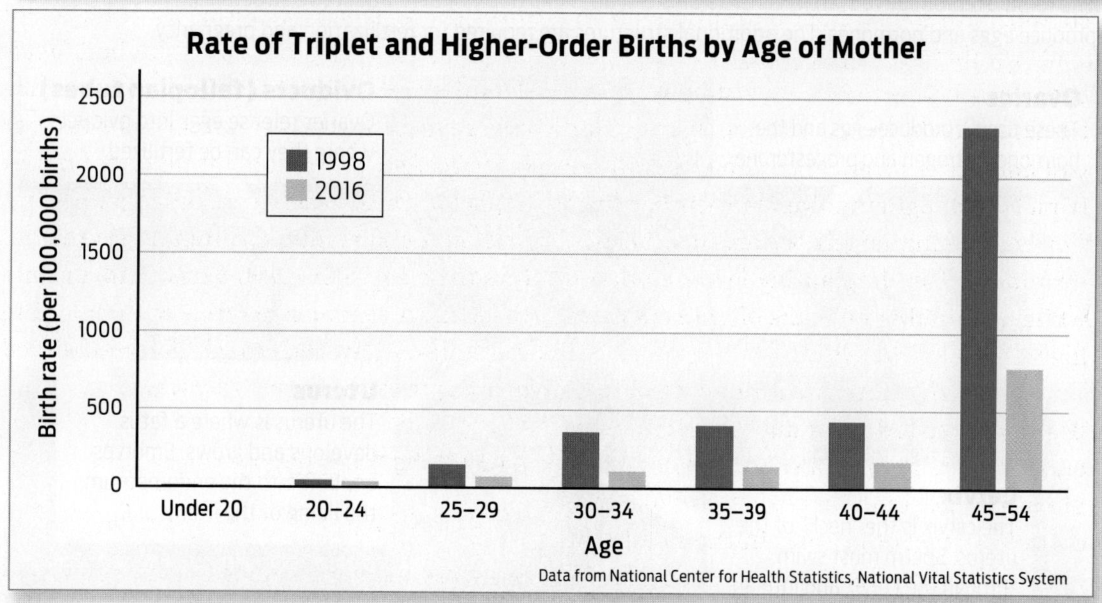

? In what year did the rate of triplet and higher-order births peak? In 2016, which age group had the highest rate of triplets (and higher-order) births?

prevented her from conceiving a child naturally. If she did, she wasn't alone. In the United States, an estimated one out of eight couples—about 7.3 million women and their partners—experience infertility, which is defined as the inability to conceive within a year or to bring a pregnancy to term. Many things can cause infertility: advanced age, infections, hormonal imbalances, chromosomal abnormalities, and physical blockage

OVARIES
Paired female reproductive organs; the ovaries contain eggs and produce estrogen and progesterone.

ESTROGEN
A female sex hormone produced by the ovaries that supports female sexual development and function.

PROGESTERONE
A female sex hormone produced by the corpus luteum of the ovary that prepares and maintains the uterus for pregnancy.

OVIDUCT
The tube connecting an ovary and the uterus in females. Eggs are ovulated into and fertilized within the oviducts.

UTERUS
The muscular organ in females in which a fetus develops.

ENDOMETRIUM
The lining of the uterus.

TESTES
Paired male reproductive organs, which contain sperm and produce androgens (primarily testosterone).

of reproductive passages. Men can suffer from a low sperm count or have abnormal sperm. In many cases, the reason for a couple's infertility remains unknown, and most couples are unaware that they have a fertility problem until they begin trying to conceive a child and can't.

According to the ASRM, modern medicine can offer treatment to the 90% of infertile couples for whom a cause of the infertility can be identified. But fertility isn't an exact science, and treatment isn't always effective. Human reproduction involves numerous organs and hormones, and communication among them is a highly orchestrated process; even in the best circumstances, successful pregnancies require a bit of lucky timing.

The female reproductive system consists of two **ovaries** and additional structures that are required for fertilization and pregnancy. The ovaries contain eggs and produce **estrogen** and **progesterone,** the major female sex hormones. Each month, one ovary typically releases one egg into a tubelike adjacent organ called the **oviduct,** also known as the fallopian tube. The egg travels within the oviduct to the **uterus,** an elastic muscular compartment where a fetus may develop should an egg be fertilized by sperm. Lining the uterus is a tissue, called the **endometrium,** which becomes enriched with blood vessels to support a potential pregnancy **(INFOGRAPHIC 31.2).**

In males, paired glands called **testes** (or testicles) produce sperm. The testes are

INFOGRAPHIC 31.2

Female Reproductive System

The female reproductive system consists of the ovaries and additional reproductive structures. The ovaries produce eggs and hormones. The additional structures are required for fertilization and pregnancy.

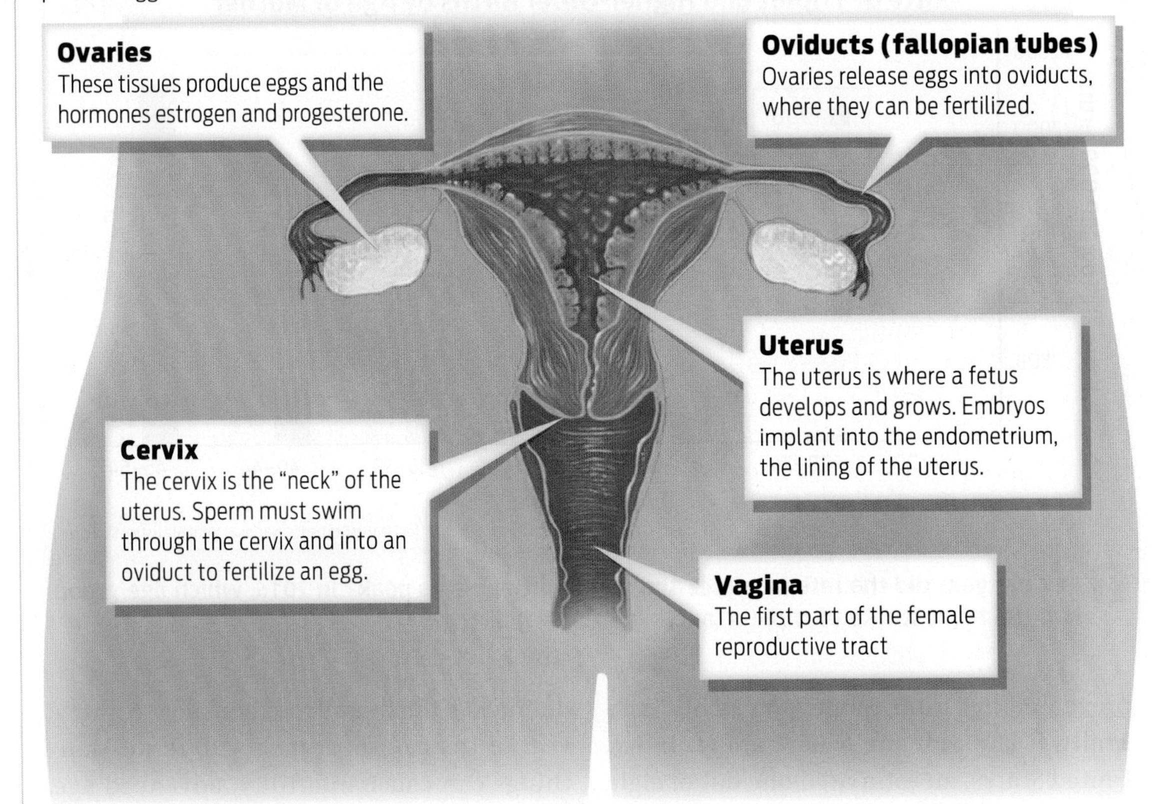

? **Where does fertilization, if it occurs, take place? Where will the embryo develop?**

INFOGRAPHIC 31.3

Male Reproductive System

Paired testes produce sperm and hormones. Sperm travel through a series of ducts and are ejaculated through the urethra.

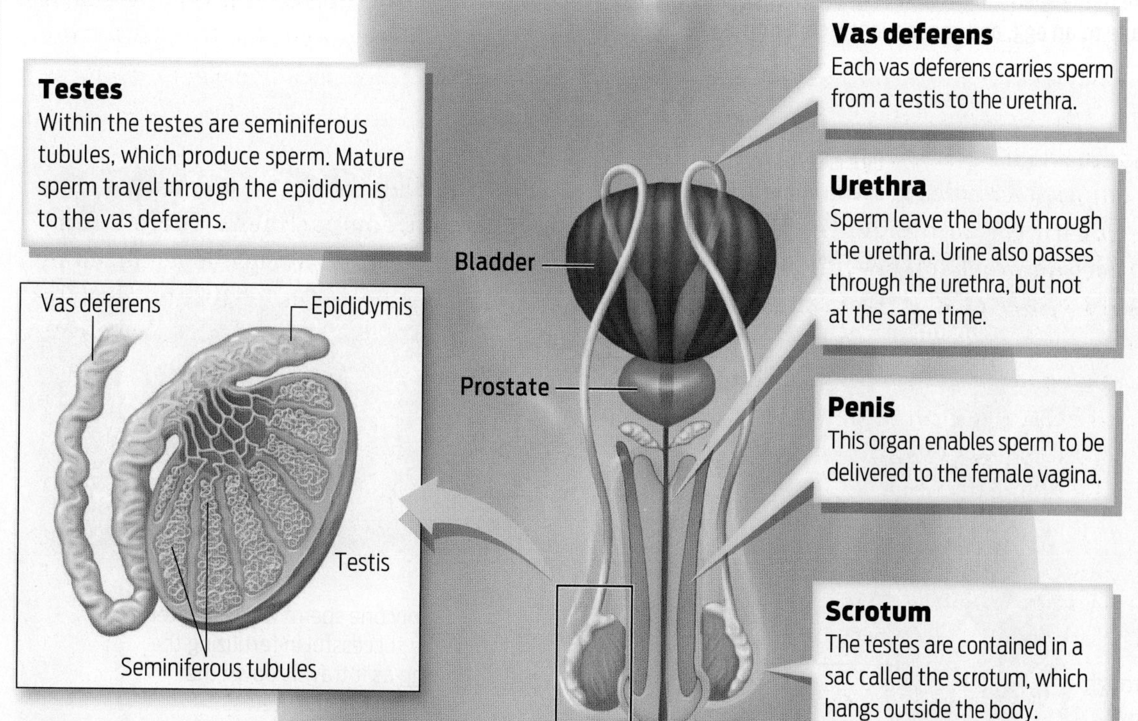

? **Describe the path that sperm take from the seminiferous tubules to the end of the urethra.**

contained in a sac of skin called the **scrotum** that hangs outside the body (an arrangement that keeps the testes slightly cooler than body temperature, ensuring proper sperm development). The testes produce **testosterone,** the primary male sex hormone. Each testis contains tightly coiled **seminiferous tubules** within which sperm develop. Remarkably, each testis contains approximately 250 meters of seminiferous tubules. Sperm travel through the seminiferous tubules and enter the **epididymis,** a system of tubes where the sperm mature and are stored until ejaculated. From the epididymis, sperm travel through paired tubes called the **vas deferens** and exit the body through the **urethra,** which ends at the tip of the penis. Along the way, the prostate and other glands add fluid to the sperm that helps the sperm survive in the female reproductive system. This fluid contains the sugar fructose as an energy source, bases that help buffer the acidic pH in the vagina and protect the sperm DNA from being damaged, and mucus that helps keep the sperm mobile. The mixture of ejaculated sperm and accompanying fluid is called **semen** **(INFOGRAPHIC 31.3).**

During sex, when a man ejaculates through the penis into a woman's **vagina,** sperm swim up the female reproductive tract, through the opening into the uterus called the **cervix,** and through the uterus into the oviducts. The oviducts are where **fertilization**—the fusion of egg and sperm—occurs. Only a single sperm will

SCROTUM
The sac in which the testes are held.

TESTOSTERONE
The primary male sex hormone, which stimulates the development of masculine features and plays a key role in sperm development.

SEMINIFEROUS TUBULES
Coiled structures that constitute the bulk of the testes and in which sperm develop.

EPIDIDYMIS
A system of tubes in which sperm mature and are stored before ejaculation.

VAS DEFERENS
Paired tubes that carry sperm from the testes to the urethra.

URETHRA
A tube that connects the bladder to the genitals and carries urine out of the body. In males, the urethra travels through the penis and also carries sperm.

SEMEN
The mixture of fluid and sperm that is ejaculated from the penis.

VAGINA
The first part of the female reproductive tract, extending to the cervix; also known as the birth canal.

CERVIX
The opening or "neck" of the uterus, where sperm enter and babies exit.

FERTILIZATION
The fusion of an egg and a sperm; the resulting cell is called a zygote.

INFOGRAPHIC 31.4

Fertilization Occurs in the Oviduct

During intercourse, sperm ejaculated from the penis enters the female reproductive tract. Sperm must swim through the tract into the oviducts to fertilize an ovulated egg. While many sperm may make it to an oviduct, only one will actually fertilize an egg. Blockages in the female reproductive tract can impede sperm passage to an egg, and consequently compromise fertilization.

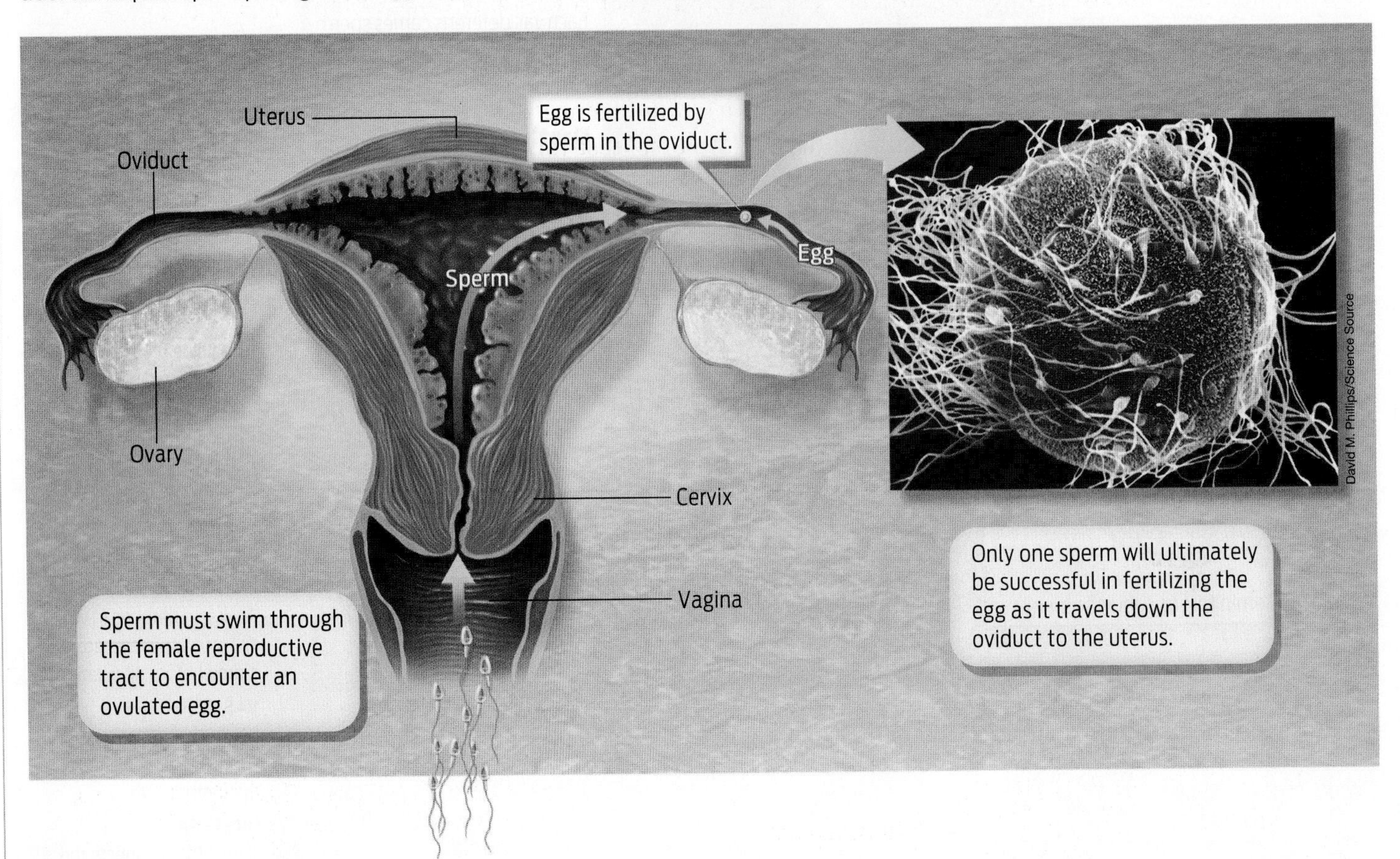

? What structure do sperm have to pass through from the vagina to the uterus? Once sperm are in the uterus, is their journey complete?

be successful in fertilizing an egg **(INFOGRAPHIC 31.4)**.

Assisted Reproduction

▸ In vitro fertilization

That's the way conception normally happens—but several things can go wrong along the way. For example, physical damage to the reproductive organs can prevent fertilization. In some cases, a woman's oviducts can be blocked or damaged, preventing eggs from entering the uterus or sperm from getting into the oviduct, where fertilization normally takes place. The most common cause of blocked tubes is pelvic inflammatory disease, which can be caused by sexually transmitted diseases such as chlamydia, a bacterial infection. An untreated infection can cause scar tissue to build up in the oviducts, blocking the passage of egg and/or sperm. In interviews, Suleman said that she suffered from fibroids—benign tumors that grow in muscle—that left her oviducts scarred.

In men, obstructions in the vas deferens or epididymis can block sperm transport. Varicose veins in the testicles and sexually transmitted bacterial infections such as chlamydia or gonorrhea can also block the male reproductive tubes.

Fertility specialists can test for physical blockages and, in some cases, surgically correct them. When surgery isn't an option, in vitro fertilization (IVF) is often recommended. In IVF, hormones are administered to a woman to promote egg development, and then eggs are extracted from her ovaries through a needle inserted through the vagina. Sperm are extracted from the man's ejaculate or, in cases of physical blockage, from his epididymis. The sperm and eggs are combined outside the body in a petri dish to allow the sperm to fertilize the eggs. The fertilized eggs begin to divide, and the resulting **embryos** are then inserted into the woman's uterus in hope that at least one will develop into a fetus **(INFOGRAPHIC 31.5)**.

IVF has been used to help infertile couples conceive children since 1978, when Louise Brown, the first "test-tube baby" conceived

EMBRYO
An early stage of development; an embryo forms when a zygote undergoes cell division.

INFOGRAPHIC 31.5

In Vitro Fertilization: How It Works

In vitro fertilization involves extracting eggs and sperm and combining them outside the body to allow fertilization. The resulting embryos are then inserted back into a woman's uterus in hope that at least one of them will implant into the uterus and grow into a fetus.

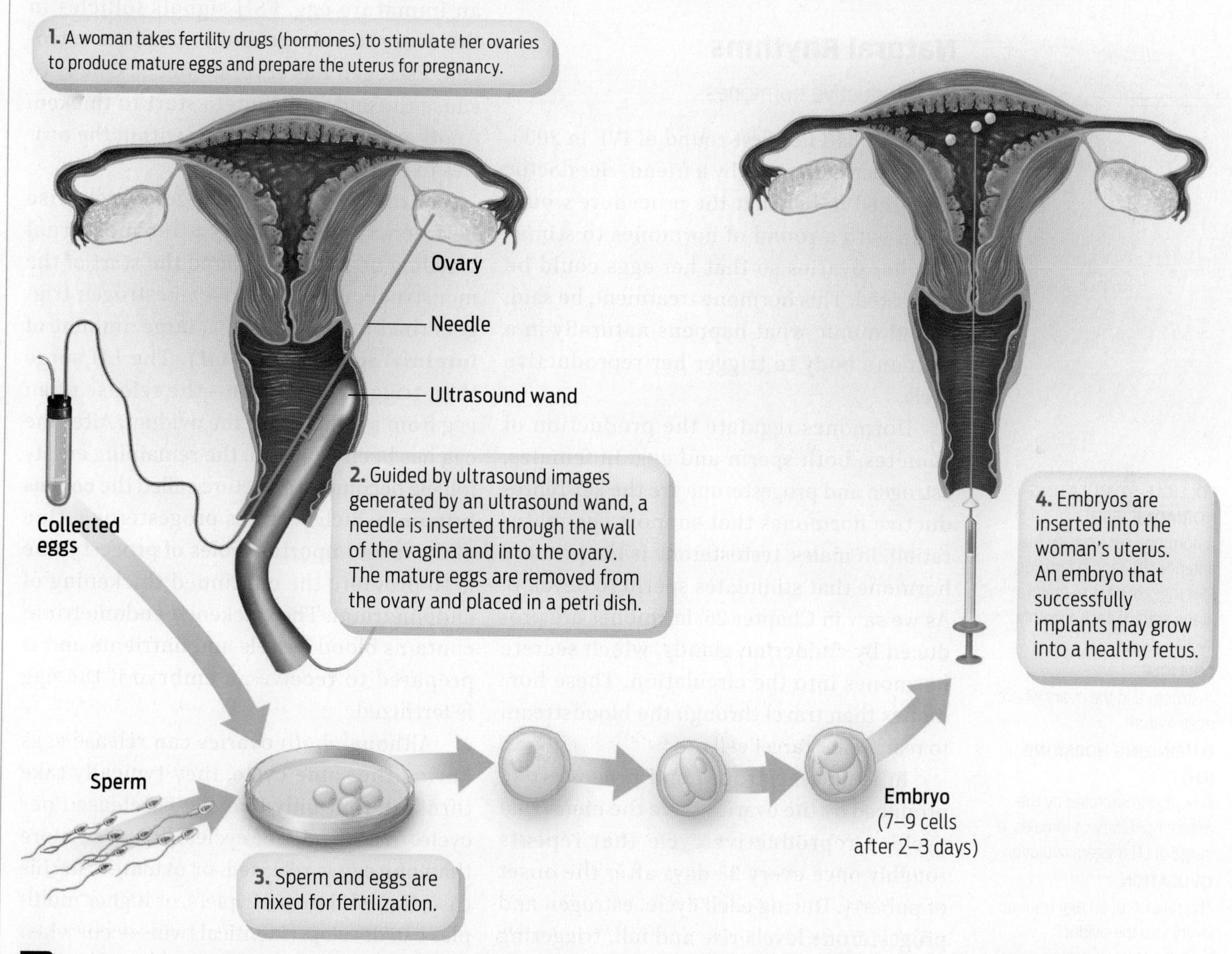

? In IVF, which processes occur in their normal locations? Which occur in other places? (Hint: Think about egg production, fertilization, and embryo implantation.)

with IVF, was born (see **Milestone 8: Pandora's Dish**). In addition to cases of blocked passages, IVF may also be recommended in cases of low sperm number or abnormal sperm, and even when the cause of infertility can't be determined. In such cases, the term "infertility" can actually be a misnomer. Many couples with defective sperm or unexplained infertility can still conceive a child naturally—it just may take longer. But because no one can predict how long it might take to achieve a successful pregnancy and because fertility decreases with age, IVF makes conception more likely by bringing sperm and egg together artificially.

Natural Rhythms

▶ Reproductive hormones

Suleman had her first round of IVF in 2000, using sperm donated by a friend. Her doctor explained to her that the procedure would begin with a round of hormones to stimulate her ovaries so that her eggs could be harvested. This hormone treatment, he said, would mimic what happens naturally in a woman's body to trigger her reproductive cycle.

Hormones regulate the production of gametes, both sperm and egg. In females, estrogen and progesterone are the key reproductive hormones that support egg maturation. In males, testosterone is the primary hormone that stimulates sperm to develop. As we saw in Chapter 26, hormones are produced by endocrine glands, which secrete hormones into the circulation. These hormones then travel through the bloodstream to reach their target cells.

In females, estrogen and progesterone produced by the ovaries drive the menstrual cycle, a reproductive cycle that repeats roughly once every 28 days after the onset of puberty. During each cycle, estrogen and progesterone levels rise and fall, triggering the ovaries to release an egg and prepare a woman's uterus for pregnancy should an egg be fertilized.

The brain's hypothalamus controls levels of estrogen and progesterone in the body, so it is the ultimate regulator of fertility in females. The hypothalamus works closely with the anterior pituitary, which sits just below the hypothalamus in the brain. The hypothalamus secretes hormones that act on the anterior pituitary, causing it to produce two hormones of its own. These hormones, called follicle-stimulating hormone and luteinizing hormone, travel through the bloodstream and directly stimulate the ovaries.

In women, **follicle-stimulating hormone (FSH)** acts on structures in the ovaries called **follicles,** each of which contains an immature egg. FSH signals follicles in the ovary to enlarge and to produce estrogen. Estrogen has several effects. One is to cause the endometrium to start to thicken. Another is to stimulate eggs within the ovaries to mature.

In most women, estrogen levels rise between 10 and 14 days after menstrual bleeding begins (considered the start of the menstrual cycle). This rise in estrogen triggers the brain to release a large amount of **luteinizing hormone (LH).** The LH surge then triggers **ovulation**—the release of an egg from a follicle into the oviduct. After the egg has been ovulated, the remaining empty follicle becomes a structure called the **corpus luteum,** which secretes progesterone. One of the most important roles of progesterone is to promote the continued thickening of endometrium. The thickened endometrium contains blood vessels and nutrients and is prepared to receive an embryo if the egg is fertilized.

Although both ovaries can release eggs during the same cycle, they typically take turns, so that only one egg is released per cycle. In about 1% of cycles, however, more than one egg is released, or ovulated. In this case, fraternal twins, triplets, or higher multiples can develop. (Identical twins occur when a single egg is released and fertilized and then splits into two embryos early in embryonic development.)

FOLLICLE-STIMULATING HORMONE (FSH)
A hormone secreted by the anterior pituitary. In females, FSH triggers eggs to mature at the start of each monthly cycle.

FOLLICLE
A structure in the ovary where eggs mature.

LUTEINIZING HORMONE (LH)
A hormone secreted by the anterior pituitary. In females, a surge of LH triggers ovulation.

OVULATION
The release of an egg from an ovary into the oviduct.

CORPUS LUTEUM
The structure in the ovary that remains after ovulation. It secretes progesterone.

INFOGRAPHIC 31.6

Hormones Regulate the Menstrual Cycle

A complex interplay of hormones from the hypothalamus, anterior pituitary gland, and ovaries drives the monthly female reproductive cycle.

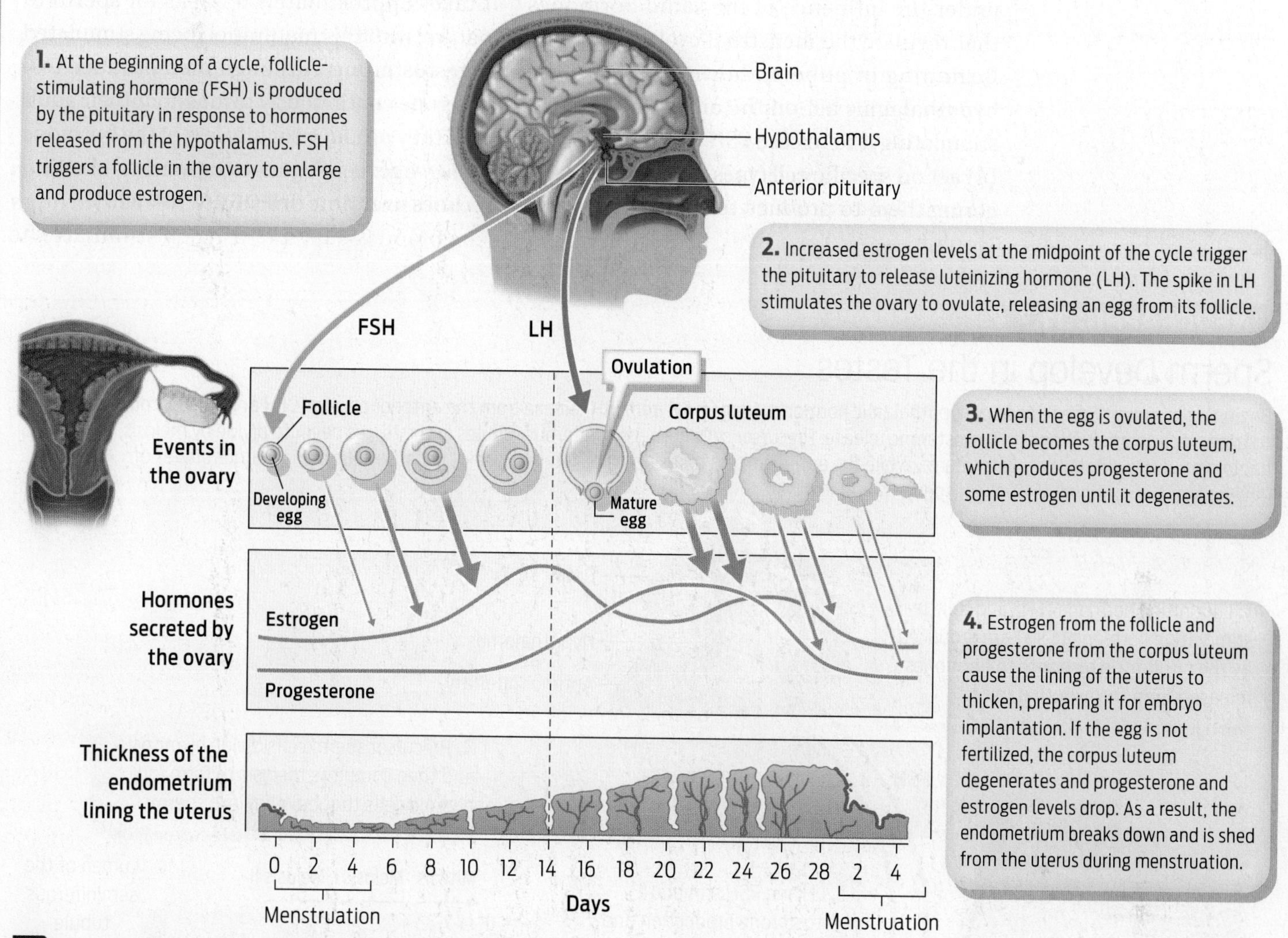

? For each of the four reproductive hormones (LSH, FH, estrogen, and progesterone), state where it is produced and what effect it has on its target.

Ovulation presents a crucial time window during which a woman can become pregnant. Sperm must swim through the cervix and uterus and into the oviduct containing the released egg to fertilize it. Because sperm can survive in the female reproductive tract anywhere from 3 to 7 days, a woman can become pregnant even if she has sex before she ovulates. Sperm can, in effect, wait in the oviduct for an egg to be ovulated. Once the egg leaves the oviduct and enters the uterus, however, the odds that it will be fertilized are extremely small.

If an egg is not fertilized within 24 hours of ovulation, it is no longer viable. The corpus luteum degenerates at about day 26 of the cycle, progesterone levels drop, and the uterine lining sloughs off in the **menstruation** that follows **(INFOGRAPHIC 31.6)**.

A woman's hormonal cycles continue until the time of menopause, which in American women occurs at about age 51. At this time, the ovaries stop responding to follicle-stimulating hormone and luteinizing hormone. As a result, women stop ovulating and stop having monthly reproductive

MENSTRUATION
The shedding of the uterine lining (the endometrium) that occurs when an embryo does not implant.

cycles—which means they are no longer able to conceive.

Men do not have a monthly hormone cycle, but the male gametes (sperm) develop under the influence of the same hormones that regulate the menstrual cycle in women. Beginning in puberty, hormones from the hypothalamus act on the anterior pituitary, stimulating it to release FSH and LH. FSH and LH act on specific cell types in the testes, triggering them to produce testosterone, which is essential for sperm production. The seminiferous tubules house precursor sperm cells that go through meiosis (see Chapter 11) and specialization to produce mature sperm cells. It takes approximately 6 weeks for sperm to mature, with this maturation being stimulated by testosterone. Although men produce testosterone continuously throughout their adult lives, they produce slightly less of the hormone as they age; consequently, sperm production declines over time **(INFOGRAPHIC 31.7)**.

INFOGRAPHIC 31.7

Sperm Develop in the Testes

Beginning at puberty the release of hypothalamic hormones triggers LH and FSH release from the anterior pituitary. LH and FSH act on the testes to trigger sperm production and testosterone release. Precursor cells in the seminiferous tubules in the testes begin to divide by meiosis, and the haploid products differentiate into sperm. While the entire process takes approximately 6 weeks, because cells are in various stages of development, a continuous supply of sperm is produced.

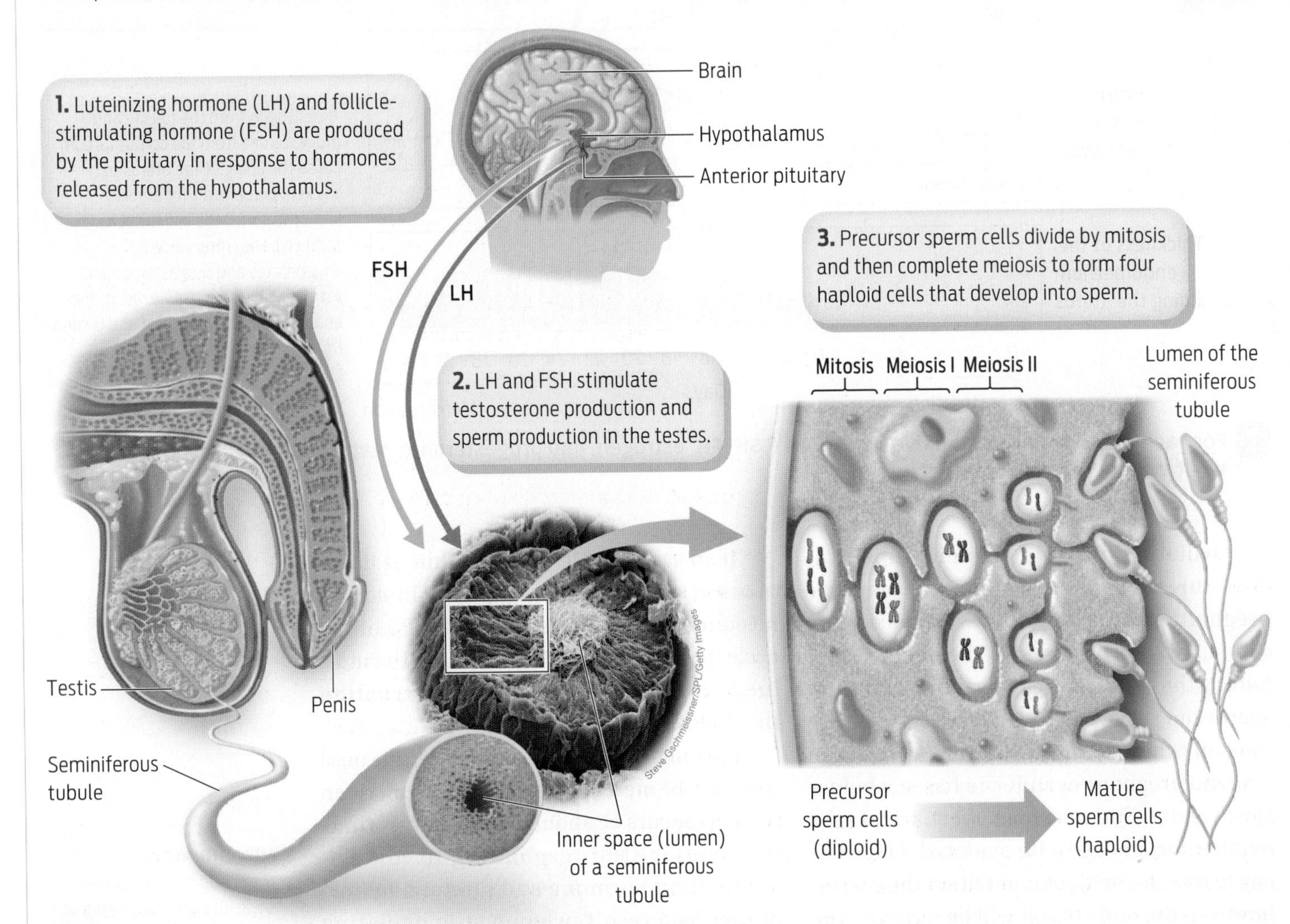

? Where are the hormones that trigger sperm production produced?

INFOGRAPHIC 31.8

Hormones Support Pregnancy

Estrogen and progesterone support the implanted embryo as it develops. Early in pregnancy, the embryo secretes human chorionic gonadatropin (hCG), which signals the corpus luteum to continue to produce these hormones. The placenta, once it has formed, takes over estrogen and progesterone production.

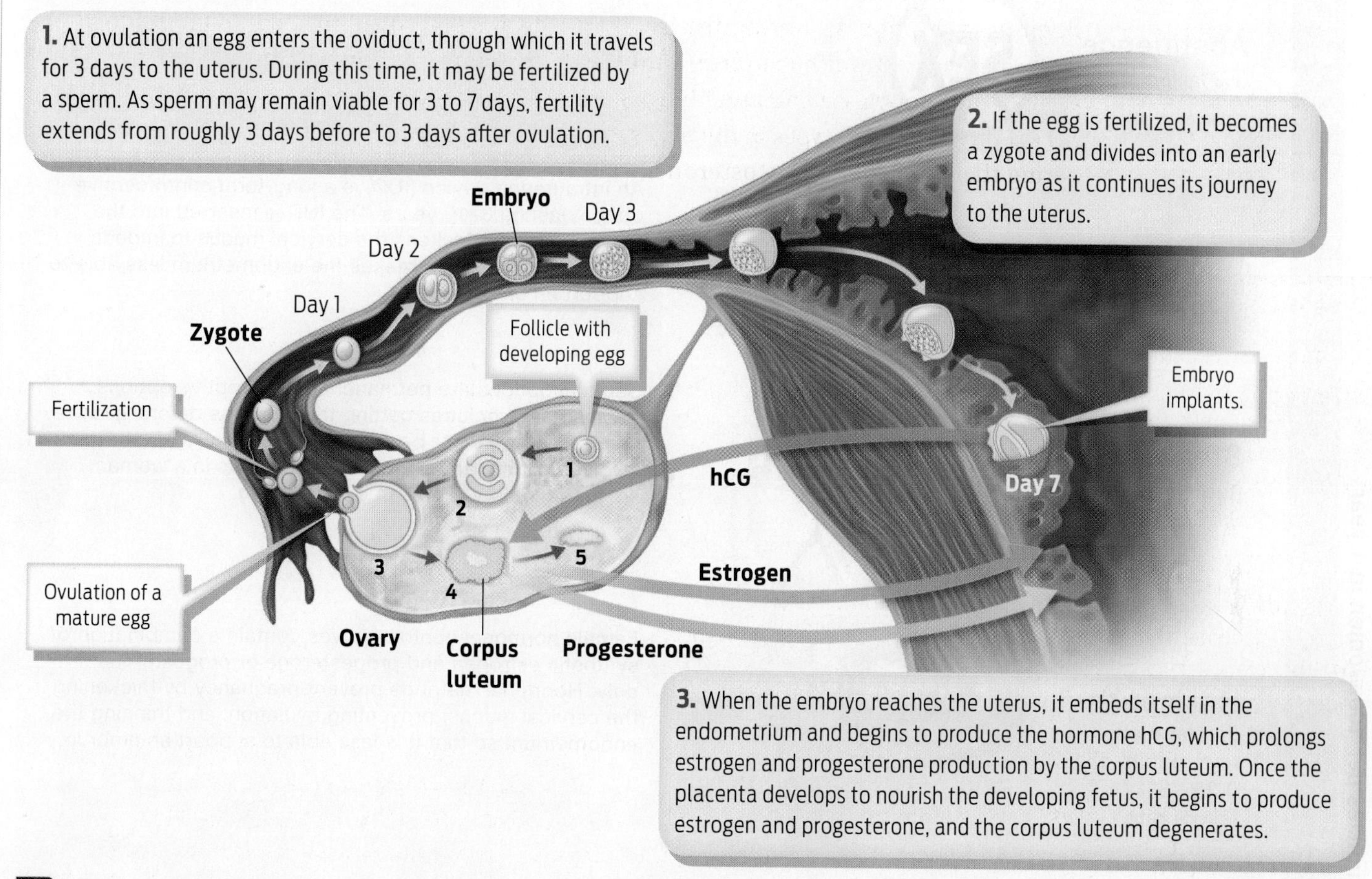

? **What makes hCG? Why is it important during early pregnancy?**

Family Planning

▶ Pregnancy and contraception

If a healthy sperm meets a healthy egg, the result may be fertilization, and pregnancy will have begun. A fertilized egg is called a **zygote**. As the zygote travels to the uterus, it begins dividing and developing into an embryo. Typically, the embryo implants in the endometrium of the uterus about 1 week after the egg is fertilized. Once implanted in the uterus, the embryo secretes a hormone called **human chorionic gonadotropin (hCG).** This hormone signals the corpus luteum to continue producing progesterone, which supports the thickening endometrium. In effect, hCG tells the reproductive system that pregnancy has begun; this hormone is also what most pregnancy tests measure to detect a pregnancy.

Once the embryo implants, tissues from the embryo and maternal endometrium interact to form the **placenta,** a disc-shaped structure that provides nourishment and support to the d eveloping **fetus,** as the embryo is now called. The placenta delivers oxygen, nutrients,and other key molecules like antibodies from the mother that help protect the embryo against infections. In addition, it eventually takes over from the corpus luteum the task of producing estrogen and progesterone **(INFOGRAPHIC 31.8).**

ZYGOTE
A fertilized egg.

HUMAN CHORIONIC GONADOTROPIN (HCG)
A hormone produced by an early embryo that helps maintain the corpus luteum until the placenta develops.

PLACENTA
A structure made of fetal and maternal tissues that helps sustain and support the embryo and fetus.

FETUS
After the eighth week after fertilization, the embryo is referred to as a fetus.

TABLE 31.1 Contraception

Pregnancies per 100 women in 1 year: <1 (top) → 10–20 → 85 (bottom)

METHOD	DESCRIPTION
Abstinence 0% failure rate SEX Photomac/FeaturePics.com	No sexual intercourse
Intrauterine device 0.2%–0.8% failure rate Saturn Stills/Science Source	An intrauterine device (IUD) is a long-term contraceptive option, lasting 3–10 years. The IUD is inserted into the uterus, where it thickens the cervical mucus to impede sperm movement and makes the endometrium less able to support an embryo.
Sterilization surgery (tubal ligation, vasectomy): 0.5%–0.15% failure rate Tek Image/Science Source	Surgical options are permanent contraceptive options. A vasectomy includes cutting the male vas deferens, so sperm can no longer be ejaculated. A tubal ligation involves cutting and tying off the oviducts in a woman, so sperm cannot reach an ovulated egg.
Hormones (implant, shot, ring, patch, pill): 0.3%–8% failure rate crankyT/iStockphoto moodboard/Superstock	Female hormonal contraceptives contain a combination of synthetic estrogen and progesterone or progesterone only. Hormonal methods prevent pregnancy by thickening the cervical mucus, preventing ovulation, and thinning the endometrium so that it is less able to support an embryo.
Barriers 2%–21% failure rate Superstock Jenny Swanson/iStockphoto	The male and female condom, the diaphragm, and the cervical cap prevent sperm from entering the uterus and are typically used with spermicidal foams or jellies.
Rhythm method and withdrawal 3%–27% failure rate	The rhythm method involves avoiding intercourse around the time a woman ovulates, and withdrawal of the penis before ejaculation.
No contraception 85% failure rate	Sexual intercourse without any method of contraception

Data from Trussel, J. (2007). In Hatcher, R. A., et al., *Contraceptive Technology,* 19th revised ed. New York: Ardent Media.

Because hormones play such a crucial role in pregnancy and reproduction, many types of **contraception** are designed to interfere with the normal female hormone cycle. Most birth control pills contain both estrogen and progesterone at levels that prevent the anterior pituitary from releasing follicle-stimulating and luteinizing hormones. In the absence of these hormones, ovulation does not take place, and consequently a woman taking this pill does not release eggs. Progesterone in birth control pills prevents successful pregnancy in other ways, too: it thickens the cervical mucus, blocking sperm from entering the uterus and oviducts, and it also reduces endometrial thickening—a process that is necessary to support an embryo (**TABLE 31.1**).

What Can Go Wrong

▸ Causes of infertility

Given how important hormones are to egg maturation and ovulation, it's not surprising that hormonal imbalances are a common cause of female infertility. Even slight irregularities in levels of luteinizing hormone and follicle-stimulating hormone can prevent the ovaries from releasing eggs. Specific causes of such hormonal imbalances include injury, tumors, excessive exercise, and starvation. Some medications can interfere with ovulation, and some studies have shown that stress can lower fertility, as can poor nutrition.

Some women experience polycystic ovary syndrome, a condition defined by the presence of multiple ovarian cysts. The cysts impair ovulation and are associated with the production of excessive amounts of **androgens** (the "male" sex hormones, typically present in lower amounts in females). This syndrome is one of the most common hormonal disorders, affecting an estimated 10% of women of reproductive age. In addition to impairing ovulation, polycystic ovary syndrome is associated with irregular menstrual cycles, diabetes, and obesity.

One of the most common causes of infertility in women is advanced maternal age (typically defined as 35 years or older). Infertility due to advanced age is especially common in Western industrialized societies, where women may delay trying to have children for a variety of reasons. Fertility in women peaks around age 25 and declines thereafter. According to the Centers for Disease Control and Prevention, by age 40, nearly half of all women have difficulty conceiving. Age-related declines in fertility occur in part because both the quantity and quality of a woman's eggs decline with age. At birth, a baby girl has about 1 million follicles in her ovaries. By puberty, she has about 300,000 follicles. Of these, only about 300 will ever ovulate an egg.

Men, too, experience declines in fertility as they age and testosterone levels fall, but their decrease in fertility usually does not happen as quickly or as dramatically as in women. Men produce sperm throughout their lives, so they can still father children in their 50s, 60s, and 70s. However, many men suffer from other

CONTRACEPTION
The prevention of pregnancy through physical, surgical, or hormonal methods.

ANDROGEN
A class of sex hormones, including testosterone, that is present in higher levels in men than in women and causes male-associated traits like deep voice, growth of facial hair, and defined musculature.

Professors Pietro M. Motta & Sayoko Makabe/Science Source

Enlarged ovarian follicles form cysts protruding from a polycystic ovary.

problems, such as erectile dysfunction and testicular varicose veins, that can interfere with sperm reaching their target.

INTRAUTERINE INSEMINATION (IUI)
A form of assisted reproduction in which sperm are injected directly into a woman's uterus.

About a third of all cases of infertility are caused by reproductive impairments in women, and about a third are caused by impairments in men. In the remaining cases, the impairments affect both men and women or the cause of infertility cannot be determined **(INFOGRAPHIC 31.9)**.

Modern Families

▸ Types of assisted reproduction

While Suleman's case made for shocking headlines and generated a strong backlash among

INFOGRAPHIC 31.9

Causes of Infertility

There are multiple causes of infertility in both women and men that disrupt the normal function of reproductive tissues.

In Women

Blockages
Passages may become blocked or disabled in both male and female reproductive systems because of tissue scarring, infection, cancer, or abnormal tissue growth.

Cysts, fibroids, polyps
Each of these is a type of abnormal growth that may block passages or interfere with normal function of the tissue.

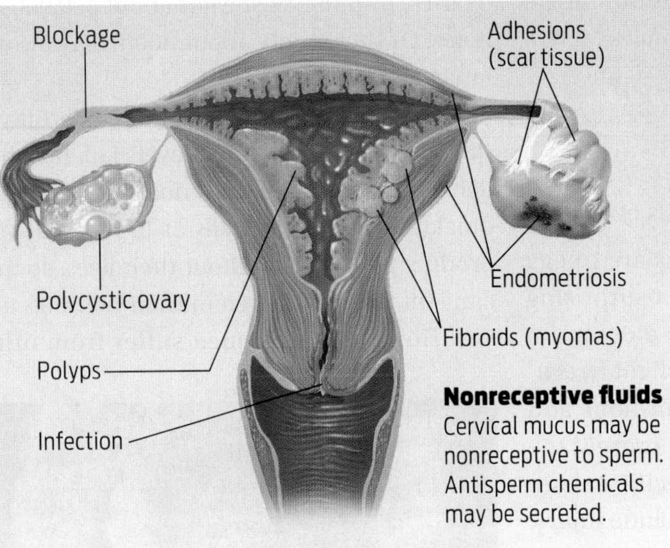

Nonreceptive fluids
Cervical mucus may be nonreceptive to sperm. Antisperm chemicals may be secreted.

Nonfunctional ovaries
Ovaries may fail to ovulate because of a variety of factors, including hormonal imbalances, genetic abnormalities, undeveloped ovarian tissue, endometriosis, and cancer.

Endometriosis
The tissue that lines the uterus grows abnormally and invades other tissues in the pelvic region. The wayward tissue irritates the nerve endings of these organs and interferes with their function.

In Men

Blockages
Passages may become blocked or disabled in both male and female reproductive systems because of tissue scarring, infection, cancer, or abnormal tissue growth.

Erectile dysfunction
Genetic abnormalities, neurological problems, hormonal imbalances, and physical blockages inhibit the ability of blood to flood the penis tissue to support an erection.

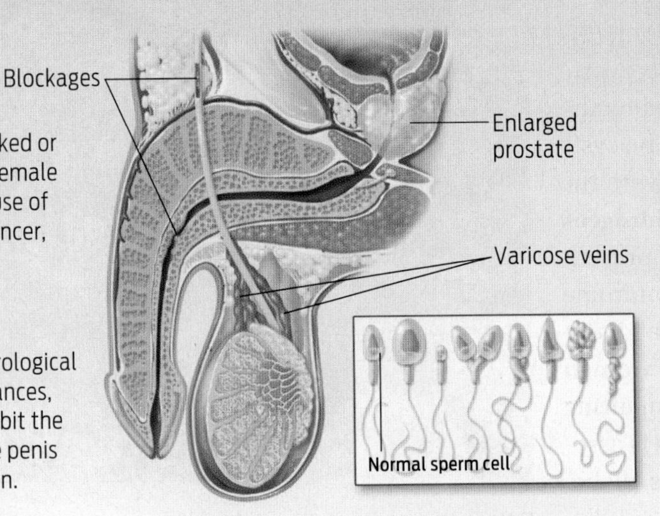

Prostatitis
Sperm pass through and receive fluid from the prostate on their way into the urethra. An enlarged prostate can block the passage of semen.

Testicular varicose veins
Valves in the veins that keep blood flowing in one direction deteriorate, causing blood to back up and pool. These enlarged veins can interfere with sperm production and transport.

Sperm abnormalities
Men may have low numbers of healthy sperm and/or physically abnormal sperm.

? **Identify at least two specific places in the male and female reproductive tracts where blockages may lead to infertility.**

the public, the lessons of the case extend well beyond this one woman's unusual story. Increasingly, people with all kinds of fertility difficulties are turning to assisted reproduction for help conceiving a child. Lesbian and gay couples, too, are seeking help from fertility specialists in their quest to become parents. Approximately 2% of all infants born in the United States every year are conceived using some form of assisted reproductive technology. Multiples are often an unintended consequence of the technologies used.

Kate Gosselin and six of her children.

© TLC/Courtesy: Everett Collection

Assisted reproduction takes a number of forms. Besides IVF, another popular method is **intrauterine insemination (IUI),** in which sperm are injected directly into the uterus. This approach, also known as artificial insemination, is used most often when infertility can be traced to low sperm count, or to sperm that are "slow swimmers." It is also commonly used when a woman is trying to conceive with donor sperm or if the cause of infertility cannot be determined. Ultrasound technology is often used to monitor ovulation, so the insemination can be performed when the chances are highest that one or more eggs have been released into the oviduct.

Commonly, both IVF and IUI start with a course of fertility drugs administered in the weeks before the procedure. These drugs usually contain FSH and LH, the hormones that stimulate ovarian follicles to develop and eggs to ovulate. The treatment works well for promoting follicle development and for controlling the precise time of ovulation, but there is a catch: fertility drugs almost always cause multiple eggs to develop in multiple follicles. This effect is desirable during IVF, in which the clinician can control the number of eggs that are fertilized outside the body and implanted into the uterus. But this advantage turns into a liability with IUI. Since millions of sperm are injected into the uterus during insemination, it's difficult to control the number of eggs that are fertilized. Multiple births result when more than one egg is fertilized (each by a different sperm), leading

©Joshua A. Bickel/The Columbus Dispatch-USA TODAY NETWORK

Mia and Rozanno McGhee tried to become pregnant for many years. After using fertility drugs, Mia gave birth to sextuplets in 2010.

to more than one embryo implanting in the uterus and developing into a fetus.

That's what happened to Kate Gosselin, who, with her husband Jon, starred in the TLC television reality series *Jon & Kate Plus 8* from 2007 to 2011. The couple first had twins and then later sextuplets, all of whom were conceived through intrauterine insemination **(INFOGRAPHIC 31.10)**.

Risky Business

▶ Fetal development and potential complications

In the United States, only 17 states currently mandate that health insurance policies must offer coverage for infertility treatment, and even those policies often do not cover the more expensive treatments, like IVF. As a

INFOGRAPHIC 31.10

Assisted Reproductive Technologies Can Result in Multiple Births

A hazard of assisted reproduction is a high probability of multiple births. Babies born as multiples are more likely to be born underweight and premature, putting them at risk for a variety of serious health conditions.

In Vitro Fertilization (IVF)

A woman is given fertility drugs to stimulate egg development and maturation. Eggs are then retrieved and fertilized in a dish. If multiple embryos are injected, more than one of them may implant in the uterus.

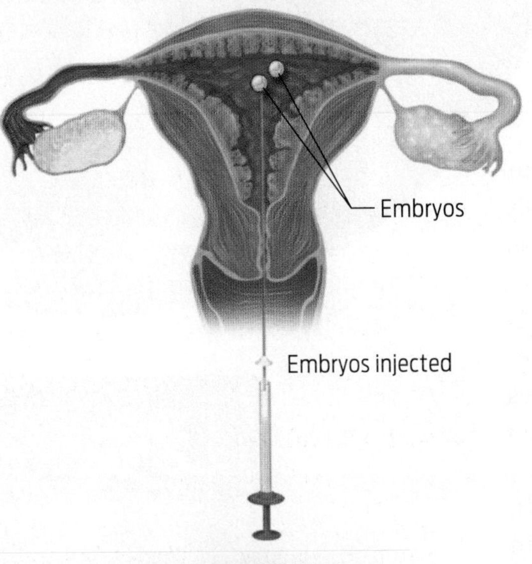

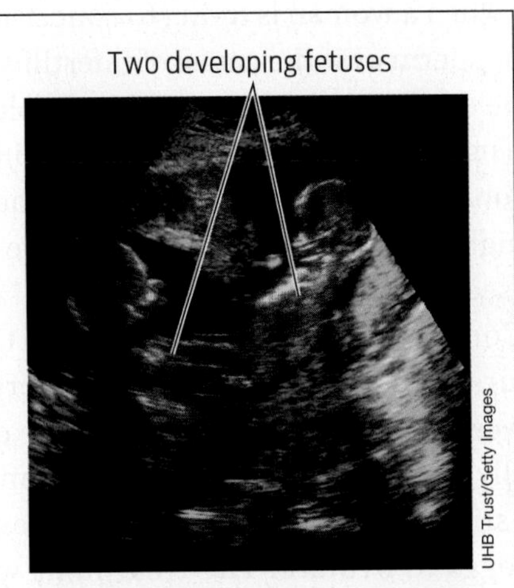

Multiple births: 41% of births by IVF are twins; 2.4% are triplets or more.

Intrauterine Insemination (IUI)

When a woman has been given fertility drugs to stimulate ovulation, several eggs may mature and ovulate simultaneously. Sperm injected into the uterus may fertilize multiple eggs and create multiple embryos.

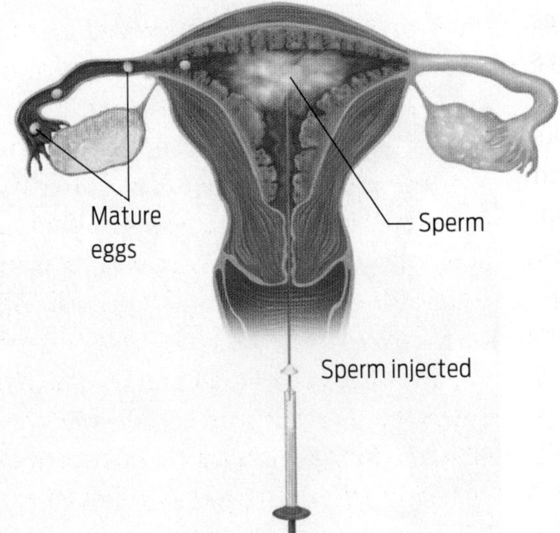

? Why is it easier to reduce the chance of multiples with IVF than with IUI?

consequence, many patients must pay out of pocket for fertility-related services. Because the costs of treatment are so high, many doctors find themselves under pressure to be aggressive with treatment with patients who have limited resources. Therefore, financial considerations can dictate a course of treatment, even when the treatment is unlikely to be successful or may be too successful, as it proved to be in Suleman's case.

A single insemination treatment (including fertility drugs and monitoring) can cost as much as $1,000—steep, but less than one-tenth the cost of an average cycle of in vitro fertilization. Many couples may choose insemination over in vitro fertilization simply because it is less expensive, even though the procedure has a low success rate. On average, only 5% to 15% of all insemination treatments result in a live birth. But younger women and women who take fertility drugs tend to have higher success rates. The multiple ovulations stimulated by fertility drugs increase the odds of a successful pregnancy by insemination.

IVF has a higher success rate—about 30% for women aged 35, according to ASRM—but if the woman must undergo multiple rounds before becoming pregnant, the $10,000–$15,000 price tag for each round can be an obstacle. This financial burden can cause couples to hedge their bets by insisting that the doctor implant more embryos at one time than is typically recommended. The more embryos that are transferred into a woman's uterus during each round of IVF, the higher the odds of a pregnancy become.

Medical details revealed during a court case brought by the California Medical Board in 2009 against Suleman's doctor, Michael Kamrava, show that Kamrava created 14 embryos and implanted a dozen of them in Suleman—six times the recommended number for a woman her age. Eight embryos survived, and the babies were delivered 9 weeks prematurely by C-section (cesarean section). The Associated Press reported that during the hearing Kamrava said he regretted implanting the 12 embryos and "would never do it again."

Kamrava further stated that Suleman was adamant about using all 12 embryos, even though he suggested implanting only 4. "She just wouldn't accept doing anything else with those embryos. She did not want them frozen, she did not want them transferred to another patient in the future," he said, according to the AP story. Kamrava said that he consented only after Suleman agreed to have a fetal abortion if necessary to reduce risk. After the implantation, however, he only heard from Suleman after the birth of her octuplets.

After a lengthy investigation, the California Medical Board revoked Kamrava's medical license. The Board claimed that he was negligent not only in Suleman's case but also in the cases of two other women who suffered serious medical complications because of aggressive fertility treatments.

"If nothing else, these high-profile cases have served as a wake-up call."

—Barbara Collura

All pregnancies carry risks, but multiple-birth pregnancies carry greater risks. One of the most common is premature birth. Pregnancy is typically divided into three trimesters of 3 months each. Each trimester has unique developmental milestones and potential associated complications. A full pregnancy lasts 40 weeks (counting from the time of the last menstrual period). Babies who are born more than 3 weeks before their due date are considered premature. According to the March of Dimes, 50% of twins and 90% of triplets are born prematurely, as are virtually all quadruplets and quintuplets. When babies are born prematurely, their organs may not be fully developed, which can lead to medical problems. In particular, their lungs are often immature, requiring the babies to be hooked up to mechanical breathing ventilators after birth. These treatments sometimes scar the lungs, leaving these children prone to asthma, pneumonia, chronic lung disease, and other respiratory disorders for the rest of their lives. And because their brains aren't fully developed, premature babies are susceptible to brain hemorrhages and to developmental difficulties, including learning disabilities **(INFOGRAPHIC 31.11)**.

More than three-fourths of all triplets and higher-order multiples born in the United States are attributable to artificial reproductive technology. To reduce the incidence of multiples and prevent the associated health problems, in 1998 the ASRM began recommending limits on the number of embryos transferred; it has since adjusted its initial recommendations downward. Current guidelines recommend transferring no more than two embryos in women younger than 35

INFOGRAPHIC 31.11

Stages of Prenatal Development

Pregnancy is divided into three trimesters. A normal pregnancy lasts 38 weeks from the time of fertilization, which is 40 weeks from the last menstrual period.

First trimester: Development of tissue layers and vital organs

The first trimester includes the embryonic stage of development and the early fetal stage. The embryo and fetus grow rapidly and critical organs develop.

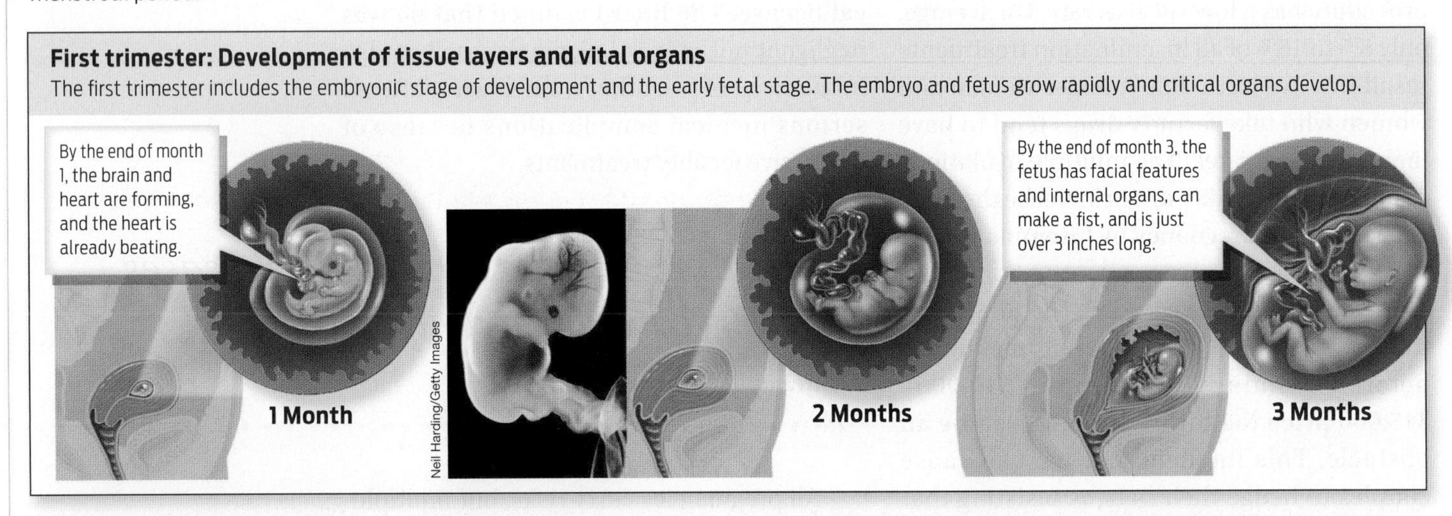

Second trimester: Growth and sex organ development

During the second trimester, the fetus continues to grow and develop features, including external genitalia.

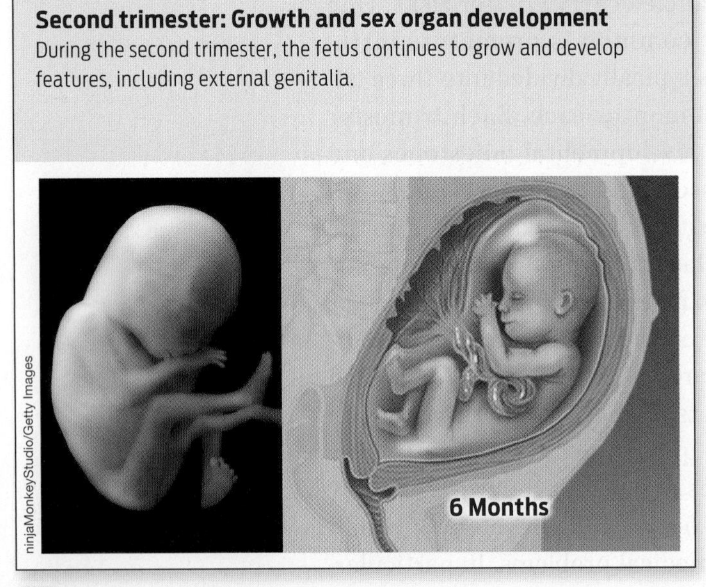

Third trimester: Weight gain and organ system development

During the third trimester the fetus becomes fully developed and continues to gain weight. A fetus is technically "full term" at 35 weeks after fertilization (or 37 weeks after the last menstrual period), but most pregnancies last 38 weeks after fertilization (or 40 weeks after the last menstrual period).

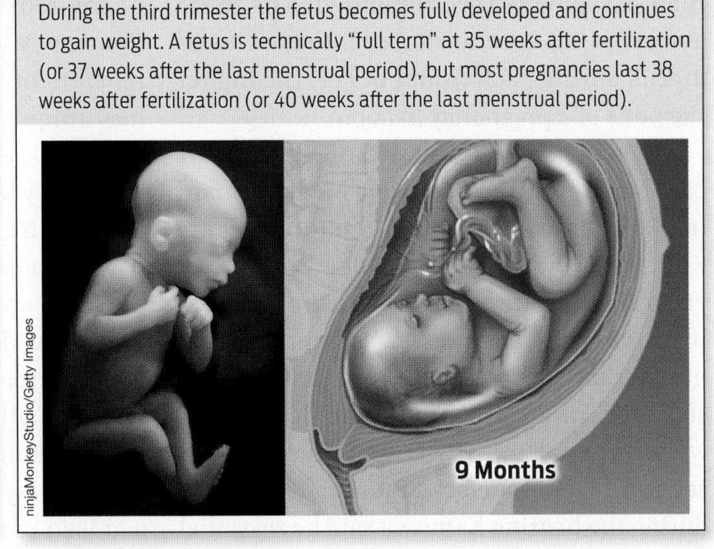

? **What is the earliest trimester during which the sex of the baby can be determined by ultrasound? Explain your answer.**

years of age and no more than five embryos in women older than 40—older women are allowed more embryos because as a woman ages the success rate tends to drop. These guidelines appear to have helped: in 2016, 1.1% of all infants born via assisted reproduction were triplets, compared to 2.4% in 2012 and 6.4% in 2003. But assisted reproduction still results in a high number of twins—in 2016, 30% of births from IVF alone were twins, compared to 3% of all births.

Many other countries, including Canada, the United Kingdom, Sweden, Germany, and Australia, heavily regulate the fertility business. Because the United States does not, the risk remains that doctors may act against the ASRM guidelines, increasing the risk of poor outcomes for women and their babies.

Moreover, as long as fertility treatment continues to place a financial burden on couples, fertility doctors will face pressure from patients to give them the most they can get for their money, says Barbara Collura, executive director of RESOLVE, an advocacy group supporting infertile couples. Broad health insurance coverage would eliminate the cost factor. The result might be that couples would be able to forgo treatments such as insemination that are characterized by high risks and low success rates, and skip directly to IVF when appropriate.

MOVI Inc./MEGA

Nadya Suleman with some of her 14 children.

"If nothing else, these high-profile cases have served as a wake-up call," says Collura. The community of health care workers is examining its own procedures and methods because it would rather self-regulate than have regulation imposed on it from the outside, she adds. "After Suleman, the community is really looking into how [the birth of octuplets] happened and how it can prevent it from happening again." ■

CHAPTER 31 Summary

Driving Question 1 What is the anatomy of the male and female reproductive systems, and how does the anatomy contribute to the function of the reproductive system?

- The female reproductive system consists of paired ovaries, which produce eggs and the hormones estrogen and progesterone, and accessory structures that enable fertilization and support pregnancy.
- The male reproductive system consists of paired testes, which produce sperm and the hormone testosterone, and accessory structures that produce seminal fluids and permit the delivery of sperm to the egg.
- Fertilization of an egg by a sperm occurs in an oviduct. Only one sperm can fertilize an egg.

Driving Question 2 What hormones are involved in reproduction, and how do they work in the reproductive systems of males and females?

- The monthly menstrual cycle in females and production of sperm in males are coordinated by a complex balance of hormones that is controlled by the brain.
- Hormones from the hypothalamus trigger the anterior pituitary to release follicle-stimulating hormone (FSH) and luteinizing hormone (LH). In females, these hormones stimulate eggs to develop and the ovary to secrete estrogen and progesterone. In males, FSH and LH trigger the production of sperm and testosterone.
- Ovulation, the monthly release of an egg from an ovarian follicle, is triggered by a spike in production of LH. Estrogen and progesterone stimulate eggs to develop and the endometrium to thicken and prepare for a possible pregnancy.

- Upon fertilization, the zygote divides and travels to the uterus, where it implants in the nutrient-rich endometrium. If an egg is not fertilized, progesterone levels fall and the endometrium sloughs off during menstruation.
- There are many approaches to contraception, including placing barriers between sperm and eggs and manipulating reproductive hormones.

Driving Question 3 What are the different types of assisted reproduction, and how do they work?

- Infertility has many causes, including blocked passageways caused by infection and scar tissue, chromosomal abnormalities, hormonal deficiencies, and advanced age.
- Assisted reproduction involves artificially bringing sperm and egg together, either inside the body (in IUI) or outside the body (in IVF).
- Fertility drugs increase the number of eggs that mature and are ovulated by a female at one time. Multiple pregnancies result when sperm fertilize more than one available egg.

More to Explore

- TED Radio Hour. (2019). The biology of sex: https://www.npr.org/programs/ted-radio-hour/852195850/the-biology-of-sex
- Arney, K. (2019). Everything you ever wanted to know about the evolution of sex (but were too afraid to ask): https://geneticsunzipped.com/news/2019/8/15/evolution-of-sex
- Richardson, S. S. (2013). *Sex Itself: The Search for Male and Female in the Human Genome.* Chicago: University of Chicago Press.
- Fausto-Sterling, A. (2020). *Sexing the Body: Gender Politics and the Construction of Sexuality.* New York: Basic Books.
- Okwerekwu, J. A. (2016). Where have all the triplets (and quadruplets, and quintuplets) gone? https://www.statnews.com/2016/04/28/triplets-multiple-births/

CHAPTER 31 Test Your Knowledge

DRIVING QUESTION 1 What is the anatomy of the male and female reproductive systems, and how does the anatomy contribute to the function of the reproductive system?

By answering the questions below and studying Infographics 31.2, 31.3, and 31.4, you should be able to generate an answer for this broader Driving Question.

Know It

1. Sperm develop in the
- **a.** epididymis.
- **b.** vas deferens.
- **c.** seminiferous tubules.
- **d.** urethra.
- **e.** penis.

2. Why can untreated pelvic inflammatory disease lead to infertility?
- **a.** because it prevents ovulation
- **b.** because it scars and blocks the oviducts
- **c.** because it scars and blocks the cervix
- **d.** because it interferes with estrogen production by the ovaries
- **e.** because it interferes with FSH and LH production by the anterior pituitary gland

3. Describe the relationship between the uterus and the cervix, and between the uterus and the endometrium.

Use It

4. List the structures that sperm must pass through to reach and fertilize an egg. Begin with the seminiferous tubules.

5. A woman can become pregnant if she has intercourse 5 days before ovulating or on the day she ovulates, but not generally more than 24 hours after ovulation. What does this suggest about sperm and egg?
- **a.** Sperm can survive for up to 6 days in the female reproductive tract.
- **b.** Fertilization can occur in the ovary.
- **c.** Human eggs cannot survive for very long after ovulation.
- **d.** all of the above
- **e.** both a and c

6. A friend tells you that her boyfriend has received a diagnosis of gonorrhea, a sexually transmitted infection. She isn't worried for herself because she doesn't have any symptoms of infection. What can you tell her about the invisible risks of an untreated sexually transmitted bacterial infection?

DRIVING QUESTION 2 What hormones are involved in reproduction, and how do they work in the reproductive systems of males and females?

By answering the questions below and studying Infographics 31.6, 31.7, and 31.8, you should be able to generate an answer for this broader Driving Question.

Know It

7. What is the source—testes, ovaries, anterior pituitary gland, or embryo—of each of the following hormones?

Luteinizing hormone (LH) ________

Follicle-stimulating hormone (FSH) ________

Testosterone ________

Estrogen ________

Progesterone ________

Human chorionic gonadotropin (hCG) ________

8. The hormone hCG is an indicator of pregnancy; it also

a. signals the corpus luteum to keep producing progesterone.

b. triggers ovulation.

c. acts on the anterior pituitary gland, causing it to release a surge of LH.

d. acts on the endometrium, causing it to thicken.

e. attracts sperm.

9. Which of the following hormones is/are produced by the anterior pituitary in males?

a. FSH

b. LH

c. testosterone

d. hCG

e. FSH and LH

f. FSH, LH, and testosterone

Use It

10. Which of the following would most directly cause reduced levels of estrogen production?

a. an anterior pituitary tumor that increases secretion of LH

b. an increase in hypothalamus hormones that target the anterior pituitary

c. an anterior pituitary tumor that increases secretion of FSH

d. a decrease in hypothalamus hormones that target the anterior pituitary

e. anterior pituitary damage that prevents synthesis and release of FSH

11. In an episode of a popular "ripped from the headlines" TV crime series, a blood sample from a crime scene was found to have extremely low levels of FSH and LH. From this information, detectives determined that the blood came from a prepubescent girl, not a woman of reproductive age. Explain how they reached this conclusion.

12. As discussed in this chapter, oral contraceptives (such as the combination birth control pill, which contains both estrogen and progesterone) are designed to block ovulation in women. As males do not ovulate, a male hormonal contraceptive would have to target sperm development. Why would blocking testosterone secretion or action be effective in terms of contraception? What would be a likely undesired consequence of this type of male contraception?

Interpreting Data

Apply Your Knowledge

13. Emergency contraception—that is, contraception following intercourse—works by delaying or preventing ovulation. To be effective, it must be taken before LH levels start to rise.

a. Will emergency contraception be effective in preventing a pregnancy after unprotected intercourse on the day that a woman ovulates?

b. Different forms of emergency contraception have different efficacy rates, expressed as the percent reduction in the number of pregnancies that would have otherwise occurred. For example, if 80 women out of 1,000 become pregnant after unprotected intercourse, and an emergency contraceptive is 75% effective, then one would expect the use of the emergency contraceptive to reduce the number of pregnancies by 75% of 80 pregnancies, to 20 pregnancies. Predict the number of pregnancies with each of the following emergency contraceptives (given 80 of 1,000 pregnancies as a baseline): the pill marketed as Plan B (progestin—a synthetic progesterone-only contraceptive), 89% effective; combined progestin and estrogen, 74% effective.

DRIVING QUESTION 3 What are the different types of assisted reproduction, and how do they work?

By answering the questions below and studying Infographics 31.1, 31.5, 31.9, and 31.10, you should be able to generate an answer for this broader Driving Question.

Know It

14. Which of the following could interfere with ovulation?

- **a.** blocked oviducts
- **b.** chronically low levels of LH
- **c.** excessive production of cervical mucus that blocks the cervix
- **d.** presence of sperm in the oviduct
- **e.** low levels of hCG

15. Compare and contrast in vitro fertilization (IVF) and intrauterine insemination (IUI).

Use It

16. Assume that an array of diagnostic methods are available to you, including blood tests to determine hormone levels and ultrasound to visualize internal structures. What results might confirm each of the following infertility-associated conditions? Be specific.

- **a.** a blocked epididymis
- **b.** polycystic ovary syndrome
- **c.** menopause
- **d.** oviduct scarring

17. Why does IUI create a higher risk of multiple births than IVF?

Mini Case

Apply Your Knowledge

18. A young couple has been trying to have a baby for more than a year, but so far they have not had any luck. Analysis of the man's semen reveals a normal sperm count and no evidence of high rates of abnormal sperm. A physical exam and ultrasound reveal blockages in both of the woman's oviducts (fallopian tubes). From this information, describe two forms of fertility treatments (including assisted reproduction) that the couple could consider, and two that would not be helpful. For each, describe what is involved in the treatment, and why it would or would not be a helpful strategy for this couple.

Bring It Home

Apply Your Knowledge

19. From the perspective of a fertility specialist, how would you respond to a congressional representative proposing increased regulation of fertility clinics? To make a convincing argument, include both pros and cons, medical and scientific considerations, and a description of the patient population whom this specialist serves.

PANDORA'S Dish

The power, promise, and politics of stem cells

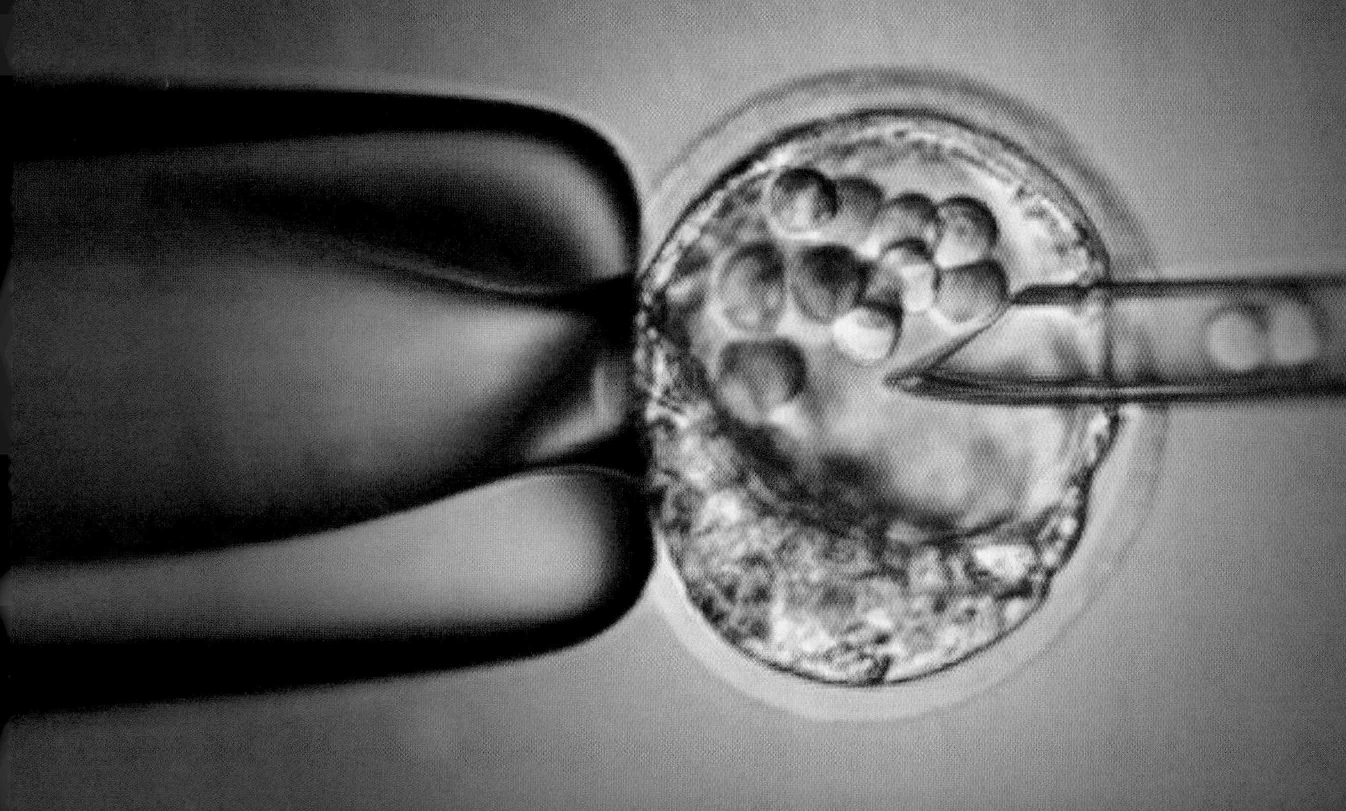

ANNE-CHRISTINE POUJOULAT/AFP/Getty Images

DRIVING QUESTIONS

1 What are stem cells, and how do different types of stem cells differ from one another?

2 How do cells become more specialized?

3 How can different types of stem cells be used in research and medicine?

Louise Brown recalls the hate mail her family received. One letter was doused in blood. Another package contained a tiny plastic fetus, a broken test tube, and a "test tube baby warranty card."

Louise's parents, Lesley and John Brown, were the first people in the world to conceive a child through in vitro fertilization (IVF). This procedure involves bringing sperm and egg together outside the body (see Chapter 31). It's a common medical procedure now, but in 1977, when the Browns used it, it seemed more like science fiction.

So novel and misunderstand was IVF that, when Louise was born on July 25, 1978, newspapers greeted her as the world's first "test tube baby"—a nickname she's carried her whole life. Paparazzi flocked to the family's home in Bristol, England. Lesley couldn't take her daughter out in a stroller without being mobbed by reporters.

"It was menacing and scary," Louise wrote in her memoir, *My Life as a Test Tube Baby,* published in 2015. "Imagine how worrying this was for mum."

From today's perspective, it can be hard to fathom the controversy that surrounded Louise's birth. (To be clear: a test tube was never involved; Louise grew to term inside her mother's womb.) Millions of children around the world have been born through IVF, and in some countries as many as 3% of all births are conceived this way.

But when Lesley and John sought help from doctors for their infertility, IVF was still largely untested and unproven. Many people—scientists included—feared that babies born this way would have serious medical problems. Ethicists and moralists worried that the procedure would lead to "assembly-line fetuses grown in test tubes," as one newspaper put it.

It was only when Louise proved to be perfectly healthy, and other children were born this way—to very grateful parents—that the controversy began to die down.

Yet in many ways, Louise Brown's birth was indeed a world-changing development. Not only did it help to solve the problem of infertility for millions of people, but it also birthed an entirely new field of biological research: stem cell biology.

Typically, more embryos are created than are used by parents visiting fertility clinics. Through this surplus of human embryos, IVF made it possible for scientists to study the earliest stages of human development. From these extra embryos, they would eventually isolate and grow embryonic stem cells (ESCs). These flexible precursor cells can generate every cell type in

Ben Birchall—PA Images/Getty Images

Daniel Leal-Olivas/Getty Images

Louise Brown, born in 1978 through IVF, became known as the world's first "test tube baby."

the human body, from retinal cells in the eyes to skin cells on the toes.

Scientists think ESCs hold the key to regenerative medicine—treatments that repair or regrow damaged organs. But the powerful cells have generated equally potent controversy. That's because, to obtain them, researchers have to destroy embryos. Scientists have sought ways around this problem, with varying degrees of success. But the ethical questions surrounding stem cell biology show no signs of abating. In the near future, some applications of the science could make IVF seem old-fashioned.

How Cells Grow Up

▸ Stem cells and cell differentiation

What most intrigues scientists about embryonic stem cells is their natural ability to divide repeatedly and generate essentially any cell type in the body. Most cells in an adult organism can only give rise to more cells of the same type, if they can divide at all. But ESCs have almost unlimited potential to divide and produce different cell types. This quality—which is unique to cells present in developing embryos—makes them attractive as a source of cells to repair damaged or diseased tissues.

Scientists draw distinctions between the relative flexibility, or potency, of different cells to generate other cell types. Cells found in the earliest stages of an embryo are described as totipotent: these cells can give rise to every cell type in the body, as well as those making up tissues outside the embryo, such as the placenta.

Embryonic stem cells, which come from 5-day-old embryos, are considered pluripotent: they can give rise to all cell types in the body, but not to cells in tissues outside the embryo. At 5 days old, the embryo is still just a hollow ball of cells, termed a blastocyst; ESCs are obtained from the inner cell mass of the blastocyst.

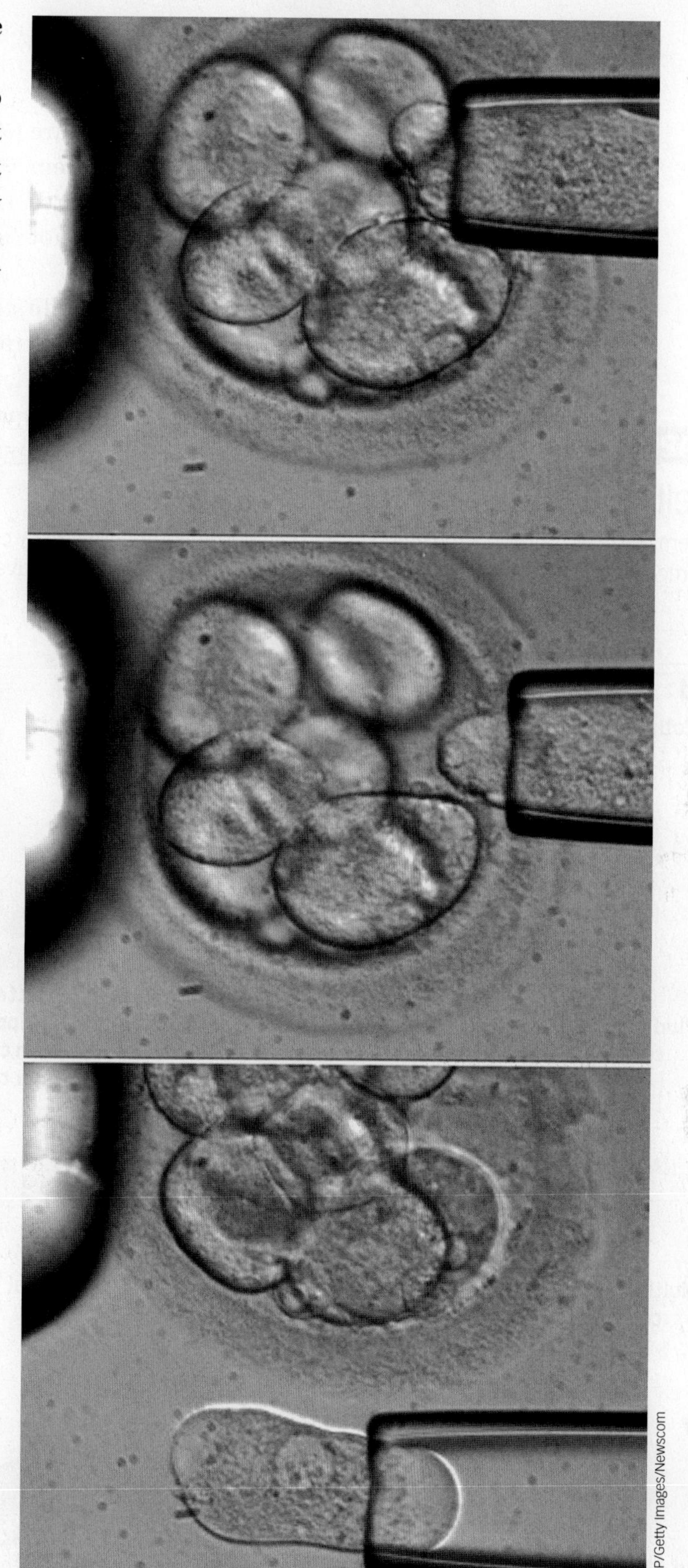

AFP/Getty Images/Newscom

Harvesting embryonic stem cells from an eight-cell embryo.

Some adult tissues also contain stem cells. These cells are multipotent: they can give rise to only a few possible types, typically cells of the tissue in which they are found. For example, blood-forming stem cells found in the bone marrow can produce different types of blood cells, but not brain or heart cells.

Stem cells in adult tissues provide replacements for cells that have reached the end of their life span or are damaged in an injury. Adult stem cells can continue to supply cell replacements throughout a lifetime because they are never used up. When a stem cell divides, one of the daughter cell "grows up" into a specific cell type, while the other daughter cell remains a stem cell and can divide again.

The process by which a cell grows up to have a defined identity, like a neuron or

INFOGRAPHIC M8.1

Stem Cells and Cell Differentiation

Embryonic stem cells are found in early embryos and can develop into almost any cell type, making them attractive tools for regenerative medicine. As cells become more specialized (differentiated), they have less ability to develop into different cell types.

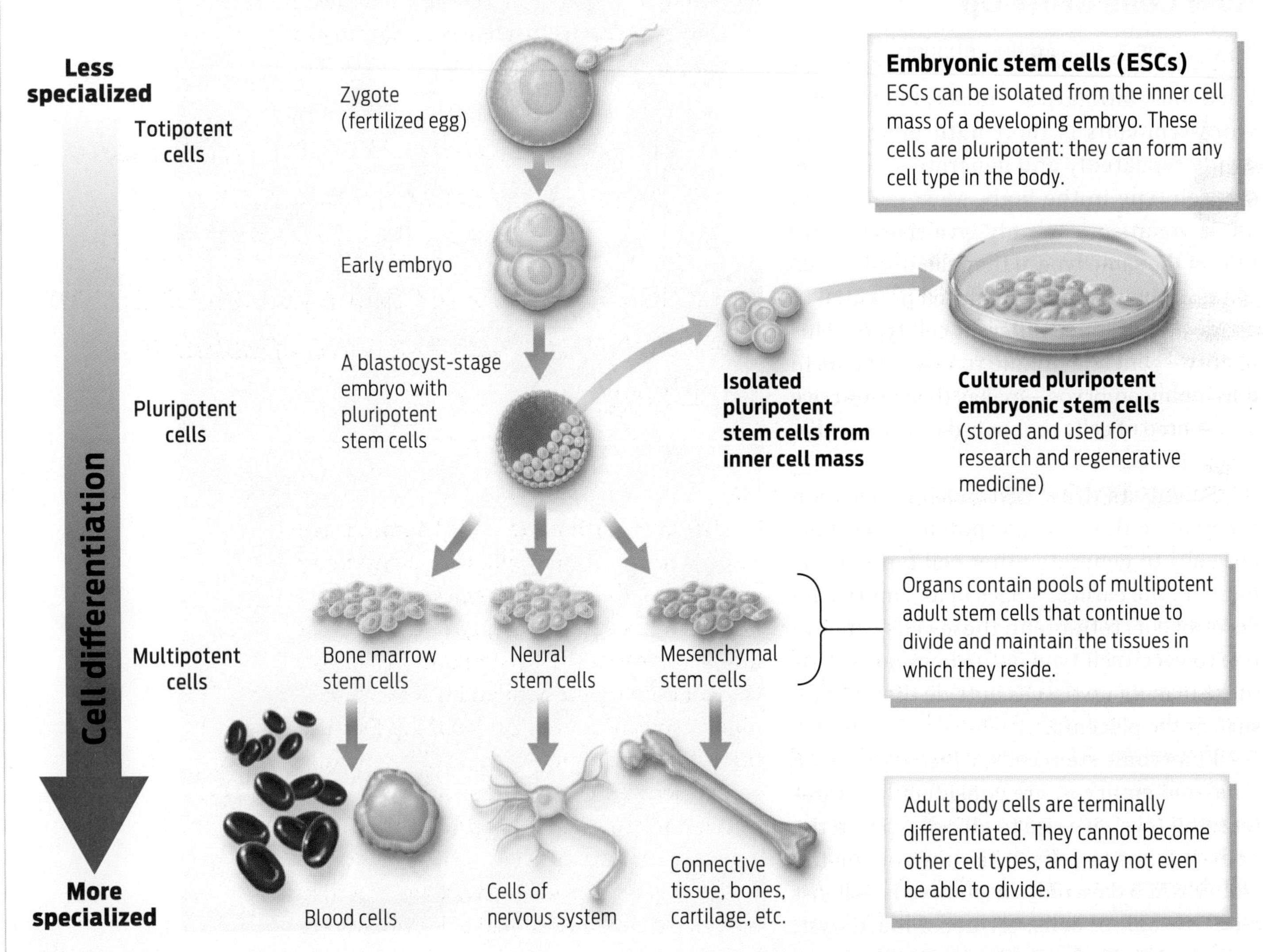

? **Characterize each of the following as multipotent, pluripotent, or totipotent: zygote, bone marrow stem cell, embryonic stem cell.**

a heart cell, is called cell differentiation. Like potency, cell differentiation falls along a spectrum—in this case, from less to more differentiated.

You can think of the difference between an undifferentiated cell and a differentiated cell as the difference between a teenager who hasn't picked a career yet and a doctor with advanced degrees who specializes in cardiology. One has a lot of potential and could follow many different career paths, while the other is highly committed to one path. Similarly, cells must choose early on in development which path they will follow.

Stem cells remain undifferentiated throughout their life. By contrast, most cells in adult tissues are terminally differentiated. This means they have very little potential to divide and differentiate into other cell types. In fact, some, like neurons or cardiac (heart) muscle cells, may never divide at all **(INFOGRAPHIC M8.1)**.

The idea behind regenerative medicine is that embryonic stem cells could be used to produce a limitless supply of pluripotent daughter cells that can differentiate into the specific types of cells needed to heal a damaged tissue. ESCs might even be used to create an entire organ.

Turning Back the Clock

▸ Cloning, gene expression, induced pluripotent stem cells

Once a cell has differentiated, it does not typically go in the other direction—back toward becoming an undifferentiated cell. Is that because it can't? Do specialized cells permanently lose the ability to become another cell type? Is their genetic material altered in some way so that it no longer contains the information necessary to change the cell's identity? Many scientists in the 1950s thought the answer to these questions was likely "Yes."

The first person to answer the question definitively was John Gurdon, a developmental biologist at Oxford University in England. In 1962, he scooped the nucleus out of a cell taken from a tadpole's intestine and slipped it into a frog egg whose own nucleus had been removed. (Recall that the nucleus contains a cell's genes, while the cytoplasm contains proteins and other molecules.) The egg with the donated nucleus developed into a healthy tadpole.

This landmark experiment showed that specialized cells (like the intestinal cells used in this experiment) contain all the genetic information in their nuclei necessary to make a complete frog. Cells do not lose this genetic information as they differentiate, as had previously been thought. Moreover, Gurdon's experiment showed that the egg's cytoplasm provides the signals that tell a cell how to use its genetic material.

You can think of the difference between an undifferentiated cell and a differentiated cell as the difference between a teenager who hasn't picked a career yet and a doctor with advanced degrees who specializes in cardiology.

Gurdon's technique, called somatic cell nuclear transfer, has enabled scientists to create animals from adult cells without having to join sperm and egg. Perhaps most famously, in 1997, Ian Wilmut at the University of Edinburgh in Scotland used this technique to create a sheep named Dolly. Because the offspring animals produced by somatic cell nuclear transfer are genetically identical to the nucleus donor, this procedure is also known as cloning.

Somatic cell nuclear transfer can also be used to generate cloned embryonic stem cells for use in regenerative medicine. In this case, the nucleus from a cell taken from a patient is inserted into a donor egg to form an embryo. The embryonic stem cells are removed from the embryo, which is never implanted. The embryonic stem cells can then be used to generate replacements for a patient's damaged or diseased cells **(INFOGRAPHIC M8.2).**

Gurdon's frog experiment showed that adult cells contain the genetic information necessary to make other cell types, even if they usually do not do so. But then how do cells remember what type of cell they are? Why doesn't a neuron one day decide to become a heart cell?

INFOGRAPHIC M8.2

Somatic Cell Nuclear Transfer Produces Cloned Embryonic Stem Cells

Somatic cell nuclear transfer, or cloning, involves replacing the nucleus of an egg with a nucleus from a specialized cell, creating an embryo that is genetically identical to the donor cell. John Gurdon used this technique to show that specialized cells contained all the genetic information necessary for the development of an embryo and eventually an adult organism. Ian Wilmut famously used this technique to clone Dolly the sheep.

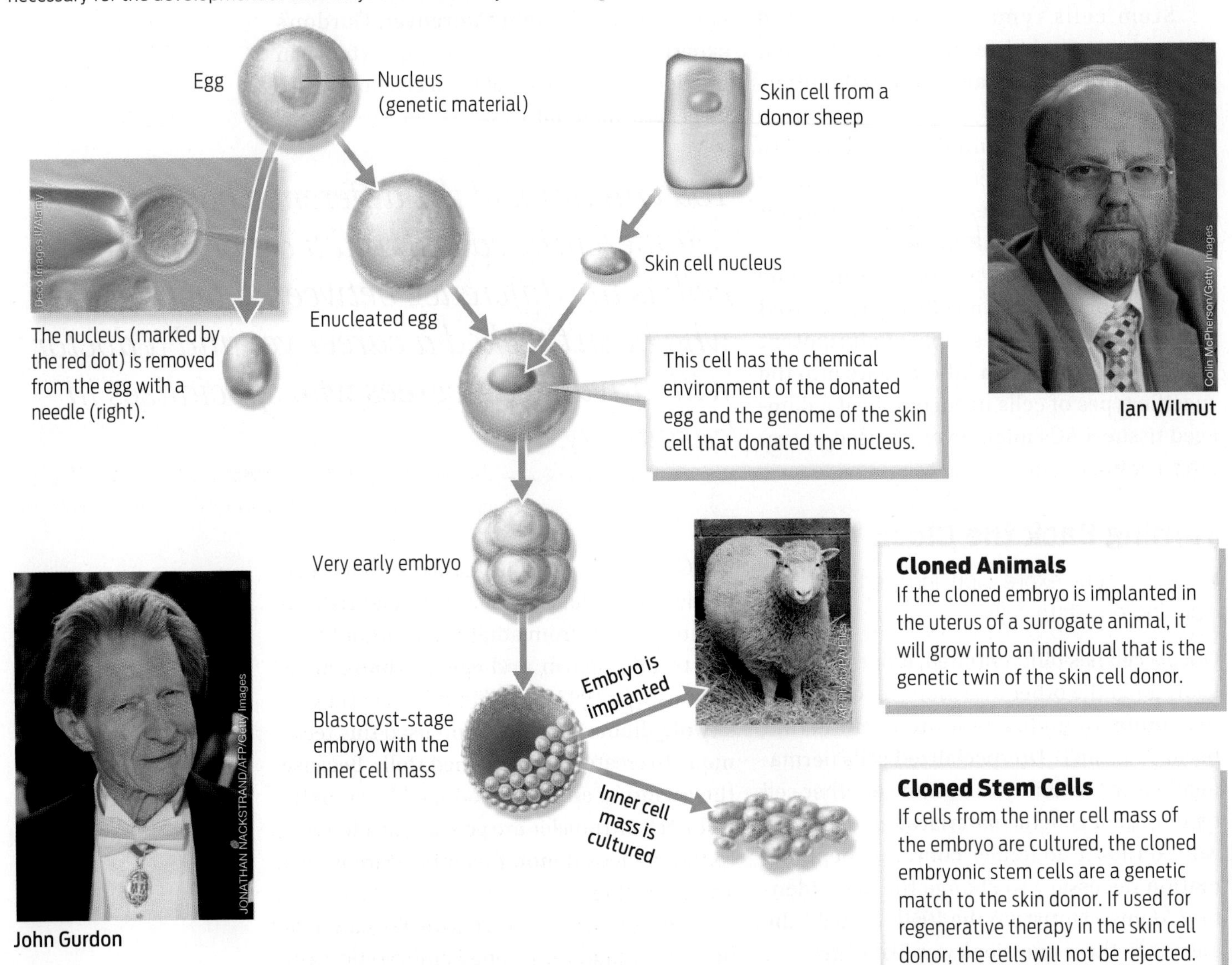

? A nucleus is removed from an egg taken from a Labrador retriever, and is replaced by a nucleus taken from a skin cell of a chihuahua. The egg is implanted in the uterus of a greyhound. What kind of puppy will the greyhound give birth to?

INFOGRAPHIC M8.3

Differentiated Cells Express Different Sets of Genes

Every cell in your body has the same genes, or genome. What distinguishes one cell type from another is the pattern of gene expression and, consequently, the proteins each cell makes. A muscle cell makes a different set of proteins than a B cell, a type of immune-system cell. Muscle cells, for example, express large amounts of actin and myosin proteins, which help muscles contract, whereas B cells express high levels of antibody proteins, which help the body fight infections.

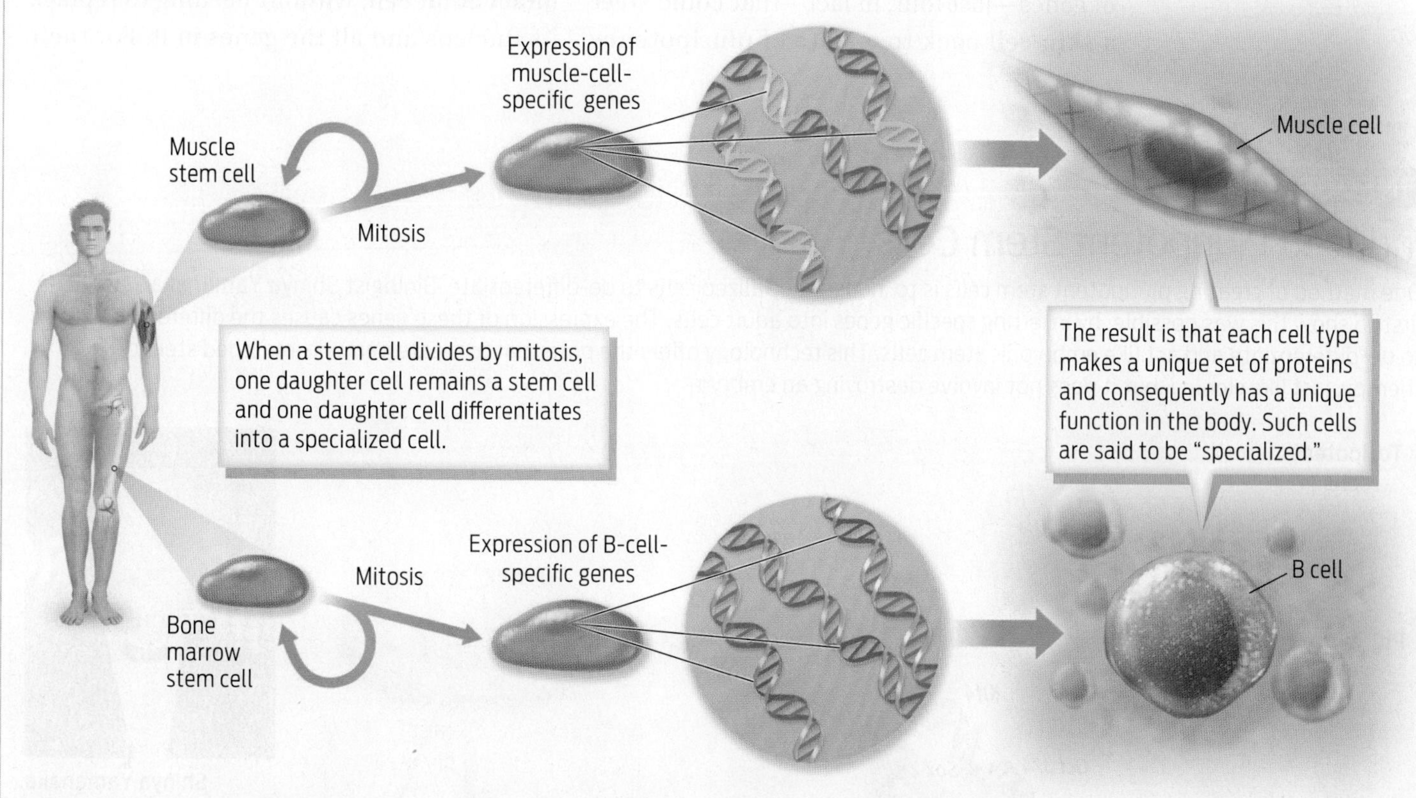

? A specific form of myosin called myosin heavy-chain beta is found primarily in cardiac (heart) muscle. Which of the following cell types would you expect to have the gene encoding myosin heavy-chain beta: embryonic stem cells, bone marrow stem cells, B cells, cardiac muscle cells?

It would be a few years before scientists figured out the answer to this question. They now know that the difference between cell types comes down to which genes are turned on in those cells. While all cells contain the same genes, not all of these genes are turned on in every cell. What distinguishes one cell type from another are the proteins made from actively expressed genes. Muscle cells, for example, make a protein called myosin that allows muscle fibers to contract. Immune cells make proteins called antibodies that protect us from disease. What specifies a cell's identity is the particular combination of genes that are turned on (or off) in that cell **(INFOGRAPHIC M8.3)**.

If what distinguishes a differentiated cell from an undifferentiated cell is the genes that are turned on in each, might it be possible to cue an adult cell to become a stem cell by turning on or off certain genes in that cell?

That was the question that biologist Shinya Yamanaka at Kyoto University began investigating in the mid-2000s. He first looked for genes that are turned on in a stem cell but turned off in a skin cell (a type of differentiated cell). From these genes, he identified a long list of potential genes that he thought might be most important to maintaining the cell's status as a stem cell. Then, Yamanaka began transferring copies of these genes into skin cells, using viruses as vectors. He noted those genes that, when expressed, were able to alter the characteristics of the skin cell. Eventually, in 2006, he identified a cocktail of genes—just four, in fact—that could steer a skin cell back to a state of pluripotency when they were turned on. He called these cells induced pluripotent stem cells (iPSCs) **(INFOGRAPHIC M8.4)**.

More than 40 years after Gurdon's discovery in frogs, Yamanaka had shown it was possible to turn back the clock on an intact adult cell, without needing to replace its nucleus and all the genes in it. For their

INFOGRAPHIC M8.4

Induced Pluripotent Stem Cells

One method of creating pluripotent stem cells is to induce specialized cells to de-differentiate. Biologist Shinya Yamanaka was the first to show this was possible, by inserting specific genes into adult cells. The expression of these genes causes the differentiated cells to de-differentiate and act like embryonic stem cells. This technology offers the potential to create immune-matched stem cells for therapy, just like cloning, but it does not involve destroying an embryo.

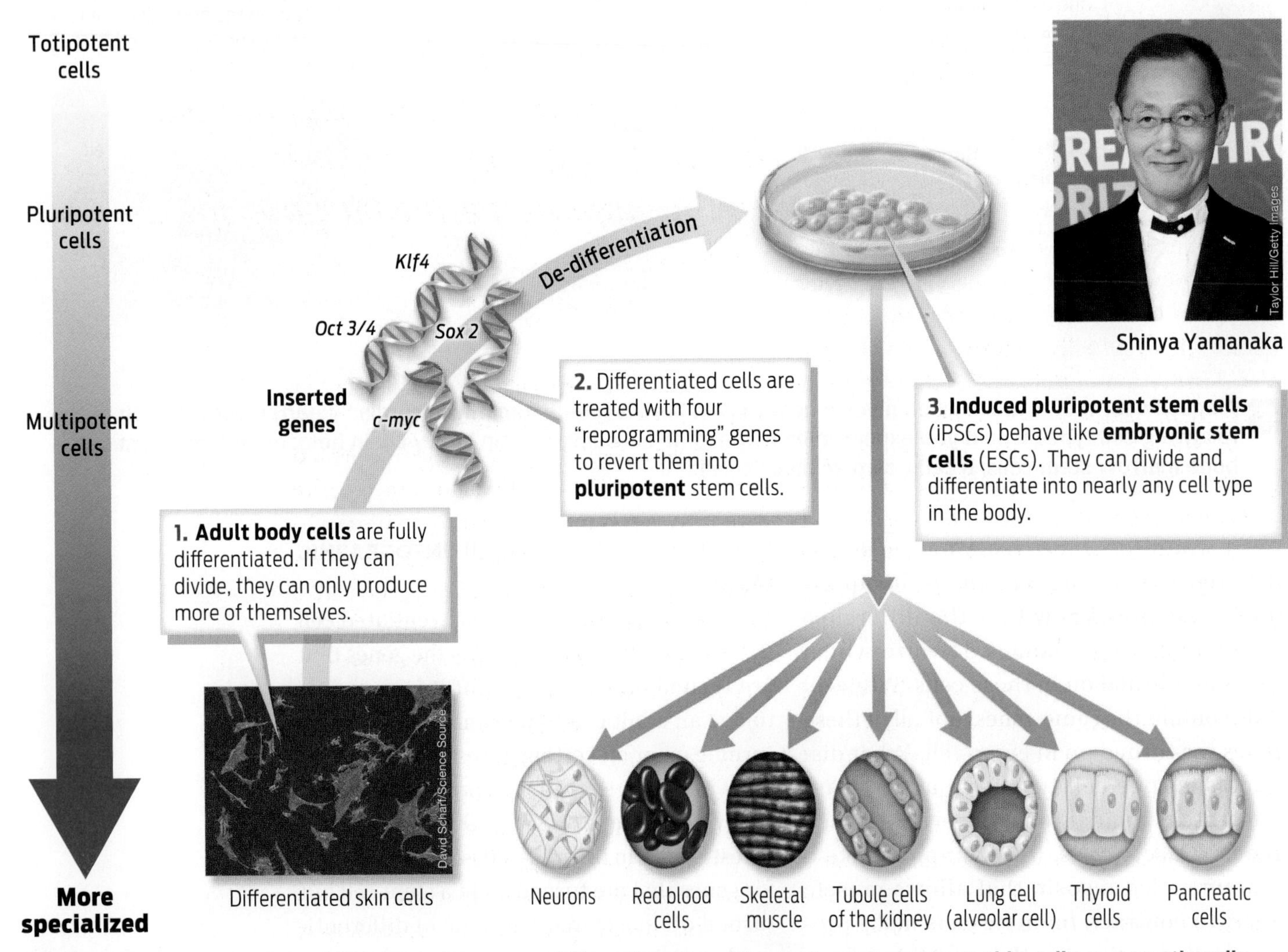

? Characterize each of the following cell types as differentiated, multipotent, or pluripotent: skin cells, pancreatic cells, iPSCs.

work, separated by decades, the two scientists shared the 2012 Nobel Prize in Physiology or Medicine.

Matters of Life and Death

▶ Stem cell uses and controversies

If embryonic stem cells already have the ability to differentiate into many cell types, why bother trying to create pluripotent cells from adult cells?

This question goes to the heart of the ethical debate surrounding the use of embryonic stem cells in research and medicine. ESCs come from embryos—embryos that have the potential to become future people, if they were implanted in a woman's womb. The ESCs that currently exist for research and medical purposes all came from embryos originally obtained from fertility clinics. Many people oppose experimentation on ESCs because isolating those ESCs involves destroying a human embryo, and therefore a potential life.

On the other side of the debate are many other people, including many scientists, who point out that unused embryos in fertility clinics will eventually be thrown away, which also destroys potential life. So why not make use of them to lessen human suffering?

There is no one right answer to this question, only different views of what is ethical according to one's moral principles. But the intensity of the debate eventually forced Congress to act to regulate the use of human embryos in research.

In 1995, Congress passed the Dickey-Wicker Amendment, which prohibits federal funds from being used to create embryos for the purpose of research. The amendment was signed into law by President Bill Clinton in 1996. Left open was the question of whether federal funds could be used to fund research on embryos that were left over from fertility treatments. Scientists used such embryos to isolate and grow the first human ESCs in 1998.

Soon thereafter, in 2001, President George W. Bush issued a ban on the use of federal funds to create ESCs. His decision limited research to ESC lines that already existed at the time of the ban—those in which "the life-and-death decision has already been made," he stated. Just a handful of workable ESC lines existed at that time.

To get around this moral conundrum while still allowing the science of regenerative medicine to progress, scientists sought ways to create pluripotent stem cells that didn't come from embryos. Yamanaka's breakthrough in creating stem cells derived from specialized adult cells (iPSCs) was welcomed by everyone involved in the debate.

In 1995, Congress passed the Dickey-Wicker Amendment, which prohibits federal funds from being used to create embryos for the purpose of research.

And indeed, stem cells derived from adult cells have proved incredibly useful. With these cells, scientists have created so-called diseases in a dish—cells that contain the genetic abnormalities of diseases. Researchers can then test potential drugs on cells modeling these conditions. iPSCs have also been used to develop cell therapies to replace cells lost in diseases such as Parkinson's disease, which results from the destruction of specific cell types in the brain. In some cases, the genetic defects causing the disease may be edited out before the cells are returned to the patient. iPSCs can even be used as the starting point to grow artificial organs. In each case, because iPSCs are derived from the specific person to be treated, they will not be rejected by the immune system, which is a big plus in their favor compared

INFOGRAPHIC M8.5

Applications of Induced Pluripotent Stem Cells

Induced pluripotent stem cells (iPSCs) have the potential to advance our understanding and treatment of human disease and assisted reproduction. iPSCs from an individual with a tissue injury or disease can be used to generate specialized differentiated cells that can replace cells lost in the injury or due to the disease. Therapeutic drugs can be tested on these specialized cells in the lab. For people with a genetic disease, the mutation can be corrected in iPSCs by gene editing. The corrected iPSCs can be used to generate specialized differentiated cells to be used in treatment. iPSCs can also be used to generate gametes, used in assisted reproduction.

Disease Modeling and Drug Testing

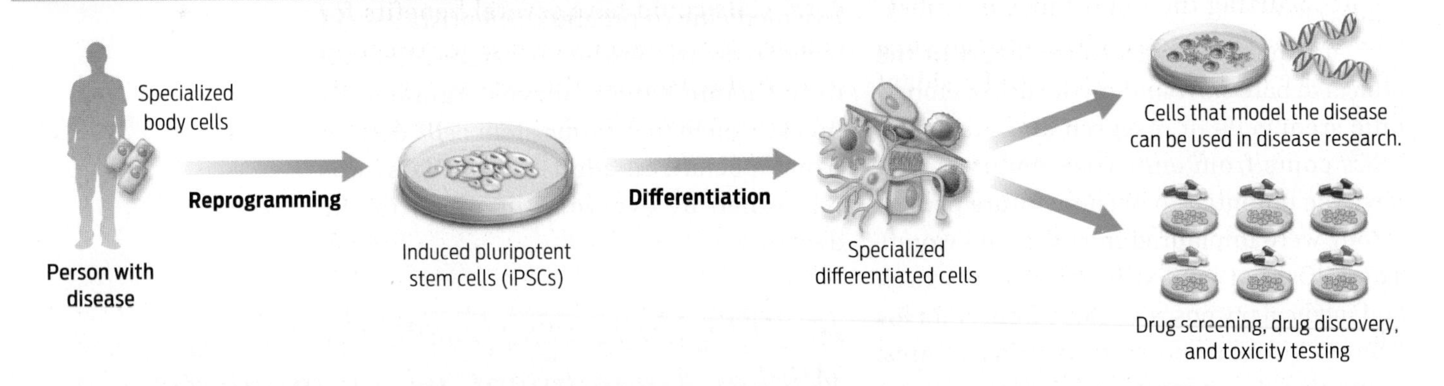

Cell Therapy

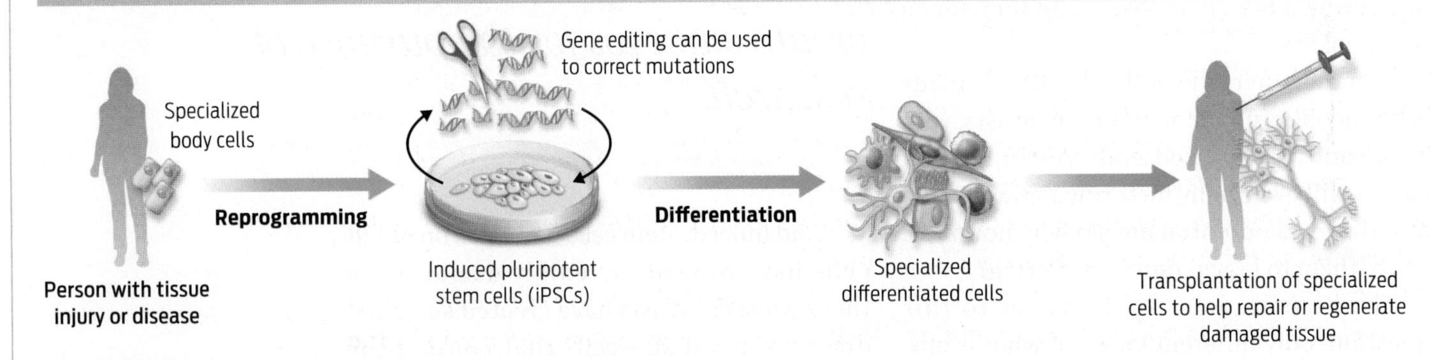

In Vitro Gametogenesis

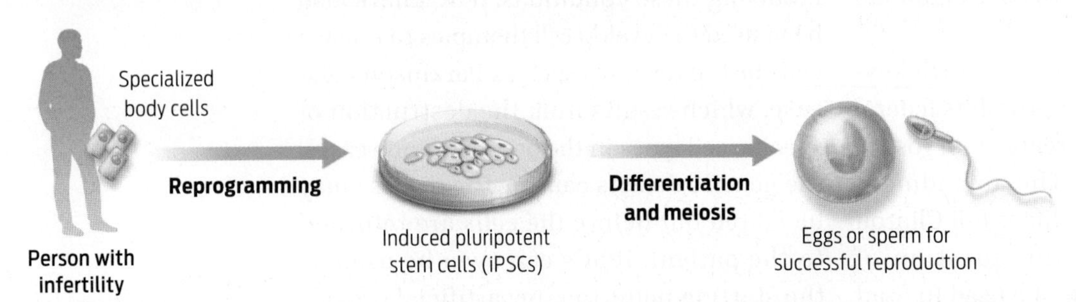

? A person has an inherited form of Alzheimer's disease. Which iPSC applications would be beneficial to this person? Explain your answer.

to replacement cells derived from embryonic stem cells (**INFOGRAPHIC M8.5**).

Unfortunately, stem cells derived from adult cells have some drawbacks when compared with embryonic stem cells. They are not quite as flexible in their potential to differentiate into different cell types as ESCs, and they have a few unexpected traits, such as a tendency to form cancer, that may limit their use in regenerative medicine.

Recognizing the importance of embryonic stem cells for research and coming to a different ethical conclusion than President Bush, President Barack Obama in 2009 loosened the restrictions on federal funding of ESC research. The new guidelines allow researchers to use federal funds to conduct research on new ESC lines, but they still forbid researchers to use federal money to derive ESCs. Scientists who are interested in creating new ESC lines must obtain private or state funding. California is a state leader in this type of funding.

Reproduction without Sex?

Considering the controversy that greeted Louise Brown's birth, it is perhaps not surprising that of all the potential uses of stem cells, the one that raises the most ethical questions is in vitro gametogenesis—creating a sperm or an egg from adult cells, such as a skin or hair cell.

If an adult cell can be reprogrammed to become a stem cell, can that stem cell be converted into a sperm or an egg? In principle, yes. It's a bit trickier because both sperm and eggs are haploid, whereas a skin or hair cell is diploid. A stem cell (either an ESC or iPSC) would need to be prodded to undergo meiosis in addition to differentiating into the germ cells. But nothing prevents that from being a realistic objective. In fact, researchers have already achieved success with this technique in mice. Many hurdles, technical and moral, still stand in the way of using the technique in humans, but that is clearly a goal.

The ability to convert stem cells to germ cells could have several benefits for couples. Some couples cannot make sperm or eggs due to infertility or cancer treatment; they would be able to produce germ cells from adult cells obtained from their body. Same-sex couples could produce offspring that are related to both parents. (Any embryos made by in vitro fertilization of the "generated" gametes would still need to be implanted into a woman's uterus to grow.)

Other possible uses are more sinister. For example, this technique could potentially be used to generate germ cells from someone without their consent. A skin cell retrieved from a celebrity's soda can, for example, might be used to create a baby the celebrity didn't even knew they had. Even single-parent children might one day be possible.

Will these additional forms of reproductive technology become common in the future? They very well could be. It will be up to bioethicists, lawyers, and voters to decide what reproductive future we want, before the technologies become a reality. ■

More to Explore

- The Hastings Center: https://www.thehastingscenter.org/briefingbook/stem-cells/
- Brown, L. (2015). *My Life as the World's First Test-Tube Baby.* Bristol: Bristol Books.
- Cohen, I. G., and Pearlman, A. (June 5, 2019). Creating eggs and sperm from stem cells: The next big thing in assisted reproduction? https://www.statnews.com/2019/06/05/creating-eggs-sperm-stem-cells/
- Rehm, J. (October 11, 2018). Healthy mice from same-sex parents have their own pups. *Nature*. https://www.nature.com/articles/d41586-018-06999-6

MILESTONES IN BIOLOGY 8 Test Your Knowledge

1. Put the following cell types in order, from the broadest to the narrowest potential:
   __ Multipotent adult stem cells
   __ Pluripotent embryonic stem cells
   __ Totipotent cells from an early embryo

2. Embryonic stem cells are best described as
   a. multipotent.
   b. pluripotent.
   c. totipotent.
   d. all of the above

3. To cure type 1 diabetes, specific pancreatic cells need to be replaced. Which cell type would be the most appropriate to generate the replacement cells—ESCs or multipotent blood stem cells? Explain your answer.

4. A muscle cell produces high levels of myosin; it does not produce antibodies. If the muscle cell is induced to become an iPSC, and the iPSC is coaxed to specialize into a B cell, will the B cell produce antibodies? Explain your answer.

32 Immune System

THE FORGOTTEN PLAGUE

After nearly a century, scientists learn what made the 1918 influenza virus so deadly

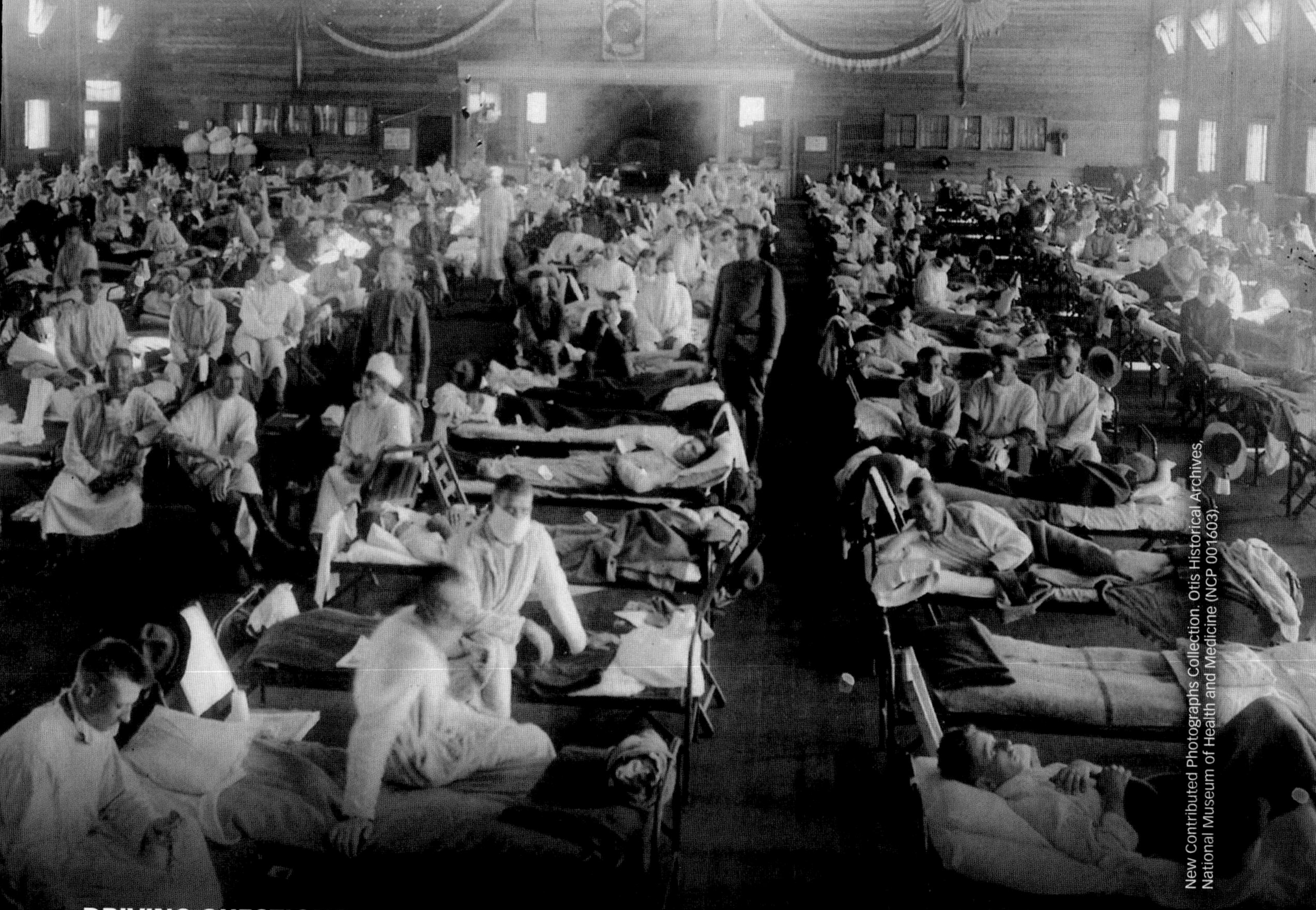

New Contributed Photographs Collection. Otis Historical Archives, National Museum of Health and Medicine (NCP 001603).

DRIVING QUESTIONS

1 What is the structure of a virus, and how do viruses cause disease?

2 Why is nonspecific immunity important even though another system provides both specificity and memory?

3 What is adaptive immunity, and how does vaccination rely on adaptive immunity?

4 What are the features of the influenza virus that allow it to cause worldwide outbreaks?

Time Life Pictures/Getty Images

Police in Seattle, Washington, keeping the peace during the 1918 flu pandemic.

AP Photo/Seth Wenig

Police in New York City wearing masks during the 2020 COVID-19 pandemic.

In the fall of 1918, World War I was coming to an end—but another, more insidious threat was surfacing. In communities around the world, a deadly illness was gaining momentum, spreading like wildfire from person to person.

At first the disease seemed much like a bad case of influenza, or "flu." High fevers and body aches were common. But some patients developed more severe symptoms, such as coughing so intense that they spat up blood. Many died so quickly after falling ill that doctors began to suspect they were witnessing something far worse than ordinary flu.

An army physician stationed at Camp Devens in Massachusetts wrote to a friend that soldiers would rapidly "develop the most vicious type of pneumonia that has ever been seen" and later would "struggle for air until they suffocated." A physician from Philadelphia wrote that his patients "died struggling to clear their airways of a blood-tinged froth that sometimes gushed from their nose and mouth."

Between 1918 and 1920, the mysterious disease swept around the globe. Had the world not been at war, the scourge might never have spread so far. Troops carried the illness with them across Europe and back to the United States, infecting relatives, friends, and strangers along the way.

The result was one of the deadliest **pandemics,** or global epidemics, in human history. According to historian John Barry, the 1918 flu pandemic killed more people in 1 year than the Black Death killed in the entire 14th century; it killed more people in 25 weeks than AIDS killed in 25 years.

PANDEMIC
A worldwide outbreak of disease; a global epidemic.

The 1918 flu pandemic killed more people in 1 year than the Black Death killed in the entire 14th century; it killed more people in 25 weeks than AIDS killed in 25 years.

By the time it was over, between 50 and 100 million people around the world had perished—more than the total number killed in World War I **(INFOGRAPHIC 32.1)**.

Why this particular strain of flu was deadlier than any pandemic that came before (or after), scientists of the time couldn't say. For decades after the pandemic had subsided, scientists thought they would never be able to explain the flu's severity, since the virus itself was lost to history. But recent discoveries have made it possible to peer into the viral past, and scientists have begun to uncover what made the 1918 virus such a relentless killer.

INFOGRAPHIC 32.1

Pandemic Threats throughout History

There have been many deadly outbreaks caused by viruses and bacteria. To date, the Spanish flu pandemic remains the deadliest, having caused at least 50 million deaths.

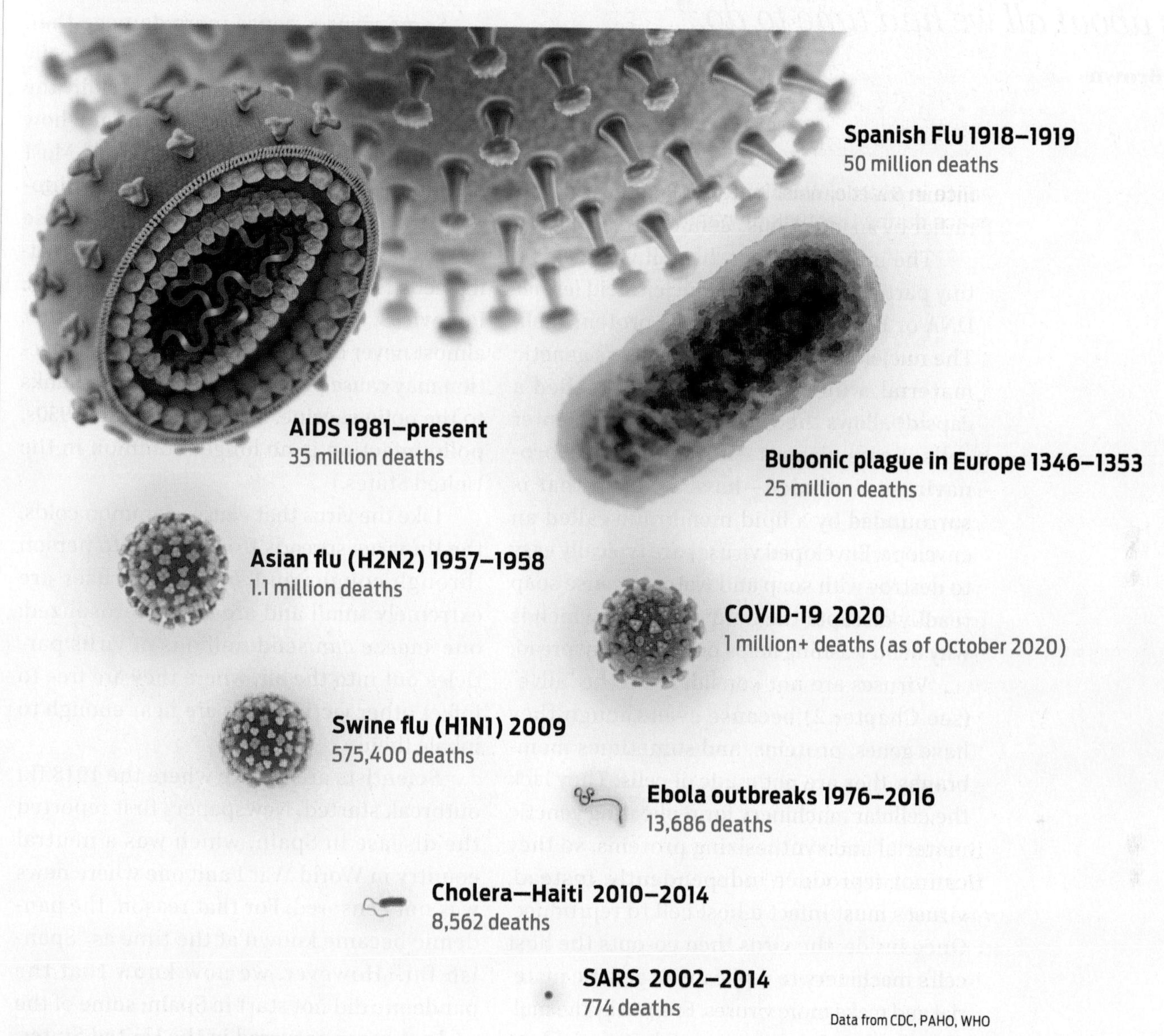

? **If the world population in 1918 was approximately 1.8 billion people, what percentage of the population was killed by the Spanish flu?**

Silent Invader

▶ Virus structure and replication

Each year in the United States, between 12,000 and 60,000 people die from complications of flu. Most of these deaths strike young children and the elderly, whose bodies are less able to fight off the infection. The 1918 flu, by contrast, struck down people in the prime of life: people age 20 to 40 were the hardest hit. The death toll in this age group was so severe that it reduced the average U.S. life expectancy in 1918 by 12 years.

At the time, doctors did not know what caused influenza. Some suspected that it was caused by a bacterium, as were many other known diseases, such as cholera and tuberculosis. But research in 1933 showed that

"We didn't take temperatures; we didn't even have time to take blood pressure. We would give them a little hot whisky toddy; that's about all we had time to do."

—Josie Brown

influenza is caused not by a bacterium, but rather by an infectious agent called a **virus.**

The influenza virus, like all viruses, is a tiny particle made up of a nucleic acid (either DNA or RNA) surrounded by a protein shell. The nucleic acid serves as the virus's genetic material, while the protein shell (called a capsid) allows the virus to bind to and enter cells. Some viruses—influenza and coronaviruses included—have a capsid that is surrounded by a lipid membrane called an envelope. Enveloped viruses are typically easy to destroy with soap and water because soap readily disrupts the lipid envelope, which is why hand washing helps prevent their spread.

Viruses are not considered to be "alive" (see Chapter 2) because even though they have genes, proteins, and sometimes membranes, they are not made of cells. They lack the cellular machinery for replicating genetic material and synthesizing proteins, so they cannot reproduce independently. Instead, viruses must infect a host cell to reproduce. Once inside, the virus then co-opts the host cell's machinery to replicate its genetic material and make more viruses. Eventually the multiplied viruses burst from and destroy the host cell **(INFOGRAPHIC 32.2).**

Unlike cells, which rely on DNA as their hereditary material, viruses can store their genetic information in the form of either DNA or RNA (see Chapter 2). RNA viruses include the influenza virus; coronaviruses like those that cause the common cold, severe acute respiratory syndrome (SARS), and COVID-19; the viruses that cause measles and mumps; and those responsible for AIDS and polio. Diseases caused by DNA viruses include hepatitis B, chickenpox, and herpes. Each type of virus has a distinctive protein shell with a characteristic shape. The flu virus, for example, has a spherical capsid surrounded by a lipid envelope with knob-like proteins jutting out of it.

Some viruses cause more damage than others. The severity of a viral illness typically depends on how quickly and successfully the body's defenses rally to kill the virus and how well the infected tissue can repair itself. Most people recover completely from the rhinovirus that causes the common cold because the respiratory tract contains rapidly dividing cells that quickly replace damaged ones. Poliovirus, however, attacks nerve cells that almost never divide, which is why polio infection may cause permanent paralysis. (Thanks to the polio vaccine, introduced in the 1950s, polio infection is no longer common in the United States.)

Like the virus that causes common colds, the flu virus spreads from person to person through coughs and sneezes. Viruses are extremely small and are easily aerosolized: one sneeze can send millions of virus particles out into the air, where they are free to infect other victims who are near enough to inhale them.

Scientists aren't sure where the 1918 flu outbreak started. Newspapers first reported the disease in Spain, which was a neutral country in World War I and one where news was not censored. For that reason, the pandemic became known at the time as "Spanish flu." However, we now know that the pandemic did not start in Spain; some of the earliest cases occurred in the United States, among soldiers housed in close quarters. Such U.S. military bases were also where Spanish flu first demonstrated its deadly power.

"We didn't have the time to treat them," recalled Josie Brown, a nurse stationed at a naval base in Illinois in 1918. "We didn't take temperatures; we didn't even have time to take blood pressure. We would give them a little hot whisky toddy; that's about all we had time to do."

VIRUS
A noncellular infectious particle consisting of a nucleic acid (DNA or RNA) surrounded by a protein shell.

A Microscopic Battleground

▶ Pathogens and immune defenses

Some early clues to what made the 1918 flu infection so deadly came from examining bodies of the deceased. Autopsies performed at the time showed that the infection caused extensive damage to the lungs. In severe cases, the lungs looked torn apart, as if ravaged by shrapnel. The lungs, it seems, were the central battleground of the war between the influenza virus and the **immune system** of its human host.

Like an army, the immune system defends the body from invaders—principally infectious agents such as viruses, bacteria, and parasites that enter the body and cause disease. Infectious agents that cause disease are called **pathogens.** In addition, the immune system plays an important role in helping our bodies heal from injuries, and even protects us as against cancer, which can be thought of as treasonous cells. The immune system basically reacts against anything it encounters as foreign, or "nonself," and tries to destroy it **(INFOGRAPHIC 32.3).**

IMMUNE SYSTEM
A network of cells and tissues that acts to defend the body against infectious agents and helps to heal injuries.

PATHOGENS
Infectious agents, including certain viruses, bacteria, fungi, and parasites, that may cause disease.

INFOGRAPHIC 32.2

Viruses Infect and Replicate in Host Cells

Viruses can replicate only within a host cell. Viral genes direct the host cell to synthesize new viral particles. Ultimately, viral replication kills the host cell either by causing it to burst or by depleting the cell's resources.

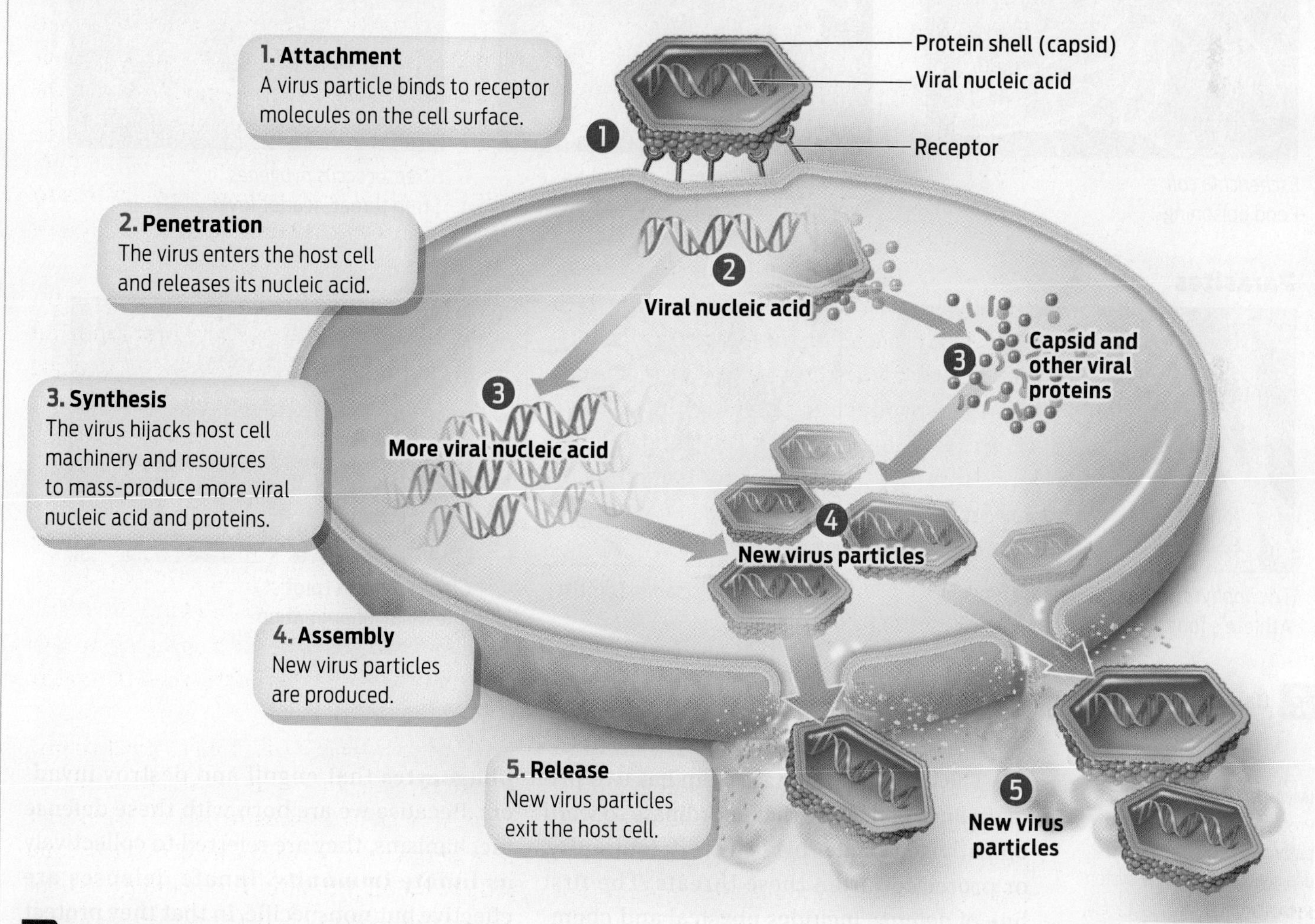

? **If a cell does not have receptor molecules to which a specific virus can bind, can that cell be a host for that virus?**

INFOGRAPHIC 32.3

The Immune System Defends against a Variety of Pathogens

The immune system can protect the body against many specific pathogens. Viruses, bacteria, and parasites can all be detected as foreign, triggering a variety of defenses.

Viruses

Cynthia Goldsmith/CDC Public Health Image Library

Influenza virus
Flu

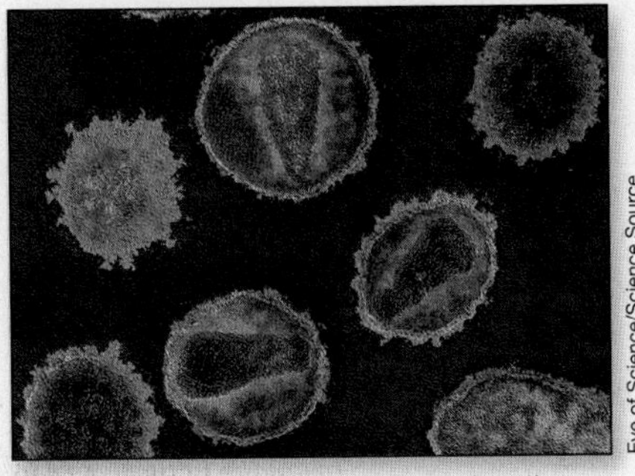

Eye of Science/Science Source

Human immunodeficiency virus (HIV)
AIDS

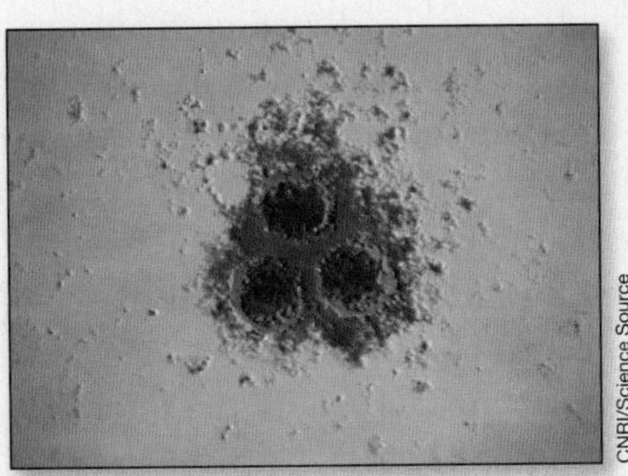

CNRI/Science Source

Varicella zoster virus
Chicken pox

Bacteria

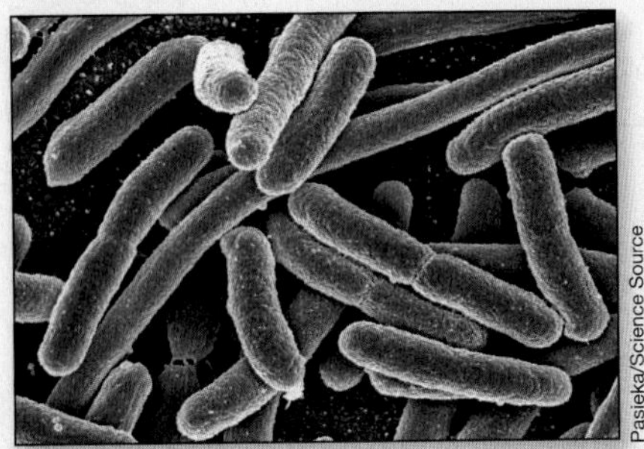

Pasieka/Science Source

Escherichia coli
Food poisoning

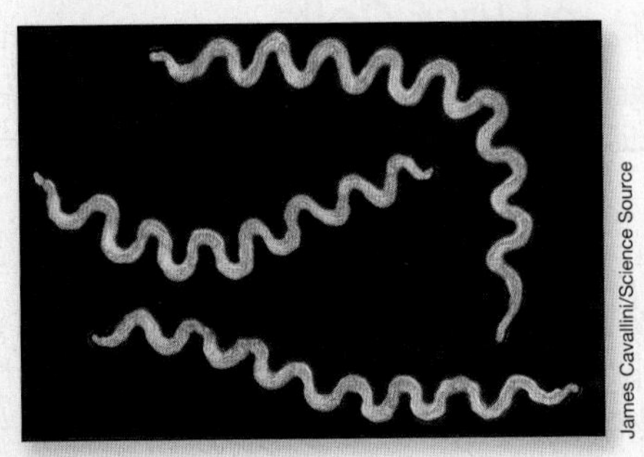

James Cavallini/Science Source

Treponema pallidum
Syphilis

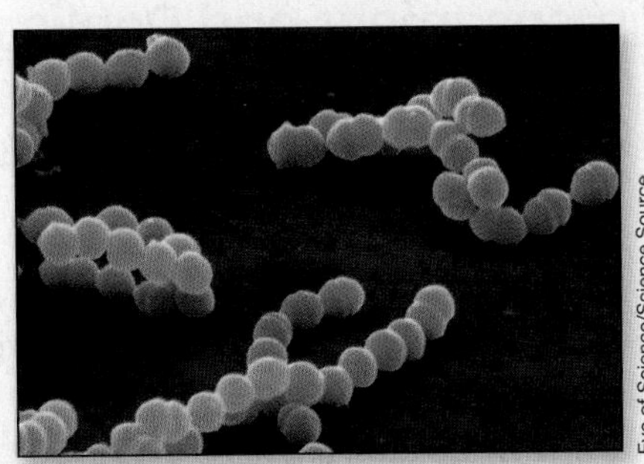

Eye of Science/Science Source

Streptococcus pyogenes
Strep throat, scarlet fever

Parasites

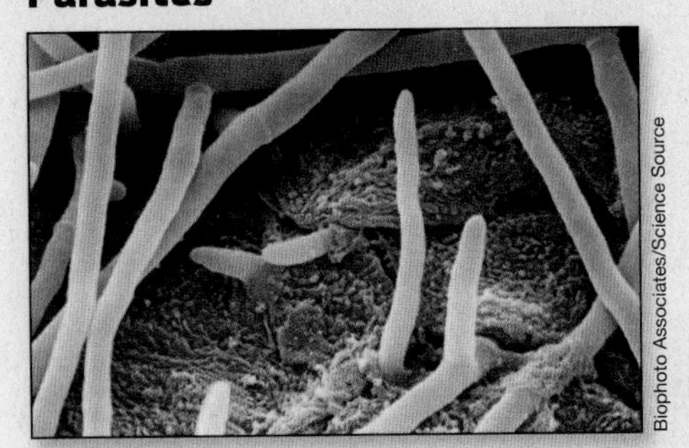

Biophoto Associates/Science Source

Trychophyton mentagrophytes (fungus)
Athlete's foot

NIAID

Plasmodium falciparum (protist)
Malaria

Dr. Stan Erlandsen/CDC/Public Health Image Library

Giardia lamblia (protist)
Intestinal inflammation

? **Can eukaryotic organisms cause disease in humans?**

IMMUNITY
Protection from a given pathogen conferred by the activity of the immune system.

PHAGOCYTE
A type of white blood cell that engulfs and digests pathogens and debris from dead cells.

The human immune system has two primary lines of defense that coordinate to ward off pathogens and provide us with **immunity,** or protection from these threats. The first line of defense includes physical and chemical barriers such as the skin and mucous membranes as well as white blood cells called **phagocytes** that engulf and destroy invaders. Because we are born with these defense mechanisms, they are referred to collectively as **innate immunity.** Innate defenses are effective but nonspecific, in that they protect us against a wide variety of invaders but do not target any one in particular. Phagocytes,

INFOGRAPHIC 32.4

Innate and Adaptive Immunity

Invading pathogens encounter a variety of defenses. Physical and chemical barriers are a first line of defense. Pathogens that breach these barriers encounter nonspecific cellular defenses. These early lines of defense are known as innate immunity, and are always present and ready to protect against a wide range of pathogens. Adaptive immunity, which relies on the activity of specialized white blood cells, is the next line of defense. Adaptive immunity mounts a unique defense against each specific pathogen it encounters.

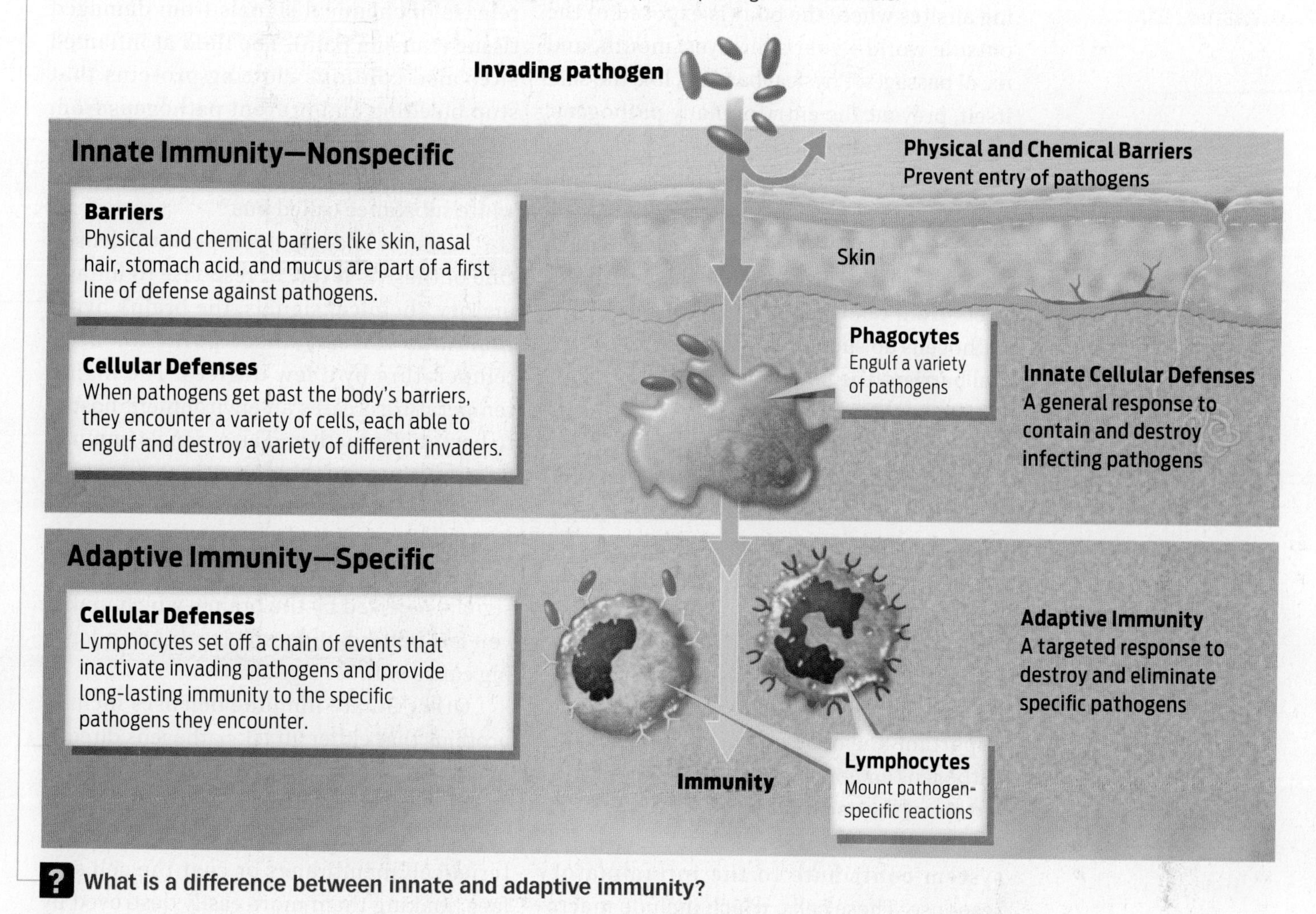

? **What is a difference between innate and adaptive immunity?**

for example, may detect a particular carbohydrate molecule found in many bacterial cell walls, but cannot distinguish among different bacteria. Innate immune defenses are always present and require little or no time to become active.

The second line of defense is **adaptive immunity,** which includes the coordinated actions of specialized white blood cells called **lymphocytes** that take aim at specific pathogens. Adaptive immunity takes time to develop, but it has three main advantages over innate immunity. First, the adaptive immune system is *specific*: it is able to recognize particular pathogens—one type of virus, for example—even if it has never encountered that pathogen before. This feature is important because pathogens evolve quickly, and many have found ways to evade our innate defenses. Second, the adaptive immune system is *diverse*: it can mount an immune response to essentially any threat that comes its way. Finally, the adaptive immune system exhibits *memory*: once you have chickenpox, the activity of the adaptive immune system means you won't get it again. But even with an adaptive immune system, we still need our innate defenses to be in good working order; in fact, the two systems work together to protect us from illness **(INFOGRAPHIC 32.4).**

INNATE IMMUNITY
Nonspecific defenses, such as physical and chemical barriers and specialized white blood cells, that are present from birth and require little or no time to become active.

ADAPTIVE IMMUNITY
A protective response, carried out by lymphocytes, that confers long-lasting immunity against specific pathogens.

LYMPHOCYTE
A specialized white blood cell of the immune system. Lymphocytes are important in adaptive immunity.

First Responders

▶ Innate immunity

The innate immune system starts defending at sites where the body is exposed to the outside world—your skin, eyes, mouth, and nasal passages. Physical barriers, like the skin itself, prevent the entry of many pathogens. Enzymes in saliva destroy some pathogens; nasal hair and the mucus that lines the throat trap others. Under the layer of mucus is a layer of cells with hairlike projections called cilia that sweep away pathogens to prevent them from taking hold in the throat. When pathogens do successfully breach these physical barriers, the body tries to flush them out with more fluid: runny noses, watery eyes, coughs, and sneezes are the body's way of expelling the invaders.

When pathogens overcome these physical and chemical defenses, they trigger additional innate responses. One of the most important is **inflammation,** a suite of reactions that generates redness, heat, swelling, and pain at the affected site ("inflammation" means "setting on fire"). Inflammation is an important line of defense that kills invading pathogens and prevents them from spreading further in the body.

The phagocytes of the innate immune system contribute to the inflammatory response. These cells, which include macrophages ("big eaters"), surround and engulf the invaders in a process called phagocytosis. They also help clean up debris from dead cells and play an important role in promoting an adaptive immune response (a topic to which we'll return later in this chapter).

What triggers inflammation? Tissues that have been damaged by injury or by infection release chemicals such as **histamine** that cause blood vessels to expand and leak fluid into surrounding tissues. These chemical signals also attract phagocytes from the blood and surrounding tissue to the region. The redness, heat, swelling, and pain associated with inflammation are the result of increased blood flow to the area (causing redness and heat), accumulation of fluid and immune cells from the leaky blood vessels (causing swelling), and release of chemical signals from damaged tissue (causing pain). The fluid at inflamed sites also contains clotting proteins that stop bleeding and prevent pathogens from spreading to neighboring tissues. Together, this collection of materials forms a milky white substance called pus.

The body has other nonspecific defenses. One of these is **fever.** In response to inflammatory chemical signals, the brain's hypothalamus raises the set point for body temperature by a few degrees. The higher temperature creates an environment hostile to bacterial growth and also signals immune cells to ramp up their reactions. (Aspirin, ibuprofen, and acetaminophen are anti-inflammatory drugs that help reduce fever and pain by interrupting the inflammatory signals received by the brain—which makes you feel better but doesn't treat the underlying cause of those signals.)

Other innate immune defenses include proteins that either attack pathogens directly or prevent them from reproducing. For example, defensive proteins in the blood called **complement proteins** punch holes in bacterial cell membranes or coat the cell surface, making them more easily destroyed by phagocytes **(INFOGRAPHIC 32.5).**

Though the innate immune system provides some protection against influenza, occasionally virus particles manage to evade the body's physical and chemical barriers and take up residence in the upper respiratory tract: the mouth, the nose, and the throat. Within minutes, these virus particles attach to epithelial cells that line the respiratory tract and slip inside the cells. The virus then hijacks the host cells' cellular machinery to replicate its own genetic material. Generally, about 10 hours after the virus invades a cell, the cell begins to release newly synthesized viral particles, sending out between 1,000 and

INFLAMMATION
An innate immune defense that is activated by infection or tissue damage; characterized by redness, heat, swelling, and pain.

HISTAMINE
A molecule released by damaged tissue and during allergic reactions that promotes inflammation.

FEVER
An elevated body temperature.

COMPLEMENT PROTEINS
Proteins in blood that help destroy pathogens by coating or puncturing them.

INFOGRAPHIC 32.5

Some Important Features of Innate Immunity

Innate immunity is always "on" and ready to contribute to defenses. Innate immunity does not recognize specific pathogens, but relies on physical, cellular, and chemical defense strategies.

Physical barriers

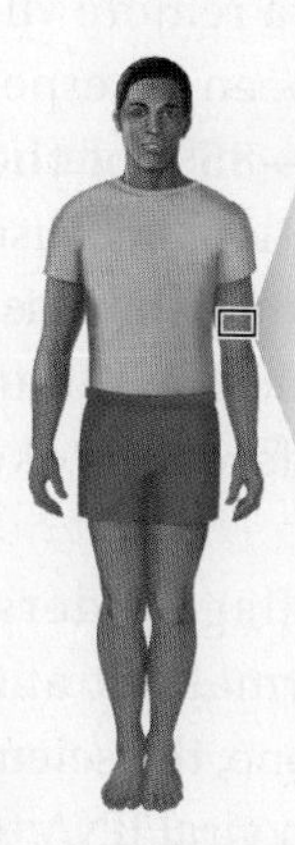

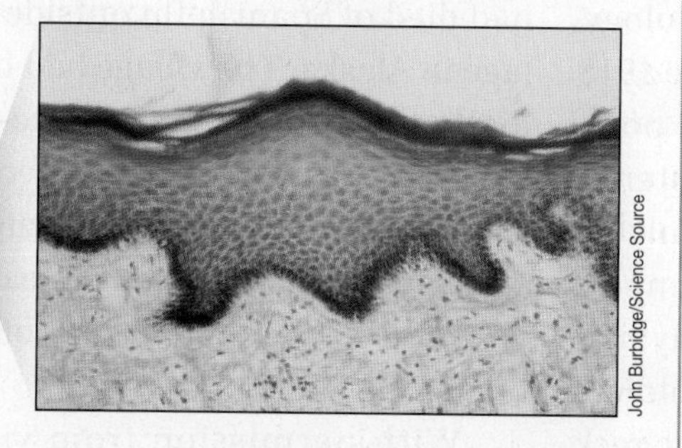

Skin (on the outside of the body) and mucous membranes (lining the inside of the body) have layers of tightly packed cells that prevent pathogens from entering the body. The mucus that coats mucous membranes traps foreign substances.

Chemical defenses

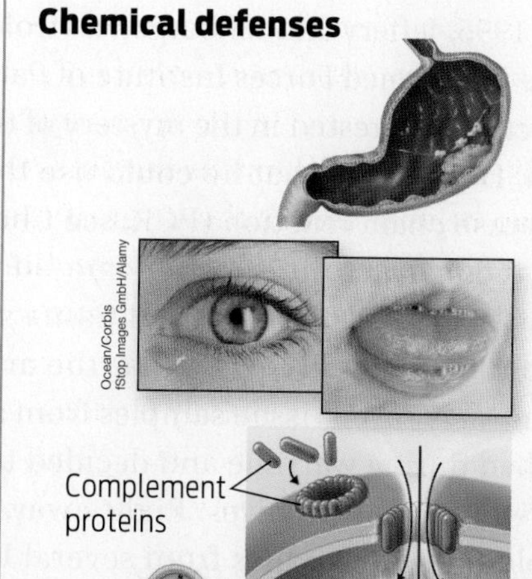

Acid in the stomach kills many of the microorganisms that we ingest.

Tears and **saliva** contain an enzyme that breaks down the cell walls of bacteria, causing them to burst.

Complement proteins in blood punch holes in bacterial cells or coat the cells' surface, making them more easily destroyed by phagocytes.

Fever is induced by chemical signals acting in the brain in response to infection. Fever inhibits bacterial growth and stimulates the immune system.

Phagocytes

Phagocytes ingest and destroy pathogens. Phagocytes also trigger inflammation and adaptive immune responses.

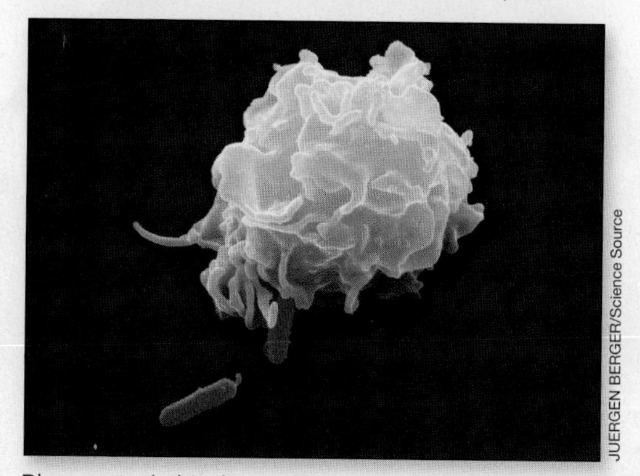

Phagocyte (white) engulfs several pathogens (red).

Inflammation

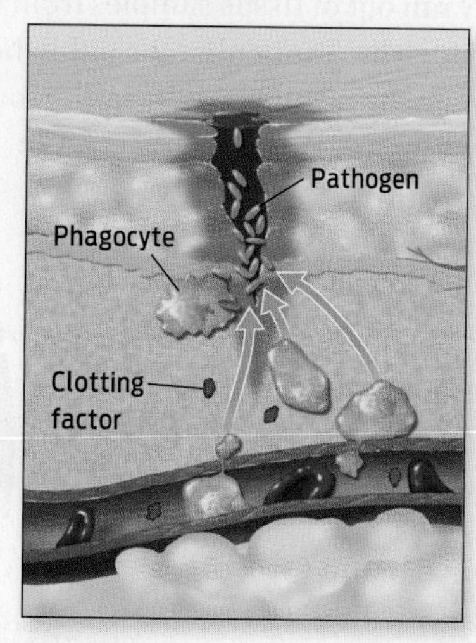

1. Pathogens get past physical and chemical barriers.
2. Pathogens and damaged cells release histamines and other molecules that increase blood flow and attract white blood cells to infected areas.
3. Blood vessels leak, causing surrounding tissue to swell with fluid that contains clotting factors and white blood cells.
4. Phagocytes ingest pathogens and trigger an adaptive response. Clotting reactions contain the infection.

? What innate defenses does an ingested pathogen encounter as it passes through the digestive tract?

10,000 viruses capable of infecting other cells. Eventually the infected cells die, weakening the respiratory tract.

Normally our bodies are able to mount a sufficient immune response to contain the infection, so a typical course of flu lasts only about a week and is not generally fatal. Why the 1918 flu was so deadly, researchers at the time could not say. More perplexing still was why the virus tended to kill people in the prime of their lives—those with the healthiest immune systems. It would be more than 75 years before researchers could even begin to answer those questions.

Viral Time Capsule

In 1995, Jeffery Taubenberger, a virologist at the U.S. Armed Forces Institute of Pathology, became interested in the mystery of the 1918 flu. He realized that he could use the polymerase chain reaction (PCR; see Chapter 7) to reconstruct the viral genome—if only he had access to some original tissue samples from the dead. He knew that the army had routinely saved tissue samples from soldiers killed during wartime and decided to check the institute's archives. Right away, he was able to locate tissues from several U.S. servicemen who had died of flu in 1918. Over the next several months, Taubenberger and his colleagues used PCR to recover viral RNA from these samples; they were then able to sequence a few of the virus's genes. But the results were incomplete, and the team eventually ran out of tissue samples from which to obtain virus. Taubenberger published these preliminary results in 1997 in the journal *Science*.

Later that year, Johan Hultin, a Swedish pathologist and physician, came across Taubenberger's paper and realized he had a solution to the problem of finding more tissue samples. Back in the 1950s, while doing research for his dissertation on the flu, Hultin had discovered a mass grave of people who had died of Spanish flu outside a remote village in Alaska. The village had been hit especially hard by the pandemic—85% of the villagers had died from the infection. Because the dead were buried in permafrost, their tissues might still be intact, he reasoned. Hultin wrote to Taubenberger and offered to take him to the site.

With permission from village elders, the scientists dug into the permafrost and exhumed several bodies. From one, the scientists were able to obtain enough viral RNA to eventually reconstruct the complete 1918 flu virus sequence—all eight of its genes.

From these gene sequences, scientists at the U.S. Centers for Disease Control and

Sample blocks containing lung and brain tissue from soldiers who died of the 1918 flu.

James King-Holmes/Science Source

Karl Mondon/KRT/Newscom

From his apartment in San Francisco, Dr. Johan Hultin recalls his two adventures to Brevig, Alaska, where he unearthed tissues tainted with the 1918 flu virus from a gravesite in a remote village.

Prevention (CDC) were able to reconstruct the virus in a lab (taking extreme precautions to keep it isolated in a secure containment area) and test it on mice. By substituting alleles of genes in a normal flu virus with those from the 1918 flu virus and comparing the severity of each composite virus, scientists were eventually able to determine what made the 1918 flu virus so deadly: alleles of four genes allowed the virus to penetrate deep into the lungs and cause infection there. While less lethal flu viruses replicate primarily in the upper respiratory tract, viruses with these four alleles cruise past the mouth, nose, and throat and invade the lungs. Replication of the virus in the epithelial cells lining the lungs killed these cells, and triggered a massive inflammatory response.

The inflammatory response is generally a good thing: it is a very effective way to contain and destroy invaders. But the balance between defense and destruction is delicate. In unusual cases, the inflammatory response can go into overdrive and destroy the very organ it is trying to save. This is what researchers now think happened with the 1918 flu. The inflammatory response within infected individuals was so massive that it damaged their lung tissue, ultimately causing some people to suffocate from fluid buildup in the lungs. This aggressive immune response helps explain why those with the strongest immune systems—people age 20 to 40—were the ones who suffered most.

Another important factor in explaining the large death toll is what came next. Because influenza destroyed the protective layer of cells lining the lungs and upper respiratory tract, bacteria and other pathogens had unimpeded entry into the lungs. Once inside, they grew like mad. Because antibiotics had not yet been discovered (see Chapter 3), doctors had no way of treating these infections. Bacterial pneumonia was responsible for at least half of all deaths during the 1918 pandemic, according to Taubenberger.

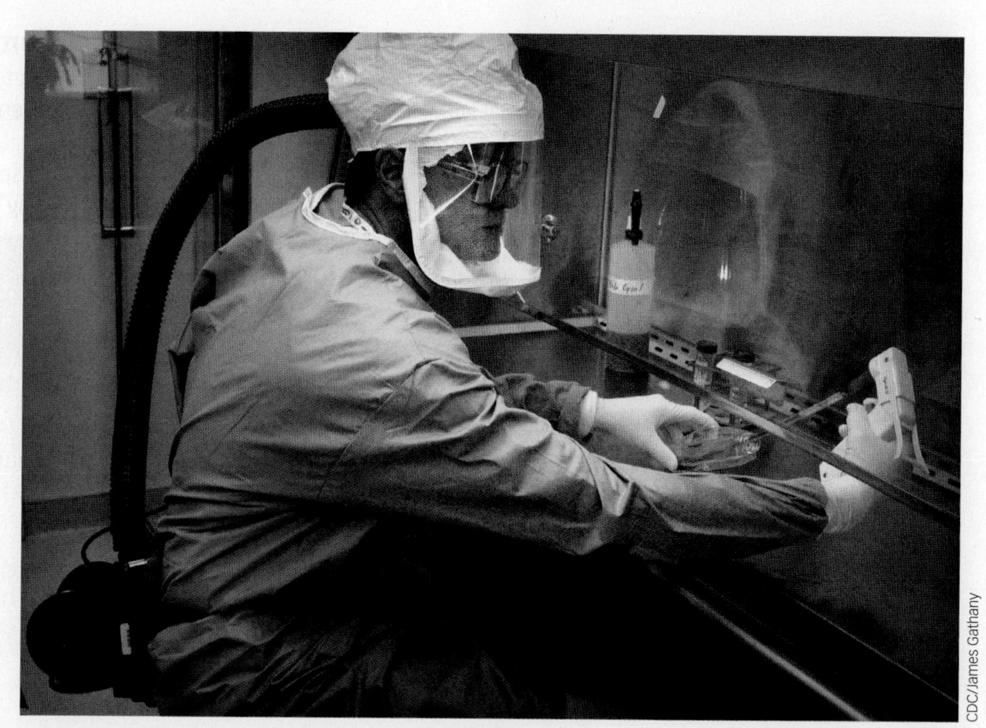

CDC/James Gathany

A CDC researcher working with samples of recreated 1918 flu virus.

Immunological Memory

▸ Adaptive immunity

Not everyone who became infected with the 1918 flu died, despite its virulence. In fact, while there were approximately 50 to 100 million deaths, experts estimate that some 525 million people were infected—about 30% of the entire world population. How were so many people able to fight the infection while others died? Of those who became infected and then recovered, some may have had partial immunity from an earlier influenza infection. Such long-lasting immunity is conferred by the adaptive immune system.

Whereas the innate immune system is always ready to fight, the adaptive immune system must be primed over time. From birth

A U.S. Public Health Service ad that appeared in newspapers in 1918.

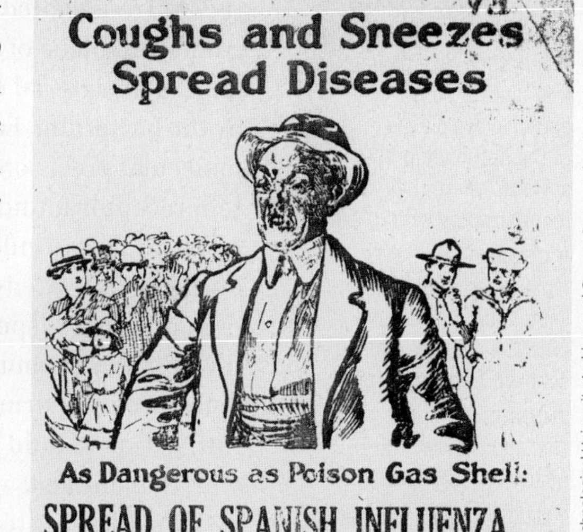
Coughs and Sneezes Spread Diseases

As Dangerous as Poison Gas Shells

SPREAD OF SPANISH INFLUENZA MENACES OUR WAR PRODUCTION

Michigan Technological University Archives and Copper Country Historical Collections

Normal lung
A normal lung is clear on an x-ray, having plenty of space that can fill with air.

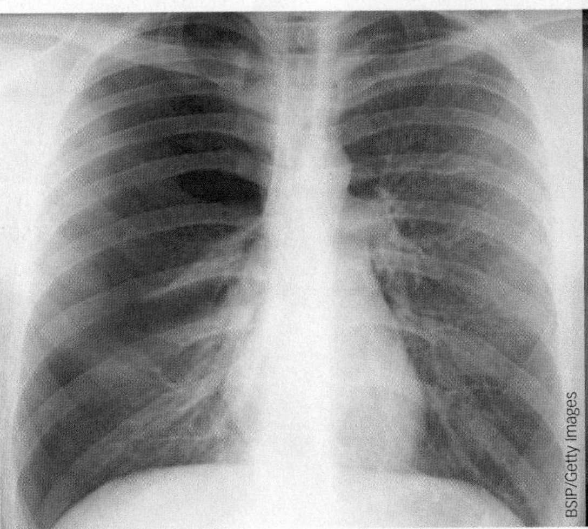

Pneumonia lung
An infected lung is cloudy on an x-ray. It is filled with pathogens, fluid, inflammatory cells, and debris, making it difficult to breathe. Bacterial lung infections were responsible for at least half of all deaths resulting from the 1918 flu pandemic.

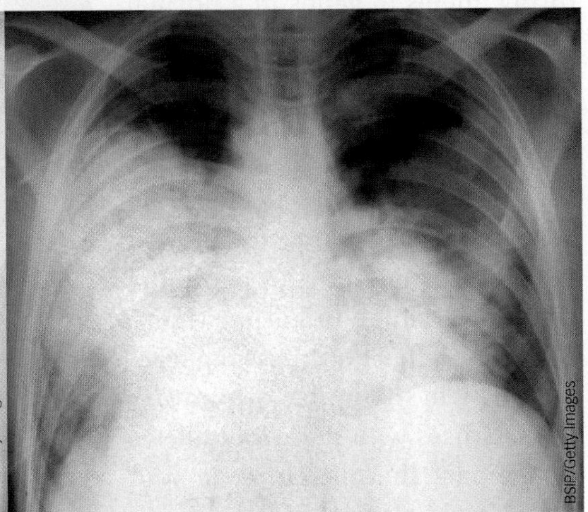

into adulthood, our bodies are continually assaulted by pathogens. With repeated exposure, our adaptive immune system develops a memory of every pathogen we encounter that gets past our innate defenses. Should we confront the same pathogen again, this immunological memory helps our bodies fight off infection before it can take hold.

To respond to and remember specific pathogens, our adaptive immune system must be able to recognize them—to distinguish one from another. The particular identifying detail of a pathogen that lymphocytes recognize is called an **antigen.** An antigen could be a piece of a viral protein, a component of a bacterial cell wall, or a toxin made by the bacterium. Each antigen matches with molecular precision the exact shape of a protein receptor found on a lymphocyte. When a lymphocyte finds an antigen "match," it becomes activated—that is, it turns on specific immune response genes and begins to divide. You can think of an activated lymphocyte as being "armed" and ready to spring into action should it encounter an intruder with that antigen.

The phagocytes of innate immunity we encountered earlier play an important role in adaptive immunity: the antigens that activate lymphocytes are remnants of pathogens that phagocytes have swallowed and chewed up. When these cells are done chewing, they post little bits of the invaders—antigens—on their surface, where they present them to lymphocytes. This process is called antigen presentation, and the cells that do this vital work are called **antigen-presenting cells (APCs).**

Lymphocytes have receptors on their surface that can bind to antigens on the surface of APCs. A lymphocyte with a receptor that matches the antigen perfectly will bind to it, and that interaction then activates the lymphocyte. Each lymphocyte recognizes only one antigen, but since our bodies can produce billions of unique lymphocytes, each with a unique receptor, chances are good that at least one will match. This specificity is the basis of adaptive immunity **(INFOGRAPHIC 32.6).**

Two types of lymphocytes play crucial roles in adaptive immunity: B cells and T cells. Like all blood cells, B and T cells develop from stem cells made in the bone marrow. Some immature lymphocytes develop in the bone marrow and become **B cells** (think "B" for "bone"). Other immature lymphocytes migrate from the bone marrow to the **thymus,** a gland in the chest, where

ANTIGEN
A specific molecule (or part of a molecule) to which immune receptors bind and against which an adaptive immune response is mounted.

ANTIGEN-PRESENTING CELL (APC)
A cell, such as a phagocyte, that digests pathogens and displays pieces of the pathogen on its surface, where they can be recognized by lymphocytes.

B CELLS
White blood cells that mature in the bone marrow and produce antibodies during an adaptive immune response.

THYMUS
The organ in which T cells mature.

T CELLS
White blood cells that mature in the thymus and play several roles in adaptive immunity.

INFOGRAPHIC 32.6

Lymphocytes Are Activated by Antigen-Presenting Cells

Phagocytes are antigen-presenting cells that play an important role in activating lymphocytes, arming them against specific invaders. Phagocytes ingest pathogens, digest them into small pieces, and then display these fragments on their surface. Once a phagocyte has displayed an antigen on a surface display molecule, a lymphocyte can recognize it and bind to it. Ultimately the lymphocyte is activated and can contribute to an adaptive response.

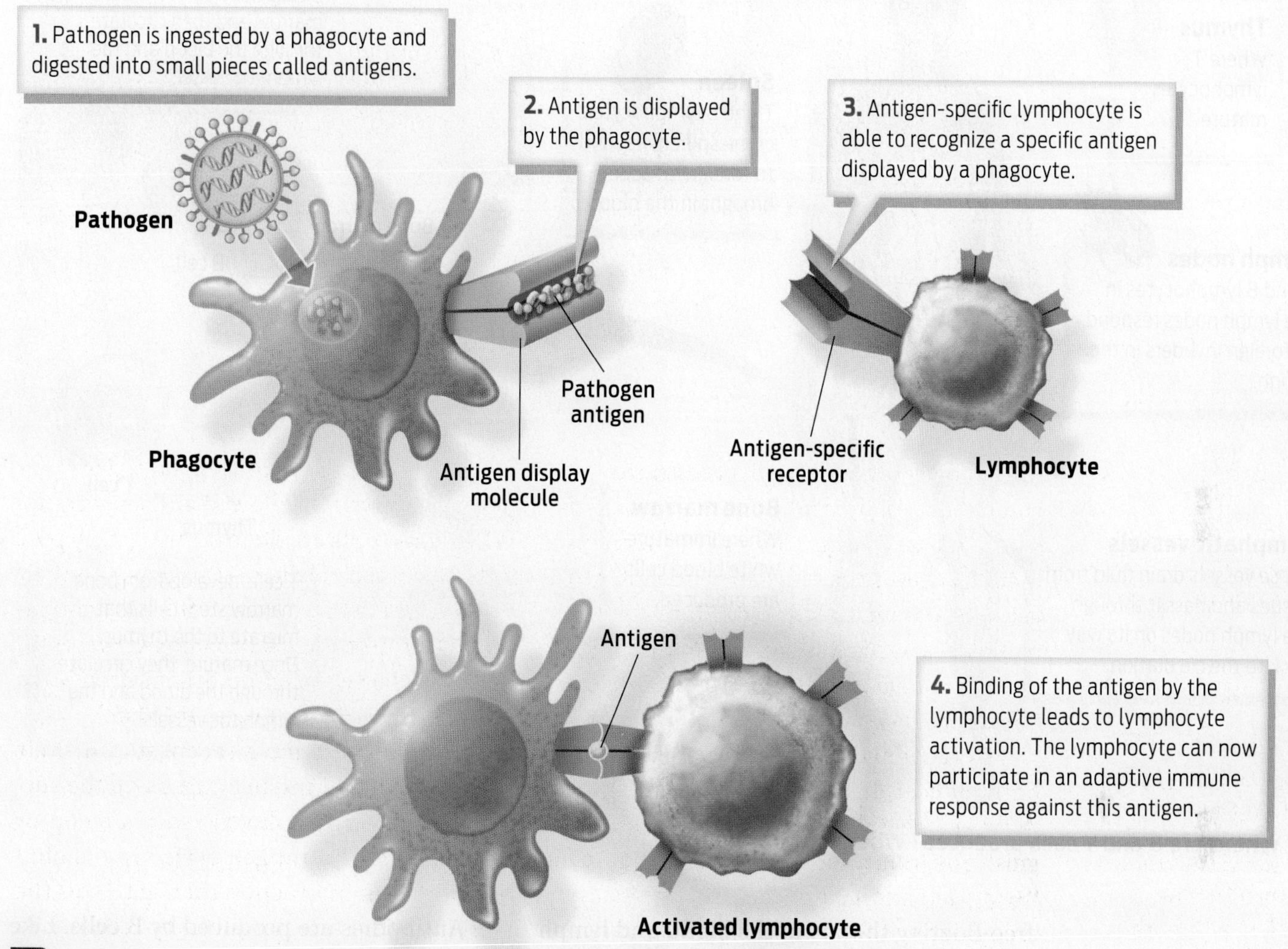

? **Describe the role of phagocytes in recruiting lymphocytes to an adaptive response.**

they develop into **T cells** (think "T" for "thymus"). Both B cells and T cells eventually travel to the **lymph nodes** and other organs of the **lymphatic system,** where they monitor body fluids for signs of infection.

The lymphatic system is connected by a set of vessels (the lymphatic vessels) that weave between blood capillaries and drain fluid (lymph) from tissues. Lymph contains white blood cells and sometimes pathogens picked up from tissues. Lymph travels from tissues to lymph nodes, where antigen presentation between phagocytes and lymphocytes occurs. From there, lymph returns to the blood circulation through the veins. In this way, the lymphatic system and the blood circulatory system are connected. Lymphocytes continually circulate through both the blood and the lymph as they patrol the body looking for invaders **(INFOGRAPHIC 32.7).**

B and T cells cooperate to produce two main types of adaptive immunity. One type, called **humoral immunity,** targets

LYMPH NODES
Small organs in the lymphatic system where B and T cells may encounter pathogens.

LYMPHATIC SYSTEM
The system of vessels and organs that drains fluid (lymph) from the tissues and sends it through lymph nodes on its way to the circulation.

HUMORAL IMMUNITY
The type of adaptive immunity that fights free-floating pathogens in the blood and lymph.

INFOGRAPHIC 32.7

The Lymphatic System: Where B and T Lymphocytes Develop and Act

The lymphatic system is a series of vessels and organs. The lymphatic system moves lymph and cells of the immune system around the body. Lymphocytes develop, mature, and act in a variety of tissues, including the bone marrow, thymus, spleen, and lymph nodes. There are two main types of lymphocytes (B and T), each of which plays an important role in the adaptive response.

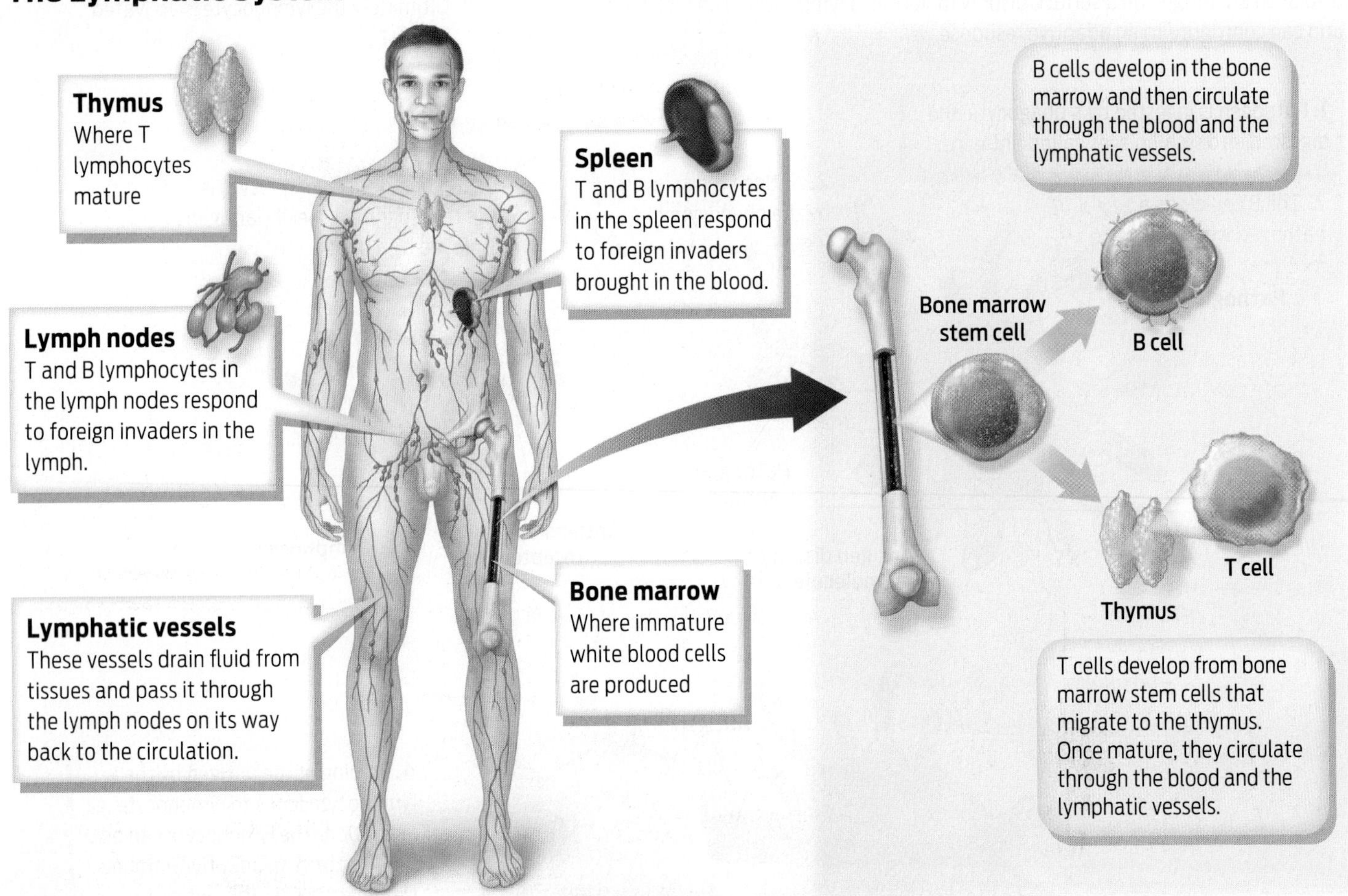

? **Where are B and T cells produced? Where do they mature?**

free-floating threats in the blood and lymph ("humoral" refers to the "humors," an old name for the liquids in the body). These threats may include virus particles, bacterial cells, and bacterial toxins.

Humoral immunity produces **antibodies**—proteins that circulate in body fluids and help fight infections by binding to specific antigens on pathogens. Antibodies fight infections in several ways. First, by binding to antigens, they may physically block the infectious organism from infecting other cells, rendering it harmless. Second, they can flag a pathogen to be digested by phagocytes. Finally, they can mark the pathogen so that complement proteins will bind to it and destroy it.

Antibodies are produced by B cells. Like all lymphocytes, B cells have receptors on their surface that recognize specific antigens. When a B cell encounters an antigen on a pathogen, it binds to it with its receptor. The B cell then internalizes the pathogen, digests it, and presents these antigens on its surface. (B cells are a type of antigen-presenting cell.) To begin producing antibodies that recognize and bind to that antigen, the B cell needs the assistance of specialized T cells called **helper T cells.** These cells also have receptors that can recognize antigens. When an activated helper T cell meets a B cell that has encountered the same pathogen, the helper T cell releases signaling molecules

ANTIBODY
A protein produced by B cells that fights infection by binding to specific antigens on pathogens.

HELPER T CELL
A type of T cell that helps activate other lymphocytes, including B cells and cytotoxic T cells.

INFOGRAPHIC 32.8

Humoral Immunity: B Cells Produce Antibodies

The humoral immune response protects the body from free-floating pathogens found in the blood and lymph. When a B cell and a helper T cell recognize the same antigen, the T cell can activate the B cell to divide and become plasma cells. Plasma cells secrete antibodies that bind to the specific antigens present only on the target pathogens. The binding inactivates and destroys these pathogens. Specific memory B cells are also produced and spring into action if the same pathogen is encountered again in the future.

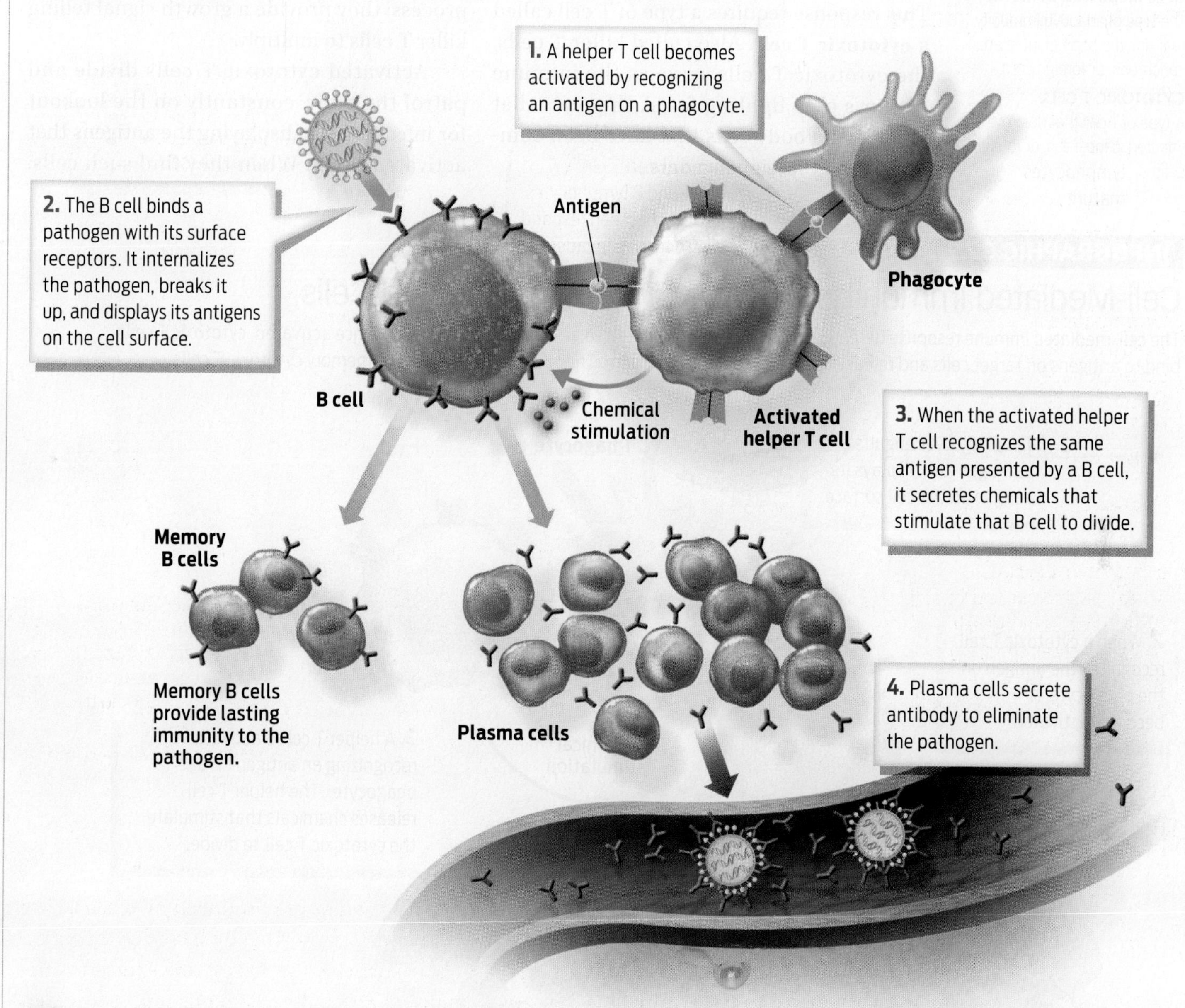

? **What molecules are produced by plasma cells? How do these molecules protect the body during an adaptive response?**

that trigger the B cell to divide. The B cell then divides repeatedly to create an army of **plasma cells**—cells that secrete many copies of an antibody specific to that particular antigen. Some of the dividing B cells will become **memory cells,** which remain in the bloodstream and "remember" the infection; they can be called on in the future to secrete antibodies in response to an infection.

Each B cell produces antibodies specific for one particular antigen. And antibodies produced against one antigen are usually ineffective against any other antigen—they are highly specific. But because the body can produce more than a billion unique B cells, there are more than a billion different kinds of potential antibodies—a very large war chest indeed **(INFOGRAPHIC 32.8).**

PLASMA CELL
An activated B cell that divides rapidly and secretes an abundance of antibodies.

MEMORY CELL
A long-lived B or T cell that is produced during an immune response and that can "remember" the pathogen.

CELL-MEDIATED IMMUNITY The type of adaptive immunity that rids the body of infected, cancerous, or foreign cells.

CYTOTOXIC T CELL A type of T cell that destroys infected, cancerous, or foreign cells.

The second type of adaptive immunity is **cell-mediated immunity,** which targets infected or altered body cells—for example, cells infected with the influenza virus. This response requires a type of T cell called a **cytotoxic T cell.** Also called killer T cells, the cytotoxic T cells serve as the immune system's elite fighting force, the ones that actually kill body cells that have been compromised by foreign invaders.

Cytotoxic T cells are activated by antigens they encounter on the surface of phagocytes that have engulfed pathogens. As with humoral immunity, helper T cells assist in the process: they provide a growth signal telling killer T cells to multiply.

Activated cytotoxic T cells divide and patrol the body, constantly on the lookout for infected cells displaying the antigens that activated them. When they find such cells,

INFOGRAPHIC 32.9

Cell-Mediated Immunity: Cytotoxic T Cells Kill Infected Cells

The cell-mediated immune response defends against infected or altered (e.g., cancerous) body cells. Once activated, cytotoxic T cells bind to antigens on target cells and release chemicals that destroy them. They also divide to produce memory cytotoxic T cells.

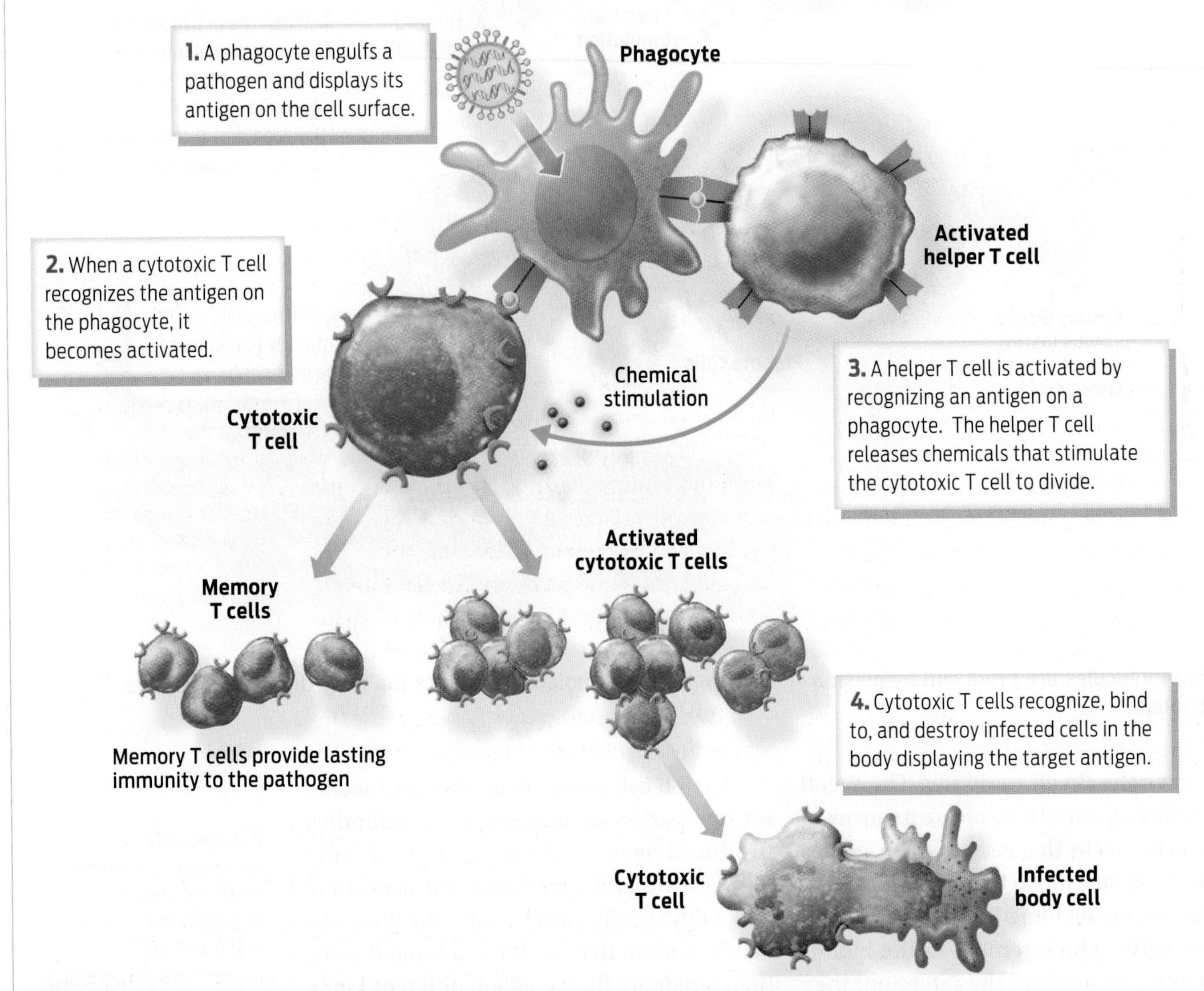

? **How does a cytotoxic T cell recognize a cell as a target for destruction?**

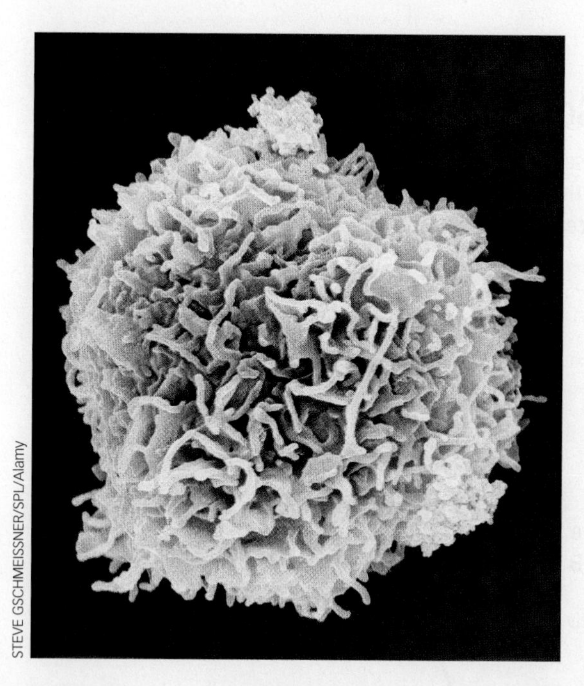
STEVE GSCHMEISSNER/SPL/Alamy

A T cell.

the cytotoxic T cells bind to the antigens and release chemicals that kill the rogue cells. As with B cells, some of the dividing cytotoxic T cells become long-lived memory cells, ready to recognize the same antigens in the future. In addition to targeting virus-infected body cells, cytotoxic T cells target foreign cells (from a transplanted organ or tissue, for example) and even cancer cells that the body recognizes as genetically altered **(INFOGRAPHIC 32.9)**.

While both the humoral and cell-mediated immune responses destroy invaders, there is a key difference between them: humoral immunity produces antibodies that bind to antigens on free-floating pathogens in blood and lymph, whereas cell-mediated immunity marshals cytotoxic T cells that bind to and destroy infected or altered cells in body tissues.

With its billions of lymphocytes, each equipped to recognize one particular antigen, the adaptive immune system enables our bodies to fight off countless numbers of pathogens. But occasionally, for reasons that aren't entirely clear, the system runs amok and immune cells become active against antigens that aren't harmful to us, or even worse, begin attacking healthy cells in the body.

When the immune system attacks harmless antigens from outside the body, like those in dust, pollen, or certain types of food, an **allergy** results. Allergies are quite common. They're often accompanied by a runny nose, watery eyes, and sneezing—all triggered by histamine, the same chemical involved in inflammation. When the immune system attacks the body's own healthy cells, a more serious **autoimmune disease** can occur. Multiple sclerosis, lupus, and rheumatoid arthritis are all autoimmune diseases, caused by a body at war with itself.

Building a Line of Defense

▶ Primary and secondary responses; vaccines and herd immunity

The advantage of adaptive immunity is that it "learns" to respond to new invaders and targets them specifically. The downside is that it takes a while for the response to kick in. First-time exposure to a pathogen that has breached our innate defenses will almost certainly cause illness, because the adaptive response takes 7 to 10 days to develop. Over time an exposed individual will recover as T and B cells are activated and antibody levels increase. This initial slow response is the **primary immune response.** As B and T cells are churned out, some of them become memory cells. These memory cells remain in the body and "remember" the infection. The next time the same pathogen is encountered, memory B and T cells become active, dividing rapidly and leading to the production of very high levels of antibodies. These cells fight the specific pathogen so quickly that the illness usually doesn't occur a second time.

ALLERGY
An immune response that is misdirected against harmless environmental substances, such as dust, pollen, and certain foods, and that causes uncomfortable physical symptoms.

AUTOIMMUNE DISEASE
A misdirected immune response in which the immune system attacks the body's own healthy cells.

PRIMARY IMMUNE RESPONSE
The adaptive immune response mounted the first time the immune system encounters a particular antigen.

Because the body can produce more than a billion unique B cells, there are more than a billion different kinds of potential antibodies—a very large war chest indeed.

INFOGRAPHIC 32.10

Memory Cells Mount a Swift Secondary Response

The adaptive immune system's primary humoral response is slow and produces low levels of antibodies. Upon subsequent exposures, memory B cells produced during the primary response respond quickly, rapidly dividing and developing into plasma cells that produce high concentrations of antibodies.

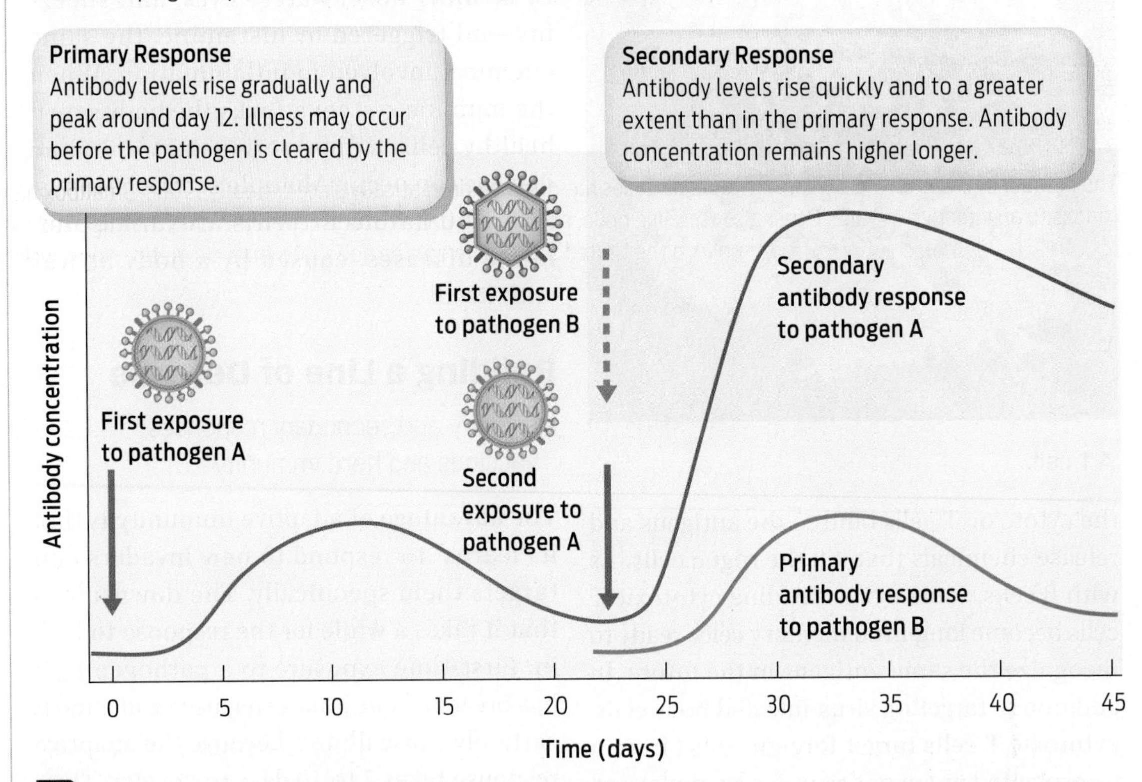

? **If a third pathogen (pathogen C) is encountered at day 22, what do you predict about the response to that pathogen?**

This rapid reaction is the **secondary immune response (INFOGRAPHIC 32.10).**

The secondary immune response is what confers immunity to a particular infection. It's also how **vaccines** work. Vaccines are essentially dead or weakened versions of a pathogen administered for the purpose of generating immunity to that pathogen. The goal is to create a primary immune response in the body without causing disease. Thus, if the pathogen is subsequently encountered naturally, the secondary response is already primed. Vaccination is like being infected with a pathogen but not actually having the disease.

It is hard to overstate the impact that vaccines have had on public health. Many once-deadly and disfiguring diseases have been relegated to the historical dustbin thanks to these medicines. Many people today cannot remember the time when polio meant confinement in an iron lung just to breathe, and often-fatal smallpox caused blisters and then left scars all over the body **(INFOGRAPHIC 32.11).**

In fact, you could say that vaccines have become victims of their own success. It's precisely because they have been so effective at reducing the burden of illness over the last hundred years that some people have begun to question whether it's really in their best interest to vaccinate their children. This movement against vaccination is sometimes called antivaccination.

The modern antivaccination push really took off in the late 1990s, in the wake of a fraudulent research paper published by a doctor in England, Andrew Wakefield. That paper, which was published in the journal

SECONDARY IMMUNE RESPONSE
The rapid and strong immune response mounted when the immune system encounters a particular antigen subsequent to the first encounter.

VACCINE
A preparation of killed or weakened pathogen that is administered to people or animals to generate protective immunity to that pathogen.

The Lancet, purported to show a link between the MMR (measles–mumps–rubella) vaccine and the development of autism in 8 of 12 children referred to a London hospital for unusual symptoms. When news outlets picked up the story, overnight millions of parents became concerned that vaccines might harm their children. In truth, Wakefield's paper was fundamentally flawed and contained falsified data; it was eventually retracted. Wakefield was ultimately accused of scientific misconduct and had his medical

INFOGRAPHIC 32.11

The Burden of Vaccine-Preventable Diseases Has Decreased Dramatically

The number of cases of vaccine-preventable diseases has declined dramatically since the early 1900s. Smallpox has been completely eradicated thanks to an effective vaccine. Other diseases, like polio, occur in the United States only when they are imported by a traveler from another country. The high vaccination rates for polio in the United States prevent the polio virus from circulating and causing disease within this country. Measles cases in the United States have doubled since 2013, and mumps cases have quadrupled, because rates of childhood vaccination for these diseases have declined.

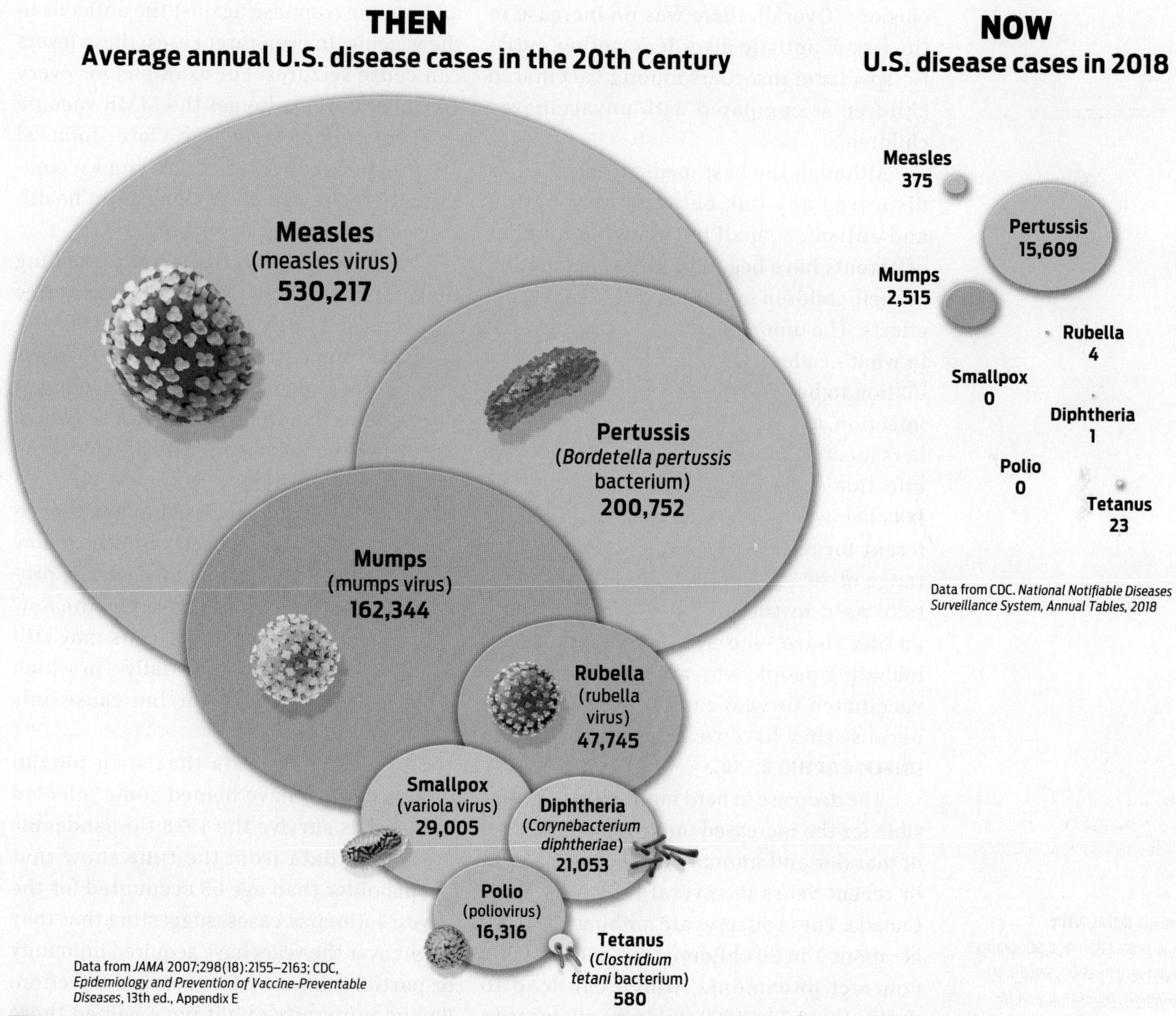

? **If there is an effective vaccine for measles, why have measles cases not dropped to near zero?**

license revoked. But the damage was done: as a result of his fraudulent paper, many reasonable people believed that there was a link between vaccines and autism.

Since Wakefield's paper was published, many large studies have looked specifically at the relationship between vaccines and autism and found no link whatsoever. For example, a study published in 2002 in the *New England Journal of Medicine* looked at all children born in Denmark between 1991 and 1998—a total of 537,303 children—and compared the rate of autism among those children who received the MMR vaccine and those who did not. The authors' conclusion: "Overall, there was no increase in the risk of autistic disorder or other autistic-spectrum disorders among vaccinated children as compared with unvaccinated children."

Although the best medical science has disproved any link between vaccination and autism, a small but growing number of parents have begun to forgo vaccination of their children out of fear of possible side effects. The unfortunate result is a decrease in what's called **herd immunity.** For a population to be adequately protected from an infection, a certain percentage of its members must be immune, either through prior infection or vaccination. If the percentage falls below a certain number (which is different for different pathogens, but in the range of 80% to 90%), then infections can take hold in the population and spread among those who are not vaccinated—including people who are too young to be vaccinated or who cannot be vaccinated because they have weak immune systems **(INFOGRAPHIC 32.12).**

The decrease in herd immunity is responsible for the increased number of outbreaks of measles and mumps that have occurred in recent years in several U.S. states and Canada. These diseases are nothing to sneeze at: about 1 in 20 children with measles will contract pneumonia, which can lead to death. About 1 in 1,000 children with measles will develop encephalitis (swelling of the brain), which can lead to convulsions, deafness, and intellectual disability. And for every 1,000 children who get measles, 1 or 2 will die. Yet this disease is largely preventable with routine vaccination.

How safe are vaccines? Remarkably safe. While use of any medicine carries some amount of risk—and vaccines are no different—there is no comparison between the risks posed by vaccines and the threats from the diseases they aim to prevent. For most people, vaccine side effects are nonexistent or very mild. In rare cases, vaccines may cause fever—a consequence of stimulating an immune response against the antigens in the vaccine. In even rarer cases, these fevers can cause seizures. For example, for every 10,000 children who get the MMR vaccine as infants, about 4 will have a fever-induced seizure during the week following vaccination; there are usually no long-term health consequences of these seizures.

Vaccines are very effective at preventing illness, but they can't absolutely guarantee that someone won't get the disease. One reason is that some strains of bacteria or viruses may be different enough from those found in the vaccines that they may still cause illness. Even then, the vaccine will still provide some measure of protection.

Similarly, people exposed to a pathogen that is similar to a pathogen with which they were previously infected may be partly protected from the disease caused by the new pathogen. Memory B and T cells may still respond, although just partially—in which case the illness may occur, but cause only mild symptoms.

Evidence suggests that such partial immunity may have helped some infected individuals survive the 1918 flu pandemic. Statistical data from the time show that people older than age 65 accounted for the fewest influenza cases, suggesting that they might over the years have acquired immunity or partial immunity from earlier infection. Partial immunity might have helped these

HERD IMMUNITY
The protection of a population from an infection, based on a certain percentage of its members being immune.

INFOGRAPHIC 32.12

Herd Immunity Protects against the Spread of Infections

When a sufficient proportion of people in a community are immune to a particular contagious pathogen (because they have been vaccinated, for example), then that infection cannot spread in that community. Low rates of vaccination lead to low rates of herd immunity and the spread of an infection in a community. High rates of vaccination protect those who cannot be vaccinated—for example, very young babies and people with cancer.

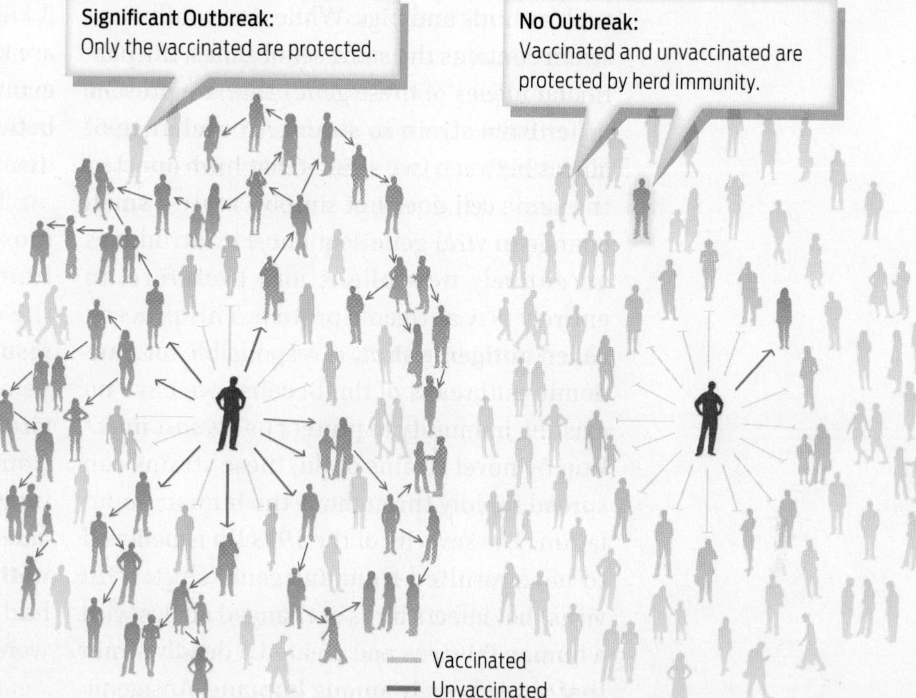

Unvaccinated Community

In a community where no one has been vaccinated, an outbreak can easily occur. One person can infect dozens of unvaccinated people, each of whom goes on to infect dozens more.

Partially Vaccinated Community

In a community where some people have been vaccinated, the infected person is able to infect multiple unvaccinated people, creating a significant outbreak of the disease. Some people (those who have been vaccinated) are protected.

Vaccinated Community

In a community where most people have been vaccinated, the infected person is surrounded by people who cannot become infected. An outbreak cannot occur. For this scenario to be effective, a high percentage of people must be immune to the disease.

? If a community is experiencing a measles outbreak, what can you infer about vaccination rates in that community?

people fight off the virus before it dug deep into their lungs.

An Evolving Enemy

▸ Antigenic shift and antigenic drift

When researchers discovered that the 1918 flu virus carried specific alleles that enabled the virus to replicate in the lungs, making it more deadly, a piece of the 75-year-old mystery was solved. But where did these alleles come from?

New influenza viruses are constantly being produced by two mechanisms: mutation and gene swapping. Influenza viruses replicate their genetic material very rapidly and don't "proofread" the replicated copies. As a result, mistakes often occur, leading to mutations. **Antigenic drift** is the gradual accumulation of mutations that cause small

ANTIGENIC DRIFT
Changes in viral antigens caused by genetic mutation during normal viral replication.

changes in the antigens on the virus surface. Antigenic drift explains why different types, or strains, of influenza can be circulating at the same time.

An influenza virus strain can also swap genetic material with other strains of influenza, including strains that infect animals such as birds and pigs. While every influenza strain contains the same set of genes, the particular alleles of these genes that are present differ from strain to strain. An exchange of alleles between two strains that have infected the same cell does not simply create a small change in viral gene sequence: it introduces an entirely new allele, and therefore an entirely new antigenic protein. This process, called **antigenic shift,** is responsible for pandemic outbreaks of flu. Because we have no existing immunity to protect us against infection by novel strains of flu, these strains can spread rapidly throughout the human population. The severity of the 1918 flu is believed to have resulted from antigenic shift: a flu virus that infects birds exchanged alleles with a human flu virus and created a deadly strain that spread easily among humans. Antigenic shift is also thought to have played a role in the emergence of "avian flu" in the late 1990s and "swine flu" in 2009. (These nicknames reflect where scientists believe some of the viral material originated.)

To keep track of the many different strains of flu viruses, scientists name them based on two proteins found their surface: hemagglutinin and neuraminidase. Hemagglutinin (H) binds to receptors on host cells and enables the virus to enter those cells. Neuraminidase (N) helps newly formed viruses exit host cells. At least 18 H types and 11 N types are found in nature, and these can occur in any combination. The flu strains that typically cause seasonal flu in humans are H1N1 and H3N2. The 1918 flu virus strain was also H1N1, but at that time it was novel, so most people had not previously encountered it and developed immunity. A novel form of H1N1 flu (dubbed "swine flu") also emerged in 2009, causing a pandemic and killing several hundred thousand people. Since the late 1990s, an avian H5N1 flu virus that sickens poultry has been sporadically infecting humans in Southeast Asia—mostly those who have direct contact with infected poultry. The disease does not spread easily between people, but it is deadly: it kills 60 percent of those infected. Scientists are keeping a close eye on H5N1, because if it evolves to become more easily transmissible between humans, it could cause a deadly pandemic **(INFOGRAPHIC 32.13).**

ANTIGENIC SHIFT
Changes in viral antigens that occur when viruses of one strain exchange genetic material with other strains.

ZOONOTIC DISEASE
An infectious disease that normally affects animals but can be transmitted to humans.

It is not a coincidence that some of the most dangerous strains of flu have come from animals like birds and pigs. Many of the deadliest pandemics in history are the result of animal viruses "spilling over" into people. Ebola viruses, HIV, and the 1918 flu virus are all examples of viruses that "jumped" from wild animals into humans. This type of jump is called a zoonosis, and the diseases caused by this phenomenon are called **zoonotic diseases**. Because our immune systems had never seen these viruses before, they were not prepared to fight them.

SARS-CoV-2, the coronavirus that causes COVID-19, is believed to be zoonotic. Evidence suggests that it jumped from bats to humans through an intermediate mammal—perhaps one sold as meat in a market in Wuhan, China, where the first documented cases of COVID-19 occurred. Bats are a source of many potential zoonotic diseases because they are teeming with viruses, yet do not themselves become sick, thanks to some quirks of their immune system. As the rapidly growing human population encroaches on wild habitats, the chance of a virus jumping from wild animals to humans becomes ever higher.

Preventing and Responding to Pandemics

Because viruses can swap genes so easily, public health officials closely monitor the types of viruses that infect both domesticated

INFOGRAPHIC 32.13

Antigenic Drift and Shift Create New Influenza Strains

Mutation and gene exchange are two mechanisms by which influenza viruses can change. Mutations that accumulate gradually can cause variations in surface antigen. This process is called antigenic drift. Different strains can also swap genes and cause surface antigens to change more dramatically. This is called antigenic shift. Drift is responsible for annual seasonal variation in influenza; shift is responsible for dramatic pandemics.

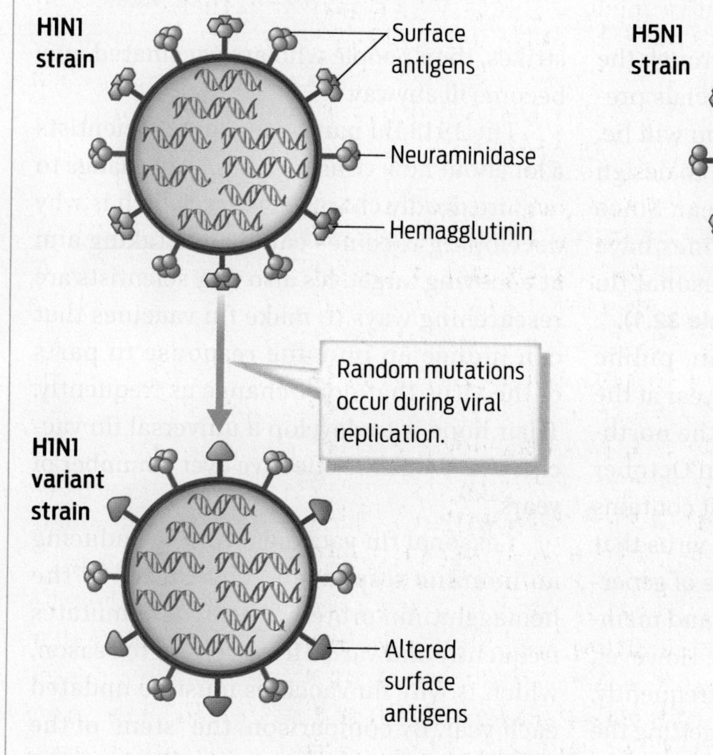

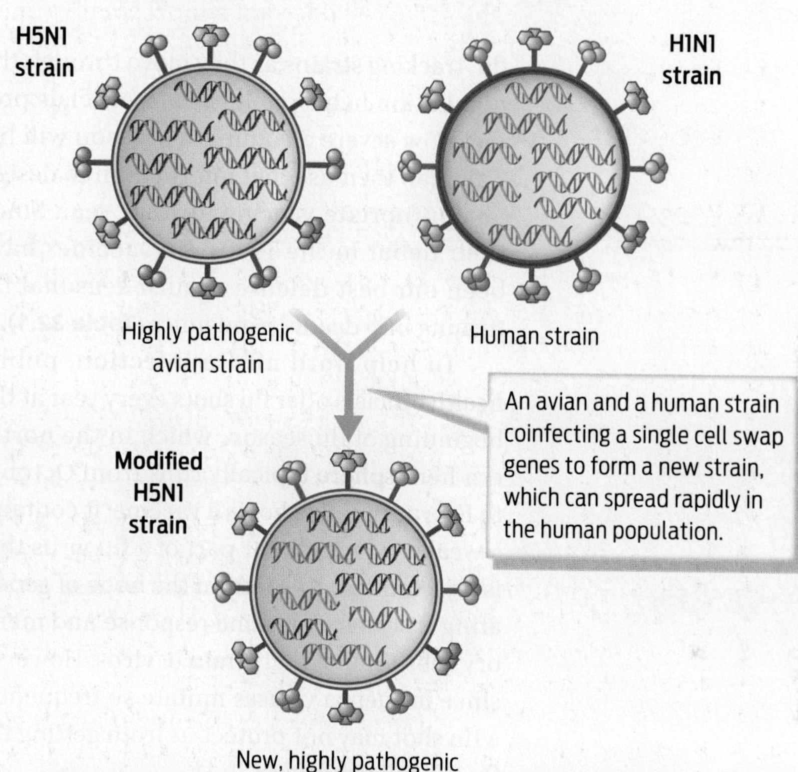

? This year's flu virus has only slightly different H and N antigens relative to last year's virus. What mechanism generated this year's strain?

and wild animals. Before 2020, many public health officials who study animal viruses had been warning that a virus with pandemic potential was likely to emerge. But governments around the world were still caught off guard and unprepared when a new viral disease, COVID-19, emerged at the end of 2019. It soon became clear that, in today's hyperconnected world, there is little to prevent a deadly virus from quickly spreading from one town or country to every country around the world. Whether COVID-19 will continue to circulate among humans seasonally, as flu does, is, at the time of this writing, still an open question.

But the ways we as a society have learned to deal with flu may provide important lessons for dealing with COVID-19. In the case of

TABLE 32.1 Known Flu Pandemics

Name	Date	Virus antigen subtype	Number of deaths
Asiatic (Russian) flu	1889–1890	H2N2	1 million
1918 (Spanish) flu	1918–1919	H1N1	20–100 million
Asian flu	1957–1958	H2N2	1–1.5 million
Hong Kong flu	1968–1969	H3N2	0.75–1 million
Swine flu	2009–2010	H1N1	150,000–575,000

flu, tracking strains as they move through the population helps public health officials predict how severe a coming flu season will be. They can then use that information to design an appropriate vaccine for that year. Since their debut in the 1940s, flu vaccines have been our best defense against seasonal flu turning into deadly pandemics **(Table 32.1)**.

To help ward off flu infection, public health officials offer flu shots every year at the beginning of flu season, which in the northern hemisphere typically runs from October to February. A flu shot is a vaccine: it contains a weakened version or part of a flu virus that is injected into the body in the hope of generating a primary immune response and memory cells against that strain of virus. However, since influenza viruses mutate so frequently, a flu shot may not protect us from getting the flu the following year.

Scientists create a new vaccine each year by tracking which strains of influenza are circulating worldwide and studying the antigens on their surfaces. These antigens are the same hemagglutinin and neuraminidase proteins that we saw help viruses enter and exit cells. Scientists then make a vaccine using a strain of flu with the antigens that seem most likely to characterize that year's virus. Sometimes, however, their predictions may be wrong. If public health officials release a vaccine against a strain of influenza with one variant of hemagglutinin, but a strain with a different variant of the antigen actually strikes, then people who are vaccinated may become ill anyway.

The 1918 flu pandemic taught scientists a lot about how viruses can quickly change to acquire deadly characteristics, which is why developing vaccines can be like taking aim at a moving target. It's also why scientists are researching ways to make flu vaccines that can induce an immune response to parts of the virus that don't change as frequently. Their hope is to develop a universal flu vaccine that would be effective over a number of years.

Seasonal flu vaccines work by inducing an immune response to the "head" of the hemagglutinin protein. This region mutates frequently and varies from season to season, which is why flu vaccines must be updated each year. By comparison, the "stem" of the hemagglutinin protein typically does not change (as fast), making it a potentially good target for a universal flu vaccine.

So far, at least two proposed universal flu vaccines have been shown in laboratory animals to provide protection against lethal strains of influenza, including H5N1 and H1N1. Clinical trials of these vaccines in humans are under way. Researchers are also working on different antiviral drugs to help reduce the impact and spread of the influenza virus. The hope is that such measures will suppress the next big flu outbreak, whenever and wherever it happens. ■

CHAPTER 32 Summary

Driving Question 1 What is the structure of a virus, and how do viruses cause disease?

- The immune system defends the body against infection by pathogens. It also helps heal injuries and fight cancer.
- Pathogens are infectious agents that can cause disease; they include certain bacteria, viruses, fungi, and parasites.
- Viruses are noncellular particles consisting of a genome of nucleic acid (DNA or RNA) contained within a protein shell that may or may not be surrounded by a lipid envelope. Viruses cannot reproduce on their own, but rather must rely on the replication machinery of the host cells they infect to make more copies of themselves.

Driving Question 2 Why is nonspecific immunity important even though another system provides both specificity and memory?

- The immune system has two main arms: one that is nonspecific (innate immunity) and one that is specific (adaptive immunity). Innate immunity is the first line of defense against invaders; adaptive immunity comes into play once innate defenses are breached.
- The innate immune system includes defenses with which we are born and which are always active: they include barriers such as skin and mucous membranes; antimicrobial chemicals in tears, saliva, and other secretions; and phagocytic cells that engulf and destroy pathogens.
- The inflammatory response, which is part of the innate immune system, is triggered by tissue infection or injury. During an inflammatory response, blood vessels swell and leak, bringing phagocytic cells and protective molecules to the area to contain the infection.
- Phagocytes are antigen-presenting cells that digest pathogens and present them on their surface where they can be "seen" by lymphocytes, a necessary step to activate the adaptive immune response.

Driving Question 3 What is adaptive immunity, and how does vaccination rely on adaptive immunity?

- Adaptive immunity is conferred by specialized lymphocytes called B and T cells that recognize antigens unique to a specific pathogen.
- B cells produce antibodies that circulate in body fluids, recognize specific antigens unique to a pathogen, and mark that pathogen for destruction. The defense provided by B cells is called humoral immunity.
- Cytotoxic T cells destroy infected, foreign, and cancer cells. The defense provided by cytotoxic T cells is called cell-mediated immunity.
- Both humoral and cell-mediated immunity require the assistance of helper T cells.
- The immune response can go awry, causing allergies and autoimmune conditions. Allergies are immune responses to intrinsically harmless substances (for example, pollen); autoimmune conditions occur when the immune system mounts a response against the body's own cells and tissues.
- At the first exposure to a particular pathogen, a primary immune response is generated that takes time to become fully effective; this primary response also produces memory cells.
- Memory cells remain in the body and become active at the time of a subsequent exposure to the same pathogen; they produce a more rapid and vigorous secondary immune response that fights the pathogen and usually prevents the associated illness.
- Vaccination elicits a primary immune response. Memory cells produced during the primary response protect against illness following subsequent exposure to the actual pathogen.

Driving Question 4 What are the features of the influenza virus that allow it to cause worldwide outbreaks?

- The adaptive immune response is highly specific for particular pathogens. If the pathogen changes, by either antigenic shift or antigenic drift, the body must mount a new primary response for each strain of the pathogen.
- The influenza virus undergoes both antigenic shift and antigenic drift, so a new flu vaccine is required every year.
- Pandemic strains of flu and other viruses often are the result of zoonoses—when a virus that infects wild or domesticated animals changes to become capable of infecting and spreading among humans who lack immunity to it.

More to Explore

- *Pandemic: How to Prevent an Outbreak* (Netflix series, 2020): https://www.netflix.com/title/81026143
- Jordan, D., for the Centers for Disease Control and Prevention. *The Deadliest Flu: The Complete Story of the Discovery and Reconstruction of the 1918 Pandemic Virus:* https://www.cdc.gov/flu/pandemic-resources/reconstruction-1918-virus.html
- Taubenberger, J. K., and Kash, J. C. (2011). Insights on influenza pathogenesis from the grave. *Virus Res* 162(1):2–7.
- Barry, J. M. (2005). *The Great Influenza: The Story of the Deadliest Pandemic in History.* New York: Penguin Books.
- Mnookin, S. (2011). *The Panic Virus: The True Story Behind the Vaccine–Autism Controversy.* New York: Simon & Schuster.
- Osaka, S. (2020). These scientists saw COVID-19 coming. Now they're trying to stop the next pandemic before it starts: https://www.motherjones.com/environment/2020/05/these-scientists-saw-covid-19-coming-now-theyre-trying-to-stop-the-next-pandemic-before-it-starts/

Chapter 32 Test Your Knowledge

Driving Question 1 What is the structure of a virus, and how do viruses cause disease?

By answering the questions below and studying Infographics 32.2, 32.3, and 32.13, you should be able to generate an answer for this broader Driving Question.

Know It

1. Which of the following is found in all viruses?

a. DNA
b. RNA
c. membranous envelope
d. protein shell
e. cell membrane

2. Explain how viruses replicate within humans.

3. Why does poliovirus cause long-lasting damage, whereas people infected by influenza virus typically make a full recovery?

Use It

4. Both viruses and bacteria can be human pathogens. Describe some key differences between them. (*Hint:* You may wish to refer to Chapter 3.)

5. Why do poliovirus, influenza virus, and HIV infections cause different symptoms?

Driving Question 2 Why is nonspecific immunity important, even though another system provides both specificity and memory?

By answering the questions below and studying Infographics 32.4 and 32.5, you should be able to generate an answer for this broader Driving Question.

Know It

6. Name three components of the innate immune system. For each, provide a brief description of how it offers protection.

7. What are phagocytes, and what do they do?

Use It

8. From what you know about innate immunity, would you predict different or identical innate responses to infections from *Escherichia coli* (a bacterium) and *Staphylococcus aureus* (another bacterium)? Explain your answer.

9. Neutropenia is a deficiency of neutrophils, a type of phagocytic cell. Neutrophils are among the "first responders" to an injury or infection. Would you expect someone with neutropenia to be able to mount an effective inflammatory response? Explain your answer.

10. Why might someone taking anti-inflammatory drugs be more susceptible than others to bacterial infections?

Driving Question 3 What is adaptive immunity, and how does vaccination rely on adaptive immunity?

By answering the questions below and studying Infographics 32.4, 32.6, 32.7, 32.8, 32.9, 32.10, 32.11, and 32.12, you should be able to generate an answer for this broader Driving Question.

Know It

11. Compare and contrast the features of innate and adaptive immunity.

12. B cells, plasma cells, and antibodies are all related. Describe this relationship, using words, a diagram, or both.

Use It

13. Anti–hepatitis C antibodies present in a patient's blood indicate

a. that the patient is mounting an innate response.

b. that the patient has been exposed to HIV.

c. that the patient was exposed to hepatitis C within the last 24 hours.

d. that the patient was exposed to hepatitis C at least 2 weeks ago.

e. that the patient has hepatitis.

14. Vaccination against a particular pathogen stimulates what type of response?

a. innate

b. primary

c. secondary

d. autoimmune

e. b and c

15. Will someone who has been exposed to seasonal influenza in the past

a. have memory B cells?

b. still be at risk for seasonal influenza next year? Why?

c. still be at risk for pandemic swine flu? Why?

16. *Staphylococcus aureus* can cause a bacterial skin infection that can become very serious.

a. Why does the body exhibit innate and adaptive responses to *S. aureus* but not to its own skin cells?

b. Will the innate response to *S. aureus* be equally effective against *Streptococcus pyogenes*, another bacterium that can cause skin infections? Explain your answer.

c. Will the adaptive response to *S. aureus* be equally effective against *S. pyogenes*? Explain your answer.

17. HIV is a virus that infects and eventually destroys helper T cells. Why do people with AIDS (that is, with advanced HIV infections) often die from infections by other pathogens?

Mini Case

Apply Your Knowledge

18. In 2008, an outbreak of measles occurred in San Diego. The first child infected in that city was a 7-year-old unvaccinated boy who had returned home from a family trip to Switzerland. He began to develop a cough, sore throat, and fever, but continued to attend school. The child was taken to his pediatrician when he developed a rash, and then was sent to the emergency room because of a very high fever. Blood tests revealed antimeasles antibodies. Eleven other children ended up developing measles; none had been vaccinated. The other cases were the 2 siblings of the first patient, 4 children in his school, and 5 children who were in the pediatrician's office at the same time as the patient. Of the 11 additional cases, 3 children were younger than 1 year.

[Data are from Outbreak of measles–San Diego, California, January–February 2008. (2008). *MMWR* 57(8):203–206.]

a. Why was the presence of antimeasles antibodies in the first case an important finding?

b. What can you infer about how easily measles spreads?

c. What does this case suggest about the importance of measles vaccinations?

d. Were all the unvaccinated children necessarily behind on their vaccination schedule? (*Hint:* Look up the recommended measles vaccination schedule on the Centers for Disease Control and Prevention website, www.cdc.gov.)

Bring It Home

Apply Your Knowledge

19. Almost 10% of the children in the school attended by the original infected child described in the Mini Case were unvaccinated because their parents had filed Personal Belief Exemptions stating that they did not want to vaccinate their children. What is your local school district's or state's policy on vaccinations for enrolled students? This information is typically available on the school district's website or the state health department's website. Based on the policies in place (and whether they permit any exemptions), do you think it is possible that a measles outbreak could occur in a local school?

Driving Question 4 What are the features of the influenza virus that allow it to cause worldwide outbreaks?

By answering the questions below and studying Infographic 32.13, you should be able to generate an answer for this broader Driving Question.

Know It

20. What is the difference between antigenic shift and antigenic drift?

21. A strain of influenza can infect and replicate in birds without causing them disease. The same strain can then be transmitted from birds to humans, causing severe illness in humans. What else would this strain need to be able to do in order to become pandemic?

Use It

22. What processes are responsible for the emergence of pandemic influenza strains such as swine flu? Explain how these strains can spread so successfully through the human population.

23. Why are people with influenza infections susceptible to bacterial pneumonia?

Interpreting Data

Apply Your Knowledge

24. The following table gives the approximate cumulative rate of influenza hospitalizations in the United States in newborns to 4-year-olds and in people 65 years and older for the 2009–2010 flu season and the 2014–2015 flu season. The flu season runs from the fall of one year through the spring of the next, and the data are reported by week (40th week of the first year to 12th week of the next).

a. Graph the cumulative incidence rates by week for each age group in each flu season.

b. Examine each season and compare and contrast the age-specific hospitalization rates in the two seasons.

Age Group and Season	Cumulative incidence rate (per 100,000)						
	Week 40	Week 44	Week 48	Week 52	Week 4	Week 8	Week 12
Birth to 4 years; 2009–2010	40	45	50	60	60	60	60
65 years and older; 2009–2010	10	18	20	20	20	20	20
Birth to 4 years; 2014–2015	0	0	1	40	48	50	50
65 years and older; 2014–2015	0	0	8	125	240	280	305

Data from Update: influenza activity–United States, 2009–10 season. (2010). *MMWR* 59(29):901–908; Influenza activity–United States, 2014–15 season and composition of the 2015–16 influenza vaccine. (2015). *MMWR* 64(21):583–590.

Answers to Infographic Questions

Chapter 1

IG1.1 Continue to refine and test the hypothesis using different experiments.

IG1.2 The independent variable is exposure to cell phone radiation. The dependent variable is development of cancer. The independent variable is intentionally changed from 0 in the control group to different levels in the experimental rats.

IG1.3 In the left panel, the percentages of affected rats range from 0% to 38%. The low (0%) is lower than the actual percentage in the total population (14%) and the high (38%) is higher than the actual percentage (14%) in the total population. In the right panel, the sample with the largest population (430 rats) gives the percentage that most closely matches the percentage in the total population.

IG1.4 Theories are hypotheses that are supported by large bodies of evidence and have not been disproved. The reporter should have stated that researchers were testing the hypothesis that cell phones cause cancer.

IG1.5 Highest energy (highest frequency) to lowest is: x-rays; ultraviolet (UV); microwaves; radio waves.

IG1.6 Cell phone subscriptions went from 125 million to 250 million. This is a doubling (100% increase). If cell phone use was a direct and immediate cause of brain cancer, then brain cancer diagnoses should have increased by the same amount—from 6.5 per 100,000 to 13 per 100,000.

IG1.7 The NTP rodent study is most similar to a randomized clinical trial in that there were different groups of rodents randomly assigned to different levels of exposure to cell phone radiation.

IG1.8 Information that may be missing includes the sample size, the population studied, and details of how data were collected and analyzed.

Chapter 2

IG2.1 It grows taller, it reproduces (by producing avocados with seeds), it maintains water balance by regulating water uptake and loss, its leaves can turn toward sunlight, and it uses the energy of sunlight to carry out photosynthesis.

IG2.2 The atomic number of magnesium is 12; it has 12 protons.

IG2.3 Every carbon in glucose has four covalent bonds. The carbon of CO_2 has four covalent bonds.

IG2.5 Phospholipids and proteins

IG2.6 You can show the oxygen atom of the bottom right water molecule forming a hydrogen bond with a hydrogen atom of another water molecule.

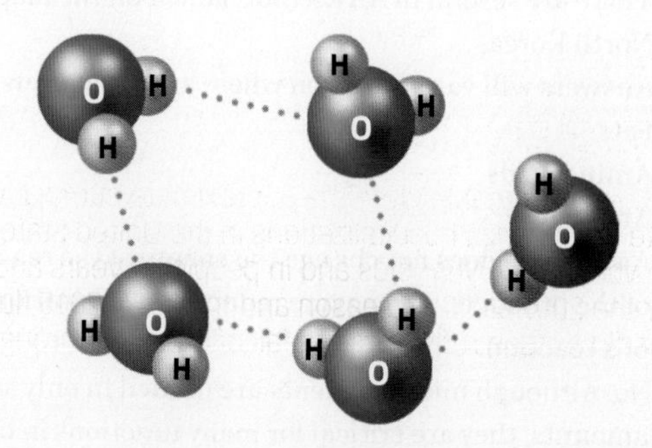

IG2.7 There are many possibilities, but in all cases, the oxygen atom of one water molecule will form a hydrogen bond with a hydrogen atom of another water molecule. Here is one possibility:

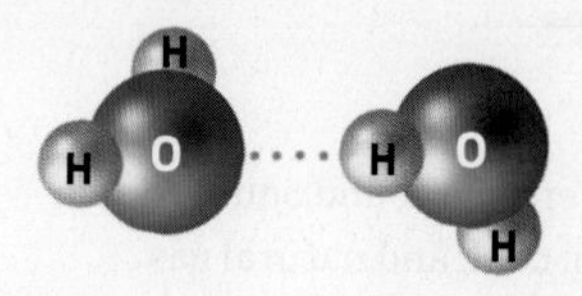

IG2.8 A sodium ion is positively charged because it has lost an electron. A chloride ion is negatively charged because it has gained an electron.

IG2.9 Because water is so crucial for life on Earth, the presence of liquid water is considered a proxy for life.

IG2.10 7.4

IG2.11 Both viruses and prions are infectious, and neither are made of cells. Viruses have genetic material enclosed in a protein shell and reproduce in infected cells. Prions do not have genetic material (they are made only of protein) and are also able to replicate in infected cells.

Chapter 3

IG3.1 There would not be any "clearing" of bacterial growth next to the fungus. The bacteria would grow well next to the fungus.

IG3.2 Diatoms, amoeba, molds, and *Elodea*

IG3.3 Eukaryotic cells have organelles, including a nucleus.

IG3.4 The cell will swell and may burst.

IG3.5 Eukaryotic ribosomes have a different structure from prokaryotic ribosomes. The antibiotic targets only prokaryotic ribosomes.

IG3.6 The hydrophobic tails

IG3.7 Diffusion does not require an input of energy; active transport does.

IG3.8 Three of the organelles present in both plant and animal cells and their functions are: the nucleus—it holds DNA; mitochondria—they convert energy from food into a useful form; the Golgi apparatus—it packages and processes proteins.

Milestone 1

IGM1.1 Circular

IGM1.2 Nonphotosynthetic bacteria

Chapter 4

IG4.1 There are several in Africa (not named on the map) and North Korea.

IG4.2 Answers will vary, based on where a student grew up.

IG4.3 Fats

IG4.4 Amino acids

IG4.5 Anabolic

IG4.6 An enzyme does not change the energy of the reactants or of the products. An enzyme reduces the activation energy of a reaction.

IG4.7 No. Although micronutrients are needed in only small amounts, they are critical for many functions in cells and organisms.

IG4.8 RUTF was more effective as a full dietary replacement than as a supplement.

IG4.9 These diet plans will be challenging to fit into a balanced diet because a balanced diet includes food from a variety of food groups.

Chapter 5

IG5.1 Canada, Venezuela, and Saudi Arabia

IG5.2 Petroleum, coal, and natural gas

IG5.3 Algal oil lipids and carbohydrates are used to produce biofuels. Algal cell wall biomass can be burned to generate electricity.

IG5.4 Kinetic energy of heat, muscle movement, and wheel movement

IG5.5 Potential energy (stored chemical energy)

IG5.6 Glucose can be used immediately as energy to power cellular functions, stored as potential energy (in molecules like starch), and used as energy to make cellular structures (such as the cell wall).

IG5.7 The carbon atoms in the glucose come from the carbon atoms in CO_2, and the oxygen atoms in the O_2 come from water (H_2O).

IG5.8 Violet

IG5.9 ATP

IG 5.10 Fossil fuels: 10; Crop-based biofuels: 9; Algae biofuels: 12. Answers regarding weighting will vary.

Chapter 6

IG6.1 Obese

IG6.2 Fat stores more than twice the number of Calories per gram as carbohydrates.

IG6.3 Approximately 5 hours and 15 minutes

IG6.4 Walking the dog and weeding the garden are NEAT activities. Studies show that increased amounts of NEAT activities can reduce fat gain in people who overeat and that lean people spend more time carrying out NEAT activities than do obese people.

IG6.5 They both require an energy source in the form of ATP.

IG6.6 Glucose from the digestion of carbohydrates is delivered from the small intestine via the circulation. Oxygen is delivered from the lungs, also via the circulation.

IG6.7 Glycolysis, the citric acid cycle, and electron transport. Glycolysis occurs in the cytoplasm; the citric acid cycle occurs in mitochondria; and the electron transport chain occurs in the mitochondria.

IG6.8 Pyruvate accepts electrons from NADH during fermentation. Pyruvate is converted to lactic acid or alcohol, depending on the organism and fermentation reaction.

IG 6.9 In approximately 1970

IG6.10 A triglyceride stores more energy per gram than glycogen. Triglycerides are stored in fat cells.

IG6.11 In photosynthesis, CO_2 is the source of carbon used to generate glucose. In cellular respiration, CO_2 is a waste product.

IG6.12 There are many possible combinations. For example, using a headset to talk while making phone calls, walking 15 minutes back and forth to the convenience store and carrying milk home, and spending an hour playing cards or working a puzzle.

Chapter 7

IG7.1 There are 46 chromosomes; each has a single DNA molecule. Therefore, there are 46 DNA molecules in the nucleus of each human cell.

IG7.2 TACGACT

IG7.3 DNA polymerase

IG7.4 DNA polymerase

IG7.5 He inherited six copies from his mother, and four copies from his father.

IG7.6 One half of the child's bands should match bands from his or her biological father.

IG7.7 $0.2 \times 0.2 \times 0.2 = 0.008$, or 1 in 125

Milestone 2

IGM2.1 A phosphate group, a deoxyribose sugar, and a nitrogenous base

IGM2.2 The phosphate groups on the outside of the helix can interact with water, as illustrated by the change in the structure of DNA in the presence of water.

IGM2.3 If a DNA molecule is 20% G, then it must also be 20% C. The remaining 60% is equally divided between A and T, giving the %A as 30%.

IGM2.4 A DNA molecule has two strands running in opposite directions, each made up of covalently bonded deoxyribonucleotides. The sugars and phosphates form the backbone of each strand, while paired bases form "rungs" in the middle of the molecule. The bases pair by hydrogen bonding.

Chapter 8

IG8.1 Strength is important to hold the sutured tissue together. Flexibility is important to allow sutured tissue to maintain its normal range of motion.

IG8.2 Because leucine is hydrophobic, it will interact with other hydrophobic amino acids in the interior of the protein, rather than lie on the surface of a protein that interacts with water.

IG8.3 The cell still has the gene encoding protein B, but that gene is not expressed in this particular cell.

IG8.4 The regulatory sequence of each gene determines where it is expressed. A milk protein gene is expressed in mammary glands, so it has a regulatory sequence for mammary gland expression. Insulin is expressed in the pancreas, so it has a regulatory sequence for expression in the pancreas. The genes have different coding sequences because the proteins they encode have different amino acid sequences, structures, and functions.

IG8.5 The product of transcription is an RNA molecule; the product of translation is a protein molecule.

IG8.6 RNA polymerase is the enzyme responsible for transcription. It adds ribonucleotides to the growing RNA molecule.

IG8.7 tRNA transports amino acids to the growing protein. One end binds to a specific amino acid, and its anticodon end forms a complementary base pair with the mRNA codon.

IG8.8 The start codon has the sequence AUG; it specifies the amino acid methionine.

IG8.9 The scientists wanted to express the spider spidroin gene in yeast, so they needed to provide a yeast regulatory sequence to drive expression in yeast.

Chapter 9

IG9.1 In sickle cell disease, the red blood cells are sickle shaped, and carry less oxygen. In people without sickle cell disease, the red blood cells are round (with a "thumbprint") and can carry adequate amounts of oxygen.

IG9.2 The mutation in the gene alters the DNA. The mRNA is altered, as is the protein.

IG9.3 Avoid exposure to mutagens, including alcohol, charred meat, and tobacco smoke.

IG9.4 The virus is the vector that delivers the normal beta-globin gene to the cells to be treated (in this case, hematopoietic stem cells).

IG9.5 Fetal hemoglobin is expressed after this gene therapy. It contains alpha and gamma subunits.

IG9.6 CRISPR uses a complementary RNA to direct a specific enzyme to the target DNA sequence.

IG9.7 It is a somatic gene therapy. Hematopoietic stem cells are somatic cells, not gametes (egg or sperm).

IG9.8 Having a single copy of the mutation protects against malaria. The mutation was selected for by the high frequency of malaria in Africa and South America.

Chapter 10

IG10.1 Breast cancer and pancreatic cancer

IG10.2 Blood cells

IG10.3 Interphase prepares the cell for division. Interphase includes G_1, S, and G_2.

IG10.4 So that mutations are not copied (during S phase) and passed on to daughter cells.

IG10.6 Most cancers are sporadic—they occur as the result of mutations that happen during the individual's life. In contrast, inherited cancers occur as the result of having inherited a mutation in a cancer-associated gene.

IG10.7 The cell will divide since it is unable to detect the DNA damage. Normally, DNA damage would be detected and either be repaired or cause the cell to die by apoptosis.

IG10.8 No. The *BRCA1* mutation would have contributed to the failure to repair other mutations, allowing cells to continue to divide even as they accumulate unrepaired mutations.

IG10.9 Chemotherapy is delivered to the whole body and can interfere with the division of both healthy and cancer cells. Targeted therapy is also delivered to the whole body, but interferes only with cancer cells.

IG10.10 Mammography

Chapter 11

IG11.1 The median survival for people with CF born between 2013 and 2017 is much higher (44 years) than for people born between 2000 and 2004 (35 years).

IG11.2 Three nucleotides differ; there is a single amino acid difference.

IG11.3 Humans are diploid because they have two complete sets of chromosomes, one inherited from the mother and one from the father.

IG11.4 A sperm is haploid; a skin cell and a zygote are both diploid.

IG11.5 The first meiotic division (meiosis I)

IG11.7 Recombination mixes up combinations of alleles of genes on a single chromosome by the exchange of segments between homologous chromosomes.

IG11.8 The CFTR membrane protein allows chloride ions to move across the cell membrane. When ions leave the cell, water can follow, keeping the mucus thin and slippery. If the CFTR membrane protein does not work properly, chloride ions are retained in cells, as is water. The extracellular mucus is then thick and sticky.

IG11.9 The probability is 50% (2 in 4).

IG11.10 These parents cannot have a child with two copies of the dominant (*T*) allele (0% probability).

IG11.11 The probability is 1 in 16.

Milestone 3

IGM3.1 Both proposed that preformed humans were present in one of the gametes.

IGM3.2 Purple flower color is dominant.

IGM3.3 Only one allele for any given gene is segregated into a single gamete.

IGM3.4 Because the *R* alleles segregate away from each other at meiosis I, and the *Y* alleles also segregate away from each other at meiosis I.

Chapter 12

IG12.1 Their biological sex would be male, because of the Y chromosome.

IG12.2 All of their children (sons and daughters) will have red–green color blindness.

IG12.3 If Irene (a carrier) married Maurice (who has hemophilia), 50% of their daughters would be predicted to have hemophilia, and 50% of their sons would be predicted to have hemophilia.

IG12.4 Only his sons will inherit this allele.

IG12.5 Thomas, Edy, Beverly, Madison, and Eston would all be predicted to have the same Y chromosome as Thomas. If Harriet had a son, he would not share his Y chromosome with his grandfather Thomas (he would share his Y chromosome with his father, and his father's father).

IG12.6 All of their children will have the mild form of the disease.

IG12.7 The AB blood type exemplifies codominance. A person with AB blood could have parents with the following blood types: A and B; AB and A; AB and B; AB and AB.

IG12.8 Only people with AB-positive blood can safely receive AB-positive blood.

IG12.9 This person could have inherited three dominant alleles, or could have inherited four dominant alleles but have a nutritionally poor diet.

IG12.10 The nurturing behavior of a rat is determined by the nurturing behavior it experienced as a pup, not by inheritance. A pup raised by a highly nurturing mother will grow up to be highly nurturing. Its identical twin, raised by a low-nurturing mother, will grow up to be low nurturing.

IG12.11 The zygote will have 44 autosomes and a Y chromosome. The egg and the zygote are aneuploid in this case.

IG12.12 There would be a total of 45 chromosomes; 44 autosomes and an X chromosome.

Chapter 13

IG13.1 Colonization by *S. aureus* is the growth of *S. aureus* in or on the body, with no disease. A MRSA infection is the growth of MRSA in or on the body that causes disease.

IG13.2 The skin of other wrestlers; shared equipment or towels; contaminated surfaces (for example, mats or benches).

IG13.3 They would not rupture because there would be no influx of water to put pressure on the weakened cell wall.

IG13.4 The daughter cells are identical to each other and to the parent cell.

IG13.5 Mutation is more likely to introduce a new allele, by mutating an existing allele. Gene transfer can introduce an entirely new gene.

IG13.6 In the absence of antibiotics, there is no difference between the fitness of antibiotic-sensitive and antibiotic-resistant variants (both have average fitness).

IG13.7 Over time, the crabs will pose a selective pressure, and a higher proportion of the blue mussels will have the allele that permits the development of thicker shells.

IG13.8 Diversifying selection will select for both lighter and darker oysters in this environment.

IG13.9 Select poultry and meat that has been raised without antibiotics; do not take antibiotics for viral infections; wash hands frequently to reduce the transmission of infections, including antibiotic-resistant infections.

Milestone 4

IGM4.1 Yes. The Lamarck model predicts that acquired traits can be passed on to offspring.

IGM4.2 Darwin did not visit North America or Asia; thus the development of his ideas was based on his observations in other parts of the globe.

IGM4.3 Darwin was influenced by substantial bodies of work ranging from 1798 to 1854.

IGM4.4 Darwin's key observations were made on his *Beagle* voyage (1831–1836), during which he visited South America (including the Galápagos Islands) and Africa. Wallace's key trip was to the Amazon (1848–1852). During their travels, they observed, collected specimens, and learned from the works of others.

Chapter 14

IG14.1 Evolution did not occur. The allele frequencies did not change over time.

IG14.2 The Highland Park and the Willow Lake populations are likely to be isolated populations.

IG14.3 Both are types of genetic drift. In the bottleneck effect, a large population is reduced in size (and genetic variation); in the founder effect, a small number of individuals (with limited genetic variation) leave a larger population.

IG14.4 A new red allele was introduced into population B.

IG14.5 A gene involved in detoxification of chemicals (like pesticides) would be most likely to vary between the two populations, given that agricultural chemicals are present in only one of the two populations.

IG14.6 If the frequency of homozygous recessives is 10% (0.1), then the frequency of the recessive allele is the square root of 0.1 (= 0.316) and the frequency of the dominant allele is $1 - 0.316 = 0.684$.

IG14.7 The pavement ant (*Tetramorium caespitum*) is present in 93.2% of medians.

IG14.8 Hybrid inviability, gametic isolation, and hybrid infertility

IG14.9 No. Differences in beak shape alone do not establish that species are distinct. Whether or not species are distinct is determined by their ability to reproduce and produce fertile offspring.

Chapter 15

IG15.1 Because flies do not have bones or other hard parts, they are most likely to be preserved if they are trapped in amber.

IG15.2 The middle toe is thicker and more prominent in *Merychippus*.

IG15.3 Approximately 511 million years

IG15.4 Lobe-finned fishes have bony rays at the tips of their thick forelimb bones. *Tiktaalik* does not have these bony rays.

IG15.5 Weight-bearing forelimbs could be used to pull *Tiktaalik* onto land and crawl on land.

IG15.6 Hindlimbs are substantially reduced in species that lived only in water.

IG15.7 The humerus

IG15.8 All except the human

IG15.9 There are 2 differences in 30 nucleotides between the close and distant relative. 2/30 = 7% difference, or 93% similarity.

Chapter 16

IG16.1 The time it takes for half of the element present to decay

IG16.2 The amount of uranium-238 decreases, and the amount of lead-206 increases.

IG16.3 It was the first demonstration that organic molecules could form from inorganic molecules, and it also showed that hypotheses regarding the origin of life could be tested.

IG16.4 Prokaryotes were dominant during the Archaen time period; they were dominant in water.

IG16.5 If brown bears evolved in New Zealand, polar bears could have diverged from brown bears in New Zealand. Because penguins evolved in New Zealand, polar bears and penguins could coexist in New Zealand and Antarctica.

IG16.6 Approximately 65 mya

IG 16.7 These animals are not closely related, so they differ in their evolutionary history. They differ in how they give birth to their young, and in where they evolved.

IG16.8 Animals

IG16.9 Supergroup

IG16.10 Bird

IG16.11 Archaea and Eukarya

Chapter 17

IG17.1 *Staphylococcus epidermidis, Staphylococcus aureus, Corynebacterium jeikeium, Propionibacterium acnes, Thaumarchaeota*

IG17.2 Size (prokaryotes are typically smaller); location of DNA (contained in a nucleus only in eukaryotes); organelles absent in prokaryotes

IG17.3 Active decay

IG 17.4 The degree of genetic relatedness and evolutionary history, based on 16S rRNA sequences

IG17.5 The sequence for 16S rRNA genes

IG17.6 Capsules and pili can be used to attach disease-causing organisms to host cells. Flagella can be used to swim to host cells (for example, through a layer of mucus in the stomach).

IG17.7 High temperature, low temperature, high salt, extremely acidic pH

IG17.8 Fiber is a food source for gut microbes, which produce beneficial compounds that suppress gut inflammation and reduce the severity of chronic disease.

Chapter 18

IG18.1 Indonesia: about 27%; Colombia: about 15%

IG18.2 Animals

IG18.3 Cell walls made of cellulose; not mobile; autotrophic

IG18.4 Mosses have no vascular system, so they can't transport water long distances within the plant; their sperm require water to survive.

IG18.5 Vascular tissue

IG18.6 Cell walls made of chitin; external digestion; heterotrophic; not mobile

IG18.7 Molds

IG18.8 They can evade predators, seek out food, and seek out mates.

IG18.9 Sand dollar—radial; parrot—bilateral; jellyfish—radial; earthworm—bilateral; octopus—bilateral

IG18.10 Asymmetrical—sponges; radial—jellyfish and starfish; bilateral—humans, birds, insects

IG18.11 Protists are not closely related to one another. They are found in many groups on the eukaryotic tree of life.

IG18.12 There was a decrease between 1988 and 1990. Deforestation then increased until 2004, followed by a decrease until 2012. Since 2012, deforestation has increased slightly.

Milestone 5

IGM5.1 It was most likely not described as a hot tea. As heating destroys the activity of the compound, a hot tea would not have been effective.

IGM5.2 No. The parasite completes a portion of its life cycle in the mosquito.

IGM5.3 *Pneumocystis jirovecii* is more closely related to humans (both are opisthokonts).

IGM5.4 Mosquito eradication programs could have eliminated the mosquito that carries the malaria parasite. Effective treatment of people who do develop malaria can also prevent the transmission.

IGM5.5 No. The nonphotosynthetic host did not contain chloroplasts (or remnants of chloroplasts) before the endosymbiosis event. Because chloroplasts (and their remnants) are targets of the antimalarial herbicide, the host would not be susceptible to this drug.

Chapter 19

IG19.1 Because humans are 99.9% similar to one another, there would be only 0.1% difference between two people, or 1 nucleotide out of 1,000.

IG19.2 Melanocytes

IG19.3 Dark skin confers more of an advantage in a high-UV environment. Dark skin protects folate from degradation, which is more of a risk in a high-UV environment, and, where the UV level is high, still allows sufficient UV exposure for vitamin D production.

IG19.4 In a low-UV environment, the predominant pressure selecting for light skin is the need to make sufficient amounts of vitamin D.

IG19.5 Mitochondrial DNA is used to trace female lineages (mothers).

IG19.6 The female's mtDNA (shown as purple here) would be passed on to all the couple's offspring.

IG19.7 There would be higher levels of genetic diversity in Australian populations relative to African populations.

IG19.8 *Australopithecus africanus* (minimal overlap), *Paranthropus boisei, Paranthropus robustus, Homo rudolfensis, Homo erectus*

IG19.9 Dark-skin alleles would be predicted to experience negative selection in a low-UV environment.

IG19.10 Light skin is a disadvantage in a high-UV environment because it cannot protect folate from destruction.

IG19.11 Blood type does not appear to have been selected by UV intensity. Based on the map, there is no correlation between UV intensity and blood type. Nor is there any correlation between blood type and skin color, suggesting that these two traits were under different selective pressures.

Chapter 20

IG20.1 Ecosystem ecology

IG20.2 No. They estimate the population size on the basis of aerial counts of a portion of the island.

IG20.3 Clumped (clumps of plants near the watering holes)

IG20.4 Limiting amounts of food; disease; limiting amounts of suitable habitat

IG20.5 As the predator (wolf) population increases, more prey (moose) are hunted and killed, reducing the prey population.

IG20.6 Many large trees suggest that the moose population is low, presumably because of a large wolf population.

IG20.7 The composition of the vegetation in their diet and their overall nutritional status

IG20.8 Widespread flooding is abiotic and density-independent.

IG20.9 Tick populations increase in warmer conditions. An abundance of ticks will lead to a reduction in the number of moose.

Chapter 21

IG21.1 Nevada, Iowa, and Louisiana all have particularly high rates of losses, and Nebraska, Indiana, and New York all have particularly low rates of losses.

IG21.2 Bees use pollen as a protein source and nectar as an energy-rich source of sugar.

IG21.3 Approximately $2.7 billion

IG21.4 Bears can be characterized as consumers, herbivores, and carnivores.

IG21.5 The difference is whether both species benefit or only one. In a commensalism, one species benefits and the other is not harmed. In a mutualism, both species benefit.

IG21.6 Killer bees are out-competing other bee species in their niche, by aggressively defending their food resources.

IG21.7 They can coexist by relying on nectar from different species of wildflowers.

IG21.8 Honeybees are important pollinators of crops that human rely on.

Chapter 22

IG22.1 Temperature and rainfall (abiotic factors) are influencing vegetation type (biotic factor).

IG22.2 Researchers can study the interactions of different species and their impact on the environment to predict which species might be keystone species. The true determination comes from studying an ecosystem after a species is lost. The loss of keystone species have large impacts on the ecosystem.

IG22.3 Nitrogen has a greater abundance in the atmosphere. The atmospheric form (N_2) is not usable to plants.

IG22.4 10% of the energy in the grass is stored in the cow; 1% of the energy stored in the grass is stored in the human.

IG22.6 Oceans and aquifers store water; groundwater returns water to lakes and oceans.

Milestone 6

IGM6.1 Typhus; malaria

IGM6.2 By eating earthworms

IGM6.3 Chemicals like DDT are retained in the bodies of organisms that ingest them (such as plankton in water). When many of these organisms are eaten by another (such as small fish), all the retained DDT ends up being retained in the consumer. At each level of the food chain, the concentration of the chemical increases (from small fish to large fish to birds of prey, such as eagles).

Chapter 23

IG23.1 If the mosquitos expand their range, dengue virus infections may occur outside of Florida.

IG23.2 One milder winter is a relatively short-term difference, so reflects a difference in weather.

IG23.3 The temperatures in the southeastern United States are decreasing, while they are increasing in New England.

IG23.4 Increasing spring temperatures are causing earlier flowering. This means that plants flower well before deer are born, so there is limited food for young deer, and their survival is reduced.

IG23.5 Bark beetles are increasing in numbers, and trees are less able to defend themselves. This means that bark beetles are killing more trees. The dead trees are a source of fuel for wildfires.

IG23.6 Greenhouse gases trap in the atmosphere some of the heat energy that is reflected off Earth's surface.

IG23.7 The most recent data show the current temperature to be approximately 0.8°C warmer than the 20th-century average. The temperature has been consistently warmer than the 20th-century average since approximately 1980.

IG23.8 Northern regions have the greatest increase in temperatures, and some areas near the southern pole have colder than average temperatures.

IG23.9 If human activities had no effect on the carbon cycle, there would be 9 fewer gigatons of CO_2 added the atmosphere each year. This would result in greater amounts of CO_2 being taken out of the atmosphere than added to the atmosphere.

IG23.10 CO_2 remained less than 300 ppm for 800,000 years. It is now at 407.4 ppm. This is an increase of 135.8% ($407.4/300 \times 100$).

IG23.11 The proportion of emissions from transportation would decrease. If the cars are recharged from the electrical grid, the emissions from electricity may increase.

IG23.12 Natural gas

IG23.13 There are many possibilities, including using LEDs, recycling whenever possible, walking or riding a bike when possible, and reducing vampire energy waste by using a power strip that is turned off each night.

Chapter 24

IG24.1 Increases in crop yield

IG24.2 Water—through roots; CO_2—through leaves; nitrogen—through roots

IG24.3 An excess of these nutrients stimulates the growth of algae. When the algae die, decomposition removes oxygen, which harms the fish.

IG24.4 Legumes naturally form root nodules with nitrogen-fixing bacteria.

IG24.5 Fruit

IG24.6 Pollination occurs at the stigma. Fertilization occurs in the ovules in the ovary.

IG24.7 One sperm fertilizes the egg to create the embryo. One sperm fuses with two nuclei in a central cell in the ovule to generate the endosperm.

IG24.8 Sexual reproduction with animal pollination

IG24.9 The increased use of *Bt* corn has reduced the number of insect pests. With fewer insect pests, there is less need for insecticide use.

IG24.10 Transgenic organisms

Chapter 25

IG25.1 Too much fertilizer will create a relatively high solute concentration in the soil, causing water to leave the roots by osmosis.

IG25.2 Both transport water and dissolved substances throughout the plant. Xylem carries water from roots to shoots. Phloem transports dissolved sugars from shoots to roots or from roots to shoots.

IG25.3 The lateral meristems—the vascular cambium and the cork cambium

IG25.4 Nitrogen

IG25.5 Many plants close their stomata on very hot, sunny, dry days to prevent water loss. Closed stomata prevent CO_2 entry, limiting the amount of photosynthesis that can take place.

IG25.6 They help capture additional wavelengths of light for photosynthesis.

IG25.7 No. Burrs can attach to the fur of any animal and do not need to be eaten to be dispersed.

IG25.8 The cells elongate.

IG25.9 Gibberellins

IG25.11 Both. Physical defense—tough exterior; chemical defense—toxins

Chapter 26

IG26.1 A tissue has different cell types, which work together to produce the function of the tissue. An organ has multiple tissue types, which work together to produce the function of the organ.

IG26.2 A body temperature of 36.5°C (97.7°F) is within normal range and will not trigger a response to restore homeostasis.

IG26.3 Shivering generates heat. We cannot generate "cool"—instead, we rely on heat dissipation through sweating and vasodilation to reduce body temperature.

IG26.4 Sweat glands and blood vessels. Sweat glands release sweat; the evaporation of sweat cools the body. Blood vessels dilate to release heat or constrict to conserve heat.

IG26.5 The pituitary

IG26.7 Insulin reduces the level of glucose (sugar) in the blood, whereas glucagon elevates the levels of glucose in the blood.

IG26.8 ADH is made in and released by the pituitary. Its target organ is the kidney. ADH causes the kidney to conserve water (by reabsorbing water).

Chapter 27

IG27.1 Asthma, cardiovascular disease or high blood pressure, arthritis

IG27.2 Salivary glands, liver, gallbladder, pancreas

IG27.3 Key events that occur in the stomach include mixing of food with stomach acid and digestive enzymes; digestion of proteins by pepsin; unfolding and loss of shape of proteins by the action of stomach acid; destruction of bacteria by stomach acid.

IG27.4 They are all accessory organs that secrete enzymes and other substances into the small intestine.

IG27.5 A person lacking villi would not be able adequately to absorb nutrients.

IG27.6 No. Fats are fully digested before leaving the small intestine.

IG27.7 Both procedures reduce the size of the stomach. Sleeve gastrectomy does not re-route the passage of food; gastric bypass does.

IG27.8 216 pounds. (In 10 years, 28% of body weight is lost; 28% of 300 pounds is 84 pounds; $300 - 84 = 216$.)

IG27.9 leptin—satiety; PYY—satiety; ghrelin—hunger

Milestone 7

IGM7.1 The islet of Langerhans cells produce insulin; the duct cells produce digestive enzymes.

IGM7.2 When the pancreas was removed, the dog was unable to produce insulin and so was unable to control its blood-sugar levels.

IGM7.3 The primary difference relates to the production (or nonproduction) of insulin. In type 1 diabetes, insulin is not produced; in type 2 diabetes, insulin is produced but cells do not respond properly to the hormone.

Chapter 28

IG28.1 38.8%

IG28.2 An artery

IG28.3 Fatty streaks are still a cause for concern even if they don't presently obstruct blood flow because they may increase in size and harden, developing into plaques that can obstruct blood flow.

IG28.4 Oxygen diffuses into the tissue (from the blood) and carbon dioxide diffuses into the blood (from the tissue). The amount of oxygen in the blood increases, and the amount of carbon dioxide in the blood decreases.

IG28.5 Red blood cells transport oxygen, white blood cells contribute to immune defenses, and platelets form a blood clot at the site of an injury.

IG28.6 Blood in a vein in the pulmonary circuit is traveling toward the heart (from the lungs). It is oxygen rich.

IG28.7 High blood pressure can cause tears in artery walls where plaque builds up and can cause plaques to rupture.

IG28.8 The extent of plaques on the lining of the aorta of those with 2 risk factors would be similar to their extent in people with 0 or 1 risk factors. The extent of plaques on the lining of the coronary arteries would be greater than in people with 0 or 1 risk factors, but not as extensive as in people with 3 or 4 risk factors.

IG28.9 Cholesterol is a lipid, so it is hydrophobic and not soluble in blood.

IG28.10 Texas and southern states have consistently had high proportions of overweight and obese people. Colorado has consistently had low proportions of overweight and obese people.

IG28.11 Texas and southern states have high rates of both overweight and obesity, as well as high rates of deaths due to heart disease.

Chapter 29

IG29.1 From the alveoli to the blood

IG29.2 In the lungs, O_2 enters the blood and CO_2 leaves the blood. In a tissue (such as muscle), O_2 leaves the blood and CO_2 enters the blood.

IG29.3 If the supplemental oxygen is sufficient to mimic the partial pressure of oxygen at the individual's home (lower) altitude, then that individual will not experience hypoxia and will not have an elevated RBC count.

IG29.4 Partial pressure at the peak of Mount Everest is 53 mmHg; at Colorado Springs, Colorado, it is 129 mmHg; at sea level, it is 159 mmHg.

IG29.5 An athlete's RBC increases in hypoxic conditions. A higher RBC is advantageous when an athlete competes at sea level because it increases the O_2-carrying capacity of the blood.

IG29.6 Air flows in.

IG29.7 The change in oxygen uptake for males and for females is very close. The females appear to have a slightly larger increase (from 64 to 66 mg/kg/min) compared to males (from 75.7 to 77 mg/kg/min). The males had a slightly larger change in race time, with a 5.8-second improvement relative to the 5.5-second improvement by females.

IG29.8 A dietary iron deficiency would translate into an inability to make heme groups of hemoglobin, thereby reducing the ability of red blood cells to transport oxygen.

IG29.9 40%

IG29.10 The kidneys

Chapter 30

IG30.1 Approximately 137 are likely to have vaped, and approximately 29 are likely to have smoked cigarettes.

IG30.2 The nicotine in the chewing tobacco can still be absorbed into the bloodstream (through the mucous membranes that line the mouth). From the bloodstream it will reach the brain.

IG30.3 The peripheral nervous system senses and responds to the mosquito.

IG30.4 The dendrites receive signals; the axon terminals transmit signals.

IG30.5 Sodium ions (Na^+) flowing into the cell trigger an action potential at a particular spot along a neuron.

IG30.6 An injury to the cerebellum could impair movement, coordination, and balance.

IG30.7 An axon terminal

IG30.8 Drug-addicted people need to take increasing amounts of the given drug to compensate for the reduced number of dopamine receptors on receiving cells.

IG30.9 The participant who recalled the greatest number of words had more dopamine reuptake receptors than the participant who recalled the fewest number of words.

Chapter 31

IG31.1 Triplet and higher births peaked in 1998. Women between 45 and 54 years of age had the highest rate of triple (and higher) births in 2016.

IG31.2 Fertilization occurs in the oviduct; the embryo develops after implantation into the lining of the uterus.

IG31.3 Sperm travel from the seminiferous tubules through the epididymis, the vas deferens, and finally the urethra.

IG31.4 Sperm pass through the cervix from the vagina to the uterus. Once in the uterus, sperm must travel to the oviducts (fallopian tubes) in order to fertilize an egg.

IG31.5 Egg production occurs in the ovary (normal location); fertilization occurs outside the body (in a dish); embryo implantation occurs in the lining of the uterus (normal location).

IG31.6 FSH is produced by the anterior pituitary and acts on the ovary, causing a follicle to develop; LH is produced by the anterior pituitary and acts on the ovary, causing ovulation; estrogen is produced by a follicle in the ovary and acts on the endometrium, causing it to thicken; progesterone is produced by the corpus luteum in the ovary and acts on the endometrium, causing it to thicken.

IG31.7 FSH and LH, which in males trigger spermatogenesis, are produced by the anterior pituitary.

IG31.8 The hormone hCG is made by the embryo. It is important in early pregnancy because it causes the corpus luteum in the ovary to continue secreting the hormones estrogen and progesterone to maintain the endometrium.

IG31.9 There are many possible blockage sites, including the oviducts and the vas deferens.

IG31.10 In IVF, the physician can control the number of embryos that are implanted. In IUI, if many eggs are ovulated, there is a chance that all of them will be fertilized and implanted.

IG31.11 The second trimester. The external genitalia develop during this trimester.

Milestone 8

IGM8.1 Zygote—totipotent; stem cell from the bone marrow—multipotent; embryonic stem cell—pluripotent

IGM8.2 A chihuahua (as that was the donor of the nucleus in the cloning procedure)

IGM8.3 Every cell in the body will have the gene for the myosin heavy-chain beta.

IGM8.4 Skin cells and pancreatic cells are differentiated; iPSCs are pluripotent.

IGM8.5 If there is no standard treatment for their form of Alzheimer's disease, then using iPSCs to model their disease could result in a new treatment. If their disease could benefit from new and healthy neurons in the brain, they could try cell therapy using corrected iPSCs to generate specialized cells (neurons).

Chapter 32

IG32.1 Approximately 2.7% (50 million/1.8 billion × 100)

IG32.2 No. A cell cannot be a host for a virus if the cell does not have the specific receptor molecules to which that virus binds.

IG32.3 Yes. Fungi and various other eukaryotic pathogens can cause disease in humans.

IG32.4 There are many differences between innate and adaptive immunity, including specificity, whether or not the defense is always present and active, and the types of cell involved.

IG32.5 Mucous membranes lining the digestive tract, saliva (with bacteria-destroying enzymes), and acid in the stomach

IG32.6 Specialized phagocytes process ingested pathogens, then display (that is, present) antigens to lymphocytes of the adaptive response. Lymphocytes bind to the displayed antigen, leading to lymphocyte activation.

IG32.7 B and T cells are both produced in bone marrow. B cells mature in bone marrow, and T cells mature in the thymus.

IG32.8 Plasma cells produce antibodies. Antibodies bind to specific antigens on pathogens, resulting in the elimination of the pathogen.

IG32.9 Cytotoxic T cells recognize target antigens from the pathogen displayed on the surface of the host cell.

IG32.10 The first response to pathogen C will be a primary response.

IG32.11 Because not every person is vaccinated, some people can still be infected.

IG32.12 A community experiencing a measles outbreak likely has low rates of vaccination against measles.

IG32.13 Slight differences are the result of antigenic drift.

Answers to End-of-Chapter Questions

See below for the answers to the "Know It" questions and multiple-choice "Use It" questions. Sign on to Achieve to see the answers to the rest of the "Use It" questions, Mini Cases, Bring It Home, and Interpreting Data problems.

Chapter 1

1. c
2. e
3. b
4. c
5. d
8. a
9. e
10. e
11. c
15. e
16. b
17. e
20. c
21. b

Chapter 2

1. c
2. Homeostasis is the maintenance of a relatively stable internal environment even when the external environment changes. It helps maintain the conditions necessary for life: many of the molecules and processes of life cannot function outside a narrow range of, for example, pH and temperature.
5. e
6. b
7. six
8. A polymer is a molecule made up of smaller and typically repeating subunits. Examples include proteins, which are made up of amino acids; complex carbohydrates, which are made up of simple sugars, or monosaccharides; and nucleic acids, which are made up of nucleotides.
9. d
11. f
13. b
14. e
17. Olive oil does not mix with water, so it is hydrophobic; it is made up of hydrophobic triglycerides. Salt will dissolve in water, so it is hydrophilic. When salt dissolves, the charged Na^+ and Cl^- ions can interact with the water molecules.
18. a
19. a. The solvent in coffee and tea is water.
b. The main solute is the dissolved sugar molecules, but there are also other compounds in coffee and tea that dissolve in the water as the beverage is prepared; these are also solutes.
c. Because the sugar dissolves in water, sugar is hydrophilic.
20. c
21. a
22. Both ionic bonds and hydrogen bonds are based on electrical attractions. Ionic bonds are strong electrical attractions between oppositely charged ions. Hydrogen bonds are weak electrical attractions between atoms with partial charges.
24. b

Chapter 3

1. The cell theory states that all living organisms are made of cells, and that all cells arise from existing cells.
2. d
3. Prokaryotic and eukaryotic
4. d
5. d
6. e
8. c. This organism cannot be an animal because animal cells do not have cell walls. Choice c is convincing evidence that the organism is a bacterium and not a plant.
9. e
10. c
11. a. In both simple diffusion and facilitated diffusion, a solute moves across a membrane from an area of higher solute concentration to an area of lower solute concentration without the input of any additional energy.
b. Active transport and facilitated diffusion both require a transport protein to move a solute across a membrane.
12. b
14. c
17. b
18. Phospholipids of the cell membrane would not be a good target for an antibiotic because both human and bacterial cells have phospholipids in their cell membranes. Thus, the proposed antibiotic would harm both bacterial and human cells.
24. a. Mitochondria are rod-shaped organelles. They are surrounded by a double membrane (they have both an inner and an outer membrane). Mitochondria are important in the reactions that extract energy from food and convert it to an immediately usable form.
b. The nucleus is a large organelle that stores the genetic instructions (DNA). It has a double membrane (the nuclear envelope) containing pores through which molecules can move into and out of the nucleus.
c. The endoplasmic reticulum is an extensive network of membrane tubes. Its critical functions include protein and lipid synthesis.
d. The chloroplast contains an internal system of stacked membrane disks surrounded by two membranes on the outside of the organelle. It is the site of photosynthesis in plants and algae.
25. a
26. d
28. c

Milestone 1

1. Mitochondria are cellular "power plants." They carry out the reactions of cellular respiration to generate usable energy. Chloroplasts are the photosynthetic organelles in plant cells. Using the energy of sunlight, they convert atmospheric carbon dioxide to organic carbohydrates (sugars).
2. Chloroplasts have their own DNA; they have their own ribosomes; they are about the same size as bacteria and replicate in the same way as bacteria; mitochondria share these traits with chloroplasts.
3. a. cyanobacteria; b. *Rickettsia*
4. c
5. a. You could not live without your mitochondria (endosymbiotic organelles). Humans require continuous cellular respiration to sustain life.
   b. If plants did not have mitochondria, they would die (much as humans would die without mitochondria). That would reduce the food supply for humans (directly or indirectly because the animals that many humans eat, in turn, eat plants). Furthermore, if plants lacked chloroplasts, they would not be able to carry out photosynthesis. They would die (which would be very bad for humans). Additionally, when plants carry out photosynthesis, oxygen is released, and humans rely on this oxygen.

Chapter 4

1. c
2. f
3. b
4. a
5. b
6. e
9. a
11. e
12. Both cofactors and coenzymes help enzymes speed up their reaction rates. Cofactors are typically inorganic metals, whereas coenzymes are organic molecules, such as vitamins.
13. b
14. c
17. a. Vitamin C is water soluble, so any excess is removed from the body in the urine. Vitamin E is fat soluble, so any excess is stored in fatty tissues in the body, possibly accumulating to toxic levels.
    b. Taking high levels of a vitamin C supplement will not lead to storage of vitamin C in the body. Any excess over what is needed is excreted from the body, dissolved in the urine.
18. e

Chapter 5

1. They are all photosynthetic—that is, they carry out photosynthesis, using the energy from sunlight and carbon dioxide from air to synthesize sugars.
2. No. Animals cannot carry out photosynthesis. However, they rely on photosynthesis, because they eat plant material, which is the product of photosynthesis, as well as material from animals that are sustained by eating plants.
3. c
4. The main photosynthetic pigment is chlorophyll, which reflects green wavelengths.
5. Photosynthetic algae rely on photosynthesis for the production of sugars that can be used for short-term energy needs or for longer-term energy storage. Animals cannot make their own sugars (or other energy-rich organic molecules) from scratch, so they must obtain them from their diet.
8. e
9. c
10. c
12. a
13. c
14. oxygen (O); carbon dioxide (I); photons (I); glucose (O); water (I)
15. a
19. g
20. e

Chapter 6

1. a
2. d
3. c
4. e
7. c
8. b
9. a
10. b
11. d
13. c
15. d
16. c
17. b
18. a
22. obese
23. b

Chapter 7

1. e
2. d
3. c
6. e
7. c
8. b
9. e
10. a
11. c
12. a
15. Lane b
16. a. Suspect B's profile matches the evidence at all markers tested. Suspect B is more than likely the source of the DNA evidence at the crime scene.
    b. Suspect B is likely to be related to the victim because suspect B shares at least one band with the victim at every allele tested. Suspect B could be the parent or child of the victim.
17. a
18. d
19. a. F; b. T; c. F; d. F; e. T

Milestone 2

1. c
2. d
3. Franklin carried out the key x-ray diffraction studies. Wilkins recognized that Franklin's data might be helpful to Watson and Crick. Watson and Crick synthesized existing information to build models that were consistent with the observed data. Franklin's experimental observations were critical in assembling the correct model, the one that accurately represented the structure of DNA.
4. The DNA stretched out in the presence of water, suggesting that the water must be interacting with and coating the molecule. As the charged phosphate groups are hydrophilic, and as the water was interacting with DNA on the surface of the DNA, the phosphate groups had to be on the outside of the molecule—a key insight.
5. The DNA double helix is arranged as two strands running in opposite directions. The phosphate groups of each strand are oriented along the external backbone of the molecule, and the nucleotide bases are oriented toward the interior of the helix. The bases from one strand pair with bases on the other strand, according to specific rules: A pairs with T and G pairs with C. The helix has a constant diameter along its length.

Chapter 8

1. b
2. e
5. This statement is not accurate. Chromosomes contain genes, and genes encode proteins. A possible rewrite: "A chromosome contains many genes. Each gene encodes one or more proteins."
6. c
7. RNA polymerase (N); ribosome (C); tRNA (C); mRNA (C) (mRNA is made by transcription in the nucleus, but it acts in translation in the cytoplasm.)
8. d
9. d
14. e
15. The vector is a "carrier" DNA molecule that carries the recombinant gene into the organism and its genome.
16. The recombinant gene would need a regulatory sequence from a spider silk gene and the coding sequence of the desired yeast gene.
20. Unlike silkworms (which produce silkworm silk in large quantities on silk "farms"), spiders cannot easily be raised in large groups (because they will eat one another). It is challenging to raise enough spiders to produce large quantities of spider silk. Transgenic technology allows large quantities of spider silk to be produced by another organism (such as yeast).
21. Spider silk has a unique combination of strength and elasticity, making it useful in the manufacture of a variety of products, including bulletproof vests, medical devices, and airbags.

Chapter 9

1. d
2. b
3. c
4. c
5. e
9. b
10. The virus naturally delivers genetic material in a way that it can be stitched into the genome of the stem cells.
11. a
15. b
16. e

Chapter 10

1. Mitosis
2. c
3. b
4. Anaphase of mitosis
5. S phase
6. b
10. b
11. e
12. In their nonmutant states, both tumor suppressor genes and proto-oncogenes are important for proper cell cycle progression. In their mutant states, both contribute to the development and progression of cancer. Normal tumor suppressor genes act to prevent the cell cycle from progressing inappropriately (e.g., when there is rampant DNA damage). When mutated, tumor suppressor genes can no longer pause the cell cycle when necessary, and cells with DNA damage may continue to divide. Normal proto-oncogenes act to promote cell division in response to appropriate signals to divide. When proto-oncogenes are mutated to oncogenes, they continuously "push" cells to divide, even in the absence of growth-promoting signals.
13. *BRCA1* is a tumor suppressor gene that encodes a protein involved in repair of DNA.
14. e
17. a
21. c
22. Surgery and radiation are generally targeted at a specific location. Surgery removes a tumor (and potentially surrounding tissue) at a specific location, and radiation is typically used to irradiate a tumor at a specific location. Chemotherapy can reach every cell of the body, including metastatic cancer cells that have spread from the original tumor.

Chapter 11

1. a
2. a. There are 46 chromosomes in each human liver cell.
   b. There are two alleles of each gene in each liver cell.
3. a. A human gamete has 23 chromosomes.
   b. There is one allele of each gene present in a gamete.
7. c
8. f
9. See Infographics 11.4 and 11.5.
10. a. Homologous chromosomes (homologues) separate. This is different from what happens in mitosis, during which sister chromatids separate.
    b. At the end of meiosis I, cells are haploid (they have one copy of each homologue). This result is different from epithelial cells that have completed mitosis, during which diploid epithelial cells divide to produce identical diploid cells.
14. b

15. a. no CF; b. no CF; c. CF

16. A person with CF has two copies of the CF-associated (recessive) alleles of the CF gene in his or her lung cells. A carrier is heterozygous, having one CF-associated (recessive) allele and one non-CF allele. A homozygous dominant person has no CF-associated alleles.

17. c

23. CF cannot be diagnosed prenatally by chromosome analysis because the mutations associated with CF are too small to be detected by that method.

Milestone 3

1. a. A tall pea plant could be homozygous dominant (*TT*) or heterozygous (*Tt*).
b. You would cross the tall pea plant with a homozygous recessive (true-breeding short) plant (*tt*). If the tall plant is true breeding (homozygous), then 100% of the offspring will be *Tt* (tall). If the tall plant is heterozygous, 50% of the offspring will be *Tt* (tall) and 50% will be *tt* (short).

2. a. If both parents are true breeding for both traits, then 100% of the offspring will be tall with purple flowers.
b. In this case, all the offspring will be homozygous dominant (*TTPP*) (as both of their parents were homozygous dominant). If all the plants are homozygous dominant, then they will be true breeding.

3. Mendel's law of segregation. The two alleles will segregate from each other at meiosis I.

4. a. Mendel's law of independent assortment would be violated, as the alleles of each gene do not appear to be assorting independently of each other.
b. These results would suggest that the *T* and *B* alleles are closely linked on one homologue, and that the *t* and *b* alleles are closely linked on the other homologue.

Chapter 12

1. c

2. Males have only one X chromosome, while females have two. This means that if a male inherits an X-linked recessive allele (from his mother's X chromosome), that is the only allele of that gene he has—there is no dominant allele on another X chromosome to mask the recessive. In this case, the male will develop the disease. If a female inherits an X-linked recessive allele on one of her two X chromosomes, she will not develop the disease if her other X chromosome has the dominant allele to mask the recessive allele.

3. No. Sons cannot inherit X-linked conditions from their fathers because sons inherit their father's Y chromosome, not the X. The son will inherit his X chromosome from his mother.

4. c

8. In incomplete dominance, heterozygotes have a phenotype that is intermediate between the phenotypes of the homozygous dominants and the homozygous recessives. In codominance, heterozygotes display traits of both alleles present.

9. If you are type A-positive, then you can donate to other A-positive people as well as to AB-positive people. If you are A-positive, you can receive type O-negative, O-positive, A-positive, and A-negative blood.

12. Many genes contribute to height. As multiple genes contribute to the phenotype, height is a polygenic trait.

13. d

14. In polygenic inheritance, multiple genes influence the phenotype. Multifactorial traits are those on which environment has an influence.

18. a. 23; b. 23; c. 46

19. Karyotype analysis can be used to detect trisomy 21 because an extra chromosome is easily visible at this level. However, cystic fibrosis is caused by mutations that change the nucleotide sequence of the gene—these changes cannot be detected by simply looking at the chromosomes.

20. c and d

Chapter 13

1. d

2. a

3. In colonization, the bacteria are growing on or in the body without causing disease. Infections are associated with disease.

4. c

5. a. A penicillin-sensitive strain of *S. aureus* will eventually burst (lyse) and die because of its weakened cell wall.
b. A resistant strain of *S. aureus* will not be affected by penicillin and will continue to grow.

6. d

10. b

11. c

15. Fitness, in an evolutionary context, describes the ability of an individual to survive and reproduce in a given environment. Individuals that are more fit leave more offspring, and thus more of their alleles in the next generation, relative to individuals that are less fit.

16. a. For exposure to penicillin, the fitness of the MRSA strain will be high (because it is resistant to this antibiotic, it will continue to divide and grow).
b. For exposure to ciprofloxacin, its fitness will be low (because it is not resistant to this antibiotic).
c. For exposure to vancomycin, its fitness will be high.

17. c

18. d

19. a

20. c

Milestone 4

1. That although the giant sloth was an extinct animal, it resembled modern-day sloths in Argentina. This observation led him to consider that the modern-day animals might be descendants of the ancient giant sloths.

2. Malthus wrote about factors (such as hunger) that would limit the growth of a population. Darwin realized that individual organisms must therefore compete for access to resources, and that any small variation that gave an individual an advantage would lead to its success over others. If the variations were successful, then the individual with those variations would survive and reproduce to a greater extent than individuals without these successful variants. Over time, these variations, or traits, would become more common in the population.

3. Wallace focused on Malthus's writings about disease and how disease limited the growth of populations. As Wallace himself was suffering from malaria, he realized that disease would eliminate the weakest members of a population, leaving the strongest (the most fit) individuals to survive and reproduce. He reasoned that the survival of the strongest individuals would lead to changes in the population, leading to adaptations, and even to new species.

4. Both Darwin and Wallace made careful observations while on their voyages. Darwin had read Lyell's work, and when he observed a bed of seashells in a cliff well above sea level and the effects of an earthquake, he truly appreciated how much the geology of Earth changed over long periods. His observation of the extinct sloth and smaller but similar modern sloths led him to consider how resemblances suggested ancestral relationships. Wallace's observations of distinct yet similar species on either side of a physical separation (e.g., a river or canyon) led him to the idea that each species was somehow related to a preexisting species. Without actually seeing these fossils, organisms, and events, it is less likely that either scientist would have been able to develop his understanding of evolution.

Chapter 14

1. 3,200 mice have 6,400 alleles, of which 2,200 are *g*. 2,200/6,400 gives an allele frequency of 0.34.
2. Population A has fewer alleles than the other populations, suggesting that it may be an isolated population.
6. e
7. d
8. c
9. a. The allele frequency for every allele except *H* is now 0. The allele frequency for *H* is 1 (100% of the alleles are *H*).
   b. As a random event affecting allele frequencies, this is an example of genetic drift—specifically, it has created a bottleneck.
14. e
15. Evolution has occurred. The allele frequencies of the *TUB* gene have changed over many generations (in this case, in response to a change in the environmental conditions).
16. Matings between relatives can produce offspring with two (detrimental) recessive alleles. Over time, inbreeding reduces the frequency of heterozygotes, and produces homozygotes that have two deleterious alleles.
18. d
19. When populations are geographically isolated, they do not exchange alleles. This means that if a mutation arises in one population and not in the other, the mutation will be present in only one of the two populations. As the number of different mutations accumulates in each isolated population and the resulting phenotypes are acted on by natural selection, the two populations could diverge enough so that they cannot successfully interbreed if they come into contact with one another.

Chapter 15

1. c
2. The newest fossils will be in the layers closest to the surface.
3. A fossil with four limbs and digits would be the fossil of an organism more recent than a "standard" fish. Thus, the "standard" fish fossil would be in deeper layers, and the four-limbed fossil would be in layers above that fish fossil.
4. This fossil seems to be similar to modern-day bony fishes. It has fins with rays, suggesting that this organism was aquatic and able to swim. There does not appear to be a distinct neck, consistent with a fish, and there could be a gill cover present, again consistent with an aquatic organism. This may have been a predatory fish, as teeth appear to be present.
8. c
9. Long and sturdy ribs (to help support the body) and pectoral fins that have wrists and can bear weight
13. The skeletal anatomy of an eagle wing and that of a human arm are very similar. All major bones are present in each, and in the same locations relative to other bones. In the human, the most distal bones (the tips of the digits) are longer and arranged in a way that permits fine manipulation of objects with hands and fingers. Birds do not need to carry out this fine-scale manipulation, and their wings are specialized for flying.
14. Middle ear bones in humans; gills in adult bony fish
15. You could make pairwise comparisons between the sequences, counting the number of nucleotide differences between them. The more differences there are, the less similar the sequences are. More-similar sequences suggest closer relationships, and less-similar sequences suggest more distant relationships. Ideally, you would compare a large number of genes before coming to a conclusion.

Chapter 16

1. They are all radioactive isotopes that decay into other elements at constant rates.
2. You would use uranium-238, which has the longest half-life of the three (4.5 billion years). Isotopes with shorter half-lives may no longer be present in very ancient samples, having completely decayed.
3. (1) the first prokaryotes (~3 billion years ago); (2) an increase of oxygen in the atmosphere (~2.5 billion years ago); (3) the first multicellular eukaryotes (~1.2 billion years ago); (4) the Cambrian explosion (~545 million years ago); (5) the first animals (~540 million years ago); (6) the Permian extinction (~248 million years ago); (7) the extinction of dinosaurs (~65 million years ago)
9. No. While they may be closely related, they may also represent convergent evolution, in which unrelated groups of organisms share common characteristics because of independent natural selection in similar environments.
10. See Infographic 16.6. Several of the continental landmasses were much closer together (e.g., North America and Eurasia). Over time, through plate tectonics, the landmasses have moved farther apart to their present locations. As the landmasses moved, organisms moved with them and were subjected to changing environments, which influenced the evolution of organisms.
14. f
15. (1) domain; (2) supergroup; (3) kingdom; (4) phylum; (5) genus; (6) species
16. d
18. c

Chapter 17

1. e
2. e

3. Babies start to acquire their microbiome during the birth process, picking up microbes from the birth canal. Microbes continue to be introduced into the microbiome through breast feeding.
6. c
7. d
8. d
9. a
10. b
15. e
16. b

Chapter 18

1. d
2. Eukaryotes are defined by having cells with membrane-bound organelles, including a nucleus.
3. They are both eukaryotes. Both have eukaryotic cells (with a nucleus and internal membrane-bound organelles).
7. Plants similar to bryophytes were probably the first plants to live on land. As they do not have a vascular system to transport water throughout their bodies, they live in damp environments. They lack adaptations that would enable them to live in other, drier environments.
8. d
12. a. Fungi are not photosynthetic; they are heterotrophs.
    b. They do not ingest their food.
    c. Fungi obtain nutrients and energy by secreting digestive enzymes onto their food source in the external environment. The food is then digested into smaller subunits, which are then absorbed by the fungi.
13. c
15. b
16. d
17. f
18. a
20. b

Milestone 5

1. a. *Plasmodium falciparum* requires the *Anopheles* mosquito in order to complete its life cycle.
   b. *Anopheles* mosquitoes are insects in the supergroup Opisthokonts and *Plasmodium falciparum* is a member of the Rhizaria supergroup.
2. *Phytophthora infestans* has been reclassified as a Stramenopila, on the basis of molecular evidence. *Pneumocystis jirovecii* is currently classified as a fungus, on the basis of genetic evidence. *Trypansoma brucei* is classified as a member of the Excavata supergroup.
3. *Pneumocystis jirovecii* was formerly classified as animal-like protist. It is now classified as a fungus. This classification has allowed the use of more appropriate drugs (antifungal drugs) to treat the pneumonia caused by this organism.
4. a. Artemisinin is directed against the parasite. Resistance is emerging.
   b. DDT is directed against the mosquito. Resistance is emerging.
   c. Quinine is directed against the parasite. Resistance is emerging.
5. As climates warm, the mosquito vector of malaria can expand its range, introducing malaria into areas that were previously free of the disease.
6. a. DDT and other insecticides kill the mosquito, so they block all steps occurring in the mosquito, as well as transmission to humans by the bite of a female mosquito.
   b. There are several points that you could target to block the life cycle and minimize symptoms in humans:
   (i) You could block step 1 (transfer of parasites into humans).
   (ii) You could block the life-cycle steps that occur in the liver, so that the parasite can never be released from the liver into the red blood cells (because once the parasite gets in the red blood cells, the symptoms occur). You could also propose blocking step 4 (reproductive life cycle continues in the mosquito), although this won't prevent the symptoms from occurring at step 3.
7. Plants are in one supergroup (Chromaveolata). Animals and fungi are in another supergroup (Opisthokonta). Protists are found across all the supergroups.

Chapter 19

1. Darker skin confers protection from UV destruction of folate, while still permitting vitamin D synthesis in a high-UV environment. Their skin color was likely dark.
2. c
3. b
5. d
7. c
9. mtEve is a single female ancestor of all modern humans. mtEve lived in eastern Africa between 200,000 and 150,000 years ago.
10. a
11.

Type of DNA	*Description*
mtDNA	a. Inherited by all children only from their mother
nuclear DNA	d. Inherited by all children from both their mother and their father
Y-chromosome DNA	c. Inherited only by sons from their father

15. e
16. (1) last common ancestor of chimpanzees and humans; (2) *Ardipithecus ramidus;* (3) *Australopithecus;* (4) *Homo erectus;* (5) *Homo sapiens*
17. A person with dark skin could have any blood type. A person with type A blood could have any skin color. There is a huge amount of variation in a trait like blood type, even in a population of people of similar skin color.

Chapter 20

1. A population is a group of interbreeding organisms of the same species living in a particular geographic area. A community includes all the organisms (and populations) in a geographic area. Populations represent a single species; communities represent multiple species.

2. e

3. Because oases are not evenly distributed, the distribution of palm trees would most likely appear to be clumped (with a clump of trees at each oasis).

7. c

8. c

9. At carrying capacity, the population growth rate is 0.

12. Shootings and traffic kills are directly related to human activities, and are thus biotic factors.

13. The drying up of the pond due to drought is an abiotic factor.

14. c

Chapter 21

1. A community is a collection of interacting populations of different species in a given geographic area. A population is a group of interbreeding organisms of the same species in a given geographic area.

2. b

3. a

7. b

8. f

9. d

10. a

13. b

14. a

17. The honey bee niche includes flowers of a particular shape or color that attracts bees and assists the transfer of pollen. The niche will contain a variety of flowering plants that flower at different times during the seasons when the bees are active. Other nectar-feeding organisms can coexist with bees by relying on flowers other than the ones the bees are attracted to.

18. d

Chapter 22

1. f

2. Keystone species are species that have a strong influence on the community without necessarily having a high abundance.

3. d

6. N_2 is abundant in the atmosphere. Ammonia is the major form of nitrogen taken up by plants.

7.

Nitrogen Conversion Process	*Organism*
$N_2 \rightarrow NH_3$	b. nitrogen-fixing bacteria
proteins $\rightarrow NH_3$	c. decomposers
$N_2 \rightarrow$ chemical fertilizer	a. nitrogen-recycling bacteria in soil

8. c

9. Phosphorus is an important component of the phospholipids that make up cell membranes, and of the nucleotides that make up DNA.

10. b

13. a

14. c

Milestone 6

1. a. DDT was intended to combat insect-borne diseases, specifically typhus and malaria, among U.S. soldiers during World War II.

b. It was very effective, saving the lives of countless soldiers.

2. a. DDT is toxic to the nervous systems of insects.

b. DDT accumulates in organisms up the food chain, and the high concentrations in top predator birds cause reproductive failures—eggshells thin to the point that the eggs are easily crushed before the chicks hatch.

3. Biomagnification is the process by which chemicals (particularly toxic chemicals) increase in concentration with each trophic level. Organisms at the lowest trophic levels have the lowest concentrations and organisms at the highest trophic levels have the highest concentrations.

4. Like DDT, PCBs are very stable in the environment and degrade very slowly. Even though they have not been used for decades, they are still present in the environment and can still magnify up trophic levels.

5. Some of the questions to consider and test include: Do the pesticides leave a residue on the plants or in the environment? Are humans, other animals, or nontarget insects harmed by exposure to these pesticides even at low levels? How long does it take for the pesticides to degrade? Do the pesticides accumulate in the tissues of nontarget insects, wildlife, or humans? If the pesticides accumulate, what are their impacts on the health and behavior of nontarget insects, wildlife, and humans, in both the short term and the long term? If there are any toxic effects, are they reversible with treatment?

Chapter 23

1. Weather refers to daily or short-term fluctuations in temperature and precipitation, whereas climate refers to long-term averages for a particular region.

2. Plants are flowering earlier in the spring; early flowering reduces the survival of young roe deer. Bark beetle larvae are able to survive warmer winters, and bark beetles have extended their northern range as those regions have warmed. Polar bears are facing starvation because of melting sea ice.

3. Possible answers:

- Trees produce less sap during droughts. With less sap, trees can't protect themselves from bark beetles, so more trees die. Dead trees provide fuel for fires.
- More bark beetles survive the warmer winters, leading to more-severe infestations of trees. Dead trees provide fuel for fires.
- Drier conditions associated with drought make fires more likely.

7. b

8. e

9. No. The greenhouse effect helps trap heat in the environment. In the absence of the greenhouse effect, the temperature of the planet would be too cold to support life as we know it.

10. Melting sea ice will not change sea levels. In contrast, melting ice caps will cause a rise in sea levels, putting low-lying cities such as Miami at risk of flooding.

13. b

14. a

15. Using *fossil fuels* for energy converts organic carbon to CO_2. Most organisms, including plants, animals, and decomposers, perform *cellular respiration*, producing CO_2 from organic food. CO_2 is released to the *atmosphere*.

CO_2 is absorbed by the *oceans.*
Plants perform *photosynthesis,* fixing CO_2 into organic molecules.
Coal and oil are *fossil fuels,* which trap carbon below the surface.

16. Burning fossil fuels (e.g., by driving cars) and eating red meat increase atmospheric CO_2 levels. Decomposition increases CO_2 levels in the atmosphere, as do forest fires.

19. d. The ideal data would be the analysis of ice cores formed in 1750. Historical temperature records do not directly measure CO_2. The Mauna Loa observatory was not recording data in 1750, and tree-ring analysis does not reveal CO_2 levels.

Chapter 24

1. N_2 is abundant in the atmosphere. Ammonia (NH_3) is the major form of nitrogen taken up by plants.

2. c

3. e

7. See Infographic 24.4. anther, D; filament, F; ovary, C; ovule, B; carpel, G; stamen, H; style, E; stigma, A

8. stamen (M); carpel (F); ovary (F); anther (M); filament (M); ovule (F); stigma (F); style (F)

9. Endosperm provides nutrients and energy for the germinating seed.

10. The *egg* is the female gamete. In angiosperms, it is found in an *ovule,* which in turn is found in the *ovary.*

12. Possible answers:

- To produce resistance to herbicides (e.g., Roundup Ready seeds)
- To produce resistance to pests (e.g., *Bt* corn)
- To increase the shelf life of produce (e.g., using CRISPR to mutate a gene involved in ripening/browning)
- To speed up the generation of a commercially viable product (relative to selective breeding) (e.g., CRISPR was used to modify groundcherries to make the fruit larger and the plant hardier)

13. *Bt* plants contain a gene from a bacterium that encodes a toxin that kills insects that attempt to eat the plant. The *Bt* plants become protected against insects. Roundup Ready plants are engineered with a recombinant gene that confers resistance to a synthetic herbicide (Roundup). These crops can be treated with Roundup—the modified crops will survive, but weeds will be killed. Farmers growing Roundup Ready crops may increase the amount of the herbicide applied to fields. Farmers must buy new seed for Roundup Ready crops every year, increasing their costs. Concerns about GM crops include unintended consequences for nontarget organisms (e.g., insects that wouldn't normally eat the crop plant) and the potential for the recombinant gene to "escape" and become incorporated into plants without detection.

14. b

15. Using a synthetic chemical fertilizer. Synthetic chemical fertilizers contain high levels of nutrients that can run off into aquatic environments. The resulting growth of algae sets off a chain of events resulting in the depletion of oxygen.

Chapter 25

1. b

2. b

3. d

4. c

5. d

7. a

9. a

10. f

11. e

12. The bright coloration of the trumpet pitcher attracts insects, a food source that can supplement the plant's nutrition. The bright coloration of the squash blossoms attracts pollinating insects that transfer pollen from one squash blossom flower to the next.

16. c

17. a

Chapter 26

1. Anatomy is the study of the structure of living organisms. Physiology is the study of how organisms function, particularly with respect to maintaining homeostasis.

2. (1) mucus-secreting cell of the small intestine; (2) layer of muscle that contributes to the function of the small intestine; (3) small intestine; (4) digestive system

3. b

4. a

7. Homeostasis is the maintenance of a relatively stable internal environment, even when the external environment changes.

8. In cold conditions, *sensors* detect the *low body temperature* and send a signal to the *hypothalamus.* The hypothalamus then sends signals to *effectors* (including *muscles* and blood vessels). The effectors exert their effects (e.g., shivering and vasoconstriction), bringing about a *normal body temperature.*

9. d

10. c

11. In the short term, Sherpas individually adapt to low O_2 levels, producing higher numbers of red blood cells in response to that environmental factor. In the longer term, the population has likely adapted. Evolutionary adaptations to high altitude and low O_2 could include increased lung volume and the capacity of individual red blood cells to carry more O_2 (because of altered hemoglobin).

14. c

15. b

16. a

17. In a dehydrated person, *blood volume* is reduced, as is *blood pressure.* Blood has a higher-than-normal solute concentration. Specific receptors detect the changes in blood pressure and volume. In response, the *hypothalamus* releases *ADH.* ADH acts on the *effector*—in this case the *kidneys,* which will reabsorb water into the bloodstream—to increase blood volume and pressure and dilute the solutes in the blood.

Chapter 27

1. (1) mouth; (2) esophagus; (3) stomach; (4) small intestine; (5) large intestine; (6) anus

2. d

3. With an expandable stomach you can eat more at one time than you can immediately process. The stomach expands to store the extra food until there is room for it in the small intestine. If the stomach were not expandable, we would have to eat small amounts of food continuously throughout the day.

4. c
6. e
9. a
10. Pepsin and salivary amylase are both digestive enzymes. Pepsin digests proteins in the stomach, an environment with an acidic pH. Salivary amylase digests carbohydrates in the mouth, an environment with a neutral pH.
11. Lipase is produced by the pancreas.
12. a
17. a
18. Both procedures dramatically reduce the size of the stomach. Sleeve gastrectomy retains the normal passage of digested food through the stomach and the small intestine, whereas gastric bypass "reroutes" digested food so that it bypasses most of the stomach, as well as the upper part of the small intestine.
19. Because gastric bypass reroutes food, bypassing most of the stomach and part of the small intestine, the digestive system not only takes in less food, but also digests and absorbs less of the food it does take in. Thus, fewer Calories and nutrients (e.g., fatty acids) are absorbed. The impact on conditions such as diabetes and cardiovascular disease can be dramatic.

Milestone 7

1. c
2. c
3. a
4. It is not possible to distinguish between type 1 diabetes and type 2 diabetes on the basis of blood-sugar levels: both types lead to elevated levels of blood sugar. To distinguish between them, one would need to monitor insulin levels. In type 1 diabetes, insulin levels would be essentially at zero at all times. In type 2 diabetes, they would be low between meals and remain high after meals, providing a critical clue that blood glucose is remaining high after meals even in the presence of a high insulin level.
5. b
6. In both types, levels of blood glucose are elevated, both immediately after a carbohydrate-rich meal and also for extended periods after the meal. In type 1 diabetes, no insulin is produced, so insulin levels will remain low or at zero at all times. In type 2 diabetes, the body can make and release insulin normally, so (at least in early stages of the disease) insulin will be low between meals, then rise after a carbohydrate-rich meal. In type 2 diabetes, blood-glucose level will not come down in response to insulin.

Chapter 28

1. blood, blood vessels, and heart
2. b
3. b
7. c
8. d
9. b
10. a
12. d
14. Different white blood cells have different specific functions, but overall, white blood cells are important for the immune response. Red blood cells carry oxygen and deliver it to cells. Platelets are important in blood clotting.
15. e
16. d
19.

Disease	*Underlying cause*
heart attack	c. blockage of a coronary artery
hypertension	a. elevated systolic and diastolic pressure
stroke	d. blockage of blood vessels in the brain
atherosclerosis	b. fatty streaks in arteries

20. There are many possibilities. Some include family history/genetics (not modifiable); diet (modifiable); exercise (modifiable); race (not modifiable); smoking (modifiable); and obesity (modifiable to a certain extent). Type 2 diabetes and hypertension are risk factors that have both modifiable and unmodifiable aspects; in both cases, proper management after diagnosis is important and modifiable.
21. The plaques of atherosclerosis can rupture, promoting blood clot formation. If these clots travel to and block a blood vessel in the brain, a stroke can occur.
22. b

Chapter 29

1. A. pharynx; B. larynx; C. trachea; D. bronchioles; E. bronchus; F. lung
2. a
3. d
6. d
7. d
10. e
11. b
12. c
16. e
17. b
18. There are many possibilities. The independent variable will be the treatment: altitude training, hypoxic chambers, or no oxygen manipulation. The dependent variable could be a variety of factors relative to athletic performance—speed, endurance, strength—and there are a variety of ways to measure these. You should also consider whether to measure any athletic advantage at sea level or at altitude. You could also consider the relative timing of the treatment and the duration of the treatment before assessing athletic advantage.

Chapter 30

1. light-detecting receptor in the eye (PNS); amygdala (CNS); pain receptor in the skin (PNS); spinal cord (CNS); thalamus (CNS)
2. movement—cerebellum; body temperature—hypothalamus (diencephalon)
5. b
6. d
7. b
8. c
9. The neurotransmitter diffuses from the signaling cell to the receiving cell, then binds receptors on the receiving cell.

10. Electrical signals travel along the length of axons, as sodium and potassium ions move across the axon membrane. Chemical signaling molecules, called neurotransmitters, are released into synapses and bind to receptors on nearby dendrites. Chemical signals communicate information between neurons and between neurons and their effector cells (other neurons, glands, muscles).
14. b
15. Receptors respond to the initial surge of dopamine by down-regulating—that is, some receptors are shut down. With fewer active receptors, the drug cannot elicit excessive feelings of pleasure unless more drug is ingested to overcome the effect of fewer active receptors.

Chapter 31

1. c
2. b
3. The uterus is the organ in which embryos and fetuses develop. The cervix is the opening into the uterus from the vagina. The endometrium is the lining of the uterus. The embryo implants into the endometrium.
7. luteinizing hormone (LH)—anterior pituitary; follicle-stimulating hormone (FSH)—anterior pituitary; testosterone—testes; estrogen—ovary; progesterone—ovary; hCG— embryo
8. a
9. e
10. e
14. b
15. In both IUI and IVF, drugs are administered to stimulate egg production in the woman. In IVF, the eggs are surgically removed, then fertilized in vitro (that is, in a dish). The embryos are placed in the woman's uterus with the hope that at least one will successfully implant. In IUI, the timing of ovulation is carefully monitored so that sperm can be inserted in the uterus just before ovulation (this procedure generally takes place in a doctor's office or a fertility clinic). Multiple eggs may be ovulated in response to the drug, and the hope is that at least one of them will be fertilized by one of the sperm inserted in the uterus.

Milestone 8

1. totipotent cells from an early embryo, pluripotent embryonic stem cells, multipotent adult stem cells
2. b
3. Embryonic stem cells would be more appropriate, because they are pluripotent and can differentiate into any cell type. Multipotent blood stem cells can only differentiate into blood cells, not pancreatic cells.
4. A muscle cell contains all the genes in the genome (including antibody genes). If a muscle cell is induced to become an iPSC, and the iPSC is coaxed to specialize into a B cell, then the B cell could express the antibody genes and thereby produce antibodies.

Chapter 32

1. d
2. Viruses must enter their specific host cell in the human, thus infecting it. They direct the host cell to replicate the virus, using host resources.
3. Poliovirus infects and damages nerve cells. Because nerve cells do not replicate, they cannot repair themselves. Influenza virus infects cells lining the respiratory tract; these cells are replaced at high frequency.
6. Skin provides a barrier to pathogen entry. Enzymes in tears and saliva digest components of a pathogen. Phagocytes ingest and destroy pathogens.
7. Phagocytes are white blood cells that engulf and destroy invaders, including bacteria.
11. Innate immunity is present since birth, always active and nonspecific. It does not have "memory." Adaptive immunity is specific for a particular pathogen and must be turned on when that pathogen is encountered—it is not always "on." Adaptive immunity exhibits "memory": it strengthens with repeated exposures to the same pathogen.
12. B cells are lymphocytes that are activated during an adaptive response. Upon activation, they become antibody-producing plasma cells specialized to act against a specific pathogen. The antibodies produced specifically bind to and inactivate their targets.
13. d
14. b
20. Antigenic shift occurs when viruses exchange genes, such that a virus can have genes from a completely different virus. A virus resulting from antigenic shift tends to be very different from either starting virus. Antigenic drift describes small changes, caused by mutations, in the viral genome.
21. To become pandemic, this virus would have to spread easily from human to human, thus spreading widely throughout the human population.

Glossary

abiotic Refers to the nonliving components of an environment, such as temperature and precipitation.

abscisic acid (ABA) A plant hormone that helps seeds remain dormant.

acclimatization The process of physiologically adjusting to an environmental change over a period of time. Acclimatization is generally reversible.

acid A substance that increases the hydrogen ion concentration of solutions, making them more acidic.

acidosis A dangerous condition in which blood is too acidic.

action potential An electrical signal that travels down a neuron, caused by ions moving across the cell membrane.

activation energy The energy required for a chemical reaction to proceed. Enzymes accelerate reactions by reducing their activation energy.

active site The part of an enzyme that binds to a substrate.

active transport The process by which solutes are pumped from an area of lower concentration to an area of higher concentration with the help of transport proteins; requires an input of energy.

adaptation The process by which populations become better suited to their environment as a result of natural selection.

adaptive immunity A protective response, carried out by lymphocytes, that confers long-lasting immunity against specific pathogens.

adaptive radiation The spreading and diversification of organisms that occur when the organisms colonize a new habitat.

adenosine triphosphate (ATP) The molecule in cells that powers energy-requiring functions.

adhesion The attraction between molecules (or other particles) and a surface.

aerobic respiration A series of reactions that occurs in the presence of oxygen and converts energy stored in food into ATP.

allele Any of the alternative versions of the same gene that have different nucleotide sequences.

allele frequency The relative proportion of an allele in a population.

allergy An immune response that is misdirected against harmless environmental substances, such as dust, pollen, and certain foods, and that causes uncomfortable physical symptoms.

altitude sickness An illness that can occur as a result of an abrupt move to an altitude with a reduced partial pressure of oxygen.

alveoli Air sacs in the lung across which gases diffuse between air and blood.

amino acids The building blocks of proteins. There are 20 different amino acids.

amniocentesis A procedure that removes fluid surrounding the fetus so as to obtain and analyze the chromosomal makeup of fetal cells.

anabolic reaction Any chemical reaction that combines simple molecules to build more complex molecules.

anatomy The study of the physical structures that make up an organism.

anchorage dependence The need for normal cells to be in physical contact with another layer of cells or a surface.

androgens A class of sex hormones, including testosterone, that are present in higher levels in men and cause male-associated traits like deep voice, growth of facial hair, and defined musculature.

anecdotal evidence An informal observation that has not been systematically tested.

aneuploidy An abnormal number of one or more chromosomes (either extra or missing copies).

angiogenesis The growth of new blood vessels.

angiosperm A seed-bearing flowering plant with seeds typically contained within a fruit.

animal A eukaryotic multicellular organism that can move and that obtains nutrients by ingesting other organisms.

annelid An invertebrate with a soft, segmented body; annelids are commonly referred to as worms.

anther The part of the stamen that produces pollen.

anthropocene The geological period of the present day, dominated by human activities.

antibiotic A chemical that can slow or stop the growth of bacteria; many antibiotics are produced by living organisms.

antibody A protein produced by B cells that fights infection by binding to specific antigens on pathogens.

anticodon The part of a tRNA molecule that binds to a complementary mRNA codon.

antigen A specific molecule (or part of a molecule) to which immune receptors bind and against which an adaptive immune response is mounted.

antigen-presenting cell (APC) A cell, such as a phagocyte, that digests pathogens and displays pieces of the pathogen on its surface, where they can be recognized by lymphocytes.

antigenic drift Changes in viral antigens caused by genetic mutation during normal viral replication.

antigenic shift Changes in viral antigens that occur when viruses of one strain exchange genetic material with other strains.

aorta The large artery that receives oxygenated blood from the left ventricle and delivers it to the body.

apoptosis A type of cell death; often referred to as cellular suicide.

aquifer An underground layer of porous rock that contains water.

archaea One of the two domains of prokaryotic life; the other is Bacteria.

arteries Blood vessels that carry blood away from the heart.

arthropod An invertebrate having a segmented body, a hard exoskeleton, and jointed appendages.

atherosclerosis A condition in which fatty deposits build up in the lining of arteries, restricting blood flow; also known as hardening of the arteries.

atom The smallest unit of an element that still retains the property of the element.

atomic number The number of protons in an atom, which determines the atom's identity.

atria The upper chambers of the heart that receive blood. In humans, the right atrium receives oxygen-poor blood from the body, and the left atrium receives oxygen-rich blood from the lungs.

autoimmune disease A misdirected immune response in which the immune system attacks the body's own healthy cells.

autosomes Paired chromosomes present in both males and females; all chromosomes except the X and Y chromosomes.

autotrophs Organisms such as plants, algae, and certain bacteria that can make their own food from inorganic starting materials (e.g., CO_2).

auxin A plant hormone that causes elongation of cells as one of its effects.

axon The long extension of a neuron that conducts electrical signals away from the cell body toward the axon terminal.

axon terminals The tips of an axon, which communicate with the next cell or cells in the pathway.

B cells White blood cells that mature in the bone marrow and produce antibodies during an adaptive immune response.

bacteria One of the two domains of prokaryotic life; the other is Archaea.

base A substance that reduces the hydrogen ion concentration of solutions, making them more basic.

benign tumor A noncancerous tumor whose cells will not spread throughout the body.

beta-globin One of the proteins that makes up hemoglobin.

bilateral symmetry The pattern exhibited by a body plan with right and left halves that are mirror images of each other.

bile salts Chemicals produced by the liver and stored by the gallbladder that emulsify fats so that they can be chemically digested by enzymes.

binary fission A type of asexual reproduction in which one parental cell divides into two.

biodiversity The number of different species and their relative abundances in a specific region or on the planet as a whole.

biofuels Renewable fuels made from living organisms (e.g., plants and algae).

biogeography The study of the distribution of organisms in geographic space.

biological species concept The definition of a species as a group whose members can interbreed to produce fertile offspring.

biology The study of life.

biome A large geographic area defined by its characteristic plant life, which in turn is determined by temperature and levels of moisture.

biotic Refers to the living components of an environment.

blood A circulating fluid that contains several types of cells and transports substances, including nutrients, gases, and hormones.

blood pressure The overall pressure in blood vessels, expressed as the systolic pressure over the diastolic pressure.

body mass index (BMI) An estimate of body fat based on height and weight.

bottleneck effect A type of genetic drift in which a population is suddenly reduced to a small number of individuals, and alleles are lost from the population.

brain The organ of the central nervous system that integrates information and coordinates virtually all functions of the body.

brain stem The part of the brain that is closest to the spinal cord and which controls vital functions such as heart rate, breathing, and blood pressure.

bronchi Two airways that branch from the trachea; one bronchus leads into each lung.

bronchioles Small airways that branch from the bronchi.

Calorie A Calorie (spelled with a capital "C") is 1,000 calories or 1 kilocalorie (kcal). The Calorie is the common unit of energy used in food nutrition labels.

calorie A calorie (spelled with a lowercase "c") is the amount of energy required to raise the temperature of 1 gram of water by 1°C.

cancer A disease in which cells divide repeatedly and without restraint, in some cases forming a tumor.

capillaries The smallest blood vessels, which are the sites of gas, nutrient, and waste exchange between the blood and tissues.

capsule A sticky coating surrounding some bacterial cells that adheres to surfaces.

carbohydrate An organic molecule made up of one or more sugars.

carbon capture Using technology to remove CO_2 from the atmosphere.

carbon cycle The movement of carbon atoms through the ecosystem as they cycle between organic molecules and inorganic CO_2.

carbon fixation The conversion of inorganic carbon (e.g., CO_2) into organic forms (e.g., sugars like glucose, $C_6H_{12}O_6$).

carbon footprint A measure of the total greenhouse gases produced by human activities.

carcinogen Any substance that causes cancer. Most carcinogens are mutagens.

cardiovascular disease (CVD) A disease of the heart or blood vessels, or both.

cardiovascular system The system that transports nutrients, gases, and other critical molecules throughout the body. It consists of the heart, blood vessels, and blood.

carnivore An organism (typically an animal) that eats animals.

carpel The female reproductive structure of a flower made up of a stigma, style, and ovary.

carrier An individual who is heterozygous for a recessive allele and can therefore pass it on to offspring without showing any of its effects.

carrying capacity The maximum population size that a given environment or habitat can support given its food supply and other natural resources.

case-control study A type of epidemiological study to assess an association between an exposure and an outcome.

catabolic reaction Any chemical reaction that breaks down complex molecules into simpler molecules.

catalysis The process of speeding up the rate of a chemical reaction (e.g., by an enzyme).

cell The basic structural unit of living organisms.

cell body The part of a neuron that contains the nucleus and most of the cell's other organelles.

cell cycle The ordered sequence of stages through which a cell progresses to divide. The stages include preparatory phases (G_1, S, G_2) and division phases (mitosis and cytokinesis).

cell cycle checkpoint A cellular mechanism that ensures that a stage of the cell cycle is completed accurately.

cell division The process by which a cell reproduces itself; cell division is important for normal growth, development, maintenance, and repair of an organism.

cell membrane A phospholipid bilayer with embedded proteins that forms the boundary of all cells.

cell theory The concept that all living organisms are made of cells and that cells are formed by the division of existing cells.

cell wall A rigid structure present in some cells that encloses the cell membrane and helps the cell maintain its integrity.

cell-mediated immunity The type of adaptive immunity that rids the body of infected, cancerous, or foreign cells.

central nervous system (CNS) The brain and the spinal cord.

central vacuole A fluid-filled compartment in a plant cell that contributes to cell rigidity by exerting turgor pressure against the cell wall.

centromere The specialized region of a chromosome where the sister chromatids are joined; it is critical for proper alignment and separation of sister chromatids during mitosis.

cerebellum The part of the brain that processes sensory information and is involved in movement, coordination, and balance.

cerebral cortex The outer layer of the cerebrum, which is involved in many advanced brain functions.

cerebrum The region of the brain that controls intellect, learning, perception, and emotion.

cervix The opening or "neck" of the uterus, where sperm enter and babies exit.

chemical energy Potential energy stored in the bonds of biological molecules.

chemical reaction A rearrangement of atoms in molecules to form different molecules by forming or breaking bonds.

chemotherapy Treatment using toxic chemicals that kill cancer by interfering with cell division.

chlorophyll The pigment present in the green parts of plants that absorbs photons of light energy during the "photo" reactions of photosynthesis.

chloroplast An organelle in plant and algal cells that is the site of photosynthesis.

cholesterol A lipid that is an important component of cell structures; it is used to make important molecules and also plays a role in heart disease.

chromosome A single, large DNA molecule wrapped around proteins. Chromosomes are located in the nuclei of eukaryotic cells.

chyme The acidic "soup" of partially digested food that leaves the stomach and enters the small intestine.

citric acid cycle A set of reactions that takes place in mitochondria and helps extract energy (in the form of high-energy electrons) from food; the second stage of aerobic respiration.

climate The long-term average of atmospheric conditions.

climate change Any substantial change in climate that lasts for an extended period of time (decades or more).

coding sequence The part of a gene that specifies the amino acid sequence of a protein. Coding sequences determine the identity, shape, and function of proteins.

codominance A form of inheritance in which the effects of both alleles are displayed in the phenotype of a heterozygote.

codon A sequence of three mRNA nucleotides that specifies a particular amino acid.

coenzyme A small organic molecule, such as a vitamin, required to activate an enzyme.

cofactor An inorganic substance, such as a metal ion, required to activate an enzyme.

cohesion The attraction between molecules (or other particles).

commensalism A type of symbiotic relationship in which one member benefits and the other is unharmed.

community A group of interacting populations of different species living in the same area.

competition An interaction between two or more organisms that rely on a common resource that is not available in sufficient quantities.

competitive exclusion principle The concept that when two species compete for resources in an identical niche, one is inevitably eliminated from that niche.

complement proteins Proteins in blood that help destroy pathogens by coating or puncturing them.

complementary Fitting together; two strands of DNA are said to be complementary in that A in one strand always pairs with T in the other strand, and G always pairs with C.

conservation of energy The principle that energy cannot be created or destroyed, but can be transformed from one form to another.

consumers Heterotrophs that eat other organisms lower on the food chain to obtain energy.

contact inhibition A characteristic of normal cells that prevents them from dividing once they have filled a space and are in contact with their neighbors.

continental drift The movement of the continents relative to one another over time.

continuous variation Variation in a population showing an unbroken range of phenotypes rather than discrete categories.

contraception The prevention of pregnancy through physical, surgical, or hormonal methods.

control group The group in an experiment that experiences no experimental intervention or manipulation.

controlled experiment An experiment in which there are at least two groups, a control group and an experimental group.

convergent evolution The process by which organisms that are not closely related evolve similar adaptations as a result of independent episodes of natural selection.

coronary arteries The blood vessels that deliver oxygen-rich blood to the heart muscle.

corpus luteum The structure in the ovary that remains after ovulation. It secretes progesterone.

correlation A consistent relationship between two variables.

covalent bond A strong interaction resulting from the sharing of a pair of electrons between two atoms.

CRISPR A genome-editing tool based on a natural defense system in bacteria.

cuticle The waxy coating on leaves and stems that prevents water loss.

cytokinesis The physical division of a cell into two daughter cells.

cytoplasm The gelatinous, aqueous interior of all cells.

cytoskeleton A network of protein fibers in eukaryotic cells that provides structure and facilitates cell movement.

cytotoxic T cell A type of T cell that destroys infected, cancerous, or foreign cells.

decomposer An organism such as a fungus or bacterium that digests and uses the organic molecules in dead organisms as sources of nutrients and energy.

dendrites Branched extensions from the cell body of a neuron, which receive incoming information.

density-dependent factor A factor whose influence on population size and growth depends on the number and crowding of individuals in the population (e.g., predation).

density-independent factor A factor that can influence population size and growth regardless of the numbers and crowding within a population (e.g., weather).

deoxyribonucleic acid (DNA) The molecule of heredity, common to all life forms, which is passed from parents to offspring.

dependent variable The measured result of an experiment, analyzed in both the experimental and control groups.

descent with modification Darwin's term for evolution, combining the ideas that all living things are related and that organisms have changed over time.

diabetes A disease characterized by chronically elevated blood-sugar levels.

diaphragm A sheet of muscle that contributes to breathing by contracting and relaxing.

diastolic pressure The pressure in arteries when the ventricles are relaxed.

diencephalon A brain region located between the brain stem and the cerebrum that regulates homeostatic functions like body temperature, hunger, thirst, and sex drive.

digestion The mechanical and chemical breakdown of food into subunits, enabling the absorption of nutrients.

digestive system The organ system that breaks down food molecules into smaller subunits, absorbs nutrients, and eliminates waste; it consists of the digestive tract and accessory organs.

diploid Having two copies of every chromosome.

directional selection A type of natural selection in which organisms with phenotypes at one end of a spectrum are favored by the environment.

dispersion pattern The way organisms are distributed in geographic space, which depends on resources and interactions with other members of the population.

diversifying selection A type of natural selection in which organisms with phenotypes at both extremes of the phenotypic range are favored by the environment.

DNA polymerase An enzyme that "reads" the nucleotide sequence of a DNA strand and incorporates complementary nucleotides into a new strand during DNA replication.

DNA profile A visual representation of a person's unique DNA sequence.

DNA replication The natural process by which cells make an identical copy of a DNA molecule.

domain The highest (most inclusive) category in the modern system of classification. There are three domains: Bacteria, Archaea, and Eukarya.

dominant allele An allele that can mask the presence of a recessive allele.

dopamine A chemical messenger that is involved in conveying a sense of pleasure in the brain.

dose–response relationship A relationship between the amount of a chemical or physical (e.g., radiation) exposure, and the risk of a specific outcome in an exposed organism.

double fertilization The process in angiosperms in which one sperm fertilizes the egg and one sperm fuses with another cell in the ovule.

double helix The spiral structure formed by two strands of DNA nucleotides held together by hydrogen bonds.

duodenum The first portion of the small intestine, where mixing of chyme and digestive enzymes occurs.

ecology The study of the interactions among organisms and between organisms and their nonliving environment.

ecosystem All the living organisms in an area and the nonliving components of the environment with which they interact.

effector A cell or tissue that responds to information relayed from a sensor.

electron A negatively charged subatomic particle with negligible mass in an atom.

electron transport chain The transfer of electrons that takes place in mitochondria and produces the bulk of ATP during aerobic respiration; the third stage of aerobic respiration.

element A pure substance that cannot be chemically broken down; each element is made up of and defined by a single type of atom.

embryo An early stage of development reached when a zygote undergoes cell division to form a multicellular structure.

emigration The movement of individuals out of a population.

endocrine system The collection of hormone-secreting glands and organs with hormone-secreting cells.

endometrium The lining of the uterus.

endoplasmic reticulum (ER) A network of membranes in eukaryotic cells where proteins and lipids are synthesized.

endoskeleton An internal body skeleton, typically made of cartilage or bone.

endosperm A part of the seed containing nutrients that is the seed's source of energy; produced when a sperm fuses with a cell in the ovule during double fertilization.

endosymbiosis The scientific theory that free-living prokaryotic cells engulfed other free-living prokaryotic cells billions of years ago, forming eukaryotic organelles such as mitochondria and chloroplasts.

energy The capacity to do work. Cellular work includes processes such as building complex molecules and moving substances into and out of the cell.

enzyme A protein that speeds up the rate of a chemical reaction.

epidemiology The study of patterns of disease in populations, including risk factors.

epididymis A system of tubes in which sperm mature and are stored before ejaculation.

epigenetics Changes in gene expression that are not based on changes in the DNA sequence.

erythropoietin (EPO) A hormone produced by the kidneys that stimulates red blood cell production.

esophagus The section of the digestive tract between the mouth and the stomach.

essential amino acids Amino acids that can't be made by the body, and so must be obtained in pre-assembled form from the diet.

essential nutrients Nutrients that can't be made by the body, and so must be obtained from the diet.

estrogens A class of sex hormones, including estradiol, that are present in higher levels in women than in men and that support female sexual development and function.

ethylene A gaseous plant hormone that promotes fruit ripening as one of its effects.

eukaryote Any organism of the domain Eukarya; eukaryotic cells are characterized by the presence of a membrane-enclosed nucleus and organelles.

eukaryotic cells Cells that contain membrane-bound organelles, including a central nucleus.

eutrophication An overabundance of nutrients in bodies of water that can cause excessive growth of algae.

evolution Change in allele frequencies in a population over time.

exoskeleton An external skeleton; in arthropods, the exoskeleton is made up of proteins and chitin.

experiment A carefully designed test, the results of which will either support or rule out a hypothesis.

experimental group The group in an experiment that experiences the experimental intervention or manipulation.

exponential growth The unrestricted growth of a population increasing at a constant growth rate.

extinction The elimination of all individuals in a species; extinction may occur over time or in a sudden mass die-off.

facilitated diffusion The process by which large, hydrophilic, or charged solutes move across a membrane from an area of higher concentration to an area of lower concentration with the help of transport proteins; does not require an input of energy.

falsifiable Describes a hypothesis that can be ruled out by data that show the hypothesis does not explain the observation.

feedback loop A pathway in which the output from an effector feeds back to a sensor and changes further output.

fermentation A series of chemical reactions beginning with glycolysis and taking place in the absence of oxygen. Fermentation produces far less ATP than does aerobic respiration.

ferns The first true vascular plants; ferns do not produce seeds.

fertilization The fusion of an egg and a sperm; the resulting cell is called a zygote.

fertilizer A substance applied to soil so as to provide one or more nutrients to plants.

fetus After the eighth week after fertilization, the embryo is referred to as a fetus.

fever An elevated body temperature.

fitness The relative ability of an organism to survive and reproduce in a particular environment.

flagella (singular: flagellum) In bacteria, long, slender appendages extending from some bacterial cells, used in movement of the cell.

folate A B vitamin also known as folic acid; folate is an essential nutrient, necessary for basic cellular processes such as DNA replication and cell division.

follicle A structure in the ovary where eggs mature.

follicle-stimulating hormone (FSH) A hormone secreted by the anterior pituitary. In females, FSH triggers eggs to mature at the start of each monthly cycle.

food chain A linked series of feeding relationships in a community in which organisms further up the chain feed on ones below.

food web A complex, interconnected set of feeding relationships in a community.

fossil fuel A carbon-rich energy source, such as coal, petroleum, or natural gas, formed from the compressed, fossilized remains of once-living organisms.

fossil record An assemblage of fossils arranged in order of age, providing evidence of changes in species over time.

fossils The preserved remains or impressions of once-living organisms.

founder effect A type of genetic drift in which a small number of individuals leaves one population and establishes a new population, resulting in lower genetic diversity than in the original population.

frameshift mutation A shift in the reading frame, such that codons start and end at an alternative position.

fruiting body A fungal structure that is specialized for the release of spores.

fungus A unicellular or multicellular eukaryotic organism that obtains nutrients by secreting digestive enzymes onto organic matter and absorbing the digested product.

gallbladder An organ that stores bile salts and releases them as needed into the small intestine.

gametes Specialized reproductive cells that carry one copy of each chromosome (i.e., they are haploid). Sperm are male gametes; eggs are female gametes.

gas exchange In the lungs, the process of taking up and releasing oxygen and carbon dioxide across capillaries.

gel electrophoresis A laboratory technique that separates fragments of DNA by size.

gene A sequence of DNA that contains the information to make at least one protein.

gene editing A way to change the sequence of a gene.

gene expression The process of using DNA instruction as to make proteins.

gene flow The movement of alleles from one population to another, which may increase the genetic diversity of a population.

gene pool The total collection of alleles in a population.

gene therapy A treatment that aims to cure, treat, or prevent human disease by replacing defective genes with functional ones.

gene transfer The process by which bacteria can exchange segments of DNA between them.

generalist An organism that can be successful in a variety of environments, often because of a varied diet.

genetic code The set of rules relating particular mRNA codons to particular amino acids.

genetic drift Random changes in the allele frequencies of a population between generations; genetic drift tends to have more dramatic effects in smaller populations than in larger ones.

genetic engineering Altering or manipulating the DNA of organisms by modern laboratory techniques.

genetically modified organism (GMO) An organism whose genome has been altered through genetic engineering techniques, often to contain a gene from another species.

genome One complete set of genetic instructions encoded in the DNA of an organism.

genotype The particular genetic makeup of an individual.

germ cells Reproductive cells of the body.

gibberellins Plant hormones that cause cell division and stem elongation.

glial cells Supporting cells of the nervous system, some of which produce myelin.

global warming An increase in Earth's average temperature.

glucagon A hormone secreted by the pancreas that causes an increase in blood sugar.

glycogen A complex animal carbohydrate, made up of linked chains of glucose molecules, that stores energy for short-term use.

glycolysis A series of reactions that breaks down sugar into smaller units; glycolysis takes place in the cytoplasm and is the first stage of both aerobic respiration and fermentation.

Golgi apparatus An organelle made up of stacked membrane-enclosed discs that packages proteins and prepares them for transport.

gonads Sex organs: ovaries in females, testes in males.

Gram-negative Describes bacteria with a cell wall that includes a thin layer of peptidoglycan surrounded by an outer lipid membrane that does not retain the Gram stain.

Gram-positive Describes bacteria with a cell wall that includes a thick layer of peptidoglycan that retains the Gram stain.

gravitropism The growth of plants in response to gravity. Roots grow downward, with gravity; shoots grow upward, against gravity.

greenhouse effect The natural process by which heat from sunlight is radiated from Earth's surface and trapped by gases in the atmosphere, helping to maintain Earth at a temperature that can support life.

greenhouse gas Any of the gases in Earth's atmosphere that absorb heat radiated from Earth's surface and contribute to the greenhouse effect—for example, carbon dioxide and methane.

growth rate The difference between the birth rate and the death rate of a given population; also known as the rate of natural increase.

gymnosperm A seed-bearing plant with exposed seeds typically held in cones.

habitat The physical environment where an organism lives and to which it is adapted.

half-life The time it takes for one-half of a sample of a radioactive isotope to decay.

haploid Having only one copy of every chromosome.

Hardy–Weinberg principle The principle that, in a nonevolving population, both allele and genotype frequencies remain constant from one generation to the next.

heart The muscular pump that generates force to move blood throughout the body.

heart attack Damage to the heart muscle resulting from the restriction of blood flow to heart tissue.

heat The kinetic energy generated by random movements of molecules or atoms.

helicase An enzyme that unwinds and unzips the DNA double helix during DNA replication.

helper T cell A type of T cell that helps activate other lymphocytes, including B cells and cytotoxic T cells.

heme groups Iron-containing structures on hemoglobin, the sites of oxygen binding.

hemoglobin A protein found in red blood cells that is specialized for transporting oxygen.

herbivore An organism that eats plants.

herd immunity The protection of a population from an infection, based on a certain percentage of its members being immune.

heterotrophs Organisms, such as humans and other animals, that obtain energy by consuming organic molecules that were produced by other organisms.

heterozygous Having two different alleles for a given gene.

high-density lipoprotein (HDL) A form of cholesterol and protein that is protective against CVD.

histamine A molecule released by damaged tissue and during allergic reactions that promotes inflammation.

homeostasis The maintenance of a relatively stable internal environment even when the external environment changes.

hominids Any living or extinct member of the family Hominidae, the great apes—humans, orangutans, gorillas, chimpanzees, and bonobos.

homologous chromosomes A pair of chromosomes that both contain the same genes. In a diploid cell, one chromosome in the pair is inherited from the mother, the other from the father.

homology Anatomical, genetic, or developmental similarity among organisms due to common ancestry.

homozygous Having two identical alleles for a given gene.

hormone A chemical signaling molecule that is released by a cell or gland and travels through the bloodstream to exert an effect on target cells.

human chorionic gonadotropin (hCG) A hormone produced by an early embryo that helps maintain the corpus luteum until the placenta develops.

humoral immunity The type of adaptive immunity that fights free-floating pathogens in the blood and lymph.

hydrogen bond A weak electrical attraction between a partially positive hydrogen atom and an atom with a partial negative charge.

hydrophilic "Water-loving"; hydrophilic molecules attract water.

hydrophobic "Water-fearing"; hydrophobic molecules repel water.

hypertension Elevated (high) blood pressure.

hypertonic Describes a solution surrounding a cell that has a higher concentration of solutes than the cell's cytoplasm.

hypha (plural: hyphae) A long, threadlike structure through which fungi absorb nutrients.

hypothalamus A master coordinator region of the brain responsible for a variety of physiological functions.

hypothermia A drop of body temperature below 35°C (95°F), which causes enzyme malfunction and eventually death.

hypothesis A tentative explanation for a scientific observation or question.

hypotonic Describes a solution surrounding a cell that has a lower concentration of solutes than the cell's cytoplasm.

hypoxia A state of low oxygen concentration in the blood.

igneous rock Rock formed from the cooling and hardening of molten lava.

immigration The movement of individuals into a population.

immune system A network of cells and tissues that acts to defend the body against infectious agents and helps to heal injuries.

immunity Protection from a given pathogen conferred by the activity of the immune system.

immunotherapy A cancer therapy that uses the immune system to recognize and destroy cancer cells.

in vitro fertilization (IVF) A form of assisted reproduction in which eggs and sperm are brought together outside the body and the resulting embryos are inserted into a woman's uterus.

inbreeding Mating between closely related individuals. Inbreeding does not change the allele frequencies within a population, but does increase the ratio of homozygous individuals to heterozygous individuals.

inbreeding depression The negative reproductive consequences for a population associated with having a high frequency of homozygous individuals possessing harmful recessive alleles.

incomplete dominance A form of inheritance in which heterozygotes have a phenotype that is intermediate between the two homozygotes.

independent assortment The principle that alleles of different genes are distributed independently of one another during meiosis.

independent variable The variable, or factor, being deliberately changed in the experimental group relative to the control group.

inflammation An innate immune defense that is activated by infection or tissue damage; characterized by redness, heat, swelling, and pain.

innate immunity Nonspecific defenses, such as physical and chemical barriers and specialized white blood cells, that are present from birth and require little or no time to become active.

inorganic Describes a molecule that lacks a carbon-based backbone and C–H bonds.

insect An arthropod with three pairs of jointed legs and a body with three segments.

insulin A hormone secreted by the pancreas that causes a decrease in blood sugar.

interphase The stage of the cell cycle in which dividing cells spend most of their time, preparing for cell division. There are three distinct subphases: G_1, S, and G_2.

intrauterine insemination (IUI) A form of assisted reproduction in which sperm are injected directly into a woman's uterus.

introduced species Species that are not native to a particular environment and that have arrived as a result of human activity.

invasive species Introduced species that do harm in their new environment.

invertebrate An animal without a backbone.

ion An electrically charged atom, the charge resulting from the loss or gain of electrons.

ionic bond A strong electrical attraction between oppositely charged ions formed by the transfer of one or more electrons from one atom to another.

isotonic Describes a solution surrounding a cell that has the same solute concentration as the cell's cytoplasm.

karyotype The chromosomal makeup of cells. Karyotype analysis can be used to detect chromosomal disorders prenatally.

keystone species Species on which other species depend, and whose removal has a dramatic impact on the ecosystem.

kidney An organ involved in osmoregulation, filtration of blood to remove wastes, and production of several important hormones.

kinetic energy The energy of motion or movement.

large intestine The last organ of the digestive tract, in which remaining water is absorbed and solid stool is formed.

larynx The part of the airway between the pharynx and the trachea; also known as the voice box.

light energy A type of electromagnetic radiation that includes visible light.

lignin A stiff strengthening agent found in the inner cell wall of plant stems.

limbic system A set of brain structures that is stimulated during pleasurable activities and which is involved in addiction.

lipase A fat-digesting enzyme active in the small intestine.

lipids Organic molecules that generally repel water.

liver An organ that aids digestion by producing bile salts that emulsify fats.

logistic growth A pattern of growth that starts off fast and then levels off as the population reaches the carrying capacity of the environment.

low-density lipoprotein (LDL) A form of cholesterol and protein in the blood that contributes to cardiovascular disease.

lung The major respiratory organ, the site of gas exchange between air and the blood.

luteinizing hormone (LH) A hormone secreted by the anterior pituitary. In females, a surge of LH triggers ovulation.

lymph nodes Small organs in the lymphatic system where B and T cells may encounter pathogens.

lymphatic system The system of vessels and organs that drains fluid (lymph) from the tissues and sends it through lymph nodes on its way to the circulation.

lymphocyte A specialized white blood cell of the immune system. Lymphocytes are important in adaptive immunity.

lysosome An organelle in eukaryotic cells that is filled with enzymes that can degrade worn-out cellular structures.

macromolecules Very large organic molecules that make up living organisms; they include carbohydrates, proteins, and nucleic acids.

macronutrients Nutrients, including carbohydrates, proteins, and fats, that organisms must ingest in large amounts to maintain health.

malignant tumor A cancerous tumor whose cells can spread throughout the body.

malnutrition A medical condition resulting from a lack of essential nutrients in the diet. Malnutrition is often, but not always, associated with starvation.

mammals Members of the class Mammalia; all members of this class have mammary glands and a body covered with hair.

mammogram An x-ray of the breast.

mass extinction An extinction of between 50% and 90% of all living species that occurs relatively rapidly.

matter Anything that takes up space and has mass.

meiosis A type of cell division that generates genetically unique haploid gametes.

melanin A pigment produced by a specific type of skin cell that gives skin its color.

memory cell A long-lived B or T cell that is produced during an immune response and that can "remember" the pathogen.

menstruation The shedding of the uterine lining (the endometrium) that occurs when an embryo does not implant.

meristem A plant tissue consisting of stem cells that can divide and contribute to the growth of the plant.

messenger RNA (mRNA) The RNA copy of an original DNA sequence made during transcription.

metabolism All biochemical reactions occurring in an organism, including reactions that break down food molecules and reactions that build new cell structures.

metastasis The spread of cancer cells from one location in the body to another.

microbiome A community of microbes at a particular location (e.g., on a person's skin or in the gut).

micronutrients Nutrients, including vitamins and minerals, that organisms must ingest in small amounts to maintain health.

mineral An inorganic chemical element required by organisms for normal growth, reproduction, and tissue maintenance; examples include calcium, iron, potassium, and zinc.

missense mutation A point mutation that changes the amino acid sequence of the encoded protein.

mitochondria (singular: mitochondrion) Membrane-bound organelles responsible for important energy-conversion reactions in eukaryotes.

mitochondrial DNA (mtDNA) The DNA within mitochondria; it is inherited solely from the mother.

mitosis The segregation and separation of replicated chromosomes during cell division.

mitotic spindle The microtubule-based structure that separates sister chromatids during mitosis.

molecule Atoms linked by covalent bonds.

mollusk An invertebrate with a soft, unsegmented body enclosed in a hard shell.

monocrop A single crop species grown in the same field over many seasons.

monomer One chemical subunit of a polymer.

monosaccharide The building block, or monomer, of a carbohydrate.

motor neurons Neurons that control the contraction of skeletal muscle.

multifactorial inheritance An interaction between genes and the environment that contributes to a phenotype or trait.

mutagen Any chemical or physical agent that can damage DNA by changing its nucleotide sequence.

mutation A change in the nucleotide sequence of a DNA molecule.

mutualism A type of symbiotic relationship in which both members benefit; a "win–win" relationship.

mycelium A spreading mass of interwoven hyphae that forms the often subterranean body of multicellular fungi.

myelin A fatty substance that insulates the axons of neurons and facilitates rapid conduction of action potentials.

NAD$^+$ An electron carrier. NAD$^+$ can accept electrons, becoming NADH in the process.

natural selection The greater survival and reproduction of individuals with certain traits in a particular environment that leads to a change in allele frequencies in a population over time.

NEAT Non-exercise activity thermogenesis; the amount of energy expended in everyday activities.

nerve A bundle of specialized cells that transmits information.

nervous system The collection of organs that sense and respond to information, including the brain, spinal cord, and nerves.

neurons Specialized cells of the nervous system that generate electrical signals in the form of action potentials.

neurotransmitter A chemical signaling molecule released by a neuron to transmit a signal to a neighboring cell.

neutron An electrically uncharged subatomic particle in the nucleus of an atom.

niche The space, environmental conditions, and resources that a species needs to survive and reproduce.

nitrogen cycle The movement of nitrogen atoms as they cycle between different molecules in living organisms and the environment.

nitrogen-fixing bacteria Bacteria that convert gaseous nitrogen to ammonia, a form of nitrogen usable by plants.

nonadaptive evolution Any change in allele frequencies that does not by itself lead a population to become more adapted to its environment; the mechanisms of nonadaptive evolution are mutation, genetic drift, and gene flow.

nondisjunction The failure of chromosomes to separate accurately during cell division; nondisjunction in meiosis leads to aneuploid gametes.

nonrenewable resources Natural resources that are not replenished at the same speed at which they are consumed and are therefore considered finite; fossil fuels are an example.

nonvascular plant A plant that lacks vascular tissue to transport water and nutrients through the plant body.

nuclear envelope The double membrane surrounding the nucleus of a eukaryotic cell.

nucleic acids Organic molecules made up of linked nucleotide subunits; DNA (deoxyribonucleic acid) and RNA (ribonucleic acid) are examples.

nucleotides The building blocks of DNA. Each nucleotide consists of a sugar, a phosphate group, and a base. The sequence of nucleotides (As, Cs, Gs, Ts) along a DNA strand is unique to each person.

nucleus (atomic) The dense core of an atom.

nucleus (eukaryotic) The organelle in eukaryotic cells that contains the genetic material.

nutrient cycling The movement of the atoms of nutrients as they cycle between different molecules in living organisms and the environment.

nutrients Components in food that the body needs to grow, develop, and repair itself.

obesity Having more body fat than is considered healthy.

oncogene A mutated and overactive form of a proto-oncogene. Oncogenes drive cells to divide continually.

organ A structure made up of different tissue types working together to carry out a common function.

organ system A set of cooperating organs within the body.

organelles The membrane-bound compartments of eukaryotic cells that carry out specific functions.

organic (agriculture) Describes a way of growing crops that conforms to several regulations, among them that synthetic pesticides must not be used.

organic (chemistry) Describes a molecule with a carbon-based backbone and at least one C–H bond.

osmolarity The concentration of solutes in blood and other bodily fluids.

osmoregulation The maintenance of relatively stable volume, pressure, and solute concentration of bodily fluids, especially blood.

osmosis The diffusion of water across a membrane from an area of lower solute concentration to an area of higher solute concentration.

ovaries Paired female reproductive organs; the ovaries contain eggs and produce estrogen and progesterone.

ovary (plant) The structure at the base of the pistil that contains the ovules.

oviduct The tube connecting an ovary and the uterus in females. Eggs are ovulated into and fertilized within the oviducts.

ovulation The release of an egg from an ovary into the oviduct.

ovule The part of a flower that develops into a seed after fertilization.

paleontologist A scientist who studies ancient life by examining the fossil record.

pancreas An organ that secretes the hormones insulin and glucagon as well as digestive enzymes.

pandemic A worldwide outbreak of disease; a global epidemic.

parasitism A type of symbiotic relationship in which one member benefits at the expense of the other.

partial pressure The proportion of total air pressure contributed by a given gas.

pathogens Infectious agents, including certain viruses, bacteria, fungi, and parasites, that may cause disease.

pedigree A visual representation of the occurrence of phenotypes across generations.

peer review A process in which independent experts read scientific studies before they are published to ensure that the authors have appropriately designed and interpreted the study.

pepsin A protein-digesting enzyme that is active in the stomach.

peptidoglycan The macromolecule found in all bacterial cell walls that gives the cell wall its rigidity.

peripheral nervous system (PNS) All the nervous tissue outside the central nervous system that transmits signals from the central nervous system to the rest of the body.

peristalsis Coordinated muscular contractions that force food down the digestive tract.

pesticide A substance that is toxic to pests, organisms that can damage crops or farm animals.

pH A measure of the concentration of hydrogen ions, H^+, in a solution.

phagocyte A type of white blood cell that engulfs and digests pathogens and debris from dead cells.

pharynx The throat.

phenology The study of cyclic life events such as plant flowering and animal migration and how these are influenced by climate and seasonal changes.

phenotype The visible or measurable features of an individual.

phloem Plant vascular tissue that transports sugars throughout the plant.

phospholipid A type of lipid that forms the cell membrane.

phosphorus cycle The movement of phosphorus atoms as they cycle between different molecules in living organisms and the environment.

photons Packets of light energy, each with a specific wavelength and quantity of energy.

photosynthesis The process by which plants and algae harness the energy of sunlight to make sugar from carbon dioxide and water.

phototropism The growth of the stem of a plant toward light.

phylogenetic tree A branching diagram of relationships showing common ancestry.

phylogeny The evolutionary history of a group of organisms.

physiology The study of the way a living organism's physical parts function.

pili (singular: pilus) Short, hairlike appendages extending from the surface of some bacteria that enable them to adhere to surfaces.

pituitary gland An endocrine gland in the brain that secretes many important hormones.

placebo A fake treatment given to control groups to mimic the experience of the experimental groups.

placenta A structure made of fetal and maternal tissues that helps sustain and support the embryo and fetus.

plant A multicellular eukaryote that has cell walls, carries out photosynthesis, and is adapted to living on land.

plaque Deposits of cholesterol, other fatty substances, calcium, blood clotting proteins, and cellular waste that accumulate inside arteries, limiting the flow of blood.

plasma cell An activated B cell that divides rapidly and secretes an abundance of antibodies.

plate tectonics The theory that the continents are part of large sections, or plates, that sit atop Earth's mantle and that move around and collide due to heat convection currents in the underlying mantle.

platelets Fragments of cells involved in blood clotting.

point mutation A mutation that alters a single DNA nucleotide.

polar molecule A molecule in which electrons are not shared equally between atoms, causing a partial negative charge at one end and a partial positive charge at the other. Water is a polar molecule.

pollen Small, thick-walled plant structures that contain cells that develop into sperm.

pollen tube A hollow tube that grows from a pollen grain after pollination and transports the male gametes to the egg.

pollination The transfer of pollen from male to female plant structures so that fertilization can occur.

polygenic trait A trait whose phenotype is determined by the interaction among alleles of more than one gene.

polymer A molecule made up of individual subunits, called monomers, linked together in a chain.

polymerase chain reaction (PCR) A laboratory technique used to replicate, and thereby amplify, a specific DNA segment.

population A group of organisms of the same species living and interacting in a particular area.

population density The number of organisms per unit area.

population genetics The study of the genetic makeup of populations and how the genetic composition of a population changes.

potential energy Stored energy.

prebiotics Components of food (like fiber) that promote the growth of beneficial microbes.

predation An interaction between two organisms in which one organism (the predator) feeds on the other (the prey).

primary growth Plants growing taller and roots growing longer as a result of cell division in apical meristems.

primary immune response The adaptive immune response mounted the first time the immune system encounters a particular antigen.

prion An infectious agent composed only of protein.

probiotics Living bacteria intended to be consumed and introduced into the microbiome (e.g., yogurt or capsules containing bacteria).

producers Autotrophs (photosynthetic organisms) that obtain energy directly from the sun and form the base of every food chain.

progesterone A female sex hormone produced by the corpus luteum of the ovary that prepares and maintains the uterus for pregnancy.

prokaryote A (typically) unicellular organism whose cell lacks internal membrane-bound organelles and whose DNA is not contained within a nucleus.

prokaryotic cells Cells that lack internal membrane-bound organelles.

protein A macromolecule made up of repeating subunits called amino acids, which determine the shape and function of a protein. Proteins play many critical roles in living organisms.

protist A eukaryote that cannot be classified as a plant, animal, or fungus; usually unicellular.

proton A positively charged subatomic particle in the nucleus of an atom.

proto-oncogene A gene that codes for a protein that helps cells divide normally.

pulmonary circuit The circulation of blood between the heart and the lungs.

pulse The detectable force of blood entering arteries, which can be felt in the neck or wrist.

Punnett square A diagram used to determine probabilities of offspring having particular genotypes, given the genotypes of the parents.

radial symmetry The pattern exhibited by a body plan that is circular, with no defined left and right sides.

radiation therapy The use of ionizing (high-energy) radiation to treat cancer.

radioactive isotope An unstable form of an element that decays into another element by radiation—that is, by emitting energetic particles.

radiometric dating The use of radioactive isotopes as a measure for determining the age of a rock or fossil.

randomized clinical trial A controlled medical experiment in which subjects are randomly chosen to receive either an experimental treatment or a standard treatment (or a placebo).

recall bias A type of error resulting from inaccurate recollection or reporting of past events.

receptor A molecule on or in a cell that binds to a specific signaling molecule, allowing the signaling molecule to exert an effect on that cell.

recessive allele An allele that reveals itself in the phenotype only if a masking dominant allele is not present.

recombinant gene A genetically engineered gene that contains portions of genes not naturally found together.

recombination An event in meiosis during which maternal and paternal chromosomes pair and physically exchange DNA segments.

red blood cells (RBCs) Blood cells specialized for transporting oxygen throughout the body.

regulatory sequence The part of a gene that determines the timing, amount, and location of protein production.

relative dating Determining the age of a fossil from its position relative to layers of rock or fossils of known age.

renewable resources Natural resources that do not run out or can be replenished easily; solar power, wind power, and sustainably harvested timber are examples.

reproductive isolation Mechanisms that prevent mating (and therefore gene flow) between members of different species.

resource partitioning The use of different resources by different species in a given area, enabling them to divide a niche.

respiratory surface A surface across which oxygen enters the blood and carbon dioxide leaves the blood.

respiratory system The organ system that allows us to take in oxygen and unload carbon dioxide.

ribosome The cellular machinery that assembles proteins during translation.

risk factor A behavior, exposure, or other factor that increases the probability of developing a disease.

RNA polymerase The enzyme that carries out transcription. RNA polymerase copies a strand of DNA into a complementary strand of mRNA.

root The belowground parts of a plant, which anchor it in the soil and absorb water and nutrients.

root nodule An enlargement on the root of a plant that contains nitrogen-fixing bacteria.

Rubisco The enzyme responsible for the first step of carbon fixation.

salivary glands Glands that secrete enzymes into the mouth to break down macromolecules in food. One such enzyme is salivary amylase, which digests carbohydrates.

sample size The number of experimental subjects or the number of times an experiment is repeated. In human studies, sample size is the number of participants.

saturated fat A fat, such as butter, that is solid at room temperature; often found in animal products.

science The process of using observations and experiments to draw conclusions based on evidence.

scientific theory An explanation of the natural world that is supported by a large body of evidence and has never been disproved.

scrotum The sac in which the testes are held.

secondary growth Plants growing wider as a result of cell division in lateral meristems.

secondary immune response The rapid and strong immune response mounted when the immune system encounters a particular antigen subsequent to the first encounter.

secondary metabolites Chemicals produced by plants that are not directly involved in growth or reproduction but that help protect the plant by their impacts on other organisms.

sedimentary rock Rock formed from the compression of layers of particles eroded from other rocks.

seed coat The hardy outer covering of a seed that protects the developing embryo.

semen The mixture of fluid and sperm that is ejaculated from the penis.

semiconservative DNA replication is said to be semiconservative because each newly made DNA molecule has one original DNA strand and one new DNA strand.

seminiferous tubules Coiled structures that constitute the bulk of the testes and in which sperm develop.

sensor A specialized cell that detects specific sensory input like temperature, pressure, or solute concentration.

sensory neurons Cells that convey information from both inside and outside the body to the central nervous system.

sex chromosomes Paired chromosomes that differ between males and females. Females have XX, and males have XY.

shoot The aboveground parts of a plant: the stem and photosynthetic leaves.

short tandem repeats (STRs) Sections of a chromosome in which short DNA sequences are repeated.

silent mutation A point mutation that does not change the amino acid sequence of the encoded protein.

simple diffusion The movement of small, uncharged solutes across a membrane from an area of higher concentration to an area of lower concentration without the aid of transport proteins; does not require an input of energy.

sister chromatids The two identical DNA molecules that result from the replication of a chromosome during the S phase.

small intestine The organ in which the bulk of chemical digestion and absorption of food occurs.

solute A dissolved substance.

solution The mixture of solute and solvent.

solvent A substance in which other substances can dissolve. Water is a good solvent.

somatic cells Nonreproductive cells of the body.

specialist An organism that requires a specific resource or habitat to be successful.

speciation The genetic divergence of populations, leading over time to reproductive isolation and the formation of new species.

spinal cord A bundle of nerve fibers, contained within the bony spinal column, that transmits information between the brain and the rest of the body.

sporadic Cancers that are caused by non-inherited (acquired) mutations.

spores (fungal) Fungal cells that are resistant to drying out and can be dispersed to new locations as part of sexual or asexual reproduction.

stabilizing selection A type of natural selection in which organisms near the middle of the phenotypic range of variation are favored by the environment.

stamen The male reproductive structure of a flower, made up of a filament and an anther.

staple crop A crop that is eaten in large quantities and provides most of the energy and nutrients in the human diet.

starch A complex plant carbohydrate made of linked chains of glucose molecules; a source of stored energy.

statistical significance A measure of confidence that the results obtained are "real" and not due to chance.

stigma The sticky "landing pad" for pollen on the pistil.

stomach An expandable muscular organ that stores and mechanically breaks down food. Specific enzymes in the stomach digest proteins.

stomata (singular: stoma) Pores on leaves that permit the exchange of oxygen and carbon dioxide with the air and allow water loss.

stool Solid waste material eliminated from the digestive tract.

stroke A disruption in blood supply to the brain.

style The tubelike structure that leads from the stigma to the ovary.

substrate A molecule to which an enzyme binds and on which the enzyme acts.

symbiosis A relationship in which two different organisms live together, often interdependently.

synapse The site of transmission of a signal between a neuron and another cell; includes the axon terminal of the signaling neuron, the space between the cells, and receptors on the receiving cell.

synaptic cleft The physical space between a neuron and the cell with which it is communicating.

systemic circuit The circulation of blood between the heart and the rest of the body.

systolic pressure The pressure in arteries at the time the ventricles contract.

T cells White blood cells that mature in the thymus and play several roles in adaptive immunity.

targeted therapy A cancer therapy that is specific for cancer cells and not harmful to normal cells.

taxonomy The identification, naming, and classification of organisms on the basis of shared traits.

testable Describes a hypothesis that can be supported or rejected by carefully designed experiments or observational studies.

testes Paired male reproductive organs, which contain sperm and produce androgens (primarily testosterone).

testosterone The primary male sex hormone, which stimulates the development of masculine features and plays a key role in sperm development.

tetrapod A vertebrate animal with four true limbs—that is, jointed, bony appendages with digits. Mammals, amphibians, birds, and reptiles are tetrapods.

thermoregulation The maintenance of a relatively stable internal body temperature.

thigmotropism The response of plants to touch and wind.

thymus The organ in which T cells mature.

tissue An organized collection of cells working to carry out a specific function.

trachea A large airway between the larynx and the lungs.

trans fat An unhealthful fat made from vegetable oil that has been chemically altered to make it solid at room temperature; often found in snack foods.

transcription The first stage of gene expression, during which cells produce molecules of messenger RNA (mRNA) from the instructions encoded within genes in DNA.

transfer RNA (tRNA) A type of RNA that transports amino acids to the ribosome during translation.

transgenic Refers to an organism that carries one or more genes from a different species.

translation The second stage of gene expression, during which mRNA sequences are used to assemble the corresponding amino acids to make a protein.

transpiration The loss of water from plants by evaporation, which powers the transport of water and nutrients through a plant's vascular system.

transport proteins Proteins involved in the movement of molecules and ions across the cell membrane.

triglycerides A type of lipid found in fat cells that stores excess energy for long-term use.

trisomy 21 Having an extra copy of chromosome 21; also known as Down syndrome.

trophic level The feeding level of an organism, reflecting its position in a food chain.

tumor A mass of cells resulting from uncontrolled cell division.

tumor suppressor gene A gene that codes for a protein that monitors and checks cell cycle progression. When these genes mutate, tumor suppressor proteins lose normal function.

turgor pressure The pressure exerted by the water-filled central vacuole against the plant cell wall, giving a stem its rigidity.

unsaturated fat A fat, such as olive oil, that is liquid at room temperature; often found in plants and fish.

urethra A tube that connects the bladder to the genitals and carries urine out of the body. In males, the urethra travels through the penis and also carries sperm.

uterus The muscular organ in females in which a fetus develops.

vaccine A preparation of killed or weakened pathogen that is administered to people or animals to generate protective immunity to that pathogen.

vagina The first part of the female reproductive tract, extending to the cervix; also known as the birth canal.

vas deferens Paired tubes that carry sperm from the testes to the urethra.

vascular plant A plant with tissues that transport water and nutrients through the plant body.

vascular system A system of tube-shaped vessels that transports water and nutrients throughout an organism's body.

vasoconstriction The reduction in diameter of blood vessels, which helps to retain heat.

vasodilation The expansion in diameter of blood vessels, which helps to release heat.

vector A DNA molecule used to deliver a recombinant gene to a host cell.

veins Blood vessels that carry blood toward the heart.

ventilation The process of moving air in and out of the lungs.

ventricles The lower chambers of the heart that pump blood away from the heart. In humans, the right ventricle pumps oxygen-poor blood to the lungs, and the left ventricle pumps oxygen-rich blood to the body.

vertebrate An animal with a bony or cartilaginous backbone.

vestigial structure A structure inherited from an ancestor that no longer serves a clear function in the organism that possesses it.

virus A noncellular infectious particle consisting of a nucleic acid (DNA or RNA) surrounded by a protein shell.

vitamin An organic molecule required in small amounts for normal growth, reproduction, and tissue maintenance.

vitamin D A fat-soluble vitamin required to maintain a healthy immune system and to build healthy bones and teeth. The human body produces vitamin D when skin is exposed to UV light.

water cycle The continuous movement of water on, above, and below Earth's surface.

weather Local atmospheric conditions over a short period of time.

white blood cells Cells involved in the body's immune response.

wood Hard, secondary xylem tissue found in the stem of a woody plant.

X-linked trait A phenotype determined by an allele on an X chromosome.

xylem Plant vascular tissue that transports water from the roots to the shoots.

Y-chromosome analysis The comparison of sequences on the Y chromosomes of different individuals to examine paternity and paternal ancestry.

zoonotic disease An infectious disease that normally affects animals but can be transmitted to humans.

zygote A diploid cell that is capable of developing into an adult organism. It is formed when a haploid egg is fertilized by a haploid sperm.

Index

Page numbers followed by f indicate figures; by t indicate tables. Page numbers in bold indicate definitions.